EARTH + SPACE
SCIENCE

Hendrix • Thompson • Turk

ACKNOWLEDGMENTS

Grateful acknowledgment is given to the authors, artists, photographers, museums, publishers, and agents for permission to reprint copyrighted material. Every effort has been made to secure the appropriate permission. If any omissions have been made or if corrections are required, please contact the Publisher.

PHOTOGRAPHIC CREDITS

Front cover: ESA/NASA
Back cover: NASA
Locator globe: kathykonkle/DigitalVision Vectors/Getty Images
Acknowledgments and credits continue on page 891.

ISBN: 978-0-357-11362-2
Printed in the United States of America
Print Number: 04
Print Year: 2023

Copyright © 2022 Cengage Learning, Inc.

ALL RIGHTS RESERVED. No part of this work covered by the copyright herein may be reproduced or distributed in any form or by any means, except as permitted by U.S. copyright law, without the prior written permission of the copyright owner.

"National Geographic", "National Geographic Society" and the Yellow Border Design are registered trademarks of the National Geographic Society ® Marcas Registradas

For product information and technology assistance, contact us at Customer & Sales Support, 888-915-3276

For permission to use material from this text or product, submit all requests online at
www.cengage.com/permissions

Further permissions questions can be emailed to permissionrequest@cengage.com

National Geographic Learning | Cengage
200 Pier 4 Blvd., Suite 400
Boston, MA 02210

National Geographic Learning, a Cengage company, is a provider of quality core and supplemental educational materials for the PreK-12, adult education, and ELT markets. Cengage is a leading provider of customized learning solutions with employees residing in nearly 40 different countries and sales in more than 125 countries around the world. Find your local representative at NGL.Cengage.com/RepFinder.

Visit National Geographic Learning online at NGL.Cengage.com.

BRIEF CONTENTS

Detailed Contents iv
About the Authors xiv
Program Reviewers xvi
Featured Explorers xvi
Lab Safety xviii

UNIT 1 INTRODUCTION TO EARTH AND SPACE SCIENCE
Chapter 1 Nature of Science and Engineering 4
Chapter 2 Earth Systems and Cycles 32

UNIT 2 EARTH MATERIALS AND GEOLOGIC TIME
Chapter 3 Minerals 66
Chapter 4 Rocks 94
Chapter 5 Geologic Time 124
Chapter 6 Mineral and Energy Resources 158

UNIT 3 INTERNAL PROCESSES
Chapter 7 Plate Tectonics 192
Chapter 8 Earthquakes and Earth's Structure 226
Chapter 9 Volcanoes 262
Chapter 10 Mountains 292
Chapter 11 The Seafloor 320

UNIT 4 SURFACE PROCESSES
Chapter 12 Weathering, Erosion, and Deposition 354
Chapter 13 Glaciers and Deserts 394
Chapter 14 Soil Resources 438

UNIT 5 THE HYDROSPHERE
Chapter 15 Fresh Water 472
Chapter 16 Oceans and Coastlines 496
Chapter 17 Water Resources 536

UNIT 6 THE ATMOSPHERE
Chapter 18 The Atmosphere 574
Chapter 19 Energy Balance in the Atmosphere 606
Chapter 20 Weather 636
Chapter 21 Climate and Climate Change 672

UNIT 7 EARTH AND SPACE
Chapter 22 Evolving Models of the Sky 708
Chapter 23 Our Solar System 736
Chapter 24 Stars 768
Chapter 25 Galaxies and the Universe 794

Appendices 816
Glossary 837
Glosario 853
Index 871

TABLE OF CONTENTS

About the Authors xiv
Program Reviewers xvi
Featured Explorers xvi
Lab Safety xviii

UNIT 1

INTRODUCTION TO EARTH AND SPACE SCIENCE 2

CHAPTER 1

NATURE OF SCIENCE AND ENGINEERING 4
■ **EXPLORERS AT WORK** Andrés Ruzo 6
CASE STUDY Mapping Geothermal Energy 8
1.1 The Nature of Science and Engineering 9
   Data Analysis Distinguish Between Correlation and Causation 12
   Minilab Puzzle: Law Versus Theory 14
1.2 Mapping Earth's Systems 19
   Data Analysis Develop a Topographic Profile 26

TYING IT ALL TOGETHER Geothermal Energy Resource Planning 27
Chapter Summary 29
Chapter Test 30

CHAPTER 2

EARTH SYSTEMS AND CYCLES 32
■ **EXPLORERS AT WORK** Allyson Tessin 34
CASE STUDY The Yellowstone Volcano: Major Eruptions Can Affect Earth's Systems 36
2.1 Earth's Spheres 37
   Minilab Model a Cloud in a Bottle 42
2.2 Earth's Systems 43
2.3 Earth's Life-Sustaining Cycles 54
   Data Analysis Graphing the Carbon Cycle 57
TYING IT ALL TOGETHER A Super Disruption 60
Chapter Summary 61
Chapter Test 62

iv TABLE OF CONTENTS

UNIT 2

EARTH MATERIALS AND GEOLOGIC TIME 64

CHAPTER 3

MINERALS 66
- **EXPLORERS AT WORK** Katlin L. Bowman 68

CASE STUDY Rising Atmospheric and Ocean Mercury Concentrations Linked to Small-Scale Gold Mining 70

- 3.1 What Are Minerals? 71
  - Minilab Observe Minerals 72
- 3.2 Composition and Structure of Minerals 73
- 3.3 Physical Properties of Minerals 75
- 3.4 Mineral Classes and the Rock-Forming Minerals 79
  - Data Analysis Abundance and Distribution of Silicate Minerals 82
- 3.5 Hazardous Rocks and Minerals 85

TYING IT ALL TOGETHER The Costs of Mining 90

Chapter Summary 91

Chapter Test 92

CHAPTER 4

ROCKS 94
- **EXPLORERS AT WORK** Kenneth Warren Sims 96

CASE STUDY Devils Tower National Monument, Wyoming 98

- 4.1 Rocks and the Rock Cycle 99
  - Minilab Rock Cycle Model 101
- 4.2 Igneous Rocks 104
- 4.3 Sedimentary Rocks 108
- 4.4 Metamorphic Rocks 115
  - Data Analysis Metamorphism of Mudstone 118

TYING IT ALL TOGETHER Formation of Devils Tower 120

Chapter Summary 121

Chapter Test 122

TABLE OF CONTENTS

CHAPTER 5

GEOLOGIC TIME 124
■ **EXPLORERS AT WORK** Sarah Carmichael ... 126
CASE STUDY Giant Dragonflies? 128
5.1 Mass Extinctions in Earth's History 129
5.2 Relative Age 133
5.3 Interpreting the Rock Record 139
5.4 Absolute Age 144
    Minilab Radioactivity and Half-Life 146
    Data Analysis Think About Radiometric Testing 147
5.5 The Geologic Timescale 150
TYING IT ALL TOGETHER Reconstructing Earth's Past 154
Chapter Summary 155
Chapter Test 156

CHAPTER 6

MINERAL AND ENERGY RESOURCES 158
■ **EXPLORERS AT WORK** Saleem Ali 160
CASE STUDY Strategically Important Minerals 162
6.1 Mineral Resources 163
6.2 Nonrenewable Energy Resources 172
6.3 Renewable Energy Resources 177
    Minilab Biofuel 180
6.4 Conservation and Energy for the Future ... 181
    Data Analysis Analyze Patterns in U.S. Energy Consumption 184
TYING IT ALL TOGETHER Mining Manganese Nodules 186
Chapter Summary 187
Chapter Test 188

UNIT 3

INTERNAL PROCESSES 190

CHAPTER 7

PLATE TECTONICS 192
■ **EXPLORERS AT WORK** D. Sarah Stamps 194
CASE STUDY The Splitting of a Continent 196
7.1 Early Hypotheses 197
    Data Analysis Interpret Temperature and Pressure Graphs 201
7.2 The Theory of Plate Tectonics 205
7.3 Plate Movement 211
    Minilab Convection 213
    Minilab Isostasy 216
7.4 Effects of Plate Tectonics 217
TYING IT ALL TOGETHER Rifting Apart 222
Chapter Summary 223
Chapter Test 224

CHAPTER 8

EARTHQUAKES AND EARTH'S STRUCTURE ... 226
■ **EXPLORERS AT WORK** Afroz Ahmad Shah ... 228
CASE STUDY Mexico City Earthquakes, 1985 and 2017 230
8.1 Earthquakes and Seismic Waves 231
    Minilab Build a Simple Seismograph 238
    Data Analysis Analyze Patterns of P and S Waves 239
8.2 Earthquakes and Tectonic Plate Boundaries 241
    Data Analysis Map Earthquakes in Real Time 244
8.3 Earthquake Damage 246
8.4 Studying Earth's Interior 253
TYING IT ALL TOGETHER Earthquake Vulnerability 258
Chapter Summary 259
Chapter Test 260

CHAPTER 9

VOLCANOES 262
- **EXPLORERS AT WORK** Arianna Soldati 264

CASE STUDY Volcanic Activity: Past, Present, and Future 266

9.1 Magma 267
   Minilab Soda Bottle Volcano 271
9.2 Magmatic Bodies and Volcanic Formations 274
9.3 Volcanic Eruptions 281
   Data Analysis Exploring Volcanic Risk in the United States 286

TYING IT ALL TOGETHER Communities Near Volcanoes 288
Chapter Summary 289
Chapter Test 290

CHAPTER 10

MOUNTAINS 292
- **EXPLORERS AT WORK** Cory Richards 294

CASE STUDY The Story of a Mountain Chain ... 296

10.1 Geologic Structures and Tectonic Stress 297
10.2 Mountains and Mountain Ranges 304
   Minilab Fault-Block Mountain Models 310
10.3 Continental Collision: The Himalaya 311
   Data Analysis Compare Elevations 314

TYING IT ALL TOGETHER Chains of Change 316
Chapter Summary 317
Chapter Test 318

CHAPTER 11

THE SEAFLOOR 320
- **EXPLORERS AT WORK** Marcello Calisti 322

CASE STUDY Probing the Depths 324

11.1 Earth's Oceans 325
   Data Analysis Compare Seafloor Data 327
11.2 Studying the Seafloor 328
11.3 Features of the Seafloor 331
   Minilab Hawaiian Hot Spot 339
11.4 Continental Margins 342

TYING IT ALL TOGETHER Robotic Exploration of the Seafloor 348
Chapter Summary 349
Chapter Test 350

TABLE OF CONTENTS

UNIT 4
SURFACE PROCESSES 352

CHAPTER 12
WEATHERING, EROSION, AND DEPOSITION 354
- **EXPLORERS AT WORK** Cynthia Liutkus-Pierce 356
- **CASE STUDY** Flooding from Hurricane Katrina 358
- **12.1** Mechanical Weathering 359
  - Minilab Weathering by Plant Roots 364
- **12.2** Chemical Weathering 365
  - Minilab Weathering of Iron 368
- **12.3** Stream Erosion and Deposition 369
  - Data Analysis Blackfoot River Discharge over Time 371
- **12.4** Mass Wasting 382
- **TYING IT ALL TOGETHER** Flood Control and Management 389
- Chapter Summary 390
- Chapter Test 391

CHAPTER 13
GLACIERS AND DESERTS 394
- **EXPLORERS AT WORK** Erin Pettit 396
- **CASE STUDY** A Glacier to Watch 398
- **13.1** Formation of Glaciers 399
- **13.2** Glacial Movement 401
- **13.3** Glacial Erosion 405
- **13.4** Glacial Deposits 409
- **13.5** Glaciation and Glaciers Today 414
  - Data Analysis Graphing Glacial Melt 415
- **13.6** Why Do Deserts Exist? 417
  - Data Analysis Which Is the Desert? 419
- **13.7** Water and Deserts 421
- **13.8** Wind and Deserts 428
  - Minilab Simulating Wind Erosion 431
- **TYING IT ALL TOGETHER** Ice and Wind Erosion 432
- Chapter Summary 433
- Chapter Test 434

CHAPTER 14

SOIL RESOURCES 438
■ **EXPLORERS AT WORK** Jerry Glover 440
CASE STUDY Soil Erosion and Human Activities 442
14.1 Soil Formation 443
    Data Analysis Effects on Soil Biodiversity 448
14.2 Soil Classification 449
    Minilab What Is in the Soil? 452
14.3 Soil Erosion and Desertification 458
    Minilab Modeling Soil Erosion 460
    Data Analysis Exploring Earthworm Bioremediation 465
TYING IT ALL TOGETHER Combating Soil Depletion and Pollution in the Chesapeake Bay Watershed 466
Chapter Summary 467
Chapter Test 468

UNIT 5
THE HYDROSPHERE 470

CHAPTER 15

FRESH WATER 472
■ **EXPLORERS AT WORK** Joe Cutler 474
CASE STUDY The Great Lakes Ecosystem 476
15.1 Hydrologic Cycle 477
15.2 Lakes and Wetlands 478
    Minilab Lake Turnover 482
    Data Analysis Compare Wetland Gains and Losses 485
    Minilab Role of Wetlands 486
15.3 Groundwater 487
TYING IT ALL TOGETHER Great Lakes Connections 492
Chapter Summary 493
Chapter Test 494

CHAPTER 16

OCEANS AND COASTLINES 496
- **EXPLORERS AT WORK** Caroline Quanbeck 498

CASE STUDY Coral Reef Killers 500

16.1 Ocean Geography and Seawater Composition 501

16.2 Tides and Sea Waves 505

16.3 Ocean Currents 509

    Minilab Motion in the Ocean 515

16.4 Seacoasts 516

16.5 Beaches 522

16.6 Life in the Sea 528

    Data Analysis Comparing World Fisheries 529

TYING IT ALL TOGETHER Sea Change 532

Chapter Summary 533

Chapter Test 534

CHAPTER 17

WATER RESOURCES 536
- **EXPLORERS AT WORK** Steve Boyes 538

CASE STUDY Is Sustainable Irrigation Possible? 540

17.1 Water Supply and Demand 541

    Data Analysis Global Water Use 542

17.2 Dams and Diversion 547

17.3 Groundwater Diversion and Depletion 554

17.4 Water Pollution 559

    Minilab Point Source Pollution Models 563

TYING IT ALL TOGETHER Sustainable Water Management 567

Chapter Summary 568

Chapter Test 569

UNIT 6

THE ATMOSPHERE 572

CHAPTER 18

THE ATMOSPHERE 574
- **EXPLORERS AT WORK** Katey Walter Anthony 576

CASE STUDY Manure, Cold Seeps, and Fire Ice 578

18.1 Earth's Early Atmospheres 579

18.2 Life, Iron, and the Evolution of the Modern Atmosphere 581

    Minilab Carbon Dioxide, Photosynthesis, and Oxygen 584

    Minilab Oxygen in the Atmosphere 585

18.3 Atmospheric Pressure and Temperature ...586

18.4 Air Pollution 591

    Data Analysis Air Pollution 593

TYING IT ALL TOGETHER Reducing Air Pollution 600

Chapter Summary 601

Chapter Test 602

CHAPTER 19

ENERGY BALANCE IN THE ATMOSPHERE 606
- **EXPLORERS AT WORK** Stefanie Lutz 608

CASE STUDY Melting Glaciers in Greenland ... 610

19.1 Incoming Solar Radiation 611

19.2 The Radiation Balance 617

    Minilab Simulating the Greenhouse Effect 618

19.3 Energy Storage and Transfer 620

    Minilab Modeling Atmospheric Convection and Wind 621

19.4 Geographic Factors 624

    Data Analysis Hours of Sunlight Per Day 627

TYING IT ALL TOGETHER Antarctica's Glacier Retreat 631

Chapter Summary 632

Chapter Test 633

CHAPTER 20

WEATHER 636
■ **EXPLORERS AT WORK** Anton Seimon 638
CASE STUDY Even More Powerful Storms ... 640
20.1 Moisture, Temperature, and Air 641
20.2 Clouds and Precipitation 645
   Data Analysis Record Data and Design a Game 646
20.3 Pressure, Wind, and Fronts 649
   Minilab Make a Barometer 653
   Minilab Compare Two Coriolis Effect Models 654
20.4 Storms 659
   Data Analysis Compare Storm Statistics ... 666
TYING IT ALL TOGETHER Explaining Weather Patterns 667
Chapter Summary 668
Chapter Test 669

CHAPTER 21

CLIMATE AND CLIMATE CHANGE 672
■ **EXPLORERS AT WORK** Paul Miller 674
CASE STUDY Ice Shelf Disintegration 676
21.1 Global Climate 677
21.2 Climate Types 680
   Data Analysis Compare Climographs 683
21.3 Historical Climate Change....................... 686
   Minilab Energy Exchanges in Atmospheric Models 690
21.4 Climate Change Today 692
21.5 Consequences of Climate Change 695
TYING IT ALL TOGETHER Consequences of Climate Change 701
Chapter Summary 702
Chapter Test 703

TABLE OF CONTENTS **xi**

TABLE OF CONTENTS

UNIT 7
EARTH AND SPACE 706

CHAPTER 22
EVOLVING MODELS OF THE SKY 708
- **EXPLORERS AT WORK** Brendan Mullan 710
- **CASE STUDY** Galilean Moons and Models of the Universe 712
- **22.1** Patterns in the Sky 713
- **22.2** Evolving Models of the Universe 715
- **22.3** The Role of Gravity 719
  - **Data Analysis** Quantifying the Effect of Gravity 723
  - **Minilab** Simulating Gravity 724
- **22.4** Tools of Modern Astronomy 727
  - **Data Analysis** Evolution of the Telescope 730
- **TYING IT ALL TOGETHER** Gravity in Exoplanetary Systems 732
- Chapter Summary 733
- Chapter Test 734

CHAPTER 23

OUR SOLAR SYSTEM 736
■ **EXPLORERS AT WORK** Bethany Ehlmann ... 738
CASE STUDY Small Steps and Giant Leaps 740
23.1 The Solar System: A Brief Overview 741
  Data Analysis Compare Planetary Data 743
23.2 The Terrestrial Planets and Earth's Moon 744
23.3 The Jovian Planets 752
23.4 Jovian Moons 757
23.5 Dwarf Planets, Minor Planets, and Comets 760
  Minilab Micrometeorites 763
TYING IT ALL TOGETHER Distant Worlds 764
Chapter Summary 765
Chapter Test 766

CHAPTER 24

STARS 768
■ **EXPLORERS AT WORK** Munazza Alam 770
CASE STUDY Journey to the Sun: Parker Solar Probe 772
24.1 The Birth of Stars 773
24.2 The Sun 775
  Minilab Build a Spectroscope 778
24.3 The Life and Death of Stars 780
  Data Analysis Estimating the Lifetime of the Sun 782
24.4 Extreme Stellar Remnants 785
TYING IT ALL TOGETHER We Are Made of Star-Stuff 789
Chapter Summary 790
Chapter Test 791

CHAPTER 25

GALAXIES AND THE UNIVERSE 794
■ **EXPLORERS AT WORK** Knicole Colon 796
CASE STUDY Sgr A* 798
25.1 The Milky Way and Other Galaxies 799
25.2 The Big Bang 803
  Data Analysis Quantify the Expansion Rate of the Universe 805
  Minilab Big Bang Balloon 806
25.3 Quasars, Dark Matter, and the Fate of the Universe 808
TYING IT ALL TOGETHER Quasars and the Evolution of the Universe 812
Chapter Summary 813
Chapter Test 814

APPENDIX 1 Reference Tables 816
  Periodic Table of the Elements 816
  International System of Units (SI) 818
  Important Minerals 820
  Identifying Minerals 822
  Common Rocks 824
  Geologic Timescale 826
  Common Map Symbols 827
APPENDIX 2 Climate 830
  Köppen Classification System 830
  Map of Köppen Climate Types 832
  Climographs 834
GLOSSARY 837
GLOSARIO 853
INDEX 871
ACKNOWLEDGMENTS 891

TABLE OF CONTENTS **xiii**

ABOUT THE AUTHORS

MARC S. HENDRIX

Marc S. Hendrix is a Professor of Geology at the University of Montana in Missoula, Montana. Growing up in Gettysburg, Pennsylvania, Marc developed an early love of geology in the 1970s while working as a field assistant for his father, a biology professor at Gettysburg College. Marc received a B.A. in geology from Wittenberg University in Springfield, Ohio, in 1985 and an M.S. in geology and geophysics from the University of Wisconsin—Madison in 1987. In 1992, he graduated with a Ph.D. in applied earth sciences from Stanford University, where he conducted research on the geologic record of mountain building and ancient climate in western China. Afterward he worked at Stanford as a postdoctoral researcher, analyzing the geologic history of Mongolia.

Marc joined the faculty at the University of Montana in 1994, where he has developed a field-based research program focused on the geology of the northern Rocky Mountains. Marc and his students have published a variety of technical papers on the geology of North America, Asia, and Africa. In 2011, Marc authored and illustrated the book, *Geology Underfoot in Yellowstone Country*, and in 2014, he coauthored the introductory earth sciences textbook, *Earth2*, with Gray Thompson and Jon Turk. In 2012, Marc cofounded AIM Geoanalytics, a Missoula-based geologic consulting company that provides technical services to the energy industry. Marc has served as an expert witness in the field of geology, and he continues to travel broadly to conduct geologic research. He currently lives in Missoula with his wife and two sons.

ABOUT THE AUTHORS

GRAHAM R. THOMPSON

Dr. Thompson is a geologist and mountaineer, and is now Professor Emeritus of Geology at the University of Montana. He has earth science degrees from Bates College, Dartmouth College, and Case Western Reserve University, and did postdoctoral work at the Sorbonne in Paris. Gray has published numerous research papers in scientific journals and coauthored more than a dozen textbooks with his good friends, Turk and Hendrix.

As a geologist, Gray's research interests include sedimentary basin analysis, with emphasis on clay mineral reactions during burial and petroleum generation. He has taught university courses to undergraduate and graduate students, often with Marc Hendrix, in introductory geology, mineralogy, petrology and fieldwork-based geological mapmaking. His current research focuses on mapping and interpreting the geology of Montana's numerous mountain ranges, particularly the Beartooth Mountains near Yellowstone National Park.

As a mountaineer, Gray has made numerous first ascents in many mountain ranges around the world, several with Jon Turk. Many of those climbs have been featured in popular climbing magazines.

JONATHAN TURK

Jon Turk earned his Ph.D. in chemistry in 1971, working on a theoretical problem with a long-term eye toward developing cheap, efficient, organic semiconductors for solar electric panels. In the same time period, in honor of Earth Day 1, Jon coauthored the first environmental science textbook in North America, *Ecology, Pollution, Environment*.

Over the next decades, Jon kayaked around Cape Horn and across the North Pacific, mountain biked across the northern Gobi in Mongolia, and made numerous first ski descents and first climbing ascents in North America and central Asia. During this time, he climbed frequently with Gray Thompson. Gray introduced Jon to the wonders of geology, often turning climbing expeditions into prolonged, informal field trips. Together, Jon and Gray wrote numerous geology and earth science textbooks, eventually teaming up with Marc Hendrix to produce the current text.

In 2012, at the age of 65, Jon was nominated by National Geographic as one of the ten top adventure athletes of the year. Today, in his early 70s, he lives in the Bitterroot Mountains in western Montana with his wife, Nina. He continues to write, ride his mountain bike, and ski.

PROGRAM REVIEWERS

Michele Gephart
Oscar Smith High School
Chesapeake, Virginia

Erin Graves
Herculaneum High School
Herculaneum, Missouri

Michael Jabot, Ph.D.
SUNY at Fredonia
Fredonia, New York

Jim Lindsey
Franklin Central High School
Indianapolis, Indiana

Pradip C. Misra
Bagdad Middle and High School
Bagdad, Arizona

Seyi Okuneye
Metropolitan Expeditionary
Learning School
Queens, New York

Julie Olson
Mitchell High School
Mitchell, South Dakota

Susan Pike, Ph.D.
Dover High School
Dover, New Hampshire

Alicia Pressel
Creekside High School
Saint Augustine, Florida

Kyle Tredinnick
Zoo Academy
Omaha, Nebraska

Sara Young
Waubonsie Valley
High School
Aurora, Illinois

NATIONAL GEOGRAPHIC

FEATURED EXPLORERS

Munazza Alam
Astronomer
National Geographic
Grantee

Saleem Ali
Environmental Planner,
Researcher, Educator
National Geographic
Emerging Explorer

Katlin Bowman
Oceanographer
National Geographic
Grantee

Steve Boyes
Conservation Biologist
National Geographic
Fellow

Marcello Calisti
Roboticist
National Geographic
Grantee

Sarah Carmichael
Geochemist
National Geographic
Grantee

Knicole Colón
Research
Astrophysicist
National Geographic
Grantee

Joe Cutler
Ichthyologist,
Freshwater
Conservationist
National Geographic
Explorer

Bethany Ehlmann
Planetary Geologist
National Geographic
Emerging Explorer

Jerry Glover
Agricultural Ecologist
National Geographic
Emerging Explorer

FEATURED EXPLORERS

Cynthia Liutkus-Pierce
Sedimentary Geologist
National Geographic
Grantee

Stefanie Lutz
Microbiologist,
Ecologist
National Geographic
Grantee

Paul Miller
Composer, Multimedia
Artist, Writer
National Geographic
Emerging Explorer

Brendan Mullan
Astrophysicist,
Educator
National Geographic
Emerging Explorer

Erin Pettit
Glaciologist
National Geographic
Emerging Explorer

Caroline Quanbeck
Geologist
National Geographic
Grantee

Andrés Ruzo
Geothermal Scientist,
Conservationist
National Geographic
Explorer

Anton Seimon
Tornado Scientist,
Geographer
National Geographic
Grantee

Afroz Ahmad Shah
Earthquake Geologist
National Geographic
Grantee

Kenneth Warren Sims
Isotope Geologist
National Geographic
Grantee

Arianna Soldati
Volcanologist
National Geographic
Grantee

D. Sarah Stamps
Geophysicist
National Geographic
Grantee

Allyson Tessin
Geologist,
Biogeochemist
National Geographic
Grantee

Katey Walter Anthony
Aquatic Ecosystem
Ecologist
National Geographic
Emerging Explorer

xvii

Working in a laboratory can be an exciting experience. But the laboratory can also be a dangerous place if safety rules are not followed. You must be responsible for your safety and the safety of others.

General Safety Rules

1. Read the directions several times before you begin. Follow the directions exactly as they are written. If you have questions about any part of the procedure, ask your teacher for assistance.

2. Never perform any activities that are not authorized by your teacher. Do not work in the laboratory without teacher supervision.

3. Use the safety equipment provided for you. Safety goggles and a lab apron should be worn whenever you are working with chemicals or heating any substance.

4. Never eat or drink in the lab. Never inhale chemicals. Do not attempt to taste any chemicals or materials.

5. Take care not to spill any material in the lab. If a spill occurs, immediately ask your teacher about the proper cleanup procedure. Do not pour chemicals or other substances into the sink or trash container.

6. Tie back long hair to keep it away from chemicals, burners, and other lab equipment.

7. Remove or tie back articles of loose clothing and jewelry that can hang down and come in contact with chemicals or flames. If a fire should break out in the room or if your clothing catches fire, smother it with a fire blanket or roll on the ground. NEVER RUN.

8. Know the locations of the safety showers and eyewashes and know how to use them. See Table 1 for appropriate first-aid responses.

First Aid in the Laboratory

Report any accident or injury, no matter how minor, to your teacher.

TABLE 1 First Aid Procedures

Injury	Response
Cuts and Abrasions	Stop any bleeding by applying direct pressure. Call your teacher immediately.
Foreign Matter in Eye	Use eyewash to flush with plenty of water. Call your teacher immediately.
Fainting	Leave the person lying down. Loosen any tight clothing. Call your teacher immediately. Keep people away.
Poisoning	Note the suspected substance. Call your teacher immediately.
Spills on Skin	Flush with large amounts of water or use the safety shower. Call your teacher immediately.

Working in the Laboratory

1. Always use heat-resistant (Pyrex) glassware for heating.

2. Use materials only from properly labeled containers. Read labels carefully before using chemicals.

3. Use glycerin or a lubricant when inserting glass tubing or thermometers through rubber stoppers. Protect your hand with a paper towel or a folded cloth.

4. Store backpacks, coats, and other personal items away from the immediate lab bench area.

Cleaning Up the Laboratory

1. After working in the lab, clean up your work area and return all equipment to its proper place.

2. Turn off all water and burners. Check that you have turned off the gas jet as well as the burner.

3. Dispose of chemicals and other materials as directed by your teacher. Do not use the sink to discard matches or any other solid material.

4. Wash your hands thoroughly before leaving the lab.

SAFETY AND CLEANUP SYMBOLS

The following safety and cleanup symbols are used throughout the chapter investigations and in certain minilabs. In general, it is good practice to wear safety goggles and a lab apron at all times when you are in the lab.

Safety Symbols			
	Chemical safety The corrosives symbol indicates that caustic substances are used. Knowledge of how to handle the chemicals being used is required.		**Glassware safety** The glassware symbol indicates that breakable glass objects are used.
	Electrical safety The electrical safety symbol indicates the possibility of danger from electrical shock or burn.		**Lab clothing** The apron symbol indicates that clothing protection is needed.
	Eye safety The safety goggles symbol is used to indicate that eye protection is needed.		**Sharps and scissors safety** The sharp objects symbol indicates that tools are used that can easily puncture or cut skin.
	Fire safety The flammables symbol is used to indicate that heating (via hot plate or Bunsen burner) is used.		**Skin safety** The gloves symbol is used to indicate that hand safety (protective gloves) is required.
Cleanup Symbols			
	Cleanup The wash hands symbol is used to indicate that personal hygiene is required before leaving the lab.		**Disposal** The waste disposal symbol is used to ensure that chemicals (hazardous waste) are disposed of properly.

UNIT 1 | INTRODUCTION TO EARTH AND SPACE SCIENCE

Chapter 1 **Nature of Science and Engineering**

Chapter 2 **Earth Systems and Cycles**

Where might you find scientists and engineers working in Earth and space sciences? You might find one in a remote jungle in Peru on the banks of a mysterious boiling river, hot enough to cook any animal that falls in. Or you might look for an Arctic lake where you can set methane gas aflame as it seeps up from below the ice. Perhaps you might peer into the night sky to find the International Space Station, where scientists have a panoramic view of Earth's most magnificent features. In this unit, you will learn what Earth and space scientists and engineers do. You will also look at some of the complex interactions among the rocky, watery, airy, and living realms of our planet.

Niccolò Segreto, an Italian glaciologist, explores bubblelike ice formations inside a glacier near Zermatt, Switzerland.

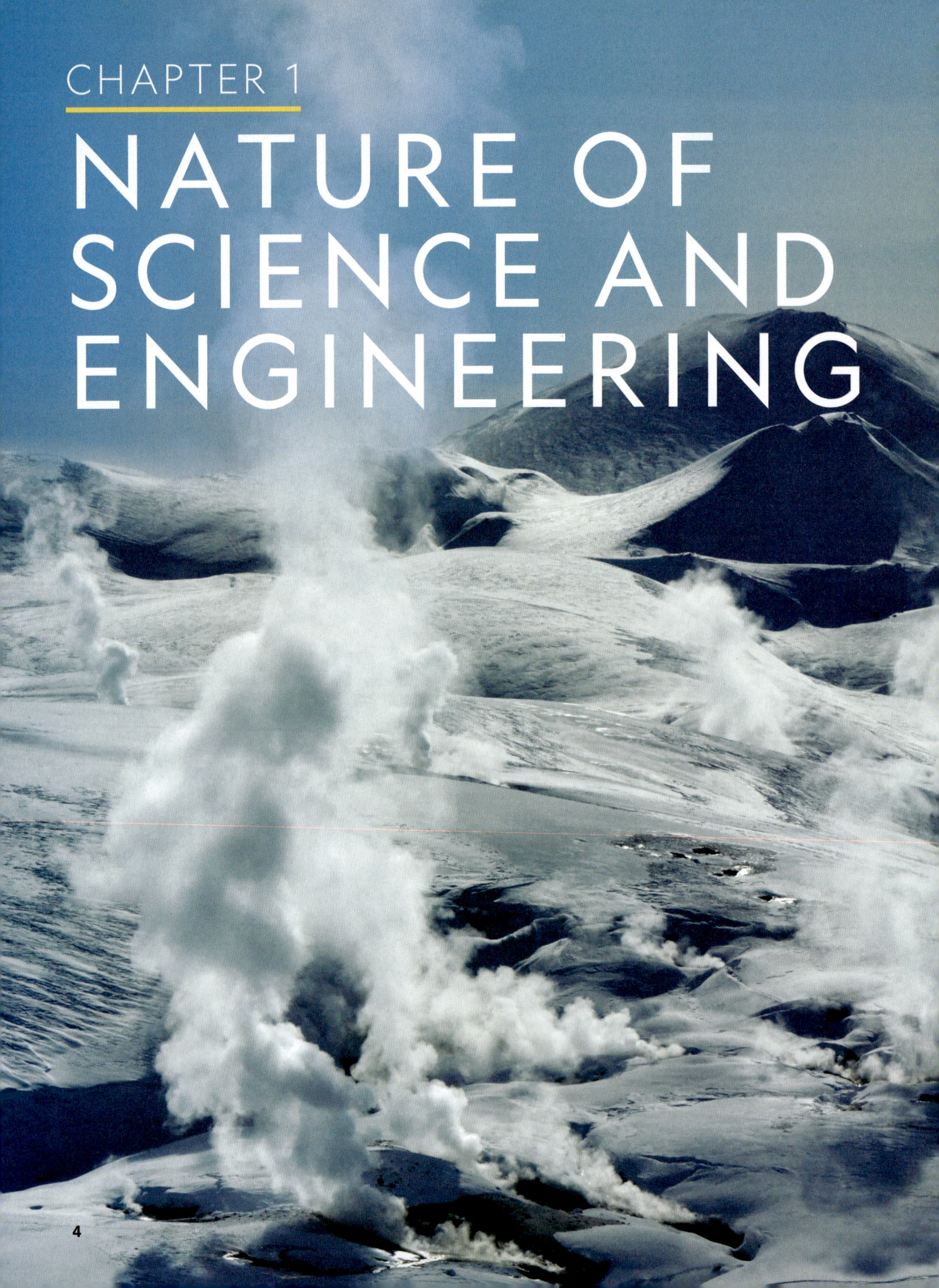

CHAPTER 1
NATURE OF SCIENCE AND ENGINEERING

The Land of Fire and Ice sounds like the title of a popular new fantasy novel. But it is also a common nickname for Iceland, a geologic oddity that has fascinated scientists for centuries. Iceland is a volcanic island that formed because of the action of two tectonic plates, massive slabs of Earth's crust that are pulling away from each other. Magnificent glaciers cover more than 10 percent of Iceland's surface. Yet in the Reykjadalur Valley, steam and boiling mud rise from the ground. The thermal pools in the Reykjadalur are replenished by rainfall and melting glaciers. Molten rock, called magma, below the surface heats up large reservoirs of underground water to temperatures as high as 400° Celsius (752° Fahrenheit). For Earth scientists, Iceland provides an up-close view of the complex processes that shape our dynamic planet. For engineers, Iceland's geothermal activity gives an opportunity for the development of reliable and sustainable clean energy.

KEY QUESTIONS

1.1 In what ways do the goals of science and engineering intersect?

1.2 How does the work of scientists and engineers contribute to the mapping of Earth's systems?

Steam rises from the natural thermal pools in Reykjadalur, Iceland. The name Reykjadalur means "steam valley" in Icelandic.

EXPLORERS AT WORK

THE WORLD IS MY CLASSROOM

WITH NATIONAL GEOGRAPHIC EXPLORER ANDRÉS RUZO

When Andrés Ruzo talks about his childhood, it is easy to understand why he became a geothermal scientist—someone who studies heat energy contained in rock and fluids beneath Earth's crust. Geothermal heat escapes at Earth's surface via land features such as hot springs, geysers, volcanoes, and fumaroles (openings near volcanoes that emit hot sulfurous gases). As a boy, Ruzo personally encountered some of these features in Nicaragua at his family's coffee farm atop the Casita Volcano.

Ruzo also grew up in Peru and remembers his grandfather telling him stories and legends, including one about a mysterious "boiling river" hidden in the Amazon. Ruzo earned an undergraduate degree in geology and become interested in geothermal energy. He had helped map out geothermal areas in the United States that could be tapped for clean, renewable energy (see Case Study) and hoped to do the same for Peru.

His curiosity led him to ask geologic experts whether such a "boiling river" could actually exist in the Peruvian jungle. Amazingly, Ruzo's aunt, who had worked with native Amazonian communities in the jungle when she was young, told him that she had seen this boiling river. One day she led him into the jungle to meet the powerful shaman, or holy man, who protected it. Ruzo's aunt explained that without the shaman's blessing, he would not be permitted to see the river because it was considered sacred. After two years of searching, he was finally able to confirm the river's existence.

The Boiling River of the Amazon is the world's largest documented thermal river. It flows hot for more than 6 kilometers. At its widest, the river is 30 meters across. Its deepest point is 4.5 meters. The hottest temperature Ruzo has measured at the Boiling River was 99.1°C. Ruzo and other scientists want to explain the cause of this incredible geothermal feature without a nearby volcano.

Ruzo's research is still ongoing. What is certain is that the Boiling River of the Amazon is a rare geothermal system. To understand it properly, Ruzo is traveling the world to study similar geologic sites that may give clues about how the Boiling River works. "As a geoscientist, the world is my classroom," Ruzo explains."

Recently, Ruzo traveled to Iceland, which has more than 30 active volcano systems that provide an abundant source of geothermal energy. Geothermal energy accounts for about one-third of electricity and some 90 percent of heating and hot water supplies in Iceland. Iceland even has its own thermal river called the Reykjadalur, which means "steam valley." Even though the Reykjadalur is next to a volcano, Peru's Boiling River is still bigger and hotter. Ruzo hopes to learn more about how such rivers form as well as their geologic, environmental, cultural, and economic importance. Perhaps most of all, he wants to protect them.

THINKING CRITICALLY

Infer How might collaborating with experts in government, conservation, and energy production be important to Ruzo's scientific research?

▲ Andrés Ruzo uses a thermal camera to take temperature measurements at a geothermal feature in Iceland.

◀ Ruzo's research takes him to the legendary Boiling River in the central Peruvian Amazon. He works closely with indigenous people to study and protect the river.

CASE STUDY
MAPPING GEOTHERMAL ENERGY

Below Earth's surface lies a clean and renewable source of energy that is always present, day or night. Geothermal energy is thermal energy produced by natural processes inside Earth's core and mantle. This energy can be accessed by drilling into Earth and using water to absorb the energy. Steam and hot water are then driven up to the surface (Figure 1-1). In general, the amount of available geothermal energy increases with depth. In some areas, there is sufficient geothermal energy just 50 to 100 meters underground to heat a home. Farther below the surface, there is enough heat to generate megawatts of electricity—enough to power a city.

Unfortunately, tapping into these large heat reserves can be challenging because they are not always the same depth below Earth's surface. In some regions, accessing geothermal sources would require drilling down many kilometers, a prospect that is expensive and difficult. In other regions, rich geothermal resources are located relatively close to the surface. For example, the island of Iceland lies above a huge, accessible geothermal resource that makes geothermal energy a feasible resource for electrical power generation.

The United States also has some areas with enormous geothermal potential. The challenge has been finding them. As a result, geothermal mapping has become an important first step for developers seeking promising sites for geothermal power plants. Working with a team at Southern Methodist University, Andrés Ruzo developed a comprehensive, detailed geothermal map of the United States. To create the map, Ruzo and his team members collected temperature data from thousands of oil, gas, mining, and water wells across the United States. They found that higher geothermal energy potential was correlated with higher temperatures in wells, and the opposite for colder wells. By noting the latitude and longitude coordinates of each site and correlating those with temperature values at specific depths, the team generated a series of layered maps. The result was a geothermal energy map (Figure 1-2).

Before Ruzo's team produced the geothermal map, it had been known for many years that states like Oregon, Idaho, and Nevada had great potential for geothermal energy development. However, the team's maps revealed that the states of Texas, Louisiana, and Arkansas had a surprising number of geothermal resources as well. Although these states have long been major sources of fossil fuel production, their geothermal potential may one day help them become a powerhouse of renewable energy.

As You Read Learn how scientists construct and use various kinds of maps. Consider the kinds of data that are needed to create maps, and how these data can be obtained. Determine what kinds of maps are most useful in the field of Earth science and learn how to use them.

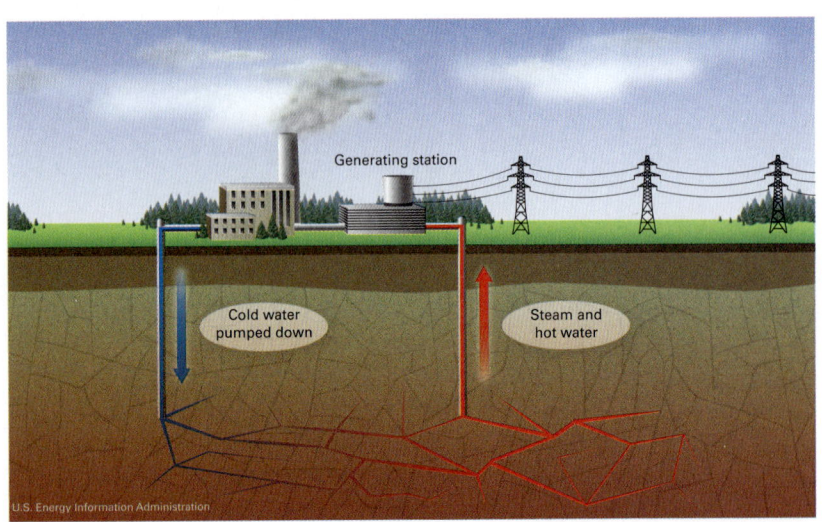

▶ **FIGURE 1-1**
One method of obtaining electricity from geothermal energy is to pump cold water below ground, where it absorbs heat from Earth's interior before being piped back to the surface as steam or hot water that can be converted into steam. The steam turns large generator turbines, which produce the electricity.

🌐 **FIGURE 1-2**
Go online to view a map of geothermal resources in the conterminous United States.

1.1 THE NATURE OF SCIENCE AND ENGINEERING

Core Ideas and Skills
- Compare the process of scientific inquiry with the engineering design process.
- Differentiate between correlation and causation.
- Describe the purpose of models in science and engineering.
- Explain the differences between hypotheses, theories, and laws.

KEY TERMS

science	law
data	engineering
model	trade-off
hypothesis	criteria
experiment	constraint
peer review	optimization
theory	

The Nature of Science

From biology and chemistry to Earth and space sciences, all branches of science start with curiosity about how things work. **Science** is both a systematic set of practices that investigate the physical world through observation and experimentation, and an organized body of knowledge based on testable explanations and predictions. The nature of science refers to the common characteristics, values, limitations, and assumptions found in all branches of science (Table 1-1).

Science is a way of knowing. Many other ways of knowing involve personal or spiritual beliefs, emotions, and personal perceptions of the world. Science is distinguished by its reliance on empirical (observable) evidence, logical arguments, and critical thinking. A key premise of science is that the universe is governed by natural laws that operate in a consistent manner over space and time. Understanding these laws allows scientists to make sense of phenomena and to make predictions about natural events that occur under specific conditions.

Like all ways of knowing, science has limitations. Science limits itself to addressing questions about the natural world and the material world; it does not attempt to answer all questions. Science generally attempts to observe and explain how things work, not necessarily how they *should* work. In other words, scientific arguments rely on empirical evidence, which is evidence that can be observed, measured, or collected. It steers

TABLE 1-1 The Nature of Science

Science is a way of knowing.
Scientific knowledge assumes an order and consistency in natural systems.
Scientific knowledge is based on empirical evidence.
Scientific models, laws, mechanisms, and theories explain natural phenomena.
Science addresses questions about the natural and material world.
Science is a human endeavor.
Scientific investigations use a variety of methods.
Scientific knowledge is open to revision in light of new evidence.

FIGURE 1-3
A geologist examines 250-million-year-old fossilized tracks of reptiles on a ledge in Grand Canyon National Park, Arizona.

LESSON 1.1

clear of questions that involve ethics, values, belief systems, or opinions because none of those can be empirically observed. Although scientific knowledge can be used to inform important decisions that impact people, most decisions are not based on science alone.

In spite of its focus on empirical evidence, science is a human endeavor that is influenced by the social and cultural backgrounds of scientists themselves. The accumulation of scientific knowledge is only possible because of the curiosity, creativity, innovation, and critical thinking of people from many different nations and cultures. As a human endeavor, the needs of society drive scientific research and affect its findings.

Scientists use a diverse range of tools and methods to carry out scientific investigations and collect **data**, or information, about the world. As new tools, methods, and technologies are developed, scientists often make discoveries that shed light on their previous explanations and understandings. As a result, scientific knowledge evolves over time as scientists refine their theories and explanations. When new evidence is uncovered, scientists must be open to revising their thinking.

checkpoint How does society affect scientific research?

What Is Earth Science?

Imagine walking through Iceland's Reykjadalur Valley and observing steam rise from the ground. This phenomenon would likely spark your curiosity, and you would probably look around for a logical explanation for the steam. Perhaps you would look for a nearby manhole cover or construction equipment to determine whether the steam was coming from human activity. Once you ruled that out, you might do research into the geology of the region to explain this steam. This process of using logic, critical thinking, and evidence to explain an observation is similar to the processes scientists use to understand and explain the world around us.

Earth science is the study of Earth and its evolution through time. There are several major branches of Earth science (Table 1-2); each branch contains numerous scientific fields. The scientists in any given field often work closely with scientists in other fields. For example, Iceland is a geologically unique location of interest to scientists working in many fields of Earth science. A volcanologist in Iceland who is studying a certain

TABLE 1-2 Major Branches of Earth and Space Sciences

Geology	The study of Earth's internal processes and landforms
Hydrology	The study of Earth's groundwater and surface water systems
Atmospheric Science	The study of Earth's atmosphere
Environmental Science	The study of the relationships between the physical and chemical components of Earth's environments
Astronomy	The study of objects and phenomena above Earth's atmosphere

volcano might collaborate with a hydrogeologist who is an expert on the behavior of groundwater and hot springs near the volcano. An atmospheric scientist such as a climatologist can study the emissions from Iceland's volcanoes and how they affect Iceland's weather and climate. Environmental scientists and biogeologists study the interactions between systems, such as how volcanic activity affects ecosystems or human transportation systems.

Astronomy, the study of objects and phenomena beyond planet Earth, connects closely to the other branches of Earth science. To better understand Earth's systems, scientists must also understand the planet's formation and its place in the universe. Planetary astronomers use what we know of Earth's geologic and atmospheric processes to better understand other planets and moons, looking for similarities and differences in their composition and formation. Because Earth is the only planet known to contain liquid water and support life, these comparisons can help identify other worlds that may be habitable. Space scientists have worked with Earth scientists to identify training sites for astronauts. Neil Armstrong and other astronauts from the Apollo missions (the first successful landings of humans on the moon in the 1960s and 1970s) trained in Iceland because its unusual terrain is similar to the natural landforms and structures on the moon. Solar physicists study the surface of the sun and monitor how solar activity impacts the atmosphere and life on Earth.

checkpoint Why does a better understanding of Earth's systems help planetary astronomers?

Scientific Inquiry

Studies of Icelandic volcanoes have identified a relationship between climate and volcanic activity. Over time frames of hundreds of years, high average temperatures in Iceland are associated with high volcanic activity in Iceland. This observation is an example of a positive correlation, which occurs when two variables change in a similar way (both variables increase or both decrease) over the same time period. A negative correlation, the opposite of a positive correlation, is when two variables change in opposite ways (one increases and the other decreases). For example, the land area covered by glacial ice in Iceland is negatively correlated with volcanic activity. A decrease in glacial coverage occurs with an increase in volcanic activity.

Scientists have investigated various explanations in order to identify the *causes* of the correlations between temperature, glacial ice, and volcanism in Iceland. For example, some scientists have hypothesized that climate-driven melting of glacial ice during warm periods causes Earth's surface to rise vertically, or rebound, because the crust is no longer pushed downward by the large mass of ice. The rebound, in turn, allows hot mantle rock in the area to flow to shallower depths.

As in all branches of science, it is important to distinguish between correlation and causation. Some correlations occur because there is indeed a cause-effect relationship between the two variables; but other correlations occur even when there is no causal relationship. In other words, correlation does not reliably imply causation. To show causation, scientists must demonstrate that there is a mechanism for one variable to effect change in another variable.

Scientific inquiry often involves identifying correlations and other patterns, and then investigating and testing possible causes that might explain the correlation. In conducting scientific investigations, scientists use a variety of methods that revolve around a set of common practices summarized in Figure 1-4. The practices are not a linear, step-by-step method for conducting research. Instead, they are a set of activities that can be undertaken in any order and adapted to suit

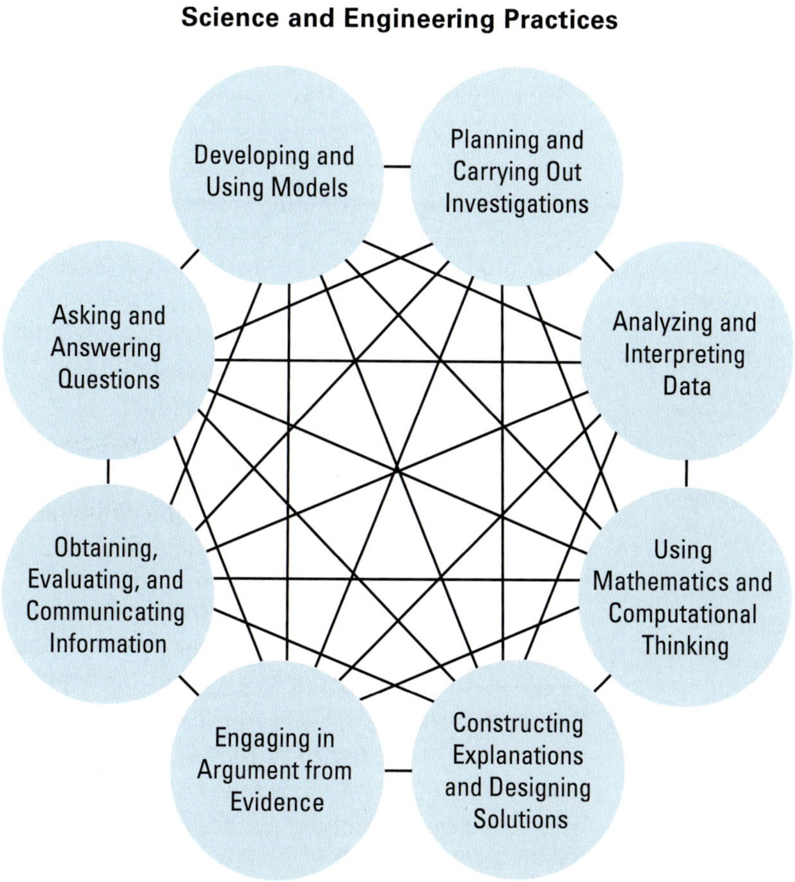

Science and Engineering Practices

- Developing and Using Models
- Planning and Carrying Out Investigations
- Analyzing and Interpreting Data
- Using Mathematics and Computational Thinking
- Constructing Explanations and Designing Solutions
- Engaging in Argument from Evidence
- Obtaining, Evaluating, and Communicating Information
- Asking and Answering Questions

FIGURE 1-4
Scientific inquiry follows a set of eight practices that are employed continuously and often in different order. New discoveries often lead to new questions and unexpected new lines of inquiry.

Source: National Research Council. 2012. *A Framework for K-12 Science Education: Practices, Crosscutting Concepts, and Core Ideas.* Washington, DC: The National Academies Press.

DATA ANALYSIS Distinguish Between Correlation and Causation

One popular belief is that animal behavior is influenced by phases of the moon. To determine if there may be any evidence to support these legends, a group of scientists collected data from a hospital in England on patients bitten by animals. Using the dates when animal-bite patients were treated over a three-year period, they compiled the following data for each phase of the moon. Use the information in the table to answer the questions that follow.

Computational Thinking

1. **Represent Data** Construct a graph that can be used to look for evidence of any correlations between phases of the moon and number of animal bites treated.

2. **Pattern Recognition** Examine your graph. Do you observe a correlation in the data? Is the correlation positive or negative? Explain.

Critical Thinking

3. **Evaluate** Based on the data presented here, can you conclude that any particular phase of the moon is the cause for greater or fewer animals biting humans? Explain.

4. **Critique** How is the study weakened by the fact that the data represent numbers of people treated for animal bites rather than the numbers of people bitten by animals?

5. **Design** Would it be possible to design a research study to gather evidence for or against a causal relationship between moon phases and the number of animals biting humans? How could you design a study to gather more correlational evidence?

TABLE 1-3 Patients Treated for Animal Bites during Different Phases of the Moon

Phase of Moon	Lunar Days	Number of Bites Treated
New Moon	14, 15	110
Waxing Crescent	16, 17, 18	137
Waxing Crescent	19, 20, 21	150
First Quarter	22, 23, 24	163
Waxing Gibbous	25, 26, 27	201
Full Moon	28, 29, 1	269
Waning Gibbous	2, 3, 4	155
Waning Gibbous	5, 6, 7	142
Last Quarter	8, 9, 10	146
Waning Crescent	11, 12, 13	148

the needs of a particular investigation. This reflects an understanding that the process of science does not have a definitive start and end point. It is an ongoing cycle of discovery, testing, evaluation of new evidence, retesting, and new discoveries that sometimes lead scientists in new directions.

checkpoint What is the difference between causation and correlation?

Models in Science and Engineering

Models are particularly important in science and engineering. A **model** is any representation that serves as a tool for thinking about or visualizing natural phenomena, or to develop solutions to problems. Good models are designed and constructed so that they can be shared, revised, and improved. Models are a powerful tool for observing or analyzing complex problems, and for studying phenomena or systems that are too small, too large, too fast, or too slow to observe directly. Examples of models include explanatory text, diagrams, graphs, physical replicas, computer simulations, and mathematical representations.

Scientists mainly use models to explain phenomena and make predictions, whereas engineers use models to design and test solutions. For example, scientists have developed computer models that simulate Earth's atmosphere. Using data from storm behavior in the past, scientists can make predictions about storm behavior in the future and help people plan for severe weather. During the design phase of an engineering project, engineers may draw scale models of systems or tools that are proportional to the measurements of the full design. Later, they generally build a prototype that can be tested against the criteria. During and after testing, the prototype is evaluated for any problems or design weaknesses. Often, a prototype must be refined and improved many times until it meets the criteria for success. A commonly recognized example of this iterative process is the work of the Wright Brothers, who refined their prototype of an airplane, modifying the wings, the air frame and other parts of the plane, until they achieved success with first flight.

Models in science are used to explain phenomena and are consistent with all available evidence. But models also have limitations and are subject to change or abandonment when new evidence is found. One famous example is the study of Earth's place in the universe. The published telescopic observations of Galileo forced the scientific community to reconsider the long-held model that Earth was the center of everything.

Scientists often evaluate the merits and limitations of a model. Sometimes, for a model to be simple and understandable, it needs to be an approximation of a phenomenon. This occasionally requires information to be excluded. Trade-offs, where one aspect is accurately portrayed while another is given less priority, are common. For example, the sizes of and distances between planets span such a large range that it is difficult to create a scale model in a classroom that is accurate with respect to both size and distance. If one model accurately shows the distances between planets, the sizes of the planets would be so small that they would not be clearly visible. If another model shows the sizes of the planets in relation to each other, their scale distances need to be approximated or the model would not fit in the classroom.

Often, scientists make different types of models about a single phenomenon and engineers make different types of models about a single solution. Models of different types can be used to approach a science explanation or engineering solution in different ways. For example, physicists can describe certain properties of light using a model where light behaves like particles, yet other properties of light can only be modeled if light is considered to act like waves. Solutions to complex engineering problems often require that models be created for different-size scales or for separate parts of the solution. The best solution is then a combination of the best models for each part.

Models are clear, shareable understandings about how things work. Scientists and engineers can revise and develop them to become more and more accurate representations of a system, phenomenon, or solution. In Earth science, models can be used to explain phenomena and make predictions, even anticipate and solve problems. The horizontal and vertical motion of the ground surface during an earthquake is well-documented and studied by scientists. Scientific simulations can predict the severity of this motion in different scenarios. Engineers can use these predictions to develop earthquake-resistant buildings that can withstand the forces involved.

checkpoint What do scientists do when new evidence conflicts with a model they have already developed?

Scientific Theories and Laws

Scientific inquiry typically builds on previous scientific knowledge. However, because scientific inquiry is ongoing, the scientific body of knowledge is always changing and evolving.

Scientific investigation often starts out with a **hypothesis**, a possible explanation for a phenomenon that can be tested to determine whether it is supported by evidence. One way to test a hypothesis is by conducting an **experiment**, a controlled procedure that produces data. The data can be analyzed to look for relationships that provide evidence to support or refute a hypothesis. Experiments generally involve two or more variables that are believed to be related. An *independent variable* is a variable that is intentionally controlled or manipulated during an experiment in order to observe and measure changes in other variables, called *dependent variables*. Researchers' observations and measurements provide empirical (observable) evidence that can be used to support, refute, or refine a hypothesis. Typically, many repeated tests are needed to confirm a hypothesis. Scientific investigations must be validated through the process of **peer review**. During peer review, scientists in the same field or a related field review the research of others. This process gives these scientists an opportunity to evaluate the validity of a study's design, data, analysis, and conclusions.

Whereas a hypothesis is an explanation that must be tested and verified, a scientific **theory** is a broader explanation that is already supported by a large body of evidence. A theory typically explains a wide range of phenomena and may incorporate many different hypotheses. Some aspects of a theory may be difficult to demonstrate or may not yet be completely understood, but a theory provides a framework for understanding that can be tested and refined as new evidence becomes available. In fact, scientists often make and test new hypotheses with the goal of refining, clarifying, or challenging parts of a theory.

LESSON 1.1

MINILAB Puzzle: Law Versus Theory

Materials

envelope containing 100 pieces of a 1,000-piece jigsaw puzzle
poster paper
marker

Procedure

1. With your group, discuss and define each of the following terms: *evidence, hypothesis, theory, scientific law.* Record the definitions on a sheet of poster paper.
2. Remove one puzzle piece from the envelope. Examine and use the piece to discuss ideas about what the completed puzzle will show. As a group, decide on one idea and record it.
3. Remove 10 more puzzle pieces. Use the information to refine your idea about the completed picture. Record refinements to your idea.
4. Remove 15 more puzzle pieces and repeat step 3, recording any changes or refinements to your idea.
5. Remove 20 more puzzle pieces, and repeat step 3 to develop a final statement about what the completed puzzle picture will show. Record your final statement.

Results and Analysis

1. **Identify** How can you use the terms *evidence, hypothesis,* and *theory* to explain the process you carried out during this exercise?
2. **Relate** How does this exercise model the development of scientific theories?
3. **Conclude** Was the concept of a scientific law modeled during this exercise? Explain.

Critical Thinking

4. **Evaluate** Review your definitions from step 1 of the procedure. Revise any that you feel were inaccurate or could use improvement.
5. **Apply** In the past, people believed that the sun, the moon, and the planets revolved around Earth. How do we think about these ideas today? Use the results of this exercise to explain why people thought about Earth as the center of the universe and the process that changed that thinking.

Unlike theories, scientific **laws** are descriptions of phenomena that are universally accepted as true. Laws, which are often expressed mathematically, describe or predict a range of natural phenomena that can be assumed to always exist or occur under given conditions. Laws are based on repeated experimental observations, and they have never been shown to be false under the specified conditions. Laws do not try to explain the causes of observed phenomena, they simply describe observations consistently. Newton's laws of motion are considered laws because they can be reliably used to predict the behavior of moving objects.

How Theories Develop Scientific theories often develop from many lines of scientific inquiry and change as new evidence becomes available. For example, most scientists today agree that dinosaurs, along with many other land and marine organisms, died off during a mass extinction event that occurred roughly 65 million years ago. This event, known as the K–Pg extinction, marked the end of the Cretaceous period and the beginning of the Tertiary period. But what exactly caused the K–Pg extinction is the focus of much ongoing research.

According to one theory, a large asteroid collided with Earth, wiping out most of Earth's large vertebrates in a short period of time. Another theory attributes the K–Pg extinction to the Deccan Traps, massive volcanoes in present-day India that altered Earth's climate and poisoned its ecosystems by pouring toxins and greenhouse gases into the atmosphere over hundreds of thousands of years. Still other theories hold that both the asteroid impact *and* the Deccan Traps contributed to the mass extinction, or that perhaps the asteroid even caused or worsened the volcanic eruptions.

All of these theories have grown out of evidence gathered by scientists across many different fields over several decades. Interestingly, the initial spark came in the 1970s from a geologist named Walter Alvarez who was studying plate tectonics—a field seemingly unrelated to dinosaurs and mass extinctions. Alvarez and his team were studying a rock formation in Italy when they made an interesting find. Between two layers of limestone was a thin, one-centimeter layer of clay that would soon become known as the K–Pg boundary because it dated to 65 million years ago, coinciding with the K–Pg extinction event. Alvarez found that the limestone layers above and below the clay boundary had microscopic fossils of single-celled organisms called foraminifera, but the clay layer had no fossils at all. Alvarez also noticed that the foraminifera in the rock layers above the clay boundary were less diverse and smaller than those in the rock layers below. Within several years, other scientists had discovered this same phenomenon in other locations around the world. This discovery prompted Alvarez to ask some new questions:

- How long did the clay layer take to form, and what evidence could the clay provide about the events that led to its formation?
- What caused the foraminifera to change so dramatically, and was it the same thing that caused the extinction of the dinosaurs?
- Did whatever caused the change in the clay layer happen suddenly or gradually?

Alvarez turned to his father, Luis, a renowned physicist, to help answer these questions (Figure 1-5A). They devised a method to calculate how long it took the clay layer to form by analyzing levels of iridium in the rock layers. Iridium is a metal that is rare in Earth's crust but commonly found in meteors, which scatter iridium dust across Earth at a relatively constant rate. Alvarez expected that the amount of iridium in the clay rock layer would help answer the question of how long it took the layer to deposit. However, the levels of iridium turned out to be more than 30 times higher than expected. This led to new questions:

- What could have caused such unusually high levels of iridium in this rock layer?
- Could the high levels of iridium in this rock layer be related to the mass extinction?

The Alvarezes proposed that the high levels of iridium must have come not from meteors but from a much larger extraterrestrial source, such as an asteroid. They also suggested that an asteroid large enough to scatter iridium all over Earth's surface must have been large enough to cause major changes in Earth's environments, enough to result in a mass extinction.

Many scientists were initially skeptical of Alvarez's arguments, which led to a flurry of new research in many fields. Biologists and paleontologists examined the fossil record to determine whether other species besides dinosaurs and marine animals had also gone extinct at the K–Pg boundary. Physicists and geologists scoured Earth for evidence left by the physical impact of an asteroid. Chemists analyzed atomic and molecular evidence to determine whether other chemicals

A

B

FIGURE 1-5
(A) Luis and Walter Alvarez pose by the K–Pg boundary between layers of rock in Bottaccione Gorge, Italy. This thin, clay rock layer holds clues to the causes of a mass extinction that occurred some 65 million years ago. (B) A massive, 180-kilometer-wide crater buried in the sediments of the Yucatán peninsula and the Gulf of Mexico suggests that a massive asteroid struck Earth 65 million years ago, likely contributing to the mass extinction event.

commonly found in asteroids were present in the K–Pg clay rock.

Some scientists pointed to the Deccan Traps in India as an alternative explanation for the high levels of iridium. Because iridium is believed to be present in high levels in Earth's core, they reasoned that a period of severe volcanic activity could have been responsible for releasing large amounts of iridium onto Earth's surface.

In the early 1990s, new evidence emerged again when scientists found Chicxulub, a 180-kilometer-wide crater that was formed 65 million years ago in Mexico's Yucatan Peninsula (Figure 1-5B). This discovery seemed to support Alvarez's theory that an asteroid impact had caused the K–Pg extinction. But it did not necessarily rule out other possible causes. Many questions remain about the causes of the K–Pg extinction. Even today, many scientists search for answers to those questions by seeking out the K–Pg boundary layer wherever it might be preserved and testing new hypotheses. No doubt this will lead to new discoveries that will further expand our scientific knowledge and lead to a more complete understanding of Earth's geologic history.

checkpoint What is the difference between a theory and a law?

Engineering Design

Engineering is a field that applies scientific knowledge, mathematics, and human innovation to design solutions to problems. Whereas the primary goal of science is to explain phenomena in the natural world, the goal of engineering is to design systems and processes that help solve problems. Engineers and scientists rely on one another's work to advance their fields of study. Engineers use scientific knowledge and mathematics to find solutions to problems, and scientists use tools, devices, and processes developed by engineers to gather more accurate data.

Like scientists, engineers rely on many of the same scientific practices shown in Figure 1-4, but with a greater emphasis on defining problems and designing solutions. The three components of the engineering design process are defining problems, developing solutions to these problems, and optimizing the design solution (Figure 1-6).

Defining Engineering Problems
Defining a problem means describing it as clearly and in as much detail as possible. The problems must be

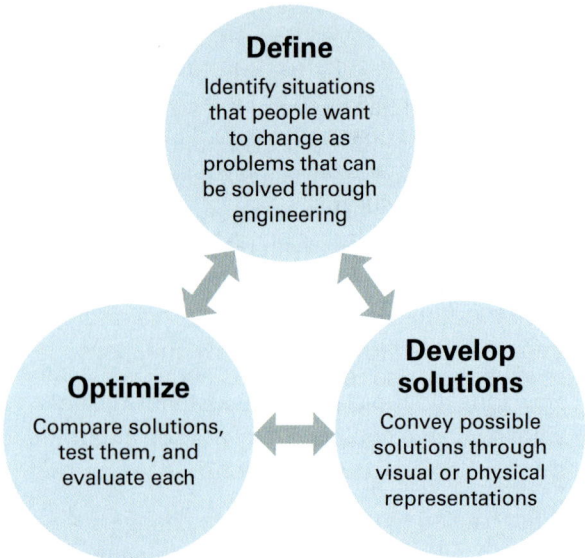

Source: National Research Council. 2012. *A Framework for K-12 Science Education: Practices, Crosscutting Concepts, and Core Ideas.* Washington, DC: The National Academies Press.

FIGURE 1-6
The engineering design process is an iterative process characterized by three basic components: defining problems, developing solutions, and optimizing solutions.

focused. It is important for engineers to carefully consider all aspects of a problem, such as "What needs or desires will solving this problem satisfy?", "What goals would a successful solution to the problem achieve?", and "What are the possible **trade-offs**, or compromises, that a solution might require if all the goals cannot be achieved at the same time?"

Defining a problem also involves identifying **criteria** and **constraints**. Criteria are the standards that a solution must meet to be considered successful. Constraints are factors that will limit the design process and the possible solutions to the problem. Political, financial, environmental, social, and ethical considerations can all act as constraints that place limits on possible solutions to a problem.

For example, engineers in Iceland face some unique challenges for meeting the island nation's energy needs. Iceland is geographically isolated from much of the world and lacks its own reserves of coal and other fossil fuels, but Icelanders need energy to power homes, businesses, automobiles, and other technology essential to maintaining their standard of living. Importing fossil fuels is expensive because of the island's remote location, and the burning of fossil fuels also creates undesirable pollution. These are all aspects of an engineering problem with many possible solutions.

Defining the problem requires establishing clear and measurable criteria, such as how much energy needs to be generated and how many homes and buildings need to be powered. Because these criteria are quantifiable, they can be used to evaluate whether a particular design meets the standards for success. The constraints in this scenario include Iceland's geography, geology, and access to natural resources. These are constraints within which all possible solutions must operate. Economic, political, and environmental considerations also represent constraints because some solutions might be too expensive, unpopular, unethical, or detrimental to the environment to gain the support of Icelanders.

Developing and Optimizing Solutions

Once a problem has been defined, engineers brainstorm many possible solutions and then decide which solutions they will try to develop. This process typically involves making value judgments, as engineers weigh each possible solution against the constraints and criteria. It may also involve breaking large problems down into smaller problems that can be solved individually.

To design and test a solution, engineers apply many of the same practices used in scientific inquiry: developing and using models; planning and carrying out investigations; analyzing and interpreting data; using mathematics and computational thinking; engaging in argument from evidence; and obtaining, evaluating, and communicating information. Before any solution is implemented, it goes through **optimization**, a systematic process of testing, refining, and re-testing in order to maximize the benefits of the solution and eliminate flaws. Optimization may also mean deciding which trade-offs are acceptable and which criteria are most important. For example, if limiting the cost of a solution is the most important criterion, engineers may decide to compromise the quality of a product by using less expensive materials. If quality is the most important criterion, engineers may decide to accept higher costs if they are necessary to achieve the highest-quality product.

The processes of defining, developing, and optimizing solutions may be lengthy, sometimes lasting decades. In Iceland, engineers decided to pursue the development of geothermal power as a solution to the island's energy problem. In contrast to its lack of natural fossil fuel deposits, Iceland's unique geology makes it one of the best places on Earth to tap into the heat energy below Earth's surface. As early as 1947, scientists began conducting research at Nesjavellir, Iceland, to determine whether it was possible to produce enough electricity from geothermal activity. Research at the site continued between 1965 and 1986, and construction on the plant finally began in 1987. Today, the plant is one of the largest geothermal plants in the world (Figure 1-7).

checkpoint What is the difference between a criterion and a constraint?

FIGURE 1-7
The Nesjavellir Geothermal Power Station is located near volcanoes in Iceland.

TABLE 1-4 Crosscutting Concepts in Science and Engineering

Crosscutting Concept	Description
Patterns	A pattern repeats in a predictable way. Patterns can be observed and used to prompt research questions, identify engineering problems, analyze data, or refine solutions.
Cause and Effect	Every phenomenon has a cause. Cause-and-effect relationships can be investigated by scientists and engineers through controlled tests and experiments. In engineering, the goal is often to design a system that causes certain effects.
Scale, Proportion, and Quantity	Both scientists and engineers need to consider how changes in time, size, or quantity can affect a natural process or a solution to a problem.
Systems and System Models	A system is made up of parts that interact to form a complex whole. Systems have boundaries, inputs and outputs, and feedback mechanisms. Positive feedbacks intensify change within a system while negative feedbacks reduce or limit change. System models of natural and artificial systems can be used to gather information about a system or to test effects on a system.
Energy and Matter	Systems consist of matter and energy. Energy and matter cannot be created or destroyed but can change forms as they flow through a system. Observable phenomena in Earth systems are the result of the flow of energy and the cycling of matter through and between systems.
Structure and Function	In both natural and artificial systems, the shape and structure of an object relates to its function. In Earth systems, sometimes the structure of an object or feature can be used to describe or explain the processes that formed it. In other cases, the opposite is true—observing a process helps scientists predict or understand the structures or features that form as a result of the process.
Stability and Change	Both natural and artificial systems are subject to change. The investigation of the factors that cause change and instability is crucial to understanding systems and for designing solutions to problems.

Source: National Research Council. 2012. *A Framework for K-12 Science Education: Practices, Crosscutting Concepts, and Core Ideas.* Washington, DC: The National Academies Press.

Crosscutting Concepts

Science and engineering depend on each other. Engineers use scientific knowledge to identify criteria and constraints and to propose potential solutions to problems, and scientists use tools designed by engineers to conduct scientific research. Scientists and engineers also recognize concepts that apply across all domains of science—called crosscutting concepts—to help them analyze, interpret, and make sense of their observations, data, and results. These concepts are described in detail in Table 1-4.

checkpoint Explain how the crosscutting concepts relate to scientists and engineers.

1.1 ASSESSMENT

1. **Contrast** Describe some similarities and differences between science and engineering.
2. **Contrast** Differentiate between a hypothesis, a theory, and a law.
3. **Describe** What role has new evidence played in developing the theory that an asteroid was responsible for the K–Pg mass extinction event?
4. **Explain** Select one scientific practice and one crosscutting concept and explain how each was applied by scientists and engineers to develop geothermal power in Iceland.

Critical Thinking

5. **Synthesize** Describe two examples of scientific or engineering models used in everyday life and explain how each can be used to represent a phenomenon, make predictions or address a problem.
6. **Apply** Explain one engineering advancement that would lead to further scientific understanding of the extinction of the dinosaurs.

1.2 MAPPING EARTH'S SYSTEMS

Core Ideas and Skills
- Explain how maps are used to represent data.
- Differentiate between various types of maps and how they are used.
- Analyze and interpret data represented on a topographic map.
- Explain how science and engineering contribute to the mapping of Earth and other planets.

KEY TERMS

isoline   topography

Maps as Representations of Data

In a general sense, a map is a visual representation of data that illustrates some portion of the world or universe as humans understand it. To create a map, people collect information, or data, and decide how to represent those data visually. The concept of mapping is not new. In fact, people have been making maps for thousands of years. However, today's maps are much more sophisticated and accurate than those of the distant past—and even the recent past—because scientists and engineers have developed better tools and technologies for collecting and representing data.

For example, the earliest surviving "world map" is the *Imago Mundi*, a Babylonian map that dates back to 700–500 B.C.E. Drawn on a clay tablet, *Imago Mundi* showed Babylon and the surrounding Euphrates River valley, bounded by a "Salt Sea" (Figure 1-8A). Today, we recognize this region as part of the modern-day country of Iraq, a small part of the world. But to the Babylonians, this map represented their world based on the limited data they were able to gather and their limited means of modeling those data.

Like the *Imago Mundi,* ancient maps were often drawn from personal experience or secondhand information from travelers. But maps improved as people developed better and more precise tools for navigation. Many ancient cultures used the patterns of the sun, the stars, and prevailing winds to determine direction. However, by the 12th century C.E., sailors in China and Europe had already started to use magnetic compasses to navigate in the four cardinal directions: north, south, east, and west. In the centuries that followed, improved sailing and navigating techniques allowed explorers to sail to distant parts of the world that were previously difficult or impossible to reach. As they discovered new continents, islands, and seas, they used these new data to construct better maps for navigation. Most modern maps still use compass symbols (compass rose) to indicate the four cardinal directions.

FIGURE 1-8A
Go online to view the oldest surviving map of the world.

FIGURE 1-8B
This interactive road map uses global positioning system (GPS) technology to help with navigation. Modern maps are the product of advancements in science and engineering.

checkpoint What was one important innovation that led to a more complete world map?

Map Projections

One of the most important innovations of cartography (mapmaking) was the use of a coordinate system to describe the position of a location on a map. Coordinate positions given in latitude and longitude essentially apply a grid to Earth's surface. The grid is fixed at Earth's North Pole and South Pole, those points on Earth's surface through which Earth's rotational spin axis projects. On Earth, lines of latitude run east-west. The latitude at the Equator is 0 degrees, and latitude increases to 90 degrees at each of the poles. Longitude lines run north-south and extend from pole to pole. The Greenwich prime meridian and its antimeridian, commonly called the international date line, divide Earth into

LESSON 1.2 **19**

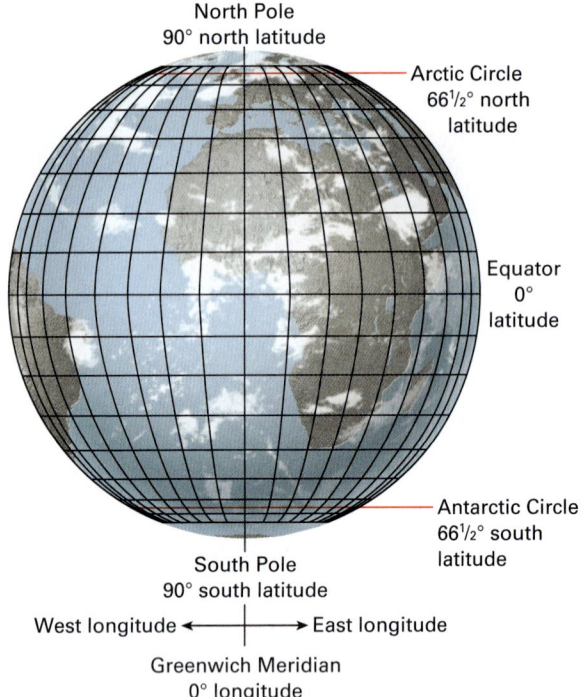

FIGURE 1-9
This diagram shows the lines of latitude and longitude on Earth's surface. Latitude and longitude allow Earth scientists and others to identify a location on a spherical Earth.

two hemispheres: the Eastern Hemisphere and Western Hemispheres (Figure 1-9).

The location of any point on Earth's surface can be expressed in degrees latitude north or south of the Equator and degrees longitude east or west of the prime meridian. Figure 1-10 shows how the latitude and longitude for a point of interest are measured. Both latitude and longitude are measured in degrees and both correspond to the angle formed between the point of interest and the Equator (for latitude) or prime meridian (for longitude). On Earth, degrees can involve large distances. Degrees are further broken down into smaller units of minutes and seconds to provide more exact locations on the globe. One degree of latitude is about 111 kilometers, one minute about 1.85 kilometers, and one second about 30.8 meters. Degrees, minutes, and seconds of longitude vary by location because longitudinal lines converge at the poles. For example, at 38 degrees north latitude, one degree of longitude is about 87.6 kilometers. At 68 degrees north latitude, the same one degree of longitude is about 41.6 kilometers. Modern GPS devices can typically pinpoint the latitude and longitude of a specific location down to the second.

Creating a Two-Dimensional Map of Earth

Have you ever tried to peel the skin off an orange in one piece? One way to do it is by piercing the top of the orange and peeling the skin in sections. If you flattened this peel out, it would have an irregular shape, and there would be gaps in the peel. A similar problem occurs when cartographers try to project Earth's curved surface onto a flat map. Gaps left in the map would suggest that there are gaps in Earth's surface where there are none; but filling in the gaps results in some distortion of size, shape, direction, or distance.

Different methods of map projections are used to address the problem of representing a three-dimensional Earth on a two-dimensional plane. Mercator projections (Figure 1-11)

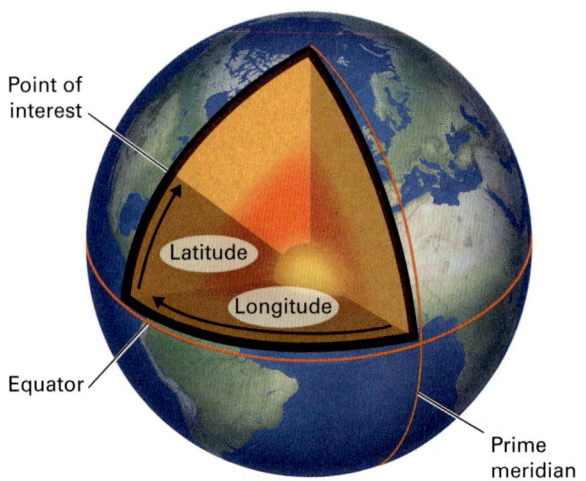

FIGURE 1-10
Latitude is a measurement of the angle produced by a point of interest and the Equator. Longitude is a measurement of the angle made by the point of interest and the prime meridian.

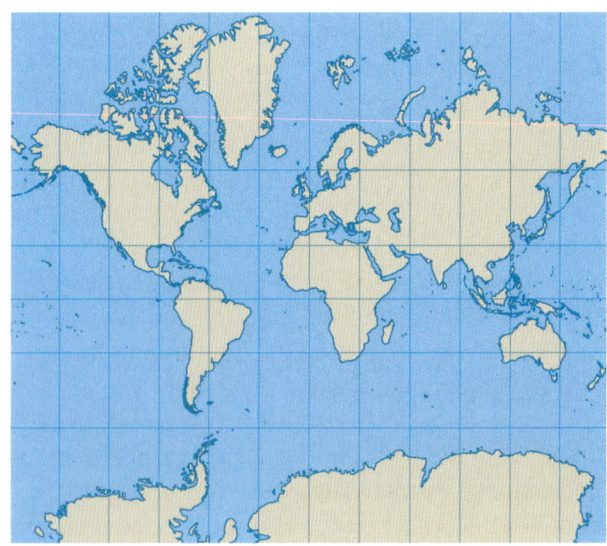

FIGURE 1-11
On Mercator projections, lines of longitude do not converge at the poles, and lines of latitude are not equidistant from one another.

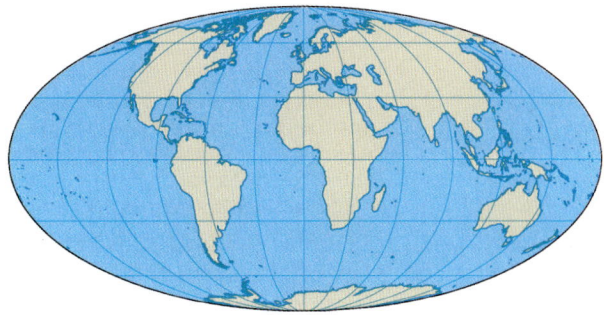

FIGURE 1-12
Mollweide projections show accurate proportions of area, but they distort shape and direction.

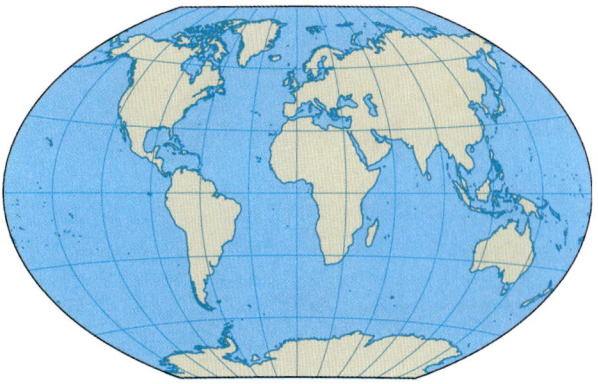

FIGURE 1-13
Winkel tripel distortion maps minimize distortion of area, distance, and direction.

were first introduced by Flemish geographer Gerardus Mercator in 1569; they portray the world as a grid with latitude and longitude lines forming 90-degree angles. The chief disadvantage of Mercator projections is that they distort sizes and shapes, especially near the poles. Lines of longitude appear to remain parallel rather than converging near the poles, and lines of latitude appear to become farther apart rather than remaining equally spaced near the poles. As a result, landmasses like Greenland and Antarctica appear much larger on Mercator maps than they are on the globe.

Mollweide projections (Figure 1-12) represent Earth as an ellipse. Lines of longitude converge at the poles, but the prime meridian is inaccurately represented as being approximately half the length as the Equator. These projections show more accurate proportions of the area of landmasses, but the shapes of the continents and direction are distorted. Scientists use Mollweide maps for representing global distributions because they show accurate proportions of area.

Winkel tripel projections (Figure 1-13) minimize distortion in three map elements: area, distance, and direction. Although there is still some distortion, the amount of distortion is less as compared to other map projections. On Winkel tripel projections, lines of latitude and longitude appear curved and nonparallel.

checkpoint How can you use a coordinate system to identify locations on a modern map?

Mapping Technology

The different types of maps available today are due to advancements in science and engineering. By determining the latitude and longitude of many places, mapmakers can create very accurate maps of Earth. However, a two-dimensional coordinate system cannot convey much information about the vertical direction. This includes much of the information that interests Earth scientists and geologists, such as the elevation or shape of landforms or the depths of the ocean. Fortunately, recent advances in technology have greatly improved scientists' ability to collect different types of data that can be used to map Earth's geologic features in three dimensions.

One such technology is the global positioning system (GPS), a network of satellites and receivers that collects various types of data used to make maps. The United States military launched the first of many satellites for the GPS network in 1978, and the system became operational in 1993. The satellites in the GPS network communicate with radio receivers on Earth to accurately pinpoint the position of the receiver using a process called trilateration (Figure 1-14).

FIGURE 1-14
Go online to view a diagram of how GPS works.

GPS can pinpoint a receiver's position with accuracy and determine its altitude and track its motion over time. In fact, the most accurate GPS satellites and receivers can pinpoint a location within centimeters. Since GPS can track the positions of moving objects, it is used to create traffic maps and directional maps. GPS can also determine the receiver's altitude, which can be used to add elevation information to maps. Geologists use GPS devices to map and track ground deformation near volcanoes, which can be used to evaluate the risk of an eruption. Earth scientists also use GPS to track the motions of tectonic plates, motions along fault lines, and the

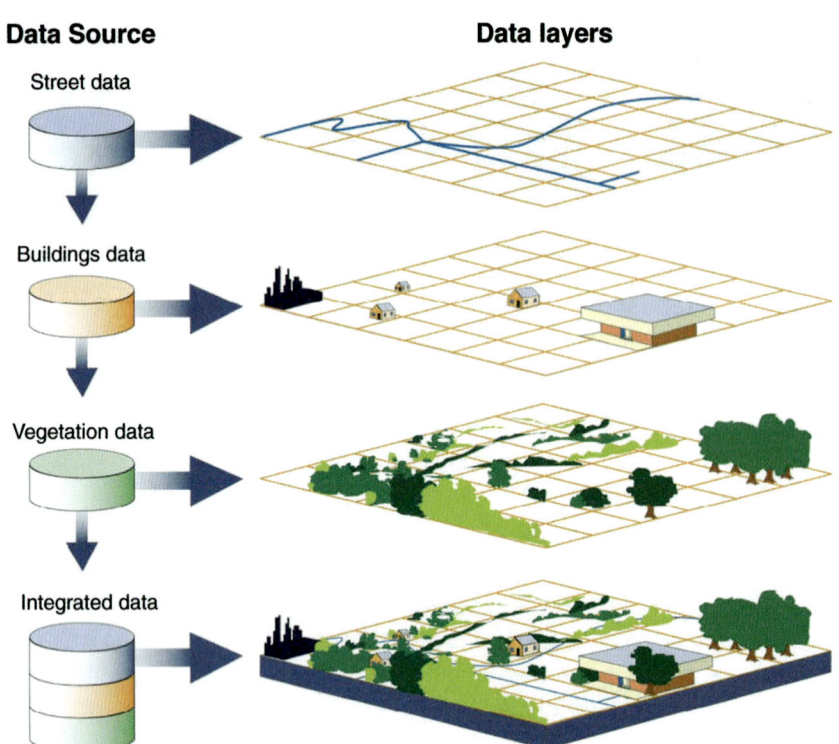

FIGURE 1-15
Geographic information systems (GIS) capture different types of spatial data, or data that relate to locations. Each data set is mapped out on a grid. The data layers are stacked on top of each other to make a map that shows all the data sets together, each in their correct location.

Source: U.S. Government Accountability Office

gradual downward motion of soil in areas that are prone to landslides.

Geographic information systems (GIS) are computer systems that capture, compile, and layer different types of spatial data and use these data to make maps. Data may be input from many sources. For example, GPS data can be used to map landforms, buildings, and streets of a region. Time-lapse imagery produced by satellites can be used to show vegetation type and density. Population and census data can be input from government records. Each type of information is layered together to create a single map that displays all the data, and this map can be used to identify and explore the relationships among variables. Figure 1-15 illustrates how a GIS turns different types of data into maps. The maps created with GIS are used for many purposes, including planning for natural disasters, engineering new buildings or structures, or making business decisions.

checkpoint How is GIS useful in mapmaking?

Maps as Models

Maps used for scientific purposes need to provide accurate scale models of Earth's features. The scale of a map refers to proportions of distance. For example, a distance of 1 centimeter on a map may represent a distance of 1 kilometer, 7 kilometers, or some other distance on the ground, depending on the scale of the map. The orientation of the cardinal directions (north, south, east, and west) on maps is provided by a compass rose symbol or lines of longitude and latitude. On a static map, the scale and compass rose are generally labeled. On computerized maps, users can often adjust the scale, zooming in and out and change the orientation by rotating the map, to examine different parts of the map in greater or lesser detail.

Many maps use **isolines**, or contour lines, to mark all points on the map that share the same value. Weather maps like the one shown in Figure 1-16 use isolines to indicate all the points on the map with a specific atmospheric pressure, temperature, or precipitation total. Topographic maps use contour lines to mark regions that share the same elevation.

Maps that give data represented by symbols or colors require a map key, or legend. The key defines what the symbols and colors mean. Maps that represent data but that are missing a map key cannot be interpreted and provide no useful information. Some of the most common types of maps are shown in Figure 1-16.

checkpoint What do the isolines on a weather map indicate?

Relief Map The physical characteristics of a region are shown on a relief map. These maps have shading to show different landforms, such as mountains.

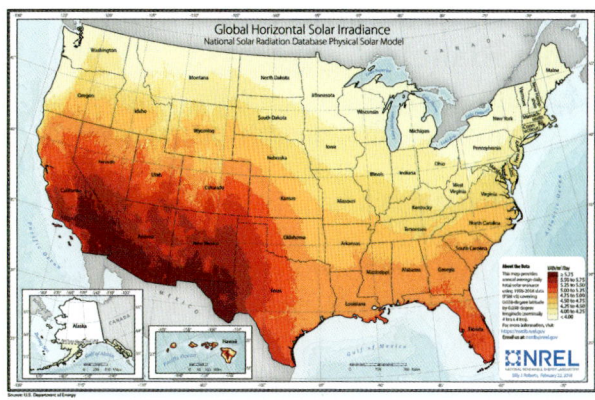

Resource Map The locations and distribution of natural resources are shown on a resource map. This map shows the availability of solar resources, which can be used to generate electricity.

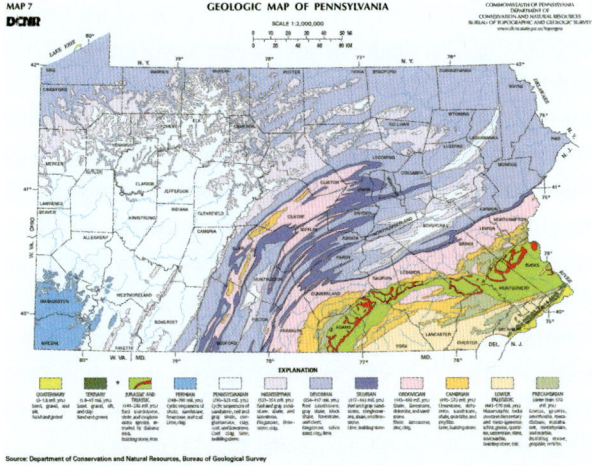

Geologic Map Uses different colors to contrast different rock types, ages, formations, or geologic structures in a region.

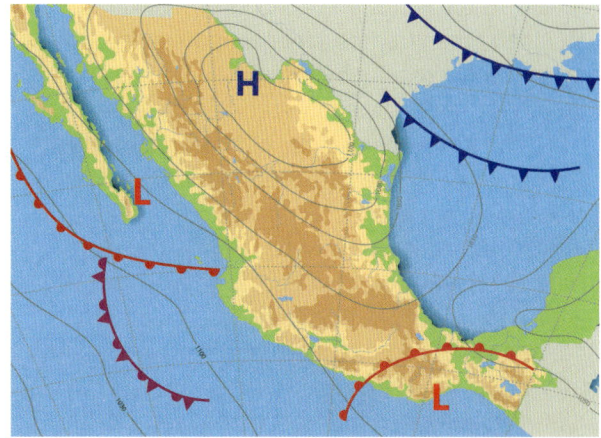

Weather Map Shows locations and properties of air masses and other atmospheric conditions to forecast changes in weather. The gray isolines indicate regions of the same pressure.

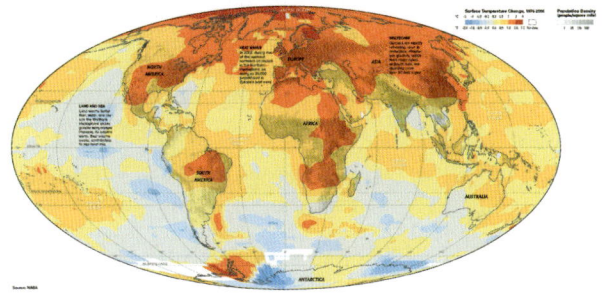

Climate Map Shows the climate characteristics of different regions.

Topographic Map Represents landforms by using contour lines that connect points of equal elevation.

FIGURE 1-16
Use of color and scale can vary greatly depending on the information to be communicated by maps. Many maps include a key indicating what the symbols or colors on the map represent. Without a key, maps that provide data cannot be interpreted.

Mapping Landforms on Earth and Other Planets

Topographic maps (Figure 1-17) have special importance for Earth scientists. **Topography** is the study of landforms and other structures on Earth's surface. Topographic maps display elevation data that are critical for studying these features. The elevation data used to create topographic maps can be obtained from devices called altimeters that determine the elevation of a point relative to sea level. Some altimeters calculate elevation by measuring atmospheric pressure, which varies systematically with elevation. Other altimeters determine elevation by using GPS data or by using radar or lasers.

Contour lines connect points of equal elevation on a topographic map. Since each contour line represents a specific elevation, they never intersect with each other. Contour lines that are close together reflect steep topography, whereas lines that are farther apart show gradual changes.

The contour interval is the change in elevation between each isoline on a topographic map. On some maps, not every contour line is labeled with the elevation, but labels will appear at regular intervals on index contour lines. For example, in Figure 1-17, every fifth contour line is labeled, and these index contour lines appear thicker than the contour lines between them. Since there is a 200-foot elevation difference between index contour lines, we can calculate a 40-foot difference between each of the smaller contour lines.

Topographic maps can also help one visualize the shapes of landforms. Closed contour lines indicate the location of a hill or peak. Where topographic contour lines cross a valley stream, they typically form V's that point upslope. The tip of the V is located where the stream crosses the topographic contour line (Figure 1-17).

Topographic maps have many uses for both Earth scientists and engineers. Earth scientists use them to help evaluate the risk for certain natural disasters, such as landslides and floods, or to predict changes in topsoil due to wind or water erosion. Engineers use topographic maps to design roadways, buildings, bridges, and other structures.

Advances in science and engineering have also led to the development of new tools to map the ocean floor. Sonar technology uses the reflection of sound waves off objects to create seafloor maps. First used during World War I, sonar technology releases sound waves from a boat or submarine; a receiver on the vessel interprets the reflected sound waves when they bounce off the seafloor or an underwater structure. Differences in the sound wave return times are used to estimate depths and map the shape of the seafloor and underwater objects. A similar technology called LIDAR (Light Detection and Ranging) creates maps based on measurements from reflected light waves. LIDAR devices are often carried by aircraft and satellites.

Sensors on the seafloor can also provide data that are used to map the seafloor and to monitor changes. The sensors are generally attached to

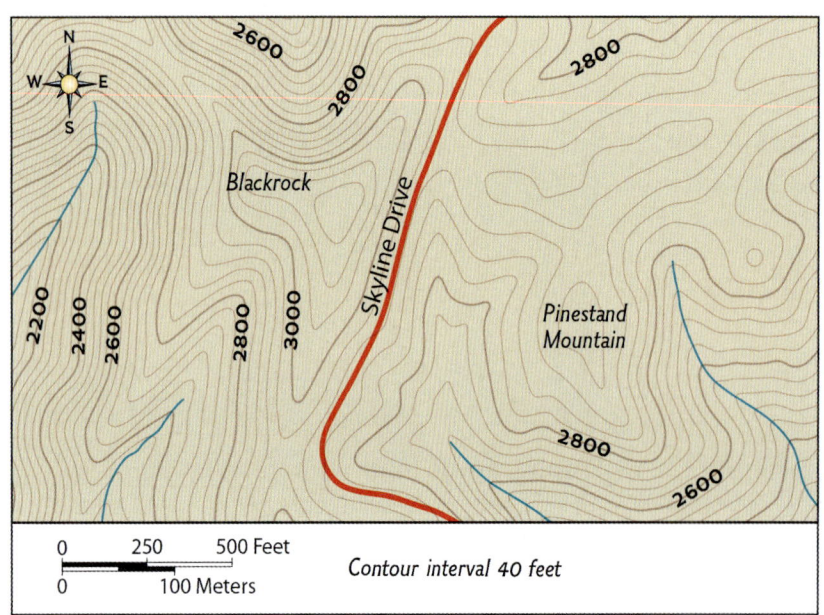

◀ **FIGURE 1-17**

This topographic map shows an area straddling a road in Shenandoah National Park, Virginia. Topographic contour lines illustrate elevation changes. The contour lines surrounding Pinestand Mountain represent the vertical topography of the mountain. The two blue lines closest to Pinestand Mountain are streams that are flowing down the mountainside. Notice that each topographic contour line forms a V that points upslope where the stream crosses it. This geometric relationship exists because the topographic lines are lines of equal elevation, whereas the mapped streams are flowing downhill through local valleys on the side of Pinestand Mountain.

Source: USGS

buoys that are anchored to the seafloor, and they are positioned in a network in order to collect data over a large area. Once installed, the sensors transmit data regularly with minimal human involvement. Underwater sensors are especially important along underwater fault lines, where they are used to detect changes in the water column and transmit warning signals for possible tsunamis.

Satellite altimetry is a more recent technique used for mapping the seafloor. This technology detects local variations in the sea's surface level in order to infer the locations of mountains on the seafloor. Satellites map the level of the sea's surface by emitting a radar pulse and measuring the time it takes for the signal to bounce back to the satellite. Mountains on the seafloor have a strong gravitational pull that attracts sea water toward them; this causes the surface of the water above the mountains to dip slightly, sometimes depressing the sea level by as much as 10 centimeters. Similarly, the water above deep ocean trenches bulges slightly because these depressions in the ocean floor exert less gravitational force on the water above. Thus, sea level data can be collected and then translated into topographical maps of the seafloor (Figure 1-18).

FIGURE 1-18
Go online to view a map of the seafloor that was created using satellite altimetry.

Advances in engineering technology have provided scientists with tools to map other planets and space objects. In 2011, scientists mapped the surface of Venus primarily using data collected during an 11-year mission by NASA's Magellan craft (Figure 1-19). Magellan mapped more than

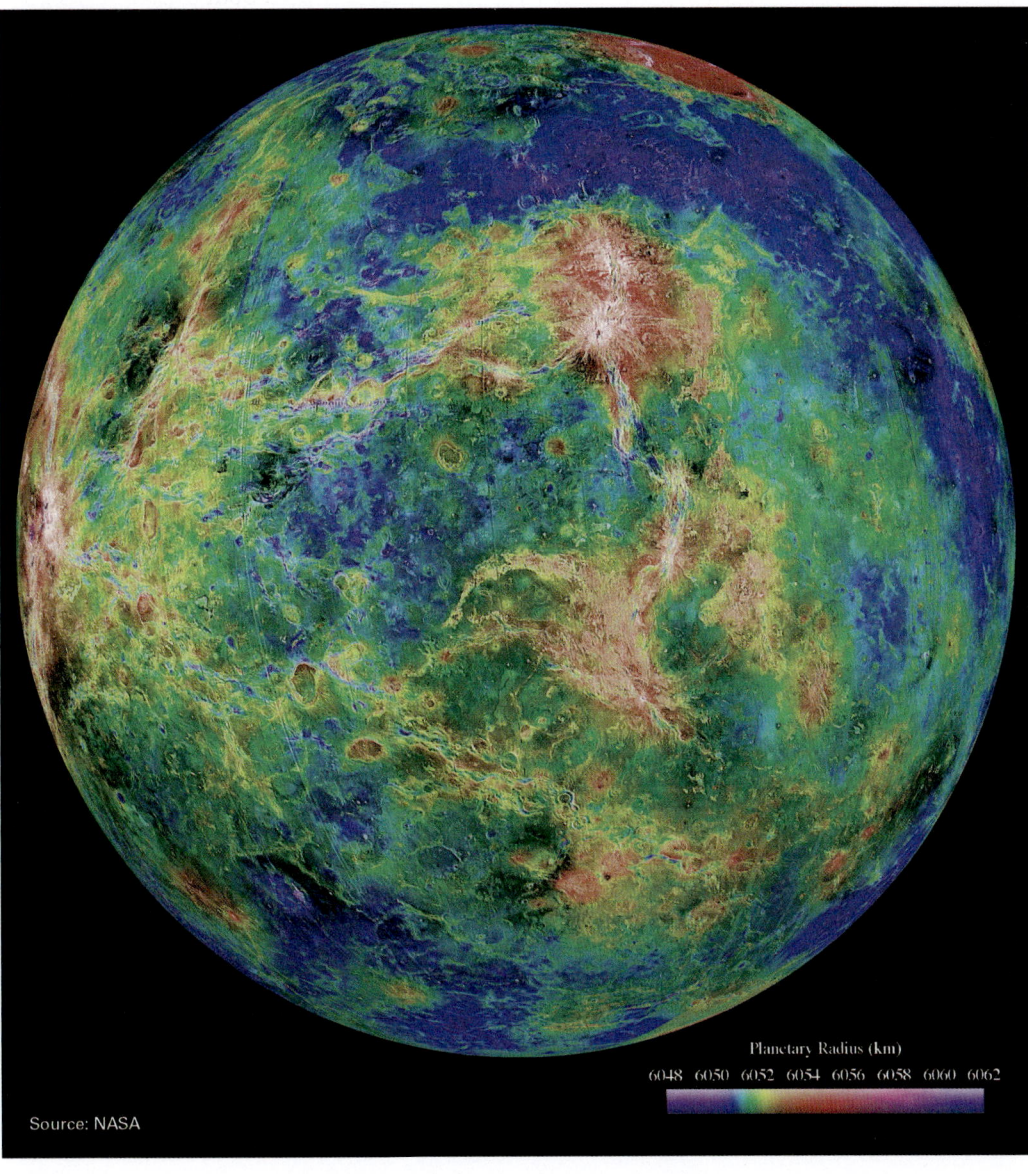

FIGURE 1-19
This map of the surface of Venus was created with data from NASA's Magellan spacecraft and National Astronomy and Ionosphere Center's Arecibo Observatory in Puerto Rico.

DATA ANALYSIS Develop a Topographic Profile

A topographic profile is a cross-sectional view along a line within a topographic map. You can think of this as making a slice vertically down through Earth along the line, pulling away the two halves, and looking at the exposed cut from the side. Construct a topographic profile of the region marked by the line (called a transect) from A to B in Figure 1-20.

i. Download and print the "Topographic Profile Handout" or use a photocopy of Figure 1-20.

ii. Cut out the graph and place it on the map. Align the graph's horizontal axis along the transect from A to B on the map.

iii. Mark a point on the graph's horizontal axis at each place where a contour line crosses the transect on the map. (You should have about 30 points.)

iv. From the map, determine the elevation at each point that you have marked on the horizontal axis of your graph. (Start from the map's 500-meter contour line and use 10 meters as the contour interval.)

v. For each point on your horizontal axis, plot another point directly above it reflecting the elevation of that point.

vi. Draw a smooth line through the points.

1. **Describe** How would you describe the terrain of the transect from A to B?

2. **Conclude** Where are the highest and lowest points along the transect from A to B? Was it easier to see these points using the topographic map or the topographic profile?

Computational Thinking

3. **Analyze Data** Water flowing along the surface of the ground, called runoff, always flows downhill. Choose the highest point along the transect from A to B. Predict the path of runoff flowing eastward from this point and then repeat for runoff flowing westward. Do both paths of runoff eventually lead to Lower Saranac Lake? Construct a topographic profile to support your claim.

4. **Represent Data** Suggest a transect for drawing a topographic profile that shows the least amount of elevation change between Lily Pad Pond and Fish Creek.

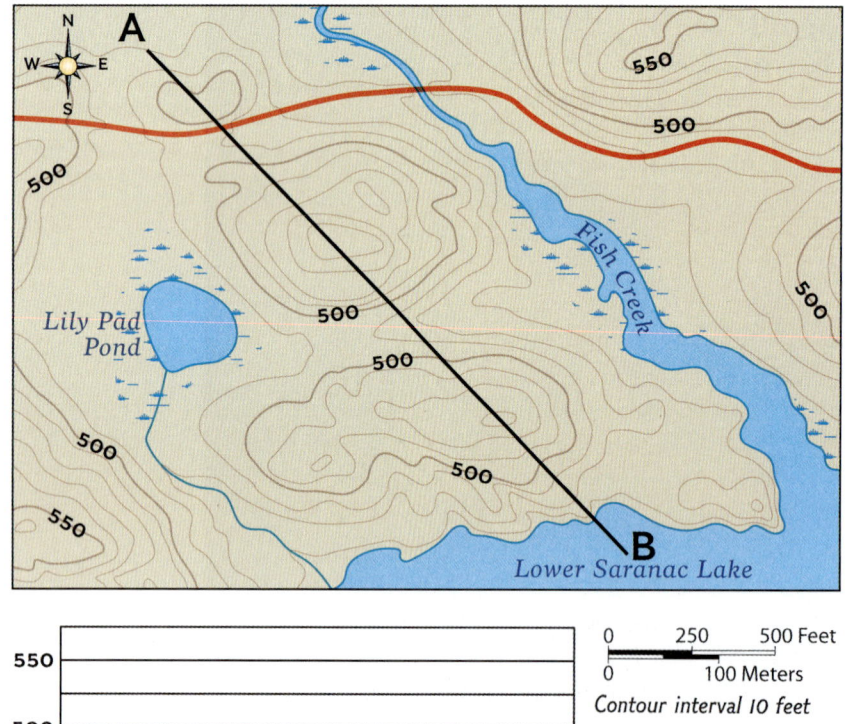

◀ **FIGURE 1-20**
This topographic map of an area near Saranac Lake, New York, shows Fish Creek flowing from the north into Saranac Lake at the south end of the map.

Source: USGS

98 percent of Venus's surface using radar and also collected data about the gravitational field of Venus. Additional data from Puerto Rico's Arecibo Observatory helped fill in gaps in the Magellan data in order to complete the map. Analysis of data from Magellan revealed that 85 percent of the surface of Venus is covered by lava flows. This analysis suggests that Venus has a history of volcanic activity.

checkpoint How do Earth scientists and engineers use topographic maps?

1.2 ASSESSMENT

1. **Summarize** Explain how a coordinate system is used to map locations on Earth.
2. **Identify** Describe at least three kinds of data that can be represented on a map.
3. **Summarize** Explain how science and engineering contribute to the mapping of Earth's systems.
4. **Contrast** What is the difference between a topographic relief map and a topographic contour map? Give one example of how a scientist or an engineer would use each type of map.

Critical Thinking

5. **Critique** Identify some problems or obstacles that limit the kinds of maps people are currently able to make.

TYING IT ALL TOGETHER
GEOTHERMAL ENERGY RESOURCE PLANNING

In this chapter, we discussed how scientists and engineers apply the practices of scientific inquiry and engineering to build scientific knowledge and develop solutions for problems. We also explored how scientists and engineers have developed and used various technologies to collect and map data related to Earth's systems. The Case Study described the efforts of scientists to map the geothermal resources in the United States. Such geothermal resource maps are now being used to make decisions about where to build geothermal power plants that can generate enough electricity for an entire region.

Most geothermal power plants are built over geothermal reservoirs that are 150°C (302°F) and within 1.5 to 3 kilometers from the surface. Engineers drill down to reach these reservoirs and install pipes to bring hot water upward. At the surface, the water expands to form steam, where it is used to drive turbines that generate electricity. High-resolution geothermal maps provide important information about where geothermal resources are located and their capacity for electric power generation. Decision-makers can consider this information along with other data, such as construction and maintenance costs and potential environmental impacts, before deciding when and where to build a new geothermal plant.

Figure 1-21 shows a map created by researchers seeking to identify promising sites for geothermal power production. The map shows underground temperatures in the Anadarko Basin, which covers parts of Oklahoma, Texas, and Kansas. To create the map, the researchers used temperature data collected from deep inside existing oil and natural gas wells. These wells vary in depth, so the map shows temperatures at different depths.

(Continued on next page.)

TYING IT ALL TOGETHER
GEOTHERMAL ENERGY RESOURCE PLANNING

Use information from the table and the map provided to answer the following questions.

1. At what temperature does water boil? Why do you think geothermal resources need to have temperatures greater than 150°C to be used for electricity generation?

2. Is it reasonable to consider any places shown on this map as potential sites for building new geothermal power plants? Use evidence and reasoning to explain your answer.

3. Based on the map, is there any correlation between temperature and depth? If so, explain what kind of correlation it is and how you know.

4. Geothermal resources can be used for applications other than electricity generation. Resources that are at least 110°C can be brought to the surface as a heat source for drying and curing cement slabs, which are used in the construction of buildings. Based on this map, are there areas in Oklahoma that are suitable for geothermal cement-curing facilities?

5. What information is not provided on the geothermal map that would be important for decision-makers to consider before planning development of electricity production facilities? Could this information be obtained from other types of maps? Explain.

FIGURE 1-21 ▼
Contour lines and colors on the map indicate temperatures at various depths in the Anadarko Basin. Use the key (upper right) to interpret these contours. The depth profile (upper left) shows the depths from which temperature data were taken along the transect from A to A'.

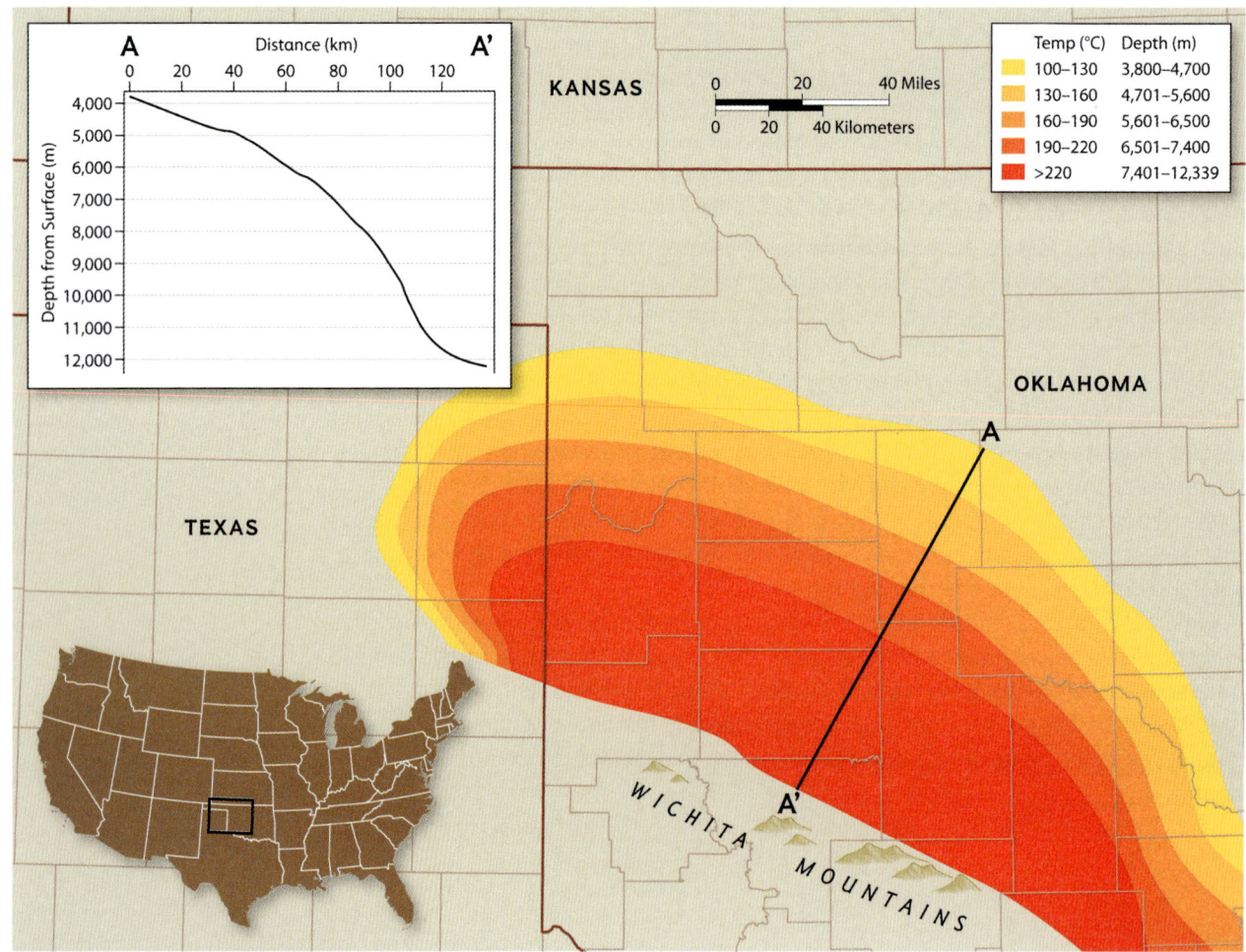

Source: National Renewable Energy Laboratory

CHAPTER 1 SUMMARY

1.1 THE NATURE OF SCIENCE AND ENGINEERING

- Science is a way of knowing and investigating the physical world through observation and experimentation. All scientific disciplines are grounded in a set of common understandings, known collectively as the nature of science (Table 1-1).

- Earth science is the study of Earth and its place in the universe. Branches of Earth science include geology, hydrology, atmospheric science, environmental science, and astronomy.

- Scientific inquiry involves identifying patterns and correlations and investigating possible causes following a set of eight practices collectively known as the science and engineering practices (Figure 1-4).

- Scientific laws are descriptions of phenomena that are universally accepted as true under given conditions.

- Scientific theories provide a framework for understanding a wide range of phenomena based on a large body of evidence. Theories may be revised, refined, or challenged as new evidence is discovered.

- Engineering applies scientific knowledge, mathematics, and human innovation to develop solutions to problems. The engineering design process incorporates many of the same practices as scientific inquiry, but with focus on defining problems and developing, testing, and optimizing solutions within a set of constraints.

- The seven crosscutting concepts emphasize the interdependence of science and engineering. They are: patterns; cause and effect; scale, proportion, and quantity; systems and system models; energy and matter; structure and function; and stability and change (Table 1-4).

1.2 MAPPING EARTH'S SYSTEMS

- All maps are visual representations of data. Historically, maps have always been limited both by our ability to collect data and by our ability to represent three-dimensional data in a two-dimensional format.

- The coordinate system uses degrees of latitude and longitude to assign a position to every location on Earth's surface relative to the Equator and the prime meridian.

- Different methods of map projections (Mercator, Mollweide, Winkel tripel, and others) use various strategies to represent Earth's curved surface on a flat, two-dimensional map. All projections result in some distortion of size, shape, direction, or distance.

- The GPS satellite network and geographic information systems have greatly advanced the ability of scientists and cartographers to collect and map numerous layers of data.

- Maps are scale models of Earth's features. Maps commonly used by Earth scientists include resource maps, geologic maps, weather maps, climate maps, and topographic maps.

CHAPTER 1 ASSESSMENT

Review Key Terms
Select the key term that best fits the definition. Not all terms will be used, and no term will be used more than once.

constraint	model
criteria	optimization
data	peer review
engineering	science
experiment	theory
hypothesis	topography
isoline	trade-off
law	

1. a statement describing or predicting a natural phenomenon that has never been shown to be false under stated conditions
2. a procedure followed to collect data and/or observations for the purpose of testing a hypothesis
3. an explanation of a phenomenon that is generally accepted to be true based on a great deal of repeated and independent experimentation
4. a tentative explanation for a natural phenomenon that can be tested to establish whether evidence exists to support it
5. information gathered during a scientific experiment
6. a line on a map that connects areas having the same elevation, depth, temperature, pressure, or other stated feature
7. a factor that limits the possible solutions that could be developed to solve a problem
8. the application of scientific and mathematical ideas to design solutions to problems
9. the process of allowing scientists to critique the experimental designs, results, and conclusions of other scientists
10. a description of the physical features of landforms in an area
11. the standards that must be met by an engineering design in order for the solution to be considered successful
12. the study of natural phenomena to better understand how the natural world works and to predict outcomes of natural events
13. a physical, diagrammatic, mathematical, or other representation of a system used to describe the system or explain how it works

Review Key Concepts
Answer each question on a separate sheet of paper to demonstrate your understanding of key concepts from the chapter.

14. What distinguishes a scientific explanation of a natural phenomenon from a nonscientific explanation?
15. What is the difference between correlation and causation as they apply to the interpretation of results from scientific investigations?
16. After completing a scientific study, a team of scientists writes a description of the study, including their hypotheses, data and observations, results, and conclusions. They then submit their study to a scientific journal for publication. Before publishing the study, however, the journal sends copies of the report to several scientists not involved in the study. What is this process called, and why it is an important step in the scientific process?
17. Describe a situation in which scientists might revise or even discard a theory.
18. Explain how criteria and constraints are involved in the engineering design process and give an example of each.
19. Give an example of how a mathematical model might be used in an engineering design.
20. Explain why all maps are representations of data.
21. What is a coordinate system and why is it useful in making maps?
22. How do Mollweide projections of Earth compare with Mercator projections?
23. What do the abbreviations GPS and GIS refer to and how do they help cartographers create new maps?
24. How do two-dimensional topographic maps convey information about three-dimensional landforms?
25. How did advances in technology enable scientists to begin mapping the seafloor?

Think Critically
Write a response to each question on a separate sheet of paper. Use concepts from the chapter to support your reasoning.

26. A scientist develops a hypothesis to explain a phenomenon and then designs an experiment to test the hypothesis. After running the experiment and analyzing the data, the scientist concludes that the data do not support her hypothesis. What would be a next step for the scientist, who wants to continue to pursue this area of research?

27. Scientists are sometimes asked to testify as expert witnesses in a court of law. During the court proceedings, a lawyer may ask a scientist, "What is your expert opinion on this matter?" Does this mean that opinions are important in scientific work? Explain.

28. People sometimes use the word "theory" in everyday conversation. For example, every day for several days, a man notices small holes in his lawn that were not present the day before. The man says, "My theory is that a skunk comes at night to dig for grubs and worms." How is the man's use of the word "theory" different from the scientific use of this word?

29. A budget consultant has suggested reducing the number of satellites in the Global Positioning System to save money. How might this affect the accuracy of maps that rely on GPS, and why? How would you respond to the consultant?

30. Some maps are used to represent areas that undergo constant change. How might a scientist studying glacier response to climate change use such a map, and what tools would be available for the scientist to create it?

31. How is a topographic map similar to and different from a weather map?

PERFORMANCE TASK

Interpreting Maps

The same area can be mapped in many different ways depending on the information to be conveyed. For this reason, it can be helpful to consult more than one map of an area when working on a project that requires various pieces of information.

In this performance task, you will use both a topographic map and an aerial map of the same land area to determine whether a municipal airport could be upgraded so that it can service larger planes.

1. Follow the instructions below to access topographic and areal maps of the area surrounding Malvern Municipal Airport in Malvern, Arkansas.
   A. Access the U.S. Geological Survey's National Map tool at the following URL: https://viewer.nationalmap.gov/advanced-viewer/
   B. In the Search box, enter "Malvern Municipal Airport."
   C. In the menu bar, select "Basemap Gallery."
   D. In the Basemap Gallery, select "Topographic."
   E. Use the Zoom controls (+ and – buttons) to zoom in and out. Drag the map in any direction to explore the area around Malvern, but stay within a 10-mile radius of the city. (Look for the scale bar, which will change with the zoom.)
   F. Return to the Basemap Gallery to switch the view to other types of maps of the same area. For this activity, you will mainly need the "Topographic" and "USGS Imagery Only" maps. However, you may also explore other maps in order to answer the questions that follow.

   At what elevation is the municipal airport?

2. The runway at the municipal airport can accommodate small planes but is too short to allow larger planes to land and take off safely. Could this runway be extended to 1,500 meters so that it can accommodate larger planes?

3. If the town of Malvern decided to build another airport at a different location, would there be an area within the range of these maps where an airport with a 1,500-meter runway could be built? Explain where this site is located and why it would be a good site for a new airport.

4. Would building an airport at the site selected above require rerouting of roads? What other structures, either natural or manmade, would be affected by building at this site?

5. Would either the existing airport or the new airport be at risk from flooding? Are there any man-made structures in the region of the map at risk from flooding? Explain your reasoning.

CHAPTER 2
EARTH SYSTEMS AND CYCLES

Grand Prismatic Spring is a favorite destination for visitors to Yellowstone National Park. The hot spring is known for its range of colors; oranges, yellows, and greens encircle deep blue waters. It is one of the park's many geologic features formed by heat beneath Earth's surface.

Grand Prismatic Spring, a hot spring in Yellowstone National Park, appears as something like a rainbow. Earth's four major systems, or spheres, are all on dramatic display. Heat from beneath Earth's surface warms rock and soil that are part of the geosphere. The root *geo-* refers to the planet Earth. The near-boiling water, heated by the hot rock, is part of the hydrosphere. The root *hydro-* refers to water. The crisp winter air makes up the atmosphere. The root *atmos-* refers to gases in the air. The colors in the spring represent different species of microscopic, single-celled organisms that can survive in the scalding water. The organisms are part of the sphere of living things—the biosphere. The root *bio-* refers to life. Each sphere may appear to be separated from the others, but they overlap and interact in endless ways that cause changes in land, water, air, and the evolution of life on Earth.

KEY QUESTIONS

2.1 What makes up each of Earth's four main systems, or spheres?

2.2 What factors cause changes in Earth's systems?

2.3 How does the cycling of water, carbon, and nitrogen demonstrate interactions within and among Earth's spheres?

EXPLORERS AT WORK

STUDYING MARINE CARBON CYCLES AND CLIMATE CHANGE

WITH NATIONAL GEOGRAPHIC EXPLORER ALLYSON TESSIN

The Arctic Ocean, known for its icy waters and vast sea ice, is nonetheless one of Earth's most rapidly warming regions due to climate change. In the last 30 years, scientists have measured a dramatic decline in both the thickness and extent of Arctic summer sea ice as air and water temperatures rise. Some models predict the Arctic Ocean will be free of summer ice by 2050. How might changes at the ocean surface impact the function of the seafloor ecosystem? This question concerns Dr. Allyson Tessin, a geologist, polar researcher, and biogeochemist at the University of Southern Mississippi. Tessin researches effects of climate change on the cycling of nutrients and carbon in the Arctic marine environment.

Oceans are vital to the global carbon cycle—the cyclic movement of carbon in various chemical forms from the environment to organisms and back again. Overall, oceans take up more carbon dioxide (CO_2) from the atmosphere than they release back to it. Because organic carbon (carbon incorporated into organisms) that gets buried in marine sediments may be removed from the ocean-atmosphere system for a very long time, oceans help stabilize Earth's climate. But how will their function be impacted if climate change increases temperatures and atmospheric CO_2? That trend is predicted if human activities continue releasing CO_2 and other gases that trap heat in the atmosphere.

To answer this question, Tessin investigates sedimentary records from the late Cretaceous period (approximately 86 million years ago). This period had elevated atmospheric CO_2 and a warm climate. The Arctic Ocean was free of sea ice in summer. Tessin says this period is "among the best examples of a greenhouse climate within the last 100 million years of geologic history."

During the late Cretaceous, the ocean flooded the western interior of North America, forming the Cretaceous Western Interior Seaway. This shallow seaway extended from the Arctic Ocean to the Gulf Coast. Tessin's analyses of biological, chemical, and geologic characteristics of carbon-rich seaway sediments help her better understand links between warmer, high-CO_2 climate conditions and processes of marine carbon burial.

Tessin also investigates the effects of retreating sea ice on the present-day Arctic Ocean and is documenting rapid chemical and biological changes occurring within the seafloor environment. Tessin hopes her research of seafloor conditions past and present will lead to more accurate predictions of the impacts of climate change on oceans.

THINKING CRITICALLY

Apply How can you link global climate change to an increase in carbon buried in ocean sediments?

▲ Allyson Tessin (left) and her team examine sediment samples from the ocean floor. By analyzing the carbon content in the samples, Tessin can get an idea of ancient climates and how they have changed.

◀ Tessin's team lowers core sampler equipment into the Barents Sea, located in the Arctic Ocean off the northern coasts of Russia and Norway. Sediment deposits made over millions of years are collected from holes drilled into the ocean floor.

CASE STUDY
THE YELLOWSTONE VOLCANO: MAJOR ERUPTIONS CAN AFFECT EARTH'S SYSTEMS

Yellowstone National Park is a carnival of geologic phenomena. Visitors can see some of Earth's biggest geysers, walk on bridges over boiling mud, spot fossilized trees, and may even feel the ground shake. The geology of Yellowstone is dominated by the structures and rocks associated with a very large, active volcano under the park. It includes an enormous caldera, or volcanic depression, that formed during a gigantic eruption about 640,000 years ago and is roughly 30 kilometers across (Figure 2-1). That eruption produced roughly 1,000 cubic kilometers of volcanic ash—enough to bury the state of Texas under nearly 1.5 meters of ash! The Yellowstone volcano also produced two earlier catastrophic eruptions at 1.3 million and 2.1 million years ago, respectively.

All three of these Yellowstone eruptions disrupted natural cycles and caused profound changes in Earth's systems. They began with explosive outbursts of volcanic ash, chunks of rock, and bits of magma from a surface opening called a vent. Ash was likely blown more than 25 kilometers into the atmosphere, blocking out sunlight across much of the region for months or possibly years.

As hot volcanic material fell back to the ground, it formed superheated mixtures of gas, lava, and rock debris that flowed toward low-lying areas such as river valleys. These so-called pyroclastic flows can race across the landscape at speeds of more than 100 kilometers per hour. They are so hot and destructive that nothing in their path survives.

If another major Yellowstone eruption occurred today, it would probably kill thousands of people, covering towns and cities beneath meters of ash. Pyroclastic flows would alter the course of rivers and streams. Airborne particles would cloud the upper atmosphere and spread across the globe, possibly cooling the climate and disrupting ecosystems worldwide.

The probability of a major Yellowstone eruption occurring in any given year is tiny. Far more likely to occur are less-catastrophic hazards such as earthquakes, landslides, and hydrothermal explosions. Hydrothermal explosions occur in regions where water deep underground has been heated past the boiling point. This superheated water cannot boil because the weight of the water above it confines it under high pressure. However, a reduction of pressure enables the superheated water to flash to steam, fragmenting overlying rock and propelling it upward along with water, steam, and mud. Today, some geologists are concerned that global warming could decrease precipitation in the Yellowstone region, thereby lowering the regional water table. The resulting decrease in pressure on superheated water underground may lead to more frequent hydrothermal eruptions.

As You Read Think about gradual changes and catastrophic disruptions to Earth's cycles and systems. Consider landforms, water sources, or environments near you. Are any of them a result of a catastrophic event such as one of the Yellowstone eruptions? How can you research the answer?

FIGURE 2-1
This map shows the locations of the three Yellowstone calderas that have formed over the past 2.1 million years. The yellow border shows the boundary of Yellowstone National Park.

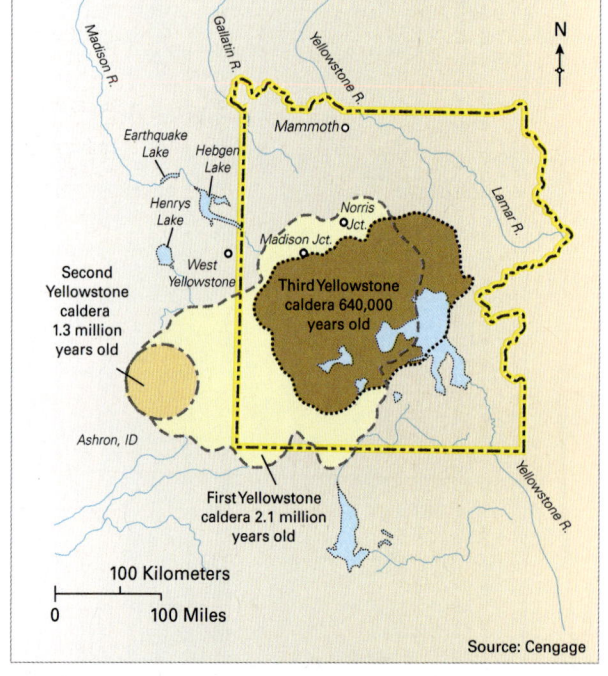

2.1 EARTH'S SPHERES

Core Ideas and Skills
- Identify and describe the components of the geosphere, hydrosphere, atmosphere, and biosphere.

KEY TERMS

geosphere	lithosphere
hydrosphere	tectonic plate
atmosphere	groundwater
biosphere	cryosphere
mantle	

The Four Interconnected Spheres

Imagine walking along a sandy beach as a storm blows in from the sea. Wind whips the ocean into whitecaps, while large waves crash onto shore. Blowing sand stings your eyes as gulls overhead frantically beat their wings en route to finding shelter. In minutes, sea spray has soaked your clothes. A hard rain begins as you hurry back to your vehicle. During this adventure, you have experienced the four major spheres of Earth. The beach sand underfoot is the surface of the **geosphere**, the rocky part of Earth that includes both molten and solid rock. The rain and sea are parts of the **hydrosphere**, the watery and icy part of our planet. The wind belongs to the **atmosphere**, the gaseous layer above Earth's surface. Finally,

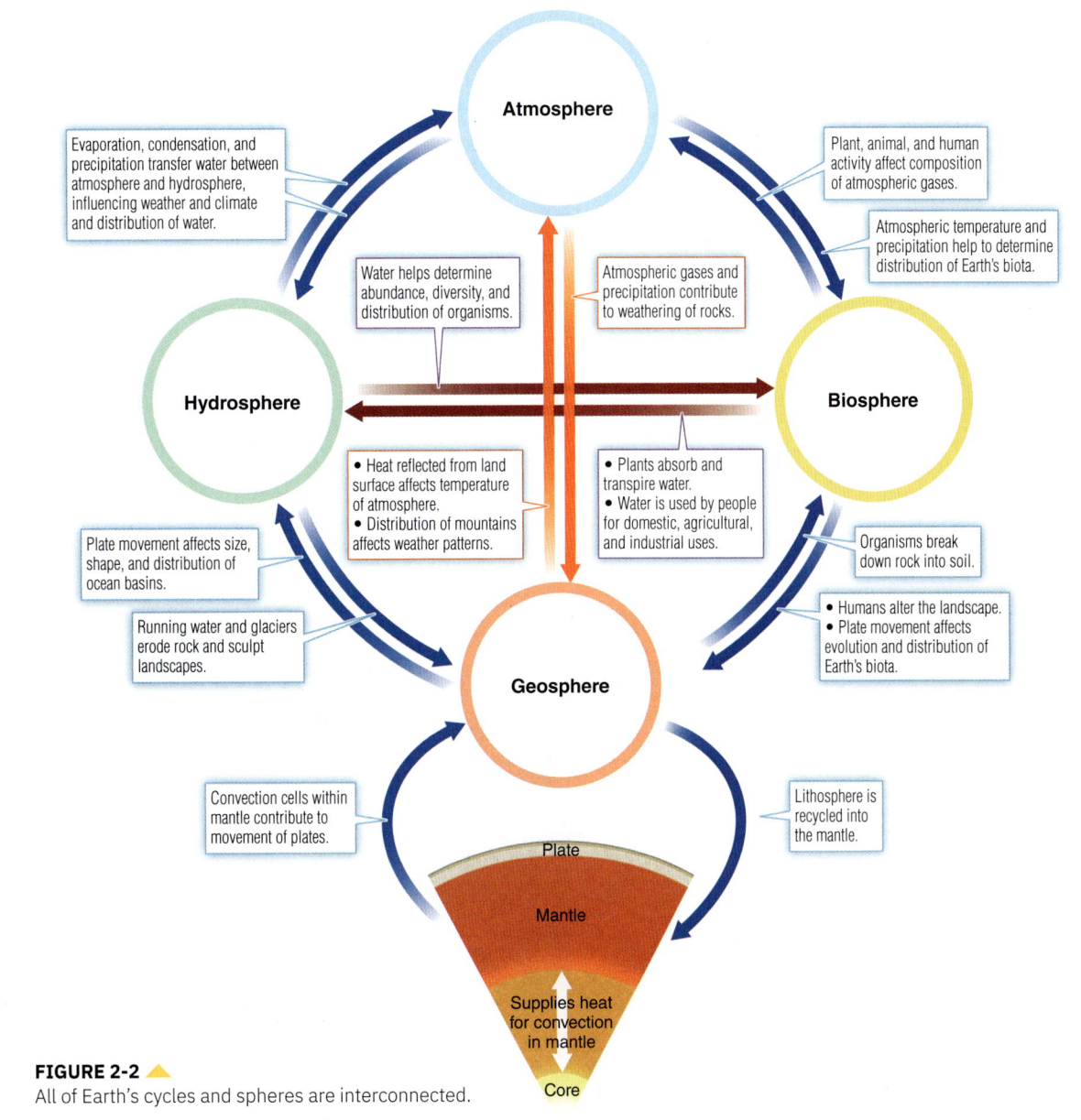

FIGURE 2-2
All of Earth's cycles and spheres are interconnected.

you, the gulls, the beach grasses, and all other forms of life are parts of the **biosphere**, the realm of organisms.

Components of each sphere are in motion. In this book, we explore phenomena that result as energy forces set matter in motion and how these motions affect the planet we live on. Figure 2-2 shows possible interactions among the spheres.

checkpoint What makes up each of Earth's major spheres?

The Geosphere

Our solar system formed from a frigid cloud of dust and gas rotating slowly in space. The sun turned on as gravity pulled material toward the swirling center. At the same time, rotational forces spun material in the outer cloud into a thin disk. Eventually, small grains of matter within the disk stuck together to form fist-sized masses. These planetary "seeds" then accreted to form rocky clumps, which grew to form larger bodies, called *planetesimals*, a hundred

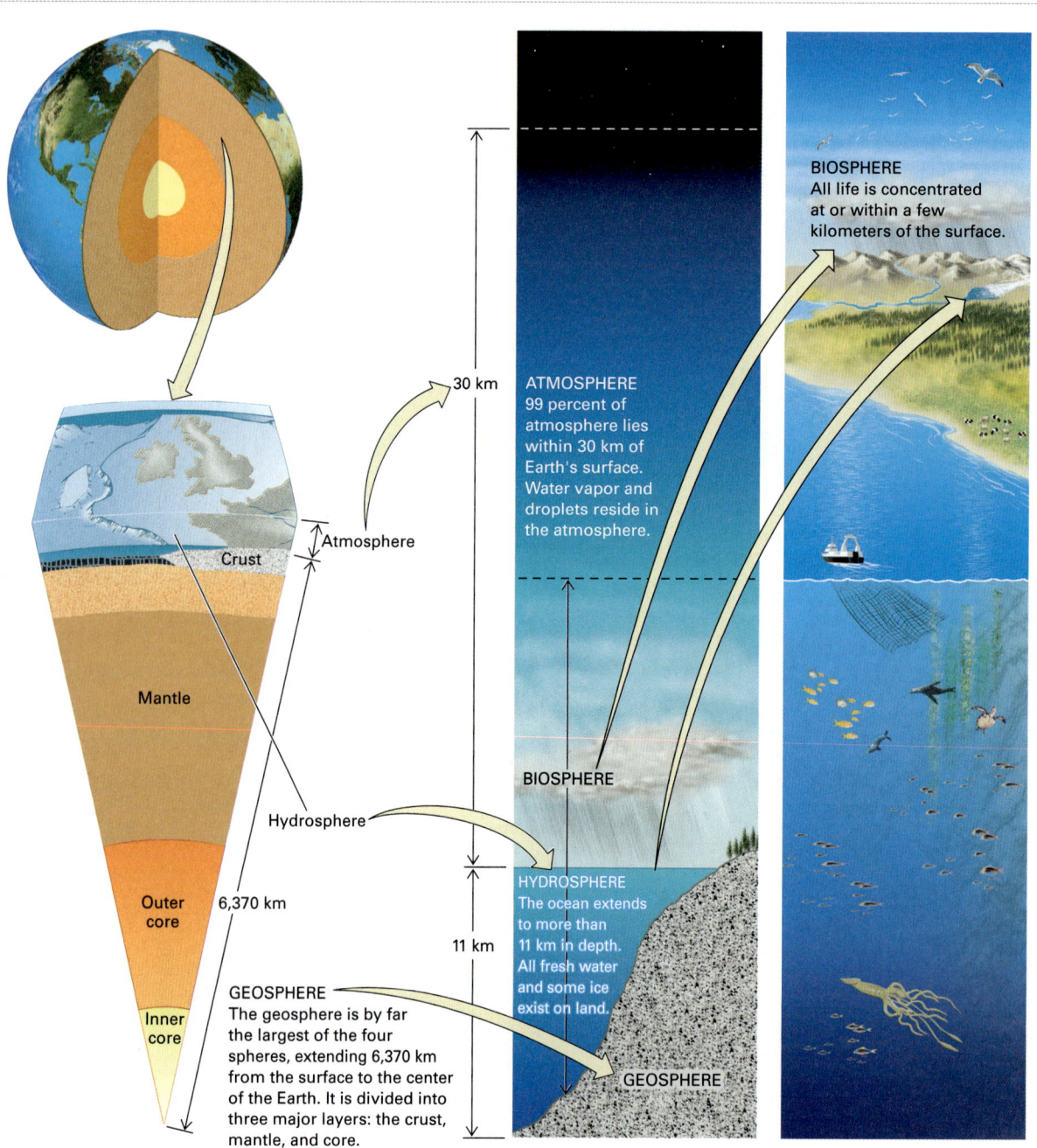

FIGURE 2-3

The geosphere is the largest component of Earth. Much of it is surrounded by the hydrosphere, the biosphere, and the atmosphere. Near Earth's surface, elements of the spheres are combined together and interact.

38 CHAPTER 2 EARTH SYSTEMS AND CYCLES

to a thousand kilometers in diameter. Finally, the planetesimals consolidated to form the planets.

As Earth formed, gravity caused the rocky chunks and planetesimals to accelerate and slam together at high speeds. Particles heat up when they collide, so the early Earth warmed as it formed. Later, asteroids, comets, and more planetesimals crashed into the surface, generating additional heat. At the same time, radioactive decay heated Earth's interior. These three processes caused the early Earth to become so hot that much of the planet melted as it formed.

Within the molten Earth, the denser materials sank toward the center, while the less dense materials floated toward the top, creating a layered structure. Today the geosphere consists of three major layers: a dense metallic core, a less dense rocky **mantle**, and an even less dense surface crust (Figure 2-3).

The temperature of modern Earth increases with depth. At its center, Earth is estimated to be about 6,000°C—about as hot as the sun's surface. The core is composed mainly of iron and nickel.

The outer core is molten metal. However, the inner core, although hotter yet, is solid because the great pressure compresses the metal to a solid state.

The mantle surrounds the core and lies beneath the crust. The physical characteristics of the mantle vary with depth. From its upper surface to a depth of about 100 kilometers, the outermost mantle is relatively cool and hard. Below a depth of 100 kilometers, however, rock making up the mantle is so hot that it is soft, plastic (able to deform permanently), and flows very slowly. Deeper in the mantle, pressure increases and the rock becomes solid again.

The crust is the outermost layer of rock extending from the ground surface or bottom of the ocean to the top of the mantle. The crust ranges from as little as 4 kilometers thick beneath the oceans to as much as 75 kilometers thick beneath the continents. Even a casual observer sees that the crust includes many different rock types: some are soft, others are hard, and they come in many colors, as shown in Figure 2-4.

FIGURE 2-4
Earth's crust is made up of different kinds of rock. (A) Sandstones, siltstones, limestones, and mudstones make up the Grand Canyon, as seen from the South Rim. These sedimentary rocks are relatively soft and crumbly and are layered horizontally. (B) The granite of Baffin Island in the Canadian Arctic is gray and hard.

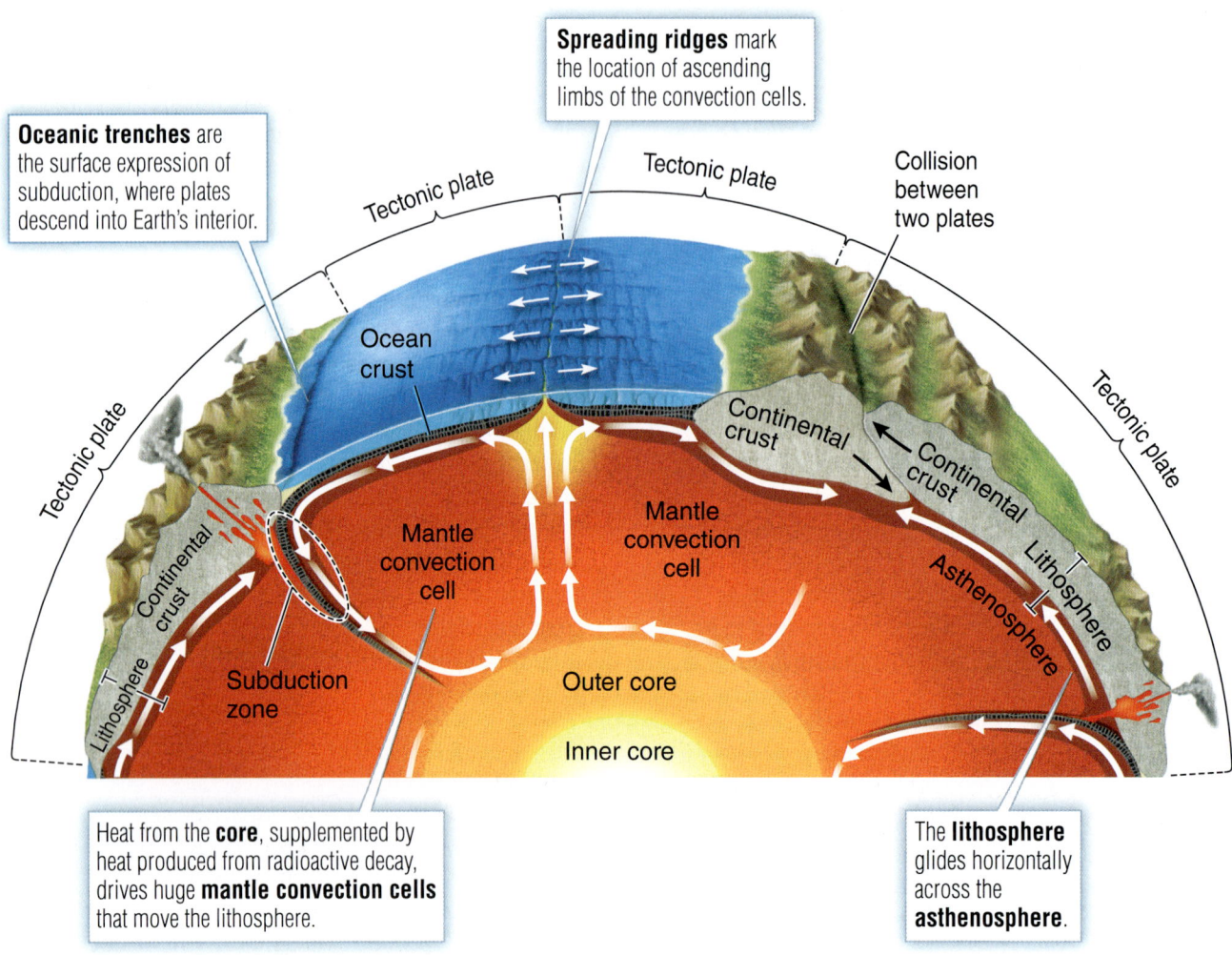

FIGURE 2-5
The lithosphere is composed of the crust and the uppermost mantle. It is a 100-kilometer-thick layer of rock that floats on the underlying plastic mantle. The lithosphere is broken into eight major segments, called tectonic plates, that glide horizontally over the plastic mantle at rates of a few centimeters per year. In the drawing, the thickness of the mantle and the lithosphere are exaggerated to show detail.

The relatively cool and hard rock of the uppermost mantle is similar to that of the crust. Together, the crust and uppermost mantle make up the **lithosphere**, which averages about 100 kilometers thick. According to the theory of plate tectonics, developed mostly in the 1960s, the lithosphere is divided into eight major and several smaller segments called **tectonic plates**. These plates float on the relatively hot, plastic mantle rock beneath and move horizontally with respect to each other (Figure 2-5). You will learn more about plate tectonics in Chapter 7.

checkpoint Describe Earth's lithosphere.

The Hydrosphere

The hydrosphere includes all of Earth's water, which circulates among oceans, continents, glaciers, **groundwater** (water stored beneath Earth's surface in soil and rock), and the atmosphere. Figure 2-6 shows the proportion of water in each of these areas. Oceans cover 71 percent of Earth and contain 97.5 percent of its water. Ocean currents transport heat across vast distances, altering global climate.

The part of the hydrosphere that consists of water in its frozen state, or ice, is the **cryosphere**. About 1.8 percent of Earth's water is frozen in ice caps, glaciers, and permanent snow. Although glaciers cover about 10 percent of Earth's land surface today, they covered much greater portions of the globe as recently as 18,000 years ago. During this glacial period, nearly all of Canada, the British Isles, and Scandinavia were covered with ice that was 4 kilometers thick. In North America, the ice was so massive that it caused the underlying lithosphere to sag beneath its weight. Hudson Bay in Canada exists because the glacial ice melted faster than the lithosphere was able to bounce back. Today, parts of the Hudson Bay region are undergoing over 10 cm of uplift per year as the lithosphere continues to recover from the weight of the ice.

Only about 0.64 percent of Earth's total water exists on the continents as a liquid. Although this is a small proportion, fresh water is essential to much of life on Earth. Lakes, rivers, and clear, sparkling streams are the most visible reservoirs of continental water, but they constitute only 0.01 percent of Earth's water. In contrast, groundwater—which occurs in pores, spaces within soil and rock of the upper few kilometers of the geosphere—is much more voluminous and accounts for 0.63 percent of Earth's water. Only a minuscule amount of water, 0.001 percent, exists in the atmosphere. But because it is so mobile, this atmospheric water profoundly affects both the weather and the climate of our planet.

checkpoint About how much of Earth's water exists as liquid fresh water?

The Atmosphere

The atmosphere is a mixture of gases: mostly nitrogen and oxygen, with smaller amounts of argon, carbon dioxide, and other gases. It is held to Earth by gravity and thins rapidly with altitude. Ninety-nine percent of the atmosphere is concentrated in the first 30 kilometers, but traces of atmospheric gas occur as far as 10,000 kilometers above Earth's surface.

The atmosphere supports life. Animals need oxygen and plants need both carbon dioxide and oxygen. In addition, the atmosphere supports life indirectly by regulating climate. Air serves as both a blanket and a filter, retaining heat at night and shielding us from direct solar radiation during the day. Wind transports heat from the Equator toward the poles, cooling equatorial regions and warming temperate and polar zones.

checkpoint How does the atmosphere support living things, such as animals and plants?

The Biosphere

The biosphere is the zone that living things inhabit. It includes the uppermost geosphere, the hydrosphere, and the lower parts of the atmosphere. Sea life concentrates near the surface, where sunlight is available. Plants also grow on Earth's surface, with roots penetrating as far as a few meters into the soil. Animals live on the surface,

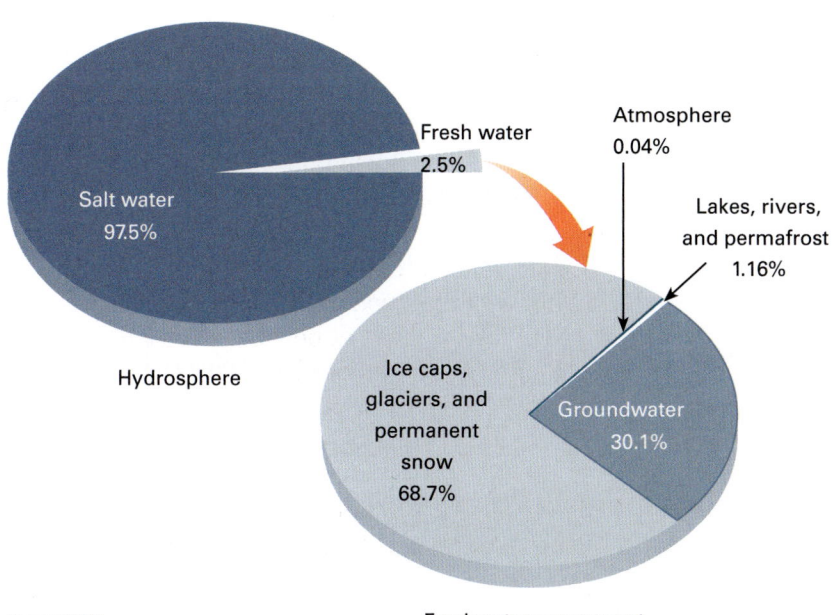

FIGURE 2-6
The oceans contain most of Earth's surface water. Most fresh water is frozen in glaciers. Most available fresh water is stored as groundwater.

LESSON 2.1

MINILAB Model a Cloud in a Bottle

Materials

100 mL hot water
1 L clear plastic water or soda bottle
safety goggles
wooden matches (about 100)

Procedure

1. Add hot tap water to the bottle. Close the cap and lay the bottle on its side. The water level should be below the cap. If it is not, pour some out.
2. With the cap closed, swirl the water around so it coats the sides of the bottle. Lay the bottle on its side and remove the cap.
3. Put on your safety goggles. Light the match, blow it out, and then hold the smoking match inside the mouth of the bottle.
4. When the match stops smoking, drop it in the bottle and close the cap tightly.
5. Holding the bottle upright, squeeze it hard with both hands and hold for two seconds. Then release your grip and observe what happens. Repeat several times.

Results and Analysis

1. **Observe** Describe what you see inside the bottle as it is squeezed and released.
2. **Infer** What is causing the change you see in the bottle?
3. **Predict** Explain what you would see if you could zoom in and see the particles within the bottle.

Critical Thinking

4. **Apply** Describe what parts of the model are similar to the interactions of Earth's geosphere, hydrosphere, and atmosphere.
5. **Critique** How could this model be modified to more accurately represent the formation of clouds in Earth's atmosphere?
6. **Synthesize** Predict how the interactions within these systems could also affect the biosphere.

---

fly a kilometer or two above it, or burrow a few meters underground. Large populations of bacteria live under glacial ice and in rock to depths as great as four kilometers. Some organisms live on the ocean floor, including entire thriving ecosystems supported by mineral-rich, super-heated water spewing from vents on the ocean floor. A few windblown microorganisms drift at heights of 10 kilometers or more. Despite these extremes, the biosphere is a very thin layer at Earth's surface.

From seemingly barren glaciers and deserts to colorful coral reefs and rain forests, life has taken hold virtually everywhere on Earth. Biodiversity, or biological diversity, is the variety of life in Earth's biosphere. Species vary in the genetic material they contain, the ecosystems in which they live, and their contributions to ecosystem processes such as energy flow and cycling of nutrients. To date, scientists have identified about two million species on Earth. They estimate the actual number most likely ranges from seven million to ten million and may be as many as 100 million.

Plants, animals, and other organisms are clearly affected by Earth's environment. Organisms take in gases from the atmosphere, require water from the hydrosphere, and thrive in a relatively narrow temperature range. Terrestrial organisms ultimately depend on soil, which is part of the geosphere. But organisms also alter the atmosphere through processes such as respiration and photosynthesis. Living things also contribute organic matter to the geosphere when they die.

checkpoint How are some organisms in Earth's biosphere affected by Earth's environment?

2.1 ASSESSMENT

1. **Describe** What are the characteristics of each of Earth's layers within the geosphere?
2. **Compare** How does the amount of salt water on Earth compare to the amount of fresh water?
3. **Recall** How does the atmosphere support life on Earth?

Critical Thinking

4. **Apply** Give three examples of interactions among all four of Earth's spheres.

2.2 EARTH'S SYSTEMS

Core Ideas and Skills
- Explain how systems are made of smaller subsystems and give examples.
- Explain how matter and energy enter and leave systems but are conserved.
- Describe how gradual and catastrophic changes to Earth's spheres have occurred throughout geologic time.
- Describe the effects of human interaction with Earth's systems.

KEY TERMS

system
ecosystem
cycle
gradualism
uniformitarianism
catastrophism
threshold effect

feedback mechanism
positive feedback mechanism
negative feedback mechanism
biodiversity
Anthropocene

Systems and Cycles

A **system** is a combination of interacting parts that form a complex whole. For example, the human body is a system, composed of bones, nerves, muscles, and a variety of specialized organs. Each organ is separate, yet all the organs interact to form a living human. For example, the heart pumps oxygen-rich blood to the brain, and the brain helps regulate the beating of the heart.

Systems are driven by the flow of matter and energy. For example, a person ingests food, which contains both matter and chemical energy, and inhales oxygen. Waste products are released through urine, feces, sweat, and exhaled breath. Some energy is used for growth and repair of tissues, respiration, and motion. The remainder is released as heat or stored as fat.

A single system may be composed of many smaller ones. For instance, the human body contains hundreds of millions of bacteria, each of which is its own system. Many of these bacteria are essential to the functioning of human life processes such as digestion of food.

In addition, humans are part of their local **ecosystem**, which is defined as a complex community of organisms and their environment functioning as an ecological unit in nature. Therefore, to understand the human body system, we must study smaller systems (e.g., bacteria) that exist within the body, while also exploring how humans interact with their larger ecosystems (Figure 2-7).

But we're not finished yet. Individual ecosystems interact with climate systems, ocean currents, and other Earth systems. We can conclude that among the great variety of Earth's systems, the size of systems varies dramatically, large systems contain numerous smaller systems, and systems interact with one another in complex ways.

FIGURE 2-7
Animals, including humans, interact with their larger systems. In this African savanna ecosystem, a variety of organisms interact with living and nonliving things in the environment.

LESSON 2.2   43

As we have learned, Earth is composed of four major systems: geosphere, hydrosphere, atmosphere, and biosphere. Each of these large systems is subdivided into a great many interacting smaller ones. For example, a single volcanic eruption is part of a system. Energy from deep within Earth melts rock, forming magma. Some of this magma escapes during the eruption, along with volcanic gases that react chemically with surface materials. But this volcanic eruption is driven by the distribution and movement of heat within Earth's interior, which is also a system. Volcanic ash and certain gases spewed skyward during the eruption can affect local weather and cool Earth's climate, thereby becoming part of these systems. Heat from the eruption can also rapidly melt glaciers growing near the summit of the volcano, affecting the local hydrologic system. In this book, we will study systems of all sizes and illustrate many of the complex interactions among them.

The atmosphere, the hydrosphere, and the biosphere are ultimately powered by the sun. Wind is caused by uneven solar heating of the atmosphere. Ocean waves are driven by the wind. Ocean currents move in response to wind or differences in water temperature or density. Luckily for us, Earth receives a continuous influx of solar energy, and it will continue to receive this energy for another five billion years or so. In contrast, Earth's interior is powered by the decay of radioactive elements and by residual heat from the formation of the planet. We will discuss these sources of heat in later chapters.

Fundamental to the study of Earth systems are several cycles of energy and materials. A **cycle** is a sequential process or phenomenon that returns to its beginning and then repeats itself over and over. During the course of these cycles, both matter and energy are always conserved. They never simply disappear, although either may continuously change form. For example, water evaporates from the ocean into the atmosphere, falls to Earth as rain or snow, and eventually flows back to the oceans. Ultraviolet radiation that we cannot see enters Earth's atmosphere from the sun and heats Earth's surface, which in turn emits heat. In this chapter we will examine the hydrologic cycle, the nitrogen cycle, and the carbon cycle. Later, we will examine the rock cycle (Chapter 4) and take a deeper dive into the water cycle (Chapters 15–17). We will also explore the critical role that energy and transformations of energy from one form to another play in Earth's oceans (Chapters 11 and 16) and atmosphere (Chapters 18–19).

Because matter exists in so many different chemical and physical forms, most materials occur in all four of Earth's major spheres—geosphere, hydrosphere, atmosphere, and biosphere. Water, for example, is chemically bound into clays and other minerals as a component of the geosphere. It is the primary component of the hydrosphere and exists in the atmosphere as vapor and clouds. Water is also essential for the survival of all living organisms. Salt, an essential component of life, is another example of a material that occurs in all four spheres. Thick layers of salt occur as chemical sedimentary rocks. Large quantities of salt are dissolved in the oceans. Salt aerosols are suspended in the atmosphere. As we can see, all the spheres continuously exchange matter and energy. In our study of Earth systems, we categorize the four separate spheres and numerous material cycles independently, but we also recognize that Earth materials and processes are all part of one integrated system (Figure 2-2).

checkpoint Explain how water changes as it cycles through different spheres.

Time and Rates of Change

Early scientists had no way of measuring the magnitude of geologic time. However, modern geologists have learned that certain radioactive materials in rocks can be used as clocks to record the passage of time. Geologists used these "clocks" and other clues embedded in Earth's crust, the moon, and meteorites fallen from the solar system to estimate Earth's age. They concluded that Earth formed about 4.6 billion years ago.

The newly formed Earth was vastly different from our modern world. There was no crust as we know it today, there were no oceans, and the diffuse atmosphere was vastly different from the modern one. There were no living organisms.

No one knows exactly when or how the first living organisms evolved, but we know that life existed at least as early as 3.8 billion years ago. That is 800 million years after the planet formed. For the following 3.3 billion years, life evolved slowly. Although some multicellular organisms developed, most of the biosphere consisted of single-celled organisms. Organisms rapidly became more complex, abundant, and varied about 541 million years ago. The dinosaurs flourished between 225 million and 65 million years ago. Earth's history

FIGURE 2-8
The geologic timescale is represented graphically. Note the great length of time before multicellular organisms became abundant about 541 million years ago.

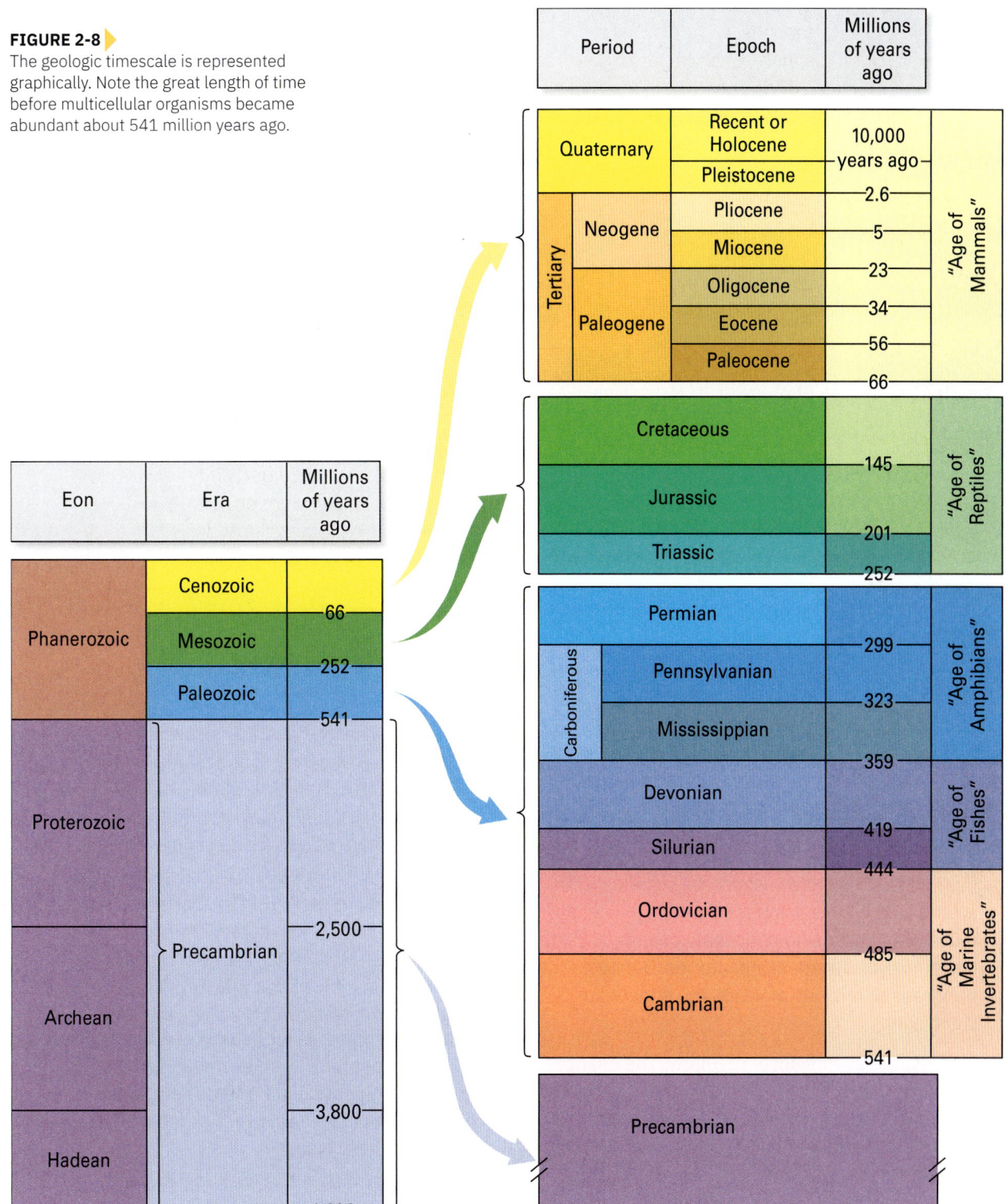

is summarized in Figure 2-8. You will learn more about geologic time in Chapter 5.

Geologists routinely talk about events that occurred millions or even billions of years ago. For example, about 1.7 billion years ago, the granite now forming Mount Rushmore cooled from molten rock and crystallized. About a half billion years ago, the Appalachian Mountains began to form in the eastern part of the North American plate. About 150 million years ago, a blanket of mud containing the remains of tiny plankton accumulated in deep water off the western coast of North America. That

LESSON 2.2   **45**

FIGURE 2-9
(A) The north end of the Golden Gate Bridge, shown here, was built on rock that was deposited as mud in the deep ocean about 150 million years ago. (B) The mud hardened to form these layers of rock. The sharp folds in the rock layers were produced by tectonic forces related to the western boundary of the North American plate.

sediment has since hardened to form one of the rock units on which the Golden Gate Bridge was built (Figure 2-9).

There are two significant consequences of the vast span of geologic time:

1. Events that occur slowly become significant. If a continent moves a few centimeters a year, the movement causes no noticeable change of Earth systems over decades or centuries. But over hundreds of millions of years, the effects are significant.

2. Improbable events occur regularly. The chances are great that a large meteorite won't crash into Earth tomorrow, or next year, or during the next century. But during the past 500 million years, several catastrophic impacts have occurred, and they will probably occur sometime in the future.

In the late 1700s, Scottish geologist James Hutton deduced that sandstone forms when rocks slowly decompose to sand, the sand is transported to lowland regions, and the grains cement together. This process occurs step by step over many years. Hutton's conclusions led him to formulate the principles now known as **gradualism** and **uniformitarianism**. The principle of gradualism states that geologic change occurs over long periods of time, by a sequence of almost imperceptible events. Uniformitarianism states that geologic processes operating today also operated in the past. Therefore, scientists can explain events that occurred in the past by observing slow changes occurring today. Sometimes this idea is summarized in the statement, "The present is the key to the past."

However, not all geologic change is gradual. William Whewell, a geologist in the 19th century, argued that geologic change was sometimes rapid and catastrophic. Earthquakes and volcanoes are examples of rapid, catastrophic events. Whewell argued for the principle of **catastrophism**—that occasionally huge catastrophes have altered the course of Earth history.

Today geologists know that both gradualism and catastrophism are correct. Over the great expanses of geologic time, slow, relatively uniform processes alter Earth and its systems. In addition, infrequent catastrophic events radically modify the path of slow change.

Gradual Change in Earth History
A good example of a slow but continuous Earth process is the movement of tectonic plates. This occurs too slowly to be observed directly without sensitive

instruments. Over geologic time, the movement of tectonic plates alters the shapes of ocean basins. The movement forms lofty mountains and plateaus, generates earthquakes and volcanic eruptions, and affects our planet in many other ways.

Catastrophic Change in Earth History

Although the chance that you will experience a large earthquake this year is very small, it nonetheless exists. In fact, deadly earthquakes occur each year somewhere on the planet. The largest recorded earthquake in the United States occurred in Alaska on September 2, 1964. This event caused significant loss of property and caused 139 human deaths. More recent catastrophic earthquakes include the 2004 Sumatra-Andaman earthquake and tsunami in Indonesia, the 2011 Tōhoku earthquake and tsunami in Japan, and large earthquakes that occurred i n Pakistan in 2005, China in 2008, and Haiti in 2010. Though each was devastating, even these catastrophes are minuscule in comparison with events that have occurred in the past.

When geologists study the 4.6 billion years of Earth history, they find abundant evidence of catastrophic events that are highly improbable in a human lifetime or even in human history. Clues preserved in rock and sediment indicate that giant meteorites have smashed into our planet. They vaporized portions of the crust, spreading dense dust clouds over the sky, and leaving large craters

FIGURE 2-10
Meteor Crater in Arizona formed from the impact of a meteorite.

(Figure 2-10). Geologists have suggested that some meteorite impacts have almost instantaneously driven millions of species into extinction. As another example, when the continental glaciers receded at the end of the last ice age, huge floods occurred (Figure 2-11). In addition, catastrophic volcanic eruptions have changed environmental conditions for life around the globe.

checkpoint Give an example of a gradual and a catastrophic change in Earth's history.

FIGURE 2-11
These giant current ripples in western Montana were formed by a catastrophic flood that occurred during sudden draining of a large glacial lake about 15,000 years ago. The floodwaters flowed from left to right. Notice the truck near the center of the photo for scale.

LESSON 2.2 47

CONNECTION TO ART

FIGURE 2-12
This depiction of dinosaurs and their habitat is based on fossil evidence. For instance, fossil remains indicate that *Triceratops* (left) had three horns, and *Parasaurolophus* (middle) had a long, bony crest on its head. Features such as color, however, largely are inferences based in part on animals that live in similar habitats today.

Scientific Illustration

Scientists must use clues from the past, such as fossil evidence, and study animals and plants today to make inferences about organisms that lived in the distant past. They might rely on scientific illustrators and animators to create images of these living things. Artists use their talent and training to create visual representations of how ancient plants, animals, or other organisms might have looked. The challenge is to accurately base their art on scientific concepts and evidence.

Threshold Effects and Feedback Mechanisms

Earth systems often change in ways that are difficult to predict. Two mechanisms that contribute to the challenges and complexities facing Earth science today are threshold and feedback effects.

Threshold Effects Consider a single rainstorm in Southern California. If the rain is gentle enough and it has not rained much recently, all of the rain falling will soak into the soil. If the rain lasts long enough or becomes heavy enough, a point will be reached where the soil becomes saturated and water begins to flow overland. Once runoff begins to flow across already-saturated soil, a very different situation exists, especially on hill slopes where the overland flow runs together. There, even small additional increases in rainfall can cause the saturated soil to begin to move downslope as a landslide. Each year, landslides in Southern California are a serious threat, although widespread landsliding occurs only during especially wet years. The landslide shown in Figure 2-13 occurred in March 1995, following an unusually wet spring. Part of the slope moved again in 2005, killing 10 people.

FIGURE 2-13 ▼
This landslide in Southern California formed during a threshold event in 1995. Heavy rains soaked into the ground. The ground became saturated to the point where it began to move downslope.

A **threshold effect** occurs when the environment initially changes slowly (or not at all) in response to a small disturbance in a system. But after the threshold is crossed, an additional small disturbance causes rapid and dramatic change. In the example above, the initial rainfall simply soaked into the soil. Once the soil became saturated and additional rain began to flow across the surface, the saturated soil could no longer maintain its slope and the landslide began.

If we don't understand a system thoroughly, it is easy to be deluded into a false sense of complacency. Imagine, in the case just described, that we didn't know about landslide hazards. If we observed that normal amounts of rainfall simply soaked into the ground or produced a muddy stream, we might conclude that a landslide was not possible. But once the threshold level of saturation is crossed, the land will move.

Populations of plants, animals, and people also respond to thresholds. In 1944, the U.S. Coast Guard released 29 reindeer onto St. Matthew Island in the Bering Sea. Because there were no natural predators on the island, the reindeer population increased dramatically. In 1957, wildlife biologist Dave Klein visited the island and counted 1,350 reindeer. Most of the reindeer appeared well fed and healthy. During that visit, Klein noticed some local overgrazing and trampling of lichens (the main food source for the reindeer). He returned in 1963 and counted 6,000 reindeer. But this time the average reindeer size and the ratio of adults to yearlings were down. Klein also noticed obvious signs of overgrazing. The reindeer herd was eating more food than was growing and was rapidly heading toward a threshold. When Klein next visited St. Matthew Island, in 1966, he was greeted by hundreds of bleached skeletons and counted only 41 emaciated survivors, a population drop of over 99 percent in three years or less.

FIGURE 2-14 ▼
(A) A population curve for reindeer on St. Matthew Island shows a marked population increase and crash. (B) A population curve for humans since around 1700 BCE shows a long period of stable population with a sudden rise in recent time. For most of human history, there were fewer than a half-billion people on Earth. In mid-2019, about 7.7 billion people inhabited our planet.

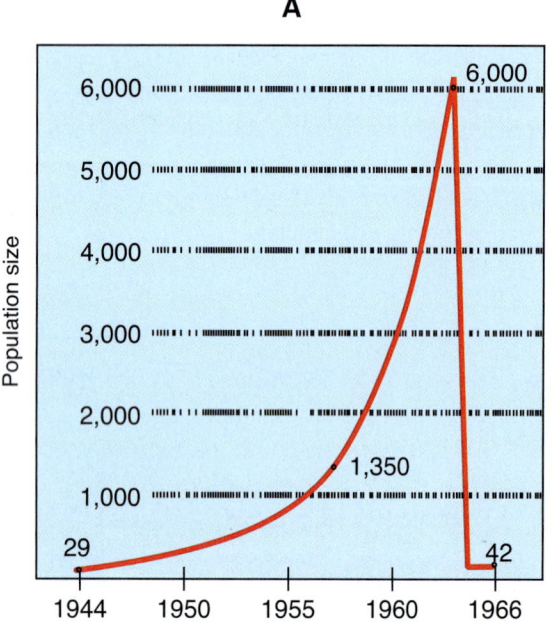

Assumed population of the St. Matthew Island reindeer herd. Actual counts are indicated on the population curve.

Source: Geophysical Institute, University of Alaska Fairbanks

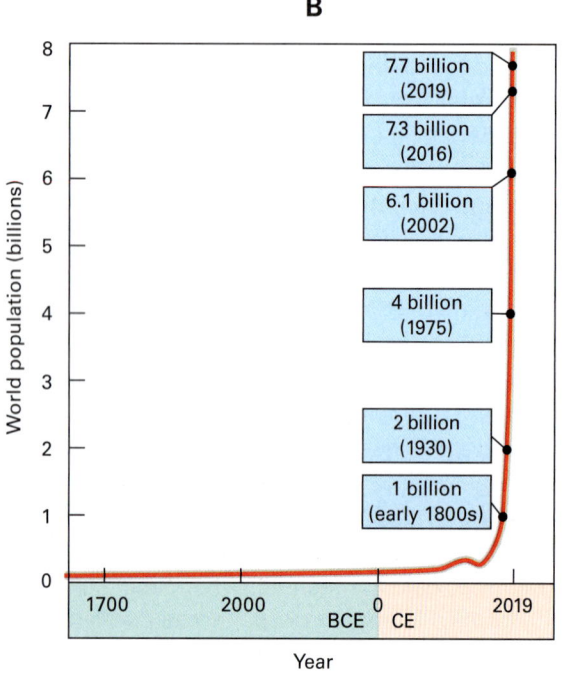

Source: Geophysical Institute, University of Alaska Fairbanks

Figure 2-14A shows the population curve for the St. Matthew Island reindeer over the 24-year period of the study. In comparison, the world population of humans follows a similar growth curve over the last few decades as shown in Figure 2-14B.

Feedback Mechanisms
A **feedback mechanism** occurs when a change or disturbance in one component affects a different component of a system. In a **positive feedback mechanism**, the original effect is enhanced or amplified, leading to an even greater effect. For example, as Earth warms, the rate at which organic matter in soil decomposes also increases. The increased rate of decomposition releases more carbon dioxide into the atmosphere. The additional carbon dioxide gas causes Earth's climate to warm further, promoting even faster rates of decomposition. In a **negative feedback mechanism**, the original effect is reduced or lessened. For example, as Earth warms, more water can be held in the atmosphere, leading to an increase in cloud cover. Clouds reflect sunlight energy, which can reduce the amount of light striking the ground. Less energy is absorbed by Earth, and atmospheric warming may be slowed. Although no one fully understands the eventual ramifications of these feedback loops for Earth's climate, Chapter 21 explores what scientists do know.

checkpoint Describe one threshold event and one feedback mechanism.

Humans and Earth Systems
Throughout history, humans have had an impact on Earth's systems, including the biosphere. Earth has numerous biomes, or communities of plants, animals, and other organisms that survive in particular environments. **Biodiversity** includes the genetic variations within populations and variation of populations within ecosystems that make up Earth's biosphere. Over time, healthy biomes tend to have increased biodiversity. When environments are disturbed, biodiversity can decrease, and organisms can become extinct. Reduced biodiversity affects humans. With less variety, food supplies such as crops are more vulnerable to pests and disease. An area planted with a single crop can be wiped out by conditions or disease that affects one type of organism, while an area with more variety may only lose a relatively small number of organisms. Factors that affect biodiversity can result in extinction of organisms.

CAREERS IN EARTH AND SPACE SCIENCES

Geobiologist

Geobiology is a relatively young field of science. A geobiologist studies connections between the physical, nonliving spheres, and the biosphere, both in the present and in the past. A geobiologist uses methods for studying Earth materials and biology, focusing on finding evidence demonstrating how living things and Earth's physical characteristics have iinteracted and evolved together. One of the greatest examples in Earth's history shows the result of interactions between microorganisms and the atmosphere. Current evidence indicates that Earth's early atmosphere was devoid of oxygen. Photosynthetic organisms evolved and produced oxygen as a waste product, greatly influencing the atmosphere's composition. This type of interaction between Earth and life processes through Earth's geologic history is particularly suited for investigation by geobiologists.

Although extinction of species occurs naturally on Earth, human population growth has reduced biodiversity. In mid-2009, there were 6.7 billion people on Earth. Farms, pasturelands, and cities covered 40 percent of the land area of the planet. Humans had diverted half of all the fresh water on the continents and controlled 50 percent of the total net terrestrial biological productivity. As the land has been domesticated and the waters manipulated, other animal and plant species have perished. The species extinction rate today is 1,000 times more rapid than it was before the Industrial Revolution. In fact, species extinction hasn't been this rapid since a giant meteorite smashed into Earth's surface 65 million years ago, leading to the demise of the dinosaurs and many other species. We are presently living through the largest documented extinction on Earth since the age of the dinosaurs.

Besides accelerating the rate of extinction, human activities are causing other changes to Earth's systems. Emissions from fires, factories, electricity-generating plants, and transportation machines have raised the atmospheric carbon dioxide concentration to its highest level in 420,000 years. Since the Industrial Revolution, Earth's mean annual temperature has increased by 0.8°C. This increase contributes to major changes in Earth's climates, as we will discuss in depth in Chapter 21.

Much of the climate-changing emissions come from the most developed nations on Earth. Roughly 20 percent of the human population lives in wealthy industrialized nations where food, shelter, clean drinking water, efficient sewage disposal, and advanced medical attention are all readily available. In contrast, as of 2016, about 767 million people worldwide survived on less than two dollars per day. As of 2018, 821 million were malnourished.

Yet no matter where we live and regardless of whether we are rich or poor, our survival is interconnected with the flow of energy and materials among Earth systems. We affect Earth, and at the same time Earth affects us. It is the most intimate and permanent of marriages. But what lies ahead in this complex relationship between humans and Earth's systems? Predictions for the future are, by their very nature, uncertain. More than 200 years ago, in 1798, Reverend Thomas Malthus argued that the human population was growing exponentially, while food production was increasing much more slowly. As a result, humankind was facing famine, accompanied by "misery and vice." But technological advancements have led to an exponential increase in food production, and today food is plentiful in many regions. At the same time, however, serious shortages exist, especially in parts of Africa.

Uncontrolled human population growth can deplete Earth's resources. Carrying capacity is the maximum population of a species in an ecosystem that can be supported by available resources. No population can increase indefinitely. Limiting factors influence population growth and density. Examples of limiting factors include availability of sufficient food and water sources. Human population growth and accompanying destruction of habitat for living space and agriculture, pollution, overharvesting, introduction of nonnative species, and consumption of more natural resources than the planet can sustain are factors that limit populations and reduce biodiversity.

FIGURE 2-15
ON ASSIGNMENT National Geographic Explorer and photographer Joel Sartore is the founder of the National Geographic Photo Ark, a project that uses photography to inspire people to help save species at risk before it is too late. Many, though not all, of the species pictured in the Photo Ark are at risk of extinction because of human activities. Pictured here are some of the primates featured in the Photo Ark. "It is folly to think we can destroy one species and ecosystem after another and not affect humanity," Joel remarks. "When we save species, we're actually saving ourselves."

52 CHAPTER 2 EARTH SYSTEMS AND CYCLES

LESSON 2.2

Scientists are concerned about our future. John Holdren is a professor at the John F. Kennedy School of Government. He argues, "The problem is not that we are running out of energy, food, or water but that we are running out of environment—that is, running out of the capacity of air, water, soil, and biota to absorb, without intolerable consequences for human well-being, the effects of energy extraction, transport transformation and use."

Three broad trends are currently facing humanity:

1. The human population is large and continues to increase—although the rate of increase is slowing. As of September 2019, the U.S. Census Bureau estimated the world population at 7.6 billion. According to the United Nations, world population is projected to reach 9.8 billion in 2050, and 11.2 billion in 2100.

2. At the same time, extreme poverty is decreasing. In India and China, with one-third of the world's population, economic output and standards of living are rising rapidly.

3. When people have money, they consume more resources. As both population and per capita consumption rise, pollution and pressure on Earth's resources also continue to rise.

Because of our huge population and the technologies we have developed, humans have significant impacts on Earth systems. Scientists give names to different periods of geologic time based on organisms and events that have major influence on the planet. Some scientists argue that in Earth's timescale, we have entered a new time period. They refer to this period as the **Anthropocene**, from *anthropo,* for "human," and *cene,* for "new." In recent human history, people have caused great changes in the atmosphere and the environment, drastically accelerating extinctions of plants, animals, and other organisms (Figure 2-15). Humans are making permanent changes to Earth's systems that previously had not been possible.

Jared Diamond, a Pulitzer-Prize-winning author, argues that we are in a "horse race" with the environment. One horse, representing increased human consumption and the resulting environmental degradation and resource depletion, is galloping toward an unsustainable society with a possbile catastrophic collapse. The other horse, powered by improved technology, increased awareness, and declining rates of population growth, leads us toward a happy, healthy, sustainable society. Which horse will win?

checkpoint How have humans affected the natural rate of species extinction?

2.2 ASSESSMENT

1. **Define** What is a system?
2. **Explain** What does it mean to say that matter and energy in Earth's cycles are conserved? Give an example.
3. **Compare** What is the difference between gradualism and catastrophism? Give an example of each type of change.
4. **Summarize** Describe a threshold effect and how it affects the system in which it occurs.

Critical Thinking

5. **Critique** Do you think that Anthropocene is a good term to apply to the present period in which humans live? Explain.

2.3 EARTH'S LIFE-SUSTAINING CYCLES

Core Ideas and Skills
- Describe the significance of the hydrologic cycle and how water cycles through Earth's spheres.
- Describe the significance of the carbon cycle and how carbon cycles through Earth's spheres.
- Describe the significance of the nitrogen cycle and how nitrogen cycles through Earth's spheres.

KEY TERMS

hydrologic cycle	nitrogen cycle
carbon cycle	

There are countless systems on Earth. Many of the systems can be described as cycles in which materials move from the biosphere, through Earth's other spheres, and back to the biosphere. These systems are vital to life on Earth. Many elements can be found in living organisms, but the six most common are carbon, hydrogen, nitrogen, oxygen, phosphorus, and sulfur. You can remember these

FIGURE 2-16

In the water cycle, the sun and gravity cause water to be continually cycled through Earth's spheres. Water can take countless paths through this cycle.

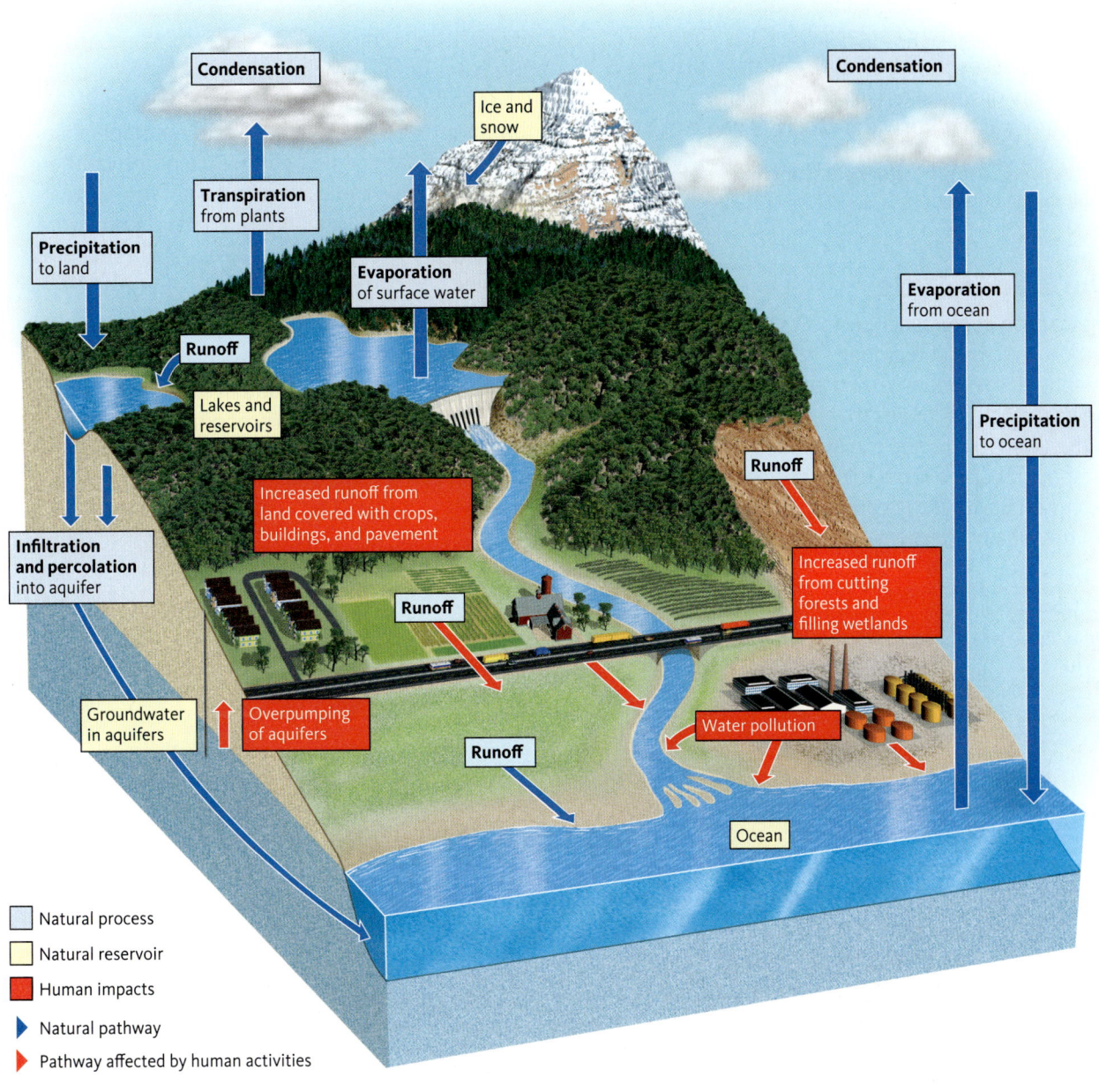

by noting that the first letter of each element forms the acronym CHNOPS. These elements combine chemically in different ways to form the substances that make up living organisms. The organisms die and decay, allowing the elements to combine in other ways as they cycle through Earth's spheres, and back to the biosphere. Let's look at some important cycles that contain carbon, hydrogen, nitrogen, and oxygen.

The Hydrologic Cycle

Organisms in Earth's biosphere are dependent on water, a compound containing hydrogen and oxygen. The **hydrologic cycle**, or water cycle, describes the continual cycling of water through Earth's spheres (Figure 2-16). Through evaporation, condensation, and precipitation, water is continually recycled over and over again. Reservoirs of water in Earth's spheres include the atmosphere, oceans, ice caps and glaciers, lakes and rivers, and groundwater. Liquid water is essential for life on Earth. As water cycles through the biosphere, it is taken in and given off by living things. Water allows for the transport of nutrients through organisms so they can be used for energy. Water is vital for carrying wastes out of organisms. Many chemical reactions necessary for life can only take place in the watery environment in living organisms. You will learn more about the water cycle in Chapter 15.

checkpoint Why is water necessary for the survival of living things on Earth?

The Carbon Cycle

The **carbon cycle** describes how the element carbon is exchanged among the atmosphere, biosphere, hydrosphere, and geosphere. Carbon is the basis for substances that make up living things. In the atmosphere and the hydrosphere, carbon compounds play a large role in the balance of energy on Earth. If the carbon cycle is disrupted, it can drastically affect life on Earth.

Carbon in the Atmosphere Carbon circulates among the atmosphere, the hydrosphere, the biosphere, and the geosphere and is stored in each of these reservoirs. Carbon exists in the atmosphere mostly as carbon dioxide (CO_2), and in smaller amounts as methane (CH_4). Although only 0.1 percent of the total carbon near Earth's surface is in the atmosphere, this reservoir plays an important role in controlling atmospheric temperature. Carbon dioxide and methane absorb infrared radiation and heat the lower atmosphere. If either compound is removed from the atmosphere, the atmosphere cools. If either one is released into the atmosphere, the air becomes warmer.

Carbon in the Biosphere Carbon is the fundamental building block for all organic tissue. Plants extract carbon dioxide from the atmosphere and build their body parts predominantly from carbon and hydrogen. This process occurs both on land and in the sea. Most of the aquatic fixation of carbon is conducted by microscopic phytoplankton. Therefore, healthy terrestrial and aquatic

FIGURE 2-17
In this representation of the carbon cycle, numbers show the size of the reservoirs in billions of tons of carbon.

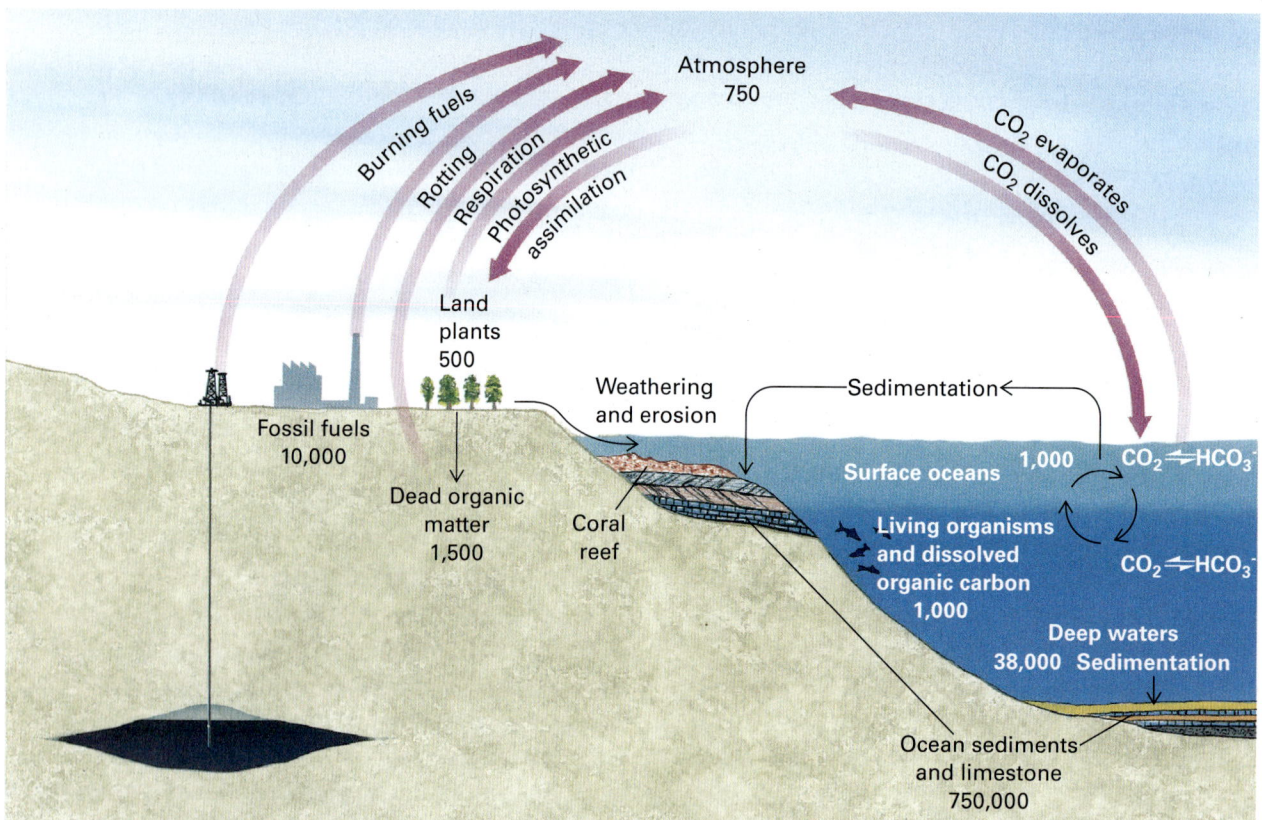

Source: U. Seigenthaler and J. L. Sarmiento, "Atmospheric Carbon Dioxide and the Ocean." Nature 365: 119, 1993.

DATA ANALYSIS Graphing the Carbon Cycle

Carbon cycles through Earth's atmosphere by way of natural and human-made processes. The graph (Figure 2-18) and table (Table 2-1) show the changes in the concentration of atmospheric carbon dioxide from 1960 through 2013 and 2014 through June 2018, respectively. The red zig-zag line in the graph represents seasonal variations in atmospheric carbon dioxide. The black line represents the mean carbon dioxide levels, corrected for seasonal variations.

1. **Identify** What is the trend for atmospheric carbon dioxide since 1960?

2. **Represent Data** Graph the data in the table.

Computational Thinking

3. **Analyze Data** Does the data between 2014 and 2018 support continuation of the trend shown in the graph provided?

4. **Recognize Patterns** Does the shape of the trend line change between 2014 and 2018? Predict how many years it will take for the atmospheric concentration of carbon dioxide to reach 420 ppm.

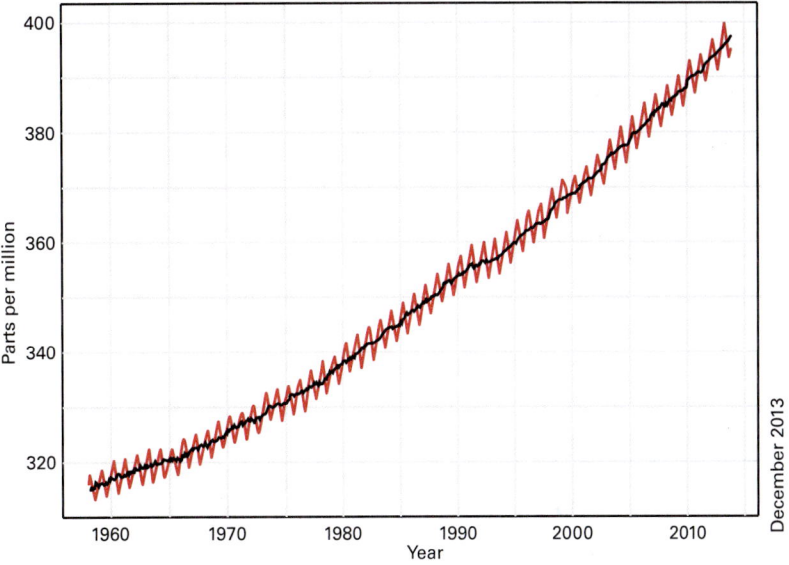

FIGURE 2-18
Atmospheric CO_2 at Mauna Loa Observatory, 1960–2013, is shown here.

Source: Scripps Institution of Oceanography, NOAA Earth System Research Laboratory

TABLE 2-1 Atmospheric CO_2 at Mauna Loa Observatory, 2014–2018

Year	Quarter	Atmospheric CO_2 (ppm)	Year	Quarter	Atmospheric CO_2 (ppm)
2014	1	398.54	2016	2	407.33
2014	2	401.47	2016	3	402.58
2014	3	397.17	2016	4	403.20
2014	4	397.41	2017	1	406.62
2015	1	400.60	2017	2	409.20
2015	2	403.35	2017	3	405.21
2015	3	399.29	2017	4	405.19
2015	4	400.10	2018	1	408.56
2016	1	403.85	2018	2	410.76

Source: Scripps Institution of Oceanography, and NOAA Earth System Research Laboratory

ecosystems play a vital role in removing carbon from the atmosphere.

Most of the carbon is released back into the atmosphere by natural processes, such as respiration, fire, or rotting. These are all part of the carbon cycle (Figure 2-17). However, at certain times and places, organic material does not decompose completely and is stored as fossil fuels—coal, oil, and gas. In this way, plants transfer carbon from the biosphere to rocks of the upper crust.

Carbon in the Hydrosphere
Carbon dioxide dissolves in seawater. Most of it then reacts to form bicarbonate, HCO_3^- (commonly found in your kitchen as baking soda or bicarbonate of soda, $NaHCO_3$), and carbonate, CO_3^{2-}.

The amount of carbon dioxide dissolved in the oceans depends in part on the temperature of the atmosphere and the oceans. When seawater warms, it releases dissolved carbon dioxide into the atmosphere, causing warming of the atmosphere. In turn, this warming further heats the oceans, causing more carbon dioxide to escape. Warmth in the atmosphere causes seawater to evaporate. The additional water vapor in the air also absorbs infrared radiation. Clearly, such a feedback mechanism can escalate.

Carbon in the Crust and Upper Mantle
As shown in Figure 2-18, the atmosphere contains about 750 billion tons of carbon. In contrast, the crust and upper mantle contain 1,000 times as much. The upper geosphere, combined with the hydrosphere and the biosphere, contain almost 800 trillion tons of carbon. If only a minute fraction of the carbon in the geosphere, the hydrosphere, and the biosphere is released, the atmospheric concentration of carbon dioxide can increase dramatically and severely affect climate.

checkpoint How do rising levels of carbon dioxide in the atmosphere affect Earth's climate?

The Nitrogen Cycle
The **nitrogen cycle** describes the five main processes that occur as nitrogen moves through the biosphere, atmosphere, hydrosphere, and geosphere. Nitrogen gas is the most abundant element in Earth's atmosphere. As a gas, nitrogen (N_2) is inert. Through the process of nitrogen fixation, nitrogen gas is converted by various bacteria into ammonium (NH_4^+). Next, through nitrification, bacteria combine ammonium with oxygen. The resulting nitrites (NO_2^-) and nitrates (NO_3^-) can be used by plants and other photosynthetic organisms to build tissues. Animals and other organisms that consume plant or other animal material use these nitrogen compounds to make new proteins and nucleic acids necessary for growth and survival. This process is called nitrogen uptake, or assimilation. Nitrogen mineralization occurs when organisms such as bacteria and fungi break down dead organisms. In this process, much of the nitrogen is converted to ammonium. Finally, through the process of denitrification, certain bacteria convert materials such as nitrates and nitrites into nitrogen gas which then enters the atmosphere (Figure 2-19).

Humans can have negative impacts on the environment at different points in the nitrogen cycle. One example is the use of high-nitrogen fertilizer in agriculture. If excess nitrogen from farm fields is washed into streams, rivers, lakes, and other bodies of water, excess growth of algae and other organisms can occur. When these organisms die and decay, oxygen can be depleted from the water, harming or killing organisms that rely on oxygen for respiration.

checkpoint What is one negative impact humans can have on the nitrogen cycle?

FIGURE 2-19

The nitrogen cycle is shown below. Through five processes, nitrogen is converted to different materials as it cycles through Earth's spheres.

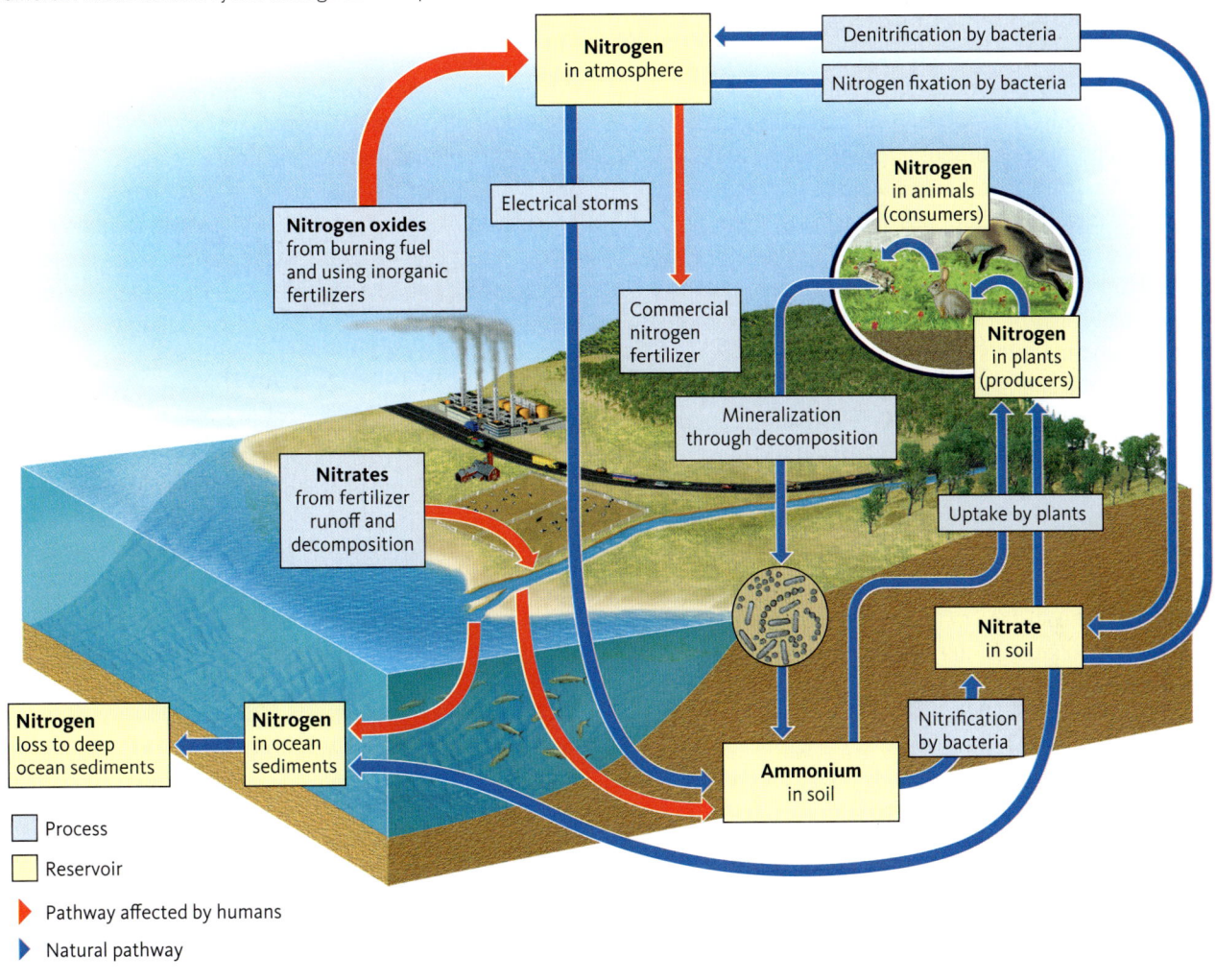

2.3 ASSESSMENT

1. **Recall** What are the main processes in the water cycle?
2. **Explain** How are the water cycle, the carbon cycle, and the nitrogen cycle important to the survival of living organisms in the biosphere?
3. **Describe** What are some negative impacts humans can have on the nitrogen cycle?

Critical Thinking

4. **Apply** Explain how rising levels of carbon dioxide in the atmosphere and oceans can cause a positive feedback mechanism.

TYING IT ALL TOGETHER
A SUPER DISRUPTION

In this chapter, you learned about Earth's major spheres, systems, and cycles. The Case Study examined prehistoric eruptions of the Yellowstone volcano. Such large-scale natural disasters help reveal the interconnectedness of Earth's systems. Because energy and matter continually cycle through Earth's spheres, a major geologic event can dramatically disrupt these cycles, shifting the delicate balance among Earth's spheres and causing major environmental changes.

In the case of the most recent Yellowstone super eruption, the local geosphere underwent the most immediate changes. The initial explosion ejected 1,000 cubic kilometers of material. The surrounding landscape was altered through the formation of a large caldera. The ejection of ash, gas, rock, and magma left the surface material unsupported, causing the collapse of the ground surface. Ash and other pyroclastic materials flowed outward from the eruption, flattening vegetation, smothering living things, and covering surrounding areas with a blanket of ash locally up to more than 200 meters (660 feet). Beds of ash from this eruption have been found as far away as California, Montana, and Louisiana.

The super eruption affected the atmosphere on an even broader scale. Enormous quantities of ash, sulfuric acid, and other gases and particulate matter lingered in the atmosphere for decades, reducing the amount of sunlight reaching Earth's surface and causing a "volcanic winter" that lasted for years. Changes in the atmosphere led, in turn, to changes in the hydrosphere. Because the volcanic ash and gases reduced the amount of solar radiation that reached Earth's surface, the ocean absorbed less energy from the sun. As a result, the ocean's surface temperature dropped. Airborne ash and gases led to acidic precipitation that contaminated freshwater sources across the globe. All of these changes had a profound effect on the biosphere. Sudden environmental changes severely limited organisms' access to clean air, water, and food, likely causing large die-offs and possibly extinctions.

If the Yellowstone volcano erupted today, it could have similarly devastating effects on Earth's spheres.

Projections suggest that another eruption would result in catastrophic levels of damage to cities within 500 kilometers of the eruption, including Casper, Wyoming, and Billings, Montana, which have a combined population of more than 150,000 people. Depending on the size of the eruption, these nearby cities could be covered by up to a meter of ash. For perspective, just 20 centimeters of ash can cause buildings to collapse.

As grim as that sounds, it is far from the worst-case scenario. Scientists theorize that a super eruption in Siberia about 252 million years ago contributed to a mass extinction that wiped out 96 percent of marine species and nearly two-thirds of all land species on Earth.

Think about natural disasters you have read or heard about. Research another historic natural disaster (for example, the Mount Tambora eruption of 1816; the Krakatau eruption of 1883; the Toba eruption; the 1900 Galveston, Texas, hurricane; or the 1500 BCE Stroggli earthquake and tsunami) and its short- and long-term effects on Earth's spheres.

Respond to the following questions and statements based on your research.

1. Describe the natural disaster. Was it a result of processes within Earth's atmosphere, geosphere, hydrosphere, or biosphere?

2. Construct a flow chart that shows how each of Earth's different spheres was affected during and after the natural disaster.

3. How do scientists study similar natural disasters? In what ways are scientists and engineers trying to limit the effects of future disasters?

4. Design a presentation you might give to government officials explaining measures for preventing long-term damage as a result of a future disaster like the one you researched.

FIGURE 2-20
Go online to view comparisons of amounts of magma ejected by various eruptions of the Yellowstone and other volcanoes.

CHAPTER 2 SUMMARY

2.1 EARTH'S SPHERES

- Earth consists of four spheres: geosphere, hydrosphere, atmosphere and biosphere.
- The geosphere is composed of solid and molten rock.
- The hydrosphere is mostly ocean water. Most of Earth's fresh water is locked in glaciers. Most of the liquid fresh water lies in groundwater reservoirs; streams, lakes, and rivers account for only 0.01 percent of the planet's water.
- The atmosphere is a mixture of gases, mostly nitrogen and oxygen. Earth's atmosphere supports life and regulates climate.
- The biosphere is the thin zone that life inhabits.

2.2 EARTH'S SYSTEMS

- A system is composed of interrelated, interacting components that form a complex whole. All of Earth's systems continuously exchange matter and energy.
- Earth is about 4.6 billion years old. Life formed at least 3.8 billion years ago. Abundant multicellular life evolved about 542 million years ago.
- The principle of gradualism states that geologic change occurs over a long period of time by a sequence of almost imperceptible events. In contrast, the principle of catastrophism postulates that geologic change occurs mainly during infrequent catastrophic events.
- A threshold effect occurs when the environment initially changes slowly (or not at all) in response to a small disturbance, but after the threshold is crossed, an additional small disturbance causes rapid and dramatic change.
- A feedback mechanism occurs when a small initial disturbance in one component affects a different component of Earth systems, amplifying the original effect, which impacts the system even more and leads to an even greater effect.
- Uncontrolled human population growth can deplete Earth's resources. Human population growth and accompanying destruction of habitat, introduction of nonnative species, and consumption of more natural resources than the planet can sustain are factors that limit populations and reduce biodiversity.
- Due to our population and technological advances, humans have become a significant engine of change in Earth systems. Some scientists argue that in Earth's time scale, we have entered a new epoch. This time period is known as the Anthropocene.
- In recent human history, people have caused great changes in the environment, such as drastically accelerated extinctions of plants, animals, and other organisms, pollution, and changes to the atmosphere. Humans are making permanent changes to Earth's systems.

2.3 EARTH'S CYCLES SUPPORT LIFE

- Life in Earth's biosphere depends on a number of cycles.
- The hydrologic cycle describes the continual movement of water through Earth's spheres. Through evaporation, condensation, and precipitation, water is continually recycled. Reservoirs of water in Earth's spheres include oceans, ice caps and glaciers, lakes and rivers, groundwater, and the atmosphere. The water cycle is driven by energy from the sun and gravity.
- The carbon cycle describes the cycling of carbon through Earth's spheres. Carbon cycles as carbon dioxide and methane in the atmosphere and hydrosphere, as carbonate and bicarbonate ions in organic tissues, shells, and skeletons in the biosphere, and as fossil fuels, limestone, and other rocks in the geosphere.
- The nitrogen cycle describes the cycling of nitrogen through Earth's spheres. Nitrogen gas in our atmosphere is transformed and recycled by biological processes. Humans cause disruption to the nitrogen cycle by adding excess nitrogen to the system.

CHAPTER 2 ASSESSMENT

Review Key Terms
Select the key term that best fits the definition. Not all terms will be used, and no term will be used more than once.

Anthropocene	hydrologic cycle
atmosphere	hydrosphere
biodiversity	lithosphere
biosphere	negative feedback mechanism
carbon cycle	
catastrophism	nitrogen cycle
cryosphere	positive feedback mechanism
cycle	
ecosystem	system
feedback mechanism	tectonic plate
geosphere	threshold effect
gradualism	uniformitarianism
groundwater	

1. the rigid outer layer of Earth, which includes the crust and uppermost mantle
2. living organisms in an area that interact with each other and the nonliving environment
3. the principle that states that the processes affecting geologic changes on Earth today are the same as they were in the past
4. the process by which carbon moves through different parts of the biosphere, geosphere, hydrosphere and atmosphere
5. the gaseous layer above Earth's surface
6. a phenomenon in which a system responds slowly to an initial perturbation but then begins changing rapidly once the perturbation reaches a certain level
7. large segment of the crust and uppermost mantle that floats on the relatively hot, plastic rock beneath
8. includes glaciers, ice caps, permanent snow, and all of Earth's frozen water
9. the variety of living organisms on Earth or in an ecosystem
10. water found below Earth's surface within soil and rock
11. the principle that states that geologic change occurs over long periods of time due to small events
12. the principle that states that large catastrophic events shape geologic change
13. a mechanism whereby a system stays in balance; as one factor increases, it creates effects that then lead to the original factor decreasing
14. a sequential process or phenomenon that returns to its beginning and then repeats itself over and over
15. the movement of water through Earth's spheres via precipitation, evaporation, and condensation.

Review Key Concepts
Answer each question on a separate piece of paper to demonstrate your understanding of key concepts from the chapter.

16. In what ways do the hydrosphere and the atmosphere overlap?
17. In what ways can the cryosphere affect the geosphere?
18. Describe ways that the atmosphere supports the biosphere, both directly and indirectly.
19. How can a volcanic eruption impact the entire Earth?
20. Use examples to describe the threshold effect.
21. How does a growing human population affect biodiversity?
22. Identify a feedback mechanism related to the dissolving of carbon dioxide in Earth's oceans and rising global temperatures. Is the feedback mechanism you identified positive or negative? Explain why.
23. What role do bacteria play in the nitrogen cycle?

Think Critically
Write a response to each question on a separate sheet of paper. Use concepts from the chapter to support your reasoning.

24. Compare and contrast gradualism and catastrophism. Which do you think has a bigger effect on Earth? Why?
25. Review the story of the reindeer on St. Matthew Island. How did being on an island affect the reindeer population?
26. Compare and contrast a system found within the human body, such as the circulatory system or digestive system, to an ecosystem. What characteristics make each of these a system? What is the benefit of using a systems approach to studying both the human body and complex ecosystems?
27. Review the environmental to-do list at the end of Lesson 2.2. Which of these steps do you think you can take action on now? How could you take action? Which of these steps do you think will be most challenging for the human population as a whole? Why?

PERFORMANCE TASK

Modeling the Carbon Cycle

The carbon cycle moves carbon through the geosphere, hydrosphere, atmosphere, and biosphere. Human impact on the carbon cycle is quite pronounced due to the burning of fossil fuels. Another significant human impact on the carbon cycle is the production of cement. One of the steps in cement production is the heating of limestone. As the limestone breaks down, it releases carbon dioxide.

In this performance task, you will create your own model of the carbon cycle that includes cement production, following the directions below. You'll use this to predict the effect of cement production on Earth's systems.

1. Sketch the beginnings of your carbon cycle.
    A. Include:
        a. Carbon moving to living things through photosynthesis
        b. Carbon being released to the atmosphere through respiration and decomposition
        c. Carbon being released to the atmosphere through burning of fossil fuels
        d. Carbon being absorbed and released by the oceans
        e. Carbon building up as sediment in oceans
    B. Label the hydrosphere, atmosphere, geosphere, and biosphere in your sketch.
    C. Use the following table to add quantities to your diagram.

Carbon Reservoir	Quantity of Carbon (billions of tons)
Atmosphere	750
Land Plants	500
Fossil Fuels	10,000
Oceans	40,000
Sediments and Limestone	750,000

Source: Data from U. Siegenthaler and J.L. Sarmiento, "Atmospheric Carbon Dioxide and the Ocean," *Nature* 365 (September 9, 1993): 119.

   D. Include in your diagram, the movement of carbon from ocean sediments (limestone) to the atmosphere through the human production of cement.

   Use your diagram to answer the questions that follow.

2. Based on your model, how will cement production affect carbon levels in the atmosphere?
3. Explain how cement production will affect the biosphere, hydrosphere, and geosphere.
4. It is estimated that carbon emissions due to burning fossil fuels are about 20 times higher than emissions from cement production. How could you modify your diagram to reflect this?
5. Use the following table to add quantifying labels to the carbon emissions from fossil fuels and cement production in your diagram.

Carbon transfer	Quantity (billions of tons per year)
Fossil fuels to atmosphere	9.5
Limestone sediments to atmosphere	0.5

Sources: Boden, T.A., Marland, G., and Andres, R.J. (2017). Global, Regional, and National Fossil-Fuel CO2 Emissions. Carbon Dioxide Information Analysis Center, Oak Ridge National Laboratory, U.S. Department of Energy, Oak Ridge, Tenn., U.S.A. doi 10.3334/CDIAC/00001_V2017

With the features of your carbon cycle complete, create a final version that shows how carbon cycles through the geosphere, hydrosphere, atmosphere, and hydrosphere. Use it to model and quantify the effect of cement production and the burning of fossil fuels.

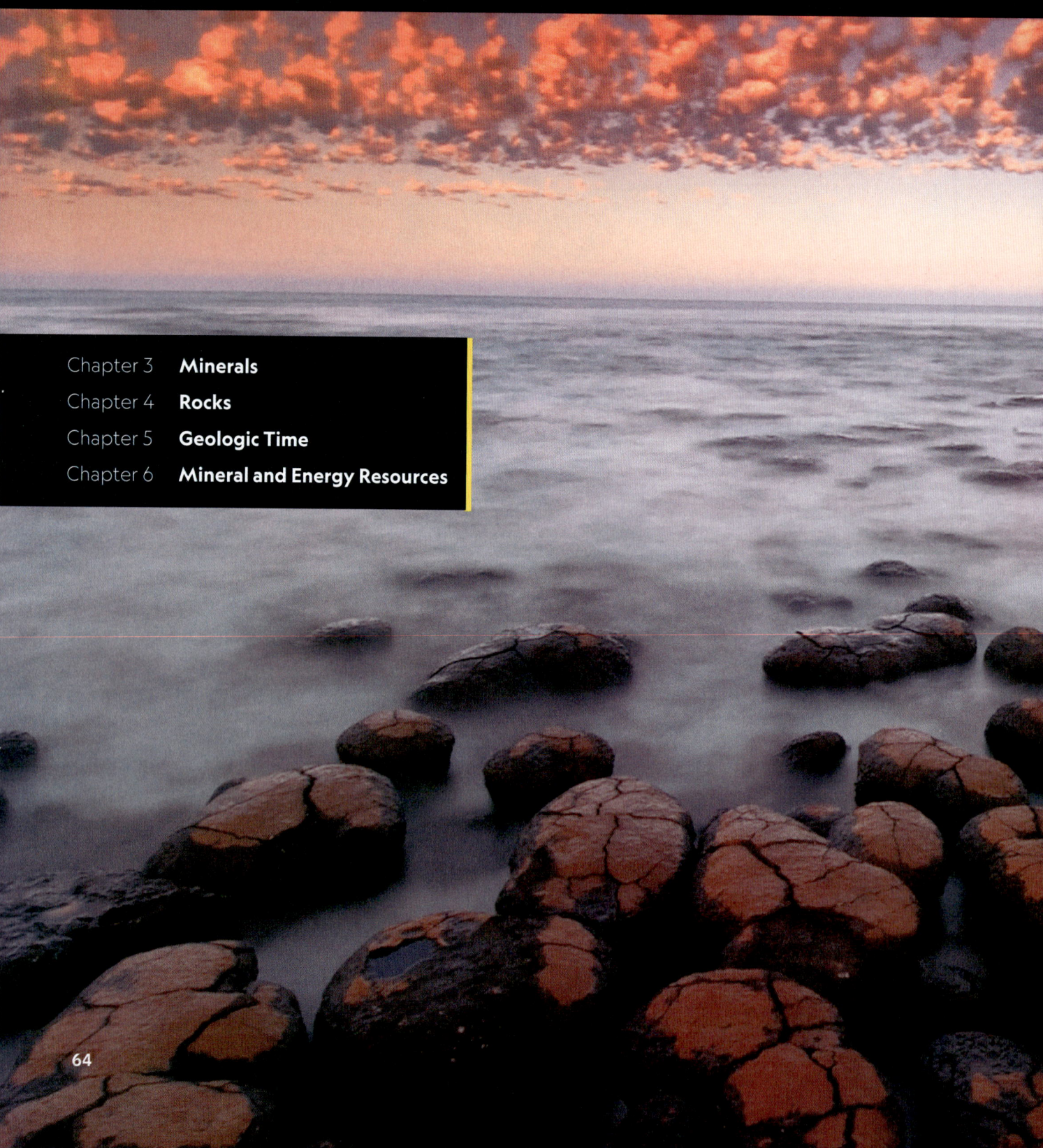

UNIT 2: EARTH MATERIALS AND GEOLOGIC TIME

Chapter 3 **Minerals**
Chapter 4 **Rocks**
Chapter 5 **Geologic Time**
Chapter 6 **Mineral and Energy Resources**

In Western Australia, a mist rises among rust-colored, rounded, rocky objects. Called stromatolites, these rocky structures were formed by one of Earth's oldest types of living organisms, cyanobacteria. Once considered extinct, the stromatolite-forming microbes are the modern representatives of ancient cyanobacteria that thrived about 3.7 billion years ago—a time so distant in the past it is practically incomprehensible to humans. Such "living fossils" provide a glimpse of how Earth might once have appeared. In this unit, you will learn about Earth's incredible history and explore the materials that make up rocks and minerals and how we use them.

The fossilized ancestors of these living stromatolites in Shark Bay, Western Australia, go back billions of years, putting this family of cyanobacteria in one of the longest known continuous lineages on Earth.

CHAPTER 3
MINERALS

Blue liquid fills pools in the white terraces of Pamukkale, a landform in Turkey. The water in the hot springs is saturated with dissolved carbon dioxide gas and the ions of calcium and bicarbonate.

How would you describe this strange landform? You may not be able to tell, but the liquid in these blue pools is pleasingly warm. The white solid material surrounding the pools is travertine, a type of limestone formed from the hot springs. The limestone is made almost entirely of a single mineral called a carbonate. Minerals are the building blocks of Earth's geosphere. Read on to find out about the many minerals that make up the geosphere and how they interact with Earth's other systems.

KEY QUESTIONS

3.1 What distinguishes minerals from rocks and organic matter?

3.2 What are the chemical and structural properties of minerals?

3.3 What physical properties are analyzed to identify minerals?

3.4 What are different classes of minerals and which are rock-forming minerals?

3.5 How does human interaction with minerals affect health and the environment?

EXPLORERS AT WORK

TRACKING MERCURY

WITH NATIONAL GEOGRAPHIC EXPLORER KATLIN L. BOWMAN

Dr. Katlin Bowman is a chemical oceanographer who has been to ports of call across the globe. She's sailed to Tahiti, Ecuador, and the North Pole. She's even traveled to the seafloor inside a deep-ocean submersible. Bowman's destinations are not all far from home. She works in a lab at the University of California Santa Cruz and conducts research in San Francisco Bay. Bowman's extensive time at sea is helping her track levels of toxic mercury contamination in the world's marine environments.

Mercury is a naturally occurring element that exists as a liquid, a solid, or a gas depending on where it is found. Most mercury that enters the ocean is deposited from the atmosphere. Some of the gaseous mercury in Earth's atmosphere is released by natural sources such as volcanoes. But increasingly, gaseous mercury is also being released through human activities, including fossil fuel combustion, such as burning coal for electricity generation. Freshwater and marine ecosystems can also be contaminated by mercury if it has been used in gold-mining operations and made its way into local waters (Case Study).

Bowman and other scientists are concerned about mercury's effects on living organisms. Certain bacteria in the environment can transform mercury into a much more toxic form called methylmercury. Once methylmercury is ingested, it accumulates in tissues and becomes increasingly concentrated as it moves up the food chain. People who regularly consume fish and shellfish contaminated by methylmercury may suffer damage to the nervous system, brain, kidneys, and lungs. Bowman says it is not clear why microorganisms take up mercury and return it to the environment in a chemically altered form, and she is working to identify species that are able to do so.

Bowman is also researching the role of plastic pollution in the mercury cycle. This may be another way in which mercury enters food webs. She's focusing on microplastics, tiny pieces of plastic measuring less than 5 millimeters in length. These bits of plastic float just below the water's surface and have been observed in oceans worldwide. Bowman says mercury tends to stick to microplastic particles. She also says microplastics tend to be ingested by sea animals that mistake them for food. The indigestible plastic either kills the animals or exposes them to toxic mercury. Bowman is analyzing samples of the stomach contents and tissues from fish to determine whether those that ingest more microplastics have higher concentrations of mercury in their bodies.

It also turns out that microplastics provide ideal surfaces for microbial growth. Bowman hypothesizes that microorganisms living on contaminated microplastics may be those that can produce methylmercury. She is growing bacteria on microplastics collected from San Francisco Bay to find out. Bowman hopes her research will provide evidence for the need to minimize plastic pollution entering oceans and other ecosystems. "There is no way to remove mercury that is already contaminating our oceans and land, the only solution is to decrease inputs."

THINKING CRITICALLY

Infer Bowman says even the most remote oceans are contaminated by mercury and gaseous mercury may remain in the atmosphere for up to one year. How might this fact explain Bowman's first statement?

Katlin Bowman (left) and Gretchen Swarr (right) analyze mercury in seawater onboard a research vessel in the Pacific Ocean. Bowman is working to identify microorganisms that change mercury into methylmercury, a form that can accumulate in marine food webs.

Bowman uses special equipment such as a manta net to collect microplastics from the ocean surface.

CASE STUDY
RISING ATMOSPHERIC AND OCEAN MERCURY CONCENTRATIONS LINKED TO SMALL-SCALE GOLD MINING

Mercury is a highly toxic metallic element that occurs naturally in some mineral compounds and inorganic salts. Over the past two centuries, human activities have increased mercury concentrations in the atmosphere by more than 400 percent and in surface ocean waters by more than 200 percent. Efforts to curb mercury production have been challenging. Over the past thirty years, large-scale industrial emissions have significantly declined. Yet mercury concentrations in the air and oceans are still rising. How is this possible, and what does it mean for the environment?

Part of the problem is the staying power of mercury contamination. The mercury released from mining and coal combustion over the past 150 years gets re-emitted as it recycles through Earth's spheres. (See Lesson 3.5, Here to Stay: Mercury.) But researchers have discovered a relatively new source of mercury pollution that is driving global levels up: artisanal small-scale gold mining (ASGM).

ASGM is a subsistence practice often carried out in mineral-rich, impoverished regions. Individual miners or small groups use mercury to separate fine gold particles from ore (Figure 3-1). The United Nations Industrial Development Organization estimates that there are at least 10 million artisanal miners operating in 70 countries. Many of the most active locations are biologically diverse and sensitive tropical forests.

Worldwide, artisanal miners use an estimated 1,400 metric tons of mercury annually. A third of that mercury is emitted into the air during gold processing. The rest makes its way into soil and surface waters. The United Nations Environmental Programme estimates that ASGM accounts for about 35 percent of new global mercury emissions and more than 800 metric tons of mercury discharged to surface waters annually.

These statistics are alarming because mercury is a potent biological toxin. Mercury contamination causes physical abnormalities, reduced reproduction, behavioral changes, and death in many animals and plants. Wetland and marine species suffer the most harm because elemental mercury is converted to a more toxic form of methymercury in aquatic habitats.

Countries associated with ASGM include Brazil, Columbia, Guyana, Indonesia, Ghana, and Peru—yet these countries are also home to ecologically vulnerable areas. In the Amazon rain forest, an artisanal gold rush threatens species such as the critically endangered giant river otter. Ghana, a West African country and the second largest producer of cocoa in the world, has also seen a recent surge in ASGM activity. There, mercury emissions and contaminated runoff have stunted the growth of cocoa trees and damaged fruits. Left unchecked, ASGM could hurt the environment for centuries and endanger thousands of species.

As You Read Cell phones, batteries, roads, and buildings are all made from mineral resources. What is the environmental sustainability of these mineral resources? What actions could you take to make your consumption of mineral resources more sustainable?

FIGURE 3-1
With his bare hands, an artisanal miner presses excess mercury from gold extracted at an illegal mining operation. Artisanal mining uses large quantities of toxic mercury to liberate gold from mined ores, contaminating air, land, and ocean resources and threatening biodiversity.

3.1 WHAT ARE MINERALS?

Core Ideas and Skills
- Describe the characteristics and uses of minerals.
- Explain how the mining of minerals affects humans positively and negatively.
- Explain the differences between minerals, rocks, and organic matter.

KEY TERMS
mineral

What Is a Mineral?

Minerals are the building blocks of rock. Therefore, the geosphere is composed of minerals. Minerals make up soil and rock found at Earth's surface, as well as the deeper rock of the crust and mantle. It is not sufficient to study minerals in isolation from the rest of the planet. Rather, we can learn more by observing the ways in which minerals interact with other Earth systems.

Most minerals in their natural settings are harmless to humans and other species. In addition, many resources that are essential to modern life are produced from rocks and minerals: iron, aluminum, gold, and all other metals; concrete; fertilizers; coal; and nuclear fuels. Both oil and natural gas are recovered from natural reservoirs in rock. The discovery of methods to extract and use these Earth materials has profoundly altered the course of human history. The Stone Age, the Bronze Age, and the Iron Age are historical periods named for the rock and mineral resources that dominated the development of civilization during those times. The industrial revolution occurred when humans discovered how to convert the energy stored in coal into useful work. The automobile age occurred because humans found vast oil reservoirs in shallow rocks of Earth's crust and built engines to burn the refined fuels. The computer age relies on the movement of electrons through wafer-thin slices of silicon, germanium, and other materials that are also crystallized from minerals.

However, some minerals are naturally harmful. For example, uranium-bearing minerals are radioactive, while sulfide-bearing minerals can dissolve and release harmful elements including

FIGURE 3-2
Each of the differently colored grains in this close-up photo of granite is a different mineral. The top of the U.S. penny is for scale.

lead and arsenic into the environment. Some common minerals become harmful to humans when they are crushed to dust during mining, quarrying, road building, and other activities. In this chapter, we will consider the nature of minerals and then return to a discussion of minerals as natural resources and environmental hazards.

Pick up any rock and look at it. You may see small, differently colored specks, such as those you see in the close-up photo of granite in Figure 3-2. Each speck is a mineral. In the photo, the white grains are the mineral feldspar, the black ones are biotite, and the glassy-gray ones are quartz. Most rocks contain two to five abundant minerals plus minor amounts of several others (Figure 3-3). Few

FIGURE 3-3
The Bugaboo Mountains of British Columbia are made of granite. Granite is a rock that contains different minerals.

LESSON 3.1

rocks are made of only one mineral.

Although we can define minerals as the building blocks of rocks, such a definition does not tell us much about the nature of minerals. More precisely, a **mineral** is a naturally occurring inorganic solid with a definite chemical composition and a crystalline structure. Chemical composition and crystalline structure are the two most important properties of a mineral. They distinguish any mineral from all others. Before discussing these two properties, however, let's briefly consider the other properties of minerals.

checkpoint What is a mineral?

Inorganic and Naturally Occurring

A synthetic diamond can be identical to a natural one, but it is not a true mineral, according to our definition, because a mineral must form by natural processes. Like diamonds, most other gems that occur naturally can be manufactured by industrial processes. Natural gems are more highly valued than manufactured ones. For this reason, jewelers should always tell their customers whether a gem is natural or artificial by prefacing the name of a manufactured gem with the term *synthetic*.

Organic substances are composed mostly of carbon that is chemically bonded to hydrogen. Inorganic compounds do not contain carbon–hydrogen bonds. Although organic compounds can be produced in laboratories and by industrial processes, plants and animals create most of Earth's organic material because plant and animal tissue is organic. When these organisms die, they decompose to form other organic substances. For example, coal, oil, and natural gas form by the decay of plants and animals. Because of their organic properties and lack of a definite chemical composition, none of these substances is a mineral.

MINILAB Observe Minerals

Materials
1:1 solution of Epsom salt
300 mL chilled purified water
salt
small paintbrush or toothpick
600-mL clean glass beaker
tablespoon
glass slide
thermometer
microscope
large bowl with enough crushed or cubed ice to submerge half of the beaker

Procedure

Part A.

1. Using a clean paintbrush or toothpick, spread a small amount of Epsom salt solution on a glass slide.
2. Observe the slide at 40X under a microscope. Record your observations.

Part B.

1. Pour 300 mL chilled purified water into the glass beaker.
2. Place the beaker in a bowl. Add ice to the bowl until the level of the ice is higher than the level of water in the beaker. Avoid spilling any ice into the beaker.
3. Sprinkle a couple of tablespoons of salt onto the ice. Take care not to get any of the salt in the beaker of water.
4. Insert a thermometer into the beaker of water. When the temperature of the water reaches −2°C, remove the thermometer. Then slowly and carefully remove the beaker from the bowl of ice.
5. Drop a small piece of ice into the beaker and watch closely. Record your observations.

Results and Analysis

1. **Identify** In Part A, your teacher supplied you with a saturated solution of Epsom salt ($MgSO_4$) and water (H_2O). Which component of this solution is a mineral? How do you know?
2. **Explain** What process allowed the crystals to form on the slide in Part A?
3. **Compare** How is the frozen water in Part B similar to the salt crystals you observed in Part A? How is it different?

Critical Thinking

4. **Evaluate** Is frozen water a mineral? Is distilled water a mineral? Explain your reasoning.
5. **Apply** Compare the physical or chemical changes involved in the formation of the ice crystals and $MgSO_4$ crystals.

All minerals are also solids, which disqualifies oil and natural gas. On the other hand, ice *is* a mineral, but not liquid water or water vapor.

Some material that organisms produce is not organic. For example, limestone, one of the most common sedimentary rocks, commonly contains the shells of clams, snails, and similar marine organisms. Shells, in turn, are made mostly of the mineral calcite. Although organisms produce calcite, the calcite is a mineral—an inorganic solid that formed naturally and has a definite chemical composition and crystalline structure.

checkpoint Why are fossil fuels not considered minerals?

3.1 ASSESSMENT

1. **Explain** How are minerals related to rocks?
2. **Identify** What two properties distinguish any mineral from all other minerals?
3. **Make Comparisons** Summarize the basic differences between organic and inorganic substances.

Critical Thinking

4. **Evaluate** People consider naturally occurring gems more valuable than manufactured gems even though they are made of the same substances. Do you think this is a reasonable consideration? Explain why or why not.

3.2 COMPOSITION AND STRUCTURE OF MINERALS

Core Ideas and Skills
- List examples of minerals that are made of different elements.
- Describe the crystalline structure of minerals.

KEY TERMS
element	anion
ion	crystal
cation	crystal face

Chemical Composition of Minerals

Minerals are the fundamental building blocks of rocks, but what are minerals made of? Minerals and all other Earth materials are composed of chemical elements. An **element** is a fundamental component of matter that cannot be broken into simpler particles by ordinary chemical processes. Most common minerals consist of a small number of different chemical elements—usually two to five.

A total of 88 elements occur naturally in Earth's crust. However, eight elements—oxygen, silicon, aluminum, iron, calcium, sodium, potassium, and magnesium—make up more than 98 percent of the crust (Table 3-1). Each element is represented by a one- or two-letter symbol, such as O for oxygen and Si for silicon. In nature, most chemical elements have either a positive or negative electrical charge. For example, oxygen has a charge of negative 2 (−2) and silicon has a charge of plus 4 (+4). An atom with an electrical charge, whether it is positive or negative, is called an **ion**. A positively charged ion is a **cation**; a negatively charged ion is an **anion**. Ions with opposite charges are attracted to each other, like the positive end of a magnet attracts the negative end.

Recall that a mineral has a definite chemical composition. A substance with a definite chemical composition is made up of elements that are combined in definite proportions. Therefore, the composition of a mineral can be expressed as a chemical formula, written by combining the

TABLE 3-1 The Eight Most Abundant Chemical Elements in Earth's Crust

Element	Symbol	Percentage of Crust by Number of Atoms
Oxygen	O	62.55
Silicon	Si	21.22
Aluminium	Al	6.47
Sodium	Na	2.64
Calcium	Ca	1.94
Iron	Fe	1.92
Magnesium	Mg	1.84
Potassium	K	1.42
	Totals	100.00

symbols of the individual elements with the number of each element within the molecular unit that forms the mineral crystal. A few minerals, such as gold and silver, consist of only a single element. Their chemical formulas, respectively, are Au (the symbol for gold) and Ag (the symbol for silver). Most minerals, however, are made up of two to five essential elements. For example, the formula of quartz is SiO_2. It consists of one atom of silicon (Si) for every two atoms of oxygen (O). Quartz from anywhere in the universe has that exact same composition. If it had a different composition, it would be some other mineral. The compositions of some minerals, such as quartz, do not vary by even a fraction of a percent. The compositions of other minerals vary slightly, but the variations are limited.

The 88 elements that occur naturally in Earth's crust can combine in many ways to form many different minerals. More than 3,500 minerals are known. However, the eight abundant elements commonly combine in only a few ways. As a result, only nine rock-forming minerals (or mineral "groups") make up most rocks of Earth's crust. Rock-forming minerals will be described later in the chapter, in Lesson 3.4.

checkpoint In SiO_2, for every atom of silicon, how many atoms of oxygen are there?

Crystalline Structure of Minerals

Every mineral has a crystalline structure. Therefore, every mineral is a crystal. A **crystal** is any solid element or compound whose atoms are arranged in a regular, periodically repeated three-dimensional pattern, like a wallpaper pattern that extends into three dimensions. The mineral halite (common table salt) has the composition NaCl. This formula means that there is one sodium ion (Na^+) for every chlorine ion (Cl^-). Figure 3-4 shows both exploded and realistic views of the ions in halite. In both images, the sodium and chlorine ions alternate in orderly rows and columns intersecting at right angles. This orderly, repetitive arrangement of atoms is the crystalline structure of halite.

Most minerals initially form as tiny crystals that grow as layer after layer of atoms (or ions) are added to their surfaces. A halite crystal might grow, for example, as salty seawater evaporates from a tidal pool. At first, a tiny mineral grain might form, similar to the sketch of halite in Figure 3-4. This model shows a halite crystal containing 125 atoms; it would be only about one-millionth of a millimeter long on each side. It would be far too small to see with the unaided eye. As evaporation continued, more and more sodium and chlorine ions would precipitate onto the corners, edges, and faces of the growing crystal.

A **crystal face** is a flat surface that develops if a crystal grows freely in an uncrowded environment,

FIGURE 3-4 ▼
NaCl, or halite, consists of an orderly arrangement of sodium and chlorine ions. The crystal model in (A) is exploded so that you can see into it. The ions are actually closely packed, as represented in (B).

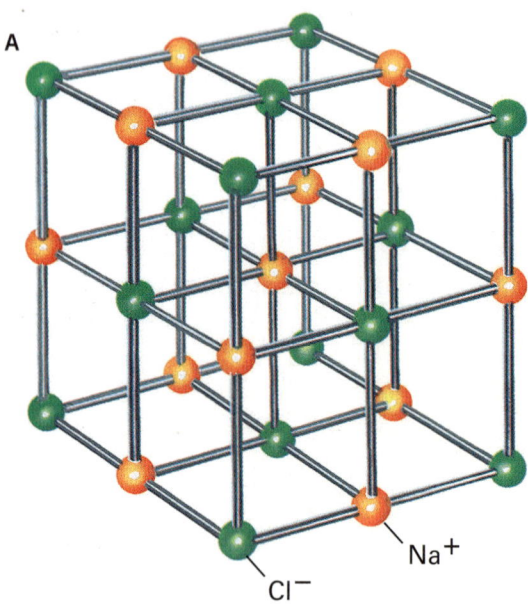

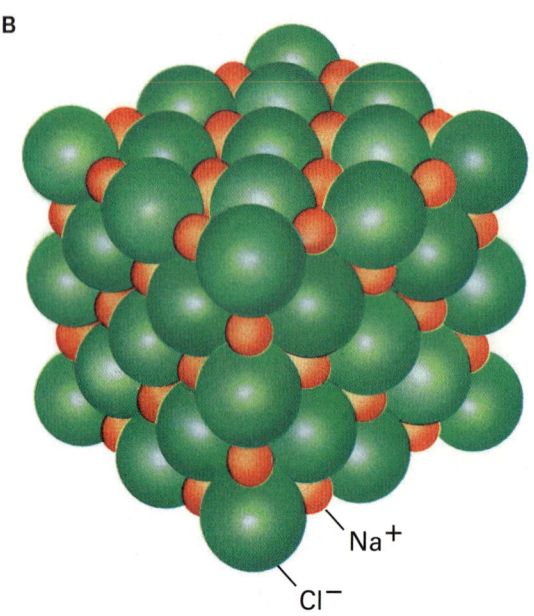

FIGURE 3-5
Photomicrograph of a thin slice (0.03 mm) of granite

such as halite growing in a pool of evaporating seawater. When a mineral grows freely like this, it commonly forms a symmetrical crystal with perfectly flat faces that reflect light like a mirror. In nature, however, mineral crystals often impede the growth of adjacent crystals. For this reason, minerals rarely show perfect development of crystal faces. Figure 3-5 is a photomicrograph (a photo taken through a microscope) of a thin slice of granite in which the crystals fit like pieces of a jigsaw puzzle. This interlocking texture developed because some crystals grew around others as molten magma cooled and solidified.

checkpoint What is a crystal face of a mineral?

3.2 ASSESSMENT

1. **Summarize** What is a substance with a definite chemical composition made up of? Why is this significant to determine what mineral the substance is?
2. **Explain** Describe how crystals grow to develop into minerals.
3. **Explain** In nature, why do crystals rarely grow to develop symmetrical crystal faces?

Computational Thinking

4. **Analyze Data** Oxygen is by far the most abundant element in Earth's crust. Refer to Table 3-1. Use the percentages to calculate about how many times more abundant oxygen is in Earth's crust than the other elements shown.

3.3 PHYSICAL PROPERTIES OF MINERALS

Core Ideas and Skills
- Distinguish the physical properties of minerals and how they are measured.
- Describe several methods for the identification of minerals.

KEY TERMS

Mohs' scale	cleavage
streak	fracture
luster	specific gravity
crystal habit	

Mineral Identification

How does a geologist identify a mineral in the field? Chemical composition and crystal structure distinguish each mineral from all others. For example, halite always consists of sodium and chlorine in a one-to-one ratio, with the atoms arranged in a cubic fashion. But if you pick up a crystal of halite, you cannot see the atoms. You could identify a sample of halite by measuring its chemical composition and crystal structure using laboratory procedures. But such analyses are expensive and time-consuming. Instead, geologists can identify minerals by visual recognition and then confirm the identification with simple tests.

Most minerals have distinctive appearances. Once you become familiar with common minerals, you will recognize them just as you recognize any familiar object. For example, an apple looks like an apple, even though apples come in many colors and shapes. In the same way, quartz looks like quartz to a geologist. Some minerals, however, look like others, so their physical properties must be examined further to make a correct identification. For example, halite can look like calcite, quartz, and several other minerals, but halite tastes salty and scratches easily with a knife blade. These two characteristics distinguish it from the other minerals. Geologists commonly use properties such as crystal habit, cleavage, fracture, hardness, specific gravity, color, streak, and luster to identify minerals.

checkpoint What are the physical properties that geologists commonly use to identify minerals?

FIGURE 3-6
Pyrite, or fool's gold, has a metallic luster.

Hardness

Hardness is the resistance of a mineral to scratching and is one of the most commonly used properties for identifying a mineral. It is easily measured and is a fundamental property of each mineral because it is controlled by the bond strength between the atoms in the mineral. Geologists commonly gauge hardness by attempting to scratch a mineral with the tip of a pocketknife. If the metal scratches the mineral, the mineral is softer. If the knife tip cannot scratch the mineral, the mineral is harder.

To measure hardness more accurately, geologists use a scale based on 10 minerals, numbered 1 through 10 (Table 3-2). Each mineral is harder than those with lower numbers on the scale—so 10 (diamond) is the hardest and 1 (talc) is the softest. The scale is known as the **Mohs' scale**, after Friedrich Mohs, the Austrian mineralogist who developed it in the early 19th century.

As you can see from Table 3-2, Mohs' scale shows that a mineral scratched by quartz but not by orthoclase would have a hardness between 6 and 7. Because the minerals of Mohs' scale are not always handy, it is useful to know the hardness values of common materials. A fingernail has a hardness of slightly more than 2, a pocketknife blade slightly more than 5, window glass about 5.5, and a steel file about 6.5. If you practice with a knife and the minerals of Mohs' scale, you can develop a "feel" for minerals with hardness values of 5 and under by seeing how easily the blade scratches them.

When testing hardness, it is important to determine whether the mineral has actually been scratched by the object or the object has simply left a trail of its own powder on the surface of the mineral. To check, rub away the powder trail and feel the surface of the mineral with your fingernail for the groove of the scratch. Fresh, unweathered mineral surfaces must be used in hardness measurements because weathering often produces a soft rind on minerals.

checkpoint How can hardness of minerals be determined?

Color, Streak, and Luster

Color is the most obvious property of a mineral, and it is often used in identification. But color can be unreliable because small amounts of

TABLE 3-2 Mohs' Scale of Mineral Hardness

Hardness	Mineral	Common Objects with Similar Hardness
1	Talc	
2	Gypsum	Fingernail (2.5)
3	Calcite	Copper Penny (3.5)
4	Fluorite	
5	Apatite	Glass (5.5)
6	Orthoclase	Stell file (6.5)
7	Quartz	Streak plate (7)
8	Topaz	
9	Corundum	
10	Diamond	

FIGURE 3-7
(A) Prismatic quartz grows as elongated crystals. (B) Massive rosy quartz shows no characteristic shape.

chemical impurities and imperfections in crystal structure can dramatically alter color. For example, corundum (Al_2O_3) is normally a cloudy, translucent, brown or blue mineral. The addition of a small amount of chromium can convert corundum to the beautiful, clear, red gem known as ruby. A small quantity of iron or titanium turns corundum into the striking blue gem called sapphire. Quartz also occurs in many colors, including white, clear, black, purple, and red, as a result of tiny amounts of impurities and minor defects in the perfect ordering of atoms.

Streak refers to the color of the fine powder of a mineral. It is observed by rubbing the mineral across a piece of unglazed porcelain known as a "streak plate." Many minerals leave on the plate a streak of powder with a diagnostic color. Streak is commonly more reliable for identification than the color of the mineral itself. For example, the mineral hematite (iron oxide) can occur both as a dull red mineral or in a shiny black form that closely resembles black mica, but both types will leave the same red powder on a streak plate.

Luster is the manner in which a mineral reflects light. A mineral with a metallic look, irrespective of color, has a metallic luster. Pyrite is a yellowish mineral with a metallic luster (Figure 3-6). As a result, it looks like gold and is commonly called fool's gold. The luster of nonmetallic minerals is usually described by words such as *metallic, glassy, pearly, earthy, greasy,* and *resinous*.

checkpoint Why is streak commonly more reliable for identifying a mineral than color?

Crystal Habit, Cleavage, and Fracture

Crystal habit refers to the characteristic shape of an individual crystal and the manner in which crystals grow together. If a crystal grows freely, it develops a characteristic shape that the arrangement of its atoms controls. Some minerals occur in more than one habit. For example, Figure 3-7A shows quartz with a prismatic (pencil-shaped) habit, and Figure 3-7B shows a massive quartz in which numerous individual microscopic crystals intergrow to produce the habit.

FIGURE 3-8
Mica has a single, perfect cleavage plane.

Cleavage is the tendency of some minerals to break along flat surfaces. The surfaces are planes of weak bonds in the crystal. Some minerals, such as the micas, have one set of parallel cleavage planes (Figure 3-8). Others have two, three, or even four different sets, as shown by the three photos in Figure 3-9. In some cases, the expression of the cleavage is excellent. For example, you can peel sheet after sheet from a mica crystal as if you were peeling layers from an onion. In other cases, cleavage is not well-developed. Many minerals, such as quartz, have no cleavage at all because they have no planes of weak bonds. The number of cleavage planes, the expression—or quality—of cleavage, and the angles between cleavage planes all help in mineral identification.

A flat surface created by cleavage and a flat crystal face can appear to be very similar. However, a cleavage surface is repeated when a crystal is broken, whereas a crystal face is not.

Fracture is the manner in which a mineral breaks other than along planes of cleavage. Many minerals fracture into characteristic shapes. For instance, a conchoidal fracture creates smooth, curved surfaces (Figure 3-10). It is characteristic of quartz and olivine. Some minerals break into splintery or fibrous fragments; and most of them fracture into irregular shapes.

checkpoint Compare and contrast the cleavage of mica and calcite.

Specific Gravity and Other Properties

Specific gravity is the weight of a substance relative to that of an equal volume of water. If a mineral weighs 2.5 times as much as an equal volume of water, its specific gravity is 2.5. You can estimate a mineral's specific gravity simply by lifting a sample in your hand. If you practice with known minerals, you can develop a feel for specific gravity. Most common minerals have specific gravities of about 2.7. Metals have much higher specific gravities; for example, lead is 11.3, silver is 10.5, and copper is 8.9. Gold has the highest specific gravity of all minerals at 19.

FIGURE 3-9
Some minerals have more than one cleavage plane. (A) The mineral microcline (a feldspar) shows two well-developed cleavage planes that intersect at right angles. (B) Calcite has three sets of cleavage planes that do not intersect at right angles. (C) Fluorite has four cleavage planes. Each cleavage face has a parallel counterpart on the opposite side of the crystal.

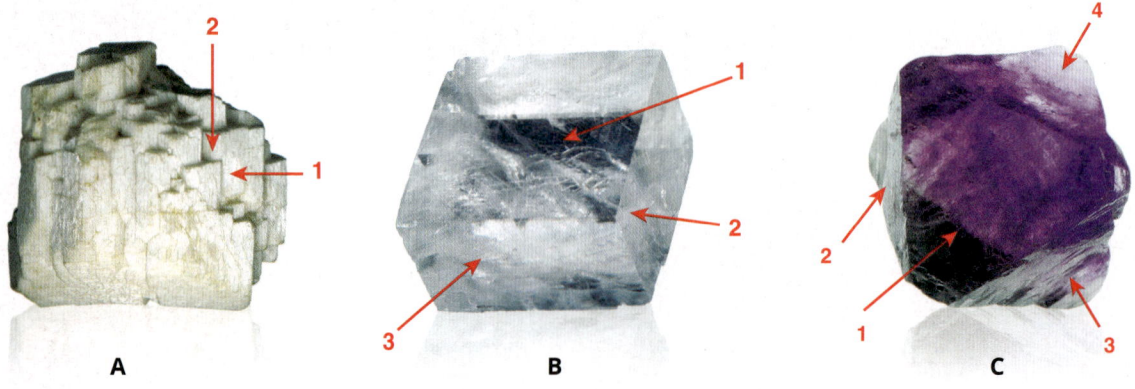

A B C

FIGURE 3-10
Some minerals without cleavage break along smoothly curved surfaces, called conchoidal fractures. This sample is quartz.

Properties such as reaction to acid, magnetism, radioactivity, fluorescence, and phosphorescence can be characteristic of specific minerals. Calcite and some other carbonate minerals dissolve rapidly in acid, releasing visible bubbles of carbon dioxide gas. The mineral magnetite displays an obvious, diagnostic attraction to magnets. Carnotite, a uranium-bearing mineral, emits radioactivity that can be detected with a device called a scintillometer. Fluorescent materials emit visible light when they are exposed to ultraviolet light. Phosphorescent minerals continue to emit light after the external stimulus ceases.

checkpoint What is specific gravity?

3.3 ASSESSMENT

1. **Identify** List the physical properties of minerals that are most useful for identification.
2. **Distinguish** Why is color often an unreliable property for mineral identification?
3. **Explain** Why do some minerals have cleavage whereas others do not?

Critical Thinking

4. **Design** Propose a set of diagnostic tests to identify a small set of minerals. Determine the order of tests and the properties you would analyze. List any tools you might need to complete the evaluations.

3.4 MINERAL CLASSES AND THE ROCK-FORMING MINERALS

Core Ideas and Skills

- Explain how systems are made of smaller subsystems and give examples.
- Explain how matter and energy enter and leave systems but are conserved.
- Describe how gradual and catastrophic changes to Earth's spheres have occurred throughout geologic time.
- Describe the effects of human interaction with Earth's systems.

KEY TERMS

silicate	native element
carbonate	silicate tetrahedron
sulfide	ore

Mineral Groups

Geologists classify minerals according to their chemical elements (Table 3-3). For example, the **silicates** contain silicon and oxygen, the carbonates carbon and oxygen, and the **sulfides** sulfur (without oxygen). The **native elements** are a small class of minerals, including pure gold and silver, that consist of only a single element.

Although more than 3,500 minerals are known to be in Earth's crust, only a small number (between 50 and 100) are common or valuable. Nine rock-forming minerals make up most of the crust; they are composed primarily of the eight most common elements in Earth's crust (Table 3-1).

The nine rock-forming minerals are feldspar, quartz, pyroxene, amphibole, mica, clay, olivine, calcite, and dolomite. They fall into two of the mineral groups shown in Table 3-3. Calcite and dolomite are carbonates, while the other seven groups are silicates.

checkpoint To which two mineral groups do the nine rock-forming minerals belong?

LESSON 3.4

TABLE 3-3 Important Mineral Groups

Group	Example Members	Economic Uses
Oxides (O^{2-} anion)	Hematite, Magnetite, Corundum, Chromite	Ores of iron and chromium, gemstones
Sulfides (S^- anion)	Galena, Sphalerite, Pyrite, Chalcopyrite, Bornite, Cinnabar	Ores of lead, zinc, copper, and mercury
Sulfates (SO_4^{4-} anion)	Gypsum, Anhydrite, Barite	Plaster, drilling mud
Native elements	Gold, Copper, Diamond, Sulfur, Graphite, Silver, Platinum	Electronics, jewelry, gemstones, chemicals, pencils, dry lubricant, photography, catalysts
Halides (halogens such as Cl^- and F^-)	Halite, Fluorite, Sylvite	Common salt, steel making, fertilizers
Carbonates (CO_3^{2-} anion)	Calcite, Dolomite, Aragonite	Portland cement
Hydroxides (OH^- anion)	Limonite, Bauxite	Ores of iron and aluminum, pigments
Phosphates (PO_4^{3-} anion)	Apatite, Turquoise	Fertilizer, gemstones
Silicates ($SiO4_4^{4-}$ anion)	Quartz, Feldspar, Mica, Amphibole, Pyroxene, Olivine, Clay	Glass, ceramics, paints, drilling mud, construction materials, gemstones, paper

Rock-Forming Minerals

Silicates Silicates make up about 92 percent of Earth's crust. They are abundant for two reasons. First, silicon and oxygen are the two most plentiful elements in Earth's crust. Second, silicon and oxygen combine easily.

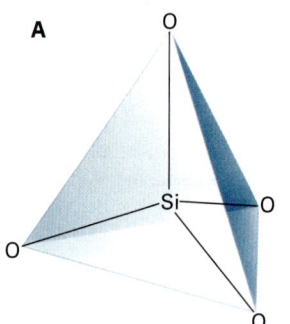
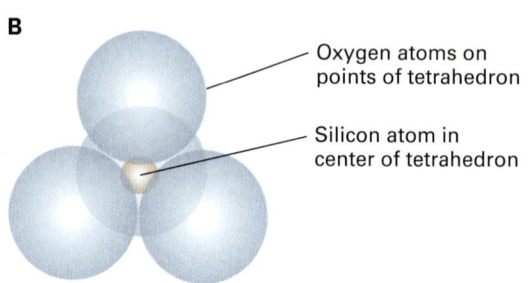

FIGURE 3-11 The silicate tetrahedron consists of one silicon atom surrounded by four oxygen atoms. It is the fundamental building block of all silicate minerals. (A) A schematic representation is shown. (B) A proportionally accurate model is shown.

To understand silicate minerals, remember that:

1. Every silicon atom in a silicate mineral surrounds itself with four oxygen atoms. The bonds between silicon and its four oxygen atoms are very strong.

2. The silicon atom and its four oxygen atoms form a pyramid-shaped structure, called the **silicate tetrahedron**. A tetrahedron (plural, tetrahedra) is simply a pyramid shape with four triangular faces. In the silicate tetrahedron, a silicon atom is at the center, and four oxygen atoms form the four corners (Figure 3-11). The silicate tetrahedron is the fundamental building block of all silicate minerals.

3. Most silicate tetrahedra combine with additional elements in Earth's crust. Quartz is the only common silicate that contains just silicon and oxygen.

4. Silicate tetrahedra commonly link together by sharing oxygen atoms to form chains, sheets, or three-dimensional networks as shown in Figure 3-12.

The silicate minerals fall into five groups, based on five ways in which tetrahedra link together (Figure 3-12). Each group contains at least one of the seven rock-forming silicates.

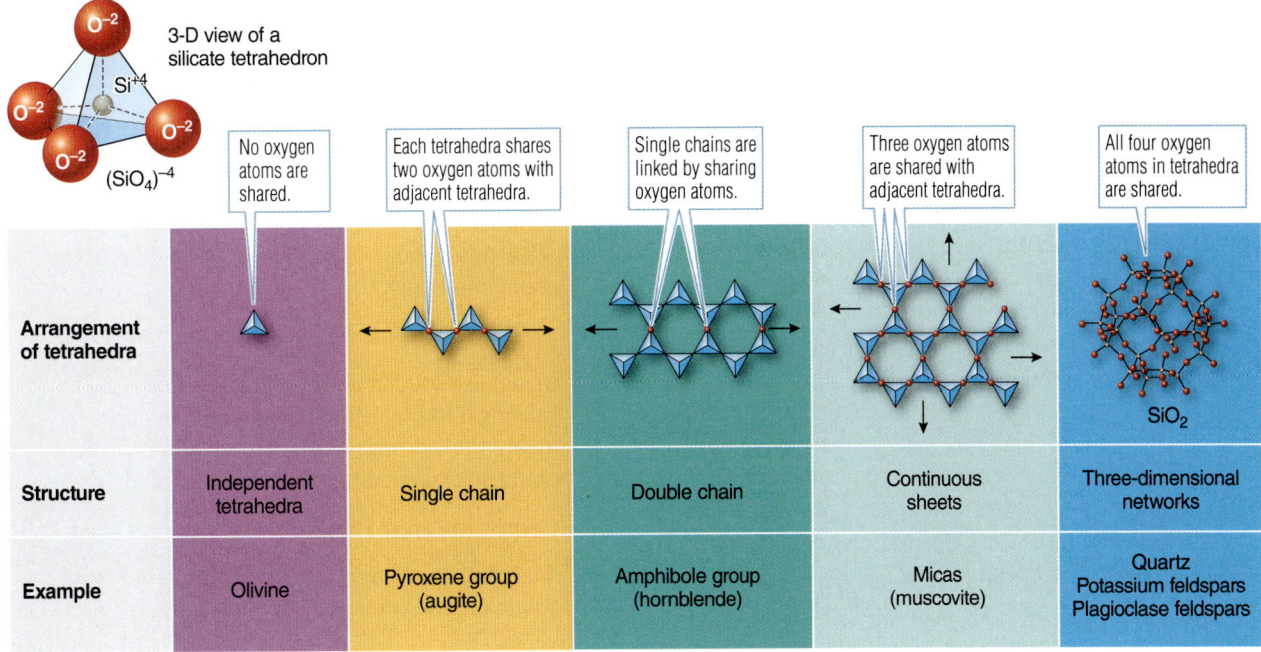

FIGURE 3-12
The five silicate structures are based on sharing of oxygen atoms among the silicate tetrahedra.

Figure 3-13 shows that feldspar alone makes up roughly half of Earth's crust. Feldspars are divided into two main groups: plagioclase feldspar contains calcium, and alkali feldspar contains potassium or sodium instead of calcium. Orthoclase is a common alkali feldspar and is abundant in granite and other rocks of the continents. Rocks of the oceanic crust contain about 50 percent plagioclase.

Quartz comprises 12 percent of crustal rocks. It is widespread and abundant in most continental rocks but rare in oceanic crust and the mantle. Quartz is the only common silicate mineral that contains no cations other than silicon. Quartz is pure SiO_2.

Pyroxene makes up another 11 percent of the crust. The basalt rock of oceanic crust is made up almost entirely of two minerals, plagioclase feldspar and pyroxene.

Basalt commonly contains 1 or 2 percent olivine. But olivine is otherwise uncommon in Earth's crust. However, basalt makes up a large proportion of mantle rocks, so olivine is an abundant and important mineral.

Amphibole, mica, and clay are the other rock-forming silicate minerals. Amphiboles and micas are common in granite and many other types of continental rocks but are rare in oceanic crust. Clay forms when rain and atmospheric gases chemically decompose other silicate minerals. Therefore, clay is common in soils. Clay minerals are especially common in mudstone, which is the most abundant of all sedimentary rocks and is described in Chapter 4.

Carbonates Carbonate minerals are much less common overall than silicates. But many sedimentary rocks are formed from carbonate minerals. These now cover large regions of every continent, as described in the next chapter. The shells and other hard parts of many marine organisms—including clams, snails, corals, and even certain types of plankton and green algae—are made of carbonate minerals. The hard pieces

FIGURE 3-13
More than 92 percent of Earth's crust is composed of silicate minerals. Feldspar alone makes up about 50 percent of the crust, and pyroxene and quartz constitute another 23 percent.

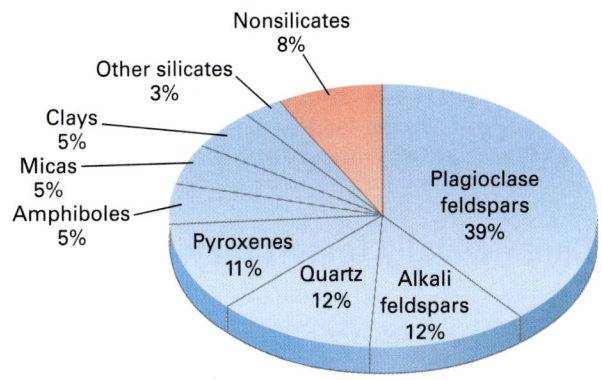

Source: Klein, C. (2002). *The 22nd edition of the manual of mineral science.* New York, NY: John Wiley and Sons.

LESSON 3.4

DATA ANALYSIS: Abundance and Distribution of Silicate Minerals

Silicate minerals are abundant in Earth's crust but are not equally distributed because of differences in how parts of the crust are formed. Continental crust is older, more exposed to weathering, and predominantly composed of granitic rocks. Basalt and gabbro make up the majority of the oceanic crust. Basalt is also found on volcanoes and lava fields.

Sandstone forms from eroded silicates and is a major component of the sedimentary cover of both continental and oceanic crusts. Mudstone, the most abundant of all sedimentary rocks, consists mostly of clay minerals that have become stone. Mudstone can also contain minor amounts of quartz, other minerals, and organic material. Use this information with Table 3-4 to answer the questions.

1. **Identify** Which silicate is relatively evenly distributed across these five rock types?
2. **Infer** At which location would you most likely find quartz crystals: a sandy beach or along the slope of a volcano? Why?
3. **Contrast** How is the overall mineral composition of Earth's oceanic crust different from that of the continental crust?

Critical Thinking

4. **Generalize** Mudstone has a planar cleavage, causing it to break into thin sheet-like pieces. Relate the cleavage of mudstone to the cleavage of the silicate minerals that make up mudstone.

Computational Thinking

5. **Analyze Data** What class of silicates is the most abundant and widely distributed?

TABLE 3-4 Circles of Four Different Sizes Representing the Relative Abundance of Silicate Minerals in Common Rock Types

Silicate Group \ Common Rock Types	Basalt	Gabbro	Granite	Sandstone	Mudstone
Plagioclase	●	●	○	·	
Alkali-Feldspar			●	·	
Quartz			●	●	·
Pyroxenes	●	●			
Amphiboles			·		
Micas	·	·	·		
Clays				·	●

accumulate to form the abundant sedimentary rock called limestone. Limestone is comprised of calcite ($CaCO_3$). It is a rock type mined as a raw ingredient of cement. Another very common carbonate mineral is dolomite ($CaMg(CO_3)_2$), which forms a rock also called dolomite, or dolostone.

checkpoint Why are silicates the most common type of mineral in Earth's crust?

Commercially Important Minerals

Many minerals are commercially important even though they are not abundant. Our industrial society depends on metals such as iron, copper, lead, zinc, gold, and silver.

An **ore** is a substance from which minerals, metals, or other elements can be profitably mined or recovered. A few metal ores, such as native gold and native silver, are composed of a single element. Most metals, however, are chemically bonded to other elements. Iron is commonly bonded to oxygen. Copper, lead, and zinc are commonly bonded to sulfur to form sulfide minerals. Galena (shown in Figure 3-14) is the most abundant sulfide minerals and is an important lead ore.

FIGURE 3-14
Go online to view an example of galena.

Industrial minerals are not metal ores, fuels, or gems, but they have economic value nonetheless. Halite is mined for table salt. Gypsum is mined for

FIGURE 3-15
Yellow native sulfur is forming today in the vent of the Ollague volcano on the Chile-Bolivia border.

plaster and drywall. Apatite and other phosphate-containing minerals are sources of the phosphate fertilizers used in modern agriculture. Calcite is a raw ingredient of cement. Native sulfur—used to manufacture sulfuric acid, insecticides, fertilizer, and rubber—is mined from the craters of dormant and active volcanoes, where it is deposited from gases emitted from vents like the one in Figure 3-15.

A gem is a mineral that is prized primarily for its rarity or beauty. Some gems, such as diamonds, are also used industrially. Depending on its value, a gem can be either "precious" or "semiprecious." Precious gems include diamond, emerald, ruby, and sapphire (Figure 3-16). Several varieties of quartz—including amethyst, agate, jasper, and tigereye—are semiprecious gems. Garnet, olivine, topaz, turquoise, and many other minerals can also occur as attractive semiprecious gems (Figure 3-17).

⊕ **FIGURE 3-16**
Go online to view an image of a sapphire.

⊕ **FIGURE 3-17**
Go online to view an image of topaz.

When you look at a lofty mountain or a steep cliff, you might not immediately think about the tiny mineral grains that form the rocks. Yet minerals are the building blocks of Earth. In addition, some minerals provide the basic resources for our industrial civilization. As we will see throughout this book, Earth's surface and the human environment change as minerals react with the air and water, with other minerals, and with living organisms.

checkpoint What are some commercially important minerals, and how are they used?

3.4 ASSESSMENT

1. **Define** List the rock-forming minerals.
2. **Classify** Which minerals are rock-forming silicates, and why are there so many silicates?
3. **Distinguish** Provide examples of minerals that are not metal ores but are valued as industrial minerals.
4. **Classify** List some precious gems and some semiprecious gems.

Critical Thinking

5. **Infer** If most metals are bonded to other elements, what can you infer about methods for accessing a metal for industrial use?

LESSON 3.4

ON ASSIGNMENT
FIGURE 3-18 Diamond is the hardest mineral, but it can be cut to make a dazzling gem. A laser was used to make the engraving on the edge of the diamond. This image was taken by National Geographic photographer Cary Wolinksy.

3.5 HAZARDOUS ROCKS AND MINERALS

Core Ideas and Skills

- Identify rocks and minerals that are hazardous to mine, and explain why they are hazardous to people.
- Describe the risks of exposure to radon gas and how exposure can be kept to a minimum.
- Describe how the element mercury can be released into the environment and the consequences of exposure to mercury.

KEY TERMS

carcinogenic

Most rocks and minerals in their natural states are harmless to humans and other organisms. That is not surprising. All organisms evolved among the minerals and rocks that make up Earth, and species would not survive if they were poisoned by their surroundings. A few rocks and minerals, however, are harmful to human health. Asbestos is one such mineral; it is a powerful carcinogen—a cancer-causing substance. Some rocks and minerals emit radon gas, a radioactive carcinogen. Other rocks and minerals contain toxic elements such as mercury, lead, arsenic, uranium, and sulfur.

In nature, most environmentally hazardous rocks and minerals are buried beneath the surface. Here, they are inaccessible to plants and animals. Weathering and erosion expose them slowly so they do little harm. Most minerals that contain toxic metals or other elements such as lead, mercury, and arsenic are also relatively insoluble. In most natural environments, they weather so slowly that they release the toxic materials in low concentrations. However, if pollution controls are inadequate, mining, milling, and processing metals from ore can concentrate and release hazardous natural materials at greatly accelerated rates, poisoning humans and other organisms.

Mineral Extraction and Human Health

Most common minerals such as feldspar and quartz are harmless in their natural states in solid rock. However, if these minerals are ground to dust, they can enter the lungs and cause serious and even fatal inflammation and scarring of the lung tissue. This condition is called silicosis. In advanced cases, the lungs become inflamed and may fill with fluid. On-the-job exposure to silica dust occurs in several occupations and hobbies. These include mining, stonecutting, quarrying, building and road construction, working with abrasives, and sandblasting. Intense exposure to silica may produce silicosis in a year or less. But it usually takes at least 10 or 15 years of exposure before symptoms develop. Silicosis has become less common since the Occupational Safety and Health Administration (OSHA) instituted regulations requiring the use of protective equipment that limits the amount of silica dust inhaled. Coal dust was inhaled by coal miners in large quantities before federal laws required dust reduction in coal mines. Breathing in coal dust can cause a disease called black lung.

Asbestos is an industrial name for a group of minerals that crystallize as long, thin fibers. One type is a sheet silicate mineral called chrysotile. This mineral has a crystal structure and composition similar to that of the micas and forms tangled, curly fibers (Figure 3-19). The other type of asbestos includes four similar varieties of amphibole. These minerals crystallize as straight, sharply pointed needles. Chrysotile has been the more valuable and commonly used type of asbestos because the fibers are flexible and tough and can be woven into fabric. It was considered to be a less harmful form because some believed its curly fibers would be more easily expelled from the lungs. However, recent studies show that all forms of asbestos are **carcinogenic**, or have the potential to cause cancer.

FIGURE 3-19
Chrysotile asbestos is believed by some to be a less potent carcinogen than other forms of asbestos. But it is still being actively removed from public buildings.

Asbestos is commercially valuable because it is flameproof, chemically inert, and extremely strong. For example, chrysotile fibers are eight times stronger than a steel wire of equivalent diameter. Asbestos has been used to manufacture brake linings, fireproof clothing, insulation, shingles, tiles, pipes, and gaskets, but now is allowed only in brake pads, shingles, and pipes.

In the early 1900s, asbestos miners and others who worked with asbestos learned that prolonged exposure to the fibers caused an irreversible respiratory disease called asbestosis. Later, in the 1950s and 1960s, studies showed that asbestos also causes lung cancer and other forms of cancer. One reason so much time passed before scientists recognized the cancer-causing properties of asbestos is that the disease commonly does not develop until decades after exposure.

Although it is not clear how asbestos fibers cause cancer, it seems that the shapes of the crystals play an important role. Statistical studies also show that the sharp, pointed amphibole asbestos is a more potent carcinogen than the flexible chrysotile fibers, although both types cause cancer. In response to growing awareness of the health effects of asbestos, the Environmental Protection Agency (EPA) banned its use in building construction in 1978. However, the ban did not address the issue of what should be done with the asbestos already installed. In 1986, Congress passed a ruling called the Asbestos Hazard Emergency Response Act, requiring that all schools be inspected for asbestos. Public response has resulted in hasty programs to remove asbestos from schools and other buildings, at a cost of billions of dollars. But what is the real level of hazard?

Most asbestos in buildings is the less potent chrysotile, woven into cloth or glued into a tight matrix, and often the surface has been further stabilized by painting. Therefore, the fibers are not free to blow around. The levels of airborne asbestos in most buildings are no higher than those in outdoor air. Some scientists argue that asbestos insulation, despite being carcinogenic, poses no health danger if left alone. But when the material is removed, it is disturbed and asbestos dust escapes. Not only are workers endangered, but airborne asbestos can persist in a building for months after completion of an asbestos-removal project.

checkpoint What are some hazards of breathing in particles of minerals as they are extracted?

Hazards of Radon

Radon is one of a series of elements formed by the radioactive decay of uranium. Uranium occurs naturally in small concentrations in several minerals and in all types of rock. But it concentrates in two abundant rocks—granite and mudstone. It is also found in soil that formed from granite and mudstone, as well as in construction materials made from those rocks, such as aggregate or concrete. Radon is itself radioactive, and it decays quickly into other radioactive elements. Because of their radioactivity, radon and its decay products are carcinogenic. And because it is a gas, we can breathe radon into our lungs.

Radon seeps from the ground into homes and other buildings. Here it becomes concentrated. Indoor radon exposure causes an estimated 5,000 to 20,000 cancer deaths per year among Americans. The risk of dying from radon-caused lung cancer in the United States is about 0.4 percent over a lifetime. That is much greater than the risk of dying from cancer caused by asbestos, pesticides, or other air pollutants. It is nearly as high as the risk of dying in an auto accident, from a fall, or in a fire at home.

Not all Americans are exposed to equal amounts of radon. Some buildings contain very low concentrations of the gas; others have high concentrations. The variations in concentration are due to two factors: geology and home ventilation. Geology is important because some types of rocks, such as granite and mudstone, contain high concentrations of uranium and radon, while others contain relatively little.

Radon forms through the slow radioactive decay of uranium in bedrock, soil, or construction materials. It can seep into the basement of a home and circulate throughout the house. Radon concentrations are highest in poorly ventilated homes built on granite or mudstone, or built on soil derived from these rocks. Radon levels can also be high in building constructed with concrete and concrete blocks containing granite or mudstone. The highest home radon concentrations ever measured were found in houses built on a rock formation called the Reading Prong. This uranium-rich body of granite extends from Reading, Pennsylvania, through northern New Jersey and into New York. The air in one home in this area contained 700 times more radon than the EPA "action level." That is the concentration at which the Environmental Protection Agency recommends that action be taken to reduce the amount of radon in indoor air.

People should ask two questions in regard to radon hazards: "What is the radon concentration in my home?" and "If it is high, what can be done about it?" Because radon is radioactive, it can be measured inexpensively with a simple detector available at most hardware stores or from local government agencies. If the detector indicates excessive radon, several solutions are possible for lowering the radon level. One solution is to extend a ventilation duct either from the basement or from below the basement floor to the outside of the house. The ventilation duct prevents radon from accumulating in or below the basement by venting it directly to the outdoors. Another solution is to pump outside air into the house to keep indoor air at a slightly higher pressure than the outside air. This positive pressure prevents gas from seeping from soil or bedrock into the basement. Despite these solutions for lowering radon levels, it is impossible to avoid exposure to radon completely because it is everywhere—in outdoor air as well as in homes and other buildings. But it is relatively easy and inexpensive to minimize exposure and avoid a significant cause of cancer.

checkpoint Why do people test radon levels in buildings?

Acid Mines and Heavy Metals

Sulfide ore minerals are combinations of metals with sulfur. Some contain arsenic or other toxic elements in addition to the metals. These minerals are mined for their metals, which are essential to modern industrial societies. Such metals include lead, zinc, copper, cadmium, and mercury.

Mining and refining sulfide ore minerals can create serious air and water pollution problems. When these minerals are mined or refined without adequate pollution control, sulfur escapes into streams, groundwater, and the atmosphere. It then forms hydrogen sulfide and sulfuric acid. Waterborne sulfides poison aquatic organisms. In addition, atmospheric sulfur compounds contribute to acid precipitation.

The sulfuric acid that forms from the weathering of sulfide ore minerals can cause other minerals from the rock in and around the mining site to be dissolved. Dissolution of these minerals releases the metals they contain into the water. Once dissolved, the metals can quickly become part of the groundwater and surface water systems if not contained. Lead, zinc, and copper are several of the metals that are common pollutants in areas affected by acid mine drainage (Figure 3-20). Arsenic is another common pollutant.

checkpoint What effects do waterborne and airborne sulfides have on the environment?

FIGURE 3-20
The Rio Tinto (translated "Red River") in Spain owes its name to the color resulting from acid mine drainage. The Rio Tinto flows past a large open pit mining site. Since about 3000 B.C.E., gold, silver, and copper have been extracted and processed here. The orange and yellow colors of the rocks in this acidic stream are produced by a coating of iron oxide.

FIGURE 3-21
Nineteenth-century hard-rock miners in Montana commonly used mercury to extract gold and silver from ore.

Here to Stay: Mercury

Mercury that was used to extract gold and silver from ores in Montana during the 19th century (Figure 3-21) now reaches across the food chain. One of the most challenging environmental problems associated with gold and silver smelting is the introduction of mercury into the environment. Mercury, symbolized Hg, is a toxic metal that affects the brain, kidneys, and lungs. Mercury is the only native metal that is in the liquid state at room temperature. In this form, mercury exists as a native element with no charge (Hg^0). Yet mercury also can exist as a positively charged ion (Hg^{2+}) capable of forming complexes with inorganic and organic compounds. Once released, the chemically reactive nature of mercury results in it being quickly dispersed into the environment.

Because it exists as a liquid metal under normal conditions, mercury can be used to recover tiny flecks of gold from crushed rock or sediment. When the crushed rock or sediment is mixed with liquid mercury, the gold metal and mercury metal stick together. This mixture is easily separated from the other minerals in the crushed rock. Once separated from the other minerals, the mercury-gold mixture is heated in a crucible. The heat drives off the mercury as a vapor, while leaving the gold behind.

The mercury vapor disperses quickly into the atmosphere, often by clinging to tiny dust particles that also are released by the smelting process. The mercury-laden dust and vapor enter the global atmospheric cycle. Eventually it combines with precipitation to settle on plants, soil, and lake and river surfaces, where it enters the base of the food chain.

Recent environmental investigations in western Montana have used the osprey, a fish-eating bird of prey, as a means of pinpointing specific river drainages in which mercury contamination exists (Figure 3-22). The contamination is a legacy of the mining and smelting that occurred during the 19th and early 20th centuries. Osprey exclusively eat fish. The fish feed on aquatic insects that in turn eat mercury-contaminated microorganisms in the streams. With each step in the food chain, the mercury is concentrated, or biomagnified. The osprey, at the top of the food chain, can contain very high levels of the metal in their body tissue. It is therefore not surprising that those stream drainages in western Montana that have hosted long histories of gold mining and smelting are inhabited by osprey with the highest mercury levels in their blood.

The release of toxic elements such as lead, mercury, cadmium, and arsenic into the

CONNECTIONS TO HISTORY

Mercury Madness

Have you ever heard the phrase "mad as a hatter?" It is a 19th-century expression referring to someone who exhibits bizarre behavior. In the 19th century, mercury was used to prepare felt for making hats. At that time, little was known about the negative health effects of mercury exposure. The physical symptoms caused by exposure to mercury can include trembling, mood swings, and antisocial behavior. The mercury that hatters were exposed to could have affected their nervous systems, causing them to tremble and appear insane. This provides circumstantial evidence to link the phrase "mad as a hatter" to the hat-making industry.

---

environment from mining and smelting activities is a global problem. Areas with historic mining are far likelier to have high mercury concentrations. But mercury has been detected even in the sediment of isolated alpine lakes in the western United States. Many scientists have conclude d that mercury is carried in the atmosphere from distant industrial sources, particularly in Asia. Areas formerly considered pristine now contain detectable mercury from atmospheric fallout.

Mercury is chemically reactive and can combine with negatively charged organic molecules. Because of this, mercury tends to occur naturally in sediments rich in organic matter. Coal, a rock containing more than 50 percent organic carbon, is used worldwide as a source of energy. Unfortunately, the increasingly widespread burning of coal since the advent of the industrial revolution has released much mercury into the atmosphere.

This problem is likely to grow in the future as the demand for world energy increases. Coal is plentiful and a relatively cheap source of energy. As recently as 2018, six or seven full trainloads of coal passed daily from coal mines in Montana and Wyoming westward to the coast. There, the coal is loaded on barges and shipped overseas to Asia. Ironically, much of the mercury contained in these coal trains is blown back to North America as air pollution after the coal is burned in Asia. The consumption of large quantities of coal is providing much of the energy needed for global economic growth. But the burning of the coal comes with serious environmental challenges, including the release of mercury, that must be considered by current and future generations.

checkpoint How does mercury undergo the process of biomagnification?

3.5 ASSESSMENT

1. **Summarize** In what industrial environments does solid rock that has been ground to dust impact humans, and how have government regulations attempted to protect workers?

2. **Explain** What properties of asbestos have made it commercially valuable?

3. **Explain** How can mining release harmful or poisonous materials from rocks that were harmless in their natural environment?

Critical Thinking

4. **Apply** Do some research to find out what level of radon is considered acceptable in homes. Make a plan for testing the radon level in your home.

FIGURE 3-22
Mercury that occurs in the streams and lakes is concentrated in aquatic insects, fish, and osprey, a bird that eats fish exclusively. Scientists analyze the mercury levels in the blood of osprey. They can then identify drainages affected by mercury pollution from historic mining activities.

LESSON 3.5

TYING IT ALL TOGETHER
THE COSTS OF MINING

In this chapter, you learned about minerals and their importance to human societies. You also learned that human extraction and processing of mineral resources is often associated with significant environmental pollution and a range of human health problems.

Some pollution pathways, such as mercury from ASGM, may contaminate systems for hundreds to thousands of years. To cut back on mercury releases, the international community has worked to adapt proven industrial-scale mining processes to ASGM. These alternative methods take advantage of the physical and chemical properties of gold and associated minerals to isolate gold from waste rock without using mercury.

Mineral ores provide many metals other than gold that you use every day. Cell phones, for instance, are made up of more than two dozen metals and other elements extracted from mineral ores. Often, the metals of interest are bonded to other elements within the crystalline structure of the mineral. Processing and refining of these minerals breaks these bonds to isolate the economically important elements. These processes are extremely energy-intensive and sometimes release dangerous compounds and elements into the environment.

Work individually or with a group to complete the steps below.

1. Research one of these ores commonly used to produce metals for cell phone components: *hematite*, *chalcopyrite*, *cuprite*, *galena*, *pyromorphite*, *sperrylite*, *monazite*, or *loparite*. Use online resources to gather information about the ore: its physical and chemical properties; methods used in extracting, processing, and refining the ore; target metals and other elements; and known environmental impacts. Respond to the following questions based on your research.

2. Describe the mineral ore you researched. What are its physical properties?

3. What elements are bonded together in the ore? What class of minerals does it belong to?

4. What metals are extracted from the ore? How are these metals used? Are there other uses for the mineral?

5. Describe the geologic occurrence of the ore. Is it widely distributed? Does it commonly occur in association with other minerals?

6. What methods are used to extract, process, and refine the ore? Does extraction, processing, or refining of the mineral ore you researched negatively impact the environment? How?

Discuss your findings as a class. Are there some metal ores that can be mined to extract the same metal with less environmental impact? What characteristics limit mining of some ores?

FIGURE 3-23
Unlike artisanal small-scale gold mining (ASGM), large-scale industrial gold mining operations do not typically use mercury to separate gold from ore. However, all forms of mineral extraction have environmental consequences. Pictured here is an open-pit gold mine in Australia.

CHAPTER 3 SUMMARY

3.1 WHAT ARE MINERALS?

- A mineral is a naturally occurring, inorganic solid with a definite chemical composition and a crystalline structure.
- Minerals are the substances that make up rocks. Most rocks contain two to five abundant minerals plus minor amounts of other minerals.
- Every mineral consists of chemical elements bonded together in definite proportions, so its chemical composition can be represented as a chemical formula.

3.2 COMPOSITION AND STRUCTURE OF MINERALS

- Every mineral has a crystalline structure—an orderly, periodically repeated arrangement of its atoms—and therefore every mineral is a crystal.
- The shape of a crystal is determined by the shape and arrangement of its atoms.
- Every mineral is distinguished from others by its chemical composition and crystalline structure.
- Eight elements make up more than 98 percent of Earth's crust. They are oxygen, silicon, aluminum, iron, calcium, sodium, potassium, and magnesium.

3.3 PHYSICAL PROPERTIES OF MINERALS

- Most common minerals are easily recognized and identified visually.
- Identification of minerals is aided by observing a few physical properties, including crystal habit, cleavage, fracture, hardness, specific gravity, color, streak, and luster.

3.4 MINERAL CLASSES AND THE ROCK-FORMING MINERALS

- Nine rock-forming minerals make up most of Earth's crust. They are feldspar, quartz, pyroxene, amphibole, mica, clay, olivine, calcite, and dolomite.
- The first seven on this list are silicates; their structures and compositions are based on the silicate tetrahedron, in which a silicon atom is surrounded by four oxygen atoms.
- Two carbonate minerals, calcite and dolomite, are also sufficiently abundant to be considered rock-forming minerals.
- Ore minerals, industrial minerals, and gems are important for economic reasons.

3.5 HAZARDOUS ROCKS AND MINERALS

- Rocks and minerals such as feldspar and quartz become harmful only when they are crushed to dust during mining, road building, and other activities.
- The group of minerals known as asbestos causes cancer. Asbestos is commercially valuable and still used today in brake pads, shingles, and pipes.
- Radon is a carcinogenic gas produced by the radioactive decay of uranium. Radon seeps from the ground into homes and other buildings where it may become concentrated.
- When sulfide ore minerals are extracted without adequate controls, escaped sulfur leads to the release of other pollutants such as arsenic, lead, zinc, and copper.
- Mercury, which has been used in gold mining since the 19th century and is released in large quantities by coal-fired power plants, is a highly toxic metal. Mercury contamination can persist for hundreds to thousands of years.

CHAPTER 3 ASSESSMENT

Review Key Terms
Select the key term that best fits the definition. Not all terms will be used, and no term will be used more than once.

amphibole	ion
anion	luster
carbonate	mineral
carcinogenic	Mohs' scale
cation	native element
cleavage	silicate
crystal	silicate tetrahedron
crystal face	specific gravity
crystal habit	streak
element	sulfide
fracture	

1. a regularly repeating arrangement of atoms in solid form
2. an inorganic solid found in nature that has a definite chemical composition and crystal structure
3. fundamental unit of matter that cannot be divided by chemical means
4. an atom that has either lost or gained electrons to become electrically charged
5. a class of minerals that all contain silicon and oxygen
6. the color of a fine powder form of a mineral
7. a class of minerals that all contain sulfur and at least one other element
8. the characteristic shape of a single mineral crystal or crystal group
9. description of the way light interacts with the surface of a mineral
10. a ranking of minerals according to their resistance to scratching
11. a type of rock-forming silicate mineral that, along with mica, is common in granite and other continental rocks but rare in oceanic crust
12. the weight of a substance relative to an equal volume of water
13. an arrangement of four oxygen atoms around a silicon atom, forming the fundamental building block of all silicate minerals
14. a flat surface that develops as a crystalline solid grows freely in an uncrowded environment
15. the manner in which a mineral breaks other than along flat surfaces to form characteristic shapes

Review Key Concepts
Answer each question on a separate sheet of paper to demonstrate your understanding of key concepts from the chapter.

16. In terms of composition and structure, what do all minerals have in common and how can they differ?
17. Describe how the atoms that make up a crystalline mineral are arranged.
18. Explain why ice is considered a mineral but liquid water is not.
19. Explain what streak, luster, and cleavage are and how tests for these qualities help identify unknown minerals.
20. How is the specific gravity of a mineral found and how is it used in identifying an unknown mineral?
21. Explain how minerals are classified according to the elements that comprise them. Give examples of three classes of minerals and describe their chemical makeup.
22. Which class of minerals is most abundant on Earth? How many groups is this class divided into and what is the basis for these different groups?
23. Explain the difference between ore minerals and industrial minerals and give examples of each.
24. What are gems and how are they distinguished from other minerals?
25. What is acid mine drainage and why is it a concern?
26. How are mining practices responsible for the mercury poisoning of osprey, large birds of prey native to parts of the United States?
27. How can grinding of certain minerals lead to conditions that threaten human health?

Think Critically
Write a response to each question on a separate sheet of paper. Use concepts from the chapter to support your reasoning.

28. If you attended a rock and mineral show, you would be able to view many interesting specimens of both rocks and minerals. How would you be able to tell the difference between specimens that are rocks and specimens that are minerals?
29. Why is it useful to have a wide variety of characteristics that can be used to identify an unknown mineral instead of just a few? Give an example to support your answer.

30. Use Table 3-2 to predict what would happen in each of the following scenarios.
    a. A piece of quartz is rubbed against a piece of topaz.
    b. A piece of fluorite is rubbed against a copper penny.
    c. An unknown mineral that scratches a steel file is rubbed against apatite.
    d. An unknown mineral that is scratched by fluorite is rubbed against window glass.
31. Diamond and graphite are both native element minerals consisting of carbon. Diamond has a hardness of 10 on the Mohs' scale, whereas graphite has a hardness of between 1 and 2. How can you explain the difference in hardness of these two minerals?
32. Heavy metals found in minerals pose little to no health risk to living things in their natural state. How do humans create conditions so that heavy metals in minerals become a significant health risk to themselves and to other living things?
33. Describe two hazards that can be present in buildings that arise from mineral sources. What minerals are involved and how do each result in a health hazard?

PERFORMANCE TASK

Mineral Field Guide

Imagine that you are a mineralogist who writes field guides to educate the general public about minerals. Field guides can be published in many different formats, but any good field guide must present accurate information clearly and concisely using text and images. Field guides also generally have a specific focus. For example, a field guide may focus on a particular class of minerals, minerals found in a particular region, or minerals used for a specific commercial purpose.

In this performance task, you will create your own mineral field guide on a focus of your choice.

1. Develop your field guide.
   A. Choose a focus. You may use any of the examples listed below or, with your teacher's approval, choose another area of interest.
      - A specific group of minerals
      - Local minerals
      - Minerals found in a particular area, region, national park, etc.
      - Minerals used for a specific commercial purpose

      Use Table 3-3 from the text to help you narrow down your focus.
   B. Identify the minerals you will include in your field guide, and then conduct research on each mineral. Gather information about each mineral's physical properties, chemical composition, and crystal structure.
   C. Create your field guide in one of the following formats: picture book, slide show, or poster. Include both images and descriptions of the minerals that appear in your guide. Include a list of the references you used to gather the information you will be presenting.
   D. Plan and deliver an oral presentation to your classmates summarizing the most important information from your field guide.

   When you have completed your field guide, use it to answer the questions that follow.

2. Which classes of minerals are represented in your field guide?
3. In terms of physical and chemical characteristics, how are the minerals in your field guide similar to and different from one another? Compare and contrast at least three minerals with respect to three or more of the following characteristics: cleavage, fracture, color, specific gravity, hardness, streak, luster, crystal shape, and reaction with acid.
4. What are some uses for any of the minerals included in your field guide? Are there any safety issues related to using these minerals?

CHAPTER 4
ROCKS

The lithosphere, Earth's thinnest layer, holds all the rocks with which we are familiar. From the pebbles on the edge of a river to the towering mountains that scrape the sky, rocks are everywhere. Rocks are what we think of when we think of the nonliving part of Earth. Yet they are not everlasting. Rocks go through significant change over geologic timescales and in different environments within the crust. Some rock-changing processes take place at the surface and involve interactions with the atmosphere and biosphere. Other processes take place at great depth as a result of increased temperature and pressure, including the outright melting of rocks into magma.

KEY QUESTIONS

4.1 What processes change one type of rock into another?

4.2 What are the two types of igneous rock?

4.3 What characteristics do all sedimentary rocks have in common? What distinguishes them?

4.4 What processes transform metamorphic rocks?

Over millions of years, layers of sediment may build up and harden into sedimentary rock, such as this sandstone that forms The Wave at Kaleidoscope Ridge in Arizona. Sandstone forms as tiny grains of sand are cemented together into solid rock.

EXPLORERS AT WORK

GOING TO VOLCANIC EXTREMES

WITH NATIONAL GEOGRAPHIC EXPLORER KENNETH WARREN SIMS

When it comes to studying Earth's volcanic activity, Dr. Kenneth Warren Sims likes to be where the action is. In fact, Sims has stood along the rims of active volcanoes around the world, including those in Antarctica, Ecuador, Italy, Hawai'i, Iceland, the Democratic Republic of the Congo, and Nicaragua. He has peered into their steaming depths—and then climbed down into them.

Sims has scooped up samples from bubbling lakes of lava. He has braved noxious fumes and sweltering temperatures to collect volcanic gases. He has taken to the sea, descending to the ocean floor aboard submersibles to study underwater volcanoes that form the world's most extensive mountain chain, the Mid-Ocean Ridge system.

Sims's expertise as a climber enables him to explore volcanoes more closely than most volcanologists. He started climbing as a boy and worked as a professional mountain guide and instructor before studying to become a geologist and a volcanologist. Sims is currently a professor at the University of Wyoming. In addition to leading research expeditions, Sims gives lectures on geology and volcanology to audiences of all ages. He feels it is important for the public to understand his research and the questions he is exploring.

Some questions Sims explores are immediate in nature, such as how to better predict volcanic eruptions. Others touch on processes that occur over long timescales. For example, volcanoes play a critical role in the rock cycle—the physical and chemical processes that change Earth's rocks and recycle them over millions of years. Volcanic eruptions transport magma from within Earth's crust onto its surface. There, the magma becomes lava that cools and crystallizes into igneous rock, one of three main types of rock on Earth.

Sims and his team are researching a type of igneous rock called mid-ocean-ridge basalt (MORB) that covers the entire ocean floor. Sims uses crewed and remotely-controlled submersibles to collect rock samples from different mid-ocean ridges, regions where the seafloor is split into parts that move away from each other. This motion (measured in millimeters per year) causes magma to repeatedly erupt through the seafloor and cool into new rock.

As new rock forms, it displaces older rock, pushing it further from the ridge; this is called seafloor spreading. The rock furthest from the ridge is recycled as it collides with an adjacent tectonic plate and is pushed down into the mantle where it melts into magma. Sims is quantifying the timescales involved in these processes to better understand how Earth has evolved.

As Sims explains, volcanism is one of the most fundamental of all Earth processes and there is much yet to be learned. Whether that means visiting volcanoes thousands of meters above sea level or below, Sims is up for the challenge.

THINKING CRITICALLY

Explain Why does seafloor spreading occur at mid-ocean ridges?

Dr. Kenneth Warren Sims collects a sample below the first terrace inside the crater of Mount Nyiragongo, an active volcano in the Democratic Republic of the Congo.

Sims climbs out of the crater as activity in the lava lake increases.

97

CASE STUDY
DEVILS TOWER NATIONAL MONUMENT, WYOMING

Rising majestically from the grasslands and Ponderosa pines of northeastern Wyoming, the rock structure known as Devils Tower inspires wonder and curiosity. Several Native American tribes consider the landform sacred and have passed down various narratives about its origins. For example, oral histories of the Crow tribe, who call the tower "Bear Lodge," recount how two girls were saved from a bear when the Great Spirit caused the rock to rise up beneath them. The bear could not climb the rock but continued to claw at its sides, leaving ridges all along the outer edge.

Such stories developed to provide an explanation for the strikingly distinctive geologic features of Devils Tower. The tower is made up of many smaller columns—most of them hexagonal—bunched together into one larger column like a bundle of sticks. These tightly packed columns give the tower a ridged outer surface. Made up of light-colored rock, the tower rises 264 meters from the base to the summit. The base measures about 305 meters in diameter, and the summit about 84 meters.

In contrast to the vertical columns that make up Devils Tower, the rock surrounding the landform occurs in horizontal layers of various colors and thicknesses. A deep-red layer known as the Spearfish formation is the oldest visible layer of rock in the area (Figure 4-1). It is made up mostly of iron-rich red sandstone and maroon siltstone. Above this is a layer of lighter-colored rock called the Gypsum Springs formation. It consists of white gypsum, a very common type of rock containing calcium and sulfur. Above the Gypsum Springs formation are several layers that make up the Sundance formation. The main layers, or "members," of this formation are the Stockade Beaver mudstone and Hulett sandstone members. The former is mostly gray and gray-green mudstone, with pockets and streaks of sandstone, limestone, and red mudstone. The latter is yellow and has a very fine grain size.

How the rock of Devils Tower emerged in the middle of this landscape is a matter of some debate. Most geologists agree that Devils Tower formed about 40 to 50 million years ago, but there is little consensus as to how the tower rose to such a great height and obtained such a peculiar shape. As you will explore at the end of this chapter, several competing hypotheses propose various roles that volcanic activity, erosion, and other forces may have had in shaping Devils Tower.

As You Read Consider how rocks form and change over geologic time. What processes play a role in the formation of landforms like Devils Tower? Think about other interesting landforms you have seen and the forces that may have shaped them.

FIGURE 4-1
Devils Tower National Monument in Wyoming looms large above the surrounding area, inviting observers to wonder how it formed. Layers of rock surrounding Devils Tower provide a striking visual contrast to the vertical rise of the tower.

4.1 ROCKS AND THE ROCK CYCLE

Core Ideas and Skills
- Explain what the rock cycle is.
- Give an example of each of the processes involved in the rock cycle.
- Describe the three types of rock, how they are identified, and how they are classified.

KEY TERMS

magma	precipitation
igneous rock	metamorphic rock
weathering	texture
sedimentary rock	rock cycle
lithification	

Earth is solid rock to a depth of almost 3,000 kilometers, where the mantle meets the liquid outer core. Even casual observation reveals that rocks are not all alike. The great peaks of the Sierra Nevada in California are hard, strong granite. The red cliffs of the Utah desert are soft sandstone. The top of Mount Everest is limestone containing the fossil remains of small marine invertebrates. The fossils tell us that this limestone formed in the sea.

What forces lifted the rocks atop Mount Everest to the highest point on Earth? Where did the vast amounts of sand in the Utah sandstone come from? What created the granite of the Sierra Nevada? These questions are about the processes that formed and changed rocks. In this chapter, we will study rocks, what they are made of, and how they form.

The Rock Cycle and Earth Systems

A rock is a solid collection of one or more minerals. Based on how the rocks form, geologists define three main categories: igneous rocks, sedimentary rocks, and metamorphic rocks.

Earth's interior is hot and dynamic. The high temperature that exists within a few hundred kilometers of the surface can cause solid rock to melt, forming a molten liquid called **magma**. Because the liquid magma is less dense than the surrounding solid rock, the magma rises slowly toward Earth's surface. As it rises, the magma cools. **Igneous rock** forms when cooling magma solidifies.

All rocks may seem permanent and unchanging over a human lifetime, but this is an illusion created by our short observational time frame. Over geologic time, water and air attack rocks at Earth's surface through the process called **weathering**. Weathering breaks rocks down into smaller and smaller particles. The particles, including gravel, sand, clay, and all other fragments weathered and eroded from rock, are transported away from the site of weathering by streams, glaciers, wind, and gravity. Eventually the particles accumulate in layers of loose, unconsolidated sediment. Sand on a beach and mud on a lake bottom are examples of sediment. **Sedimentary rock** forms when sediment particles become compacted by pressure and cemented, gradually forming into solid rock. This process is called **lithification**.

Weathering processes also form dissolved ions such as sodium, calcium, and chloride. (Ions are atoms or molecules with a positive or negative electrical charge.) Weathering of solid rock causes some minerals in the rock to change chemically or dissolve, freeing up positive and negative ions. The ions are transported away from the site of weathering by streams and groundwater.

Most of the dissolved ions formed by weathering eventually are carried to the sea. There, marine organisms such as clams, oysters, and corals extract dissolved calcium from seawater. They combine the calcium with carbonate, which forms from atmospheric carbon dioxide dissolved in seawater. The organisms use the calcium and carbonate ions to form their shells. After

CONNECTIONS TO ART

Lapidary Art

The art of cutting and polishing stone is called lapidary art. The earliest humans began developing tools and weapons from rocks. As colored minerals and crystals became more prevalent, those skills eventually were applied to creating adornments used in homes, on clothing, and as jewelry for people, pets, and labor animals such as horses and elephants. In the 1950s, lapidary art became a popular hobby in the United States. Two types of stonework are common: cabochons—opaque stones polished into smooth shapes—and faceted transparent stones such as emeralds, diamonds, and other gemstones.

the organisms die, the remains of those shells accumulate. During burial, in a process called **precipitation**, minerals can crystallize out of the water left in the poor spaces between shell fragments, gradually cementing the shell fragments together into limestone. Thus, limestone forms by direct interactions among the geosphere, hydrosphere, biosphere, and atmosphere.

A **metamorphic rock** forms when heat, pressure, or hot water alters any preexisting rock. For example, when rising magma intrudes Earth's crust, it heats the rock around it. Although the heat may not be sufficient to melt rock surrounding the magma, it can bring about other changes. For example, the mineral crystals in the heated rock may grow or change their internal arrangement. Such alterations in the size, shape, and arrangement of mineral crystals within the rock are changes in the rock's **texture**. In addition, temperature and pressure within Earth's crust may cause other types of minerals that are stable in those conditions to grow. Thus, metamorphism can change both the texture and the mineral types within a preexisting rock. Unlike igneous rocks, however, metamorphic rocks form without completely melting.

No rock is permanent. Instead, all rocks undergo processes that change them from one of the three rock types to another. This continuous process is called the **rock cycle** (Figure 4-2). For example, as sediment accumulates and is buried, it typically cements together to form a sedimentary rock. If sedimentary rock at the bottom of the accumulation is buried deeply enough, the rising temperature and pressure will cause it to undergo changes in texture and mineral composition. In other words, it converts to a metamorphic rock. With additional burial and temperature increase, the metamorphic rock can melt, forming magma. The magma will then rise in the crust, slowly cool, and solidify to become igneous rock. Millions of years later, movement of Earth's crust might raise the igneous rock to the surface, where it will weather to form sediment. Rain and streams will then wash the sediment into a new basin, renewing the cycle.

The rock cycle does not follow a set order and can take many different paths. For example, all

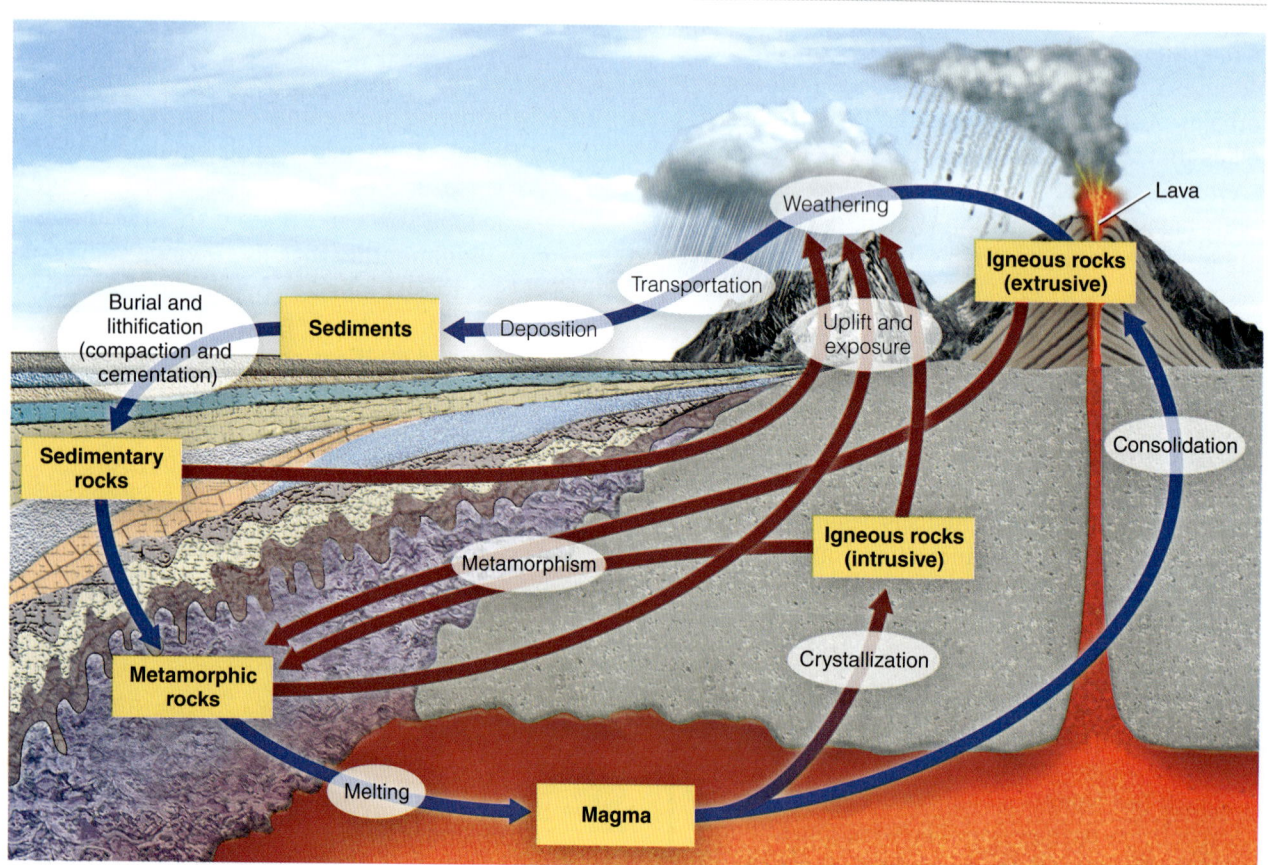

FIGURE 4-2
The rock cycle shows that rocks change over time. The arrows show the paths rocks can follow as physical processes change them.

MINILAB Rock Cycle Model

Materials

safety goggles
crayons in various colors
manual pencil sharpener
butter knife
candle
multi-purpose lighter
small bowl
heat-resistant surface
2 rectangular sheets heavy-duty aluminum foil, approximately 23 cm × 33 cm
2 wooden blocks, approximately 9 cm × 4 cm × 2 cm
tongs

CAUTION: Use safety goggles. Tie back long hair. Do not touch hot objects. Monitor the flame at all times.

Procedure

1. Choose four different colored crayons. Use the pencil sharpener to create four piles of crayon shavings, keeping colors separate.
2. On one sheet of aluminum foil, spread one color of shavings to create a base layer about 0.5 cm thick. Then spread another color of shavings on this base layer. Repeat this with the other colors.
3. Fold the aluminum foil over the crayon shavings to make a small sealed packet. Place the small packet between two wooden blocks and press the two blocks together. Open the foil and observe the shavings.
4. Use the second sheet of aluminum foil to create a small bowl. You may wish to use a small bowl as a mold for your foil bowl. Break the pressed crayon shavings up into the bowl.
5. Light the candle. Use tongs to hold the foil bowl over the candle. Adjust the distance between the candle flame and the bottom of the bowl so the crayon shavings melt slowly without sputtering. Be careful to avoid touching the flame and hot wax.
6. When the crayon shavings are melted, set the bowl on a surface to cool. Extinguish the candle.
7. After the melted crayon shavings have cooled, they should form one piece of crayon wax. Use the butter knife to slice this wax piece into two pieces.

Results and Analysis

1. **Observe** Describe what you observed after steps 2, 3, and 7.
2. **Connect** How did your actions in the lab model the physical processes of the rock cycle?

Critical Thinking

3. **Evaluate** What properties of the crayon shavings made them a good choice for this model? Explain.
4. **Compare** What were the main differences between your model and the rock cycle?
5. **Compare** How did the energy sources in your model compare to the energy sources in the rock cycle?

---

three rock types can melt to form magma and eventually cool into an igneous rock. Similarly, all three rock types can be weathered to form sediment or can be metamorphosed, or altered, through changes in texture and mineral composition. The term "rock cycle" simply expresses the idea that rocks are not permanent but change over geologic time.

The rock cycle illustrates several types of system interactions. These interactions are between rocks, the atmosphere, the biosphere, and the hydrosphere. Rain and air, aided by acids and other chemicals secreted by plants, decompose solid rocks to form large volumes of clay and other sediment. During these processes, water reacts chemically with the clay and can be incorporated into the clay mineral crystals. Thus, these processes transfer water from the atmosphere and the hydrosphere to the solid minerals of the geosphere. More rain then washes the clay and other sediment into streams, which carry it to a sedimentary basin where the clay is deposited. Recall from Chapter 2 that solar energy drives the hydrologic cycle. So, sunlight evaporates moisture to form rain, which in turn feeds flowing streams. But these same processes are also part of the rock cycle, illustrating once again that all Earth processes interact.

The rock cycle is also driven by Earth's internal heat. For example, when a sedimentary basin sinks under the weight of additional sediment, the deeper layers become heated and metamorphosed by Earth's heat. The same heat may melt the rocks to produce magma. That hot magma may rise upward and perhaps even erupt onto Earth's surface from a volcano. In this way, heat is transferred from Earth's interior to the atmosphere.

FIGURE 4-3
ON ASSIGNMENT In Hawai'i, National Geographic photographer Patrick Kelly snapped a photo of lava falling more than 18 meters into the ocean, explosively cooling and forming new rock as it hit the water.

Throughout this chapter, we emphasize the interactions between Earth's spheres. They illustrate how Earth systems continuously exchange both energy and material so that Earth functions as a single, integrated system.

checkpoint What is the link between weathering and sedimentary rocks?

4.1 ASSESSMENT

1. **Define** What are the three main categories of rocks and what distinguishes them?
2. **Describe** Summarize the rock cycle. How is each type of rock transformed into each other type of rock?

Critical Thinking

3. **Evaluate** Is the claim "All igneous rocks are younger than all sedimentary or metamorphic rocks" valid? Why or why not?
4. **Recognize Patterns** In what ways are the rock cycle and the water cycle similar? In what ways are they different?

4.2 IGNEOUS ROCKS

Core Ideas and Skills
- Describe how igneous rocks form from magma deep within Earth.
- Define how igneous rocks are identified.
- Distinguish between the formation process and appearance of extrusive and intrusive igneous rocks.

KEY TERMS

intrusive igneous rock　　　extrusive igneous rock

Formation and Features of Igneous Rocks

If you were to drill a well deep into the middle of one of Earth's continents, you would find that the temperature within the crust rises about 25°C for every kilometer of depth. In the mantle between depths of 100 and 350 kilometers, the pressure and temperature are such that rocks in some areas melt to form magma. Unlike ice, which *decreases* in volume when it melts to form water, rocks *increase* in volume when they melt. When a rock melts to form magma, it expands by about 10 percent. Melted rock is less dense than the solid rock around it and therefore rises as it forms. When magma rises, it enters the cooler environment near Earth's surface, where it solidifies into igneous rock.

In addition, unlike ice, which melts completely above 0°C, rocks do not melt completely at a single temperature because each mineral in the rock has its own unique melting temperature. For example, quartz melts at about 500°C. The mineral olivine can reach temperatures in excess of 1,100°C. Magma forming from melting rocks commonly has a different composition than the original rock. Only those minerals with the lowest melting temperature melt at a given temperature, leaving behind those minerals with higher melting temperatures. The fact that different minerals melt and solidify, or crystallize, at different temperatures results in much of the variety of igneous rock types found on Earth.

checkpoint Why don't rocks melt at one single temperature?

Types of Igneous Rocks

Some igneous rocks form when magma solidifies within Earth's crust; other igneous rocks are created when magma erupts onto the surface. When magma solidifies, it usually crystallizes to form minerals. As magma cools, different minerals crystallize at different times. The size, shape, and arrangement of mineral crystals in an igneous rock are referred to as the rock's texture. Although some igneous rocks consist of mineral crystals that are too small to be seen with the unaided eye, other igneous rocks are made up of thumb-size, or even larger, crystals. Table 4-1 compares crystal sizes of different igneous rocks.

Extrusive Rocks When magma rises all the way through the crust to erupt onto Earth's surface, it forms **extrusive igneous rock**. Lava is magma that flows from a crack or volcano onto Earth's surface. The term also refers to the rock that forms when lava cools and becomes solid.

After lava erupts onto the relatively cool surface, it solidifies in a few days to a few years. Crystals form but do not have much time to grow. As a result, many volcanic rocks consist of crystals too small to be seen with the unaided eye. Basalt is a common, very finely crystalline volcanic rock (Figure 4-4A).

If erupting lava encounters glacial ice or cold seawater, it may solidify within a few hours. Because the magma hardens so quickly, the atoms have no time to align themselves to form crystals. As a result, the atoms are frozen into a random chaotic pattern, as happens in volcanic glass (Figure 4-4B).

If magma rises slowly through the crust before erupting, some crystals may grow while most of the magma remains molten. If this mixture of magma and crystals then erupts onto the surface, the magma solidifies quickly, forming porphyry, a rock with large crystals embedded in a fine-grained matrix (Figure 4-4C).

TABLE 4-1 Igneous Rock Textures Based on Crystal Size

Name of Texture	Crystal Size
Glassy	No mineral crystals
Very finely crystalline	Too fine to see with unaided eye
Finely crystalline	Up to 1 millimeter
Medium crystalline	Between 1 and 5 millimeters
Coarsely crystalline	More than 5 millimeters
Porphyry	Relatively large crystals in a finely crystalline matrix

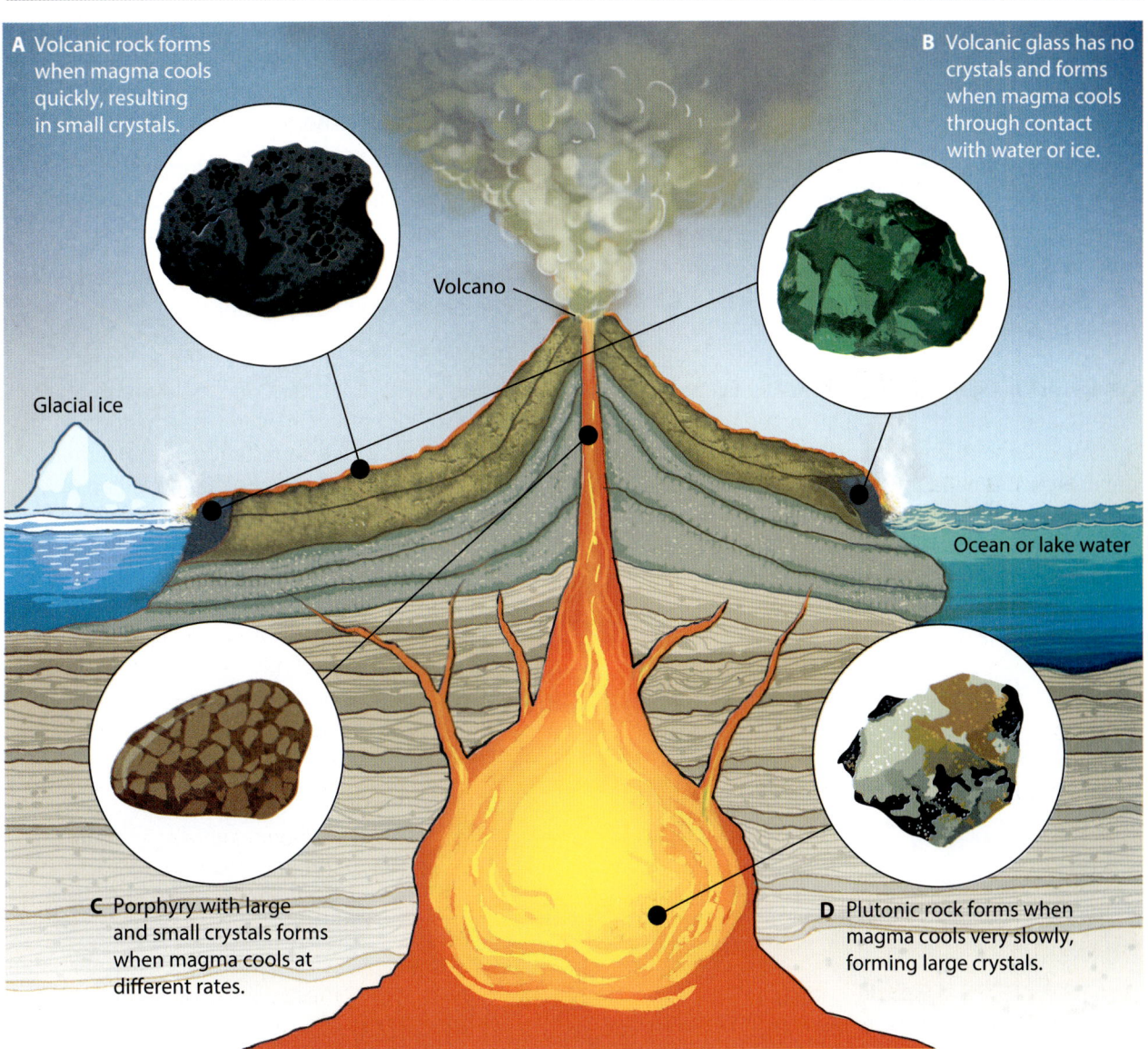

FIGURE 4-4
Differences in igneous rock textures result from differences in the length of time magma has to cool before becoming solid igneous rock.

Intrusive Rocks When magma solidifies within the crust, without erupting to the surface, it forms an **intrusive igneous rock**, also called plutonic rock (Figure 4-4D). The rock overlying the magma insulates it like a thick blanket, providing hundreds of thousands, or even millions, of years for the magma to crystallize. As a result, most intrusive igneous rocks are medium-to-coarsely crystalline.

Geologists commonly use the terms *basement rock, bedrock, parent rock,* and *country rock*. Bedrock is the solid rock that lies beneath soil. It can be igneous, metamorphic, or sedimentary. Parent rock is any original rock before it is changed by weathering, metamorphism, or other geologic processes. The rock that is already in an area and is cut into by intrusive igneous rock is called country rock. Basement rock is the igneous and metamorphic rock that lies beneath the thin layer of sediment and sedimentary rocks covering much of Earth's surface. It forms the "basement" of the crust.

Common Igneous Rocks Geologists have different names for igneous rocks based on their minerals and texture. Take for example two rocks consisting mostly of feldspar and quartz. Medium or coarsely crystalline igneous rock made up of these minerals is called granite, the most abundant rock in continental crust (Figure 4-5A). The crystals in granite are clearly visible. Many are a millimeter across, and some may be much larger. When igneous rock with these same minerals is very finely crystalline, it is called rhyolite (Figure 4-5B). The same magma that solidifies slowly within the crust to form granite can also erupt onto Earth's surface to form rhyolite. If the magma cools quickly through contact with water or ice, obsidian with no crystal structure forms (Figure 4-5C).

Like granite and rhyolite, most common igneous rocks are classified in pairs. The two members of a pair contain the same minerals but have different crystal sizes. The crystal size depends mainly on whether the rock is extrusive (volcanic, finely crystalline) or intrusive (plutonic, coarsely crystalline).

The specific combination of minerals that make up an igneous rock determines the rock's composition. Igneous rock compositions are classified along a range from those that contain the most silica to those that have little silica but are rich in other minerals such as iron.

- **Felsic** Silica-rich igneous rocks are called *felsic*. Felsic rocks contain more than 65 percent silica. Examples include granite and rhyolite.
- **Intermediate in Composition** Igneous rocks with intermediate levels of silica are called *intermediate in composition*. Examples include andesite and diorite.
- **Mafic** Igneous rocks that contain roughly 50 percent silica are called *mafic*. Examples are basalt and gabbro.
- **Ultramafic** *Ultramafic* rocks, such as peridotite, contain less than 50 percent silica.

Understanding composition differences in igneous rocks is important. The differences relate to the geologic setting in which magma formed and cooled. This provides important clues to earlier parts of Earth's history. Magma composition plays a direct role in how powerful a volcanic eruption can be. The largest, most catastrophic volcanic eruptions involve felsic magmas.

As discussed, granite contains mostly feldspar and quartz. It is found nearly everywhere on the

A   B   C

FIGURE 4-5
Igneous rocks made of the same minerals look very different because of crystal size. (A) Granite contains coarse individual crystals of quartz, feldspar, and mica. (B) Rhyolite contains fine crystals. (C) Obsidian contains no crystals.

FIGURE 4-6
Obsidian Cliff in Yellowstone National Park is made mostly of volcanic glass formed at the leading edge of a flow of lava from the Yellowstone volcano that erupted 500,000 years ago.

FIGURE 4-7
Outside the city of Seattle is Mount Rainier, an andesitic volcano that is part of the Cascade Mountains.

continents as basement rock, beneath the soil and younger sedimentary rocks that cover it. Granite is hard and resistant to weathering. It forms steep, sheer cliffs in many of the world's great mountain ranges. Mountaineers prize granite cliffs for the steepness and strength of the rock.

The geologically active Yellowstone volcano contains felsic magma located only a few kilometers below the surface. Over the past 2.1 million years, the Yellowstone volcano has produced three catastrophic eruptions. Numerous flows of rhyolite have spread across the volcano, filling in its central crater. The longest flows are tens of kilometers in length. Several of those flows have an outer edge of obsidian that formed by rapid cooling of the rhyolite by water or glacial ice (Figure 4-6).

Basalt is the finely crystalline igneous rock that makes up most of the oceanic crust. It consists of approximately equal amounts of feldspar and pyroxene. Gabbro has the same minerals as basalt but consists of larger crystals. Gabbro is uncommon at Earth's surface, although it is abundant in deeper parts of the oceanic crust.

Andesite is a volcanic rock intermediate in composition between basalt and granite. It is commonly gray or green and consists of feldspar and dark minerals (usually biotite, amphibole, or pyroxene). It is named for the Andes Mountains, where andesite is abundant. Because it is volcanic, andesite is typically very fine-grained. Diorite is the plutonic equivalent of andesite. It forms from the same magma as andesite and often underlies andesitic mountain chains such as the Andes.

Andesite is most common in volcanic chains called volcanic arcs because they typically form an arclike pattern on geologic maps. Volcanic arcs are developed where part of Earth's crust sinks beneath another part and is slowly melted in the process. The Cascade Mountains of the Pacific Northwest are a volcanic arc. They are made mostly of andesite and diorite (Figure 4-7).

Peridotite, an ultramafic igneous rock that makes up most of the upper mantle, is rare in Earth's crust. It is coarse-grained and composed of olivine and small amounts of pyroxene, amphibole, or mica. Komatiite is the finely crystalline equivalent of peridotite. It is also rare in Earth's crust. Modern volcanoes do not erupt ultramafic magma because it is too dense to reach the surface. Before reaching the surface, magma must pass through a much more felsic crust with minerals that melt at low temperatures. More than 4 billion years ago, however, the first crust was made of ultramafic rock. As a result, komatiites formed abundantly. Today they are preserved only in a few places.

checkpoint What are the mineral composition differences between felsic, mafic, and ultramafic igneous rock?

4.2 ASSESSMENT

1. **Explain** What factors are important in a rock's texture?
2. **Define** Compare and contrast the two types of igneous rock.
3. **Distinguish** What are the most common igneous rocks, and what makes them different?

Critical Thinking

4. **Synthesize** Describe a situation where both extrusive and intrusive igneous rocks might be found at the same location.

LESSON 4.2   **107**

4.3 SEDIMENTARY ROCKS

Core Ideas and Skills
- Explain the process of precipitation and how it is related to sedimentary rocks.
- Describe the characteristics that all sedimentary rocks have in common.
- Name the four major types of sedimentary rocks and how they can be identified.

KEY TERMS

deposition
clastic sedimentary rock
bioclastic sedimentary rock
organic sedimentary rock
chemical sedimentary rock
carbonate rock

A

B

FIGURE 4-8
(A) Rounded gravel has been deposited along the Flathead River in Montana. The gravel is derived from the physical weathering and breakdown of bedrock, followed by transport and rounding of the sediment by the river. (B) Dissolved ions derived from weathered bedrock are transported into Death Valley and then concentrated through evaporation.

Sedimentary rocks make up only about 5 percent of Earth's crust. However, because they form on Earth's surface, sedimentary rocks are widely spread in a thin veneer over underlying igneous and metamorphic rocks. As a result, they cover about 75 percent of continents.

Formation and Features of Sedimentary Rocks

Over geologic time, the atmosphere, the biosphere, and the hydrosphere weather rock, breaking it down to gravel, sand, silt, clay, and ions (positively or negatively charged atoms or molecules) dissolved in water. Weathering of rocks occurs by both chemical and physical processes. Glaciers, flowing water, gravity, and wind all erode the rock, transport the rock fragments downslope, and deposit them at lower elevations (Figure 4-8A). The last step in this process is called **deposition**.

Dissolved ions from weathering are transported downslope, commonly all the way to the sea. Here the ions are concentrated. Marine organisms such as clams, snails, certain kinds of green algae, and corals use some of these ions to form shells and other hard parts. After the organisms die, these hard parts accumulate to form limestone. Other ions can react chemically to produce a solid salt through the process of precipitation. For example, the mineral halite (rock salt) precipitates from lakes that form on the floor of Death Valley (Figure 4-8B).

Nearly all sedimentary rocks contain sedimentary structures. These physical features develop during or shortly after deposition of the sediment. Most sedimentary structures form through the interaction of moving water and loose grains of sediment. For example, when a stream moves sand along its bed, the size of the particles that can be moved is limited by the strength of the flow. Only those particles small enough to be moved by the flow move. Larger particles are left behind. In this way, sediment is sorted by size as the water interacts with it. Environments with high levels of wave energy, such as beaches, typically produce sediments that are well-sorted. There is little size variation among sediment particles on most beaches.

Winds typical in many desert environments also move sand grains. Desert winds as well as beach waves organize sediment grains into sedimentary structures. Examples of sedimentary structures include sand dunes in a desert and sediment ripples in a stream. Many sedimentary structures form little by little, as wind or water currents sort and move the grains into piles. Because sedimentary rocks form over long periods of time on Earth's surface, their structures contain clues about environmental conditions when the rocks were formed.

The most obvious and common sedimentary structure is bedding, or stratification (Figure 4-9). Bedding forms because sediment accumulates layer by layer, with short pauses in sediment deposition. These pauses are represented by the surfaces that separate individual beds.

FIGURE 4-9
Go online to see an example of stratification where different layers of sediment were deposited at different times.

Ripple marks are small ridges and troughs that are similar to dunes. They form by organized movement and sorting of loose sediment grains by a flow. Ripples are common in shallow streams and lake shorelines. Ripple marks are also commonly preserved in ancient sandstone. Their external shape is preserved along bedding planes (Figure 4-10A). Ripple marks can form by currents moving in a single direction, as in a stream. They can also form by currents moving back and forth, as in a shallow lake where sand on the bottom is pushed forward and backward by waves on the lake surface.

Mud cracks are irregular polygon-shaped cracks that form when mud shrinks as it dries out (Figure 4-10B). They indicate that the mud dried after being deposited. Mud cracks are common in river floodplains. Mud deposited by a flood eventually dries out when the floodwaters recede.

Graded beds are a commonly occurring sedimentary structure. They form when a flow that carries sediment slows down. The size of

FIGURE 4-10
(A) Ancient ripple marks are preserved in sandstone. (B) Mud cracks, common in river floodplains, form along the edge of a small pond in Montana. (C) Graded bedding shows the coarsest sediment at the base and the finest sediment at the top. (D) Burrows in dolomite from western Wyoming were made by ancient shrimp burrowing into the seafloor to evade predators. Tracks, trails, burrows, and borings left behind by ancient organisms are called trace fossils.

TABLE 4-2 Clastic Sediment Particle Types and Sizes

Sediment Particle	Clay*	Silt	Sand	Granule	Pebble	Cobble	Boulder
Diameter (mm)	<0.004	0.004-0.063	0.063-2.000	2.00-4.0	4-64	64-256	>256

the sediment that is deposited decreases over time (Figure 4-10C). That is, the graded bed has the coarsest sediment at its base and the finest sediment at its top. Graded beds can be produced by a flood that wanes through time. Graded beds are also very common in deep marine environments where mixtures of sediment and water currents move downslope, coming to rest as the current slows down and eventually stops.

Trace fossils are sedimentary structures that result from the activities of plants and animals, but that do not include the bodily remains of the organism itself (Figure 4-10D). For example, tracks, trails, and burrows are trace fossils, whereas ancient shells, bones, and teeth are not.

checkpoint Name and describe the most common sedimentary structures.

Clastic and Bioclastic Sedimentary Rocks

Sedimentary rocks are broadly divided into four categories:

1. **Clastic sedimentary rock** is composed of particles of weathered rocks that have been transported, deposited, and lithified. The generic term *clastic* refers to any rocks that are composed of fragments of older rocks. The fragments themselves, such as sand grains and pebbles, are called *clasts*. This category includes conglomerate, sandstone, and mudstone. Clastic sedimentary rock makes up about 85 percent of all sedimentary rock.

2. **Bioclastic sedimentary rock** is composed of broken shell fragments and similar remains of living organisms. The fragments are transported like clastic particles, but they have a biological origin. Many limestones are formed from broken shells and thus are bioclastic sedimentary rocks.

3. **Organic sedimentary rock** consists of the lithified remains of plants or animals. Coal is an organic sedimentary rock that contains such a high percentage of decomposed and compacted plant remains that the rock itself will burn.

4. **Chemical sedimentary rock** forms by direct precipitation of minerals from solution. Rock salt, for example, forms when halite precipitates from evaporating seawater or salty lake water.

Clastic Sedimentary Rocks

Clastic sediment is called gravel, sand, silt, or clay, in order of decreasing particle size. Table 4-2 shows the main sediment particle sizes and their corresponding diameters. As clastic particles ranging in size from boulders to silt tumble downstream, their sharp edges are worn off and become rounded. Finer silt and clay do not round effectively because they are so small, and have so little mass, that water and wind cushion them as they are buffeted along.

If you hold a pile of sand in your hand and dribble water onto it, the water will soak into the empty zone between the sand grains. This empty zone is called pore space (Figure 4-11). Commonly, sand and similar sediment have about 20 to 40 percent pore space.

As sediment accumulates and is buried, it is compressed by the weight of the overlying layers.

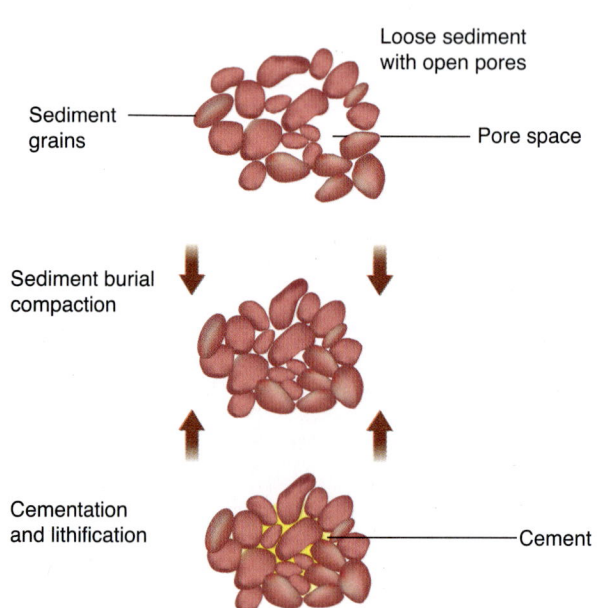

FIGURE 4-11
Pore space is the open space between loose sediment grains. Compaction squashes the grains together, reducing the pore space. Cement fills some of the remaining pore space, lithifying the sediment.

The compression partially collapses the pore spaces between sediment grains, forcing water out. This process is called compaction.

As sediment is buried and compacted, groundwater slowly circulates through the remaining pore space. The water commonly contains dissolved ions that can precipitate in the pores. This process forms cement that binds the clastic grains firmly together into a hard rock. Calcite, quartz, and iron oxides are the most common cements in sedimentary rocks.

Conglomerate is lithified gravel (Figure 4-12A). Each clast in a conglomerate is usually much larger than the individual mineral grains in the clast. In many conglomerates, the clasts may be fist-size or larger. They typically retain most of the characteristics of the parent rock. If enough is known about the geology of an area where conglomerate is found, it may be possible to identify approximately where the clasts originated. For example, a granite clast probably came from nearby granite bedrock.

The next time you walk along a gravely streambed, look carefully at the gravel. You will probably see sand or silt trapped among the larger clasts. In a similar way, most conglomerates contain fine-grained sediment among the large clasts.

Sandstone consists of lithified sand grains (Figure 4-12B). The sand forms from the physical and chemical breakdown of preexisting rock. For example, weathering of granite typically produces sand-size grains of quartz, feldspar, and other minerals. Streams, wind, glaciers, and gravity all can carry sand downslope. Sand grains carried in streams and by wind repeatedly undergo collisions as they bounce along. These impacts wear the sharp edges off each grain, causing the grains to become rounded. Eventually, the sand grains accumulate to form a deposit of sand. Over time, the deposit compacts and lithifies to form sandstone. Quartz is so hard that it resists physical and chemical breakdown. Many sandstones consist predominantly of rounded quartz grains.

Mudstone is a clastic sedimentary rock that consists mostly of tiny clay minerals, silt, and organic particles (Figure 4-12C). Mudstone can vary widely in color, ranging from red to green to black. Mudstone is a dark-colored sedimentary rock due to the abundance of organic matter. As it is buried, mudstone can convert into oil and gas, which rise toward the surface through pore spaces in the rock.

Bioclastic Sedimentary Rocks

Carbonate rocks are bioclastic sedimentary rocks. These

A

B

C

FIGURE 4-12
Sedimentary rocks show a variety of sediment particle size. From large to small: (A) conglomerate, (B) sandstone, and (C) mudstone.

are primarily made up of the carbonate minerals aragonite, calcite, and dolomite. Limestone is made primarily of calcite and aragonite, two minerals with the same composition but with a different crystalline structure. Rocks rich in dolomite are called simply dolomite.

Dissolved calcium ions are released into water during the chemical weathering of limestone. Carbonate ions form when carbon dioxide gas

LESSON 4.3 **111**

from the atmosphere dissolves in water. Seawater contains large quantities of both dissolved calcium and carbonate ions. Clams, snails, corals, and many other marine organisms convert dissolved calcium and carbonate ions into shells and other hard body parts. When these organisms die, waves and ocean currents break the shells into fragments. These pieces can range in size from boulders to clay-size particles. A rock formed by lithification of such sediment is called bioclastic limestone (Figure 4-13A). It forms by both biological and clastic processes.

All limestones are calcareous, meaning they contain the mineral calcite. The calcite mineral grains are often too small to see with the unaided eye or a hand lens. Field and laboratory geologists typically apply a small drop of acid to the surface of a rock to test whether it is calcareous. If the rock is calcareous, the drop of acid will fizz. The acid rapidly reacts with the calcite in the rock to produce carbon dioxide gas. If the drop of acid does not react, the rock lacks much calcite.

Many limestones are bioclastic. They contain the skeletal debris from broken-down shells and other marine organisms with hard body parts. Many marine organisms, such as clams and snails, produce a hard skeleton that breaks down upon death. These include corals, sea urchins, and several kinds of green algae. The diversity and abundance of marine organisms with hard skeletal parts made of calcite or aragonite is remarkable. They have provided paleontologists with much information about changes to marine ecosystems through geologic time.

Perhaps no environment has undergone a better documented set of ecological changes through time than that of the reef. A reef is a wave-resistant structure of carbonate rock formed from carbonate-forming organisms such as corals. In today's shallow marine reefs, corals are common (Figure 4-13B). About 100 million years ago, a type of clam was the dominant reef-building organism. But 350 million years ago, a class of organisms called echinoderms was the main reef-building organism. Other reef-building organisms of the geologic past have no modern surviving counterparts.

Many environments where limestones form are in shallow water that is almost saturated with calcium carbonate. Limestones commonly contain sand-size grains called ooids. Ooids form from the precipitation of aragonite and calcite on the surface of a single sand grain (Figure 4-13C). Cross sections of ooids show concentric structures from the layer-by-layer precipitation of minerals.

Although reefs and reef-building organisms produce large pieces of skeletal debris, many limestones are much finer-grained. Individual

A

B

C

FIGURE 4-13
Common forms of limestone include (A) bioclastic limestone made of shell debris, (B) reef limestone found in corals, and (C) oolitic limestone made of sand-sized ooids.

FIGURE 4-14
Chalk, formed from the remains of calcareous plankton, make up the White Cliffs of Dover on the southern coast of England.

fossils making up the grains are too small to see without a microscope. For example, chalk is a very fine-grained, soft, white bioclastic limestone. Chalk is made of the fossilized skeletons of plankton, tiny organisms that float through the ocean and produce a calcium carbonate skeleton. When plankton die, their remains sink to the bottom. Their skeletal debris accumulates to form chalk. The White Cliffs of Dover are an example of accumulated chalk (Figure 4-14).

checkpoint What is the difference between bioclastic and organic sedimentary rock?

Organic and Chemical Sedimentary Rocks

Organic Sedimentary Rocks Organic sedimentary rocks form by lithification of the remains of plants and animals. Coal is perhaps the most obvious example of an organic sedimentary rock (Figure 4-15). Coal contains so much organic matter that the rock will burn. It forms from the

FIGURE 4-15
Coal comes in different varieties depending on the carbon content, from the lowest concentration, (A) peat, to the highest, (B) anthracite. Anthracite, a metamorphic rock, is formed over millions of years when peat is buried and compacted.

accumulation of peat in swamps and similar environments where dead plants accumulate faster than they can be broken down by decay. Peat is buried and compacted by overlying sediments. It is lithified and over time turns into anthracite, the most carbon-rich form of coal.

Chemical Sedimentary Rocks Some common elements such as calcium, sodium, potassium, and magnesium dissolve during chemical weathering.

LESSON 4.3

FIGURE 4-16
(A) Bedded chert is a chemical sedimentary rock formed by the slow accumulation of plankton in deep marine water. These beds of chert have been folded by tectonic compression. (B) A nodule of chert is dark gray in this limestone rock. The chert consists of tiny quartz crystals.

They are carried by groundwater and streams downslope. These elements form chemical sedimentary rocks. Most streams eventually reach the sea, but some are landlocked and end instead in a lake with no outlet. The Great Salt Lake in Utah is an example of a lake in which water escapes only by evaporation or downward seepage. When the water evaporates, salts remain behind and the lake water becomes saltier. Chemical sedimentary rocks form when evaporation concentrates the salts so they precipitate out from the water. The same process can occur if ocean water is trapped in coastal or inland basins where it can no longer mix with the open sea.

A more common chemical sedimentary rock is chert, which is composed of very finely crystalline quartz. Chert comes in many different colors and is one of the earliest geologic natural resources. Flint, a dark gray-to-black variety of chert, was frequently used by humans for arrowheads, spear points, scrapers, and other tools chipped to hold a sharp edge.

Chert typically forms in two varieties. Bedded chert occurs as sedimentary layers. Nodular chert is found as irregularly shaped lumps called nodules within other sedimentary rocks (Figure 4-16A). Microscopic examination of bedded chert often shows that it contains the remains of tiny marine organisms whose skeletons are composed of silica instead of calcium carbonate. In other cases, silica-rich volcanic ash from a distant eruption can settle onto the sea surface. The ash slowly sinks to the bottom to become a layer of silica-rich mud that lithifies to become bedded chert. In contrast to bedded chert, nodular chert (Figure 4-16B) forms when silica-rich groundwater soaks through limestone and forms nodules.

checkpoint What are four elements commonly found in chemical sedimentary rock?

4.3 ASSESSMENT

1. **Identify** What percentage of Earth's crust is made up of sedimentary rock, and how are these rocks distributed on the land?
2. **Relate** Compare and contrast bioclastic and organic sedimentary rocks.
3. **Describe** How is sediment of different sizes moved by a stream?
4. **Contrast** What is the main difference between clastic and bioclastic sedimentary rock?

Critical Thinking

5. **Evaluate** How does sediment transported at a great distance compare to sediment transported a short distance?

4.4 METAMORPHIC ROCKS

Core Ideas and Skills
- Describe the connection between minerals and metamorphic rocks.
- Define metamorphic grade and give examples of rocks that are linked by metamorphism.

KEY TERMS

metamorphism foliation
metamorphic grade

A potter forms a delicate vase from clay. She places the soft piece in a kiln and slowly heats it to 1,000°C. As the temperature inside the kiln rises, the clay minerals decompose. Atoms from the clay then recombine to form new minerals that make the vase strong and hard. The breakdown of the clay minerals, the growth of new minerals, and the hardening of the vase all occur without melting the solid materials.

Formation and Features of Metamorphic Rocks

Metamorphism (from the Greek words for "changing form") is a process that transforms rocks and minerals through rising temperature and pressure or through changing chemical conditions. Metamorphism occurs in solid rock, like the transformations in the vase as the potter fires it in her kiln. Small amounts of water and other fluids speed up the metamorphic mineral reactions, but the rock remains solid as it changes. Metamorphism can change any type of parent rock: sedimentary, igneous, or another metamorphic rock.

A mineral that does not decompose or change in other ways, no matter how much time passes, is a "stable" mineral. A stable mineral can become unstable when environmental conditions change. Three types of environmental change affect mineral stability and cause metamorphism: rising temperature, rising pressure, and changing chemical composition. For example, when the potter placed clay in the kiln and raised the temperature, the clay minerals decomposed because they became unstable. The atoms from the clay then recombined to form new minerals that were stable at the higher temperature.

Similarly, if hot water seeping through bedrock carries new chemicals to a rock, those chemicals may react with the rock's original minerals. Different minerals that are stable in the new chemical environment may form. Metamorphism occurs because each mineral is stable only within a certain range of temperature, pressure, and chemical environment. If temperature or pressure rises above that range, or if chemicals are added or removed from the rock, the rock's original minerals may decompose. The mineral's components recombine to form new minerals that are stable under the new conditions.

The **metamorphic grade** of a rock is the intensity of metamorphism that formed it. Temperature is the most important factor in metamorphism. Because temperature increases with depth in Earth, a general relationship exists between depth and metamorphic grade. Low-grade metamorphism occurs at shallow depths, less than 10 kilometers beneath the surface, where temperature is no higher than 300°C to 400°C. Medium-grade conditions, where temperatures are between 400°C and 600°C, exist at depths between about 10 and 40 kilometers. High-grade conditions are found deep within the continental crust and in the upper mantle, 40 to 55 kilometers below Earth's surface. The temperature here is 600°C to 800°C, close to the melting point of rock. High-grade conditions can develop

CAREERS IN EARTH AND SPACE SCIENCES

Paleontologist
Paleontologists study fossils of plants and animals that lived on Earth up to billions of years ago. Their research involves many science disciplines including geology, chemistry, biology, physics, and anthropology. Paleontologists discover more about the history of life, the process of evolution, and even extinction events by analyzing fossils found in rocks at collection sites all around the world. They plan and direct the search for fossil samples in not only rocks but also in lakes, soil, and ice. They preserve specimens in the laboratory and publish journal articles to share their work with others.

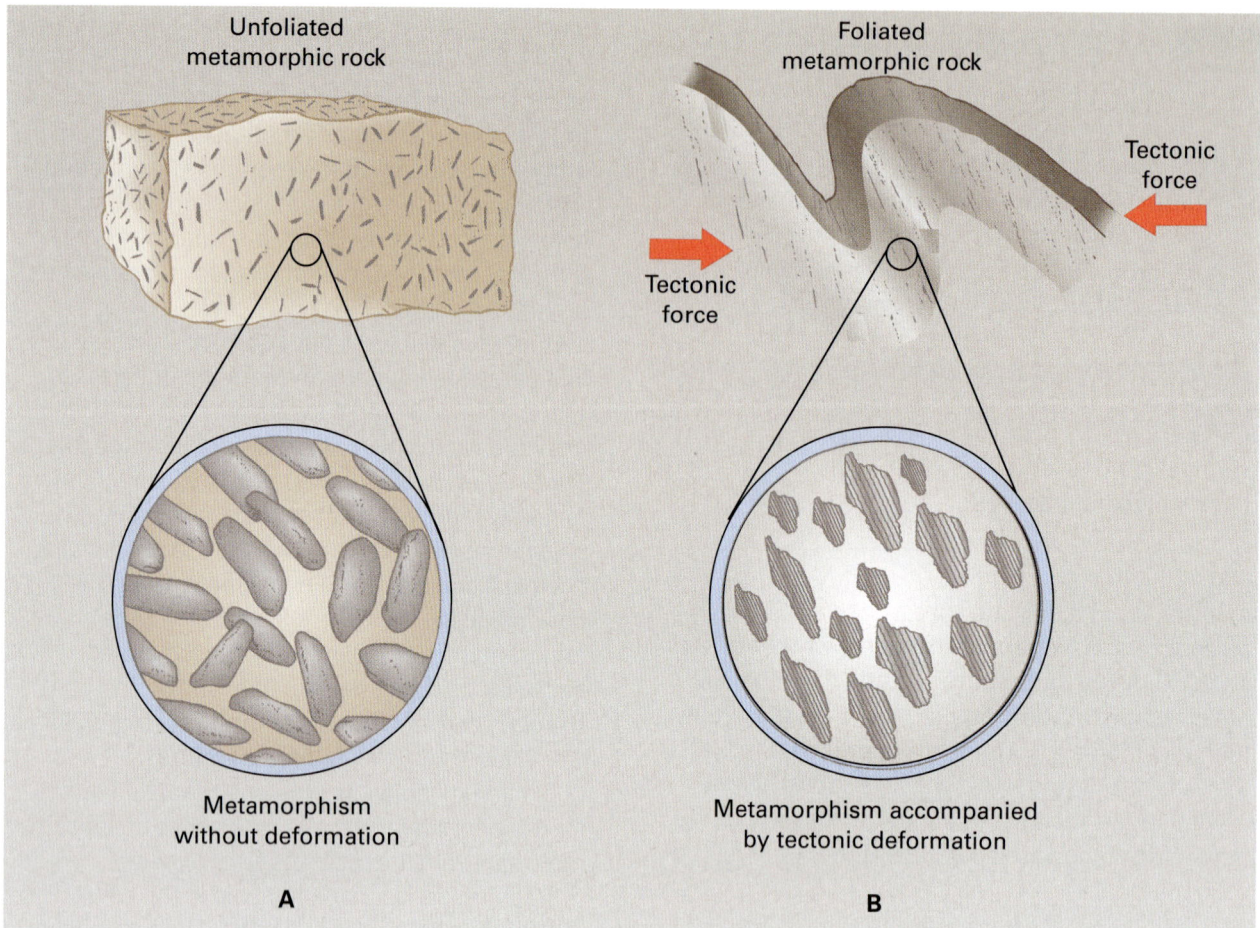

FIGURE 4-17
A diagram depicts the differences between foliated and unfoliated metamorphic rock. (A) Unfoliated rock results from metamorphism without deformation with random orientations of mica minerals. (B) Metamorphism with deformation forces the micas to align in foliated rock.

at shallower depths, however, in rocks adjacent to hot magma. For example, metamorphic rocks are forming today beneath Yellowstone Park, where hot magma lies close to Earth's surface. Metamorphism commonly alters both the texture and mineral content of a rock.

As a rock undergoes metamorphism, most mineral grains grow larger. The shapes of the grains may also change. For example, fossiliferous limestone is a sedimentary rock, but if it is subject to high temperature and pressure, the small calcite crystals that make up both the fossils and the cement between them will recrystallize into larger crystals. In the process, the fossils will be slowly destroyed and the rock will transform into marble, a metamorphic rock composed of calcite but with a texture consisting instead of large, interlocking crystals.

Micas are common metamorphic minerals that form when many different parent rocks undergo metamorphism. Micas are shaped like pie plates. When metamorphism occurs without deformation or a change in shape, the micas grow with random orientations (Figure 4-17A). However, when tectonic force squeezes rocks as they are heated during metamorphism, the rock deforms into folds. When rocks are folded as mica crystals are growing, the micas grow with their flat surfaces perpendicular to the direction of the squeezing. This parallel alignment of micas (and other minerals) produces the metamorphic layering called **foliation** (Figure 4-17B).

The foliation layers can range in thickness from a fraction of a millimeter to a meter or more. Metamorphic rocks commonly will break parallel to foliation planes to form thin, parallel cleavage.

Although metamorphic foliation and the thin cleavage that results can resemble sedimentary bedding (stratification), the two types of layering are different in origin. Foliation results from

FIGURE 4-18
Two photos show the results of metamorphism: (A) un-metamorphosed mudstone and (B) high-grade metamorphosed mudstone, also called gneiss.

the alignment of metamorphic minerals during metamorphism. Stratification develops because sediments are deposited layer by layer.

When a parent rock contains only one mineral, metamorphism transforms the rock to have a coarser texture. Limestone converting to marble is one. Both rocks consist of the mineral calcite, but their textures are very different. Another example is the metamorphism of quartz sandstone to quartzite.

Metamorphism of a parent rock containing several minerals usually forms a rock with new and different minerals and a new texture. For example, a typical mudstone (Figure 4-18A) contains large amounts of clay as well as quartz feldspar and several other minerals. When heated, some of those minerals decompose. Their atoms recombine to form new minerals such as mica, garnet, and a different kind of feldspar. Figure 4-18B shows a rock called gneiss that was formed when metamorphism altered both the texture and the minerals of mudstone. If fluids alter the chemical composition of a rock, new minerals form.

checkpoint What two factors affect metamorphic grade?

Types of Metamorphism

Rising temperature, rising pressure, and changing chemical environment can cause metamorphism. In addition, deformation caused by the movement of Earth's crust causes foliation and strongly affects the texture of a metamorphic rock. Four different geologic processes create these changes.

Contact Metamorphism Contact metamorphism occurs where hot magma intrudes cooler rock of any type. The highest-grade metamorphic rocks form at the contact point, closest to the magma. Lower-grade rocks develop farther away. A metamorphic "halo" around a body of igneous rock can range in width from less than a meter to hundreds of meters. Contact metamorphism commonly occurs without deformation. As a result, the metamorphic minerals grow with random orientations and the rocks develop no foliation.

Burial Metamorphism Burial metamorphism results from the burial of rocks in a sedimentary basin. Younger sediment may bury the oldest layers to depths greater than 10 kilometers in a large basin. Over time, temperature and pressure increase within the deeper layers until burial metamorphism begins. Burial metamorphism is occurring today in the sediments underlying many large deltas, including the Mississippi River Delta. Like contact metamorphism, burial metamorphism occurs without deformation. The rocks are unfoliated, and minerals grow with random orientations.

Regional Metamorphism Regional metamorphism occurs where major crustal movements build mountains and deform rocks across broad regions. These metamorphic rocks

DATA ANALYSIS Metamorphism of Mudstone

The metamorphic grade of a rock is the intensity of metamorphism that formed it. Figure 4-19A shows the metamorphic grade of a rock in different conditions. The blue line traces the path of increasing temperature and pressure with depth in Earth's continental crust.

Figure 4-19B shows how mudstone changes as it is exposed to different conditions. As mudstone is metamorphosed, it undergoes changes in texture and mineral content. Low-grade metamorphism changes mudstone to slate, a dull, finely textured rock harder than mudstone. Slate transforms to phyllite, a shiny rock with foliated surfaces. Schist has crystals big enough to see with the unaided eye. Gneiss is coarse in texture and includes individual layers of light and dark-colored minerals. Migmatite rock typically shows well-developed folds.

FIGURE 4-19
(A) A graph compares the temperature and pressure at different depths. (B) A graph demonstrates the effects of metamorphism by high temperature and high pressure on mudstones ranging from slate (low metamorphic grade) to migmatite (high metamorphic grade).

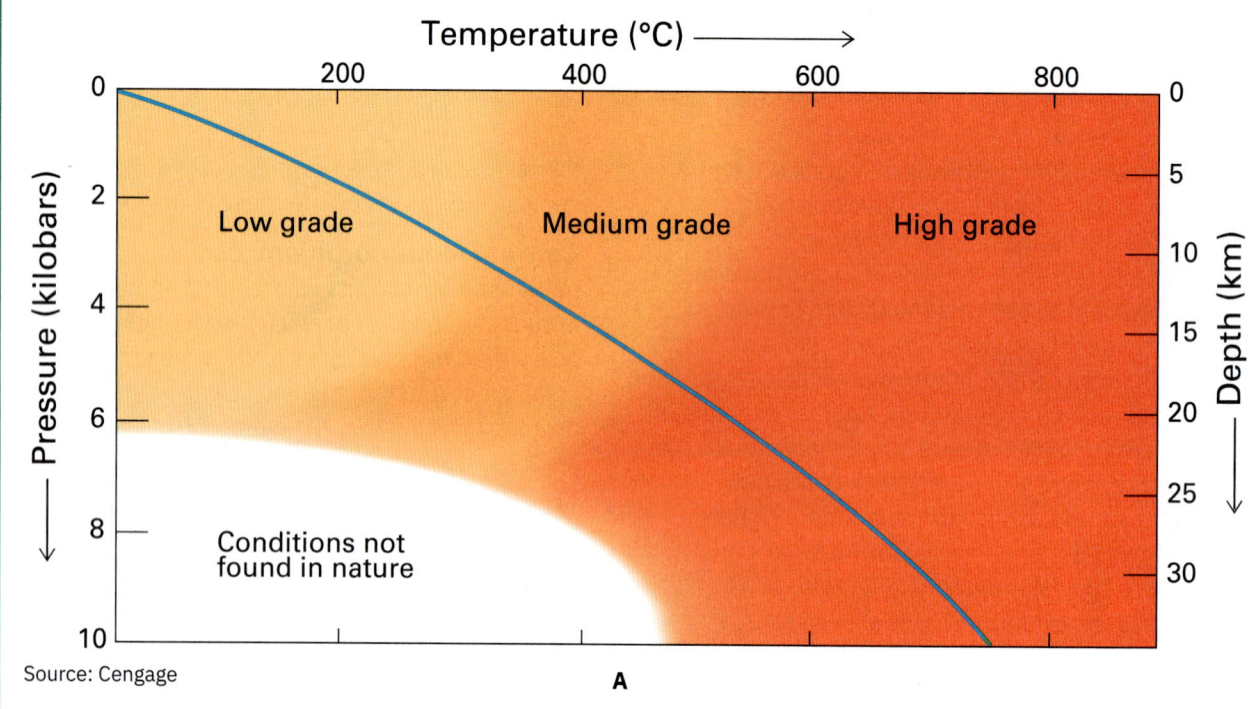

Source: Cengage

are deformed and heated at the same time. Such rocks are the most common and widespread type of metamorphic rocks.

Magma rises and heats large portions of the crust in places where tectonic plates converge. The high temperatures cause new metamorphic minerals to form throughout the region. As it forces its way upward, the magma also deforms the hot, plastic country rock. At the same time, the movements of the crust squeeze and deform rocks. As a result of all these processes, regionally metamorphosed rocks are strongly foliated. Regional metamorphism produces zones of foliated metamorphic rocks tens to hundreds of kilometers across. For example, large portions of the Appalachian Mountains in the eastern United States have undergone regional metamorphism and exhibit well-developed foliation.

Hydrothermal Metamorphism Water is a chemically active fluid because it dissolves rocks and minerals. If the water is hot, it dissolves minerals even more rapidly. Hydrothermal metamorphism occurs when hot water and ions dissolved in the hot water react with a rock to change its chemical composition.

checkpoint Which types of metamorphism result in foliated metamorphic rock?

1. **Recognize Patterns** Describe the correlation between depth and temperature.
2. **Recognize Patterns** Describe the correlation between depth and pressure.
3. **Infer** How does the energy available for metamorphism change as depth changes?
4. **Contrast** Distinguish between the processes that transform rocks close to Earth's surface with those that occur deep within Earth.

Computational Thinking

5. **Represent Data** Create a diagram that shows the depths at which you are likely to find different metamorphic grades of mudstone. Label the temperature and pressures in your model.
6. **Generalize Patterns** In Yellowstone Park, hot magma lies closer to the surface than it does in most places on Earth. How would you need to adjust both Figure 4-9A and Figure 4-9B to represent metamorphism in the rocks under Yellowstone?

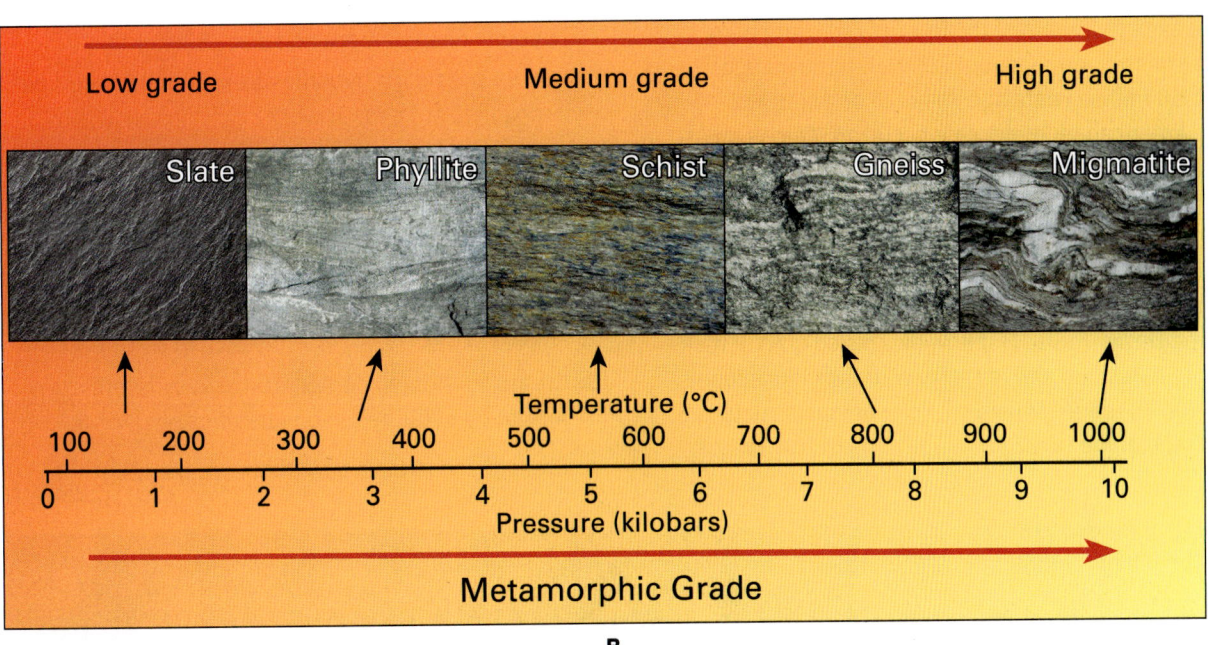

Source: Cengage

B

4.4 ASSESSMENT

1. **Explain** In the absence of drastic changes in temperature and pressure, what conditions speed up the metamorphism of a rock?
2. **Define** What is a stable mineral and what environmental factors can cause it to become unstable?
3. **Describe** Under what conditions is metamorphism considered low-grade, medium-grade, and high-grade?
4. **Distinguish** What are the four types of metamorphism and in what geologic settings do they occur?

Critical Thinking

5. **Recognize Patterns** Explain how sedimentary rocks and certain metamorphic rocks might display similar patterns.

TYING IT ALL TOGETHER
FORMATION OF DEVILS TOWER

In this chapter, you learned about the processes that drive the rock cycle. Although rocks change very slowly from a human perspective, Earth's systems interact over geologic time to form new rock and transform old rock. This results in the rock cycle, in which rocks transform among the three types of rock: igneous, sedimentary, and metamorphic. You also examined how surface processes, such as weathering, and various forms of metamorphism can cause chemical and physical changes in rocks over time.

In the Case Study, you learned about Devils Tower, a unique column of rock that rises above a landscape consisting primarily of layered rock. The oldest exposed layers of the layered rock date back to the Triassic period, about 200 million years ago. About 50 million years ago, major changes to the North American tectonic plate occurred. Across the northern Rocky Mountains, pressure in Earth's crust bent the rock into new shapes, forming uplifts that include the Black Hills, where Devils Tower is located. Magma pushed up from below the layered rock in the Black Hills area, cooling and forming the rock that has become Devils Tower. But how did this column of rock come to rise so far above the surrounding landscape? What forces are responsible for the unique structure and shape of the rock?

In this activity, you will investigate hypotheses proposed by geologists to explain how Devils Tower took its present form. Work with one or two other students to complete the steps below.

1. Find several reliable sources on the formation of Devils Tower.

2. With your partner or group, work to identify and compare different explanations offered in your sources for the formation and shaping of Devils Tower. Use the following focus questions to guide your research.

    - How do the hypotheses compare? What are the points of agreement and disagreement?
    - How does each hypothesis explain the observed changes at Devils Tower?
    - How does each hypothesis explain the formation of Devils tower in terms of the processes and rock types of the rock cycle?
    - What evidence supports each hypothesis, and what evidence presents possible weaknesses of each hypothesis?

3. Create a model to demonstrate the different formation hypothesis you identified.

4. Present and explain your model to the class.

FIGURE 4-20
(A) Devils tower looms over the landscape in Wyoming. (B) A rock climber scales one of the columns of this massive geologic feature.

CHAPTER 4 SUMMARY

4.1 ROCKS AND THE ROCK CYCLE

- Geologists divide rocks into three groups depending upon how the rocks formed. Igneous rocks solidify from magma.
- Sedimentary rocks form from clay, sand, gravel, and other sediment that accumulates at Earth's surface.
- Metamorphic rocks form when any rock in the solid state is altered by temperature, pressure, or an influx of hot water.
- The rock cycle summarizes processes by which rocks continuously transform among the three main types of rock: igneous, sedimentary, and metamorphic.
- Rock cycle processes exchange energy and materials with the atmosphere, the hydrosphere, and the biosphere.

4.2 IGNEOUS ROCKS

- Extrusive (or volcanic) igneous rocks are finely crystalline rocks that solidify from magma that has erupted onto Earth's surface.
- Intrusive (or plutonic) igneous rocks are medium-to-coarsely crystalline rocks that solidify within Earth's crust.
- Granite and basalt are the two most common igneous rocks.

4.3 SEDIMENTARY ROCKS

- Sediment forms by the weathering of rocks and minerals.
- Sediment includes all solid particles such as rock and mineral fragments, organic remains, and precipitated minerals.
- Sediment is transported by streams, glaciers, wind, and gravity; is deposited in layers; and eventually is lithified to form sedimentary rock.
- Mudstone, sandstone, and limestone are the most common kinds of sedimentary rock.

4.4 METAMORPHIC ROCKS

- When a rock is heated, when pressure increases, or when hot water alters its chemistry, its minerals and texture change in a process called metamorphism.
- Contact metamorphism affects rocks heated by a nearby igneous intrusion.
- Burial metamorphism alters rocks as they are buried deep within Earth's crust.
- In regions where tectonic plates converge, high temperature, deformation from rising magma, and plate movement all combine to cause regional metamorphism.
- Hydrothermal metamorphism is caused by hot solutions soaking through rocks.
- Slate, schist, gneiss, and marble are common metamorphic rocks.

CHAPTER 4 ASSESSMENT

Review Key Terms
Select the key term that best fits the definition. Not all terms will be used, and no term will be used more than once.

bioclastic sedimentary rock	lithification
carbonate rock	magma
chemical sedimentary rock	metamorphic grade
	metamorphic rock
clastic sedimentary rock	metamorphism
	organic sedimentary rock
deposition	precipitation
dike	rock cycle
extrusive igneous rock	sedimentary rock
foliation	texture
igneous rock	weathering
intrusive igneous rock	

1. liquid rock
2. parallel layering of minerals in a rock
3. rock formed when molten rock solidifies
4. formation of a solid from ions previously dissolved in a liquid
5. rock formed from the remains of living organisms
6. rock made from fragments of weathered rocks
7. rock formed when lava cools
8. the breakdown of rock into smaller pieces
9. a measurement of the intensity of change in a rock caused by heat, pressure, or chemical conditions
10. a description of the size, shape and arrangement of minerals in a rock
11. an example of a bioclastic rock with high amounts of calcite and/or dolomite.

Review Key Concepts
Answer each question on a separate sheet of paper to demonstrate your understanding of key concepts from the chapter.

12. Use a model to describe changes that occur to rocks and how these changes lead to the rock cycle.
13. What is the main observable difference between extrusive and intrusive igneous rocks? What causes the difference?
14. Compare and contrast the conditions under which granite, rhyolite, and obsidian form from magma.
15. What is a clast, and what environments or conditions can cause clasts to have a rounded shape?
16. Describe the steps in formation of sedimentary rocks.
17. Describe the key differences between clastic and bioclastic rock. Give examples of each.
18. Give two examples of how precipitation can result in the formation of a rock.
19. Make a chart or bulleted list that summarizes the conditions and results of the four types of metamorphism.
20. Compare and contrast stratification and foliation.

Think Critically
Write a response to each question on a separate sheet of paper. Use concepts from the chapter to support your reasoning.

21. Is the rock pictured igneous, sedimentary, or metamorphic? What evidence can you use to determine the type?

22. Is the rock pictured igneous, sedimentary, or metamorphic? What evidence can you use to determine the type?

23. Compare and contrast metamorphism and the melting of rock. How are they similar and different in terms of their causes and outcomes?
24. Explain why igneous rock found at the surface of Earth is not always located near a volcano.
25. What kinds of rocks—igneous, metamorphic, or sedimentary—would most likely contain fossils? Explain your reasoning.
26. Describe why peat is classified as sedimentary rock but anthracite is classified as metamorphic rock.

PERFORMANCE TASK

Rocks in U.S. National Parks

Impressive geologic formations are important features at many of our national parks. In this performance task, you will identify the different types of rocks found at our national parks and explore how they vary across the country.

1. Explore the rock types at different national parks. Review the information on igneous, metamorphic, and sedimentary rock, and visit the Geodiversity Atlas pages for the national parks listed on the page for each rock type. Select one park as the focus of your investigation, or investigate another park assigned by your teacher.

   Characterize the types of rock that are found in the park. Identify and describe the processes responsible for the formation of these rocks.

2. Working with other students in the class, collect the data from your investigations and add it to a map of the United States. Use labels and symbols to indicate the location of each park and the kinds of rocks found there. When the map is complete, analyze the distribution of rock types and describe any patterns you observe.

   Compare the geology of different regions of the U.S. and use your knowledge of the rock cycle to infer which processes have most recently played a role in producing each region's rocks.

CHAPTER 5
GEOLOGIC TIME

The 600-meter high cliffs around Western Brook Pond in Canada's Gros Morne National Park are made primarily of the metamorphic rock gneiss and sedimentary rocks, such as limestone and sandstone. The rocks found across the park range in age from hundreds of millions to billions of years old.

The landscape of Gros Morne National Park in Newfoundland and Labrador represents a unique opportunity for geologists to study a diverse collection of rocks and rock formation processes in one location. The rocks in this United Nations Educational, Scientific and Cultural Organization (UNESCO) World Heritage Site include samples that are among the oldest in the world, dating back to 3.8 billion years ago. These rocks formed during Earth's early history, the Precambrian, which spans from formation 4.6 billion years ago until approximately 541 million years ago.

The study of geologic time helps us reconstruct the time line along which Earth's historical events occurred. Read on to learn about the basic principles geologists use to place geologic events into a time sequence; that is, what the relative ordering of multiple events was in time. This chapter also introduces the concept of absolute dating, in which geologists use isotopes to determine the numerical age of a geologic unit such as a granite pluton or an ancient lava flow.

KEY QUESTIONS

5.1 How did the recognition of abrupt mass extinctions change our understanding of how Earth changes over time?

5.2 What are the principles of relative age and how do they differ from those of absolute age?

5.3 What information do the boundaries between sedimentary rock layers provide about environmental conditions over time?

5.4 How is radiometric dating used to determine the age of rocks?

5.5 What is the geologic timescale?

EXPLORERS AT WORK

CONSTRUCTING ROCK RECORDS

WITH NATIONAL GEOGRAPHIC EXPLORER SARAH CARMICHAEL

Sarah Carmichael is a geochemist who studies the chemistry of rocks and minerals. As part of her research, Dr. Carmichael analyzes the geochemical makeup of ancient rocks to better understand mass extinctions. A mass extinction is an event characterized by the complete die-off of a high number of species in a relatively short period of geologic time. There have been five such mass extinctions in the past. Carmichael is primarily interested in mass extinction pulses from 375 million to 350 million years ago (late Devonian period) that wiped out 70 percent of Earth's marine species and decimated coral reefs.

Carmichael is analyzing the rock record from this period to determine the environmental conditions at the time of the mass extinction, as well as how some species were able to survive it. She says the causes for what happened are poorly understood, but she hypothesizes that the mass extinction was climate-driven and global in scope. Carmichael and her collaborators also assert that the marine species survived because they were isolated in the open ocean. If this hypothesis is correct, then information from research like hers could help predict how the rapid climate change today will impact marine mammals and ocean geochemistry.

What scientists do know is that the late Devonian was a time when Earth's climate was very unstable, The chemical composition of its atmosphere was changing. The mass extinction was associated with a severe loss of oxygen from the oceans. Yet, some marine species did survive, and Carmichael wants to know why. However, she finds previous studies about the mass extinction's causes, scope, and impacts problematic. They are limited and do not represent global oceanic conditions at that time.

In 2018, Carmichael led the Devonian Anoxia, Geochemistry, Geochronology, and Extinction Research group (DAGGER), a team of scientists through Appalachian State University in North Carolina, where Carmichael is a professor, to western Mongolia. She says that western Mongolia is a key site for two reasons. First, it was formed from isolated volcanic island chains, and therefore its ancient sediments likely represent the oceanic conditions in the late Devonian. Second, it is part of the region where most marine animals survived and evolved into much of today's marine life.

Carmichael's analysis of ancient ocean sediments in western Mongolia supports her team's hypothesis that mass extinction was global in scope and that oceanic isolation was a factor in the survival of marine species. She conducts research in Mongolia, Southeast Asia, Central Asia, and Europe. As Carmichael explains, "Understanding where and how marine animals can survive rapidly changing oceanic conditions and oxygen loss during past mass extinctions is a vital tool for predicting how marine ecosystems will respond to a predicted future biodiversity crisis."

THINKING CRITICALLY

Compare and Contrast How is Sarah Carmichael's research similar to and different from the work of Allyson Tessin featured in Chapter 2?

Meet National Geographic Explorer Sarah Carmichael and learn how mass extinctions tell us about the geologic past. Find out about key concepts tied to geologic time, including fossils and rock layers.

◀ Carmichael and her team gathered fossils of echinoderms and different shells. The team wants to determine not only which species survived the late Devonian mass extinctions, but also how they survived. Sediments in which the fossils are preserved also tell a great deal about the environment in which the animals lived.

127

CASE STUDY
GIANT DRAGONFLIES?

Gigantic insects, several feet long, might seem like the stuff of horror films or far-fetched science fiction. But based on the fossil record, they are actually the stuff of Earth's past.

In 1940, Frank Carpenter, a Harvard fossil specialist, was working with his colleagues to excavate and collect fossilized insects in Noble County, Oklahoma. Among the approximately 5,000 insect fossils they discovered was a large fossil of *Meganeuropsis americana*, a distant relative of the modern dragonfly. Modern dragonflies are sizable insects with a typical wingspan of about 10 centimeters (4 inches). But the *Meganeuropsis* fossil Carpenter discovered had a wingspan of about 75 centimeters (2.5 feet), larger than that of a red-tailed hawk (Figure 5-1).

The area of Oklahoma where Carpenter's *Meganeuropsis* was found was once a tropical wetland, which at intervals became covered in seawater. The sediments that are deposited in such submerged area commonly contain the remains of prehistoric plants and animals that died and were buried. Over time, the sediment lithified to form sedimentary rock, and the organic remains became fossils. By dating the rock in which the fossils were found, scientists have determined that *Meganeuropsis* lived about 290 million to 248 million years ago, during the Permian period.

Noble County, Oklahoma, is not the only location in which *Meganeuropsis* fossils are preserved. The Grand Canyon area is also known for its abundant fossils, including *Meganeuropsis* and a number of other marine and land animals. Hermit Formation, one of many sedimentary rock layers exposed in the Grand Canyon, contains *Meganeuropsis* fossils. About 280 million years ago, *Meganeuropsis* individuals died and were buried in mud and silt associated with ancient streams in the area.

Scientists can learn about the lifestyle of *Meganeuropsis* from the features preserved in fossils. It was a powerful predator with forceful jaws and teeth for securing large prey. It had spiny legs and was agile in the air, as dragonflies are today. The fossils also contain clues about the nature of Earth's atmosphere during the Permian period. Some researchers hypothesize that higher levels of oxygen in the atmosphere during the Permian period allowed for the evolution of large insects, such as *Meganeuropsis*. Widespread forests that developed during the Carboniferous Period, which immediately preceded the Permian, drove down atmospheric carbon dioxide and drove up atmospheric levels of oxygen. (Remember that plants take up carbon dioxide and release oxygen during photosynthesis.) When a mass extinction wiped out the majority of life on Earth, including *Meganeuropsis,* oxygen levels dropped, ensuring that insects today would not grow to the large proportions of the past.

As You Read Consider how organisms have changed over very long periods of time and how the fossil record helps us reconstruct Earth's prehistoric past. Think about how evidence of changes that occurred in the geosphere and the biosphere can also reveal information about other spheres of Earth, such as the atmosphere.

FIGURE 5-1
An artist's vision of the *Meganeuropsis americana* portrays the size of this early dragonfly species against tree trunks in an imagined forest.

5.1 MASS EXTINCTIONS IN EARTH'S HISTORY

Core Ideas and Skills
- Examine our changing understanding of mass extinctions and their causes.

KEY TERMS

mass extinction   anoxic

In the 19th century, geologists researching in North America and Europe found thick sequences of sedimentary rock layers containing an abundance of fossils preserved in them. The geologists also found layers of rock containing few or no fossils lying above the fossil-rich rocks. Even higher in the sequence of rock layers, they found fossils again. The fossils in the higher rock layers were of organisms that were very different from the ones found in the older, lower rock layers. Most surprising, many of the most abundant fossilized organisms in the lower layers never appeared again in the younger rocks. Those organisms simply had disappeared from the surface of Earth forever, as though suddenly extinguished.

Changing Scientific Understanding

Today we know through the fossil record that more than once, a sudden, catastrophic event has abruptly decimated life on Earth, causing a **mass extinction** in which many types of organisms simply were wiped out. Following each extinction event, new life-forms slowly emerged, and these new life-forms recolonized the planet.

Many geologists of the 20th century did not accept the idea that life on Earth underwent catastrophic extinctions that affected the entire planet. Instead, they suggested that perhaps the rocks containing the evidence of more gradual changes in Earth's life-forms simply had been destroyed by erosion—or were never deposited. According to this reasoning, evidence for the gradual decline of species that became extinct and the slow emergence of new species was simply missing from the rock record. Or it had not yet been found. Following the scientific viewpoint of the time, scientists concluded that the extinctions of old life-forms and the emergence of new ones occurred slowly, as a result of gradually changing conditions.

As travel became easier in the later 20th century and more of Earth's rock record could be reached and studied by scientists, many searched for fossiliferous rocks of the appropriate ages to fill in the gaps. Finding such rocks would provide evidence for gradual extinctions. However, these rocks were never found. Instead, the scientists found an abundance of geologic evidence showing that near-instantaneous mass extinctions occurred at least five times in the geologic past. In each instance, life in the forms as it existed was decimated (Figure 5-2).

FIGURE 5-2
Go online to view an image of the geologic timescale.

The most dramatic extinction occurred about 252 million years ago, near the end of the Permian period. At that time, 90 percent of all species in the oceans suddenly died out. On land, about two-thirds of reptile and amphibian species and 30 percent of insect species vanished, including the giant *Meganeuropsis* described previously. Author Douglas Erwin claims it was "the closest life has come to complete extermination since its origin."

The death of most life-forms at the end of the Permian period left huge ecological voids in the biosphere. Ocean ecosystems changed as new organisms emerged in an environment relatively free of predators and competition. On land, terrestrial animals including dinosaurs slowly appeared and proliferated. Plants, too, underwent major changes that led to the evolution of the first flowering plants. About 65 million years ago, at the end of the Cretaceous period, another catastrophic extinction wiped out up to 50 percent of Earth's organism groups, including the dinosaurs. Small mammals survived this disaster, facing a new world free of the efficient predators that had hunted them. Grasses also evolved after the Cretaceous extinction event, leading to the establishment of widespread grassland ecosystems. Eventually, humans evolved.

We consider three hypotheses for mass extinctions. Each of these hypotheses involves large-scale interactions among Earth's four major systems. The geosphere, the atmosphere, the hydrosphere, and the biosphere all were involved in each type of mass extinction. The trigger that started the radical changes among the four Earth systems and caused the mass extinction is the main difference between each hypothesis.

checkpoint Explain how the discovery of new fossil evidence in the 20th century forced geologists to revise previous theories about the history of life on Earth.

Extraterrestrial Impacts

The father-and-son team of Walter and Luis Alvarez conducted research in the late 1970s. They were studying sedimentary rocks deposited at the time dinosaurs and so many other life-forms were becoming extinct. In one layer, the Alvarez duo found abundant dinosaur fossils. Just above it, they found very few fossils of any kind and no dinosaur fossils. Between these two rock layers they found a thin, sooty layer of clay. They brought samples of the clay to the laboratory and found that it contained high concentrations of the element iridium. This discovery was surprising because iridium is rare in Earth rocks. They wondered where it came from?

Although rare in Earth's crust, iridium is abundant in meteorites. In a paper published in 1980, Walter and Luis Alvarez suggested that 65 million years ago, a meteorite roughly 10 kilometers in diameter hit Earth. The meteorite strike had the explosive energy many thousands of times greater than that of today's entire global nuclear arsenal. The collision vaporized both the meteorite and Earth's crust at the point of impact. It formed a plume of dust and hot gas that rose into the upper atmosphere and circled the globe. This thick, dark cloud blotted out the sun, significantly reducing the amount of solar energy that reached Earth's surface for up to several years. Particularly hard-hit were microscopic algae and other tiny photosynthesizing organisms in the oceans. Such algae and organisms were unable to survive under the reduced levels of solar energy reaching the surface. The widespread die-off of photosynthesizing organisms in the oceans greatly reduced the base of the food chain. The reduced solar energy reaching the surface resulted in less vegetation on land too. A reduction in vegetation eliminated all dinosaurs and many reptiles. Analysis of the fossil record has suggested that nothing with a body mass greater than 25 kilograms (about the size of a large dog) survived.

The now-famous 1980 paper by the Alvarez team created a stir among geologists. The paper cited compelling evidence for the impact but did not identify the actual impact site. At about the same time, geophysicist Glen Penfield was analyzing magnetic surveys of the Yucatan Peninsula for a Mexican oil company. Penfield discovered the presence of a large, symmetrical semicircle in the Gulf of Mexico on the northwest side of the peninsula. Obtaining data from an older gravity survey of the peninsula, Penfield noticed that a second semicircle was present onshore and that the two semicircles together formed a large, circular structure about 70 kilometers across. Although Penfield and his coworkers suspected that the circular structure might be the "missing" impact site, they were not able to secure any rock samples from the area to examine them for evidence of an impact. The oil company did not allow Penfield's team to publish the geophysical data behind their discovery. It was not until 1991 that a team of scientists (among them Penfield) had amassed sufficient data, including rock samples from old oil wells in the area, to publish the first paper describing the site as an impact structure. Since then, many analyses of the impact site and surrounding areas have suggested that an impact occurred 65 million years ago, coinciding with the Cretaceous extinction event.

Although much evidence in support of the Yucatan site as a major impact structure has been published since 1991, its relation to the mass extinction at the end of the Cretaceous period is still the subject of vigorous debate. Careful analyses of fossils preserved in sediment cores recovered from the crater have indicated that the impact actually occurred about 300,000 years prior to the extinction event. In addition, other possible large impact sites of late Cretaceous age have been described in the North Sea, the Ukraine, and India (Figure 5-3). A possible impact crater more than 500 kilometers in diameter—the biggest on Earth—has been reported in India.

For many years, the Yucatan impact site was considered the preliminary event that caused the Cretaceous terminal extinction event. More recent findings cast doubt on that simple explanation. Some scientists have suggested that widespread volcanism in India, combined with severe greenhouse conditions at the end of the Cretaceous, caused significant environmental stress that reduced the population sizes of many species (or eliminated them entirely) prior to the Yucatan impact. These scientists have suggested that the impact itself did not catastrophically kill a healthy, thriving biosphere. Rather, the meteorite pushed an already very-stressed biosphere over the threshold

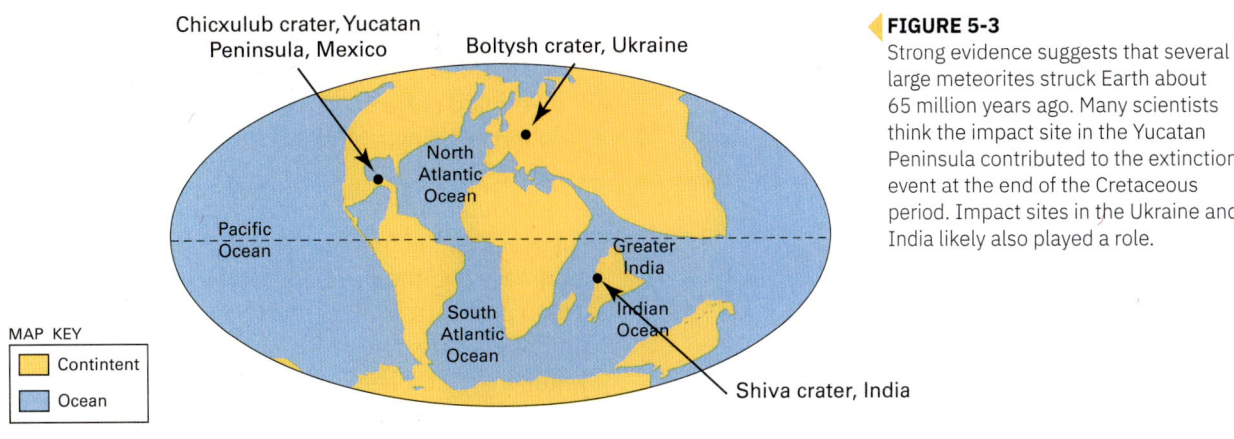

FIGURE 5-3
Strong evidence suggests that several large meteorites struck Earth about 65 million years ago. Many scientists think the impact site in the Yucatan Peninsula contributed to the extinction event at the end of the Cretaceous period. Impact sites in the Ukraine and India likely also played a role.

that led to a major extinction. The extraterrestrial impact hypothesis remains a popular one. However, reports of other possible major impact sites, and uncertainty about the role that factors such as volcanism may have played, keep scientific interest fixed on the other cause or causes of the Cretaceous extinction event.

checkpoint What event did geophysicist Glen Penfield discover evidence for while surveying the Yucatan Peninsula?

Volcanic Eruptions and Mass Extinctions

A volcanic eruption ejects gas and fine ash into the atmosphere. The volcanic ash acts as an "umbrella" that reflects sunlight back into space. This causes the climate to cool. One volcanic gas, sulfur dioxide, forms small particles called aerosols that also reflect sunlight and cool Earth. Carbon dioxide is another gas emitted by volcanoes. It is a greenhouse gas that can cause warming of Earth's atmosphere. Therefore, volcanoes erupt materials capable of causing both atmospheric cooling and warming. In many eruptions, the cooling overwhelms the warming effects. This cooling can occur rapidly, even within a few weeks or months of the eruption.

Scientists have noted that some mass extinctions coincided with unusually high rates of volcanic activity. For example, massive flood basalts erupted onto Earth's surface in Siberia roughly 252 million years ago—at the same time that Permian extinction occurred. According to one hypothesis, explosive volcanic eruptions blasted massive amounts of volcanic ash and sulfur aerosols into the air. The ash and aerosols spread throughout the upper atmosphere, blocking out the sun. Earth's atmospheric and oceanic temperatures plummeted, plants withered, and animals starved or froze to death. If this hypothesis is correct, the Permian extinction (and perhaps other extinction events) resulted from a classic Earth-systems interaction. A volcanic eruption occurring within the geosphere altered the composition of the atmosphere, leading to dramatic consequences for atmospheric and marine temperatures, and for life.

checkpoint What are two gases released during volcanic eruptions and how do they affect the temperature of the atmosphere?

Supercontinents and Earth's Carbon Dioxide Budget

In modern oceans, cold polar seawater sinks to the seafloor and flows as a deep-ocean current toward the Equator. This current transports oxygen from the surface to the deep-ocean basins and forces cold water to rise near the Equator, where it is warmed. The current mixes both heat and gases throughout the oceans.

Surface marine organisms that photosynthesize absorb carbon dioxide that has dissolved in ocean water from the atmosphere. The organisms use this carbon to produce organic tissue. When they die, they settle to the seafloor, trapping the carbon in sediment. Decomposers in the deep ocean consume some of the settling organic matter, converting carbon back to carbon dioxide. The carbon dioxide then returns to the atmosphere as modern ocean currents mix the deep and shallow seawater. Any carbon that is deposited on the seafloor and buried before it is consumed is ultimately removed from the atmosphere.

Ocean currents are partially controlled by the positions of the continents. Recall that the continents move slowly. Several times in Earth history, all the continents have joined together

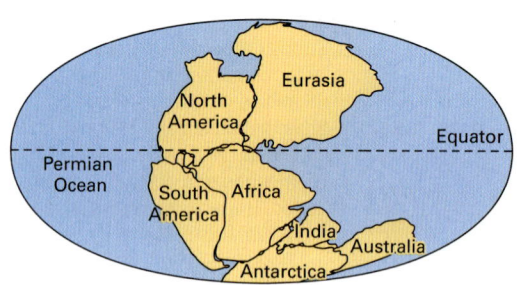

FIGURE 5-4
This map shows an assembly of all continents into the supercontinent Pangaea.

to form one giant supercontinent. One such supercontinent, called Pangaea, developed during the Permian period. The formation of Pangaea also resulted in a single global ocean called the Permian Ocean (Figure 5-4).

Computer models and the rock record both suggest that the merging of continents into a single landmass and the development of a single global ocean prevented mixing of deep and shallow water. As ocean water stagnated, two major changes in water chemistry began to take place in some parts

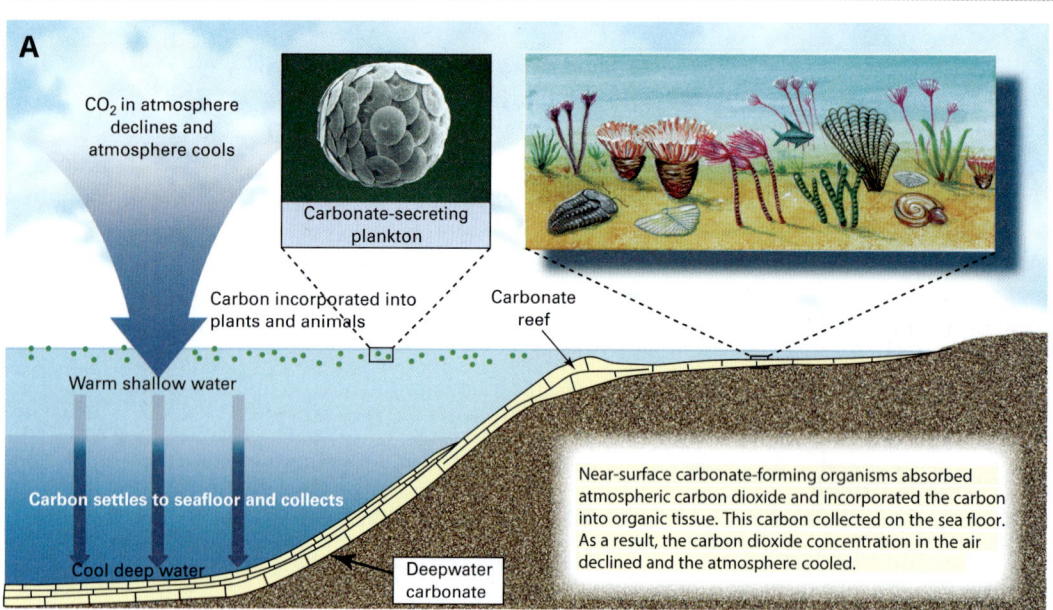

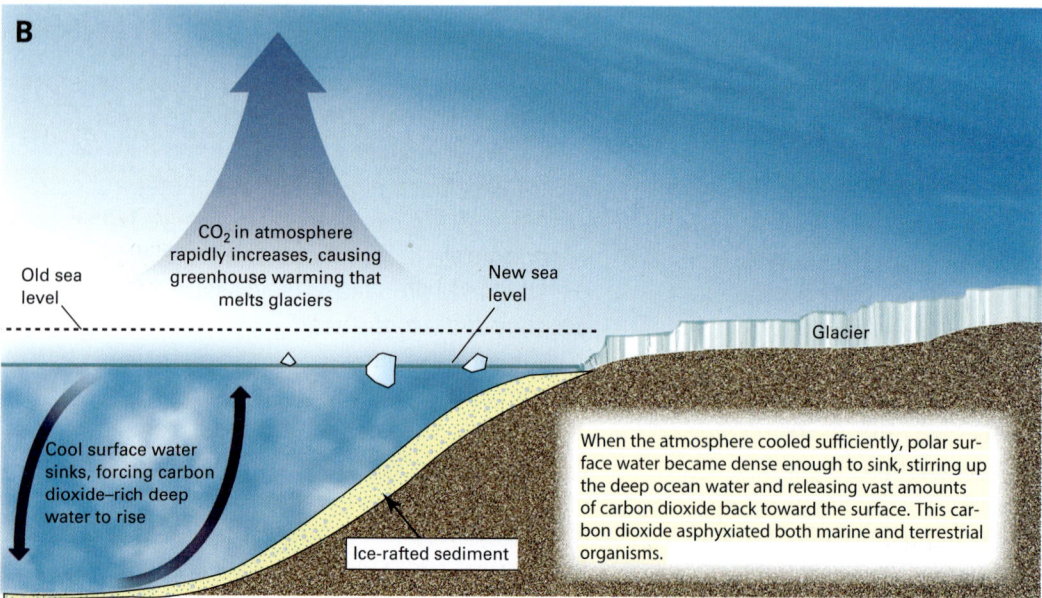

FIGURE 5-5
(A) Global cooling in the late Permian period was caused by a decline in CO_2 and carbon removal from the atmosphere. (B) After polar surface water restarted the deep-ocean currents, vast amounts of CO_2 were released into the atmosphere, resulting in the Permian extinction.

of the Permian Ocean. These conditions are called *anoxia* and *euxinia*.

As organic matter sank and was decomposed by oxidizing bacteria, Permian seawater lost most or all of its dissolved oxygen. Oxidizing bacteria require oxygen to decompose organic matter, so eventually the available oxygen was used up, resulting in **anoxic** waters. There was no measurable dissolved oxygen in the ocean. Once the oxygen was gone, the oxidizing bacteria died off while sulfate-reducing bacteria continued to decompose the available organic matter. Sulfate-reducing bacteria convert sulfate dissolved in the seawater into sulfide. This conversion produces the by-product hydrogen sulfide, a colorless, poisonous gas. In some parts of the Permian Ocean, the water became euxinic; it was so full of dissolved hydrogen sulfide that it began to form tiny bubbles.

The anoxic and euxinic conditions that developed in some parts of the Permian Ocean were deadly to nearly all species of multicellular marine life living near or on the seafloor. Organisms living near the surface were less affected because oxygen from the atmosphere could still mix with the water from wave energy to keep the water chemistry at the surface from becoming anoxic. Some surface plankton were able to thrive. They continued to remove carbon dioxide from the atmosphere and supply carbon to the bottom waters that remained anoxic and euxinic. Without ocean currents mixing the deep and shallow seawater, their removal of carbon dioxide from the atmosphere caused a decrease in greenhouse warming and resulted in global cooling. Some scientists have suggested that this global cooling led directly to the formation of continental ice sheets and the lowering of the global sea level.

When water cools, it becomes denser. The late Permian global cooling eventually reached a threshold at which polar surface water became denser than the deep water. At this point, the cool, dense surface water sank, displacing the carbon dioxide–rich deep water, which rose toward the sea surface. Massive amounts of carbon dioxide were released into the atmosphere. According to this hypothesis, the high levels of atmospheric carbon dioxide asphyxiated much life both in the seas and on the continents and caused the Permian extinction event (Figure 5-5).

checkpoint What were the two changes in the chemistry of the deep Permian Ocean caused by the lack of deep-ocean currents?

5.1 ASSESSMENT

1. **Explain** How did early geologists' understanding of the fossil record differ from what later geologists understood about the rock layers?
2. **Summarize** How have more recent research findings affected the acceptance of the Yucatan impact site hypothesis?

Critical Thinking

3. **Predict** If ocean currents are partially controlled by the positions of the continents, is another mass extinction millions of years from now unavoidable? Explain your reasoning.
4. **Synthesize** Choose one mass extinction event discussed in Lesson 5.1 and one hypothesis for what caused it. Use the hypothesis to explain the specific cause-and-effect interactions among each of Earth's spheres that triggered the mass extinction.

5.2 RELATIVE AGE

Core Ideas and Skills
- Compare methods for measuring geologic time.
- Analyze why geologists use a set of principles to determine relative age.

KEY TERMS

absolute age	dike
relative age	principle of included fragments
principle of original horizontality	principle of faunal succession
principle of superposition	geologic formation
principle of crosscutting relationships	

While most of us think of time in terms of days or years, Earth scientists commonly refer to events that happened millions or billions of years ago. Earth is approximately 4.6 billion years old. Yet our humanlike ancestors originated only some 5 to 7 million years ago, and humans (*Homo sapiens*) originated just 200,000 to 300,000 years ago. Recorded history is only a few thousand years old. How do scientists measure the ages of rocks and events that occurred millions or billions of years ago, long before recorded history or even before human existence?

Relative and Absolute Age

Scientists measure geologic time in two ways. **Absolute age** is age in numeric years. For example, the dinosaurs became extinct about 65 million years ago. The Permian extinction occurred 252 million years ago. The Yellowstone volcano first erupted 16.5 million years ago. Absolute age tells both the order in which events occurred and the amount of time that has passed since they occurred. Read more about how absolute age is determined in Lesson 5.4.

Relative age determination relies on the order in which events occurred and is based on a few simple principles. For an event to affect a rock, for example, the rock must exist first. Therefore, the rock must be older than the event. This principle seems obvious, yet it is the basis of much geologic work. Sediment normally accumulates in horizontal layers. However, some sedimentary rocks are folded and tilted (Figure 5-6). We deduce that the folding and tilting occurred after the sediment accumulated. The order in which rocks and geologic features formed can usually be interpreted by such observation and logic.

checkpoint What is the chronology of the folding or tilting of rock in relation to the rock sediment?

Principles of Relative Age

Absolute age measurements have become common only since the second half of the 20th century. Prior to that time, geologists used field observations to determine relative ages. Even with today's sophisticated laboratory instrumentation, most field geologists routinely make relative age determinations based on five simple principles.

The **principle of original horizontality** is based on the observation that sediment usually accumulates in horizontal layers. If sedimentary rocks lie at an angle or if they are folded, as shown in Figure 5-6, we can infer that tectonic forces folded them after the rocks formed.

⊕ **FIGURE 5-6**
Go online to view an example of how tectonic forces uplifted and folded layers of limestone.

The **principle of superposition** states that in an undeformed sequence of layered sedimentary rocks, the oldest layers are at the bottom and the youngest layers are at the top. For example, in Figure 5-7, the sedimentary layers become progressively younger, from 1 to 5. Two notable exceptions to this principle occur if tectonic forces have turned the layers of sedimentary rock upside down or have caused an older layer to be shoved up and over a younger layer.

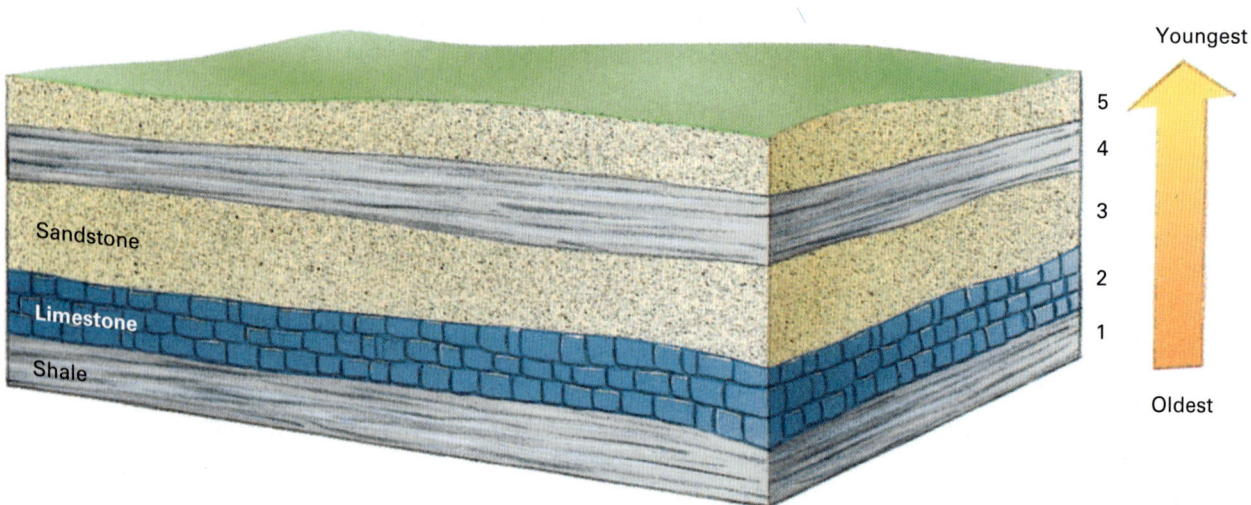

FIGURE 5-7 ▲
In a sequence of sedimentary beds, according to the principle of superposition, the oldest bed is the lowest, and the youngest is on top. The arrow shows the direction of time, from oldest to youngest.

FIGURE 5-8
An igneous intrusion cuts across conglomerate in northwestern Wyoming. By the principle of crosscutting relationships, the dike must be younger than the conglomerate.

The **principle of crosscutting relationships** is based on the fact that a rock must first exist before anything can happen to it. Figure 5-8 shows an igneous intrusion—called a **dike**—cutting across a conglomerate. Clearly, the dike must be younger than the conglomerate.

Figure 5-9 shows an illustration of sedimentary rock layers intruded by three granite dikes. Dike 2 cuts into dike 1 and dike 3 cuts into dike 2. So dike 2 is younger than dike 1, and dike 3 is younger than dike 2. Dike 3 is the youngest and dike 1 is the oldest. The sedimentary rocks must be older than all three dikes.

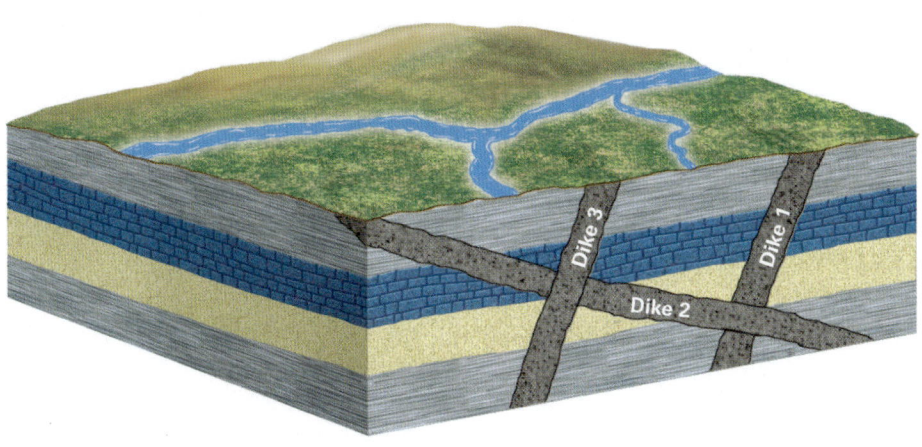

FIGURE 5-9
Three granite dikes cut sedimentary rocks. By the principle of crosscutting relationships, the dikes cut through the rock layers in the order 1, 2, 3, making dike 1 the oldest and dike 3 the youngest. The sedimentary rocks must be older than all three dikes.

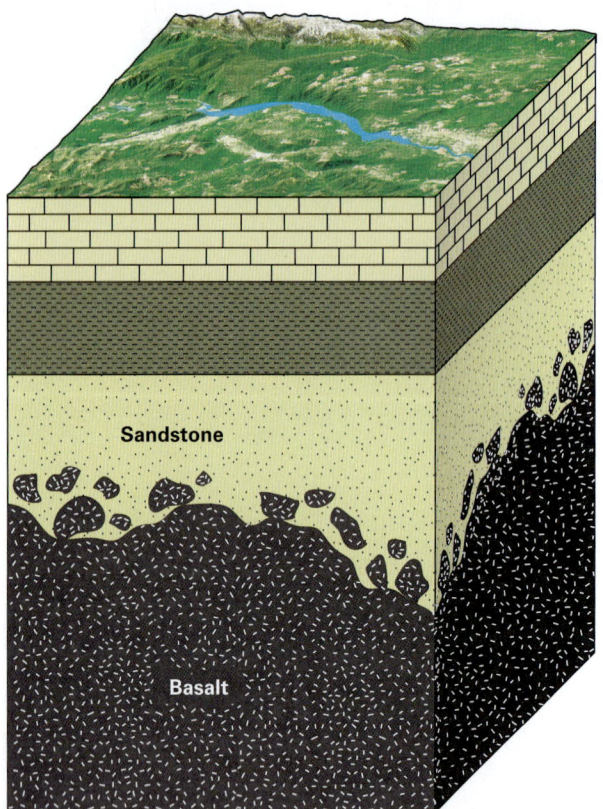

FIGURE 5-10
A rock first has to exist before pieces of it can be broken off and incorporated into another rock. In the example shown, pieces of the basalt have been eroded and deposited in sand that later lithified to sandstone. By the principle of included fragments, the basalt is older.

Another foundational idea is that a rock must first exist before pieces of it can be broken off and incorporated into another geologic unit. Therefore, according to the **principle of included fragments**, a conglomerate that contains clasts, or fragments, of granite must be younger than the age of the granite. Similarly, pieces of basalt incorporated into a layer of sandstone must be older than the sandstone (Figure 5-10).

The theory of evolution states that life-forms have changed throughout geologic time. Species emerge, persist for a while, and then become extinct. Fossils consist of the remains and other traces of prehistoric life preserved in sedimentary rocks. Fossils are useful in determining relative ages of rocks because different animals and plants lived at different times in Earth history. For example, trilobites lived from about 521 to 252 million years ago. Rocks containing trilobite fossils cannot be older than about 521 million years (Figure 5-11).

The principle of superposition tells us that sedimentary rock layers become younger from bottom to top. If the rocks formed over a long time, different fossils will appear and then vanish from upper layers, in the same order in which the organisms evolved and then became extinct. The first dinosaurs appear about 240 million years ago, well after the last trilobites had become extinct. Rocks containing dinosaur bones must be younger than those containing trilobite remains.

The **principle of faunal succession** states that species succeeded one another through time in a definite and recognizable order. This principle also states that the relative ages of sedimentary rocks can therefore be recognized from their fossils. For example, a layer of sedimentary rock containing trilobite fossils will always be older than a layer of sedimentary rock with dinosaur bones, because trilobites evolved and went extinct before the dinosaurs evolved (Figure 5-12).

CAREERS IN EARTH AND SPACE SCIENCES

Museum Conservator

Museums of natural history exhibit and display fossils and other objects of natural history. Exhibits come to life with the efforts of teams of people, including museum scientists, curators, designers, and educators. Specimens, such as fossils, are preserved and stored in museum collections.

A museum conservator is responsible for caring for specimens, keeping each in good condition for continued use. Conservators must learn about an artifact's or specimen's material and origin. Then, preservation treatments are often applied to prevent decay. Conservators monitor light, temperature, humidity, and air pollutants to attempt to control damage or wear.

Most conservators of natural-history artifacts and specimens have a degree in museum conservation. The coursework may include anthropology, paleontology, and studies in technology and chemistry, as related to caring for specimens. Students usually complete internships in hopes of finding a job at a natural history museum. A love for preserving the past and careful attention to detail are good qualities to have in this career. This kind of work can be filled with surprising and amazing discoveries. By preserving specimens that might tell about life billions of years ago, a conservator contributes to the understanding of Earth's history.

FIGURE 5-11
Trilobites are preserved in Burgess Shale on the side of Mount Stephen in Yoho National Park in British Columbia, Canada. The shale, a type of mudstone made almost entirely of clay-sized mineral grains, is about half billion years old and showcases preserved Cambrian sea life.

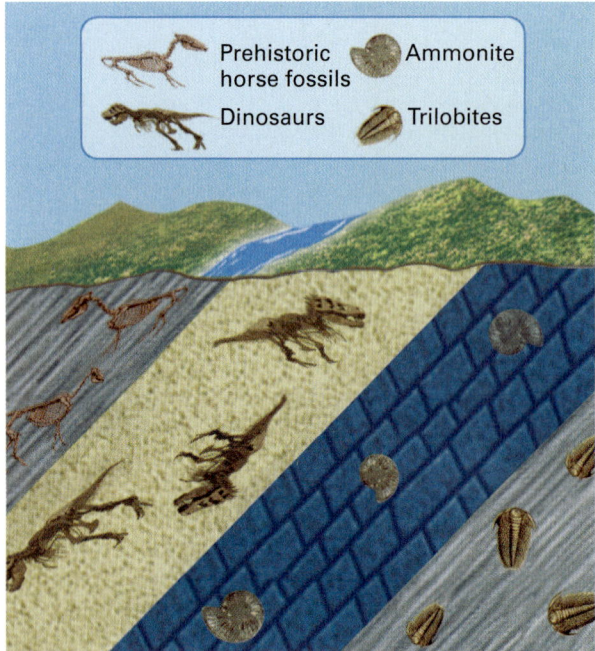

FIGURE 5-12
In a tilted bed, four sedimentary rock layers containing four different fossils represent different relative ages since the organisms existed at separate times with no overlap. The oldest rock layer, the one containing trilobite fossils, was deposited before the layer containing ammonites. The layer with the ammonites was deposited before the layer with the dinosaur fossils. The last layer to be deposited contains fossils of prehistoric horses and is the youngest.

checkpoint Which principle explains that the oldest layers of rock are at the bottom of an undeformed rock sequence?

Interpreting Geologic History from Fossils

Paleontologists study fossils to understand the history of life and its evolution. The oldest-known fossils are traces of bacteria-like organisms that lived about 3.5 billion years ago. A much younger fossil consists of the frozen and mummified remains of a Bronze Age man, recently found in a glacier near the Italian-Austrian border.

Fossils also allow geologists to interpret ancient environments because most organisms thrive in specific environments and do not survive in others. For example, all modern species of coral live in salt water environments. Corals do not survive in fresh or even brackish (only slightly salty) water. Yet today, fossil corals are abundant in the rocks of many mountain environments. Fossil corals have been found in the Canadian Rockies, the European Alps, and the Tian Shan of western China. Each of these mountain ranges are hundreds of kilometers or more away from the nearest ocean. The fossil corals tell us that the exposed rocks in these mountains were once submerged beneath marine waters. Today, geologists understand that tectonic processes raised the marine rocks and their fossils to their present altitudes.

Although some individual fossils, such as corals, can be used to identify ancient environments, geologists more often use the entire fossil assemblage in a rock **geologic formation** to interpret its environment at the time it was deposited. For example, if the collection of fossils in a rock sequence includes numerous different species and a wide variety of trace fossils, geologists might infer that the environment was widely suitable for life. Shallow marine environments in tropical or temperate waters are such environments. Their rich ecosystems support a diversity of life-forms. On the other hand, if only one or two fossil types occur in a sedimentary rock sequence and the trace fossil diversity also is low, geologists might infer that the environment of deposition was stressful and only a few types of specialized organisms could survive. As we will see in Chapter 13, deserts represent such a hostile environment of deposition.

checkpoint What information would a fossil coral give if found in a desert environment?

5.2 ASSESSMENT

1. **Describe** How are layers of rocks identified using the principle of superposition?
2. **Explain** How does the principle of original horizontality rely on observation and logic?
3. **Relate** Explain how the principle of faunal succession connects prehistoric life to the age of sedimentary rocks.

Computational Thinking

4. **Simulate** Fossil corals have been found in the Canadian Rockies, the European Alps, and the Tian Shan of western China. Explain how you would depict a model showing how one of these mountain environments could have fossil corals. The model could be a series of images or a three-dimensional model with multiple parts. Describe what the model would show and how you might create it.

5.3 INTERPRETING THE ROCK RECORD

Core Ideas and Skills
- Examine how geologic correlation can provide information about the nature of unconformities.
- Compare different types of unconformities.

KEY TERMS
conformable	rock record
unconformity	correlation
disconformity	index fossil
angular unconformity	key bed
nonconformity	

The rock layers of the Grand Canyon reveal nearly 2 billion years of Earth history. The exposed layers are like the rungs of a ladder descending into the depths of time.

The Grand Canyon
The walls of the Grand Canyon are composed of sedimentary rocks lying on older igneous and metamorphic rocks (Figure 5-13). Their ages range from about 200 million years to nearly 2 billion years. The principle of superposition tells us that the deepest sedimentary rocks are the oldest, and the rocks become younger as we climb up the canyon walls. However, no principle assures us that the rocks formed continuously from 2 billion to 200 million years ago. The rock record may be incomplete. Suppose that no sediment were deposited for a period of time, or that erosion removed some sedimentary layers before younger layers accumulated. In either case, a gap would exist in the rock record. We know that in an undeformed layered sequence of sedimentary rock, any one layer is younger than the layer below it, but without more information we do not know how much younger.

FIGURE 5-13
Go online to view an image of the Grand Canyon.

checkpoint For which principle of relative age is the Grand Canyon a clear example?

Unconformities
Layers of sedimentary rocks are considered **conformable** if they were deposited without detectable interruption. An **unconformity** is the contact boundary between two rock bodies and represents an interruption in deposition, usually of long duration. Some rock may have been eroded during the time frame represented by the unconformity, but eventually deposition of sediment resumed and buried the unconformity. The missing rock record may involve billions of years.

Commonly, geologists identify three types of unconformities: disconformities, angular unconformities, and nonconformities. In a **disconformity**, the sedimentary layers above and below the unconformity are parallel (Figure 5-14).

Disconformities such as the sandstone layers shown in Figure 5-14 are relatively easy to recognize in a good exposure of sedimentary rocks. However, many disconformities are much more subtle and difficult to recognize without detailed investigation of the rock texture, composition, sedimentary structures, and fossil content.

Layers of sediment that accumulate slowly on parts of the seafloor can all consist of similar-looking mud. In such mud-rich sediment

FIGURE 5-14
These layers of sandstone in Tanzania were deposited in a deep lake, possibly by large floods. Separating the five individual flood deposits are disconformities, indicated by red arrows. Each disconformity was formed by a long pause between successive floods during which no sediment was deposited.

sequences, disconformities that may represent tens or hundreds of thousands of years can be difficult to distinguish without detailed study of the mud itself. Figure 5-15 illustrates an example of a disconformity formed by a break in deposition on a shallow marine sandbar.

In some cases, disconformities are identified by determining the ages of rocks, using methods based on fossils and absolute dating. Read more about absolute dating in Lesson 5.4.

In an **angular unconformity**, tectonic activity has uplifted and tilted older sedimentary rock layers, and erosion has worn down the older layers before younger sediment accumulated. Figure 5-16 illustrates this process.

While disconformities may also be uplifted and tilted by tectonic forces (as shown in Figure 5-12), the rock layers are parallel. At an angular unconformity, younger layers accumulate at a different angle above older layers that have been tilted. Figure 5-17 shows an example of an angular unconformity in the Grand Canyon.

Finally, a **nonconformity** is an unconformity in which sedimentary rocks lie on igneous or metamorphic rocks. Perhaps the best widespread example of a nonconformity is the unconformity that separates metamorphic and igneous rocks, which make up the basement rocks of Earth's

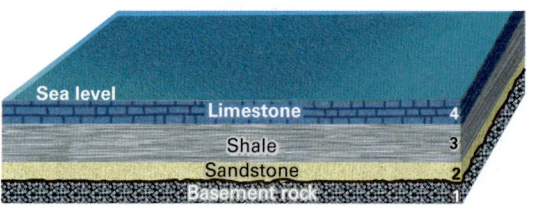

A Sediment is deposited in horizontal layers and lithifies to form rock.

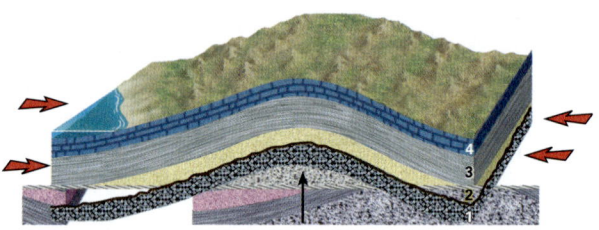

B Rocks are compressed, uplifted, and tilted.

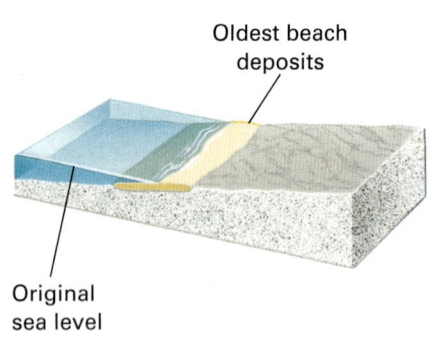

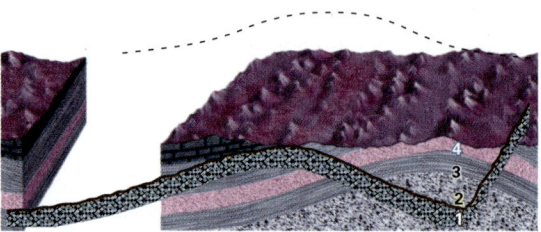

C Erosion exposes folded rocks.

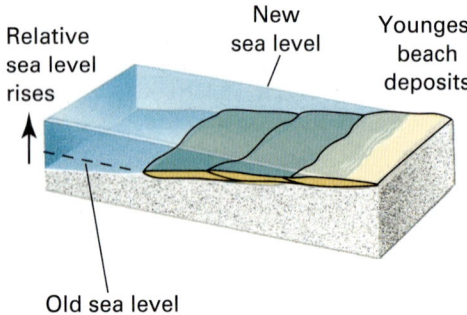

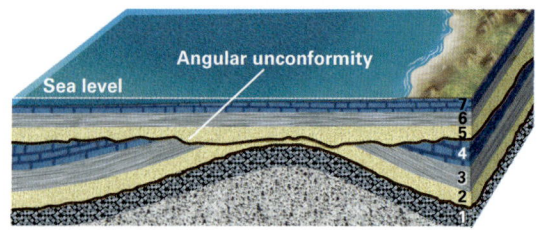

D Rocks sink below sea level and new rocks are deposited.

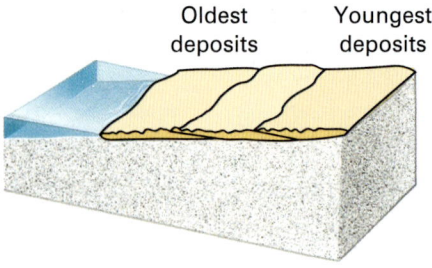

FIGURE 5-15
A disconformity is created when there is a pause in the deposition of sediment, sometimes accompanied by minor erosion. When a river channel shifts its position, the sandbar formed at its mouth no longer receives sediment and begins to erode. Commonly, organisms will burrow into this stable surface before the river channel shifts back and deposition resumes on the bar.

FIGURE 5-16
An angular unconformity develops when older sedimentary rocks are folded and partially eroded before younger sediment accumulates.

FIGURE 5-17
The Grand Canyon provides an example of an angular unconformity where flat-lying Cambrian sedimentary rocks rest on top of tilted and partially eroded Precambrian sedimentary rocks. The top of the angular unconformity is highlighted by the dashed white line.

continents and the overlying sedimentary rocks deposited on top of it. Because it is so widespread and represents such a significant break in the geologic record, this nonconformity commonly is called "The Great Unconformity."

checkpoint Which type of unconformity reveals that tectonic activity tilted sedimentary rocks?

Correlation

Ideally, geologists would like to develop a continuous history for each region of Earth by interpreting rocks that formed in that place throughout geologic time. However, there is no single place on Earth where rocks were continuously formed and preserved. Consequently, in any one place, the **rock record**—that is, the rocks that currently exist on Earth and contain the record of Earth's history—is full of unconformities. To assemble as complete and continuous a record as possible, geologists gather geologic evidence from many locations and match rocks of the same age in a process called **correlation** (Figure 5-18).

The correlation of rock of the same age over great distances is at the heart of reconstructing Earth's ancient systems and recognizing when and how these changes affected the surface of Earth. Correlating rock sequences from one region on Earth to another is complicated by the great variety of environments that existed at any single time in Earth's geologic past. The drastic changes that Earth's systems have undergone over the past 4.6 billion years add more complication.

In order to understand what even a small part of Earth's surface looked like during the geologic past, geologists must be able to correlate rocks that formed in different environments that existed at the same point in time. For example, the East Coast of North America includes the Appalachian Mountains and a very gently sloping, mostly flat coastal plain. It includes the shoreline environment and the shallow marine settings offshore. Beyond the coast are environments of the deeper ocean. Each of these environments is very different. Yet all exist today—at the same point in geologic time. The primary aim of correlation is the recognition of time-equivalence between different sequences of physical sedimentary rock.

If you follow a single sedimentary bed from one place to another, then it is clearly the same layer in both places. It is natural to suppose that because the layer can be physically traced from one place to another that the layer must be the same age everywhere. This assumption is almost never the case. Although a layer may be physically traced, the result reflects a set of environments that changed in space and through time. For example, many rivers will shift laterally (horizontally) over time as one bank erodes and the opposite bank builds out. The result is a layer of sandstone that was not deposited

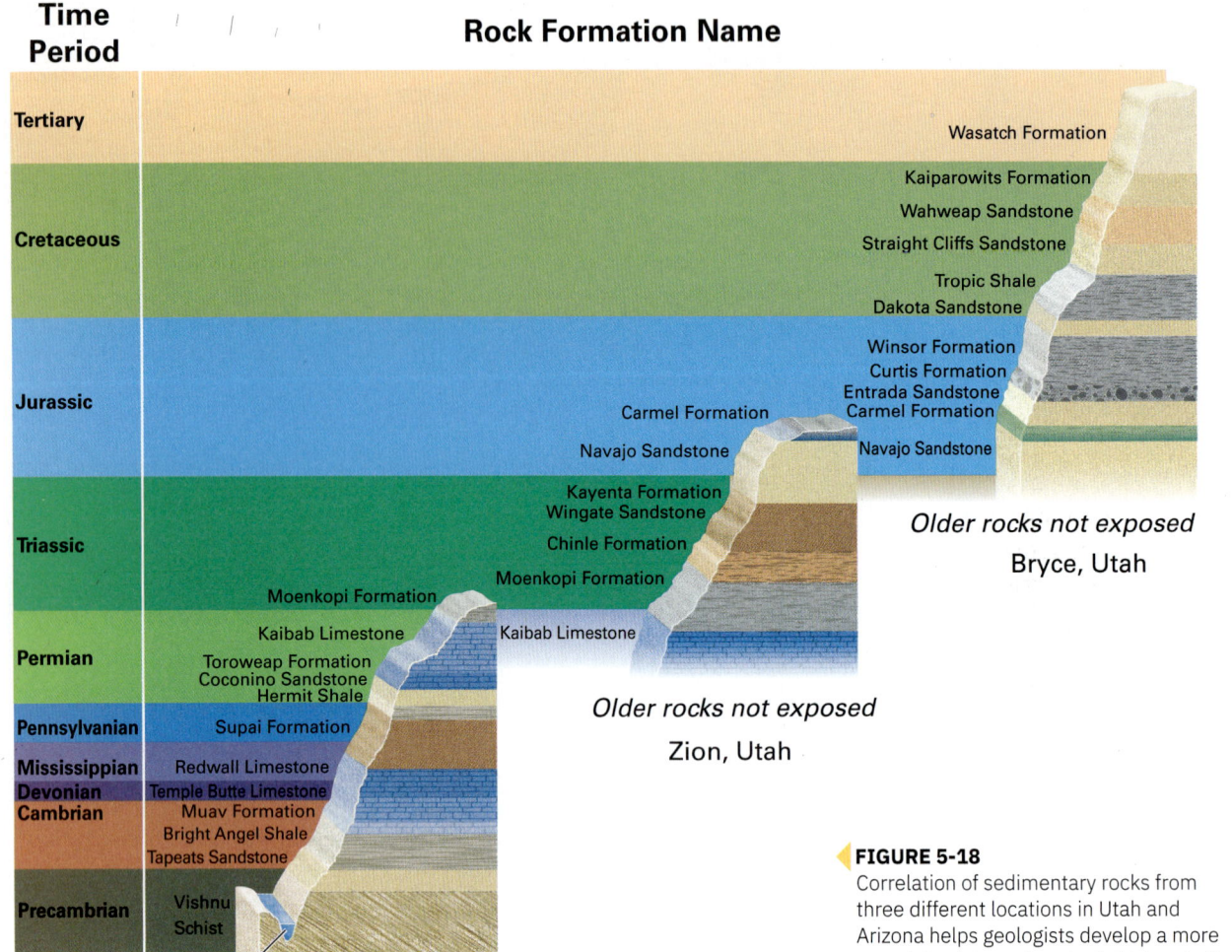

FIGURE 5-18
Correlation of sedimentary rocks from three different locations in Utah and Arizona helps geologists develop a more complete picture of geologic history in the region. The rock units are called geologic formations.

all at once, but instead grew laterally as the river slowly migrated (Figure 5-19).

In a larger scale example, suppose you are attempting to trace a layer of sandstone deposited on a beach. If the sea level rose, the beach would migrate inland. However, the beach deposits closer to the edge of the continent would be older than the beach deposits that are farther inland. Although the entire layer would be physically continuous, it would not be the same age everywhere (Figure 5-20). Over a few hundred meters, the age difference within the sandstone layer may be very unimportant, but over hundreds of kilometers it could vary by millions of years.

To construct a record of Earth history and a geologic timescale, Earth scientists use several types of evidence to correlate rocks of the same age. **Index fossils** are a very useful tool for correlation because they can precisely indicate the ages of sedimentary rocks. The best index fossils are abundantly preserved in rocks, are geographically widespread, existed as a species or genus for only a relatively short time, and are easily identified in the field. Tiny marine organisms make some of the best index fossils because they spread rapidly and widely throughout the seas. And they evolved quickly. The shorter the time that a species existed, the more precisely the index fossil reflects the age of a rock. In many cases, the presence of a single type of index fossil is sufficient to establish the age of a rock. More commonly, Earth scientists use an assemblage of several fossils to date and correlate rocks (Figure 5-21).

Another correlation tool is the key bed. A **key bed** is a thin, widespread, easily recognized sedimentary layer that was deposited rapidly and simultaneously over a wide area. Many volcanic eruptions eject great volumes of fine, glassy volcanic ash into the atmosphere. Wind carries the ash over great distances before it settles. Some historic ash clouds

FIGURE 5-19
One bank of a river channel erodes while the opposite builds out. As the river migrates over hundreds or thousands of years, the channel leaves behind a layer of sandstone. The sandstone is one physically continuous layer, but it does not have the same age.

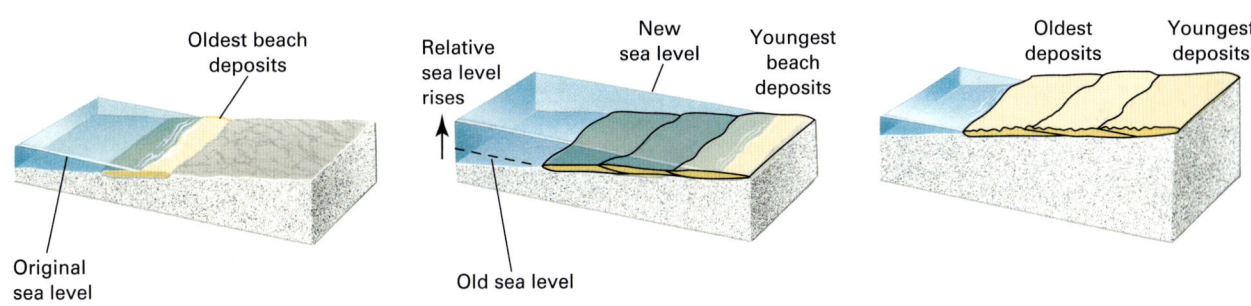

FIGURE 5-20
As sea level rises slowly through time, a beach migrates inland. When sea level falls, a semi-continuous blanket of sand can be correlated. The sand deposits are not the same age everywhere.

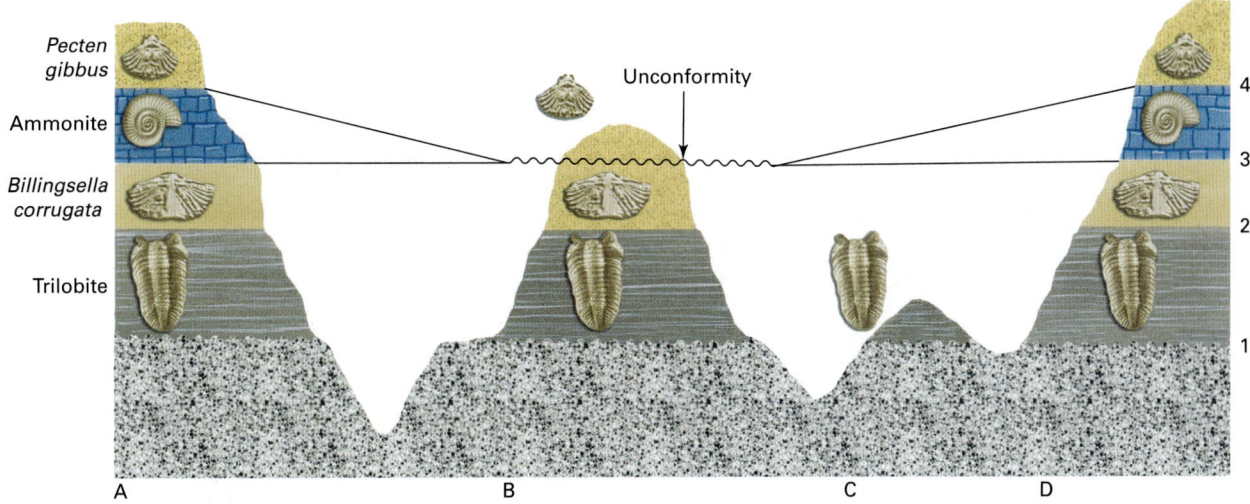

FIGURE 5-21
Index fossils demonstrate age equivalency of sedimentary rocks from widely separated locations. Sedimentary beds containing the same index fossils are interpreted to be of the same age. In this illustration, the fossils show that at locality B, an unconformity exists because layer 4 lies directly on top of layer 2. Either layer 3 was deposited and eroded away before layer 4 was deposited, or it was never deposited. At locality C, all layers above layer 1 are missing, either because of erosion or because they were never deposited.

LESSON 5.3

have encircled the globe. When the ash settles, it forms a depositional layer, and the glassy fragments within the ash layer typically crystallize to form clay. Because such volcanic eruptions occur at a precise point in time, the ash deposit is the same age everywhere. A recent key bed is the thin layer of white volcanic ash shown in Figure 5-22 from western Montana. The layer of ash shown in the photo came from an eruption that occurred about 11,200 years ago from Glacier Peak, an active volcano northeast of Seattle, Washington.

FIGURE 5-22
Go online to view an example of a key bed.

checkpoint How does the fossil of a species or genus that existed for only a short time help date the age of a rock?

5.3 ASSESSMENT

1. **Explain** What are possible causes of gaps in the rock record?
2. **Contrast** Explain the differences between disconformities, angular unconformities, and nonconformities.
3. **Recall** What is the purpose of correlation?
4. **Relate** Describe the characteristics of ancient organisms that geologists consider to be very effective index fossils.
5. **Explain** In order to use a key bed as a correlation tool, what other information must be known?

Critical Thinking

6. **Analyze** The rock layers shown in Figure 5-18 are given proper names to distinguish them as unique sedimentary structures. Layers with names that include the word "formation" consist of a mix of rock types but those with names that include specific rocks, such as sandstone or shale, are made predominantly of that one rock type. For each time period, determine the number of sedimentary layers and compile a list of the rock types present. What other information is needed and what additional questions would you ask when using this data representation to compare the deposition that occurred during the different time periods?

5.4 ABSOLUTE AGE

Core Ideas and Skills
- Explain the method of radiometric dating.
- Analyze data for some of Earth's oldest dated rock.

KEY TERMS

isotope   radiometric dating
half-life

Geologists have a challenging task—they are attempting to measure the absolute age of events that occurred before history was recorded. Measuring absolute age relies on a process that occurs at a constant rate, such as Earth revolving around the sun every 365.26 days, and some way to keep a cumulative record of that process, such as marking a calendar each time a year passes. Measurement of time with a calendar, a clock, an hourglass, or any other device depends on these two factors.

Isotopes

Scientists have found a natural process that occurs at a constant rate and accumulates its own record: the radioactive decay of the elements that are present in many rocks. Therefore, many rocks have built-in calendars.

To begin to understand radioactivity, we need a basic understanding of the atom. An atom consists of a small, dense nucleus surrounded by a cloud of negatively charged electrons. A nucleus consists of positively charged protons and neutral particles called neutrons. All atoms of any specific element have the same number of protons in the nucleus. However, the number of neutrons may vary. **Isotopes** are atoms of the same element with different numbers of neutrons. For example, all isotopes of carbon contain 6 protons, but only carbon-12 also has 6 neutrons. Carbon-13 has 7 neutrons, and carbon-14 has 8 neutrons.

Many isotopes are stable so they do not change with time. Carbon-12 and carbon-13 are two such isotopes. Over a million years, all of the atoms in a sample of carbon-12 or carbon-13 remain unchanged. However, other isotopes are unstable, or radioactive. Given time, their nuclei spontaneously decay, forming a different isotope and commonly a different element. For example, carbon-14, the one unstable isotope of carbon, spontaneously decays to form nitrogen-14.

A radioactive isotope such as carbon-14 is known as a parent isotope. An isotope created by radioactivity, such as nitrogen-14, is called a daughter isotope.

Many common elements, including carbon and potassium, consist of a mixture of radioactive and nonradioactive isotopes. With time, the radioactive isotopes decay, but the nonradioactive ones do not. Some elements, such as uranium, consist only of radioactive isotopes. The total amount of uranium on Earth slowly decreases as it decays to other elements, such as lead.

checkpoint What causes radioactive isotopes to form a different isotope over time?

Radioactivity and Radiometric Dating

If you watch a single atom of carbon-14, when will it decompose? This question cannot be answered, because any particular carbon-14 atom may or may not decompose at any time. Each atom has a certain probability of decaying at any time. Averaged over time, half of the atoms in any sample of carbon-14 will decompose in about 5,730 years. The **half-life** is the time it takes for half of the atoms in a radioactive sample to decompose. The half-life of carbon-14 is 5,730 years. Therefore, if 1 gram of carbon-14 were placed in a container, 0.5 gram would remain after 5,730 years, 0.25 gram after 11,460 years, and so on. Each radioactive isotope has its own half-life; some half-lives are fractions of a second, and others are measured in billions of years.

Two properties of radioactivity are essential to the dating rocks. First, the half-life of a radioactive isotope is constant. It can be measured in the laboratory and is unaffected by geologic or chemical processes, so radioactive decay occurs at a known, constant rate. Second, as a parent isotope decays, its daughter accumulates in the rock. The longer the rock exists, the more the daughter isotope accumulates. **Radiometric dating** is the process of determining the absolute ages of rocks, minerals, and fossils by measuring the relative amounts of parent and daughter isotopes.

Figure 5-23 shows the relationships between age and relative amounts of parent and daughter isotopes. At the end of one half-life, 50 percent of the parent atoms have decayed to daughter. At the end of two half-lives, the mixture is 25 percent

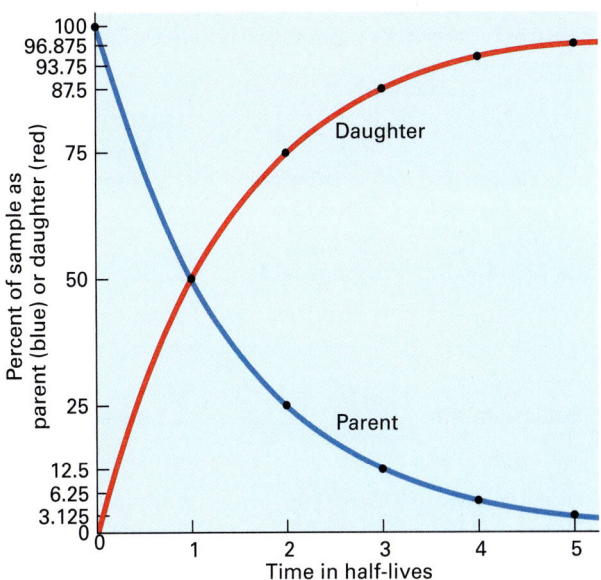

FIGURE 5-23

As a radioactive parent isotope decays to its daughter isotope, the proportion of parent decreases (blue line) and the amount of daughter increases (red line). The half-life is the amount of time required for half of the parent to decay to daughter. At time zero, when the radiometric calendar starts, a sample is 100 percent parent. By measuring the proportions of parent and daughter in a rock, the rock's age in half-lives can be obtained. Because the half-lives of all radioactive isotopes are well-known, it is possible to convert age in half-lives to age in years.

parent and 75 percent daughter. To determine the age of a rock, a geologist measures the proportions of parent and daughter isotopes in a sample and compares the ratio to a similar graph. Consider a hypothetical parent–daughter pair having a half-life of a million years. If we determine that a rock contains a mixture of 25 percent parent isotope and 75 percent daughter, the figure shows that the age is two half-lives, or in this case 2 million years.

If the half-life of a radioactive isotope is short, that isotope gives accurate ages for young materials. For example, the half-life of 5,730 years for carbon-14 dating allows age determinations for materials younger than about 50,000 years. It is useless for older materials, because by 50,000 years, the amount of carbon-14 that remains in the sample is too small to measure accurately.

The opposite limitation exists with isotopes that have long half-lives. Such isotopes provide accurate ages for old rocks, but not enough daughter isotope accumulates in young rocks to be measured. For example, rubidium-87 has a half-life of 47 billion years. In a geologically short period of time—10 million years or less—so little of its daughter isotope has accumulated that it is impossible to measure accurately. Therefore, rubidium-87 is not useful for rocks younger than about 10 million years. The

TABLE 5-1 The Most Commonly Used Isotopes in Radiometric Age Dating

Isotope (Parent)	Isotope (Daughter)	Half-Life of Parent (Years)	Effective Dating Range (Years)	Minerals and Other Materials That Can Be Dated with This Isotope
Carbon-14	Nitrogen-14	5,730 ± 30	100–50,000	Anything that was once alive: wood, other plant matter, bone, flesh, or shells; also, carbon in carbon dioxide dissolved in groundwater, deep layers of the ocean, or glacial ice
Potassium-40	Argon-40 Calcium-40	1.3 billion	50,000–4.6 billion	Muscovite Biotite Hornblende Whole volcanic rock
Uranium-238 Uranium-235 Thorium-232	Lead-206 Lead-207 Lead-208	4.5 billion 710 million 14 billion	10 million–4.6 billion	Zircon Uraninite
Rubidium-87	Strontium-87	47 billion	10 million–4.6 billion	Muscovite Biotite Potassium feldspar Whole metamorphic or igneous rock

MINILAB Radioactivity and Half-Life

Materials
paper cup   100 pennies

Procedure
1. Place the pennies into the paper cup. Give the cup a gentle shake to mix. Then dump out the pennies onto a table or desk.
2. Remove the pennies that land tails up and set them aside. These are no longer in use for the rest of the lab.
3. Count the pennies showing heads and record this number in a data table as Trial 1. Return these pennies to the cup.
4. Repeat this process until you run out of pennies or have completed seven trials. Count the number of pennies that land with heads up for each trial and record the number in your data table. Then use those pennies for the next trial.

Results and Analysis
1. **Interpret Data** What do your data show? How do your data compare with the data of other students?
2. **Model** How does the lab model the process of radioactive decay?

Critical Thinking
3. **Evaluate** Do the pennies in this lab accurately model half-life? What other objects could be substituted? Explain your reasoning.
4. **Connect** How did this lab help you understand the process of radiometric dating?
5. **Predict** Suppose you performed this lab with 500 pennies instead of 100. How would this affect the results?

radioactive isotopes that are most commonly used for dating are summarized in Table 5-1.

Advances in absolute age dating have developed as technology in general has advanced. Today, for example, it is possible to determine with accuracy and precision the age of individual grains of sand collected from an ancient sandstone. By isolating mineral grains from the sandstone, geologists can measure the age of each grain, forming an age spectrum, a histogram of ages from all the individual grains measured. The age spectrum tells geologists the ages of the rocks that were eroded to form the sandstone. Along with information about the composition of the grains—that is, what types of minerals and rock fragments are represented—age spectra provide a direct link to the original source rock from which the minerals and rock fragments were eroded. In addition, by applying the principle

DATA ANALYSIS Think About Radiometric Testing

Earth is approximately 4.6 billion years old. Due to the rock cycle, in which rock is continually being recycled into newer rock, pieces of Earth's original crust are few. Yet there are several places on Earth where very ancient rock can be found. Table 5-2 shows data about some of Earth's oldest identified rock. Use the data in the table as well as Figure 5-23 and Table 5-1 to answer the questions.

1. **Infer** What can you infer about the presence of life on Earth from the data about Greenland's ISB?
2. **Analyze** Compare the data in Table 5-1 showing the most commonly used isotopes in radiometric dating with the information in Table 5-2. Which isotopes were originally present (parent isotopes) in Greenland's ISB? Why do you think scientists examined so many isotopes to date the ISB?
3. **Infer** Samarium-146 is an extinct radioactive isotope that decays to neodymium-142. Although some samarium-146 was present on Earth after the formation of the solar system, all of it decayed to neodymium-142 during Earth's first 500 million years. How do you think this knowledge helped scientists date the rock of the Canadian Shield?

Computational Thinking

4. **Generalize Patterns** Why did the presence of significant amounts of lead-206 in Australia's Jack Hills lead scientists to conclude it was very old? About what ratio of uranium-238 to lead-206 must have been in the rock when researchers examined it?

TABLE 5-2 Examples of Earth's Oldest Identified Rock

Location where rock was found	Age arrived at by radiometric testing	Isotopes examined	Other information
Greenland's Isua supracrustal belt (ISB)	between 3.7 billion and 3.8 billion years	lead-206 lead-207 rubidium-87 strontium-87	Ratio of carbon isotopes is consistent with either biological origin or decomposition of carbonate rocks.
Australia's Jack Hills	4.4 billion years	lead-206 lead-207	Zircon crystals were examined.
Canadian Shield, Nunavik, Quebec	4.3 billion years	samarium-146 neodymium-142	Samples taken from basalt encased in granite

of included fragments to ages measured from individual grains of sandstone, geologists are able to determine the maximum depositional age of the sandstone. The grains must first have existed before they were eroded, transported, and deposited in the layer of sandstone that was collected by the geologist for analysis.

checkpoint How do we know that radioactive decay occurs at a constant rate?

5.4 ASSESSMENT

1. **Recall** What natural process allows rocks to be dated radiometrically?
2. **Explain** Why is a radioactive isotope with a short half-life, such as carbon-14, ineffective for older materials?
3. **Infer** Radiometric dating of a sample of sandstone produces an age spectrum that consists of a population of young individual grains that are 10 million years old and an older population of grains that are dated at 500 million years. What can you conclude about the age of the sandstone?

Computational Thinking

4. **Analyze Data** The isotope uranium-235 decays into the isotope lead-207. Suppose you have three samples of rock. Sample A contains three lead-207 atoms for every atom of uranium-235. Sample B contains equal numbers of uranium-235 and lead-207 atoms. Sample C contains 124 lead-207 atoms for every uranium-235 atom. Uranium-235 has a half-life of 710 million years. Rank the samples by age. Explain your reasoning.

FIGURE 5-24
ON ASSIGNMENT Photographer Brian Skerry has been taking his camera underwater for many years, capturing ocean images for *National Geographic*. Often, Skerry photographs marine life. However, on a 2018 trip to Yellowstone National Park, the objective was to snap images of a unique landform. In this photo, Brett Seymour, a diver and photographer for the National Park Service, gets close to an 8-meter tall underwater spire. The spire was formed 11,000 years ago by now dormant hydrothermal vents in Yellowstone Lake, the largest high-altitude lake in North America, sitting at 2,350 meters above sea level.

5.5 THE GEOLOGIC TIMESCALE

Core Ideas and Skills
- Interpret the geologic timescale through periods, eons, and eras.
- Compare fossils from different eras.

KEY TERMS	
geologic timescale	epoch
eon	age
era	Precambrian
period	Phanerozoic eon

The Scale

No single locality exists on Earth where a complete sequence of rocks formed continuously throughout geologic time. However, geologists have analyzed and correlated rocks from many localities around the world to create the **geologic timescale**, a formal subdivision of Earth time based on numerical dates (Figure 5-25). The geologic timescale is frequently revised as more rock units are dated radiometrically and more fossil assemblages are described around the world.

Just as a year is subdivided into months, months into weeks, and weeks into days, geologic time is split into smaller intervals. The units are named, just as months and days are. The largest time units are **eons**, which are divided into **eras**. Eras are

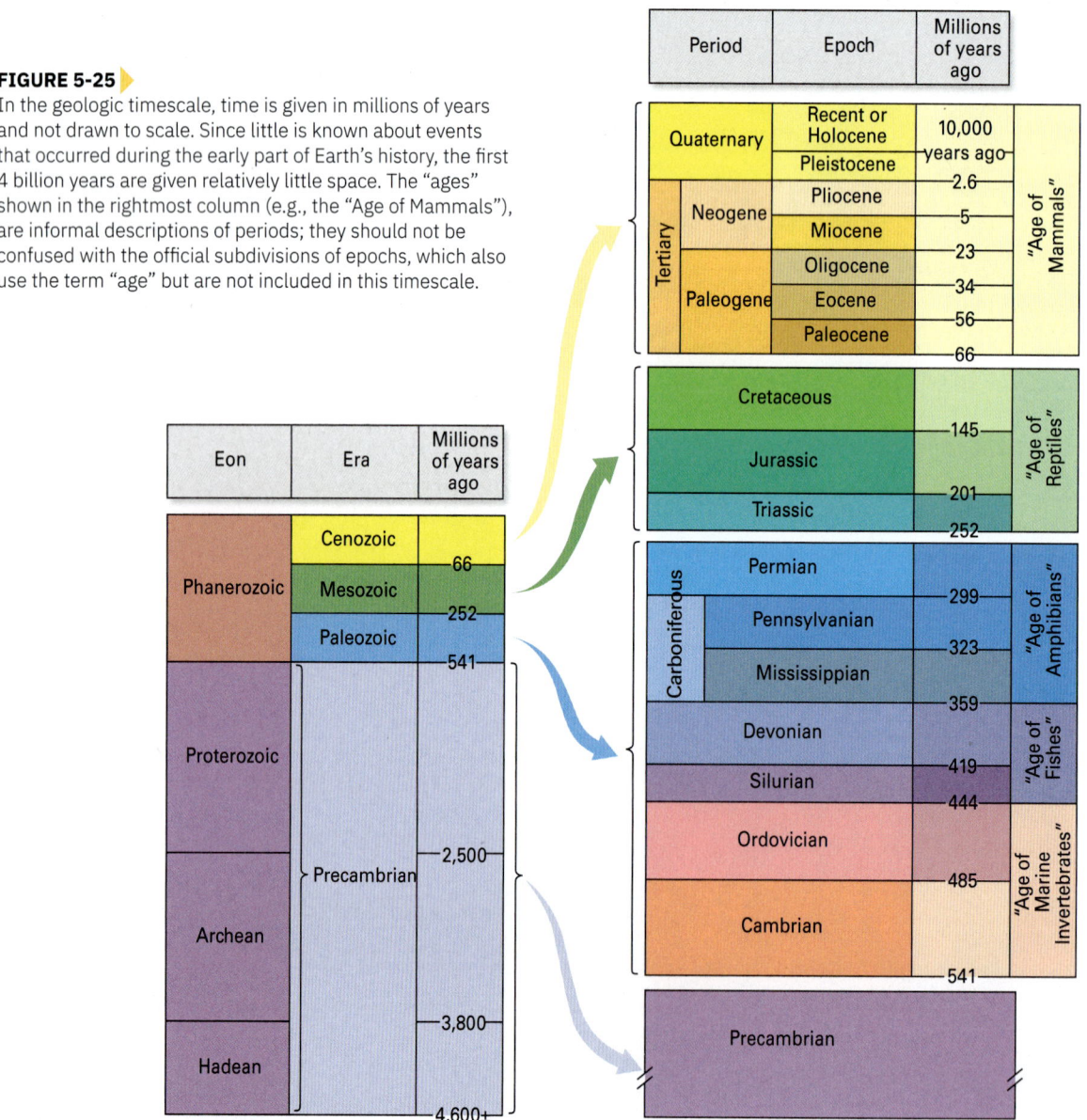

FIGURE 5-25
In the geologic timescale, time is given in millions of years and not drawn to scale. Since little is known about events that occurred during the early part of Earth's history, the first 4 billion years are given relatively little space. The "ages" shown in the rightmost column (e.g., the "Age of Mammals"), are informal descriptions of periods; they should not be confused with the official subdivisions of epochs, which also use the term "age" but are not included in this timescale.

subdivided, in turn, into **periods**, which are further subdivided into **epochs** and, even further, into **ages**.

The geologic timescale was originally constructed on the basis of relative age determinations. When geologists developed radiometric dating, they added absolute ages to the timescale. Geologists commonly use the timescale to date rocks in the field. Imagine that you are studying sedimentary rocks in the walls of the Grand Canyon. If you find an index fossil or a key bed that has already been radiometrically dated by other scientists, you know the age of the rock you are studying and you do not need to send the sample to a laboratory for radiometric dating. For example, the ash bed in Figure 5-22 does not need to be dated because it is recognized as a key bed that corresponds to an eruption of Glacier Peak that occurred in 11,200 B.C.E.

checkpoint What causes revisions or additions to the geologic timescale?

The Earliest Eons of Geologic Time: Precambrian Time

The earliest eons—the Hadean, Archean, and Proterozoic eons—together they constitute a time interval of 4 billion years—nearly 90 percent of Earth's history (Figure 5-25). These early eons are commonly referred to with the informal term **Precambrian** because they preceded the Cambrian period, when fossil remains became very abundant.

The Hadean eon (Greek for "beneath the earth") is the earliest time in Earth history and ranges from the planet's origin 4.6 billion to 3.8 billion years ago. Only a few Earth rocks are known that formed during the Hadean eon. No fossils of Hadean age are known, making it impossible to subdivide the Hadean eon based on fossils.

There are few fossils among the rocks of the Archean eon (Greek for "ancient"), and they are not preserved well enough to allow for finely tuned subdivision of this eon that spanned from 3.8 billion to 2.5 billion years ago. The fossil record does indicate that life began on Earth during the early Archean eon, and geologists recently have described fossilized algal structures, called stromatolites, in rocks as old as 3.45 billion years in Western Australia.

More diverse groups of fossils have been found in sedimentary rocks of the Proterozoic eon (Greek for "earlier life"), 2.5 billion to 541 million years ago. The most complex are multicellular and have different kinds of cells arranged into tissues and organs. A few types of Proterozoic shell-bearing organisms have been identified, but shelled organisms did not become abundant until the Paleozoic era. Because it does preserve more of a fossil record, the Proterozoic eon is further subdivided into the Paleoproterozoic, Mesoproterozoic, and Neoproterozoic.

checkpoint In which eon did life on Earth begin?

The Phanerozoic Eon

Sedimentary rocks of the **Phanerozoic eon**, which covers the most recent 541 million years of geologic time, contain abundant fossils. Additionally, sedimentary rock and sediments of Phanerozoic age are very widely exposed on all of Earth's continents. In North America, Phanerozoic sediments and sedimentary rocks make up nearly all surface exposures from the Florida Keys to the San Juan Islands of Washington State, and from Southern California to the tip of Cape Cod, Massachusetts. Because they are widely exposed and form much of the ground immediately under our feet, Phanerozoic sedimentary rocks in North America and Europe have long been studied.

The number of species with shells and skeletons dramatically increased in the Phanerozoic eon. The total number of individual organisms preserved as fossils increased as did the total number of species preserved as fossils. In rocks of the earliest Phanerozoic time and younger, the most abundant fossils are the hard, tough shells and skeletons.

Subdivision of the Phanerozoic eon into three eras (Figure 5-25) is based on the most common types of life during each era. Sedimentary rocks that formed during the Paleozoic era (Greek for "old life") contain fossils of early life-forms, including invertebrates, fishes, amphibians, reptiles, ferns, and cone-bearing trees. The Paleozoic era ended about 252 million years ago when the terminal Permian mass extinction wiped out 90 percent of marine species, two-thirds of reptile and amphibian species, and 30 percent of insect species.

In North America, rocks of Paleozoic age are widespread. Almost all of the bedrock geology of the New England states is Paleozoic in age. In New York State, for example, only the bedrock of the Adirondack Mountains is not Paleozoic. Paleozoic bedrock also occurs widely across the mid-continent and in surrounding states, in the Appalachian Mountains, and within the Rocky Mountains (Figure 5-26). In contrast, Paleozoic

FIGURE 5-26
Bluffs along the Upper Iowa River in Winneshiek County, Iowa, near the town of Decorah, are formed from Paleozoic limestone.

FIGURE 5-27
The Durdle Door is a natural limestone arch stretching into the North Sea on the Jurassic Coast of Dorset, England. This area contains exposures of Triassic rock, Jurassic rock, and Cretaceous rock.

rocks are relatively uncommon in California, Oregon, and Washington.

Because it occurs widely, Paleozoic rocks have been used in many industrial applications historically. For example, almost all of the limestone building stone quarried in the mid-continent region is Paleozoic in age, including the limestone from Indiana used to face all Federal buildings in the area called "Federal Triangle" in Washington, D.C. Rocks of Paleozoic age contain high-quality coal in the Appalachians, and oil and gas in the mid-continent and Rocky Mountain regions. Sand of Paleozoic age from the mid-continent is used heavily today to develop oil wells.

The Mesozoic era is most famous for the dinosaurs that roamed the land. Mammals and flowering plants also evolved during this era. The Mesozoic era ended 65 million years ago with another mass extinction that wiped out the dinosaurs.

In the United States, Mesozoic sedimentary rocks and sediments occur at the surface along the Atlantic coastal plain, the lower Mississippi Valley, and across large portions of the Great Plains and Rocky Mountain regions. During Mesozoic time, a marine seaway formed in North America, and the seaway—called the "Cretaceous Interior Seaway"—stretched from the Gulf of Mexico to the Arctic Ocean. Sediments deposited in the seaway formed sedimentary rocks belonging to the youngest of the formally named marine sequences.

During the Cenozoic era (Greek for "recent life"), mammals and grasses became abundant. Humans have evolved and lived entirely in the Cenozoic era.

In the United States, almost all of the unconsolidated sediment that occurs at the surface is Cenozoic in age, including widespread glacial deposits across the northern states. Sedimentary rock and sediment of Cenozoic age covers the entire surface of Florida and Louisiana, and all of the geology of the Hawaiian Islands is Cenozoic in age. Cenozoic sediments also occur widely across other southeastern and Gulf Coast states, in a large swath across the Great Plains, within the broad valleys of the Rocky Mountain states, and across far western states.

The eras of Phanerozoic time are subdivided into periods, the time unit that geologists most commonly use. Some of the periods are named after special characteristics of the rocks formed during those periods. For example, the Cretaceous period is named from the Latin word for chalk

(creta), after chalk beds of this age in Africa, North America, and Europe.

Other periods are named for the geographic localities where rocks of that age were first described. For example, the Jurassic period is named for the Jura Mountains of France and Switzerland. The Cambrian period is named for Cambria, the Roman name for Wales, where rocks of this age were first studied.

Another reason why details of Phanerozoic time are better known than those of Precambrian time is that many of the older rocks have been deformed, metamorphosed, and eroded (Figure 5-27). It is simple probability that the older a rock is, the greater the chance that metamorphism or erosion has destroyed the rock or the features that record Earth's history.

checkpoint In what era did mammals become abundant?

5.5 ASSESSMENT

1. **Define** What is the geologic timescale, and how could it potentially change in the future?
2. **Contrast** Explain why there are fewer subdivisions of Precambrian time than of Phanerozoic time.

Critical Thinking

3. **Evaluate** How does the construction of the geologic timescale reflect the elements of the nature of science?

TYING IT ALL TOGETHER
RECONSTRUCTING EARTH'S PAST

In this chapter, you learned how geologists piece together clues from Earth's rocks to reveal the basic elements of Earth's history—a history that began billions of years before humans existed. Earth's history is full of volcanic activity, the movement of continents, and the emergence and extinction of species. Evidence of these events is preserved in fossils and layers of rock that have formed over millions of years. Geologists have applied their understanding of the rock cycle, the fossil record, and radiometric dating to reconstruct the geologic timescale.

In the Case Study, we examined how the age of rock in which fossils are found can provide clues about long-extinct organisms and their relationship to similar organisms that are alive today. The principle that we can apply our knowledge of Earth's current processes in order to understand its distant past was made popular by Scottish geologist Sir Charles Lyell (1797–1875) in his book *Principles of Geology*. As part of his research, Lyell studied the rock of the Paris Basin, an area of sedimentary rock that extends across a large part of France. Lyell examined the fossils in the rock layers and determined that the Tertiary period could be further divided into three distinct ages he called the Eocene, Miocene, and Pliocene. Lyell's analysis would later be revised and refined by other geologists using evidence from many other locations.

Follow the steps below, using what you have learned to think critically about how the study of geology helps reconstruct Earth's history.

1. Form a discussion group with two or three other students.

2. With your group, discuss the following questions:
   - What types of data does the geologic record provide? What kinds of questions can this data help us answer?
   - How are geology and evolutionary biology linked?
   - What have you learned about the nature of scientific research and scientific evidence from this chapter?

3. If necessary, do additional research to answer questions that arise during your discussion.

4. Summarize in writing the most important ideas and questions that result from your discussion.

5. Share these main ideas with the class.

FIGURE 5-28
These snail fossils, called turritellid gastropods, are preserved in sedimentary rocks that are about 13 million years old in the Paris basin of France. This particular species is now extinct.

CHAPTER 5 SUMMARY

5.1 MASS EXTINCTIONS IN EARTH'S HISTORY

- At least five catastrophic mass extinctions have occurred in geologic history.

- The greatest past extinction annihilated 90 percent of all marine species, two-thirds of reptile and amphibian species, and 30 percent of insect species.

- Many scientists regard the current loss of species due to human activities and modern climate change to be a sixth major mass extinction.

5.2 RELATIVE AGE

- Scientists measure geologic time in two ways. Relative age measurement refers only to the order in which events occurred. Absolute age is measured in years.

- Determinations of relative time are based on geologic relationships among rocks and the evolution of life-forms through time.

- The criteria for relative dating are summarized in a few principles: the principle of original horizontality, the principle of superposition, the principle of crosscutting relationships, the principle of included fragments, and the principle of faunal succession.

5.3 INTERPRETING THE ROCK RECORD

- Layers of sedimentary rock are conformable if they were deposited without major interruptions.

- An unconformity represents a major interruption of deposition and a significant time gap between formation of successive layers of rock.

- In a disconformity, layers of sedimentary rock immediately above and below the unconformity are parallel.

- An angular unconformity forms when layers of rock are tilted and partially eroded prior to deposition of the upper beds.

- In a nonconformity, sedimentary layers lie on top of an erosion surface developed on igneous or metamorphic rocks.

- Correlation is the process of establishing the age relationship of rocks from different locations on Earth by comparing characteristics of the layers or the fossils found in those layers.

- Index fossils and key beds are important tools in time correlation—the demonstration that sedimentary rocks found in different geographic localities formed at the same time.

5.4 ABSOLUTE AGE

- Absolute time is measured by radiometric dating, which relies on the fact that radioactive parent isotopes decay to form daughter isotopes at a fixed, known rate that is expressed by the half-life of the isotope.

- Rocks are dated by measuring the ratio of parent and daughter isotopes to determine how many half-lives have elapsed, and then multiplying this number of half-lives by the number of years represented in a single half-life.

5.5 THE GEOLOGIC TIMESCALE

- The major units of the geologic timescale are eons, eras, periods, and epochs.

- The Phanerozoic eon, the most recent 541 million years of geologic time, is finely and accurately subdivided because sedimentary rocks deposited at this time are often well-preserved and they contain abundant well-preserved fossils.

- Precambrian time is only coarsely subdivided because fossils are scarce and poorly preserved and the rocks that formed during that time are often altered.

CHAPTER 5 ASSESSMENT

Review Key Terms
Select the key term that best fits the definition. Not all terms will be used, and no term will be used more than once.

absolute age
age
angular unconformity
anoxic
conformable
correlation
dike
disconformity
eon
epoch
era
geologic formation
geologic time scale
half-life
index fossil
isotope
key bed
mass extinction
nonconformity
period
Phanerozoic eon
Precambrian
principle of crosscutting relationships
principle of faunal succession
principle of included fragments
principle of original horizontality
principle of superposition
radiometric dating
relative age
rock record
unconformity

1. the idea that in undisturbed layers of sediment or sedimentary rock, the age becomes progressively younger from bottom to top; younger layers always accumulate on top of older layers
2. an approach to determining the timing of geologic events based on the order in which they occurred
3. each of two or more atoms of the same element that have different numbers of neutrons
4. the idea that a rock or feature must first exist before anything can happen to it
5. the process of establishing the age relationship of rocks or geologic features from different locations on Earth
6. the largest unit of geologic time, longer than an era
7. a thin, widespread, easily recognized sedimentary layer that can be used to date rock because it was deposited rapidly and simultaneously over a wide area
8. a sudden, catastrophic event during which a significant percentage of all life-forms on Earth become extinct
9. the time it takes for half of the atoms of a radioactive isotope in a sample to decay
10. a term referring to all of geologic time before the Paleozoic era, encompassing approximately the first 4 billion years of Earth's history
11. the shortest period of geologic time, shorter than an epoch
12. the principle that a rock unit must first exist before pieces of it can be broken off and incorporated into another rock unit
13. an interruption in the deposition of sediment in which the sedimentary layers above and below the interruption are parallel
14. the most recent 541 million years of geologic time, including the present, represented by rocks that contain evident and abundant fossils
15. a fossil that dates the layers where it is found because it came from an organism that is abundantly preserved in rocks, was widespread geographically, and existed as a species or genus for only a relatively short time

Review Key Concepts
Answer each question on a separate sheet of paper to demonstrate your understanding of key concepts from the chapter.

16. Identify three events that scientists think have contributed to mass extinctions in Earth's history. Describe how each event could have led to a loss of species on Earth.
17. What evidence in the fossil record supports the idea that mass extinctions occurred at least five times in Earth's history?
18. Why is it usually reasonable to assume that older sedimentary rock layers are found below younger sedimentary rock layers? Are there any exceptions to this rule?
19. What are the relative ages of rocks A, B, C, D, E, F, and G in the figure? Explain your reasoning.

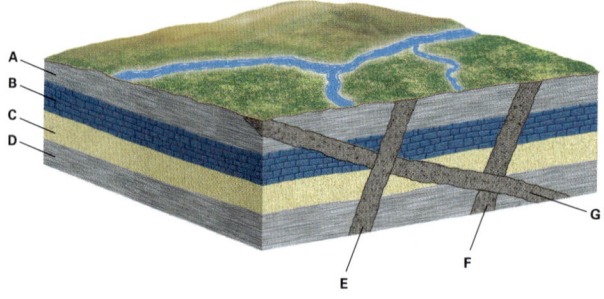

20. Explain how scientists use the principle of faunal succession to determine the age of rock.
21. What are three different types of unconformities and how do they differ?
22. How do time correlation and lithographic correlation differ? Why does lithographic correlation not imply a time correlation?
23. How do index fossils and key beds help scientists determine the age of rock?
24. Explain how scientists can use the radioactive decay of unstable isotopes to estimate the age of rock.
25. Potassium-40 decays to argon-40 with a half-life of 1.3 billion years. Suppose a sample of rock has 50 g of argon-40. How much potassium-40 was in the rock 1.3 billion years ago? Explain how you know.
26. How are eons, eras, periods, epochs, and ages related to each other?
27. The Phanerozoic eon covers the most recent 541 million years, and the geologic record of this period is abundant in fossils. How did living things on Earth change between the Paleozoic era and the Mesozoic era over the course of the Phanerozoic eon?

Think Critically

Write a response to the question on a separate sheet of paper. Use concepts from the chapter to support your reasoning.

28. Some scientists argue that Earth is currently experiencing its sixth mass extinction. If such an extinction occurs, predict what evidence scientists of the distant geologic future will find in the fossil record of the current Holocene era. Explain your reasoning.
29. Geologists have found fossils of marine creatures such as corals in the Canadian Rocky Mountains. As a result, they hypothesize that the area was once submerged beneath marine waters. Is this a reasonable conclusion? Why or why not?
30. Why does the principle of superposition apply only to sedimentary rock and not to igneous and metamorphic rock?
31. When humans build cities, they significantly alter the surface of the land. How do you think human cities will appear in the geologic record millions of years from now?
32. Identify some of the challenges faced by scientists looking to construct a more detailed timeline of Precambrian history.

PERFORMANCE TASK

Regional Geologic History

The ground you stand on documents the history of past events in your neighborhood. For example, fossils in the rock tell the story of living things that have lived in the past.

In this performance task, you will research the rock and fossils found in your neighborhood or another region identified by your teacher and construct a timeline and narrative that represent the regional geologic history.

1. Build a narrative of how things have changed over geologic time.
    A. Research the fossils that have been found in your assigned region. Use resources that allow you to view examples or replicas. Make a short list of fossils found in your assigned area.
    B. Research a particular fossil or group of fossils from your list, as directed by your teacher. For each fossil, identify which species was identified and where it was found. Then, using this information, research the fossil and the rock in which it was found. Prepare a short report describing what you have learned. Include a description of the organism, the way in which the fossil was formed, the habitat in which the organism lived, and its geologic age. Evaluate the rock layer in which the fossil was found in light of this information. Are the properties of the rock consistent with the environment in which the fossilized organism lived?
    C. As a class, construct a geologic timeline of the region based on the fossil record. When you add your fossil or fossil group to the timeline, explain to the class how you were able to determine fossil age based on evidence from fossilized biologic features or the surrounding rock.
    D. Based on the timeline that the class has constructed, write a short geologic history of the region. How has the region changed over time?

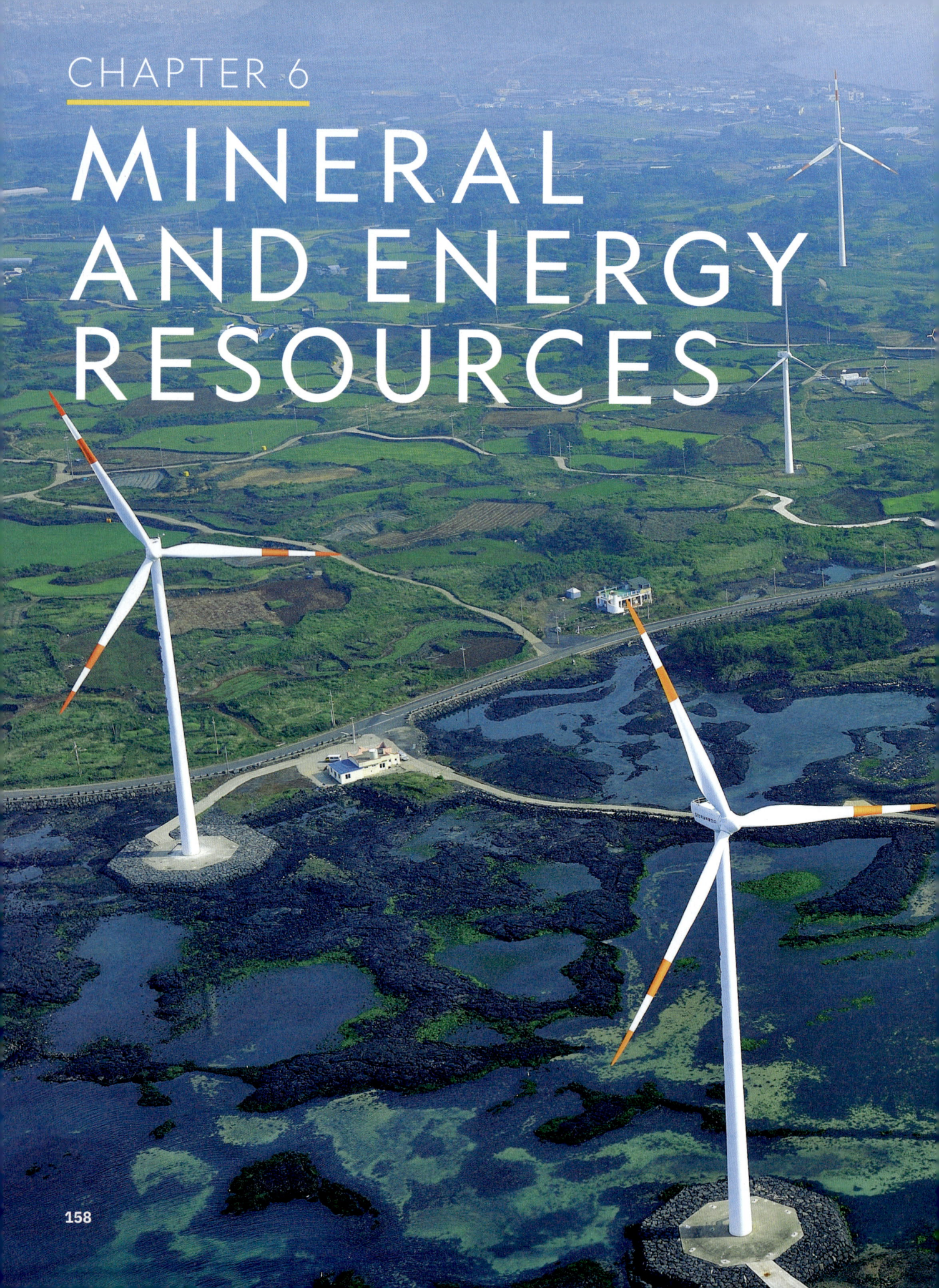

CHAPTER 6
MINERAL AND ENERGY RESOURCES

Humans are consumers of Earth's natural resources. We need materials to manufacture the products we use and to produce the energy that powers our lives. The ways humans obtain and replenish these natural resources are hotly debated topics. Predictions of global supply and demand are uncertain. Extracting and consuming natural resources can be harmful to the environment and ecosystems. Most people agree that using clean and naturally renewable resources, such as converting wind energy into electricity, are actions that can decrease human impact on Earth. Understanding the ways in which humans obtain and use geologic resources will help you make the best economic, technological, and social decisions for the future of humankind and the planet.

KEY QUESTIONS

6.1 What are the differences between ore, ore deposits, and mineral reserves?

6.2 What are Earth's nonrenewable resources?

6.3 What are Earth's renewable resources?

6.4 Why is conservation important when considering future energy sources?

A road connects multiple wind turbines on Jeju Island, South Korea. The island's provincial government wants to be carbon-neutral by 2030, largely by relying on wind energy and electric vehicles.

EXPLORERS AT WORK

MANAGING GLOBAL MINERAL RESOURCES

WITH NATIONAL GEOGRAPHIC EXPLORER SALEEM ALI

Finding solutions to environmental problems such as climate change, pollution, and habitat loss is incredibly complex. These issues are often international in scope. They may involve long-term Earth processes that people find difficult to fully understand. Too often, solutions are hindered by an "us-versus-them" mentality. Environmental groups, corporations, government agencies, indigenous people, and other stakeholders typically regard each other as opponents rather than collaborators.

Enter Dr. Saleem Ali. Dr. Ali believes environmental problems can only be solved when these opposing factions can find ways to communicate and cooperate. Ali helps them do just that as an environmental planner and professional mediator. Ali works in a number of capacities. He teaches and conducts research at several universities in the United States and Australia. He is a member of the United Nations International Resource Panel. And he is a well-published author and consultant.

Much of Ali's work focuses on the mineral sector and ways to make extracting minerals more sustainable. Ali says minerals and metals are the ultimate resource. Of particular importance are rare earth metals, a group of 17 metals that are essential for manufacturing high-tech products ranging from cell phones and laptops to hybrid and electric cars and solar panels.

Ali explains that as the world transitions from fossil fuels to renewable energy, demand is going to grow for minerals used in green technology. These include copper for wires and electric motors, lithium and cobalt for rechargeable batteries, and small amounts of rare earth metals such as indium and gallium for solar cells. According to Ali, little attention is being paid to how mineral supplies are going to keep up with demand. He says the same type of international dialogue taking place around environmental issues such as climate change, biodiversity, and sustainability should also be taking place around global mineral exploration. Otherwise, says Ali, we run the risk that mineral supply shortages will derail many environmental goals.

Ali calls for global coordination of mineral development. Such coordination should include sharing geoscience data to help predict possible shortfalls in supply, maintaining an inventory of recyclable metals, developing new mineral resources, and promoting responsible mining practices. Part of the challenge will be reaching international agreements, which countries seem increasingly reluctant to sign. And part of the challenge will be overcoming widespread public mistrust for mining operations due to their associated environmental and social abuses.

Ali is well-positioned to help governments, industries, and community members negotiate the future of mineral development. The work will be critical: Ali predicts the global mineral supply must double or triple in the next few decades in order to meet the predicted demand.

THINKING CRITICALLY

Analyze Dr. Ali says environmental issues often become a "jobs-versus-environment" debate. But he says this argument is too simplistic. The solution can never be to choose one or the other. Do you agree? Why or why not?

▲ Dr. Saleem Ali (right) consults Nic Bullivant (left), former Head Ranger of Cairngorms National Park, near Glensanda granite quarry in northern Scotland.

◀ Ali poses with the Great Rare Earth Rock monument near the city of Baotou in the Bayan Obo mining region of China. The monument plaque reads " The Home Town of Rare Earths Welcomes You."

CASE STUDY
STRATEGICALLY IMPORTANT MINERALS

If you use a mobile phone, computer, or any device with rechargeable batteries, then you depend on strategically important minerals. Certain minerals are needed to manufacture electronics, medical equipment, and other widely used products. These minerals are sometimes classified as strategically important because maintaining a steady supply is necessary for a country's economic stability. Shortages can create major headaches for manufacturers because strategically important minerals typically have no practical substitutes. Many strategically important minerals must be obtained from other countries because they are found only in certain places on Earth. As a result, supply can be affected by international affairs all over the world. Broken trade agreements, economic sanctions, and political unrest can all threaten to interrupt the supply chain of strategically important minerals.

In the United States, many of the minerals considered strategically important also happen to be rare earth metals. The term "rare earth metal" does not refer to the scarcity of the metal. The rare earth metals are a set of 17 specific metals that have unique properties and applications. Many industrial technologies rely on rare earth metals. Some examples include lasers, computer drives and memory chips, rechargeable batteries, and superconductors.

Acquiring rare earth metals can be difficult. They do not commonly concentrate in mineral ores and are expensive to refine. Mining them can have negative environmental consequences, such as increased erosion; pollution of air, water, and soil; and habitat destruction leading to loss of biodiversity. Rare earth metals are mined in a number of locations, such as Australia, Canada, and Africa.

In addition to the rare earth metals, the United States considers several other minerals strategically important. For example, niobium is a metal used to manufacture certain types of high-strength steel as well as the superconducting magnets used in particle accelerators (Figure 6-1). The United States relies solely on imports to supply all of its niobium needs. Niobium is mined in only a few countries, with Brazil producing about 90 percent of the world's supply. Manganese is another strategically important mineral that is critical to the production of high-strength steel. While the metal itself is not rare, quality manganese ores are concentrated in a limited number of locations, such as South Africa, Brazil, and Ukraine. Because niobium and manganese, along with a number of other metals, are critical to U.S. defense and economy, finding new sources is a priority.

As You Read Consider the challenges associated with mineral extraction and the importance of minerals in various kinds of manufacturing. Think about the diverse ways that minerals impact your daily life, your community, the country, and international relationships.

🌐 **FIGURE 6-1A**
Go online to view Niobium, considered a strategically important mineral in the United States, has applications in steel manufacturing and powerful particle accelerators. For scale, the blue cube is 1 cm on a side.

◀ **FIGURE 6-1B**
The large hadron collider is a particle accelerator that uses superconductors made with niobium.

6.1 MINERAL RESOURCES

Core Ideas and Skills
- Compare and contrast mineral resources and energy resources.
- Describe the natural processes that result in mineral resources.
- Distinguish between mineral resources and mineral reserves.

KEY TERMS

mineral resource	nonrenewable resource
energy resource	mineral reserve

Since humanlike animals emerged five to seven million years ago, our use of geologic resources has become more sophisticated. Early prehistoric people relied on sticks and rocks before learning how to use flint and obsidian to make more effective weapons and tools. They used natural pigments to create elegant art on cave walls. About 8000 B.C.E., people learned to shape and fire clay to make pottery. Archaeologists have found copper ornaments in Turkey dating from 6500 B.C.E. About 1,500 years later, Mesopotamian farmers used copper tools. Today, geologic resources provide iron for steel, silicon for making computer chips, and gasoline to power cars.

Types of Natural Resources

We use two types of geologic resources: **mineral resources** and **energy resources**. Mineral resources include all useful rocks and minerals. Many mineral resources are naturally concentrated by processes that involve interactions among the geosphere, atmosphere, and hydrosphere. Humans have mined and refined these resources to create the industrial world that has altered our planet. The primary energy resources of the early 21st century are coal, petroleum, and natural gas. These formed from prehistoric plant and animal remains that were altered by internal Earth processes.

Mineral resources include both metal ore and nonmetallic minerals. Recall from Chapter 3 that ore is rock sufficiently enriched in one or more minerals to be mined profitably. Geologists usually use the term "ore" along with the name of the metal, for example, iron ore or silver ore.

Nonmetallic mineral resources are rocks or minerals that are not metals, such as salt, building stone, sand, and gravel. When we think about "striking it rich" from mining, we usually think of gold. The U.S. Geological Survey (USGS) estimated that the value of all the gold mined in the United States in 2016 was $8.5 billion. But even more money was made that year from mining and selling sand, gravel, and crushed stone. Sand and gravel are mined from streams, glacial deposits, sand dunes, and beaches. Crushed stone is quarried from non-weathered bedrock. These nonmetallic resources are combined with a mixture of crushed limestone and clay to make concrete. Reinforced with steel, concrete is used to build roads, bridges, and buildings. Concrete is one of the basic building materials of the modern world. In addition, many buildings are faced with stone, usually granite or limestone. Marble, slate, sandstone, and other rocks used for building are also mined from quarries cut into bedrock (Figure 6-2).

FIGURE 6-2
Granite stone is removed from a quarry in Georgia.

There are many important minerals that are fundamental to the industries that produce products we use daily. Some of these minerals are familiar, such as iron, lead, copper, aluminum, silver, and gold. Other minerals are less well-known, such as tungsten (used in light bulb filaments) and borax (used in soaps and antiseptics).

All mineral resources are **nonrenewable resources**, meaning we use them up at a much faster rate than natural processes create them. Many mineral resources, such as gold, can be recycled. Gold mines exist all over the world and on every continent except Antarctica. The gold has accumulated in different types of deposits in different locations over geologic time. Most individual gold deposits take hundreds of thousands or millions of years to form.

If all the gold that has ever been mined by humans could be combined into a single giant cube, it would be 21 meters tall and weigh more than 170,000 metric tons. Although it might seem that most gold has already been found, the USGS estimates that roughly 52,000 more metric tons exist in gold deposits that are not yet discovered. Gold will be mined for many years to come.

Gold is easily recycled through melting. In fact, almost all the gold mined throughout history is still being used today. Your grandfather's gold watch may include some gold mined by the Romans 2,000 years ago. The recent increase in the use of gold in technology changes this long tradition of recycling. In many technology applications, the amount of gold used is so small that recycling is expensive and impractical. As a result, for the first time in human history, gold is being consumed and not recycled. However, gold represents an extreme example because it is so easily recycled and has held its value for so long. In almost all other cases, mineral ore extracted from Earth is consumed at higher rates than gold. For example, much of the iron that is extracted from Earth and used to make steel is not recycled. It is simply consumed and ends up in a landfill.

checkpoint What is the difference between mineral resources and energy resources?

Ore and Ore Deposits

If you pick up any rock and send it to a laboratory for analysis, the report will probably show that the rock contains iron, aluminum, and other valuable metals. However, the natural concentrations of these metals are often so low that the cost to extract them would be much more than the income gained by selling the refined metals. In certain locations, however, natural geologic processes have concentrated metals many times above their normal amounts. Table 6-1 shows a comparison between the natural concentration of a metal and the concentration required to operate a commercial mine. In some cases, metals are enriched up to 100,000 times their natural concentration at the location of a commercial mine.

Successful exploration for new ore deposits requires an understanding of the processes that concentrate metals. For example, platinum concentrates in certain types of igneous rocks. Therefore, if you were exploring for platinum, you would focus your search in areas where igneous rock is common. Except for magmatic processes that occur deep within the crust, the natural processes that concentrate ore minerals all involve interactions of rocks and minerals of the geosphere with water from the hydrosphere.

TABLE 6-1 Comparison of Natural and Enriched Concentrations of Metals

Element	Natural Concentration in Crust (% by Weight)	Natural Concentration Required to Operate a Commercial Mine (% by Weight)
Aluminum	8.0	24.0
Iron	5.8	40.0
Zinc	0.0082	3.0
Uranium	0.00016	0.2
Gold	0.0000002	0.0008
Mercury	0.000002	0.2

Source: Cengage

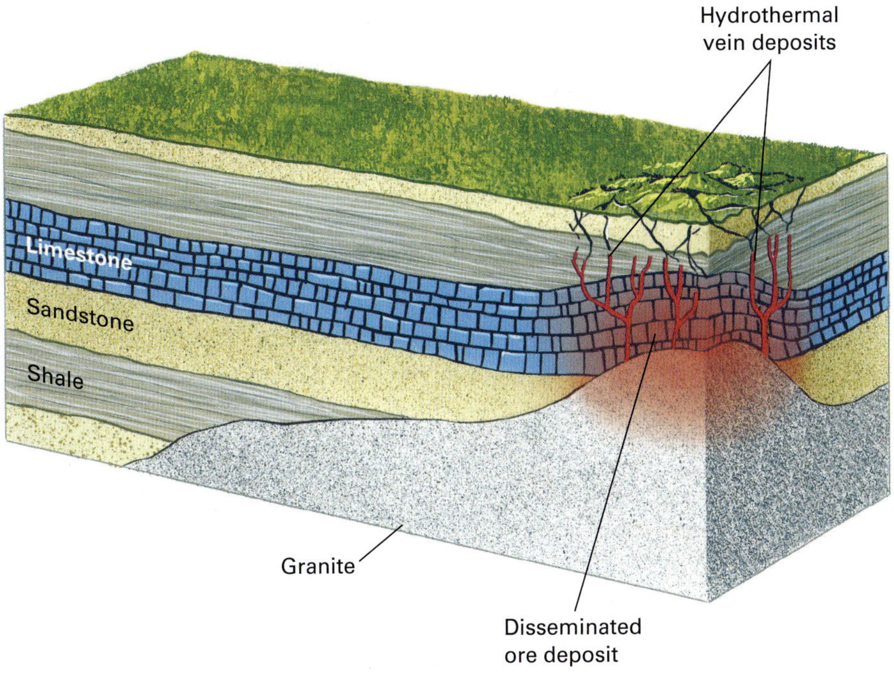

FIGURE 6-3 Hydrothermal vein deposits form in rock fractures.

Magmatic Processes Magmatic processes form mineral deposits as liquid magma solidifies to form igneous rock. These processes create many metal ores, some gems, and certain nonmetallic mineral deposits.

As some large bodies of intrusive igneous rock cool, especially those high in magnesium and iron, they solidify in layers, with each layer containing different minerals. Some layers may contain rich ore deposits. Minerals that crystallize at high temperatures crystallize first as the magma cools. Since most minerals are denser than magma, crystals that form early during cooling sink to the bottom of a magma chamber. These denser minerals accumulate in the bottom layers of igneous rock.

Hydrothermal Processes Hydrothermal processes involve interactions between hot water or steam and rocks or minerals. These processes are probably responsible for the formation of more ore than all other processes combined. Hot water dissolves metals from rock or magma, forming a metallic solution. The metallic solution can seep through cracks or through permeable rock. Then they precipitate to form an ore deposit.

Most hydrothermal water contains salts that greatly increase the water's ability to dissolve minerals. Therefore, hot, salty, hydrothermal water is a powerful solvent, capable of dissolving and transporting metals. Hydrothermal water comes from three sources.

1. Granitic magma gives off hydrothermal water as it solidifies. Many ore metals do not fit neatly into the crystal structures that form when granitic magma is cooling, so they become concentrated in the hydrothermal waters.

2. Groundwater, especially near volcanic areas, can seep deep into Earth's crust. It becomes heated and circulates through the rock in the crust, dissolving ore metals. These elements later precipitate elsewhere in concentrated form.

3. In the oceans, hot, young basalt at a mid-ocean ridge heats seawater as the water seeps into cracks in the seafloor.

Some metals have very low natural concentrations. However, hydrothermal water travels through vast volumes of rock, dissolving the metals and carrying them elsewhere. Where the hydrothermal waters encounter changes in temperature, pressure, or chemical conditions, the metals can be deposited to form a local ore deposit.

A hydrothermal vein deposit forms when metals dissolved in hydrothermal water precipitate in a rock fracture (Figure 6-3). Hydrothermal veins range from less than a millimeter to several meters in width. The gold or silver in a single vein can be worth several million dollars. The same hydrothermal solutions may also soak into pores in country rock near the vein. This can create a disseminated deposit, one that is spread over a large area and is less concentrated. Because they commonly form from the same solutions, rich ore veins and disseminated deposits are often found together. The history of many mining districts is one in which early miners dug shafts and tunnels to follow the rich veins. After the veins were exhausted, later miners used huge power shovels to dig down to the low-concentration ore deposits and mine them from an open pit.

In regions of the seafloor near mid-ocean ridges and submarine volcanoes, seawater circulates through the hot, fractured oceanic crust. The hot seawater dissolves metals from the rocks. As it rises through the upper layers of oceanic crust, the seawater cools. Metals precipitate to form submarine hydrothermal deposits.

The metal-rich solutions can be seen today as jets of black water called black smokers (Figure 6-4). These solutions spout from fractures and vents in the Mid-Ocean Ridge. The black color is caused by precipitation of fine-grained metal sulfide minerals as the solutions cool upon contact with seawater. The precipitating metals form rich ore deposits near the hot-water vent.

Sedimentary Processes Gold is denser than any other mineral. Therefore, if you swirl a mixture of water, gold dust, and sand in a gold pan, the gold sinks to the bottom fastest. This separation also occurs in nature. Many streams carry silt, sand, and gravel with a small amount of gold. The gold settles fastest when the water current slows down. Over years, currents agitate the sediment and the dense gold works its way into cracks and crevices in the streambed. Grains of gold concentrate in gravel as well as in cracks in the bedrock of the streambed. This forms a placer deposit (Figure 6-5). The prospectors who rushed to California in the gold rush of 1849, for example, mined placer deposits of gold.

Groundwater dissolves minerals as it seeps through soil and bedrock. In most environments, groundwater eventually flows into streams and then to the sea. Some of the dissolved ions, such as sodium and chloride, make seawater salty. In deserts, however, lakes develop with no outlet to the ocean. Water flows into the lakes and can escape only by evaporation. As the water evaporates, the dissolved salts concentrate until they precipitate to form evaporite deposits. The composition of the salt and specific salt minerals that form depend on the composition of dissolved ions. Deposits in desert lakes include sodium chloride (table salt), borax, sodium sulfate, and sodium carbonate. These salts are used in the production of paper, soap, and medicines, and for the tanning of leather.

Several times during the past 500 million years, shallow seas covered large regions of North America and all other continents. At times, those seas were

FIGURE 6-4
A black smoker at Sully Vent in the northeast Pacific Ocean pumps metal-rich solutions into the ocean.

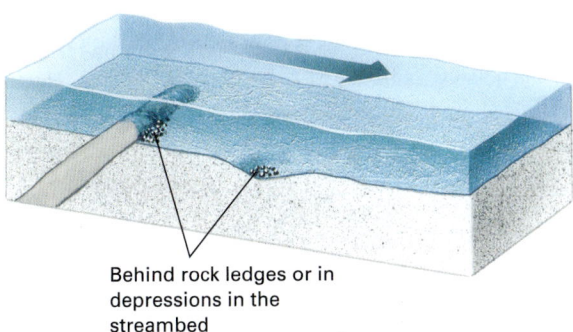

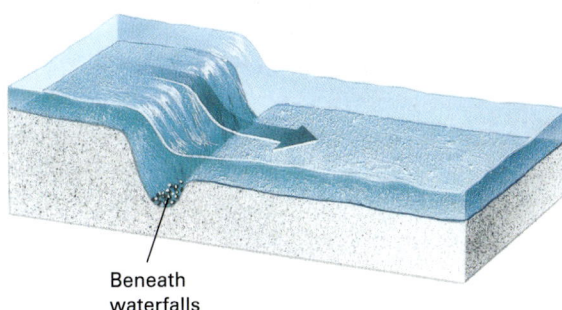

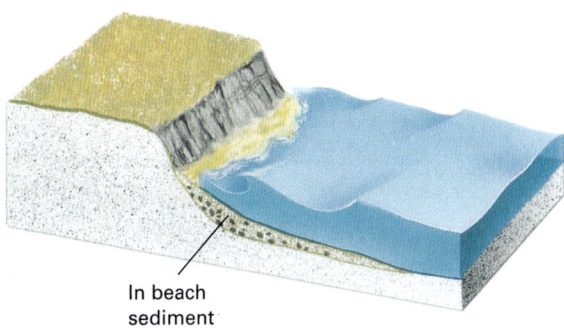

FIGURE 6-5
Placer deposits form where water currents slow down and deposit high-density minerals.

FIGURE 6-6
A two-billion-year-old banded iron specimen shows alternating iron-rich and silica-rich layers.

FIGURE 6-7
This spheroidal texture is typical of bauxite, which is aluminum ore deposited by intense tropical weathering of aluminum-rich rocks. The pen tip is pointing to concentric layers within a spheroid that has been broken.

so weakly connected to the open oceans that water did not circulate freely between seas and the oceans. Evaporation concentrated the dissolved salts. Periodically, new seawater from the open ocean would replenish the shallow seas, providing a new supply of salt. Thick salt beds, formed in this way, underlie nearly 30 percent of North America. Table salt, gypsum (used to manufacture plaster and sheetrock), and potassium salts (used in fertilizer) are mined extensively from these deposits.

Most of the world's supply of iron is mined from sedimentary rocks called banded iron formations composed of alternating iron-rich and silica-rich layers (Figure 6-6). These iron-rich rocks precipitated from the seas between 2.6 and 1.9 billion years ago.

Weathering Processes In environments with high rainfall, water dissolves and removes some ions from soil and rock near Earth's surface. This process leaves behind other ions in the soil that form residual ore deposits. Both aluminum and iron are not easily dissolved in water. Bauxite, the principal source of aluminum, forms as a residual deposit (Figure 6-7). Most bauxite deposits form in warm and rainy tropical or subtropical environments where chemical weathering occurs rapidly. Bauxite ores are common in Jamaica, Cuba, Guinea, Australia, and parts of the southeastern United States.

checkpoint Match each mineral deposit process (magmatic, hydrothermal, sedimentary, and weathering) with a rock type (igneous, sedimentary, metamorphic).

FIGURE 6-8
A worker inspects a machine excavator in a coal seam passage of an underground mine.

FIGURE 6-9
Bingham Canyon copper mine in Utah is the largest excavation in the world.

Mineral Reserves and Mining

Mineral resources are all occurrences of a mineral, including those only thought to exist, that have present or potential future value. **Mineral reserves**, on the other hand, are the known supply of ore in the ground that can be mined for profit.

Mining depletes mineral reserves by decreasing the amount of ore remaining in the ground. Mineral reserves may increase in two ways. First, geologists may discover new mineral deposits, adding to the known amount of ore. Second, mineral deposits where the metal is not concentrated enough to be mined for profit can become profitable. If the price of a metal increases or improvements in mining or refining technology reduce the costs, the mineral reserve of that metal increases.

For example, in 1970, world copper resources were estimated to be 1.6 billion metric tons. World reserves of copper were estimated at only 280 million metric tons. By 2010, however, improved mining techniques and rising copper prices resulted in the production of about 400 million metric tons of copper. Improved mining technology, rising prices for copper, and increasing demand for copper has continued to drive the exploration and production of new copper sources.

Miners extract both ore and coal from underground mines and surface mines. Large underground mines contain kilometers of interconnected passages that follow ore veins or coal seams (Figure 6-8). The lowest levels of the mine may be several kilometers deep. A surface mine is a hole excavated in Earth's surface. Most modern coal mining is done by large power shovels that extract coal from huge surface mines. The largest human-made surface mine on Earth is the open-pit copper mine at Bingham Canyon, Utah (Figure 6-9). The mine is 4 kilometers in diameter and 1.2 kilometers deep, and it can be seen with the unaided eye from space. Since its beginning in 1873, the mine has produced about 13 trillion kilograms of copper, along with significant amounts of gold, silver, and molybdenum.

Although underground mines do not directly disturb the land surface, some abandoned mines collapse. Occasionally buildings have sunken into collapsed mines. Over 8,000 square kilometers of land in central Appalachia have settled into underground coal mine shafts.

Mining also creates huge piles of waste rock that must be removed to get at the ore or coal. Some waste rock is left over after the refining of the ore. If the waste piles are not treated properly, oxygen from the atmosphere will combine with sulfur in the waste rock to form sulfuric acid. The acid partially dissolves the bits of loose rock, releasing toxic elements such as arsenic into the local surface and groundwater systems.

checkpoint How and why do mineral reserves change over time?

6.1 ASSESSMENT

1. **Explain** What are the two types of geologic resources and why do we need both?
2. **Define** From which of Earth's systems are mineral resources obtained?
3. **Describe** Explain why commercial mines in specific locations extract some mineral resources and not others.
4. **Distinguish** Different geologic processes form specific mineral resources. Distinguish among three such geologic processes and the ores associated with them.

Critical Thinking

5. **Apply** Refer to Table 6-1 to answer the question. What would happen to the natural concentration of aluminum required to operate a commercial mine if the demand for aluminum dramatically increased? Explain your reasoning.

FIGURE 6-10
ON ASSIGNMENT National Geographic photographer Robb Kendrick documented miners lifting coal out of a mine shaft two tons at a time in Meghalaya, India.

6.2 NONRENEWABLE ENERGY RESOURCES

Core Ideas and Skills
- Compare and contrast the pros and cons of different fossil fuels.
- Explain the main difference between fossil fuels and nuclear fuels.

KEY TERMS

fossil fuel	hydraulic fracturing (fracking)
coal	nuclear fuel
petroleum	
natural gas	

Coal, petroleum, and natural gas are called **fossil fuels** because they are formed from the remains of plants and animals. Fossil fuels are nonrenewable. When a lump of coal or a liter of oil (petroleum) is burned, the energy is used and dissipates, and for all practical purposes thereafter, is lost. Our fossil fuel supply diminishes over time.

Coal

Coal is a sedimentary rock made mostly of organic carbon that will burn without refining. Coal power plants burn about 92 percent of the coal consumed in the United States. Coal power plants produce less than half of the nation's electricity and that number is dropping. Although it is easily mined and abundant in many parts of the world, coal emits air pollutants that can be removed only with expensive control devices. Mercury, in particular, is released into the atmosphere mainly through coal power plants. Despite these drawbacks, coal is an abundant resource, with widespread availability projected to last beyond the 21st century.

In North America, large quantities of coal formed between 360 and 286 million years ago, and again between 146 and 56 million years ago. During this time, warm, humid swamps covered broad areas of low-lying land. When plants die in forests and grasslands, organisms consume some of the plant matter. Chemical reactions with oxygen and water decompose the remains. As a result, little organic matter accumulates except in topsoil. In some swamps and bogs, however, plants grow and die rapidly. Newly fallen vegetation quickly buries older plant remains. The new layers prevent atmospheric oxygen from penetrating into the deeper layers. Decomposition stops before it is complete, leaving brown, partially decayed plant matter called peat.

Plant matter is composed mainly of carbon, hydrogen, and oxygen and contains large amounts of water. During burial, rising pressure expels the water. Chemical reactions release most of the hydrogen and oxygen. As a result, the proportion of carbon increases until coal forms (Figure 6-11). The grade of coal and the heat that can be recovered by burning different grades of coal varies considerably depending on the carbon content. Table 6-2 compares the amount of heat that can be produced by different grades of coal.

checkpoint Why does coal obtained from greater depths have greater heat value than coal obtained from the surface?

TABLE 6-2 Classification of Different Grades of Coal by Color, Composition, and Heat Value (British Thermal Units per Mass of Fuel)

Type	Color	Water (%)	Other Volatiles and Noncombustible Compounds (%)	Carbon (%)	Heat Value (kJ/kg)
Peat	Brown	75	10	15	1,400–2,400
Lignite	Dark brown	45	20	35	3,300
Bituminous (soft coal)	Black	5–15	20–30	55–75	5,700
Anthracite (hard coal)	Black	4	1	95	6,700

Source: Cengage

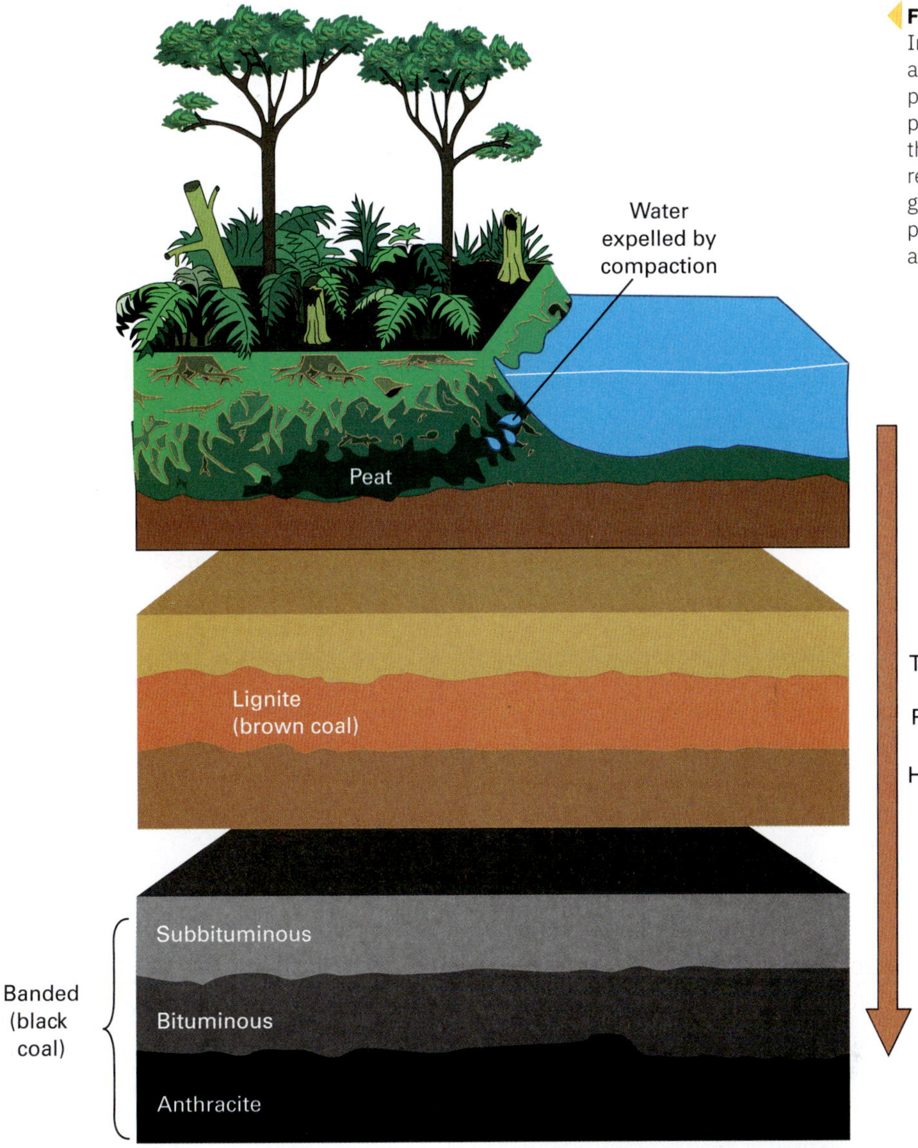

FIGURE 6-11 Increasing pressure and heat acting on plant matter over long periods of time raises the carbon content, resulting in different grades of coal—from peat to anthracite. (See also Figure 4-15.)

Petroleum

The word **petroleum** comes from the Latin for "rock oil" or "oil from the earth." In North America, the first commercial oil well was drilled in 1859 in Titusville, Pennsylvania, ushering in a new energy age. Petroleum is called crude oil when first pumped from the ground. Crude oil is made up of thousands of different chemical compounds and ranges widely in consistency and color. Some are brown, waxy substances that are solid at room temperature but liquid at higher temperatures. Some are yellowish or nearly clear liquids that resemble refined gasoline. Most are rather thick and dark colored. Once recovered from a well, crude oil is refined to produce propane, gasoline, heating oil, and other fuels (Figure 6-12). Petroleum is also used to manufacture plastics, nylon, and other useful materials.

FIGURE 6-12
An oil refinery converts crude oil into useful products such as gasoline.

LESSON 6.2

Petroleum forms from plant matter that accumulates in mud found in swamps, lakes, and marine waters. Over millions of years, the mud is buried to depths of a few kilometers. Rising temperature and pressure convert the mud into mudstone. The organic matter turns into a solid that breaks down chemically, releasing small organic molecules that form liquid petroleum. If it is not trapped along the way, the oil will migrate to the surface, forming a natural oil seep. The La Brea Tar Pits in downtown Los Angeles is perhaps the most famous example of a natural oil seep (Figure 6-13). Between 40,000 and 8,000 years ago, hundreds of species of organisms became trapped in tar formed from the La Brea oil seeps, died, and were preserved.

checkpoint Where does petroleum come from?

Natural Gas

Natural gas is an energy resource that forms when crude oil is heated above 100°C during burial. Organic molecules break down to form methane. Many petroleum reservoirs contain a layer of gas-saturated rock above the heavier liquid petroleum.

Natural gas is used without refining for home heating and cooking. It is also used for fueling large electricity generating plants. Because natural gas contains few impurities, it releases few pollutants such as mercury or sulfur when it burns. However, as with all fossil fuels, combustion of natural gas releases carbon dioxide, a greenhouse gas. This fuel is more energy efficient, produces fewer pollutants, and is less expensive to produce than petroleum.

checkpoint What is the connection between natural gas and crude oil?

FIGURE 6-13
Museum staff and volunteers dig through tar at Pit 91 of the La Brea Tar Pits, a natural oil seep, to find bones and other items buried there.

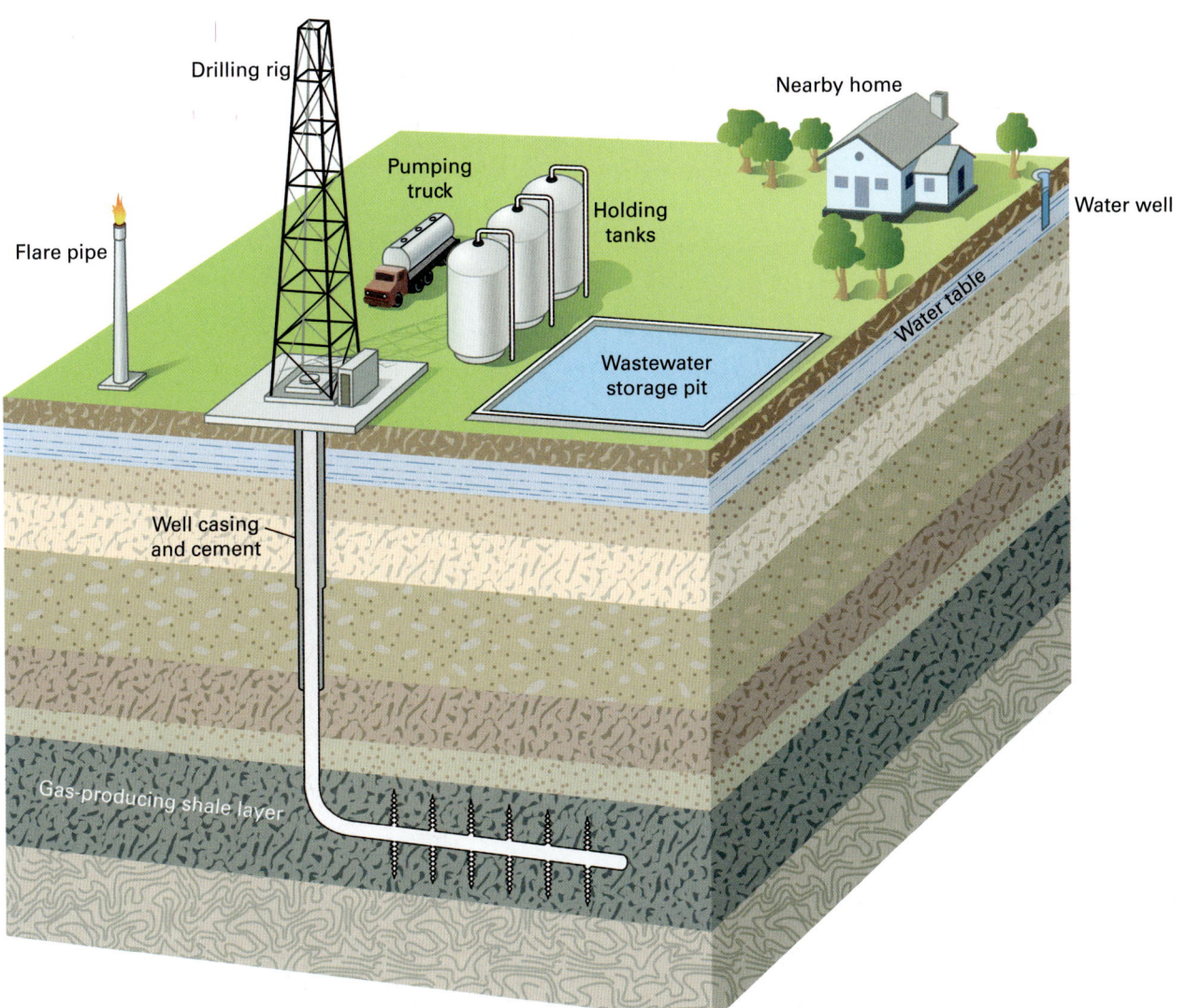

FIGURE 6-14
Most oil wells in the U.S. involve a vertical section and a horizontal section that is drilled into the target layer. This layer is then hydraulically fractured as large volumes of water and sand are pumped down the well. The fractures release oil and gas that is trapped in the layer and allow it to flow to the well where it is produced.

Nontraditional Fossil Fuel Reservoirs

Some natural gas and oil is trapped in dense rock layers, and the oil and gas is difficult or impossible to access through conventional vertical drilling. **Hydraulic fracturing**, also known as fracking, is the high-pressure injection of water and chemicals into subsurface rock formations (Figure 6-14). The high water pressure fractures the rock, allowing it to release some of the gas and oil contained in it. New techniques such as horizontal drilling have created additional access to these unconventional fossil fuel reservoirs. Fracking is controversial because of environmental and health concerns such as the large volume of toxic wastewater it produces. In many cases wastewater is injected into deep disposal wells. The disposal water can cause faults in the subsurface to slip, generating an earthquake.

checkpoint What fractures during the process of hydraulic fracturing?

Nuclear Fuels and Reactors

Nuclear fuels are radioactive isotopes that produce heat through fission reactions, in which nuclei are split. The heat is used to generate electricity in nuclear power plants. Uranium is the most commonly used nuclear fuel. Although it is relatively abundant, uranium is nonrenewable, like all mineral resources.

▸ **FIGURE 16-15**
The nuclear power plant at Three Mile Island in central Pennsylvania used two cooling towers (left). The two cooling towers behind them were permanently shut down following the 1979 accident. The entire plant shut down in September 2019.

FIGURE 6-16 ▾
A map and bar graph compare the number of nuclear reactors per state and their net summer nuclear energy capacity in megawatts of electricity in 2017.

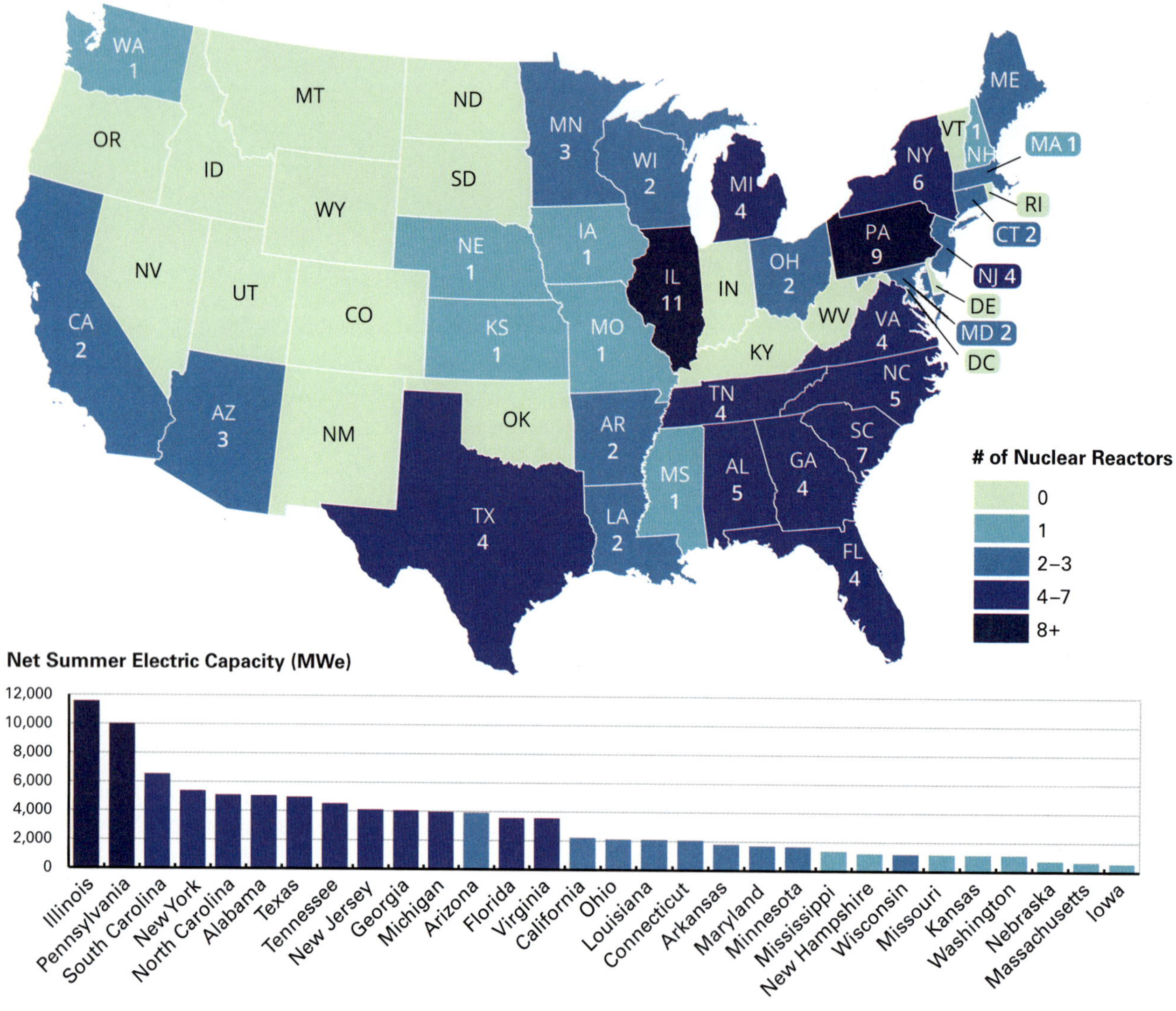

# of Nuclear Reactors
- 0
- 1
- 2–3
- 4–7
- 8+

Net Summer Electric Capacity (MWe)

Source: U.S. Energy Information Administration (EIA) data for January, 2017.

176   CHAPTER 6   MINERAL AND ENERGY RESOURCES

Every step in the mining, processing, and use of nuclear fuel produces radioactive waste. The waste discarded during the mining of uranium is radioactive. Two useless radioactive nuclei are produced when a single uranium nucleus splits inside a nuclear reactor. This material must be discarded. After several months of use in a reactor, the concentration of useful uranium in the fuel rods is too low to generate sufficient heat, and the fuel rods are replaced. Old fuel rods accumulate as radioactive waste.

In the early 1970s, the nuclear industry in the United States was growing rapidly. Many energy experts predicted that nuclear energy would dominate the generation of the country's electric energy. These predictions have not been realized. Four factors have led to the decline of the nuclear power industry.

1. Construction of new reactors in the United States has become so costly that electricity generated by nuclear power is more expensive than that generated by coal power plants.
2. After major accidents at Three Mile Island (Figure 6-15) in the United States (1979), Chernobyl in Ukraine (1986), and Fukushima Daiichi in Japan (2011), many people have become concerned about safety.
3. Serious concerns remain about the safe disposal of nuclear wastes.
4. The demand for electricity has risen less than expected since 1990.

Figure 6-16 shows the number of nuclear power plants per state and their energy production capacity. It is interesting to note that most of the operating nuclear plants in the United States as of 2017 are located east of the Mississippi River or on the river itself. More energy production is needed in the more densely populated eastern part of the United States. The relatively arid west of the country has easier and more abundant access to other alternative energy sources, especially solar, wind, and geothermal.

checkpoint What are the by-products of nuclear energy production?

6.2 ASSESSMENT

1. **Define** What are nonrenewable energy resources and why are they classified by this term?
2. **Summarize** Why has the construction of new nuclear power plants in the United States decreased since the 1990s?
3. **Explain** What is one key similarity and one key difference between energy resources such as uranium and mineral resources such as gold?

Computational Thinking

4. **Analyze Data** Rank the top 10 states based on nuclear energy production.

6.3 RENEWABLE ENERGY RESOURCES

Core Ideas and Skills
- Describe the different sources of renewable energy.
- Compare and contrast the pros and cons of different renewable energy resources.

KEY TERMS

renewable resource
solar cell
geothermal energy
hydroelectric energy
biomass fuel
biomass energy

Solar, wind, geothermal, hydroelectric, and wood are examples of **renewable resources**. Natural processes replenish renewable resources even as we use them. The amount of energy produced today by renewable resources is small compared to that produced by nonrenewable resources. However, renewable resources have the potential to supply all of our energy needs. As the prices of conventional fossil fuels have risen along with worldwide energy demand, some renewables have become more economical. Except for biomass fuels, renewable energy sources emit no carbon dioxide and therefore do not contribute to global warming.

LESSON 6.3 177

Solar Energy

Current technologies allow us to use solar energy in three ways: passive solar heating, active solar heating, and electricity production by solar cells.

A passive solar house is built to absorb and store the sun's heat directly. In active solar heating systems, solar thermal collectors absorb the sun's energy and use it to heat water. Pumps then circulate the hot water through radiators to heat a building. The inhabitants also may use the hot water directly for washing and bathing.

A **solar cell** (photovoltaic cell, or PV) produces electricity directly from sunlight. A modern solar cell contains a semiconductor, a device that can conduct electrical current under some conditions but not others. Sunlight energizes electrons in the semiconductor, producing an electric current.

Figure 6-17 shows an installation of industrial-scale solar arrays in China. Although solar power still accounts for less than 1.5 percent of the total energy consumed annually in the United States, solar energy is our most abundant resource worldwide. Photovoltaic (PV) cell production is the fastest-growing segment of the energy industry. Arrays of PV cells are now competitive with electricity costs during peak demand times in many desert areas, especially those installed for single-family homes. PV arrays are also cost-effective for electricity needs far from existing power lines.

checkpoint Why is a PV cell considered a semi conductor?

Wind Energy

In the United States, wind energy grew 1,000 percent in the first decade of the 21st century, making up about 6 percent of the country's total energy demand. Wind energy production has grown rapidly because construction of wind generators is cheaper than building new fossil fuel power plants. Wind energy is also clean and abundant. Gigantic wind farms like the one shown at the beginning of this chapter generate electricity in many countries. The main drawbacks to wind energy include its inconsistency, the visibility of their structures (which some people view as unsightly), and the noise generated by the blades. In addition, the best places to place a wind farm do not always have large nearby human populations. Transmitting the electricity generated by the wind farm to fulfill energy needs has high economic and environmental costs.

checkpoint What are the main drawbacks of wind energy?

FIGURE 6-17
A photovoltaic power plant in China's Shanxi Province covers the landscape. In recent years, China has vigorously developed and utilized solar energy by building photovoltaic power plants across entire mountainsides.

FIGURE 6-18
The geothermal power station in Nesjavellir, Iceland, delivers steam and hot water by way of long pipelines as far as Reykjavik, 40 kilometers away.

Geothermal Energy

Energy extracted from Earth's internal heat is called **geothermal energy** (Figure 6-18). Geothermal plants collect underground steam from thermal regions. This steam is used to spin turbines that generate electricity. Naturally hot groundwater can also be pumped to the surface to generate electricity. Natural groundwater also can be used directly to heat or cool homes and other buildings through the installation of heat exchange pumps. The United States is the largest producer of geothermal electricity in the world, but geothermal electricity generation accounts for less than 0.5 percent of the total energy produced in this country.

In the United States, most geothermal plants are located in the western states because the region is more geologically active. The oldest, and also presently the largest, steam-driven geothermal plant is located at The Geysers, north of San Francisco, California.

Relative to other renewables such as wind and solar, geothermal energy has a larger capacity and can be used 24 hours a day, seven days a week. Because the wind does not blow all the time and the sun does not shine all the time, those energy sources have less capacity than geothermal.

checkpoint What advantage does geothermal energy have over solar energy and wind energy?

Hydroelectric Energy

If a river is dammed, the energy of water dropping downward through the dam can be harnessed to turn turbines that produce electricity called **hydroelectric energy**. Although hydroelectric energy is renewable, it is entirely dependent on the amount of water available in dammed up rivers. Recent climate change has caused many of the large reservoirs in the western United States to have low water levels. Inadequate water level decreases the ability of hydroelectric power plants to produce electricity. In addition, the construction of dams and the formation of reservoirs destroy wildlife habitats. Agricultural land, towns, and migratory fish populations are also negatively affected. Large dams are also expensive to build, and few suitable sites remain.

For these reasons, the United States is unlikely to increase its production of hydroelectric energy, which currently accounts for about seven percent of the total energy produced in this country. Many historic dams, some with hydroelectric facilities, are being removed.

checkpoint How does a hydroelectric power plant produce electricity?

MINILAB Biofuel

Materials

safety goggles
1 peanut
2 large metal paper clips
2 test tubes
water (at least 200 mL)
test tube stand
glass saucer
forceps
thermometer
multi-purpose lighter
graduated cylinder
stopwatch
balance
1 tree nut, such as cashew or almond

CAUTION: Use safety goggles. Tie back long hair. Do not touch hot objects. Monitor the flame at all times.

Procedure

1. Measure and record the mass of the tree nut and the mass of the peanut.
2. Measure 100 mL of room temperature water and add it to a test tube. Record the temperature of the water.
3. Secure the test tube to the test tube stand. Place the saucer where it will catch any falling ash.
4. Use forceps to hold the tree nut 5 mm under the bottom of the test tube. Use the lighter to ignite the nut. The nut should burn on its own; relight if it stops burning.
5. Allow the nut to burn for 1 minute. Extinguish the flame. Measure and record the temperature of the water.
6. Measure and record the total mass of the burned nut and any ash that fell into the saucer. The ash must be included in the mass because it is a product of the combustion.
7. Repeat steps 2–5 using the peanut and the second test tube.

Results and Analysis

1. **Calculate** For each fuel (nut), calculate the change in water temperature per gram of fuel. Make a table to record your results.
2. **Conclude** Use evidence to explain which fuel contains the most stored energy.
3. **Explain** How does this lab demonstrate the use of energy resources? Are the resources renewable or nonrenewable? Why?

Critical Thinking

4. **Apply** What are possible real-world uses of biofuels? What constraints affect the use of biofuels?
5. **Compare** How do the risks and benefits of using biofuels and fossil fuels compare?

Biomass Energy

Biomass fuels are plant-based materials that can provide energy. The burning of wood as a source of heat is familiar example. Today, industries burn agricultural products to produce steam and generate **biomass energy**. Additionally, biomass from oil-rich plants such as canola can be used as transportation fuel. Research is currently being directed toward the production of liquid fuels and other chemicals.

Biomass energy can be produced domestically, thereby creating local jobs and reducing the need to import foreign oil. However, production of biofuels is not always efficient. Biomass energy makes up less than 2 percent of the energy production in the United States. In some cases, more energy is used in the production and processing of biomass fuels than can be extracted. In addition, burning of biomass produces carbon dioxide and releases pollutants into the atmosphere.

checkpoint What is a common example of a biomass fuel?

6.3 ASSESSMENT

1. **Define** What are renewable energy resources and why are they classified by this term?
2. **Summarize** Describe the advantages and disadvantages of renewable energy sources as compared to nonrenewable energy sources.
3. **Compare** Rank the renewable energy sources by the amount of electricity they provide in the United States.

Critical Thinking

4. **Evaluate** Describe the areas in the United States where the different types of renewable energy sources would be most efficient.

6.4 CONSERVATION AND ENERGY FOR THE FUTURE

Core Ideas and Skills

- Describe how energy consumption and production from renewable and nonrenewable sources are changing in the United States and globally.
- Identify effective solutions for preserving and conserving energy resources.
- Compare the energy production in different sectors (renewable and nonrenewable) in the past to the current production as well as the predicted future production.

KEY TERMS

hydrocarbon

Energy and Environmental Conservation

The quickest and most effective way to decrease energy consumption and to prolong the availability of fossil fuels is to conserve energy (Figure 6-19). Policies to improve energy efficiency are more cost-effective than building new power plants. Such policies help to reduce air pollution and dependence on oil imports while saving money for consumers and industry.

Energy conservation has produced dramatic results in the United States. The cost of producing energy fell from about 13.5 percent of gross domestic product (GDP) in 1980 to about 6 percent of GDP in 2000. Higher global energy prices since 2000 have caused spending on energy production to rise again. Clearly, conservation is a critical component of keeping the total spending for energy in the country from continuing to rise. Some energy experts have suggested that if people in

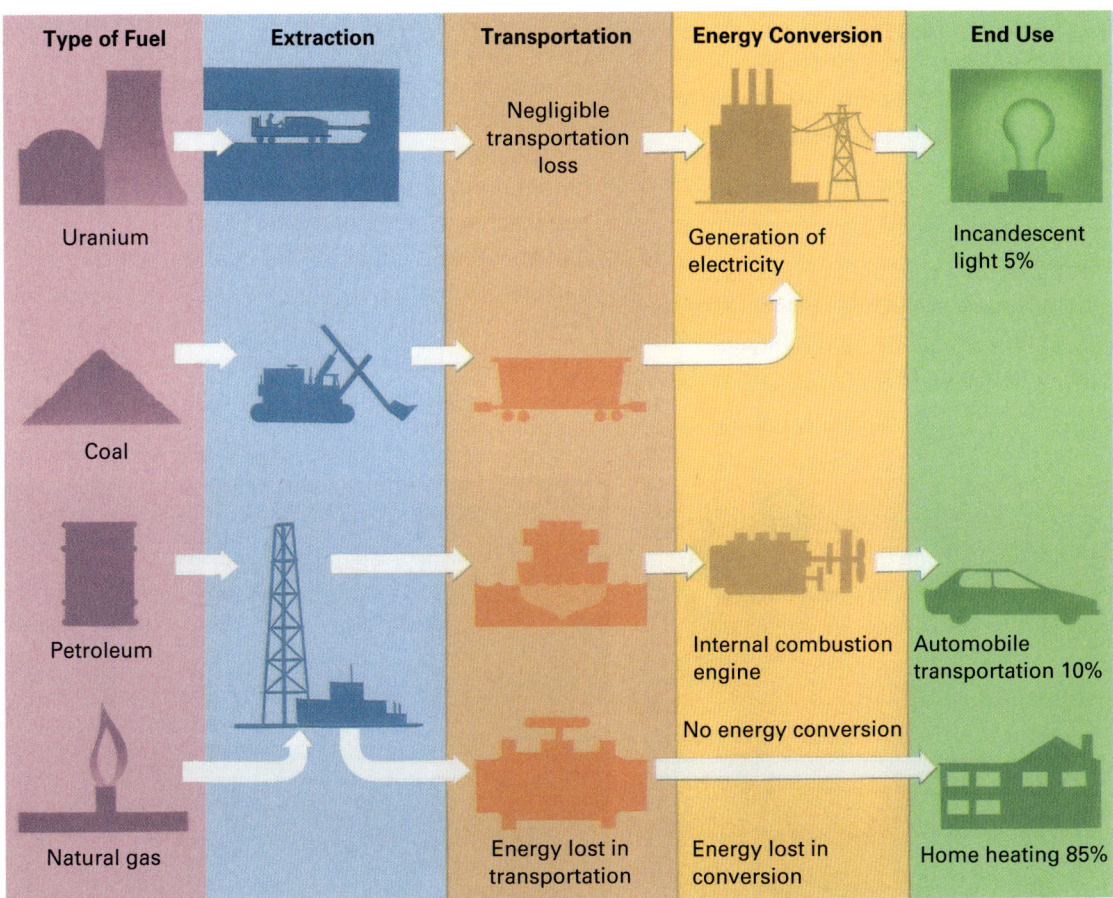

FIGURE 6-19
A flowchart shows the end-use efficiencies of common energy-consuming systems. Home heating represents the only energy-consumption system that is even remotely efficient, and even there, 15 percent of energy is wasted. Energy to produce incandescent lighting is 95 percent wasted. Energy for automobile transportation is 90 percent wasted.

industrialized nations use more efficient equipment and develop more efficient habits, these nations could conserve as much as half of the energy they currently consume.

Energy use in the United States falls under three categories: buildings, industry, and transportation. Two types of conservation strategies can be applied in each of those categories. Technical solutions involve switching to more efficient devices. Social solutions involve people's choices and decisions to use existing energy systems more efficiently.

Technical Solutions In 2010, residential and commercial buildings consumed about 41 percent of all the energy produced in the United States. Most of that energy was used for heating, air-conditioning, and lighting.

Significant energy savings are possible in all aspects of energy consumption in buildings. As one example, lighting accounts for about 20 percent of the average U.S. home's electricity bill. Because incandescent lighting is about 95 percent inefficient in energy consumption, savings in that area alone are potentially great. A fluorescent bulb consumes one-fourth as much energy as a comparable incandescent bulb. Fluorescent bulbs can last 10 times longer too. New solid-state technology promises further advances in energy-efficient lighting. For example, light-emitting diodes (LED) are lights that are illuminated by the movement of electrons through a semiconductor. There is no filament as with incandescent lights. LEDs release almost no heat, so they are much more efficient, last much longer, and use far less energy. Today, LEDs are used in clock radios, jumbo TVs, and many other applications. According to the U.S. Department of Energy, widespread switching to LED lighting technologies (Figure 6-20) over the next 15 years could save the equivalent of 44 years worth of annual electrical output by large power plants.

Industry consumes about 31 percent of the energy used in the United States. In general, conservation practices are cost-effective. Many companies are taking advantage of the fact that saving energy is profitable, although industry still wastes great amounts of energy.

For example, about two-thirds of the electricity consumed by industry drives electric motors for machinery and tools. Most motors used in industry are inefficient because they run only at full speed and are slowed by brakes to perform their tasks. Replacing older electric motors with variable-speed motors would save vast amounts of electricity. Such replacement has been slow.

Hybrid cars use a small, fuel-efficient gasoline engine combined with an electric motor. They consume less gas and produce less pollution per mile than gasoline engines. Current models of hybrid cars achieve fuel efficiencies ranging from 30 to 50 miles per gallon. They produce as much as 90 percent fewer harmful emissions than a comparable gasoline engine. Using hybrids and other energy-efficient vehicles, American motorists could achieve a 50 percent or greater increase in fuel economy, the equivalent of about one-third of current oil imports.

Another solution is the electric car. Battery-only and gasoline-electric hybrid cars have seen a recent increase in popularity and availability. If that trend

A

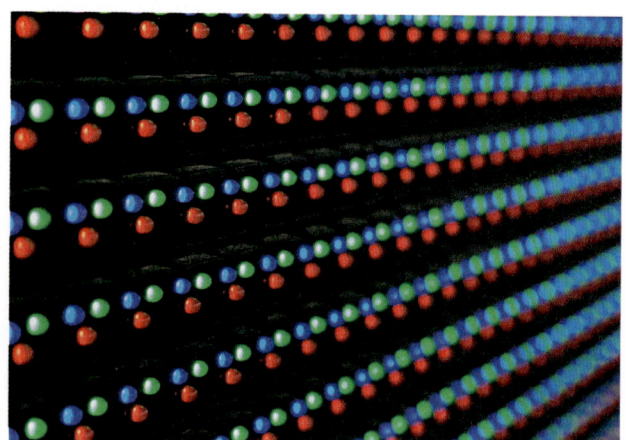

B

FIGURE 6-20
(A) Artificial lighting has progressed from candle to tungsten bulb to fluorescent bulb to LED bulb (left to right). (B) LED screen displays are quickly becoming a popular lighting source.

continues, perhaps we can further reduce the cost of and dependence on petroleum. Challenges associated with battery-powered vehicles include the need to recharge the battery frequently. This draws energy off the electric grid, which means that the energy must be generated elsewhere in a power plant and delivered to the recharging station via transmission lines. In addition, the availability of key mineral resources needed to build the batteries, especially the element lithium, is a challenge.

Social Solutions Social solutions involve altering human behavior to conserve energy. Energy-conserving actions can be used in buildings, industry, and transportation. Some result in inconvenience to individuals. For example, if people choose to carpool rather than drive their own car, they save fuel but inconvenience themselves by coordinating their schedule with their carpool companions. Small and lightweight, fuel-efficient vehicles are available, but they will make an impact only if people make the social decision to use them. People argue that this decision comes at a great cost because light vehicles make the driver and passengers more vulnerable in the case of an accident. Studies have shown that lighter, more agile vehicles, with better turning capacity and more effective braking, are actually safer than heavier SUVs.

An example of a modest change in lifestyle that could have a major positive impact is decreasing the role that plastics play in our society. Plastics are used in a wide array of products such as straws and cup lids, children's toys, modern fabrics, car parts, and even satellite dishes. Our society consumes plastics at a voracious rate and the consequences are only recently being brought to light.
All plastics are made from **hydrocarbons**, compounds of hydrogen and carbon from organic sources. Plastics are not grown, nor are they mined. Ultimately all plastics are synthesized from crude oil through chemical engineering.

The gigantic volume of plastics that are consumed globally has had a significant negative impact on the global environment. Much of the consumed plastic ends up as waste that must be disposed. A large amount finds its way into the environment as litter. Disposing of plastic through incineration generates a variety of toxic chemicals that are released into the atmosphere. Plastic that ends up in the environment as litter has accumulated in famous giant collections, such as the Great Pacific garbage patch, at certain spots in

CONNECTIONS TO ENVIRONMENTAL SCIENCES

Planet or Plastics?

Today, awareness of the negative consequences of using so much plastic in our society is increasing. For example, a recent multiyear campaign called "Planet or Plastic?" was launched by National Geographic and Sky Media to raise public awareness of the problem of single-use plastics. The goal is to encourage people to decrease their use of products made from plastics that cannot be recycled and are simply thrown away or end up in landfills or the ocean. It is through awareness campaigns like this and changes in mindset and behavior that humans will be able to reverse some of the negative consequences of our "throw-away" society.

FIGURE 6-21
Plastic pollution floats in the ocean.

the middle of the ocean. Plastics break down into microplastics when exposed to the environment. Microplastics have a large negative impact on the ocean ecosystem. They can fill the digestive system of small organisms that mistake the floating bits of plastic as food and eat it.

checkpoint What are one technological and one social action that can be taken to conserve energy?

The Future of Renewable and Nonrenewable Energy Resources

According to a 2018 report by the U.S. Energy Information Administration, the mix of energy resources both produced and consumed by the United States will change in the next thirty years. The consumption of natural gas will grow the most during this period. The development of non-

LESSON 6.4 183

DATA ANALYSIS Analyze Patterns in U.S. Energy Consumption

Energy used in the United States comes from a combination of fossil fuel, nuclear, and renewable sources. Figure 6-22A shows how each source contributed to the total energy consumption each year from 2000 through 2016. Figure 6-22B shows positive and negative change for each type of energy source from 2015 to 2016.

1. **Describe** Use Figure 6-22A to summarize how U.S. energy consumption changed from 2000 through 2016. Which energy sources increased? Which energy sources decreased?
2. **Infer** What do you think accounts for the changes, and lack of change, you described?

Computational Thinking

3. **Recognize Patterns** Compare Figure 6-22A and Figure 6-22B. How do the data in Figure 6-22B support the trends you observe in Figure 6-22A?
4. **Generalize Patterns** Based on the two graphs, offer a statement that relates the use of other fuel sources to the use of coal.

Data Challenge

Go to the Data Analysis in MindTap to complete the data challenge.

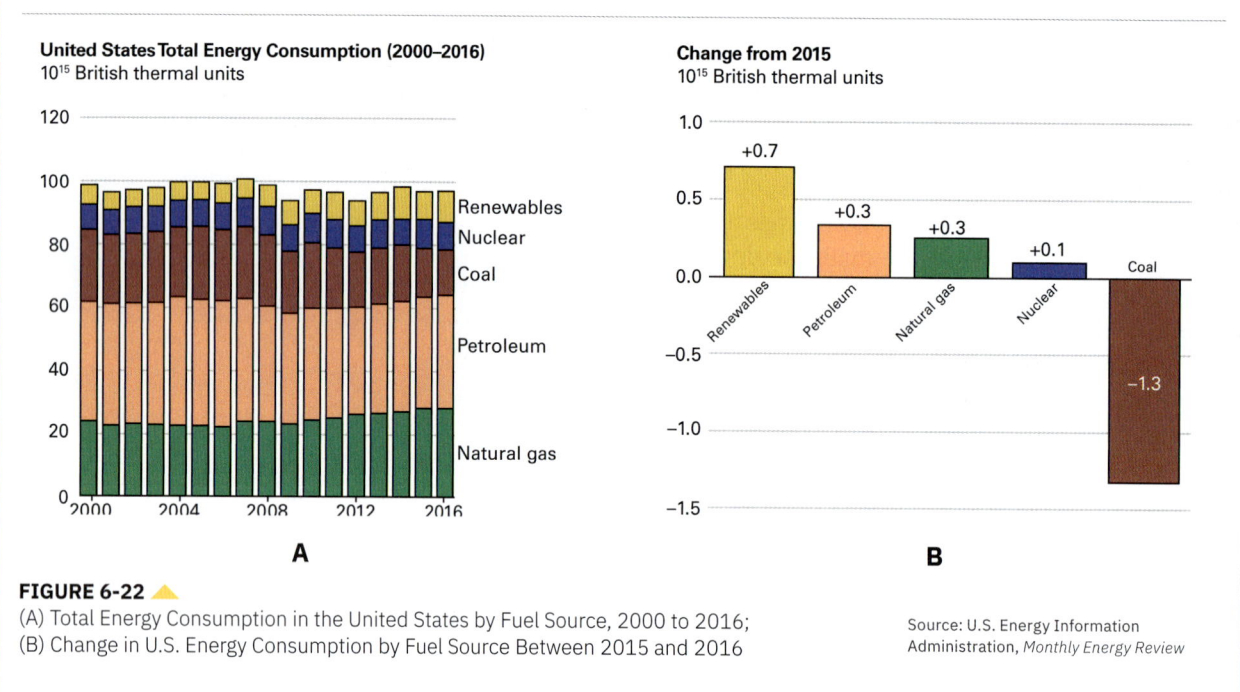

FIGURE 6-22
(A) Total Energy Consumption in the United States by Fuel Source, 2000 to 2016;
(B) Change in U.S. Energy Consumption by Fuel Source Between 2015 and 2016

Source: U.S. Energy Information Administration, *Monthly Energy Review*

hydroelectric renewable energy resources such as solar and wind is projected to be the fastest-growing segment of the energy economy. The percentage of overall energy produced in the United States by non-hydroelectric renewable resources will still remain small compared to that produced by natural gas.

The global picture of energy production and consumption over the next several decades is projected to be somewhat different from that of the United States. According to the *International Energy Outlook* published in 2017, between 2015 and 2040, world energy consumption is projected to increase by 28 percent. More than half the increase is attributed to growth in Asia, especially China and India, where strong economic growth drives increasing energy demand.

Global energy consumption is expected to increase for all major fuel types except coal, with renewables growing at the fastest rate. Global consumption from natural gas, petroleum, and nuclear are all projected to grow through 2050, despite the slowdown of nuclear reactor construction in the United States. Even with the projected decline in energy produced from coal, however, global energy production is projected to be dominated by petroleum and natural gas.

checkpoint Why is energy consumption expected to increase over the next few decades?

Energy for the 21st Century

Is it possible to completely change global energy production from fossil-fuel based to renewable-energy based? In 2001, the United Nations Intergovernmental Panel on Climate Change (IPCC) concluded that this is possible with renewable energy systems that exist without any new technology breakthroughs needed. In other words, they suggested that if we focus on developing renewable energy facilities, the global economic system could absorb a drastic decline in petroleum production without massive energy disruptions.

A year later, in 2002, a group of energy experts published a rebuttal proposing the opposite conclusion. Using almost the same phrases, with the simple addition of the word *not*, they argued that existing renewable energy technology could not support energy demand. Their basic counterargument was that global energy consumption is huge and renewable sources have low power output. Unfortunately, we do not have the available land, nor could we quickly build the required infrastructure to replace fossil fuels.

Almost a decade later, the IPCC again argued that close to 80 percent of world energy supplies could be met by renewable energy sources, as long as they are backed by policies that encourage their development. To stress the importance of the need to move away from fossil-fuel based energy that releases greenhouse gases and other pollutants, IPCC published a summary of the risks posed by continued increases in global temperature. These included: (1) negative impacts on unique and already threatened systems such as coral reefs; (2) an increase in extreme climate and weather events, for example stronger more dangerous tropical cyclones; (3) greater impacts on poor and developing countries rather than industrialized countries; (4) severe economic impacts worldwide and (5) large-scale impacts such as climate instability and disruptions to global systems such as ocean currents.

When experts disagree, it is difficult for the public to evaluate the merits of the contradictory arguments. But whoever is right, it is clear that if global energy demand significantly exceeds supply, the world will suffer economically. On the other hand, if we continue to base our future energy economy on fossil fuels and fail to decrease the impact on the environment, we will push Earth's systems past the point at which the ongoing mass extinction will ultimately involve us. We can only hope that human ingenuity will combine with economic and political commitment to develop alternative energy resources before these catastrophes become reality.

checkpoint Why is switching to renewable energy so important to the environment?

6.4 ASSESSMENT

1. **Explain** Describe two steps meant to prolong the availability of nonrenewable energy resources.
2. **Summarize** What are some of the ways energy is being conserved in the transportation industry?
3. **Compare** Rank the energy resources by expected growth in the United States over the next few decades. Which are expected to grow fastest? Which are expected to decline?

Critical Thinking

4. **Evaluate** Construct your own argument for or against replacing nonrenewable energy sources with renewable energy sources.

TYING IT ALL TOGETHER
MINING MANGANESE NODULES

You may recall that minerals such as manganese, niobium, and rare earth metals can be classified as strategically important for a country if they are widely used in manufacturing and their continuous supply is important to the economy. One of the keys to maintaining a steady supply of strategically important minerals is finding diverse sources. When production of a mineral is controlled by just a few countries, political choices and trade negotiations can threaten other nations' supply. Finding alternative sources allows demand to shift to another source if acquiring the mineral from one source becomes problematic.

One option for diversifying the sourcing of strategically important minerals is large-scale mining of manganese nodules from the deep seafloor. Much of the Pacific Ocean floor is covered with manganese nodules (Figure 6-23). Shaped something like a potato, these rocks range from golf-ball size to bowling-ball size and are rich in manganese and other metals. At least 60 different elements have been identified in manganese nodules, several of which are metals with significant commercial applications. Scientists believe these metals are probably introduced into seawater by volcanic activity at mid-ocean ridges. Specialized bacteria and algae on the seafloor precipitate minerals from seawater, producing a thin film that is deposited on the seafloor. This process eventually forms a small nodule that continues to grow as more layers are added.

In this activity, you will research manganese nodules and the potential for obtaining manganese and rare earth metals from them. Work with one or two other students to complete the steps below.

1. Find several reliable sources on manganese nodules.

2. Develop three to four research questions to focus your research. The following are a few suggestions:
   - What metals can manganese nodules contain? What are their uses?
   - What steps do people need to take to successfully gather and use the manganese nodules?
   - How do the locations of manganese nodules affect the cost of obtaining them?
   - What are the advantages and disadvantages of mining manganese nodules?
   - How might increasing demand or rising prices of rare earth metals affect the costs and benefits of mining manganese nodules?
   - How might large-scale mining of manganese nodules from the ocean floor affect the environment?

3. Create a presentation or slideshow using your research findings, and present it in person or online.

A B

FIGURE 6-23
(A) Manganese nodules cover large portions of the seafloor. These are from the central North Pacific Ocean at a depth of 5,000 meters. (B) Close-up of a manganese nodule from the South Pacific Ocean shows the average size compared to a penny.

CHAPTER 6 SUMMARY

6.1 MINERAL RESOURCES

- Useful rocks and minerals are called mineral resources; they include both nonmetallic mineral resources and metals.
- All mineral resources are nonrenewable.
- Ore is rock sufficiently enriched in one or more minerals to be mined profitably; geologists usually use the term to refer to metallic mineral deposits.
- Four types of geologic processes concentrate elements to form ore.
- Magmatic processes form ore as magma solidifies.
- Hydrothermal processes transport and precipitate metals from hot water.
- Sedimentary processes form placer deposits, evaporite deposits, and banded iron formations.
- Weathering removes easily dissolved elements from rocks and minerals, leaving behind residual ore deposits such as bauxite.
- Mineral reserves are the known quantity of ore that can profitably be extracted from a deposit.
- Metal ores and coal are extracted from underground mines and surface mines.

6.2 NONRENEWABLE ENERGY RESOURCES

- Fossil fuels (coal, oil, and natural gas) are important nonrenewable energy resources.
- Peat converts to coal when it is buried and subjected to elevated temperature and pressure.
- Petroleum forms from the remains of organisms that settle to the ocean floor or lake bed and are incorporated into rock.
- Natural gas forms in rock or in an oil reservoir subjected to high temperature, and consequently many oil fields contain a mixture of oil with natural gas floating above the heavier liquid petroleum.
- Nuclear power is expensive, and questions about the safety and disposal of nuclear wastes have diminished its future importance as an energy resource in the United States.

6.3 RENEWABLE ENERGY RESOURCES

- Solar, wind, geothermal, hydroelectric, and biomass fuels are renewable sources of energy.
- Alternative energy resources currently supply a small fraction of our energy needs but have the potential to provide abundant renewable energy.

6.4 CONSERVATION AND ENERGY FOR THE FUTURE

- The quickest and most effective way to decrease energy consumption and to prolong the availability of fossil fuels is to conserve energy.
- There are technical and social solutions to reducing energy consumption.
- Energy production and consumption is shifting from nonrenewable sources to renewable sources.

CHAPTER 6 ASSESSMENT

Review Key Terms
Select the key term that best fits the definition. Not all terms will be used, and no term will be used more than once.

biomass energy	mineral reserve
biomass fuel	mineral resource
coal	natural gas
energy resource	nonrenewable resource
fossil fuel	nuclear fuel
geothermal energy	petroleum
hydraulic fracturing	renewable resource
hydrocarbon	solar cell
hydroelectric energy	

1. any naturally occurring substance or phenomenon that can be used to generate useful heat, light, and/or electricity
2. any useful substance or phenomenon that can be replenished by natural processes faster than it is consumed
3. any naturally occurring inorganic solid that is useful to humans and can be reasonably extracted at a profit now or in the future
4. energy produced by the burning of organic material
5. energy resource that forms when crude oil is heated above 100°C during burial
6. a radioactive substance that is used to generate electricity
7. the known amount of mineral deposits that can be profitably extracted from Earth
8. heat extracted from Earth's interior and used to generate heat and electricity at the surface.
9. any organic, combustible compound that is composed of only hydrogen and carbon atoms
10. process by which fluids are injected into a dense rock in order to expand naturally occurring fractures so that trapped natural gas and oil can flow more freely
11. a substance of plant or animal origin that can be burned or processed to generate energy
12. any fuel derived from ancient organic remains and formed by geological processes
13. a device that converts sunlight to electricity
14. any resource that cannot be replenished by natural processes as quickly as it is consumed
15. power generated by converting the energy of flowing water to electricity
16. a complex mixture of different organic compounds that can be refined to manufacture gasoline, plastics, nylon, and other materials

Review Key Concepts
Answer each question on a separate sheet of paper to demonstrate your understanding of key concepts from the chapter.

17. Describe how human use of minerals has changed over time.
18. Describe the two processes through which metal ore deposits are formed.
19. Compare and contrast mineral resources and mineral reserves. Give an example of how each can increase or decrease.
20. Identify two types of mineral resources and give examples of everyday goods made from each type.
21. Uranium is abundant and able to produce enormous amounts of energy by nuclear fission. Why then is nuclear energy not a primary energy source in the United States?
22. Compare the advantages and disadvantages of solar, wind, and geothermal power.
23. Explain the disadvantages of using wind and sunlight to replace all fossil fuels in the production of useful power.
24. Summarize how the use of renewable and nonrenewable resources for energy production is predicted to change in the future.

Think Critically
Write a response to each question on a separate sheet of paper. Use concepts from the chapter to support your reasoning.

25. Describe at least one technological solution or one social solution not mentioned in the text that can help decrease the use of fossil fuels.
26. The Arabian Peninsula is a mostly desert region with some of the largest conventional oil and natural gas fields in the world. What can geologists deduce about the geologic history of this region based on the abundant fossil fuel resources available?

PERFORMANCE TASK

Evaluate the Energy Balance of Power Generation Systems

An energy balance analysis compares the energy input of a system to the system's output, or the amount of useful energy it produces. All power generation requires energy inputs. For example, the extraction and processing of coal requires energy. Transporting coal by rail or truck to power plants also requires energy. Similarly, the production of solar and wind energy requires energy to mine minerals, convert them to useful materials, and manufacture them into solar cells and wind turbines. Energy is also required to operate and maintain a power plant once it is constructed. The table summarizes the energy inputs included in the analysis that you will carry out.

One way to express the energetic costs and gains is the Energy Return On Investment (EROI). This measure is a ratio of power output to the sum of energy inputs. For example, a system with an EROI of 50 means that the system's energy output is 50 times greater than the energy input. To calculate the EROI of a system, divide the expected lifetime energy output of the system by the sum of all of the energy inputs. A system with a high EROI is much more efficient than one with a lower EROI.

In this performance task, you will evaluate the EROI for five different power sources.

Category	Energy Inputs
Fuel	Mining and manufacture of materials consumed in drilling and mining of fuel
	Extraction, processing and transport of fuel
Power plant construction	Mining and manufacture of materials consumed in constructing facilities
	Fuels consumed in constructing facility and access roads
Power plant operations	Energy used during operation and maintenance of facilities

Materials
calculator paper
pen or pencil

1. Use the data in the table to calculate the EROI for each power source. Record your answers and use them to answer the questions that follow.
2. Which power system is most energy-efficient? Which is least efficient? Describe how you determined your answers.
3. How does EROI relate to environmental impacts?
4. Compare the costs and benefits of fossil fuel systems and renewable energy.
5. Explain how changes in resource availability and technological advancements may affect power system EROIs.

Power Source	Fuel Cycle (GWh)	Operation and Maintenance (GWh)	Construction and Decommissioning (GWh)	Lifetime Power Output
Coal	4,028	2,056	573	190,875
Natural Gas	7,250	71	139	215,250
Nuclear	5,222	1,917	1,444	643,200
Wind	0	0	4	60
Solar thermal	0	0	603	2,250

Source: Weissbach et al. 2013. Energy intensities, EROIs, and energy payback times of electricity generating power plants. *Energy*. doi: 10.1016/j.energy.2013.01.029

UNIT 3 | INTERNAL PROCESSES

Chapter 7 **Plate Tectonics**
Chapter 8 **Earthquakes and Earth's Structure**
Chapter 9 **Volcanoes**
Chapter 10 **Mountains**
Chapter 11 **The Seafloor**

Earth's surface, on which all life depends, is continually and imperceptibly changing. It is also reshaped abruptly and without warning from within. Earthquakes impact communities unexpectedly. Some volcanoes erupt regularly while others become active seemingly without warning. The seafloor contains geologic features as diverse as—and more extreme than—those on land. In this unit, you will learn about physical processes that occur within Earth, resulting in the transformation of the planet's surface.

Pictured nine months after a new volcanic island broke through the surface of the western Pacific Ocean in 2013, the eruption on Nishinoshima continues, and the island grows.

CHAPTER 7
PLATE TECTONICS

If you could time-travel 250 million years into the future, you would find a very different arrangement of land on Earth than exists today. Scientists created this map of Pangaea Proxima, the expected configuration of continents based on data of past and present rates of change in Earth's surface. Earth's landmasses are in constant motion. Continents push together and split apart. Mountains form. Oceans combine. Land is pushed down into Earth's interior, and magma rises up and forms new land. Usually these motions are too slow for us to notice. But every tremor in the ground and every expulsion of lava from a volcano serves as a reminder that while we walk atop crust that is usually stable, molten rock continuously churns deep beneath us. The land up here is not as still as it may seem.

KEY QUESTIONS

7.1 What evidence led to the theory of plate tectonics?

7.2 How do tectonic plates move relative to one another?

7.3 What causes tectonic plates to move?

7.4 How do tectonic movements affect Earth's surface and climate?

EXPLORERS AT WORK

MEASURING TECTONIC MOVEMENTS

WITH NATIONAL GEOGRAPHIC EXPLORER D. SARAH STAMPS

Dr. D. Sarah Stamps and her team have studied the motions of tectonic plates in Madagascar, Tanzania, Kenya, and Uganda. Stamps is an expert on the tectonic activity of the East African Rift System and has studied these tectonic motions to improve our understanding of plate tectonics in East Africa. "Our work entails measuring Earth's surface motions with millimeter precision," says Stamps.

Movements within Earth's crust provide clues as to how Earth's tectonic plates break apart, slide against, and crash into each other. The immense forces behind these movements of Earth's crust usually work slowly and are not commonly noticed in our everyday lives. But events such as earthquakes and volcanic eruptions clue us in to the effects of the forces causing deformation of Earth's surface.

Stamps uses computational modeling to understand the physical processes driving Earth's volcanoes, earthquakes, and non-seismic deformation. "We tackle questions such as: How and why does the Earth move and deform? Why do continents break apart? How do volcanoes influence continental rift initiation?"

Stamps is a geophysicist in the Department of Geosciences at Virginia Tech. Geophysicists measure Earth using physical quantities such as gravity, rotation, or motion. Stamps's research is based on using GNSS/GPS (Global Navigation Satellite System/Global Positioning System) technology to study minute movements of the crust. You may have used a GNSS/GPS-enabled smartphone to monitor your location as it changes during a car trip or to find specific locations. GNSS/GPS relies on measurements using radio signals from satellites in orbit around Earth to help pinpoint an exact location on Earth's surface. A GNSS/GPS-enabled smartphone is typically accurate to within a few meters of your actual location. By using GNSS/GPS and state-of-the-art sensors and cybertools, Stamps's research team can measure movements with *millimeter-level* accuracy.

With new technologies, surface measurements can now be made in real time. Stamps is using these measurements to monitor the active volcano Ol Doinyo Lengai in Tanzania. She is the U.S. leader of the Tanzania Volcano Observatory along with Dr. Elifuraha Saria of Tanzania and Dr. Kang-Hyeun Ji from South Korea.

Stamps hopes the results of her research will provide the local people with advance warning of volcanic eruptions. Another reason she studies Ol Doinyo Lengai is to understand the influence of the volcano on tectonic movements during continental breakup. Perhaps the volcano promotes slip on nearby faults, or the nearby faults may influence volcanic eruptions.

THINKING CRITICALLY

Evaluate How might real-time measurements of the tiniest movements of Earth's tectonic plates benefit people living near volcanoes?

▸ Dr. D. Sarah Stamps installs a high-precision GNSS (Global Navigation Satellite System) sensor in Uganda to monitor movements in Earth's crust.

◂ The photo shows the inside of a GPS box located at the base of the active volcano Ol Doinyo Lengai in Tanzania. Stamp's team uses this device to investigate changes in the shape and position of the ground surface due to tectonic forces associated with the volcano. Her team also uses it to monitor the volcano for changes in the rate and direction of ground movement.

CASE STUDY
THE SPLITTING OF A CONTINENT

FIGURE 7-1
This panoramic view of the East African Rift Valley includes the Rift Valley escarpment on the left and the tip of Lake Natron at the far right.

Looking down from the slopes of the Ol Doinyo Lengai in northern Tanzania, Sarah Stamps could take in a view of the East African Rift Valley like the one shown in Figure 7-1. This valley is actually the eastern branch of the much larger East African Rift System, which consists of multiple rifts, or fissures, running for thousands of kilometers through East Africa. West of Lake Victoria, the western branch of the rift system extends from Uganda southward all the way through Tanzania to Lake Malawi. The lakes in this region are each long and narrow, much like Lake Turkana in Kenya. Many are bordered by tall cliffs and have deep floors, some below sea level.

Each of these features is a product of the splitting and pulling apart of the African continent. As you can see in Figure 7-2, there are three plates that are moving apart in East Africa: (1) the Nubian Plate, (2) the Somalian Plate, and (3) the Arabian Plate. If the rifting continues, the trench could eventually widen enough to form an extended ocean basin separating the Somalian Plate from the rest of the African continent.

As You Read Explore models of Earth's layers to understand how and why plates move. Talk about the effects of plate movement and ask questions about landforms that exist where you live or that have been part of your experience. Find out whether any of them may have resulted from movements in Earth's surface.

🌐 **FIGURE 7-2**
Go online to view a map of East Africa showing how plates are moving apart from each other.

7.1 EARLY HYPOTHESES

Core Ideas and Skills
- Explain the bases of Wegener's continental drift hypothesis.
- Describe how the discovery of Earth's layers supported the concept of moving tectonic plates.
- Explain how data revealed by new technology led to the hypothesis of seafloor spreading.

KEY TERMS

plate tectonics
continental drift
asthenosphere
Mid-Ocean Ridge
normal magnetic polarity
magnetic reversal
seafloor spreading

About five billion years ago, a ball of dust and gas, one among billions in the universe, collapsed into a slowly spinning disc. The inner portion of the disc condensed to form our sun, while the outer parts coalesced to form planets orbiting the sun. Our Earth is one of those planets.

Earth began to form as particles of dust and gas were drawn together by gravity and began to collide. These collisions caused the coalescing particles to become hotter and hotter. Frozen crystals of carbon dioxide, methane, and ammonia melted as the spinning mass—early Earth—slowly heated up. Eventually, ice melted. The young planet grew hotter as asteroids, comets, and other space debris bombarded its surface. Additional heat was released by the decay of radioactive isotopes within its interior. By about 4.6 billion years ago, the planet became hot enough that it melted. Then, as the bombardment slowed down and radioactive isotopes decayed and became less abundant, much of Earth's heat radiated into space and our planet began to cool and solidify.

Today, although Earth's surface has cooled to temperatures that support living organisms, the interior remains hot, both from heat left over from the early melting event and from continued decay of radioactive isotopes. Consequently, Earth becomes hotter with depth. At its center, Earth is close to 7,000°C, similar to the temperature of the sun's surface. This internal heat causes earthquakes, volcanic eruptions, mountain building, and continual movements of the continents and ocean basins. These effects, in turn, profoundly affect our environment—Earth's atmosphere, hydrosphere, and biosphere. Earth's internal heat engine and its effects are described in the theory of **plate tectonics**, a simple theory that provides a unifying framework for understanding the way Earth works and how Earth systems interact to create our environment. The term "tectonics" is taken from the Greek *tektonikos*, meaning "construction."

Like most great scientific theories, the plate tectonics theory developed incrementally over many years, building on earlier observations, hypotheses, and theories. The story illustrates how a scientific theory evolves through the accumulation of evidence and how scientists rely on the work and discoveries of earlier scientists.

Continental Drift Hypothesis

Although the theory of plate tectonics was not developed until the 1960s, it was foreshadowed early in the 20th century by a young German scientist named Alfred Wegener. Wegener noticed that the African and South American coastlines on opposite sides of the Atlantic Ocean seemed to fit as if they were adjacent pieces of a jigsaw puzzle (Figure 7-3). He realized that the apparent

FIGURE 7-3 ▼

The African and South American coastlines appear to fit together like adjacent pieces of a jigsaw puzzle on Wegener's reconstruction map. Several of the unique rock types correlated between the two continents are shown. These include areas of Precambrian stable crust (green), Precambrian and Cambrian mountain belts (blue lines), and the Cape and Sierra de la Ventana Fold Belts of Paleozoic age (orange). The darker brown regions are the continental shelves, which are the actual edges of the continents.

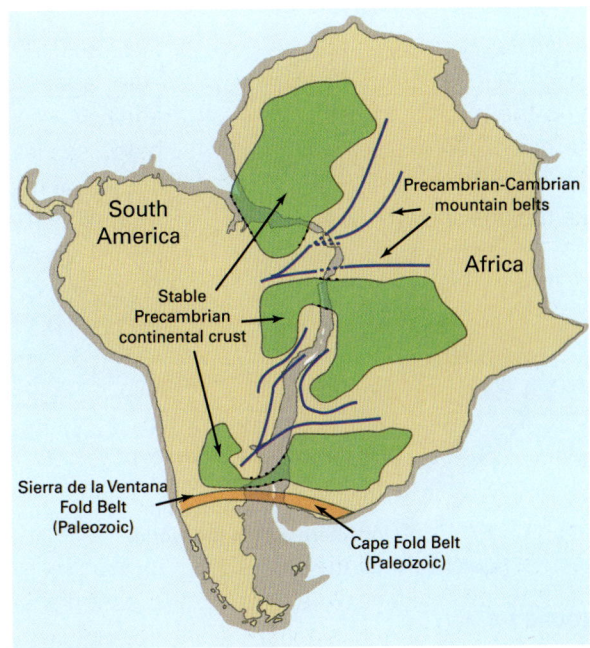

Source: Cengage

fit suggested that the continents had once been joined together and had later separated by thousands of kilometers to form the Atlantic Ocean.

Wegener was not the first to make this observation, but he was the first scientist to pursue it with additional research. Studying world maps and making paper cutouts of each continent that he could move around, Wegener realized that not only did the continents on both sides of the Atlantic fit together, but other continents, when positioned correctly, also fit like pieces of the same jigsaw puzzle (Figure 7-4). On his map, all the continents joined together into one huge landmass he called Pangaea, from the Greek root words for "all lands." The northern part of Pangaea is commonly called Laurasia and the southern part Gondwana.

Wegener understood that the fit of the continents alone did not prove that a single landmass had existed. He began seeking additional evidence in 1910 and continued work on the project until his death in 1930.

One line of evidence Wegener found in support of his hypothesis is the occurrence of uncommon rock types or distinctive sequences of rocks that are identical on both sides of the Atlantic Ocean. When he plotted the distinctive rocks on a map of Pangaea, those currently on the east side of the Atlantic were continuous with their counterparts on the west side (Figure 7-3). For example, the deformed rocks of the Cape Fold Belt of South Africa are similar to rocks found in the Sierra de la Ventana Fold Belt of Argentina. Plotted on a map of Pangaea, the two sequences of rocks appear as a single, continuous belt.

Using fossil evidence to support the existence of Pangaea, Wegener compiled information regarding locations of certain fossilized plant and animal species that would not have been able to survive long oceanic crossings. Today these fossils are found in Antarctica, Africa, Australia, South America, and India, all of which are separated by wide oceans. However, when Wegener plotted the same fossil localities on his reconstruction of

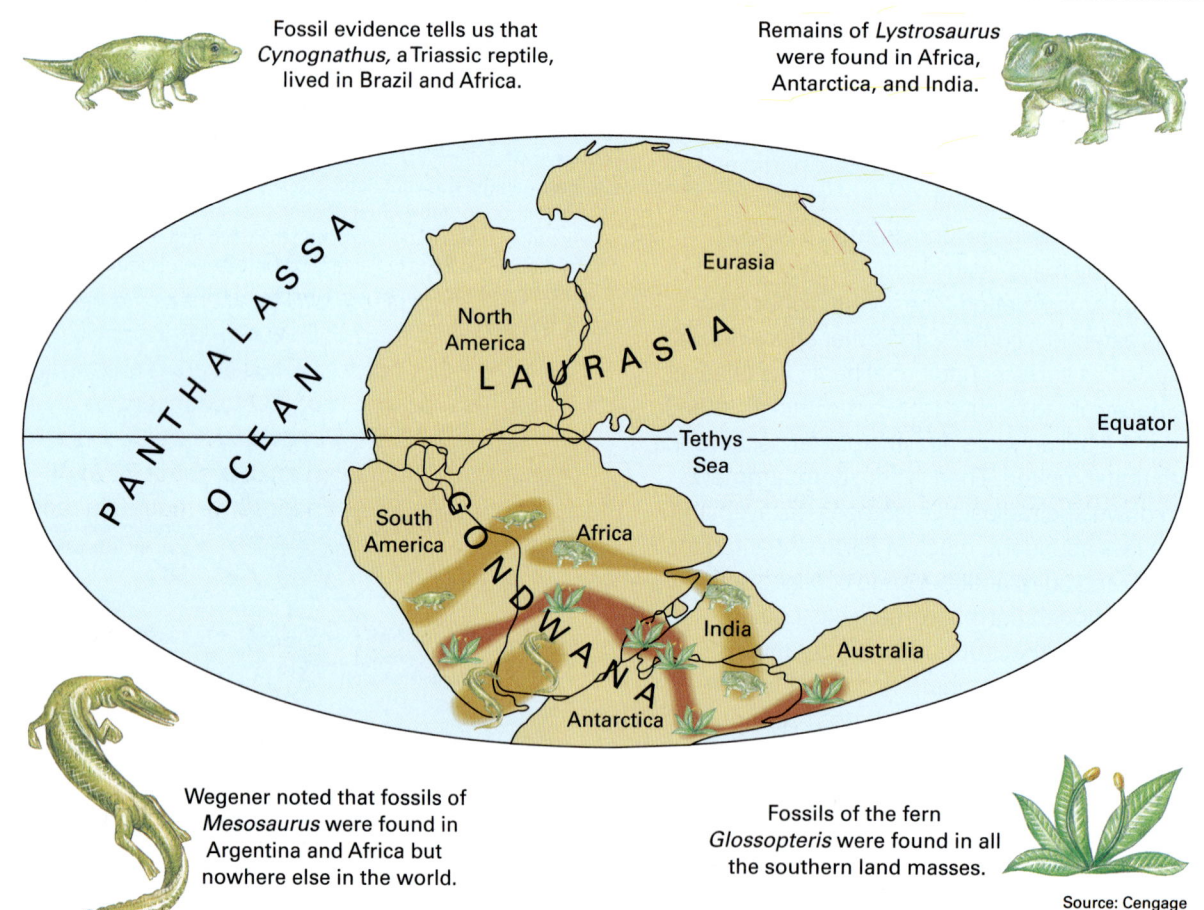

FIGURE 7-4
Geographic distributions of plant and animal fossils on Wegener's map indicate that a single landmass, called Pangaea, existed between about 300 million and 200 million years ago.

Pangaea, he found that they all occurred in the same region (Figure 7-4). Wegener deduced that rather than migrating across the wide oceans that presently separate the different fossil locations, each species had evolved and spread over a portion of Pangaea *before* the landmass broke apart.

Wegener also cited evidence from sedimentary rocks known to form in specific climate zones to support the existence of Pangaea. Glaciers and gravel deposited by glacial ice, for example, form in cold climates and are therefore typically found at high latitudes and high altitudes. Sandstones that preserve the structures of desert sand dunes form where deserts are common, near latitudes 30° north and south. Coral reefs and coal swamps thrive in near-equatorial tropical climates. Each rock type reflects an ancient environment characteristic of a specific latitude.

Wegener plotted 250-million-year-old glacial deposits on a map showing the present-day distribution of continents. If the continents had not moved, the glacial deposits would have to have formed in tropical and subtropical climate zones. In Wegener's reconstruction of Pangaea, the glaciers clustered neatly near the South Pole. Coral reefs and coal both occurred in equatorial positions, as they might today. Additionally, desert environments existed around 30° north, similar to the distribution of deserts today.

Wegener's concept of a single landmass that broke apart to form the modern continents is called **continental drift**. Wegener first presented the framework of his hypothesis in 1912 and published a more thorough treatment in 1915 in the first edition of his book, *The Origin of Continents and Oceans*. Wegener's hypothesis was not well-received at the time. He did not provide any alternative explanations, and his field-based evidence was not quantitative. Additionally, Wegener did not provide any convincing explanation regarding how the continents move. Although Wegener did suggest two possible mechanisms, both were proved impossible by physicists.

Much of the hypothesis of continental drift is similar to modern plate tectonics theory. Modern evidence indicates that the continents were once joined together much as Wegener portrayed them in his map of Pangaea. Today, most geologists recognize Wegener's contributions as foundational to the development of the current theory of plate tectonics.

checkpoint What lines of evidence supported Wegener's continental drift hypothesis?

Discovery of Earth's Layers

After Wegener died and his hypothesis was mostly forgotten, geologists discovered that Earth has layers. They did this by studying the energy released by earthquakes, which travels through Earth as waves. Geologists learned that both the speed and the direction of these waves change abruptly at certain depths, as the waves pass through Earth. They soon realized that these changes reveal that Earth is a layered planet. To consider the theory of plate tectonics, you first need to understand Earth's layers: the crust, mantle, lithosphere, asthenosphere, and core.

The Crust
The crust is the outermost and thinnest layer (Figure 7-5, Table 7-1). Because it is cooler than the layers below, the crust consists of hard, strong rock. Crust beneath the oceans differs from that of continents. Oceanic crust is between 4 and 7 kilometers thick and composed mostly of dark, dense basalt. In contrast, the average thickness of continental crust is about 20 to 40 kilometers, although under some mountain ranges

TABLE 7-1 Composition, Depth, and Other Properties of Earth's Layers

	Layer	Composition	Depth Range (km)	Properties
Lithosphere and crust	Oceanic crust	Basaltic	0–7	Cool, hard, and strong
	Continental crust	Granitic	0–70	
	Remainder of lithosphere	Silicates	Down to 100	
Mantle	Asthenosphere	Mainly peridotite; mineralogy varies with depth	100–350	Hot, weak, and 1 to 2% melted
	Remainder of mantle		350–2,900	Hot, under pressure, plastic
Core	Outer core	Iron and nickel	2,900–5,150	Liquid
	Inner core		5,150–6,370	Solid

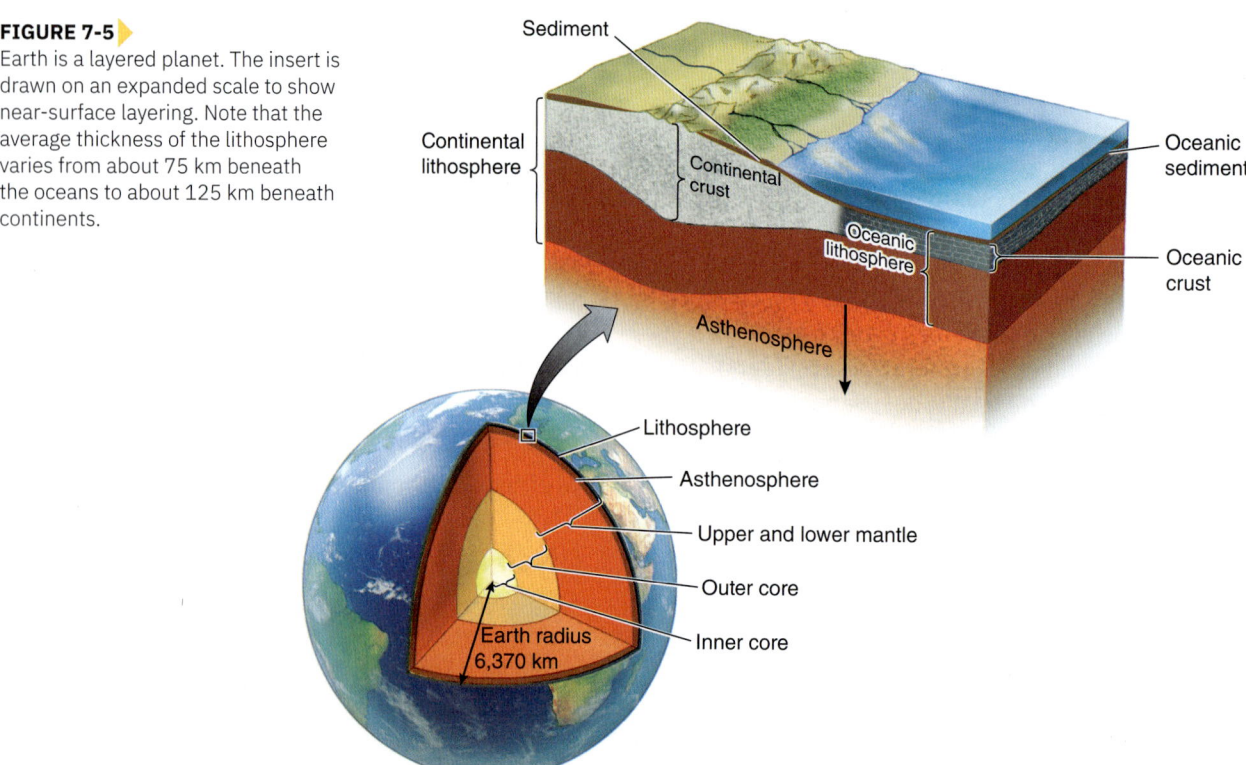

FIGURE 7-5
Earth is a layered planet. The insert is drawn on an expanded scale to show near-surface layering. Note that the average thickness of the lithosphere varies from about 75 km beneath the oceans to about 125 km beneath continents.

Source: Cengage

it can be as much as 70 kilometers thick. The overall composition of continental crust is granitic, making it less dense than oceanic crust.

The Mantle The mantle lies directly below the crust. It is almost 2,900 kilometers thick and makes up 80 percent of Earth's volume. The mantle is composed mainly of peridotite, a rock that is denser than continental or oceanic crust.

Although the chemical composition may not vary much throughout the entire mantle, temperature and pressure increase with depth. Figure 7-6A shows that the temperature at the top of the mantle is near 1,000°C and that near the mantle/core boundary it is about 3,300°C. These changes cause the strength of mantle rock to vary with depth. The differences in strength create layering. Figure 7-6B shows that internal pressure also increases with depth.

Most people understand that increasing temperature will eventually melt a rock. Less obvious, however, is the fact that high pressure *inhibits* melting because rock expands by about 10 percent when it melts. High pressure makes it more difficult for a rock to expand and therefore impedes melting. If the combined effects of temperature and pressure are close to—but just below—a rock's melting point, the rock remains solid but loses strength, so it becomes weak and plastic. In such a weakened state, rock can flow slowly, similar to the way cold honey spills from a jar. At that point, if temperature rises and pressure decreases, the rock will begin to melt.

The Lithosphere Figure 7-6 shows that the uppermost mantle is cool and its pressure is low, conditions similar to those in the crust. Both factors combine to produce hard, strong rock similar to that of the crust. Recall from Chapter 2 that the outer part of Earth, including both the crust and the uppermost mantle, make up the lithosphere. The average thickness of the lithosphere is about 100 kilometers but ranges from about 75 kilometers beneath ocean basins to about 125 kilometers under the continents (Figure 7-5, Table 7-1). The lithosphere, then, consists mostly of the cold, strong uppermost mantle; the crust is just a thin layer of buoyant rock forming the top of the lithosphere.

The Asthenosphere At a depth varying from 75 to 125 kilometers beneath Earth's surface, the temperature and pressure conditions are close to the melting point of mantle rock. As a result, at this depth the mantle abruptly loses strength relative to the overlying rock and becomes weak and plastic (Figure 7-6). About 1 to 2 percent of the rock melts, although the rest remains solid. This weak, plastic, and partly molten character extends to a depth of about 350 kilometers, where increasing pressure overwhelms

DATA ANALYSIS Interpret Temperature and Pressure Graphs

At the center of Earth, the temperature is close to 7,000°C. For reference, the temperature of an oven baking a cake is about 175°C and a heated steel bar turns cherry red at 746°C.

1. **Summarize** Use Figure 7-6A to summarize the temperature ranges in the mantle, outer core, and inner core. Which has the largest range of temperature?

Computational Thinking

2. **Recognize Patterns** Examine Figure 7-6B. To what degree does pressure increase or decrease with respect to depth? Explain your response by referencing the graph's line and values.

3. **Generalize Patterns** Based on Figure 7-6A, offer a statement that correlates temperature with depth.

4. **Represent Data** Express the relationship between the thickness of the continental lithosphere, the asthenosphere, and the mantle using a three-term ratio.

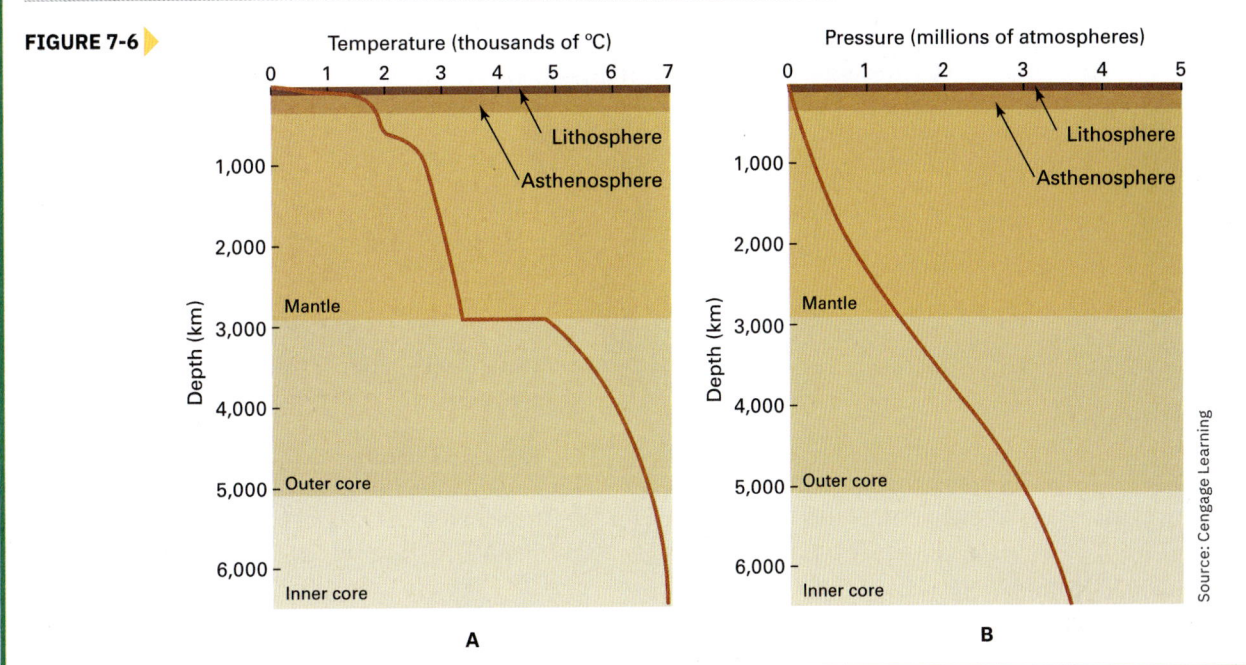

FIGURE 7-6

temperature and the rock becomes stronger again. This layer of weak mantle rock extending from about 100 to 350 kilometers deep is the **asthenosphere** (from the Greek for "weak layer").

The average temperature in the asthenosphere is about 1,800°C, although the temperature increases with depth as it does in other Earth layers. Pressure in the asthenosphere rises from about 35,000 atmospheres near the top to about 118,000 atmospheres at the base. One atmosphere is equal to the weight of Earth's atmosphere at sea level, which is about 14.6 pounds per square inch. That means the pressure near the top of the asthenosphere is equivalent to more than 500,000 pounds of weight per square inch. Pressure at the base of the asthenosphere is about 1.7 million pounds per square inch.

If you apply force to a plastic solid, it deforms slowly. The soft, plastic rock of the asthenosphere behaves in this way, relative to the strong, hard lithosphere that lies on top of it. The lithosphere is not rigidly supported by the rock beneath it, but instead floats on the soft, plastic asthenosphere. The concept of a floating lithosphere is important to our understanding of plate tectonics and Earth's internal processes.

At the base of the asthenosphere, the increasing pressure overcomes the effect of rising temperature, and the strength of the mantle increases again (Figure 7-6). Although the mantle below 350 kilometers is stronger than the asthenosphere, it is not as strong as the lithosphere, but rather is plastic and capable of flowing slowly over geologic time.

LESSON 7.1

FIGURE 7-7
In Iceland's Thingvellir National Park, a diver explores a large vertical fracture. This region forms the boundary between the North American and Eurasian plates, which are being pulled apart.

The Core The core is the innermost of Earth's layers. It is a sphere with a radius of about 3,470 kilometers, about the same size as Mars, and is composed largely of iron and nickel. The pressure in the outer core is intense. However, the rock is still molten because of the extremely high temperature. Temperatures approach 7,000°C in the inner core and the pressure is 3.5 million times that of Earth's atmosphere at sea level. This extreme pressure compresses the inner core to a solid.

checkpoint How does pressure make it possible for the asthenosphere to flow?

Seafloor Spreading Hypothesis

Studies of the seafloor that began in the mid-1900s led to another important line of evidence that supports the theory of plate tectonics. As they mapped the seafloor, oceanographers discovered the largest mountain chain on Earth, now called the **Mid-Ocean Ridge** system. It circles the planet like the seam on a baseball. One branch of this huge submarine mountain range, called the Mid-Atlantic Ridge, lies directly in the middle of the Atlantic Ocean (Figure 7-7).

Oceanic crust is composed mostly of basalt, an igneous rock rich in iron. As basaltic lava cools and forms solid rock, the iron-rich mineral crystals in the basalt operate like weak magnets. The magnetic fields of these minerals align parallel to Earth's magnetic field. The basalt preserves a record of the orientation and strength of Earth's magnetic field at the time the rock cools.

By towing devices called magnetometers behind their research vessels, oceanographers were able to detect and record magnetic patterns in the basalt forming the deep-ocean floors. Figure 7-8 shows the magnetic orientations of seafloor rocks near a part of the Mid-Atlantic Ridge southwest of Iceland. In this figure, purple stripes represent basalt with a magnetic orientation parallel to Earth's current magnetic field, called **normal magnetic polarity**. The stripes form a symmetric pattern of normal and reversed polarity about the axis of the ridge.

Why do the seafloor rocks have alternating normal and reversed polarity, and why is the pattern symmetrically distributed across the Mid-Ocean Ridge? In the mid-1960s, three scientists—Frederick Vine, Drummond Matthews and Lawrence Morley—proposed an explanation for these odd magnetic patterns on the seafloor. They knew that other scientists had been studying the magnetism

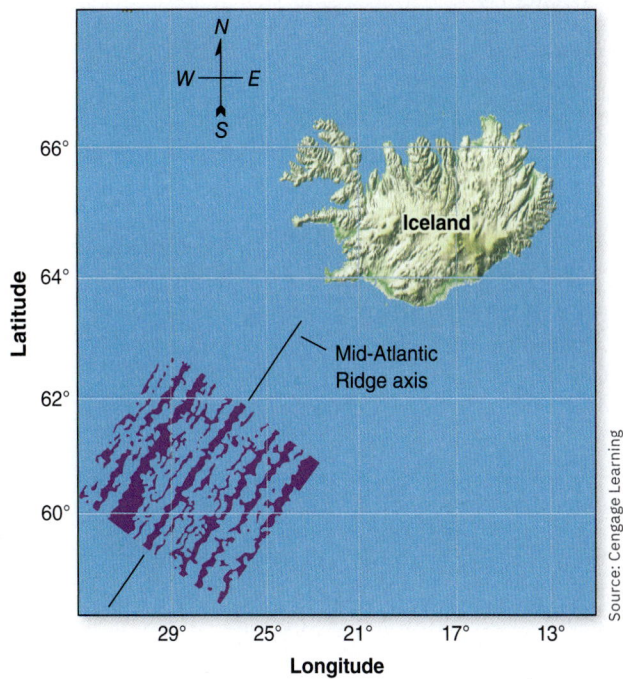

FIGURE 7-8
The Mid-Atlantic Ridge runs through Iceland. Magnetic orientation of seafloor rocks near the ridge is shown in the lower-left portion of the map. Alternating purple and blue stripes show normal and reversed magnetic polarity in the rocks.

preserved in layers of basalt and discovered that Earth's magnetic field has reversed its polarity on the average of every 500,000 years during the past 65 million years. The data from Hawai'i indicated that when a **magnetic reversal** of Earth's field occurs, the north magnetic pole becomes the south magnetic pole, and vice versa.

Vine, Matthews, and Morley suggested that the symmetrical magnetic stripes they observed in the seafloor were produced by the continuous spreading of newly formed oceanic crust away from the Mid-Ocean Ridge, like two conveyor belts moving away from each other (Figure 7-9). They recognized that the seafloor and oceanic crust become older with increasing distance from the ridge axis. New basalt lava rises through cracks that form at the ridge axis as the two sides of the seafloor separate. As the lava cools and solidifies, the basalt records the strength and orientation of Earth's field. Because Earth's field periodically reverses, the magnetism preserved in the basalt of the ocean floor acquires a striped pattern.

At the same time as these seafloor magnetic patterns were being detected and explained, oceanographers discovered that the layer of mud overlying the seafloor basalt in most parts of the oceans typically is thinnest at the Mid-Ocean Ridge

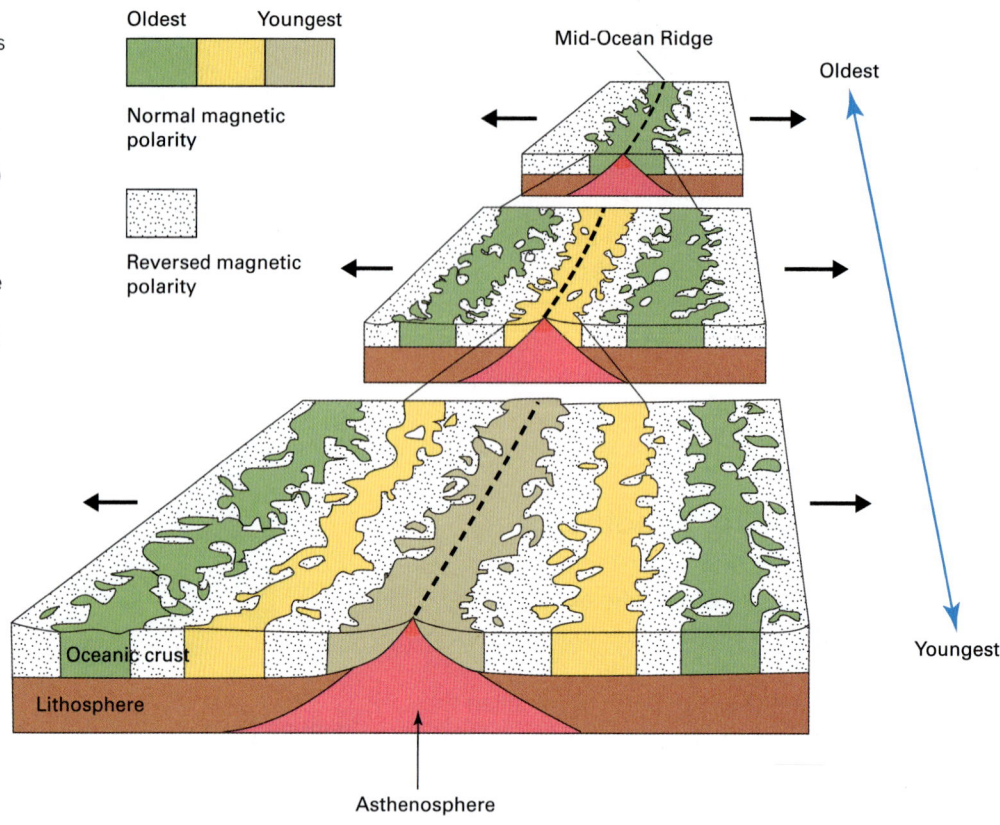

FIGURE 7-9 As new oceanic crust cools at the Mid-Ocean Ridge, it acquires the magnetic orientation of Earth's field. Notice the alternating stripes of normal (colored) and reversed (dotted) polarity. The pattern of reversals is a record of Earth's magnetic field. The three frames show the evolution of the spreading center through time from oldest to youngest.

and becomes progressively thicker with distance from the ridge. They reasoned that if mud settles onto the seafloor at the same rate everywhere, the mud layer would be thinnest over the newest part of the seafloor. Because oceanic crust is progressively older with increased distance from the ridge axis, more time has elapsed for mud to accumulate, so the layer of mud becomes progressively thicker.

The oceanographers also found that fossils in the deepest layers of mud overlying basalt are very young at the ridge axis but become progressively older with increasing distance from the ridge. This discovery, too, indicated that the seafloor becomes older with increasing distance from the ridge axis.

Symmetrical magnetic patterns and similar mud age and thickness trends were quickly discovered along other parts of the Mid-Ocean Ridge system and in other ocean basins. Therefore, the hypothesis of **seafloor spreading** was proposed as a general model for the origin of all oceanic crust. In a very few years, the seafloor spreading hypothesis became the basis for development of the much broader theory of plate tectonics.

checkpoint How did stripes of magnetic polarity support the seafloor spreading hypothesis?

7.1 ASSESSMENT

1. **Use Evidence** Explain how Wegener supported his continental drift hypothesis with fossil evidence.
2. **Explain** Why were Wegener's ideas largely dismissed until the 1960s?
3. **Describe** How does Earth's interior structure lend support to the concept of a floating lithosphere?
4. **Relate** Explain how technology led to the hypothesis of seafloor spreading.

Computational Thinking

5. **Abstract Information** Use one or more analogies to describe how Earth's layers result in a moving lithosphere.

Critical Thinking

6. **Synthesize** Use the findings of Vine, Matthews, and Morley to defend the claim that it is important for geologists to have a background in physical science.

7.2 THE THEORY OF PLATE TECTONICS

Core Ideas and Skills
- Describe the modern theory of plate tectonics.
- Differentiate the three types of plate boundaries.
- Infer how a mountain range formed based on its location along plate boundaries.
- Use plate tectonics to explain ages of crustal rocks

KEY TERMS

plate boundary	transform boundary
divergent boundary	continental rifting
convergent boundary	subduction

Plate Boundaries

Like many great unifying scientific ideas, the plate tectonics theory is simple. Briefly, it states that the lithosphere is a shell of hard, strong rock that floats on the hot, plastic asthenosphere (Figure 7-10). The lithosphere consists of eight large (and several smaller) segments called tectonic plates (Figure 7-11). They are also called lithospheric plates, or simply plates, the terms are interchangeable. The tectonic plates slide slowly over the asthenosphere at rates ranging from less than 1 to about 16 centimeters per year. Continents and ocean basins make up the upper parts of the lithospheric plates, so as the plates slide over the asthenosphere, the continents and oceans move with them.

A **plate boundary** is a fracture that separates one plate from another. Neighboring plates can move relative to one another at these boundaries in three ways, shown by the insets in Figure 7-11. At a **divergent boundary**, two plates move apart from each other; at a **convergent boundary**, two plates move toward each other; and at a **transform boundary**, two plates slide horizontally past each other.

The great forces generated at plate boundaries build mountain ranges, cause earthquakes, and produce many of Earth's volcanoes. In contrast, the interior portions of plates usually are tectonically quiet because they are further from the zones where two plates interact.

checkpoint What are the three types of plate boundaries?

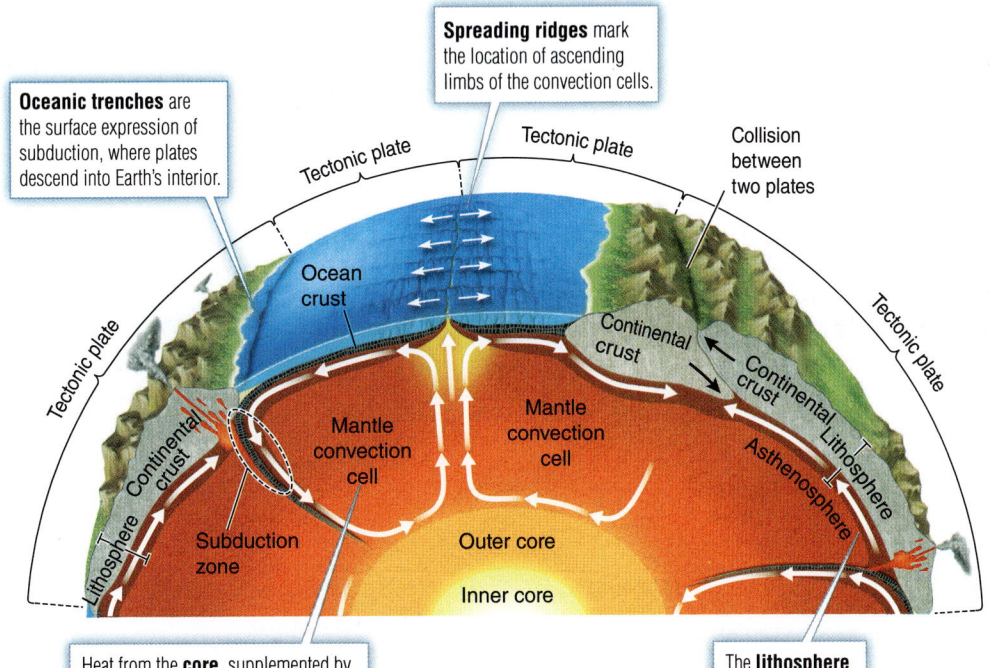

FIGURE 7-10
A cutaway view of Earth shows that the lithosphere glides horizontally across the asthenosphere at rates of a few centimeters each year. The top of the lithosphere includes the crust. Below the crust, the asthenosphere and the deeper mantle circulate in convection cells. In this illustration, the circulating cells involve the entire mantle. Some geologists have suggested that mantle convection may involve two layers of shallow and deep convection. The thickness of the lithosphere is exaggerated here for clarity.

LESSON 7.2 **205**

Divergent Plate Boundaries

At a divergent plate boundary, also called a spreading center, two plates spread apart from one another, as shown at the center of Figure 7-11. The underlying asthenosphere rises upward to fill the gap between the separating plates. As it rises, the decrease in pressure causes the hot asthenosphere to melt and form magma. As this magma continues to rise, it cools to form new crust. Most of this activity occurs at divergent plate boundaries within the ocean basins, but it also can occur between two continental plates that are rifting apart, as in East Africa. As the asthenosphere rises closer to Earth's surface between two separating plates, it cools, gains mechanical strength, and transforms into new lithosphere. In this way, new lithosphere continuously forms at a divergent boundary.

At a divergent boundary, the rising asthenosphere is hot, weak, and plastic. Only the upper 10 to 15 kilometers cools enough to gain the strength and hardness of lithosphere rock. As a result, the lithosphere rock, including the crust and the upper few kilometers of mantle rock, can be as little as 10 or 15 kilometers thick at a spreading center. But as the lithosphere spreads, it cools from the top downward and thickens (Figure 7-12).

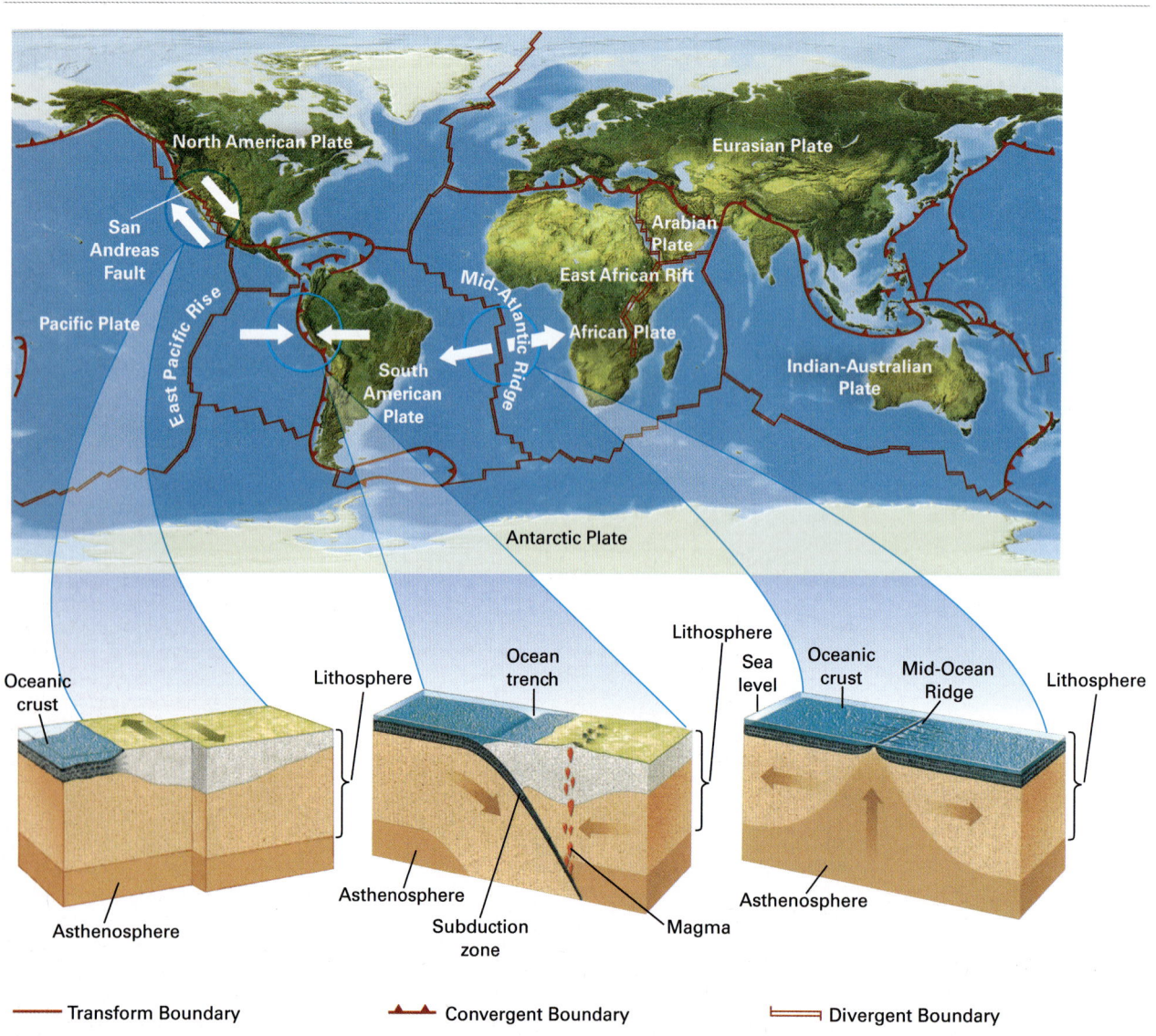

FIGURE 7-11
Earth's lithosphere is broken into eight large tectonic plates, called the African, Arabian, Eurasian, Indian-Australian, Antarctic, Pacific, North American, and South American plates. The white arrows show how the plates move in different directions. The three different types of plate boundaries are shown below the map.

206 CHAPTER 7 PLATE TECTONICS

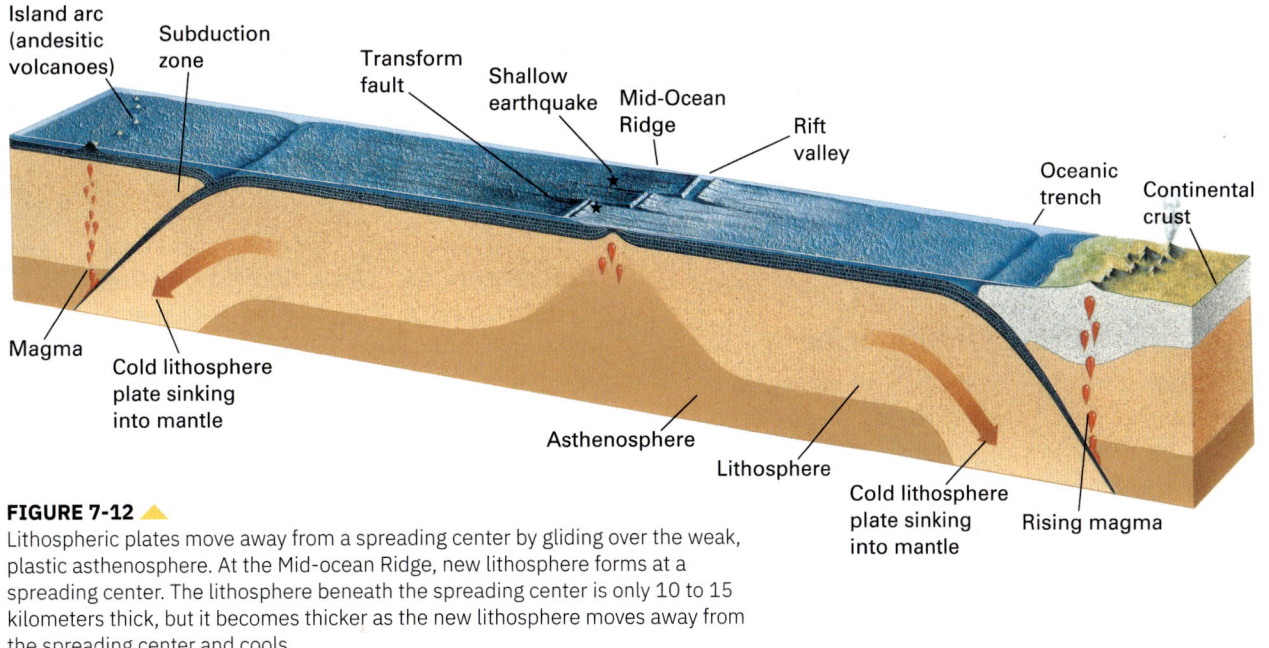

FIGURE 7-12
Lithospheric plates move away from a spreading center by gliding over the weak, plastic asthenosphere. At the Mid-ocean Ridge, new lithosphere forms at a spreading center. The lithosphere beneath the spreading center is only 10 to 15 kilometers thick, but it becomes thicker as the new lithosphere moves away from the spreading center and cools.

As it spreads outward and cools, the new lithosphere also thickens because the boundary between cool rock and hot rock migrates downward. Consequently, the thickness of the lithosphere increases as it moves away from the spreading center. Think of ice freezing on a pond. On a cold day, water under the ice freezes and the ice becomes thicker. In a similar fashion, the cooling lithosphere thickens to about 75 kilometers beneath ocean basins and to about 125 kilometers beneath continents.

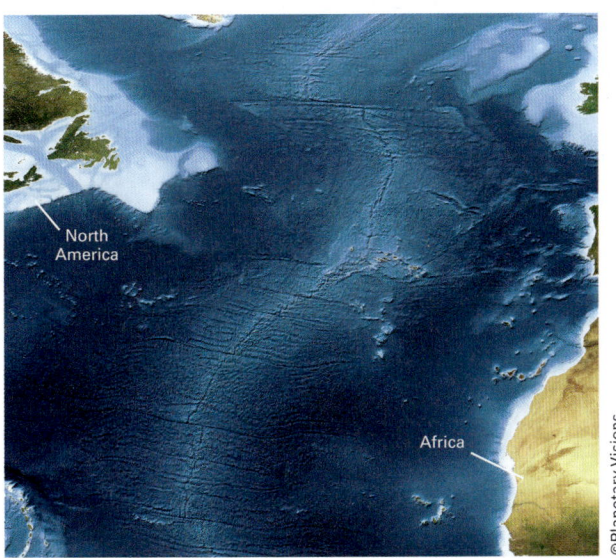

FIGURE 7-13
The Mid-Atlantic Ridge lies in the middle of the Atlantic Ocean, halfway between North America and South America to the west, and Europe and Africa to the east.

The Mid-Ocean Ridge: Rifting in the Oceans
New lithosphere at an oceanic spreading center is hotter and therefore less dense than older lithospheric rock farther away from the divergent boundary. Therefore, it floats to a higher level, forming the undersea mountain chain called the Mid-Ocean Ridge system (Figure 7-9). But as lithosphere migrates away from a spreading center, it cools and becomes denser. As a result, the lithosphere sinks into the plastic asthenosphere (Figure 7-12), causing the depth of the sea floor to increase with distance from the Mid-Ocean Ridge.

Splitting Continents: Rifting in Continental Crust
A divergent plate boundary can split apart continental crust in a process called **continental rifting**. A rift valley develops in a continental rift because continental crust stretches and fractures. Basaltic magma from the base of the crust typically wells up through the fractures. Some of the basaltic magma reaches the surface to form volcanoes. As the basalt magma works its way through the granitic continental crust, it partially melts it to form additional magma that is rhyolitic in composition. Typically, both basalt magma and rhyolite magma erupt from evolving continental rifts.

As the continental crust continues to stretch and fracture, and as more basalt intrudes through or erupts onto it, the surface of the rift begins to sink. As the process continues, eventually the continental crust separates into two sides with a zone of much thinner basaltic crust in between. Continued rifting

LESSON 7.2 **207**

along the same zone causes the two sides of the continent to continue to move apart. New oceanic crust forms in the middle of the new ocean basin. With time, a narrow oceanic gulf like the Red Sea can spread to become a major ocean.

checkpoint What geologic features form at divergent plate boundaries?

Convergent Plate Boundaries

At a convergent plate boundary, two lithospheric plates move toward each other. Not all lithospheric plates are made of equally dense rock. Where two plates of different densities converge, the denser plate sinks into the mantle beneath the less dense one. This sinking process is called **subduction**. This process is shown on both the far left side and far right side of Figure 7-12. A subduction zone is a long, narrow belt where a lithospheric plate is sinking into the mantle. On a worldwide scale, the rate at which old lithosphere sinks into the mantle at subduction zones is equal to the rate at which new lithosphere forms at spreading centers. In this way, Earth maintains a global balance between the creation of new lithosphere and the destruction of old lithosphere.

Plate convergence can occur (1) between a plate carrying oceanic crust and another carrying continental crust, (2) between two plates carrying oceanic crust, and (3) between two plates carrying continental crust.

Convergence of Oceanic and Continental Crusts Recall that oceanic crust is generally denser than continental crust. In fact, the entire lithosphere beneath the oceans is denser than continental lithosphere. When an oceanic plate converges with a continental plate, subduction occurs and the denser oceanic plate plunges into the mantle beneath the edge of the continent. As a result, many subduction zones are located at continental margins.

Today, oceanic plates are subducting beneath the western edge of South America; along the coasts of Oregon, Washington, and British Columbia; and at several other continental margins shown in Figure 7-11. When the descending plate reaches the asthenosphere, large quantities of magma are generated. Magma rises through the lithosphere of the overriding plate; much of the magma

TABLE 7-2 Characteristics and Examples of Plate Boundaries

Type of Boundary	Types of Plates Involved	Topography	Geologic Events	Modern Examples
Divergent	Ocean–ocean	Mid-Ocean Ridge	Seafloor spreading, shallow earthquakes, rising magma, volcanoes	Mid-Atlantic Ridge
	Continent–continent	Rift valley	Continents torn apart, earthquakes, rising magma, volcanoes	East African Rift
Convergent	Ocean–ocean	Island arcs and ocean trenches	Subduction, deep earthquakes, rising magma, volcanoes, deformation of rocks	Western Aleutians
	Ocean–continent	Mountains and ocean trenches	Subduction, deep earthquakes, rising magma, volcanoes, deformation of rocks	Andes
	Continent–continent	Mountains	Deep earthquakes, deformation of rocks	Himalayan Mountains
Transform	Ocean–ocean	Major offset of Mid-Ocean Ridge axis	Earthquakes	Offset of East Pacific Rise
	Ocean–continent	Linear, deformed mountain ranges	Earthquakes, deformation of rocks	Northern portion of San Andreas Fault
	Continent–continent	Linear, deformed mountain ranges	Earthquakes, deformation of rocks	Southern San Andreas Fault

reaches the surface, where it erupts from a chain of volcanoes that form parallel to the subduction zone. The Andes—a chain of volcanic mountains—formed as a result of the subduction of the Nazca plate beneath the South American plate (Figure 7-14).

FIGURE 7-14 Go online to view a map showing the Peru-Chile trench, where the Nazca plate is subducting beneath the South American plate.

The oldest seafloor rocks on Earth are only about 200 million years old because oceanic crust is being destroyed continuously where it subducts and is melted. In contrast, rocks older than four billion years are found on continents because subduction consumes little continental crust.

Convergence of Oceanic Crust Newly formed oceanic lithosphere is hot, thin, and of relatively low density, but as it spreads away from the Mid-Ocean Ridge, it becomes older, cooler, thicker, and denser. The density of oceanic lithosphere increases with its age. When two oceanic plates converge, the older, denser one subducts into the mantle. Oceanic subduction zones are common along the southwest edge of the Pacific Ocean Plate as well as the northwest edge (Figure 7-11), forming the Aleutian Islands in Alaska.

Convergence of Continental Crust If two converging plates carry continents, the relatively low density of the continental lithosphere prevents either plate from subducting deeply into the mantle. Continental lithosphere does not normally sink into the mantle at a subduction zone for two reasons. First, continental crust generally is too thick to subduct. Second, continental crust resists subduction for the same reason a log does not sink into a lake: It has lower density than the material beneath it.

Rather, when two plates with continental lithosphere collide, they crumple against each other and form a huge mountain chain. The Himalaya, Alps, and Appalachians all formed as a result of continental crust convergence (Figure 7-15 and Figure 7-16).

checkpoint What geologic features form at convergent plate boundaries?

Transform Plate Boundaries

A transform plate boundary forms where two plates slide horizontally past one another as they move in opposite directions (Figure 7-11). This type of boundary can occur in both oceans and continents and can result in frequent earthquakes. California's San Andreas Fault is a transform boundary between the North American plate and the Pacific plate.

Table 7-2 summarizes characteristics and examples of the three types of plate boundaries.

checkpoint What is a transform boundary and what type of geologic event occurs there?

7.2 ASSESSMENT

1. **Summarize** Give three main points of the theory of plate tectonics.
2. **Describe** How did the Mid-Atlantic Ridge form?
3. **Compare and Contrast** How are the geologic events that occur at transform boundaries similar to and different from those at convergent boundaries? What accounts for the differences?
4. **Explain** Use the concept of density to explain why subduction is more likely to occur at the convergence of an oceanic plate and a continental plate than at the convergence of two continental plates.

Critical Thinking

5. **Infer** The Caucasus Mountains in western Asia extend along the boundary between the Eurasian and Arabian tectonic plates. Describe what type of plate boundary likely formed the mountains and how they formed.
6. **Apply** The Rio Grande Valley is a rift valley that formed roughly 30 million years ago in what is now South Texas. Explain why rock samples found near the Rio Grande Valley are younger than those located farther away.

FIGURE 7-15
The summit of Gokyo Ri in eastern Nepal is renowned for its spectacular views of the Himalaya, a vast mountain range resulting from the convergence of the Eurasian and Indian plates. Ten of the planet's 14 peaks above 8,000 meters occur in this mountain system.

FIGURE 7-16
The Indian plate continues to move northward at a rate of about 4 to 6 centimeters each year, forcing the Himalaya upward at a rate of about 1 centimeter per year.

7.3 PLATE MOVEMENT

Core Ideas and Skills
- Explain how density within Earth's interior influences the movement of tectonic plates.
- Model the process of mantle convection.
- Determine past and future positions of land masses based on present data.
- Apply the concept of isostasy to plate movement and crustal thickness.

KEY TERMS

convection
mantle plume
hot spot
supercontinent
isostasy

After geologists had developed the theory of plate tectonics, they began to ask, "*Why* do the great slabs of lithosphere move?" Research has shown that subduction can continue slowly all the way to the core–mantle boundary, to a depth of 2,900 kilometers. At the same time, hot rock rises from the deep mantle toward the surface to replace the lithosphere lost to subduction.

Why Plates Move

Convection is the transfer of thermal energy resulting from the circulating flow of fluids (gases or liquids) in response to heating and cooling. The process of mantle convection continually stirs the entire mantle as rock that is hotter than its surroundings rises toward Earth's surface and old plates that are colder than their surroundings sink into the mantle. In this way, the entire mantle–lithosphere system slowly circulates in cells that carry rock from as deep as the core–mantle boundary toward Earth's surface and then back into the deepest mantle (Figure 7-10).

A pot of soup on a hot stove illustrates the process of convection. Rising temperature causes most materials, including soup (or rock), to expand. When soup at the bottom of the pot is heated by the underlying stove, it becomes warm and expands. It then rises because it is less dense than the soup at the top. When the hot soup reaches the top of the pot, it flows along the surface until it cools and sinks (Figure 7-17).

Although the circulation of soup in a hot pot provides a good example of convection, the

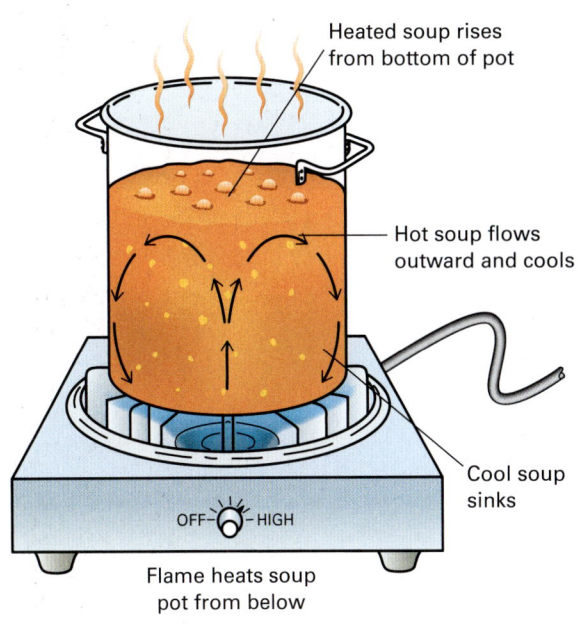

FIGURE 7-17
Soup convects when it is heated from the bottom of the pot.

circulation of the mantle is much more complex. The mantle is not shaped like a pot and the circulation of the mantle is driven by three processes: heat from the core below, radioactive decay of unstable isotopes within the mantle, and the cooling of the upper mantle in direct contact with the lithosphere.

Today, the specifics of mantle convection are not well understood, although there is general agreement that mantle convection and plate tectonics are part of the same system and that mantle convection is the main mechanism for transport of heat away from Earth's interior to the surface. What is less certain is the structure of the circulating mantle rock. For example, upwelling of hot mantle beneath mid-ocean ridges is quite shallow and not related to deep-mantle circulation. In this case, it appears to be the diverging motion of the lithospheric plates that causes upwelling of mantle from shallow depths, and not the other way around.

Two other processes, shown in Figure 7-18, may facilitate the movement of tectonic plates. Notice that the base of the lithosphere slopes downward from a spreading center; the grade can be as steep as 8 percent, steeper than most paved highways. Calculations show that even if the slope were less steep, gravity would cause the lithosphere to slide away from a spreading center over the soft, plastic asthenosphere at a rate of a few centimeters per year. This downslope sliding of the lithosphere away

LESSON 7.3 **211**

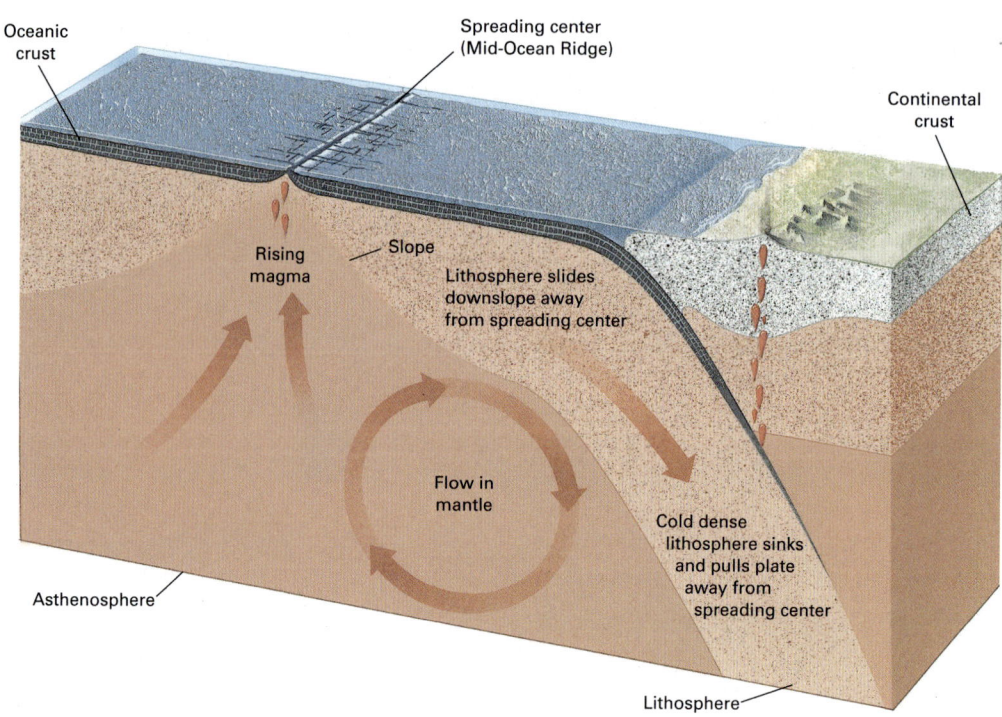

FIGURE 7-18 New lithosphere glides downslope away from a spreading center. At the same time, the old, cool part of the plate sinks into the mantle at a subduction zone, pulling the rest of the plate along with it. (The steepness of the slope at the base of the lithosphere is exaggerated in this figure.)

from a spreading center is called "ridge push" and may contribute to the movement of plates.

As the lithosphere moves away from a spreading center, it cools and becomes denser. Eventually, old lithosphere may become denser than the asthenosphere below. Consequently, it can no longer float on the asthenosphere and sinks into the mantle in a subduction zone, pulling the trailing plate along with it in a phenomenon referred to as "slab pull." Both ridge push and slab pull contribute to the movement of a lithospheric plate as it slides over the asthenosphere.

checkpoint How does mantle convection affect the movement of tectonic plates?

Mantle Plumes and Hot Spots

In contrast to the huge, ridge-shaped mass of mantle that rises beneath a spreading center, a relatively small column of plastic mantle rock that is hotter than surrounding rock often rises up in a **mantle plume**. Some mantle plumes appear to rise from great depths in the mantle, probably because small zones of rock near the core–mantle boundary become hotter and more buoyant than surrounding regions of the deep mantle. Others form as a result of heating in shallower portions of the mantle.

As pressure decreases in a rising mantle plume, rock melts to form magma. The rising heat and magma produce a **hot spot** in the upper mantle, which in turn heats the overlying lithosphere, forming a volcanic center. The Hawaiian Island chain is an example of a volcanic center over a hot spot. As demonstrated by the destructive lava flows on the Island of Hawai'i in 2018, the hot spot is active and poses a constant hazard (Figure 7-19). The mantle plume beneath the Hawaiian Islands originates deep in the mantle, far from any plate boundary and below the level of lateral plate motion. The volcanic center lies in the middle of the Pacific tectonic plate. New islands have formed as the plate moves northwest over the hot spot. Kauai is the oldest of the islands, and Hawai'i is the newest.

Some researchers have suggested that the mantle consists of two primary layers, with each layer undergoing convection. The shallower layer, located above 660 kilometers in depth, behaves dynamically and is characterized by relatively rapid convection. Below 660 kilometers, convection is more sluggish. This two-layered mantle model explains why the chemical composition of basalts from mid-ocean ridges is different from those that erupt at hot spots such as Hawai'i. Basalt erupting at mid-ocean ridges is part of the shallow convection system, where the mantle is well mixed. In contrast, the basalts erupting in Hawai'i are more primitive and come from a mantle plume of mantle welling up from the deeper convection system.

checkpoint How do the Hawaiian islands form?

MINILAB Convection

Materials
safety goggles
aluminum pie pan
three pennies
hot plate
food coloring
pearlescent liquid soap
tablespoon
water
graduated cylinder

Procedure
1. Place three pennies next to each other (not stacked) on the center of the hot plate.
2. Place the pie pan on top of the pennies on the hot plate.
3. Add two tablespoons of liquid soap to the pan.
4. Measure and pour 200 mL of water into the pan. Use the spoon to gently stir the solution.
5. Add two drops of food coloring to the center of the solution.
6. Put on your safety goggles. Turn the hot plate to a medium setting.

CAUTION: Do not touch the surface of the hot plate.

Results and Analysis
1. **Observe** Describe the general pattern of motion you observe.
2. **Infer** What purpose do the pennies serve in this model?
3. **Predict** Explain what you would expect to see if you could view a cross-section of the pan from the side.

Critical Thinking
4. **Critique** How could this model more accurately represent convection within Earth's mantle?
5. **Apply** Describe the parts of the model that are similar to subduction zones.

FIGURE 7-19
A burst of lava from a fissure on Kilauea volcano threatens a home during the May 2018 eruption on the island of Hawai'i.

LESSON 7.3

FIGURE 7-20

The arrangement of landmasses is continually changing. When you view Earth through the lens of geologic time, the illusion of permanence disappears. The next frame in this time sequence is presented at the beginning of this chapter.

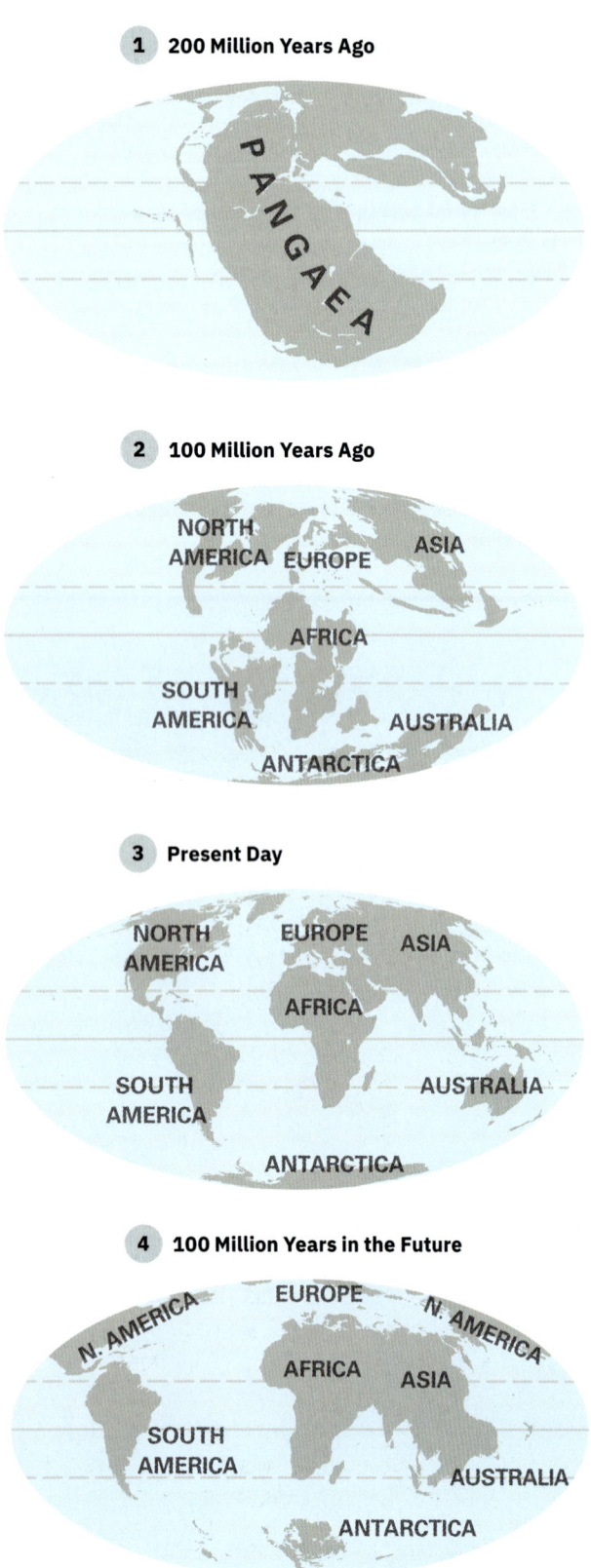

Supercontinents

By analyzing the composition and magnetism of rock samples formed at different times during Earth's history, scientists have pieced together a story of continents merging and separating. Prior to two billion years ago, large continents as we know them today may not have existed. Instead, many—perhaps hundreds—of small masses of continental crust and island arcs similar to Japan, New Zealand, and the islands of the southwest Pacific Ocean probably dotted a global ocean basin. Then, between 2 billion and 1.8 billion years ago, tectonic plate movements brought these microcontinents together to form a single landmass called a **supercontinent**. After a few hundred million years, this supercontinent, called Nuna, developed rifts and broke into fragments. The fragments then separated, each riding away from the others on its own tectonic plate.

About one billion years ago, the fragments of continental crust reassembled, forming a second supercontinent, called Rodinia. In turn, this continent fractured and the fragments came together into a third supercontinent about 300 million years ago. For perspective, that was about 70 million years before the appearance of dinosaurs. This third supercontinent is Alfred Wegener's Pangaea, which began to break apart about 235 million years ago in the late Triassic period (Figure 7-20).

The tectonic plates have continued their slow movement to create the mosaic of continents and ocean basins that shape the map of the world as we know it today. They will continue to shape it into the future. One recent model of future plate motions points to another supercontinent, Pangaea Proxima, existing about 250 million years from now. (See the Chapter Opener.)

checkpoint What aspects of Wegener's continental drift hypothesis are consistent with the theory of plate tectonics?

Isostasy

If you have ever used a small boat, you may have noticed that the boat settles in the water as you step into it and rises as you step out. Likewise, if a large mass is added to the lithosphere, the underlying asthenosphere flows laterally away from that region to make space for the settling lithosphere.

But how is mass added or subtracted from the lithosphere? One process that adds and removes

FIGURE 7-21

The weight of an ice sheet causes continental lithosphere to sink in response to the added burden. Notice that thicker ice will depress the lithosphere to a greater degree.

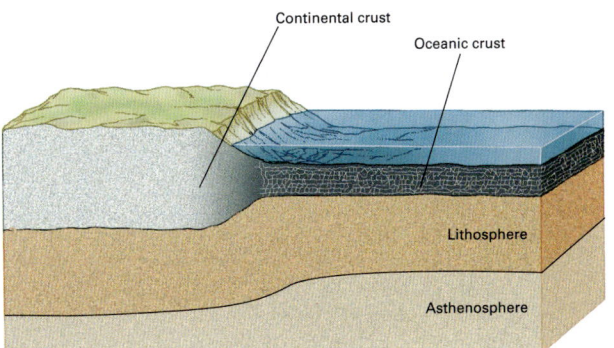
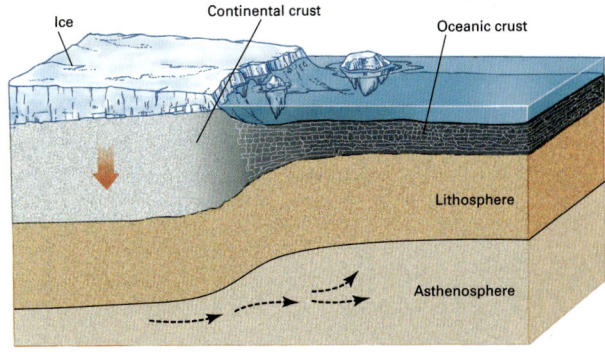

mass is the growth and melting of large glaciers. When a glacier grows, the weight of ice forces the lithosphere downward. The asthenosphere moves laterally away from the depressed region. For example, the Hudson Bay region of Canada was depressed during the last glaciation by an ice sheet about 3,000 meters thick. When a glacier melts, the continent rises, or rebounds. The rate of rebound is slower than the rate at which the ice melts. The surface formerly below thick glacial ice can remain depressed for thousands of years after all the ice is gone. The Hudson Bay region of Canada remains below sea level, although it is slowly rising at a rate of about one centimeter per year. The underlying asthenosphere is slowly flowing back into the region once covered by glaciers.

The Great Lakes, located near the former southern margin of the ice sheet, are also slowly rebounding following the melting of glacial ice. The rebound is faster on the northern side of the lakes because the ice was thicker there. Similarly, geologists in Scandinavia have discovered ice-age beaches that are tens of meters above modern sea level. The beaches formed when glaciers had depressed the Scandinavian crust to sea level, but they now lie well above that elevation because the land rose after the ice melted.

The equilibrium that results in the lithosphere floating on the asthenosphere is called **isostasy**, and the vertical movement in response to a changing burden is called isostatic adjustment (Figure 7-21).

An iceberg demonstrates an additional effect of isostasy (Figure 7-22). A large iceberg has a high peak, but its base extends deeply below the water's surface. The lithosphere behaves similarly. Continents rise high above sea level, and

FIGURE 7-22

(A) Due to isostasy, icebergs that rise higher above sea level also extend deeper below sea level. (B) Similarly, the lithosphere extends deeper into the asthenosphere beneath the mass of high mountains than it does under lower-elevation regions.

A

B
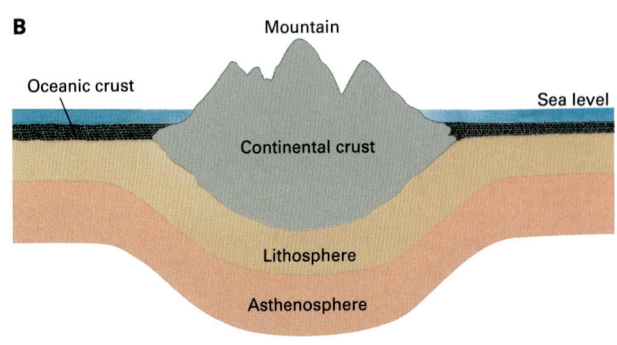

the lithosphere beneath a continent has a "root" that extends as much as 125 kilometers into the asthenosphere.

In contrast, most ocean crust lies approximately 5 kilometers below sea level, and oceanic lithosphere extends only about 75 kilometers into the asthenosphere. For similar reasons, high mountain ranges have deeper roots than do low plains, just as the bottom of a large iceberg is deeper than the base of a small one.

checkpoint Explain how isostatic adjustment affects crust thickness.

MINILAB Isostasy

Materials
metric ruler
felt tip pen
one thick block of wood: 8 cm × 15 cm × 4 cm
one thin block of wood: 8 cm × 15 cm × 2 cm
aquarium or small glass tank with water

Procedure

1. Place the thicker block of wood in the water. Measure how much of the block is below water and how much is above water. Record your measurements.
2. Determine the volume of the wood below the water: multiply the depth of the block below the water by the length and width of the block. The volume of the wood below the water is the same as the volume of the water displaced. The weight of the floating wood block is equal to the weight of the water that is displaced.
3. Mark on the glass the position of the top and bottom of the thick block using a felt tip pen.
4. Place the thin block on top of the thick block. Observe how the position of the thick block changes.
5. Mark the position of the top of the thin block. Measure the height of the wood above the water and the depth of wood below the water.
6. Determine the volume of the wood above and below the water.

Results and Analysis

1. **Identify** What part of the model represents Earth's lithosphere? What part of the model represents Earth's asthenosphere?
2. **Describe** How did the volume (and weight) of the wood below water and above water change after you placed the thin block on the thick block?
3. **Infer** What process or processes does placing the thin block of wood on the thick block of wood represent?
4. **Predict** Explain what would happen if the wooden blocks were placed in a liquid of greater density.

Critical Thinking

5. **Critique** Describe how you would change this model to more accurately represent the vertical movement of Earth's lithosphere.

7.3 ASSESSMENT

1. **Compare and Contrast** Describe the similarities and differences between ridge push and slab pull.
2. **Sequence** List the steps of volcano formation from a mantle plume of magma.
3. **Define** What is a supercontinent?
4. **Evaluate** What evidence would help you determine whether a volcano formed by hot spot or plate movement?

Critical Thinking

5. **Abstract Information** How is a mountain range like the tip of an iceberg?

7.4 EFFECTS OF PLATE TECTONICS

Core Ideas and Skills

- Explain how tectonic activity results in volcanoes, earthquakes, and mountains.
- Give an example of how plate tectonics affects Earth's climate.
- Deduce ways in which plate tectonics affects living things.

The movements of tectonic plates generate volcanic eruptions and earthquakes, which help shape Earth's surface. They also build mountain ranges and change the global distributions of continents and oceans. Tectonic activities strongly affect our environment in other ways by impacting global and regional climate, the atmosphere, the hydrosphere, and the biosphere.

Volcanoes

Most of Earth's volcanoes result from plate movements. At a divergent boundary (spreading center), hot asthenosphere oozes upward to fill the gap left between the two separating plates. Portions of the rising asthenosphere melt to form basaltic magma, which erupts onto Earth's surface. The Mid-Ocean Ridge is in part a chain of submarine volcanoes.

Volcanoes are common in continental rifts as well. For example, the longest continuously erupting lava lake in the world occurs at the volcano Erta Ale, located in Ethiopia in the northern portion of the East African Rift System. Numerous other large volcanoes are scattered down the East African Rift System for more than 1,000 kilometers (Figure 7-2).

Today, a large portion of western interior North America is also undergoing continental rifting with accompanying volcanism. There, the rifting and volcanoes are spread out over a larger region than in the narrower East African Rift System. The ongoing rifting in western North America has produced a geologic province called the Basin and Range Province. The Basin and Range reaches from north of the Canadian border to south of the Mexican border and from Utah to the Eastern Sierra Nevada of California (Figure 7-23). The Basin and Range is so named because the extension has broken the crust into numerous tilted segments.

FIGURE 7-23 ▼

Much of the western interior of North America is undergoing active rifting, forming a region called the Basin and Range Province. The rifting causes the crust to break into blocks that tilt. The high points form mountain ranges and the low points, called basins, fill with sediment eroded from the nearby mountains. The Cenozoic volcanic fields contain 65 million year old volcanic rocks.

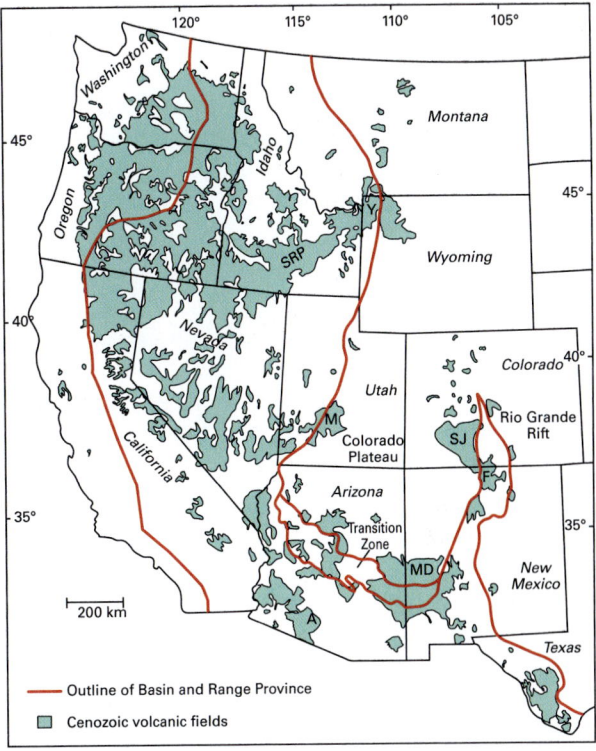

Source: Cengage

The high-elevation parts of each segment form isolated ranges. As each range erodes, it produces an abundance of sediment that fills in the lower-elevation basins.

Like in continental and oceanic rift systems, huge quantities of magma also form in the descending lithosphere of a subduction zone. Some of the magma solidifies within the crust, and some erupts at Earth's surface from a chain of volcanoes that forms parallel to the subduction zone. In North America, the Cascade Range in the Pacific Northwest and Canada is one such chain of volcanoes. The Aleutian Islands of Alaska, the Andes Mountains of South America, and the volcanic islands of the Caribbean and western Pacific are all examples of volcano chains formed in this manner. Volcanoes will be discussed in greater detail in Chapter 9.

checkpoint Describe the effects of rifting in North America.

FIGURE 7-24
ON ASSIGNMENT National Geographic photographer Carsten Peter snapped this shot of a caver standing on the remnants of a lava pond. The cavers are exploring Hawai'i's largest known lava tube, Kazumura, beneath the eastern slope of Kilauea. The lava tube was created after an eruption about 600 years ago.

Earthquakes and Mountain Building

Earthquakes are common at all three types of plate boundaries but are generally uncommon within the interior of a tectonic plate. As a result, changes in the land surface resulting from earthquakes are most common near plate boundaries. Earthquakes concentrate at plate boundaries simply because these are zones where one plate slips past another.

The slippage of one plate relative to its neighbor is rarely smooth and continuous. Rather, stress builds up along the plate boundary and is stored in the rock as energy. Eventually, the stress causes the rock to suddenly break, releasing the stored energy and causing rock on one side of the break to lurch violently past rock on the other. An earthquake is the vibration of the rock due to this sudden movement and rapid release of energy. During an earthquake, the ground surface can be broken. Earthquakes also can cause major landslides and even change the course of rivers and streams. Earthquakes will be discussed in greater detail in Chapter 8.

Great chains of volcanic mountains form at mid-ocean ridges because large amounts of magma form in these zones, and the new, hot lithosphere rises to a high level. Along subduction zones, long, linear mountain chains form as rock from the descending oceanic lithosphere melts and some of the magma ascends to the surface. If two continents collide at a convergent plate boundary, the ground surface will rise for the same reason a mound of bread dough thickens when compressed from both sides. Such continent–continent collisions thrust great masses of rock upward, creating huge mountain chains such as the Himalaya, the Alps, and the Appalachians. Mountain building will be discussed in greater detail in Chapter 10.

checkpoint Describe how mountains form at subduction zones.

How Plate Tectonics Affects Earth's Climate

The tectonic movements of the continents and the opening and closing of ocean basins have profoundly altered Earth's oceanic and atmospheric systems. Today's ocean currents carry warm water from the Equator toward the Poles and cool water from polar regions toward the Equator. These currents warm the polar regions and cool the tropics. Similarly, prevailing winds transport heat and moisture over the globe.

Changes in these ocean currents and wind patterns have had far-reaching consequences for regional climates. For example, widespread glaciation across Antarctica began occurring between 38 and 28 million years ago as seafloor spreading caused the continent to drift farther away from the southern tips of South America, Africa, and Australia. These tectonic movements isolated Antarctica over the South Pole. They also established the Antarctic Circumpolar Current, a strong west-to-east ocean current that continuously flows in a clockwise direction around the continent. As it formed, this current prevented warmer water from the southern Atlantic, Pacific, and Indian Oceans from reaching Antarctica, effectively putting the continent into a deep freeze that continues today.

Tectonic movements have also altered the composition of the atmosphere and oceans, producing major changes in global climate. For example, some scientists have proposed that the Cenozoic uplift of the Tibetan Plateau in Asia—the largest such uplift in the world—caused a strengthening of the Asian monsoon that led to Earth's recent glaciations. According to this idea, the additional monsoonal rainfall and widespread exposure of new rock as the plateau was uplifted significantly increased the rate of chemical weathering of silicate minerals, including those

---

CAREERS IN EARTH AND SPACE SCIENCES

Engineering Geologist

Engineering geologists are scientists who analyze the geologic factors involved in an engineering project. They assess construction sites for projects ranging from housing developments and skyscrapers to bridges and tunnels. They study the risk of hazards such as earthquakes, floods, and landslides. They also test rock and soil stability to ensure the ground can withstand the demands of the project and will remain safe if conditions change. Students of engineering geology take classes in math, physics, and Earth science. Engineering geologists spend part of their time indoors working in an office and the remainder of their time outdoors investigating sites.

FIGURE 7-25
Death Valley in southern California holds the record for the hottest recorded surface temperature on Earth. Its hot, dry climate is the result of tectonic forces. The valley is walled by steep mountain ranges and a valley floor that is 86 meters below sea level. Hot air rises but cools only slightly before sinking and reheating, creating super-heated air masses that blow through the valley during summer months.

containing calcium. Once released, the calcium ions combined with carbon dioxide (CO_2) molecules from the atmosphere to form calcite. In this way, CO_2 (a greenhouse gas) was removed from the atmosphere, causing Earth to cool and contributing to the development of major glaciations during late Cenozoic time.

With these changes in environments and climates come changes in the ecosystems supported by them. Tectonic processes also affect the biosphere. Streams and drainage patterns must respond to changes in slope direction and altered distributions of precipitation. Lakes may dry up, or new lakes may form. Ultimately, plants and animals may die or migrate away, to be replaced by new species that are adapted to the new climatic conditions.

checkpoint How do tectonic movements affect climate?

7.4 ASSESSMENT

1. **Apply** In 1980, Mount St. Helens, a volcano in the Cascade Range in Pacific Northwest state of Washington, erupted. Describe the tectonic activity that caused the eruption.

2. **Explain** Give an example that explains how tectonic activity affects climate.

Critical Thinking

3. **Apply** How can tectonic activity cause air pollution?

4. **Synthesize** In the process of geographic isolation, new species develop when a population is separated into groups that can no longer interbreed. Describe a scenario in which tectonic activity could lead to speciation.

TYING IT ALL TOGETHER
RIFTING APART

In this chapter, you learned about the theory of plate tectonics and plate movement, including continental rifting. The Case Study focused on Africa's Great Rift Valley, where giant cracks in Earth are prevalent (Figure 7-1).

These fissures are the result of continental rifting along the continental plate boundaries. The eastern part of the African continent is comprised of two different tectonic plates: the African (or Nubian) Plate to the west and the Somali Plate to the east. The two plates lie along a divergent plate boundary. A third plate, the Arabian plate, is pulling away from Africa to the northeast.

East Africa has several rift valleys that are characterized by long, narrow, deep lakes. Lake Turkana, Lake Malawi, and the Gulf of Aden all formed as a result of continental rifting. The rifting process likely caused areas of land to sink and form depressions that filled with water. The Red Sea also formed as a result of continental rifting along the boundary between the Arabian and African plates. If continental rifting continues in this region for tens of millions of years, an extended ocean basin could eventually form, separating the Somali Plate from the African continent.

Work individually or with a group to complete the steps below.

1. Select one of the following questions to research:

- Besides fissures, what other kinds of geologic events are typical in the East African Rift System? Identify and describe at least three recent examples and explain the connections between these events and plate tectonics.
- How does plate tectonics affect the climate and the biosphere in the Great Rift Valley?
- What types of plate boundaries are found closest to where you live? Identify at least three geologic events and/or formations related to these plate boundaries and explain the connections.
- How does plate tectonics affect the climate and the biosphere in the area where you live?

2. Find at least three reputable sources of information to help you address your research question. Make sure your sources contain enough information to help you fully address the research question. Before proceeding any further, check with your teacher to make sure your sources are credible and relevant.

3. Once your sources have been approved, gather information that will help you address your research question. Use your findings to make a hypothesis about the ways that tectonic processes such as mantle convection, ridge push, and slab pull contribute to the geologic events or formations in the region you chose to research. Explain your reasoning.

4. Write a report summarizing your findings in your own words. Cite your sources.

5. Find a classmate or group who selected the same research question you did. Compare your findings and explain any significant differences.

CHAPTER 7 SUMMARY

7.1 EARLY HYPOTHESES

- Alfred Wegener's hypothesis of continental drift, proposing that Earth's continents were once joined together in a single landmass before drifting apart, foreshadowed the theory of plate tectonics.
- The theory of plate tectonics provides a unifying framework for much of modern geology.
- Earth is a layered planet with the lithosphere and crust as the outermost cool and hard layer.
- The mantle extends from the base of the lithosphere to the core and includes the plastic asthenosphere.
- The core is mostly iron and nickel and consists of a liquid outer layer and a solid inner sphere.
- The hypothesis of seafloor spreading is the general model for the origin of all oceanic crust.

7.2 THE THEORY OF PLATE TECTONICS

- Plate tectonics theory is based on the concept that the lithosphere floats on the asthenosphere.
- The lithosphere is segmented into eight major tectonic plates that move relative to one another by gliding over the asthenosphere.
- Three types of plate boundaries exist: divergent, convergent and transform.
- Most of Earth's major geologic activity such as volcanoes, earthquakes, and mountain building occur near plate boundaries. Interior parts of tectonic plates are geologically stable.
- Tectonic plates move horizontally at rates that vary from one to 16 centimeters per year.
- Plate movements carry continents across the globe, cause ocean basins to open and close, and impact climate and the distribution of plants and animals.

7.3 PLATE MOVEMENT

- Mantle convection and movement of tectonic plates occurs because the mantle is hot, plastic, and capable of flowing.
- The entire mantle, from the top of the core to the crust, convects in huge cells. Horizontally moving tectonic plates are at the uppermost portions of convection cells.
- Convection occurs because (1) the mantle is hottest near its base, (2) new lithosphere glides downslope away from a spreading center, and (3) the cold leading edge of a plate sinks into the mantle and drags the rest of the plate along.
- Mantle plumes are relatively narrow rising columns of mantle rock hotter than surrounding rock.
- A hot spot is created when a mantle plume reaches the surface forming a volcano where it breaks through the lithosphere.
- Supercontinents may assemble, split apart, and reassemble every few hundred million years.
- The concept of balance between gravity and buoyancy that causes the lithosphere to float on the mantle at different elevations is called isostasy.
- When weight is added to or removed from Earth's surface, the lithosphere sinks or rises. This vertical movement in response to changing burdens is called isostatic adjustment.

7.4 EFFECTS OF PLATE TECTONICS

- The movements of tectonic plates generate volcanic eruptions, earthquakes, and mountain ranges and change the global distributions of continents and oceans.
- Tectonic activities strongly affect our environment in other ways—impacting global and regional climate, the atmosphere, the hydrosphere, and the biosphere.
- The tectonic movements of the continents and the opening and closing of ocean basins through rifting and subduction profoundly alter Earth's oceanic and atmospheric systems.
- Ocean currents warm polar regions and cool the tropics, while winds transport heat and moisture over the globe.
- Changes in ocean currents and wind patterns have far-reaching consequences for regional climate.

CHAPTER 7 ASSESSMENT

Review Key Terms
Select the key term that best fits the definition. Not all terms will be used, and no term will be used more than once.

asthenosphere	Mid-Ocean Ridge
continental drift	normal magnetic polarity
continental rifting	
convection	plate boundary
convergent boundary	plate tectonics
divergent boundary	seafloor spreading
isostasy	subduction
magnetic reversal	supercontinent
mantle plume	transform boundary

1. a theory stating that the lithosphere is segmented into several plates that move about relative to one another by floating on and sliding over the upper mantle
2. the hypothesis that segments of oceanic crust are separating at the Mid-Ocean Ridge
3. the hypothesis proposed by Alfred Wegener that Earth's continents were once joined together and later split and drifted apart
4. the upward and downward flow of fluid material in response to density changes produced by heating and cooling, which occurs slowly in Earth's mantle and much more quickly in the oceans and atmosphere
5. a relatively small rising column of mantle rock that is hotter than surrounding rock
6. the concept of balance between gravity and buoyancy that causes the lithosphere to float on the asthenosphere at different elevations
7. the process in which two lithospheric plates of different densities converge and the denser one sinks into the mantle beneath the other
8. a change in Earth's magnetic field in which the north magnetic pole becomes the south magnetic pole and vice versa, which has occurred on average every 500,000 years over the past 65 million years
9. a magnetic orientation the same as that of Earth's current magnetic field
10. an undersea mountain chain that forms at the boundary between divergent tectonic plates within oceanic crust
11. the portion of the upper mantle just beneath the lithosphere, extending from about 100–350 kilometers below the surface and consisting of weak, plastic rock where magma may form

Review Key Concepts
Answer each question on a separate sheet of paper to demonstrate your understanding of key concepts from the chapter.

12. Use a model to explain the role of convection in the movement of tectonic plates.
13. Identify the similarities and differences between the lithosphere and the asthenosphere.
14. Explain the evidence for seafloor spreading that was discovered using magnetometers in the mid 1900s.
15. Describe how mud and fossil deposits support the hypothesis of seafloor spreading.
16. Explain how the different properties of basaltic and granitic rock result in rift valleys.
17. Identify each type of plate boundary shown.

A

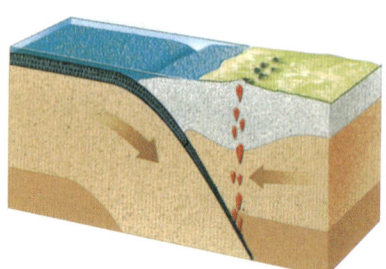

B

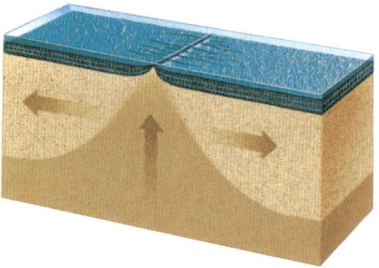

C

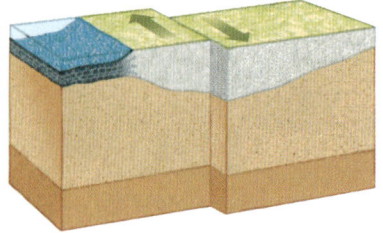

Think Critically

Write a response to each question on a separate sheet of paper. Use concepts from the chapter to support your reasoning.

18. Wegener understood that the fit of the continents alone did not prove that a supercontinent had existed. Explain how each additional piece of evidence supported his hypothesis of continental drift.
19. Describe the similarities and differences in how mountains and volcanoes form.
20. Does the number and shape of Earth's plates change over time? Explain.
21. Explain how tectonic forces affect climate. Include one example of how tectonics affect climate over a short period of time and one example of how tectonics affect climate over millions of years.
22. Plate tectonics is described as a "unifying theory." What does it unify, and how?

PERFORMANCE TASK

Create a Scientific Illustration

Suppose you are a scientist tasked with explaining to the public how a natural phenomenon has occurred on Earth's surface. The phenomenon is a result of plate tectonics, such as a chain of mountains, a volcanic eruption, a basin, or an earthquake. For this task, you will create a scientific illustration of Earth's layers and use it to explain a phenomenon that has occurred on Earth's surface. A scientific illustration is a two-dimensional model that can be used to explain or predict actual scientific phenomena.

TABLE 7-3 Depths and Densities of Earth's Layers

	Average Depth from Earth's Surface to Base of Layer (kilometers)	Average Density (g/cm³)
Crust	70	2.6
Lithosphere	125	3.3
Mantle	2,900	4.5
Outer Core	5,150	11.1
Inner Core	6,370	12.5

Materials
- ruler
- compass
- waterproof fine line pen
- watercolor paper
- watercolor paints
- colored paper scraps
- scissors
- glue stick

1. Construct and use a model.

    A. Make an accurate cross-section illustration of Earth's layers. Start with a pencil sketch before adding permanent ink and color.
    - Your illustration must be drawn to scale. That is, it must show relative thicknesses of Earth's actual layers. Use the information in Table 7-1 to guide you. The radius of Earth is 6,370 kilometers.
    - Your illustration must include the following information (and may include more):
        ○ Key showing the scale; for example, "Key: 1 cm = X" (Replace X with quantity represented by each centimeter in the illustration.)
        ○ Names of layers
        ○ Primary composition of each layer (for example, "basaltic")
        ○ General properties of each layer (for example, "hot, weak")
        ○ At least one convection cell

    B. Use your illustration to explain an actual event or feature that has occurred on Earth as a result of plate tectonics. You may use text, arrows, labels, and captions in your explanation. Explain what happened, how it happened, and roughly when it occurred.

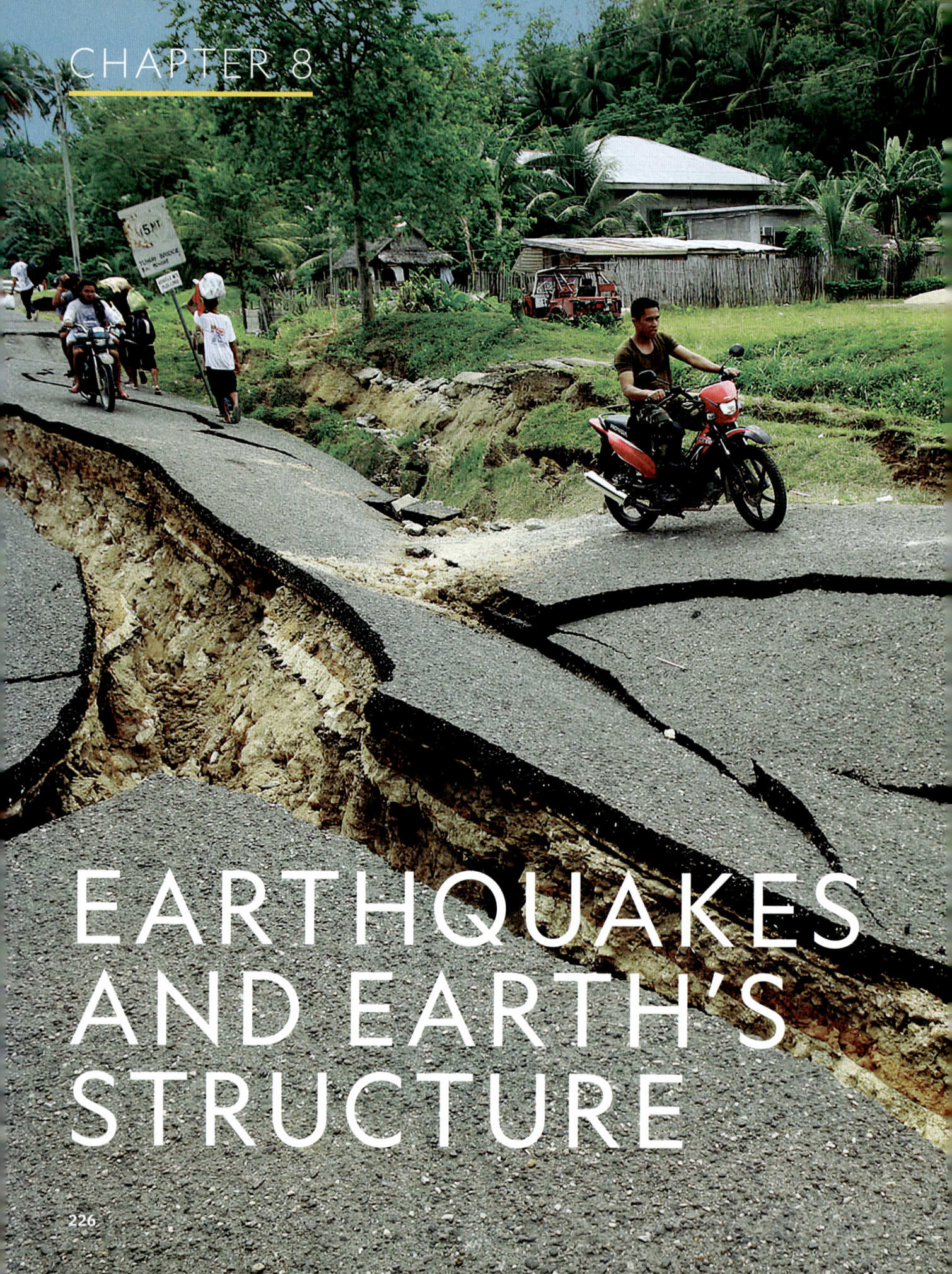

CHAPTER 8

EARTHQUAKES AND EARTH'S STRUCTURE

An earthquake caused this severe road damage in La Libertad in the central Philippines when it struck on February 7, 2012.

Knowledge about plate tectonics and the identification of significant faults allows scientists to predict the locations most at risk for future earthquakes. Yet, it hasn't been possible to predict exactly when an earthquake will occur nor mitigate its effects. In 2012, an earthquake shook the island of Negros in the Philippines. The tremors were felt within a 112-kilometer radius. Some roads, like the one pictured here, split apart as the underlying rock fractured. Like many earthquakes, it triggered a tsunami and landslide, which destroyed homes and buildings. Threats posed by earthquakes are a regular part of life in the Philippines, which has recovered from the effects of countless past quakes. Residents continue to cope with the daily risk of yet another major earthquake and its aftereffects.

KEY QUESTIONS

8.1 How is energy stored in rock and then released by an earthquake?

8.2 How do earthquakes that originate at different types of plate boundaries differ?

8.3 How do the properties of rock and soil affect seismic waves?

8.4 What do earthquakes reveal about Earth's interior?

EXPLORERS AT WORK

MAPPING EARTHQUAKE-CAUSING FAULTS

WITH NATIONAL GEOGRAPHIC EXPLORER AFROZ AHMAD SHAH

Spanning some 2,400 kilometers, the mountains of the Himalaya run across the nations of India, Pakistan, Afghanistan, China, Bhutan, and Nepal. The towering peaks of this "abode of snow"—which include the world-famous Mt. Everest—inspire a sense of awe in locals and tourists alike. Yet, for geoscientist Dr. Afroz Ahmad Shah, these magnificent mountains inspire something more: a sense of urgency.

Shah, an assistant professor of structural geology at the Universiti Brunei Darussalam, is on a mission to identify major earthquake hazards in Jammu and Kashmir, a state in northern India that lies in the northwestern Himalaya. Jammu and Kashmir were among the areas impacted by a magnitude 7.6 earthquake in 2005 that devastated parts of Pakistan, India, and Afghanistan. The earthquake triggered landslides and caused extensive damage, claiming more than 80,000 lives. In 2015, Nepal experienced a magnitude 7.5 earthquake.

Shah worries that government officials in Jammu and Kashmir have done very little in response to those recent quakes. He stresses that the area is due for one or more earthquakes that could be even more powerful. Shah's strategy for avoiding a repeat of 2005 is threefold: Develop an earthquake scenario map to identify where earthquakes are most likely to occur, educate local people about earthquake preparedness, and advocate for stricter earthquake-resistant building codes.

The Himalaya range began forming between 40 million to 50 million years ago with the collision of two continental plates, the Indian plate (which underlies most of India and some portions of Pakistan) and the much larger Eurasian plate. Mountains formed over millions of years as the Eurasian plate buckled and was lifted up and over the Indian plate. The plates continue to collide today, which means the Himalaya continue to rise by more than 1 centimeter each year. It also means this region is prone to earthquakes.

According to Shah, three active major faults run through the area around Jammu and Kashmir. Geologists are not sure how long tectonic strain has been building up in the region, or how much of it has been released in past earthquakes. Shah is studying satellite imagery and seismic data from past earthquakes to help answer these questions. He is also identifying new areas of fault activity. He will use the information to create an up-to-date map of faults throughout the region.

People who live in areas prone to earthquakes must focus on preparing for them. One way to do so is to build earthquake-resistant structures. Shah is urging the government to invest in training for engineers and to develop better communication with the public about earthquake hazards and preparedness. Shah says, "It is never too late to turn and to think of a safer future, where people can live with earthquakes."

THINKING CRITICALLY

Evaluate Consider Dr. Shah's approaches to reducing the effects of earthquakes in Jammu and Kashmir. Which approach is most effective, and why? How might its success be measured?

▲ Dr. Afroz Ahmad Shah works with a team, taking samples from exposed sediment in the Leh district of Jammu and Kashmir, India.

◀ Dr. Shah and a student examine a rock sample extracted from one of the newly exposed fault surfaces in northern India.

CASE STUDY
MEXICO CITY EARTHQUAKES, 1985 AND 2017

On September 19, 1985, a magnitude 8.0 earthquake struck Mexico, leaving approximately 10,000 dead, 30,000 injured, and thousands more homeless. Even though the quake was centered about 400 kilometers west of Mexico City, it was one of the most devastating earthquakes to ever hit the area. A total of 3,000 buildings had to be demolished, and an additional 100,000 were left seriously damaged (Figure 8-1).

In the years following this catastrophe, the Mexican government took bold steps to fortify the country against future earthquakes. Engineers and seismologists assessed the structural damage. They determined that despite the fact that Mexico experiences about 90 quakes per year of magnitude 4.0 or greater, many of the collapsed buildings were poorly designed and ill-suited to withstand earthquakes. Over time, Mexico implemented stricter building codes.

By the early 1990s, Mexico had also developed an early warning system to alert city residents of impending earthquakes. Sensors to detect seismic activity were installed along Mexico's western coast, where many earthquakes originate. Loudspeakers were placed in schools, parks, and other public areas in cities across the country. When the sensors detect an earthquake occurring, they send a signal that sounds alarms across the country. Although shock waves from an earthquake travel quickly, such alarms can sometimes give residents precious minutes to get to safety before the shaking starts.

In 2017, more than three decades after the 1985 quake, Mexico's improved infrastructure was put to the test when two more large earthquakes struck within a period of twelve days. On September 8, a magnitude 8.2 earthquake struck just off the Pacific coast in southern Mexico. Although it was even stronger than the 1985 quake, it killed 98 people—far fewer than the 10,000 killed in 1985. A short time later, a magnitude 7.1 quake struck in central Mexico. Although it was about ten times less powerful, it struck closer to Mexico City, resulting in a death toll of at least 360 people.

By 2017, Mexico City was clearly better prepared than it had been in 1985. Whereas the 1985 quake toppled many high-rise buildings, the 2017 quakes mostly damaged buildings that were built prior to 1985 (Figure 8-2). In addition, millions of residents received alerts on their mobile phones minutes before the quake arrived. As such, it would be easy to credit improvements in infrastructure for the lower death tolls. However, no two earthquakes are the same, and geologic factors beyond human control can also affect the damage caused by a particular earthquake.

As You Read Evaluate the causes of earthquakes and how they are measured. What factors determine how much damage a quake causes? What steps should scientists and engineers take to help limit earthquake damage?

⊕ **FIGURE 8-1**
Go online to view a high-rise buildings were severely damaged in Mexico City after the magnitude 8.1 earthquake in 1985.

◀ **FIGURE 8-2**
Most of the buildings that were damaged after the two quakes in 2017 were built before 1985.

8.1 EARTHQUAKES AND SEISMIC WAVES

Core Ideas and Skills
- Explain the physical processes that lead to an earthquake.
- Differentiate between body waves and surface waves.
- Explain the process of recording a seismogram.
- Compare and contrast different earthquake evaluation scales.
- Summarize how geologists use a travel-time curve.

KEY TERMS

earthquake
fault
seismic wave
focus
epicenter
body wave
surface wave

primary wave (P wave)
secondary wave (S wave)
seismograph
Mercalli scale
Richter scale
moment magnitude

As you learned in Chapter 7, Earth's tectonic plates slide slowly over the soft asthenosphere, about as fast as your fingernails grow. Friction often locks the two sides of a plate boundary together. Rocks stretch and compress across the boundary as forces build, but otherwise nothing major happens for long periods of time. This can range from decades to thousands of years. Then, suddenly, the two sides of the boundary snap free, releasing all at once the stored energy that had built up over time and causing an earthquake.

An **earthquake** is a sudden motion or trembling of Earth caused by the abrupt release of elastic energy that is stored in rocks. The interiors of plates move at a steady rate, but motion is constrained at plate boundaries, causing rocks there to slowly accumulate energy from ongoing plate movements. Once the threshold level of energy has accumulated, it is released as an earthquake.

Anatomy of an Earthquake

Rock appears rigid, but when you apply enough stress to a rock, it can deform. When stress is applied to a rock, the rock can deform in one of the three ways: elastically, by fracturing, or plastically.

Under small amounts of stress, rock deforms elastically. If the stress is removed, the rock returns to its original size and shape. A rubber band deforms elastically when you stretch it. The energy used to stretch the rubber band is stored in the elongated rubber band. When the stress is removed, it immediately springs back and releases the stored elastic energy. In the same way, an elastically deformed rock will spring back to its original shape and release its stored energy when the force is removed.

However, like a rubber band, every rock has a limit beyond which it cannot deform elastically. Under certain conditions, an elastically deformed rock may suddenly fracture. When large masses of rock in Earth's crust deform elastically and then fracture, vibrations are formed that travel through Earth and are felt as an earthquake.

Under other conditions, when its elastic limit is exceeded, a rock continues to deform, like the bending of a steel nail. This behavior is called plastic deformation. A rock that has deformed plastically keeps its new shape when the stress is released and, consequently, does not store the energy used to deform it. Therefore, earthquakes do not occur when rocks deform plastically.

The photo at the beginning of this chapter shows an example of asphalt pavement that fractured during an earthquake. The elastic limit of the rock under the pavement was exceeded, so it suddenly ruptured. In contrast, Figure 8-3 shows a rock outcrop that was deformed plastically by forces deep within Earth's crust. Parallel rock layers were bent and twisted into an S-shaped fold over time while the rocks were hot and under pressure.

A fracture becomes a **fault** when rock on either side of the fracture ruptures and the two sides move

FIGURE 8-3
This rock deformed plastically when stressed.

LESSON 8.1

relative to one another. Most movement of crustal rock occurs due to slippage along established faults. This movement occurs because the friction binding the two sides of the fault is weaker than the rock itself. As additional tectonic stress then builds and strain begins to accumulate in the rock again, the rock is more likely to move along the existing fault than to crack somewhere else and create a new fault. However, new faults can form if the orientation and curvature of the existing faults can no longer most efficiently release stress that builds in the rock.

Many faults develop as part of a larger group of faults that together release much of the strain that accumulates in a region. For example, thousands of separate normal faults exist today in the North American Basin and Range Province. Many of these faults form the geologic boundary separating each mountain range from its adjacent basin. The faults develop through time as the crust extends, tilting large blocks of the crust. As extension continues, the first faults can no longer accommodate the stretching. New faults develop with different orientations and the old faults are no longer active.

Tectonic plates move at rates between 1 and 16 centimeters per year. Friction prevents the plates from slipping past one another continuously while the edges remain locked and immobile. Elastic deformation strains the rock, causing stored energy to build. When the strain reaches a critical threshold, rock suddenly fractures. All at once the built-up stored energy is released (Figure 8-4).

Many earthquakes occur along established faults as strain is suddenly released. For example, the San Andreas Fault in Southern California (Figure 8-5) lies along a tectonic plate boundary that has slipped many times in the past. It will certainly slip again in the future. The boundary includes numerous individual faults, some of which are active and some of which are not. Although much of the movement across the plate boundary is associated with the San Andreas, other faults also contribute and some have significant earthquake

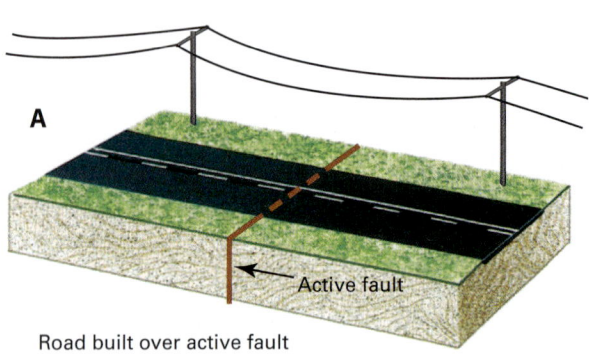

Road built over active fault

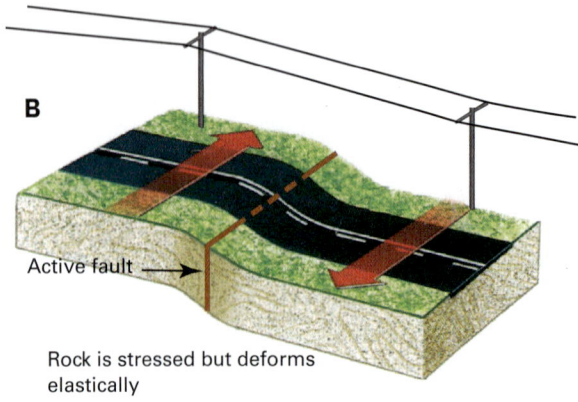

Rock is stressed but deforms elastically

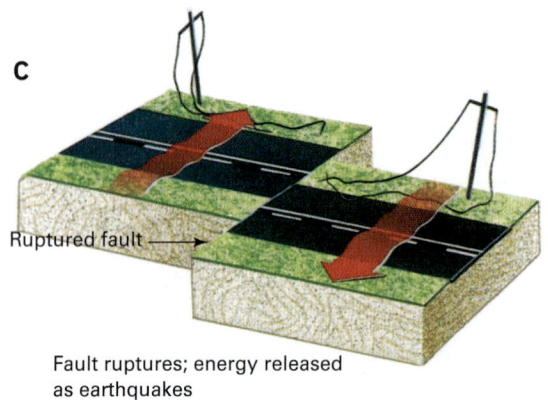

Fault ruptures; energy released as earthquakes

FIGURE 8-4
(A) A road is built across an active fault. (B) The rock stores energy when it is stressed by a tectonic force. The stress causes elastic deformation and stored energy builds. (C) At a critical point, the rock fractures, releasing the stored energy and returning the rock to its unstressed shape on either side of the fault.

FIGURE 8-5
California's San Andreas Fault, the source of many earthquakes, is the boundary between the Pacific plate and the North American plate.

risk associated with them. For example, the UC Berkeley football stadium is bisected by the Hayward Fault, which in places is slipping a rate of 5 millimeters per year. The Hayward Fault last generated an earthquake with an estimated 7.0 magnitude in 1868 (Figure 8-6).

🌐 FIGURE 8-6
Go online to view a satellite image of Hayward Fault, an active fault in the Bay Area.

The United States Geological Survey (USGS) considers the Hayward Fault to be the most likely to generate a magnitude 7.0 or greater earthquake within the next 30 years (31 percent probability). For comparison, the San Andreas Fault, located across the San Francisco Bay Area from the Hayward Fault, is estimated to have only a 21 percent chance of generating a magnitude 7.0 earthquake over the same time frame (Figure 8-7).

During an earthquake, rock can move from a few centimeters to several meters. A single earthquake makes only small changes in the topography or geography of a region. But over tens of millions of years, the relentless motion of plates, accompanied by hundreds of thousands of earthquakes, changes the surface of the planet. In this way, mountains are raised, continents are split apart, and new ocean basins are formed.

checkpoint What are the differences between elastic rock deformation and plastic rock deformation?

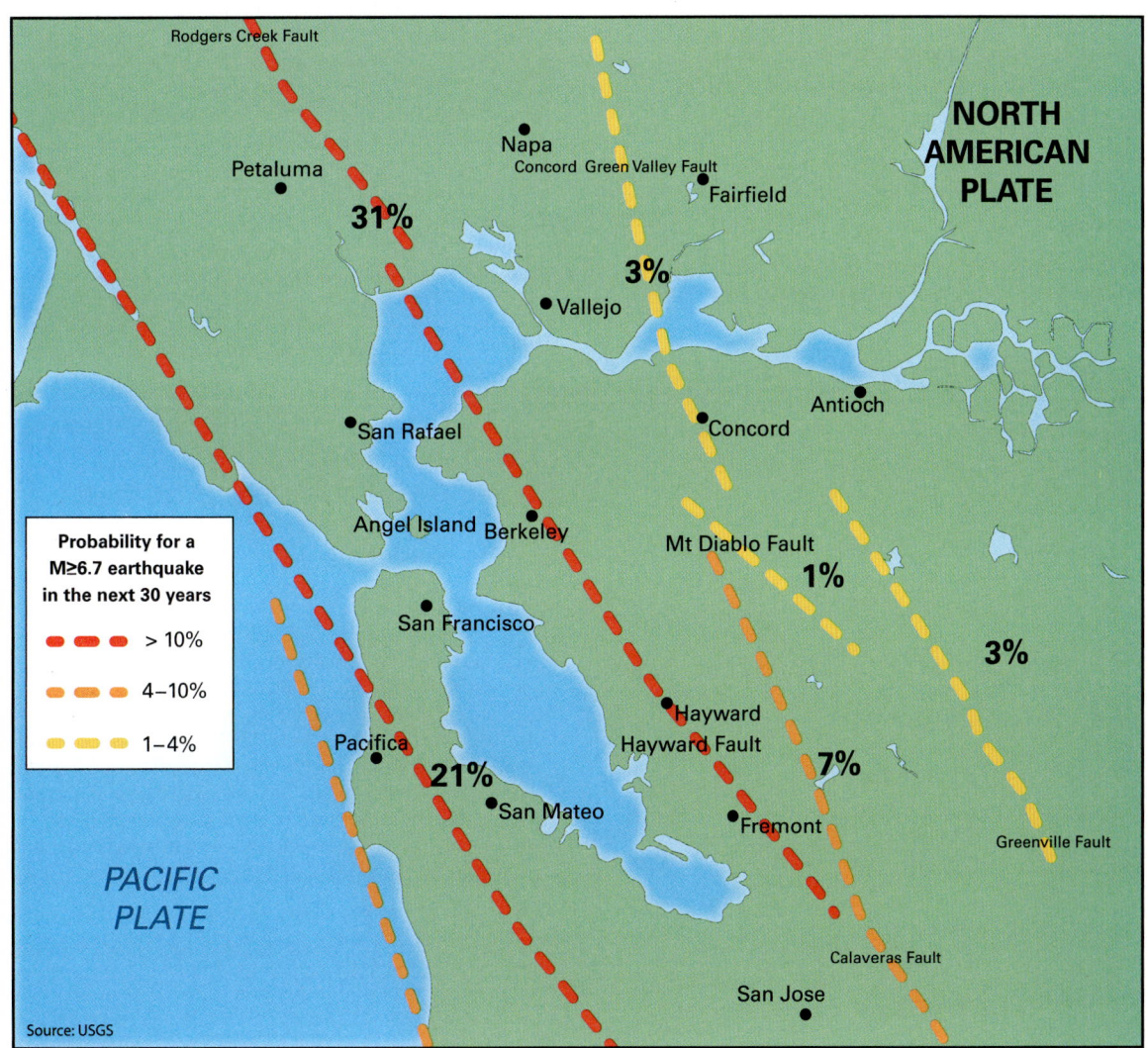

FIGURE 8-7
The San Francisco Bay Area straddles the boundary between the North American and Pacific plates. It is crisscrossed with individual faults that together accommodate the total slip across the plate boundary. Estimates show the likelihood of 7.0 or greater magnitude in the next 30 years. The fault with the most risk in the Bay Area is the Hayward Fault (31 percent) with the San Andreas Fault second at 21 percent.

Earthquake Waves

A wave transmits energy from one place to another. A drumbeat travels through air as a sequence of waves. The sun's heat travels to Earth as waves. And a tap travels through a watermelon in waves. Waves that travel through rock are called **seismic waves**. Earthquakes and explosions produce seismic waves. Seismology is the study of earthquakes and the nature of Earth's interior based on evidence from seismic waves.

The initial rupture point, where abrupt movement creates an earthquake, typically lies below the surface at a point called the **focus**. The point on Earth's surface directly above the focus is the **epicenter**. An earthquake produces two main types of seismic waves. **Body waves** travel through Earth's interior and carry some energy from the focus to the surface. **Surface waves** then radiate from the epicenter and travel along Earth's surface, somewhat like swells on the surface of the sea (Figure 8-8).

Body waves can be of two main types. **Primary waves (P waves)** travel fast and are the first or "primary" seismic waves to reach an observer. P waves are compressional waves that cause the rock to undergo alternating compression and expansion (Figure 8-9). Consider a long spring, such as in a toy, with one end attached to a wall. If you stretch and then rapidly push the end of the spring, a compressional wave travels back and forth along its length. Each segment of the spring is compressed and then stretched as the wave travels. An ant sitting on the spring would move toward, then away from, the wall. Therefore, the wave vibrates in the direction in which it propagates. P waves travel through air, liquid,

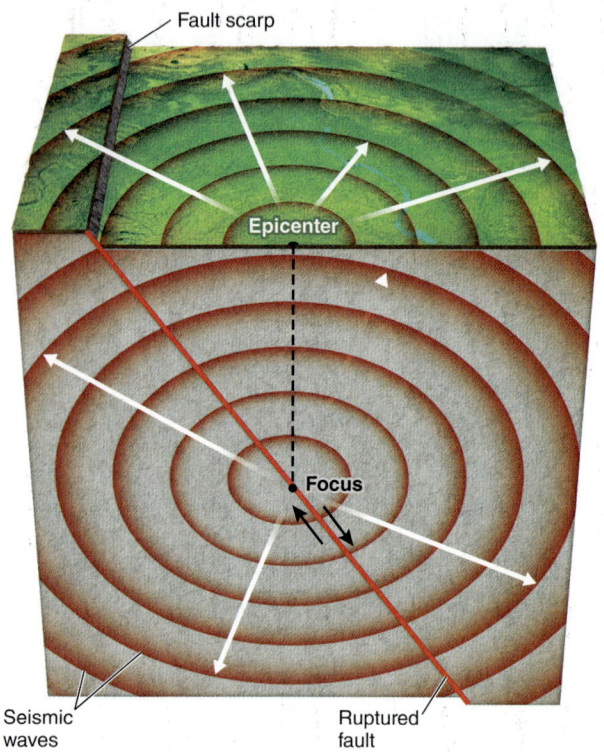

FIGURE 8-8
Body waves, generated by abrupt movement on a fault, radiate outward from the focus of an earthquake.

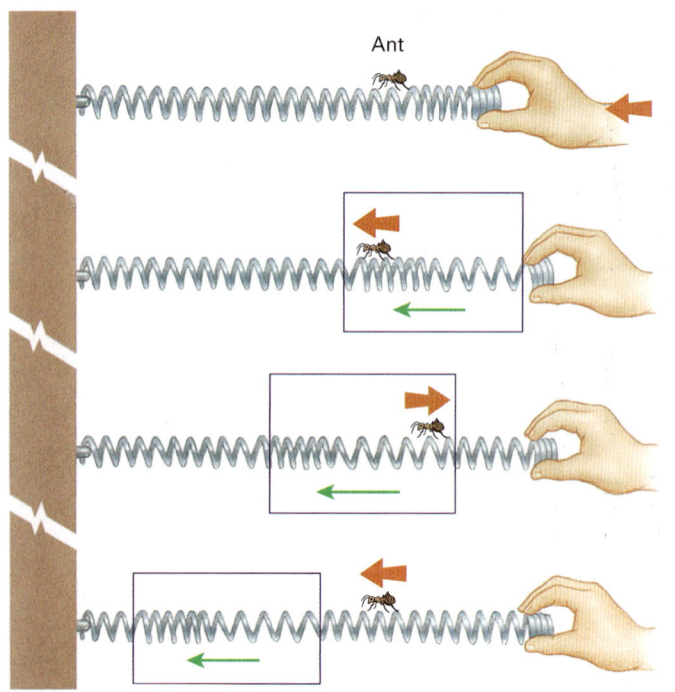

FIGURE 8-9
A spring is used to demonstrate P waves. A P wave is created when the direction of vibration (orange arrow) is parallel to the direction in which the wave travels (green arrow). (A) The hand pushes the end of the spring toward the wall, initiating the P wave. (B) The compressed end of the wave reaches an ant sitting on the spring, causing it to move toward the wall. (C) The stretched part of the wave reaches the ant, causing it to move away from the wall. (D) The ant returns to its original position relative to the wall after the wave has passed by.

and solid material. If a person is underwater in a bathtub, and he or she taps the sides of the tub, P waves can be heard.

P waves travel at speeds between 4 and 7 kilometers per second in Earth's crust and about 8 kilometers per second in the uppermost mantle. For comparison, the speed of sound in air is only 0.34 kilometers per second, and the fastest jet fighters fly about 0.96 kilometers per second.

Another type of body wave, **Secondary waves (S waves)**, are slower than P waves and thus are the "secondary" waves to reach an observer. Also called shear waves, S waves have a shearing motion that can be illustrated by tying a rope to a wall, holding the opposite end, and giving the rope a sharp up-and-down jerk (Figure 8-10). Although the wave travels or propagates parallel to the rope, each segment of the rope moves at right angles to the rope length. An ant sitting on the rope would move up and down as the wave passes by. A similar motion in an S wave produces shear stress in a rock—stress that acts in parallel but opposite directions. S waves vibrate at right angles to the direction in which the wave is moving and travel between 3–4 kilometers per second in the crust.

🌐 **FIGURE 8-10**
Go online to view illustrations that demonstrate S waves.

Unlike P waves, S waves move only through solids. Because molecules in liquids and gases are only weakly bound to one another, they slip past each other and thus cannot transmit an S wave.

Surface waves travel more slowly than body waves. Surface waves undulate across the ground like the waves that ripple across the water when you throw a rock into a calm lake, although the actual wave mechanism is different. Two types of surface waves occur simultaneously: an up-and-down rolling motion called Rayleigh waves and a side-to-side vibration called Love waves (Figure 8-11). Both waves are named after the

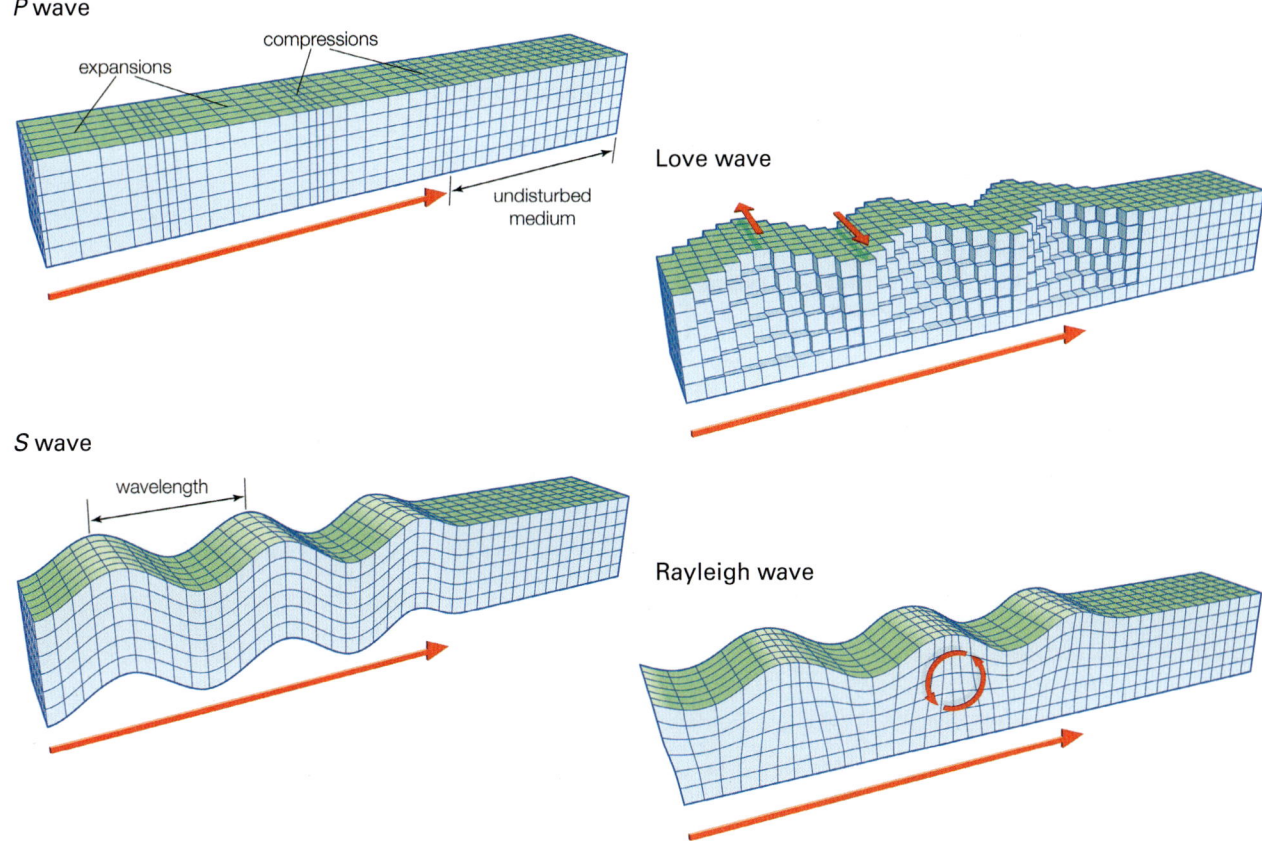

FIGURE 8-11 ▲
This diagram shows how each wave type affects the surface of Earth. P waves compress and stretch the rocks in the direction of travel. S waves shear the rock up and down, perpendicular to the direction of travel. Love waves cause a back-and-forth shear that is restricted to the region near the ground surface. Rayleigh waves also compress and stretch the near surface while shearing it perpendicular to the direction of travel.

scientists who first described each wave type. During an earthquake, Earth's surface rolls like ocean waves and writhes from side to side like a snake.

checkpoint What is the difference between surface waves and body waves?

Measurement of Seismic Waves

A **seismograph** is a device that records seismic waves. In early seismographs, a heavy weight was suspended from a spring (Figure 8-12). A pen attached to the weight was aimed at the zero mark on a piece of graph paper. The graph paper was mounted on a rotating drum that was attached firmly to bedrock. During an earthquake, the graph paper jiggled up and down, but inertia kept the weight and its pen stationary. As a result, the paper moved up and down beneath the pen, while the rotating drum recorded this earthquake motion over time. This physically written record of Earth vibration is called a seismogram (Figure 8-13). Modern seismographs use a coil fixed to a pendulum that is suspended in a magnetic field. When an earthquake causes the pendulum to move within the magnetic field, it creates an electric current that is amplified. A computer records this electric current as data.

The USGS hosts a large earthquake-monitoring network, which makes information from thousands of deployed seismographs around the world available immediately. Basic information provided by the monitoring network includes such data as an earthquake's location, magnitude, and the depth of the earthquake.

Seismologists are able to use information from the worldwide network of seismographs to study the details of specific earthquakes, especially the largest ones. One such study is called finite fault modeling, which provides a description of the dynamics of the faulting event that caused the earthquake. Included in the finite fault model are estimates of the total slip distance that occurred on the fault during the earthquake, how much surface area on the fault was involved, and how long the slipping lasted. The USGS has finite fault model results for all the largest earthquakes to occur over the past several years. Figure 8-14 shows an example of the finite fault model for the devastating March 9, 2011, Tōhoku earthquake in Japan.

checkpoint How do modern seismographs create a seismogram?

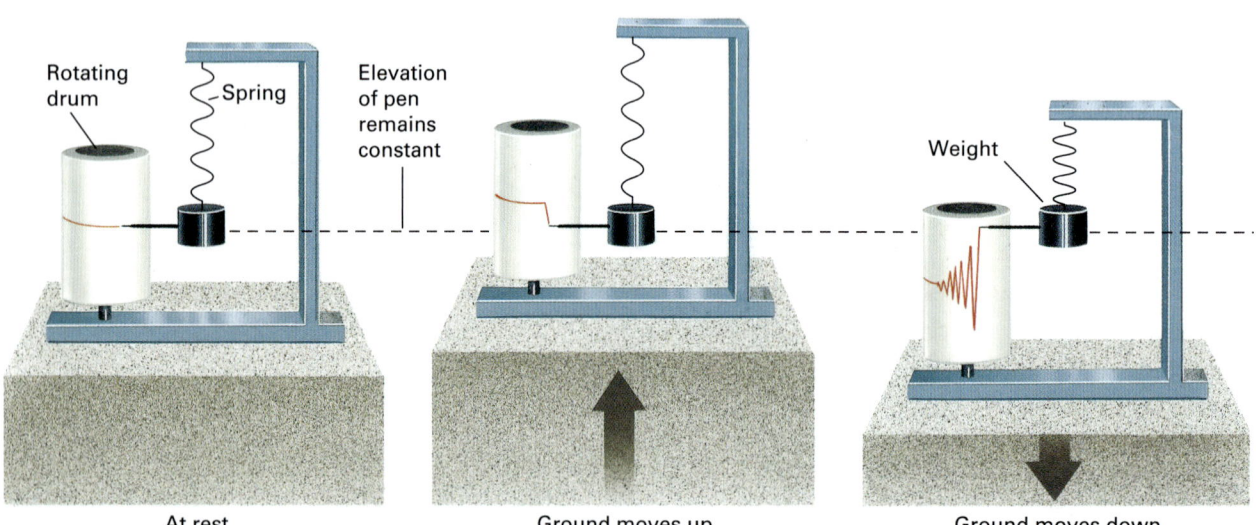

FIGURE 8-12
In early seismographs, the pen draws a straight line across the rotating drum when the ground is stationary. When the ground rises abruptly during an earthquake, it carries the drum up with it. But the spring stretches, so the weight and pen hardly move. Therefore, the pen marks a line lower on the drum. Conversely, when the ground sinks, the pen marks a line higher on the drum. During an earthquake, the pen traces a jagged line as the drum rises and falls.

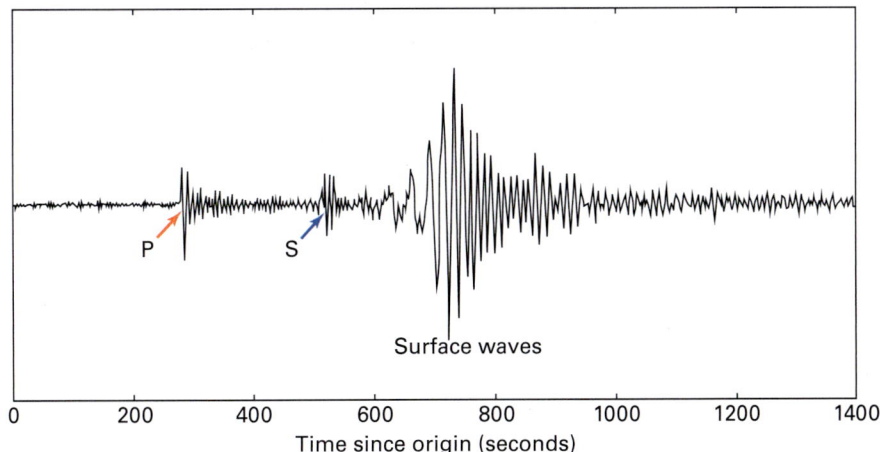

FIGURE 8-13
This seismogram was recorded at a station in Belarus from a 5.0 magnitude earthquake in the North Atlantic. The arrows mark the first arrival of P waves (red) and S waves (blue).

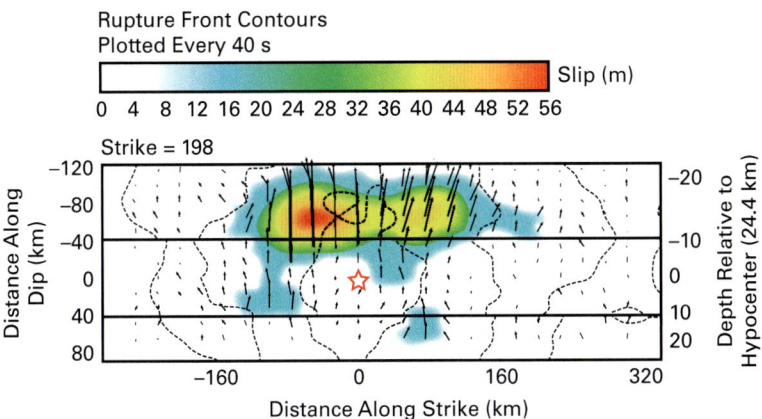

FIGURE 8-14
This figure shows finite fault modeling results for the March 9, 2011, 9.0 magnitude Tōhoku earthquake in Japan. The colors correspond to the total slip distance for the fault that ruptured to cause the earthquake. As shown by the color key at the top of the figure, the warmest colors correspond to more than 50 meters of slip during the earthquake. The horizontal and vertical scales represented by the cross section suggest that this area of the fault surface that moved measured tens of kilometers both laterally and vertically.

Source: USGS

Measurement of Earthquake Strength

Over the past century, geologists have devised several scales to express the strength of an earthquake. Before seismographs became widespread, geologists evaluated earthquakes using the **Mercalli scale**, which measures the intensity of an earthquake based on structural damage. On the Mercalli scale, an earthquake that destroyed many buildings was rated as more intense than one that destroyed only a few. However, this system did not accurately measure the energy released by a quake because structural damage also depends on distance from the focus, the rock or soil beneath the structure, and the quality of construction.

In 1935, Charles Richter devised the **Richter scale** to express the amount of energy released during an earthquake. Richter magnitude is calculated using the height of the largest earthquake body wave recorded on a specific type of seismograph. The Richter scale is more quantitative than earlier intensity scales, but it is not a precise measure of earthquake energy.

A sharp, quick jolt would register as a high peak on a Richter seismograph. However, a very large earthquake can shake the ground for a long time without generating extremely high peaks. And so, a strong earthquake can release a huge amount of energy that is not recorded in the height of a single seismograph peak and is not adequately expressed by Richter magnitude.

Modern equipment and methods enable seismologists to measure the amount of movement and the surface area of a fault that moved during a quake. The product of these two values and the shear strength of the faulted rock allow them to calculate the **moment magnitude**. Most seismologists now use moment magnitude rather than Richter magnitude, because it more closely reflects the total amount of energy released during an earthquake. An earthquake with a moment magnitude of 6.5 has an energy about equal to the mechanical energy released by the atomic bomb that was dropped on the Japanese city of Hiroshima at the end of World War II.

MINILAB Build a Simple Seismograph

Materials
large, sturdy, corrugated cardboard box, roughly 25 cm tall, 10 cm deep, and 10 cm wide
1 rectangular piece of cardboard (larger than cereal box base)
8-10 oz. Styrofoam cup with lid
fine-tipped marker or pen
scissors
ruler
packing or shipping tape
roll of adding machine or similar paper
piece of string, 60 cm long
wooden craft stick
gravel (100 g)

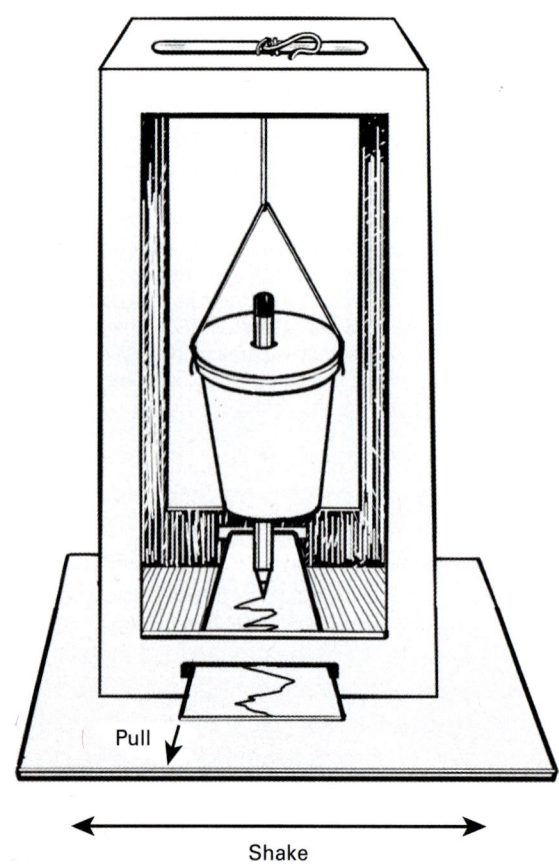

Procedure

1. Cut the front and back sides out of the cardboard box, leaving a 2.5-cm border.
2. In the middle of both bottom edges, cut a slot wide enough to slide the adding machine paper through.
3. Make a small hole in the center of the top of the box.
4. Make a hole in the center of the lid and another in the center of the bottom of the cup.
5. With the scissors, make two small holes near the upper ridge on both sides of the cup.
6. Push the marker or pen through the top and bottom holes in the cup, making sure the point is facing downward.
7. Fill the cup halfway with gravel.
8. Thread the string through the two holes on the sides of the cup.
9. Hang the cup so that the marker lightly touches the paper by bringing the ends of the string together and threading them through the hole in the top of the cereal box; tie the ends of the string to the craft stick.
10. Tape the bottom of the box to the piece of cardboard, securing it firmly.
11. Without moving the box, slowly pull the adding machine paper through the slots at a steady rate. Observe the markings on the paper and record your observations.
12. Now simulate an earthquake by quickly moving the cardboard bottom side to side in short, repetitive motions. Meanwhile, have a partner continue to slowly and steadily pull the adding machine paper through the slots. Observe the markings on the paper and record your observations.
13. Repeat step 12 at least three times. Vary the intensity of the shaking with each trial.

Results and Analysis

1. **Observe** Describe the general patterns of motion you observe on the seismogram.
2. **Infer** How does the size of the peaks relate to the magnitude of an earthquake?
3. **Predict** What would you expect the seismogram to look like if your seismograph was at a station close to the epicenter of an earthquake? How would it differ if the earthquake occurred farther away from the seismograph?

Critical Thinking

4. **Critique** How could this model be improved to more accurately represent the effects of an earthquake? What were some of the challenges you faced when building your seismograph?
5. **Apply** Describe the information that can be measured based on a single seismogram reading. Describe the information that cannot be determined from a single seismogram reading.

DATA ANALYSIS: Analyze Patterns of P and S Waves

The distance from a seismic station to an earthquake is determined using a travel-time curve. Refer to Figure 8-15. Seismograms from different stations are fit to the curve by aligning the arrival of P waves (red arrows) and S waves (blue arrows) to match the two curves on the graph. The distance to the epicenter from each seismic station can then be read off the x-axis of the graph. Once distances from three different stations are determined, you can draw circles on a map around each seismic station. The radius of each circle should be equal to the distance from the epicenter to that station. The point where the three circles intersect is the location of the earthquake epicenter.

1. **Identify** How long does it take for a primary wave to travel 3,000 km?
2. **Identify** How long does it take for a secondary wave to travel 3,000 km?
3. **Evaluate** Which seismic station is 3,000 km from the epicenter?
4. **Infer** A P wave arrives at 2:00 p.m. If the S waves arrive at 2:05 p.m., approximately how far away was the earthquake's epicenter?
5. **Predict** What do you think the time difference would be between the arrival of P and S waves at 5,000 km from the epicenter?

Computational Thinking

6. **Generalize Patterns** How does the distance from the epicenter affect the interval between the first arrival of P waves and the first arrival of S waves?
7. **Analyze Data** Where is the earthquake's epicenter?
8. **Recognize Patterns** Approximate the time difference between the first arrival of P waves and the first arrival of S waves for a seismic station located 1,600 km from the epicenter.

On both the moment magnitude and the Richter scales, the energy of the quake increases by a factor of about 30 for each whole number increment on the scale. Therefore, a magnitude 6 earthquake releases roughly 30 times more energy than a magnitude 5 earthquake.

Moment magnitude depends on the strength of rocks because a strong rock can store more energy before it fractures than a weak rock can. The two largest earthquakes ever measured occurred in Chile in 1960 and Alaska in 1964. These quakes had moment magnitudes of 9.6 and 9.2—respectively, about 25,000 times and 7,500 times, respectively, greater than the energy released by the Hiroshima bomb.

checkpoint How are body waves used to calculate a value on the Richter scale, and how is this type of calculation potentially a false representation of earthquake strength?

Locating the Source of an Earthquake

A simple technique used during an electrical storm can estimate the distance between you and the place where the lightning strikes. After the flash of a lightning bolt, count the seconds that pass before hearing the thunder. Although the electrical discharge produces thunder and lightning simultaneously, light travels much faster than sound. Light reaches us virtually instantaneously. Sound travels much more slowly through the atmosphere at about 340 meters per second. If the time interval between the lightning flash and the thunder is 1 second, then the lightning struck 340 meters away.

The same principle is used to determine the distance from a recording station to both the epicenter and the focus of an earthquake. Recall that P waves travel faster than S waves and that surface waves are slower yet. If a seismograph happens to be close to an earthquake epicenter, the different waves will arrive in rapid succession for the same reason the thunder and lightning come close together when a storm is close. On the other hand, if a seismograph is located far from the epicenter, the S waves arrive after the P waves arrive. The surface waves are even farther behind.

Geologists use a travel-time curve to calculate the distance between an earthquake epicenter and a seismograph. A travel-time curve is constructed using P-wave and S-wave arrival times from multiple stations at different locations for an earthquake with a known epicenter and occurrence time. Geologists use the resulting graph to measure the distance between a recording station and an earthquake whose epicenter is unknown.

Travel-time curves were first constructed using data obtained from natural earthquakes. However,

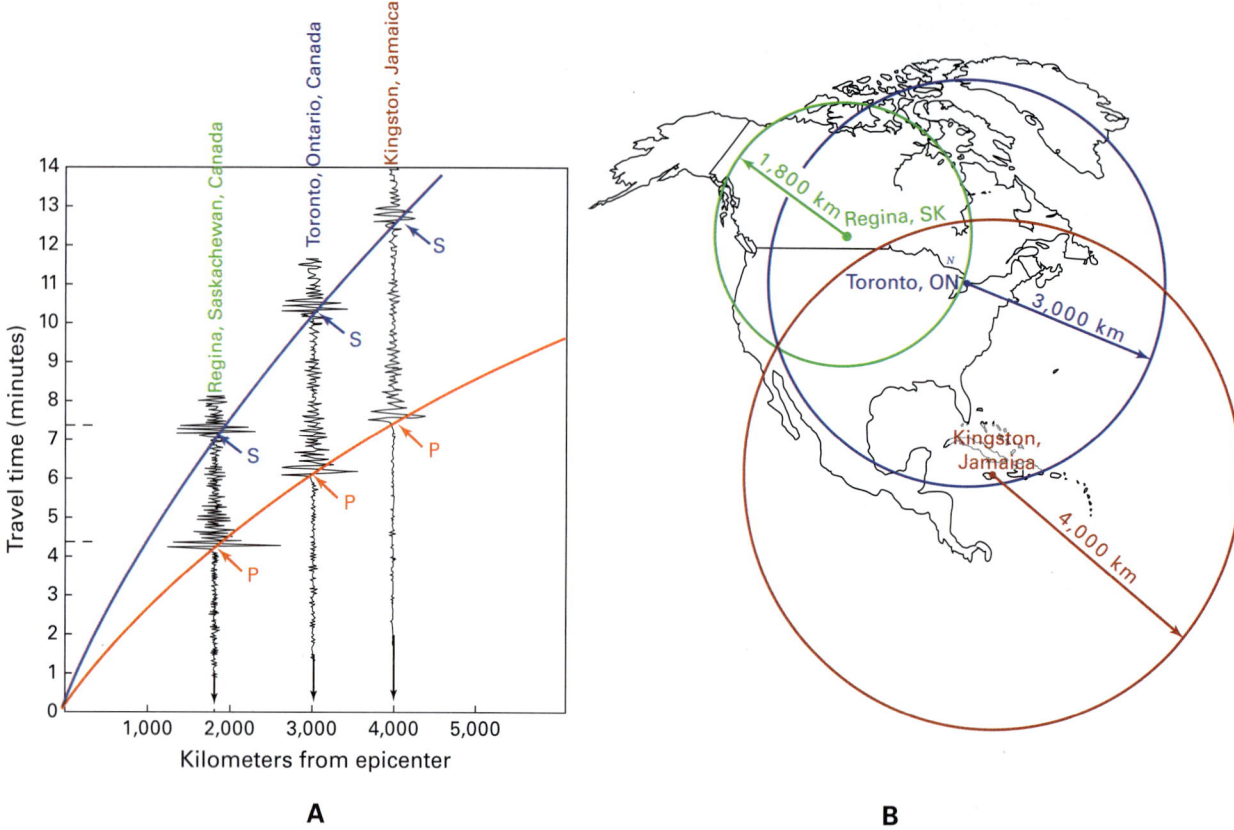

FIGURE 8-15
(A) A travel-time curve shows the expected arrival time of seismic waves based on distance from the epicenter. (B) The distance can be mapped, and the epicenter can be located by triangulation. The epicenter of the earthquake is located at the point where all three circles intersect.

scientists do not always know precisely when and where an earthquake occurs. In the 1950s and 1960s, geologists studied seismic waves from atomic bomb tests to improve the travel-time curves. Geologists analyzed atomic bomb tests because they could verify both the location and timing of the explosions.

Figure 8-15 illustrates how the location of the epicenter of an earthquake is determined. Using seismograms from three or more seismic recording stations, a seismologist plots the first arrival times for P and S waves on a travel-time curve. The corresponding value on the *x*-axis of the graph is the distance between the epicenter and each seismic station. The seismologist then plots a circle around each of the seismic station locations on a map. The radius of each circle is the linear distance to the epicenter from that station. The one point on the map where the three (or more) circles intersect is the location of the earthquake's epicenter.

checkpoint What information does a travel-time curve provide, and how is it calculated?

8.1 ASSESSMENT

1. **Explain** When is a fault considered active, and for how long does it keep that status?
2. **Define** Explain the relationship between an earthquake's focus and epicenter.
3. **Summarize** What information generated from a network of seismographs is included in a finite fault model?
4. **Describe** How does a modern seismograph record seismic waves?
5. **Distinguish** How does the measure of moment magnitude differ from the value measured by the Richter scale?

Critical Thinking

6. **Generalize** How does ongoing earthquake monitoring contribute to the body of knowledge about earthquake patterns? How does this work help build scientific knowledge and protect communities?

8.2 EARTHQUAKES AND TECTONIC PLATE BOUNDARIES

Core Ideas and Skills
- Recount the history and patterns of the San Andreas Fault.
- Explain the movements of the Pacific and North American plates.
- Describe the characteristics of a Benioff zone.
- Use the New Madrid Fault zone to explain earthquakes at plate interiors.

KEY TERMS

strike-slip fault
fault creep
Benioff zone

Before the theory of plate tectonics was developed, geologists recognized that earthquakes occur frequently in some regions and infrequently in others, but they did not understand why. Modern geologists know that most earthquakes occur along plate boundaries, where tectonic plates diverge, converge, or slip past one another. The area in the Pacific known as the Ring of Fire (Figure 8-16) marks the boundaries of several plates. Ninety percent of the world's earthquakes take place in this region.

Earthquakes at a Transform Plate Boundary: The San Andreas Fault Zone

The populous region from San Francisco to San Diego, California, straddles the San Andreas Fault zone, a transform boundary between the Pacific plate and the North American plate. The fault zone ranges in width from less than 1 kilometer to

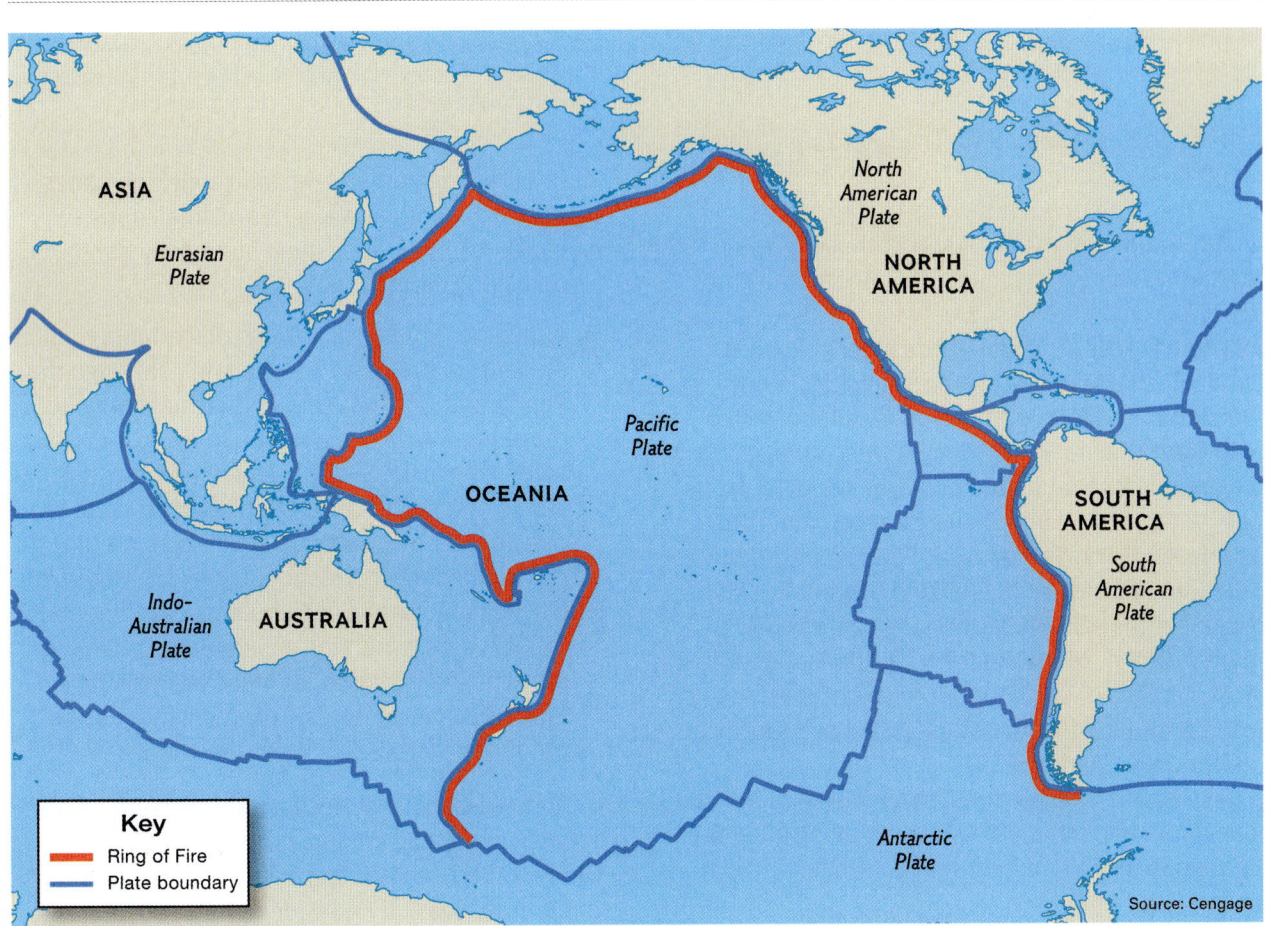

FIGURE 8-16
The Ring of Fire follows the boundaries of tectonic plates. Most earthquakes occur in this zone.

LESSON 8.2 **241**

FIGURE 8-17
The horizontal displacement shown in this field is an example of a strike-slip fault. The land on either side of the fault move horizontally relative to each other.

FIGURE 8-18
This scene of destruction captured in 1906 after the San Francisco earthquake shows a glimpse of the damage caused by the earthquake and fires that swept throughout the city.

several tens of kilometers. The fault zone is made up of numerous individual faults. However, almost all of the slip between the two plates occurs on only a handful of major faults. Of these, the San Andreas Fault is the biggest, accommodating up to 90 percent of total slip between the two plates in some places. The San Andreas Fault itself is vertical, but the rocks on opposite sides move horizontally. A fault of this type is called a **strike-slip fault** (Figure 8-17). Plate motion stresses rock adjacent to the fault, and numerous smaller faults are generated as stored energy accumulates and is released. The San Andreas Fault zone comprises the San Andreas Fault itself and the smaller faults around it that distribute the strain across the broader zone. In the past few centuries, hundreds of thousands of earthquakes have occurred in this zone, including the infamous 1906 San Francisco earthquake (Figure 8-18).

The Pacific and North American plates move past one another in three different ways along different segments of the San Andreas Fault zone:

1. Along some portions of the fault zone, rocks slip past one another at a continuous, snail-like pace referred to as fault creep. This movement occurs without violent and destructive earthquakes because the rocks slip continuously and slowly.

2. In other segments of the fault zone, the plates pass one another in a series of small hops, causing a number of small but mostly non-damaging earthquakes.

3. Along the remaining portions of the fault, friction prevents slippage across the fault, although the Pacific and North American plates continue to move relative to one another. In this case, rock across the locked fault segment deforms and stores energy. Because the plates move past one another at an average of 3.5 centimeters per year, energy corresponding to 3.5 meters of elastic deformation accumulates over a period of 100 years. When the accumulated energy exceeds the frictional resistance across the fault, the fault ruptures. This resulting rupture produces a large, destructive earthquake.

Plates move slowly and silently every day. Rocks stretch, bend, and compress, unseen beneath our feet. Geologists understand the mechanisms of earthquakes and know for certain that more destructive quakes will occur along the San Andreas Fault (Figure 8-19). But no one knows when or where the "Big One" will strike.

checkpoint Describe the motion in a strike-slip fault and how smaller faults are connected.

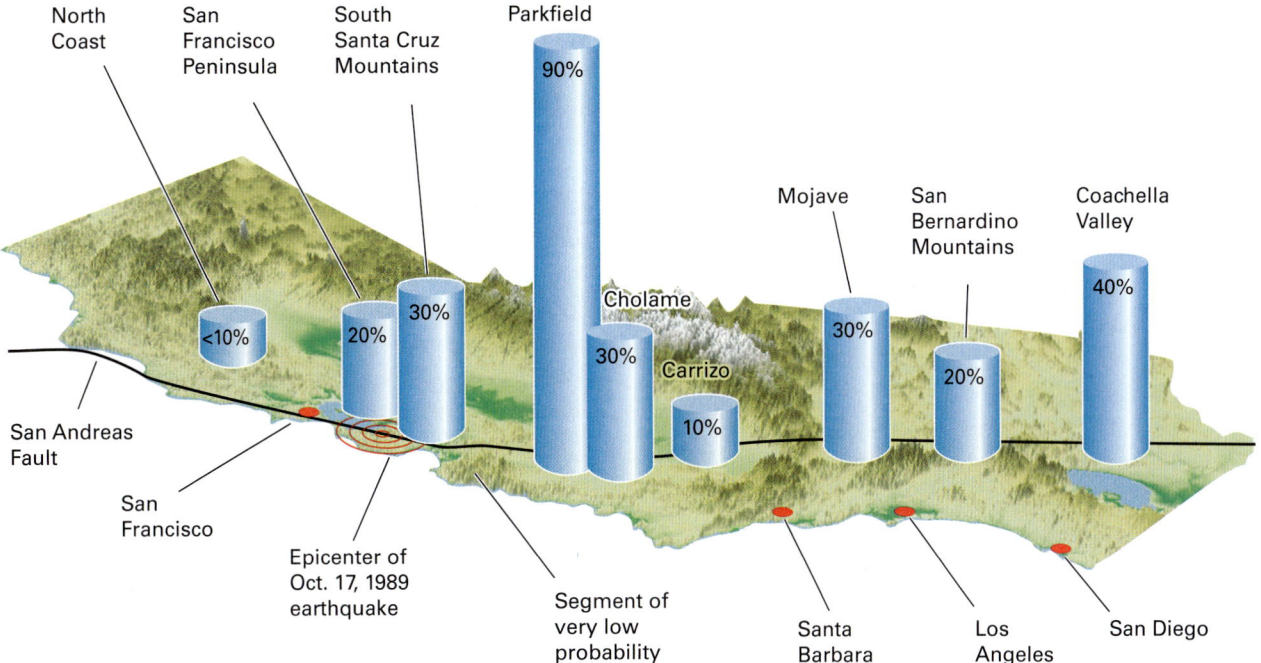

FIGURE 8-19
The illustration shows an earthquake hazard map of the San Andreas Fault zone. Percentages indicate the probability of an earthquake with a magnitude greater than 5 in the next 30 years.

Earthquakes at Convergent and Divergent Plate Boundaries

Recall from Chapter 7 that at a subduction zone, a relatively cold, rigid lithospheric plate dives beneath another plate and slowly sinks into the mantle. In most places, the subducting plate slips past the plate above it with intermittent jerks, giving rise to numerous earthquakes. The earthquakes occur mostly along the upper part of the sinking plate, where it scrapes past the opposing plate. Additional earthquakes occur within the downward-moving slab as it is bent and extended upon being pulled into the mantle. Collectively, these earthquakes form the **Benioff zone,** named after Hugo Benioff, the geologist who first recognized it (Figure 8-20). Many of the world's strongest earthquakes occur in subduction zones, including all five of the largest recorded earthquakes.

When two converging plates both carry continental crust, neither can sink deeply into the mantle. Instead, the crust buckles up to form high mountains, such as the Himalaya. Rocks fracture or slip during this buckling process, generating frequent earthquakes. The 2005 earthquake along the India–Pakistan border was generated by the slow convergence of the subcontinent of India with Asia.

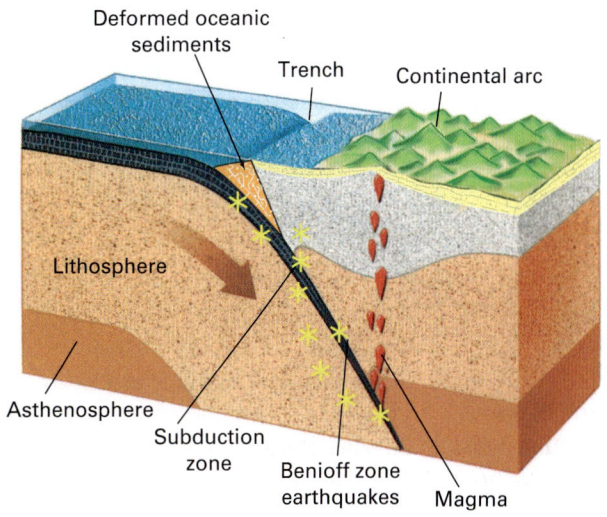

FIGURE 8-20
A descending lithospheric plate generates magma and earthquakes in a subduction zone. Earthquake foci, shown by stars, concentrate mostly along the upper portion of the subducting plate, called the Benioff zone.

Earthquakes frequently shake the Mid-Ocean Ridge system as a result of faults that form as two plates separate. Blocks of oceanic crust drop downward along most mid-ocean ridges, forming a rift valley in the center of the ridge. Only shallow earthquakes occur along the Mid-Ocean Ridge,

LESSON 8.2 **243**

because the asthenosphere there rises to depths as shallow as 10 to 15 kilometers and is too hot and plastic to fracture.

checkpoint Where do most earthquakes occur in a subduction zone?

Earthquakes in Plate Interiors

No major earthquakes have occurred in the central or eastern United States in the past 100 years, and no current lithospheric plate boundaries are known in these regions. However, the largest historical earthquake sequence in the contiguous 48 states occurred near New Madrid, Missouri. In 1811 and 1812, three shocks with estimated moment magnitudes between 7.3 and 7.8 altered the course of the Mississippi River and rang church bells 1,500 kilometers away in Washington, D.C. These earthquakes occurred within a zone of active extensional tectonics called the New Madrid Fault zone, in the lower Mississippi River Valley region.

Earthquakes in plate interiors are not as well understood as those at plate boundaries, but modern research is revealing some clues. Tectonic forces stretched North America in Precambrian time. As the continent pulled apart, rock fractured to create two huge fault zones that crisscross the continent like a giant *X*. Although the fault zones failed to develop into a divergent plate boundary, they remain weaknesses in the lithosphere. New Madrid lies at a major intersection of the faults—at the center of the *X*. As the North American plate slides over the asthenosphere, it may pass over irregularities, or "bumps," in that plastic zone, causing slippage and earthquakes along the deep faults.

DATA ANALYSIS Map Earthquakes in Real Time

The USGS monitors seismic activity and seismic hazards around the world. Visit the interactive real-time earthquake map at the USGS website. The interactive map shows all earthquakes that have occurred in the past 24 hours around the world. To the left of the map, you can see the time, location, and magnitude of each earthquake. Click on each earthquake for more detailed information, such as the depth of the quake and the latitude and longitude of the epicenter. On the right, the earthquakes are mapped and you can zoom in to various areas to see each location in detail. The settings menu in the upper right corner allows you to change the time frame (earthquakes within the past 7 days or 30 days) or limit the map to earthquakes above a certain magnitude.

1. **Sequence** Rank the magnitudes of the ten largest earthquakes in the past seven days. Use the latitude and longitude coordinates to identify the location of each earthquakes.

2. **Identify** Identify the depth at which each earthquake occurred.

3. **Infer** What type of plate boundary is found at each earthquake's location? Use Table 8-1 to help you answer this question.

Computational Thinking

4. **Recognize Patterns** Do the earthquakes occur with the same frequency throughout the world or are they concentrated in certain areas? Explain your answer.

Data Challenge

Go to the Data Analysis in MindTap to complete the data challenge.

TABLE 8-1 Types of Plate Boundaries with Examples

Plate Boundary	Location Examples
Divergent Plate Boundaries	Mid-Atlantic Ridge (Ocean–Ocean) East African Rift (Continent-Continent)
Convergent Plate Boundaries	Western South America (Ocean-Continent) Aleutians (Ocean–Ocean) Himalaya (Continent–Continent)
Transform Plate Boundaries	San Andreas Fault

FIGURE 8-21

This map shows earthquake hazards in the United States as of 2014, when the most recent long-term risk assessment was conducted by the USGS. The predictions are based on records of frequency and Mercalli magnitudes of historical earthquakes. Peak acceleration refers to how fast the ground surface would move up and down during an earthquake as a fraction of 1 g—the gravitational acceleration at Earth's surface.

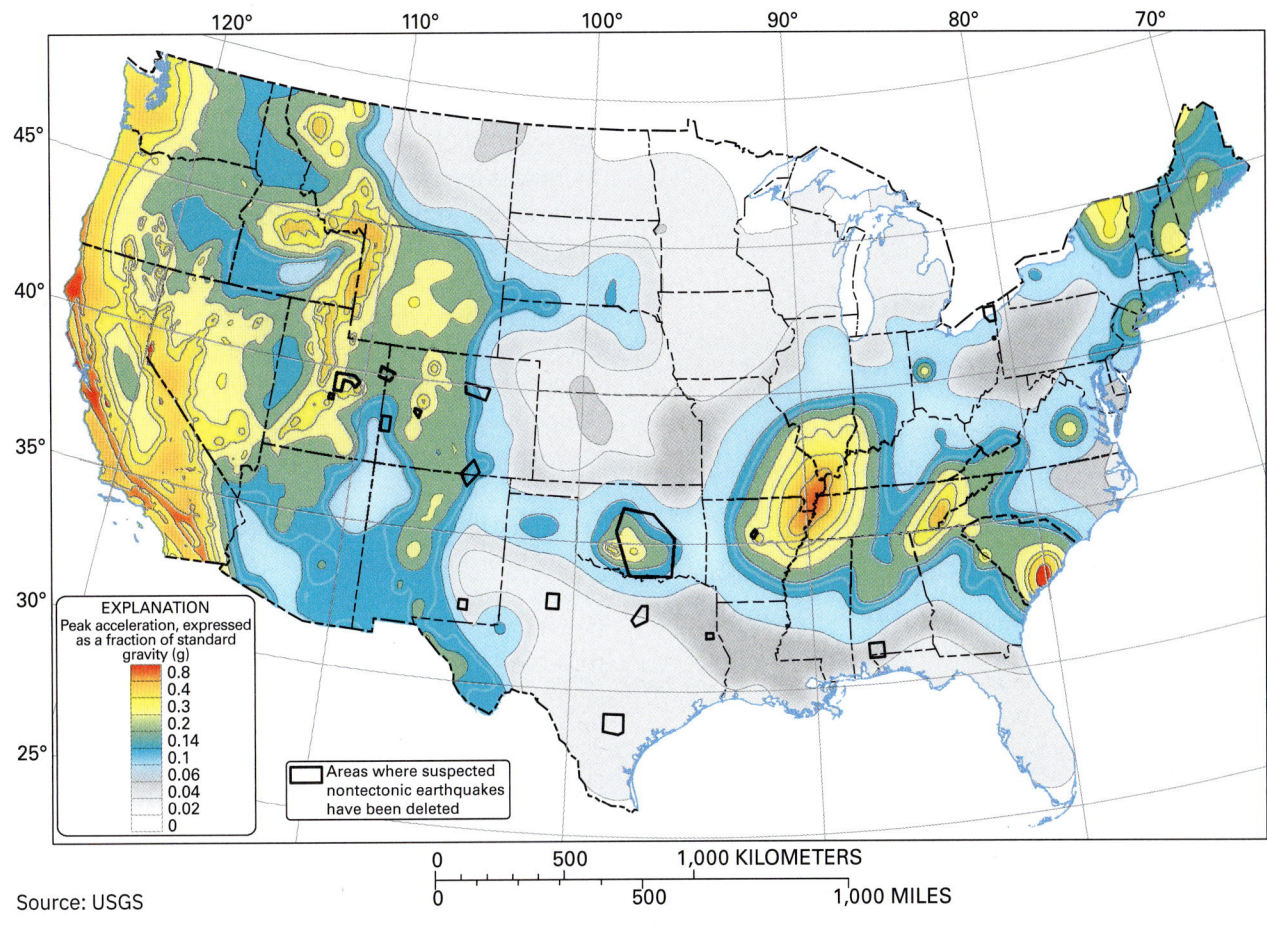

Source: USGS

In 2005, researchers reported that the rocks adjacent to the New Madrid Fault are currently deforming about as fast as rocks adjacent to active tectonic plate boundaries. This deformation indicates that additional earthquakes are likely there in the future, and the USGS has assigned a high risk of future seismicity to the region as shown on the seismic hazards map (Figure 8-21).

checkpoint What movement of the North American plate causes earthquakes in the area of New Madrid, Missouri?

8.2 ASSESSMENT

1. **Summarize** What pattern did modern geologists confirm about where earthquakes frequently occur?

2. **Explain** What causes the movement of the Pacific and North American plates to potentially result in a large destructive earthquake every hundred years?

3. **Compare** How do earthquakes at plate interiors differ from earthquakes at convergent plate boundaries?

4. **Define** What is a Benioff zone? At what type of tectonic boundary does it occur?

Critical Thinking

5. **Evaluate** Consider the significant zones for earthquake activity: San Andreas Fault zone, New Madrid Fault zone, and Mid-Ocean Ridge system. Make an evaluation of earthquake history and severity in these areas or zones. Present your information in a chart.

LESSON 8.2

8.3 EARTHQUAKE DAMAGE

Core Ideas and Skills
- Analyze factors other than earthquake magnitude that help determine the extent of earthquake damage.
- Connect earthquakes with tsunamis and summarize characteristics of a tsunami.
- Recount significant earthquakes that included tsunamis.
- Compare and contrast short-term and long-term predictions.

KEY TERMS

liquefaction   foreshock
tsunami

Most earthquake fatalities and injuries occur when falling structures crush people. Structural damage, injury, and death depend on the magnitude of the quake, its proximity to population centers, rock and soil types, topography, and the quality of construction in the region.

Influence of Rock and Soil

In many regions, bedrock lies at or near Earth's surface, and buildings are anchored directly to the rock. Bedrock vibrates during an earthquake, and buildings may fail if the motion is violent enough. However, most bedrock returns to its original shape when the earthquake is over. If structures can withstand the shaking, they will survive. Therefore, bedrock forms a desirable foundation in earthquake-prone areas.

In many places, structures are built on sand, silt, or clay. Sandy sediment and soil commonly settle during an earthquake. This displacement tilts buildings, breaks pipelines and roadways, and fractures dams. To avert structural failure in such soils, engineers drive steel or concrete pilings through the sand to the bedrock below. These pilings anchor and support the structures, even if the ground beneath them settles.

Mexico City provides one example of what can happen to clay-rich soils during an earthquake. The city is built on a high plateau ringed by even higher mountains. European settlers built the modern city on water-soaked, clay-rich, sediment deposited in an ancient former lake. On September 19, 1985, an earthquake with a moment magnitude of 8.1 struck about 500 kilometers west of the city. Seismic waves shook the wet clay beneath the city and reflected back and forth between the bedrock sides and bottom of the basin, just as waves in a bowl of jelly bounce off the sides and bottom of the bowl. The reflections amplified the waves, which destroyed more than 500 buildings and killed between 8,000 and 10,000 people. Meanwhile, there was comparatively little damage in Acapulco, a city closer to the epicenter but built on bedrock.

If soil is saturated with water, the sudden shock of an earthquake can cause the grains to shift relative to one another. When this occurs, the pressure in the pore water filling the spaces between grains increases. The pore pressure may rise enough to temporarily suspend the grains in the water, so that they are no longer held together by frictional forces. In this case, the soil loses its shear strength and behaves like a fluid in a process called **liquefaction**. A familiar example of liquefaction occurs when you tap your foot on saturated sand at the beach. The forces produced by your tapping foot cause compression waves to travel into the sand, liquefying it, and producing a slurry of sand and water that flows over your toes. When soils liquefy on a hillside due to earthquake shaking, they can be powerful enough to rip buildings off their foundation and carry them downslope. Landslides are common effects of earthquakes.

checkpoint How did seismic waves affect Mexico City's clay-rich sediment when an earthquake struck in 1985?

Construction Design and Earthquake Damage

Structural failure is caused by a variety of factors, including soil type and distance from the epicenter. But the tremendous differences in loss of human life between poor, less-developed countries and richer, developed countries are due to one simple factor: quality of construction. In addition to differences in the ability to withstand shaking associated with earthquakes, building materials differ in their susceptibility to fires caused by rupturing of electrical or gas lines during the quake.

Some common framing materials used in buildings, such as wood and steel, bend and sway during an earthquake, but they resist failure. However, brick, stone, adobe (dried mud), and other masonry products are brittle and likely to fail during an earthquake. Although masonry can be reinforced

FIGURE 8-22
The map shows the location of the 2004 Sumatra-Andaman earthquake, the third largest seismic event ever recorded.

with steel, in many regions of the world people cannot afford such structural reinforcement.

Loss of human life from earthquakes is an example of the interface between natural systems and human ones. Forces deep within Earth, beyond human control, drive earthquake frequency and magnitude. But earthquake mortality is closely connected to political and economic systems.

checkpoint What is the benefit of using wood and steel as framing materials in buildings?

Tsunamis

When an earthquake occurs beneath the sea, part of the seafloor rises or falls. Water is displaced in response to the seafloor movement, forming a wave or a series of waves. Sea waves produced by an earthquake are often called tidal waves, but they have nothing to do with tides. Therefore, geologists call them by their Japanese name, **tsunami**.

In the open sea, a tsunami is so flat that it is barely detectable. Typically, the crest may be only 1 to 3 meters high. Successive crests may be more than 100 to 150 kilometers apart. However, a tsunami may travel at 750 kilometers per hour. When the wave approaches the shallow water near shore, the base of the wave drags against the bottom and the water stacks up, increasing the height of the wave. The rising wall of water then flows inland. A tsunami can flood the land for as long as five to 10 minutes.

Today, the central part of the Indo-Australian plate is subducting beneath the islands of Sumatra and Java (Figure 8-22). On December 26, 2004, a fault rupture spanning a distance of 1,500 kilometers occurred. It is the longest fault rupture ever recorded. Like a giant piece of paper being torn over a period of 10 minutes, the fault rupture propagated from south to north at a velocity of about 1.2 kilometers per second. The resulting magnitude 9.2 Sumatra-Andaman earthquake was the third-largest seismic event ever recorded and caused the seafloor to move westward about 6 meters and upward about 2 meters. This tremendous displacement of rock triggered a massive tsunami that radiated in all directions. An estimated 286,000 people were killed. Survivors reported that, moments prior to the deadly wave, coastal water retreated and exposed dry mud in ocean bays. Then the wave raced inward, rearing upward as much as 24 meters, as high as a 10-story building, and inundating some areas 30 meters above sea level. Although mortality was highest in Sumatra, people died in coastal areas as far away as Port Elizabeth, South Africa, 8,000 kilometers from the epicenter.

An even larger tsunami occurred on March 11, 2011, when an earthquake with a moment magnitude of 9.0 struck off the eastern coast of Japan. The 2011 Tōhoku earthquake was the largest ever reported from Japan and the fourth largest on Earth since the beginning of seismic instrumentation. The earthquake resulted from a fault rupture where the Pacific plate is subducting beneath the northern Japanese island of Honshu at a rate of about 8 to 9 centimeters per year. The rupture was only about 250 kilometers long, roughly half of that normally needed to produce an earthquake of this magnitude. The rupture was only one-sixth the length of the rupture that caused the 2004 Sumatran-Andaman earthquake. However, the displacement across the ruptured fault in 2011 was estimated to be more than 50 meters in some places—more than many seismologists expected.

The 2011 Tōhoku earthquake lasted between three and five minutes and caused northern Japan to lurch about 4 meters closer to North America. This sudden change in the distribution of mass caused Earth's rotational axis to shift between 10 and 25 centimeters and the Earth to rotate slightly faster, effectively shortening the length of a day by 1.8 millionths of a second. Ultralow-frequency sound waves from the earthquake were even detected by satellites.

The Tōhoku earthquake produced a devastating tsunami that began to sweep ashore about an hour after the ground shaking and was estimated to be as high as 39 meters high in places. The tsunami washed as far as 10 kilometers inland, inundating more than 550 square kilometers. Entire towns were overrun by the water, as were more than 100 tsunami evacuation sites. The tsunami traveled all the way across the Pacific Ocean (Figure 8-23).

Recent analysis of the fault rupture zone through oceanic drilling and analysis of data from the dense cluster of seismometers located in and around Japan have shown that the very large size of the tsunami resulted from the manner in which the fault

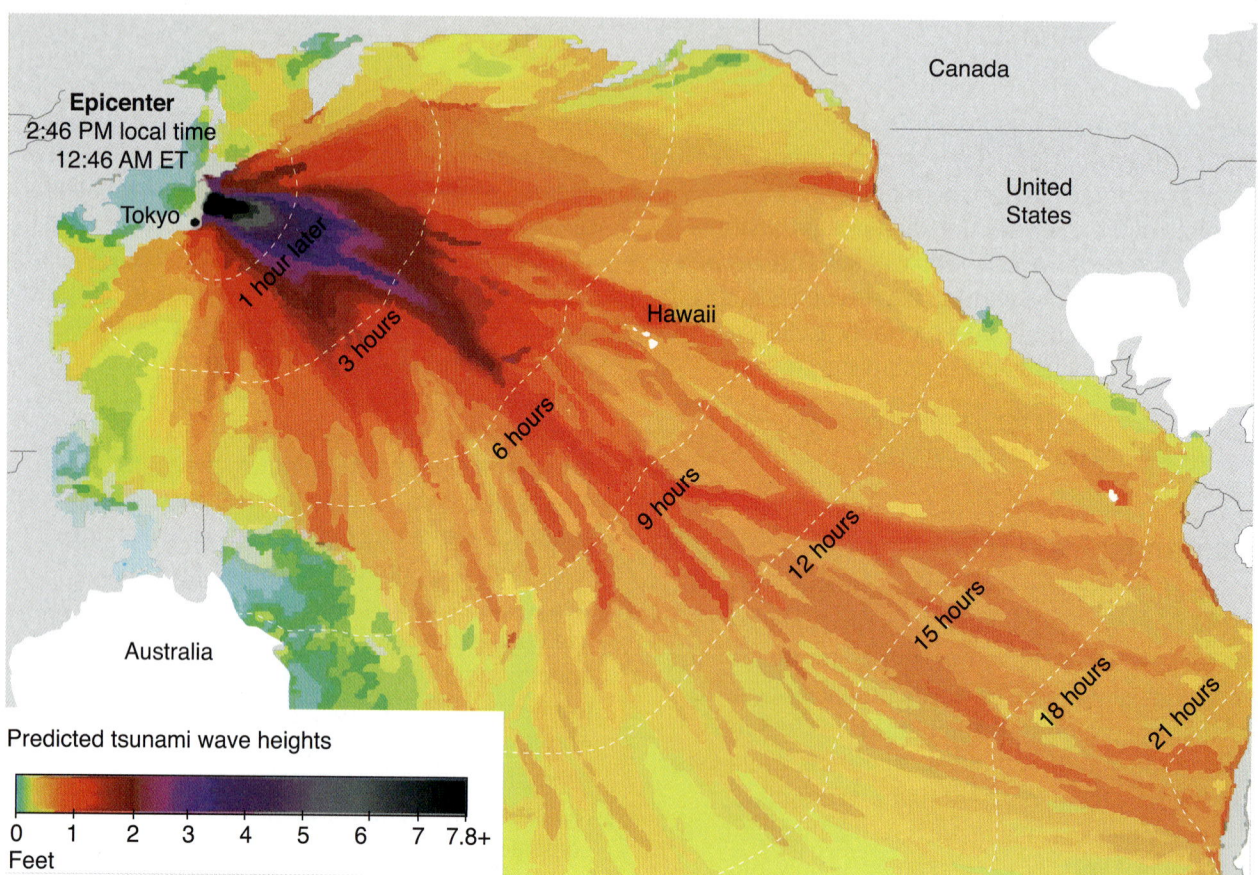

FIGURE 8-23
The map shows the path of the tsunami triggered by the 2011 Tōhoku earthquake in Japan. The tsunami traveled across the entire Pacific Ocean, taking more than 21 hours to reach the coast of Chile. The thin strip of red along the west coasts of the Americas indicate that the wave heights increased there due to the shallowing water depths.

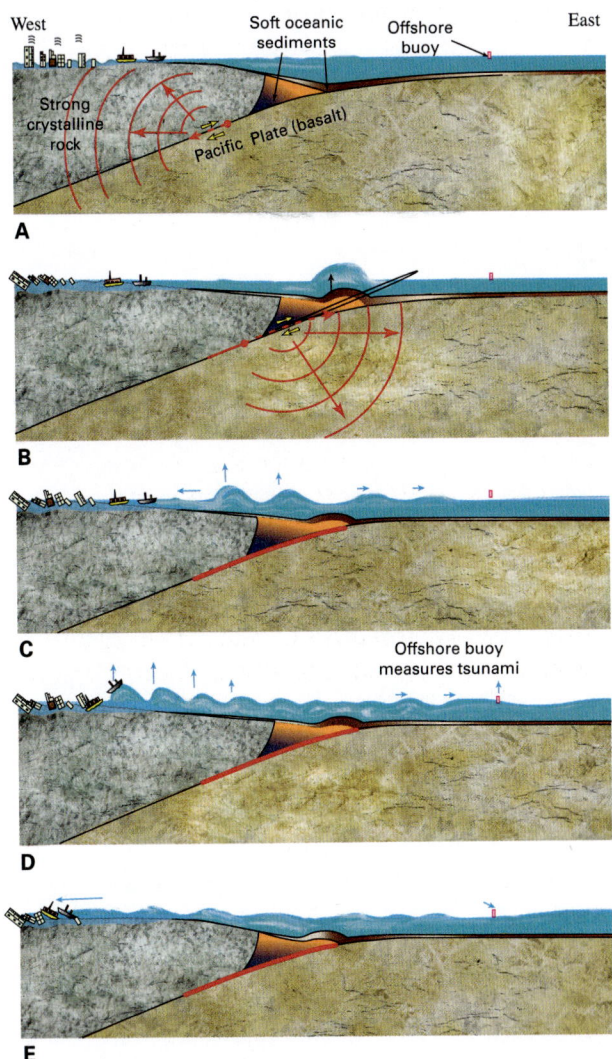

FIGURE 8-24
The illustration shows the evolution of the tsunami triggered by the 2011 Tōhoku earthquake in Japan. (A) The initial fault rupture started about 32 kilometers below the seafloor (red dot). Energy propagated west causing severe ground-shaking. (B) The fault rupture violently punched upward through the soft sediment near the seafloor and the deep water overlying it. (C) Tsunami waves traveling shoreward grew larger as the bottom shallowed. (D) As they approached the shoreline, the tsunami waves became even larger. (E) The tsunami rushed onshore.

broke (Figure 8-24). The fault rupture began along the plate boundary about 32 kilometers below the seafloor. During the first 40 seconds, the fault plane broke westward and downward, through relatively strong rock. Then, to the surprise of many geologists and geophysicists, the rupture continued eastward and upward from its starting point for another 35 seconds. As this fault break grew shallower and began to involve relatively soft sediments near the seafloor, its size increased dramatically, violently deforming the sediments, punching the seafloor upward, and triggering the large tsunami.

The 2011 Tōhoku earthquake and tsunami killed more than 15,000 people and left an additional 2,600 people missing and 6,000 injured. The earthquake also caused a dam to fail, set widespread fires including one at an oil refinery, and caused the third-largest nuclear accident in history. The nuclear accident resulted when water from the tsunami flooded backup generators needed to power circulation pumps that cooled nuclear reactors at the Fukushima Daiichi nuclear power plant. The lack of cooling water caused the reactors to overheat and melt down. In turn, several explosions released radioactivity into the surrounding environment.

checkpoint What were the sources of the two earthquakes described in the lesson, the 2004 Sumatra-Andaman earthquake and the 2011 Tōhoku earthquake?

Earthquake Predictions

Long-Term Predictions Long-term earthquake prediction recognizes that earthquakes have recurred many times along existing faults and will probably occur in these same regions again. For example, in the United States, although some faults exist in plate interiors (as in the New Madrid Fault zone), the most active faults are at plate boundaries along the West Coast and within the actively extending Basin and Range Province (Figure 8-26).

Long-term prediction tells us *where* earthquakes are likely to occur. This information is useful because it allows engineers to establish strict building codes in earthquake-prone regions. However, long-term prediction gives only a vague idea of *when* the next earthquake will strike. The last earthquake near New Madrid struck in 1812. Given an average 300-year interval, one might expect a quake in the year 2100, but such an analysis could be in error by hundreds of years.

Geophysicists have shown that the distribution of earthquakes in time cannot be statistically distinguished from random. However, recent analyses of large earthquakes suggest that they appear to be grouped in time in a synchronous manner, happening at the same time. The cause is not known. One interesting possibility posed by the authors of the study is that slight slowdown of Earth's rotation causes internal stress to build up because the lithosphere slows down at a different speed than other parts of Earth's interior. As a result of differences in the rate of slowdown, every few

FIGURE 8-25
ON ASSIGNMENT National Geographic photographer Michael Yamashita captured the effects of the 2011 earthquake in Japan in this image of the heavily flooded town of Ishinomaki. The town of Ishinomaki was one of the places where the 2011 earthquake lowered land elevation. A reduction in elevation can lead to frequent flooding. Yamashita has traveled across Asia, taking photographs that convey the rich history, both cultural and natural, of many places. Earthquakes are part of Japan's natural history. Such geologic events leave indelible effects on the people of Japan, a place that is geographically connected to active earthquake faults in the Ring of Fire.

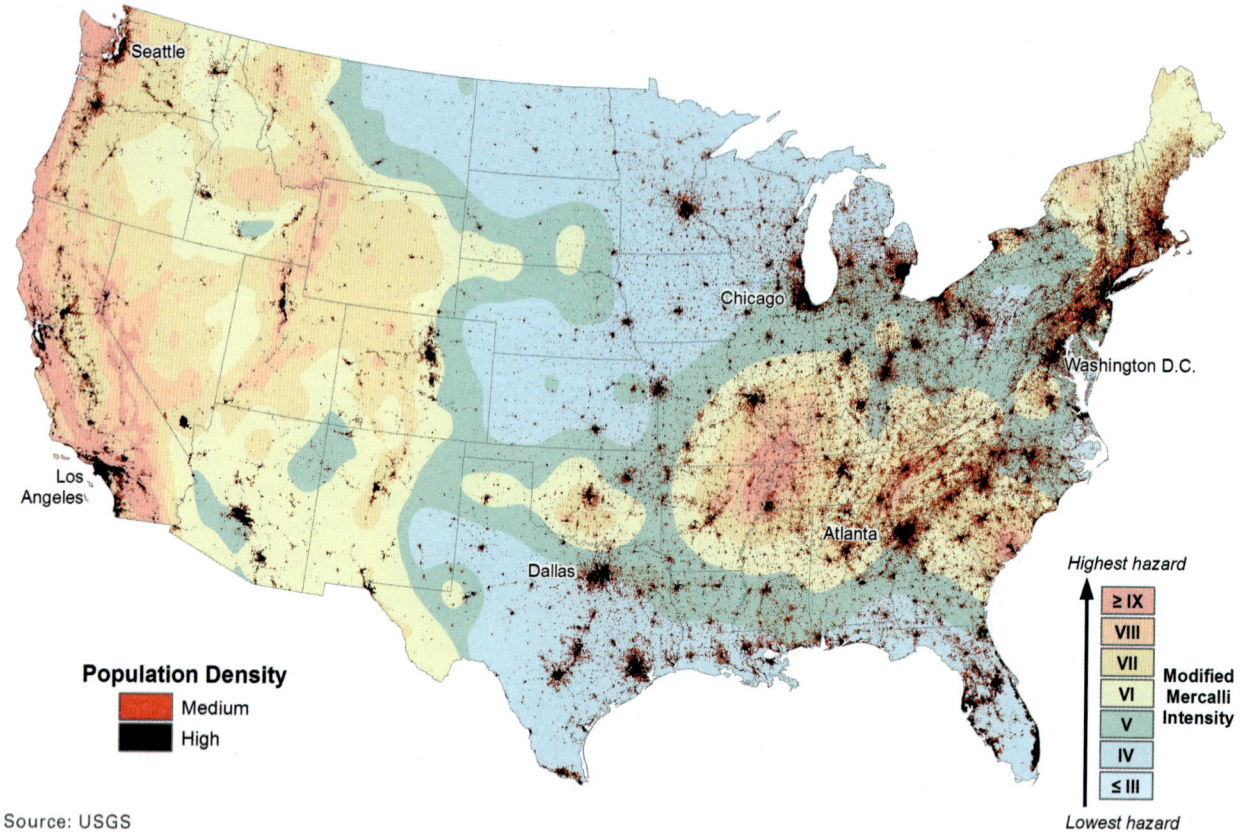

FIGURE 8-26
The map shows population density and the intensity of potential earthquake ground shaking that has a 2 percent chance of occurring in the next 50 years.

decades Earth has a cluster of large earthquakes to relieve the stress that has accumulated.

Short-Term Predictions Short-term predictions are forecasts that an earthquake may occur at a specific place and time. Short-term prediction depends on signals that immediately precede an earthquake.

Foreshocks are small earthquakes that precede a large quake by a few seconds to a few weeks. The cause of foreshocks can be explained by a simple analogy. If you try to break a stick by bending it slowly, you may hear a few small cracking sounds just before the final snap. Foreshocks are not a reliable tool for short-term prediction because they do not precede all earthquakes. Of those selected for study, foreshocks preceded only about half of the earthquakes. In addition, some swarms of small shocks thought to be foreshocks were not followed by a large quake.

Another approach to short-term earthquake prediction is to measure changes in the land surface near an active fault zone. Seismologists monitor unusual earth movements with tiltmeters and laser surveying instruments to detect distortions of the crust that may precede a major earthquake. This method has successfully predicted some earthquakes, but in other instances the predicted quakes did not occur or quakes occurred that had not been predicted.

Other types of signals can be used in short-term prediction. When rock is deformed prior to an earthquake, microscopic cracks may develop as the rock approaches its rupture point. In some cases, the cracks release radon gas previously trapped in the rock. The cracks may fill with water and cause the water levels in wells to fluctuate. Air-filled cracks do not conduct electricity as well as solid rock does, so the electrical conductivity of rock also decreases as the microscopic cracks form.

Over the past few decades, short-term prediction has not been reliable. Some geologists continue to search for reliable indicators for short-term earthquake prediction, but many geologists have concluded that this goal will remain elusive.

checkpoint What is usually the most reliable type of information given in long-term earthquake predictions?

CAREERS IN EARTH AND SPACE SCIENCES

Earthquake Engineer

Both civil engineers and structural engineers can specialize in engineering for earthquakes. With their expertise, earthquake engineers plan structures using earthquake design principles. They analyze seismic hazards and the dynamics of structure. Such engineers can plan structures that minimize the effects of ground shaking. For example, building foundations in seismically active regions might include springs that absorb much of the shock before it reaches the building itself. Earthquake engineers also study damage from past earthquakes to determine building and construction codes. Structural engineers who specialize in earthquake engineering helped plan and direct the update and the extension to the San Francisco-Oakland Bay Bridge. The original bridge sustained damage after the 1989 Loma Prieta earthquake. New construction has replaced and extended the bridge. The bridge now features a seismically resistant design intended to protect it from future earthquakes.

8.3 ASSESSMENT

1. **Identify** In addition to earthquake magnitude, name four other factors that affect earthquake damage.
2. **Explain** How do bedrock foundations help maintain buildings in earthquake-prone areas?
3. **Summarize** What is liquefaction, and how can liquefaction add to the destruction caused by an earthquake?
4. **Relate** Describe the effects of the 2004 fault rupture near Sumatra and Java, including the change in the seafloor and the tsunami's effects. Explain the connection between changes beneath Earth and effects on land and water.

Critical Thinking

5. **Collect Data** Review the different pieces of data shared about the 2004 and 2011 earthquakes in Sumatra and Japan. Present key values in a chart and offer comparisons between the two earthquakes. Consider rupture length, magnitude, tsunami wave height, and geographically affected area.

8.4 STUDYING EARTH'S INTERIOR

Core Ideas and Skills
- Discuss principles of seismic wave behavior.
- Explain the significance of the Mohorovičić discontinuity.
- Summarize the body of knowledge that exists about Earth's core.
- Discuss the connection between the liquid outer core and Earth's magnetic poles.

KEY TERMS

Mohorovičić discontinuity

Discovery of the Crust–Mantle Boundary

Earth's deepest borehole, located in northern Russia, extends to a depth of 12 kilometers, or about one-third through the crust. Despite the lack of deeper boreholes, scientists have learned a remarkable amount about Earth's structure by studying the behavior of seismic waves. The principles necessary for understanding the behavior of seismic waves include the following points.

1. In a uniform, homogeneous medium, a wave radiates outward in concentric spheres and at constant velocity.
2. The velocity of a seismic wave depends on the nature of the material that it travels through. Seismic waves travel at different velocities in different types of rock, varying with the rigidity and density of that rock.
3. When a wave passes from one material to another, it bends, or refracts, and sometimes reflects back. Boundaries between Earth's layers refract and reflect back seismic waves.
4. P waves are compressional waves and can travel through gases, liquids, and solids. S waves are shear waves and travel only through solids.

In 1909, Croatian seismologist Andrija Mohorovičić discovered that seismic waves passing through the upper mantle travel more rapidly than those passing through the shallower crust. By analyzing earthquake arrival-time data from many different seismographs, he identified the boundary between the crust and the mantle. Today, this boundary is called the **Mohorovičić discontinuity** (moh-ho-ro-vi-chich), or simply the *Moho*, in honor of its discoverer.

The Moho lies at a depth ranging from 4 to 70 kilometers. As you have learned, oceanic crust is thinner than continental crust, and continental crust is thicker under mountain ranges than it is under plains. Although the present-day Moho has never been sampled directly, the Moho probably corresponds to a change in rock type that affects the velocity of earthquake waves.

checkpoint Why do seismic waves travel at different velocities?

The Structure of the Mantle

The mantle is almost 2,900 kilometers thick and makes up about 80 percent of Earth's volume. Much of our knowledge of the composition and structure of the mantle comes from seismic data. Seismic waves speed up abruptly at the crust–mantle boundary (Figure 8-27). Seismic waves slow down again when they enter the plastic and partially melted asthenosphere at a depth between 75 and 125 kilometers. At the base of the asthenosphere, 350 kilometers below the surface, seismic waves speed up again because increasing pressure overwhelms the temperature effect and the mantle becomes stronger and more rigid.

At a depth of about 660 kilometers, seismic wave velocities increase again because pressure is great enough there to produce denser minerals. The zone where the change occurs is called the 660-kilometer discontinuity. The base of the mantle lies at a depth of 2,900 kilometers. Recent research has indicated that the base of the mantle, at the core–mantle boundary, may be so hot that despite the tremendous pressure, rock in this region is partially liquid.

checkpoint How do seismic waves move differently within the asthenosphere, and why?

Discovery of the Core

Scientists use seismic evidence to infer the characteristics of the core. Using a global array of seismographs, seismologists can detect direct P and S waves up to 105° from the focus

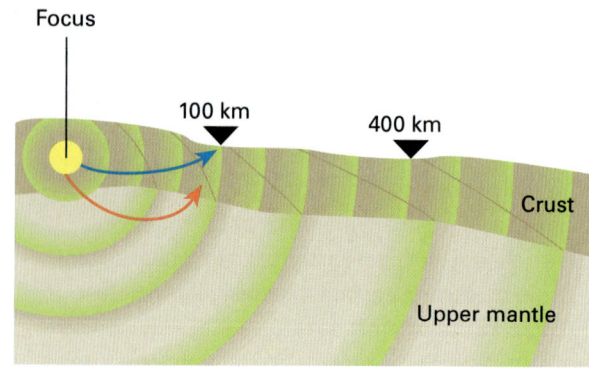

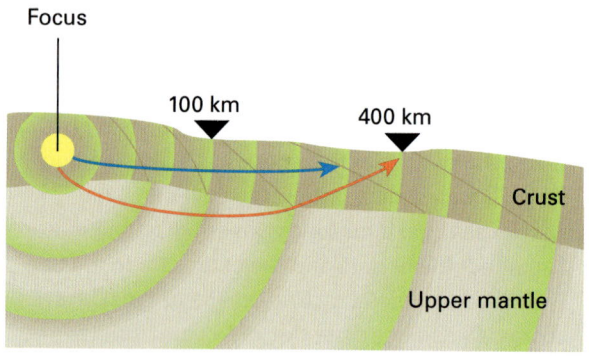

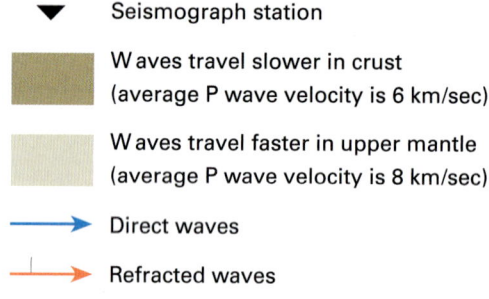

FIGURE 8-27
This illustration shows the velocities of P waves in the crust and the upper mantle. Generally, the velocity of P waves increases with depth.

of an earthquake. Between 105 and 140° is a "shadow zone" where no direct P waves arrive at Earth's surface. This shadow zone is caused by a discontinuity, which is the mantle–core boundary. When P waves pass from the mantle into the core, they are refracted as shown in the (Figure 8-28). The refraction deflects the P waves away from the shadow zone.

No S waves arrive beyond 105°. Their absence in this region shows that they do not travel through the outer core. Recall that S waves are not transmitted through liquids, so the failure of S waves to pass through the outer core indicates that it is liquid.

Refraction patterns of P waves indicate that another boundary exists within the core

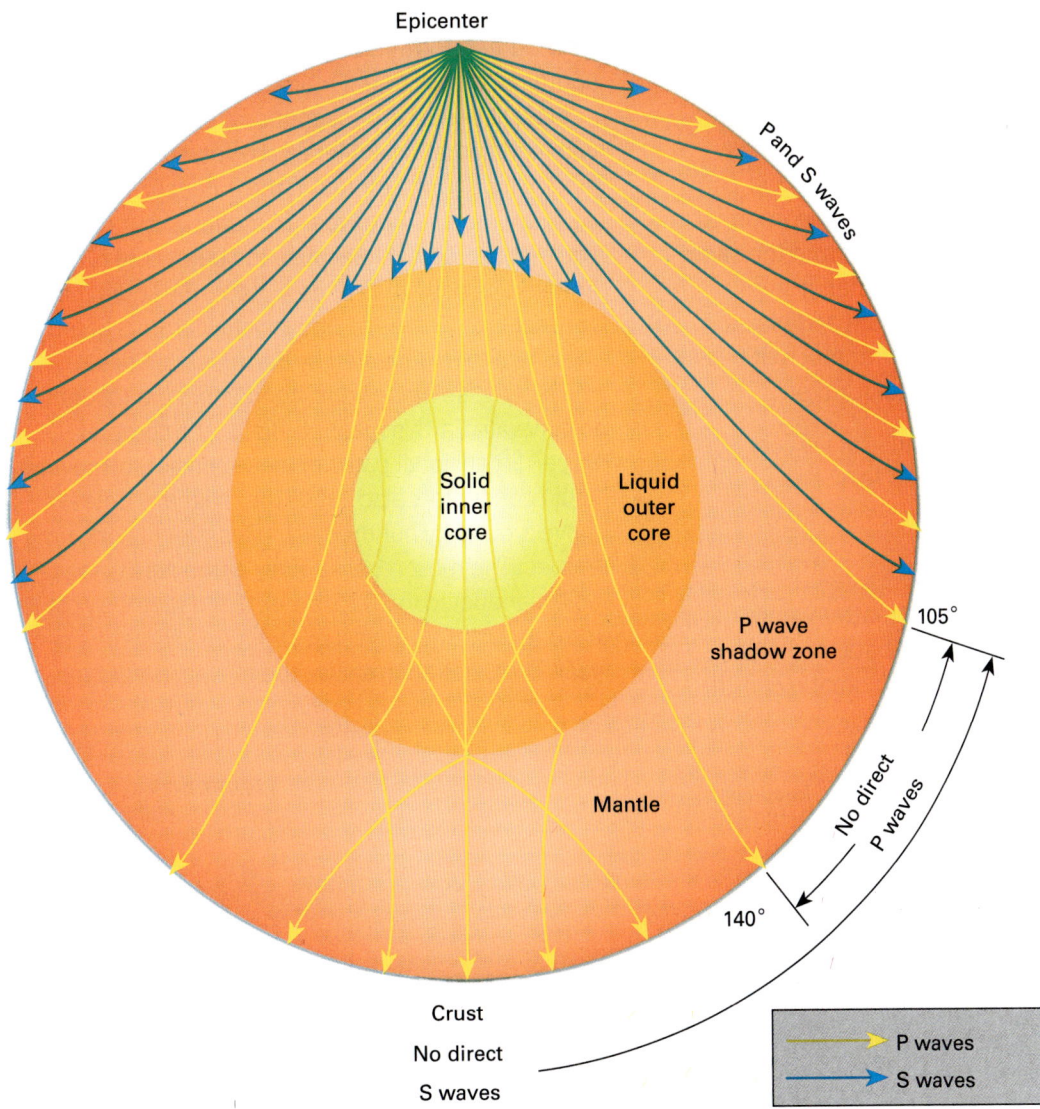

FIGURE 8-28
This illustration shows a cross section of Earth showing paths of seismic body waves. The waves bend gradually because of increasing pressure with depth. The body waves also bend sharply where they cross major layer boundaries in Earth's interior. Note that S waves do not travel through the liquid outer core and therefore are observed only within an arc of 105° from the epicenter, creating an S-wave shadow elsewhere. P waves are refracted sharply at the core–mantle boundary, so there is a P-wave shadow from 105° to 140°.

(Figure 8-28). It is the boundary between the liquid outer core and the solid inner core. Although seismic waves tell us that the outer core is liquid and the inner core is solid, other evidence tells us that the core is composed of iron and nickel.

Data shows seismic waves travel at different speeds in different directions through the inner core. Different rates of speed reveal that the inner core is not homogeneous. One series of measurements suggests that the inner core is rotating at a significantly faster rate than the mantle and crust. Some researchers have proposed that the solid inner core may convect just as the solid mantle does.

The overall density of Earth is 5.5 grams per cubic centimeter (g/cm^3), but both crust and mantle have average densities less than this value. The density of the crust ranges from 2.5 g/cm^3 to 3.0 g/cm^3, and the density of the mantle varies from 3.3 g/cm^3 to 5.5 g/cm^3. Because the mantle and crust account for slightly more than 80 percent of Earth's volume, the core must be very dense to account for the average density of Earth.

Calculations show that the density of the core must be 10 g/cm³ to 13 g/cm³, which is the density of many metals under high pressure.

Many meteorites are composed mainly of iron and nickel. Astronomers think that meteorites formed at about the same time that Earth did and that they reflect the composition of the primordial solar system. Because Earth formed from the same material as meteorites and other solar system objects, scientists conclude that iron and nickel must be abundant on Earth and that much of this iron and nickel exists as the metallic core.

checkpoint What is the composition of the core and how does the core's density support this?

Earth's Magnetism

Early navigators learned that no matter where they sailed, a needle-shaped magnet balanced on a point would align itself in an approximately north–south orientation. From this observation, they learned that Earth has a north magnetic pole and a south magnetic pole (Figure 8-29).

Most likely, Earth's magnetic field is generated within the outer core. Metals are good conductors of electricity, and the metals in the outer core are liquid and very mobile. Two types of motion occur in the liquid outer core: (1) Because the outer core is much hotter at its base than at its surface, the liquid metal convects. (2) The rising and sinking currents of molten metal are then deflected by Earth's spin. These convecting, spinning liquid conductors are thought to generate Earth's magnetic field. The complex, three-dimensional paths of these constantly moving liquid conductors also cause the magnetic poles to move around by hundreds of miles over historic time frames. The most recent survey of the north magnetic pole, conducted in 2015, determined that the pole is moving north-northwest at about 55 kilometers per year. The illustration shows the movements of the

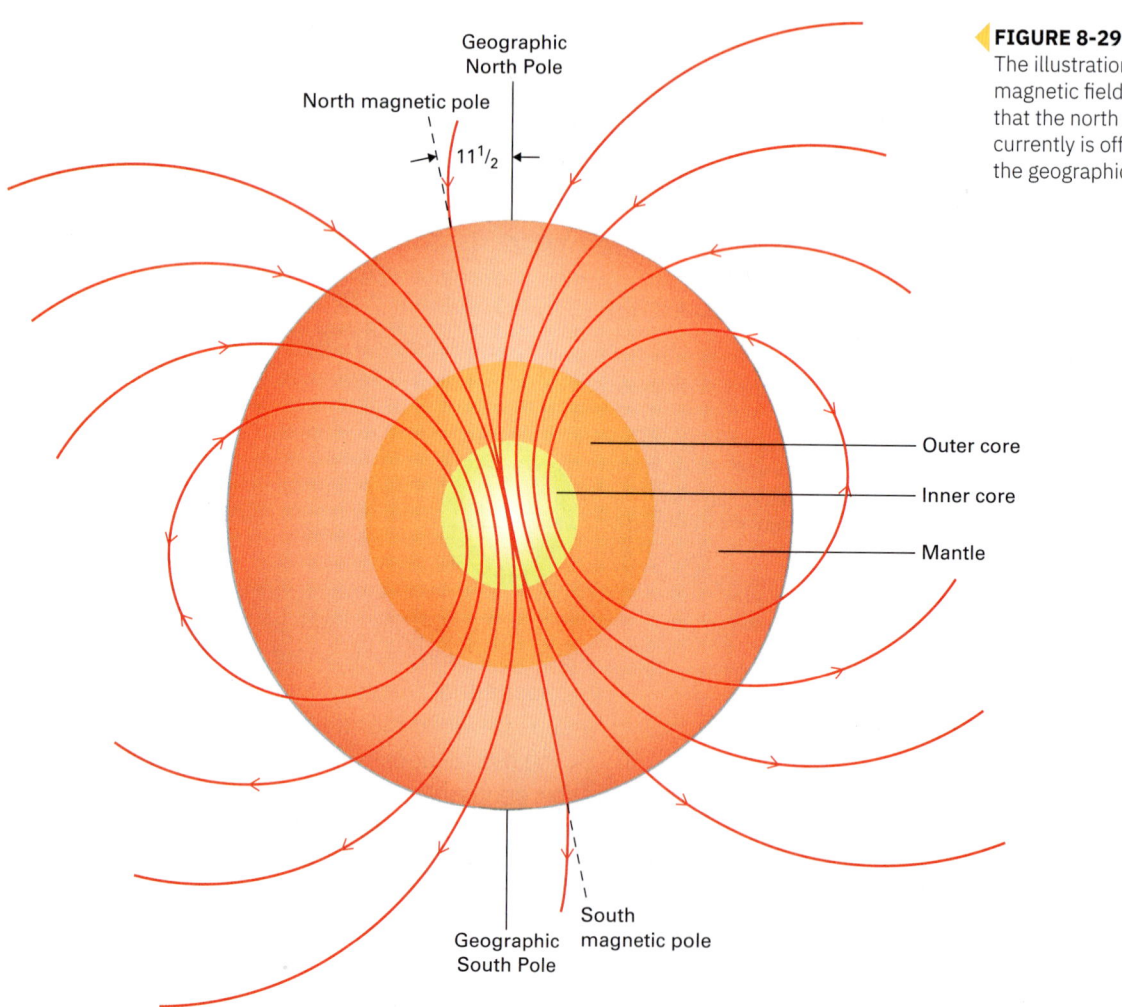

FIGURE 8-29
The illustration shows the magnetic field of Earth. Note that the north magnetic pole currently is offset 11.5° from the geographic North Pole.

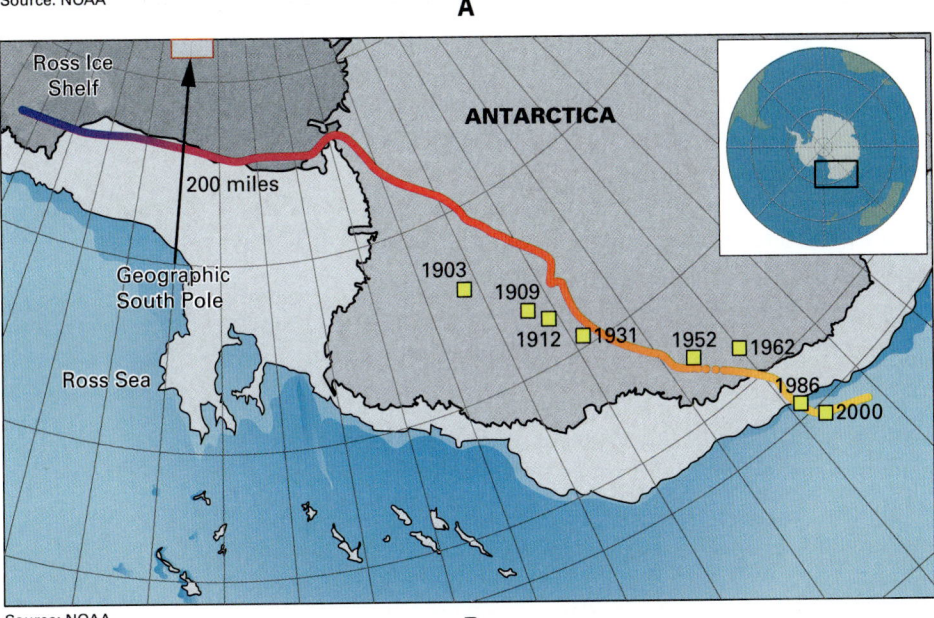

FIGURE 8-30
(A) Map A shows changes in the location of the north magnetic pole. (B) Map B shows changes in the location of the south magnetic pole over the past hundred years. The colored paths on both maps show the location of each magnetic pole between 1590 (blue end) and 2020 (yellow end), as calculated by computer modeling. The yellow squares in each map correspond to the observed location of each magnetic pole according to surveys conducted during the year next to each square. On both maps, the magnetic pole is far from the geographic pole, where the lines of longitude shown in the thin gray lines converge.

magnetic north and south magnetic poles over the past century (Figure 8-30).

checkpoint How do Earth's metals generate Earth's magnetic field?

8.4 ASSESSMENT

1. **Recall** What method did Andrija Mohorovičić use to arrive at the discovery of the boundary now named in his honor, the Mohorovičić discontinuity?
2. **Infer** What evidence enables us to infer that the inner core is not homogeneous?

Critical Thinking

3. **Synthesize** Using what you read about the structure of the mantle, explain the relationship between depth, pressure, and seismic waves.
4. **Infer** What inference can you make about the findings of the 2015 survey of the magnetic north pole?

LESSON 8.4 **257**

TYING IT ALL TOGETHER
EARTHQUAKE VULNERABILITY

In this chapter, you learned about the causes of earthquakes, the behavior of seismic waves, and some of the geologic and human factors that determine the amount of damage inflicted by earthquakes. In the Case Study, we compared three massive earthquakes that shook Mexico City in 1985 and 2017. While all three resulted in structural damage and human casualties, there were notable differences in the amount and type of damage.

The geography of Mexico City explains why it is so vulnerable to earthquakes. It is positioned near the boundaries of three tectonic plates—the North American, Cocos, and Pacific plates—with numerous subduction zones just off the Pacific coast. Certain areas of Mexico—including Mexico City—are prone to earthquake damage. Even though Mexico City is hundreds of kilometers from the coast, the soil composition of the city can amplify seismic waves.

Because the city was built on an ancient lakebed, its soil is a mixture of wet clay and sand that is prone to settling and liquefaction during an earthquake.

During the 1985 quake, shaking lasted nearly three minutes in Mexico City, causing rockslides, landslides, and openings in the ground. Numerous buildings collapsed as the ground settled, and ruptured gas lines caused fires and explosions. Since then, structural upgrades and Mexico's state-of-the-art early warning system have improved the city's ability to withstand major earthquakes. However, many of Mexico's impoverished rural areas do not benefit from the same level of preparedness. Remote villages still lack building codes or building inspectors, and poorly constructed buildings and roads abound. Since large earthquakes will remain a threat for the foreseeable future, it is critical for Mexico to continue improving its infrastructure to protect all of its people.

Research another country that has experienced the devastating effects of earthquakes, such as Indonesia, India, or Japan.

Use the questions below to guide your research. Then create a presentation to report your findings.

1. Why is the country prone to earthquakes? Describe any plate boundaries or faults in the area. What is the soil composition and how does it affect the outcome of earthquakes?

2. Compile a time line of earthquakes that have affected the country.

3. Based on the historical record, what are your short-term and long-term predictions for future earthquakes in the area?

4. Describe the structural failings that have occurred due to recent earthquakes in the country.

5. What factors have affected the amount of damage caused by major earthquakes in the country? What efforts have people made to prepare for or limit the effects of future earthquakes?

CHAPTER 8 SUMMARY

8.1 EARTHQUAKES AND SEISMIC WAVES

- An earthquake is a sudden motion or trembling of Earth caused by the abrupt release of slowly accumulated energy in rocks.
- Most earthquakes occur along tectonic plate boundaries.
- Earthquakes occur either when the energy accumulated in rock exceeds the friction that holds rock along a fault, or when the energy exceeds the strength of the rock and the rock breaks.
- Seismology is the study of earthquakes and the nature of Earth's interior based on evidence from seismic waves.
- An earthquake starts at the initial point of rupture, called the focus, typically lying below Earth's surface.
- The location on Earth's surface directly above the focus is the epicenter.
- Seismic waves include body waves, which travel through the interior of Earth, and surface waves, which travel on the surface.
- Surface waves travel more slowly than both types of body waves, P and S waves.

8.2 EARTHQUAKES AND TECTONIC PLATE BOUNDARIES

- Earthquakes are common at all three types of plate boundaries.
- The San Andreas Fault zone is an example of a strike-slip fault along a transform plate boundary.
- Along some portions of the fault zone, rocks slip past one another at a continuous, snail-like pace called fault creep.
- In other regions, friction prevents slippage until elastic deformation builds and is eventually released in a large earthquake.
- Subduction zone earthquakes occur along the Benioff zone when the subducting plate slips suddenly.
- Earthquakes occur in plate interiors along old faults.

8.3 EARTHQUAKE DAMAGE

- Earthquake damage is influenced by rock and soil type, construction design, and the likelihood of fires, landslides, and tsunamis.
- Long-term earthquake prediction is based on the observation that most earthquakes occur on preexisting faults at tectonic plate boundaries.
- Short-term prediction is based on occurrences of foreshocks, release of radon gas, and changes in the land surface, and the water table.

8.4 STUDYING EARTH'S INTERIOR

- Earth's internal structure and properties are known by studies of earthquake wave velocities and of the refraction and reflection of seismic waves as they pass through Earth.
- The boundary between the crust and the mantle, called the Mohorovičić discontinuity (or the Moho), lies at a depth ranging from 4 to 70 kilometers.
- Oceanic crust is thinner than continental crust, and continental crust is thicker under mountain ranges than it is under plains.
- The mantle is almost 2,900 kilometers thick and composes about 80 percent of Earth's volume. The upper portion of the mantle is part of the hard and rigid lithosphere.
- Beneath the lithosphere, the asthenosphere is plastic and partially melted.
- Beneath the asthenosphere, the mantle is stronger and less plastic; at the 660-kilometer discontinuity, the mineral composition of the mantle changes.
- Wave and density studies show that the outer core is liquid iron and nickel and the inner core is solid iron and nickel.
- Flowing metal in the outer core generates Earth's magnetic field.

CHAPTER 8 ASSESSMENT

Review Key Terms
Select the key term that best fits the definition. Not all terms will be used, and no term will be used more than once.

Benioff zone	Mohorovičić discontinuity
body wave	moment magnitude
earthquake	P wave
epicenter	Richter scale
fault	S wave
fault creep	seismic wave
focus	seismograph
foreshock	strike-slip fault
liquefaction	surface wave
Mercalli scale	tsunami

1. a sudden motion or trembling of Earth caused by the abrupt release of slowly accumulated elastic energy in rocks
2. body wave formed by alternating compression and expansion of rock parallel to the direction of wave travel
3. a vertical plate boundary across which rocks on opposite sides move horizontally
4. a measure of earthquake intensity that expresses the amount of energy released
5. the initial rupture point of an earthquake, typically located below Earth's surface
6. a geologic process in which a soil or sediment loses its shear strength and acts like a fluid during an earthquake
7. a seismic wave that radiates from an earthquake's epicenter and travels along Earth's surface
8. body wave that causes shearing motion in rock as the wave vibrates perpendicular to the direction of travel
9. the point on Earth's surface directly above the initial rupture point of an earthquake
10. any elastic wave that travels through rock, produced by an earthquake or explosion
11. a system for measuring of earthquake intensity based on destructive power and effects on buildings and people
12. a large, destructive sea wave produced by an undersea earthquake or volcano
13. an instrument that records seismic waves
14. a continuous, slow movement of solid rock along a fault, resulting from a constant stress acting over a long time

Review Key Concepts
Answer each question on a separate sheet of paper to demonstrate your understanding of key concepts from the chapter.

15. Explain what causes earthquakes.
16. Describe what a travel-time curve is and how it is used.
17. What type of fault is the San Andreas Fault? How do the rocks along the San Andreas Fault move?
18. Explain why earthquakes occur in a Benioff zone.
19. Describe how an earthquake might affect a city with soil consisting primarily of uncompacted loose sand, silt, or clay.
20. Explain how a tsunami forms.
21. Explain the difference between the Mohorovičić discontinuity and the 660-kilometer discontinuity.
22. Explain how S waves interact with Earth's outer core and why this is significant.
23. Explain how a seismograph is designed and how this design helps it record the waves of an earthquake.
24. Explain what causes foreshocks and how foreshocks can be used to predict the occurrence of an earthquake.
25. Explain what convection is and how it is responsible for generating Earth's magnetic field.

Think Critically
Write a response to each question on a separate sheet of paper. Use concepts from the chapter to support your reasoning.

26. Two earthquakes with the same moment magnitude strike two different cities. After analyzing the damage, seismologists determine that one city suffered a magnitude VI earthquake on the Mercalli scale whereas the other suffered a magnitude X earthquake. Give some reasons that may account for the differences.
27. Along some parts of the San Andreas Fault zone, rocks continuously slip past each other at a slow, steady rate. At other places, friction between rocks prevents slippage across the fault. Which parts of the fault would produce the largest earthquakes? Explain why.

28. Suppose you are an architect designing a housing development that will be built near the San Andreas Fault. Describe how the region's geology would affect your choices for building materials. What other strategies would you use to limit the potential for structural damage caused by earthquakes?

29. A P wave from an earthquake suddenly changes direction as it moves from a layer of basalt to clay. Explain why this occurs.

30. Compare current methods of long-term and short-term earthquake prediction in terms of their reliability and importance. Are long-term or short-term predictions more reliable? Which is currently more important in terms of preventing damage and loss of life? Make a claim and defend your reasoning.

PERFORMANCE TASK

Illustrate Release of Energy

Scientists can use the moment magnitude to measure the magnitude of earthquakes. The moment magnitude tells scientists how much energy is released by an earthquake. Its numerical value is calculated by the logarithm of the amplitude of the wave recorded on a seismometer.

In this performance task, you will research the severity and consequences of five recent earthquakes and use a model to compare them and make predictions.

Materials
computer with internet access
graph paper
pen or pencil

1. Use the USGS website, which can be found at https://earthquake.usgs.gov, to search for recent earthquakes. From the home page, select "Earthquakes" and then "Search Earthquake Catalog." Use the menus to identify recent earthquakes with magnitudes between 5.0 and 7.0. Select five of these earthquakes to research. Then search the internet for reports on the amount of damage caused by each earthquake.
    A. Record your findings in a three-column table. You might label the columns: "Earthquake Location and Date," "Magnitude," and "Summary of Damage."
    B. Draw a bar graph with a logarithmic y-axis scale showing the relative magnitude of each earthquake. Remember that a magnitude 7.0 earthquake is roughly 30 times as powerful as a 6.0 earthquake.
2. Choose a city near you and predict how it would be affected by a magnitude 5.0 earthquake and a magnitude 7.0 earthquake. Use your visual to compare the power of the two quakes, but also consider other factors that might affect the amount of damage, such as the rock and soil types in the city, its topography, and the quality of construction.

CHAPTER 9
VOLCANOES

The island country of Japan is connected to the longest chain of volcanoes on Earth—the Ring of Fire. Japan has more than a hundred active volcanoes. Mount Sakurajima in southern Japan has been consistently active since 1955. Records of this volcano's activity go as far back as the eighth century. Today the volcano spews ash in giant clouds that citizens in the nearby city of Kagoshima can regularly view.

Lightning accompanies volcanic eruptions when the ash cloud reaches great heights and the ash becomes electrified. Observations of lightning help scientists study the intensity of a volcanic eruption. Volcanologists look for patterns in the frequency and quantity of lightning strikes. As many as 500 lightning strikes have been recorded during a single volcanic eruption.

KEY QUESTIONS

9.1 What forms magma and how does it behave under different conditions?

9.2 How can volcanic activity change the landscape?

9.3 What are the characteristics of different types of eruptions and lava flows?

Mount Sakurajima is an active volcano in Japan. This eruption occurred on February 5, 2016. Turbulence in the ash cloud, which reached a height of 5 kilometers, caused the buildup of static electricity that resulted in lightning. In the foreground, glowing lava flows down the volcano's slopes.

EXPLORERS AT WORK

LOOKING AT LAVA FLOWS

WITH NATIONAL GEOGRAPHIC EXPLORER ARIANNA SOLDATI

Arianna Soldati calls her work as a volcanologist the "ultimate frontier of exploration on planet Earth." Soldati says that's because volcanoes build new land when they release molten magma from within Earth's crust. The magma—referred to as lava once it reaches the surface—eventually cools and hardens to become igneous rock. To stand on "newly erupted land," as Soldati calls it, is to stand on a place that literally did not exist on Earth the day before—a pristine new frontier. Volcanic activity can build up landscapes, extend coastlines, form underwater mountains, and even create islands.

Soldati's research focuses on volcanoes that produce lava flows rather than those that erupt violently and spew volcanic ash high into the atmosphere. Specifically, she is interested in basaltic lava flows, which are streams of partially molten rock that gently ooze or spill from a volcano and subsequently move downhill away from it. Some flows can move quickly and travel long distances before they cool to the point where they stop flowing. Along the way they destroy anything in their path, including human-built structures.

Being able to predict the path a lava flow will take during an eruption, how fast it will move, and how far it will travel could reduce hazards posed by effusive volcanic eruptions. Toward that end, Soldati works in the field and the lab to study factors that influence the behavior of basaltic lava flows. She collects much of her data from three sites: an active volcano in Guatemala, an active volcano on the island of La Réunion in the Indian Ocean, and an ancient lava field in California produced by some 40 extinct volcanoes. She has collected samples from present-day eruptions as well as samples of dark basaltic rock produced by previous flows.

Soldati studies the physical properties of the lava produced by these volcanoes because characteristics such as temperature, gas content, and mineral content affect how easily lava flows, its thickness and shape, and its overall dynamics. She also studies the terrain around each site to understand how lava flows over steep terrain versus gently sloping terrain, over flat areas, or through narrow channels. She compares flows of different sizes and durations. All this information helps her identify different patterns of lava movement.

Soldati has found that people have misconceptions about volcanoes. In fact, she says, adults often lack an opportunity to meet scientists to discuss science topics. That is why she launched the Science on Wheels program during her time as a Ph.D. student at the University of Missouri-Columbia. The program sends teams of graduate students into rural communities across Missouri to speak with adult audiences about their research. The presentations describe how students' research connects to everyday life. Soldati hopes these discussions re-engage adults with science. And she hopes to expand the program to other states.

THINKING CRITICALLY

Infer Arianna Soldati and graduate students visit communities to teach adults about volcanoes. What geographical areas might she and her program, Science on Wheels, prioritize to address misconceptions about volcanoes?

Arianna Soldati points out a type of lava, pahoehoe, while researching at Piton de La Fournaise, La Réunion.

Soldati stands in the lava ropes.

CASE STUDY
VOLCANIC ACTIVITY: PAST, PRESENT, AND FUTURE

In 79 C.E., the Roman cities of Pompeii, Herculaneum, and several neighboring villages were famously destroyed by the eruption of Mount Vesuvius in modern-day Italy. During the eruption, a pyroclastic flow streamed down the flanks of the volcano, burying the cities under 5 to 8 meters of hot ash. When archaeologists excavated Pompeii nearly 17 centuries later, they found plaster casts of inhabitants trapped by the sudden pyroclastic flow as they attempted to flee (Figure 9-1).

Nearly two millennia later, Mount Nyiragongo in central Africa has wreaked similar havoc on the city of Goma in the Democratic Republic of Congo. Goma is built atop the hardened lava flows from Nyiragongo's numerous eruptions. Nyiragongo has a large caldera at its summit that contains the world's largest lava lake. This lava periodically drains through fissures in the volcano's outer flanks. When this occurs, lava flows across the land at frighteningly high speeds. An eruption in 1977 killed hundreds of people, as did another in 2002 that flooded Goma's city center with lava and left about 200,000 homeless.

Nyiragongo and Vesuvius remain two of the world's most active and potentially dangerous stratovolcanoes. Yet the threat of eruptions has not deterred millions of people from living in their shadows. Vesuvius has erupted about three dozen times since 79 C.E. Yet the port city of Naples, Italy, lies less than 10 kilometers away, and some 3 million people live near the volcano. An eruption in 1906 prevented the 1908 Olympics from being held in Naples. The most recent eruption occurred in 1944 during World War II. That eruption forced the evacuation of an American bomber squadron and destroyed 88 of their planes. Goma, established in 1941, is home to about a million residents and refugees in a war-torn region near Congo's border with Rwanda. It is located on the southern fracture zone of Nyiragongo. After the widespread destruction of the 1977 and 2002 eruptions, the city was rebuilt on both of the lava flows. It is now the site of a United Nations peacekeeping mission that helps maintain political stability of the region.

Geologists consider both Vesuvius and Nyiragongo to be at high risk for future eruptions. With so many people nearby, the results could be catastrophic. To help mitigate this risk, observatories were established to provide a base of operations for scientists to monitor the volcanoes for signs of an impending eruption. With enough advance warning, residents of Naples and Goma may have time to escape and avoid a repeat of the sad fate of Pompeii.

As You Read Investigate the processes that contribute to volcanic activity. How does the behavior of magma produce volcanoes? What are the different kinds of volcanoes? What types of landforms do they create? Consider why people continue to live near active volcanoes and how scientists can mitigate the risks to life near an active volcano.

FIGURE 9-1
Plaster casts show two of the inhabitants of Pompeii who perished in the 79 C.E. eruption of Mount Vesuvius.

9.1 MAGMA

Core Ideas and Skills
- Analyze the process of magma formation.
- Describe magma behavior in a mantle plume and a subduction zone.
- Differentiate between the two main types of magma.
- Compare the effects of silica and water on magma behavior.

KEY TERMS

volcano
superheating
pressure-release melting
partial melting

In Chapter 4, we learned that rocks melt in certain environments to form magma. This process is one example of the constantly changing nature of rocks described by the rock cycle. Why does rock melt, and in what environments does magma form? To examine the effects of magma that reaches the Earth's surface as lava, first consider how temperature and pressure can alter solid rock. The processes involved in magma formation can eventually result in the formation of a volcano. A **volcano** is a hill or a mountain formed from lava, ash, and rock fragments ejected through a volcanic vent.

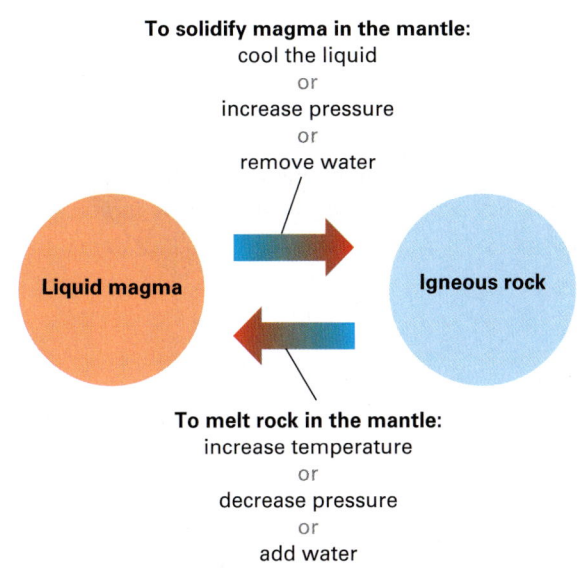

FIGURE 9-2
Cooling, increased pressure, and water loss all solidify magma to form igneous rock. Increasing temperature, the addition of water, and decreased pressure all melt rock to form magma.

Processes That Form Magma

Recall that the asthenosphere is the layer in the upper mantle that extends from a depth of about 100 kilometers to 350 kilometers. In the asthenosphere, the combined effects of temperature and pressure are such that one or two percent of the mantle rock is molten. Although the majority of the asthenosphere is solid rock, it is so hot and so close to its melting point that large volumes of rock can melt with relatively small changes in temperature, pressure, or the volume of water present. An increase in temperature will melt a hot rock (Figure 9-2). Oddly, however, increasing temperature is the least important cause of magma formation in the asthenosphere.

Recall that a mineral is a naturally occurring inorganic solid with a definite chemical composition and a crystalline structure. When a mineral melts, the bonds between the atoms are broken, causing the atoms to become disordered and move freely, taking up more space than the solid mineral. Magma occupies about 10 percent more volume than the rock that melted to form it (Figure 9-3). This behavior is opposite to that which occurs when ice melts to form water. In that case, the water is more dense and occupies a smaller volume than the equivalent amount of ice. That's why ice cubes float. In contrast, cubes of solid igneous rock would sink if tossed into a pool of completely molten magma of the same composition. If a rock is heated to its melting point on Earth's surface, it melts because there is little pressure to keep it from

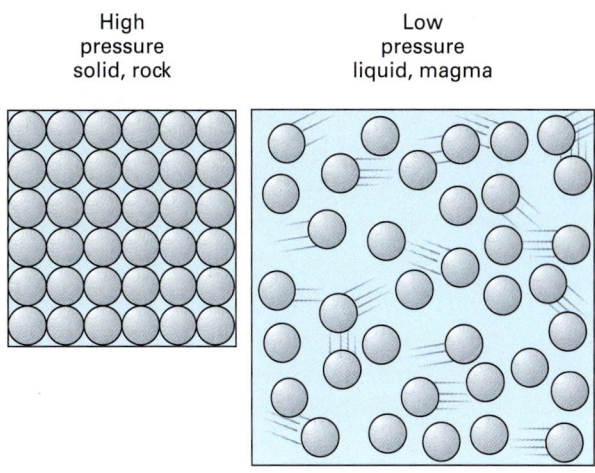

FIGURE 9-3
When most minerals melt, the volume of the resulting magma increases. As a result, high pressure favors the dense, orderly atomic arrangement of a solid mineral. Low pressure favors the random, less dense atomic arrangement of molecules in liquid magma.

LESSON 9.1 **267**

expanding. The temperature in the asthenosphere is hot enough to melt rock, but the high pressure prevents the rock from expanding, so it does not melt. The process in which a rock is heated above the melting point but does not melt is called **superheating**. The additional volume required for the rock to melt is not available because of the high pressure in the asthenosphere, so the rock remains mostly solid. However, if the pressure were to decrease, large volumes of superheated rock making up the asthenosphere would melt.

Melting caused by decreasing pressure is called **pressure-release melting** or decompression melting. A rock containing small amounts of water generally melts at a lower temperature than an otherwise-identical dry rock. Consequently, the addition of water to rock that is near its melting point can cause the rock to melt. Certain tectonic processes add water to the hot rock of the asthenosphere to melt it and form magma.

checkpoint Why does a rock under great pressure remain solid even though it is heated above its melting point?

Environments of Magma Formation

Magma forms abundantly in three tectonic environments: spreading centers, mantle plumes, and subduction zones. Let us consider each environment to see how rising temperature, decreasing pressure, and the addition of water can melt rock to create magma.

Spreading Centers
As lithospheric plates separate at a spreading center, the hot, plastic asthenosphere wells upward to fill the gap. As the asthenosphere rises, the surrounding pressure drops and pressure-release melting forms magma with a basaltic composition. Because the magma has lower density than the surrounding rock, it rises buoyantly toward the surface.

Most of the world's spreading centers lie in the ocean basins, where they form the Mid-Ocean Ridge system (Figure 9-4). The rising basaltic magma is injected into the spreading center where it solidifies to form new oceanic crust. Some magma erupts onto the seafloor. Once formed, the new oceanic crust then drifts away from the spreading center on both sides, riding atop the separating tectonic plates. Nearly all of Earth's oceanic crust is created in this way at the Mid-Ocean Ridge system. In most places, the ridge lies beneath the sea.

In a few places such as Iceland, the ridge rises above sea level, and basaltic magma erupts onto Earth's surface. Some spreading centers, such as the East African Rift or the North American Basin and Range Province, occur in continental crust.

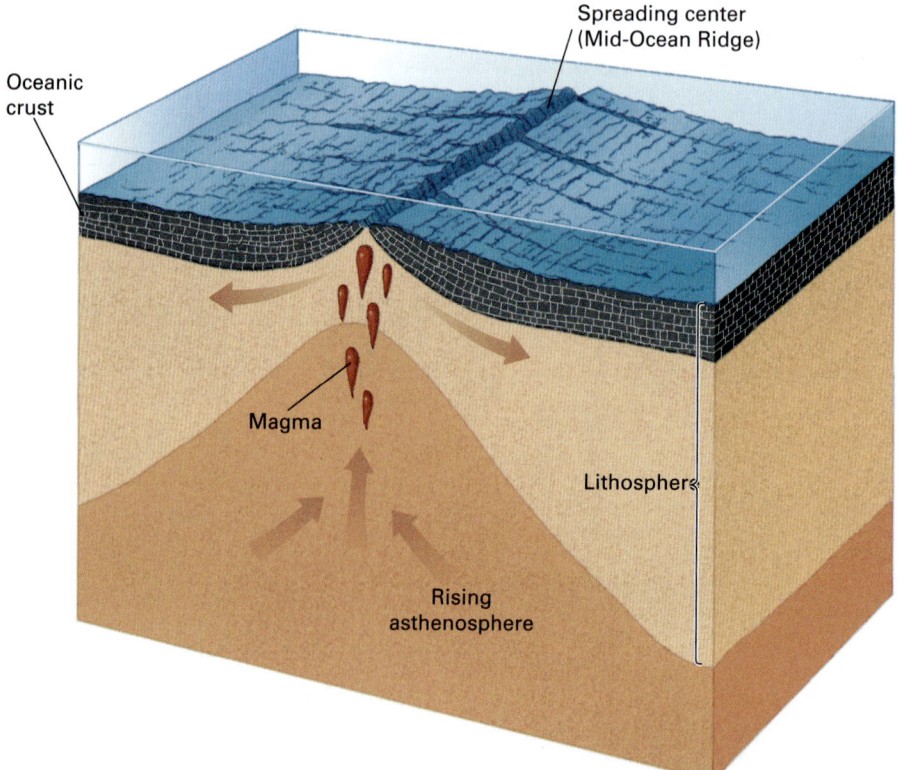

FIGURE 9-4 Pressure-release melting produces magma beneath a spreading center, where hot asthenosphere rises to fill the gap left by the two separating tectonic plates.

Thinning of the crust decreases the pressure on the hot asthenosphere below it. Basaltic magma forms and erupts onto the surface at continental rifts. Melting of the continental crust by the rising basaltic magma also produces felsic magma that erupts in continental rift settings.

Mantle Plumes Recall that a mantle plume is a rising column of hot, plastic rock that originates within the mantle. The plume rises because it is hotter than the surrounding mantle and, consequently, is less dense and more buoyant. As a plume rises, pressure-release melting forms magma, which continues to rise toward Earth's surface.

Because mantle plumes form below the lithosphere, they commonly occur within tectonic plates rather than at a boundary. For example, the Yellowstone Volcano—responsible for the volcanic activity, geysers, and hot springs in Yellowstone National Park—results from a shallow mantle plume that lies far from the nearest plate boundary. If a mantle plume rises beneath oceanic crust, volcanic eruptions build submarine volcanoes and volcanic islands. For example, the Hawaiian Islands are a chain of volcanoes that formed over a long-lived mantle plume beneath the Pacific Ocean (Figure 9-5).

Subduction Zones In a subduction zone, the addition of water, decreasing pressure, and heat from friction all combine to form huge quantities of magma. A subducting plate is covered by water-saturated ocean sediment. The upper portions of the underlying basalt also are saturated with water. As the wet rock and sediments dive into the hot mantle, the heated water ascends into the hot asthenosphere directly above the sinking plate.

As the subducting plate descends, it drags plastic asthenosphere rock down with it. Rock from deeper in the asthenosphere then flows upward to replace the sinking rock. Pressure decreases as this hot rock rises.

Friction generates heat in a subduction zone as the descending plate scrapes past the overriding plate. The addition of water, pressure release, and frictional heating combine to melt asthenosphere rocks in the zone where the subducting plate passes into the asthenosphere. The addition of water is probably the most important factor producing melting in a subduction zone. Frictional heating is the least important (Figure 9-6).

The subduction process leads directly to the formation of volcanoes. The volcanoes of the Pacific Northwest, the granite cliffs of Yosemite, and the Andes Mountains are all examples of volcanic and plutonic rocks formed through subduction. As noted in Chapter 8, the Ring of Fire in the Pacific Ocean is a chain of subduction zones at convergent plate boundaries (Figure 8-16).

checkpoint What happens if a mantle plume forms under oceanic crust?

Magma Types

Basalt and Basaltic Magma Basalt and granite are the most common igneous rocks. Basalt makes up most of the oceanic crust, and granite is the most abundant rock in continental crust. Basaltic magma forms by the melting of the asthenosphere. But the asthenosphere is peridotite. Basalt and peridotite are quite different in composition. Peridotite contains about 40 percent silica (SiO_2), but basalt contains about 50 percent.

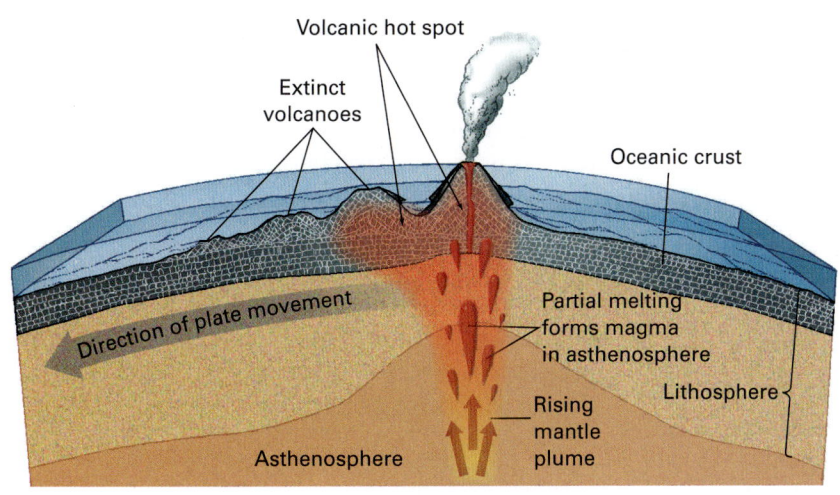

FIGURE 9-5
Pressure-release melting produces magma in a rising mantle plume. The magma rises to form a volcanic hot spot.

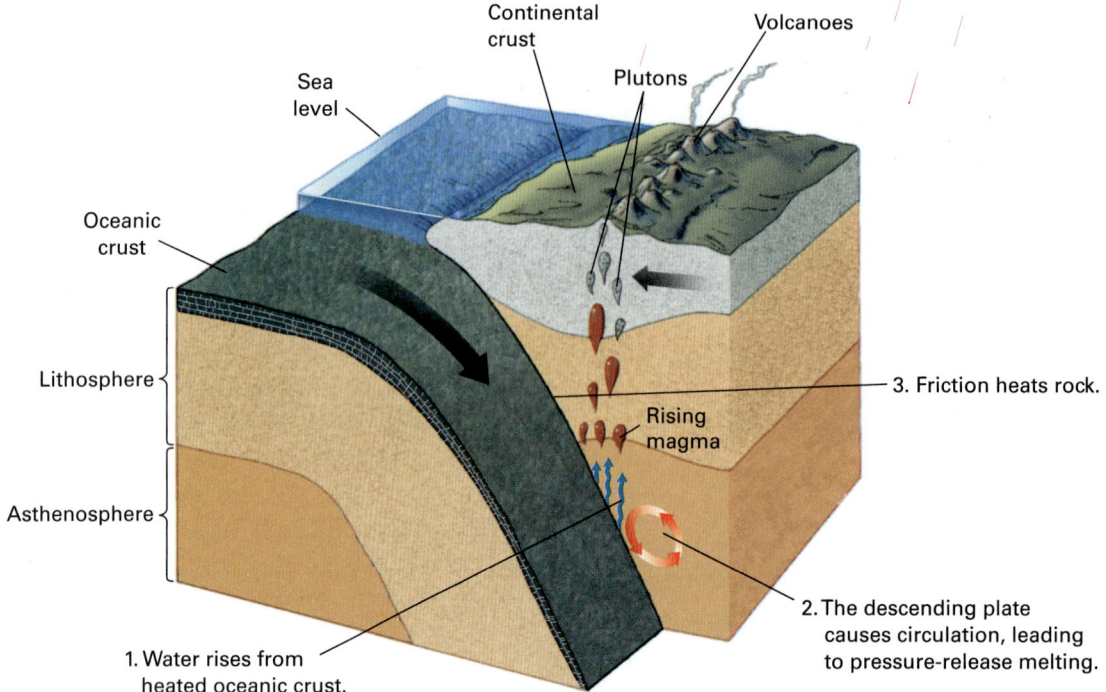

FIGURE 9-6
Three processes melt the asthenosphere to form magma at a subduction zone: (1) Geothermally heated water rises. (2) Circulation in the asthenosphere above the subducting plate causes decreasing pressure on hot mantle rock. (3) Friction heats rocks in the subduction zone.

Peridotite contains considerably more iron and magnesium than basalt, so how does it melt to create basaltic magma? Why does the magma have a composition different from that of the rock that melted to produce it?

It is a general rule that a mixture of two or more minerals will begin to melt at a temperature lower than the melting point of any one of the minerals in its pure state. Peridotite consists mainly of the minerals olivine and pyroxene, with small amounts of calcium feldspar. In one set of experiments designed to simulate the melting of mantle peridotite, pure olivine melted at 1,890°C, pure pyroxene melted at 1,390°C, and pure calcium feldspar melted at 1,550°C. However, peridotite rock consisting of all three minerals began to melt at 1,270°C.

In the process of **partial melting**, a silicate rock only partly melts and the composition of the magma is richer in silica than the original rock. When mantle peridotite melts, the basaltic magma is about 20 percent richer in silica than the peridotite. Because the new basaltic magma is less dense than the peridotite rock, the magma begins to rise toward Earth's surface. When rocks made of silicate minerals begin to melt, the most silicate-rich minerals melt first while the minerals with less silica remain solid. The result is a partial melt that is richer in silica than the original parent rock.

Granite and Granitic Magma Granite contains more silica than basalt and therefore melts at a lower temperature than basalt, typically between 700°C and 900°C. Basaltic magma that forms beneath a continent and then rises into the continental crust will cause the crust to partially melt. Because the lower continental crust is already hot, a small volume of basaltic magma that comes into contact with it can melt a large volume of lower continental crust to form granitic magma. Typically, the granitic magma rises a short distance and then solidifies within the crust to form granitic bodies. Most granitic bodies solidify at depths between about 5 and 20 kilometers. In continental rift zones where the crust has thinned, some granitic magma reaches the surface where it erupts as rhyolite. Granite forms by this process in a continental rift zone and through a mantle plume rising beneath a continent.

checkpoint What causes basaltic magma to rise toward Earth's surface?

Magma Behavior

Once magma forms, it rises toward Earth's surface because it is less dense than surrounding rock. As it rises, two changes occur:

- Magma cools as it enters shallower and cooler levels of Earth.
- Pressure drops because the weight of overlying rock decreases.

Cooling tends to solidify magma but decreasing pressure tends to keep it liquid. So, does magma solidify or remain liquid as it rises toward Earth's surface? The answer depends on the type of magma. Basaltic magma commonly remains liquid and rises to the surface to erupt from a volcano or flow onto the seafloor at the Mid-Ocean Ridge. In contrast, granitic magma usually solidifies within the crust.

The contrasting behavior of granitic and basaltic magmas is a consequence of their different compositions. Granitic magma contains about 70 percent silica, whereas the silica content of basaltic magma is only about 50 percent. In addition, granitic magma generally contains up to 10 percent water, but basaltic magma contains only one to two percent water.

TABLE 9-1 Composition of Magma Types

Typical Granitic Magma	Typical Basaltic Magma
70% silica; up to 10% water	50% silica; 1% to 2% water

Effects of Silica

Because of its higher silica content, granitic magma contains longer silicate chains than does basaltic magma. The long chains become tangled, causing the magma to become stiff, or viscous. It rises slowly because of its viscosity and has ample time to solidify within the crust before reaching the surface. In contrast, basaltic magma, with its shorter silicate chains, is less viscous and flows more easily. Because of its fluidity, it rises rapidly to Earth's surface.

MINILAB Soda Bottle Volcano

Materials
safety goggles
1-L bottle of plain seltzer or soda water
paper towels or rags

Procedure

1. Carefully place the bottle on the table. Avoid shaking or disturbing it.
2. Observe the bottle and its contents. Describe the appearance of the contents and any visible bubbles; the cap; and the firmness of the bottle (you may touch it gently). Record your observations.
3. Shake the bottle vigorously for 30 seconds, being careful to point the cap away from yourself and other people.
4. Place the shaken bottle back on the table. Observe how the appearance and feel of the bottle have changed or remained the same. Record your observations.
5. Shake the bottle once more for 30 seconds. Then pointing the bottle away from yourself and others, carefully unscrew the cap.
6. Observe what happens when you uncap the bottle. Record what you see and hear.
7. Clean up any spilled liquid immediately.

Results and Analysis

1. **Describe** How do your observations of the bottle before and after it was shaken differ?
2. **Predict** What do you think would happen to the size of the "eruption" if the bottle were shaken for 10 seconds instead of 30 seconds and then opened? What if it were shaken for more than 30 seconds? Explain your reasoning.

Critical Thinking

3. **Relate** What does the carbonated water represent in this model?
4. **Relate** Describe how parts of the soda bottle are similar to the structure of a volcano.
5. **Critique** How could this model be modified to more accurately represent a volcanic eruption?

FIGURE 9-7
ON ASSIGNMENT A team of volcanologists on an expedition into the Nyiragongo volcano in the Democratic Republic of the Congo rest at base camp. On this expedition in July 2010, National Geographic photographer Carsten Peter photographed the team as they entered the crater containing the volcano's lava lake. Volcanologists Dario Tedesco and National Geographic Explorer Kenneth Sims were there to collect a fresh sample from the lava lake. In carrying out this extremely dangerous task, Dr. Sims provided important information to better understand the dynamics of this volcano.

LESSON 9.1

Effects of Water A second difference between the two magmas is that granitic magma contains more water than basaltic magma does. Water lowers the temperature at which magma solidifies. If dry granitic magma solidifies at 700°C, the same magma with 10 percent water may not become solid until the temperature drops below 600°C.

Water tends to escape as steam from hot magma. But deep in the crust where granitic magma forms, high pressure prevents the water from escaping. As the magma rises, the pressure decreases and water escapes. Because the magma loses water, its melting temperature rises and it solidifies. Water loss causes rising granitic magma to solidify within the crust. Because basaltic magmas have only 1 to 2 percent water to begin with, water loss is relatively unimportant. As a result, rising basaltic magma usually remains liquid all the way to Earth's surface, erupting as lava through volcanic vents.

checkpoint How does the silica composition of basaltic magma affect the type of flow it has compared to granitic magma?

9.1 ASSESSMENT

1. **Describe** Under what conditions can rocks become superheated?
2. **Summarize** Describe the process through which nearly all of Earth's oceanic crust was created.
3. **Explain** Why do mantle plumes commonly form within tectonic plates instead of at plate boundaries?
4. **Distinguish** Describe two key differences between granitic and basaltic magma.
5. **Explain** Why does partial melting result in magma that is richer in silica than the original rock in its solid form?

Critical Thinking

6. **Synthesize** Consider a sample of magma within Earth's mantle. Describe the processes that transform the liquid magma to solid rock.

9.2 MAGMATIC BODIES AND VOLCANIC FORMATIONS

Core Ideas and Skills
- Identify and compare volcanic formations, including batholiths, plutons, and lava plateaus.
- Compare the characteristics of the different types of volcanoes and eruptions.

KEY TERMS

pluton	flood basalt
batholith	lava plateau
sill	vent
vesicle	crater
columnar joint	shield volcano
pyroclastic rock	cinder cone
volcanic ash	stratovolcano
cinder	

Plutons and Igneous Intrusions

The subduction process leads directly to the formation of large plutons and volcanoes. A **pluton** is any large mass of intrusive igneous rock. The volcanoes of the Pacific Northwest, the massive cliffs of Yosemite, and the Andes Mountains are all examples of volcanic and plutonic rocks formed through subduction (Figure 9-8).

Many granitic plutons measure tens of kilometers in diameter. How can such a large mass of viscous magma rise through solid rock?

FIGURE 9-8
Most of California's Sierra Nevada is made of granitic plutons, including these mountains in Yosemite National Park.

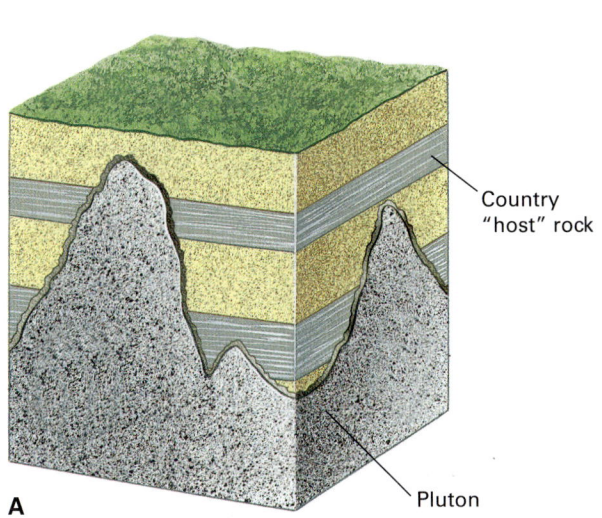
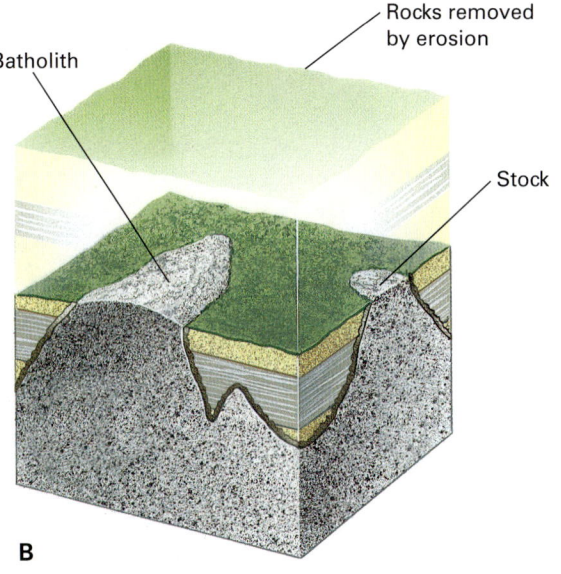

FIGURE 9-9
(A) A pluton is any large mass of intrusive igneous rock. (B) A batholith is a pluton with more than 100 square kilometers exposed at Earth's surface. A stock is similar to a batholith but has a smaller surface area.

If you place oil and water in a jar, screw the lid on, and shake the jar, oil droplets disperse throughout the water. When you set down the jar, the droplets coalesce to form larger blobs, which rise toward the surface, easily displacing the water as they ascend. Granitic magma rises in a similar way, except that the process is slower because it rises through solid rock. Granitic magma forms near the base of continental crust, where surrounding rock is hot and plastic. As the magma rises, it pushes aside the plastic country rock, which then slowly flows back to fill in behind the rising blobs of granitic magma (Figure 9-9).

After a pluton forms, tectonic forces may push that part of the crust upward, and erosion may expose parts of the pluton at Earth's surface (Figure 9-10). A **batholith** is a pluton exposed across more than 100 square kilometers of Earth's surface. An average batholith is about 10 kilometers thick, although a large one may be 20 kilometers thick. A stock is similar to a batholith but is exposed across less than 100 square kilometers.

A large body of magma engulfs or pushes country rock aside as it rises (Figure 9-11). In contrast, a smaller mass of magma may flow into a fracture or between layers in country rock. Dikes, discussed in Chapter 5, cut across sedimentary layers or other features in country rock and range from less than a centimeter to more than a kilometer thick. In some cases, a dike may be more resistant to weathering

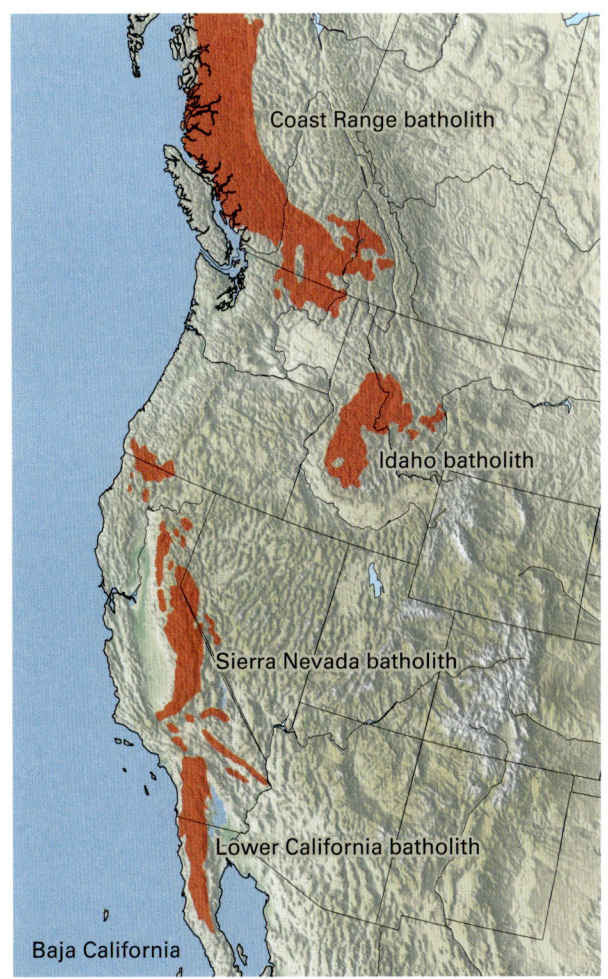

FIGURE 9-10
The large batholiths in western North America form high mountain ranges.

LESSON 9.2

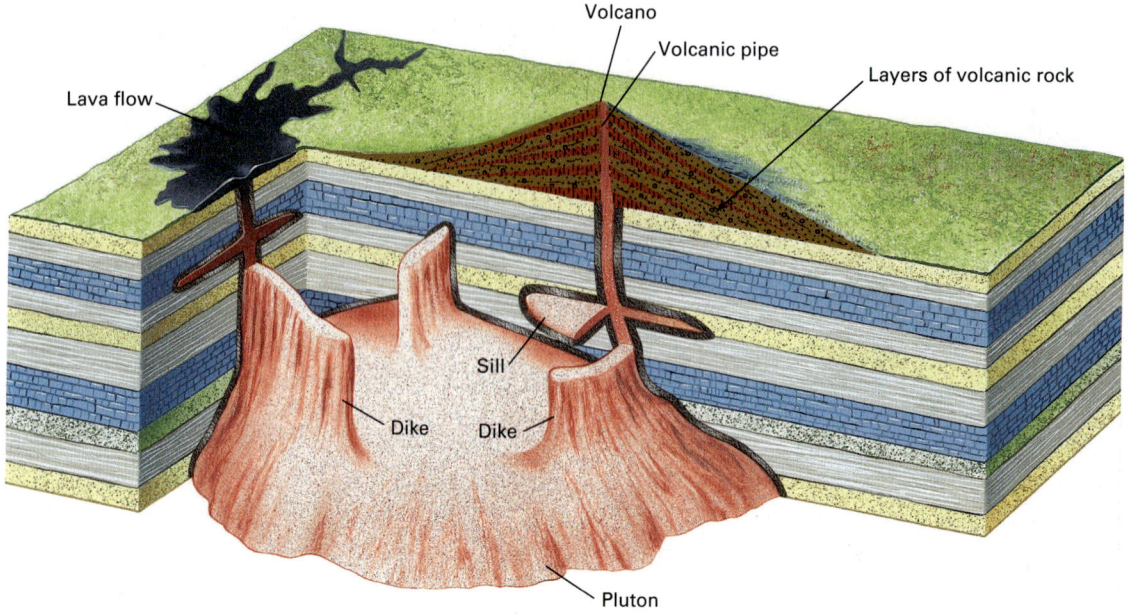

FIGURE 9-11
A large magma body may solidify within the crust to form a pluton. Some magma may rise to the surface to form volcanoes and lava flows; some intrudes country rock to form dikes and sills.

than surrounding rock. As the country rock erodes, the dike is left standing on the surface (Figure 9-12).

🌐 **FIGURE 9-12**
Go online to view (A) a large dike near Shiprock, New Mexico, that has been left standing after softer sandstone rock eroded and (B) Purcell sill visible on a mountainside in Glacier National Park.

Magma that flows between layers of country rock forms a sheet-like layer called a **sill**. Like dikes, sills vary in thickness from less than a centimeter to more than a kilometer and may extend for tens of kilometers in length and width.

checkpoint How does a pluton become a batholith?

Lava and Pyroclastic Rocks

The material erupted from volcanoes creates a wide variety of rocks and landforms. Many islands, including the Hawaiian Islands, Iceland, and most islands of the southwestern Pacific Ocean, were built entirely through volcanic eruptions.

Lava Lava is magma that flows onto Earth's surface; the word also describes the rock that forms when the magma solidifies. Lava with low viscosity may continue to flow as it cools and stiffens, forming smooth, glassy-surfaced, wrinkled, or "ropy" ridges. This type of lava is called pahoehoe (pronounced "puh-HOY-hoy"), from the Hawaiian meaning "smooth" or "polished" (Figure 9-13A). If the viscosity of lava is higher, its surface may partially solidify as it flows. The solid crust breaks up as the deeper, molten lava continues to move, forming aa (pronounced "ah-ah") lava, with a jagged, rubbled, broken surface (Figure 9-13B). When lava cools, escaping gases such as water and carbon dioxide form bubbles in the lava. If the lava solidifies before the gas escapes, the bubbles are preserved in the rock as holes called **vesicles**.

As hot lava cools and solidifies, it shrinks. The shrinkage pulls the rock apart, forming cracks that grow as the rock continues to cool. In Hawai'i, geologists have observed this phenomenon while watching fresh lava cool; when a solid crust measuring only 0.5 centimeter thick formed on the surface of the glowing liquid, five- or six-sided cracks develope. As the lava continued to cool and solidify, the cracks spread downward through the flow. Such cracks, called **columnar joints**, are regularly spaced and intersect to form five- or six-sided columns when viewed in cross section (Figure 9-14).

Pyroclastic Rock If a volcano erupts explosively, it may eject a combination of hot gas, liquid magma, and solid rock fragments. A rock formed from this material is called a **pyroclastic rock** (from *pyro*, meaning "fire," and *clastic*, meaning "particles"). The smallest particles, called

FIGURE 9-13
(A) Lava flows from a fissure near several homes in Pahoa, Hawai'i, in May 2018. The eruption of Kilauea lasted five months and was the longest and most destructive eruption on the island in decades. (B) A burned out vehicle stands as an example of the destructive power of the lava flow that also destroyed hundreds of homes and businesses.

A

B

volcanic ash, consist of tiny fragments of glass that formed when liquid magma exploded into the air along with small pieces of volcanic rock and mineral crystals. **Ciders** are volcanic fragments that vary in size from 4 to 32 millimeters.

The gentlest, least explosive type of volcanic eruption occurs when magma is so fluid that it flows from cracks in the land surface called fissures and flows over the land like water. Basaltic magma commonly erupts in this manner because of its low viscosity. Fissures and fissure eruptions vary greatly in scale. In some cases, lava pours from small cracks on the flank of a volcano. Fissure flows of this type are common on Hawaiian and Icelandic volcanoes.

In other cases, however, fissures extend for tens or hundreds of kilometers and pour thousands of cubic kilometers of basaltic lava onto Earth's surface. A fissure eruption of this type creates a **flood basalt**, which covers the landscape like a flood. It is common for many such fissure eruptions to occur in rapid succession and to create a **lava plateau**, covering thousands of square kilometers.

The Columbia River plateau in Washington, Oregon, and Idaho is a lava plateau containing 350,000 cubic kilometers of basalt. The lava is up to 3,000 meters thick and covers 200,000 square kilometers. The Columbia River basalts formed as a series of eruptions that began as early as

A

B

FIGURE 9-14
(A) Columnar joints form Devils Postpile National Monument in California. (B) This top view of Devil's Postpile National Monument shows the geometric shape of the columns.

LESSON 9.2

TABLE 9-2
Characteristics of Different Types of Volcanic Features

Type of Volcanic Feature	Physical Form	Size	Type of Magma	Activity	Example
Basalt plateau	Flat to gentle slope	100,000 to 1,000,000 km² in area; 1 to 3 km thick	Basalt	Formed by gentle fissure eruptions	Columbia River plateau
Shield volcano	Slightly sloped	Up to 9,000 m high	Basalt	Gentle; some lava fountains	Hawai'i
Cinder cone	Moderate slope	100 to 400 m high	Basalt	Ejections of pyroclastic material	Hawai'i
Composite volcano (Stratovolcano)	Alternate layers of flows and pyroclastics	100 to 3,500 m high	Variety of types of magmas and ash	Often violent	Vesuvius (Italy); Mount St. Helens; Aconcagua (Argentina)
Caldera	Circular depression, sometimes with steep walls	Less than 40 km in diameter	Granite	Formed by a violent cataclysmic explosion; potential for violent eruption remains	Yellowstone volcano; San Juan Mountains

FIGURE 9-15
Hot gas rises from the Marum Volcano on Vanuatu Island in the South Pacific.

17.4 million years ago, peaked about 16 million years ago, and essentially ended about 12 million years ago. The individual flows are between 15 and 100 meters thick and some flowed for hundreds of kilometers. Much of the basalt erupted from a series of dikes where the states of Washington, Oregon, and Idaho share a border.

checkpoint Compare what happens during a highly explosive eruption and a less explosive eruption where magma travels through fissures.

Volcano Types

Volcanoes differ widely in shape, structure, and size (Table 9-2). Lava and rock fragments commonly erupt from an opening called a **vent**, located in a **crater**, a bowl-like depression at the summit of the volcano that was itself created by volcanic activity (Figure 9-15). Lava or pyroclastic material may also erupt from a fissure on the flanks of the volcano.

Shield Volcanoes Fluid basaltic magma often builds a gently sloping mountain called a **shield volcano**. The sides of a shield volcano generally slope away from the vent at angles between 6 and 12 degrees from horizontal. Although their slopes are gentle, shield volcanoes can be enormous (Figure 9-16). The height of Mauna Kea volcano in Hawai'i, measured from its true base on the seafloor to its top, is 10,200 meters (33,476 feet) making it the tallest mountain the world, exceeding the height of Mount Everest by more 1,200 meters (4,000 feet).

FIGURE 9-16
Go online to view a snow covered shield volcano in Iceland, Skjaldbreiour, which translates to "broad shield". The lava bed of the volcano covers an area of 200 square kilometers.

Shield volcanoes, such as those of Hawai'i and Iceland, erupt regularly, but the eruptions are normally gentle and rarely life threatening. Lava flows occasionally overrun homes and villages. But the flows advance slowly enough to give people time to evacuate.

Beginning in May 2018, Kilauea, the largest volcano on the Island of Hawai'i, began a series of eruptions to the surprise of many volcanologists. The initial eruption began in a subdivision of homes on the east side of the island where the steaming ground cracked opened and began to spew lava. High levels of toxic sulfur dioxide gas and lava fountains more than 90 meters (300 feet) tall were observed. Three days later, over a dozen eruptive fissures had opened. After several days, four new fissures began to erupt several miles away. This new series of fissures produced slow-moving lava that overran everything in its path to the shoreline. When the lava encountered seawater, the steam mixed with volcanic gases and particulates in a hazardous mixture called volcanic haze or laze. Within a few days, the lava began to build a small peninsula in the ocean. Twice during this period of volcanism, the volcano's summit miles away erupted violently, sending clouds of ash 9 kilometers (more than 5.5 miles) into the air.

A couple of weeks later, a second set of eruptions produced low viscosity pahoehoe lava. The lava poured across the land surface in fast-moving, spectacular streams. A month after the initial eruptions, the lava flow burried the village of Kapoho and obliterated a subdivision popular with tourists. By July 2018, more than 700 homes were destroyed, along with several sections of highway, a park, and a school (Figure 9-17).

FIGURE 9-17
Go online to view (A) a satellite map of Mount Kilauea and (B) a map showing the areas affected by the August 2018 lava flows on the Island of Hawai'i.

Cinder Cone Volcanoes A **cinder cone** is a small volcano composed of pyroclastic fragments (Figure 9-18). A cinder cone forms when large volumes of gas accumulate in rising magma. When the gas pressure builds sufficiently, the entire mass erupts explosively. This explosion hurls cinders, ash, and molten magma into the air. The particles

FIGURE 9-18
This image taken in Hawai'i Volcanoes National Park features several cinder cones in the sunlit foreground. Beyond the fog on the horizon is Mount Kilauea, a shield volcano.

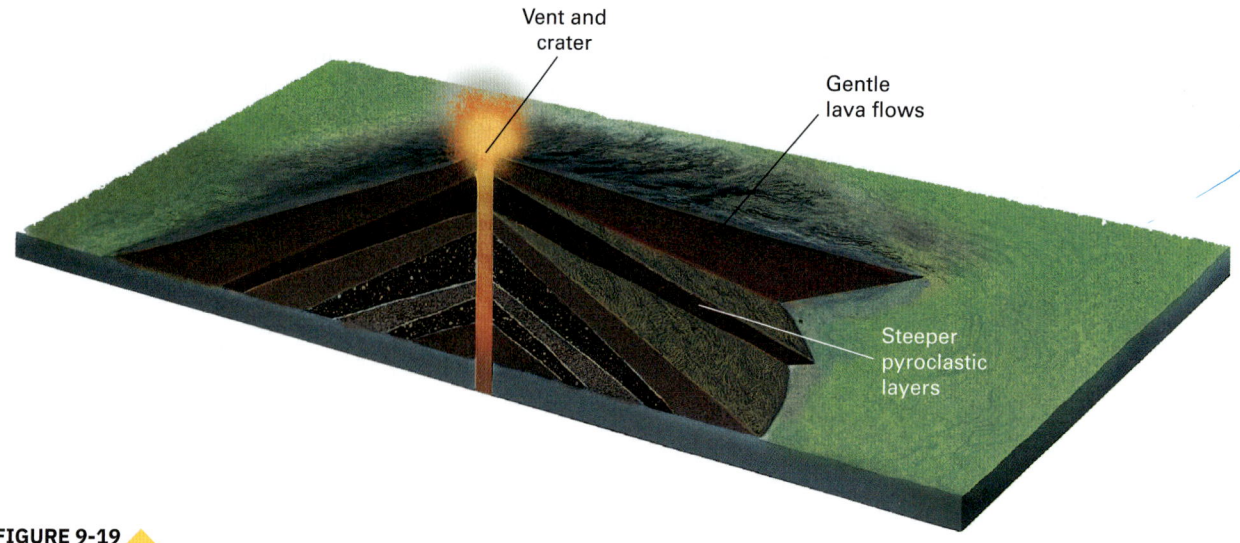

FIGURE 9-19
A composite cone consists of alternating layers of lava and loose pyroclastic material.

then fall back around the vent, to accumulate as a small mountain of pyroclastic debris. A cinder cone is usually active for only a short time, because once the gas escapes, the driving force behind the eruption is gone.

Cinder cones usually are symmetrical and can be steep (about 30°), especially near the vent, where ash and cinders pile up. Most are less than 300 meters high, although a large one can be up to 700 meters high. A cinder cone erodes easily and quickly because the pyroclastic fragments are not cemented together.

Composite Cone Volcanoes A **stratovolcano**, sometimes called a composite cone, forms over a long period of time as a sequence of lava flows and pyroclastic eruptions. The hard lava covers the loose pyroclastic material and protects it from erosion (Figures 9-19 and 9-20).

Many of the highest mountains of the Andes and some of the most spectacular mountains of western North America are composite cones. Repeated eruptions are a trademark of a composite volcano. Mount St. Helens, in the state of Washington, erupted dozens of times in the 4,500 years preceding its most recent eruption in 1980. Mount Rainier, also in Washington, has been dormant in recent times but could become active again and threaten nearby populated regions.

checkpoint Describe the typical impact of a shield volcano eruption.

9.2 ASSESSMENT

1. **Make Connections** What is the connection between a batholith, a stock, and a pluton?
2. **Distinguish** Describe one effect that flowing lava may have on Earth's surface. Interpret that effect in the context of lava behavior.
3. **Describe** Explain fissures and describe the conditions that create a lava plateau.
4. **Summarize** How was the 2018 eruption of the Kilauea volcano, a shield volcano in Hawai'i, different from the typical eruptive impact of shield volcanoes?

Computational Thinking

5. **Recognize Patterns** Although the intensity of eruptions varies, each type of volcano has patterns associated with it. Use Table 9-2 to compare the characteristics of shield volcanoes, cinder cones, and stratovolcanoes.

FIGURE 9-20
Mount Hood in Oregon is a composite cone volcano.

9.3 VOLCANIC ERUPTIONS

Core Ideas and Skills
- Explain a pyroclastic flow and its possible effects.
- Compare historical data from significant eruptions and understand the effects of eruptions on ecosystems and people.

KEY TERMS
pyroclastic flow
ash-flow tuff
caldera

Pyroclastic Flows and Calderas

Pyroclastic Flows Under certain conditions granitic magma rises to Earth's surface, where it erupts violently. Granitic magmas generally have a high water content. Decreasing pressure allows the dissolved water to form a frothy, pressurized mixture of gas and liquid magma as hot as 900°C. As the mixture rises to within a few kilometers of Earth's surface, it fractures overlying rocks and explodes through the fractures.

A large and violent eruption can blast a column of pyroclastic material 10 to 12 kilometers into the sky. A cloud of fine ash may rise even higher into the upper atmosphere. The force of material streaming out of the magma chamber can hold the column up for hours or even days.

When much of the gas has escaped from the upper layers of magma, the eruption ends. The airborne column of ash, rock, and gas then falls back to Earth's surface, spreading over the land and funneling downstream to valleys beyond (Figure 9-21). Such a flow is called a **pyroclastic flow**.

FIGURE 9-21
Go online to view a pyroclastic flow that descended the slopes of the Soufrière Hills volcano on the Caribbean island of Montserrat in 2010.

FIGURE 9-22
Ash-flow tuffs formed in Smith Rock State Park in Bend, Oregon. A pyroclastic flow forms ash-flow tuffs when the flow stops.

LESSON 9.3

FIGURE 9-23

(A) When granitic magma rises to within a few kilometers of Earth's surface, it stretches and fractures overlying rock. Gas separates from the magma and rises to the upper part of the magma body. (B) The gas-rich magma explodes through fractures, rising as a vertical column of hot ash, rock fragments, and gas. (C) When the gas is used up, the column collapses and spreads outward as a high-speed pyroclastic flow. (D) Because so much material has erupted from the top of the magma chamber, the roof collapses to form a caldera.

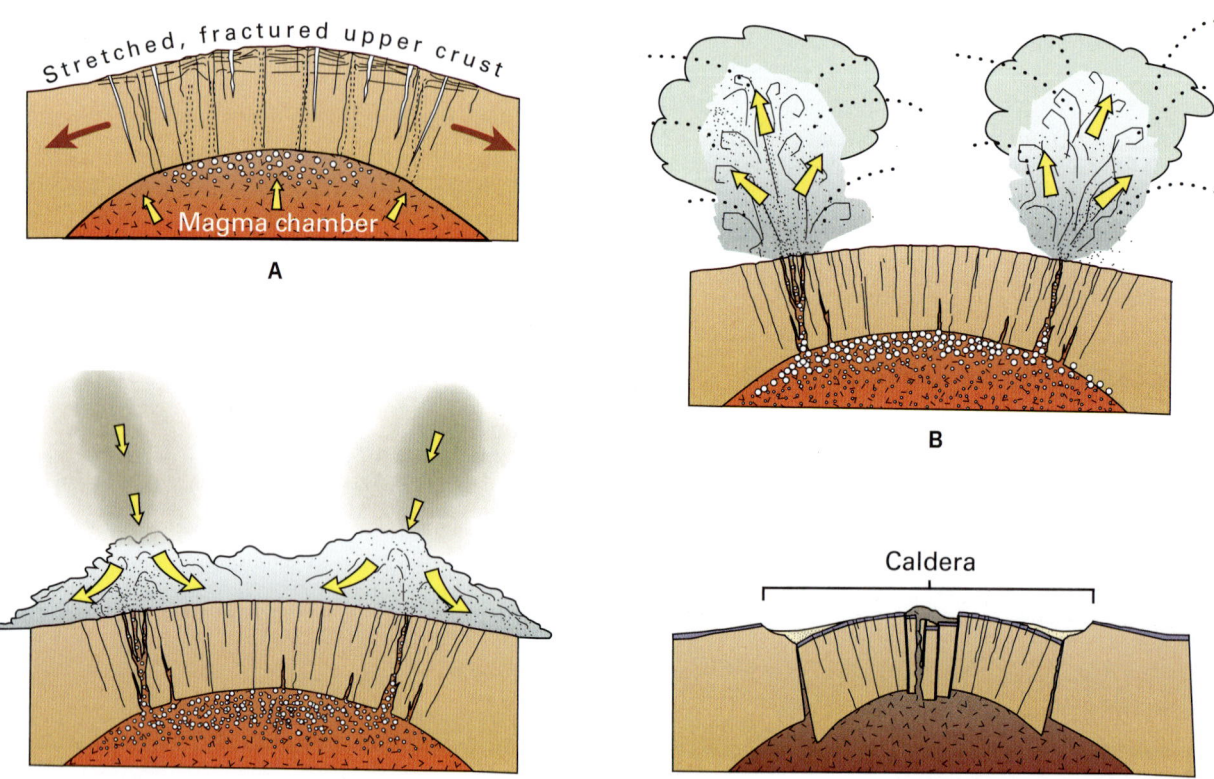

When a pyroclastic flow stops, most of the gas escapes into the atmosphere, leaving behind a chaotic mixture of volcanic ash and rock fragments called **ash-flow tuff** (Figure 9-22). If the deposit is thick enough, the ash at the bottom of the pile begins to compact and may partially melt from the residual heat, causing the ash to fuse together and form a welded tuff. Typically, only the lower portion of a pyroclastic flow becomes welded; the upper portion usually remains a relatively porous accumulation of ash particles. Welded tuffs form several prized sport-climbing areas in the western United States because of their strength.

Calderas After the gas-charged magma erupts, the roof of the magma chamber can collapse into the space the magma had filled. Typically, the collapsing roof forms a circular depression (Figure 9-23), called a **caldera**. A large caldera may be 40 kilometers in diameter and have walls a kilometer high. Some calderas fill up with volcanic rocks from later eruptions; others maintain the circular depression and steep walls.

We usually think of volcanic landforms as mountain peaks, but the topographic depression of a caldera is an exception. Calderas, ash-flow tuffs, and related rocks occur over a large part of western North America. The Yellowstone volcano, Crater Lake in Oregon, and Long Valley in California are well-known examples. Take another look at the map of Yellowstone National Park (Figure 2-1), which shows several calderas.

checkpoint What are the characteristics of a caldera?

Risk Assessment: Predicting Volcanic Eruptions

Scientific records include data on the major known volcanic disasters since the year 1500 C.E. The potential for such disasters in the future makes a volcanic eruption one of the greatest of all geologic hazards. It also makes risk assessment and prediction of volcanic eruptions an important part of modern science.

Approximately 1,300 active volcanoes are recognized globally, and nearly 6,000 eruptions have occurred in the past 10,000 years. These figures do not include the numerous submarine volcanoes of the Mid-Ocean Ridge system. Many volcanoes have erupted recently, and we are certain that others will erupt soon. How can geologists predict an eruption and reduce the risk of a volcanic disaster?

Regional Prediction Risk assessment for regional predictions is based both on the frequency of past eruptions and on potential violence. However, regional predictions based on the concentration of volcanoes in an area can only estimate probabilities and cannot be used to determine exactly when a particular volcano will erupt or the intensity of a particular eruption.

Short-Term Prediction In contrast to regional predictions, short-term predictions attempt to forecast the specific time and place of an impending eruption. They are based on instruments that monitor an active volcano to detect signals that the volcano is about to erupt. The signals include changes in the shape of the mountain and surrounding land, earthquake swarms indicating movement of magma beneath the mountain, increased emissions of ash or gas, increasing temperatures or changing compositions of nearby hot springs, and any other signs that magma is approaching the surface.

In 1978, two U.S. Geological Survey (USGS) geologists, Dwight Crandall and Don Mullineaux, noted that Mount St. Helens had erupted more frequently and violently during the past 4,500 years than any other volcano in the contiguous 48 states. They predicted that the volcano would erupt again before the end of the 20th century.

In March 1980, about two months before the great May eruption, puffs of steam and volcanic ash rose from the crater of Mount St. Helens, and swarms of earthquakes occurred beneath the mountain (Figure 9-24). This activity convinced other USGS geologists that Crandall and Mullineaux's prediction was correct. In response, they installed networks of seismographs, tiltmeters, and surveying instruments on and around the mountain.

In the spring of 1980, the geologists warned government agencies and the public that Mount St. Helens showed signs of an impending eruption. The U.S. Forest Service and local law enforcement officers quickly evacuated the area surrounding the mountain, averting a much larger tragedy.

checkpoint Why was it possible for the public to prepare in time for the 1980 Mount St. Helens eruption?

Volcanic Eruptions and Global Climate

A volcanic eruption can profoundly affect the atmosphere, the climate, and living organisms. An eruption can be an excellent example of

---

CONNECTION TO HISTORY

Historic Crop Failure and Revolution

The 1783 Laki crater eruption in Iceland altered weather patterns in Iceland and Europe. In Iceland, violent thunderstorms and hailstorms killed cattle and destroyed crops. The crop failure resulting from the reduced solar energy and extreme weather events are estimated to have killed about 24 percent of the human population there. In Europe, the summer of 1783 was more like a winter, with the sun remaining a pale ghost in the sky or a strange blood-red color. The cold summer temperatures were followed by an extremely harsh winter in 1783 to 1784, and for several years afterward, the destruction of crops and livestock brought about famine and poverty. Historians believe this poverty contributed directly to the French Revolution, which started in 1789. French citizens struggled with poverty and hunger due to the successive crop failures.

FIGURE 9-24
Clouds of ash pour out of Mount St. Helens the day of its epic eruption, May 18, 1980. Geologists at the U.S. Geological Survey accurately predicted the eruption, and the area was mobilized for evacuation.

interactions among systems. For instance, the 1991 eruptions of Mount Pinatubo in the Philippines produced the greatest ash and sulfur clouds in the latter half of the 20th century. Satellite measurements show that the total solar radiation reaching Earth's surface declined by two to four percent after the Pinatubo eruptions. The following two years, 1992 and 1993, were a few tenths of a degree Celsius cooler than the temperatures of the previous decade. Temperatures rose again in 1994, after the ash and sulfur settled (Figure 9-25).

Another example of a volcanic eruption that affected climate occurred in 1783 in Iceland. A largely nonexplosive eruption of the Laki crater occurred during June of that year. The eruption lasted nearly nine months and produced a bluish haze of sulfur aerosols across Iceland. The aerosols subsequently spread across Europe. The haze

FIGURE 9-25 ▼
The figures represent data collected by NASA's Earth Radiation Budget Satellite for the 1991 eruption of Mount Pinatubo in the Philippines. Shaded areas reflect the presence of aerosols in the atmosphere before, at the time of, and following the eruption. Hot colors (reds) indicate high aerosol concentrations, whereas cool colors (blues) indicate low concentrations.

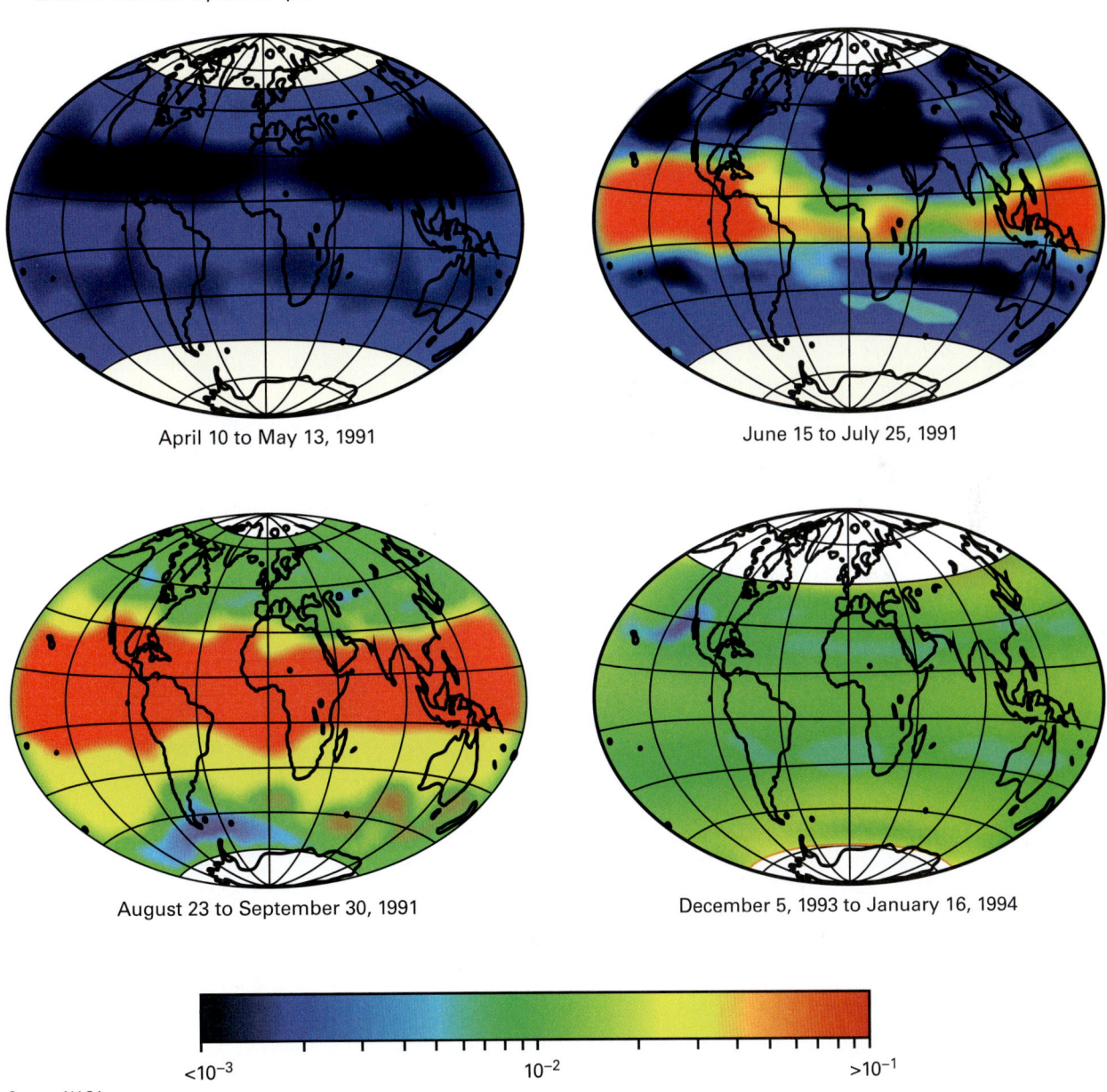

Source: NASA

DATA ANALYSIS Exploring Volcanic Risk in the United States

The United States is the third-most volcanically active country in the world, following only Indonesia and Japan. The Volcanic Threat Assessment scores U.S. volcanoes and assigns threat levels. The score is based on exposure risk to people, property, and infrastructure, and the danger of natural volcanic hazards. USGS monitors volcanoes and provides timely warnings of volcanic acitivity. Refer to Figure 9-26 to examine the volcanic threat assessment data.

Not all of these volcanoes pose the same level of threat to humans. The USGS monitors many volcanoes and assigns each volcano a Volcanic Threat Assessment score ranging from "very low" to "very high." This helps the USGS prioritize the risk reduction efforts near volcanoes that pose the most danger to large populations.

Computational Thinking

1. **Generalize Patterns** Which geographic areas of the United States have the greatest concentration of volcanoes? The least?

2. **Identify** For the states with volcanoes, create a ranking of the states and territories from most to fewest number of volcanoes.

3. **Analyze Data** What percent of all U.S. volcanoes are located in the five states or territories with the most volcanoes?

4. **Represent Data** Create a pie chart that represents the percentage of U.S. volcanoes in each threat level category.

5. **Interpret** Of all U.S. volcanoes, what percentage are considered a "very high" or "high" threat? What does this mean?

Figure 9-26
The U.S. Geological Survey analyzes volcanic threats.

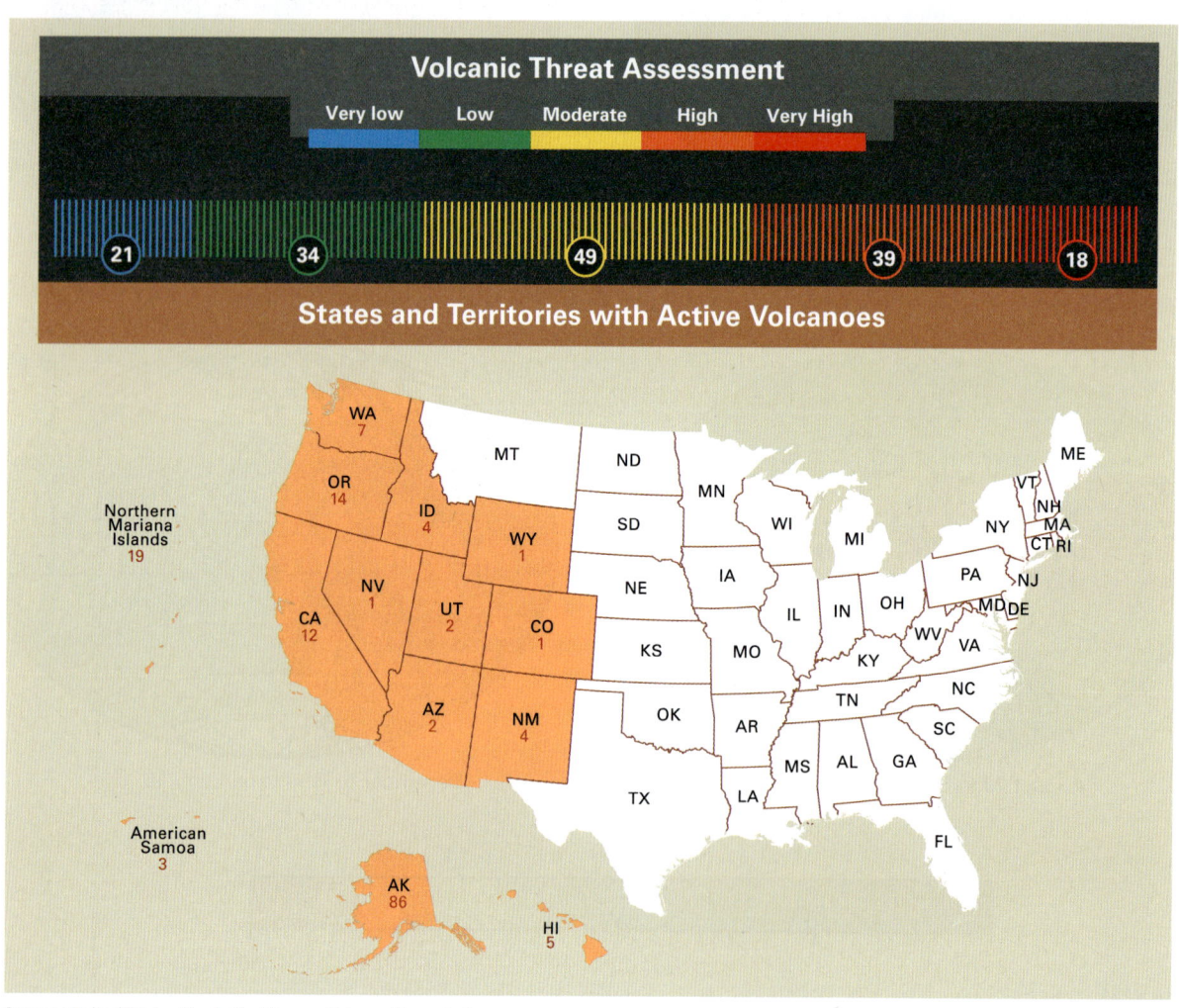

Source: United States Geological Survey Volcano Hazards Program

obscured the sun, significantly reducing the solar energy reaching the surface.

Studies of global temperatures before and after major volcanic eruptions shows a positive correlation between global cooling and volcanic eruption. The correlation supports meteorological models showing that high-altitude dust and sulfur aerosols reflect sunlight and cool the atmosphere. The primary impact of volcanic eruptions on the climate comes from the conversion of sulfur dioxide (SO_2) to sulfuric acid (H_2SO_4), which condenses to form sulfur aerosols. These aerosols reflect radiation back into space, cooling Earth's surface. Aerosols also absorb heat that radiates from the Earth, thereby warming the stratosphere. The sulfur aerosols also change the types of chlorine and nitrogen molecules in the upper atmosphere. Chlorine monoxide (ClO) is produced through reaction with the sulfur aerosols, and in turn destroys ozone.

Recent eruptions such as Pinatubo and Laki have been minuscule compared with some in the more distant past. About 248 million years ago, at the end of the Permian period, for example, 90 percent of all marine species and two-thirds of reptile and amphibian species died suddenly in the most catastrophic mass extinction in Earth history. This extinction event coincided with a massive volcanic eruption in Siberia that disgorged a million cubic kilometers of flood basalt onto Earth's surface, forming a great lava plateau. The eruption must have released massive amounts of ash and sulfur compounds into the upper atmosphere, leading to cooler global climates. Many geologists think Earth cooled enough to cause, or at least contribute to, the Permian mass extinction.

checkpoint How can a volcanic eruption cause global cooling?

9.3 ASSESSMENT

1. **Explain** What happens to rising granitic magma as pressure decreases?
2. **Explain** What causes a welded tuff?
3. **Compare** How does a caldera differ from most other volcanic landforms?
4. **Summarize** How do vast quantities of volcanic ash and sulfur gases affect solar radiation?

Computational Thinking

5. **Generalize Patterns** Examine the maps in Figure 9-25. What pattern can you identify?

TYING IT ALL TOGETHER
COMMUNITIES NEAR VOLCANOES

In this chapter, you learned about the formation and behavior of magma and how it creates various landforms. You also explored how plate tectonics influences volcanic activity and the various ways that scientists monitor volcanoes and assess the risk of future eruptions.

In the Case Study, we examined two active stratovolcanoes: Mount Vesuvius in Italy and Nyiragongo in the Congo. Mount Vesuvius is the site of the world's oldest volcanic observatory, built on the slopes of the mountain in 1841. Geologists studying seismic data beneath the volcano have detected a sill of magma extending outward for at least 400 square kilometers. By comparison, scientists only began monitoring Nyiragongo recently. The Goma Volcano Observatory (GVO) was funded in 1986 to monitor Nyiragongo and another nearby volcano. Geologists are studying the parasitic cones surrounding Nyiragongo to gain a better understanding of the volcano's history, which may lead to better predictions regarding the flow of lava for future eruptions.

Though some volcanoes present a catastrophic threat to humans, they are also responsible for creating habitable land. One reason people have historically continued to repopulate places like Naples (Figure 9-27) and Goma after volcanic eruptions is for their fertile volcanic soil. Volcanoes also play a large role in the formation of mountains, plutons, and lava plateaus that shape continental landscapes, as well as the formation of island chains. These landforms provide a variety of environments that support various ecosystems, including those in which humans thrive.

Research other notable landforms or features that were formed by volcanic activity. Some examples include Multnomah Falls in the Columbia River plateau; Crater Lake in Oregon; plutons in the Sierra Nevada mountain range; the Kilauea volcano in Hawai'i; geysers, hot springs, or batholiths at Yosemite National Park; or the basaltic cinder cones in Uinkaret Volcanic Field near the Grand Canyon. Then answer the following questions.

Respond to the following questions and statements based on your research.

1. Describe how the landform or feature was formed. What type of volcano or volcanic activity created the landform or feature?

2. If you chose to research an active landform, explain what action is still occurring both above and below ground. If it is not active, describe when the last volcanic activity occurred.

3. Describe the type of magma and its behavior related to the formation of the landform or feature.

4. Explain how partial melting likely contributed to the composition of the rock in the landform.

5. Are scientists monitoring volcanic activity in this area? Do they predict new eruptions in the near future? If so, what are the risks associated with an eruption?

6. Consider how a scientist studying the landform would share his or her research. Design a presentation to educate the public about the landform. Be sure to include information about the name of the landform or feature, the type of magma and magma behavior that composed it, and the process by which it was formed. Also explain any potential for future volcanic activity in the area and propose practical steps that can be taken to reduce risk.

CHAPTER 9 SUMMARY

9.1 MAGMA

- Rocks of the asthenosphere partially melt to produce basaltic magma as a result of three processes: rising temperature, pressure-release melting, and addition of water.
- The processes that produce magma occur beneath spreading centers, in mantle plumes, and in subduction zones to form both volcanoes and plutons.
- Basalt makes up most of the oceanic crust, and granite is the most abundant rock in continents.
- Basaltic magma forms by partial melting of mantle peridotite.
- Granitic magma forms when basaltic magma rises into and melts granitic rocks of the lower continental crust.
- Earth's earliest continents were probably formed by partial melting of the original peridotite crust to form basalt, and then by further partial melting of the basalt to form granite.
- Basaltic magma usually erupts in a relatively gentle manner onto Earth's surface from a volcano.
- In contrast, granitic magma typically solidifies within Earth's crust. When granitic magma does erupt onto the surface, it often does so violently.
- These contrasts in behavior between the two types of magma are caused by differences in silica and water content.

9.2 MAGMATIC BODIES AND VOLCANIC FORMATIONS

- A pluton is any intrusive mass of igneous rock.
- A batholith is a pluton with more than 100 square kilometers of exposure at Earth's surface.
- A dike and a sill are both sheet-like bodies of volcanic rock. Dikes cut across layering in country rock, and sills run parallel to layering.
- Magma may flow onto Earth's surface as lava or may erupt explosively as pyroclastic material.
- Fluid lava forms lava plateaus and shield volcanoes.
- A pyroclastic eruption may form a cinder cone.
- Alternating eruptions of fluid lava and pyroclastic material from the same vent create a stratovolcano.
- When granitic magma rises to Earth's surface, it may erupt explosively, forming ash-flow tuffs and calderas.

9.3 VOLCANIC ERUPTIONS

- Volcanic eruptions are common near subduction zones, spreading centers, and hot spots over mantle plumes, but are rare in other tectonic settings.
- Eruptions on a continent are often violent, whereas those that occur on oceanic crust are gentle. Such observations form the basis for predicting regional volcanic hazards.
- Violent eruptions called pyroclastic flows can blast an airborne column of ash, rock, and gas high into the atmosphere.
- Short-term predictions of a volcanic eruption are made on the basis of earthquakes caused by magma movements, swelling of a volcano, increased emissions of gas and ash from a vent, and other signs that magma is approaching the surface.
- Large volcanic episodes affect the atmosphere, climate, and living organisms.

CHAPTER 9 ASSESSMENT

Review Key Terms

Select the key term that best fits the definition. Not all terms will be used, and no term will be used more than once.

ash-flow tuff	pressure-release melting
batholith	pyroclastic flow
caldera	pyroclastic rock
cinder cone	shield volcano
cinder	sill
columnar joint	stratovolcano
crater	superheating
flood basalt	vent
lava plateau	vesicle
partial melting	volcanic ash
pluton	volcano

1. a large pluton, exposed across more than 100 square kilometers of Earth's surface
2. the process in which a small amount of a silicate rock transitions to a liquid, forming magma that is more silica-rich than the original rock
3. a small volcano, typically less than 300 meters high, made up of loose pyroclastic fragments blasted out of a central vent; usually active for only a short time
4. increasing the temperature of a substance above a phase-transition point without the transition occurring, such as when solid rock is heated beyond its melting point without melting
5. a volcanic rock formed when a pyroclastic flow solidifies
6. transition of superheated rock from solid to liquid caused by a decrease in pressure and expansion of rock volume
7. small fragment of glassy, pyroclastic rock measuring 4 mm to 32 mm in size
8. a bowl-like depression at the summit of a volcano, created by the accumulation of volcanic material around the vent
9. an opening in a volcano, typically in the crater, through which lava and rock fragments erupt
10. small hole in rock formed when the lava solidified before a bubble of gas or water could escape
11. large circular depression created by the collapse of the magma chamber after an explosive volcanic eruption
12. regularly spaced cracks that develop in lava flows, grow downward starting from the surface, and form five- or six-sided structures
13. rock particles of various sizes formed from liquid magma and rock fragments that were ejected explosively from a volcanic vent
14. a large, gently sloping volcanic mountain formed by successive flows of basaltic magma
15. a sheetlike layer of igneous rock, parallel to the grain or layering of country rock, that forms when magma is injected between layers

Review Key Concepts

Answer each question on a separate sheet of paper to demonstrate your understanding of key concepts from the chapter.

16. Rocks in the asthenosphere are under high pressure. How does this pressure prevent the rocks from changing state?
17. Compare magma formation in spreading centers, mantle plumes, and subduction zones. How are these processes similar and different?
18. Explain how plutons are formed from granitic magma.
19. Basaltic magmas tend to rise to the surface and erupt from a volcano before cooling into solid rock. Granitic magmas, on the other hand, tend to solidify before reaching the surface. What properties of basaltic and granitic magmas can account for this difference in behavior?
20. Compare and contrast batholiths and stocks.
21. Rising magma can form different rock features as it cools. What structures are labeled A, B, and C in the figure? How can you differentiate between each of these features?

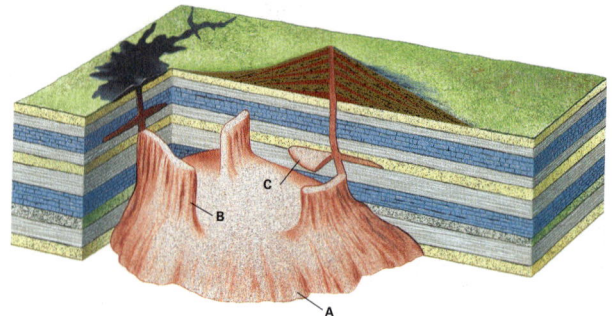

22. The Giant's Causeway in Northern Ireland is famous for its rock features. What are these features called? What processes were most likely responsible for the structures shown in the photo?

23. Compare and contrast the formation of cinder cone and composite cone volcanoes.
24. Describe how welded tuffs form from pyroclastic flow.

Think Critically
Write a response to each question on a separate sheet of paper. Use the concepts from the chapter to support your reasoning.

25. Suppose that a massive volcano in Indonesia begins erupting tomorrow and continues to release large amounts of volcanic material for the next century. What impact would this likely have on global climates? Explain your reasoning.
26. Basaltic magma is composed of 45 percent to 55 percent silica with high iron, magnesium, and calcium content. Andesitic magma is composed of 55 percent to 65 percent silica with intermediate iron, magnesium, and calcium content. Based on the composition of basaltic and andesitic magma, what can you conclude about the temperature and viscosity of andesitic magma compared to basaltic magma?
27. Io is a volcanically active moon orbiting Jupiter. It has a much thinner atmosphere than Earth. How do you think the lower atmospheric pressure would affect explosive volcanic eruptions on Io?
28. Some researchers have proposed a controversial technique called stratospheric aerosol injection (SAI) as a possible solution for climate change. SAI involves releasing large volumes of aerosol particles into the stratosphere. How is SAI similar to volcanic eruptions? What effects would you expect SAI to have on global climate?

PERFORMANCE TASK

Volcanic Effects

Volcanoes form as a result of ruptures in the crust that allow magma, ash, gases, and rock fragments to escape from below the surface. Volcanoes can have a significant impact on the areas that surround them. For example, the debris produced by eruptions creates new hills and mountains.

In this performance task, work with your classmates in groups of three or four to research an active volcano. Then draw a map to illustrate how the volcano has affected the region around it.

1. Use the Smithsonian Institute's list of Holocene volcanoes to search for active volcanoes (http://volcano.si.edu/list_volcano_holocene.cfm).
   A. Identify a volcano that has had active lava flow in the past century. Using research from the Internet, describe the type of volcano and the lava flow that is occurring there.
   B. Then research the environment around the volcano. How did the environment change during its last eruption? In your group, discuss how features of the environment could have arisen as a result of volcanic activity in the area.
   C. Next, research human activity that occurs around the volcano. How did the last volcanic eruption affect human activity in the area? In your group, discuss how human activity in the region has been shaped by volcanic activity.
   D. With your group, draw a map of the volcano and the region around the volcano. On your map, use symbols and a legend to identify 6–8 sites associated with environmental features or human activity that have been affected by volcanic activity. Then create a graphic organizer to accompany your map. Your graphic organizer, such as a table or chart, should summarize how each site on your map has been influenced by volcanic activity.

CHAPTER 10
MOUNTAINS

Tectonic forces at plate boundaries have produced mountains on every continent and in the ocean. The rugged terrain and sheer size of a mountain can fill onlookers with a sense of awe. As an example, Denali in the Alaska Range rises 6,190 meters (20,310 feet) above sea level.

Throughout time, humans have lived near mountains, adapting to the climate and geography of their region. For some people, however, a first glimpse of a mountain stirs a connection, and a new course is set to visit, study, or even climb one. Whether a traveler, geologist, or climber, knowing the principles of mountain formation offers a common starting point to understanding the processes that have shaped the highest landforms on Earth.

KEY QUESTIONS

10.1 How does tectonic stress form geologic structures?

10.2 What are the processes that produce mountains?

10.3 How do continental collisions create mountain chains?

Denali in Alaska is the highest mountain peak in North America.

EXPLORERS AT WORK

TO THE SUMMIT

WITH NATIONAL GEOGRAPHIC EXPLORER CORY RICHARDS

Alpinist and photographer Cory Richards will tell you mountain climbing saved his life. As a struggling teen, Richards began to climb mountains as a way to leave his troubles down at street level. In the years that followed, climbing became a profession for Richards. It inspired him to take a class in photography and to carry a camera along with his climbing gear. Richards quickly became an accomplished mountaineer, scaling peaks in North America, the Alps, and the Himalaya.

Then mountain climbing nearly stole Cory Richards's life. In February 2011, Richards and his team were hit by an avalanche on Gasherbrum II (G2) in Pakistan. Richards spent an hour squeezed under the snow, literally trying to swim his way out. All survived, and for Richards there was even a silver lining: a jumpstart to his photography career, thanks to pictures he took on G2. One of them, a selfie taken after the avalanche, made the cover of *National Geographic* magazine.

Today, Richards is a world-renowned photographer and storyteller who has climbed the world's most remote and dangerous mountains, including the iconic Mount Everest. Richards has made three treks up the world's tallest peak. The first was in 2012, when Richards and fellow mountaineer Conrad Anker tackled Everest as part of a National Geographic expedition. Their climb was to commemorate the upcoming 50th anniversary of the first Everest summit. Richards was just short of the summit when he had to be evacuated due to difficulty breathing.

Richards returned to Everest in 2016 with Adrian Ballinger. They attempted to reach the summit without the aid of supplemental oxygen, all the while documenting their progress on social media. Richards reached the summit but Ballinger had to turn back. The rest of the world also missed Richards's summit—his phone died. In 2017 both men returned to Everest and both summited, although Richards had to use supplemental oxygen. He was, however, able to get a few photos at the top.

Richards says climbing Everest was about just that: climbing in one of Earth's largest and harshest mountain environments. It was also about telling an authentic story. It was *not* about doing something heroic. In fact, Richard calls his sport dangerous and even self-indulgent, words that become especially significant as record numbers of people attempt to climb Everest. Few are experienced mountaineers.

National Geographic created a 50th anniversary Everest video that features interviews with several members of the 1963 expedition team as well as contemporary climbers and sherpas. All say Everest is no longer about traditional mountaineering. They say for many climbers, it's about a bucket list, and for many sherpas, it's about a paycheck. They wonder, What do we want Everest to look like in 50 years? How should people interact with it? How can it remain sacred and iconic? What does it symbolize, if not exploration?

Richards says, "As we look at [Everest] as a microcosm for just simple human acts, I think it becomes a much more powerful symbol. Because it is no longer exploration. The symbol that it has the potential to be now is how to be a good steward and a good human being."

THINKING CRITICALLY

Infer What inferences can you make about the statement that Everest can be a symbol for how to be a good steward and a good human being?

Cory Richards was named National Geographic's Adventurer of the year in 2012 Here he pauses during a climbing expedition in Khumbu valley, Nepal.

Richards snapped this shot of a fellow climber crossing a crevasse in the Khumbu icefall in Nepal. An aluminum ladder supports the climber.

CASE STUDY
THE STORY OF A MOUNTAIN CHAIN

Explorer Cory Richards climbs the Himalaya in India, Nepal, and China for the thrill of adventure and the beauty of high-altitude vistas (Explorers at Work). Like thousands of mountaineers, Richards is drawn to the Himalaya for their alluring topography and the chance to conquer their highest peaks.

Although much smaller in terms of topographic relief and peak elevation, the Appalachian Mountains of North America once were similar to the Himalaya. So how and why do these two ranges look very different today? The simple answer is time. The Himalaya are actively forming today through the tectonic collision of the Indian plate with the Eurasian plate. In contrast, tectonic forces that produced the Appalachians ceased about 250 million years ago.

Three tectonic collisions formed the Appalachian Mountains. The first collision took place about 444 million years ago as an island arc piled onto the eastern margin of North America at a subduction zone. This collision provided the material that later formed the Blue Ridge Mountains (Figure 10-1).

Then, about 350 million years ago, continued subduction caused another major, similar event. A small continent known as Avalonia collided with the eastern continental margin of North America.

Finally, from 325 to about 260 million years ago, a major collisional orogen formed. (An *orogen* is a belt of rocks that is deformed during mountain building.) South America and Africa together collided into the eastern and southern margin of North America. This huge continent-continent collision resulted in the Pangaean supercontinent (Lesson 7.3). It also formed a mountain range that stretched from present-day eastern Canada, southward through New England and the mid-Atlantic states, then westward across Alabama, Arkansas, and Oklahoma to western Texas. At that time, the Appalachian range was about twice as long as the modern Himalayan range. Its peaks likely were as high as those of the Himalaya today.

Active tectonic building of the Appalachian Mountains ended with the break-up of Pangaea, starting about 220 million years ago. Since that time, the Appalachians have been eroding. That erosion has greatly reduced their size. It also has caused rocks that were formed deep within the active orogen to be exposed at the surface.

Scientists can study the geology of one range to better understand that of the other. The Himalaya and Appalachians both formed through continent-continent collisions. With access to rocks that were once deep within the Appalachian orogen, those seeking to understand the geology of the Himalaya can turn to the geologic roots of the Appalachian range by examining the exposed rocks that were once deep within its orogen. Conversely, scientists seeking to understand the tectonic history of the Appalachians can turn to the Himalayan range to learn about the processes that underlie an ongoing continent-continent collision.

As You Read Investigate various ways mountains form. Which mountain chains are more like the Himalaya or the Appalachian range? What geologic forces affected the size and structure of different mountain ranges?

FIGURE 10-1
A panoramic view of the Blue Ridge Mountains, a section of the Appalachian Mountain range, reveals rounded peaks from millions of years of erosion.

10.1 GEOLOGIC STRUCTURES AND TECTONIC STRESS

Core Ideas and Skills
- Explain the factors that control how a rock responds to stress.
- Identify and make connections between geologic structures.

KEY TERMS

limb	graben
axial surface	horst
anticline	reverse fault
syncline	thrust fault
slip	joint
normal fault	

Tectonic Stress

Continents move at rates between less than 1 and about 16 centimeters per year, about as fast as your fingernails grow. This movement is too slow to feel or sense in any ordinary way. Try to imagine the immense forces involved as continent- or ocean-size slabs of rock slowly grind past one another or collide in slow motion. Chapters 8 and 9 explained how tectonic activity generates earthquakes and volcanic eruptions. In this chapter, we will consider how rock crumples and buckles under tectonic forces to form mountains.

Stress is a force directed against an object. Stress is conventionally measured as force per unit area, or pressure. When a rock is stressed, the rock may deform elastically or plastically—or rock may simply break by brittle fracture, as described in Chapter 8.

Four factors control how a rock responds to stress.

1. **Characteristics of the rock** Think of a quartz crystal, a gold nugget, and a rubber ball. If you strike quartz with a hammer, it shatters. That is, it fails by brittle fracture. In contrast, if you strike the gold nugget, it deforms plastically. It flattens and stays flat. If you hit the rubber ball, it deforms elastically and rebounds immediately, sending the hammer back toward you. Initially, all rocks react to stress by deforming elastically by a slight amount. Near Earth's surface, where temperature is relatively low, different types of rocks behave differently to stress. Granite and quartzite tend to fracture in a brittle manner. Other rocks, such as mudstone, limestone, and marble, tend to deform more plastically.

2. **Temperature** The higher the temperature, the greater the tendency of a rock to deform plastically. It is difficult to bend an iron bar at room temperature, but if the bar is heated to a red-hot temperature, it becomes plastic and bends easily.

3. **Pressure** Like temperature, higher pressure promotes plastic deformation. Both temperature and pressure increase with depth. Thus, deeply buried rocks have a greater tendency to bend and flow under stress than do shallow rocks.

4. **Time** Stress applied slowly, rather than suddenly, also favors plastic behavior. For example, over a hundred years, marble park benches in New York City have sagged plastically under their own weight. In contrast, rapidly applied stress, such as the blow of a sledge hammer, would cause the same marble park benches to fracture.

checkpoint What is geologic stress, and in what ways can a rock respond?

Folds

Tectonic movement creates tremendous stress that deforms rocks. A geologic structure is any feature produced by rock deformation. Tectonic stress creates three types of geologic structures: folds, faults, and joints.

A fold is a bend in rock. Some folded rocks display little or no fracturing, indicating that the rocks deformed in a plastic manner (Figure 10-2).

FIGURE 10-2
This image shows plastic deformation in several tight folds.

LESSON 10.1

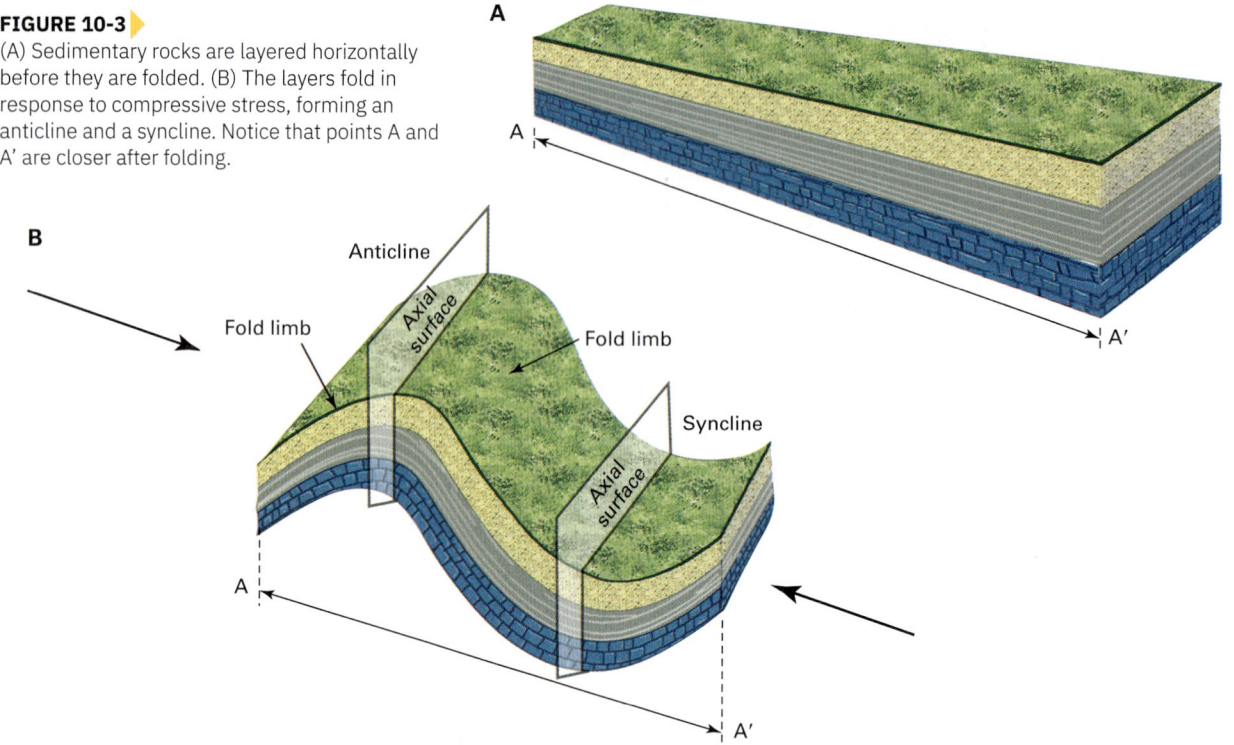

FIGURE 10-3
(A) Sedimentary rocks are layered horizontally before they are folded. (B) The layers fold in response to compressive stress, forming an anticline and a syncline. Notice that points A and A' are closer after folding.

In other cases, folding occurs by a combination of plastic deformation and brittle fracture. Folds formed in this manner exhibit many tiny fractures.

If you push together two ends of a long, slender piece of clay, the clay deforms into a sequence of folds. This demonstration illustrates three characteristics of folds.

1. Folding usually results from compression.

2. Folding always shortens the horizontal distances in rock. Notice in Figure 10-3 that the distance between the two points A and A' is shorter in the folded rock than it was before folding.

3. A fold usually occurs as part of a group of many similar folds, as seen in Figure 10-2.

Figure 10-3 shows the basic parts of a fold. The sloping sides of a fold are called the **limbs**. The **axial surface** of a fold is an imaginary surface that passes through all the points of maximum curvature on the fold. A fold arching upward is called an **anticline**. A fold arching downward is called a **syncline**.

FIGURE 10-4
Go online to view an example of an anticline and its topographic expression.

Folds may be *symmetric, asymmetric, overturned,* or *recumbent* (Figure 10-5). These types are differentiated based on the orientation of their axial surface and the facing direction of the two limbs. In a symmetric fold the axial surface is vertical. In an asymmetric fold the axial surface is tilted. Overturned folds also have a tilted axial surface, but one of the limbs faces down. In a

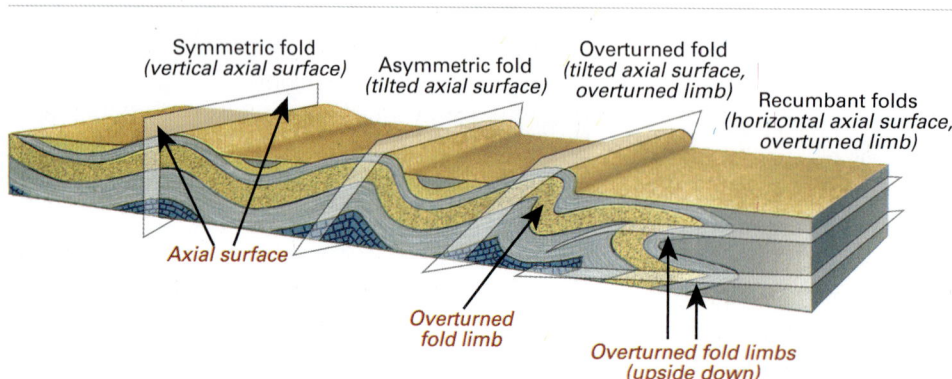

FIGURE 10-5
Folds are classified on the basis of their axial surface orientations and the facing direction of the two limbs. Folds can be symmetrical (vertical axial surface) or asymmetrical (tilted axial surface). A fold with one tilted axial surface and one limb facing downward (upside down) is overturned.

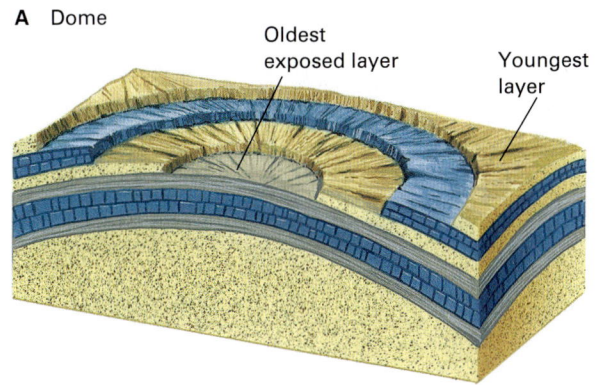

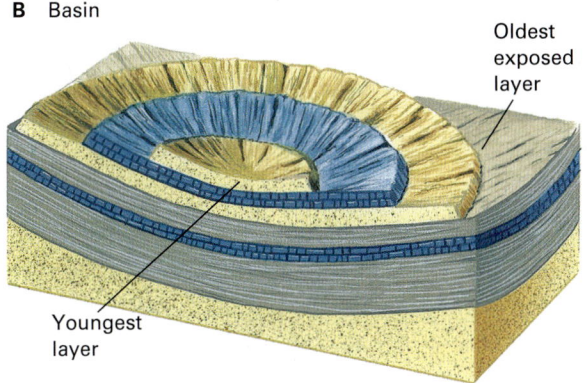

FIGURE 10-6
(A) Sedimentary layering dips away from a dome in all directions. (B) Layering dips toward the center of a basin.

recumbent fold, the axial surface is horizontal and one limb is completely upside down.

Folding on a larger scale forms domes and basins. A circular or elliptical anticlinal structure is called a dome (Figure 10-6A). A dome resembles an upside-down bowl. Sedimentary layering dips away from the top of a dome in all directions. A basin resembles a right-side-up bowl (Figure 10-6B). Domes and basins range in size from a few kilometers to hundreds of kilometers in diameter. Large domes and basins form by the sinking or rising of continental crust in response to tectonic forces. Large structural basins include the Michigan Basin, which covers much of the state of Michigan, and the Illinois Basin, which covers much of Illinois.

checkpoint What are the four common types of folds?

Faults and Joints

Like folds, faults and joints are geologic structures caused by rock deformation. Recall that a fault is a fracture along which rock on one side has moved relative to rock on the other side. **Slip** is the

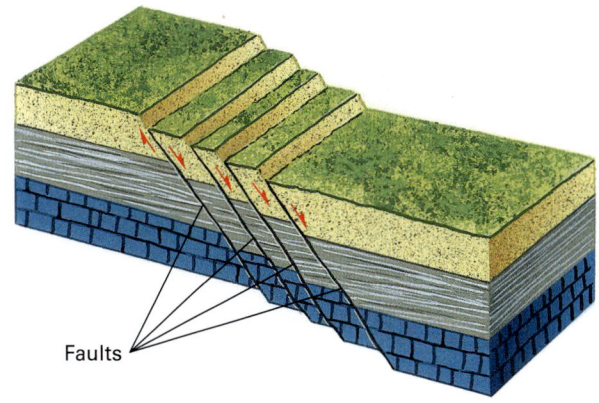

FIGURE 10-7
Faults with a large slip commonly move along numerous closely spaced faults, forming a fault zone.

distance that rocks on opposite sides of a fault have moved. Movement along a fault may be gradual or sudden. Some faults occur as single fractures in rock. Most, however, consist of numerous closely spaced fractures within a fault zone (Figure 10-7). Fault zones can range in thickness from less than a centimeter to over a kilometer. Over time, rock may slide hundreds of meters or many kilometers along a large fault zone.

Rock moves repeatedly along many faults and fault zones because tectonic forces commonly persist in the same place over long periods of time (such as at a tectonic plate boundary). In addition, once a fault forms, it is easier for rock to move again along the same fracture than for a new fracture to develop nearby. Many faults are not vertical but dip at an angle and therefore have an upper and a lower side. The block above an inclined, or angled, fault is known as the "hanging wall" while the block underneath is the "footwall." If you could stand or climb between the two blocks of crust, the side that supported your weight would be the footwall.

Normal Faults A **normal fault** forms where tectonic forces cause Earth's crust to stretch, pulling it apart and causing normal faults to form. In each normal fault, the hanging wall slips downward against the footwall. Notice in Figure 10-8 that the horizontal distance between points A and A' on opposite sides of the fault is greater after normal faulting occurs.

A wedge or keystone-shaped block of rock that has dropped downward between pairs of normal faults is called a **graben** (Figure 10-9). The word *graben* comes from the German word for

LESSON 10.1

FIGURE 10-8
(A) A normal fault accommodates an extension of the crust. The hanging wall slips down a normal fault. (B) This image shows a normal fault within layers of mudstone and sandstone.

FIGURE 10-9
Two normal faults have offset the same layers of rock resulting in a graben. The fault movement has tilted the fault block in the middle, rotating it counterclockwise.

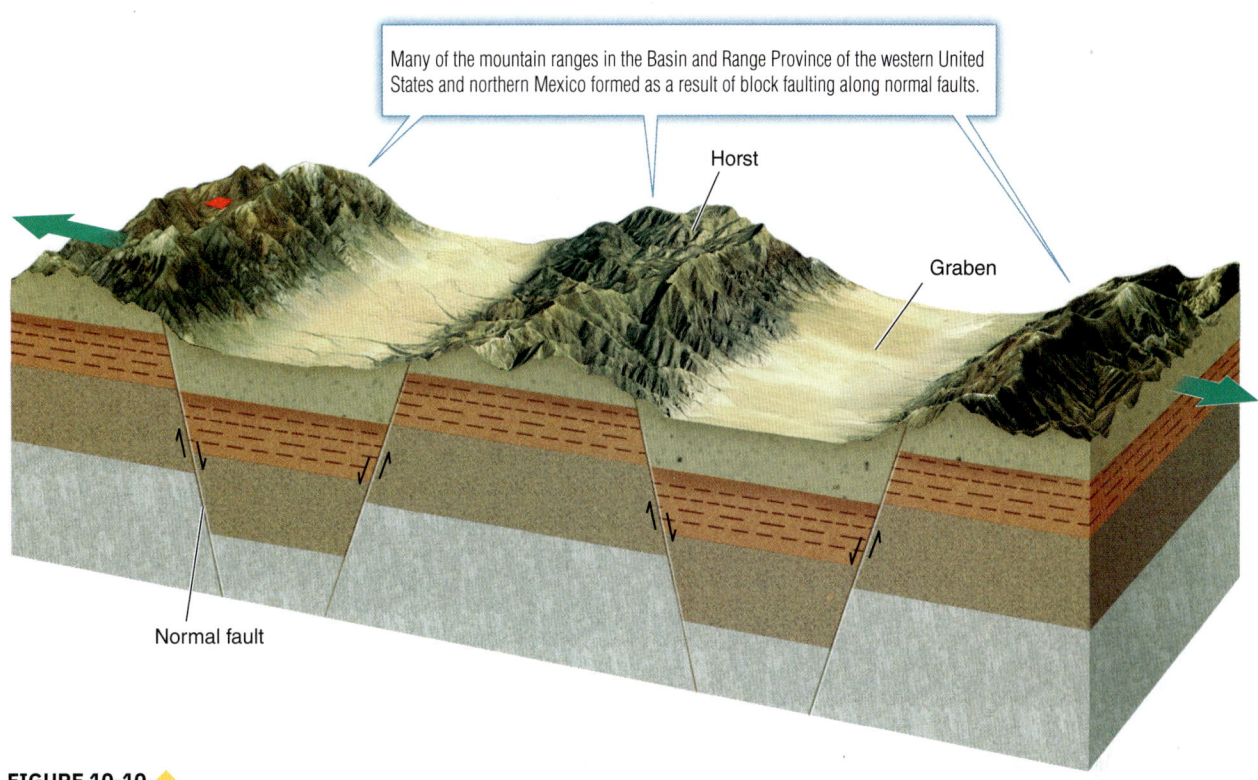

FIGURE 10-10
Horsts and grabens commonly form where tectonic forces stretch the crust over a broad area, generating a basin-and-range topography.

"ditch or valley." (Think of a large block of rock settling downward to form a valley.) If tectonic forces stretch the crust over a large area, many normal faults may develop. This allows numerous grabens to settle downward between the faults. A **horst** is the block of rock between two down-dropped grabens that appears to have moved upward relative to the grabens (Figure 10-10).

Normal faults, grabens, and horsts are common where the crust is rifting at a spreading center, such as the Mid-Ocean Ridge and the East African Rift zone. These features are also found where tectonic forces stretch a single plate, as in the Basin and Range Province of parts of western North America.

In contrast, horizontal compressive forces may fracture rock to produce a **reverse fault** in which the hanging wall moves upward relative to the footwall. In Figure 10-11, the distance between points A and A' is shortened by the faulting.

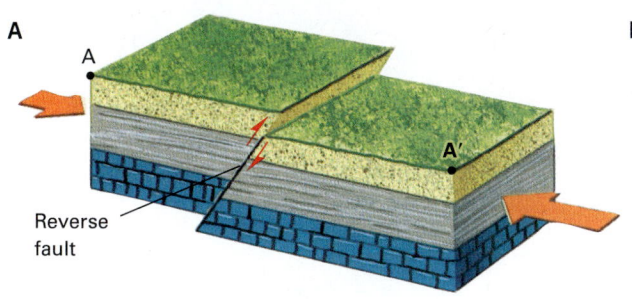

FIGURE 10-11
(A) A reverse fault accommodates the compression of the crust. The hanging wall slides up a reverse fault. (B) This photo shows a reverse fault in the Appalachian Mountains in northeastern Tennessee. The rock above the fault on the right is sandstone while the rock below the fault on the left is limestone.

Thrust Faults A **thrust fault** is a special type of reverse fault that is nearly horizontal (Figure 10-12A). In some thrust faults, the rocks of the hanging wall have moved many kilometers over the footwall. For example, all of the rocks of Glacier National Park in northwestern Montana slid 50 to 100 kilometers eastward along one or more thrust faults to their present location. One of these faults is the Lewis Overthrust (Figure 10-12B). These thrust faults formed from about 180 million to 55 million years ago as compressive tectonic forces built the mountains of western North America. During this time, widespread thrust faulting moved large slabs of rock, some even larger than that of Glacier Park, from west to east in a zone extending from Alaska to Mexico.

Faults and Plate Boundaries Each of the three types of plate boundaries—divergent, transform, and convergent—produces different tectonic stresses. Each type of plate boundary has characteristic folds and faults. At divergent plate boundaries, stretching produces normal faults and grabens. A transform plate boundary is a strike-slip fault zone. Recall from Chapter 8 that a strike-slip fault is one in which the fracture is vertical, or nearly so. Rocks on opposite sides of the fracture move horizontally past each other. The famous San Andreas Fault is the main fault within a zone of many strike-slip faults that collectively form the boundary between the Pacific plate and the North American plate. Near a convergent plate boundary, compression commonly produces large regions of folds, reverse faults, and thrust faults. Folds and thrust faults are common in the mountains of western North America, the Appalachian Mountains, the Alps, and the Himalaya (Case

A B

FIGURE 10-12
(A) A thrust-fault is a low-angle, nearly horizontal reverse fault. (B) The Lewis Overthrust in Montana is a major thrust fault. Precambrian rocks from the hanging wall of the fault have been displaced up and over much younger Cretaceous rocks. The Cretaceous rocks form the footwall of the fault.

FIGURE 10-13
Joint fractures are common near Earth's surface. Here they can be seen along the shore of George Lake in Ontario, Canada.

Study), all of which formed at convergent boundaries.

Joints Unlike a fault, a **joint** is a fracture across which the rock on both sides does not slip (Figure 10-13). In Chapter 9, we discussed columnar joints that form in basalt as it cools. Tectonic forces also create joints. Most rocks near Earth's surface are jointed, but joints become less abundant with depth. At deeper levels in the crust, rocks become more plastic and less prone to fracturing.

Joints and faults are planes of weakness in otherwise strong rock. Dams constructed in jointed rock often leak because water seeps into the joints and flows around the dam through the fractures. You can commonly observe seepage caused by such leaks in canyon walls downstream from a dam.

checkpoint What is the difference between a normal fault and a reverse fault?

10.1 ASSESSMENT

1. **Identify** What four factors control how a rock responds to tectonic stress?
2. **Explain** Give two reasons why rock tends to move repeatedly along the same faults instead of forming new faults.
3. **Contrast** What are the differences between folds, faults, and joints?
4. **Explain** Why would you not find horsts and grabens in the fault zone of a reverse fault?

Critical Thinking

5. **Apply** Imagine a complex geologic setting in which major folding occurred millions of years ago. The topography is hilly, with synclines at higher elevations than anticlines. Identify the type of force that caused the folding. Then offer a hypothesis for why the topography does not match the structure of the folds.

10.2 MOUNTAINS AND MOUNTAIN RANGES

Core Ideas and Skills
- Understand the key processes in orogeny, or mountain building.
- Describe how island arcs form.
- Explain how a continental arc, or Andean margin, forms.

KEY TERMS
orogeny
underthrusting
island arc
subduction complex
forearc basin
continental arc
foreland basin

Mountain Building

Tectonic forces have created mountains at each of the three types of tectonic plate boundaries. The world's largest mountain chain, the Mid-Ocean Ridge system, forms a network of divergent plate boundaries beneath the ocean. Mountains also rise at divergent plate boundaries on land. Mount Kilimanjaro and Mount Kenya, two volcanic peaks near the Equator, lie along the East African Rift. Other ranges, such as the California Coast Ranges, formed at transform plate boundaries. Finally, the volcanic island arcs of the southwestern Pacific Ocean, Alaska's Aleutian Islands, the Cascade Mountain Range, and the Andes Mountains all formed along convergent plate boundaries.

Orogeny refers to the process of mountain building and includes all the processes associated with divergent, transform, and convergent plate boundaries. The belt of rocks that is deformed in an orogeny is called an *orogen*.

FIGURE 10-14
ON ASSIGNMENT Journalist Paul Salopek has been walking across the globe as part of the National Geographic sponsored Out of Eden Walk. Salopek began his journey in 2013 and plans to cover 33,796 kilometers (21,000 miles). Salopek has traversed a variety of terrain, which includes mountainous regions. National Geographic photographer John Stanmeyer helped document Salopek's walk through Wadi Musa in Jordan. The Out of Eden Walk aims to connect culture and geography, bringing a greater understanding to the story of human migration.

Plate boundaries are nearly linear or slightly curved, so mountains most commonly occur as linear or slightly curved ranges and chains. For example, the Andes extend in a narrow band along the west coast of South America. The Appalachians form a gently curving uplift along the east coast of North America and parts of Arkansas and Oklahoma.

Recall from Lesson 7.3 that *isostasy* is a state of gravitational equilibrium between the lithosphere and the asthenosphere (Figure 7-22). The elevation at which a tectonic plate "floats" depends on its thickness and density. Where tectonic plates converge, the lithosphere becomes thicker, causing continental mountain ranges to rise isostatically.

Several processes can affect the thickness of the lithosphere in mountainous regions, causing isostatic adjustment of the crust.

- In a subduction zone, the descending slab generates magma. The magma rises to cool within the overlying lithosphere and forms plutons or erupts onto the surface to form volcanic rocks. Both the plutons and volcanics thicken the lithosphere by adding volumes of new mass to it.

- Magmatic activity heats the lithosphere above a subduction zone, causing it to expand and become thicker.

- Compressive tectonic forces squeeze the crust horizontally and increase its thickness by folding and faulting.

- In a region where two continents collide, one continent may be forced beneath the other. This process, called **underthrusting**, can double the thickness of continental crust in the collision zone. For example, underthrusting of the Indian-Australian plate under the Eurasian plate has caused the Himalayan region to rise isostatically. Underthrusting also occurs in subduction zones to form a subduction complex (described in the next section).

- In continental collision zones where crustal thickening is extreme, partial melting of the middle crust and intense compression can cause plastically deforming rock to push outward and upward.

FIGURE 10-15
NASA's Terra satellite took this image of the Himalaya in Bhutan, which support numerous glaciers and glacial lakes. The glaciers are the long snake-like features that project away from the center of the range. Several small glacial lakes appear as dark-colored patches at the tips of some of the glaciers.

As mountain peaks rise, streams, glaciers, and landslides erode them. The sediment is carried into adjacent valleys. Initially, when the mountains erode, they become lighter and rise isostatically. Imagine a canoe or a small boat rising as you step out of it. Eventually, erosion wins over isostatic rebound. The Appalachians are an old range where erosion is now wearing away the remains of peaks that may once have been the size of the Himalaya. Conversely, the Alps in Switzerland are younger mountains with sharper peaks.

checkpoint What process can as much as double the thickness of continental crust in a collision zone?

Island Arcs

An **island arc** is a volcanic mountain chain that forms where two plates carrying oceanic crust converge. The convergence causes the older, colder, and denser plate to sink into the mantle beneath the other plate. This creates a subduction zone and an oceanic trench. Magma forms in the subduction zone and rises to build submarine volcanoes. Typically, these volcanoes grow above sea level, creating an arc-shaped volcanic island chain that runs parallel to the trench and exists on the overriding plate (Figure 10-16).

A layer of sediment a half kilometer or more thick commonly covers the oldest basaltic crust of the deep seafloor. As the two plates converge, some of the sediment is scraped off the subducting slab and plastered against the side of the overriding plate. Occasionally, slices of basalt from the oceanic crust, underwater mountains called seamounts, and even the upper mantle are scraped off and mixed in with the seafloor sediment. The process is like a bulldozer scraping soil from bedrock, occasionally knocking off a chunk of bedrock along with the soil. Rock and sediment added in this way to the overriding plate forms a **subduction complex**. A subduction complex is characterized by abundant folds, faults, and fractures that resulted from the scraping and compression.

Growth of the subduction complex occurs by underthrusting, in which sediment and rock are added to the bottom of the complex, forcing it to grow upward. Underthrusting thickens the crust, leading to isostatic uplift of the entire complex. Sediment eroded from the volcanic arc and transported toward the sea can be impounded, or trapped, by the rising subduction complex, forming a sedimentary basin called a **forearc basin**.

Island arcs are abundant in the Pacific Ocean, where convergence of oceanic plates is common.

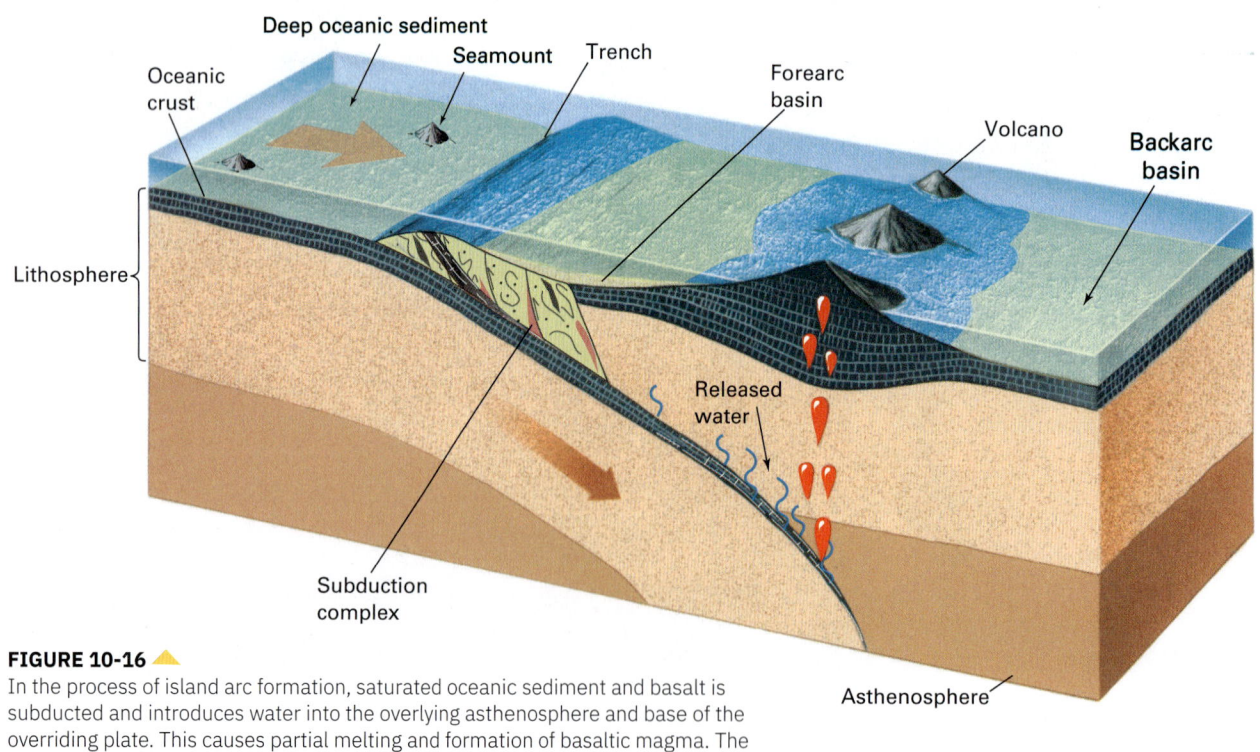

FIGURE 10-16
In the process of island arc formation, saturated oceanic sediment and basalt is subducted and introduces water into the overlying asthenosphere and base of the overriding plate. This causes partial melting and formation of basaltic magma. The magma rises and erupts, building submarine volcanoes that eventually grow above sea level to form a volcanic mountain chain.

FIGURE 10-17
The Cordillera Real mountains are a range within the Bolivian Andes. Two of the highest peaks in the northern section of the range are Illampu peak (6,368 meters) at left and Ancohuma peak (6,427 meters), which appears beneath the moon.

The western Aleutian Islands and most of the island chains of the southwestern Pacific are island arcs.

checkpoint Where are island arcs mainly found?

Subduction at a Continental Margin

The Andes are the world's second highest mountain chain, with 49 peaks above 6,000 meters (Figure 10-17). They rise from the Pacific coast of South America, starting nearly at sea level, and extend through seven South American countries. Igneous rocks make up most of the Andes. The chain also contains folded sedimentary rocks on both sides of the mountains. The Andes are a good example of a **continental arc**, formed by subduction of oceanic lithosphere beneath continental lithosphere. This type of plate margin is also called an Andean margin.

Subduction along the west coast of South America has a long geologic history that some geologists think extends as far back as the Paleozoic era, when South America was part of Gondwana. As the process of subduction proceeded, sediment and rock was scraped off the descending plate and piled onto the western edge

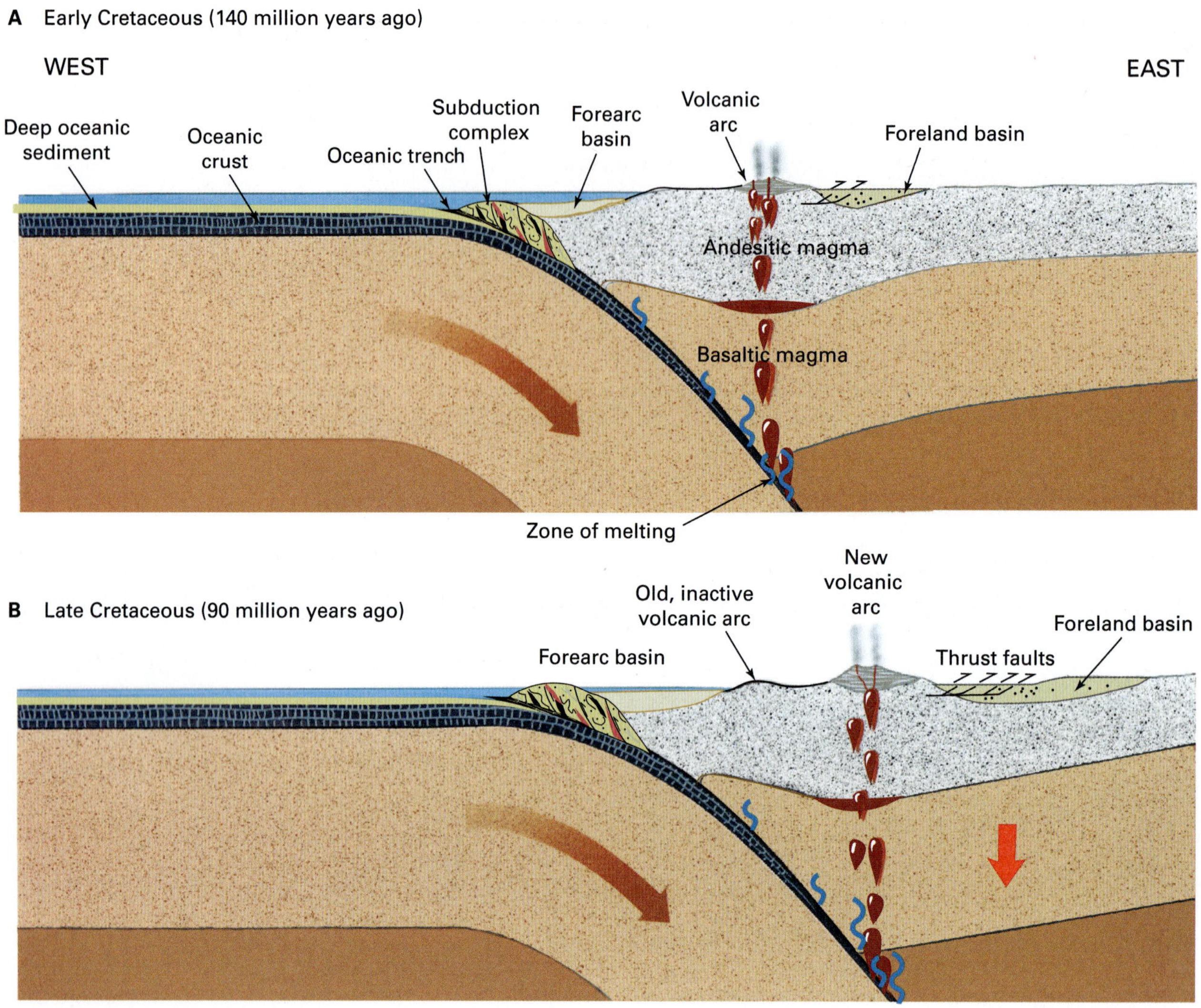

A Early Cretaceous (140 million years ago)

B Late Cretaceous (90 million years ago)

FIGURE 10-18
(A) Subduction of oceanic lithosphere beneath the western side of the South American plate has been ongoing for at least 140 million years. Basaltic magma from the descending slab rose into the overlying crust and partially melted it, forming andesitic and granitic magma. Eruption of this magma formed a continental arc on the overriding plate. (B) As subduction continued, the size of the subduction complex and the width of the forearc basin grew. Meanwhile, the rising magma migrated eastward to form new volcanoes.

---

of the continent, forming a subduction complex (Figure 10-18). Meanwhile, the basalt of the descending slab and water contained in the oceanic sediments were injected into the lithosphere of the overriding South American plate, generating huge volumes of magma. This caused the lithosphere to partially melt, forming additional magma that rose toward the surface. The rising magma formed and arc of plutons and volcanoes along the entire western length of the continent.

The rising magma also heated and thickened the crust. The crust rose isostatically and formed a large, linear mountain range parallel to the subduction zone. This range forms the Andes Mountains. As subduction took place, much sediment from the mountains was shed westward. A forearc basin formed between the arc and the subduction complex. With time, the subduction complex grew and the arc shifted to the east, causing the forearc basin to widen. Meanwhile, compression associated with the converging plates initiated a series of folds and thrust faults east of the Andes Mountains. Movement along the thrust faults caused the sections between the faults to stack up. The weight of the stack caused the crust on the east side of the mountains to flex downward, producing a **foreland basin**, which captured sediment that eroded off the east side of the mountains.

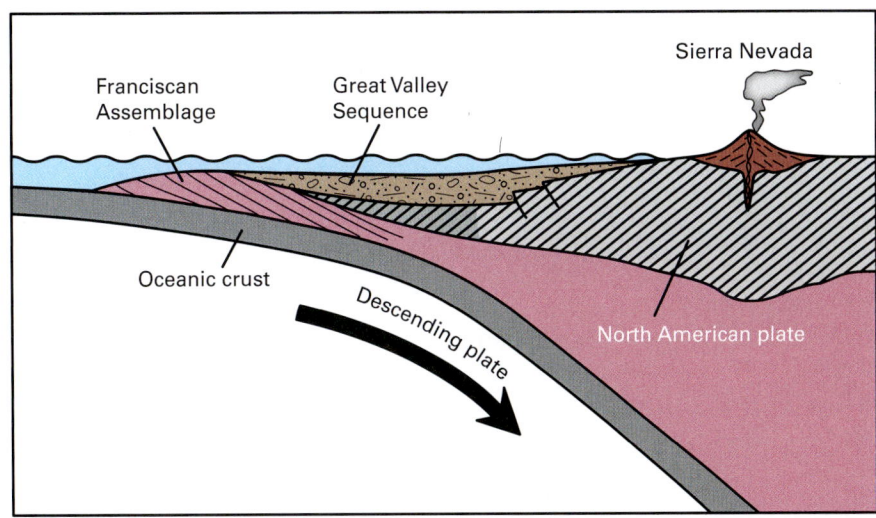

FIGURE 10-19
A cross section depicts the western margin of North America as it looked in California during the Cretaceous. This is before the evolution of the San Andreas Fault. Oceanic lithosphere of the Farallon plate subducted under North American continental lithosphere. The Sierra Nevada was an active continental arc. The Franciscan Assemblage is a subduction complex that is presently exposed in much of California's coastal ranges.

CONNECTIONS TO HUMAN GEOGRAPHY

Women Porters of the Peruvian Andes

Every mountaineering expedition requires a team of people. Supplies, including food and gear, are necessary at each stage of the ascent. Porters are key members of a climbing team, ensuring that supplies reach every camp along the way. A porter's labor is compensated very meagerly, about $72 for each four-day trek. Yet, this amount is a reliable source of income for those who can tolerate the physical demands of the job.

Until recently, women have been excluded from this earning potential. Many porter agencies dismissed women as incapable or too weak to do the work. Women mostly earned money from selling handicrafts to tourists.

Currently, at least one expedition group is trying to change their hiring practices. Some women are working as porters for climbing expeditions, carrying packs with gear over difficult terrain. They set up camps and ascend the mountainside with professional climbers. Working as porters is a means to better income and career development. With enough experience, porters can become guides. A guide earns an even higher income and has opportunities to earn certifications in English language, history, and navigation. Replacing handicraft sales with work as porters or guides creates a new mountain legacy for women in the Andes.

FIGURE 10-20
Peruvian female porters climb an Andean trail. Regulations require women to carry around 15 kilograms (33 pounds) while men can carry about 20 kilograms (44 pounds).

MINILAB Fault-Block Mountain Models

Materials
modeling clay
2 hard, dry cookies
protective gloves
utility knife

Procedure

1. Construct a model of a fault block using modeling clay. Use the utility knife to form the angles of the fault blocks in relation to one another.
2. Push the two clay blocks toward each other to model a fault. Observe the results.

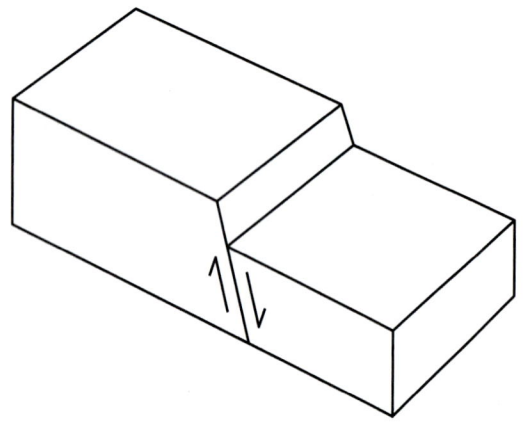

3. Construct a second model of a fault block using two hard cookies. Use the utility knife to carefully cut or scrape the cookies to form angles at the fault boundary similar to those in your clay model.
4. Push the two fault cookies toward each other to model a fault. Observe the results.

Results and Analysis

1. **Observe** Describe how the fault blocks move vertically after a thrust occurs.
2. **Infer** What type of surface feature might be formed by the movement at the faults you modeled?
3. **Predict** Explain what you would see in the rock layers of that surface feature.

Critical Thinking

4. **Apply** Describe how the models represent a fault. What process drives movement in a fault? How is that represented in each model?
5. **Critique** Explain which model is a better representation of how mountains form at faults.

The California coast offers another example of subduction at a continental margin. Coastal ranges expose a large subduction complex that was built during the Mesozoic era, when the Farallon plate subducted beneath western North America (Figure 10-19). Although now completely subducted, the Farallon plate transported oceanic crust eastward toward the convergent margin for tens of millions of years. The longevity of this ancient subduction zone caused the growth of a large subduction complex that incorporated many different elements of marine geology, including bits of oceanic crust and deepwater sediments. Today, much of San Francisco, including the Golden Gate Bridge, is built on rocks belonging to the ancient subduction complex.

checkpoint Why do the Andes have sedimentary rock on both sides of the mountains?

10.2 ASSESSMENT

1. **Explain** What needs to happen to the lithosphere for an area to rise isostatically?
2. **Summarize** What processes lead to isostatic adjustment in a subduction zone?
3. **Summarize** What processes lead to isostatic adjustment where two continents converge?
4. **Compare and Contrast** Explain the similarities and differences between forearc basins and foreland basins.

Critical Thinking

5. **Generalize** The Himalaya continue to rise, yet their overall elevation has not changed. Give one possible explanation for this observation.

10.3 CONTINENTAL COLLISION: THE HIMALAYA

Core Ideas and Skills
- Describe the geologic events that led to the formation of the Himalaya.
- Analyze and visualize data to compare the altitudes of different mountains.

The Himalaya are the world's highest mountain chain (Figure 10-21). The chain separates China from India, includes parts of Nepal, Bhutan, and Sikkim, and has the world's highest peaks. If you were to stand on the southern edge of the Tibetan Plateau and look southward, you would see the high peaks of the Himalaya. Beyond this great mountain chain lie the rain forests and hot, dry plains of India. If you had been able to stand in the same place 200 million years ago and look southward,

FIGURE 10-21
The Himalaya first began to rise as part of the Andean-style volcanic arc. The modern range resulted from continent-to-continent collision between the Indian and Asian plates. The Ama Dablam mountain is a peak in the Himalaya range of eastern Nepal. The main peak has an altitude of 6,812 meters.

you would have seen only ocean. At that time, India was located south of the Equator, separated from southern Asia by thousands of kilometers of open ocean. The Himalaya had not yet begun to rise.

Evolution of Earth's Highest Peaks

Formation of an Andean-Type Margin

About 80 million years ago, a triangular piece of lithosphere that included present-day India had split off from a large mass of continental crust near the South Pole. It began drifting northward toward Asia as fast as 20 centimeters per year. Geologically speaking, this is very rapid. As this Indian plate started to move, its leading edge, consisting of oceanic crust, began to subduct beneath the Eurasian plate's southern margin. This subduction formed an Andean-type continental margin, complete with a continental arc, along the southern edge of Tibet. Volcanoes erupted and plutons with granitic magma rose into the overlying continental crust (Figure 10-22).

Continent–Continent Collision

By about 50 million years ago, subduction had consumed all the oceanic lithosphere between the continents on the Indian and Eurasian plates. The two landmasses began to collide, and the continental lithosphere of India began to subduct beneath the Eurasian continent.

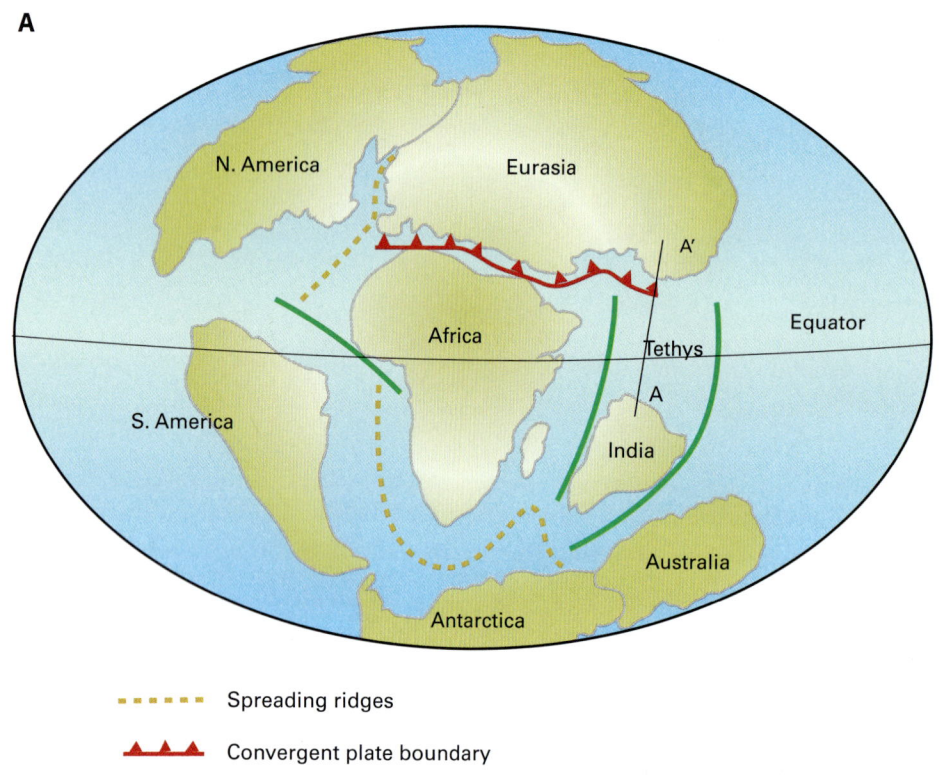
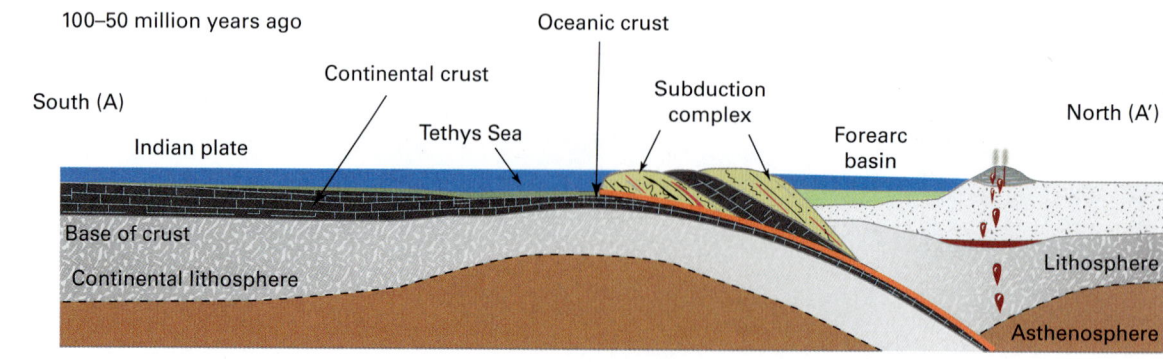

FIGURE 10-22
(A) About 150 million years ago, the landmass known as Gondwana began to break apart through tectonic rifting. At the same time, a convergent plate boundary along the southern margin of Eurasia provided for northward subduction of oceanic lithosphere beneath that plate. (B) A cross section shows the subduction complex and active volcanoes on the Asian continent. Notice points A and A' on the map, which roughly correspond to the South and North labels in the cross section.

312 CHAPTER 10 MOUNTAINS

As India subducted beneath modern-day Asia, arc volcanism associated with previous subduction of oceanic crust stopped. Thick sequences of sediment that had accumulated on India's northern continental shelf were scraped off the descending plate. These sediments were pushed into great folds separated by major thrust faults.

The intense compression and the thickening of the lithosphere that resulted caused widespread melting of the middle crust (Figure 10-23). As the compression continued, this plastic rock in the core of the orogen was squeezed out—southward and upward—like toothpaste from a giant tube. These southward-extruded rocks, called the Greater Himalayan Sequence, are bounded below by a major thrust fault and above by a major normal fault. The highest peaks of the Himalaya occur in the Greater Himalayan Sequence.

The indentation of the Eurasian plate by the Indian continent not only compressed Tibet in a north-south direction but also acted like a giant wedge on the Asian continent. This wedge-like action pushed large blocks of lithosphere north of the collision zone along huge strike-slip faults.

The continent–continent collision also caused structural reactivation of earlier-formed faults across much of Central Asia. This happened despite the fact that many of these faults are hundreds

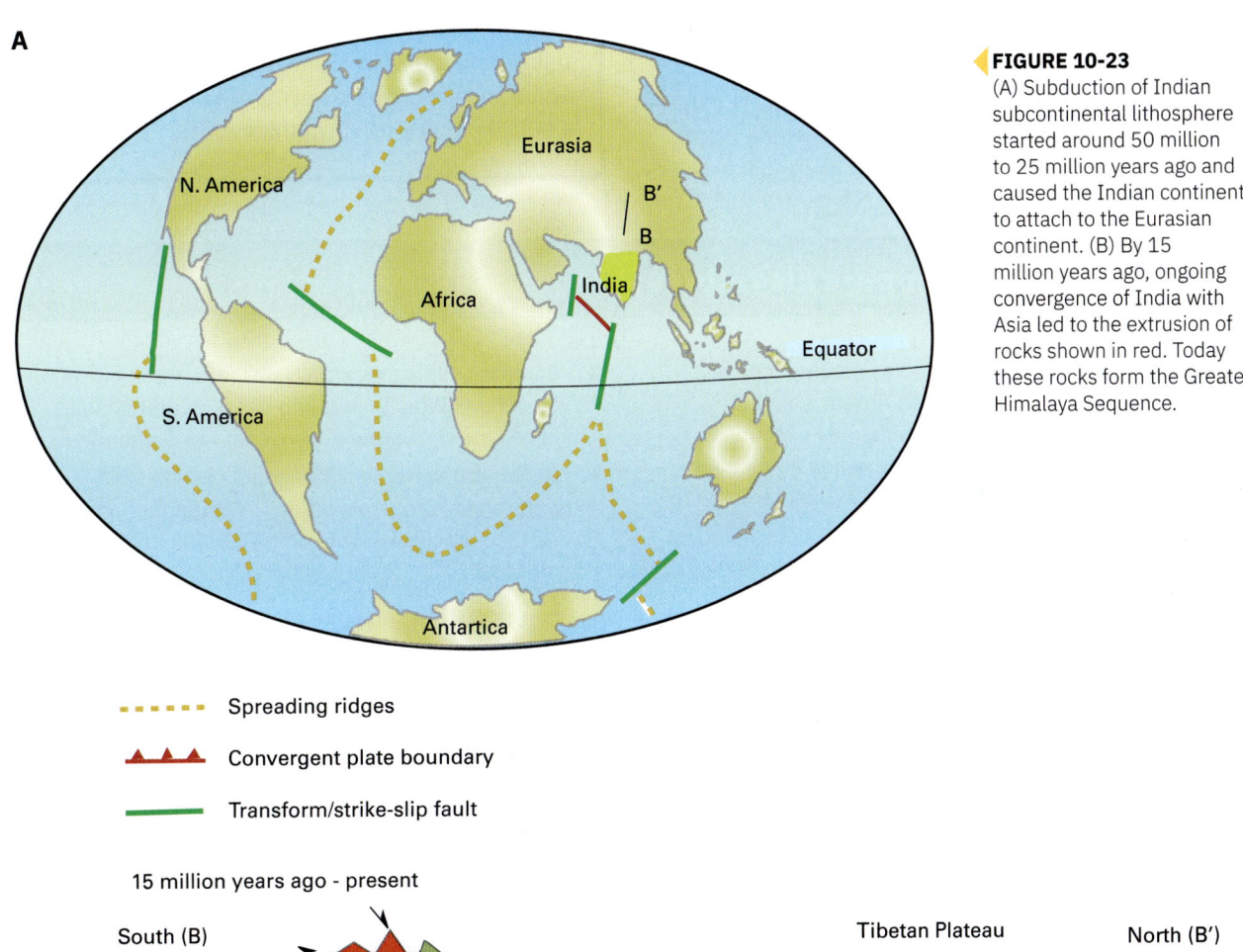

FIGURE 10-23
(A) Subduction of Indian subcontinental lithosphere started around 50 million to 25 million years ago and caused the Indian continent to attach to the Eurasian continent. (B) By 15 million years ago, ongoing convergence of India with Asia led to the extrusion of rocks shown in red. Today these rocks form the Greater Himalaya Sequence.

of kilometers north of the Indian-Eurasian plate boundary. Deformation across this collective, complex network of faults by the ongoing collision has created the tall, linear mountain ranges and vast sedimentary basins of Central Asia.

The Himalaya Today Today's Himalaya contain igneous, sedimentary, and metamorphic rocks. Many of the sedimentary rocks contain fossils of marine organisms that lived in the shallow sea of the Indian continental shelf. Volcanic rock and some plutonic rock formed when the range was an Andean margin. Large volumes of plutonic rocks also formed by midcrustal melting (Figure 10-23). Rocks of all types were metamorphosed by the tremendous pressure and heat generated during subduction and continent–continent collision.

Estimates of the convergence rate between India and Asia suggest that it slowed significantly when the continental lithosphere began to subduct. Even so, present-day estimates of the convergence rate range from about 3 to 5 centimeters per year.

The underthrusting of Indian continental lithosphere beneath Tibet and the squashing of Tibet have greatly thickened the continental lithosphere under the Himalaya and the Tibetan Plateau to the

> **CONNECTIONS TO LIFE SCIENCES**
>
> **Mountain Habitats**
> Mountains affect plant and animal habitats. Cool climates prevail at high elevations, even at the Equator. As a result, plant and animal communities change rapidly with elevation. For example, if you climbed one of Colorado's 4,000-meter mountains, you would reach alpine tundra around 3,400 meters. To find an equivalent alpine tundra environment at low elevation, you would have to travel to the Canadian Barren Lands 3,000 kilometers north of Colorado.

DATA ANALYSIS Compare Elevations

Alaska's Denali, formerly known as Mount McKinley, is the highest mountain in North America. Asia's Himalaya include the world's tallest peak, Mount Everest. Table 10-1 provides information about each mountain's size and approximate age. Use the table to answer the questions.

Computational Thinking

1. **Analyze Data** Calculate the overall height of both mountains. Which mountain's summit reaches a higher elevation?

Critical Thinking

2. **Design** Draw a model that illustrates the information in Table 10-1.

3. **Compare** Which mountain's base is at the higher elevation? Which mountain has a greater change in elevation between its base and its summit?

4. **Generalize Patterns** What could explain the difference in total elevations? What effect do surface processes have on the overall height of a mountain?

Data Challenge

Go to the Data Analysis in MindTap to complete the data challenge.

TABLE 10-1 Elevation and Age Characteristics for Mount Everest and Denali

Mountain	Change in Elevation from Base to Summit	Elevation of Base above Sea Level	Approximate Age
Mount Everest	3,700 m	5,150 m	50 million years
Denali	5,500 m	690 m	25 million years

Source: USGS

north. Consequently, the region floats isostatically at high elevation. Even the valleys lie at elevations of 3,000 to 4,000 meters above sea level, and the Tibetan Plateau has an average elevation of 4,000 to 5,000 meters. One reason the Himalaya contain all of Earth's highest peaks is simply that the entire plateau lies at such a high elevation. From the valley floor to the summit, Mount Everest is actually smaller than Alaska's Denali. Mount Everest rises about 3,300 meters from base to summit, whereas Denali rises about 4,200 meters. The difference in elevation of the two peaks results from the fact that the base of Mount Everest is at about 5,500 meters, but Denali's base is at 2,000 meters. (See Data Analysis: Compare Elevations.)

The Himalayan chain is only one example of a mountain chain built by a collision between two continents. The Appalachian Mountains formed when eastern North America collided with Europe, Africa, and South America between 470 million and 250 million years ago. The European Alps formed during collision between northern Africa and southern Europe beginning about 30 million years ago. The Ural Mountains, which separate Europe from Asia, formed by a similar process about 250 million years ago.

checkpoint How is it possible that sedimentary rocks in the Himalaya contain fossils of marine organisms?

10.3 ASSESSMENT

1. **Summarize** Explain why the view from the southern edge of the Tibetan Plateau would have looked dramatically different 200 million years ago.
2. **Explain** What caused the Tibetan Plateau region of the Himalaya to float isostatically at a high elevation?
3. **Compare** How does a caldera differ from most other volcanic landforms?
4. **Summarize** How do vast quantities of volcanic ash and sulfur gases affect solar radiation?

Critical Thinking

5. **Synthesize** Refer to the map of Pangaea Proxima at the beginning of Chapter 7. The map is a model of the next supercontinent some 250 million years into the future. Identify a new mountain chain on this map. Use concepts from this chapter to describe how the mountain chain may have formed, such as by continent–continent collision, as a continental arc, or through a combination of processes.

TYING IT ALL TOGETHER
CHAINS OF CHANGE

In this chapter, you learned about the geologic forces that form and shape mountain ranges over millions of years (Figure 10-24). In the Case Study, we examined the Appalachians and the Himalaya, two mountain ranges at very different stages in their geologic history. While the major plate tectonic forces that formed the Appalachians have stalled, the current colliding of the Indian and Eurasian plates continues to fold and shape the Himalaya. At the same time, surface forces continue to shape both mountain chains.

Mountain climbing is a journey across geologic time, each footstep guided by the ancient tectonic forces thrusting Earth upward and the relentless surface forces that wear it back down. Although the changes that shape mountains occur slowly from a human perspective, scientists worry that climate change is increasing the pace of erosion. A warmer, drier climate in the Appalachians may lead to lower levels of precipitation, more wildfires, and more extreme weather events that cause flooding or drought. Though it is hard to predict the specific effects of climate change, collectively all of these changes can modify sedimentation and increase erosion.

Research another mountain range, such as the Alps, Andes, Atlas Mountains, Great Dividing Range, Rocky Mountains, Sierra Nevada, or Ural Mountains. Then answer the questions below.

1. Identify the location and estimated age of the mountain range. Find out what geologic processes contributed to its formation and current features.

2. Construct a map of the mountain range, identifying major peaks and the processes that contributed to the current shape of each mountain.

3. Explain how the height of peaks may have changed due to erosion and climate change.

4. Organize and present information that could be used for a travel brochure or website for tourists or mountaineers. Include information about major peaks and other attractions of the mountain range you studied. Summarize the current research about the geologic history and future of the chain.

FIGURE 10-24
Mountain chains throughout the world vary in height, length, and age. Nonetheless, every continent features a mountain chain, including the Transantarctic Mountains in Antarctica (not pictured on this map).

CHAPTER 10 SUMMARY

10.1 GEOLOGIC STRUCTURES AND TECTONIC STRESS

- When stress is applied to rocks, the rocks may deform elastically or plastically, or they may break by brittle fracture.
- The characteristics of the rock, and the temperature, pressure, and rate at which stress is applied, all affect how a given rock will respond to stress.
- Folds form when rocks are compressed.
- A fault is a fracture along which rock on one side has moved relative to rock on the other side.
- Normal faults are usually caused when rocks are pulled apart; reverse and thrust faults are caused by compression; and strike-slip faults form where blocks of crust slip horizontally past each other along vertical fractures.
- A joint is a fracture where the rock on either side has not moved.

10.2 MOUNTAINS AND MOUNTAIN RANGES

- Mountains form when the crust thickens and rises isostatically.
- If two converging plates carry oceanic crust, an island arc forms from volcanic activity.
- The Andes Mountains formed at a continental margin near a convergent plate boundary where an oceanic plate is subducting beneath a continental plate.
- The Andes Mountains are an example of a continental arc; continental arcs are also called Andean margins.
- Both island arcs and continental arcs form from volcanic activity and are characterized by subduction complexes and sedimentary rocks deposited in forearc and foreland basins.

10.3 CONTINENTAL COLLISION: THE HIMALAYA

- The Himalaya began with the formation of an Andean-type continental margin along the southern margin of the Eurasian plate. Later, the Indian plate collided with present-day Tibet, producing compressional forces that created the mountain range.
- Mountain ranges formed by continent–continent collisions are dominated by vast regions of folded and thrust-faulted sedimentary and metamorphic rocks and by earlier-formed plutonic and volcanic rocks.
- Mountain ranges formed by continent–continent collisions include the Himalaya, the Appalachian Mountains, the Alps, and the Ural Mountains.

CHAPTER 10 ASSESSMENT

Review Key Terms
Select the key term that best fits the definition. Not all terms will be used, and no term will be used more than once.

anticline	limb
axial surface	normal fault
continental arc	orogeny
forearc basin	reverse fault
foreland basin	slip
graben	subduction complex
horst	syncline
island arc	thrust fault
joint	underthrusting

1. word that means "mountain building"
2. an imaginary surface passing through all the points of maximum curvature of a fold in rock
3. a volcanic mountain chain produced as one oceanic plate subducts under another oceanic plate
4. a fault in which the hanging wall moves upward relative to the footwall
5. either of the two sides of a fold
6. a synclinal structure that forms as rock is pushed onto an area, causing the underlying crust to collapse downward
7. the mass of rock that is scraped from the upper layers of the subducting slab in a subduction zone and added to the overriding plate
8. a continental margin characterized by subduction of an oceanic plate beneath a continental plate; also called an Andean margin
9. slab of rock that appears higher than neighboring slabs that have dropped due to faulting
10. a fold in rock in which the oldest layers of rock are pushed into the middle of the fold, leaving the younger layers of rock on the outside
11. a nearly horizontal break in rock with movement of the hanging wall of one slab up over the footwall of the other
12. action in which one continental plate is forced under another
13. a fault in which the hanging wall moves downward relative to the footwall
14. slab of rock that has dropped down between two normal faults
15. a fold in rock in which the youngest layers of rock are pushed into the middle of the fold, leaving the older layers of rock on the outside

Review Key Concepts
Answer each question separate sheet of paper to demonstrate your understanding of key concepts from the chapter.

16. Explain two fundamental ways that rock can respond to stress. How does temperature influence which of these responses is more likely to happen?
17. Compare how rocks respond to stress applied rapidly as opposed to stress applied over long periods of time. Explain the effects.
18. What is slip, and how does it compare in normal and reverse faults?
19. How are geologic domes and basins similar and how do they differ?
20. What is a recumbent fold, and how is it formed?
21. Why does weathering expose older rocks in the center of a dome?
22. Why are normal faults, grabens, and horsts common in areas where rifting or extension of the crust is occurring?
23. Explain the difference between a fault and a joint.
24. Explain why the Andes Mountains consist of high volumes of igneous rock.
25. The city of San Francisco is built on top of an ancient subduction complex called the Franciscan Assemblage. How did this ancient subduction complex form?
26. Explain why Mount Everest in Tibet is higher in elevation than Denali in Alaska, yet the vertical distance from base to summit for Denali is almost a kilometer greater than that of Mount Everest.

Think Critically
Write a response to each question on a separate sheet of paper. Use concepts from the chapter to support your reasoning.

27. What type of rock deformations would you expect to see at a convergent boundary and why? Describe the observations you could make at a convergent boundary and explain your reasoning.
28. If you were examining folds in rock, how would you distinguish a symmetric fold from an asymmetric fold? Which do you think would be more commonly observed and why?

29. What evidence exists in the Andes Mountains that supports the conclusion that this mountain chain formed as the result of oceanic plate subduction under the South American plate? How does this evidence support the conclusion?

30. Geologists have concluded that an island arc formed in between the Indian and Eurasian continental plates long before the Himalaya began forming. However, this island arc is no longer present in the location where it was thought to have formed close to the Eurasian plate. How can you use what you know about the orogeny of the Himalaya to explain this change?

31. Compare the orogeny of the Andes and the Himalaya. Explain how they are similar and how they differ.

PERFORMANCE TASK

Present-Day Mountain Building

The mountain chains that we see on Earth today were built over long periods of time as immense forces exerted by tectonic plates pushed and pulled different areas of Earth's crust. These same forces are operating today to continue to shape existing mountains and build new ones.

In this performance task, you will choose one of three types of orogeny, or mountain building, and make an illustrated model to show the sequence of events involved in your chosen type of mountain formation. You will then use your illustration to answer questions about real-world examples of mountain formation.

1. Choose one of the types of orogeny listed below and develop an illustrated model that shows the sequence of events leading to mountain formation. Your illustrated model should have multiple frames (at least four), with each frame depicting a different stage in the mountain-building sequence. Name the type of orogeny involved on your model and provide labels as needed to help others understand various parts of your model.

   Types of orogeny:
   - Subduction at oceanic plates
   - Subduction at a continental margin
   - Continental collision

   For each frame in your model, write an explanatory paragraph to summarize the events depicted at that stage in the mountain building sequence. For each frame, explain what happened and what caused the event to happen.

Use your model to answer the questions that follow.

2. How important are faulting, folding, and magmatic activity in the orogenic process you illustrated?

3. What type of plate boundary (divergent, convergent, or transform) is involved in the type of orogeny you illustrated, and how does this affect the orogenic process?

4. Give a real-world example of a mountain chain that formed through the mechanism that you illustrated in your model. Where is this mountain chain located and how does its location help provide support for your claim of its particular type of orogeny?

5. Choose one of the other types of orogeny and compare it to the type you illustrated in your model. How are they similar and how do they differ? Discuss at least two similarities and two differences.

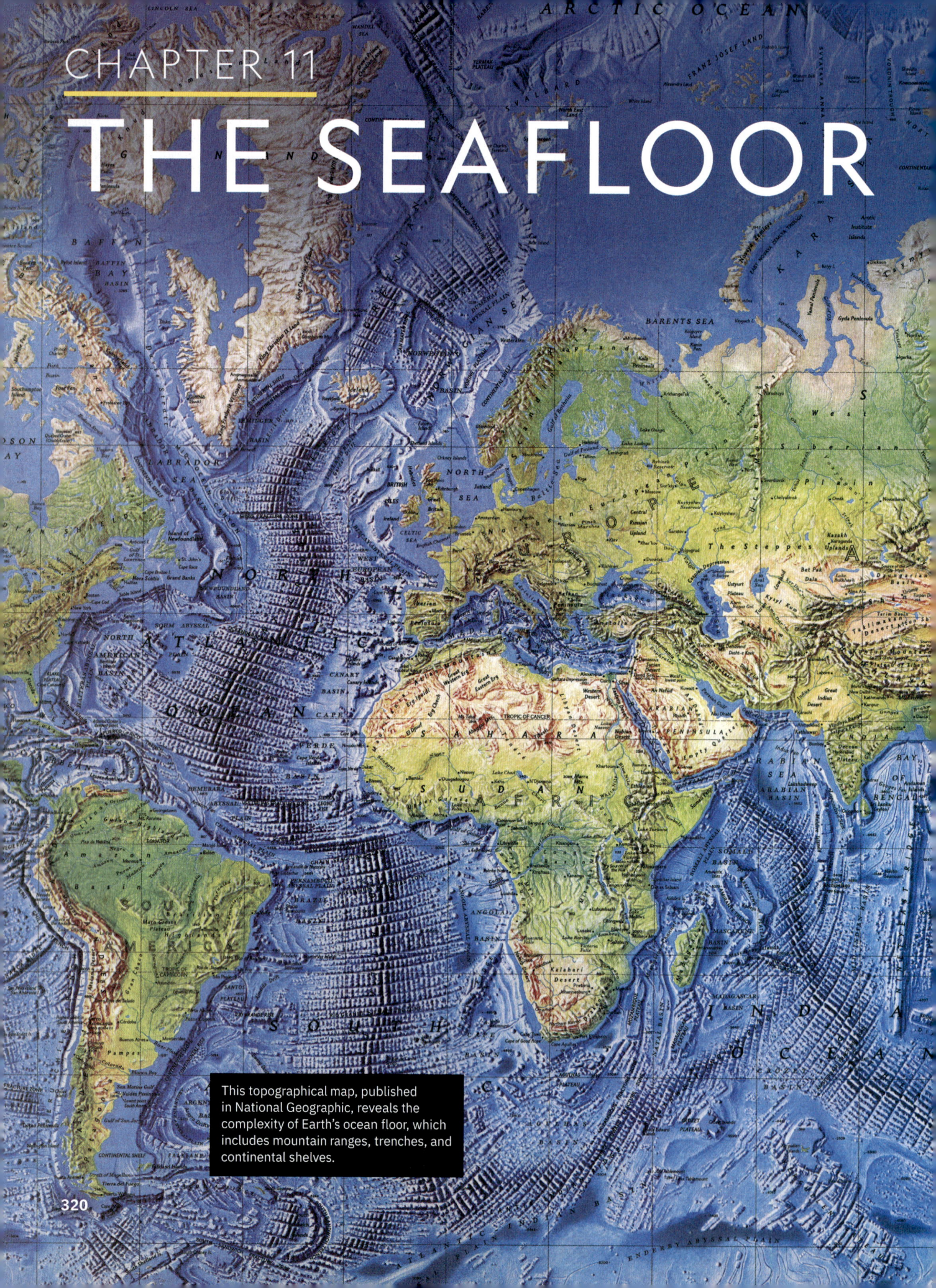

CHAPTER 11
THE SEAFLOOR

This topographical map, published in National Geographic, reveals the complexity of Earth's ocean floor, which includes mountain ranges, trenches, and continental shelves.

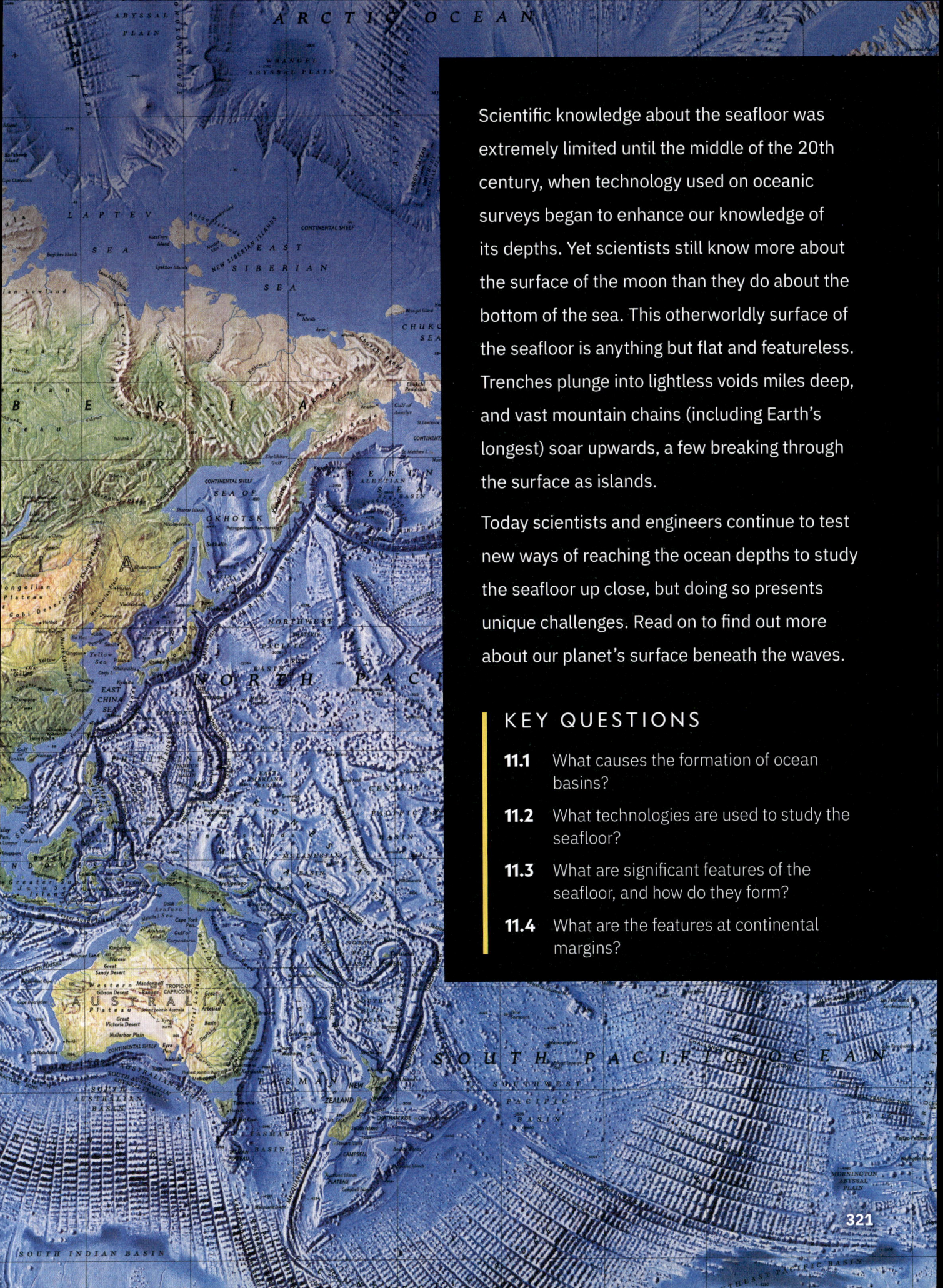

Scientific knowledge about the seafloor was extremely limited until the middle of the 20th century, when technology used on oceanic surveys began to enhance our knowledge of its depths. Yet scientists still know more about the surface of the moon than they do about the bottom of the sea. This otherworldly surface of the seafloor is anything but flat and featureless. Trenches plunge into lightless voids miles deep, and vast mountain chains (including Earth's longest) soar upwards, a few breaking through the surface as islands.

Today scientists and engineers continue to test new ways of reaching the ocean depths to study the seafloor up close, but doing so presents unique challenges. Read on to find out more about our planet's surface beneath the waves.

KEY QUESTIONS

11.1 What causes the formation of ocean basins?

11.2 What technologies are used to study the seafloor?

11.3 What are significant features of the seafloor, and how do they form?

11.4 What are the features at continental margins?

EXPLORERS AT WORK

FINDING BIOINSPIRATION
WITH NATIONAL GEOGRAPHIC EXPLORER MARCELLO CALISTI

At first glance, Marcello Calisti's underwater robot looks a little like a dog. It moves a little like a dog too, gingerly picking up each of its limbs to step, hop, and climb over uneven terrain. Dr. Calisti's robot is a prototype for a project called SILVER, which stands for Seabed-Interaction Legged Vehicle for Exploration and Research. Calisti says the goal of the SILVER project is to "provide an underwater robot that safely, precisely, repeatedly, and with a simple user interface could take high-resolution images, move inside underwater structures, collect samples, withstand disturbances, and continuously monitor a target site." In short, Calisti aims to develop a new generation of underwater robots that will vastly improve marine exploration and photography.

Calisti works as a roboticist and assistant professor at the BioRobotics Institute of Scuola Superiore Sant'Anna in Italy. Calisti specializes in developing robots that can navigate underwater using legged locomotion rather than traditional propellers. In other words, they are designed to walk rather than swim. Calisti's prototypes represent a new generation of robots called soft robots.

Soft robots differ from traditional robots, which are designed to perform very precise movements millions of times, with very specific inputs. Those robots are made of stiff materials such as metal, gears, motors, and joints. They operate in very specific environments and require a lot of computing power. Calisti explains that soft robots reflect a "bioinspired approach." They are made of flexible materials that are able to bend, shape, twist, and deform. They tend to require less computing power to control and operate. They also are designed to deal with more uncertain environments and gently interact with their surroundings, including manipulating objects.

Soft robots often mimic the behavior and function of natural organisms. Calisti says it makes sense to look to nature for inspiration, as natural solutions can be applied to human structures and machines. For example, SILVER may look doglike, but Calisti says its limbs mimic those of crabs as they crawl and pick their way across the seabed. SILVER uses springs to hop, as well.

Calisti and his team field-tested the SILVER prototype by sending it to explore a shipwreck off the coast of Italy's island of Elba. The robot's mission was to capture exterior images of the wreck as well as images from inside the ship. The robot successfully walked the site, producing high-quality videos and long-exposure pictures.

Calisti also was a member of a team that designed a soft robot to move and manipulate objects in much the same way as an octopus does. PoseiDRONE has arms like those of an octopus, and it was able to crawl and swim. Calisti currently is working to develop a six-legged underwater robot that may someday be able to scuttle across the seabed, locating and cleaning up plastics and microplastics.

In time, Calisti's soft robots could be used to study sea life and to collect samples from the seafloor. Calisti hopes they will aid "geologists, photographers, filmmakers, underwater archaeologists, and whoever wants to explore the depth of the oceans."

THINKING CRITICALLY

Evaluate Choose one of the applications Calisti suggests for the underwater robots he is developing and explain how soft robots might be useful in that field of work.

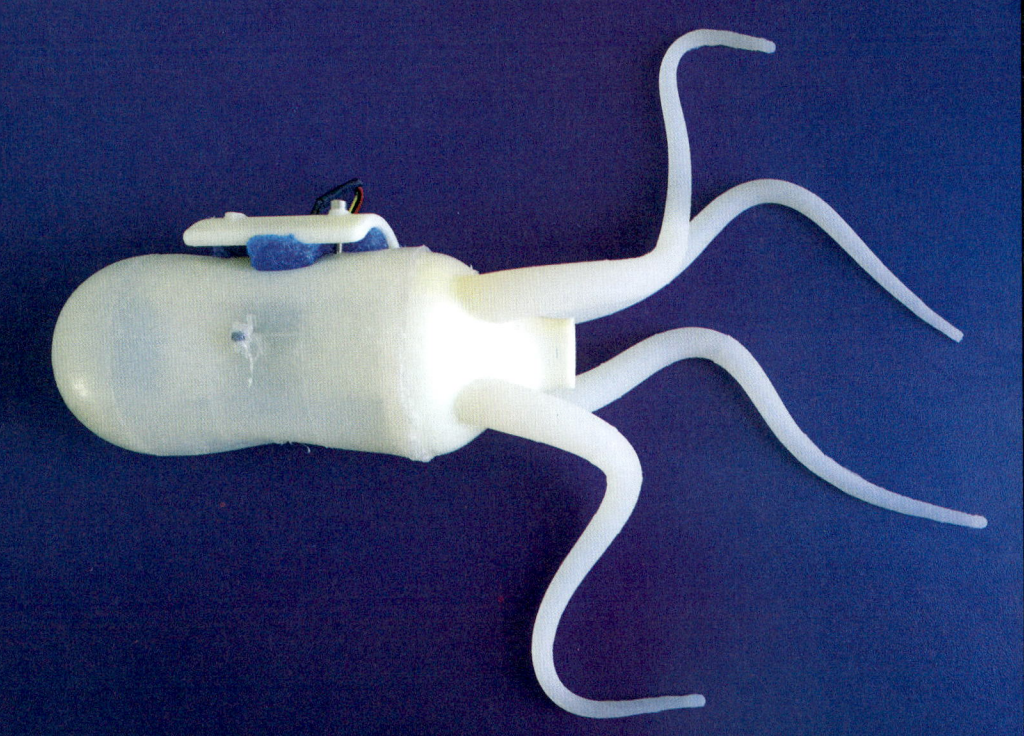

▲ Marcello Calisti examines part of the robot developed for the SILVER project. This robot walks on the seafloor with four legs. The top portion of the robot is shown in this photograph.

◀ Dr. Calisti developed a soft robot with arms like an octopus. It's called PoseiDRONE.

323

CASE STUDY
PROBING THE DEPTHS

On March 11, 2011, the magnitude 9.1 Tōhoku Earthquake struck off the eastern coast of Japan. The quake originated at a subduction zone beneath the seafloor, triggering a devastating tsunami that killed more than 15,000 people. The tsunami also caused the meltdown of three nuclear reactors at the Fukushima Daiichi power plant.

More than a year later, Japanese scientists boarded a research vessel, the *Chikyu*, and began drilling into the seafloor some 250 kilometers east of Japan (Figure 11-1). By studying the seafloor near the epicenter of the quake, the scientists hoped to learn more about earthquakes, landslides, and tsunamis.

Drilling into the seafloor is no small task, especially in waters that are 5.9 kilometers (3.6 miles) deep. Scientists aboard the *Chikyu* used a deep-sea drilling rig equipped with a drill pipe 7.7 kilometers in length, a world record. On April 25, 2012, they successfully extracted a cylindrical core of sediment and rock from the seafloor measuring 850 meters long. Using a technology called logging while drilling, they were able to take precise measurements of the temperature, pressure, radioactivity, and conductivity of the sediment at the same time the core was being drilled. After extracting the sediment core, the scientists installed an array of temperature and pressure sensors into the drill hole. These sensors were able to measure how much frictional heat was still dissipating from the rupture more than a year after the earthquake. This data helped scientists more accurately calculate the amount of energy released by the quake. The *Chikyu's* logging-while-drilling expedition was not the only effort to gather seafloor data after the Tōhoku quake. A separate team of German and Japanese scientists deployed an unmanned robotic vehicle, or autonomous underwater vehicle, called the MARUM-SEAL to study how the earthquake affected the seafloor. Armed with advanced sonar technology, the MARUM-SEAL constructed detailed maps of the seafloor in the quake zone. By comparing these post-earthquake maps with previous mappings of the seafloor, scientists were able to determine how much the seafloor had shifted along the fault lines.

As You Read Consider the different techniques scientists use to gather information about the seafloor. Think about how seafloor features form and change over time. What does the geology of the seafloor reveal about Earth's history?

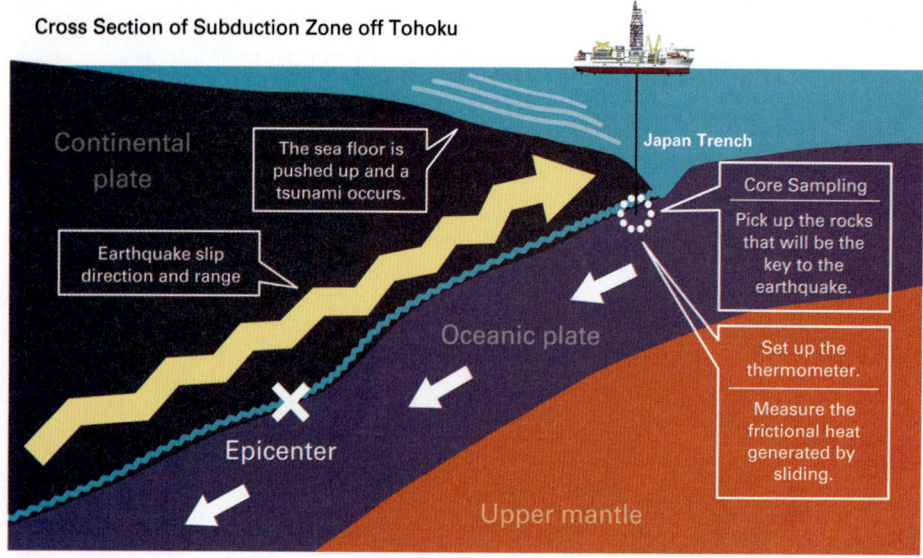

FIGURE 11-1
Go online to view an example of a research vessel outfitted with a deep-sea drilling rig.

FIGURE 11-2
A cross section of the subduction zone off Tōhoku shows where the *Chikyu* took the core sample and measurements.

11.1 EARTH'S OCEANS

Core Ideas and Skills
- Describe the leading hypothesis of the origin of Earth's water.
- Explain why Earth has ocean basins.

KEY TERMS

bolide   ocean basin

Origin of the Ocean

The primordial Earth, heated by the impacts of colliding planetesimals and the decay of radioactive isotopes, was molten, or near molten. The sky, without an atmosphere, was black. There were no oceans, no life. Today we consider Earth in terms of four spheres: the geosphere, the hydrosphere, the atmosphere, and the biosphere.

Try to abandon the idea of Earth's four spheres and think of only two kinds of Earth compounds: compounds that are volatile (substances that can evaporate rapidly and escape into the atmosphere easily) and compounds that are not volatile. Water and carbon dioxide are good examples of volatile compounds. Most scientists agree that the surface of primordial Earth consisted mainly of rock, partially molten rock, and a few volatile compounds. How, then, did our thick atmosphere, ample hydrosphere, and rich biosphere form? One leading hypothesis states that the raw materials came from the outer solar system.

Recall that our region of space heated up as dust, gas, and planetesimals collided to become planets. At the same time, the sun began to produce energy that radiated outward to heat the inner solar system even more. As a result, most of primordial Earth's volatile compounds essentially boiled off and were swept into the cold outer regions of the solar system.

As they blew away from the sun, volatiles entered a cooler region beyond Mars. Most of the volatiles were captured by gravitational attraction by the outer planets—Jupiter, Saturn, Uranus, and Neptune. Some continued their journey toward the outer fringe of the solar system (Figure 11-3). There, beyond the orbits of the known planets, volatiles from the inner solar system combined with residual dust and gas to form comets. A comet's nucleus has been compared to a dirty snowball because it is made mainly of ice and rock. However, other compounds not common in snowballs are found in comets as well. These include frozen carbon dioxide, ammonia, and simple organic molecules.

Astronomers calculate that the early solar system was crowded with comets, meteoroids, and asteroids—space debris left over from planetary formation. Much of this material contained volatile compounds. Some large pieces of space debris crash explosively into planets; these are called **bolides**. Early in the formation of the solar system, numerous bolides crashed into Earth. These bolides delivered and deposited a large mass of volatile compounds to the inner solar system. While falling space debris added less than one percent to Earth's total mass, it imported roughly 90 percent of its modern reservoir of volatiles.

Upon entry and impact, the frozen volatiles in the bolides vaporized, releasing water vapor, carbon dioxide, ammonia, simple organic molecules, and other volatile compounds. As the planet cooled and atmospheric pressure increased, the water vapor condensed to liquid, forming the first oceans. The light molecules transported to Earth in bolides also provided gases that formed the atmosphere and the raw materials for life. (The formation and evolution of the atmosphere is discussed in Chapter 18.)

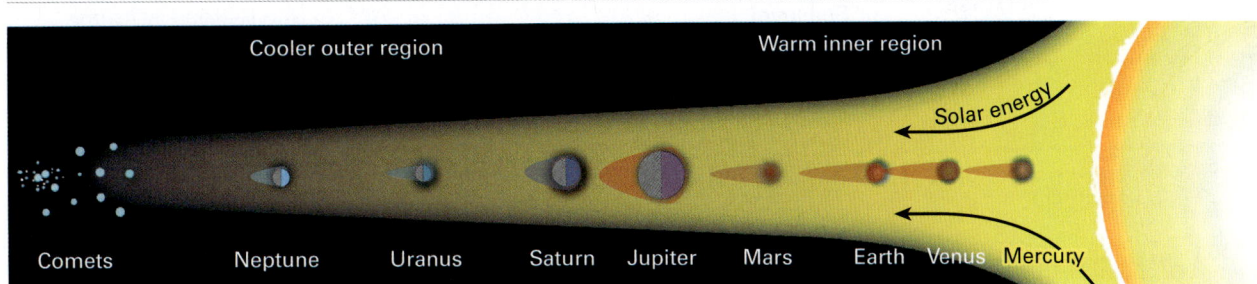

FIGURE 11-3
As the solar system formed, volatiles boiled away from the inner planets. Solar energy blew volatiles out to the cooler regions beyond Mars. The inner four planets were left with little or no atmosphere or oceans.

At least some, and probably most, of the compounds necessary to produce the hydrosphere, atmosphere, and biosphere traveled to Earth from outer regions of the solar system. The water that fills Earth's oceans came from interplanetary space.

Later in Earth's history, impacts from outer space blasted rock and dust into the sky, causing mass extinctions and killing large portions of life on Earth. (See Chapter 5 for more on mass extinctions.) Ironically, extraterrestrial impacts may have provided the raw materials for the oceans, the atmosphere, and for life. But they later caused mass extinctions.

checkpoint From where did the volatile materials that make up Earth's hydrosphere, atmosphere, and biosphere come?

Ocean Basins

To many people, the difference between a continent and an ocean is that a continent is land and an ocean is water. This observation is true, of course. But to a geologist, what is more important is that the rock beneath the ocean water is different from the rock beneath the land surface. An **ocean basin** is the rock surface under the ocean. The accumulation of seawater in the world's ocean basins is a result of that difference.

Modern oceanic crust is dense basalt. It varies from 4 to 7 kilometers thick. Continental crust is made of lower-density granitic rock and averages 20 to 40 kilometers in thickness. Not only are there these major differences in the crust of continents versus oceans, but the entire continental lithosphere is both thicker and less dense than oceanic lithosphere. The thick, lower-density continental lithosphere floats isostatically higher than the thinner, higher-density oceanic lithosphere. As a result, the top surface of continental lithosphere is almost entirely above sea level. The top surface of oceanic lithosphere—the seafloor—is several kilometers below sea level. Because water flows downhill, the majority of Earth's water has collected in the vast depressions formed by oceanic lithosphere. Even if no water existed on Earth's surface, oceanic crust would form deep basins and continental crust would form regions of higher elevation.

Oceans cover about 71 percent of Earth's surface. The seafloor is about 5 kilometers deep in the central parts of the ocean basins, although it is only 2 to 3 kilometers deep above the Mid-Ocean Ridge. The seafloor plunges to nearly 11 kilometers depth in the Mariana Trench (Figure 11-4).

The ocean basins contain 1.4 billion cubic kilometers of water. That is 18 times more than the volume of all land above sea level. So much water exists at Earth's surface that if Earth were a perfectly smooth sphere, it would be covered by a global ocean 2,000 meters deep.

The size and shape of Earth's ocean basins have changed over geologic time. At present, the Atlantic Ocean is growing wider at a rate of a few centimeters each year as the seafloor spreads apart at the Mid-Atlantic Ridge and as the Americas move away from Europe and Africa. At the same time, the Pacific is shrinking at a similar rate, as oceanic

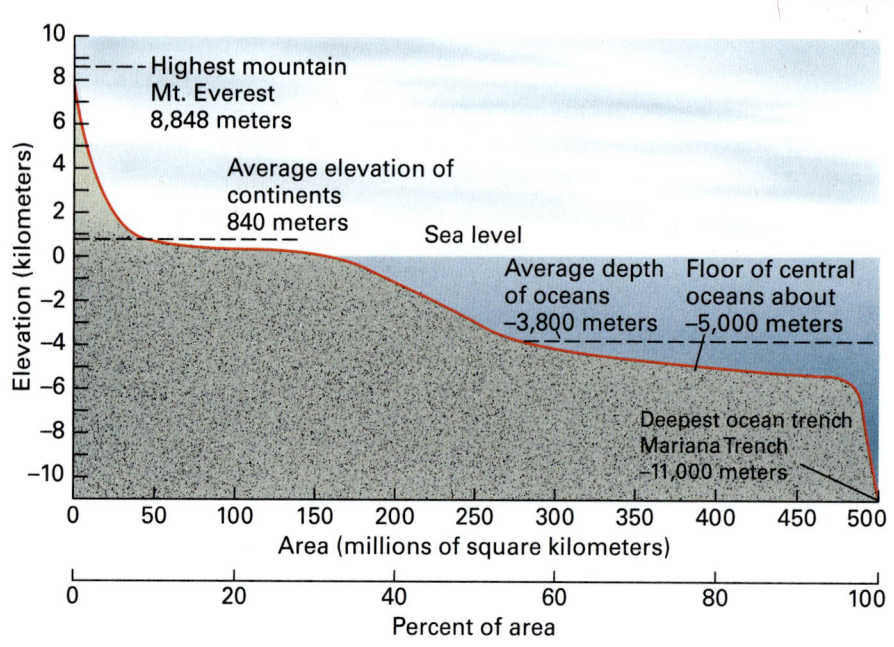

FIGURE 11-4
The vertical axis on this graph shows elevations relative to sea level. The horizontal axis shows the basic breakdown of Earth's main topographic surfaces. Roughly 29 percent, or about 150 million square kilometers, of Earth's surface lies above sea level.

DATA ANALYSIS Compare Seafloor Data

Like the surface of the continents, the elevation of the seafloor varies. The graph in Figure 11-4 shows how much of Earth's surface area lies at each elevation.

TABLE 11-1 Seafloor Trench Comparisons

Feature	Location	Deepest Point Below Sea Level (km)
Mariana Trench	Western Pacific Ocean	11.03
Tonga Trench	Southwest Pacific Ocean	10.88
Philippine Trench	Philippine Sea	10.54
Kuril-Kamchatka Trench	North of Japan	10.50
Kermadec Trench	South Pacific Ocean	10.04
Izu-Ogasawara Trench	Western Pacific Ocean	9.78

Source: National Oceanic and Atmospheric Administration

The vertical axis shows elevation relative to sea level. Positive elevation means height above sea level while negative elevation means depth below sea level. The horizontal axis shows the total area in millions of square kilometers and the percentage of Earth's total surface at each elevation. Table 11-1 shows the six deepest trenches in the world's oceans.

1. **Summarize** According to Table 11-1, where are most of the world's deepest points located? Based on what you know about the movement of Earth's plates, why is this likely true?
2. **Compare** How does the total area of Earth's surface above sea level compare with the total area of Earth's surface below sea level?
3. **Infer** What can you infer about the relative proportion of Earth's surface that has been directly explored by humans? Explain your reasoning.

Computational Thinking

4. **Analyze Data** How does the average elevation of land above sea level compare to the average depth of ocean basins?
5. **Analyze Data** How do Earth's highest peaks compare to the deepest trenches in the oceans in terms of elevation?

---

lithosphere sinks into subduction zones around its edges. In short, the Atlantic Ocean basin is now expanding at the expense of the Pacific.

Global climate is profoundly influenced by the immense volume of Earth's oceans and the fact that the oceans consist of liquid water. On average, the ocean is about 3,800 meters deep. The thermal capacity of Earth's entire atmosphere—the amount of heat the atmosphere can absorb before an increase in temperature—is equivalent only to the uppermost 2.5 meters of ocean water. That's less than 0.05 percent of the thermal capacity of Earth's oceans. Ocean currents transport heat from the Equator toward the poles, cooling equatorial climates and warming polar environments. The seas also absorb and store solar heat more efficiently than do rocks and soil. As a result, oceans are commonly warmer in winter and cooler in summer than adjacent land. Most of the water that falls as rain or snow is water that evaporated from the seas. In these and other ways, the oceans have played a major role in controlling Earth's climate and the distribution of different climate types through geologic time.

checkpoint Why does oceanic lithosphere form deep basins?

11.1 ASSESSMENT

1. **Describe** What is the current leading hypothesis for the origin of Earth's water?
2. **Explain** What is the significant difference between oceanic lithosphere and continental lithosphere, and how has that resulted in ocean basins?
3. **Explain** Why do the size and shape of Earth's ocean basins change?

Critical Thinking

4. **Generalize** What role do Earth's oceans have in regulating the global climate? Explain your answer with evidence from the text.

11.2 STUDYING THE SEAFLOOR

Core Ideas and Skills
- Compare three techniques for gathering seafloor samples.
- Describe three remote sensing technologies for mapping the seafloor.

KEY TERMS

sonar   magnetometer

Seventy-five years ago, scientists had better maps of the moon than of the seafloor. The moon is clearly visible in the night sky, and we can view its surface with a telescope. The seafloor, however, is deep, dark, and inhospitable to humans. Modern oceanographers use a variety of techniques to study the seafloor, including several types of sampling and remote sensing.

Sampling

Scientists use several devices to collect sediment and rock directly from the ocean floor.

Dredging A rock dredge is an open-mouthed steel net dragged along the seafloor behind a research ship. The dredge breaks rocks from submarine outcrops and hauls them to the surface along with whatever other sediment and organisms might be scooped up.

Piston Coring Oceanographers sample seafloor mud by lowering a piston corer to the bottom (Figure 11-5). This device consists of a steel core barrel with a sharp bit on the lower end. The weighted core barrel falls to the seafloor where it plunges bit-first into the mud. Once the barrel has stopped settling into the mud (usually less than a minute), the entire device is yanked out of the bottom sediments, lifted to the surface, and brought aboard. There, scientists extract, split, and analyze the core. The length of the core itself is limited by the strength of the barrel and its ability to penetrate downward into the sediment. Most piston cores are more than 10 meters long.

Drilling In contrast, seafloor drilling methods developed for oil exploration can recover continuous cores from the seafloor that are

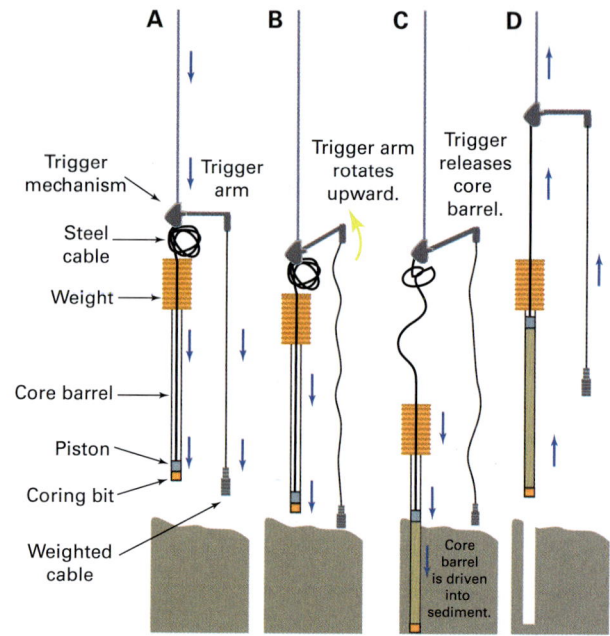

FIGURE 11-5
This illustration shows how a piston corer works. (A) The core barrel system falls to the seafloor. (B) When the weighted cable reaches the seafloor, it goes slack and the trigger arm rotates upward. (B) The trigger arm releases the core barrel which plunges into the sediment. (D) The mud core is extracted from the seafloor and pulled up to the research vessel for analysis.

hundreds of meters long. Large drill rigs are mounted on offshore platforms and on research vessels. The drill cuts cylindrical cores from sediment and rock. The cores are then brought to the surface for study. (See this chapter's Case Study.)

checkpoint Which of the sampling techniques described would be used to gather material that recently settled onto the seafloor?

Remote Sensing

Remote sensing methods do not require direct physical contact with the ocean floor. There are three main types of remote sensing technologies.

Sonar The echo sounder is an instrument commonly used to map seafloor topography. It emits a brief sound pulse from a research ship and then records the signal after it bounces off the seafloor surface and travels back up to the ship. The water depth is calculated from the time required for the sound to make the round trip. A topographic map of the seafloor is constructed as the ship steers a carefully navigated course with the echo sounder operating continuously. Modern echo sounders, called **sonar**, transmit multiple signals at

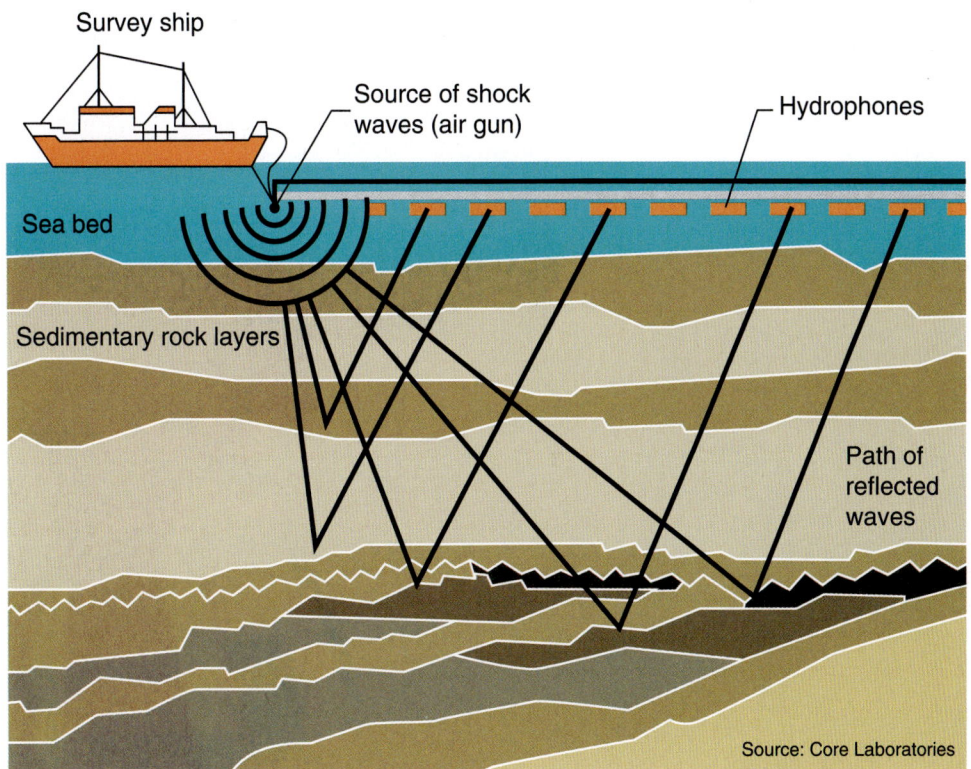

FIGURE 11-6 A seismic reflection profiler produces a visual representation of the surface and the geology of different layers below the seafloor.

a time to create more complete and accurate maps. The word *sonar* stands for **so** (und) **na** (vigation and) **r** (anging).

A seismic reflection profiler works in a similar way but uses a wide range of sound frequencies. The higher-energy signal penetrates and reflects from the boundaries of different layers of sediment and rock below the seafloor (Figure 11-6). This technique provides an image of the layering and structure of seafloor sediments, the rock layers and crust below, and major structures such as folds and faults that might be present.

Magnetometers A **magnetometer** is an instrument that measures a magnetic field. Magnetometers towed behind research ships measure the magnetism of seafloor sediment and rock. Data collected by magnetometers resulted in the now-famous discovery of symmetric magnetic stripes on the seafloor (Chapter 7). That discovery led to the seafloor spreading hypothesis and to the theory of plate tectonics shortly thereafter.

Satellites Satellite-based microwave radar instruments measure the echo of microwave pulses to detect subtle swells and depressions on the sea surface. These features reflect seafloor topography. For example, the mass of a seafloor mountain 4,000 meters high creates sufficient gravitational attraction to produce a gentle 6-meter-high swell on the sea surface directly above it. The microwave data are used to make seafloor maps.

checkpoint Which remote sensing method results in an image of the layering and structure of the seafloor, including the rock layers and crust?

11.2 ASSESSMENT

1. **Compare** What different methods have been developed for extracting samples of the seafloor?
2. **Distinguish** How is the data collected by a seismic reflection profiler different from that collected by an echo sounder?

Critical Thinking

3. **Apply** Describe a scenario in which remote sensing and sampling would be used together by scientists to make a discovery about the seafloor.
4. **Critique** Sonar and seismic reflection profilers emit sound to collect remote sensing data. Marine organisms such as dolphins and some whales rely on their hearing to communicate, navigate, and find prey. Using this information, construct an argument describing the pros and cons of the use of sound in remote sensing technologies.

LESSON 11.2

FIGURE 11-7
ON ASSIGNMENT This image shows cartographer Marie Tharp's seafloor map, which was published in the June 1968 issue of *National Geographic*. The map was the first published map to show the Mid-Atlantic Ridge. Marie Tharp's cartography supported Alfred Wegener's continental drift hypothesis. Tharp worked closely with geologist Bruce Heezen. Heezen used sonar measurements from research vessels to collect data, which Tharp and Heezen used to create this map and others.

11.3 FEATURES OF THE SEAFLOOR

Core Ideas and Skills

- Understand the processes that take place in the Mid-Ocean Ridge system.
- Explain how seafloor spreading affects global sea level.
- Describe seafloor features common at convergent and divergent plate boundaries.
- Distinguish between island arcs, seamounts, oceanic islands, and atolls.
- Summarize how sediment disperses and forms layers on the seafloor.

KEY TERMS

chemosynthesis
transform fault
oceanic trench
accreted terrane
seamount
oceanic island

guyot
atoll
terrigenous sediment
pelagic sediment
abyssal plain
pillow basalt

The Mid-Ocean Ridge System

Recall that the Mid-Ocean Ridge system is a continuous submarine mountain chain formed by the rifting apart of two oceanic plates. The Mid-Ocean Ridge system encircles the globe and has a total length exceeding 80,000 kilometers (Figure 11-8). In some places it is more than 1,500 kilometers wide. The ridge system rises an average of 2 to 3 kilometers above the surrounding deep seafloor because heat from the rifting causes rocks in and around the plate boundary to be hot. The hot rocks of the ridge system take up more volume—they are less dense—so they float higher in the underlying asthenosphere. Although the Mid-Ocean Ridge lies almost exclusively beneath the sea surface, it is Earth's longest continuous mountain chain.

The discovery of the extent of the Mid-Ocean Ridge system took place over the course of several decades. The discovery was possible only by scientific analysis of large volumes of oceanographic data, much of which was collected for military purposes. Following World War I (1914–1918), oceanographers began using early versions of echo-sounding devices to measure ocean depths. Those surveys showed that the seafloor was much more rugged than previously thought.

FIGURE 11-8 ▼
This map shows the Mid-Ocean Ridge system, which coincides exactly with divergent plate boundaries, or spreading centers, in the world's ocean basins. The spreading centers are shown in double red lines; the single red lines are transform faults.

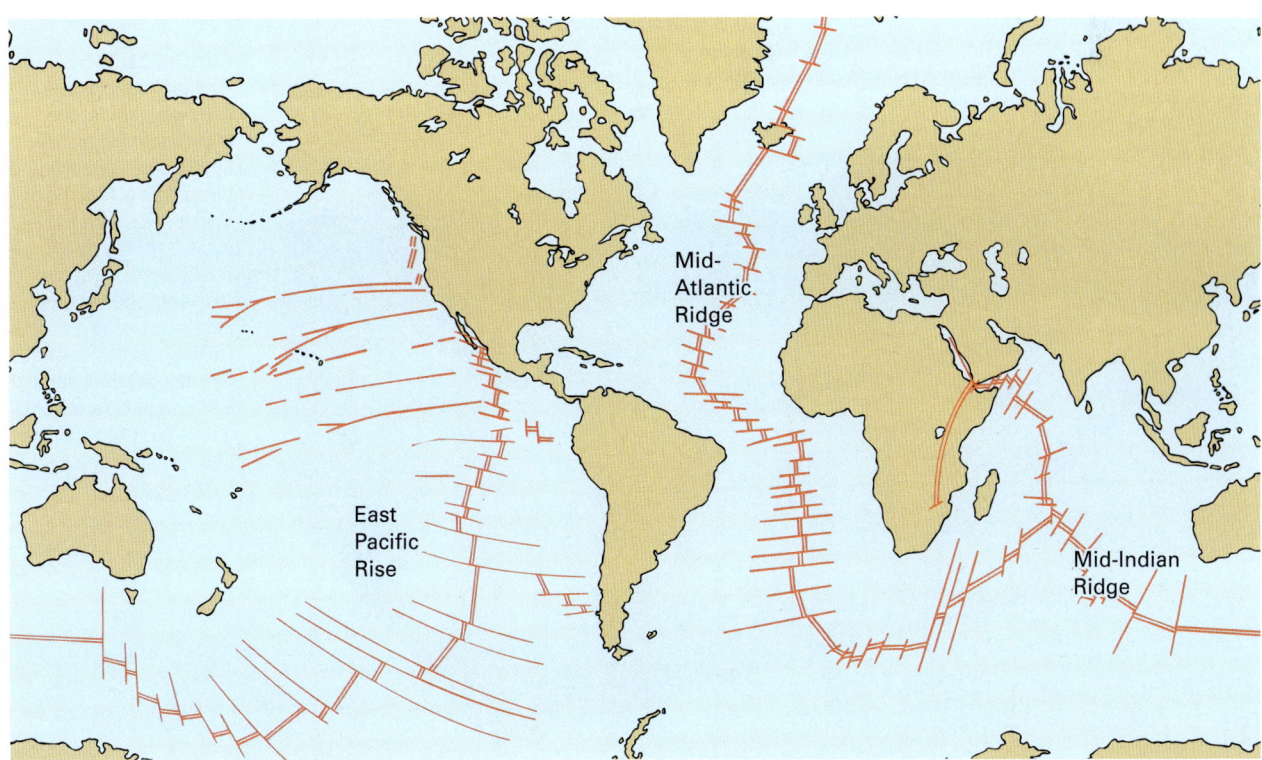

LESSON 11.3 **331**

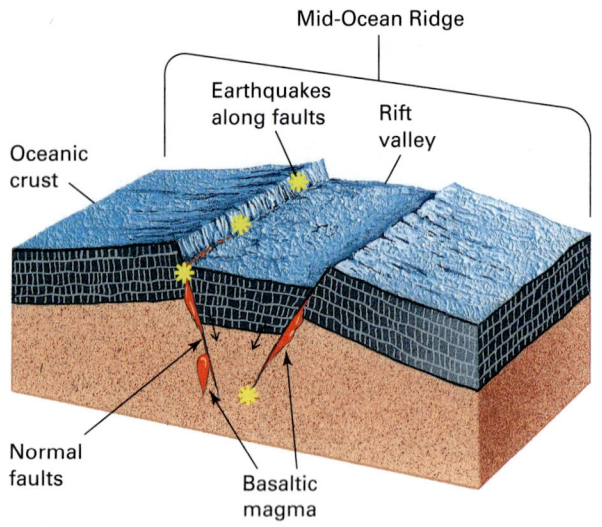

FIGURE 11-9
This is a cross-sectional view of the central rift valley in the Mid-Ocean Ridge. It shows that as the plates separate, blocks of rock drop down along the fractures to form the rift valley, bounded by normal faults. Movements across these faults cause earthquakes.

During World War II (1939–1945), naval commanders needed topographic maps of the seafloor to support submarine warfare. Those detailed maps, made with early versions of the echo sounder, were kept secret by the military. They became available to the public after peace was restored. Scientists were surprised to learn that the ocean floor has at least as much topographic diversity and relief as the continents. Broad plains, high peaks, and deep valleys form a varied and fascinating submarine landscape. In the 1950s, oceanographic surveys conducted by several nations led to the discovery that the Mid-Atlantic Ridge is just part of a great submarine mountain range, now called the Mid-Ocean Ridge system.

A rift valley is an elongated depression that develops at a divergent plate boundary (Figure 11-9). In the Mid-Ocean Ridge system, a rift valley 1 to 2 kilometers deep and several kilometers wide splits many segments of the ridge crest. Oceanographers in small research submarines have dived into the rift valley. Here they have documented gaping vertical cracks up to 3 meters wide on the rift valley floor. In some cases, extremely hot water highly concentrated with dissolved metals and metal-bearing compounds streams out of the cracks.

To understand the geologic significance of such cracks, recall that the Mid-Ocean Ridge system is a spreading center, where two lithospheric plates are moving apart from each other. The cracks form as stress at the ridge axis causes brittle oceanic crust to separate. Basaltic magma then rises through the resulting crack. It then flows onto the floor of the rift valley. This basalt becomes new oceanic crust as two lithospheric plates spread outward from the ridge axis.

The new crust (and the underlying lithosphere) at the ridge axis is warmer. Therefore, it has a

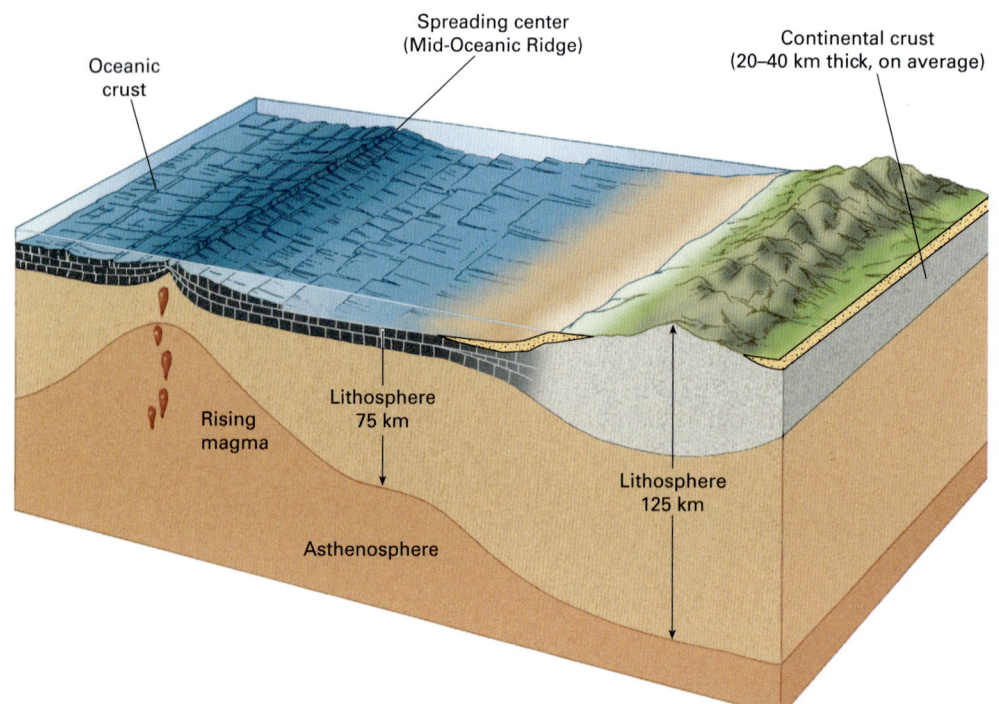

FIGURE 11-10
The seafloor sinks as it grows older. At the Mid-Ocean Ridge, new lithosphere is buoyant because it is hot and of low density. As it moves away from the ridge, the lithosphere cools, thickens, and becomes denser, causing it to sink. On average, the seafloor lies at a depth of about 5 kilometers, relative to 2 to 3 kilometers at the Mid-Ocean Ridge.

FIGURE 11-12
These red tubeworms are part of a thriving plant and animal community living near a black smoker.

relatively lower density than older crust and lithosphere located farther from the ridge axis. Its buoyancy causes it to float high above the surrounding seafloor. The buoyancy elevates the Mid-Ocean Ridge system 2 to 3 kilometers above the deep seafloor. The new lithosphere cools as it spreads away from the ridge. As a result of cooling, the lithosphere shrinks and cracks as it becomes denser and sinks to lower elevations, forming the deeper seafloor on both sides of the ridge (Figure 11-10).

Normal faults and shallow earthquakes are common along the Mid-Ocean Ridge system because oceanic crust fractures as the two plates separate. Blocks of crust drop downward along some of the seafloor cracks, forming faults that bound the rift valley.

The Mid-Ocean Ridge system is volcanically active. Here, hot rocks heat seawater as it circulates through fractures in oceanic crust. The hot water dissolves metals and sulfur from the rocks. Eventually, the hot metal-and-sulfur-laden water rises back to the seafloor surface, spouting from fractures as a jet of cloudy, black water called a black smoker (Figure 11-11). The black color is caused by precipitation of fine-grained metal sulfide minerals as the solutions cool on contact with seawater.

⊕ **FIGURE 11-11**
Go online to view an example of "smoke" emerging from a black smoker at an oceanic ridge.

These scalding, sulfurous waters are as hot as 400°C, and seemingly would produce a sterile environment in which nothing could survive. Yet, remarkably, the deep seafloor around a black smoker teems with life. At the vents, bacteria produce energy from hydrogen sulfide in a process called **chemosynthesis**. These bacteria produce energy from chemicals instead of sunlight. The chemosynthetic bacteria are the foundation of a deep-sea food web: either larger vent organisms eat them or the larger organisms live symbiotically with the bacteria. For example, instead of a digestive tract, the red tube worm (Figure 11-12) has a special organ that hosts the chemosynthetic bacteria. In return for providing a home for the bacteria, the worm receives nutrition from the bacteria's wastes. Other vent organisms in this unique food chain include giant clams and mussels, eyeless shrimp, crabs, and fish.

In addition to cracks or faults located along the axis of the mid-ocean ridges and rift valleys are hundreds of fractures, called **transform faults**. These faults cut across the rift valleys and ridges (Figure 11-13). The fractures extend through the entire thickness of the lithosphere and develop because the Mid-Ocean Ridge system consists of many short segments. Each segment is slightly offset from adjacent segments by a transform fault. Transform faults are original features of the Mid-Ocean Ridge; they form as an accommodation to Earth's spherical shape when lithospheric spreading begins.

Some transform faults displace the ridge by less than a kilometer, but others offset the ridge by hundreds of kilometers. In some cases, a transform fault can grow so large that it forms a transform plate boundary. The San Andreas Fault in California is a transform plate boundary that offsets both oceanic and continental crust.

checkpoint The Mid-Ocean Ridge is volcanically active and heats seawater as it circulates through fractures in the oceanic crust. What does the hot seawater do to the seafloor rocks, and what is the result of this?

Global Sea-Level Changes

Thin layers of marine sedimentary rocks blanket large areas of Earth's continents. The rocks tell us that those places must have been below sea level when the sediment accumulated. Some layers of marine sediments accumulated on every continent simultaneously. Those layers suggest that the

> ### CAREERS IN EARTH AND SPACE SCIENCES
>
> #### Marine Geophysicist
> A marine geophysicist works with seafloor maps and images. A background in geology allows a marine geophysicist to analyze underwater features, such as ridges and submarine volcanoes. A marine geophysicist might work with organizations such as the National Oceanic and Atmospheric Administration (NOAA). Important technology used in this career includes remote sensing equipment, which a marine geophysicist uses while working on ocean vessels in the open ocean.

flooding occurred on a global scale. Although plate tectonics could explain the sinking of individual continents or parts of continents, it does not explain how all continents could sink at once. If the land did not sink, we need to explain how sea level could rise globally by hundreds of meters, flooding all continents.

The simplest possible explanation for a dramatic rise in global sea level would be the melting of continental glaciers. Continental glaciers have advanced and melted numerous times in Earth's history. During the growth of continental glaciers, more and more water is frozen into the ice that rests on land. As a result, sea level drops. When continental glaciers melt, the water runs back into

FIGURE 11-13
Transform faults offset segments of the Mid-Ocean Ridge. Adjacent segments of the ridge may be separated by steep cliffs 3 kilometers high. Note the flat abyssal plain far from the ridge.

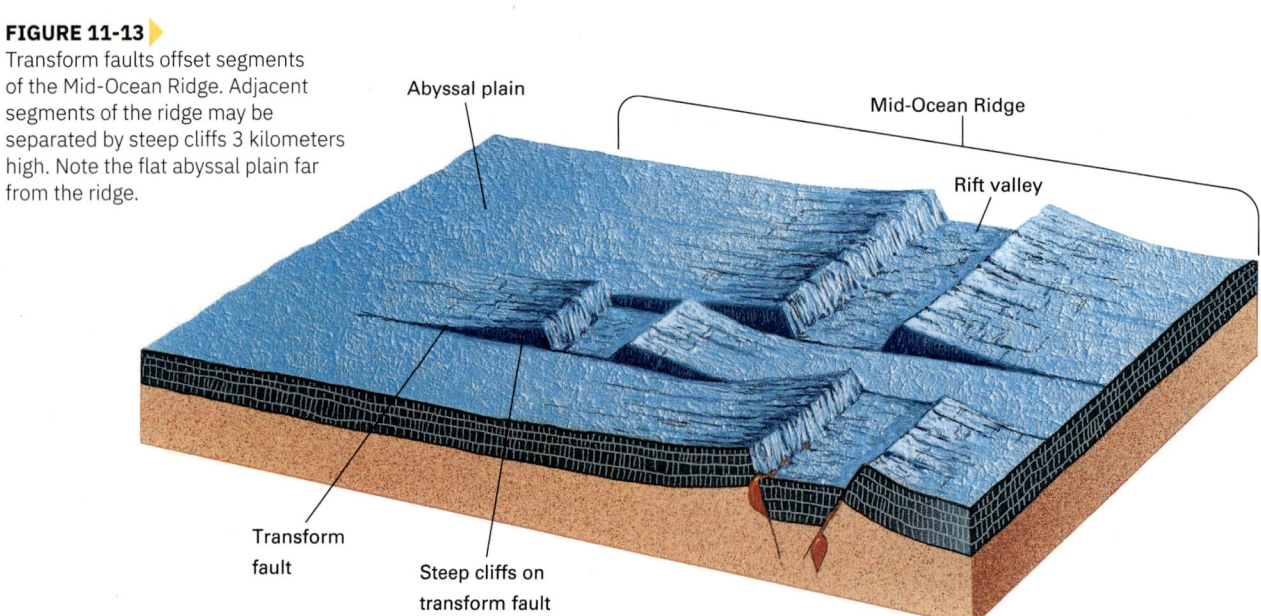

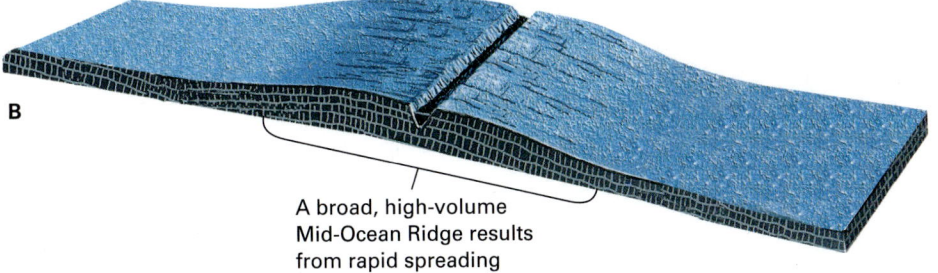

FIGURE 11-14
(A) Slow seafloor spreading creates a narrow, low-volume Mid-Ocean Ridge that displaces less seawater and lowers sea level. (B) Rapid seafloor spreading creates a wide, high-volume ridge that displaces more seawater and raises sea level.

the oceans and sea level rises. The growth and melting of glaciers during the Pleistocene epoch, for example, caused sea level to fluctuate by as much as 150 meters. But the ages of many marine sedimentary rocks on continents do not coincide with times of glacial melting. Therefore, we must look for a different cause to explain continental flooding.

Could seafloor spreading along the Mid-Ocean Ridge system explain these global sea-level changes? Recall that the new, hot lithosphere at a spreading center is buoyant, causing the Mid-Ocean Ridge system to rise above the surrounding seafloor. This submarine mountain chain displaces a huge volume of seawater. If the Mid-Ocean Ridge system were smaller, sea level would fall. If it were larger, sea level would rise.

The Mid-Ocean Ridge rises highest at its spreading center, where new lithosphere rock is hottest and has the lowest density. The elevation of the ridge decreases on both sides of the spreading center because the lithosphere cools, shrinks, and becomes more dense as it moves outward.

Now consider a spreading center where spreading is very slow (perhaps 1 to 2 centimeters per year). At such a slow rate, the newly formed lithosphere would cool before it moved far from the spreading center. As a result, the ridge would be narrow and of low volume, as shown in Figure 11-14. In contrast, rapid seafloor spreading of 10 to 20 centimeters per year would create a high-volume ridge because the newly formed, hot lithosphere would be carried farther from the spreading center before it cooled and shrank. This high-volume ridge would displace much more seawater than a low-volume ridge.

To summarize, rapid spreading produces a larger volume of hot rock and pushes aside more seawater, causing global sea level to go up. If spreading then slows down, a smaller volume of hot rock is produced, displacing less seawater and causing global sea level to fall.

Seafloor age data indicate that the rate of seafloor spreading has varied from about 2 to 16 centimeters per year over the past 200 million years. Since no oceanic crust is older than about 200 million years, it is not possible to compare the spreading rate to earlier times when extensive marine sedimentary rocks accumulated on continents.

During late Cretaceous times, between 110 million and 85 million years ago, we know that seafloor spreading was unusually rapid. That rapid spreading caused an unusually high-volume Mid-Ocean Ridge to form. As the spreading rate increased and the Mid-Ocean Ridge grew, more and more seawater was displaced. The low-lying portions of continents were gradually submerged. Geologists have found marine sedimentary rocks of late Cretaceous age on nearly all continents. These rocks indicate that late Cretaceous time was a time of abnormally high global sea level. Therefore, a process started by heat transfer deep within the geosphere profoundly affected the hydrosphere and life throughout the biosphere.

checkpoint Why do scientists rule out glacial melting as the cause of the global sea-level rise discussed in this section?

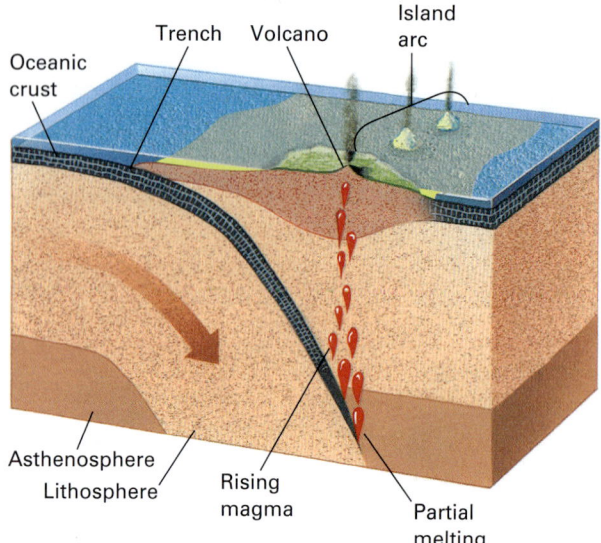

FIGURE 11-15
An oceanic trench forms at a convergent boundary between two oceanic plates. One of the plates sinks and heats, generating magma that rises to form a chain of volcanic islands called an island arc.

Oceanic Trenches and Island Arcs

In many parts of the Pacific Ocean, two oceanic plates converge. One dives beneath the other, forming a subduction zone. The sinking plate pulls the seafloor downward, forming a long, narrow depression called an **oceanic trench**. The deepest place on Earth is in the Mariana Trench, north of New Guinea in the southwestern Pacific, where the ocean floor is nearly 11 kilometers below sea level. Depths of 8 to 10 kilometers are common in other trenches.

Huge amounts of magma are generated in the subduction zone. The magma rises and erupts at the seafloor to form submarine volcanoes next to the trench. The volcanoes eventually grow to become a chain of islands called an island arc (Figure 11-15), as we learned in Chapter 10. The western Aleutian Islands are an example of an island arc. Many others occur at the numerous convergent plate boundaries in the western Pacific.

FIGURE 11-16
Go online to view an example of a portion of an island arc formed by the partial melting of a subducting plate.

If subduction stops after an island arc forms, volcanic activity also ends. The island arc may then ride passively along with another tectonic plate until it arrives at another subduction zone. However, the density of island arc rocks is relatively low, making them too buoyant to sink into the mantle. Instead, the island arc is mashed against the side of the overriding plate (Figure 11-17).

When an island arc and continent begin to slowly mash together, or "collide," the subducting plate commonly fractures on the seaward side of the island arc to form a new subduction zone. In this way, the island arc breaks away from the oceanic plate. The island arc becomes part of the continent, enlarging it. Much of the crust now

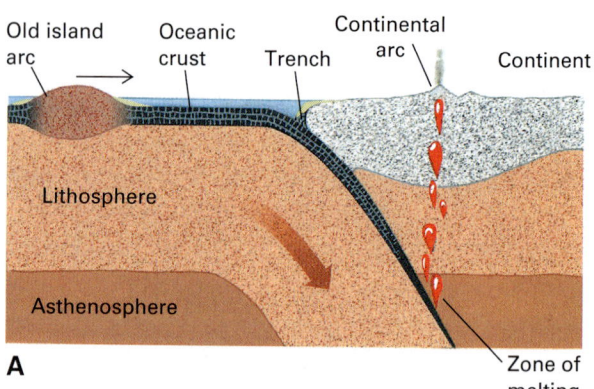

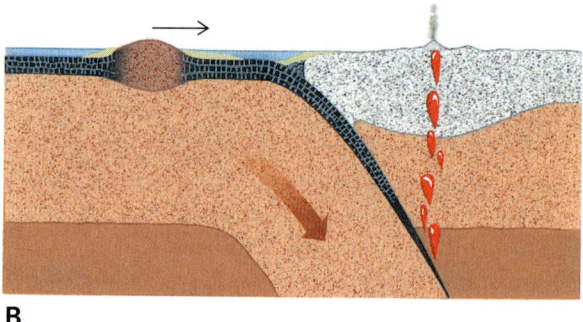

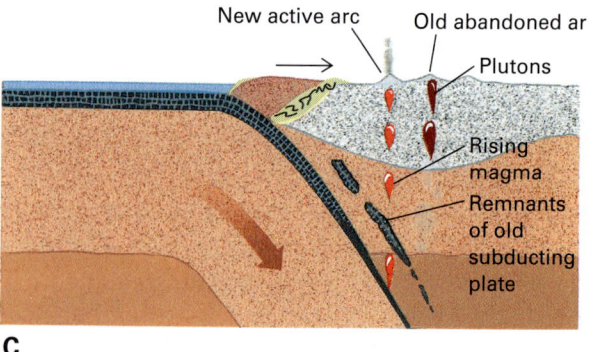

FIGURE 11-17
(A) An island arc forms on an oceanic plate that is itself being subducted beneath a continent. (B) The island arc reaches the subduction zone but the arc's low density prevents it from subducting into the mantle. (C) The island arc becomes part of the continent. The subduction zone and trench both jump to the seaward side of the island arc, thereby enlarging the continent.

Ocean-Floor Sediment

Early oceanographers believed that the oceans were 4 billion years old. If so, mud on the ancient seafloor should have been very thick, having had so much time to accumulate. In 1947, however, scientists on the U.S. research ship *Atlantis* discovered that the mud layer on the bottom of the Atlantic Ocean is much thinner than they expected. Why is there so little mud on the seafloor? The answer to this question would be a crucial piece of evidence in the development of the theory of plate tectonics.

Earth is 4.6 billion years old, and rocks as old as 4.04 billion years have been found on the North American continent. Because of its buoyancy, most continental crust remains near Earth's surface. In contrast, no parts of the seafloor are older than about 200 million years because oceanic crust forms continuously at the Mid-Ocean Ridge and then recycles into the mantle at subduction zones. As the theory of plate tectonics was emerging in the early 1960s, it became apparent that the presence of only a thin layer of mud covering the oceanic crust could be easily explained if the oldest seafloor is less than 200 million years old, not 4 billion years old as oceanographers had once advocated. This evidence supported the plate tectonics theory in its early days.

Seismic reflection profiling and seafloor drilling show that oceanic crust is made of three layers. The uppermost layer consists of sediment, and the lower two are basalt (Figure 11-24).

The uppermost layer of oceanic crust consists of two types of sediment. **Terrigenous sediment** is sand, silt, and clay eroded from the continents and delivered to deep water mainly through submarine canyons that extend outward from the continental shelf. So, most terrigenous sediment is found close to the continents.

Pelagic sediment is muddy ocean sediment made up of the skeletons of tiny marine organisms (Figure 11-25). As such, pelagic sediment can be found anywhere on the seafloor. Pelagic sediment accumulates at a rate of about 2 to 10 millimeters per 1,000 years.

Near the Mid-Ocean Ridge system there is very little to almost no sediment because the seafloor is so young. The seafloor becomes older as it spreads away from the ridge, so the sediment thickness increases with distance because it has had more time to accumulate. Accumulations are typically around four kilometers thick. The observation that the thickness of seafloor mud increases with distance from the ridge provided support for the plate tectonics theory in its early days.

Parts of the ocean floor beyond the Mid-Ocean Ridge system are flat, level, essentially featureless surfaces called **abyssal plains**. They are the flattest surfaces on Earth. Seismic profiling shows that the basaltic crust is rough and jagged throughout the ocean. On abyssal plains, however, pelagic sediment buries this rugged surface, forming smooth surfaces. If you were to remove all the sediment, you would see rugged topography similar to that of the Mid-Ocean Ridge.

Basaltic Oceanic Crust
Below the layer of sediment on the ocean floor is a layer of basalt. The layer is about 1 to 2 kilometers thick. It consists mostly of **pillow basalt**, which forms as hot magma oozes onto the seafloor. Contact with cold seawater

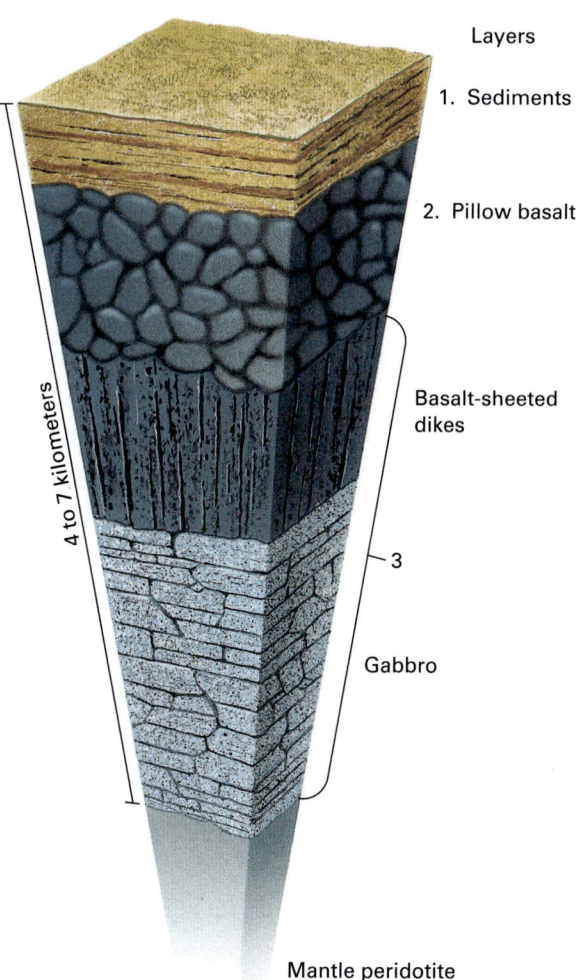

FIGURE 11-24
The illustration depicts the three layers of oceanic crust. The uppermost layer consists of sediment. The middle layer consists of pillow basalt. The deepest layer is made of vertical dikes of basalt that merge downward into gabbro. Below this lowermost layer of oceanic crust is the upper mantle.

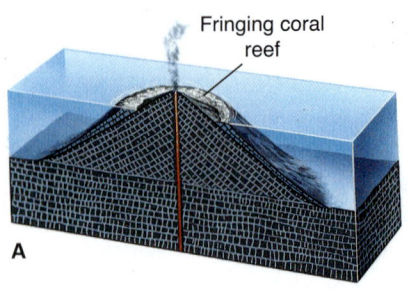
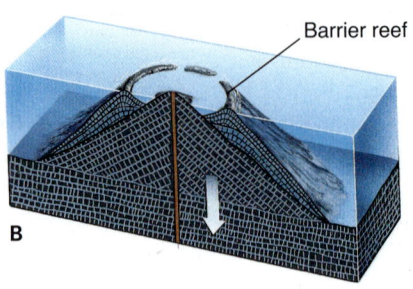
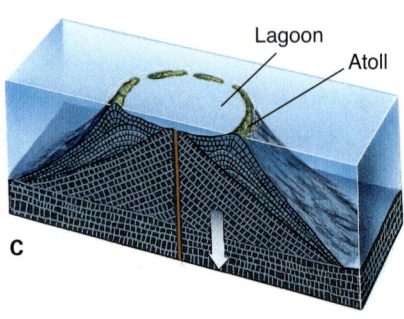

FIGURE 11-22
(A) A fringing reef grows along the shore of a young volcanic island. (B) As the island sinks, the reef continues to grow upward to form a barrier reef that encircles the island. (C) Finally, the island sinks below sea level and the reef forms a circular atoll.

If corals live only in shallow water, how did atolls form in the deep sea? Charles Darwin studied this question during his famous voyage on the *Beagle* from 1831 to 1836. He reasoned that a coral reef must have formed in shallow water on the flanks of a volcanic island. Eventually the island sank, but the reef continued to grow upward, so that the living portion always remained in shallow water (Figure 11-22). This proposal was not accepted at first because scientists could not explain how a volcanic island could sink. However, when scientists drilled into a Pacific atoll shortly after World War II and found volcanic rock hundreds of meters beneath the reef, Darwin's hypothesis was revived. It is considered accurate today, in light of our ability to explain why volcanic islands sink.

checkpoint If a seamount or oceanic island is isolated rather than part of a chain, what can we assume about its formation in contrast to a chain of oceanic islands?

MINILAB Hawaiian Hot Spot

Materials
calculator
physical map of the Hawaiian Islands
metric ruler
pencil

Procedure

1. Using the map, determine the distances between each pair of islands within the Hawaiian Island chain. Record your data in a table like Table 11-2.

TABLE 11-2 Hawaiian Island Chain Data

Island Pairs	Distance Between Island Peaks (km)	Age Difference (years)	Rate of Movement of Pacific Plate (cm/year)
Hawai'i and Maui			
Maui and Molokai			
Molokai and Oahu			
Oahu and Kauai			

2. Calculate the differences in geologic age between each of the islands. Record the data in the table.

3. For each pair of islands, calculate the rate of movement of the Pacific plate using the formula below. Add your results to the table.

Rate of movement (cm/yr) = Distance (cm)/Age difference (yr)

Results and Analysis

1. **Observe** In what direction is the Pacific plate moving? Explain your reasoning.

2. **Calculate** Find the average rate of movement of the Pacific plate.

Critical Thinking

3. **Evaluate** Has the Pacific plate always moved at the same rate? Explain.

4. **Infer** What conclusion can you draw about the relationship between the age of an island and its distance from the active hot spot?

FIGURE 11-23
Go online to see a map showing the seven inhabited Hawaiian Islands.

FIGURE 11-19 ▼
The island of Hawai'i is the youngest island in the Hawaiian Island-Emperor Seamount chain. The numbers represent ages, in millions of years, of the oldest volcanic rocks of each island or seamount.

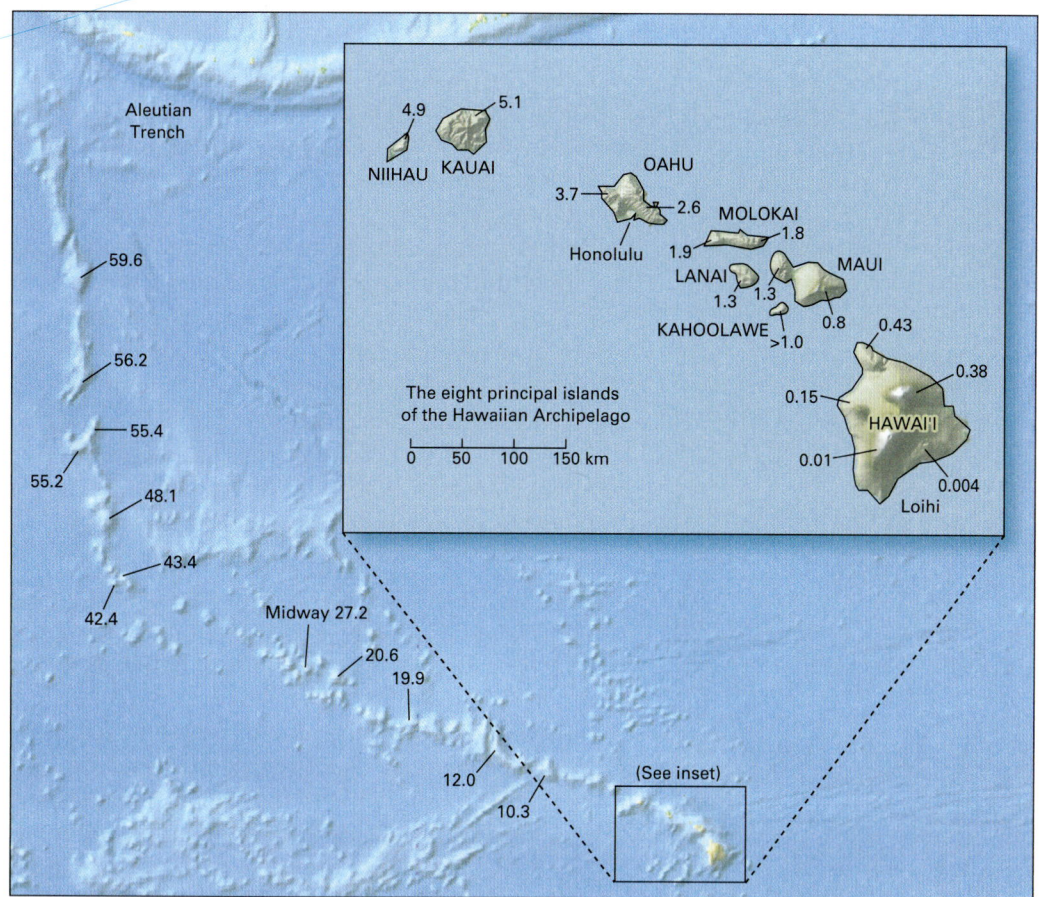

Source: Cengage

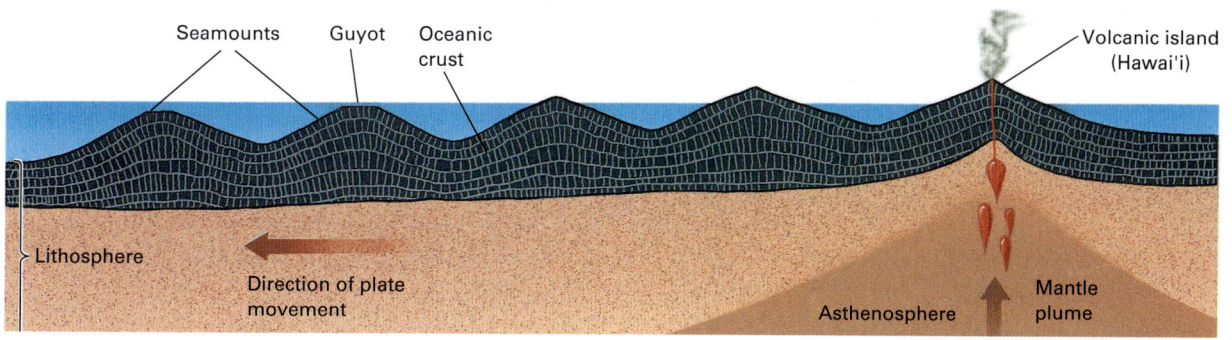

FIGURE 11-20 ▲
The Hawaiian Islands and Emperor Seamounts sink as they move away from the mantle plume currently located under the volcanically active Island of Hawai'i.

upper surface on the sinking island. A flat-topped seamount called a **guyot** (pronounced "gee-o," after Swiss-born American geologist Arnold Henri Guyot), will form.

The South Pacific and portions of the Indian Ocean are dotted with numerous other islands called atolls. An **atoll** is a circular coral reef that forms a ring of islands around a central lagoon. Atolls vary from 1 to 130 kilometers in diameter and are surrounded by the deep water of the open sea.

🌐 **FIGURE 11-21**
Go online to see how a volcanic island can become a guyot.

338　CHAPTER 11　THE SEAFLOOR

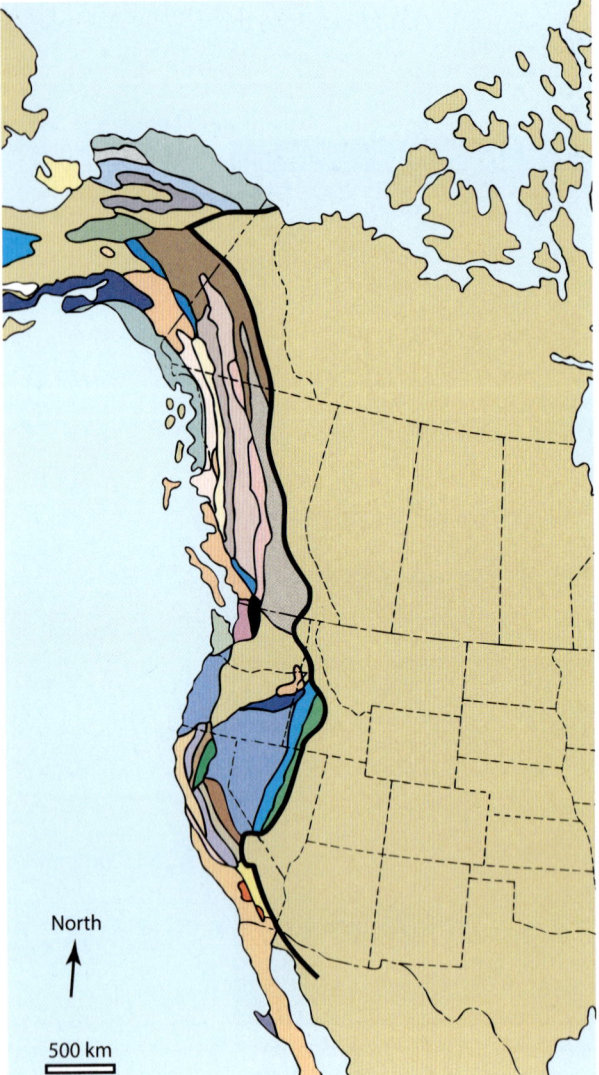

FIGURE 11-18
The accreted terranes of western North America are microcontinents and island arcs that were added to the continent from the Pacific Ocean. Each colored section on this map represents an individual terrane. Large sections of many western states and provinces in the United States and Canada have this feature.

underlying western California, Oregon, Washington, and western British Columbia was added to North America in this way between 180 million and about 50 million years ago. During that time, several island arcs formerly in the eastern Pacific Ocean slowly collided with the western margin of the North American plate. These late additions to the North American continent are called **accreted terranes** (Figure 11-18).

checkpoint Why don't old island arcs subduct with the oceanic crust at a convergent ocean–continent boundary?

Seamounts, Oceanic Islands, and Atolls

A **seamount** is a submarine mountain that rises 1 kilometer or more above the surrounding seafloor. An **oceanic island** is a seamount that rises above sea level. Both are common in all ocean basins but are particularly abundant in the southwestern Pacific Ocean. Seamounts and oceanic islands sometimes occur as isolated peaks, but they are more commonly found in chains. Dredge samples show that seamounts, oceanic islands, and the ocean floor itself are all made of basalt. Many seamounts and oceanic islands are volcanoes that formed at a hot spot above a mantle plume. And most form within a tectonic plate rather than at a plate boundary. An isolated seamount or short chain of small seamounts probably formed over a plume that lasted for only a short time. In contrast, a long chain of large islands, such as the Hawaiian Island-Emperor Seamount chain, formed over a long-lasting plume. In this case the lithospheric plate migrated over the plume as the magma continued to rise from a source beneath the lithosphere. Each volcano formed directly over the plume and then became extinct as the moving plate carried it away from the plume. As a result, the seamounts and oceanic islands become progressively younger toward the Island of Hawai'i, located at the end of the chain and currently located over the active mantle plume (Figure 11-19).

After a volcanic island forms, it begins to sink, or subside. Three factors contribute to the sinking:

1. If the mantle plume stops rising, it stops producing magma. Then the lithosphere beneath the island cools and becomes denser, and the island sinks. Alternatively, a moving plate may carry the island away from the hot spot. This also results in the cooling, contraction, and sinking of the island.

2. The weight of the newly formed volcano causes isostatic sinking.

3. Erosion lowers the top of the volcano.

These three factors gradually transform a volcanic island to a seamount over geologic time (Figure 11-20). Calculations suggest that if the Pacific plate continues to move at its present rate, the island of Hawai'i may sink beneath the sea within 10 million to 15 million years. In this case, sea waves likely will erode a horizontal

FIGURE 11-25
This colored scanning electron microscope image shows foraminifera, tiny organisms that float near the ocean surface. When these organisms die, their remains sink to the seafloor to become part of the pelagic sediment. Each of the fossils is the size of a fine sand grain.

causes newly erupting molten lava to rapidly form a solidified outer rind. Still-molten and pressurized magma immediately below the rind causes it to deform into pillow-shaped spheroids called pillows (Figure 11-26).

FIGURE 11-26
Go online to see an example of pillow basalts on the seafloor.

Beneath the pillow basalt is the thickest layer of oceanic crust. This layer consists of between 3 and 5 kilometers of basalt that did not erupt onto the seafloor. This basalt directly overlies the mantle. The upper portion consists of vertical basalt dikes that formed as basaltic magma oozing toward the surface froze in the cracks of the rift valley. The lower portion consists of gabbro, the coarse-grained equivalent of basalt. The gabbro forms as pools of magma cool slowly, insulated by the basalt dikes above them.

The pillow basalt, vertical basalt dikes, and gabbro all form at the Mid-Ocean Ridge. These rocks make up the foundation of all oceanic crust because all oceanic crust forms at a ridge axis and then spreads outward. In some places, chemical reactions with seawater have altered the basalt. The reactions form a soft, green rock called serpentinite that contains up to 13 percent water. Serpentinite is a fairly common rock type in parts of California and the Pacific Northwest. This is due to the abundance of basalt and altered basalt that represents oceanic crust associated with the accreted terranes found there. Superb outcrops of serpentinite are found on along some coasts of California (Figure 11-27).

FIGURE 11-27
Bluish serpentinite rock formed near Marshall's Beach, San Francisco, California. The rock is the result of chemical reactions between seawater and the basalt.

checkpoint What causes the smooth, featureless surface of abyssal plains?

11.3 ASSESSMENT

1. **Explain** Why is the Mid-Ocean Ridge elevated above the surrounding ocean floor?
2. **Describe** Why is the island of Hawai'i the youngest island in the Hawaiian Island-Emperor Seamount chain?
3. **Summarize** What evidence supported Darwin's hypothesis about the sinking of volcanic islands and the formation of atolls?
4. **Distinguish** Describe the similarities and differences between island arcs, seamounts, oceanic islands, and atolls.
5. **Relate** How does the accumulation of sediment change with distance from the Mid-Ocean Ridge?

Computational Thinking

6. **Generalize Patterns** Use Figure 11-8 to predict areas where the seafloor is likely youngest and where it is likely oldest. Describe any patterns you see in seafloor age relative to the continents. How well do your predictions compare to the map of the ages of the seafloor (Chapter 7 Investigation Handout 7.3)?

LESSON 11.3

11.4 CONTINENTAL MARGINS

Core Ideas and Skills
- Compare types of continental margins.
- Analyze the slope and rise of a continental shelf.
- Explain underwater erosion effects, including the formation of submarine canyons.

KEY TERMS

continental shelf	turbidity current
continental slope	passive continental margin
continental rise	
submarine canyon	active continental margin
submarine fan	

The Continental Shelf

On all continents, streams and rivers deposit sediment in coastal deltas such as the Mississippi Delta. Much of this sediment is redistributed along the coastline by ocean currents and downslope as sediment-gravity flows. These flows are mixtures of sediment and water that flow downslope as a fluid, usually as a bottom-hugging current. Much of the sediment transported downslope from the shoreline and offshore forms a shallow, gently sloping underwater surface that marks the edge of the continent and is called a **continental shelf** (Figure 11-28). As sediment accumulates on a continental shelf, the edge of the continent sinks isostatically because of the added weight. This effect keeps the shelf generally below sea level. At the same time, the delivery and accumulation of more sediment from the weathering of the nearby continent causes the outer edge of the continental shelf to build outward. This process ultimately enlarges the size of the continent itself. In this way, mass is redistributed from the continental interior, where mountains are eroded, to the continental margins, which actively grow outward due to the accumulation of sediment there.

Over millions of years, several kilometers of sediment has accumulated on the east coast of North America, forming a broad continental shelf that projects outward along the entire coast. Beginning at the shoreline, the water depth of the North American Atlantic shelf increases gradually to about 200 meters at the outer shelf edge, which in most places is more than 100 kilometers offshore. The average inclination of the continental shelf over this distance is about 0.1 degree.

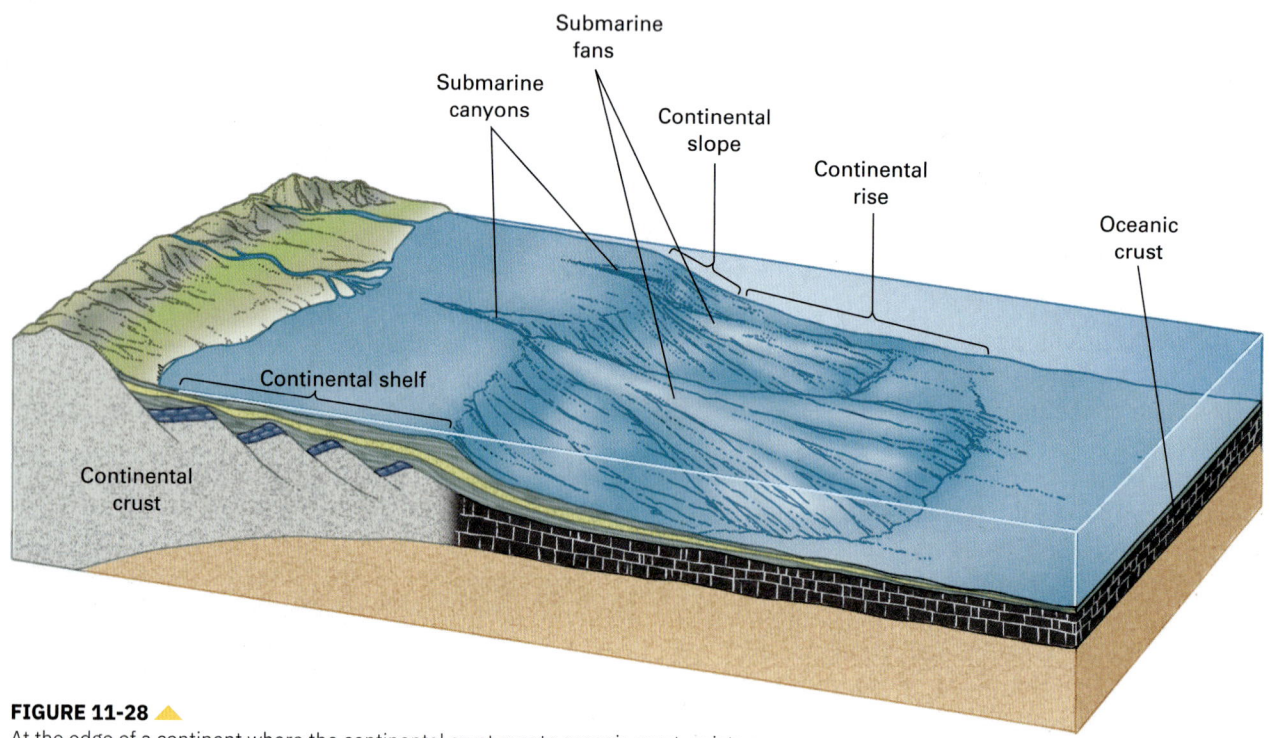

FIGURE 11-28
At the edge of a continent where the continental crust meets oceanic crust, exists a broad continental shelf, a slope, and a rise formed by accumulation of sediment eroded from the continent.

A continental shelf can be a very large feature. The shelf off the coast of southeastern Canada is about 500 kilometers wide, and parts of the shelves of Siberia and northwestern Europe are even wider.

In some places, a supply of sediment may be lacking, either because no rivers bring sand, silt, or clay to the shelf or because ocean currents do not deliver sediment to the particular region. In warm regions where terrigenous sediment is either lacking or is delivered in very small quantities, carbonate reef-building organisms such as corals, mollusks, bryozoans, calcareous algae, and numerous others can thrive. When environmental conditions are optimal, the carbonate-generating ecosystem can produce very large volumes of carbonate sediment in very short periods of time. As a result, thick beds of limestone can accumulate. Limestone accumulations of this type may be hundreds of meters thick and hundreds of kilometers across and are called carbonate platforms. The Florida Keys and the Bahamas are modern-day examples of carbonate platforms on continental shelves (Figure 11-29).

Some of the world's richest petroleum reserves occur on the continental shelves of the North Sea between England and Scandinavia, in the Gulf of Mexico, and in the Beaufort Sea on the northern coast of Alaska and western Canada. In recent years, oil companies have explored and developed these offshore reserves. Deep drilling has revealed that granitic continental crust lies beneath the sedimentary rocks, confirming that the continental shelves are truly parts of the continents despite the fact that they are covered by seawater.

At the outer edge of a shelf, the seafloor gradually steepens to an average slope of about 3

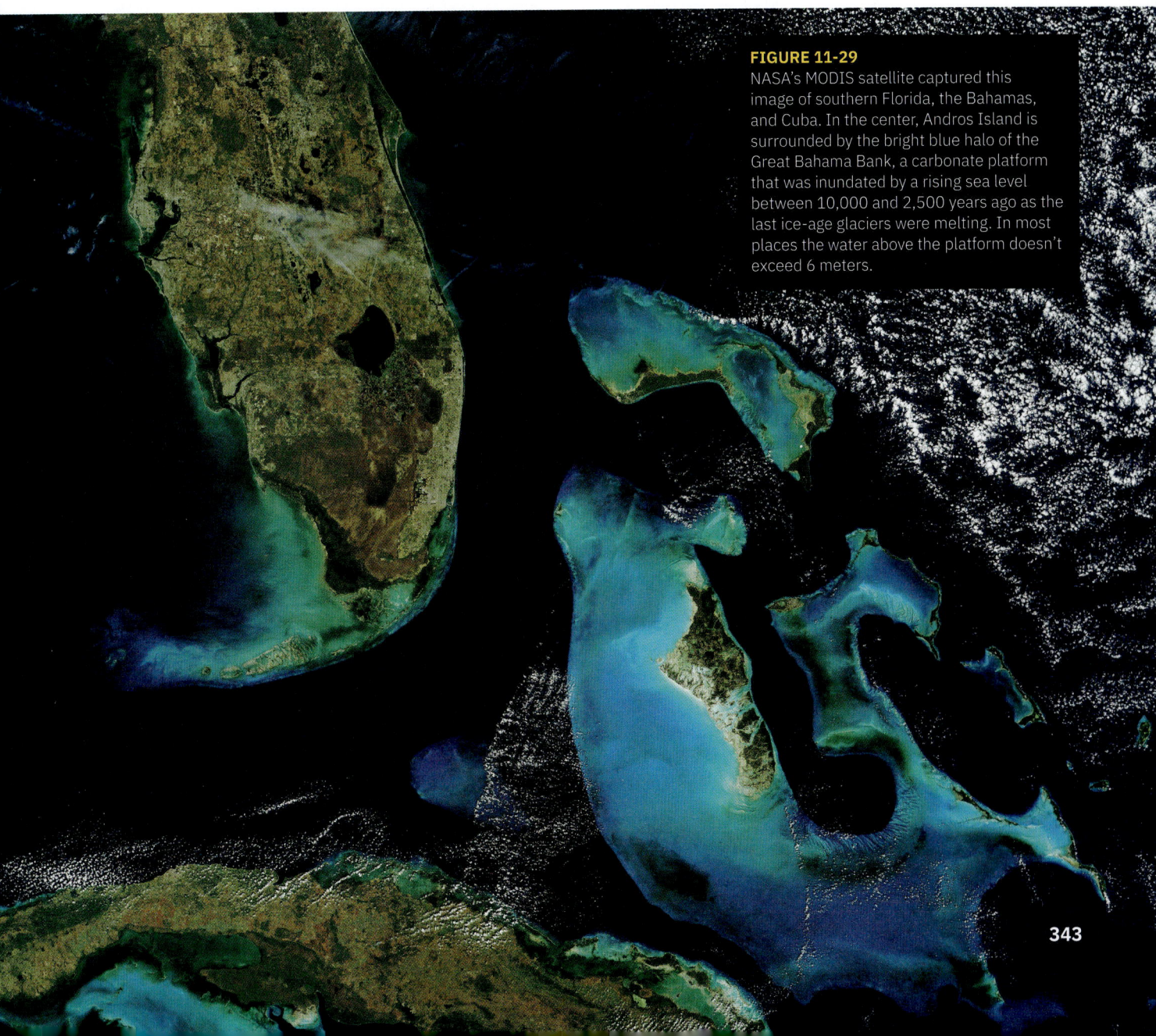

FIGURE 11-29
NASA's MODIS satellite captured this image of southern Florida, the Bahamas, and Cuba. In the center, Andros Island is surrounded by the bright blue halo of the Great Bahama Bank, a carbonate platform that was inundated by a rising sea level between 10,000 and 2,500 years ago as the last ice-age glaciers were melting. In most places the water above the platform doesn't exceed 6 meters.

degrees (though it can vary from 1 to 10 degrees) as it increases in depth from 200 meters to about 3 kilometers. This steep region of the seafloor averages about 50 kilometers and is called the **continental slope**. It is a surface formed primarily by the accumulation of mud. Its steeper angle is due primarily to thinning of continental crust where it nears the junction with oceanic crust. Offshore seismic reflection profiling shows that the sedimentary layering is commonly disrupted where sediment has slumped and slid down the steep incline.

A continental slope becomes less steep as it gradually merges with the deep ocean floor. This region, called the **continental rise**, consists of an apron of terrigenous sediment that was transported across the continental shelf and deposited on the deep ocean floor at the foot of the slope. The continental rise averages a few hundred kilometers in width. Typically, it joins the deep seafloor at a depth of about 5 kilometers. The shelf-slope-rise complex is a smoothly sloping, submarine surface on the edge of a continent formed by accumulation of sediment eroded from the continent.

checkpoint What causes the formation of carbonate platforms like the Florida Keys?

Submarine Canyons and Fans

In many places, seafloor maps show deep valleys called **submarine canyons** eroded into the continental shelf and slope. They look like submarine stream valleys. A canyon typically starts on the outer edge of a continental shelf, usually beyond the outer reaches of a major river delta on the inner shelf, and continues across the slope to the rise. At its lower end, a submarine canyon commonly leads into a **submarine fan**, a large, fan-shaped accumulation of terrigenous sediment on a continental rise (Figure 11-30).

Most submarine canyons occur downslope of a region where large rivers enter the sea. When they were first discovered, geologists thought the canyons had been eroded by rivers during the Pleistocene epoch, when accumulation of glacial ice on land lowered sea level by as much as 150 meters. However, this explanation cannot account for the deeper portions of submarine canyons that cut erosionally into the lower continental slopes at depths of a kilometer or more. These deeper parts of the submarine canyons must have formed through erosion, and a submarine mechanism must be found to explain them.

Geologists subsequently discovered that sediment gravity flows—underwater mixtures of sediment and water that flow downslope—can erode the continental shelf and slope to create or deepen the submarine canyons. Particularly efficient at eroding rock and sediment underwater are **turbidity currents**. Turbidity currents are a form of sediment gravity flow in which a turbulent mixture of sediment and seawater, which is more

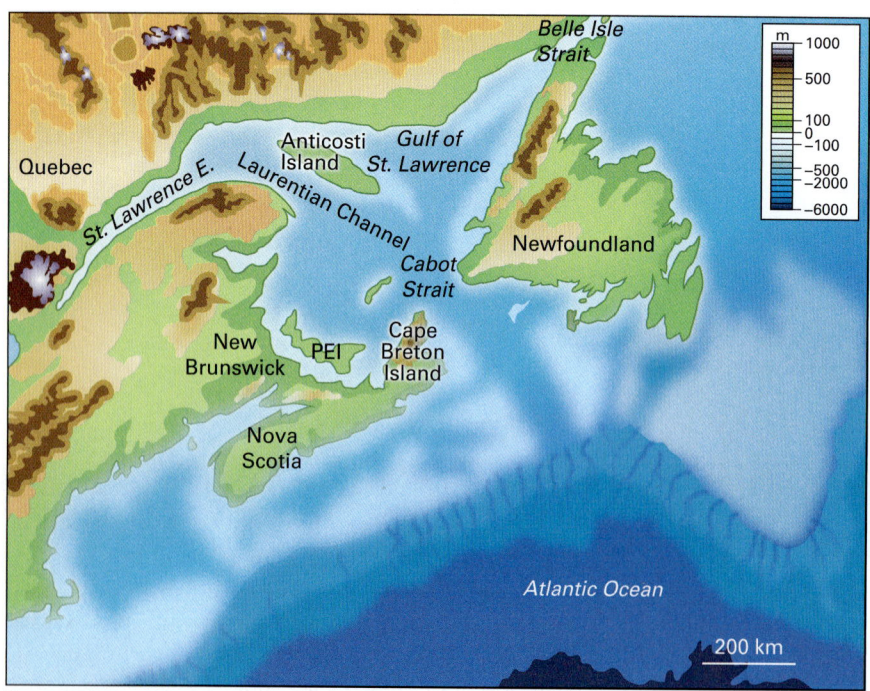

FIGURE 11-30
This map shows onshore topography and seafloor depths for parts of the United States, New Brunswick, Nova Scotia, and Newfoundland, Canada. Sediment carried down the Saint Lawrence River and across the Laurentian Channel moves through one or more submarine canyons across the continental slope to the continental rise.

FIGURE 11-31 ▼
Turbidity currents deliver a high volume of sediment to the continental rise. They are responsible for much of the erosion that occurs in submarine canyons.

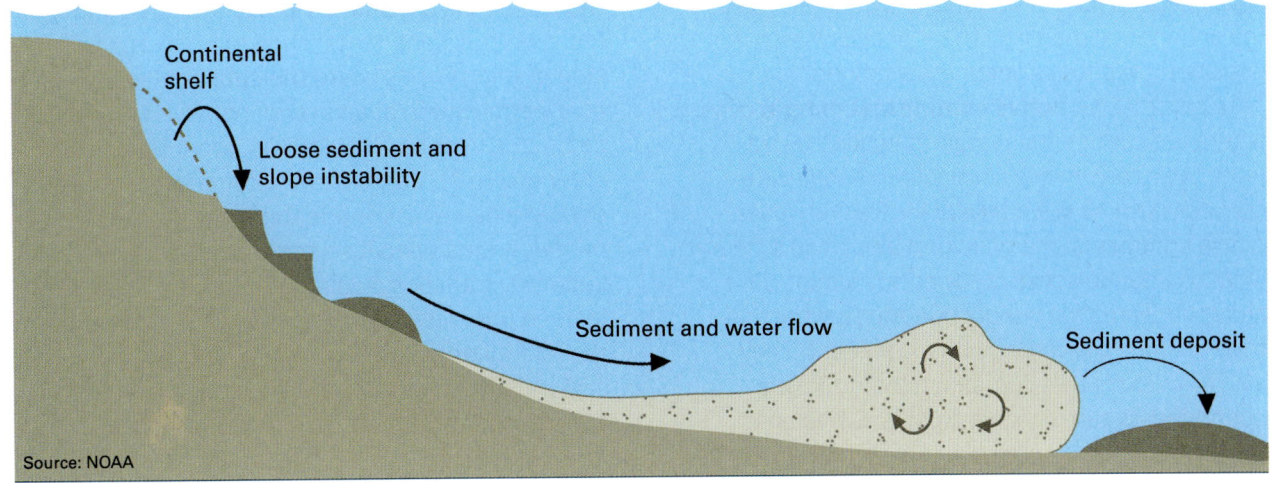

dense than the seawater alone, flows downslope along the bottom (Figure 11-31). You can create and observe a turbidity current at home by slowly pouring a cup of muddy water into a sloping basin of clear water.

Turbidity currents can be triggered by an earthquake, a large storm, or simply by the oversteepening of the slope as sediment accumulates. When the sediment starts to move, it mixes with water and flows across the shelf and down the slope as a turbulent, chaotic fluid. A turbidity current can travel at speeds greater than 100 kilometers per hour and for distances in excess of 1,200 kilometers.

Water filled with sediment traveling at such speed has tremendous erosive power. Once a turbidity current cuts a small channel into the shelf and slope, subsequent currents follow the same channel, just as a stream uses the same channel year after year. Over time, the currents erode a deep submarine canyon into the shelf and slope. Turbidity currents slow down when they reach the deep seafloor. The sediment accumulates there to form a submarine fan. Most submarine canyons and fans form near the mouths of large rivers because the rivers supply the great amount of sediment needed to create turbidity currents and other types of sediment gravity flows. Most of the largest submarine fans form on passive continental margins. Submarine fans that form on active margins typically are much smaller because the trench swallows much of the sediment. Furthermore, most of the world's largest rivers drain toward passive margins. The largest known fan is the Bengal Fan, which covers about 4 million square kilometers beyond the mouth of the Ganges River in the Indian Ocean east of India. More than half of the sediment eroded from the rapidly rising Himalaya ends up in this fan. Interestingly, the Bengal Fan has no associated submarine canyon, perhaps because the sediment supply is so great that the rapid accumulation of sediment prevents erosion of a canyon.

checkpoint What seafloor features result from turbidity currents?

Continental Margins

A continental margin is a place where continental lithosphere meets oceanic lithosphere. Two types of continental margins exist. A **passive continental margin** occurs where continental and oceanic lithosphere are firmly joined together (Figure 11-28). Because it is not a plate boundary, little tectonic activity occurs at a passive margin. Continental margins on both sides of the Atlantic Ocean are passive margins.

About 250 million years ago, all of Earth's continents were joined into the supercontinent called Pangaea. Shortly thereafter, Pangaea began to rift apart into the continents as we know them today. The Atlantic Ocean opened as the east coast of North America separated from Europe and Africa. As Pangaea broke up, the continental crust fractured and thinned near the fractures

(Figure 11-32). Basaltic magma rose at the new spreading center, forming oceanic crust between North America and Africa. All tectonic activity then focused on the continually spreading Mid-Atlantic Ridge, and no further tectonic activity occurred at the continental margins. This inactivity is why we use the term passive continental margin.

In contrast, an **active continental margin** occurs where the continental margin coincides with a plate boundary. Active continental margins are common around the edges of the Pacific Ocean. Active continental margins form along two different locations. Some active continental margins form along Andean-style subduction zones, where an oceanic plate converges with a continent. They also form along less common transform plate boundaries where oceanic crust is sliding past continental crust. This occurs along the San Andreas Fault in California.

At convergent boundaries, most of the sediment transported from a continent is swallowed up in the trench. As a result, an active margin commonly has a narrow continental shelf or none at all. The side of the trench toward the continent is the continental slope of an active margin. It typically inclines at 4 or 5 degrees in its upper part and steepens to 15 degrees or more near the bottom of the trench. The continental rise is absent or relatively small because sediment gravity flows generally transport sediment into the trench instead of across it to the ocean floor located over the subducting plate.

A trench can form wherever subduction occurs. Trenches form where oceanic crust sinks beneath

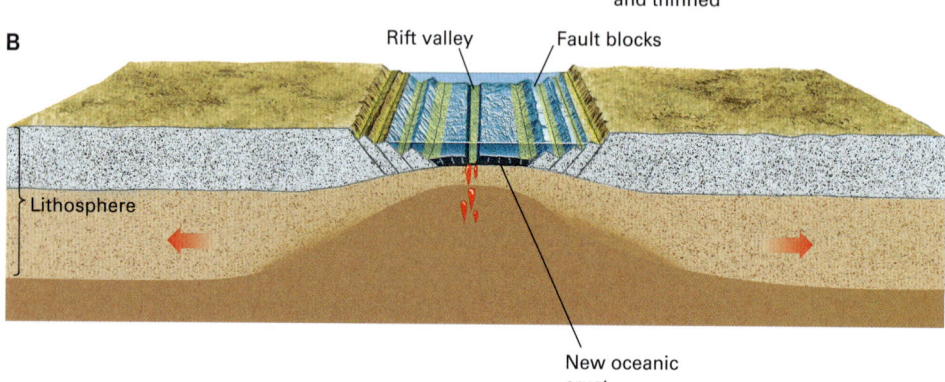
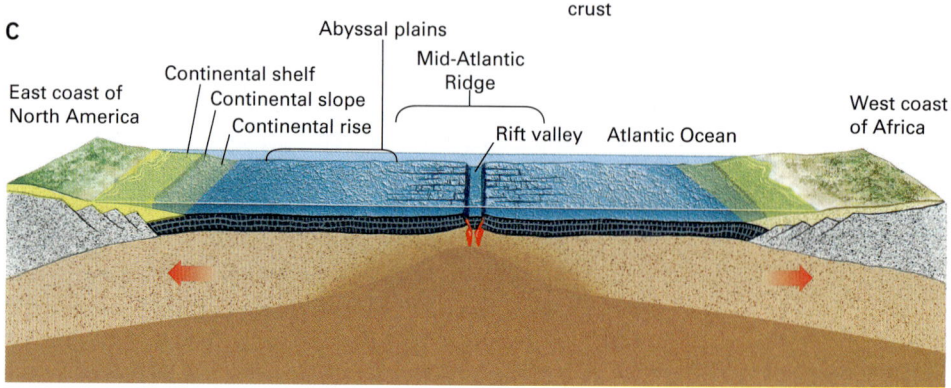

FIGURE 11-32
(A) Continental crust fractured as Pangaea began to rift. (B) Faulting and erosion thinned the crust as it separated. Rising basaltic magma formed new oceanic crust in the rift zone. (C) Sediment that eroded from the continents formed broad continental shelves on the passive continental margins of North America and Africa.

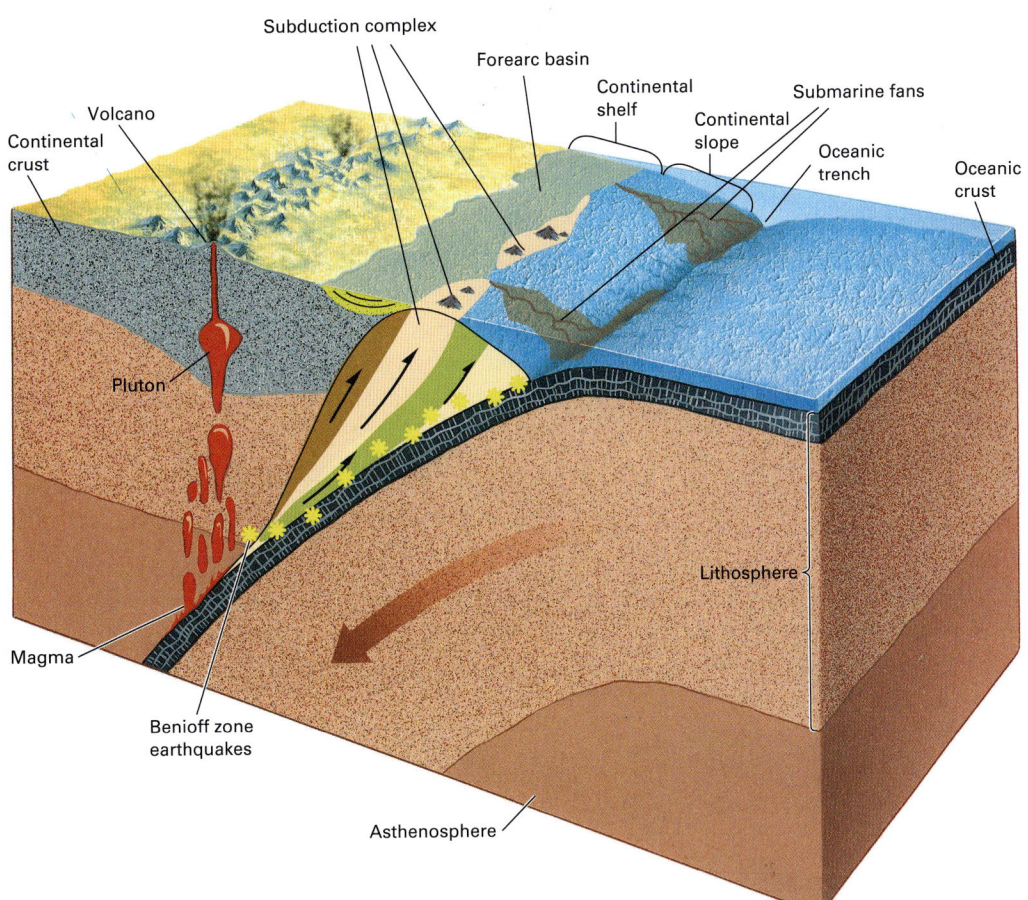

FIGURE 11-33 Along most active continental margins, an oceanic plate sinks beneath a continent, forming an oceanic trench. The continental shelf is narrow, the slope is steep, and the continental rise is small to nonexistent. Recall that Benioff-zone earthquakes are formed by release of strain as the subducting plate scrapes past the overriding plate.

the edge of a continent or where it sinks beneath another oceanic plate. However, the trench will be associated with a continental margin only when the overriding plate is made of continental lithosphere.

At convergent boundaries, the oceanic plate sinks into the mantle at descent angles ranging from about 15 degrees to more than 90 degrees (Figure 11-33). There is wide variation in the descent angle of subducting slabs around Earth's ocean basins. In most convergent margins, an oceanic trench is formed. However, in some regions the trench is filled with sediment and is not as obvious. For example, offshore of the Pacific Northwest, the Juan de Fuca plate is subducting below the North America plate. But the subduction is filled with sediment resulting from draining of the Cascades and northern Rockies. The Columbia River has been the largest deliverer of sediment to the trench. Huge volumes of water and sediment poured out of the Columbia River during and following each glaciation over the past 3 million years or so. In contrast, the Mariana Trench off the eastern coast of the Philippines includes Earth's deepest ocean depth at nearly 11 kilometers.

checkpoint Why does an active margin commonly have a narrow continental shelf?

11.4 ASSESSMENT

1. **Identify** What are the features of a continental margin? Draw a sketch and add labels to explain the differences that result between passive and active continental margins.

2. **Compare** What is the land-based equivalent of a submarine fan (Chapter 12) and how do the formation processes of both sediment accumulations compare?

3. **Explain** Where would you expect to find the deepest submarine canyons and largest submarine fans?

Computational Thinking

4. **Analyze** Use Figure 11-8 and Figure 7-11 to determine whether more continental margins are passive or active. Based on your answer, what can you conclude about the frequency of large-to-small submarine fans?

LESSON 11.4

TYING IT ALL TOGETHER
ROBOTIC EXPLORATION OF THE SEAFLOOR

In this chapter, you learned about the formation of Earth's ocean basins and the features that characterize the seafloor. You also learned about the layers of Earth's oceanic crust and the features of passive and active continental margins. Much of our current understanding of the seafloor is based on data obtained through core-drilling and remote-sensing technologies.

The Case Study explored how scientists used logging while drilling (LWD) and robotic submersibles, including the MARUM-SEAL (Figure 11-34), to investigate the effects of the 2011 Tōhoku earthquake on the Pacific seafloor. The findings of those expeditions confirm that the epicenter of the earthquake was located at a major subduction zone between the Pacific and Okhotsk plates. Analysis of the sediment core extracted from the fault revealed that the fault was lined with a very thin, weak, slippery layer of clay sediments. As stress built up along the fault, the slippery clay layer reduced the friction between the plates and allowed them to slide a great distance once the stress was released. Mapping of the seafloor by the MARUM-SEAL submersible showed a 50-meter lateral shift and 16-meter vertical shift in the seafloor as compared with previous mappings from 1999 and 2004. Further, the distribution of the lateral and vertical shifts helped to explain the many large aftershocks after the initial quake as well as the pulsating tsunami waves.

The post Tōhoku exploration efforts are just some of the ways in which scientists are using new technology to increase understanding of the ocean floor. In addition to helping us learn about past events, exploration can identify new structures on the ocean floor and provide information about how Earth is changing.

Work individually or with a group to complete the steps below.

1. Select a point of interest on the ocean floor, such as a particular seamount, trench, or other feature of the seafloor. Examples could include the Mid-Atlantic Ridge, Hawaiian Islands and Emperor Seamounts, Challenger Deep, Davidson Seamount, the Philippine Trench, or many others.

2. Select several reliable sources of information about historic and current research efforts related to your chosen point of interest.

3. Summarize the formation hypothesis (or hypotheses) for your point of interest. Identify scientific evidence that supports this hypothesis.

4. Focus your research on the following questions:
   - When and how was the seafloor feature discovered?
   - What current research efforts are underway to explore this feature?
   - What technology is being used to study or map this feature?
   - What have scientists learned about the ocean floor during their study of this feature?

5. Create an illustration of your point of interest. Include labels indicating important data points, such as elevation below sea level and from the ocean floor.

6. Develop a media-enhanced blog post about your point of interest that answers the four key questions from your research. Post to a class site.

FIGURE 11-34
The MARUM-SEAL is an autonomous underwater vehicle (AUV). In 2012, the AUV was deployed by the research vessel the *Sonne*.

CHAPTER 11 SUMMARY

11.1 EARTH'S OCEANS

- Ocean water, atmospheric gases, and the molecular building blocks of life were carried to Earth by bolides early in Earth's history.

- Oceans cover about 71 percent of Earth's surface.

- Continental lithosphere is thicker and less dense than oceanic lithosphere. As a result, continents float isostatically at relatively high elevations, whereas oceanic lithosphere sinks to relatively low elevations.

- The ocean basins are topographic depressions on Earth's surface that have filled with water to form the oceans.

11.2 STUDYING THE SEAFLOOR

- Information about the seafloor comes from direct sampling technology and remote sensing.

- Direct sampling, including dredging, coring, and drilling, allows scientists to directly study samples of the seafloor.

- Remote sensing technology uses sound, light, and magnetic fields to obtain information about the structure and characteristics of the seafloor.

11.3 FEATURES OF THE SEAFLOOR

- The Mid-Ocean Ridge system is a submarine mountain chain that extends through all of Earth's major ocean basins. It forms at spreading centers, which is where new oceanic crust is formed.

- The Mid-Ocean Ridge system supports chemosynthetic bacteria at deep-sea hydrothermal vents. These bacteria form the foundation of the deep-sea food web.

- Oceanic trenches are the deepest parts of ocean basins. A trench can form wherever subduction occurs.

- Island arcs result from chains of volcanoes formed at subduction zones where two oceanic plates collide.

- Seamounts and oceanic islands form in oceanic crust as a result of volcanic activity over mantle plumes.

- Atolls are circular coral reefs growing on sinking volcanic islands.

- Oceanic crust varies from about 4 to 7 kilometers thick and consists of three layers. The top layer is sediment, the second consists of pillow basalt, and the third contains basalt dikes on top of gabbro. The base of the third layer is the boundary between oceanic crust and mantle.

- The age of seafloor rocks increases with distance from the Mid-Ocean Ridge. No oceanic crust is older than about 200 million years because it recycles into the mantle at subduction zones.

11.4 CONTINENTAL MARGINS

- Continental margins consist of a continental shelf, a continental slope, and a continental rise formed by accumulation of terrigenous sediment.

- Passive continental margins are not associated with the boundaries of tectonic plates while active continental margins occur at plate boundaries.

- Submarine canyons are eroded into continental margins by turbidity currents and commonly lead into abyssal fans, where the turbidity currents deposit sediments on the continental rise.

- Most active continental margins occur where oceanic crust sinks into a subduction zone beneath the margin of a continent. Some active continental margins occur at a transform boundary between continental and oceanic lithosphere. All active continental margins have a narrow shelf before sloping into deeper water.

CHAPTER 11 ASSESSMENT

Review Key Terms
Select the key term that best fits the definition. Not all terms will be used, and no term will be used more than once.

abyssal plain	oceanic island
accreted terrane	oceanic trench
active continental margin	passive continental margin
atoll	pelagic sediment
bolide	pillow basalt
chemosynthesis	seamount
continental rise	sonar
continental shelf	submarine canyon
continental slope	submarine fan
guyot	terrigenous sediment
magnetometer	transform fault
ocean basin	turbidity current

1. a system detecting objects under water by emitting sound pulses and measuring their return after reflection
2. flat-topped underwater mountain formed when wave energy erodes the top of a sinking volcanic island
3. piece of space debris that typically explodes upon impact with the atmosphere
4. circular coral reef that forms a ring of islands around a central lagoon and that is bounded on the outside by the deep water of the open sea
5. sand, silt, and clay eroded from the continents and deposited on the ocean floor mainly through submarine canyons extending from the continental shelf
6. zone of the ocean floor near a continent that occurs at a convergent or transform plate boundary
7. a strike-slip fault between two offset segments of a Mid-Ocean Ridge or along a strike-slip plate boundary
8. the relatively steep (averaging 3° but varying between 1° and 10°) region of the seafloor between the continental shelf and the continental rise
9. the flat, level, largely featureless parts of the ocean floor between the Mid-Ocean Ridge and the continental rise
10. an apron of sediment at the foot of the continental slope that merges with the deep seafloor
11. a long, narrow, steep-sided depression of the seafloor formed where a subducting oceanic plate sinks into the mantle, causing the seafloor to bend downward
12. a submarine mountain, usually of volcanic origin, that rises 1 kilometer or more above the surrounding seafloor
13. zone of the ocean floor where continental and oceanic crust are firmly joined together and where little tectonic activity occurs
14. the shallow, very gently sloping portion of the seafloor that extends from the shoreline to approximately 200 meters water depth at the top of the continental slope
15. instrument that measures the strength and, in some cases, the direction of a magnetic field

Review Key Concepts
Answer each question on a separate sheet of paper to demonstrate your understanding of key concepts from the chapter.

16. The surface of the primordial Earth was composed mostly of rock and molten rock. What are two hypotheses that can explain how the hydrosphere formed on Earth?
17. How do differences between Earth's continental crust and its oceanic crust explain the formation of ocean basins?
18. Describe two devices scientists can use to collect different types of rock and sediment samples from the ocean floor.
19. Describe and compare two different remote sensing technologies that scientists can use to study the ocean floor. Explain how each method is different.
20. Describe the tectonic processes that are responsible for the formation and characteristics of the Mid-Ocean Ridge system.
21. Describe two kinds of events that can cause global sea levels to change.
22. Describe the processes responsible for the formation and sinking of a volcanic oceanic island.
23. Compare and contrast terrigenous sediment and pelagic sediment.
24. Identify and describe the three layers of the oceanic crust.
25. How are active and passive continental margins different?

26. Identify each seafloor feature labeled with a letter in the illustration.

27. Describe the processes involved in the formation of submarine canyons and submarine fans.

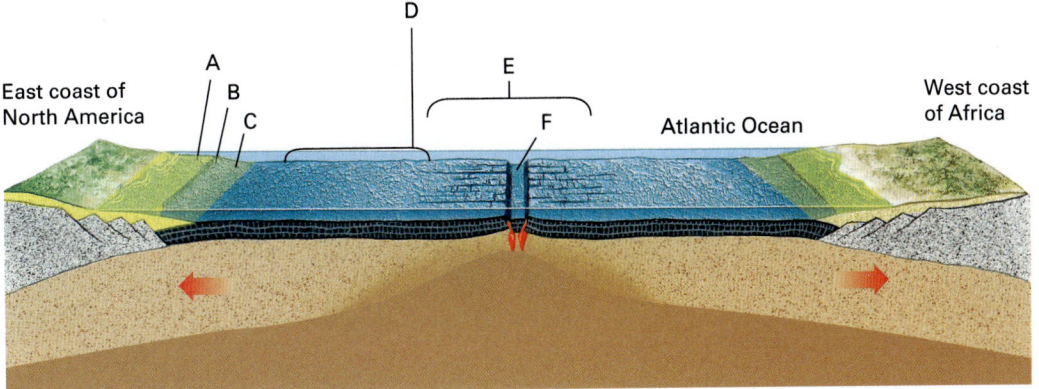

Think Critically

Write a response to each question on a separate sheet of paper. Use concepts from the chapter to support your reasoning.

28. In 2012, scientist Conel Alexander analyzed the chemical composition of comets in our solar system and found that it differs from that of rocks on Earth that formed from ancient bolides. Does this evidence support or weaken any current hypotheses about the source of Earth's water? Explain.

29. Sampling and remote sensing tools such as the rock dredge and the echo sounder are not useful for exploring the deepest parts of the seafloor. Explain why you think these tools are not suited to deep-sea exploration. Then identify any technological innovations that are being used or could be developed in the future that could advance deep-sea exploration.

30. Is the rock on volcanic islands more similar to the rock on the seafloor or the rock on the continents? Explain your reasoning.

31. Dams can be placed on rivers in order to divert water to do useful work, such as generating electricity and irrigating fields. How do you think the construction of a dam would affect the evolution of submarine canyons and fans downstream of the rivers? Explain your reasoning.

PERFORMANCE TASK

Seafloor Features Model

In this chapter, you have learned about the origin of Earth's ocean basins, the sampling and remote sensing techniques that oceanographers use to study and map the seafloor, and the major types of geologic features of the seafloor.

In this performance task, you will work with a partner to create a slideshow or a stop-motion animation video to share what you have learned with the class.

1. Review the different seafloor features you learned about in Lessons 11.3 and 11.4. With a partner, choose one of these features to model.

A. Using your textbook and other resources, research how this seafloor feature changes over time. Make sure to document how it forms, what variables are required for its formation, how it changes over time, and what factors cause these changes. Consider factors such as tectonic forces, volcanism, weathering, and erosion.

B. Create a slideshow or stop-motion animation video to model how the feature evolves over time. You may draw pictorial models or use clay to form physical models of the feature. In your model, demonstrate how the feature develops and changes. Accompanying your slideshow or animation, include text to explain how it is affected by the environment.

UNIT 4 | SURFACE PROCESSES

Chapter 12 **Weathering, Erosion, and Deposition**
Chapter 13 **Glaciers and Deserts**
Chapter 14 **Soil Resources**

Earth changes from within and from without. Fueled by the sun's energy and Earth's own gravity, powerful surface processes continually alter the planet's crust. Without the continuous interaction of Earth's spheres, its surface would look much like that of the moon: gray, barren, and cratered. Much of what makes our planet so diverse is the ice, water, wind, and organisms that break apart, wear away, and redistribute earth materials. Surface processes create precious soil too. In this unit, you will learn how surface processes change the planet.

The processes that change Earth's surface can result in formations that are majestic and exceptional, such as the eroded spires of Utah's Bryce Amphitheater in Bryce Canyon National Park.

CHAPTER 12
WEATHERING, EROSION, AND DEPOSITION

Niagara Falls, a roaring system of three waterfalls, can be a mesmerizing force of flowing, falling water. Imagine what you would you see if the 75,750-gallon-per-second flow of water were stopped. In 1969, water from the American Falls was diverted to the Canadian Horseshoe Falls. The American Falls, part of the Niagara Falls system, dwindled to a trickle. Revealed was the abrupt drop of the cliff and the gigantic pile of rocky debris shown in this photo. The boulders, the cliff, and the Niagara Falls system itself result from processes that shape Earth's landscapes: weathering, erosion, and deposition.

A trained geologist can look at a landscape and interpret how it formed. But just as you need experience and skill to read and understand a book, you need sharp observational skills and knowledge of Earth's surface processes to understand how those processes have shaped the planet's landscapes through time. What process could have brought a large boulder to an otherwise flat land? Why are the rugged, jagged peaks of the Rocky Mountains so different from the rolling, tree-covered Appalachian Mountains? The answers to questions such as these can be found in the ice, rocks, sediments, rivers, oceans, and mountains all around us. Their histories involve changes both imperceptibly slow and violently rapid. In this chapter, we explore our planet's history of wear and tear.

KEY QUESTIONS

12.1 How are the different processes of mechanical weathering alike and different?

12.2 How are the different processes of chemical weathering alike and different?

12.3 What are characteristics of streams and how do they affect Earth's surface?

12.4 Why do slides, flows, and falls occur, and how can the effects of mass wasting be predicted and avoided?

EXPLORERS AT WORK

TRACING THE FOOTPRINTS OF EARLY HUMANS

WITH NATIONAL GEOGRAPHIC EXPLORER CYNTHIA LIUTKUS-PIERCE

Dr. Cynthia Liutkus-Pierce, a geologist and professor at Appalachian State University, studies ancient sediments for clues about how humans evolved. One of her flagship studies is occuring at Engare Sero, a site containing more than 400 ancient human footprints preserved in a stretch of volcanic mudflats in northern Tanzania. The site is one of the largest of its kind. Analysis of the fossilized footprints is revealing an amazingly detailed story of the culture of early *Homo sapiens*. But the footprints are disappearing, and Liutkus-Pierce must find modern ways to preserve them.

In 2008 Liutkus-Pierce first traveled to eastern Africa to visit the site, which lies in a riverbed between a volcano and a saline lake. She counted fifty footprints and assembled a team to further excavate the site, exposing more footprints across an area about the size of a tennis court. Liutkus-Pierce concluded that the footprints formed when a rainstorm sent a flood of ash-rich mud down the flanks of a volcano and onto the riverbed. Within a few hours, a group of ancient travelers walked barefoot across the mud, which quickly dried out and captured their footprints. Soon after, another flow of muddy ash from the volcano buried the footprints, preserving them.

Based on mineral analysis of the sediment layers surrounding the footprints, Liutkus-Pierce thinks they are 19,000 to 5,700 years old. During that time, humans were hunter-gatherers. The footprints were left by about twenty people who were likely foraging for food or seeking water. Most were women, although a few children and men were among them. According to Liutkus-Pierce, such details can be used to understand gender roles, patterns of movement, and social structures.

Yet time is not on her side. The footprints are disappearing as the sediments that contain them weather away at rates between 0.10 to 0.17 millimeters per year. The footprints are located in a usually dry riverbed, but the flow of water during the rainy season gradually erodes them. Blowing sand and wind carry away bits of sediment from the edges of the prints.

In 2010, Engare Sero Footprint Project scientists made a digital copy of the site by creating a detailed 3D scan with sub-millimeter resolution. Liutkus-Pierce and local officials created a small rock wall around the footprints to divert flowing water. A fence now surrounds the site to discourage people from walking or driving over it. Ultimately, however, Luitkus-Pierce knows that the forces of erosion will win out and these treasured footprints will disappear entirely.

THINKING CRITICALLY

Evaluate Do you think natural erosion should be allowed to erase the footprints at the Engare Sero site over time? Why or why not?

Dr. Cynthia Liutkus-Pierce stands in a river channel near the volcano Ol Doinyo Lengai in northern Tanzania. In the Engare Sero Footprint Project, Liutkus-Pierce studies fossilized footprints of ancient humans.

The Engare Sero Footprint Project is conducted in an ancient riverbed near the Ol Doinyo Lengai volcano. The fossilized footprints were made by humans who walked in ash-rich mud after a rainstorm.

CASE STUDY
FLOODING FROM HURRICANE KATRINA

Hurricanes, tornadoes, and other extreme weather events have occurred throughout Earth's history. We often label such events "natural disasters," but the amount of death and destruction inflicted by a storm depends not only on its severity, but also on how many people live in the affected area. In addition, human activities can make an area more or less susceptible to damage from severe weather.

Hurricane Katrina is a case in point. In August 2005, Katrina made landfall near New Orleans, a city built on the Mississippi River Delta. As Katrina approached from the Gulf of Mexico, the storm surge sent massive volumes of water over the levees designed to protect the city. The flood waters chased 1.3 million people from their homes and caused approximately $200 billion in damages.

Decades of flood-control practices actually exacerbated the flooding along the Mississippi Delta during Katrina. A delta is a natural landform that develops gradually as sediment in a river system is deposited in the area where the river empties into the ocean. Human activity can greatly alter this process. Dams built upstream trap sediment that otherwise would reach the delta. Levees built along the river to prevent flooding also prevent the deposition of sediment that would normally be deposited on the delta during floods. Meanwhile, urban and agricultural development accelerate erosion of the delta even as the rate of sedimentation is reduced. The destruction of the natural vegetation that holds the soil in place, combined with rising sea levels and sinking coastal land, all contribute to a gradual eating away of the delta coast.

All of these factors amplified the flooding that occurred in New Orleans in 2005. Over the previous decades, the global sea level had risen while the Mississippi Delta sank and slowly eroded due to human development of the area. Between 1930 and 2012, nearly 5,000 square kilometers of the delta sank below sea level. The sinking delta surface effectively removed a set of natural buffers that helped protect the coastline from storm energy. Then, in 2005, Katrina pushed seawater across the submerged delta surface and inland up the Mississippi River channel. The resulting storm surge caused the levees around New Orleans to fail, resulting in massive flooding, destruction, and loss of life.

Since Katrina, New Orleans has rebuilt and strengthened its levees and has begun restoring wetlands. For example, the city closed a channel that allowed salty seawater to invade wetlands and kill marsh grasses and trees. Other proposals have been more controversial. Some have proposed allowing the Mississippi River to flood certain areas now protected by levees. Although this would allow more sediment to reach the delta, thereby combating coastal erosion, landowners in the floodplain could be negatively affected.

As You Read Explore how weathering, erosion, and deposition influence the shape of the land. Consider how human activities affect these natural processes, and how better land-management practices can reduce the harm caused by floods and other weather-related hazards.

FIGURE 12-1
A series of satellite images show the catastrophic flooding of the Lower Ninth Ward of New Orleans caused by Hurricane Katrina. The image on the left was taken in January 2003, before the hurricane. The second image, taken on August 31, 2005, shows the flooding after the levee failed. The third image, taken on September 21, 2005, shows the destruction after the floodwaters receded. The final image, taken in April 2011, shows more than five years of progress in rebuilding after the flood.

12.1 MECHANICAL WEATHERING

Core Ideas and Skills
- Describe the processes of weathering and erosion.
- Describe the process of mechanical weathering through pressure-release fracturing, frost wedging, abrasion, organic activity, and thermal expansion and contraction.
- Use a model to demonstrate weathering by organic activity.

KEY TERMS

erosion
mechanical weathering
chemical weathering
pressure-release fracturing
frost wedging

Five processes cause mechanical weathering: pressure-release fracturing, frost wedging, abrasion, organic activity, and thermal expansion and contraction. Two additional processes—salt cracking and hydrolysis-expansion—result from combinations of mechanical and chemical processes. You will learn about salt cracking and hydrolysis-expansion in Lesson 12.2.

Weathering, Erosion, and Deposition on Earth

As our solar system formed, multitudes of debris crashed into the planets and their moons. The impacts formed huge craters—steep, circular depressions with a surrounding blanket of debris. Earth was so hot that molten rock simply oozed inward to fill the craters. As the crust thickened, cooled, and hardened, additional bombardment formed new craters. Today, however, none of these ancient craters remain on Earth's surface. In contrast, the moon is pockmarked with craters of all ages. Why have ancient craters vanished from Earth but not the moon?

Over geologic time, tectonic processes such as mountain building, volcanic eruptions, and earthquakes continually changed Earth's surface. In addition, Earth's gravitational field is strong enough to retain its atmosphere. The atmosphere and hydrosphere continually weather and erode the surface rocks of the geosphere. The combination of tectonic activity and erosion has erased most traces of early impact craters from Earth's surface. In contrast, the smaller moon has lost most of its heat, so tectonic activity is nonexistent. The moon's small size means its gravitational force is too weak to have retained an atmosphere or significant amounts of surface water. Since neither wind nor flowing water is available to modify the moon's surface, ancient impact craters have remained there for billions of years.

Recall from Chapter 4 that the process by which solid rock is decomposed into loose gravel, sand, clay, and soil is called weathering (Figure 12-2). Weathering involves little or no movement of the broken-down rocks and minerals. The weathered material simply accumulates where it forms. However, loose soil and other weathered material offer little resistance to rain or wind and are easily eroded. **Erosion** is the removal of rock mass from an area by rain, moving water, wind, ice, or gravity. These agents of erosion may carry weathered material great distances before depositing it as sediment (Figure 12-3).

FIGURE 12-2
Go online to view the effects of weathering on the base of a sea cliff in Loutraki, Greece.

FIGURE 12-3
Weathering, erosion, and deposition—the laying down of sediment—typically occur in an orderly sequence. For example, a grain of quartz may travel thousands of kilometers over land and in water and be deposited multiple times over hundreds of years.

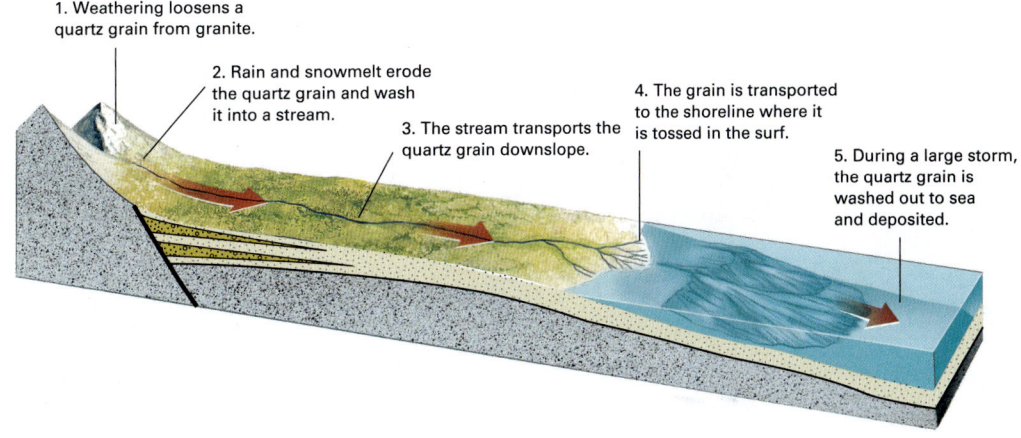

1. Weathering loosens a quartz grain from granite.
2. Rain and snowmelt erode the quartz grain and wash it into a stream.
3. The stream transports the quartz grain downslope.
4. The grain is transported to the shoreline where it is tossed in the surf.
5. During a large storm, the quartz grain is washed out to sea and deposited.

LESSON 12.1

Weathering occurs by both mechanical and chemical processes. **Mechanical weathering** (also called *physical weathering*) reduces solid rock to small fragments but does not alter the chemical composition of rocks and minerals. Think of crushing a rock with a hammer: The fragments are no different from the parent rock except that they are smaller. In contrast, **chemical weathering** occurs when air and water chemically react with rock to alter its composition and mineral content. Chemical weathering is similar to the rusting of an old truck body: The final product differs both physically and chemically from the original material (Figure 12-4). We will explore the different types of chemical weathering in the next lesson.

🌐 **FIGURE 12-4**
Go online to view an example of the chemical weathering of iron resulting in rust (iron oxide).

checkpoint What is the difference between mechanical and chemical weathering?

Pressure-Release Fracturing

Many igneous and metamorphic rocks form deep below Earth's surface. Imagine, for example, that a granitic pluton solidifies from magma at a depth of 15 kilometers. At that depth, the pressure from the weight of overlying rock is about 5,000 times that at Earth's surface. Over millions of years, tectonic forces may uplift the granite while erosional forces strip off the overlying layers of rock. As the pressure on the rock diminishes, the rock expands, but because the rock is now cool and brittle, it fractures as it expands. This process is called **pressure-release fracturing**. Many igneous and metamorphic rocks that formed at depth but now lie at Earth's surface have fractured in this manner (Figure 12-5).

FIGURE 12-5 ▲
Pressure-release fracturing contributed to the fracturing of this granite in California's Sierra Nevada.

Pressure-release fracturing is a mechanical weathering process that involves only the geosphere. All other forms of mechanical weathering involve interactions among rocks of the geosphere and the hydrosphere, the atmosphere, and the biosphere.

checkpoint How does pressure-release fracturing occur?

Thermal Expansion and Contraction

Rocks at Earth's surface are exposed to daily and yearly cycles of heating and cooling. They expand when they are heated and contract when they cool. When temperature changes rapidly, the surface of a rock heats or cools faster than its interior, causing the surface of the rock to expand or contract faster than the interior. Over time, the forces generated by this thermal expansion and contraction cause rocks to fracture and break apart.

Fire produces an extreme example of thermal expansion. A fire may heat rocks by hundreds of degrees. If you line a campfire with granite stones, for example, the rocks commonly break as you cook your dinner. In a similar manner, forest fires or brush fires occur frequently in many ecosystems, causing surface rocks to crack. In such areas, fire is an important agent of mechanical weathering by thermal expansion.

checkpoint Describe how thermal expansion and contraction weather rocks.

Frost Wedging

Water expands by seven to eight percent when it freezes. If water accumulates in a crack and then freezes, the ice wedges the rock apart in a process called **frost wedging**. During spring and autumn in a temperate climate, water freezes at night and thaws during the day. As the ice forms during the night, it pushes rock apart while at the same time holding it together. When the ice melts during the day, rock fragments loosen. For this reason, experienced mountaineers try to travel very early in the morning before ice melts and rock fragments are likelier to tumble from steep cliffs.

Over time, a pile of loose rock debris, called talus, will accumulate beneath many cliffs (Figure 12-6). These rocks fall from the cliffs mainly as a result of frost wedging over many seasons. Typically, large open pore spaces exist between the rocks.

checkpoint How does frost wedging occur?

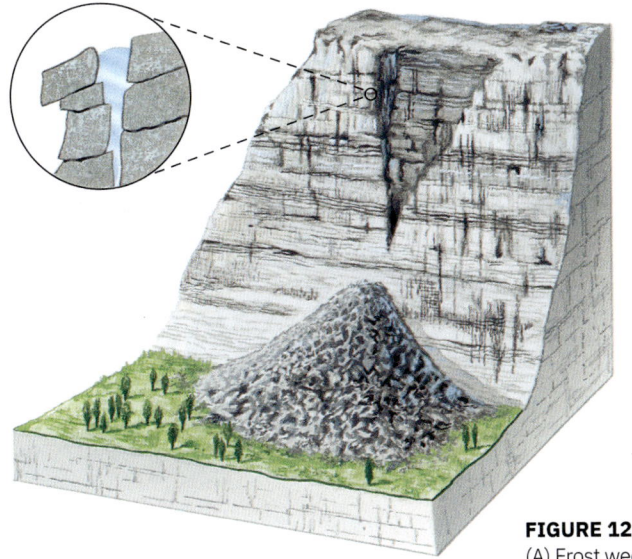

FIGURE 12-6
(A) Frost wedging dislodges rocks from a cliff and creates a talus slope. (B) Frost wedging produced this talus cone in Banff National Park, Canada. The angular blocks of rock forming the talus are porous. The stream in the center of the photograph emerges from talus at the top of the photo. It flows over bedrock for a short distance, then disappears back underground at the top of the talus cone below.

Abrasion

Rock fragments, grains of sand, and silt collide with one another when currents or waves carry them along a stream or beach. During these collisions, the sharp edges and corners of the particles are worn away and become rounded. The mechanical wearing of rock and mineral grains by friction and impact is called abrasion (Figure 12-7). Note that liquid water itself is not abrasive—it is the collisions among rock, sand, and silt in the water that does the rounding.

FIGURE 12-7
Go online to view cobbles formed by abrasion on the Pacific coast of Tasmania, Australia.

Wind-blown sand is a powerful erosional agent (Figure 12-8). Wind hurls sand and other small particles against outcrops of bedrock, sandblasting unusual shapes. Glaciers also are powerful agents of abrasion as they drag rock, sand, and silt across bedrock (Figure 12-9).

checkpoint What are some common agents of abrasion?

FIGURE 12-8
(A) Grains of sand are lifted into the air and carried in the wind during a sandstorm in Altyn Emel National Park, Kazakhstan. (B) Blowing sand sculpted and polished these unusual outcrops of weathered volcanic rock in the Libyan Desert.

FIGURE 12-9
ON ASSIGNMENT Rock frozen underneath a receding glacier carved grooves and striations into the bedrock. The bedrock was then polished by sand and small pebbles carried in the moving ice. National Geographic photographer Jason Edwards took this image while on assignment in Antarctica.

MINILAB Weathering by Plant Roots

Materials

safety goggles
protective gloves
lab apron
plaster of paris
spackling compound
sand
water
grass seeds
corn seeds (soaked overnight)
pea seeds (soaked overnight)
bean seeds
seed-starting potting soil
small paper cups
paper plate
plastic utensils (spoons, knives)

Procedure

1. Place four small paper cups on a paper plate. With a spoon, fill each cup about halfway with potting soil.
2. Plant four bean seeds in one paper cup. Lightly cover the seeds with a small amount of soil. Use a marker to label this cup "bean." Repeat with the other paper cups, planting peas, grass, and corn, each in its own cup. Water the seeds lightly.
3. Use plastic utensils to cover your seeds with about one-half inch of either plaster of paris, sand, or spackling compound. Leave a small gap in covering to allow watering.
4. Let your seeds sit for seven to ten days. Keep the soil moist by adding water carefully.
5. Examine the surface of each container. Peel off the paper cup and make additional observations.
6. Share and observe the results of other groups.

Results and Analysis

1. **Observe** Describe what you observed after seven to ten days.
2. **Compare** How did the results of other groups compare to yours?

Critical Thinking

3. **Evaluate** How did the investigation model weathering?
4. **Connect** What real examples of physical weathering from plant roots have you seen?

Organic Activity

If soil collects in a crack in bedrock, a seed may fall there and sprout. The roots work their way into the crack, expand, and may eventually widen the crack as they grow (Figure 12-10). City dwellers often see the results of this type of organic activity when tree roots raise and crack concrete sidewalks.

🌐 FIGURE 12-10
Go online to view a crack in granite widened by the roots of a tree as it grew.

Burrowing animals such as prairie dogs, moles, and badgers can indirectly contribute to weathering. As they dig, they move rock, often bringing pieces of rock to the surface. Here the rock is more exposed to the conditions that cause physical weathering as well as chemical weathering.

checkpoint Give an example of organic activity that weathers rocks.

12.1 ASSESSMENT

1. **Compare and Contrast** Describe the similarities and differences between thermal expansion-contraction and frost wedging.
2. **Explain** You observe a large bedrock surface on the ground. It appears to have a number of parallel deep scratches in the surface. What likely weathered the rock?
3. **Infer** As you walk in a city, you notice a cracked and buckled sidewalk. Next to the sidewalk is an old tree stump. Infer what caused the sidewalk to buckle.

Critical Thinking

4. **Synthesize** A large, recently exposed outcrop of rock has undergone pressure-release fracturing and is crisscrossed with small cracks. The rock is located in a natural area in a temperate climate with cool winters and hot summers. Identify which types of mechanical weathering will likely weather the rock. Explain your reasoning.

12.2 CHEMICAL WEATHERING

Core Ideas and Skills
- Describe the processes of dissolution, hydrolysis, and oxidation.
- Describe how chemical and mechanical weathering can act together.
- Model chemical weathering and apply observations to a real-world situation.

KEY TERMS

dissolution	oxidation
pH	exfoliation
hydrolysis	

Generally, rock does not change much over a human lifetime. Over geologic time, however, air and water chemically alter rocks near Earth's surface. The most important processes of chemical weathering are dissolution, hydrolysis, and oxidation. Water, acids and bases, and oxygen in the air or dissolved in water cause these processes to decompose rocks. Two additional processes—salt cracking and hydrolysis-expansion—result from combinations of mechanical and chemical processes.

Dissolution

We are all familiar with the fact that some minerals dissolve readily in water and others do not. If you put a crystal of halite (rock salt, or table salt) in water, the crystal will rapidly dissolve to form a solution. The process is called **dissolution**. Halite dissolves so rapidly and completely in water that the mineral is rare in natural, moist environments. On the other hand, if you drop a crystal of quartz into pure water, only a tiny amount will dissolve and nearly all of the crystal will remain intact.

To understand how water dissolves a mineral, think of an atom on the surface of a crystal. It is held in place because it is attracted to the other atoms in the crystal by electrical forces called chemical bonds. At the same time, electrical attractions to the outside environment are pulling the atom away from the crystal. The result is like a tug-of-war. If the bonds between the atom and the crystal are stronger than the attraction of the atom to its outside environment, the crystal remains intact. If outside attractions are stronger, they pull the atom away from the crystal and the mineral dissolves.

Rocks and minerals dissolve more rapidly when water is either acidic or basic. An acidic solution contains a high concentration of hydrogen ions (H^+). A basic solution contains a high concentration of hydroxide ions (OH^-). To express the concentration of ions conveniently, a logarithmic scale called **pH** was introduced (see Figure 16-5). Solutions with a pH of 7, such as pure water, are neutral. Acids have a pH less than 7 and bases have a pH greater than 7. Acids and bases dissolve most minerals more effectively than pure water because they provide more electrically charged hydrogen and hydroxide ions to pull atoms out of crystals. For example, limestone is made of the mineral calcite ($CaCO_3$). Calcite barely dissolves in pure water but is quite soluble in acid. If you place a drop of strong acid on limestone, bubbles of carbon dioxide gas instantly form as the calcite dissolves.

Water found in nature is never pure. Atmospheric carbon dioxide dissolves in raindrops and reacts to form a weak acid called carbonic acid. As a result, even the purest rainwater falling in the Arctic or on remote mountains is slightly acidic. This acidity can be intensified when rainwater soaks down through leaf litter on the ground. Surface water or groundwater that is acidic or basic dissolves ions from soil and bedrock and carries the dissolved material away. Groundwater also dissolves rock and can produce spectacular caverns in limestone.

Air pollution can make rain more acidic. When water vapor condenses to liquid, it requires a liquid or solid surface to do so. In the atmosphere, this surface is presented by tiny solid or liquid particles called cloud condensation nuclei. Cloud condensation nuclei are tiny particles of a liquid or solid that typically are less than 1 micrometer (1 millionth of a meter) in diameter. Air pollution creates condensation nuclei that are highly acidic. For example, the burning of coal emits tiny sulfur-rich particles that become cloud condensation nuclei and produce sulfuric acid when combined with water. More sulfur-rich particles are incorporated into the raindrops as they grow. By the time the raindrops are big enough to fall to Earth's surface, they have become acid rain. Acid rain can dissolve limestone on the facades of buildings and public works of art (Figure 12-11). Acid rain is discussed in detail in Chapter 18.

FIGURE 12-11
Go online to view an example of how acid rain has dissolved a medieval sculpture made of limestone.

checkpoint What is dissolution?

Hydrolysis

In **hydrolysis**, water reacts with one mineral to form a new mineral. Most common minerals weather by hydrolysis. For example, feldspar, the most abundant mineral in Earth's crust, weathers by hydrolysis to form clay.

In contrast to feldspar, quartz resists hydrolysis and other forms of chemical weathering because it dissolves extremely slowly and only to a small extent. It is also hard (H = 7 on the Mohs' hardness scale) and has no cleavage, so it resists abrasion. When granite weathers, the feldspar and other minerals decompose to form clay, but the unaltered quartz grains fall free from the rock. Hydrolysis has so deeply weathered some granites that quartz grains can be pried out with a fingernail at depths of several meters (Figure 12-12). The rock looks like granite but has the consistency of sand.

Because quartz is so resistant to both mechanical and chemical weathering, it is a main component in much of Earth's sand. After quartz is freed from its source rock, it is transported by water, wind, glacial ice, and gravity as sand-sized particles (Figure 12-3). Many of the quartz particles are transported to the shoreline where they are concentrated on beaches and deltas or are carried offshore by storm currents. Eventually, the sand lithifies to form sandstone.

checkpoint Describe the process of hydrolysis.

Oxidation

Many elements react with atmospheric oxygen, O_2. Iron rusts when it reacts with water and oxygen. Rusting is an example of a more general weathering process called **oxidation**. Iron is abundant in many minerals. If the iron in such a mineral oxidizes, the mineral decomposes.

Many valuable metals such as iron, copper, lead, and zinc occur as sulfide minerals in ore deposits. When these minerals oxidize during weathering, the sulfur reacts to form sulfuric acid, a strong acid. The sulfuric acid washes into streams and groundwater, where it may harm aquatic organisms. In addition, the sulfuric acid can dissolve metals from the sulfide minerals, forming a metal-rich solution that can be toxic. These reactions are accelerated when ore is dug up and exposed by mining. Acid-mine drainage is a common problem at such sites. For more on acid mines, see Chapter 3.

checkpoint What problems can occur when sulfide minerals oxidize?

FIGURE 12-12
Chemical weathering is breaking this coarsely crystalline granite (top) down to individual crystals of quartz and feldspar (bottom). The coin is 3 centimeters in diameter. The granite is found in tropical northern Queensland, Australia.

Chemical and Mechanical Weathering Acting Together

Chemical and mechanical weathering can work together, often on the same rock and at the same time. After mechanical processes fracture a rock, water and air seep into the cracks to initiate chemical weathering.

In environments where groundwater is salty, saltwater seeps through pores and cracks in bedrock. When the water evaporates, the dissolved salts crystallize. The growing crystals exert tremendous forces that loosen mineral grains and widen cracks in a process called salt cracking. Salt chemically precipitates in rock, and the growing salt crystals mechanically loosen mineral grains and break the rock apart.

Many sea cliffs show pits and depressions caused by salt cracking. This happens because spray from

FIGURE 12-13
Crystallizing salt loosened sand grains to form these depressions in sandstone. The pit at bottom right is about 15 centimeters wide.

the breaking waves brings the salt to the rock. Salt cracking is also common in deserts, where surface water and groundwater commonly contain dissolved salts (Figure 12-13).

Granite typically fractures by **exfoliation**. In this process, plates of weathered material split away like the layers of an onion (Figure 12-14). The plates may be only 10 or 20 centimeters thick near the surface, but they thicken with depth. Exfoliation fractures are usually absent below a depth of 50 to 100 meters, so exfoliation seems to be a result of exposure of the granite at Earth's surface.

Exfoliation is frequently explained as a form of pressure-release fracturing, but many geologists have argued that hydrolysis-expansion may be the main cause of exfoliation. During hydrolysis, feldspars and other silicate minerals react with water to form clay. As a result of the addition of

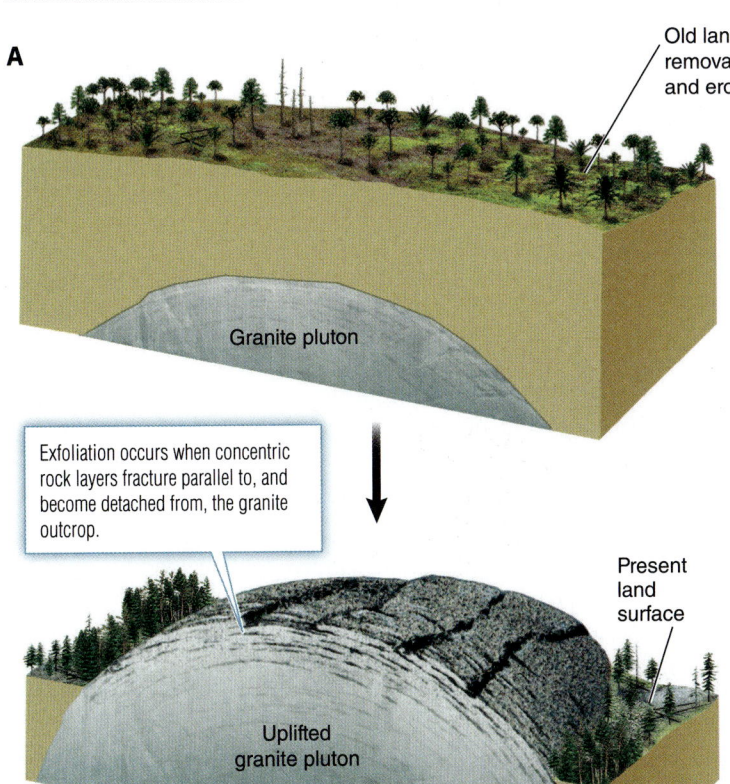

FIGURE 12-14
(A) Exfoliation occurs when concentric rock layers fracture and become detached from a granite outcrop. (B) Exfoliation has fractured this granite in Pinkham Notch, New Hampshire.

LESSON 12.2 **367**

MINILAB Weathering of Iron

Materials

2 beakers
vinegar
steel wool pad (size 0000)
thermometer
clear plastic wrap
protective gloves
safety goggles

Procedure

1. Place the steel wool in one beaker and cover with vinegar. Soak the pad in the vinegar for 2 minutes to remove the protective coating that prevents rusting. Remove the pad and tap or gently squeeze the extra vinegar out. Notice the color of the steel wool.

2. Place the steel wool pad in the second beaker. Place the thermometer into the steel wool. Cover the beaker with plastic wrap, making sure you can read the thermometer.

3. Record the temperature in the beaker. Record the temperature again after 5 minutes and again after 10 minutes.

4. Observe the color of the steel wool and note any changes that have occurred.

Results and Analysis

1. **Observe** Describe any changes you observed to the steel wool and the temperature in the beaker.

2. **Explain** Provide an explanation for the changes you observed.

Critical Thinking

3. **Predict** What would you expect to happen if you left the steel wool in the covered beaker for another day?

4. **Apply** How could you apply what you learned to a real-world situation?

---

water, clays have a greater volume than the original minerals. So, a chemical reaction (hydrolysis) forms clay, and the mechanical expansion that occurs as the clay forms may cause exfoliation. This explanation agrees with the observation that exfoliation concentrates near Earth's surface because water and chemical weathering are most abundant close to the surface.

The combined effects of mechanical and chemical weathering commonly cause coarsely crystalline igneous rocks to form spheroidal shapes (Figure 12-15). Igneous rocks exposed at the surface typically undergo pressure-release fracturing. This initially produces rocks with flat faces and sharp corners and edges. As the rocks are weathered, the corners and edges are more exposed and susceptible to hydrolysis. The corners and edges are slowly rounded. Spheroidal weathering takes place as a rock is weathered in place, without moving.

⊕ **FIGURE 12-15**
Go online to view spheroidally weathered boulders of coarsely crystalline igneous rock that cover the ground in Gettysburg National Military Park, Pennsylvania.

checkpoint What are two possible causes of exfoliation?

12.2 ASSESSMENT

1. **Explain** Why does the mineral calcite barely dissolve in water but dissolves readily in a strong acid?

2. **Explain** How does hydrolysis occur? Give an example of a mineral that undergoes hydrolysis.

3. **Explain** Why is much of Earth's beach sand composed of quartz?

4. **Explain** How can salt cause the weathering of rock in places where surface water and groundwater commonly contain dissolved salts?

Critical Thinking

5. **Synthesize** A large, recently exposed outcrop of rock has undergone pressure-release fracturing and is crisscrossed with small cracks. The rock is located in a natural area in a temperate climate with cool winters and hot summers. Identify which types of chemical weathering will likely weather the rock. Explain your reasoning.

12.3 STREAM EROSION AND DEPOSITION

Core Ideas and Skills
- Describe the factors that affect the velocity of a stream.
- Analyze data to identify and describe patterns in seasonal changes and stream discharge.
- Explain how the path of stream channels changes over time.
- Explain the processes of stream erosion, sediment transport, and deposition.
- Evaluate the effects of floods and flood control practices.

KEY TERMS

stream	base level
tributary	meander
gradient	oxbow lake
discharge	drainage basin
capacity	alluvial fan
dissolved load	delta
suspended load	bed load
bed load	floodplain
downcutting	artificial levee

Stream Flow and Velocity

Water is a primary agent of weathering and erosion—as a corrosive chemical, a physical force when it freezes, and a medium that transports rock and sediment.

Earth scientists use the word **stream** for all water flowing in a channel, regardless of the stream's size. The word *river* is commonly used for any large stream fed by smaller ones, called **tributaries**. In temperate climates, most streams run year-round, even during times of drought, because they are fed by groundwater that seeps into the streambed. Such streams are called perennial streams. In contrast, most streams in arid environments are intermittent streams that flow only when it rains.

Stream velocity is the speed of water in a stream. Three factors control stream velocity: gradient, discharge, and channel characteristics.

Gradient The steepness, or vertical drop over a specific distance, of a stream is its **gradient**. A tumbling mountain stream near the continental divide may drop 50 meters or more per kilometer. A river in the upper Great Plains, such as the Yellowstone River in eastern Montana, has a much lower gradient of about 0.5 meters per kilometer. Lower yet is the gradient of the lower Mississippi River, which has a gradient of 0.01 meters (one centimeter) per kilometer.

Discharge The volume of water flowing past a certain point along a stream over a given period of time is that stream's **discharge** at that point. It usually is expressed as cubic meters per second (m^3/sec) or cubic feet per second (cfs). The stream velocity increases when its discharge increases. As a result, a stream flows faster during a flood, even though its gradient is unchanged.

The largest river in the world by volume is the Amazon, with an average discharge of 175,000 m^3/sec (Figure 12-16). That is enough water to fill 70 Olympic-sized swimming pools every second or

FIGURE 12-16
The Amazon River is the largest river in the world in terms of discharge.

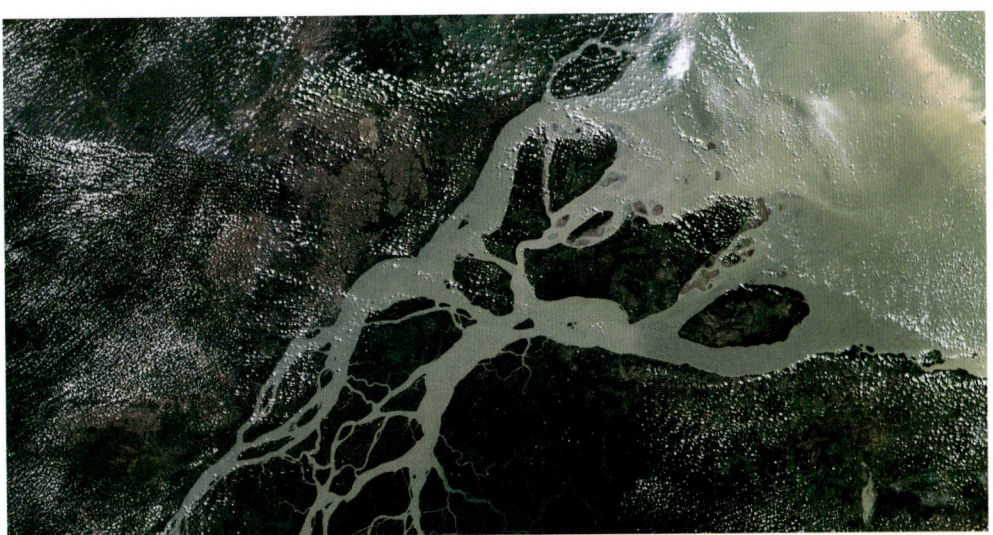

cover all of Manhattan with 18 centimeters (about seven inches) of water every minute. In contrast, the Mississippi, the largest river in North America, has an average discharge of about 16,800 m³/sec. That is enough to fill about 6.7 Olympic pools each second. In one minute of discharge, the Mississippi River would cover all of Manhattan with about 1.7 cm of water.

A stream's discharge can change dramatically from month to month or even during a single day. For example, the Blackfoot River, a mountain stream in western Montana, has a discharge of 30 to 35 m³/sec during early summer, when mountain snow is melting rapidly. During the dry season in late summer, the discharge drops to about 5 m³/sec (Figure 12-17). Intermittent streams may be dry most of the time but become the site of a flash flood during a thunderstorm.

Channel Characteristics Channel characteristics include the shape and roughness of a stream channel. The floor of the channel is called the *bed*, and the sides of the channel are the *banks*. Friction between flowing water and the stream channel slows stream velocity. Consequently, water flows more slowly near the banks than near the center of a stream. If you paddle a canoe down a straight stream channel, you move faster when you stay away from the banks. The amount of friction depends on the roughness and shape of the channel. Boulders on the banks or in the streambed increase friction and slow a stream down, whereas the water flows more rapidly if the bed and banks are smooth. Likewise, a stream that is deep and narrow will flow faster than a broad, shallow stream with the same discharge, and a stream that is straight flows faster than a stream with lots of curves.

checkpoint What three factors control stream velocity?

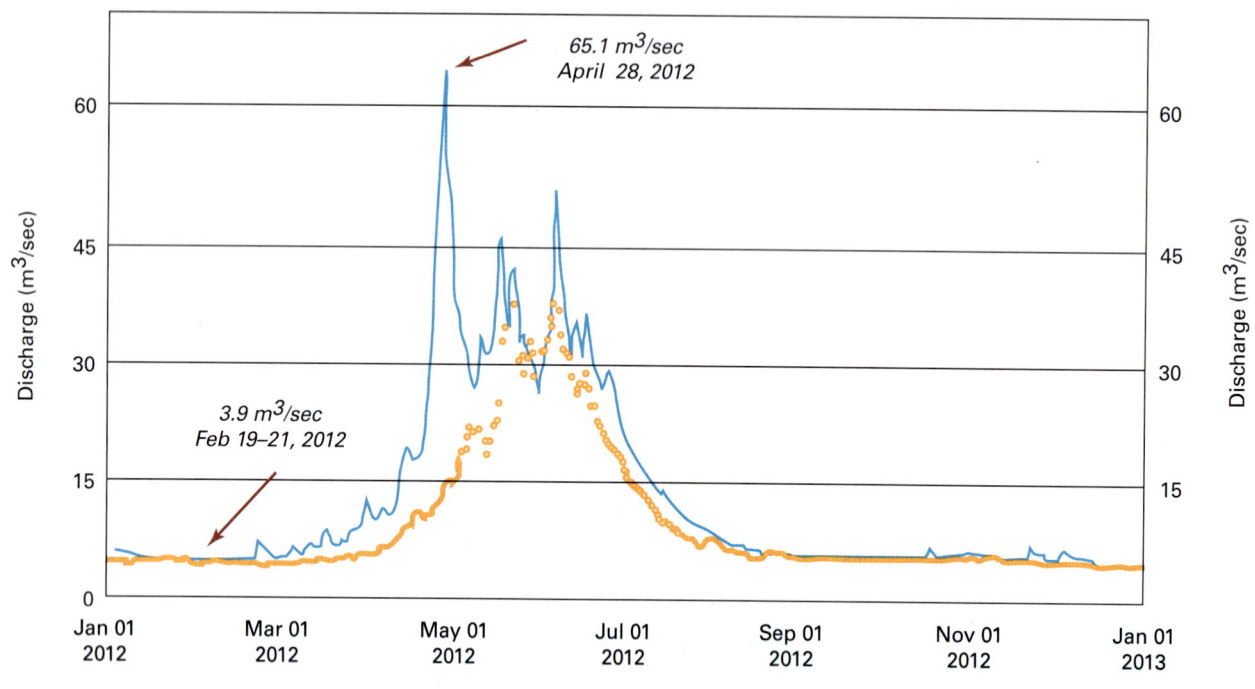

FIGURE 12-17
The 2012 hydrograph for the Blackfoot River in Montana shows that the discharge varied from a low of 3.9 m³/sec (135 cubic feet per second) in February to a high of 65.1 m³/sec (2,300 cubic feet/second) during a period of rapid snowmelt in late April. These data come from a permanent river gauge that records discharge measurements every 15 minutes and is operated by the USGS.

DATA ANALYSIS Blackfoot River Discharge Over Time

The Blackfoot River is a mountain stream in western Montana. In early summer, Montana's mountain snow melts quickly, and the area experiences a dry season in late summer. Figure 12-18 is hydrograph for the Blackfoot River from January 1, 2012, through the end of 2018. These data come from a permanent river gauge operated by the USGS that records discharge measurements every 15 minutes. Figure 12-19 is a graph of average temperature for a nearby city for the same time frame. Use the graphs to answer the questions that follow.

1. **Summarize** Describe the relationship between season and stream discharge and the relationship between season and temperature.
2. **Infer** What seasonal changes do you think affect the amount of stream discharge?

Computational Thinking

3. **Generalize Patterns** Using both Figures 12-18 and 12-19, describe the correlation between temperature and stream discharge.
4. **Analyze Patterns** Compare the graphs for June 2017 and June 2018. What do you observe? What can you conclude from your observations?

FIGURE 12-18
Blackfoot River Discharge, 2012–2018

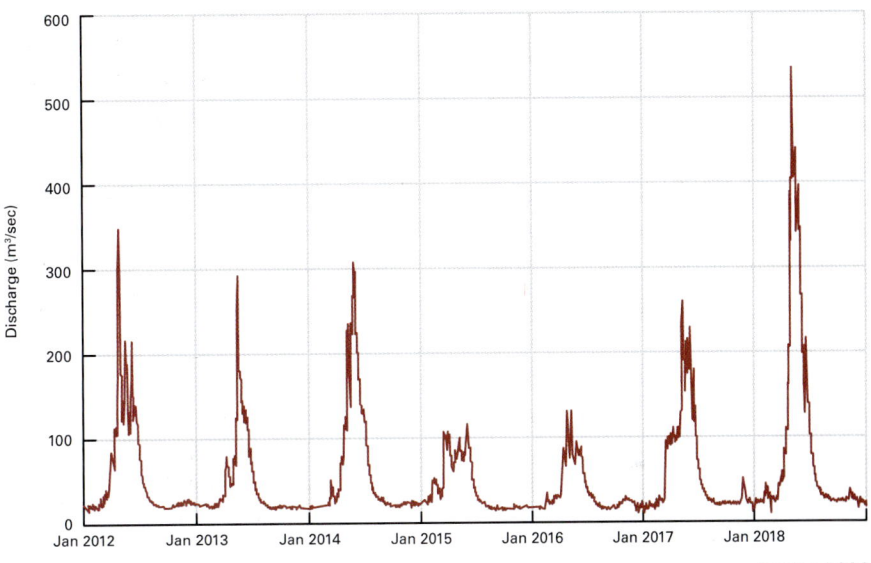

Source: USGS

FIGURE 12-19
Average Temperature, Missoula, Montana, 2012–2018

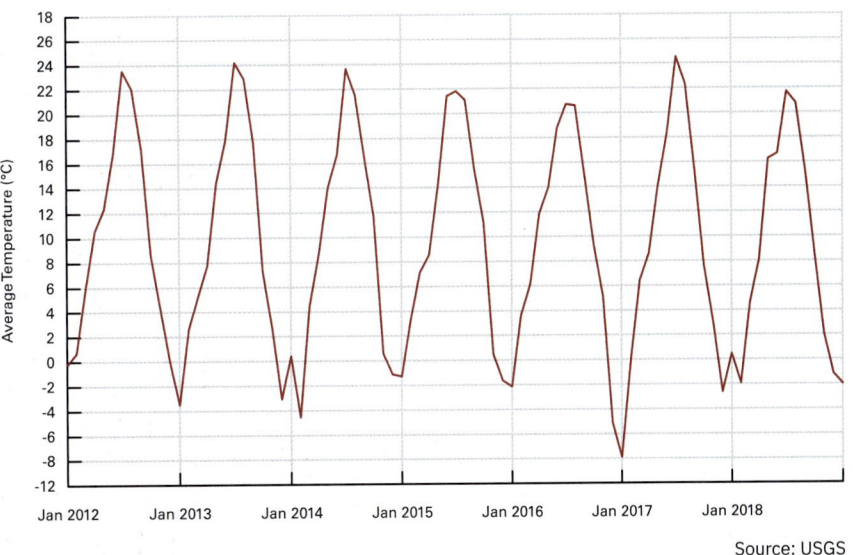

Source: USGS

Stream Erosion and Sediment Transport

Streams shape Earth's surface by eroding soil and bedrock. The flowing water carries the eroded sediment downslope. A stream may deposit some sediment in the valley through which it flows, while it carries the remainder to a lake or to the sea, where the sediment accumulates (Figure 12-20).

Stream erosion and sediment transport depend on a stream's energy. A rapidly flowing stream has more energy to erode and transport sediment than a slow stream of the same size. A stream's energy determines the size of the largest particle it can carry. A fast-flowing stream can transport cobbles and even boulders in addition to small particles. A slow stream carries only silt and clay.

The **capacity** of a stream is the total amount of sediment it can carry and is proportional to both stream velocity and discharge. Thus, a large, fast stream has a greater capacity than a small, slow one. Because the ability of a stream to carry sediment is proportional to both stream velocity and discharge, most sediment transport occurs during the few days each year when the stream is in flood. Relatively little sediment transport occurs during the remainder of the year. So flooding streams are usually muddy and dark, but the same stream may be clear during low water.

After a stream weathers soil or bedrock, it transports the products of weathering downstream in three ways: dissolved load, suspended load, and bed load.

A stream's **dissolved load** refers to all of the ions dissolved in the water. A stream's ability to carry dissolved ions depends mostly on its discharge, its pH, and the geology over which the stream flows. A stream flowing over volcanic bedrock will have a higher dissolved load than the same stream flowing over quartzite bedrock, because the volcanic rock is more chemically reactive than the quartzite. Although dissolved ions are invisible, they make up more than half the total load carried by some rivers. More commonly, however, dissolved ions make up less than 20 percent of the total load.

If you place soil with equal amounts of sand, silt, and clay in a jar of water and shake it up, the sand grains settle quickly. But the smaller silt and clay particles remain suspended in the water as **suspended load**, giving it a cloudy appearance. Clay and silt are small enough that even the slight turbulence of a slow stream keeps them in suspension. Only a rapidly flowing stream can carry sand in suspension.

During a flood, stream energy is at its highest. The rushing water can roll boulders and cobbles along the bottom as **bed load**, which is all the

FIGURE 12-20
The Yellow River in China carries more suspended sediments than any other river in the world. Its name comes from the sediment that gives the river its yellow color.

sediment that is moved, rolled, or bounced across the streambed of a channel by high-energy flow. Sand also can move as part of the bed load by rolling or bouncing along the bottom while being carried downstream by the current.

checkpoint What two factors determine the capacity of a stream?

Downcutting and Base Level

A stream can erode both downward into its bed and laterally against its banks. Downward erosion is called **downcutting** (Figure 12-21). The **base level** of a stream is the deepest level to which it can erode. Most streams cannot erode below sea level, which is regarded as the ultimate base level.

In addition to ultimate base level, a stream may have a number of local, or temporary, base levels. For example, a stream stops flowing where it enters a lake. It then stops eroding its channel because it has reached a temporary base level (Figure 12-22). A layer of rock that resists erosion may also establish a temporary local base level because it flattens the stream gradient, causing the stream to slow down and erosion to decrease. The top of a waterfall can be a temporary base level established by resistant rock. For example, Niagara Falls is formed by a resistant layer of dolomite overlying softer mudstone (Figure 12-23).

FIGURE 12-21
The Blyde River in South Africa has eroded deeply into the bedrock of the region. Here at Bourke's Luck, the river has cut downward through sandstone to form these steep canyon walls and spectacular potholes.

FIGURE 12-22
Go online to view an example of a steep mountain stream flowing down from a glacier into a temporary base in the Canadian Rockies.

FIGURE 12-23
Niagara Falls occurs at a temporary base level formed by a resistant layer of dolomite at the top of the falls.

The dolomite forms a temporary base level for the stream by resisting erosion. But the turbulent falling water undermines the soft mudstone at the base of the falls. This causes the overhanging dolomite cap to collapse and the falls to migrate upstream. In this way, Niagara Falls has retreated 11 kilometers upstream since its formation about 9,000 years ago. It is slowly eliminating the temporary base level formed by the dolomite.

A stream such as the one in Figure 12-24 erodes rapidly in the steep places where its energy is high and deposits sediment in the low-gradient stretches where it flows more slowly. Over time, erosion and deposition smooth out the irregularities in the gradient. The resulting graded stream has a smooth, concave-upward profile. Once a stream becomes graded, there is no net erosion or deposition and the stream profile no longer

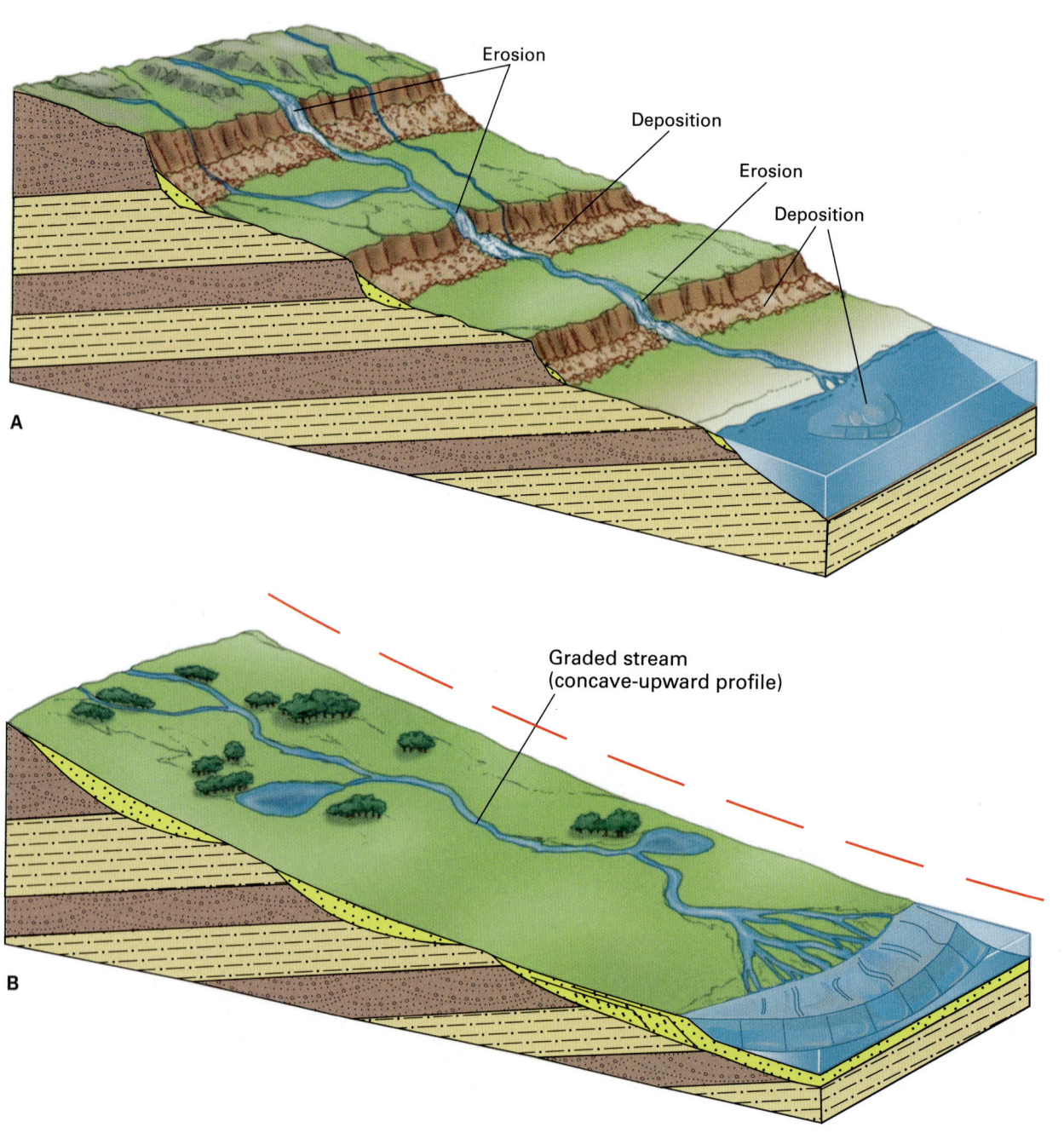

FIGURE 12-24
(A) An upgraded stream has many temporary base levels. (B) With time, the stream smooths out the irregularities and develops a graded profile, illustrated here by the red line.

374 CHAPTER 12 WEATHERING, EROSION, AND DEPOSITION

changes. An idealized graded stream such as this one does not actually exist in nature, but many streams come close.

checkpoint What is the base level of a stream?

Path of a Stream Channel

A steep mountain stream usually downcuts rapidly compared with the rate of erosion to the sides. As a result, it cuts a relatively straight channel with a steep-sided, V-shaped valley (Figure 12-25). The stream maintains its relatively straight path because it flows with enough energy to transport any material that slumps into its channel.

In contrast, a low-gradient stream is less able to erode downward into its bed. Rather, much of the stream energy is directed against the banks, causing lateral erosion. Lateral erosion occurs where a low-gradient stream forms a series of bends, called **meanders** (Figure 12-26). As a stream flows into a

FIGURE 12-25
A steep mountain stream eroded this V-shaped valley into soft mudstone in the Canadian Rockies.

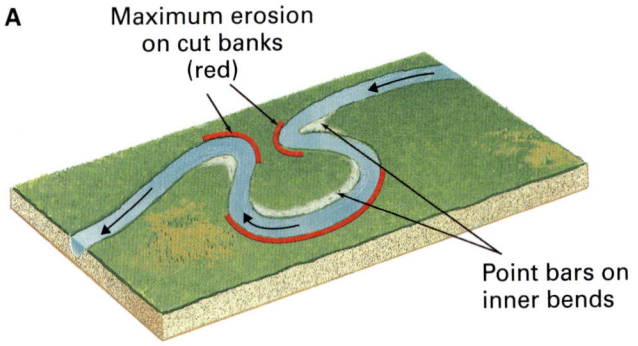

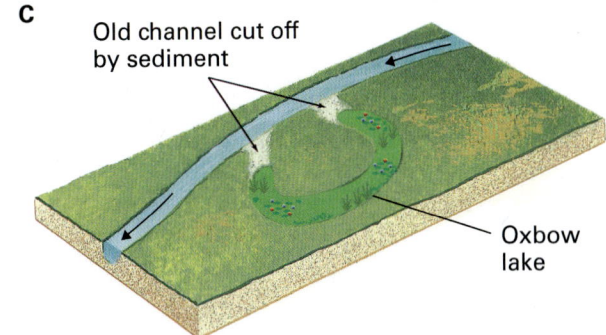

FIGURE 12-26
(A) A stream erodes the outsides of meanders (red) and deposits sand and gravel on the inside bends to form point bars. (B) Meanders and point bars form the channel of the Bitterroot River in Montana. (C) Over time, a stream may erode through the neck of a meander to form an oxbow lake that is cut off from the river and rapidly colonized by aquatic plants. (D) This oxbow lake formed through the cutoff of a meander of the upper Flathead River in Montana.

LESSON 12.3 **375**

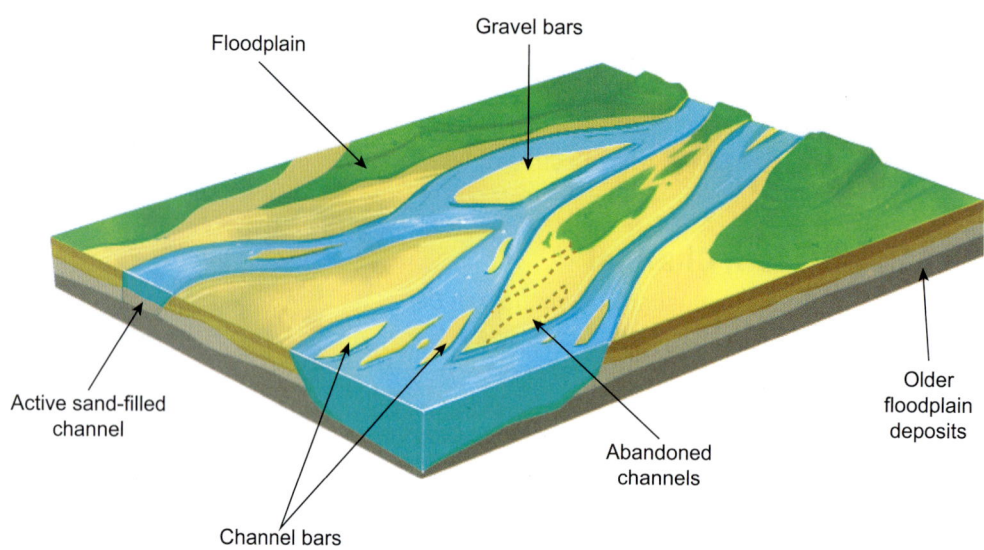

FIGURE 12-27 Braided streams form when too much sediment is available for the stream to carry. As a result, the stream piles the sediment up into bars and flows around them in multiple shallow, broad channels.

FIGURE 12-28 Joe Creek in Canada's Yukon Territory is heavily braided because melting glaciers provide more sediment than the stream can carry.

meander bend, the water pushes against the outside bank of the meander, much like your body leans against the inside of your car while going around a sharp curve at high speed. As water is pushed against the outside bank, it causes the bank to erode. The water also causes the deepest part of the channel to form beside the outside bank.

The water piling up against the outside bank causes the flow around the inner bank to be both shallower and slower. This promotes the deposition of sediment there, forming a point bar. Through time, the point bar builds farther and farther out into the river as the outside bank of the meander is eroded back. In this manner, the meander slowly migrates laterally toward its outside bank. Occasionally, such lateral migration of a stream causes it to cut across the neck of a meander, forming a short-cut that isolates the meander loop. The abandoned meander loop remains full of water and becomes an **oxbow lake**.

In contrast to meandering streams, braided streams flow in many shallow, interconnecting channels (Figures 12-27 and 12-28). A braided stream forms where more sediment is available than the stream can carry. The excess sediment accumulates in the channel, forming a series of sediment bars. During low water, the stream flows simultaneously in multiple channels that pass around the bars. During floods, the bars may be completely submerged, and the higher discharge may cause the bars to migrate downstream a short distance, while changing shape in the process. As a result, braided streams often look different from year to year.

checkpoint How does an oxbow lake form?

Drainage Basins and Drainage Divides

Only a dozen or so major rivers flow into the sea along the coastlines of the United States (Figure 12-29). Each is fed by a number of tributaries, which are in turn fed by smaller tributaries. The region drained by a single river is called a **drainage basin**. At the edge of each drainage basin is a drainage divide, an imaginary line on the ground separating adjacent drainage basins. The most famous drainage divide in North America is the Continental Divide. In the United States, the Continental Divide tends to follow a line that runs along peaks of the Rocky Mountains. Water runoff east of the Continental Divide drains into stream tributaries that ultimately lead to rivers flowing into the Atlantic Ocean. Runoff west of the Continental Divide eventually reaches the Pacific Ocean. Other important North American drainage divides separate the Columbia and Colorado River drainage basins west of the Continental Divide and the Rio Grande from the Mississippi River drainage basins east of the Continental Divide. Collectively, these four drainage basins cover three-fourths of the 48 contiguous states.

checkpoint What is the Continental Divide?

Stream Deposition

According to a model popular in the first half of the 20th century, streams erode Earth's surface and create landforms in an orderly sequence (Figure 12-30). At first, they cut steep, V-shaped valleys into mountains. Over time, the streams erode the mountains away and widen the valleys into broad plains. Eventually, the entire landscape becomes a large, featureless plain. However, if this were the only mechanism affecting Earth's surface during its 4.6-billion-year history, all landforms would have eroded to a flat plain. Why do mountains, valleys, and high plateaus still exist?

The model tells only half the story. Streams do continuously erode the landscape, wearing down mountains and widening floodplains. But at the

FIGURE 12-29
Most of the surface water in the United States flows to the sea from approximately a dozen major rivers. In this figure, the width of each river is proportional to its discharge.

Source: Cengage

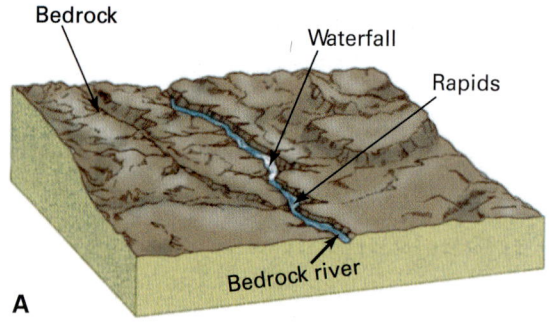
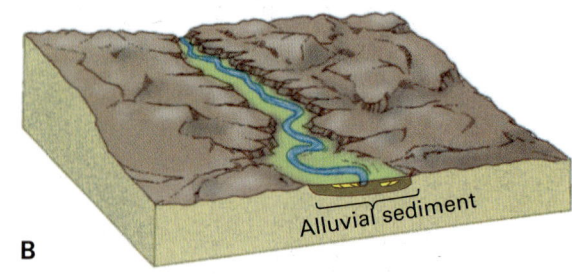
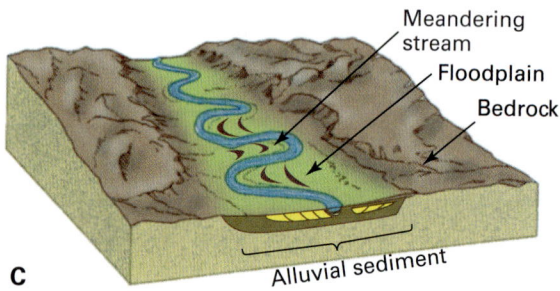
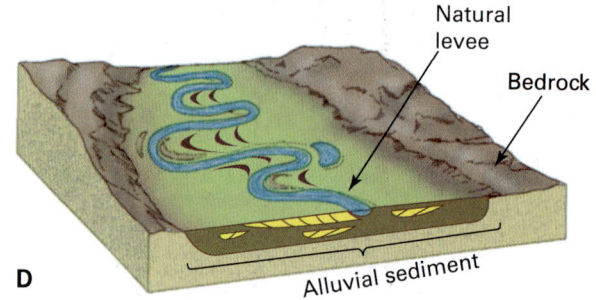

FIGURE 12-30
If tectonic activity does not uplift the land, over time streams erode the mountains and widen the valleys into broad floodplains.

same time, tectonic activity may uplift the land and interrupt this simple, idealized sequence. In this way, Earth's hydrosphere and atmosphere work together with tectonic processes in the geosphere to create landforms. Geomorphology, from the Greek words for *earth forms,* is a branch of geology that deals with the evolution of landforms through time.

Alluvial Fans and Deltas When streams slow down, they deposit their bed load first. If water slows sufficiently, the suspended load may also settle to the bottom. Sediment deposited by moving water is called alluvium.

When alluvium is deposited, new landforms are created. For example, if a steep mountain stream flows onto a flat plain, its gradient and velocity decrease abruptly. As a result, the stream deposits most of its sediment. The accumulated sediment is called an **alluvial fan**, named because the shape of the deposit resembles a handheld fan (Figure 12-31). Alluvial fans are common in many arid and semiarid mountainous regions, typically where a fault has caused a mountain range to be uplifted relative to the adjacent valley.

A stream also slows abruptly where it enters the still water of a lake or an ocean. The sediment settles out to form a **delta**, a nearly flat-topped landform with a steep frontal face. The upper surface of the delta, called the delta top, extends beyond the shoreline to the more steeply sloping, sandy delta front. Farther offshore is the muddier prodelta (Figure 12-32).

Both deltas and alluvial fans are commonly fan-shaped and resemble the Greek letter *delta* (Δ). The fan shape is produced by distributary channels. Unlike tributary channels, which merge together in a downstream direction, distributary channels split apart in a downstream direction.

FIGURE 12-31
An alluvial fan in Death Valley formed where a steep mountain stream deposited most of its sediment as it entered the flat valley. A road runs across the lower part of the fan.

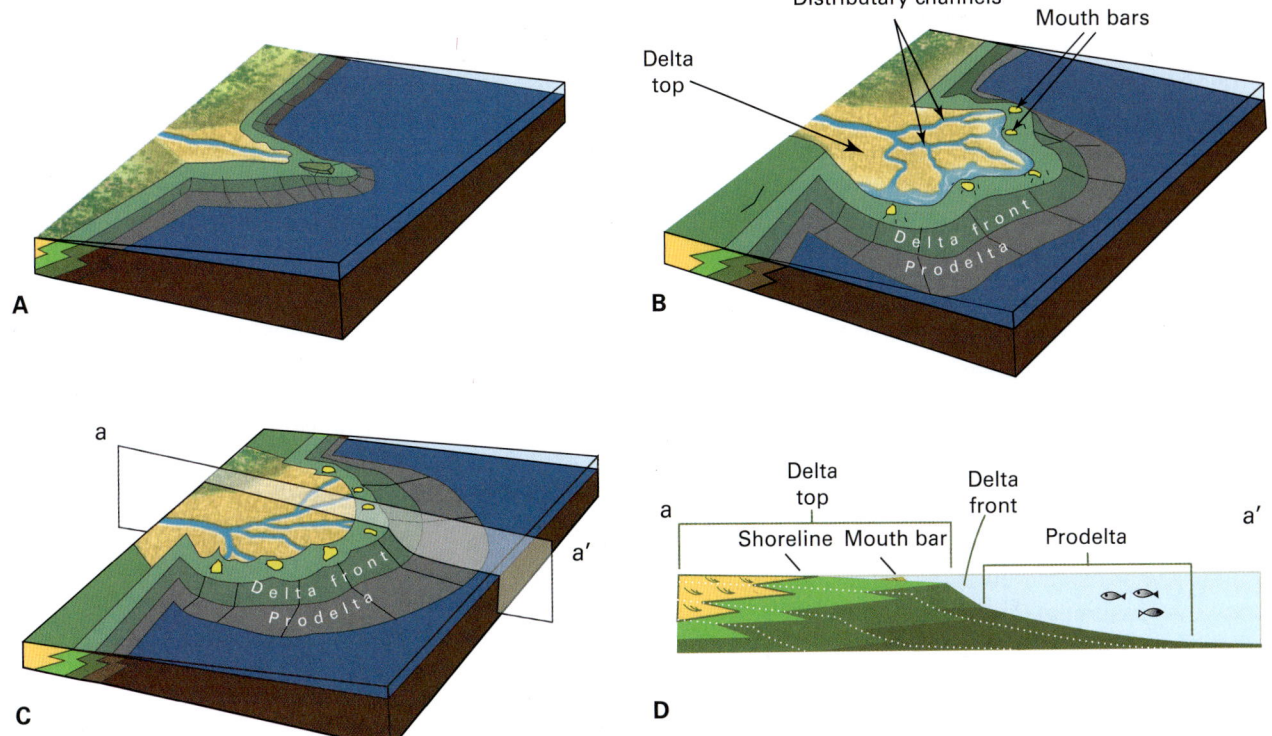

FIGURE 12-32
(A) Delta formation is initiated as a stream begins to deliver sediment to a standing water body. A mouth bar is deposited on the delta top as the ending of the stream channel. (B) Distributaries develop as sediment is spread across the delta. (C) A mature delta has numerous distributary channels. (D) This image is a cross section through a delta. The plane of cross section is shown in C.

Distributary channels on a delta carry sediment beyond the shoreline, where it accumulates in mouth bars that form where the distributary channels end (Figure 12-32). As a mouth bar grows in size, flow from the distributary channel must diverge to get around it. Eventually, the distributary channel is abandoned and the water and sediment flows down a new channel established on a different part of the delta. This process of channel movement usually takes place when the river is in flood.

As sediment is added to alluvial fans and deltas, they build outward. Most alluvial fans are relatively small, covering areas ranging from less than a square kilometer to a few square kilometers. In contrast, a large delta may cover thousands of square kilometers. Although sediment is deposited on the delta top by flooding rivers, much sediment also is deposited beyond the mouths of the distributary channels, on the delta front and prodelta.

Even though deltas cover only a small fraction of Earth's total land surface, they are environments that involve each of Earth's four spheres. The delta is composed of solid sediment, but it is a watery zone where branched distributaries meet open water and much groundwater is contained between the individual sediment grains in the delta. Because floods regularly deliver nutrient-rich sediment to deltas, they typically are fertile, and delta ecosystems are rich in natural plant and animal life.

checkpoint How are alluvial fans and deltas alike and different?

Floods and Flood Control

When rainfall is heavy or snowmelt is rapid, more water flows down a stream than the channel can hold, causing a flood. During a flood, the stream overflows onto its **floodplain**, the low-lying land adjacent to the stream. Massive floods occur somewhere in the world every year.

Deltas and floodplains have long been desirable places for humans to live. They provide natural access to transportation waterways, contain level and fertile soil for agriculture, and generally contain abundant wild game. However, flooding is a natural part of delta and floodplain formation, and all floodplains undergo flooding when enough discharge is supplied from the river. As a result, humans have developed several strategies for controlling floods so that the damage they cause can be minimized. As we will see, however, flood control strategies can cause other problems and, in some extreme cases, can make flood-related damage worse.

Flood Control Practices An **artificial levee** is a wall built along the banks of a stream to prevent rising water from spilling from the stream channel onto the floodplain. In the past 70 years, the U.S. Army Corps of Engineers has spent billions of dollars building 11,000 kilometers of levees along the banks of the Mississippi and its tributaries. Of course, levees can fail. In 2008, at least 20 levees along the Mississippi River failed. The result was widespread flooding and damage across parts of Iowa and Indiana (Figure 12-33).

FIGURE 12-33
Go online to view the results of a flood that ravaged an area in Iowa when the levees along the Mississippi failed in 2008.

Unfortunately, artificial levees create conditions that may increase both flood intensity and property damage. One factor is entirely human—the protection promised by levees encourages people to build in the floodplain. In the absence of a levee, people might decide to build on high ground that is safe from floods. But when levees are built, people are more likely to construct homes and businesses in harm's way.

Levees also can cause higher floods along nearby reaches of a river. By restricting the channel, artificial levees form a partial dam that raises the water level and increases the risk of flooding upstream (Figure 12-34).

In addition, levees prevent a stream from overflowing during small floods. Sediment that normally would be deposited on the floodplain is deposited within the channel, raising the level of the streambed (Figures 12-35). After several small floods, the entire stream may rise above its floodplain, contained only by the levees. This situation creates the potential for a truly disastrous flood. If the levee is breached during a large flood, the entire stream flows out of its channel and onto the

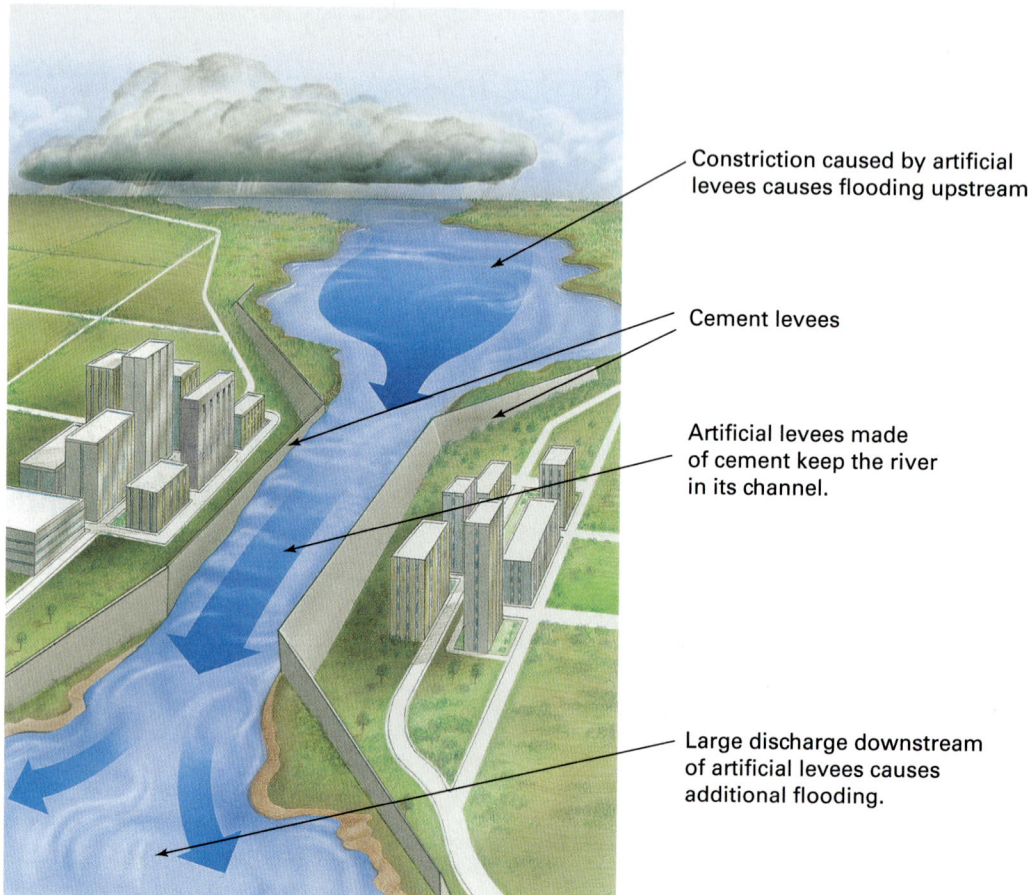

FIGURE 12-34
Artificial levees force a flooding river into a restricted channel. A partial dam is formed that raises the flood level upstream from the restriction.

Constriction caused by artificial levees causes flooding upstream.

Cement levees

Artificial levees made of cement keep the river in its channel.

Large discharge downstream of artificial levees causes additional flooding.

floodplain. As a result of artificial levees and channel sedimentation, parts of the Yellow River in China now lie 10 meters above the adjacent floodplain. As we can see, levees may solve flooding problems in the short term. But over the long term, they can cause even larger and more destructive floods.

An alternative approach to flood control is to abandon some flood-control projects and let the river spill out onto its floodplain. Of course, the question is what land should be allowed to flood. Every farmer and homeowner living on the floodplain wants to maintain the levees that protect

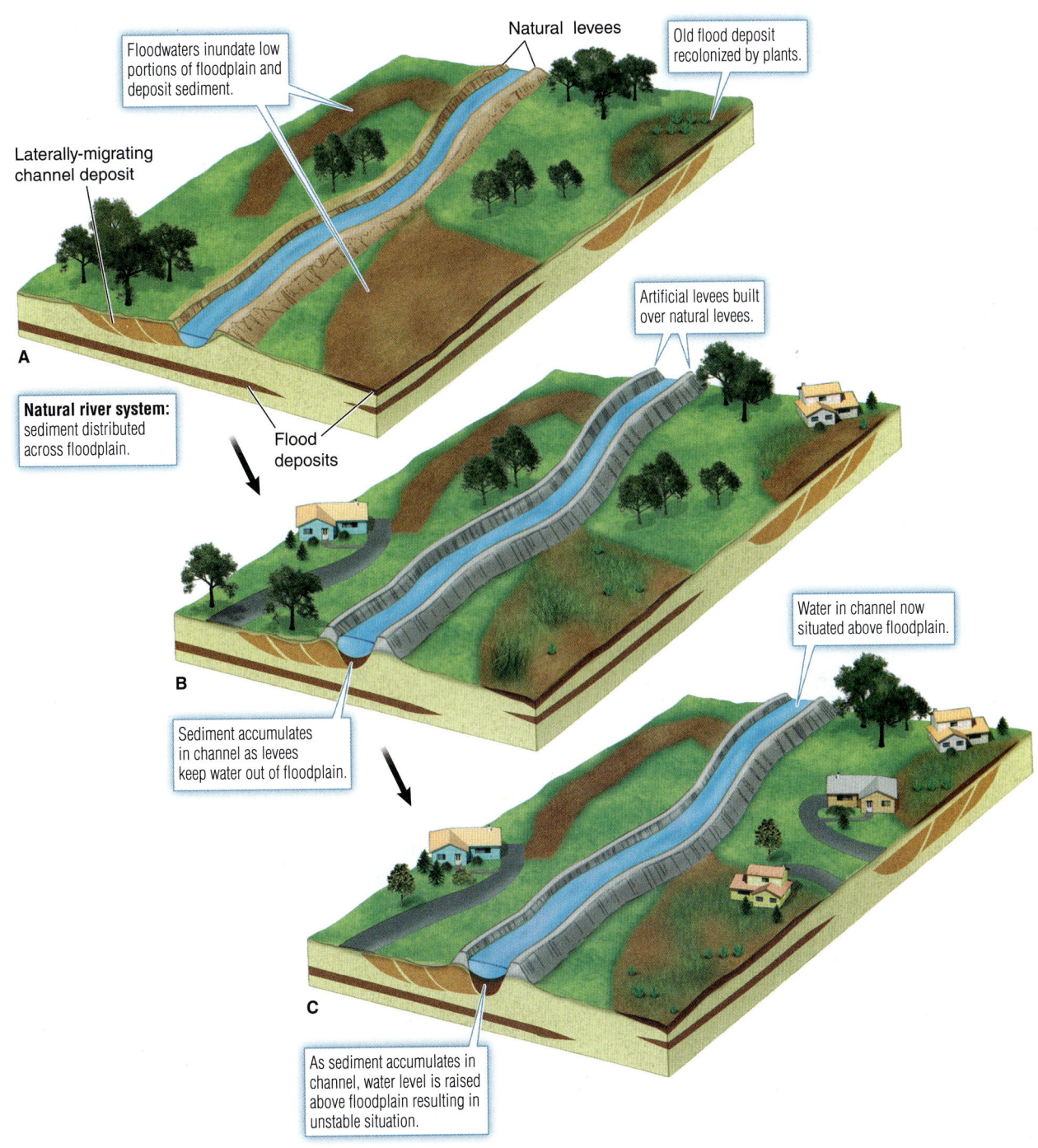

FIGURE 12-35

(A) In the natural state, a flooding stream carries sediment from the stream channel onto the floodplain. (B) Artificial levees hold water in the river channel during what would otherwise be small floods, causing sediment to accumulate there. (C) Eventually, the channel rises above the floodplain, creating the potential for a disastrous flood.

it. Currently, federal and state governments are establishing wildlife reserves in some floodplains. Because no development is allowed in these reserves, they will be permitted to flood during the next high water. However, a complete river management plan involves complex political and economic considerations.

Future of the Mississippi River Delta In the past 50 years, deposition rates in many deltas have decreased. This is partly because dams built upstream have trapped sediment that otherwise would have reached a delta. A coastal delta naturally builds upward and extends seaward by sediment deposition. At the same time, the front of the delta is eroded by ocean waves and currents. Without intermittent flooding and resultant deposition of sediment, erosion outpaces deposition. The construction of levees along distributary channels on deltas also has caused sediment to bypass the deltas rather than be deposited on them during floods. Contributing to coastal erosion is the urban and agricultural destruction of natural plant communities that normally hold the soil together. In addition, global sea-level rise causes wave energy to reach farther inland.

Other problems affect the delta. As sea levels rise, the land surface is sinking as the delta sediments compact and expel water from their pore spaces. As a result of all these processes, the Mississippi delta has shrunk. Between 1930 and 2012, nearly 5,000 square kilometers of the delta sank below sea level. One recent study estimated that by the year 2100, an additional 10,000 to 13,500 square kilometers will be submerged. That is an area roughly the size of Connecticut.

The flooding of New Orleans in 2005 is an example of both a complex systems interaction and a threshold event. (See the Case Study.) While Hurricane Katrina triggered the flood, human interference and development within a complex natural river-delta system set the stage. Over the previous decades, the global sea level rose while the delta subsided and slowly eroded. Natural buffers were removed, with no immediate ill effects. Then a catastrophic storm pushed waters inland. The levees failed, causing a huge human disaster.

checkpoint How can an artificial levee cause flooding upstream from the levee?

12.3 ASSESSMENT

1. **Explain** Where do perennial and intermittent streams tend to occur, and why?
2. **Relate** How do gradient, discharge, and channel characteristics determine stream velocity?
3. **Explain** How does a delta form as a stream enters a body of water?
4. **Explain** Why do artificial levees increase the effects of floods?
5. **Explain** How do levees and dams disrupt the natural deposition of sediment in a delta?

Critical Thinking

6. **Generalize** Based on what you know about sediment transport, explain why river deltas are typically much larger than alluvial fans.
7. **Evaluate** What are some pros and cons of different ways that people can help manage floodplains? Which do you think would be most effective?

Computational Thinking

8. **Analyze Data** About how many times greater is the average discharge of the Amazon River than that of the Mississippi River?

12.4 MASS WASTING

Core Ideas and Skills
- Identify the major causes of mass wasting.
- Define flow, slide, rockfall, slump, and creep, and explain how each of them affect the land.
- Identify effects of mass wasting and how to prevent damage from these processes.

KEY TERMS

mass wasting	slide
flow	rockslide
rockfall	angle of repose
creep	slump

FIGURE 12-36
Mass wasting falls into five categories: rockfall, slide, flow, slump, and creep.

Flow, Slide, and Rockfall

Mass wasting is erosion involving the downslope movement of earth material by gravity. *Landslide* is a general term for mass wasting and the landforms it creates.

The most rapid forms of mass wasting are flows, slides, and rockfalls (Figure 12-36). To understand the differences between these three, think of a sand castle. If one of the castle walls becomes saturated with water, the sand will flow down the face of the structure. During a **flow**, loose, unconsolidated soil or sediment moves as a fluid. Some slopes flow slowly—at a speed of 1 centimeter or less per year. In contrast, mud with high water content can flow nearly as rapidly as pure liquid water.

A slurry is a mixture of water and solid particles that flows as a liquid. Wet concrete is an example of a slurry. If heavy rain falls on unvegetated soil, rainwater can saturate the soil to form a slurry of mud and rocks called a debris flow, earthflow, or mudflow, depending on the size and sorting of the particles.

The advancing front of a debris flow, earthflow, or mudflow often forms tongue-shaped lobes. A slow-moving earthflow or mudflow travels at a rate of about 1 meter per year or slower, but others can move as fast as a car speeding along an interstate highway. A debris flow can pick up boulders and automobiles; it can also destroy houses, filling them with muddy sediment and even dislodging them from their foundations (Figure 12-37).

FIGURE 12-37
Go online to view an example of a mudslide in a village in Central Java in Indonesia.

If you undermine the base of a sand castle, the wall may fracture and a segment of the wall may slide downward. Movement of a coherent block of material along a fracture is called a **slide**. Natural slides usually move over time frames of a few seconds or minutes, although some slides may take days or weeks to finish moving.

During a **rockslide** (or rock avalanche), bedrock slides downslope over a fracture plane.

LESSON 12.4

Characteristically, the rock breaks up as it moves, and a turbulent mass of rubble tumbles down the hillside. In a large rockslide, the falling debris traps and compresses air beneath and among the tumbling blocks. The compressed air reduces friction and allows some rockslides to attain speeds of 500 kilometers per hour. The same mechanism allows a snow or ice avalanche to cover a great distance at high speed.

If you take a huge handful of sand out of the bottom of the castle, the whole castle topples. This rapid, free-falling motion of loose material is called **rockfall**. Rockfall is the most rapid type of mass wasting. If a rock dislodges from a steep cliff, it falls rapidly under the influence of gravity. Several processes commonly detach rocks from cliffs. Recall from our discussion of weathering that frost wedging can dislodge rocks from cliffs and cause rockfalls. Rockfall also occurs when flowing water or ocean waves undercut a cliff.

checkpoint Describe what happens during a flow, a slide, and a rockfall.

Creep and Slump

As the name implies, **creep** is the slow, downhill flow of rock or soil under the influence of gravity. All rock and sediment on the side of a hill or a mountain is constantly being pulled downward by gravity. Any process that causes this soil to expand and then shrink promotes the incremental downhill movement of soil particles. For example, when water in soil pores freezes or clay minerals in soil get wet, the soil expands. As it expands, soil particles are pushed outward at right angles to the slope surface. Later, when the pore water thaws, or the clay minerals dry out, and the soil shrinks, gravity pulls the soil particles directly downward, not back toward the original slope surface. Over time, this process causes soil particles to slowly creep downslope (Figure 12-38).

A creeping slope typically moves at a rate of about 1 centimeter per year, although wet soil can creep more rapidly. During creep, the shallow soil or rock layers move more rapidly than deeper material moves. As a result, anything with roots or a foundation tilts downhill.

Trees have a natural tendency to grow straight upward. As a result, when downhill creep of soil and shallow bedrock tilts a growing tree, the tree develops a J-shaped curve in its trunk, called pistol butt (Figure 12-39). People contemplating buying hillside land for a homesite should examine the trees. If the trees have pistol-butt bases, the slope is probably creeping, and creeping soil can slowly tear a building foundation apart.

A **slump** occurs when blocks of material slide downhill over a gently curved fracture in rock or loose rocky material covering bedrock. Trees remain rooted in the moving blocks. However,

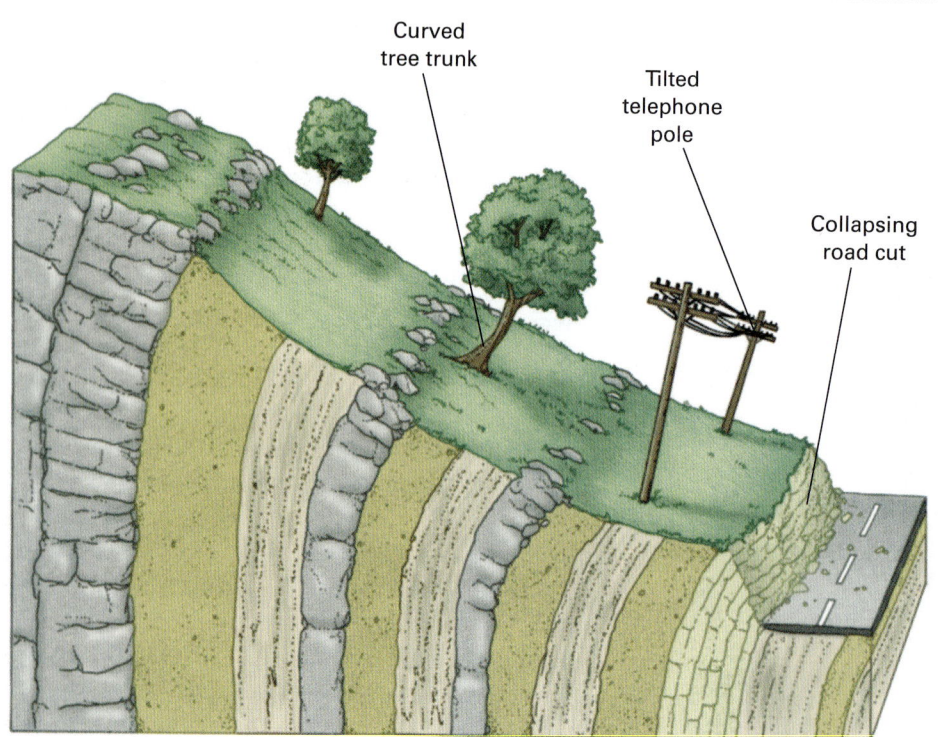

FIGURE 12-38
During creep, the land surface moves more rapidly than deeper layers, so objects embedded in rock or soil tilt downhill.

TABLE 12-1 Mass Wasting

Type of Mass Wasting	Description
Rockfall	• Material falls freely through the air. • Typically occurs in bedrock and only on steep cliffs
Slide	• Material moves downslope as consolidated blocks riding on a cushion of air. • Rockslides include segments of bedrock and can reach speeds of 200 kilometers per hour.
Flow	• Individual particles move downslope independently of one another as in a fluid. • Can occur in material with particle sizes ranging from silt to large boulders, usually in material containing a large amount of water • Mudflows can move as fast as 100 kilometers per hour.
Slump	• Blocks of material slide downslope along an upward concave curved surface. • Material travels slowly and over a short distance.
Creep	• Can occur in material that expands and contracts during freezing and thawing • The rate of movement is imperceptible, less than 1 meter per year.

FIGURE 12-39
If a hillside creeps as a tree grows, the tree trunk develops a pistol-butt shape.

because the blocks rotate on the concave fracture, trees on the slumping blocks are tilted upslope (Figure 12-40). Because of this, you can distinguish slump from creep: A slump tilts trees uphill, whereas creep tilts them downhill.

🌐 **FIGURE 12-40**
Go online to see trees tilted uphill in a slump.

It is useful to identify a slump because slumping often occurs repeatedly in the same place or appears on nearby slopes. Therefore, a slope that shows evidence of past slumping is not a good place to build a structure.

checkpoint What is the easiest way to tell whether a slope is undergoing a slump or a creep?

Effects and Risk Factors

Every year, small, rapid forms of mass wasting destroy numerous homes and farmland. Occasionally, an enormous slide or flow buries a town or city, killing thousands of people. Rapid forms of mass wasting cost billions of dollars annually. One of the most important considerations in evaluating the risks associated with mass wasting is understanding that it commonly recurs in the same area. The geologic conditions that cause mass wasting tend to be constant over a large area and for long periods of time, creating conditions for recurrence. If a hillside has slumped, nearby hills may also be vulnerable to the same type of mass wasting (Figure 12-41).

Awareness and avoidance are the most effective defenses against mass wasting. Geologists evaluate landslide probability by combining data on soil and bedrock stability, slope angle, climate, and history of slope failure in the area. They include evaluations of the probability of a triggering event such as a volcanic eruption or earthquake. Building codes then regulate or prohibit construction in unstable areas.

Imagine that you are a geologist consultant on a construction project. The developers want to build a road at the base of a hill. They wonder whether the road will be affected by slides, flows, or falls. What factors should you consider?

Steepness of the Slope
The steepness of a slope is a factor in mass wasting. If frost wedging dislodges a rock from a steep cliff, the rock tumbles to the valley below. However, a similar rock is less likely to roll down a gentle slope.

Type of Rock and Orientation of Rock Layers
If sedimentary rock layers dip in the same direction as a slope, a layer of weak rock can fail, causing layers over it to slide downslope. Imagine a hill underlain by mudstone, sandstone, and limestone oriented so that their bedding lies parallel to the slope, as shown in Figure 12-42. If the base of the hill is undercut, the upper layers of sandstone and limestone may slide over the weak mudstone. In contrast, if the rock layers dip into the hillside, the slope may be stable even if it is undercut.

Several processes can reduce the stability of a slope. Ocean waves or stream erosion can destabilize a slope, as can road building and excavation. Therefore, a geologist or engineer must consider not only a slope's stability before construction, but also how the project might alter its stability.

The Nature of Unconsolidated Materials
The **angle of repose** is the maximum slope or steepness at which loose sediment remains stable. If the slope becomes steeper than the angle of repose, the sediment slides. The angle of repose varies for different types of sediment. For example, rocks commonly tumble from a cliff and collect at the base as angular blocks of talus. The angular

FIGURE 12-41
A powerful earthquake struck the northern Japanese island of Hokkaido in 2018. It triggered several landslides that destroyed homes.

blocks interlock and jam together. As a result, talus typically has a steep angle of repose, up to 45 degrees. In contrast, rounded sand grains do not interlock and therefore have a lower angle of repose—about 30 to 35 degrees (Figure 12-43).

Water and Vegetation To understand how water affects slope stability, think again of a sand castle. Even a novice sand castle builder knows that sand must be moistened to build steep walls and towers. But too much water causes the walls to collapse. If only small amounts of water are present, the water collects only where one sand grain touches another. The surface tension and cohesion of the water binds the grains together. However, excess water fills the pores between the grains and exerts an outward pressure—called pore pressure—that pushes the grains apart. The excess water also lubricates the sand and adds weight to a slope. When some soils become saturated with water, they flow downslope, just as the sand castle collapses. In addition, if groundwater collects on an impermeable layer of clay, mudstone, or even on permafrost, it may cause overlying rock or soil to become saturated and move easily.

Roots hold soil together and plants absorb water. Therefore, a vegetated slope is more stable than a similar bare one. Many forested slopes

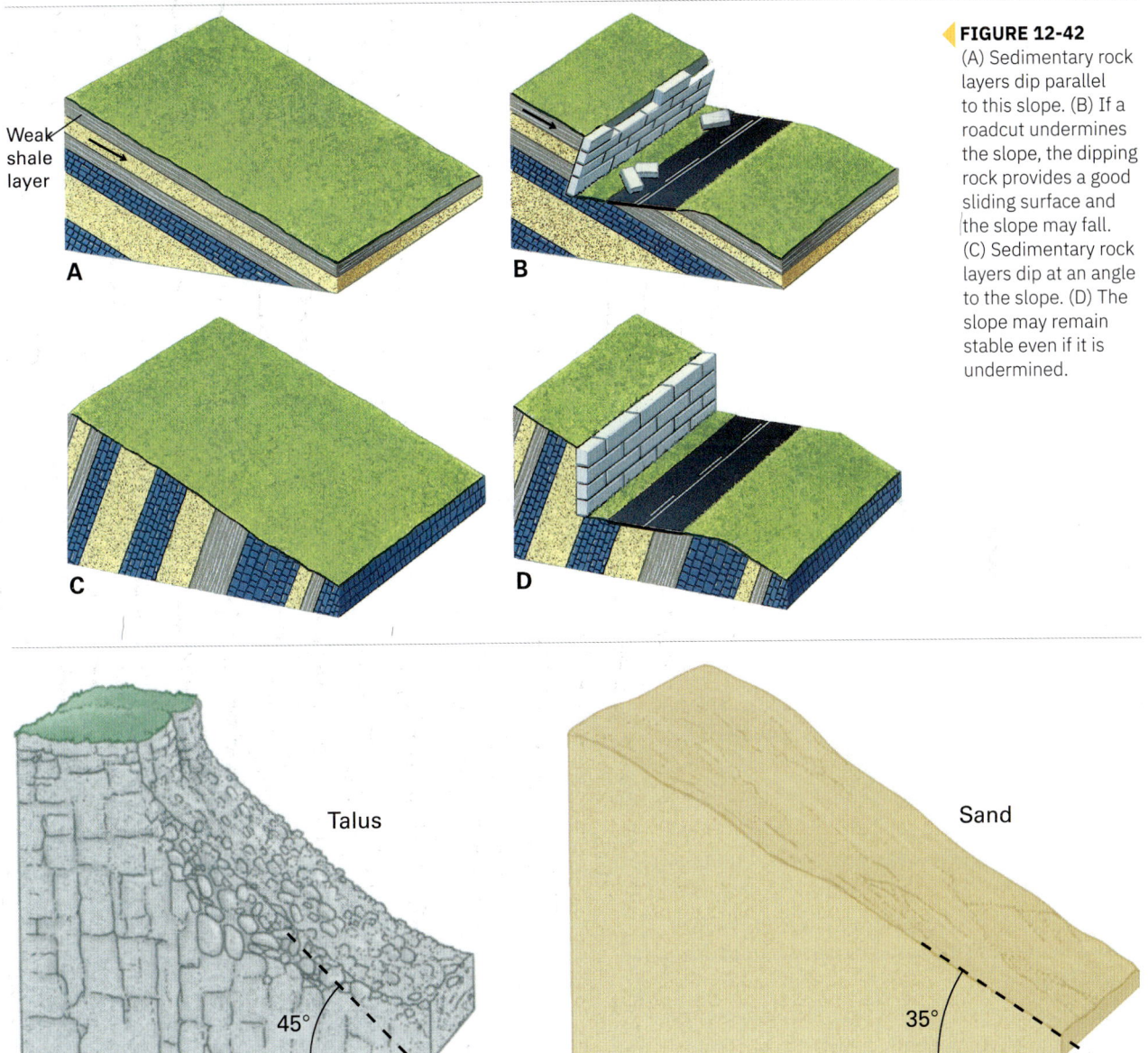

FIGURE 12-42
(A) Sedimentary rock layers dip parallel to this slope. (B) If a roadcut undermines the slope, the dipping rock provides a good sliding surface and the slope may fall. (C) Sedimentary rock layers dip at an angle to the slope. (D) The slope may remain stable even if it is undermined.

FIGURE 12-43
The angle of repose is the maximum slope at which a specific material can remain stable. Because the chunks of rock forming talus interlock and jam together, talus has a steeper angle of repose than sand, the rounded grains of which do not interlock.

FIGURE 12-44
The angle of repose depends on both the type of material and its water content. Dry sand forms low mounds, but moistened sand can form the familiar steep-sided hills and towers of sand castles. The surface tension among the grains causes them to stick together.

that were stable for centuries slid when the trees were removed during logging, agriculture, or construction.

Landslides are common in deserts and regions with intermittent rainfall. For example, Southern California has dry summers and occasional heavy winter rain. Vegetation is sparse because of summer drought and wildfires. When winter rains fall, bare hillsides often become saturated and slide. Landslides occur for similar reasons during infrequent but intense storms in deserts.

Earthquakes and Volcanoes An earthquake may cause a landslide by shaking an unstable slope, causing it to move. Saturated soils are particularly prone to movement during earthquakes. Seismic waves increase the pore pressure between soil particles, pushing them apart and causing the sediment to liquefy, making movement much more likely. If you have ever tapped with your foot on saturated sand at the beach, you've seen the same process. The energy from the tapping of your foot travels through the sand as small compression waves, increasing its pore pressure and causing the sand to liquefy.

Mass wasting is common in earthquake-prone regions and in volcanically active areas. Steep volcanoes are prone to rockslides, particularly during earthquakes. In addition, a volcanic eruption may melt the snow and ice cap at the top of the volcano. As the meltwater flows downslope, it picks up loose sediment and ash and becomes a debris flow. Many volcanic regions contain thick sedimentary areas consisting almost exclusively of debris flow deposits.

checkpoint How can vegetation affect the stability of a slope?

12.4 ASSESSMENT

1. **Compare and Contrast** How are flows, slides, and falls alike and different?
2. **Describe** What are the characteristics of a mudflow, and what damage can they cause?
3. **Summarize** What are some ways geologists evaluate the probability of a landslide?
4. **Explain** How can water and vegetation affect the stability of a slope?

Critical Thinking

5. **Apply** Suppose soil slowly creeps down a slope for months, then suddenly the soil moves quickly downslope. What conditions might have occurred to cause these movements?
6. **Recognize Patterns** Compare the speeds that can be attained by earth materials in a flow, slide, or rockfall.
7. **Evaluate** If you were planning to build a house on a slope, what conditions would you look for to ensure the building site was safe?

TYING IT ALL TOGETHER
FLOOD CONTROL AND MANAGEMENT

In this chapter, you learned about weathering, erosion, and deposition, and how these processes work together to change the landscape. You examined the causes of weathering and explored how water and wind move sediment from one location to another, resulting in new landforms. You learned how flooding and mass wasting can quickly and drastically change Earth's surface. You also looked at some methods for reducing harm caused by flooding and mass wasting.

In the Case Study, we explored how human development along the coast in the Mississippi Delta contributed to catastrophic flooding in New Orleans during Hurricane Katrina (Figure 12-45). We considered how human activity can impact the negative effects of extreme weather events such as hurricanes. Now you will expand your exploration of flood control, looking at the strategies engineers and scientists propose for controlling flooding without creating new problems in the process.

In this activity, you will investigate the various methods we have to control and manage flooding. Work with a group to complete the steps.

1. With your group, discuss how weathering, erosion, and deposition relate to flooding.

2. Together, research the main causes of flooding.

3. Investigate flood control and management in your area. Find out how flooding is currently being controlled and managed in order to reduce the negative impacts of flooding. Note any drawbacks to these methods. Pay special attention to how citizens, leaders, scientists, and engineers work together to reduce harmful flooding.

4. With your group, organize your findings and share them with others, either as a presentation or as a post to a class blog.

FIGURE 12-45
This is an aerial view of a hurricane levee in New Orleans, Louisiana. Failure of artificial levees was one cause of flooding from Hurricane Katrina.

CHAPTER 12 SUMMARY

12.1 MECHANICAL WEATHERING

- Mechanical (or physical) weathering can occur by pressure-release fracturing, frost wedging, abrasion, organic activity, and thermal expansion and contraction.
- Weathering is the decomposition and disintegration of rocks and minerals at Earth's surface. Erosion is the removal of weathered rock or soil by flowing water, wind, glaciers, or gravity.

12.2 CHEMICAL WEATHERING

- Chemical weathering is the alteration and breakdown of rock by changes in its chemistry and mineralogy. Dissolution, oxidation, and hydrolysis are among the most important chemical weathering processes.
- Chemical and physical weathering often operate together. In particular, salt cracking and hydrolysis-expansion are two processes that result from combinations of mechanical and chemical weathering.

12.3 STREAM EROSION AND DEPOSITION

- A stream is any body of water flowing in a channel. The velocity of a stream is determined by its gradient (steepness), discharge (amount of water), and channel characteristics (shape and roughness of the stream channel).
- Streams shape Earth's surface by eroding soil and bedrock. Streams transport sediment as dissolved load, suspended load, and bed load.
- Downcutting, lateral erosion, and mass wasting combine to form a stream valley. Base level is the lowest elevation to which a stream can erode its bed; it is usually sea level.
- When streams slow down, they deposit their bed load first. If water slows sufficiently, the suspended load may also sink to the bottom, but the dissolved load remains in the water until the chemical environment changes.
- A stream feeding a delta or fan splits into many channels called distributaries.
- A flood occurs when a stream overtops its banks and flows over its floodplain. Artificial levees and channels can contain small floods but may worsen large ones.

12.4 MASS WASTING

- Mass wasting is the downslope movement of earth materials by gravity and includes incremental downhill creep of soil particles as well as rapid forms of mass wasting such as landslides.
- Rapid forms of mass wasting fall broadly into three categories: flows, slides, and falls.
- Earthquakes and volcanic eruptions trigger devastating mass wasting. Proper planning and engineering can avert much damage to human habitation.

CHAPTER 12 ASSESSMENT

Review Key Terms

Select the key term that best fits the definition. Not all terms will be used, and no term will be used more than once.

alluvial fan
angle of repose
artificial levee
base level
bed load
capacity
chemical weathering
creep
delta
discharge
dissolution
dissolved load
downcutting
drainage basin
erosion
exfoliation
floodplain
flow
frost wedging
gradient
hydrolysis
mass wasting
meander
mechanical weathering
oxbow lake
oxidation
pH
pressure-release fracturing
rockfall
rockslide
slide
stream
suspended load
tributary

1. forms when a meandering stream adjusts its path and cuts off a portion of the stream
2. occurs when minerals in rock react with O_2 in the atmosphere
3. any body of water flowing through a channel
4. a wide area of sediment deposited by a stream at the base of a mountain range
5. any movement of earth material downslope by gravity
6. a stream channel that merges with another stream channel
7. the steepness of a slope, expressed as a vertical drop over a specific distance
8. the amount of water flowing in a stream over a given period of time, expressed in cubic meters per second
9. the entire region from which water flows into a single river that empties into the sea
10. the total amount of sediment a stream can carry
11. type of weathering in which a rock reacts with air, water, or other agents, altering its chemistry and mineral content
12. structure built to prevent a stream from flowing onto a floodplain
13. type of weathering that occurs due to tectonic uplift
14. type of weathering that occurs during freeze-thaw cycles
15. ions weathered from rock and transported in a stream

Review Key Concepts

Answer each question on a separate sheet of paper to demonstrate your understanding of key concepts from the chapter.

16. Explain why, unlike the moon, Earth does not have much surviving evidence from meteorite impacts that occurred millions of years ago.
17. Compare and contrast mechanical (physical) weathering and chemical weathering.
18. What factors affect a stream's capacity to transport sediments?
19. Explain what a stream's base level is and give examples of factors that might create a temporary base level for a stream.
20. Compare and contrast meandering streams and braided streams.
21. Suppose that tectonic forces did not exist. How would streams change Earth's landscape over geologic time?
22. Compare and contrast a delta and an alluvial fan.
23. A river passes through a town that has recently been built along its banks. Each of the last two springs, storms upstream have caused the river to flood homes and businesses built in the floodplain. What are two options that the town could take to prevent future flood damage?
24. Explain why the trunks of the trees in the picture curve upward in the shape of a J.

CHAPTER 12 ASSESSMENT

25. Compare and contrast flows, slides, and rockfalls.
26. Suppose one of your classmates states that "rocks along shorelines are smooth and round due to the abrasive action of water." What is wrong with this statement, and how could you expand on it to make it correct?
27. Compare and contrast hydrolysis, dissolution, and oxidation, and give an example of each.

Think Critically

Write a response to each question on a separate sheet of paper. Use concepts from the chapter to support your reasoning.

28. Identify the body of water indicated by the arrow and describe the processes that formed it.
29. What kinds of environments would be prone to weathering by abrasion? Frost wedging? Hydrolysis? Describe why using examples.
30. Discuss whether nature or human activity was a greater factor in the flooding that resulted from Hurricane Katrina. Explain your reasoning.
31. What are the advantages and disadvantages of living on a delta? Include aspects of the geosphere, biosphere, and hydrosphere in your response.
32. Describe the steps that would lead from frost wedging to rockfall.
33. Suppose you are an engineer evaluating a hillside as a possible location for a new housing development. What factors would you consider in order to determine the stability of the hillside? What would you want to see before allowing housing to be built on the hillside?

PERFORMANCE TASK

Comparing and Contrasting Mississippi and Missouri Rivers

The Mississippi and Missouri Rivers are the two longest rivers in the United States. Together, the two rivers have a drainage basin covering an area from Montana to Wisconsin. The Mississippi and Missouri meet just north of St. Louis, Missouri.

In this performance task, you will use what you know about stream flow to hypothesize which of two rivers has a greater discharge rate, then analyze data to determine whether your hypothesis is correct.

1. Look at the picture of the confluence of these two rivers. The Missouri River is known for its high sediment load. Which river do you think is the Missouri? Why?

2. Given what you know about stream flow, how do you think the Mississippi and the Missouri might differ in terms of velocity, gradient, discharge, and channel characteristics?

3. Research the elevation of the headwaters of the Missouri, the headwaters of the Mississippi, and the city of St. Louis. Which river has the greatest change in elevation as it flows to St. Louis?

4. Gradient depends on distance traveled as well as change in elevation. Which river has the steeper gradient?

5. The USGS uses river monitoring equipment to collect real-time data about the discharge rates at thousands of points across the United States. To compare the discharge of the Missouri and the Mississippi, go to https://waterwatch.usgs.gov/ and select "Current Streamflow" from the navigation menu. Find data for monitoring stations just upstream of the confluence. For example, Grafton, Illinois, provides discharge data for the Mississippi, whereas Labadie, Missouri, provides discharge data for the Missouri.

How do the discharge data compare? Do they fit your predictions based on the high sediment load of the Missouri? Why do you think discharge data might not fit your predictions?

CHAPTER 13
GLACIERS AND DESERTS

Ripples in recently deposited sand dunes in the Victoria Valley of Antarctica lie in contrast to a retreating glacier in the distance. The structure and age of sand dunes are the focus of recent studies looking at the inland impact of climate change in the Ross Sea area. The rate of dune migration has increased over the last 200 years, coinciding with the modern rise of atmospheric CO_2.

Forces of erosion shape Earth in many ways. The effects of glacial erosion can be observed in places we live, farm, and explore recreationally. Slow-moving ice scrapes the land, often yielding steep cliffs, deep lakes, and winding streams. Glacial deposits accumulate in different ways, too. In this chapter, you will learn more about the effects of glaciers and also about the formation of deserts. Like glaciers, deserts are characterized by erosion and deposition. Two extreme environments, glaciers and deserts, shape many of Earth's modern landscapes, and have left behind clues geologists use to uncover the evolution of ancient landscapes.

KEY QUESTIONS

- **13.1** What types of glaciers exist on Earth?
- **13.2** In what manner does a glacier move?
- **13.3** What erosional landforms are caused by glacial movement?
- **13.4** What types of glacial deposits occur, and how do they accumulate?
- **13.5** What are the major effects of historical glaciation and what is happening to glaciers today?
- **13.6** Where are deserts found on Earth?
- **13.7** How does water affect a desert environment?
- **13.8** How does wind affect a desert environment?

395

EXPLORERS AT WORK

AT THE BOTTOM OF THE WORLD

WITH NATIONAL GEOGRAPHIC EXPLORER ERIN PETTIT

Antarctica is one of the coldest, windiest, and driest places on Earth. In fact, Antarctica is classified as a polar desert due to the lack of annual precipitation, just 150 millimeters (6 inches), mainly in the form of snow. It also is a land of ice. About 60 percent of Earth's fresh water is frozen within the Antarctic ice sheet. Yet, even this remote environment is being impacted by climate change. So says Erin Pettit, an associate professor with Oregon State University and glaciologist who has studied glacial landscapes for more than 20 years.

Dr. Pettit says the Thwaites Glacier, which is located on the western edge of Antarctica, is rapidly melting into the ocean. A cavity two-thirds the size of Manhattan has developed beneath this disintegrating glacier, which itself is about the size of Florida. The glacier could melt completely in as little as 50 years. According to Pettit, ice melting off the Thwaites Glacier already accounts for 4 percent of global sea-level rise. If it melts completely, the global sea level will rise 0.6 meters (2 feet). That in turn will devastate coastal cities, industries, and ecosystems.

Now, government agencies from the United States and the United Kingdom are funding a $25 million, five-year study of the Thwaites Glacier. Called the International Thwaites Glacier Collaboration, the study brings together an international team of more than 60 scientists and involves eight large-scale projects aimed at better understanding and predicting ice-sheet melting. Dr. Pettit is a lead investigator for one of these projects, which focuses on how atmospheric and oceanic processes impact the glacier at the boundary where it meets the sea. The effects are twofold, explains Pettit. Not only do warmer air temperatures melt a glacier's surface, causing it to fracture, but warmer ocean waters also eat away at the glacier from below.

Pettit is using a variety of methods to take measurements from above, within, and beneath the glacier where it floats on the ocean. These include using acoustic equipment she has developed to "listen" to sounds produced by the glacier as it melts. Pettit says these sounds can range from crashes and sizzles to whooshes and groans, and they hold clues to causes and rates of melting. Pettit's team also will place sensors inside the glacier and use unmanned submarines to collect data from the ocean depths. Data will even be collected from sensors placed on seals, which are active on and under the glacier even in winter, when sea ice makes access to the glacier impossible.

Working at the bottom of the world isn't for everyone, but icy, remote, mountainous environments have always appealed to Pettit. And when she is not off on her own expeditions, she coordinates field trips for teenage girls as part of her organization Inspiring Girls Expeditions. These trips team girls up with scientists, artists, and wilderness instructors who teach them glaciology, ecology, and mountaineering skills. Pettit hopes these experiences will inspire young women to perhaps follow in her icy footsteps someday.

THINKING CRITICALLY

Evaluate What are some ways coastal cities, industries, and ecosystems might be impacted by a global sea-level rise? What preparations, if any, can be made to minimize such impacts?

Dr. Erin Pettit anchors herself in the ice.

Pettit solders wires together to connect a network of sensors that are embedded into glacier ice to measure temperature and glacial movement. One of her recent studies involves tracking how the movement of the cliffs at Taylor Glacier in Antarctica changes over a day and over the seasons.

CASE STUDY
A GLACIER TO WATCH

If you live in the United States, chances are good that you won't see a glacier when you look out the window. But even if you've never seen one, glaciers have played a key role in shaping the ground beneath your feet. Throughout Earth's history, glaciers have formed and melted many times. The advancing, receding, and melting of glaciers has helped shape continents, moving massive amounts of sediment and carving hills and valleys into the bedrock. In the United States, glaciers are responsible for many familiar landscapes. Glacial erosion sculpted Maine's rocky coastline and shaped the northern Appalachian Mountains. Further inland, glaciers carved out huge depressions in the landscape and then melted to form the Great Lakes.

Besides shaping the land, glaciers also play an important role in establishing the elevation of global sea-level. In Antarctica, scientists like Erin Pettit (see Explorers at Work) are studying the Thwaites Glacier, one of the world's largest. Currently, the Thwaites Glacier covers an area roughly the size of Florida. However, like many of the world's glaciers, the Thwaites Glacier is rapidly melting. Although most of the glacier sits above land, the ice extends out over the Amundsen Sea, where warm water is melting the ice from the bottom. This melting destabilizes a large portion of the entire ice sheet, causing the Thwaites Glacier to speed up and flow faster toward the sea. Once the ice meets the water, it can break off, forming icebergs that eventually melt and cause sea levels to rise.

Over the past 30 years, the amount of ice flowing out of the Thwaites Glacier has nearly doubled. To understand why, a team of scientists from the United States and Great Britain has launched several projects to study the glacier's evolution over time. Some studies focus on how the boundaries of the glaciers are changing. Others collect data on sediments, water depths, and rates of ice melt. By collecting this data, scientists hope to develop better models for predicting the future behavior of the glacier. Scientists estimate that if the entire Thwaites Glacier were to collapse, sea levels could rise rapidly by as much as 0.6 meters (over 2 feet), enough to devastate coastal communities worldwide. Even more concerning, the Thwaites Glacier holds back neighboring glaciers that could add another 0.6 meters (2 feet) to that total.

As You Read Learn how massive sheets of ice like the Thwaites Glacier interact with the land and the sea as they move. Discover how glaciers have shaped landscapes, coastlines, and other landforms throughout Earth's history. Find out where glaciers are currently located and study how they continue to shape the land. Also consider other processes, such as evaporation, erosion by wind and water, and the accumulation of resulting sediment, that can transform landscapes.

▸ **FIGURE 13-1**
Many of the world's glaciers are rapidly melting. In January 2019, a study led by NASA discovered an underwater cavity nearly two-thirds the size of the city of Manhattan under the Thwaites Glacier in Antarctica.

13.1 FORMATION OF GLACIERS

Core Ideas and Skills
- Explain the characteristics of glaciers.
- Compare the locations of glaciers on Earth.

KEY TERMS

glacier
continental glacier
alpine glacier
ice shelf

What Is a Glacier?

A **glacier** is a massive, long-lasting, moving accumulation of compacted snow and ice. Numerous times during Earth's history, glaciers grew to cover large parts of Earth and then melted away. Before the most recent major glacial advance, beginning about 100,000 years ago, the world was free of ice except for on high mountains and in the polar ice caps of Antarctica and Greenland. Then, in a relatively short time—perhaps only a few thousand years—Earth's climate cooled. As winter snow failed to melt completely in summer, the polar ice caps spread into lower latitudes. At the same time, glaciers formed near mountain summits, even near the Equator. They flowed down mountain valleys into nearby lowlands. When the glaciers reached their maximum size between about 24,000 and 17,000 years ago, they covered one-third of Earth's continents.

Humans lived through the most recent glaciation. In southwest France and northern Spain, humans developed sophisticated spearheads and carved body ornaments between 40,000 and 30,000 years ago. People first began experimenting with agriculture between about 12,000 and 10,000 years ago.

Glaciers form only on land, and they derive their ice from regions in which the amount of snow that falls in winter exceeds the amount that melts in summer. In most temperate regions, snow that falls in winter melts completely in spring and summer. However, in certain cold, wet environments, some of the winter snow remains unmelted during the summer and accumulates year after year. During summer, the snow crystals become rounded and denser as the snowpack is compressed and alternately warmed during daytime and cooled at night. If snow survives through one summer, it converts to rounded ice grains called firn. Mountaineers like firn because the sharp points of their ice axes and crampons sink into it easily and hold firmly. If firn is buried deeper in the snowpack, it converts to glacial ice, which consists of closely packed ice crystals (Figure 13-2).

Because ice under pressure behaves like plastic and deforms easily, mountain glaciers flow downhill. In many cases, glaciers also slide downslope along their base. Glaciers on level land flow outward in all directions along the surface. Unlike water, glacial ice is capable of moving upslope if it is being pushed from behind by ice flowing downslope from higher-elevation portions of the same glacier.

Glaciers form in two environments. Glaciers can form at all latitudes on high, snowy mountains. They also form at all elevations in the cold polar regions.

checkpoint Describe glacial ice. What characteristics does this kind of ice have?

Alpine Glaciers

Mountains are generally colder and wetter than adjacent lowlands. Near the summits, winter snowfall is deep and summers are cool. These conditions create **alpine glaciers**. With the

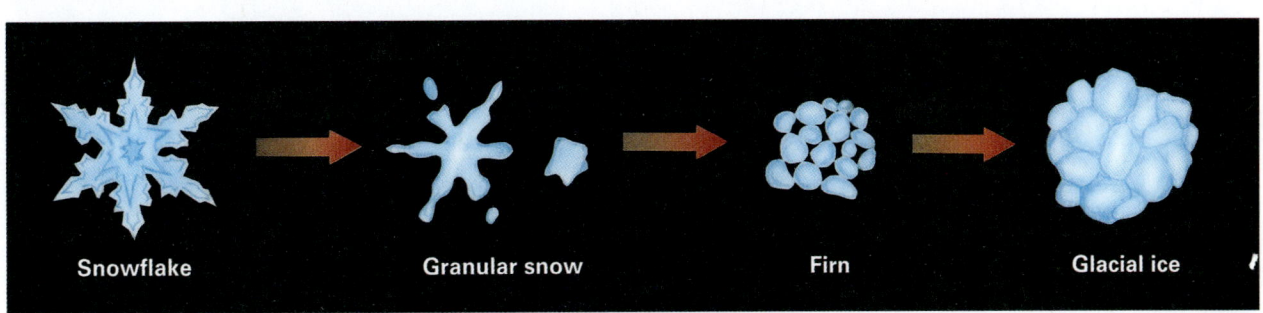

FIGURE 13-2
Newly fallen snow changes through several stages to form glacial ice.

exception of Australia, alpine glaciers exist on every continent—in the Arctic and Antarctica, in temperate regions, and in the tropics. Glaciers cover the summits of Mount Kenya in Africa and Mount Cayambe in South America, even though both peaks are near the Equator.

Some alpine glaciers flow great distances from the peaks into lowland valleys. For example, the Kahiltna Glacier, which flows down the southwest side of Denali in Alaska, is about 65 kilometers long, 12 kilometers across at its widest point, and about 700 meters thick. Most alpine glaciers are smaller than the Kahiltna. For example, Switzerland's Gorner Glacier, shown in Figure 13-3, is about 14 kilometers long, about 1.5 kilometers across at its widest point, and up to 450 meters thick.

The growth of an alpine glacier depends on both temperature and precipitation. The average annual temperature in the state of Washington is warmer than in Montana, yet alpine glaciers in Washington are larger and flow to lower elevations than those in Montana. Winter storms buffet Washington from the moisture-laden Pacific. Consequently, Washington's mountains receive such heavy winter snowfall that even though summer melting is rapid, snow generally accumulates every year. In much drier Montana, snowfall is light enough that most of it melts in the summer, and so Montana's mountains have only a few small glaciers.

checkpoint Why are mountain summits ideal for the formation of alpine glaciers?

FIGURE 13-3
The Gorner Glacier in Switzerland is the second-largest alpine glacier in the Alps.

Continental Glaciers

In polar regions, winters are so long and cold and summers so short and cool that glaciers cover most of the land regardless of its elevation. An ice sheet, or **continental glacier**, is a large accumulation of ice that spreads outward in all directions under its own weight and can cover areas of 50,000 square kilometers or more.

Today, Earth has only two ice sheets, one in Greenland and the other in Antarctica. These two ice sheets contain 99 percent of the world's ice and about three-fourths of Earth's fresh water. The Greenland sheet is more than 3.3 kilometers thick in places and covers 1.7 million square kilometers. Yet it is small compared with the Antarctic ice sheet, which blankets nearly 14 million square kilometers, roughly the size of the contiguous United States and Mexico combined. The Antarctic ice sheet covers entire mountain ranges, and the mountains that rise above its surface are islands of rock in a sea of ice. In many places, the Antarctic ice sheet extends into the sea, forming ice shelves. An **ice shelf** is a thick mass of ice that floats in the ocean but is connected to a glacier on land.

Whereas the South Pole lies in the interior of the Antarctic continent, the North Pole is situated in the Arctic Ocean. At the North Pole, only a few meters of ice freeze on the relatively warm sea surface, and the ice fractures and drifts with the currents. As a result, no ice sheet exists there.

checkpoint Where are the two continental glaciers in existence today?

13.1 ASSESSMENT

1. **Contrast** Differentiate between alpine glaciers and continental glaciers.
2. **Compare** Where are alpine glaciers and continental glaciers found today?
3. **Describe** How does seasonal precipitation affect alpine glaciers?

Computational Thinking

4. **Represent Data** Using at least two credible sources, create a data representation that displays the glacial coverage on Greenland and Antarctica.

13.2 GLACIAL MOVEMENT

Core Ideas and Skills
- Describe the effects of glacial movement and melt.
- Identify and describe features of a glacier.

KEY TERMS

basal slip	plastic flow
crevasse	terminus

Basal Slip and Plastic Flow

The rate of glacial movement varies with slope steepness, precipitation, and air temperature. In the coastal ranges of southeast Alaska, where annual precipitation is high and average temperature is relatively warm (for glaciers), some glaciers move 15 centimeters to 1 meter per day. In some instances, individual glaciers have been observed to surge at a speed of 10 to 100 meters per day. In contrast, in the interior of Alaska, where conditions are colder and drier, glaciers move only a few centimeters per day. At these rates, it takes hundreds to thousands of years for ice to flow the length of an alpine glacier.

Glaciers move by two mechanisms: basal slip and plastic flow. In **basal slip**, the entire glacier slides over bedrock in the same way that a bar of soap slides down a tilted board. Just as wet soap slides more easily than dry soap, water between bedrock and the base of a glacier accelerates basal slip.

Several factors cause water to accumulate near the base of a glacier. Earth's heat melts ice near bedrock. Friction from glacial movement also generates heat. Water occupies less volume than an equal mass of ice. As a result, pressure from the weight of overlying ice favors melting. Finally, during the summer, water melted from the surface of a glacier may seep downward to its base.

A glacier also moves by **plastic flow**, in which the ice flows as a viscous fluid. Plastic flow is demonstrated by two experiments. In one, scientists set a line of poles in the ice (Figure 13-4). After a few years, the ice moves downslope so that the poles form a U-shaped array. This experiment shows that the center of the glacier moves faster than the edges. Frictional resistance with the valley walls slows movement along the edges of

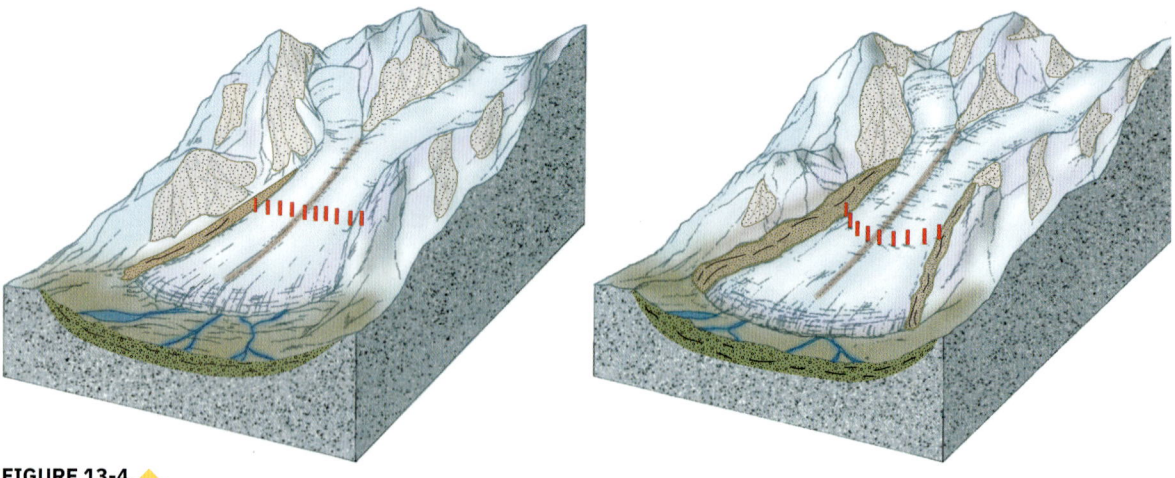

FIGURE 13-4
If a line of poles is set in a glacier, the poles near the center of the ice move downslope faster than those near the margin, demonstrating that the center of a glacier moves faster than the edges.

the glacier. Because glacial ice flows plastically, this resistance slows the edges of the glacier down relative to the center.

In another experiment, scientists drive a straight, flexible pipe downward into a glacier to study the flow of ice at depth (Figure 13-5). After several years, the entire pipe has moved downslope and become bent. This experiment demonstrates that at the surface of a glacier, the ice is brittle, like an ice cube or the ice found on the surface of a lake. In contrast, at depths greater than about 40 meters, pressure is sufficient to allow ice to deform plastically. The curvature in the pipe shows not only that ice moves plastically but also that middle levels of a glacier move faster than lower levels. The base of a glacier is slowed by friction against bedrock.

The relative rates of basal slip and plastic flow depend mainly on the steepness of the bedrock underlying the glacier, the thickness of the ice, and the temperature of the glacier. Glaciers located on steep slopes are vulnerable to more basal slip than glaciers located on flat ground. The thickness of a glacier determines the pressure at its base. Because pressure depresses the freezing/melting point of ice, a thick glacier will be more prone to develop a layer of water at its base, which will in turn promote basal sliding. Similarly, warmer temperatures at the base of a glacier will produce a layer of water that acts like a lubricant and promotes basal sliding. In contrast, if the temperature at the base of the glacier is cold enough, the glacier can freeze to the underlying substrate, thereby eliminating basal slip and causing the glacier to move entirely by internal deformation.

When a glacier flows over uneven bedrock, the deeper plastic ice bends and flows over bumps, while the brittle upper layer stretches and cracks, forming a **crevasse** (Figures 13-5, 13-6, and 13-7). Crevasses form only in the brittle upper 40 meters or so of a glacier, not in the lower plastic zone. Crevasses open and close slowly as a glacier moves. An icefall is a section of a glacier consisting of crevasses and towering ice pinnacles. The pinnacles form where ice blocks break away from the crevasse walls and rotate as the glacier moves.

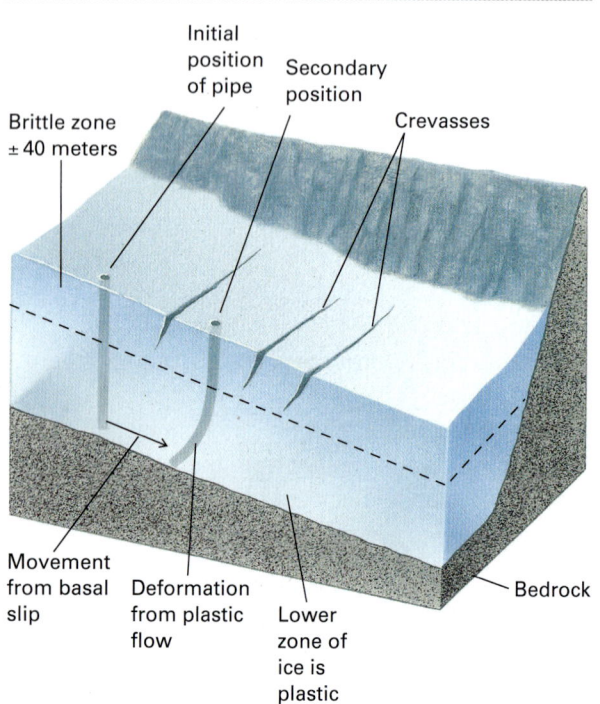

FIGURE 13-5
Notice the initial position of the pipe and the secondary position. The area above the dashed line is a brittle zone, while the lower zone is plastic.

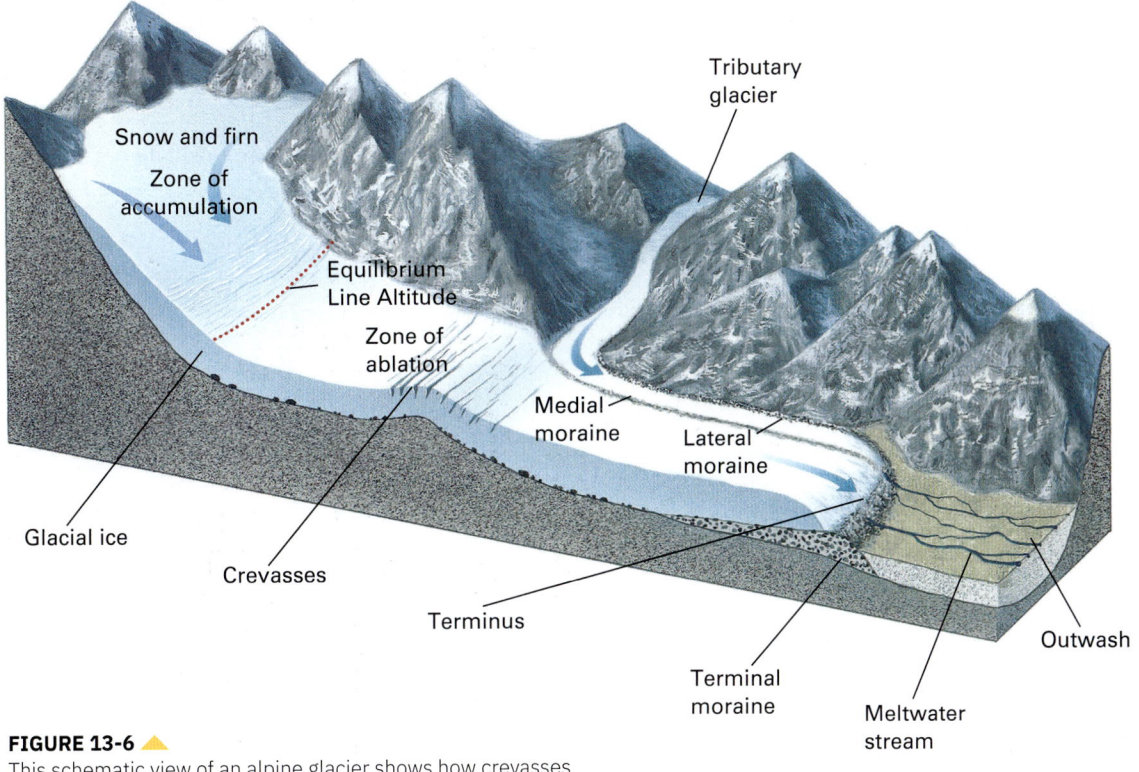

FIGURE 13-6
This schematic view of an alpine glacier shows how crevasses form in the upper, brittle zone of a glacier where the ice flows over uneven bedrock.

With crampons, ropes, and ice axes, a skilled mountaineer might climb into a crevasse. The walls are a pastel blue, and sunlight filters through the narrow opening above. The ice shifts and cracks, making creaking sounds as the glacier advances. Unfortunately, many mountaineers have been crushed by falling ice while traveling into crevasses and through icefalls.

FIGURE 13-7
Go online to view examples of crevasses in the Bugaboo Mountain Range of British Columbia.

checkpoint Why would a pipe inserted into a glacier bend over time?

The Mass Balance of a Glacier

Consider an alpine glacier flowing from the mountains into a valley. At the upper end of the glacier, snowfall is heavy, temperatures are below freezing for much of the year, and avalanches transport large quantities of snow from the surrounding slopes onto the ice. More snow falls in winter than melts in summer, and snow accumulates from year to year. This higher-elevation part of the glacier is called the zone of accumulation.

Lower in the valley, the temperature is higher throughout the year and less snow falls. This lower part of a glacier, where more snow melts in summer than accumulates in winter, is called the zone of ablation. When the snow melts, a surface of old, hard glacial ice is left behind. The equilibrium line altitude, or ELA, is the boundary between the zone of accumulation and the zone of ablation. The ELA shifts up and down the glacier from year to year, depending on weather. Ice exists in the zone of ablation because the glacier flows downward from the accumulation area. Further down the valley, the rate of glacial flow cannot keep pace with melting, so the glacier ends at its **terminus**.

Glaciers grow and shrink. For example, the zone of accumulation in an alpine glacier will grow thicker and the glacier's ELA will descend to a lower elevation if annual snowfall increases and more firn survives the summer without completely melting. At first, the glacial terminus may remain stable, but eventually it will advance farther down the valley. The lag time between a change in climate and the advance of an alpine glacier may range from a few years to several decades, depending on the size of the glacier, its rate of movement, and the

LESSON 13.2 **403**

FIGURE 13-8 Icebergs form when the terminus of a glacier reaches the sea and pieces of the ice break off. A large mass of ice is shed from the terminus of the Hubbard Glacier near Yakutak, Alaska.

magnitude of the climate change. If annual snowfall decreases or the climate warms and less firn survives the summer, the alpine glacier will grow thinner, its ELA will recede upslope, and the glacier terminus will retreat.

When a glacier retreats, its ice continues to flow downhill, but the terminus melts back faster than the rate at which ice reaches the terminus. Typically, the ELA ascends to a higher elevation as more of the glacier undergoes melting. In Glacier Bay, Alaska, glaciers have retreated more than a hundred kilometers in the past 200 years. Near the terminus, newly exposed rock is bare and lifeless. A few kilometers from the glacier, where the ground has been exposed for a few decades, lichens grow on otherwise bare rock, while mosses and dwarf fireweed can survive on thin soils formed in sheltered cracks where seabird droppings have mixed with windblown silt. Near the head of Glacier Bay, which deglaciated 200 years ago, tidal currents and ocean storms have washed enough sediment over the exposed bedrock to create a thicker soil capable of supporting a spruce-hemlock rain forest.

In equatorial and temperate regions, glaciers commonly terminate at an elevation of 3,000 meters or higher. However, in a cold, wet climate, a glacier may extend into the sea, where giant chunks of ice break off, forming icebergs (Figure 13-8).

The largest icebergs in the world are those that break away from the Antarctic ice shelf. Between the years 2000 and 2002, two plates of ice the size of Connecticut and at least one the size of Rhode Island broke free from the West Antarctic Ice Sheet and floated into the Antarctic Ocean. In 2010 a 78-kilometer-long iceberg with a surface area greater than that of Massachusetts broke off of the terminus of the Mertz Glacier in Antarctica. In early 2020 an iceberg the size of the District of Columbia broke off Pine Island Glacier in West Antarctica. Such events are becoming an annual occurrence.

checkpoint In general, what combination of events can lead to glacial retreat?

13.2 ASSESSMENT

1. **Explain** Distinguish between basal slip and plastic flow.
2. **Describe** Why are crevasses typically not more than 40 meters deep, even though many glaciers are much thicker?
3. **Compare** Describe the surface of a glacier in summer and in winter in the zone of accumulation, and in summer and winter in the zone of ablation.

Critical Thinking

4. **Generalize** What would be the result of reduced snowfall or climate warming on an alpine glacier?

13.3 GLACIAL EROSION

Core Ideas and Skills
- Explain the variation of glacial erosion.
- Describe the landforms that result from glacial erosion.

KEY TERMS

glacial striation	cirque
arête	hanging valley
fjord	

Ice Erosion

Bedrock at the base and sides of a glacier may have been fractured by tectonic forces, frost wedging, or pressure-release fracturing. Moving glacial ice will commonly dislodge and pluck individual boulders from such fractured bedrock. The plucked boulders are then transported downslope as clasts, or fragments of rock, either within or at the base of the glacier. Ice is viscous enough to pick up and carry particles of all sizes, from house-sized boulders to clay-sized grains. So, glaciers erode and transport huge quantities of rock and sediment.

Ice itself is not abrasive to bedrock, because it is too soft. However, rocks embedded in the ice scrape across bedrock, gouging deep, parallel grooves and scratches called **glacial striations** (Figure 13-9). When glaciers melt and striated bedrock is exposed, the markings show the direction of ice movement. Geologists measure the orientation of glacial striations and use them to map the flow directions of ancient glaciers.

checkpoint Is it the ice of a glacier or the rock embedded in glacier ice that causes glacial striations? Explain your answer.

Erosional Landforms

In Chapter 12, we learned that mountain streams in unglaciated regions commonly erode downward into their beds, cutting steep-sided, V-shaped valleys. A glacier, however, is not confined to a narrow streambed, but instead extends across the valley bottom and commonly fills the entire valley itself. As a result, the glacier scours the sides of the valley as well as the

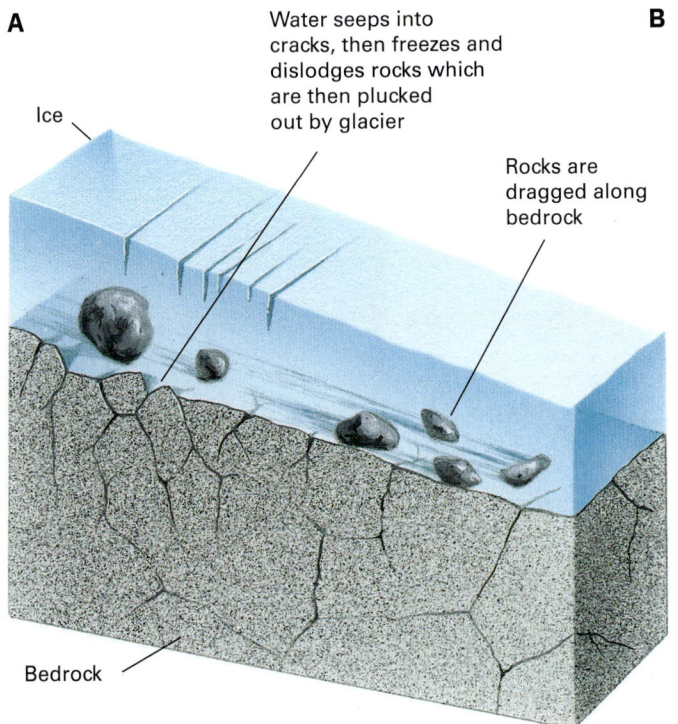

FIGURE 13-9
(A) A glacier plucks rocks from bedrock and then drags them along as clasts, abrading both the clast and the bedrock. (B) Stones embedded in the base of a glacier gouged these striations in bedrock in British Columbia.

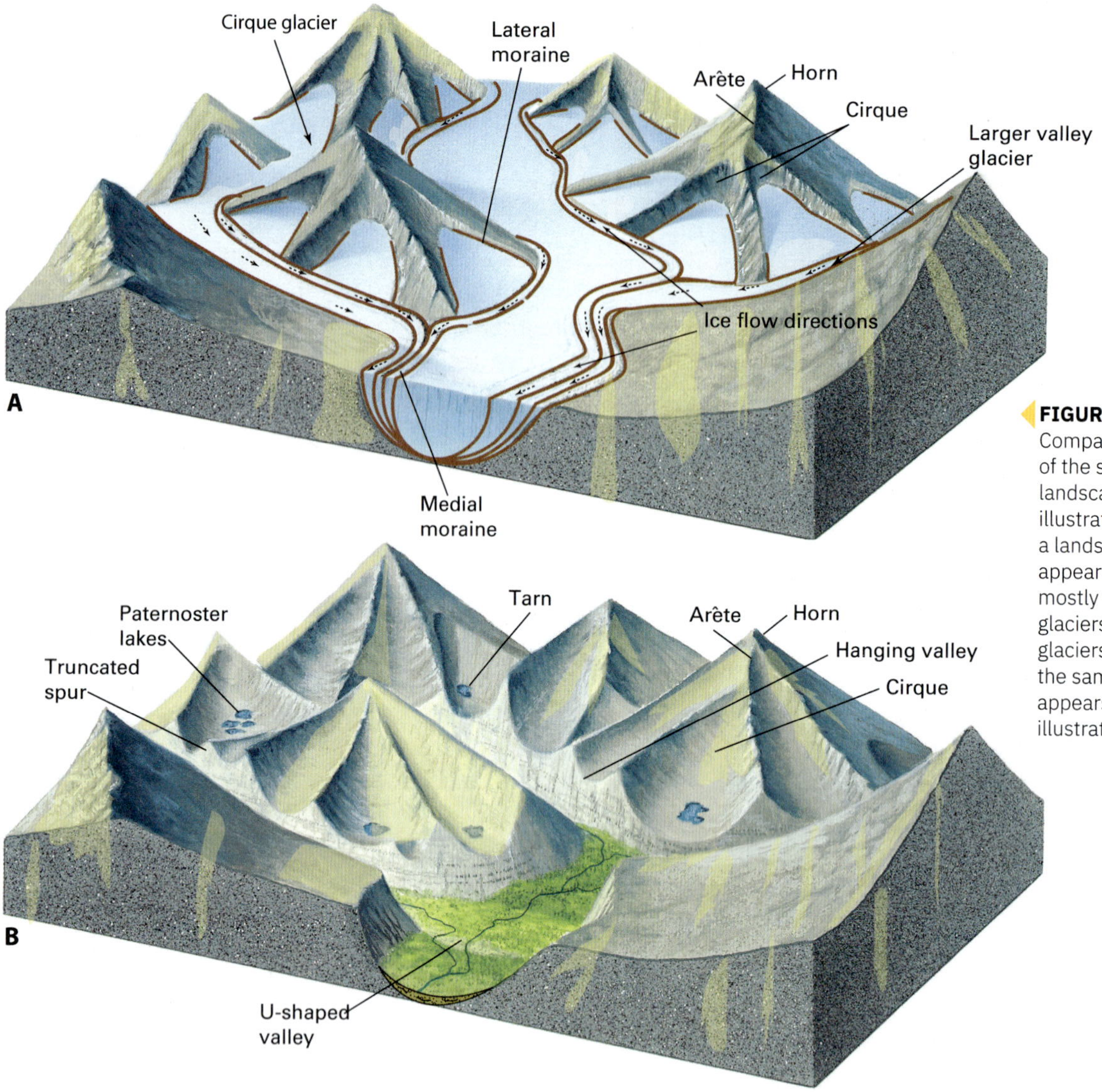

FIGURE 13-10
Compare two views of the same glacial landscape. (A) This illustration shows a landscape as it appeared when it was mostly covered by glaciers. (B) After the glaciers have melted, the same landscape appears like this illustration.

FIGURE 13-11
A U-shaped valley in Lofoten, Norway, shows the effect of glacial erosion.

406

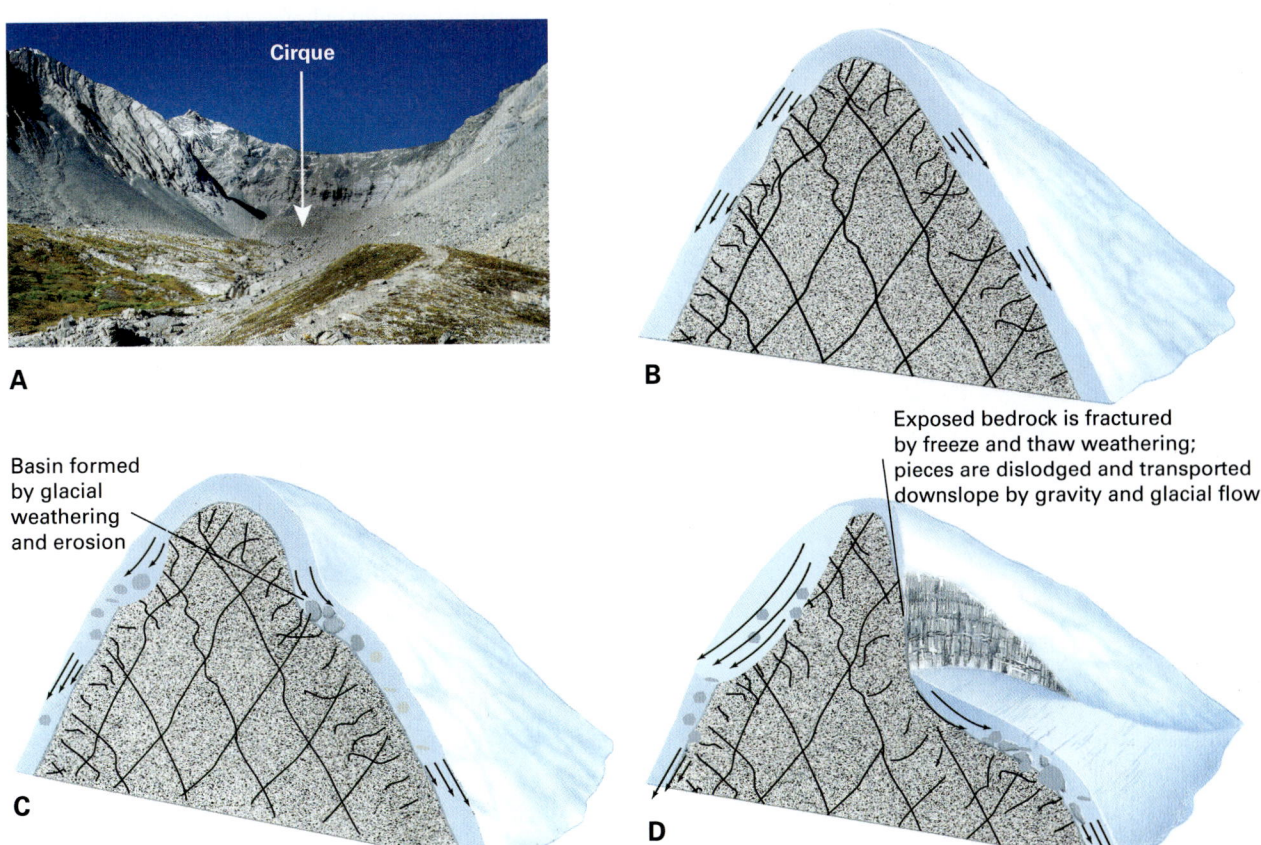

FIGURE 13-12
(A) A glacier eroded this concave cirque into a mountainside in Alberta, Canada, forming Ptarmigan Cirque. (B) To form a cirque, snow accumulates and a glacier begins to flow from the summit of a peak, shown here in cross section. (C) Weathering and glacial erosion form a small depression in the mountainside. (D) Continued glacial movement enlarges the depression. When the glacier melts, it leaves a cirque carved in the side of the peak, as in the photograph.

bottom, carving a broad, rounded U-shaped valley (Figures 13-10 and 13-11).

Steep cliffs that drop off into a scoop-shaped depression in the mountainside are called **cirques**. A small glacier at the head of the cirque reminds us of the larger mass of ice that existed in a colder, wetter time.

To understand how a glacier creates a cirque, imagine snow accumulating and a glacier forming on the side of a mountain (Figure 13-12). As the ice flows down the mountainside, it erodes a small depression that grows slowly as the glacier moves. With time and additional erosion, the depression deepens, causing the cirque walls to become steeper and higher. The glacier carries the eroded rock from the cirque to lower parts of the valley. When the glacier finally melts, it leaves behind a steep-walled, rounded cirque.

Streams and lakes are common in glaciated mountain valleys. As a cirque forms, the glacier may erode a depression into the bedrock beneath it. When the glacier melts, this depression can fill with water, forming a small lake nestled at the base of the cirque. In the valley below high cirques, one may encounter a series of paternoster lakes. The term "paternoster" refers to beads that are connected by a string. Paternoster lakes are formed as a glacier erodes a sequence of small basins that later fill with water when the glacier recedes (Figure 13-13). The lakes are connected by a stream that commonly features rapids and waterfalls.

FIGURE 13-13
Go online to see an example of a string of paternoster lakes in the Seven Rila Lakes region of Bulgaria.

If glaciers erode three or more cirques into different sides of a peak, they will create a steep, pyramid-shaped rock summit called a horn. The Matterhorn in the Swiss Alps is a famous horn

LESSON 13.3   **407**

(Figure 13-14). Similarly, two alpine glaciers flowing down opposite sides of a mountain ridge will erode both sides, forming a sharp, narrow rib of rock called an **arête**. When the glaciers melt, the arête will form a drainage divide that separates the two adjacent valleys or cirques.

A small glacial valley lying high above the floor of the main valley is called a **hanging valley**. The famous waterfalls of Yosemite Valley in California cascade from hanging valleys. A hanging valley forms where a small tributary glacier joined a much larger one. The tributary glacier eroded a shallow valley, while the massive main glacier gouged a deeper one. When the glaciers melted, they exposed an abrupt drop where the small valley joins the main valley.

Deep, narrow inlets called **fjords** extend far inland on many high-latitude seacoasts. Most fjords are glacially carved valleys that were later flooded by encroaching seas as the glaciers melted.

A continental glacier erodes the landscape just as an alpine glacier does. However, a continental glacier is considerably larger and thicker and is not confined to a valley. As a result, it covers a large area that can include entire mountain ranges. The most recent continental glacier in North America stretched from beyond the present New England coastline westward to the high plains of the Dakotas and eastern Montana. This ice sheet sculpted Maine's rocky coastline and modified the shapes of the Northern Appalachian Mountains. The ice also deposited sediment that today forms Massachusetts' famous hook-shaped peninsula and the offshore islands around Martha's Vineyard. Farther west, glacial ice carved out the depressions now occupied by the Great Lakes and deposited vast blankets of sediment across much of the upper Midwest. Even farther west, glacial erosion deepened valleys and sharpened mountain peaks and ridges in the Rocky Mountains and created the subdued lowlands of Puget Sound and the San Juan Islands in western Washington State.

checkpoint What might lead to the formation of a small lake at the base of a cirque?

13.3 ASSESSMENT

1. **Describe** How can glacial erosion create a cirque?
2. **Explain** What causes the formation of paternoster lakes?

Critical Thinking

3. **Apply** Draft a short guide that could be used to identify the formations of fjords, arêtes, and hanging valleys. Provide descriptions that include key characteristics for each feature and would be easily understood by a traveler in a region that features such formations.

FIGURE 13-14
The Matterhorn in Switzerland formed as three alpine glaciers eroded cirques into the peak from three different sides. The sharp ridge between two cirques is called an arête.

13.4 GLACIAL DEPOSITS

Core Ideas and Skills
- Summarize the processes that result in glacial deposits.
- Explain the effects of glacial deposits and identify common locations for formations.

KEY TERMS

drift	till
erratic	moraine
drumlin	outwash
esker	kame
kettle	loess

In the 1800s, geologists recognized that the large deposits of sand and gravel found in the Alps and other places had been transported from distant sources. A popular hypothesis at the time explained that this material had drifted in on icebergs during catastrophic floods. The deposits were called **drift** after this inferred mode of transport.

FIGURE 13-15
Glacial till consists of sediment deposited directly by ice and is typically unsorted.

Today we know that continental glaciers covered vast parts of the land only 10,000 to 20,000 years ago and that these glaciers (not icebergs) carried and deposited drift. Although the term *drift* is a misnomer, it remains in use. Now geologists define drift as an all-embracing term for sediments of glacial origin, no matter how the sediments were deposited.

Landforms Deposited by Glacial Ice

Till refers to sediment deposited directly by glacial ice. Ice is so much more viscous than water that it carries a wide range of particle sizes. When a glacier melts, it deposits particles of all sizes—from fine clay to huge boulders—in an unsorted, unstratified mass (Figure 13-15). Cobbles and boulders carried by a glacier often show scratches or grooves caused by the slow grinding of one clast against another during transport within the ice. However, gravel carried by a glacier is not rounded in the same way as gravel carried by a stream because the clasts in a glacier do not undergo numerous collisions during transport. Therefore, if you find rounded gravel in till, it probably became rounded by a stream before the glacier picked it up.

Erratics In some areas once covered by glaciers, one can find boulders and smaller rocks of a different type than nearby bedrock. Rocks of this type are called **erratics** (Figure 13-16). Most were transported to their present locations by a glacier. The origins of erratics can be determined by exploring the terrain in the direction the glacier came from until the parent rock is found. Erratics have often been carried hundreds of kilometers from their points of origin and provide clues to the movement of glaciers. In some cases, individual mineral grains known to come from rare types of parent rocks can be used to determine the location of the parent rocks.

FIGURE 13-16
Go online to view an example of an erratic located in Central Park in New York City.

Moraines A **moraine** is a mound or a ridge of till. Think of a glacier as a giant conveyor belt. An old-fashioned airport conveyor belt simply carries suitcases to the end of the belt and dumps them in a pile. Similarly, a glacier carries sediment and deposits it at its terminus. If a glacier is neither advancing nor retreating, its terminus may remain in the same place for years. During that time, sediment accumulates at the terminus to form a

LESSON 13.4 **409**

ridge called an end moraine (Figure 13-17). An end moraine that forms when a glacier is at its greatest advance, before beginning to retreat, is called a terminal moraine.

If warmer conditions prevail, the glacier recedes. If the glacier then stabilizes again during its retreat and the terminus remains in the same place for a sufficient amount of time, a new end moraine, called a recessional moraine, forms upslope of the terminal moraine. In some cases, multiple recessional moraines are deposited upslope of the terminal moraine.

When a glacier recedes steadily, till is deposited in a relatively thin layer over a broad area, forming a ground moraine. Ground moraines fill old stream channels and other low spots. Often this leveling process disrupts drainage patterns. Many of the swamps in the northern Great Lakes region and northern New England lie on ground moraines formed when the most recent continental glaciers receded.

End moraines and ground moraines are characteristic of both alpine and continental glaciers. An end moraine deposited by a large alpine glacier may extend for several kilometers and be so high that even a person in good physical condition would have to climb for an hour to reach the top. Moraines may be dangerous to hike over if their sides are steep and the till is loose. Large boulders are mixed randomly with rocks, cobbles, sand, and clay.

The most recent Pleistocene continental glaciers reached their maximum extent about 18,000 years ago (Figure 13-18). Their terminal moraines record the southernmost extent of those glaciers. In North America, the terminal moraines lie in a broad, undulating front extending across the northern United States. Enough time has passed since the glaciers retreated that soil has formed on the till and vegetation now covers the terminal moraines.

When an alpine glacier moves downslope, it erodes the valley walls as well as the valley floor. Additional debris falls or slides down from the valley walls and accumulates directly on mountain glaciers and their margins. Therefore, large sediment loads are carried along the lateral edges of glaciers. When the glacier retreats, this sediment

FIGURE 13-17
The image shows the terminus of an alpine glacier on Baffin Island, Canada. The sharp-crested ridge of sediment in the foreground is the end moraine; a similar ridge extends up the left side of the glacier as a lateral moraine. The two dark stripes extending up the glacier are medial moraines.

FIGURE 13-18
The map shows the maximum extent of the continental glacier in North America during the latest glacial advance, approximately 18,000 years ago. The arrows show directions of ice flow. Terminal moraines associated with the maximum advance stretch from offshore New Jersey to the Puget Sound of Washington State.

is left behind as a ridge of till called a lateral moraine (Figure 13-17).

If two alpine glaciers converge, their lateral moraines merge into the middle of the resulting larger glacier. This till forms a visible dark stripe on the surface of the ice, called a medial moraine (Figure 13-17).

Drumlins Elongated, tear-drop shaped hills, called **drumlins**, are common across the northern United States and are well exposed across parts of upstate New York, Michigan, Wisconsin, Minnesota, and Washington (Figure 13-19). The word *drumlin* is from the Old Irish for "back" or "ridge," and each one looks like a whale swimming through the ground, with its humped back in the air. An individual drumlin is typically about 1 to 2 kilometers long and about 15 to 50 meters high.

Drumlins usually occur in clusters, called drumlin fields. Drumlins typically consist of multiple layers of till that are roughly parallel to the drumlins' upper surface but that also show erosion on the upstream side. A recent study of drumlins in Iceland shows they are formed near the terminus of a glacier by multiple surges of the ice. According to this study,

FIGURE 13-19
These rolling hills in Whitman County in Washington are examples of drumlins.

FIGURE 13-20
Braided streams transport sediment downslope from glaciers, forming an outwash plain. This photo was shot looking up the valley toward the glaciers, which are visible in the background.from offshore New Jersey to the Puget Sound of Washington State.

crevasses in the ice concentrate piles of sediment directly below, along the bed of the glacier. As the glacier surges, the ice sculpts the pile of till into a teardrop shape in which the steeper upstream side is partly eroded and the till is carried to the downstream side, where it is deposited to form the more gently sloping tail of the drumlin. With each successive surge of the glacier, another layer of till is plastered to the outer edge of the drumlin and shaped by the glacial movement.

checkpoint What do all types of landforms deposited by glacial ice have in common?

Landforms Deposited by Glacial Meltwater

During summer, melting of snow and ice causes streams to form on the surface of a glacier. Many of these streams flow off the front or sides of the glacier, while others plunge into tunnels in the ice and flow downward, ultimately to become part of a subglacial stream. Because a glacier erodes great amounts of sediment, subglacial streams and outwash streams are typically laden with silt, sand, and gravel. Most of this sediment is deposited beyond the glacier terminus as **outwash** (Figure 13-20). Outwash streams carry such a heavy sediment load that they often become braided, flowing in multiple channels. Over time, outwash deposited in a valley will form a broad, gently sloping valley floor called a braidplain. Outwash deposited in front of continental glaciers forms a larger outwash plain.

Subglacial streams not only can erode the sediment bed below the glacier, but they also can carve upside-down channels into the base of the ice. These channels in the ice can become filled with well-sorted drift that forms a long, curved deposit called an **esker** when the glacier melts (Figure 13-21). Eventually, subglacial streams will reach the end of the glacier and emerge from beneath its terminus as an outwash stream.

Commonly, large pieces of glacial ice will separate from the main glacier during glacial retreat, forming stagnant ice that is no longer connected to the glacier's conveyor-belt ice delivery system. Sediment from outwash streams flowing around the stagnant ice will accumulate in a pile called a **kame**. As large blocks of stagnant ice melt, they leave behind depressions, called **kettles**, that often fill with water. Kames and kettles commonly occur together, forming topography that is typical of stagnant ice associated with a retreating glacier (Figure 13-21).

Large volumes of glacial meltwater can temporarily collect in lakes that form downslope from a glacier. Proglacial lakes form immediately in front of a retreating glacier and usually are dammed by the glacier's terminal moraine. In addition, large lakes can form in basins that do not have an outflow.

Because kames, eskers, and sediment accumulating in glacial lakes are not deposited directly by ice, they typically show sorting and sedimentary bedding, which distinguishes them from unsorted and unstratified till. In addition, transport by streams usually rounds clasts deposited in these environments, in contrast with the more angular clasts typical of glacial till.

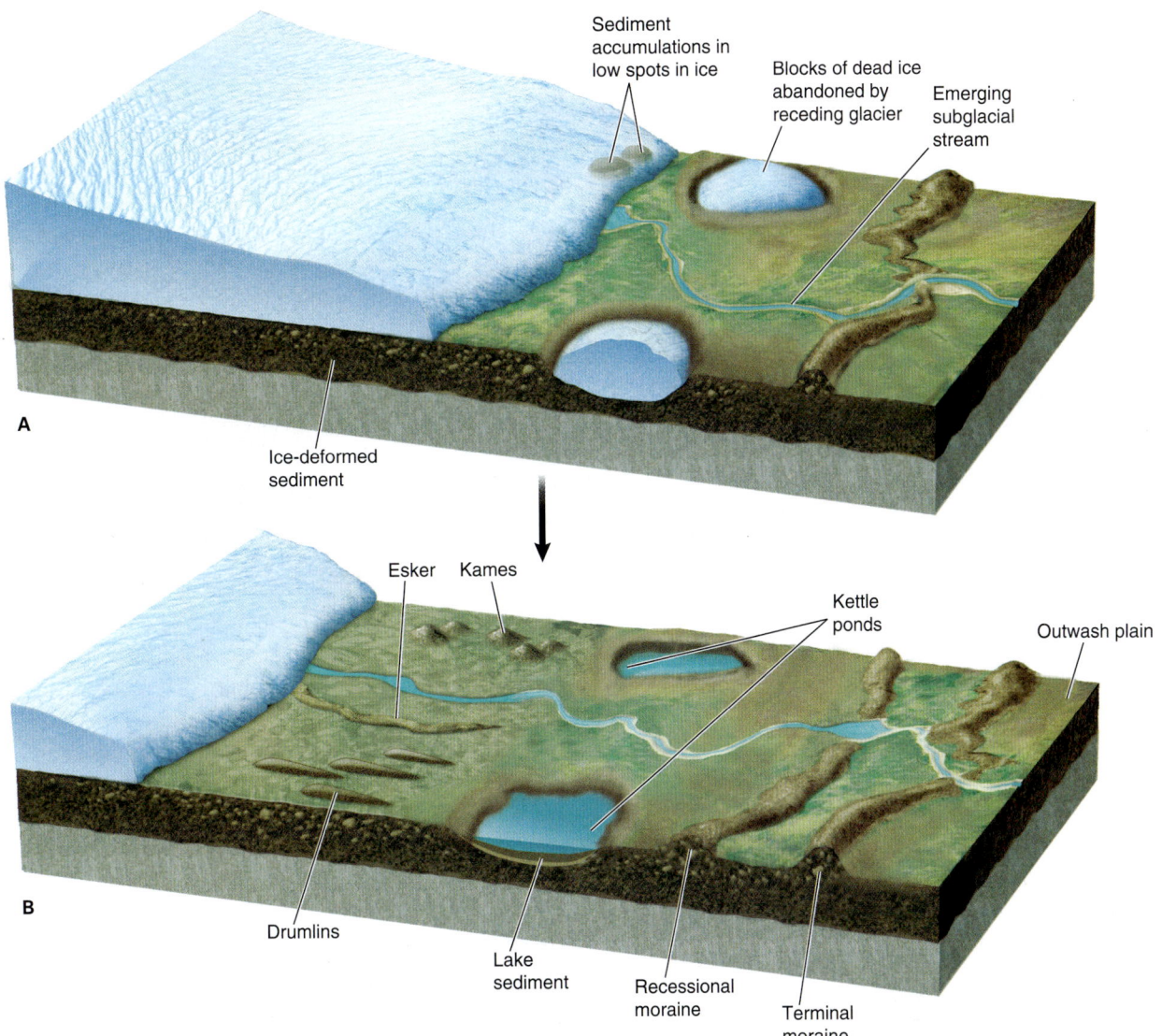

FIGURE 13-21
(A) A melting glacier causes sediment accumulations, a terminal moraine, and subglacial stream. (B) After retreat, glacial deposits are exposed, including a recessional moraine, drumlins, kames, and eskers. Sediment transported by the meltwater forms an outwash plain beyond the terminal moraine.

The deposits created by glaciers dominate much of the landscape of the northern United States. Today, terminal moraines form a broad band of rolling hills from Montana across the Midwest and eastward to the Atlantic Ocean. Drumlins dot the landscape across much of the northern Midwest. Even to the south of the terminal moraines, silt glacial deposits shape the landscape. Sediment formed by the accumulation of windblown silt, called **loess,** covers much of the northern Great Plains. Loess forms into vertical bluffs and cliffs that have weathered to form the fertile, agriculturally productive soil of North America's breadbasket.

checkpoint How does the stagnant ice from a retreating glacier ultimately form kettle lakes?

13.4 ASSESSMENT

1. **Classify** Sort the list into the landforms deposited by ice and by water: drumlins, erratics, eskers, kames, kettles, moraines, outwash plain.
2. **Compare** In what ways are moraines and eskers similar and different?
3. **Explain** Why can a glacier only have one terminal moraine but more than one of the other types of moraines?

Critical Thinking

4. **Generalize** How are crevasses in glaciers connected to the formation of drumlins?

LESSON 13.4

13.5 GLACIATION AND GLACIERS TODAY

Core Ideas and Skills
- Examine the history of the past major glaciations.
- Describe the effects of ice sheet melting on sea level.
- Summarize the great glaciations in Earth's history.
- Analyze current information about glacier melt.

KEY TERMS

glaciation   interglacial period

Periods of Glaciation

Geologists have found terminal moraines extending across all high-latitude continents. By studying those moraines, as well as lakes, eskers, outwash plains, and other glacial landforms, geologists have concluded that massive glaciers once covered large portions of the continents, altering Earth systems. A **glaciation** occurs when alpine glaciers descend into lowland valleys and continental glaciers advance across the land surface at high latitudes. During a glaciation, glaciers several kilometers thick and thousands of kilometers across can spread across the land surface. Beneath the enormous weight of ice, the continents sink deeper into the asthenosphere. The glacial ice erodes rock and soil and deposits it elsewhere, completely altering the landscape.

Geologic evidence shows that Earth has been warm and relatively ice free for at least 90 percent of the past one billion years. However, at least six major glaciations occurred during that time. Each one lasted from 2 to 10 million years and was separated in time from relatively ice-free **interglacial periods**.

The most recent major glaciations have taken place during the Pleistocene epoch and collectively are called the Pleistocene Glaciation. The Pleistocene Glaciation began about 2 million years ago in the Northern Hemisphere, although evidence of an earlier beginning has been found in the Southern Hemisphere. Earth was not glaciated continuously during the Pleistocene epoch. Instead, the climate fluctuated and continental glaciers likely grew and melted away several times (Figure 13-22). During the most recent interglacial period prior to the present day Holocene, the average temperature was about the same as it is today, or perhaps a little warmer. Then, high-latitude temperature dropped at least 15°C, causing the ice to advance. Many climate models indicate that conditions that lead to the Pleistocene Glaciation still exist and that continental ice sheets may advance once again.

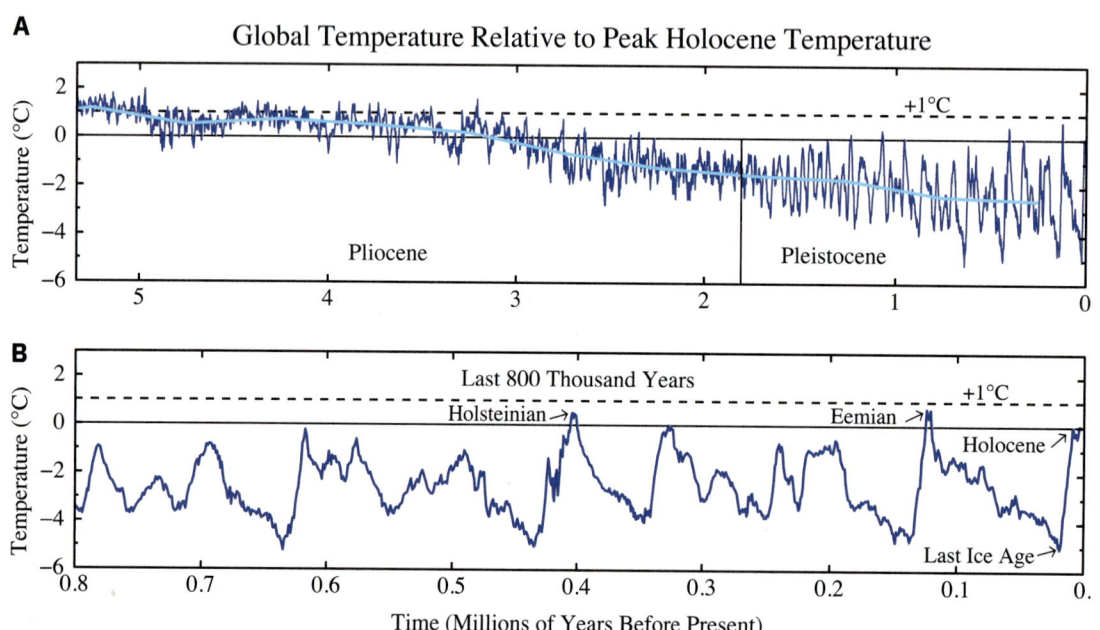

FIGURE 13-22
(A) Temperature variations from the Holocene average over the last 5.3 million years (blue line) suggest long periods of extreme cold temperature and multiple glaciations during the Pleistocene epoch. (B) A similar graph shows temperature fluctuations over the last 800,000 years. The last ice age is marked ending about 10,000 years ago.

DATA ANALYSIS: Graphing Glacial Melt

The area of a glacier can be measured using photographs taken from an airplane as it flies over the glacier. Glaciers at Glacier National Park in Montana were first measured for area in 1966 and most recently measured in 2015. Nine out of the 35 named glaciers are now considered inactive because their areas are less than 101,171 square meters (25 acres). Download "Active Glaciers", a data set of the park's 26 active glaciers to complete the questions that follow.

Computational Thinking

1. **Automate** Use a spreadsheet formula to calculate the percent area remaining for each glacier using the area values from 1966 and 2015. Round to the nearest whole number percent.

2. **Represent Data** Create a line graph comparing the change in area over time for five glaciers from the data set. Plot the glacial areas for each of the five glaciers in 1966, 1998, 2005, and 2015. Use a different color to represent each glacier and create a legend. Be sure to clearly label your axes and give your graph a title.

3. **Recognize Patterns** Describe any patterns you observe in your calculations and graph. What conclusions can you draw about how glaciers in Glacier National Park are changing?

4. **Analyze Data** Did any of the glaciers grow rather than shrink for any time period during the study?

5. **Predict** Which glaciers do you think are at highest risk of becoming inactive in the near future? Why?

At its maximum extent about 18,000 years ago, the most recent North American ice sheet covered 10 million square kilometers—most of Alaska, Canada, and parts of the northern United States. At the same time, alpine glaciers flowed from the mountains into the lowland valleys.

Sea Level Changes with Glaciation When glaciers grow, they accumulate water that would otherwise be in the oceans, and sea level falls. When glaciers melt, sea level rises again. When the Pleistocene glaciers reached their maximum extent roughly 18,000 years ago, global sea level fell to about 130 meters below its present elevation. As submerged continental shelves became exposed, the global land area increased by 8 percent (although about one-third of the land was ice covered).

When the ice sheets melted, most of the water returned to the oceans, raising sea level again. At the same time, portions of continents rebounded isostatically as the weight of the ice was removed. So, while the rising seas submerged some coastlines, others rebounded more than sea level rose, causing the sea to retreat. For example, former beaches in the Canadian Arctic now lie tens of meters to a few hundred meters above the sea.

Snowball Earth: The Greatest Glaciation in Earth's History Research suggests that at least twice, and perhaps as many as five times in late-Precambrian time between 800 million and 550 million years ago, massive ice sheets completely covered all continents and the world's oceans froze over—even at the Equator—entombing the entire globe in a 1-kilometer-thick shell of ice. This glaciation, called Snowball Earth by the researchers who discovered it, contrasts sharply with the Pleistocene glaciation, when ice covered just a third of the continents and only the polar seas froze over.

The main evidence for the Precambrian Snowball Earth is based on a unique rock called tillite. Recall that till consists of an unsorted mixture of boulders, silt, and clay that was deposited by a glacier. Pleistocene tills are made of loose sediment that can be easily dug up with a shovel. Tillite, however, is hard, solid rock that in every other respect resembles the Pleistocene tills. Simply put, tillite is till that was deposited by glaciers so long ago that it has become cemented into hard rock. Researchers have found at least two thick layers of tillite between 750 million and 580 million years old on almost every continent. Recall that continents have moved around Earth through geologic time. Other types of evidence show that some of the continents lay at the Equator when the tillites formed. Other continents were nearer to the Poles at the same time, showing that the glaciations were global in scope.

checkpoint What caused former beaches in the Canadian Arctic to lie many meters above the sea?

Earth's Disappearing Glaciers

Glaciers are now shrinking in more places and at more rapid rates than at any other time since scientists began keeping records. In 2003, during

FIGURE 13-23
(A) This 1907 photo shows Thunderbird Glacier in Glacier National Park in Montana.
(B) By 2007, glacier retreat shows a very different landscape.

a single, hot summer in Europe, 10 percent of the glacial ice in the Alps melted. Scientists point out that this accelerated loss of glacial ice coincides with an increase in atmospheric carbon dioxide levels of about 50 percent in the past century, and they suggest that the melting may be one of the first symptoms of human-caused global warming.

Alpine glaciers reflect these changes in a highly visible way. The larger glaciers in Glacier National Park, Montana, have shrunk to a third of their size since 1850, and they continue to melt away today (Figure 13-23). Many of the smaller glaciers have disappeared completely. Since 1850, the number of glaciers in the park has decreased from 150 to fewer than 25 in 2010. One computer model predicts that all glaciers will be gone from the park by 2030 if global temperatures continue to rise as predicted, though other models suggest that this prediction is overstated.

Significant loss of glacial ice volumes through melting is a global phenomenon. In some instances, ice disintegration has been particularly dramatic. For example, half of Alaska's Columbia Glacier melted between 1980 and 2011, and the glacier currently releases about 5 cubic kilometers of meltwater into Prince William Sound every year. Elsewhere, scientists predict that glaciers in the European Alps will be all but gone by 2050. Global climate changes do not occur uniformly around the globe. The north polar regions are warming faster than the average rate for all of Earth, and this warming is having a profound effect on Arctic sea ice. During the winter, Arctic sea ice covers an area about the size of the United States, though it is rapidly shrinking in size and thickness. Ice thickness measurements taken during 2011 and 2012 indicate that the summer ice pack in those years was the thinnest, least extensive on record. Scientists predict that Arctic summer sea ice will completely disappear in the next few decades. Such a loss would greatly increase the amount of solar energy that is absorbed by the Arctic Ocean, potentially accelerating global warming.

Most of Earth's ice shelves surround Antarctica. Ice shelves respond to rising temperature more sensitively than glaciers do. Since 1974, seven Antarctic ice shelves have shrunk by a total of 13,500 square kilometers. These ice shelves are described further in Chapter 21.

The melting of the Arctic and Antarctic ice has profound effects for the planet. This melting contributes to the rise in sea level, and the fresh water flowing into the ocean may ultimately alter ocean currents and global climate.

checkpoint What could happen if Arctic summer sea ice disappears?

13.5 ASSESSMENT

1. **Compare** How is sea level affected during periods of glaciation and glacial melting?
2. **Summarize** Discuss the evidence for the glaciation known as Snowball Earth.

Critical Thinking

3. **Synthesize** Consider the consequences of a Snowball Earth. Describe depositional landforms that you would or would not expect to find around Earth as a global glaciation period came to an end.

13.6 WHY DO DESERTS EXIST?

Core Ideas and Skills
- Identify the defining characteristics of a desert climate.
- Identify geographical factors that contribute to the formation of deserts.

KEY TERMS

desert   rain-shadow desert

Different regions of Earth are classified into climate types based primarily on precipitation and temperature. In turn, climate determines the communities of plants and animals that live in a region. A **desert** is any region that receives less than 25 centimeters (10 inches) of rain per year and consequently supports little or no vegetation. Most deserts are surrounded by semiarid zones that receive 25 to 50 centimeters of annual rainfall—more moisture than a true desert but less than adjacent regions.

Deserts cover 25 percent of Earth's land surface outside of the polar regions and make up a significant part of every continent. If you were to visit the great deserts of Earth, you might be surprised by their geologic and topographic variety. You would see coastal deserts along the beaches of Chile; shifting dunes in the Sahara; deep, red sandstone canyons in southern Utah; stark granite mountains in Arizona; and bitter-cold polar deserts with a few lichens clinging tenaciously to the otherwise barren rock. The world's deserts are similar only in that they all receive scant rainfall.

Rain and snow are unevenly distributed over Earth's surface. For example, Mount Waialeale, Hawaii, one of the wettest places on Earth, receives an average of 1,168 centimeters (38 feet) of rain annually. In contrast, 10 years or more may pass between rains or snowfalls in the Atacama Desert of Peru and Chile. Several factors—including latitude, mountains, and atmospheric circulation—control rainfall patterns and therefore the global distribution of deserts and semiarid lands.

Latitude

The sun shines most directly overhead near the Equator, warming air near Earth's surface. The air absorbs moisture from the equatorial oceans and rises because it is warmer, and therefore less dense, than surrounding air. Rising air cools as pressure decreases. But cool air cannot hold as much water as warm air, so the water vapor condenses and falls as rain (Figure 13-24). For this reason, vast tropical rain forests grow near the Equator.

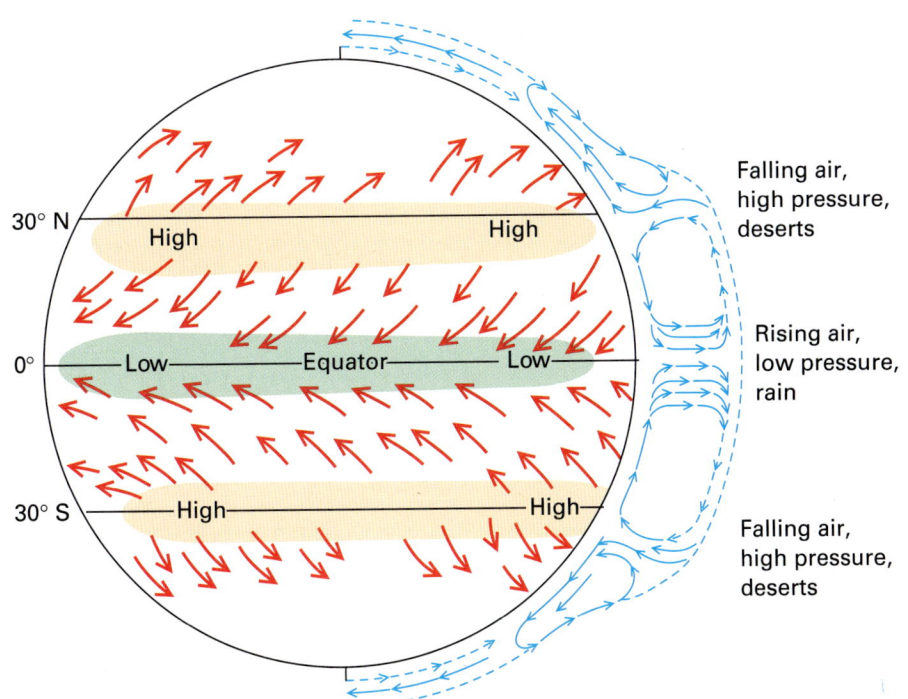

FIGURE 13-24

Falling air creates deserts at 30° north and south latitudes. The red arrows indicate surface winds. The blue arrows on the right show airflow on the surface and at higher elevations.

LESSON 13.6   **417**

This rising equatorial air, which is now drier because of the loss of moisture, flows both northward and southward at high altitudes. There the air cools, becomes denser, and sinks back toward Earth's surface at about 30° north and south latitudes. As the air falls, it is compressed and becomes warmer, which enables it to hold more water vapor. As a result, water evaporates from the land surface into the air. Because the sinking air absorbs water, the ground surface is dry and rainfall is infrequent. Consequently, many of the world's largest deserts lie at about 30° north and south latitudes (Figure 13-25).

checkpoint Why does sinking air at about 30° north and south latitudes cause the ground surface to be dry?

Mountains and Rain-Shadow Deserts

When moisture-laden air flows over a mountain range, it rises. As the air rises, it cools and its ability to hold water decreases. As a result, the water vapor condenses into rain or snow, which falls on the windward side and on the crest of the range (Figure 13-27). The drier, cool air continues down the leeward (or downwind) side of the mountain, compressing and warming as it descends. This warm, dry air creates an arid zone called a **rain-shadow desert** on the leeward side of the range.

In this way, tectonic forces, which produce mountains, affect very long-term rainfall patterns. In turn, flowing water will weather rock, forming soil and defining the local environmental conditions on which various ecosystems exist. The connection provides another example of interacting Earth systems. The building of a mountain range (a tectonic process) alters rainfall (an atmospheric process) and ultimately defines the types of ecosystems that characterize a region (the biosphere).

checkpoint Which side of a mountain is arid, the windward or leeward side?

Coastal and Interior Deserts

Because most evaporation occurs over oceans, one might expect that coastal areas would be moist and climates would become drier with increasing distance from the sea. This is generally true, but a few notable exceptions exist.

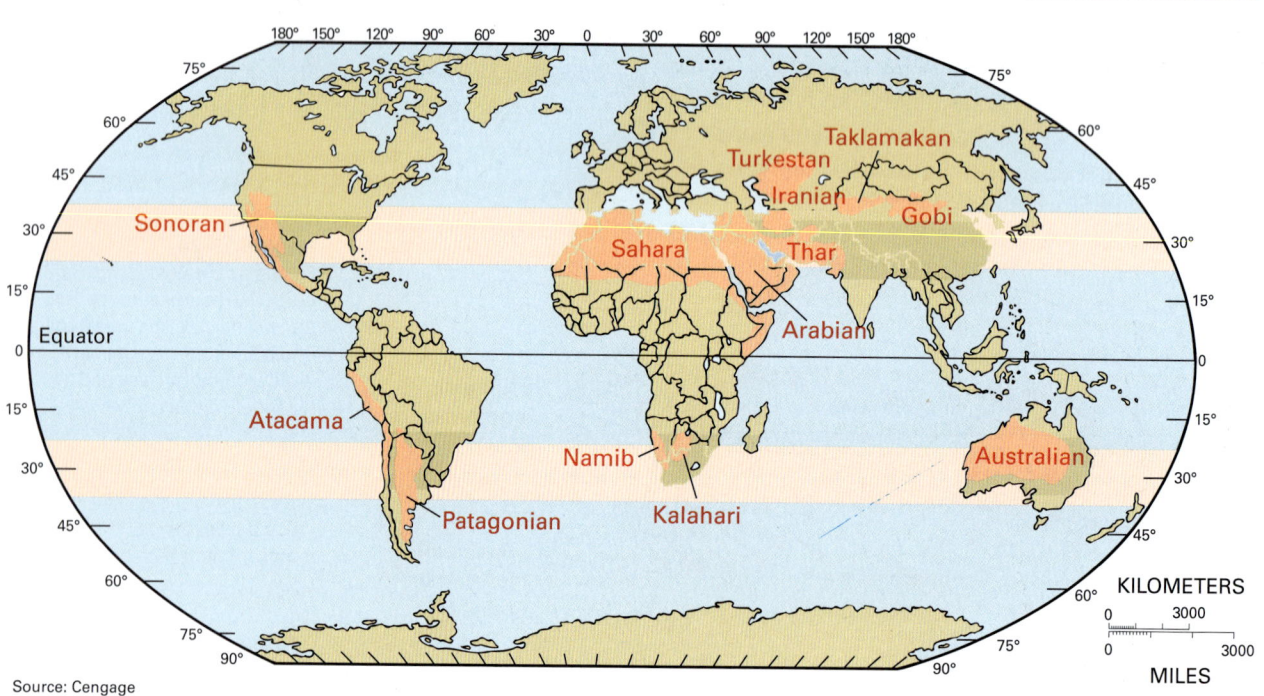

FIGURE 13-25
The major deserts of the world are concentrated at approximately 30° north and south latitudes.

DATA ANALYSIS Which Is the Desert?

A biome is a community of plants in a large area associated with a particular climate. A desert biome supports few plants because of its arid climate. The graphs below present temperature and precipitation data for three different biomes, one of which is a desert. Analyze the graphs to determine which is the desert biome and answer the questions that follow.

1. **Compare** How do the three biomes compare in terms of average temperatures throughout the year? Identify some similarities and differences.
2. **Contrast** Which biome receives the most annual precipitation? The least?
3. **Analyze** Based on the data shown in the graphs, which of the three biomes is a desert? Explain your reasoning.
4. **Predict** What kinds of vegetation would you expect to find in each biome?

Computational Thinking

5. **Recognize Patterns** Is there any correlation between temperature and precipitation? If so, does it hold true for all three biomes?

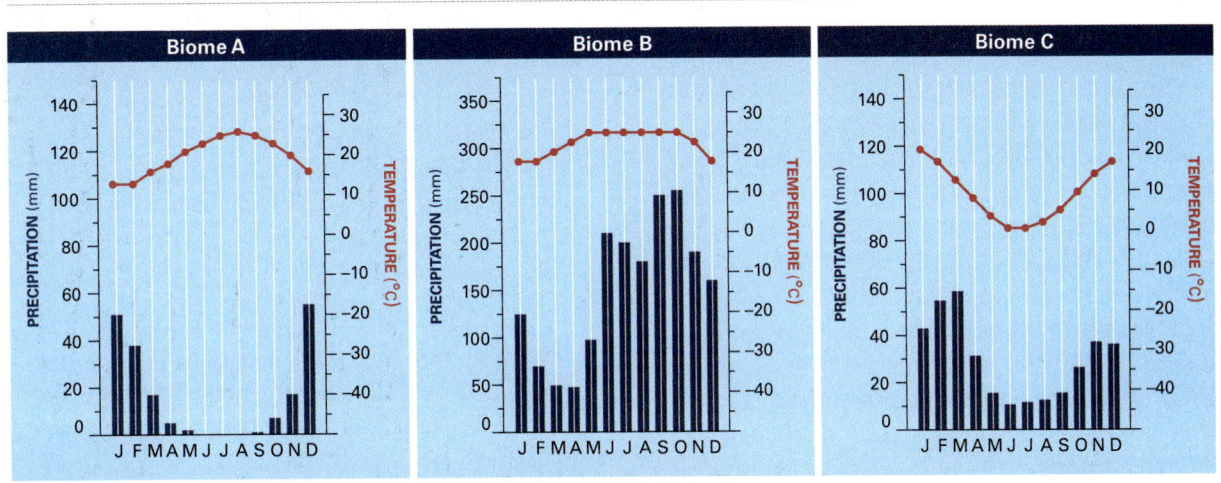

Source: NASA

FIGURE 13-26
Three graphs show the average monthly temperature (red line) in degrees Celsius and the average monthly precipitation (blue bars) in millimeters for three biomes.

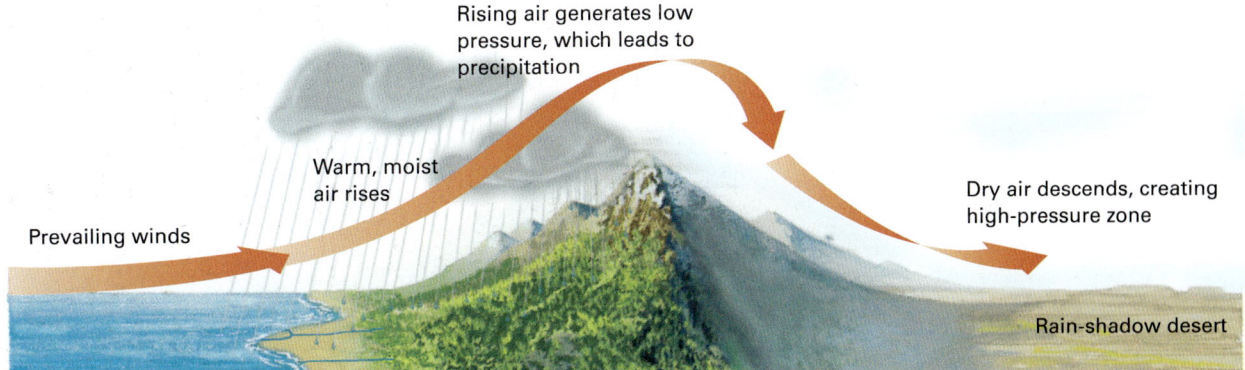

FIGURE 13-27
Warm, moist air from the ocean rises as it flows over mountains. As it rises, it cools and water vapor condenses to form rain. The dry, descending air on the leeward side absorbs moisture, forming a rain-shadow desert.

LESSON 13.6

The Atacama Desert (Figure 13-28) along the west coast of South America is so dry that portions of Peru and Chile often receive no rainfall for a decade or more. Cool ocean currents flow along the west coast of South America. The desert is situated west of the Andes Mountains, where prevailing winds flowing eastward meet the high pressure of air descending westwards from over the mountains, resulting in warm, dry air.

The Gobi is a broad, arid region in central Asia. The center of the Gobi lies at about 40° N latitude, and its eastern edge is a little more than 400 kilometers from the Yellow Sea. As a comparison, Pittsburgh, Pennsylvania, lies at about the same latitude and is 400 kilometers from the Atlantic Ocean. If latitude and distance from the ocean were the only factors, these regions would have similar climates. However, the Gobi is a barren desert and western Pennsylvania receives enough rainfall to support forests and rich farmland. The Gobi is bounded by the Tibetan Plateau to the south and the Altai and Tien Shan mountain ranges to the west, which shadow it from the prevailing winds. In contrast, winds carry abundant moisture from the Gulf of Mexico, the Great Lakes, and the Atlantic Ocean to western Pennsylvania.

In some regions deserts extend to the seashore, while elsewhere the interior of a continent can be humid. The climate at any particular place on Earth results from a combination of many factors. Latitude and proximity to the ocean are important, but complex interactions involving the direction of prevailing winds, the direction and temperature of ocean currents, and the positions of mountain ranges also control climate. These complex relationships will be explored in more detail in Chapter 21.

checkpoint Why do the Altai and Tien Shan mountain ranges create an arid region, the Gobi?

13.6 ASSESSMENT

1. **Summarize** Why are many deserts concentrated along zones at 30 degrees latitude in both the Northern Hemisphere and the Southern Hemisphere?
2. **Relate** List three factors that control global distribution of deserts.
3. **Describe** How does a rain-shadow desert form?

Critical Thinking

4. **Analyze** Compare Figure 13-25 and Figure 7-20. Based on these maps, identify where there were likely deserts 100 million years ago and where deserts might exist in 100 million years.

FIGURE 13-28
Coastal deserts, such as the Atacama Desert, provide unusual scenes of cacti growing just above the water's edge and dry slopes all around.

13.7 WATER AND DESERTS

Core Ideas and Skills
- Explain the role of water in the desert environment.
- Compare formations caused by erosion in a desert.
- Describe how two different North American deserts formed.

KEY TERMS

wash	playa
bajada	pediment
plateau	mesa
butte	spire

Although rain and snow rarely fall in deserts, water plays an important role in these dry environments. Water can reach a desert from three sources. Streams flow from adjacent mountains or other wetter regions, bringing surface water to some desert areas. Groundwater may also flow from a wetter source to an aquifer beneath a desert. Finally, rain and snow fall occasionally on deserts. So, even in the driest places on Earth, the hydrosphere and geosphere interact.

Vegetation is sparse in most deserts because of the limited water supply; much bare soil is exposed, unprotected from erosion. As a result, rain can easily erode desert soils, and flowing water is an important factor in the evolution of desert landscapes.

Desert Streams and Lakes

Large rivers flow through some deserts. For example, the Nile River flows through North African deserts and the Colorado River crosses the arid southwestern United States. Desert rivers receive most of their water from wetter, mountainous regions bordering the arid lands.

In a desert, water seeps downward from the stream into the ground. As a result, many smaller desert streams flow only for a short time after a rainstorm, or during the spring when winter snows are melting, before their water seeps below the streambed. A streambed that is dry for most of the year is called a **wash** (Figure 13-29).

During the wet season, rain and streams may fill a desert lake. Whereas some desert lakes are drained by outflowing streams, most lose water

FIGURE 13-29
This small wash in the Grand Canyon is normally dry but prone to violent flash floods, as seen here.

FIGURE 13-30
(A) The Dead Sea seashore is encrusted in salt. Water in the Dead Sea is more saline than the ocean. (B) In eastern Oregon, cracks occur in the Alvord Playa.

only by evaporation and seepage. In large desert lakes such as the Great Salt Lake in Utah and the Dead Sea in Israel and Jordan, evaporation has caused the remaining lake water to become very saline. At times, the precipitation of salt within the lake basin takes place. In smaller desert lakes, inflowing streams may cease to flow during the dry season, causing the lake to dry up completely due to evaporation and seepage. This kind of intermittent desert lake is called a playa lake, and the dry lake bed is called a **playa** (Figure 13-30).

checkpoint In a desert, what kind of lake is only present from time to time?

CONNECTIONS TO HISTORY

Mining in Death Valley

Streams and groundwater contain dissolved salts. When this slightly salty water fills a desert lake and then evaporates, the ions are left behind and salt minerals precipitate on the playa. Over many years, economically valuable mineral deposits may accumulate. Borax, a component of many detergents and cosmetics, and other valuable minerals are abundant in the evaporated deposits of Death Valley. In the 1880s, borax mining took place and large mule teams were organized to haul the ore from the valley (Figure 13-31).

FIGURE 13-31
This wagon is on display at Harmony Borax Works in Death Valley National Park. Each Harmony Borax wagon carried 10-ton loads when borax mining was in operation.

FIGURE 13-32
This alluvial fan in Mongolia has formed where an intermittent stream (dry in this photograph) passes through a small, confining canyon and emerges into a more open, gently sloping plain.

Pediments and Bajadas

When a steep, flooding mountain stream empties into a flat valley, the water slows abruptly and deposits most of its sediment at the mountain front, forming an alluvial fan. An alluvial fan is a deposit of sediment formed where a relatively steep, confined stream enters a less confined region with a lower slope angle. Streams associated with alluvial fans usually are intermittent, frequently dry, and capable of transporting large volumes of sediment and large clasts quickly. The sediment is distributed across the fan surface as the stream frequently shifts its course. Although alluvial fans form in all climates, they are particularly noticeable in deserts (Figure 13-32). A large fan may be several kilometers across and rise a few hundred meters above the surrounding valley floor.

If the mouths of several canyons are spaced only a few kilometers apart, the alluvial fans extending from each canyon may merge. A **bajada** is a broad, gently sloping depositional surface formed by alluvials fans that merge together and extend into the center of a desert valley (Figure 13-33). Typically, the fans merge incompletely, forming an irregular surface that follows the mountain front for tens of kilometers. Over millions of years,

FIGURE 13-33
The bajada in the foreground merges with a gently sloping pediment to form a continuous surface in front of these mountains in Mongolia. This basin is filling with sediment from the surrounding mountains because it has no external drainage.

the alluvial sediment forming the bajada may accumulate to a thickness of several kilometers.

A **pediment** is a broad, gently sloping erosional surface. Pediments commonly form along the front of desert mountains, because tectonic uplift results in erosion. The surface of a pediment is covered with a thin veneer of gravel that is in the process of being transported from the mountains, across the pediment, to the bajada.

Together, a pediment and bajada form a smooth surface from the mountain front to the valley center. The surface steepens slightly near the mountains. To distinguish a pediment from a bajada, you would have to dig or drill a hole. If you were on a pediment when making the hole, you would strike the eroded rock or sediment after only a few meters. On a bajada, in contrast, hundreds or even thousands of meters of recent gravel would cover the older rock or sediment.

checkpoint What formation arises when alluvial fans merge?

Two American Deserts

The Colorado Plateau The Colorado Plateau covers a broad region encompassing portions of Utah, Colorado, Arizona, and New Mexico (Figure 13-34). A **plateau** is a region of fairly flat land that has been uplifted by tectonic forces and usually is made of relatively flat-lying sedimentary or volcanic rock. During the past billion years of Earth's history, the Colorado Plateau has been alternately covered by shallow seas, deserts, and broad alluvial plains with rivers and lakes. Sediment accumulated in these environments and slowly lithified to become flat layers of sedimentary rock. Tectonic forces later uplifted the lithified flat layers of rock over a broad region, forming the Colorado Plateau. Today, much of the Colorado Plateau receives less than 25 centimeters of rainfall per year.

As the plateau rose, the Colorado River cut downward through the bedrock to form the 1.6-kilometer-deep Grand Canyon and its tributary canyons. The modern Colorado River receives most of its water from snowmelt and rain in the high Rocky Mountains east and north of the plateau.

In addition to water, the Colorado River and its tributaries transport much sediment toward the Gulf of California. Uplift of the Colorado Plateau causes these streams to erode downward into bedrock. This downcuttting continues until a resistant layer is reached. Such a resistant layer will form a temporary base level. Once the base level is reached, the stream begins to erode laterally, widening the canyon by undercutting its walls and causing the rock to collapse along vertical joints. Continued lateral erosion and removal of the sediment by the river forms a relatively level plain with flat-topped mesas and buttes that rise above it as erosional remnants that are left behind. Smaller than a plateau is a flat-topped mountain shaped like a table and called a **mesa**—Spanish for the word *table*. A **butte** is also a flat-topped mountain, smaller and more towerlike than a mesa, and characterized by steep cliff faces. Even smaller is a **spire**—a single tower with a pointy top. Mesas, buttes, and spires are all erosional

FIGURE 13-34
The Colorado Plateau and the Great Basin together make up a large desert and semiarid region of the western United States.

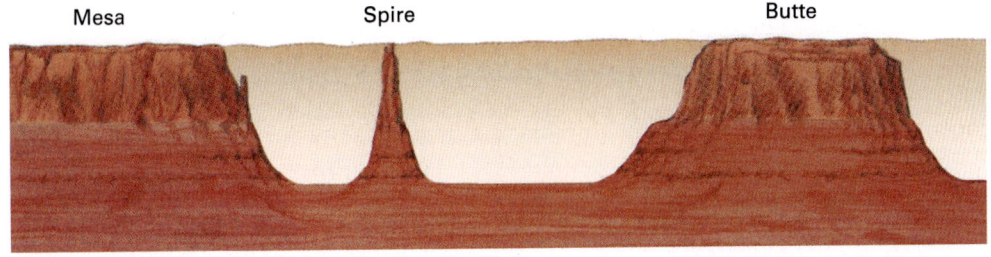

FIGURE 13-35 Buttes and spires of erosion-resistant rock remain after the lateral erosion of mesas. Some debris remains at the base of the formations, but ancient streams transported away huge volumes of sediment.

landforms that are common in the Colorado Plateau (Figure 13-35).

The flat tops of mesas and buttes usually form on a bed of sandstone or other sedimentary rock that is relatively resistant to erosion. These resistant upper layers temporarily protect the landforms from erosion, somewhat like a roof protects a building, although continued lateral erosion of the sides of the landform ultimately causes it to be removed. In many cases across the Colorado Plateau, extensive lateral erosion above a temporary base-level leaves spectacular pinnacles or spires, isolated remnants of once-continuous rock layers that the streams have all but completely eroded away. The huge volumes of sediment that are produced through this lateral erosion are transported away from the region by streams and ultimately are deposited near the shoreline as part of a delta or carried farther offshore to be deposited on a deep-water submarine fan.

Death Valley and the Great Basin Death Valley lies in the rain shadow of the High Sierra in California. The deepest part of the valley is 86 meters below sea level. It is a classic rain-shadow desert, receiving a scant 5 centimeters of rainfall per year. The mountains to the west receive abundant moisture, and during the winter rainy season and spring snowmelt, streams flow from the mountains into the valley, where a playa lake forms.

On the Colorado Plateau, the Colorado River and its tributaries carry sediment away from the desert to the Gulf of California, leaving deep canyons. In contrast, streams flow into Death Valley from the surrounding mountains, but no streams flow out. Because Death Valley has no external drainage, the valley is filling with sediment eroded from the surrounding mountains. The sediment collects to form vast alluvial fans and bajadas. Stream water flows into a playa lake that dries up under the hot summer sun.

Death Valley is just one small part of a vast desert region in the American West that has no external drainage. The Great Basin (which includes most of Nevada; the western half of Utah; and parts of California, Oregon, and Idaho) is a large desert region between the Sierra Nevada and the Rocky Mountains. Because there are no streams flowing out of the Great Basin, sediment is not removed and instead has accumulated to become thousands of meters thick.

Like the Colorado Plateau, the Great Basin currently is being uplifted, but it is also being pulled apart by ongoing tectonics. This tectonic extension causes normal faults to form, which downdrop the valleys and uplift the mountains of the Great Basin. The downdropped valleys are graben, and the uplifted mountains are horsts (Figure 10-10).

Desert streams have partially eroded the mountains and deposited sediment to form pediments, alluvial fans, and bajadas that today are very common in the Great Basin. With no streams carrying deposits out of the Great Basin, the mountains are slowly being covered by their own sediment.

checkpoint Which feature exists after the most erosion of sedimentary rock: a mesa, a butte, or a spire?

13.7 ASSESSMENT

1. **Explain** How does flowing water affect desert landscapes?
2. **Distinguish** Why are alluvial fans more conspicuous in deserts than in humid environments?
3. **Distinguish** Compare and contrast desert plateaus, mesas, and buttes.

Critical Thinking

4. **Design** Create an evaluation checklist that could be used to correctly differentiate between a pediment and a bajada.
5. **Apply** Explain how a team could study the effects of stream erosion and deposition in Death Valley.

FIGURE 13-36
ON ASSIGNMENT The spires of Monument Valley inspire awe as much for what can be seen, as for the rock we understand to have eroded. Millions of years of erosion yielded the formations in the photograph. National Geographic photographer Ira Black took this aerial photo of Monument Valley at sunset. How many sunsets and sunrises passed before the rock took the form we see here? What will be left in another million years?

13.8 WIND AND DESERTS

Core Ideas and Skills
- Examine the effects of wind in a desert environment.
- Describe the results of erosion and deposition in a desert, including the formation of dunes.

KEY TERMS	
deflation	desert pavement
saltation	dune
slip face	

Just as water from the hydrosphere plays an important role in the desert environment, moving air of the atmosphere also shapes and sculpts the desert landscape. When wind blows through a forest or across a prairie, the trees or grasses shield the soil from wind erosion. Water dampens the soil and binds particles together, further protecting it from wind erosion. In contrast, a desert commonly has little or no vegetation and rainfall, so wind erodes bare, unprotected desert soil.

Wind erosion is not limited to deserts. Wind is an important agent of erosion wherever the wind blows over unvegetated soil. Windblown dunes of sand are common along seacoasts, where salty sea spray limits plant growth, and in regions recently left bare by receding glaciers.

Wind Erosion, Transport, and Abrasion

Wind Erosion Wind erosion of loose sediment from flat, dry areas, called **deflation**, is a selective process. Wind is capable of moving only small particles, generally those sand-sized and finer. Imagine bare soil containing silt, sand, pebbles, and cobbles. When wind blows, it removes only the silt and sand, leaving the pebbles and cobbles behind to form a continuous armoring of stones called a **desert pavement** (Figure 13-37). A desert pavement prevents the wind from eroding additional sand and silt, even though this finer sediment may be abundant beneath the stony armor. Approximately 80 percent of the world's desert area is covered by a pavement of pebbles and cobbles (Figure 13-38), whereas only 20 percent is covered by sand.

⊕ **FIGURE 13-38**
Go online to view an example of stony desert pavement located in Bladensburg National Park, Australia.

Transport and Abrasion Wind abrasion is caused by sand-sized grains of sediment that are carried by the wind. When it is strong enough, wind can begin to roll and bounce sand grains across the ground surface. If the wind strengthens, some of the grains will be kicked upward then carried downwind a short distance before falling back to the surface, in a process called **saltation** (from the Latin verb *saltare*, "to dance"). The resulting trajectory of each

Wind removes surface sand

Formation of desert pavement complete—no further wind erosion

FIGURE 13-37
Wind erodes silt and sand but leaves larger rocks behind to form desert pavement.

grain is asymmetric, with a steeper ascent angle and a more shallow descent angle. When the wind is strong enough to cause many grains to saltate at one time, a saltation carpet is formed. Such windblown, saltating sand is abrasive and capable of eroding bedrock. In some cases where hard bedrock such as quartzite is exposed, windblown sand can polish the bedrock surface. Because saltation carpets form next to the ground surface, wind erosion usually is most intense close to the ground. For example, saltating sand may erode the base of a large boulder over time, resulting in a formation that looks like a large rock balancing on a thin pedestal. Eventually, such balanced rocks will collapse as the pedestal continues to be undercut.

In contrast to sand-sized particles, smaller silt and clay-sized particles can be carried by wind for much longer distances. Large storms in the Sahara, for example, can carry small particles in suspension in the atmosphere for hundreds of miles before they settle back to the surface. Skiers in the Alps commonly encounter a thin layer of silt on the snow surface resulting from wind transport of fine sediment from the Sahara across the Mediterranean Sea.

checkpoint Are more of the world's deserts covered by a pavement of pebbles or covered by sand?

Dunes

A **dune** is an asymmetric mound or ridge of wind-deposited sand (Figure 13-39). The wind commonly deposits such sand in a topographic depression or other place where the wind slows down. Dunes commonly grow to heights of 30 to 100 meters, and some giants exceed 500 meters. Most dunes are between about 100 meters and 1 kilometer across, although some dune ridges in the western Sahara are hundreds of kilometers long. The largest dune field on Earth is the Rub al-Khali ("empty quarter") in the Arabian Desert, which covers 650,000 square kilometers, an area larger than the state of California. Most dune fields cover a few square kilometers.

Dunes also form where glaciers have recently receded and along sandy coastlines. Glacial deposits consist of large quantities of bare, unvegetated sediment. A sandy beach is commonly unvegetated because sea salt prevents plant growth. As a result, both of these environments contain the essentials for dune formation: an abundant supply of sand and a windy environment with sparse vegetation.

Most dunes are asymmetrical. Sand eroded by wind from the windward side of the dune saltates up to the dune crest. From there, the sand moves down the steep leeward dune face, called the **slip face**, as a series of small, tongue-like grain flow deposits before coming to rest as an inclined layer of sand (Figure 13-39). Typically, the slip face dips about 35 degrees from horizontal—the angle of repose for dry sand. The inclined layers of sand that result from this process form cross-beds that reflect the original slope of the dune slip face.

Migrating dunes can overrun buildings and highways. For example, near the town of Winnemucca, Nevada, dunes advance across U.S. Highway 95 several times a year. Highway crews must remove as much as 4,000 cubic meters of sand (roughly half the volume of an Olympic-sized swimming pool) to reopen the road.

Engineers often attempt to stabilize dunes in inhabited areas. One method is to plant vegetation to reduce deflation and stop dune migration. The main problem with this approach is that desert dunes commonly form in regions that are too dry to support vegetation. Another solution is to build artificial windbreaks to create dunes in places where they do the least harm. For example, a fence traps blowing sand and forms a dune, thereby protecting areas downwind. Fencing is a temporary solution, however, because eventually the dune covers the fence and resumes its migration. In Saudi Arabia, dunes are sometimes stabilized by covering them with tarry wastes from petroleum refining.

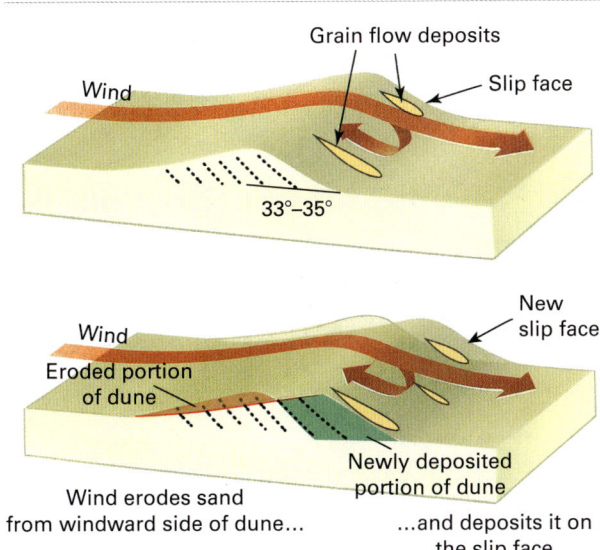

FIGURE 13-39
Sand dunes migrate in the downwind direction as sand is eroded off the upwind side of the dune and then is transported to the brink of the dune; it then avalanches down the steep slip face on the downwind side in small grain flows.

Types of Sand Dunes Wind speed and sand supply control the shapes and orientation of dunes. A barchan dune (Figure 13-40A) is one of four common types of dunes and forms in rocky deserts where there is a limited supply of sand. The center of the dune grows higher than the edges but migrates forward more slowly, causing the dune to become crescent shaped, with its tips pointing downwind. Barchan dunes are not generally connected to one another but instead migrate independently.

In environments where sand is plentiful and evenly dispersed, it accumulates in long ridges called transverse dunes (Figure 13-40B) aligned perpendicular to the prevailing wind. If sparse desert vegetation is present, the wind may form a small depression called a blowout in a bare area among the desert plants. As sand is carried out of the blowout, it accumulates in a parabolic dune, the tips of which are anchored by plants on each side of the blowout. Parabolic dunes (Figure 13-40C) are common in moist semidesert regions and along seacoasts, where sparse vegetation grows in the sand.

If two different wind directions prevail throughout the year and both are of comparable magnitude, then long, straight longitudinal dunes (Figure 13-40D) can form. In portions of the western Sahara Desert, longitudinal dunes reach 100 to 200 meters in height and are as much as 100 kilometers long.

Loess Wind can carry silt for hundreds or even thousands of kilometers and then deposit it as loess. Loess generally is composed of silt, is porous, and lacks layering. Even though loess usually is not cemented, it typically forms vertical cliffs and bluffs because the fine sediment particles stick together more than uncemented sand or gravel (Figure 13-41).

🌐 **FIGURE 13-41**
Go online to view an example of a thick loess deposit near Xian, China, into which dwellings have been dug.

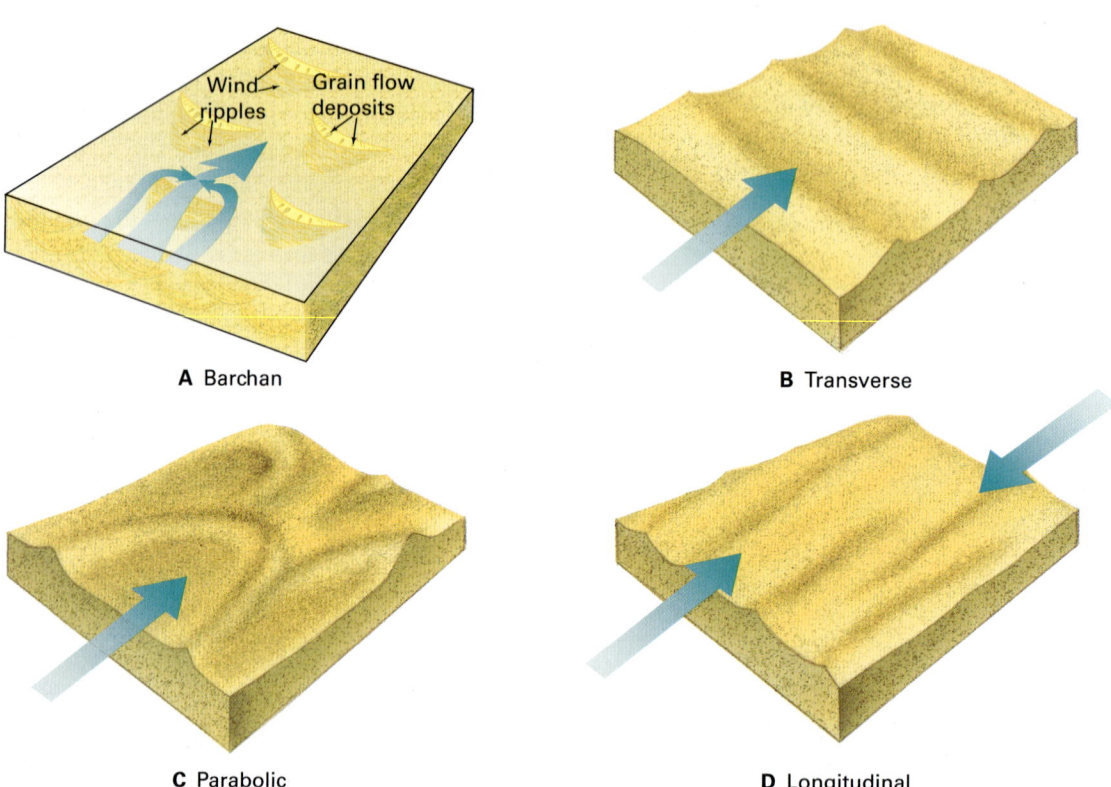

FIGURE 13-40 ▲
(A) When sand supply is limited, the tips of a barchan dune travel faster than the center and point downwind. Barchan dunes typically are covered by wind ripples on their upwind side and are characterized by a steep slip face in which small grain flows transport sand to the toe of the slip-face. (B) Transverse dunes form perpendicular to the prevailing wind direction in regions with abundant sand. (C) A parabolic dune is crescent shaped, with its tips pointing upwind. It forms where wind blows sand from a blowout, and grass or shrubs anchor the dune tips. (D) Longitudinal dunes are long, straight dunes that form when two different wind directions with comparable magnitude prevail.

MINILAB Simulating Wind Erosion

Materials

goggles
water
gravel
hair dryer
ruler
protractor
three shallow pans or containers of sand

Procedure

1. Label one of the pans "A" and set it on the table away from the other pans. Place the protractor on the table in front of pan A with the flat side facing you.

2. Rest the end of the hair dryer on the edge of the pan at a 45-degree angle. While wearing your goggles, turn the hair dryer on at the lowest setting. Without moving the hair dryer, let the stream of air blow into the pan for 30 seconds. Turn off the hair dryer and record your observations.

3. Smooth out the sand in the pan. Repeat step 3, but this time change the angle of the hair dryer to 10 degrees. Record your observations.

4. Smooth out the sand. Using the ruler, measure a distance 10 centimeters away from the pan. Position the hair dryer there and repeat steps 2 and 3, recording your observations after each trial.

5. Setting pan A aside, take a second pan of sand and label it "B." Slowly pour or sprinkle water over the sand and mix with your hands until it is moist, but not saturated. (There should be no puddles.) Repeat steps 2–4 with pan B, recording your observations after each trial.

6. Setting pan B aside, take the third pan of sand and label it "C." Sprinkle some gravel over the sand. Repeat steps 2–4 with pan C.

Results and Analysis

1. **Identify** What did the different types of sand in the pans represent? What did the hair dryer and the angles represent?

2. **Compare** How did the hair dryer affect the sand in each pan differently? Explain how this models the erosional effects of wind on land.

3. **Explain** What was the purpose of changing the angle and distance of the hair dryer? Explain how these variables affected the sand in the pans.

Critical Thinking

4. **Apply** Is wind erosion more of a problem in wet or dry climates? Use evidence from your investigation to support your answer.

5. **Predict** What do you think would happen if you added plants or vegetation into pan A? Explain your answer.

The largest loess deposits in the world, found in central China, cover 800,000 square kilometers and are more than 300 meters thick. The silt was blown from the Gobi and the Taklamakan Deserts of central Asia. The particles stick together so effectively that people have dug caves into the loess cliffs to make homes. However, in 1920 a great earthquake caused the cave system to collapse, burying and killing an estimated 100,000 people.

Thick loess deposits accumulated in North America during the most recent Pleistocene glacial advance, when continental ice sheets ground bedrock into silt. Streams carried this fine sediment from the melting glaciers and deposited much of it along their banks and floodplains. These environments were cold, windy, and devoid of vegetation, and wind easily picked up and transported the silt, depositing thick layers of loess as far south as Vicksburg, Mississippi.

Loess deposits in the United States range from about 1.5 meters to 30 meters thick. Soils formed on loess are generally fertile and support vast natural grassland ecosystems and make good farmland. Much of the rich soil of the central plains of the United States and eastern Washington State formed on loess. See Chapter 14 for more about loess and farmland.

checkpoint Which conditions make the right environment for dune formation?

13.8 ASSESSMENT

1. **Summarize** Why is wind erosion more prominent in desert environments than it is in humid regions?

2. **Sequence** Describe the evolution and shape of a dune.

Critical Thinking

3. **Synthesize** Describe a scenario where barchan dunes, transverse dunes, parabolic dunes, and longitudinal dunes all form in one area.

TYING IT ALL TOGETHER
ICE AND WIND EROSION

In this chapter, you learned about the formation of glaciers and deserts and how they can each alter the shape of Earth's surface over time. In the Case Study, you read about scientists in Antarctica studying the Thwaites Glacier, a massive, unstable ice sheet that is rapidly lurching toward the sea. Scientists consider the Thwaites Glacier an important predictor of sea-level rise because of the sheer volume of meltwater it could release into the ocean. However, it is equally fascinating to consider what new valleys, moraines, and hills the Thwaites Glacier may someday reveal if it ever fully melts.

While glaciers can take thousands of years to alter a landscape, the effects of glacial erosion are often transformational and long-lasting. The geologic forces that create deserts are equally complex and even slower to develop. Tectonic forces and mountain-building play a role in desert formation because a location's latitude and proximity to mountains can affect the amount of rainfall it receives. In addition, wind and water erosion can gradually transport soils and sediments, reshaping a desert over time.

Work individually or with a group to complete the steps below.

1. Select a desert in North America such as the Mojave Desert or the Great Basin Desert.

2. Find out basic geographic information about the desert you chose, such as its location, size, and climate.

3. Research specific features found in the desert, such as dunes, mesas, spires, or buttes. Find out more about any landforms that are particularly notable or famous.

4. With your group, discuss the geologic processes that may have contributed to the formation of the desert features you researched. Then, conduct further research to confirm your hypotheses.

5. Gather images of your selected desert and create a slide presentation that you will present to the class. Keep the text on your slides minimal but prepare a separate script with talking points that you will present as each slide is displayed. Your talking points should include in-depth information from your research.

CHAPTER 13 SUMMARY

13.1 FORMATION OF GLACIERS

- A glacier is a massive, long-lasting accumulation of compacted snow and ice that forms on land and creeps downslope or outward under the influence of gravity and its own weight.
- Alpine glaciers form in mountainous regions; continental glaciers cover vast regions.

13.2 GLACIAL MOVEMENT

- Glaciers move by two mechanisms: basal slip and plastic flow. Large cracks called crevasses can develop in a glacier's brittle upper layer.
- In the zone of accumulation, the annual rate of snow accumulation is greater than the rate of melting, whereas in the zone of ablation, melting exceeds accumulation. The equilibrium line altitude separates the two zones.

13.3 GLACIAL EROSION

- As glaciers move, embedded rocks scrape bedrock, causing glacial striations.
- Glaciers erode bedrock, forming U-shaped valleys, cirques, arêtes, hanging valleys, and fjords.

13.4 GLACIAL DEPOSITS

- Drift is rock or sediment transported and deposited by a glacier. The unsorted drift deposited directly by a glacier is called till.
- Moraines and drumlins are depositional features formed by direct ice contact.
- Sediment first carried by a glacier and then transported, sorted, and deposited by streams results in outwash plains, kames, and eskers.
- Wind-deposited silt is called loess, which typically accumulates in thick deposits.

13.5 GLACIATION AND GLACIERS TODAY

- During the past one billion years, at least six major glaciations have occurred. The most recent is the Pleistocene glaciation.
- Sea level falls when continental ice sheets form and rises again when the ice melts.
- Glaciers and sea ice are shrinking in more places and at more rapid rates now than at any other time since scientists began keeping records.

13.6 WHY DO DESERTS EXIST?

- Deserts have an annual precipitation of less than 25 centimeters. The world's largest deserts occur near 30 degrees north and south latitudes, where warm, dry, descending air absorbs moisture from the land.
- Deserts also occur on the leeward side (in rain shadows) of mountains, in continental interiors, and in coastal regions adjacent to cold ocean currents.

13.7 WATER AND DESERTS

- Desert streams are often dry for much of the year, but flash floods may occur during a rainstorm. Desert lakes that dry up periodically leave abandoned lake beds called playas.
- Alluvial fans imay be several kilometers across and rise a few hundred meters above the surrounding valley floor. Merging alluvial fans form a bajada. A pediment forms along the front of desert mountains and merges imperceptibly with a bajada.
- Streams produce erosional features that include canyons, buttes, mesas, and spires.

13.8 WIND AND DESERTS

- Desert pavement forms as wind selectively removes silt and sand, leaving larger stones at the surface.
- Sand grains are relatively large and heavy and are carried only short distances; silt can be transported great distances at higher elevations. Most of the abrasion caused by windblown particles occurs near ground level, where the heaviest grains travel.
- A dune is a mound or ridge of wind-deposited sand. Most dunes are asymmetrical, with gently sloping windward sides and steeper slip faces on the leeward sides. Winds cause dunes to migrate.

CHAPTER 13 ASSESSMENT

Review Key Terms
Select the key term that best fits the definition. Not all terms will be used, and no term will be used more than once.

alpine glacier	ice shelf
arête	interglacial period
bajada	kame
basal slip	kettle
butte	loess
cirque	mesa
continental glacier	moraine
crevasse	outwash
deflation	pediment
desert	plastic flow
desert pavement	plateau
drift	playa
drumlin	rain-shadow desert
dune	saltation
erratic	slip face
esker	spire
fjord	terminus
glacial striation	till
glaciation	wash
glacier	zone of ablation
hanging valley	zone of accumulation

1. mechanism of glacial movement in which ice within the glacier moves as a viscous fluid
2. a vast, continuous mass of ice that flows outward in all directions under its own weight and covers a large portion of a continent's area
3. region that receives less than 25 centimeters (10 inches) of rain per year
4. thick mass of ice that is floating on the ocean but connected to glaciers on land
5. groove or scratch that was gouged in bedrock by rock embedded in a moving glacier
6. the bouncing of sediment grains across a surface as they are blown by strong winds
7. broad, gently sloping depositional surface formed by merging alluvial fans and extending into the center of a desert valley
8. a rock or sediment transported and deposited by a glacier or by glacial meltwater
9. a deep inlet where a glacially carved valley has been flooded by the sea
10. sharp, narrow rib of rock formed by two alpine glaciers flowing down opposite sides of a mountain ridge
11. long, sinuous deposit of well-sorted drift formed by subglacial streams
12. higher-elevation part of an alpine glacier where snow builds up from year to year
13. asymmetric mound or ridge of wind-deposited sand
14. streambed that is dry for most of the year
15. a time when alpine glaciers descend into lowland valleys and continental glaciers advance

Review Key Concepts
Answer each question on a separate sheet of paper to demonstrate your understanding of key concepts from the chapter.

16. Describe the differences between alpine glaciers and continental glaciers.

17. Identify the zone of ablation, zone of accumulation, and terminus on the figure. What is the difference between these three areas?

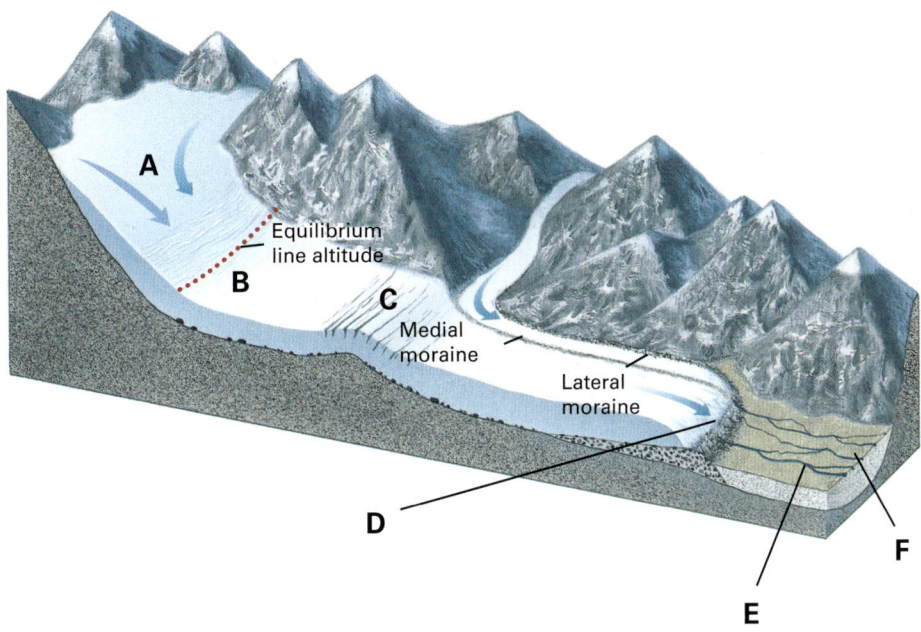

18. When a glacier moves by plastic flow, which parts of the glacier move the fastest? Which parts of the glacier move the slowest? Why?
19. How are tarns and paternosters formed by glaciers?
20. As an alpine glacier moves down a mountain, it can form cirques, arêtes, and hanging valleys. What are these features and how are they different from each other?
21. What is the difference between till and drift?
22. Identify the glacial land features in this image. Describe how each is formed.

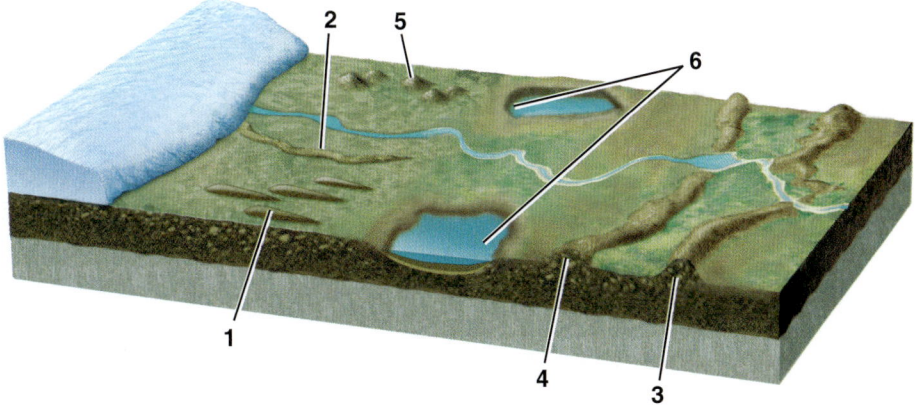

23. What is the Pleistocene Glaciation?
24. What is a desert? What are some characteristics that can differ from desert to desert?
25. Evaporation from large bodies of water, such as oceans, lakes, and rivers, adds moisture to the air. Is it possible for deserts to exist next to these sources of water? What other factors besides proximity to water affect the location of deserts? Explain your reasoning.

CHAPTER 13 ASSESSMENT   435

CHAPTER 13 ASSESSMENT

26. Explain why many desert lakes and streams are often dry for parts of the year.
27. Describe the four common types of sand dunes: barchan, parabolic, transverse, and longitudinal.

Think Critically

Write a response to each question on a separate sheet of paper. Use concepts from the chapter to support your reasoning.

28. Why do alpine glaciers tend to be smaller than continental glaciers?
29. How do you think the equilibrium altitude line varies with latitude for alpine glaciers? How strong would this relationship be? Explain your reasoning.
30. Mountain streams and alpine glaciers change the landscape in different ways. How do the effects of streams and glaciers differ? Why?
31. Despite lying along the Pacific coast, the Atacama Desert in northern Chile is one of the driest deserts on Earth. Using these maps, identify factors that contribute to this region's exceedingly dry climate.

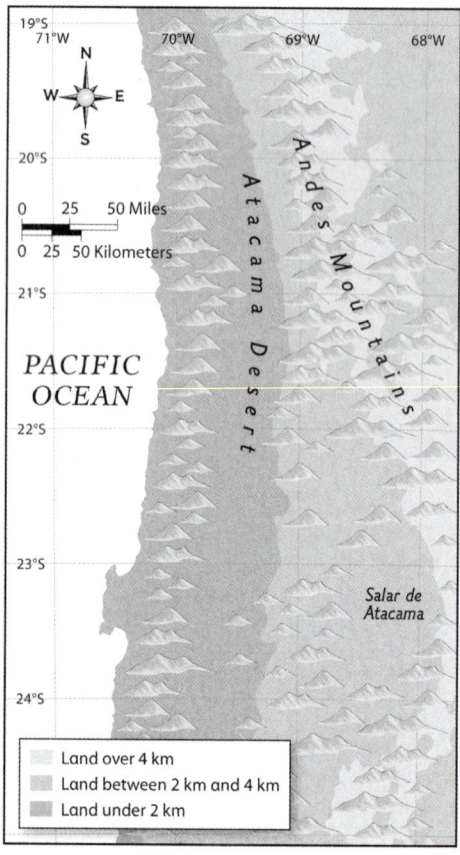

32. How is wind erosion similar to glacial erosion? How is it different?

PERFORMANCE TASK

Erosion Model

In this chapter, you have studied glaciers and deserts, two topics that may seem somewhat unrelated. However, both glaciers and deserts can cause drastic changes to the landscape through erosion and deposition. In this activity, you will further explore the erosive and depositional processes of glaciers or deserts using physical models.

1. With two or three other students, decide whether you want to study a landform created by wind, water, or glacial ice. Then, as a group, select a general type of landscape (for example, mountains, plains, plateaus, hills, or deserts) that you would like to study.

    A. Design a model to illustrate the effects of wind, water, or ice on the landscape you chose. Determine the materials you would need for your model. Consider how you would simulate wind, water, or ice in your model. Explain how the simulation would be performed and how many trials of the simulation would be necessary. Draw your design and provide a detailed narrative explaining the materials and methods you will use.

    B. At your teacher's direction, build and test your model. Record your observations in a table or graph. Use your model to answer the questions that follow.

2. Based on your model, what can you conclude about wind, water, or ice erosion and deposition? How do your results compare to the erosional or depositional effects described in the chapter?

3. What are some limitations of your model? Describe at least two limitations and propose ways to refine your model to address these limitations.

CHAPTER 14
SOIL RESOURCES

A roadcut in the Palouse region of eastern Washington State reveals soil layers.

Pick up a handful of dirt. Chances are that the soil you hold in your hand took tens or hundreds of thousands of years to transform from solid rock to the loose, crumbly material in your hand. That transformation first involved slow physical and chemical breakdown of the original rock, followed by transportation and accumulation of the resulting products.

Once accumulated, soils continue to undergo physical and chemical changes that give each soil a unique set of characteristics. In some locations, layers of sediment accumulate over time, burying older layers of soil. Elsewhere, in areas of active erosion, only a thin, superficial layer of soil is able to form.

Soils are crucial resources for the survival of humans because they provide the basis for agriculture and forestry. As you read, consider how humans interact with soil and why soil is a natural resource.

> ## KEY QUESTIONS
>
> **14.1** What is soil and how does it form?
>
> **14.2** What are the types of soil and how do soils differ from one another?
>
> **14.3** What natural and human factors affect soil quality and soil resources?

EXPLORERS AT WORK

STUDYING SOIL RESOURCES AND SUSTAINABLE FOOD SYSTEMS

WITH NATIONAL GEOGRAPHIC EXPLORER JERRY GLOVER

Earth's human population is projected to grow to 9.8 billion by the year 2050. That statistic raises a number of concerns, among them the question of how to feed all those people. Half of Earth's habitable land is already used for agriculture. Not much more is suitable for farming. Soil scientist and agroecologist Dr. Jerry Glover says that the soils in which the planet's food must be grown are the thinnest and most nutrient-depleted they have ever been. Dr. Glover works with scientists and small-scale farmers around the world. He says that protecting soil is key to making agriculture sustainable, which means taking a new approach to growing food.

Glover explains that annual crops—plants that complete their life cycles in one year and must be replanted year after year—provide more than 70 percent of our food needs. Unfortunately, says Glover, annuals have shallow root systems, which do not hold nutrients very well. Growing them requires large inputs of water, fertilizers, pesticides, and herbicides. Often, these inputs are actually lost. They seep into the ground or are washed away rather than being taken up by the plants' roots. Nutrients are also lost at harvest time. Because annual crops die, they are often removed. Wind and rain can then blow or wash away the soil.

Glover feels that shifting to perennial crops could revolutionize the agriculture industry. Perennial plants, which live for several years, can actually improve soil health because their roots often grow much deeper than those of annuals. They are better able to collect and retain water and nutrients, which they cycle back to the soil and reduce soil erosion.

Glover became interested in perennial crops after comparing the soils beneath tallgrass prairie in Kansas with soils in nearby farm fields. His research revealed that prairie soils were much healthier. Now, Glover is working with other agriculture experts who develop perennial versions of major grain crops that can produce yields needed for the world's food supply. Glover feels perennial crops can be especially important in regions with limited resources and marginal soils, such as in southern Africa. In fact, he has been working with small-scale farmers there, many of them women, on how to improve their soils by growing perennials along with other crops.

THINKING CRITICALLY

Argue How would you use evidence from what you read to persuade a community of farmers to make a shift from growing annual wheat to perennial wheat?

Dr. Jerry Glover holds a perennial grass, exposing the network of sturdy, long roots. Glover has researched perennial grains as alternatives to annual plants that are most commonly relied on as food sources. The roots of perennials do much to preserve soil fertility.

CASE STUDY
SOIL EROSION AND HUMAN ACTIVITIES

All soil originates from rock. Weathering decomposes rock, and plants and other organisms add organic material, resulting in the mixture we call soil. But soil does not continually accumulate and thicken throughout geologic time. If it did, Earth would now be covered by a layer of soil hundreds or thousands of meters thick, and rocks would not be visible at Earth's surface. Instead, natural forms of erosion combine to remove soil from the surface about as fast as it forms. Flowing water, wind, and glaciers all erode soil, and some soil simply slides downhill under the influence of gravity. Eventually, most of the eroded soil particles reach the sea, where they are deposited. Meanwhile, back on land, it can take thousands of years for weathering and organic processes to reform a soil lost to erosion. Thus, from a human perspective, soil is a nonrenewable resource. Once it is eroded, it is gone for generations.

Unfortunately, human activities have greatly accelerated soil erosion. In North America, tallgrass prairie once covered more than 65 million hectares (160 million acres) across a swath of the Great Plains stretching from southern Canada to eastern Texas. The deep roots of the native grasses served to anchor the region's thick, fertile soils. But today, as little as one percent of that grassland habitat remains, much of it having been converted for agricultural use. Plowing, grazing, and logging all accelerate soil erosion by stripping away native vegetation and disrupting the soil's surface. This exposes soil to the erosive forces of water, wind, and gravity, which can quickly carry away soils that took thousands of years to form.

Erosion of cropland in the Midwest (Figure 14-2) illustrates how intense pressures on soil resources can result in loss of productivity and biodiversity. One recent study concluded that humans' massive displacement of sediment has resulted in cropland loss at a rate 10 to 40 times faster than cropland soils are formed. This global net loss of soil resources will threaten food sustainability for future generations if left to continue.

Soil can be difficult and expensive to renew once it has become degraded. But when farmers use proper conservation measures, soil can be preserved indefinitely or even improved. Soil scientists such as Jerry Glover (see Explorers at Work) are researching new methods for slowing soil erosion, conserving water, and reducing contamination by pesticides and fertilizers. Developing more sustainable approaches to agriculture is critical to reversing the loss and degradation of the soil on which agriculture relies.

As You Read Explore how soil forms and the identify the components that distinguish different types of soils. Look for the factors that influence the development of soil and think about how human activity can alter natural processes. Find out about soil resources in your local area, and discuss how soil resources are managed in your community.

▸ **FIGURE 14-1**
Tallgrass prairies like this one once covered most of the central United States, but today, most of the land has been plowed for agriculture.

🌐 **FIGURE 14-2**
Go online to view an example of crop fields that are susceptible to erosion by wind and water.

14.1 SOIL FORMATION

Core Ideas and Skills
- Define soil and outline the importance of soil to human societies.
- Explain factors that are important to soil formation.
- Describe soil horizons.

KEY TERMS

regolith
soil
litter
humus
parent material
residual soil
transported soil
soil horizon
topsoil
leaching
eluviation
translocation
transformation

Soil may seem simple and unimportant, but it is actually very complex and essential to the survival of many organisms. Although it is not a living thing itself, soil has a dynamic, changing nature. Soil is formed through natural processes, and it can be degraded through natural processes as well. Soil can be moved from place to place by wind, ice, and liquid water, or it can be buried under tons of sediment and remain in the same place for millions of years. Soil from one area can be very different from soil found in another area—even if the two areas are only meters apart.

Soil Formation

In geologic terms, the thin layer of loose sediment that exists over top of bedrock is called **regolith**. **Soils** are the upper layers of regolith that contain organic matter and can support rooted plants. All soils are mixtures composed of three kinds of materials: inorganic matter, organic matter, and microorganisms. The relative amounts of these components can vary in different samples of soil. See Figures 14-3, 14-4, 14-5 in the next section.

The inorganic matter in soil includes minerals, water, and gases. Of these, minerals make up the largest proportion. Recall that minerals are inorganic crystalline compounds that are the building blocks of rock. Water is present in varying quantities in soils depending on conditions that will be discussed later. Gases are found within the pores, or open spaces between soil particles. The gases present in soil include nitrogen, oxygen, and carbon dioxide, which diffuse into soil from the atmosphere. Oxygen, carbon dioxide, and ammonia gases are also produced by organisms living in soil.

A small proportion of soil includes living microorganisms, such as bacteria and fungi, and the excrement or feces of animals such as earthworms and moles. Most of the organic matter in soil includes the decayed or decaying matter of dead microorganisms, plants, and animals. **Litter** is the collection of dead organic matter such as leaves, twigs, and animal bodies that lies above the surface of soil and has not yet decomposed. When litter decomposes to the extent that its origin is no longer recognizable, it is called **humus**.

The different components of soil can come together in countless ways to form soils. In the United States alone, more than 20,000 different soils have been identified and mapped. Soils are so unique that crime scene investigators often use forensic soil analysis to provide evidence of a crime. For example, they might analyze traces of soil on a suspect's shoe and compare them to soil samples from a crime scene. If the soil samples match, this could indicate that the suspect has indeed been to the crime scene.

checkpoint What is the source of organic matter in soil?

Soil-Forming Factors

Soil forms over long periods of time as weathering processes bring about changes in minerals and organic matter. Each soil has a unique origin and history of events that led to its formation, and this accounts directly for the large numbers of different soils. Five factors are most influential in the soil formation process: parent material, climate, organisms, steepness and slope, and time.

Parent Material Every soil's origin can be traced back to its **parent material**, the material from which a soil originates. Parent material can be either rock or organic matter. Soil can form from parent rock as the result of physical weathering, which breaks rock down into fine particles, or chemical weathering, which alters the chemical properties of minerals in the rock. Soil can form from organic parent material when dead plant matter accumulates in an area saturated with water for long periods of time. These conditions inhibit decomposition and allow organic material to form extensive deposits. Peat, a kind of soil that develops in swamps and bogs, is an example of a soil derived from organic parent material. To learn about how peat turns into anthracite (coal), refer to Chapters 4 and 6.

The properties of the parent material greatly influence the properties of the soil that results from it. For example, soil formed from sandstone tends to have a fine, sandy texture, whereas soil formed from weathered mudstone often has a pasty, clay-like texture.

A soil that remains near its parent rock is called a **residual soil**. By contrast, a **transported soil** is one that has been carried far from its parent material, usually by wind, water, or glaciers.

Climate Climate is the general pattern of temperature, precipitation, and wind in a region over a long period of time. Both temperature and precipitation are key factors that influence the overall style of weathering. In dry, cold regions, mechanical weathering predominates. In contrast, chemical weathering predominates in hot, humid regions. Temperature and precipitation also influence the amount and kinds of vegetation that can grow in an area. Vegetation, in turn, greatly influences the amount and chemical makeup of organic matter available for soil formation.

Organisms Soil contains organisms both living and dead. Organisms that live in soil include plants, microorganisms, fungi, worms, and insects that change the soil's chemical and physical properties and aid in the cycling of nutrients. Living plants influence soils by putting down roots that help prevent soil erosion. Larger animals such as moles, ground hogs, and prairie dogs burrow in the soil, redistributing and in some cases exposing soil, resulting in loss of moisture. Herds of elk and bison can also affect soil properties, compacting the soil by trampling and altering vegetation by grazing.

Organic matter in soils originates from living and once-living organisms. Animals deposit excrement in soil. Plants shed leaves and flowers which fall to the ground and decompose. When it dies, the plant itself falls to the ground and decomposes while the plants roots decompose underground. The decomposition of dead animals also adds organic matter to a soil. A soil's organic matter is an important source of nutrients for plant growth. In general, the more organic matter that the soil contains, the greater the soil's ability to support plant populations.

FIGURE 14-3
Vegetation and soil composition vary by climate because of differences in precipitation and temperature.

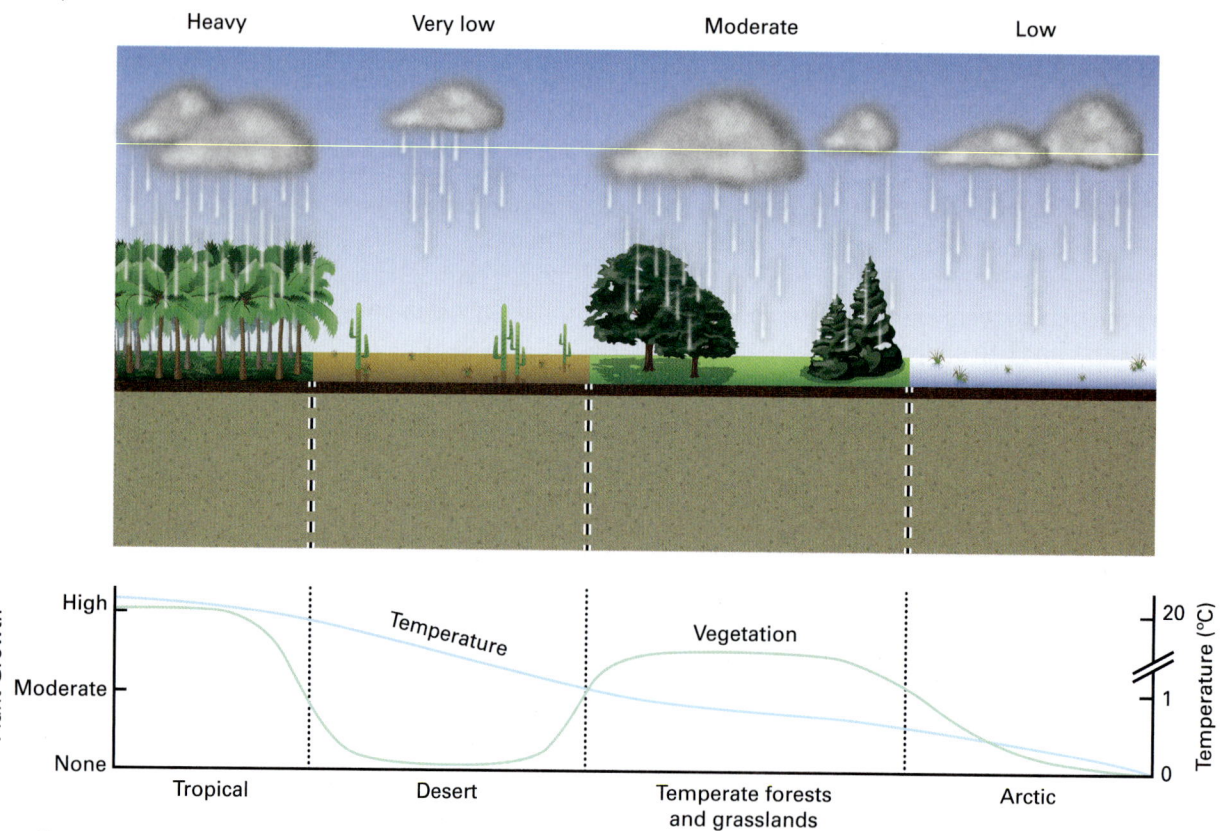

Source: Thomson Higher Education

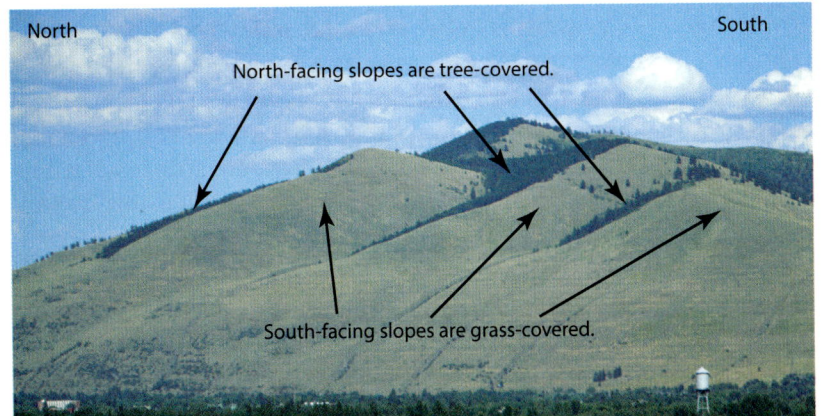

FIGURE 14-4
In this photograph from the Missoula Valley of west-central Montana, thick forests cover the cool, shady slopes with a northern aspect, but grass and sparse trees dominate the hotter, drier, south-facing exposures.

Steepness and Slope Aspect An area's relief describes the changes in relative elevation over a given distance. Areas with big, steep cliffs have high relief, whereas flat landscapes have low relief. Steepness can affect soil formation because of the role gravity plays in accelerating the flow of water over a slope. Moving water tends to cause the most erosion at the steepest parts of a slope before depositing the eroded material near the base, where the slope flattens out. As a consequence, soils formed at the tops of slopes are usually thin and less fertile than soils at the bottoms of slopes.

Relief also affects the moisture content of soils. Compared to soils on flat ground or on gentle slopes, soils on steep slopes experience greater exposure to the sun, which tends to dry them out faster. This effect can also be influenced by a slope's aspect, however. Slope aspect refers to the direction of a slope with respect to the sun. In the semiarid regions of the Northern Hemisphere, thick soils and dense forests cover the cool, shady, north-facing slopes of hills, but thin soils and grasses dominate hot, dry, southern exposures (Figure 14-4). This difference occurs because more water evaporates from the hot, sunny, southern slopes, leaving the soil drier and able to support less vegetation. The grasses that grow on south-facing slopes have shallow roots and produce little litter, which slows weathering and retards soil development. On the moist northern slopes, trees and other deep-rooted plants grow abundantly, leading to looser, thicker soil and ample leaf litter, which accelerates soil development.

Time Soils form slowly over very long periods of time, and the characteristics of a soil change gradually as it forms. Typically, a soil will have many of the same properties as its parent material earlier in its history, but it may gradually lose these properties as other factors change its composition. Figure 14-5 illustrates how a soil formed from parent rock might develop over a long period of

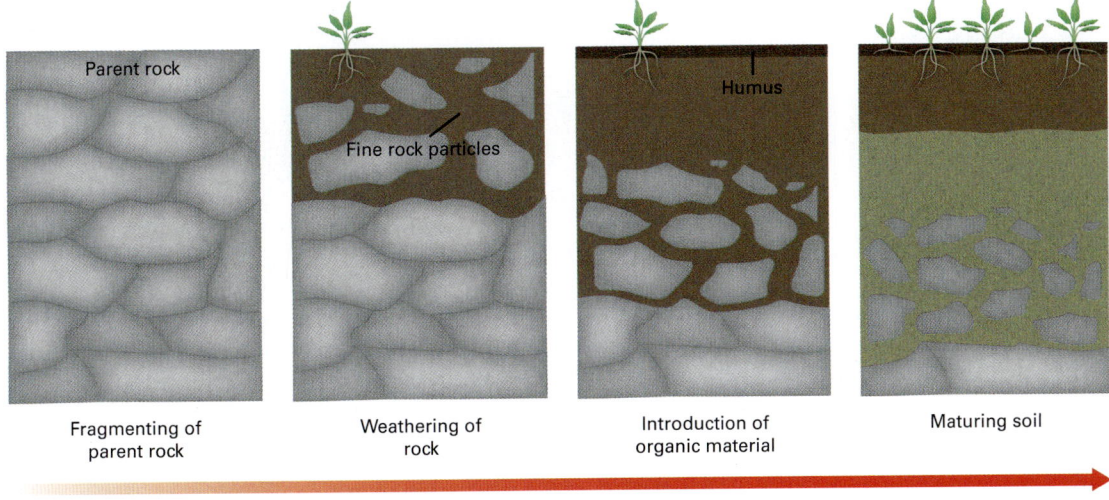

FIGURE 14-5
With time, soil forms as the result of rock weathering and an influx of organic matter.

LESSON 14.1 **445**

time, starting with the weathering and fragmenting of the parent rock. The rock gravel that is formed initially provides areas for small plants to begin growing. Some rainwater is retained in the fine gravel as more plants colonize the area and put down roots. Over time, many generations of plants live and die, contributing organic matter to the developing soil, which is broken down by bacteria and fungi that have moved in. As more organic matter accumulates, it retains more rainwater, and this combination provides a fertile site for new, larger plants to grow. These plants in turn provide new food sources and organic matter that supports new types of animals, such as earthworms, insects, and moles. Feldspar is the most common mineral at and near the Earth's surface. It weathers by hydrolysis to clay, a critical component of soil insofar as moisture retention is concerned.

Chemical weathering produces changes in the mineralogy of the soil that can have major effects on a soil's nutrient level, ability to retain water, and erodibility. In general, each stage of a soil's development changes its properties in ways that trigger additional changes. Each of the five factors discussed above (parent material, climate, organisms, relief, and time) influences the types of changes that occur as a soil develops. Because each of these factors has varied influences, there are countless ways that soils can develop, which has resulted in millions of different soils, each with unique characteristics. The end result is that no two soils are ever completely alike.

checkpoint Soil is a dynamic (always changing) substance. What factors affect the way that a particular soil changes?

Soil Horizons

Soils tend to form in layers. New material is often added on top of an existing layer, and older layers gradually become buried deeper and deeper. If you were to slice straight down through a soil and lift out a large section, you would be able to see a distinct pattern of layers called the soil profile. Soil scientists often analyze a soil's profile to determine the history of the soil's development. The horizontal layers that make up a soil's profile are called **soil horizons** (Figure 14-6). Horizons differ from one another in color, depth, and composition. Horizons

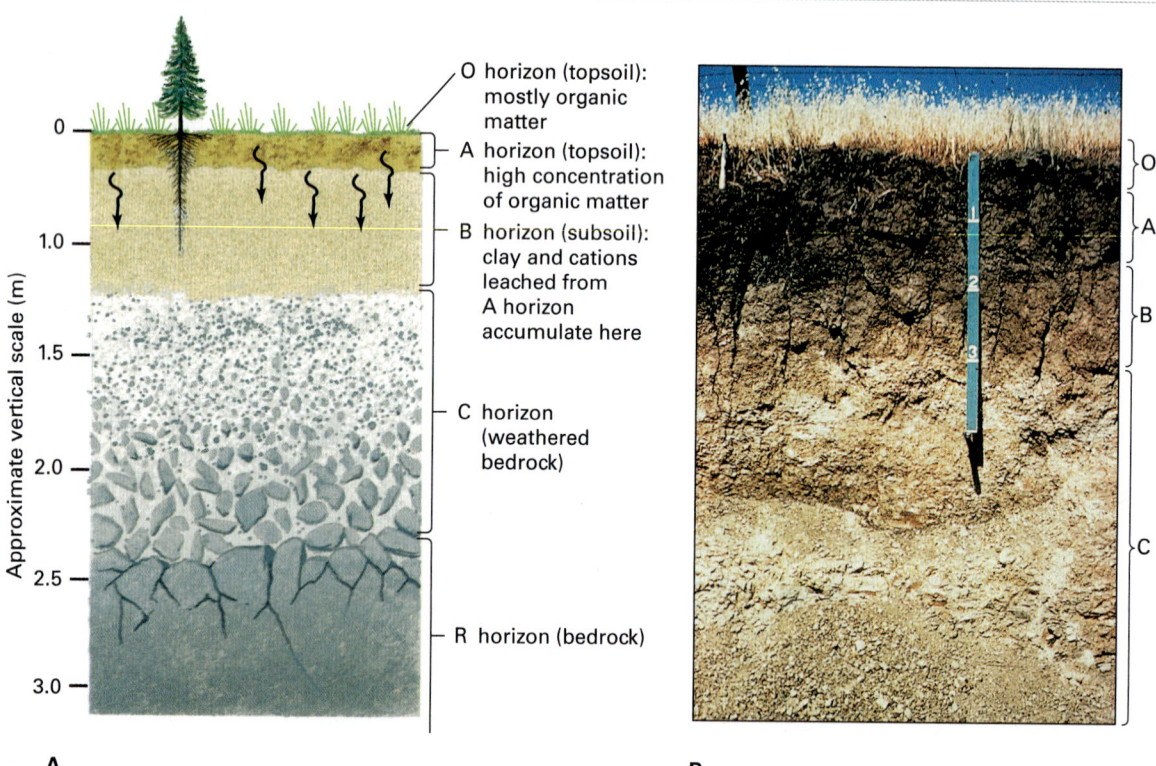

FIGURE 14-6

(A) Five different soil horizons are commonly distinguished by color, texture, and chemistry. (B) A vertical slice through the ground reveals the soil profile. In this particular soil profile, an E horizon is absent. The R horizon is the bedrock that lies below the C horizon and is not visible in the photo.

are found in the same general pattern across all soils, although not all soils have the same number of horizons. Six soil horizons can occur: O, A, E, B, C, and R. Horizons A, B, and C are found in every soil. Many soils also have horizons O, E, and R, although one or more of these may be absent.

O Horizon Not all soils have an O horizon, but if present, this layer is at the surface. The O horizon consists of partially and completely decomposed organic material, which comes from the plants and animals native to the area. For example, the litter in a forest comprises the O horizon of its soil. Soil that has been plowed for agricultural purposes lacks an O horizon. The organic matter that once made up its O horizon has been mixed into the next soil level, the A horizon.

A Horizon All soils have an A horizon. The A horizon consists mainly of minerals mixed with humus and is located either at the surface or just below the O horizon. The combined O and A horizons are often called the **topsoil**. The A horizon generally has a very dark color due to its organic-rich composition.

E Horizon Some organic and mineral matter can become dissolved or suspended in rainwater and can move out of the A horizon. As the rainwater seeps downward, it carries the dissolved and suspended substances with it into the layers below. The dissolution of minerals and ions by infiltrating water is called **leaching**, and the downward movement of leached matter through soil horizons is called **eluviation**. In some soils, so much eluviation from the A horizon occurs that it forms a subsoil horizon called the E horizon (the E is for *eluviation*). The E horizon is typically light in color because it has lost large amounts of mineral and organic content. When this matter leaches out, it leaves behind sand and other materials that do not tend to leach out. The E horizon, when present, is found between the A horizon and the B horizon.

B Horizon A B horizon is present in every soil and is sometimes referred to as the "zone of accumulation." The B horizon is the subsoil below the A and E horizons that accumulates the matter leaching out of the layers above. It typically has a much lower organic content than the A horizon but contains clay, nutrients, water, and gases that originated in the A and E horizons. Together, the O, A, E, and B horizons comprise the soil layers where most plant roots are found.

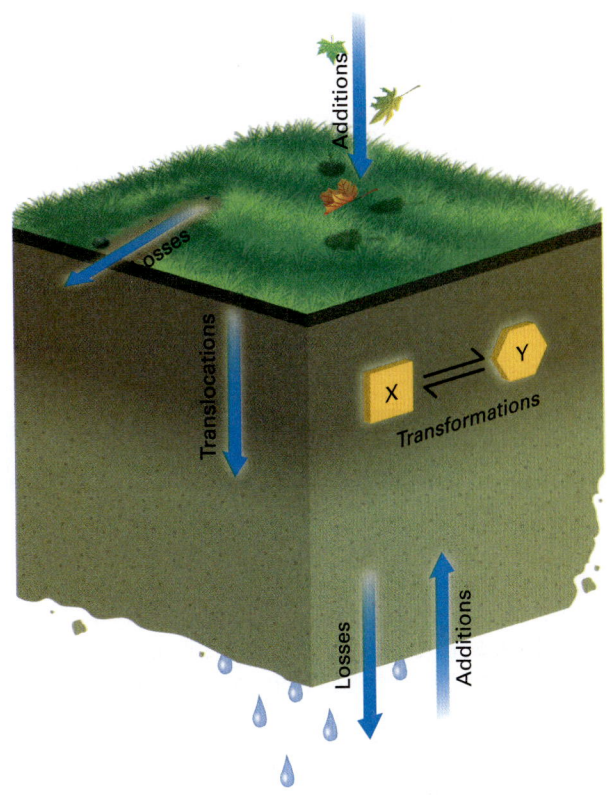

FIGURE 14-7
The four types of changes that take place in soil are additions, losses, translocations, and transformations.

C Horizon The C horizon is the layer below the B horizon. It consists of the weathered parent material that gave rise to the soil above. Although it may still undergo weathering and other changes involved in soil formation, this layer is much less active in these types of change than the layers above it.

R Horizon The R horizon is not composed of soil but is a bedrock layer underneath the C horizon.

Changes in Soil As noted earlier, soil is dynamic in that it undergoes constant change from the time of its formation from the parent material throughout its history. Figure 14-7 illustrates the four types of changes that can occur in soil: additions, losses, transformations, and translocations. Additions and losses involve the movement of matter into and out of soil. **Translocations** involve movement of matter from one part of the soil to another. **Transformations** involve physical or chemical changes to matter within the soil.

Additions, losses, and translocations occur because soils are open systems that are subject to flows of matter and energy within and through their boundaries. Additions change soil in numerous ways. For example, water readily enters soils from the surrounding environment, and the amount and flow of water into a soil can have a large impact on a soil's composition. In addition, moving water often contains suspended or dissolved matter that it may deposit on top of or within a soil. Other sources of additions include wind, animals, gases, and the sun. Wind can deposit soil particles or organic litter from elsewhere. Animals can deposit excrement or the remains of other organisms they have eaten.

Gases diffuse from the atmosphere into spaces between soil particles. Radiant energy from the sun is absorbed at the soil's surface.

Soils undergo losses as matter moves out into the surrounding environment. For example, moving groundwater can dissolve ions and other water-soluble matter in the soil, causing it to leach out of the soil along with the water. Rapidly moving water or wind can also carry away small or large quantities of soil, altering its components and structure. Plants

DATA ANALYSIS Effects on Soil Biodiversity

Mature, fertile soils typically contain a high diversity of plant, bacterial, and fungal species that contribute to their high nutrient content. Many of the world's naturally fertile soils have already been exploited for agriculture. Compare the maps to look for connections between soil biodiversity and agriculture.

1. **Compare** How does the agricultural activity of central North America compare to that of central South America?

2. **Recognize Patterns** Describe any visible correlation between percentage of area used for agriculture and level of soil biodiversity.

3. **Recognize Patterns** Describe any visible correlation between distance from the Equator and level of soil biodiversity.

Critical Thinking

4. **Infer** Identify the area in South America that has an unusual matchup for high biodiversity in soil but low land use for agriculture.

5. **Apply** Make two rough sketches of North America. Using the same color keys provided, sketch your predictions of how the maps might have looked prior to European settlement of the continent.

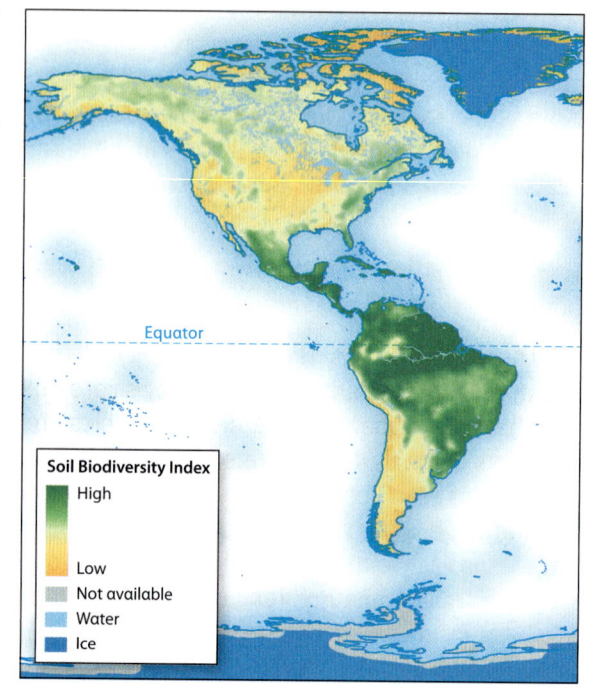

A

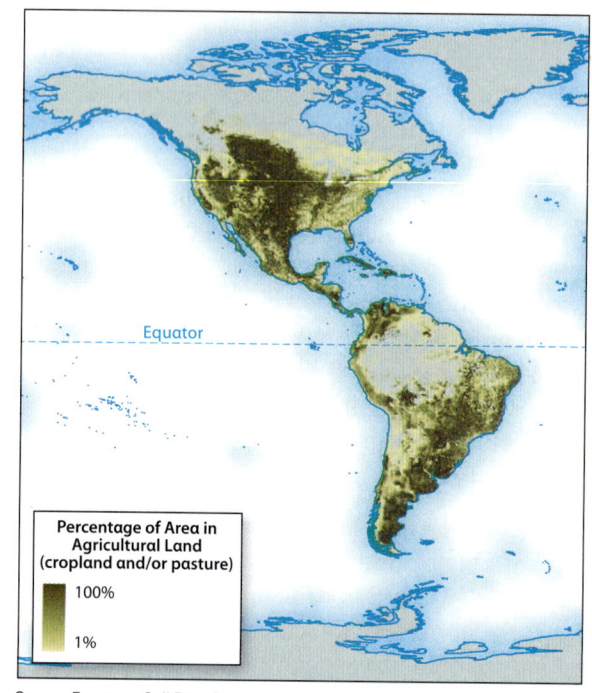

B

FIGURE 14-8
Global Biodiversity Atlas

can cause losses when they take up nutrients from the soil. Gases, including water vapor, can diffuse out of soil into the atmosphere.

Not all matter that moves through soil is lost. Translocations move matter from one part of the soil to another, which makes them essential to the formation of soil horizons. Water moving through soil is largely responsible for the translocation of ions, inorganic compounds of mineral origin, and small organic compounds from decayed matter. Even some large clay particles that do not dissolve can be suspended in water and move along with it.

Transformations include physical and chemical changes of matter within soil. For example, ice that forms then thaws in soils will slowly break down the larger mineral grains and rock fragments in the soil. Iron-bearing minerals transform when they are oxidized by oxygen in soil. Nitrogen, another gas present in soil, undergoes a chemical change when microorganisms convert it to ammonia, nitrates, and nitrites, nutrients that are essential for plant growth. This particular transformation, called nitrogen fixation, is critical for producing fertile soil.

checkpoint What is another name for the layers in a soil's profile?

14.1 ASSESSMENT

1. **Contrast** How would the top layer of soil in a forest likely differ from that of a nearby grassy meadow?
2. **Explain** Think about the soil in an area familiar to you. Construct a plausible explanation about where it came from and how it developed.
3. **Describe** Climate affects soil formation directly and indirectly through the types of plants that grow in the soil. Describe the properties of soil that you would expect to find in a hot, dry climate.

Critical Thinking

4. **Predict** Peat is a soil that forms from water-logged, plant-based organic matter. What steps do you think would occur during the formation of peat? How would these steps compare to the formation of a soil from parent rock?
5. **Synthesize** Two soils formed in the same area, from the same parent material, in the same climate, with similar vegetation, and over the same period of time. One soil is thicker than the other. What could explain this difference?

14.2 SOIL CLASSIFICATION

Core Ideas and Skills
- Describe how soil is classified based on its composition and texture.
- Predict the properties of a soil based on its particle composition.
- Explain why classifying soil is a useful enterprise.

KEY TERMS

soil texture	loam
porosity	soil order

As discussed in the previous section, no two soils are completely alike. Each soil has a unique and complex history of formation that determines its individual profile and properties. Even so, different soils can be classified and grouped into categories based on general properties that distinguish them from other soils. Many kinds of people, from soil scientists and agricultural scientists to land managers, farmers, and everyday gardeners, rely on soil classification systems. Soil classifications help people make better decisions when it comes to managing their soil and cultivating plants for food, erosion control, or ornamental value. Properties used to classify soils include predominant particle size and type, degree of fertility, age, ecological significance, and others.

Components of Soil

If you examined a soil sample from a typical Midwestern farm under the microscope, you would find three types of particles differing by size, shape, and chemical properties. The largest particles are sand grains, the next largest are silt grains, and the smallest are clay-sized particles (Figure 14-9). The relative proportions of sand, silt, and clay in a soil influence its **soil texture**.

To understand how soil texture influences the properties of a soil, consider how the porosity of a soil changes with particle size. **Porosity** is a measure of the total space between particles in a sample of soil. Sand particles are large and the pore spaces between the grains are correspondingly large. In contrast, silt grains are smaller, and the sizes of pores between the grains also are small. In unlithified sand or gravel, the network of open pore

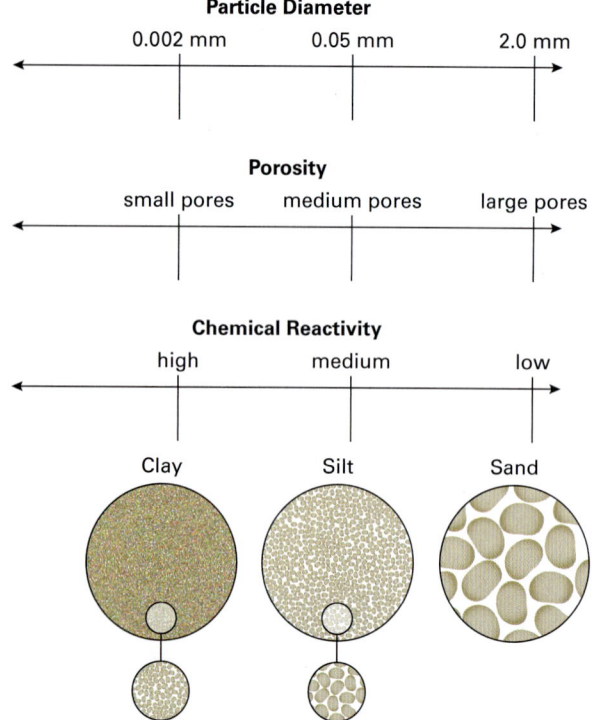

FIGURE 14-9
Soil texture, porosity, and chemical reactivity are determined by the relative amounts of three types of particles found in soil: clay, silt, and sand.

spaces between the individual sediment pieces results in porosity values that often are close to 40 percent. In contrast, unlithified mud can have porosity values of up to 90 percent. The higher porosity of unlithified mud is the result of many physical and chemical characteristics of very fine-grained sediment.

In addition to porosity, the shapes and chemical properties of soil particles also contribute to soil texture. Clay particles, the most chemically reactive of the three particle types, serve as an important source of inorganic nutrients such as calcium, magnesium, potassium, and others necessary for plant growth. Clays retain water well because the particles have a flat shape and water gets trapped in the small spaces between particles. Also, the crystal lattice of some clay minerals is expandable. When water is available, some clay minerals incorporate the water molecule into their crystal structure and swell up. Clay is also very sticky because the particles are so small and so close together. Positive and negative ions on the surface of clay particles attract one another, causing the particles to clump together.

Sand contributes very few inorganic nutrients to soil due to its chemical nature. Sand grains vary in their form from spherical to bladed to disc-shaped. They vary in roundness from very rounded to very angular. Sand particles have poor water-retention properties and little ability to stick together. Because sand particles are more massive than clay particles, electrostatic attractions between surface ions are not strong enough to make the particles stick together. These properties, along with high porosity, make sandy soils easy to plow. Sandy soils also allow incoming water to quickly reach plant roots, while draining away any excess.

checkpoint What are the three types of particles found in soils, and how are they different?

Types of Soils

Most soils contain some mixture of sand, silt, and clay, but the ratios of these particles vary. The percentage of each particle type defines the soil's texture. A soil that consists mostly of clay particles, with only small percentages of silt and/or sand, can be classified as a "clay soil," and its texture will be similar to that of pure clay. A soil that is about 50 percent clay and 50 percent sand can be classified as "sandy clay soil," and its texture will be in between that of pure sand or pure clay. A soil with a texture that is optimal for agricultural purposes, called a **loam**, has an ideal balance of sand, silt, and clay that maximizes the beneficial properties of all three. Figure 14-10 shows how a soil can be classified based on the percentages of clay, silt, and sand in its composition.

Different kinds of plants have adapted to thrive in different kinds of soils. Soil that is ideal for one kind of plant may not support another kind of plant at all. For example, some plants grow well in acidic soil, while others prefer alkaline (basic) soils. As you learned in Chapter 12, pH is the measure of H+ or OH- ions. Soils with low pH values are acidic, and the lower the value, the more acidic the soil. Soils with high pH values are alkaline, and the higher the value, the more alkaline the soil. You will learn more about pH values and the pH scale in Chapter 16.

Plants also differ in terms of their nutritional needs. Some plants require nutrients that other plants do not need, and some plants require more water than others. In addition, some plants need soils that rapidly drain water away from their roots.

Farmers and gardeners need to know whether their soil has properties suitable for the types

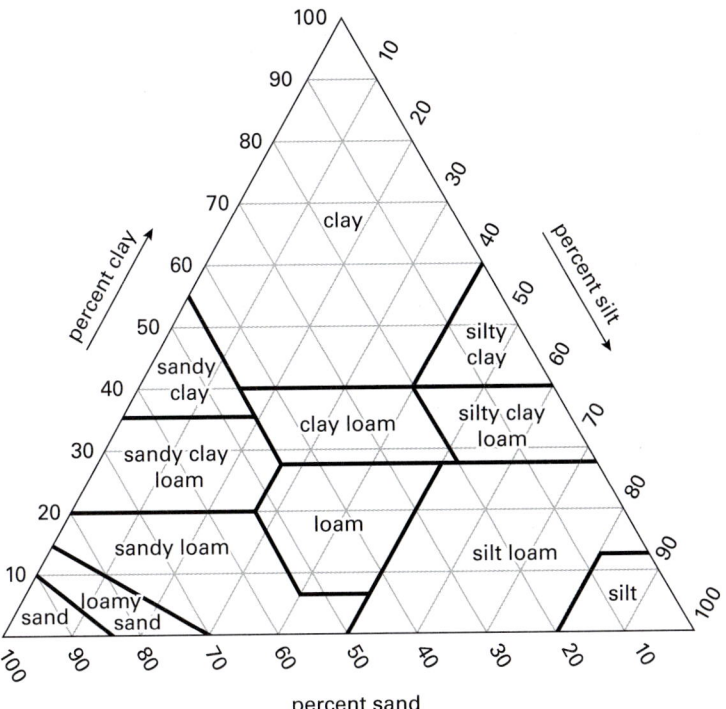

FIGURE 14-10
This triangular graph can be used to classify soils based on texture. To use the graph, find the point where the percentages of clay, silt, and sand intersect. For example, a soil containing 70 percent sand, 20 percent silt, and 10 percent clay falls inside the region labeled "sandy loam." A soil that is 40 percent sand, 40 percent silt, and 20 percent clay falls in the region labeled "loam."

of plants they want to grow. To help identify the properties of their soil that affect plant growth, they use a classification system that divides soils into six basic categories: sandy soil, clay soil, silty soil, peaty soil, chalky soil, and loamy soil. Knowing the characteristics of their soil can help farmers and gardeners avoid planting crops or other plants that their soil will not support. It can also help them decide whether to add materials such as fertilizers to their soil to make up for any nutrient deficiencies.

Sandy Soil Sandy soils contain a large percentage of sand particles. Recall that sand particles allow rapid water drainage but do not retain water or stick together. Thus, sandy soils are best for plants that need rapid and complete water drainage away from their roots and can withstand being dried out for long periods during warm months. Because they contain little clay content, sandy soils are low in inorganic nutrients. Sandy soils also tend to be very acidic.

Clay Soil Clay soils tend to retain water. Clay particles are loosely packed in soils but become much more compacted with burial beyond a few meters. Clay soils can be very fertile, but their extremely low permeability makes them retain water and inhibits the exchange of gases between the soil and atmosphere. Clay soils tend to be acidic.

Silty Soil Silty soils readily hold and release moisture to plants. They also provide reactive sites for inorganic nutrients, which makes them good for plant growth. Silty soils tend to be acidic, although not as acidic as clay soils.

Peaty Soil Peaty soils contain a much higher percentage of organic matter than other soils because they form from organic parent material. Peaty soils develop in peatlands, or bogs, which are very acidic. The high acidity slows the decomposition of dead plants in the bog. Over time, a dense mat of partially decomposed plant material accumulates in the bog. Since this waterlogged material is not exposed to air, only anaerobic microorganisms (those that can live without oxygen) can help decompose the organic matter. Peaty soils are not generally found on farms or in gardens but are often used as an additive to increase the organic matter in another soil.

Chalky Soil Chalky soils are characterized by their strongly alkaline pH, which comes from the calcium carbonate, or lime, found in these soils. Calcium carbonate originates from limestone parent material. Some plants thrive in the high alkalinity of chalky soil. For other plants, the high alkalinity impairs the ability of the roots to absorb iron, a critical nutrient.

Loamy Soil Loamy soils have the optimal properties for growing most plants. They are easily plowed, rich in inorganic nutrients, and drain well yet have good water retention characteristics.

MINILAB What Is in the Soil?

Materials

small pan or plastic container
soil sample, about 25 g
plastic cup or beaker containing water
pipette/dropper
ruler
magnifying glass, preferably 5X–10X

CAUTION Wash hands thoroughly upon completion of activity.

Procedure

1. Place the soil sample in the pan. Add a few drops of water using the pipette/dropper. Knead the soil to break up lumps. Continue adding drops of water one at a time until the moist soil can be molded with your finger.
2. Try to form the soil into a ball. Use Table 14-1 to determine or rule out soil type based on your outcome.
3. If you were able to make a ball, gently press it between your forefinger and thumb and try to form it into a ribbon. Make the ribbon as long as possible. When it breaks, measure the length of the ribbon with your ruler. Check the outcome against Table 14-1.
4. Pinch off a small amount of soil and add several drops of water. Touch the wet soil to determine the texture. Check your observations against Table 14-1.
5. Use the magnifying glass to examine the particles making up the soil.

Results and Analysis

1. **Connect** How did the procedures help you analyze the composition of soil? Why did the procedure include adding water to the sample?

Critical Thinking

2. **Apply** Which soil type was your sample? Explain how you were able to tell.
3. **Evaluate** How did the composition of the soil affect its texture?
4. **Generalize** How did your observations with the magnifying glass relate to your results?

TABLE 14-1 Exploring Soil Texture and Type

Type of soil	Can retain ball shape	Ball can be squeezed into a ribbon	Forms ribbon <2.5 cm before breaking	Forms ribbon 2.5–5 cm before breaking	Forms ribbon ≥5 cm before breaking	Feels very gritty	Feels very smooth	Feels neither gritty nor smooth
Sand	✗							
Loamy sand	✓	✗						
Sandy loam	✓	✓	✓			✓	✗	✗
Silt loam	✓	✓	✓			✗	✓	✗
Loam	✓	✓	✓			✗	✗	✓
Sandy clay loam	✓	✓	✗	✓		✓	✗	✗
Silty clay loam	✓	✓	✗	✓		✗	✓	✗
Clay loam	✓	✓	✗	✓		✗	✗	✓
Sandy clay	✓	✓	✗	✗	✓	✓	✗	✗
Silty clay	✓	✓	✗	✗	✓	✗	✓	✗
Clay	✓	✓	✗	✗	✓	✗	✗	✓

Source: U.S. Department of Agriculture

Loamy soils also tend to have large pore spaces that allow for aeration. Aeration is important for gas exchange. It allows gases from the atmosphere such as oxygen, nitrogen, and carbon dioxide to enter soil, and gases given off by plant roots, fungi, and bacteria to make their way back to the atmosphere.

checkpoint List at least two soil properties that vary between the six categories of soil.

Soil Orders

In addition to soil categories, soils have also been categorized according to age, fertility, origin of parent material, and biome location. These categories are useful in specific applications and discussions about soils.

Soil scientists at the U.S. Department of Agriculture have developed a nested hierarchy for classifying soils. This hierarchy is similar to the taxonomy used to classify living organisms. **Soil order** refers to the most general category in the hierarchy. All soils can be classified as belonging to one of twelve orders. Each order represents differences in soil-forming processes and extent of soil formation.

For example, the Mollisol order contains soils that originated from parent rock material on grasslands. Mollisol soils are characteristically very fertile and rich in both organic and inorganic nutrients as a result of their formation history. In contrast, the order Andisol contains soils that are high in volcanic ash content. Andisol soils are also very fertile as a result of their ash content, and they tend to form much more quickly than Mollisol soils.

Soil scientists have mapped the distribution of all twelve soil orders across the United States and worldwide (Figures 14-11 and 14-12). Land managers and other soil professionals rely on such maps to inform decisions about land use for farming, building construction, road construction, landfill construction, golf course management, water drainage projects, conservation projects, and many other purposes.

⊕ **FIGURE 14-12**
Go online to view a world map of soil orders.

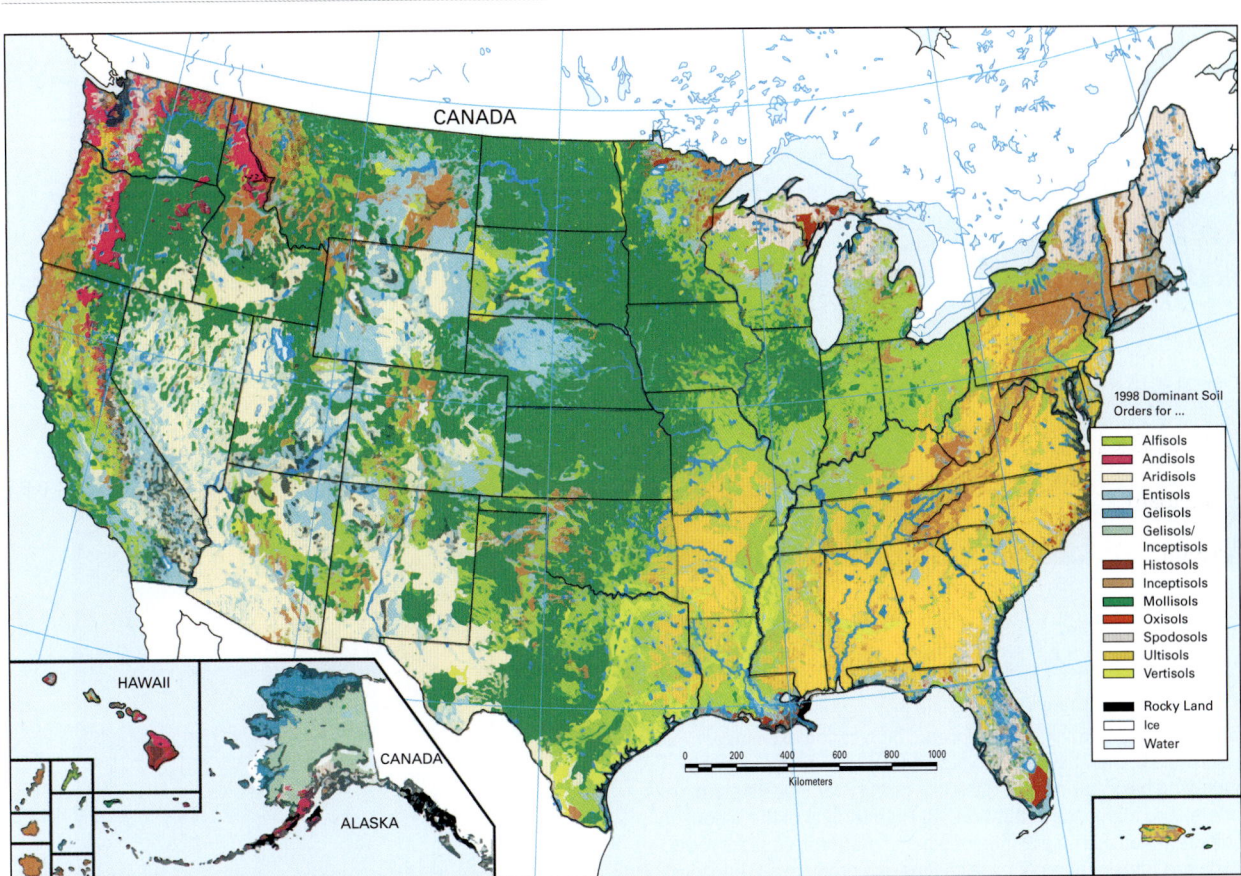

Source: U.S. Department of Agriculture

FIGURE 14-11
The colored key on this map identifies the soil orders found across the United States.

Mollisol, Entisol, and Andisol The distribution of soil components can differ from soil to soil, as illustrated in the examples shown in Figures 14-13, 14-14, and 14-15. The percentage of organic matter is highest in the soil from the Nebraska grassland prairie and lowest in the high plains soil from Montana, which is relatively high in mineral content. The soil from Mount Rainier has more moderate percentages of organic matter and mineral content.

Mollisols are the most widespread soil order in the continental United States and are extensively cultivated (Figure 14-13). Entisols have great diversity because soils not classified under any other order are classified with this soil order (Figure 14-14). Andisol is capable of supporting intensive agriculture and supports productive forests in Pacific Northwest of the United States (Figure 14-15).

⊕ **FIGURE 14-14**
Go online to view an example and diagram of Entisol soil.

⊕ **FIGURE 14-15**
Go online to view an example and diagram of Andisol soil.

checkpoint Identify some of the soil characteristics that distinguish different soil orders.

Loess

Recall that transported soils are formed from parent material that is moved by wind, water, or glaciers and deposited at another site. One type of transported soil, called loess, is formed mainly from silt deposited by wind. The silt in loess most often originates from the physical weathering of rock underneath glaciers. After the glaciers recede,

A

B

FIGURE 14-13 ▲
(A) Mollisol is a mid-latitude grassland ecosystem soil characterized by a thick, rich A horizon, high cation exchange capacity (CEC; the ability of a soil to release cations, typically by exchanging basic cations K^+, Na^+, Ca^{+2}, or Mg^{+2} for H^+ with plant rootlets) and high base saturation. (B) High fertility of the A horizon results from long-term addition of organic matter from plant roots. Downward translocation of base cations can produce a B horizon rich in calcium carbonate or other salts.

FIGURE 14-16
Much of central and eastern Washington is covered by several meters of loess. The rolling hills seen here were formed by wind deposition of silt that was later modified by erosion.

the silt becomes exposed and dries, forming a dust of fine particles. These particles are easily carried away by wind, sometimes hundreds or even thousands of kilometers. In some cases, loess can also originate from other types of weathering, such as physical weathering of sand grains in hot deserts.

Loess is a lightweight soil with a yellow or tan color. Loess can develop into good farming soil because of its mineral content and rapid drainage. Its flourlike consistency also makes it easy to plow. However, loess is sensitive to the eroding forces of wind and water, and this characteristic tends to shape loess deposits into rolling hills as shown in Figure 14-16.

checkpoint What properties of loess are beneficial to humans? Explain.

14.2 ASSESSMENT

1. **Relate** How could a gardener apply knowledge of soil classifications to decide which plants to plant in the yard?
2. **Distinguish** What tests would likely be run on a sample of soil to determine whether it is sandy, silty, loamy, chalky, peaty, or clay soil?
3. **Compare** In this section, you read about two ways to classify soil. One way classifies soil into six types and the other classifies soils into twelve soil orders. Compare and contrast these classification systems.
4. **Infer** Loess has been described as having a light, fluffy, flourlike texture. What can you infer about the particle content based on this description, and why?

Computational Thinking

5. **Represent Data** Study the map in Figure 14-11 showing the distribution of soil orders across the United States. How could you use data from this map to make a graphical representation of the frequency of each soil order? Describe the type of graph and roughly what it would show.

LESSON 14.2   455

FIGURE 14-17
ON ASSIGNMENT In Parque de la Papa, Pampallacta, Peru, a farming family harvests potatoes. Hand tilling and hand-harvesting causes less soil erosion. Such methods are more sustainable, particularly as the world's food production needs are on the rise. National Geographic photographer Jim Richardson snapped this image and others like it as he toured the globe to photograph farmers of the world. Showcasing small-scale farmers was part of an effort to highlight sustainable farming methods, often practiced on smallholder farms and on family farms. Some agricultural experts say that sustainable farming techniques practiced by smallholder farmers will contribute to the security of the global food supply.

14.3 SOIL EROSION AND DESERTIFICATION

Core Ideas and Skills
- Explain natural factors and processes affecting soil erosion.
- Describe how human activities influence the rate and extent of soil erosion.
- Define desertification and explain when and how it occurs.
- Explain what causes salinization and why it is detrimental to soil quality.
- Discuss land management practices that sustain soil resources.

KEY TERMS

soil erosion
desertification
salinization

FIGURE 14-18
The topsoil in this area has been washed away by water erosion.

Soil is in a constant state of change. Right now, some soils are forming, while others are degrading. Both natural and human factors cause loss and degradation of soil. Because healthy, fertile soil is essential for productive agriculture, humans have good reason to make sure the rate of soil degradation does not outpace the rate of formation. As the world's human population continues to grow, requiring more and more food production, it becomes increasingly important to develop sustainable land management practices. Sustainable land management aims to control soil losses and preserve soil quality.

Soil Erosion

Soil erosion is the major natural cause of soil degradation and deterioration. Recall that erosion can be a factor in soil formation, since transported soils form from particles of weathered parent rock that are carried elsewhere via erosion. By contrast, **soil erosion** is defined as the degrading action that occurs when wind, ice, or water moves through an area, carrying away the top layers of formed soil. Figure 14-18 shows an example of the effects of water erosion on soil.

Soil erosion can occur rapidly or over long periods of time. Wind storms, storm surges, hurricanes, flooding, and heavy rainfall can all cause significant erosion in a matter of hours or days. Glacial movements also cause severe soil erosion, but over hundreds or thousands of years.

Some soils are more easily eroded than others. The ability of a soil to quickly absorb water greatly influences its ability to resist erosion by heavy rain. If a soil lacks this ability, rainwater quickly saturates the topsoil. Excess water then overwhelms the mix of particles, separating them and carrying off the smaller clay and silt particles as it runs downhill. Soils with high quantities of organic matter tend to absorb water more quickly than soils with less organic matter. For these reasons, loamy soils resist erosion better than sandy, silty, or clay soils. Likewise, if a soil with B and C horizons absorbs water well, it will resist erosion better.

Another factor in preventing or slowing soil erosion is plant cover (Figure 14-19). Wherever plants grow and put down roots, soil is more resistant to erosion. During bouts of dry weather, soil with no plant cover dries out more quickly and may blow away with the wind. But roots of plants protect dry soil by holding onto it even when a drought is severe and prolonged.

Plant roots also hold onto soil particles, helping to prevent their erosion by runoff. This protective effect is observed for soils on all kinds of terrain, including steep slopes. Even dead and decaying plants, leaves, and twigs provide some protection against soil erosion. In forests, surface litter forms a protective mat, shielding the soil from heavy rain that might otherwise loosen and wash away the upper layers.

FIGURE 14-19
As plant cover increases, the rate of soil erosion by wind decreases.

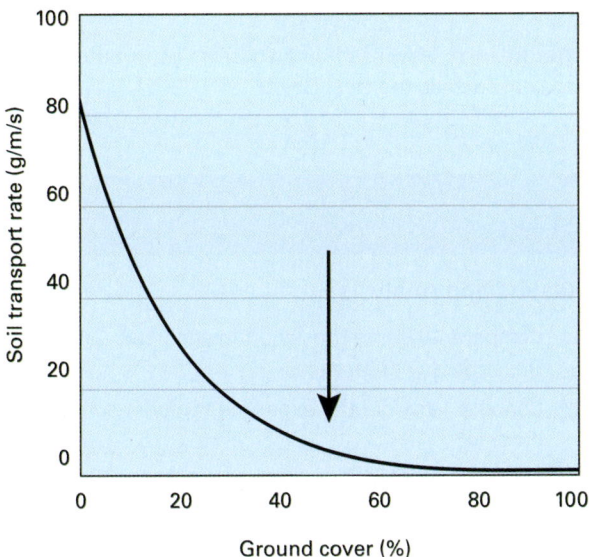

Note: The arrow shows the level of ground cover needed to control erosion.
Source: New South Wales Government Department of Planning, Industry and Environment

Human Impacts on Soil Quality Soil erosion has accelerated in places where humans have altered the landscape. In the United States as well as in other countries, people have been converting undeveloped forests, grasslands, and wetlands into farmland since the earliest human settlements. This practice continues as the need for food increases along with population. Converting wilderness to farmland involves the complete removal of native plants, followed by turnover, or tillage, of the soil to prepare it for planting crops. Although crop plants provide cover that helps reduce soil erosion, the density of plant coverage is much less than in a forest, a grassland, or a wetland. Also, the soil is vulnerable to wind and water erosion for extended periods between harvests and the next planting.

In addition to disturbing the soil structure, farming practices often involve the use of chemical pollutants that degrade soil quality. Many farms rely on chemical pesticides or herbicides to kill insects and weeds that would otherwise reduce crop yields. Over time, these substances accumulate in the soil, changing the soil's chemical properties and adversely affecting native soil organisms.

Many other human activities adversely affect soil, including development of airports, stadiums, buildings, houses, roads, parking lots, golf courses, ski slopes, and a long list of others (Figure 14-20).

All of these kinds of projects involve clearing native plants from lands that have long histories of soil formation. The soil is dug up and disturbed, sometimes removed, and often covered up. Soil quality typically declines due to increased erosion, loss of organic matter, loss of gas exchange, decline or loss of soil organism populations, and contamination by toxic chemicals.

Hard Lessons from the Dust Bowl The Dust Bowl of the 1930s served as a hard lesson on the ways human activity can amplify natural causes of erosion. In the 1920s, advances in farm machinery and high demand for wheat encouraged farmers in the Great Plains to plow large areas of grassland for crops. Plowing removed the grass that stabilized the soil across much of the plains. Some of the remaining grassland was used to graze cattle, and overgrazing led to further loss of the grass that stabilized the soil. In addition, many farmers used other practices that compromised the stability of the soil, such as burning crop residues after harvests and leaving fields bare during winter.

When a severe and long-lasting drought hit the region beginning in 1931, the conditions were suitable for large-scale soil erosion. With no rain and little organic matter to help bind its particles together, the soil turned to a fine dust as it dried out under the hot sun. Wind could easily pick up the dust and carry it away. This was the start of the Dust Bowl, an eight-year period during which massive amounts of soil were lost to wind erosion. During some dust storms, huge black clouds

FIGURE 14-20
Nearly all construction projects degrade soil quality by removing native plant cover, disturbing the natural soil structure, and increasing the soil's exposure to erosion and contamination.

traveled hundreds of miles across the country, forcing people to seek shelter indoors to avoid breathing the dust. Some storms moved so much soil that it covered the plains like a deep snow, with drifts burying barns and farming equipment (Figure 14-21). The Dust Bowl was an ecological disaster, devastating a vast region of the United States. Since it made much of the land unfit for farming, many families had no choice but to pack up and move elsewhere. Some 400,000 people were forced to leave the region.

checkpoint How has human activity contributed to soil erosion, such as the Dust Bowl?

MINILAB Modeling Soil Erosion

Materials

4 clear 2-L plastic bottles
superglue or other strong glue
latex gloves
potting soil, about 12 cups (3,000 mL) total
4 clear plastic cups
4 25-cm pieces string
sturdy cardboard, 30 cm × 50 cm
scissors or utility knife
small plant seedlings or grass seed
mulch
watering can or spray bottle
water
books or blocks

CAUTION Handle scissors or utility knife carefully. Wear gloves when applying superglue and avoid contact with skin.

Procedure

1. Using a utility knife or scissors, cut the top off each bottle lengthwise. See Minilab illustration.
2. Glue the bottles to the cardboard, side by side and oriented in the same direction.
3. Fill each bottle with the same amount of soil. To one bottle, add seeds or seedlings throughout. To a second bottle, add rows of seeds or seedlings in either a vertical or contoured (curved) arrangement, as directed by your teacher. Pat down all visible soil.
4. Add mulch to the third bottle. Add nothing to the fourth bottle.
5. Cut or punch two small holes in each plastic cup, on opposite sides just under the lip. Thread the string through and knot on each side. Hang the cups on the necks of the bottles. Refer to the Minilab illustration.
6. Once the plants are established, prop the cardboard on blocks or books so that the bottles angle downward and the cups hang over the edge of the table.
7. Using a spray bottle or watering can, add equal amounts of water to each bottle, enough so that some water runs off into the cup.

Results and Analysis

1. **Compare** Observe the runoff in each cup. Describe any differences.
2. **Connect** How did the experiment model soil erosion?
3. **Explain** What role did the mulch have in this experiment?

Critical Thinking

4. **Apply** How does soil surface affect the amount and rate of erosion?
5. **Evaluate** Compare the results obtained by groups who planted contoured rows and vertical rows. What do the differences in runoff indicate about the straight rows typical of conventional agriculture versus the contoured rows more common to sustainable agriculture?

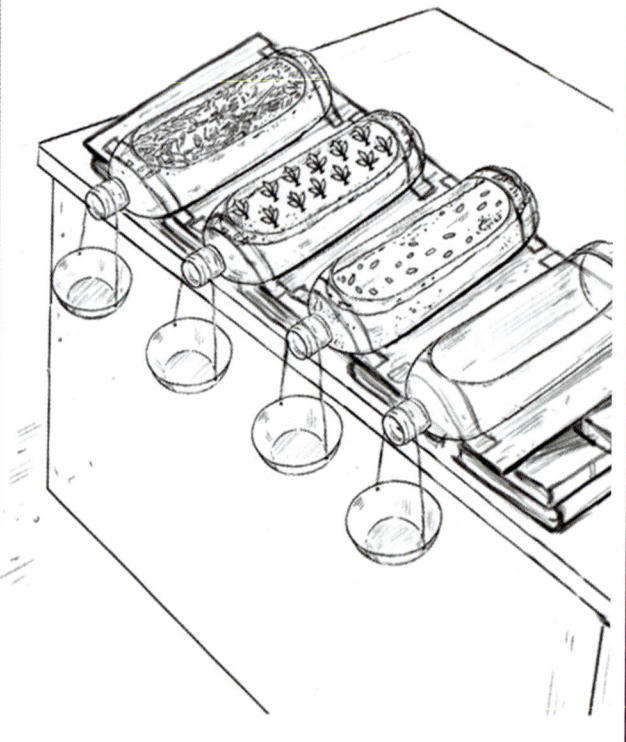

Desertification

The Dust Bowl is an extreme example of **desertification**, a type of soil degradation that occurs under very dry conditions. Desertification results in the loss of all or most of the moisture and organic matter from a soil. Without moisture and organic matter, the soil can no longer support life. Desertification can be caused by natural factors or human factors; but in the case of the Dust Bowl, human factors (poor farming practices) and natural factors (drought and wind) caused desertification.

Today, desertification is a major problem in parts of Africa and the Middle East. There, large portions of land have already turned to desert, and the conditions are right for more land to meet the same fate (Figure 14-22). The climate in these regions is dry, with little rainfall and periodic droughts. Native vegetation is sparse and organic matter is not abundant in the soil. Farming practices in these regions can contribute to the acceleration of desertification. Leaving fields bare after harvest and allowing cattle to overgraze destabilizes the soil and leads to rapid erosion.

Irrigation of crops can also contribute to desertification. Although irrigation adds moisture to the soil and helps it retain organic matter, it often results in the **salinization** of the soil over time. Unlike rainwater, which is free of dissolved

FIGURE 14-21
A dust storm buried this farming machinery in Dallas, South Dakota, during the Dust Bowl in 1936.

FIGURE 14-22 ▼
The map shows regions threatened by desertification worldwide. The red areas are the most vulnerable.

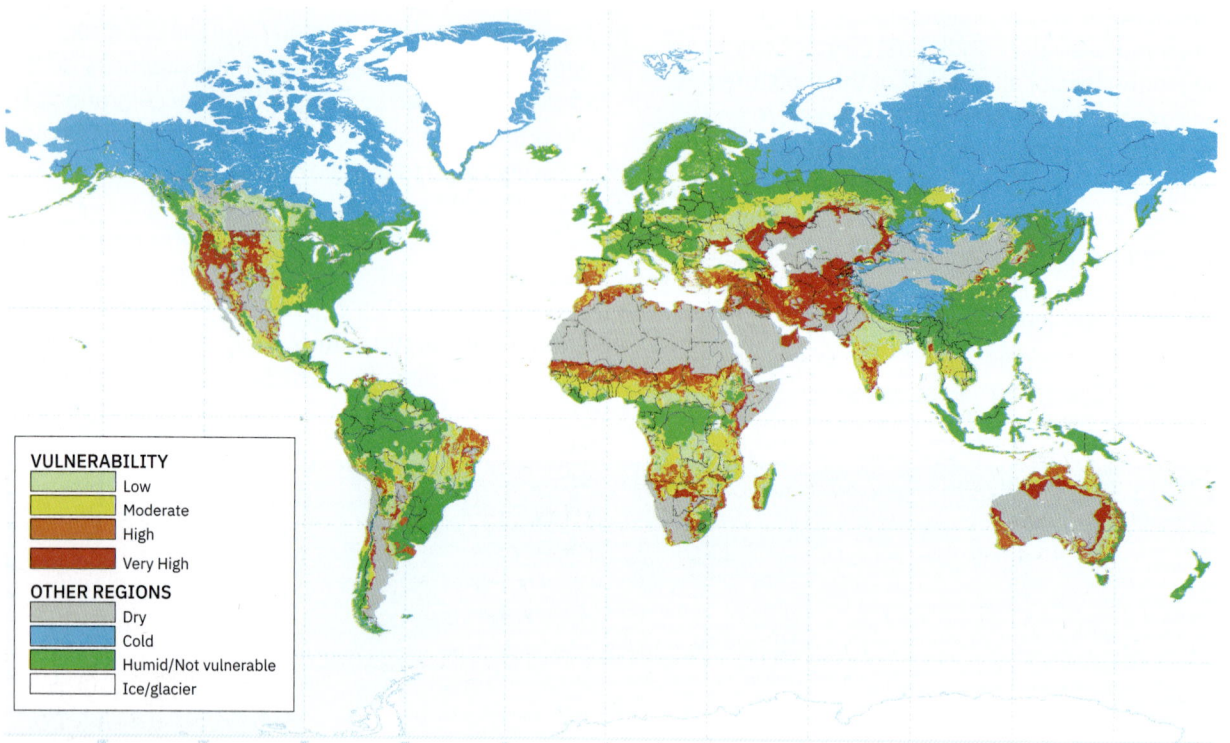

Source: U.S. Department of Agriculture

salts, most irrigation water comes from rivers, ponds, and lakes that are fed by runoff. Runoff dissolves salts as it makes its way across soil and rock. When surface water is used to irrigate crops, the dissolved salts gradually build up in the soil. Eventually, salts in the soil can reach levels that are toxic to crops as well as other plants, fungi, earthworms, and bacteria. Salinization can lead to desertification and the loss of biological productivity in a soil (Figure 14-23).

checkpoint How are salinization and desertification related?

Sustainability of Soil Resources

On average, soil forms at a rate of about 2 centimeters every 500 years. When soil is lost or degraded, it cannot be replenished in a human lifetime (or even several lifetimes), essentially making it a nonrenewable resource. Since soil is the foundation of our agricultural system as well as many terrestrial ecosystems, it is imperative that sustainable soil management practices be adopted to maintain the soil we have.

Unfortunately, social and economic pressures can make it difficult to implement and enforce sustainable soil management practices. For example, human population growth leads to higher demand for housing and food. This, in turn, leads

FIGURE 14-23 ▲
This image shows the effect of salinized soil in an irrigated vineyard in San Juan, Argentina.

462 CHAPTER 14 SOIL RESOURCES

to the clearing of more land for housing and farms. Clearing land generally involves deforestation, draining of wetlands, or conversion of grasslands, all of which degrade soil.

Economic forces can also encourage farming practices that degrade soil. High demand for a small number of crops, such as corn, results in farmers having little incentive to raise a variety of crops on their land. Instead, they raise a single crop, a practice known as monoculture, to maximize profits. Monocultures deplete soil nutrients very quickly, and the crops are more vulnerable to diseases and pests. As a result, they require more fertilizers, irrigation, herbicides, and insecticides. Raising a diversity of crops can dramatically reduce and, in some cases, completely eliminate these negative impacts.

In spite of such challenges, the current state of sustainable soil management in the United States is improving. In many ways, U.S. farming practices have made great strides since the Dust Bowl debacle. In 1935, the U.S. government established the Soil Conservation Service (SCS) to promote farming practices and sustainable land management strategies. The SCS developed local soil conservation districts where farmers could gather to learn how certain practices could prevent soil erosion. Such practices included the terracing of sloped land (Figure 14-24), establishing field drainage systems for water runoff, planting seeds using no-tillage, or no-till, methods (without turning over soil), and planting cover crops immediately after harvest. A cover crop is a plant grown primarily to prevent soil erosion and maintain the fertility of a field until the next planting season. These and other practices have since become standard, and local conservation districts continue to provide education and assistance in their own regions.

FIGURE 14-24
Constructing terraces on sloped agricultural land reduces soil erosion by slowing surface water runoff.

New technologies have been developed in more recent years to limit soil erosion, particularly in cropland. For example, thin plastic barrier sheets can be draped over soil and around vegetable plants to prevent evaporation of water. This maintains moisture within soils, reducing the need for irrigation and the risk of salinization. Advances in irrigation systems also reduce the amount of water needed to raise crops in drier climates. For example, some computerized drip irrigation systems use sensors that detect moisture levels in soil; then, they direct just the right amount of water to each individual plant. These systems reduce wasted water and prevent the accumulation of runoff that can carry soil and nutrients away.

Such innovations have great potential for improving soil management practices, but many challenges remain. Because soil composition and properties differ radically from one place to the next, there is no one-size-fits-all approach to sustainability. Nutrient availability is just one of numerous variables that determine soil quality. In light of the millions of different soils found in the world, combating soil degradation requires flexible approaches that can be adapted to soils of different composition and quality. A return to farming methods that reduce human impact on soil can prevent soil degradation (Figure 14-17). Reducing chemicals and switching from mass tilling of soil to no-till methods helps prevent nutrient loss within a given soil type (Figure 14-25).

FIGURE 14-25
Go online to view an example of a farming family that battles pests and weeds by hand.

Sustainable farming is being practiced successfully by small-scale farmers in Africa (Figure 14-26). Techniques like perennial cropping have helped to improve soil resilience and crop yields. Go online to learn more about photographer Jim Richardson's work and photography in science in Careers in Earth and Space Sciences.

checkpoint What are some old and new practices that help sustain soil health?

FIGURE 14-26
Photographer Jim Richardson proudly displays a bowl of groundnuts (peanuts) harvested and picked by the women he sits with under the shade of a tree in Mali.

14.3 ASSESSMENT

1. **Relate** Why does the increase in human population pose a threat to soil quality on Earth?

2. **Predict** Would you expect forest soil or cropland soil to be more vulnerable to erosion? Explain your reasoning.

3. **Explain** Why is loss of gas exchange detrimental to soil quality? What type of human activity can lead to this problem?

4. **Conclude** Many sustainability practices are being used routinely today to maintain soil health, so why is it important for scientists and engineers to continue to develop new sustainability practices?

Critical Thinking

5. **Apply** Suppose you have been brought in as a consultant to help farmers in a dry region combat the growing problem of soil salinization. What technological solution would you recommend to farmers in this region, and why?

DATA ANALYSIS Exploring Earthworm Bioremediation

Bioremediation involves the introduction of living organisms that can break down environmental contaminants. Earthworms process huge quantities of soil, consuming it as they burrow, and they are also highly tolerant of many toxins. This makes them excellent candidates for bioremediation of contaminated soils. The graph in Figure 14-27 shows the results of an Australian study investigating the potential for using earthworms to break down polycyclic aromatic hydrocarbons (PAHs) in soil. PAHs originate in part from the burning of wood, coal, and oil. Researchers carried out and compared the effectiveness of three 12-week treatments of PAH-contaminated soils: soil + worms + cow dung, soil + worms + kitchen waste, and soil + compost. The compost-only treatment allowed researchers to account for microbial breakdown of soil without worms present.

1. **Sequence** Based on graphed results, rank the three treatments from most effective to least effective in terms of removing PAH compounds. Were the results consistent?
2. **Conclude** Do earthworms effectively remove the studied PAHs from contaminated soil? Explain.

Computational Thinking

3. **Analyze Data** Which treatment(s) removed at least 80 percent of the PAH present, and for which compound(s)?
4. **Analyze Data** Do you think that compost only, with no worms, could effectively be used to rapidly remove the PAHs studied? Explain.
5. **Identify Patterns** If the initial soil concentration of benzo (k) fluoranthene were 4,000 mg/kg of soil, calculate how much of the contaminant would likely remain after 12 weeks, for each treatment.

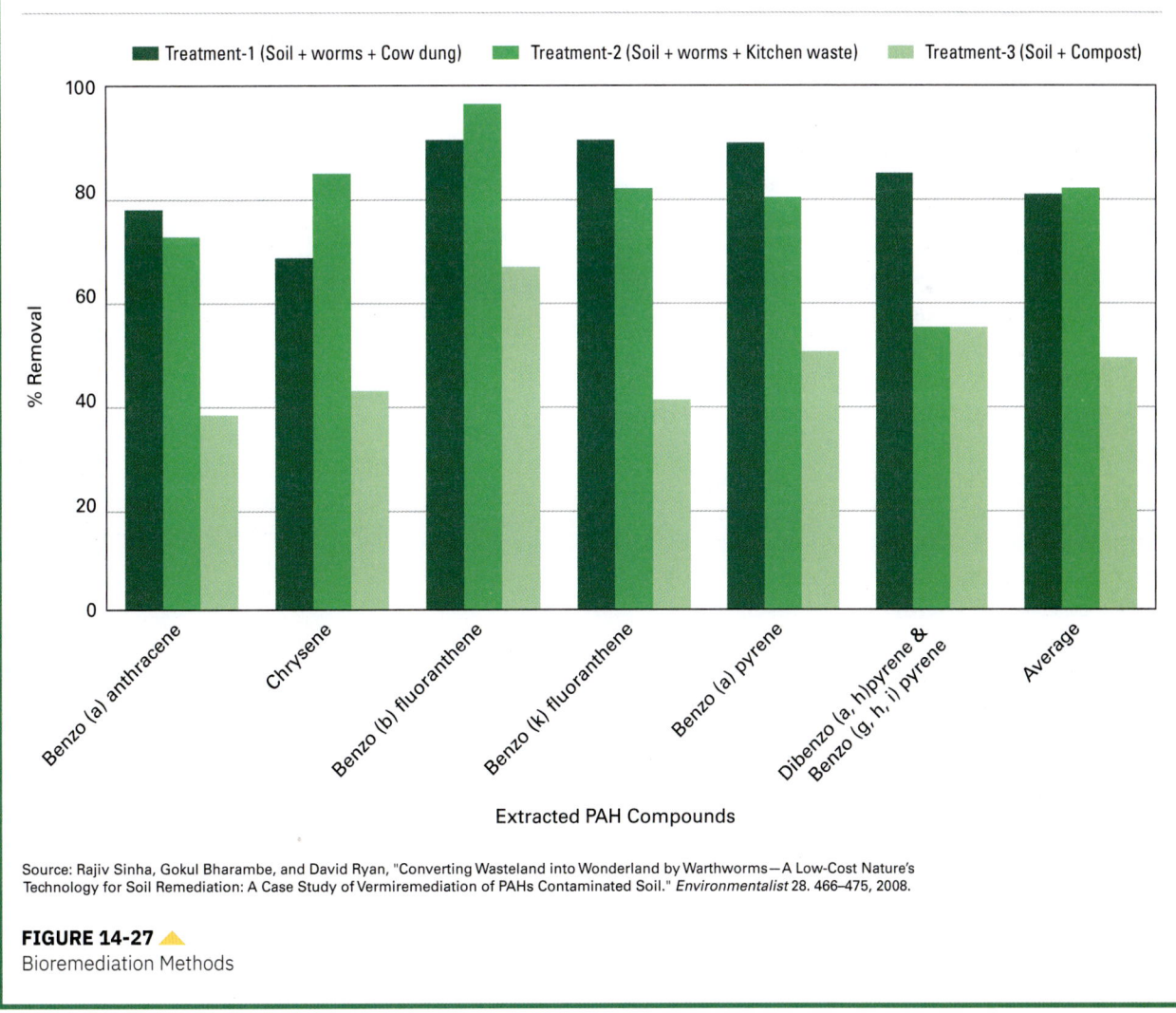

Source: Rajiv Sinha, Gokul Bharambe, and David Ryan, "Converting Wasteland into Wonderland by Warthworms—A Low-Cost Nature's Technology for Soil Remediation: A Case Study of Vermiremediation of PAHs Contaminated Soil." *Environmentalist* 28. 466–475, 2008.

FIGURE 14-27
Bioremediation Methods

TYING IT ALL TOGETHER

COMBATING SOIL DEPLETION AND POLLUTION IN THE CHESAPEAKE BAY WATERSHED

In this chapter, you learned about the formation and composition of soils, as well as the qualities that make certain kinds of soils an important natural resource. You also explored natural processes that erode and degrade soils, many of which are accelerated by human activity. Conventional farming, logging, mining, and land development all degrade soil, exposing it to erosion and often polluting it with their by-products, such as chemical toxins, pesticides, and fertilizers carried in surface runoff.

In the Case Study, we considered how agricultural development has contributed to soil depletion and degradation in the United States heartland. We also considered some of the sustainable farming practices being researched by Jerry Glover (see Explorers at Work) and other soil scientists who aim to reverse soil degradation, improve soil quality, and increase crop yields. But soil management and sustainable farming are not merely regional issues limited to the Midwest. Soil is an important resource in every inhabited region, especially those where agriculture is prevalent.

In this activity, you will investigate sustainable agriculture initiatives in another heavily farmed area. The Chesapeake Bay watershed comprises parts of six East Coast states that drain into the country's largest estuary, the Chesapeake Bay. Decades-long initiatives to improve the water quality of the Chesapeake Bay have also targeted agricultural practices. Advocates for sustainable agriculture practices in this region hope to reduce polluted runoff from entering the bay while also conserving the region's soil resources and improving crop yields.

FIGURE 14-28
Go online to view an example of chickens providing fertilizer on a Pennsylvania farm.

Work individually or with a group to complete the steps below.

1. Go online and find information about the Chesapeake Bay Foundation and its efforts to address environmental issues related to agriculture. Craft a single paragraph (3–4 sentences) summarizing how human activities contribute to soil degradation and pollution in the region.

2. Choose two of the focus questions below to extend your investigation.

   - What are five "best management practices" being implemented on farms in the Chesapeake Bay area, and what are their goals? In what way do they address soil degradation?

   - Why are farming practices being altered as part of an estuary recovery program? In what ways are these issues related?

   - Which sustainable practices are being implemented specifically on dairy farms in the region to improve soil quality and reduce polluted runoff?

   - Which specific results indicate that sustainable agriculture efforts have improved soil quality, decreased polluted runoff, and improved farming output and efficiency in the region?

   - How are the sustainable practices implemented on farms in the Chesapeake Bay similar to efforts to grow perennial plants on other U.S. farms? (See Explorers at Work.)

3. Prepare a presentation summarizing your findings. Share it within a class discussion.

CHAPTER 14 SUMMARY

14.1 SOIL FORMATION

- Soil is the layer of weathered material overlying bedrock. It consists of loose sediment, organic matter, water, gases, and microorganisms.

- Soils form and evolve over long periods of time. Factors that affect a soil's development include parent material, climate, organisms, relief, and time.

- Infiltrating water causes downward translocation of silt and clay particles, dissolved minerals, and organic matter. Over time, this leads to the development of soil horizons.

- Most translocated materials are derived from the O, A, and E horizons and deposited in the B horizon. The underlying C horizon is weathered bedrock, and the R horizon is the bedrock itself.

- Soils are dynamic, constantly changing as a result of material additions, losses, translocations, and physical and chemical transformations.

14.2 SOIL CLASSIFICATION

- Soils can be classified based on physical and chemical properties such as soil texture, porosity, and pH. These properties are largely determined by a soil's ratio of silt, clay, and sand particles, as well as its inorganic minerals and organic content.

- Sandy soils, clay soils, silty soils, peaty soils, chalky soils, and loamy soils all differ in terms of physical and chemical properties. Different plants have adapted to thrive in each kind of soil, but loamy soil has the optimal properties for growing most plants.

- Soils can also be classified by soil order, which groups all soils into twelve orders based on age, fertility, origin of parent material, and biome.

- Loess is a silty soil that is ideal for farming but also prone to erosion.

14.3 SOIL EROSION AND DESERTIFICATION

- Erosion by wind and water can degrade the soil in an area. Soils with high organic content tend to absorb water more quickly and resist erosion better than other soils.

- Plant cover and litter prevent soil erosion by anchoring soil and shielding it from erosive forces.

- Human activities such as farming, mining, and construction of roads and cities promote erosion by stripping plant cover, disturbing the soil structure, and removing soil. These activities further degrade soil by depleting organic content, inhibiting gas exchange, and adding contaminants.

- Desertification renders an area's soil unable to support life due to a near complete loss moisture and organic content. Desertification can occur because of natural factors, but human actions often compound the effects of natural causes.

- Despite social and economic pressures that lead to soil degradation, governments generally recognize the importance of conserving soil resources, actively working with scientists, engineers, farmers, and others to develop and promote sustainable solutions.

CHAPTER 14 ASSESSMENT

Review Key Terms
Select the key term that best fits the definition. Not all terms will be used, and no term will be used more than once.

desertification	salinization
eluviation	soil
humus	soil erosion
leaching	soil horizon
litter	soil order
loam	soil texture
parent material	topsoil
porosity	transformation
regolith	translocation
residual soil	transported soil

1. the downward movement of chemically dissolved minerals, clay, and nutrients from one layer of soil to another
2. the degrading action that occurs when wind, ice, or water move through an area, carrying away the top layers of formed soil
3. the matter from which a soil originates
4. a type of soil that has an ideal balance of sand, silt, and clay, maximizing the advantageous properties of all three
5. a type of soil degradation that occurs under very dry conditions and results in the loss of all or most of the moisture and organic matter
6. dark, organic component of soil that has decomposed enough so that the origin of the individual pieces cannot be determined
7. any physical or chemical change to matter within a soil
8. a measure of the total space between particles in a sample of soil
9. the collection of dead but not yet decomposed organic matter (leaves, twigs, animal bodies, animal waste) lying above the surface of the soil
10. the buildup of salts in soil to levels that are toxic to organisms living in the soil
11. one of twelve categories used to classify soils based on their common properties
12. a soil formed by the weathering of the bedrock below the soil
13. property of soil that reflects its water and air content, how quickly water drains through, and how easily it can be shoveled or plowed
14. the horizontal layers that make up a soil's profile

Review Key Concepts
Answer each question on a separate sheet of paper to demonstrate your understanding of key concepts from the chapter.

15. What is the parent material of a soil and how does the parent material influence the soil's characteristics?
16. Describe the similarities and differences between litter and humus.
17. What is the O horizon of a soil? Where would you likely find soils that have an O horizon, and where would you find soils that lack an O horizon?
18. With respect to soil formation, what is the difference between a transformation and a translocation? Use examples to compare the two.
19. What aspects of a soil determine its texture and why?
20. Why is it important to know the type of soil in an area before deciding which plants to grow in that area?
21. Summarize the two different ways that soils can be classified.
22. Why is soil erosion a threat to humans?
23. What natural and human factors led to the Dust Bowl?
24. How can irrigation practices lead to soil salinization?
25. If soil formation is an ongoing natural process, why is soil considered a nonrenewable resource?

Think Critically
Write a response to each question on a separate sheet of paper. Use concepts from the chapter to support your reasoning.

26. Suppose you are hiking through the wilderness with a group of friends. Just after dark, you realize that you have wandered off the trail and you are now lost. You know you need to head north to get back on the trail, but your compass is broken, your cell phones are dead, and it is too cloudy to navigate using the stars. You notice, however, that the vegetation on some hillsides is much different than others. How can you apply your knowledge of slope aspect and soil properties to determine which way is north?
27. Animals such as rabbits and skunks dig in soil, creating small pits as they search for food. These pits then become traps for rainwater and litter. How do you think this digging activity affects the formation and health of the soil in areas where these animals live? Explain your reasoning.

28. A homeowner had the soil from her garden tested. According to the test, she has sandy soil. The homeowner wants to plant ferns in her garden. Ferns thrive in acidic environments and need lots of moisture continuously. Would you predict that ferns will grow well in her garden? Explain your reasoning.

29. No-till farming involves drilling small holes in the ground to plant seeds. This differs from conventional planting, which involves plowing rows across a field and planting seeds in the furrows. Which planting method—no-till or plowing—would result in less soil erosion?

30. Explain why forests generally have healthy, fertile soils, and how deforestation affects the soil quality of an area.

31. Without irrigation, crops cannot survive in an arid climate, yet irrigation has enabled many arid regions to produce very high crop yields. What can you recommend to farmers in an arid climate to prevent salinization of their soil from irrigation?

PERFORMANCE TASK

Optimizing Food Production and Soil Conservation

In this performance task, you will develop and analyze a computational model using data on human populations and arable land.

The table below presents data on population and arable land for the world's five most populous countries. Combined, these countries account for 47 percent of the world population.

1. Using spreadsheet software, develop a computational model for predicting future needs for arable land and crop yield based on population increase. Your model should consider two scenarios for each country:
   - Arable land needed in 2040 if crop yield (food production per hectare) remains constant
   - Rate of increase in crop yield (food production per hectare) needed in 2040 if arable land remains constant

   Assume that the amount of arable land did not change from 2015 to 2016.

   Use your model to rank the countries in terms of arable land per capita. Which countries have the most and least arable land per person, and what are some possible reasons for any differences?

2. Based on your model, what can you conclude about the future of food production in the five most populated countries? Explain your reasoning.

3. With a group, select one of the five countries in the table and research to find out some of the principle crops grown in that country. Then, conduct additional research on one of the crops to identify and describe sustainable farming practices that can both increase crop yields and conserve the soil resources needed to grow that crop. Use the information you gathered to write a proposal for implementing the practices that you researched. Your proposal should clearly explain the practices and how they would improve crop yields and conserve soil resources.

4. Create a poster or slide presentation showing your computational model and your research findings. As you prepare your poster or slides, be sure to include the following:
   - A description of the conditions being modeled
   - A summary of the inputs and outputs used for each scenario
   - An analysis of the results from the modeling process
   - Your proposed recommendations for meeting long-term food production needs and conserving soil resources

Country	Pop. in 2015 (millions)	Estimated Pop. in 2040 (millions)	Total Arable Land in 2016 (million hectares)
China	1,397	1,417	119
India	1,309	1,605	156
United States	320	374	152
Indonesia	258	312	24
Brazil	206	232	81

UNIT 5 | THE HYDROSPHERE

Victoria Falls, originally named Mosi-oa-Tunya, which translates to "the smoke that thunders," is a legendary waterfall on the Zambezi River in Zimbabwe. The 100-meter drop in the basalt plateau has been carved out by rushing water for more than two million years. Downstream from the cascade is a series of gorges, remnants of ancient falls from an era before the river retreated. The iridescent mist from the falls can be seen up to 20 kilometers away.

More than 70 percent of Earth's surface is covered by water in solid or liquid form. A sphere containing all of Earth's water (a literal hydrosphere) would have a diameter of nearly 1,400 kilometers. Earth is the only known planet where the presence and effects of liquid water on and under the surface are readily evident. Water stores and transfers energy. It contributes to the cooling and heating of Earth systems. It is an essential resource for all organisms. In this unit, you will learn where water exists on Earth, its powerful influence on the environment, and human efforts to maintain and protect this life-sustaining resource.

Chapter 15 **Fresh Water**
Chapter 16 **Oceans and Coastlines**
Chapter 17 **Water Resources**

CHAPTER 15
FRESH WATER

Earth's surface is dominated by salt water in the oceans, yet fresh water is essential for all living organisms on the planet. Only a tiny fraction of Earth's fresh water is found flowing or stored in liquid form while the rest is frozen or buried underground. Fresh water is used for drinking, for cooking and cleaning, and in agriculture, manufacturing, and other industries. When fresh water is extracted from groundwater, lakes, or rivers, it is replenished by rain and snowmelt. It collects in lakes and wetlands that produce some of the most biodiverse ecosystems on the planet. Ecosystems that rely on fresh water are important resources for the replenishment and purification of the water that humans use.

KEY QUESTIONS

15.1 How does water change as it cycles through Earth systems?

15.2 In what bodies is fresh water stored on Earth's surface?

15.3 How does fresh water flow under Earth's surface and what effects does it have?

Frogs gather in a partially frozen freshwater pond at around 2,000 meters altitude in the French Alps.

EXPLORERS AT WORK

PROTECTING FRESHWATER BIODIVERSITY

WITH NATIONAL GEOGRAPHIC EXPLORER JOE CUTLER

Dr. Joe Cutler is an ichthyologist, a freshwater ecologist, and a self-described "certified fish head." Cutler says he has always loved fish; he was obsessed with fishing as a boy and learned all he could about fish as a teenager.

Much of Cutler's research focused on a family of fish called cichlids, and most of his fieldwork has been done in Cameroon's southwest region. Cutler first visited Cameroon as a Peace Corps volunteer after college. He visited Lake Bermin, a nearby volcanic crater lake, to check out the fish. In doing so, he actually became the first scientist to inventory fish at that lake in a quarter of a century. At that time few of Cameroon's 34 volcanic crater lakes had been well studied.

Cutler returned to Cameroon in 2015 to study fish ecology. For Cutler, these freshwater ecosystems provide a unique opportunity. Volcanic crater lakes form as fresh water fills in calderas created by volcanic eruptions. Sources of water may include precipitation, groundwater, or ice melt. Most volcanic lakes have no outflows and are essentially cut off from river systems or other bodies of water. They are self-contained habitats that are home to endemic species, fish found nowhere else in the world.

Cutler sampled nine lakes as well as 33 rivers in six months. He collected more than 3,500 fish specimens and approximately 10,000 aquatic invertebrates. Cutler spent days on foot, hacking through jungle vegetation while lugging nets, traps, chemical preservatives, and his camera. He was attacked by biting insects, infested with leeches, and struck by illness. But he was undaunted. His expedition generated more scientific data about the lakes than any other expedition had to date.

Since then, Cutler has conducted fish sampling expeditions both in Cameroon and in the Central African nation of Gabon. As part of a team from The Nature Conservancy, he is working with the Gabonese government, which plans to dam rivers to generate hydroelectric power. Cutler's team is studying the fish species in the rivers to help minimize the dams' impacts.

Cutler has collected thousands of fish specimens and identified more than 400 species, including new and endemic species. He has cataloged diverse freshwater ecosystems and pushed for their protection. He is alarmed by human impacts, including the introduction of non-native species that devastate native fish populations. But he has also met many locals who have often taught him more about local fish populations than any scientific literature. They clearly care about conserving their freshwater resources. That gives him hope for the future of the unique ecosystems he loves.

THINKING CRITICALLY

Infer Why are endemic species of fish or other organisms especially in need of protection?

Dr. Joe Cutler holds up an African pike, *Hepsetus lineata*, from the Ivindo River in Gabon, Central Africa.

CASE STUDY
THE GREAT LAKES ECOSYSTEM

From space, North America's Great Lakes are a striking feature on the face of the planet, visible as deep blue patches on the North American continent (Figure 15-1). The lakes are more than just a beautiful natural feature, however. They are a significant natural resource tied to the social, political, and economic history of the area. The five Great Lakes—Superior, Michigan, Huron, Erie, and Ontario—cover an area of 244,000 square kilometers. They hold about 20 percent of all the fresh water on Earth's surface and about 84 percent of the surface fresh water in North America. Together, they contain enough water to fill about 33 billion Olympic-size swimming pools, forming the largest system of freshwater lakes on the planet.

The Great Lakes Basin is the area around and encompassing the Great Lakes. For the most part, the borders of the Great Lakes Basin are defined by the watershed of each lake, that is, the areas that drain into the lakes themselves or the channels that connect them. European explorers first settled in the Great Lakes Basin in the 16th Century, establishing a lucrative fur trade and building towns and forts. In the centuries since then, humans have transformed the basin in order to exploit its vast natural resources. The once separate lakes were connected by canals to facilitate shipping. A robust logging industry cleared large areas of the forested northern part of the basin, providing lumber for the development of cities such as Chicago, Milwaukee, and Detroit. During the industrial revolution, iron ore shipping facilities and steel factories were built along the lakeshores. Manufacturing, fishing, and shipping industries proliferated throughout the 1900s. Today the Great Lakes Basin is home to 30 percent of Canada's population and 25 percent of its agricultural industry. On the U.S. side, the Great Lakes Basin accounts for 7 percent of U.S. farm production and 10 percent of the total population.

Human development of the Great Lakes Basin has come at a high environmental cost. The dramatic increase in population, the use of its waterways for transportation, and the rise of the manufacturing industry have endangered the health of the ecosystems of the Great Lakes. By the 1960s, Lake Erie had become so polluted that its fish began dying in large numbers, washing up on the shore. The Cuyahoga River, which flows through Cleveland and empties into Lake Erie, was so polluted that it caught fire multiple times, causing deaths and property damage. Invasive species such as the zebra mussel, sea lamprey, and alewife spread through the Great Lakes via canals and ships, nearly wiping out native species such as the lake trout and crippling the fishing industry. Development along the lakeshores has also eliminated large areas of wildlife habitat.

As You Read Explore how the water cycle and other factors are related to the water in lakes, rivers, and wetlands. Find ways that natural events and human activities affect ecosystems in and near bodies of water. Consider the importance of rivers, lakes, and wetlands and what steps can be taken to conserve them.

FIGURE 15-1
A satellite image of the Great Lakes shows (from left to right) Lake Superior, Lake Michigan, Lake Huron, Lake Erie, and Lake Ontario.

15.1 HYDROLOGIC CYCLE

Core Ideas and Skills
- Identify the changes in physical state (solid, liquid, gas) that occur as water moves among Earth's systems through the hydrologic cycle.
- Quantify the amount of fresh water and where it exists on Earth.
- Describe how water can flow or collect in different reservoirs on Earth's surface and underground.

KEY TERMS

runoff transpiration

Distribution of Water on Earth

The hydrologic cycle, or water cycle, describes the continuous circulation of water among the four spheres: the hydrosphere (or watery part of the planet), the geosphere (the solid Earth), the biosphere (life-forms in the sea, on land, and in the air), and the atmosphere. About 1.3 billion cubic kilometers of water exist at Earth's surface. Of this huge volume, 97.5 percent is salty seawater and 2.5 percent is fresh water. Although the hydrosphere contains a great volume of water, only a tiny fraction is fresh and available in streams and rivers, lakes, wetlands, and groundwater (Figure 15-2).

Water evaporates from the continents and oceans to form water vapor in the atmosphere. This vapor eventually condenses and falls back to the surface as rain, snow, sleet or hail. Most precipitation lands on the ocean, partly because it covers most of the planet. The precipitation that falls on the continents follows four paths, as illustrated in Figure 15-3:

1. Surface water flowing to the sea in streams and rivers is called **runoff**. This water may stop temporarily in a lake or wetland, but eventually it evaporates or flows to the oceans (hydrosphere).

2. Much water seeps into the ground (geosphere) to become part of a vast reservoir of groundwater. Water infiltrating underground comes from direct precipitation and from surface water sources such as rivers and lakes. Groundwater flows through pore spaces in sediment and bedrock but typically does so much more slowly than water flowing on the surface.

3. Most of the remainder of water that falls onto land evaporates back into the atmosphere. Water sucked upward through the root systems of plants also evaporates directly from plants as they respire in a process called **transpiration**.

4. A small amount of water is incorporated into the biosphere in the form of plant and animal tissue.

The water cycle refers to not only the movement of water but also energy from one part of the globe to another. Ocean currents transport huge quantities of heat from the Equator toward the poles, thus cooling the Equator and warming the higher latitudes. Evaporation is a cooling process,

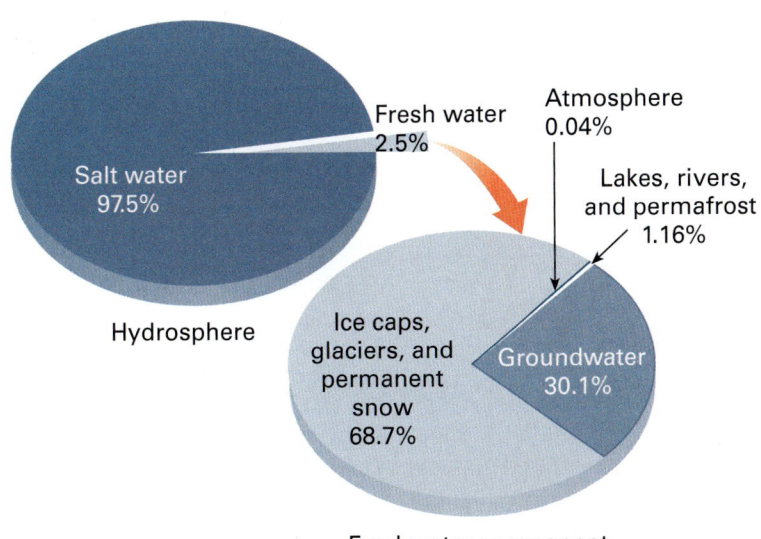

FIGURE 15-2
Two charts represent where and in what physical form water is found on Earth.

Source: USGS

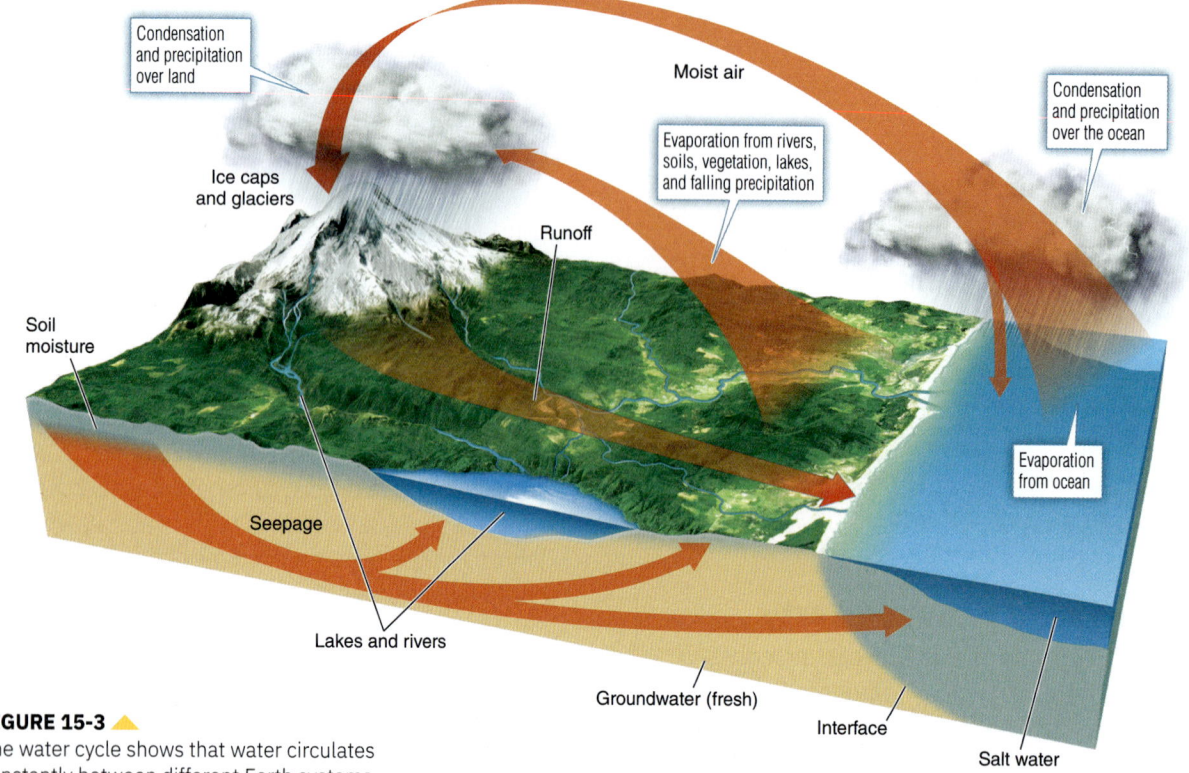

FIGURE 15-3
The water cycle shows that water circulates constantly between different Earth systems.

whereas condensation releases heat. Water vapor is a greenhouse gas, which warms the atmosphere, but clouds and sparkling glaciers reflect sunlight back out to space and thereby cool Earth. There are numerous feedback cycles, both positive and negative. This chapter describes the storage of surface water and the movement of groundwater. Other components of the water cycle are discussed in other chapters.

checkpoint Which Earth systems are involved in the water cycle?

15.1 ASSESSMENT

1. **Explain** Why is the water cycle called a "cycle"?
2. **Define** Of all the water on Earth, how much is fresh water?
3. **Distinguish** How do the physical properties of water differ in the four main reservoirs of fresh water?

Computational Thinking

4. **Analyze Data** Calculate the percent of Earth's water that is frozen in ice caps, glaciers, and permanent snow.

15.2 LAKES AND WETLANDS

Core Ideas and Skills

- Identify where on Earth's surface fresh water can accumulate.
- Describe how lake ecosystems change with the seasons.
- Summarize the importance of wetlands.

KEY TERMS

lake	eutrophic
kettle lake	thermocline
oligotrophic	turnover

Lakes

Lakes and lakeshores are attractive places to live and play. Clean, sparkling water, abundant wildlife, beautiful scenery, aquatic recreation, and fresh breezes all come to mind when we think of going to a lake. Despite their great value, however, lakes are fragile and geologically temporary.

478 CHAPTER 15 FRESH WATER

A **lake** is a large inland body of standing water that occupies a depression in the land surface. Streams flowing into the lake carry sediment, which can fill the depression in a relatively short time, geologically speaking. Soon the lake becomes a swamp, and with time the swamp fills with more sediment and vegetation and becomes a meadow or a forest with a stream flowing through it.

If most lakes fill quickly with sediment, why are these bodies of water so abundant today? Most lakes exist in places that were covered by glaciers during the latest glacial period. About 18,000 years ago, great continental ice sheets extended well south of the Canadian border, and mountain glaciers scoured alpine valleys as far south as New Mexico and Arizona. Similar ice sheets and alpine glaciers existed in the higher latitudes of the Southern Hemisphere.

The glaciers created lakes in several ways. Flowing ice eroded numerous depressions in the land surface, which then filled with water. The Great Lakes of North America and the Finger Lakes of upper New York State are examples of large lakes occupying glacially scoured depressions.

The glaciers also deposited huge amounts of sediment as they melted and retreated. Some of these great piles of glacial debris formed dams across stream valleys. When the glaciers melted, streams flowed down the valleys but were blocked by the sediment dams. Many modern lakes occupy glacially dammed valleys (Figure 15-4). In addition, large blocks of ice were left behind as the glaciers receded. When these ice blocks melted, they left behind depressions called kettles, which in many cases filled with water to become **kettle lakes**.

Most glacial lakes formed within the past 10,000 to 20,000 years, and sediment is rapidly filling them. Many of the smallest such lakes have already become swamps. In the next few hundred to few thousand years, many of the remaining lakes will fill with mud. The largest, such as the Great Lakes, may continue to exist for tens of thousands of years. But the life spans of lakes such as these are limited, and it will take another glacial episode to replace them.

Lakes also form by nonglacial means. A volcanic eruption can create a crater that fills with water, such as Crater Lake in Oregon. Oxbow lakes form in abandoned river channels. Other lakes, such as Lake Okeechobee in the Florida Everglades, form in flatlands with shallow groundwater. These types of lakes, too, fill with sediment and, as a result, exist for a limited time.

A few lakes, however, form in ways that allow them to exist beyond the span of a "normal" lake. For example, Russia's Lake Baikal is a large, deep lake lying in a depression created by active normal faults. Although rivers pour sediment into the lake, movements of the faults repeatedly deepen the basin. As a result, the lake has existed for more

FIGURE 15-4
A quiet glacial lake sits between mountain ranges in Canada's Banff National Park. The lake is dammed by a moraine at its lower end (not shown).

FIGURE 15-5
A eutrophic lake in Florida hosts a diverse range of aquatic plants and animals.

than a million years—so long that endemic species of seals, fish, and other animals have evolved in its ecosystem. A species that is *endemic* to an area exists only in that area and nowhere else in nature.

In a deep lake, sunlight is available near the surface, but nutrients are abundant mainly on the bottom. Plankton (small, free-floating organisms) growing in the surface waters are limited due to low nutrient levels. Bottom-rooted plants cannot grow in a deep lake due to lack of sunlight. Thus, deep lakes contain low concentrations of nutrients and insufficient sunlight at the bottom to sustain many aquatic organisms. The low level of biological productivity often causes these lakes to have a deep-blue color that we associate with a clean, healthy lake. Such a lake is called **oligotrophic**, meaning "poorly nourished." Oligotrophic lakes have low productivities, meaning they sustain relatively few living organisms. A lake of this type

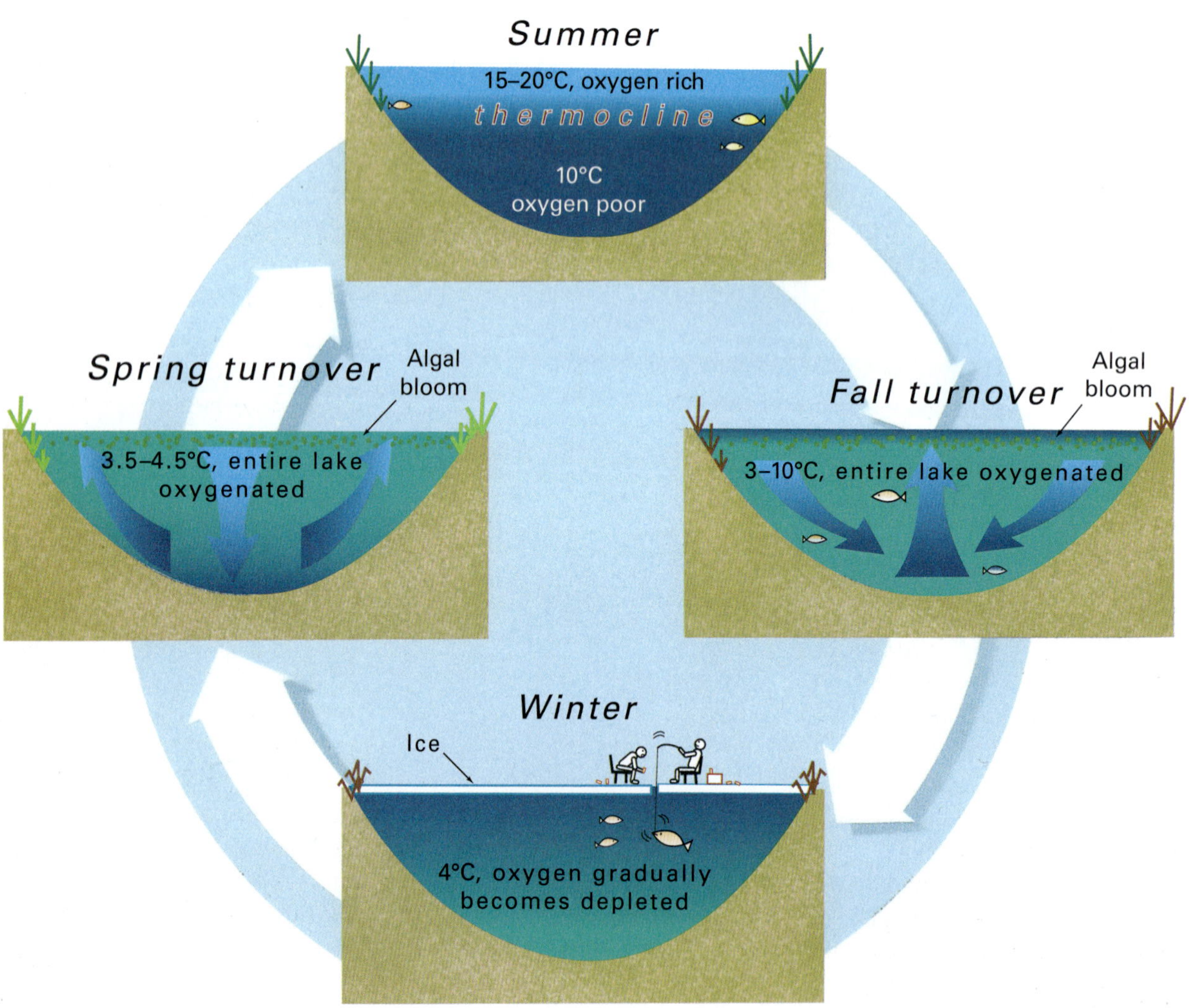

FIGURE 15-6
Oxygen and other nutrients move to different depths during lake turnover across the four distinct seasons.

CAREERS IN EARTH AND SPACE SCIENCES

Limnologist

Environmental scientists who study the ecological environments in and around inland water systems are called limnologists. A limnologist observes streams, rivers, ponds, lakes, and marshlands, and the organisms that live in those environments. Government agencies, universities, and private research companies hire limnologists to monitor the health of regional water resources.

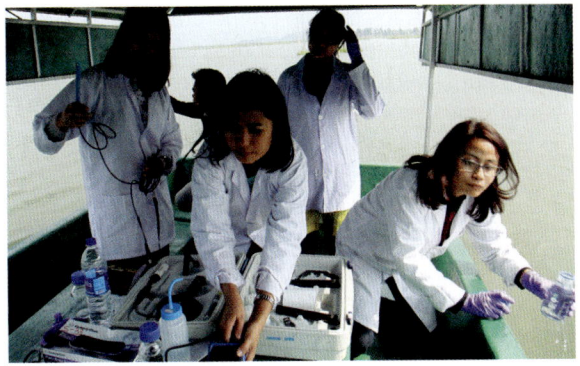

FIGURE 15-7A
A crew of limnologists analyzes the pollution level in the water of the 300-square-kilometer Loktak Lake in the eastern state of Manipur, India. The livelihoods of more than 100,000 people depend on this lake.

FIGURE 15-7B
Go online to view a limnologist taking microbial samples from a freshwater lake on Bell Island in Russia.

---

is attractive for recreation and can support limited numbers of large game fish.

As a lake fills with sediment, it becomes shallower and sunlight reaches more and more of the lake bottom. The sunlight allows bottom-rooted plants to grow. As the plants die and decay, they add nutrients to the lake water. Plankton increase in numbers, as do the numbers of fish and other organisms. The lake may become so productive that its surface is covered with a green scum of plankton or a dense mat of rooted plants. The litter from this biomass contributes to sediment filling the lake, and eventually the lake becomes a swamp.

A lake of this kind, with a high nutrient supply and high level of productivity in terms of biomass, is called a eutrophic lake (Figure 15-5). **Eutrophic** means "well nourished." Eutrophication occurs naturally as part of the life cycle of a lake. However, the addition of nutrients in the form of sewage and other kinds of pollution has greatly accelerated the eutrophication of many lakes.

If you have ever dived into a deep lake on a summer day, you have probably discovered that the top meter or so of lake water can be much warmer than deeper water. This occurs because sunshine warms the upper layer of water, making it less dense than the cooler, deeper layers. The warm, less dense water floats on the cooler, denser water. The boundary between the warm and cool layers is called the **thermocline**.

In temperate climates, colder autumn weather cools the surface water to a temperature below that of deeper water. The surface water becomes denser. Consequently, the surface water sinks and mixes with the deep water, equalizing the water temperature throughout the lake. This process is called **turnover**. In the winter, ice floats on the surface and thermal layering develops again. In spring, as ice melts on the lake, the cold surface water again is denser than deep lake waters, and spring turnover occurs. As summer comes, the lake again develops thermal layering. This seasonal process is illustrated in Figure 15-6.

Turnover in temperate lakes illustrates important interactions among the atmosphere, the hydrosphere, and the biosphere. During summer and winter when the lake water is layered, bottom-dwelling organisms may use up most or all of the oxygen in deep water. At the same time, surface organisms may deplete surface waters of dissolved nutrients. However, surface water is rich in oxygen because it is in contact with the atmosphere, and deep water may be rich in nutrients because it is in contact with bottom sediment. Turnover enriches deep water with oxygen and, at the same time, supplies nutrients to the surface water. More nutrients near the surface in spring and fall can result in an *algal bloom*—a sudden and obvious increase in the amount of floating green algae on a lake's surface.

checkpoint What is a thermocline?

LESSON 15.2

MINILAB Lake Turnover

Materials
8–10 cups crushed ice
5-gallon aquarium
cold tap water
pitcher
aquarium net
heat lamp
masking tape
7 thermometers
ruler (metric)
small electric fan
watch or clock

CAUTION: Use safety goggles. Do not allow the heat lamp or any electric appliance to come in contact with water.

Procedure

1. Measure 24 cm from the bottom of the aquarium and place a piece of masking tape on the side of the aquarium. Later, you will fill the aquarium to this mark, but do not add water yet.

2. Use masking tape to secure the thermometers at seven depths on the inside of the aquarium: 1 cm, 3 cm, 5 cm, 9 cm, 13 cm, 17 cm, 21 cm. Note: You will now be measuring *down* from the first mark you made with the tape, because depth is measured from the surface of the water downward. The thermometers should be facing outward so you can read them through the glass.

3. Use the pitcher to add cold tap water to the aquarium, filling to the 24-cm mark. Add some of the ice to the aquarium gradually, stopping when the temperature of the water near the bottom reaches 4°C. Observe what happens to the ice.

4. Once the water is still, record the temperature shown on each thermometer. Create a data table to record the temperatures.

5. Gently remove the ice from the aquarium using the net. Direct the heat lamp onto the surface of the water and turn it on. **CAUTION:** Do not allow the heat lamp to come into contact with the water.

6. In your data table, record the temperature shown on each thermometer in your data table after 5 minutes, 12 minutes, 30 minutes, and 45 minutes.

7. Turn off the heat lamp. Turn on the fan and direct it onto the surface of the water until the water mixes enough that the temperature is the same throughout. Record the temperature shown on each thermometer in your data table. **CAUTION:** Do not allow the electric fan to come into contact with the water.

8. Slowly add a layer of ice to the surface of the water, taking care to disturb the water as little as possible. Wait five minutes. In your data table, record the temperature shown on the thermometers.

Results and Analysis

1. **Observe** Describe how the temperature changed throughout the experiment.

2. **Connect** How did the heat lamp and the fan affect the temperatures of the water? What natural processes did these devices model?

3. **Explain** What did you observe about the ice in step 3? Why is this an important part of the model?

Critical Thinking

4. **Evaluate** How did the investigation model seasonal changes in a lake, including spring turnover and fall turnover?

5. **Compare** How do the seasonal changes you modeled and the properties of water affect organisms living in the lake?

Wetlands

Wetlands are known across North America as swamps, bogs, marshes, sloughs, mudflats, and floodplains. They are regions that are water soaked or flooded for all or part of the year. Some wetlands are wet only during exceptionally rainy years and may be dry for several years at a time.

Wetland ecosystems vary so greatly that the concept of a wetland defies a simple definition. All wetlands share certain properties.

- The ground is wet for at least part of the time.
- The soils support anaerobic (lacking oxygen) conditions.
- The vegetation is adapted to periodic flooding.

North American wetlands include all stream floodplains, frozen Arctic tundra, warm Louisiana swamps, coastal Florida mangrove swamps, boggy mountain meadows of the Rockies, and the immense swamps of interior Alaska and Canada (Figure 15-8).

FIGURE 15-8
North American wetlands extend from the Everglades in Florida (A) to the immense swamps of Alaska (B).

Wetlands are among the most biologically productive environments on Earth. They are important for degrading pollutants and serve to help control the effects of flooding. Two-thirds of the Atlantic fish and shellfish consumed by humans rely on coastal wetlands for at least part of their life cycles. One-third of the endangered plant and animal species in the United States also depend on wetlands for survival. More than 400 of the 800 species of protected migratory birds and one-third of all resident bird species feed, breed, and nest in wetlands. Aquatic organisms consume many pollutants and degrade them to harmless by-products. Because these organisms abound in wetlands, the ecosystems serve as natural sewage treatment systems. Wetlands also help control flooding by absorbing excess water that might otherwise overrun towns and farms.

When European settlers first arrived in North America, the continent had 87 million hectares of wetlands, not including the wetlands in Alaska. Until relatively recently, most Americans viewed wetlands as mosquito-infested, malarial swamps. This land could be farmed or developed if drained or filled but was otherwise worthless. In the mid-1800s, the federal government passed legislation known as the Swamp Land Act, which established an official policy for filling and draining wetlands in order to convert them to agricultural uses wherever possible. More than a quarter million square kilometers were officially identified by the acts as swampland available for "reclamation."

Wetlands have been lost or degraded both by humans and natural causes. Many wetlands have been drained or had their sources of water cut off by the construction of dams or dikes. Others have been

CONNECTION TO LIFE SCIENCE

The Midwest Wetlands Initiative

Wetlands produce volumes of food supporting a rich ecosystem that rivals that of rain forests and coral reefs. The high productivity is possible due to shallow waters where high levels of nutrients meet high levels of sunshine. Wetland food webs connect diverse wildlife including birds, fish, amphibians, shellfish, and insects. Two midwestern wetland areas that stand out for high levels of biodiversity, including many endemic species, are the Dixon Waterfowl Refuge in north-central Illinois and the Midewin National Tallgrass Prairie in the northeast section of the state. These environments support more than a hundred bird species, 78 percent of the mammal and reptile species in the state, and 90 percent of the amphibians. The Wetlands Initiative is an organization dedicated to education, and restoring and protecting the precious wetland resources of the Midwest.

LESSON 15.2

FIGURE 15-9
ON ASSIGNMENT National Geographic photographer Klaus Nigge snapped a photo of a sandhill crane wading with its young in the wetlands of Florida's Myakka River State Park. Florida contains about 20 percent of all the wetlands in the United States.

filled, logged, or mined. The introduction of non-native invasive species or toxic levels of pollution or nutrients continues to cause widespread wetland degradation. Natural wetland loss can also occur by sea-level rise, drought, and erosion by large storms.

According to the U.S. Environmental Protection Agency, over the past several centuries about 900,000 square kilometers of original U.S. wetlands have been degraded or destroyed. This is an area larger than Texas and Oklahoma combined.

DATA ANALYSIS Compare Wetland Gains and Losses

The extent and density of wetland areas in the United States vary from region to region due to natural conditions and human activities. Wetland density also varies over time as conditions change. The map in Figure 15-10A shows a snapshot of wetland density in three watersheds of the eastern United States in 2004. The bar graph in Figure 15-10B shows losses and gains to wetlands in the same areas from 1998–2004. Examine the map and the graph and then answer the questions.

1. **Summarize** Describe the trends shown in the graph. Which areas have the greatest gains? Which have the greatest losses? Overall, was there a net gain or a net loss of wetlands?

2. **Infer** What are possible causes of differences in the wetlands gains and losses and density of these three areas?

Computational Thinking

3. **Infer** If current trends continue, how do you think the map will change in the future?

4. **Recognize Patterns** Draw a connection between the gains and losses of wetlands and the density of wetlands in the three areas: Atlantic, Gulf, and Great Lakes.

Data Challenge

Go to the Data Analysis in MindTap to complete the data challenge.

FIGURE 15-10 ▼
(A) The map shows a comparison of wetlands density in the coastal watersheds of the Atlantic, Gulf of Mexico, and Great Lakes in 2004. (B) The graph shows wetlands gains and losses in these three watersheds between 1998 and 2004.

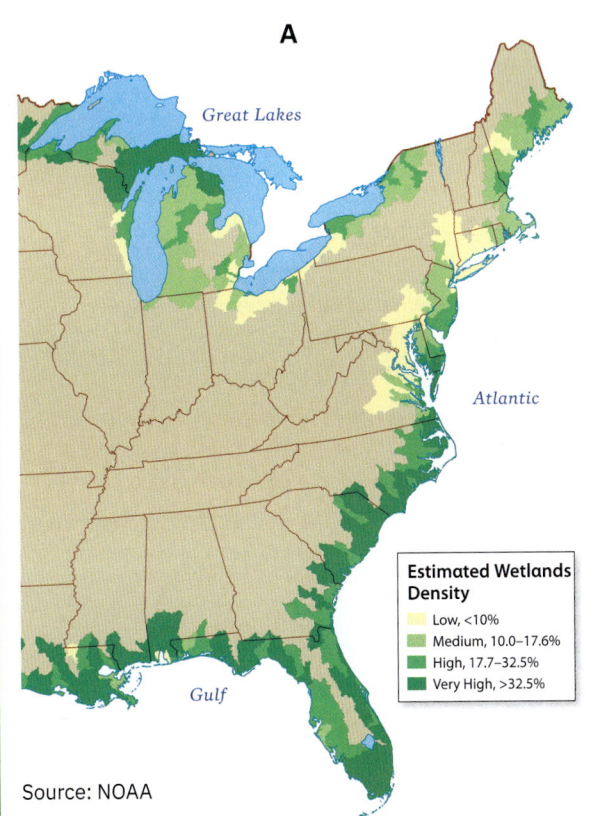

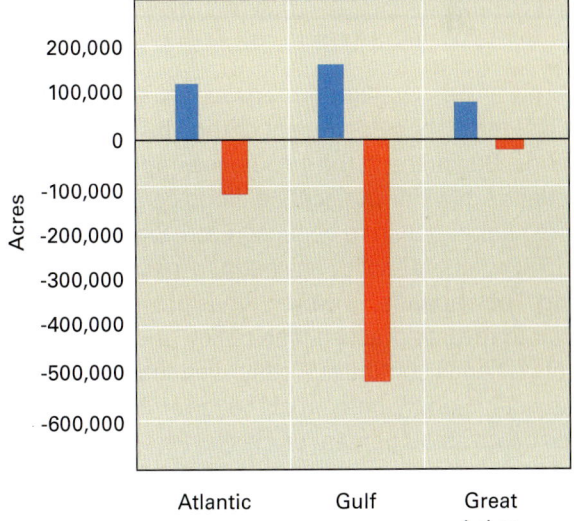

Source: NOAA

MINILAB Role of Wetlands

Materials

2 jars of muddy water, with equal parts mud and water in each
modeling clay, approximately 300–400 grams
paint roller tray
3–4 sponges
watering can
water
paper towels

CAUTION: Wash hands after completing the investigation.

Procedure

1. Press the clay along the higher end of the tray, right up to the wall of the tray. Add sponges along the inside edge of the clay. Be sure the sponges go all the way across the tray and are pushed together so there are no gaps between them. The deepest part of the tray should be empty.
2. Use the watering can to sprinkle water on the clay. Observe what happens to the water.
3. Remove the sponges and dump out the water from the tray. Leave the clay in the tray. Wipe the tray with paper towels.
4. Use the watering can to sprinkle water on the clay exactly as before, but this time without the sponges in place. Observe what happens to the water.
5. Dump out the water. Squeeze out the sponges and replace them in the tray as in step 1.
6. Pour the jar of muddy water gently over the clay. Observe the water that collects in the deep end of the tray.
7. Remove the sponges and dump out the muddy water from the tray. Leave the clay in the tray. Wipe the tray with paper towels.
8. Pour the jar of muddy water gently over the clay, but this time without the sponges in place. Observe the water that collects in the deep end of the tray.

Results and Analysis

1. **Observe** Describe what you observed in steps 2, 4, 6, and 8.
2. **Compare** Explain any differences in your observations after steps 2 and 4. Do the same for steps 6 and 8.
3. **Infer** How did the presence of the sponges affect the way the water flowed into the tray in step 2 and the quality of the water in step 8?

Critical Thinking

4. **Analyze** How did the investigation model the effect of wetlands on the flow and quality of water?
5. **Conclude** Why are wetlands important? What negative effects can occur if wetlands are eliminated from an area?

California and several upper-midwestern states have lost more than 80 percent of their historic wetlands. Beginning in the 1960s, the importance of wetlands in water purification, flood control, and wildlife habitat began to be recognized. Since the late 1980s, federal policy has been to incur "no net loss" of wetlands. However, it has proven more difficult to reverse practice than policy, and the net loss of wetlands continues. Today, wetlands make up between 6 and 9 percent of the lower 48 states and as much as 60 percent of Alaska. Preserving these wetlands along with those outside the United States will present a serious future challenge as Earth's human population continues to rise and pressure increases to develop, farm, log, or ranch these areas.

checkpoint What are the more commonly known terms for wetlands in North America?

15.2 ASSESSMENT

1. **Explain** What is the relationship between glaciers and lakes?
2. **Define** What is an oligotrophic lake?
3. **Explain** In a deep, temperate, oligotrophic lake, at what time of year would you expect algal blooms to occur? Explain why.
4. **Distinguish** How are wetlands different from lakes? How are they similar?

Critical Thinking

5. **Critique** What are the pros and cons of developing wetlands for farming or other uses?

15.3 GROUNDWATER

Core Ideas and Skills
- Identify how the porosity and permeability of soil and rock allows for water to move and collect underground.
- Name and classify bodies of rock by their capacity to store water.
- Describe the various dramatic effects groundwater has on rock and soil under and on Earth's surface.

KEY TERMS

permeability
water table
aquifer
aquitard

stalactite
stalagmite
sinkhole

In most places, if you drill a hole in the ground a few meters deep, the bottom of the hole fills with water within a few minutes. The water appears even if no rain falls and no streams flow nearby. The water that seeps into the hole is groundwater. This fresh water saturates Earth's crust from a few meters to a few kilometers below the surface.

Porosity and Permeability

Groundwater is exploited by digging wells and pumping it to the surface. It provides drinking water for more than half the population of North America and is a major source of water for irrigation and industry. However, deep wells and high-speed pumps now extract groundwater more rapidly than natural processes can replace it in many parts of the central and western United States. As a result, groundwater resources that have accumulated over thousands of years are being permanently lost. In addition, industrial, agricultural, and domestic contaminants seep into groundwater in many parts of the world. Such pollution is often difficult to detect and expensive to clean up.

Groundwater fills pores, which are small cracks and voids in soil and bedrock. The volume percentage of these open spaces is called porosity. (See Chapter 14.) In unlithified sand or gravel, the network of open pore spaces between the individual sediment clasts results in porosity values close to 40 percent. In contrast, unlithified mud can have porosity values of up to 90 percent. The higher porosity of mud results from the physical and chemical characteristics of very fine-grained sediment.

Most sedimentary rocks have lower porosities than loose sediment. For example, sandstone and conglomerate commonly have 5 to 30 percent porosity. At depths greater than 3 kilometers, mudstone and claystone are characterized by porosity values of only a few percent, much lower than the original unlithified mud. The big loss of porosity during the lithification of mud results mainly from compaction during burial. Collapsed pore spaces act like a collapsed house of cards. Minerals grow between the compacted grains, filling much of the remaining pore space. Igneous and metamorphic rocks have very low porosities (a few percent or less) unless they are fractured.

Porosity indicates the amount of water that rock or soil can hold. In contrast, **permeability** refers to how interconnected the pore spaces are. Permeability relates to the ability of rock or soil to transmit water or other fluids such as petroleum and natural gas. Water can flow rapidly through material with high permeability but flows slowly through material with low permeability.

Most materials with high porosity also have high permeability. Sand and sandstone have many relatively large, well-connected pores that allow the water to flow through the material. However, if the pores are very small, as in clay or mudstone, electrical attractions between water and soil slow the passage of water. Although unlithified mud typically has a high porosity, the electrical attraction between water and clay particles slows the passage of water, so it has low permeability and transmits water slowly.

checkpoint How is porosity different from permeability?

Water Table and Aquifers

When rain falls, much of it soaks into the ground. Water does not descend into the crust indefinitely, however. Below a depth of a few kilometers, the pressure from overlying rock closes the pores, making bedrock both nonporous and impermeable. Water accumulates above this permeability barrier, filling pores in the rock and soil. This completely wet layer of soil and bedrock above the deep, impermeable rock is called the zone of saturation. The **water table** is the top

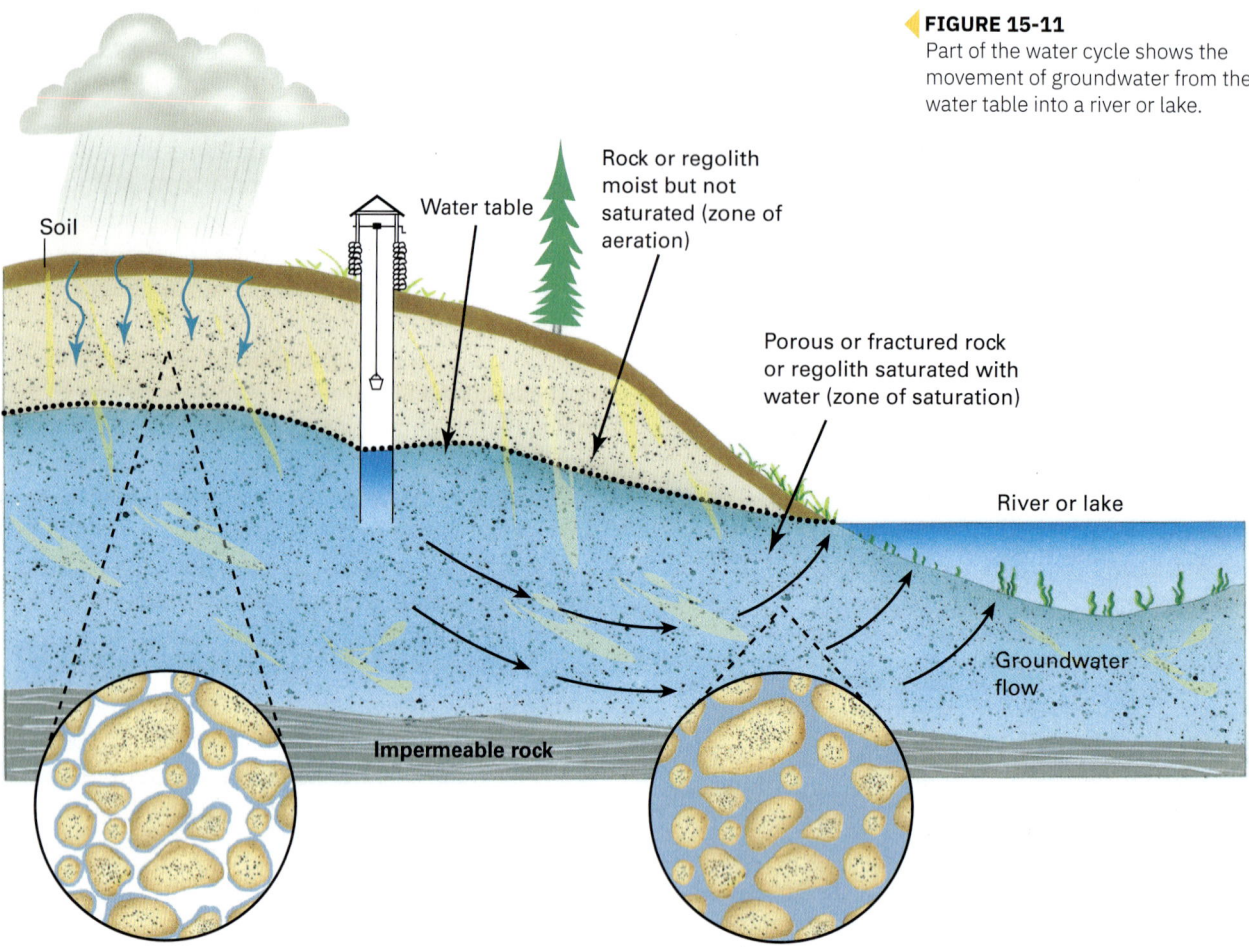

FIGURE 15-11
Part of the water cycle shows the movement of groundwater from the water table into a river or lake.

of the zone of saturation (Figure 15-11). The unsaturated zone, or zone of aeration, lies above the water table. In this layer, the rock or soil may be moist but is not saturated, and air occupies some or all of the pore space.

If you dig below the water table into the zone of saturation, you have dug a well. The water level in a well is at the level of the water table. During a wet season, rain seeps into the ground to recharge the groundwater, and the water table rises. During a dry season, the water table falls.

An **aquifer** is any body of rock or soil that can yield economically significant quantities of water. An aquifer must be both porous and permeable so that water flows into a well to replenish water that is pumped out. Sand and gravel, sandstone, limestone, and highly fractured bedrock of any kind make excellent aquifers. In contrast, mudstone, clay, and unfractured igneous and metamorphic rocks have low porosity and permeability; they are called **aquitards**.

checkpoint What is the key difference between an aquifer and an aquitard?

Groundwater Movement

Nearly all groundwater seeps slowly through bedrock and soil. Groundwater often flows at about 4 centimeters per day (about 15 meters per year), although flow rates may be much faster or slower depending on permeability. Most aquifers are like sponges rather than underground pools or streams. However, groundwater can flow very rapidly through large fractures in bedrock. In a few regions, underground rivers flow through caverns.

In general, the water table mimics its surrounding topography. Groundwater flows from zones where the water table is highest toward areas where it is lowest. Some groundwater flows roughly parallel to the sloping surface of the water table toward the valley. But groundwater also flows from zones of high pressure toward zones of low pressure. Water pressure is greatest beneath the highest part of the water table. Groundwater can flow downward beneath a hill, then sideways toward a valley, and finally upward beneath the lowest part of the valley, where a stream flows. This is how groundwater feeds a stream. It is why streams flow even when no rain has fallen for weeks or months.

A spring occurs where the water table intersects the land surface and water flows or seeps onto the surface. In some places, a layer of impermeable rock or clay lies above the main water table. Porous, saturated rock or sediment above the impermeable layer can form an aquifer. Hillside springs often flow from these aquifers. Springs also occur where fractured bedrock or cavern systems intersect the land surface.

Figure 15-12 shows a tilted layer of permeable sandstone sandwiched between two layers of impermeable shale, a type of mudstone that is made almost entirely of clay-sized mineral grains. An inclined aquifer such as the sandstone layer, bounded top and bottom by impermeable rock, is a confined aquifer. Water in the lower part of the aquifer is under pressure from the weight of water above. Therefore, if a well is drilled through the shale and into the sandstone, water rises in the well without being pumped. If pressure is sufficient, the water spurts out onto the land surface in a spring.

Like surface streams, groundwater is capable of eroding rock and leaving behind sedimentary deposits. Rainwater reacts with atmospheric carbon dioxide to become slightly acidic and capable of dissolving limestone. A cavern forms when acidic water seeps into cracks in limestone, dissolving the rock and enlarging the cracks. Mammoth Cave in Kentucky and Carlsbad Caverns in New Mexico are two famous caverns formed in this way (Figure 15-13).

Although caverns form when limestone dissolves, most caverns also contain features formed by deposition of calcite. When a solution of water, dissolved calcite, and carbon dioxide seeps through the ground, it is under pressure from water in the cracks above it. If a drop of this solution seeps into the ceiling of a cavern, the pressure decreases because the drop comes in contact with the cavern air, which is at atmospheric pressure. The high humidity of the cave prevents the water from evaporating rapidly, but the lowered pressure allows some of the carbon dioxide to escape as a gas. When the carbon dioxide escapes, the drop becomes less acidic. This decrease in acidity

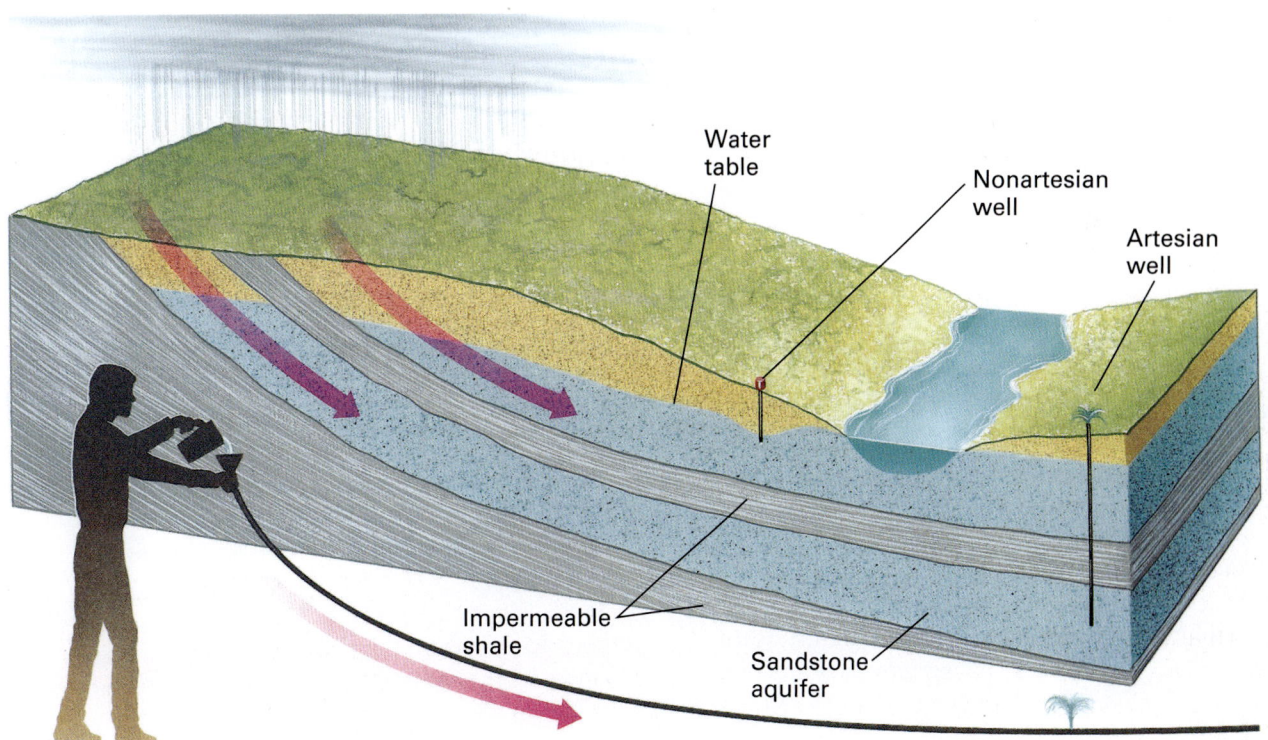

FIGURE 15-12
A confined aquifer capable of supporting different types of wells occurs where a tilted layer of permeable rock is sandwiched between layers of impermeable rock such as shale. Water rises in an artesian well without the need for a pump due to the pressure. An inset shows the same process when water is poured into a hose that contains a tiny hole.

causes some of the dissolved calcite to precipitate as the water drips from the ceiling. Over time, a beautiful and intricate stalactite (from the Greek for "drip") grows, hanging icicle-like from the ceiling of the cave.

Only a portion of the dissolved calcite precipitates as the drop seeps from the ceiling. When the drop falls to the floor, it spatters and releases more carbon dioxide. The acidity of the drop decreases further, and another minute amount of calcite precipitates. Thus, a cone-shaped **stalagmite** (from the Greek for "drop") builds from the floor upward to complement the stalactite. Because stalagmites are formed by splashing water, they tend to be broader than stalactites. As the two features continue to grow,

FIGURE 15-13
The Carlsbad Caverns in New Mexico formed when groundwater dissolved limestone, creating large open spaces underground.

FIGURE 15-14
(A) The Winter Park sinkhole opened in May 1981. It swallowed 250,000 cubic yards of soil, cars, part of an Olympic-size swimming pool, entire sections of two streets, and one three-bedroom home. (B) Today the Winter Park sinkhole is a small artificial lake in the community.

they may eventually meet and fuse together to form a column.

If the roof of a cavern collapses, a **sinkhole** forms on Earth's surface (Figure 15-14). A sinkhole can also form as limestone dissolves from the surface downward. Each year, sinkholes cause significant economic damage as buildings, roads, and other structures undergo collapse in densely populated regions underlain by limestone. A well-documented sinkhole formed in May 1981 in Winter Park, Florida. During the initial collapse, a three-bedroom house, half a swimming pool, and six Porsches in a dealer's lot all fell into the underground cavern.

Within a few days, the sinkhole was 200 meters wide and 50 meters deep, and it had devoured additional buildings and roads.

Although sinkholes form naturally, human activities can accelerate the process. The Winter Park sinkhole formed when the water table dropped, removing support for the ceiling of the cavern. The water table fell as a result of a severe drought augmented by excessive removal of groundwater by people.

checkpoint Name three structures that are formed by the movement of groundwater.

15.3 ASSESSMENT

1. **Explain** How is groundwater important for humans?
2. **Distinguish** What is the difference between porosity and permeability?
3. **Define** What are the differences between an aquifer, an aquitard, and the water table?
4. **Describe** How does groundwater create a cavern?

Critical Thinking

5. **Synthesize** Describe how the water cycle links glaciers, lakes, groundwater, and wetlands.

TYING IT ALL TOGETHER
GREAT LAKES CONNECTIONS

In this chapter, you learned about the water cycle and how it moves water continuously through the hydrosphere, the geosphere, the biosphere, and the atmosphere. You traced the path of water as it flows over the land and into streams, rivers, and lakes, and you connected the shape of the land to the direction of this water flow. You learned about drainage basins and the seasonal changes that affect the amount and temperature of water in large bodies such as rivers and lakes. You also explored the network of streams, rivers, and lakes that connect with each other and allow for the free flow of water and objects carried by water.

In the Case Study, we examined the Great Lakes Basin, an area that includes the five Great Lakes and their watersheds (Figure 15-15). We considered the history and importance of the Great Lakes as well as some ongoing threats to their ecosystems. Now you will expand your exploration of this important area, looking more closely at the connections among the lakes themselves and other bodies of water, such as rivers and oceans. The flow of water into and out of the Great Lakes affects the water levels of the lakes and the health of their ecosystems. In this activity, you will investigate these connections and create a labeled and annotated map of the Great Lakes Basin.

Work individually and with a group to complete the steps.

1. Individually, research how the Great Lakes are connected to each other and to other water systems such as rivers and the Atlantic Ocean. Investigate how these connections affect the water levels and health of the Great Lakes.

2. Write a summary of your findings in your own words. Cite your sources.

3. Form a small group with two or three other students. Compare and synthesize your findings, investigating and clarifying any significant differences.

4. With your group, create a presentation that demonstrates how the water level of the Great Lakes and connected water systems is impacted by water flowing into and out of the lakes, by different forms of human intervention, and by natural processes (such as floods, precipitation, or the water cycle). Include in your presentation an annotated map of the Great Lakes. Your annotations should include labels and short descriptions for all relevant bodies of water, arrows indicating inflow and outflow, identification of natural and human-made features, and other information about the Great Lakes system that will enhance your presentation.

5. Come together with the class. Present your group's findings and listen to other groups' presentations.

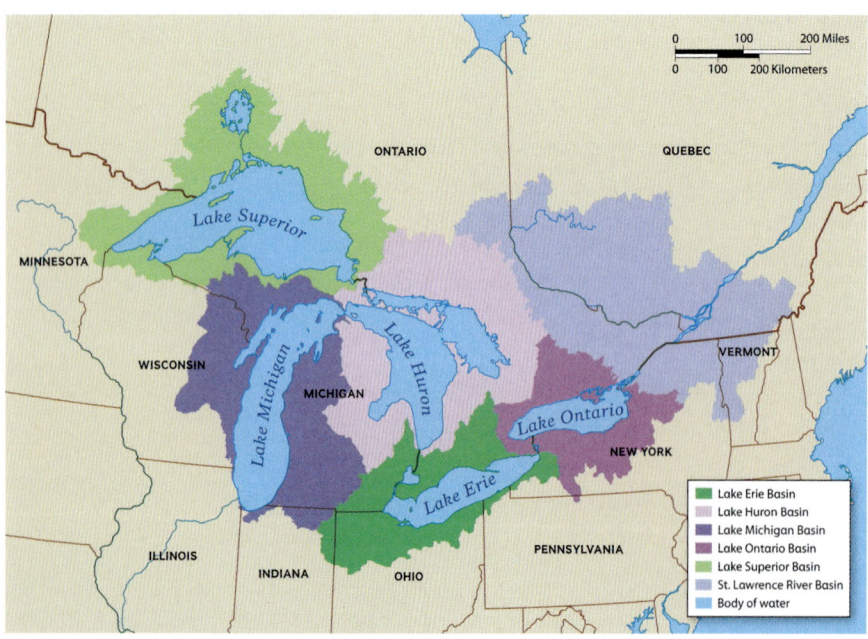

FIGURE 15-15
The Great Lakes Basin is spread across eight U.S. states and Canada.

CHAPTER 15 SUMMARY

15.1 HYDROLOGIC CYCLE

- Only about 0.63 percent of Earth's water is fresh and in liquid form. The rest is salty seawater and glacial ice.

- The constant circulation of water among the hydrosphere, the geosphere, the biosphere, and the atmosphere is called the hydrologic cycle.

- Most of the water that evaporates from the seas and from land returns to the surface as rain or snow. The precipitation that falls on the continents returns to the sea via runoff and moving groundwater, or it returns to the atmosphere via evaporation and transpiration.

15.2 LAKES AND WETLANDS

- Lakes are short-lived landforms because streams fill them with sediment.

- Recent glaciers created many modern lakes; as a result, we live in a time of unusually abundant lakes. The life history of a lake commonly involves a progression from an oligotrophic lake to a eutrophic lake to a swamp and, finally, to a flat meadow or a forest.

- Wetlands are among the most biologically productive environments on Earth.

- Wetlands are natural water purification systems, and they lessen flood effects by absorbing floodwaters.

15.3 GROUNDWATER

- Much of the rain that falls on land seeps into soil and bedrock to become groundwater.

- Groundwater saturates the upper few kilometers of soil and bedrock to a level called the water table, which separates the zone of saturation from the zone of aeration.

- Most groundwater moves slowly, about 4 centimeters per day.

- Springs occur where the water table intersects the land surface and water flows or seeps onto the surface.

- Caverns form where groundwater dissolves limestone.

- A sinkhole forms when the roof of a limestone cavern collapses.

CHAPTER 15 ASSESSMENT

Review Key Terms

Select the key term that best fits the definition. Not all terms will be used, and no term will be used more than once.

aquifer	runoff
aquitard	sinkhole
eutrophic	stalactite
kettle lake	stalagmite
lake	thermocline
oligotrophic	transpiration turnover
permeability	water table
porosity	

1. a cone-shaped rock structure formed by slow precipitation of calcite found on the floor of a limestone cave
2. the top of the zone of saturation in a layer of rock or soil
3. an area of rocks with low porosity and permeability that contains and transmits very little groundwater
4. a term describing a lake that contains few nutrients and sustains relatively few living organisms
5. any body of rock or sediment that can yield economically significant amounts of water
6. an icicle-like rock feature formed by slow precipitation of calcite found on the ceiling of a limestone cave
7. a body of standing water formed from the impression of a large piece of glacial ice, once embedded in sand and gravel, that melted after the glacier's retreat
8. the mixing of layers of water that occurs in a lake during the breakdown of thermal stratification in spring and fall
9. the movement of water from plants to the atmosphere as it evaporates through pores in the leaves
10. the ability of rock or soil to allow water or other fluids to pass through it
11. a large inland body of standing water
12. water that flows over the land in rivers and streams
13. a term describing a lake containing abundant nutrients and a high level of productivity in terms of biomass
14. the boundary between warm top waters and cold bottom waters in a lake in the temperate zone during the summer

Review Key Concepts

Answer each question on a separate sheet of paper to demonstrate your understanding of key concepts from the chapter.

15. Use a model to explain how water moves between the different systems on Earth.
16. Explain how energy changes are interconnected with the water cycle.
17. Explain the role of plants in the water cycle.
18. Explain the two factors that limit growth of organisms in a deep oligotrophic lake.
19. Explain how humans can accelerate the natural evolution of a lake.
20. Describe the physical and biological attributes of wetlands.
21. Describe three different mechanisms of lake formation.
22. Explain the importance of wetlands.
23. Why does unlithified mud, typically near the surface of the ground, have a large water capacity, and how does this change when the mud becomes lithified at greater depths?
24. Explain how aquifers and aquitards differ in terms of their porosity and permeability.
25. Explain why a stream may continue to flow even during extended periods with no rainfall.
26. What is the relationship between groundwater and caverns?
27. Describe the structures shown in the photo and how they are linked to the movement of groundwater.

Think Critically

Write a response to each question on a separate sheet of paper. Use concepts from the chapter to support your reasoning.

28. Water is abundant on Earth, yet as a resource for humans, it is increasingly scarce. Explain why, using statistics from the chapter to support your argument.

29. Explain how historical attitudes toward wetlands have affected the fate of wetlands in North America, and predict how current perceptions of wetlands may affect wetlands in the future.

30. Explain the importance of seasonal turnover to organisms living in temperate lakes.

31. A renewable resource is a natural resource that can be naturally replenished as quickly as it is used. Is groundwater a renewable resource? Defend your answer using examples from the text.

PERFORMANCE TASK

Great Lakes Invasive Species

In 2009, Great Lakes Restoration Initiative (GLRI) was established to support programs that restore the Great Lakes ecosystems. One goal of the GLRI is to lessen the impact of invasive species. Scientists and engineers developed methods to prevent the introduction of new species and control current populations. Control of invasive species often requires cooperation from the public.

In this performance task, you will develop a tool to educate the public about an invasive species in the Great Lakes. Design a presentation for an audience of sport fishermen, recreational boaters, or the general public. You should explain methods being used to control the species, their effectiveness, and any drawbacks or challenges. Explain how individuals can contribute to controlling the spread of invasive species.

1. Choose one of the following invasive species as the focus of your pamphlet or presentation: zebra mussel, Asian carp, sea lamprey, or purple loosestrife.

    A. Develop a list of questions that you plan to answer in your pamphlet or presentation. Then use your questions to guide your research on your selected topic. As you gather information, write down the answers to your questions. When all of your questions are answered, decide what information is most important to your audience and create an outline for your pamphlet. Select images that will complement the information or help you explain important concepts.

    B. Draft your pamphlet or presentation and present it to the class or a small group. After presenting the information, ask your classmates what questions they still have about your topic. Also ask your classmates for feedback to help you improve your pamphlet or presentation. Use the input from your classmates to revise your pamphlet or presentation and submit the final product to your teacher.

When you have completed your pamphlet or presentation, use it to answer the following questions.

2. Briefly summarize the information you gathered about the invasive species you researched. Include an explanation for why it is a threat to the Great Lakes ecosystem.

3. Summarize the control measures discussed in your pamphlet or presentation.

4. Evaluate the effectiveness of one of the control measures and describe alternate solutions that are being developed. You may also propose an alternate solution. Indicate the constraints such as cost, safety, and reliability, and describe potential environmental impacts.

5. Explain why informing the public about invasive species is an important part of preventing the spread of the invasive species you selected.

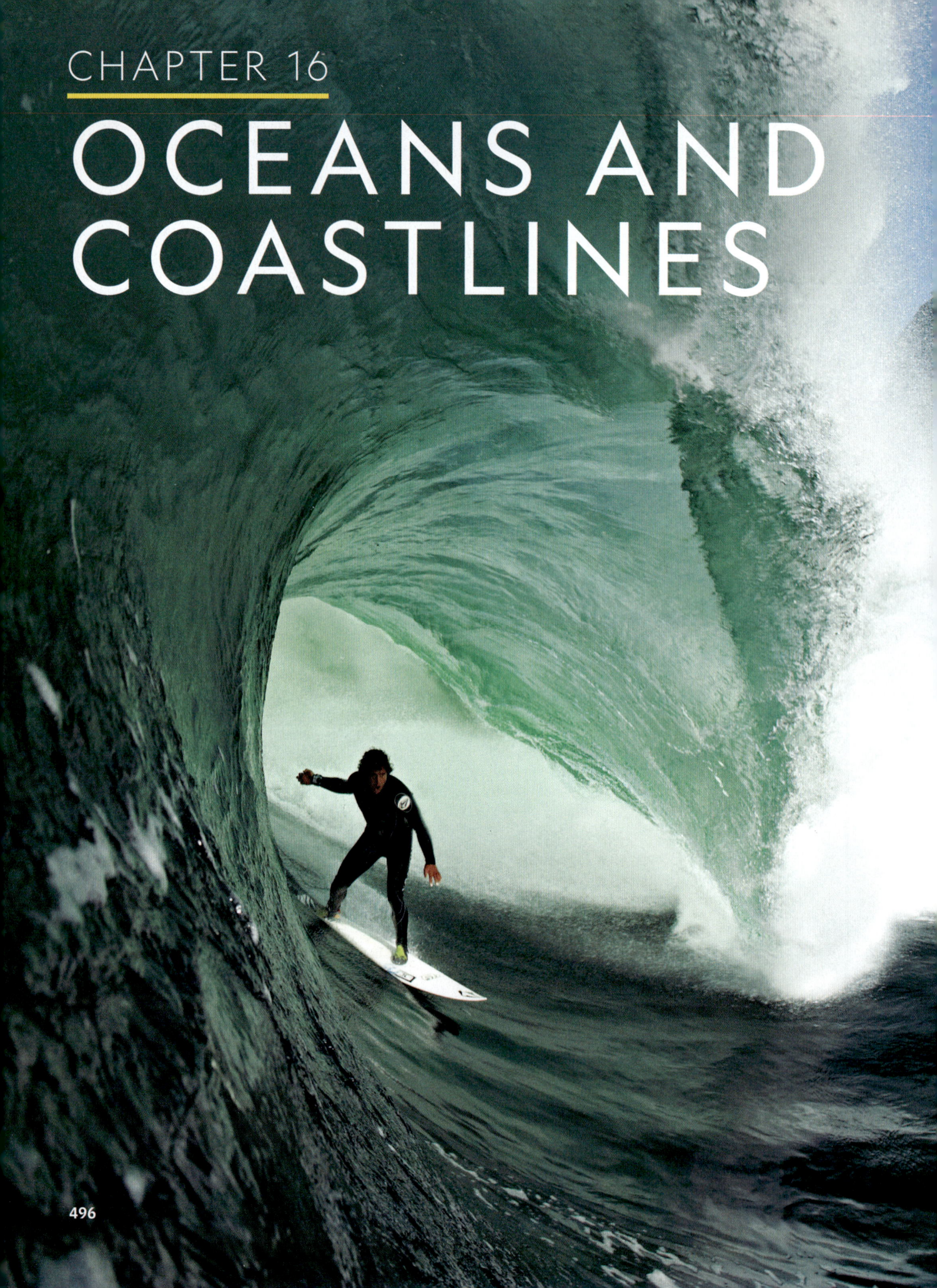

CHAPTER 16
OCEANS AND COASTLINES

Oceans shape coasts around the globe. Rough cliffs drop into the sea. Smooth beaches of soft sand surround peaceful lagoons. Migrating sand dunes slowly make their way inland, burying things in their path. All of these are generated by a dynamic system of interactions between the hydrosphere, the atmosphere, the geosphere, and the biosphere. The oceans are also affected by forces from space—gravity from the sun and the moon—and by human activities. In this chapter, you will explore the oceans, which cover about 71 percent of Earth's surface and contain about 96 percent of its water. No matter where you are on Earth, oceans have some influence on your environment.

KEY QUESTIONS

16.1 What are the characteristics of the oceans and the components of seawater?

16.2 What causes tides and sea waves?

16.3 What causes ocean currents?

16.4 How do weathering, erosion, and transport of sediment along seacoasts affect coastlines?

16.5 What are the common features of beaches and how do they form?

16.6 How do humans affect living things in the ocean, such as fisheries and coral reefs?

Near the southeastern coast of Tasmania, Australia, a huge swell begins to break, practically folding forward over itself. This is the wave that surfers seek. Riding just inside the breaking wave and using the incredible energy transferred by the water, the surfer is propelled forward. Finally, the surfer escapes the breaking wave and the ocean water crashes onto the coast with tremendous force.

EXPLORERS AT WORK

STUDYING ANCIENT CORAL REEFS

WITH NATIONAL GEOGRAPHIC EXPLORER CAROLINE QUANBECK

"Today's coastlines will be unrecognizable by the end of the century." So says Caroline Quanbeck, a geologist who conducts sea level research as part of her Ph.D. work at the University of Florida. Quanbeck is referring to the fact that Earth's global sea level is rising as polar ice sheets melt in response to a warming climate. As ocean waters encroach on shorelines around the world, many coastal communities will become uninhabitable. Coastal cities still have time to plan an organized retreat, and some may even be able to invest in adaptive solutions to slow the impacts of sea-level rise.

"Policymakers will need good guidance," notes Quanbeck. "They will need to know how high and how fast sea level will rise in their area as well as how natural coastal barriers, including coral reefs, will respond." Quanbeck's research will help provide such guidance. She's studying past sea-level changes to better predict impacts of this latest change. Specifically, she's analyzing fossils of ancient coral reefs located along the coast of Western Australia. These coral reefs formed some 125,000 years ago, during a warm period in Earth's history called the Last Interglacial. At that time, Earth's climate was similar to today's climate, yet the global sea level was 6 to 9 meters (20 to 30 feet) higher.

Quanbeck says fossil reefs are good indicators of former sea levels because corals generally grow close to the sea surface. The first step is to use the chemistry of coral skeletons to determine the age of the reef, or calculate how much time has passed since the coral grew. The age of the reef, along with a measurement of its original elevation relative to modern sea level, then provides an estimate of the height of sea level at the time the coral grew.

Measuring multiple reefs in this way yields a series of sea-level positions through time. Quanbeck is using her data to determine how quickly the sea level rose to its peak during the Last Interglacial in Western Australia, as polar ice sheets melted. That may shed light on what to expect in the next century.

Because coral reefs protect shorelines by buffering them from incoming waves, it's important to determine how reefs responded to sea-level rise in the past. So, Quanbeck is analyzing core samples, working her way up from the bottom of each one, where the oldest material generally is found, to the youngest material at the top. She records changes in characteristics such as sediment makeup, coral assemblages, and associated organisms. This tells her which species of corals might be able to keep up with the rising sea levels better than others.

Humans have dealt with rising seas before. Quanbeck notes that over the last 20,000 years, humans gradually moved their settlements inland as ice-sheet melting caused sea levels to rise more than 120 meters (about 400 feet) to current levels, which have held for the last 2,000 years. "Modern infrastructure makes moving inland much more difficult and costly than for prehistoric peoples," acknowledges Quanbeck. "But humans have successfully dealt with sea-level rise before, and we are capable of doing it again. The time to act is now."

THINKING CRITICALLY

Synthesize Review the work of Erin Pettit as described in the Explorers at Work in Chapter 13. Then explain how Caroline Quanbeck's research might benefit from Dr. Pettit's research.

Caroline Quanbeck points to the location of a fossil she extracted from an ancient coral reef that formed when the global sea level was much higher than it is today. Determining how quickly sea levels rose in the past may help us understand and make predictions about the effects of rising seas today.

Caroline Quanbeck researches fossilized coral reefs along the western coast of Australia. Because corals generally grow close to the ocean's surface, Quanbeck can get an idea of sea levels when these corals flourished over 100,000 years ago.

CASE STUDY
CORAL REEF KILLERS

The waters surrounding the Hawaiian Islands are home to numerous coral reefs that have long been a boon to the local economy. Tourists flock to Hawai'i to explore the beauty of the reefs, strapping on a snorkel mask or a SCUBA tank to revel in nature's splendor. Sadly, however, human activity is taking a devastating toll on coral reefs. Like many reefs elsewhere, Hawai'i's corals have begun to succumb to coral bleaching (Figure 16-1). In just one short year, from 2014–2015, scientists estimated that more than half of all corals in the shallows around Hawai'i had been bleached, resulting in a significant loss of biodiversity.

Coral reefs are critical not only to local economies but also to the health of marine ecosystems worldwide. Coral consists of colonies of tiny, soft-bodied organisms called polyps that host zooxanthellae algae in their tissue. Through a complex symbiotic relationship, the polyps provide the algae with a sheltered environment and nutrients needed for photosynthesis. The algae produce oxygen and nutrients that the polyps need to grow and secrete calcium carbonate. This calcium carbonate builds up over time, forming the iconic limestone structures that serve as a habitat for countless aquatic species.

Boat collisions, fishing gear, and even divers can inflict physical damage on coral reefs. But human-caused climate change poses a greater threat. In recent decades, rising concentrations of greenhouse gases in the atmosphere have led to higher ocean temperatures. Abnormally warm seawater can cause coral polyps to expel their zooxanthellae algae. The polyps lose their green pigmentation and turn white. When corals go without the algae for very long, they die. In addition to atmospheric warming, carbon-dioxide emissions have caused detrimental changes in ocean chemistry. Much of the CO_2 released by the burning of fossil fuels is absorbed by seawater. As a result, oceans have become about 30 percent more acidic over the past 200 years. Acidic seawater weakens and erodes calcium carbonate, destroying the skeleton that protects the corals and so many other reef-dwelling organisms.

In Explorers at Work, you read about Caroline Quanbeck, who is studying how prehistoric sea-level changes affected coral reef communities. Like Quanbeck, many people recognize the importance of today's fragile reef ecosystems. In 1998, the United States established a Coral Reef Task Force with the mission of protecting and conserving coral reefs from harmful activities. In Hawai'i, home to about 85 percent of coral reefs in the United States, reefs are now protected and heavily monitored by scientists. It remains to be seen whether such measures will be enough to allow coral reefs in Hawai'i and other places on Earth to rebound.

As You Read Explore relationships between the atmosphere, the biosphere, the geosphere, and the hydrosphere that affect coral reefs and other marine ecosystems. Find out what factors influence the makeup and behavior of seawater. Consider how tides, sea waves, ocean currents, and other processes shape coastlines and sustain life in the sea.

▶ **FIGURE 16-1A**
Coral reefs are large, marine ecosystems that support about one-quarter of all marine organisms. Zooxanthellae algae live in a coral polyp's tissue. Zooxanthellae can be seen as the small brown features inside the polyp. The coral polyp and algae provide benefits for each other.

🌐 **FIGURE 16-1B**
Go online to view an example of what coral looks like when it is stressed.

16.1 OCEAN GEOGRAPHY AND SEAWATER COMPOSITION

Core Ideas and Skills
- Describe characteristics of Earth's oceans and ocean basins.
- Identify common ions that make up the salts in ocean water.
- Explain how dissolved carbon dioxide affects the pH of ocean water.
- Describe the effects of rising ocean temperature and ocean acidification.

KEY TERMS

salinity
ocean acidification
reef

Earth's Oceans

All of Earth's oceans are connected, and water flows from one to another, so in one sense Earth has just one global ocean. Conversely, you may have heard the term "the seven seas," which refers to the North Atlantic, South Atlantic, North Pacific, South Pacific, Indian, Arctic, and Antarctic. However, these designations are related more closely to commerce rather than to the geology and oceanography of the ocean basins.

Geologically, there are five distinct ocean basins within the global ocean (Figure 16-2). The largest and deepest is the Pacific. It covers one-third of Earth's surface, more than all land combined, and contains more than half of the world's water. The Atlantic Ocean has about half the surface area of the Pacific. The Indian Ocean is slightly smaller than the Atlantic.

The Arctic Ocean surrounds the North Pole and extends southward to the shores of North America, Europe, and Asia. Therefore, it is bounded by land, with only a few straits and channels connecting it to the Atlantic and Pacific Oceans. The surface of the Arctic Ocean freezes in winter, and parts of it melt for a few months during summer and early fall. The Antarctic (Southern) Ocean, feared by sailors for its cold and ferocious winds, has no sharp northern boundary. The northernmost limit of the Antarctic Ocean is the zone where warm currents from the north converge with cold Antarctic water.

checkpoint Although all of Earth's oceans are connected and water flows from one to another, explain why we still name them "the seven seas."

FIGURE 16-2 ▼
Geologists and oceanographers recognize five major ocean basins: the Atlantic, Pacific, Indian, Arctic, and Antarctic. The map also identifies a few smaller seas and bays that are mentioned in this chapter.

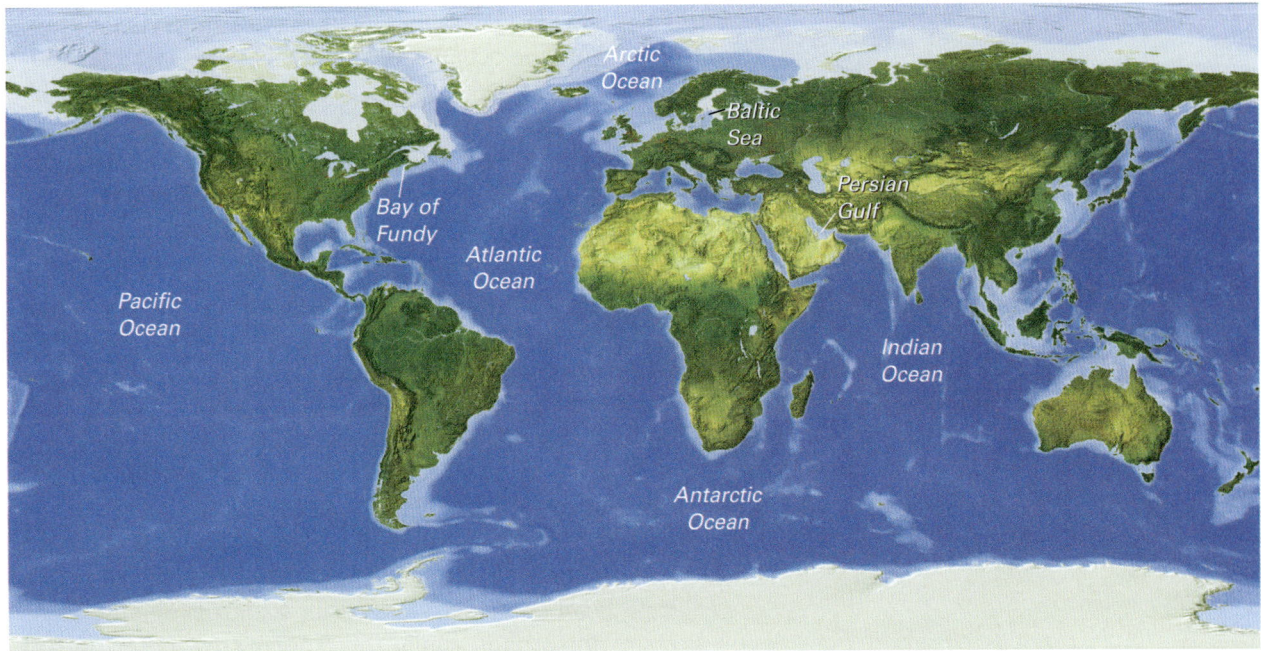

Source: Arid Ocean Maps

Salts and Trace Elements in Seawater

The **salinity** of seawater is the total quantity of dissolved salts, expressed as a percentage. In this case, the term *salts* refers to all dissolved ions, not only sodium and chloride (Figure 16-3).

Dissolved ions make up about 3.5 percent of the weight of ocean water. The six ions listed in Figure 16-3 make up 99 percent of the ocean's dissolved compounds. However, almost every other element found on land is also found dissolved in seawater, albeit in trace amounts. For example, seawater contains about 0.000000004 (4×10^{-9}) percent gold. Although the concentration is small, the oceans are large and therefore contain a lot of gold. About 4.4 kilograms of gold is dissolved in each cubic kilometer of seawater. Because the oceans contain about 1.3 billion cubic kilometers of water, about 5.7 billion kilograms of gold exists in the oceans. Unfortunately, it would be quite expensive to extract even a small portion of this amount.

The world's rivers carry more than 2.5 billion tons of dissolved salts to the oceans every year. Underwater volcanoes contribute additional dissolved ions. However, the salinity of the oceans has been relatively constant throughout much of geologic time because salt has been removed from seawater at roughly the same rate at which it has been added. When a portion of a marine basin becomes cut off from the open ocean, the water evaporates. Thick sedimentary beds of salt precipitate from the water. Additionally, large quantities of salt become incorporated into mudstone and other sedimentary rocks through the precipitation of salt minerals within these rocks.

checkpoint Why has ocean salinity remained relatively constant throughout geologic time?

Dissolved Gases

In addition to trace elements and salts, seawater also contains dissolved gases, especially carbon dioxide and oxygen. These dissolved gases exchange freely and continuously with the atmosphere. If the atmospheric concentration of carbon dioxide or oxygen rises, much of the gas quickly dissolves into seawater; if the atmospheric concentration of a gas falls, that gas exsolves (comes out of solution) from the sea and enters the atmosphere. Therefore, the seas help moderate changes in atmospheric concentrations of gases.

At the same time, the atmosphere buffers the chemistry of seawater. When carbon dioxide dissolves in seawater, the seawater becomes more acidic. In recent years, increasing concentrations of carbon dioxide gas from the burning of fossil fuels has not only caused the atmosphere and oceans to become warmer but also has caused seawater to become more acidic. That is, when carbon dioxide (CO_2) dissolves in seawater, it forms carbonic acid (H_2CO_3). As more CO_2 from the atmosphere diffuses into the oceans and converts to carbonic acid, the pH of the ocean as a whole drops (Figure 16-4). This process is called **ocean acidification**. Over the past 28 years, oceanographers have measured a drop in the pH of the oceans from about 8.1 to about 8.0. Although this small change may seem insignificant, the pH scale is not linear, and small drops in pH correspond to large increases in acidity (Figure 16-5). A pH drop from 8.1 to 8.0 indicates an increase in acidity, or the concentration of hydrogen ions, of about 50 percent.

Recent and ongoing ocean acidification has very seriously affected marine ecosystems. Corals, many kinds of algae, shellfish, foraminifera, and other organisms that make hard body parts out of calcium carbonate are hard hit because calcium carbonate dissolves under acidic conditions. As the oceans acidify, these organisms spend more of

FIGURE 16-3
Six common ions form most of the salts in seawater.

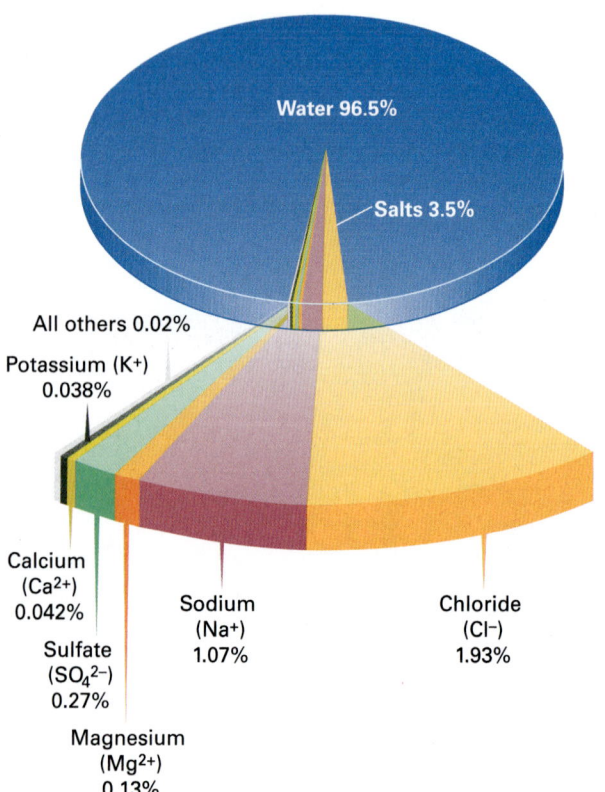

Water 96.5%
Salts 3.5%
All others 0.02%
Potassium (K^+) 0.038%
Calcium (Ca^{2+}) 0.042%
Sulfate (SO_4^{2-}) 0.27%
Magnesium (Mg^{2+}) 0.13%
Sodium (Na^+) 1.07%
Chloride (Cl^-) 1.93%

their metabolic energy trying to produce their hard body parts. They become stressed. As the acidity increases, eventually a point is reached where the organism cannot maintain its calcium carbonate production and it dies.

Carbonate reef environments are especially negatively affected by recent increases in seawater acidity and temperature. A **reef** is a wave-resistant ridge or mound built by corals, oysters, algae, or other marine organisms.

The recent loss of coral reef ecosystems is particularly alarming. From 2016 to 2018, unusually warm waters killed half of the coral reefs in Australia's Great Barrier Reef. These reefs no longer exist as a reef ecosystem. Unfortunately, they likely will never recover. Left behind are vast

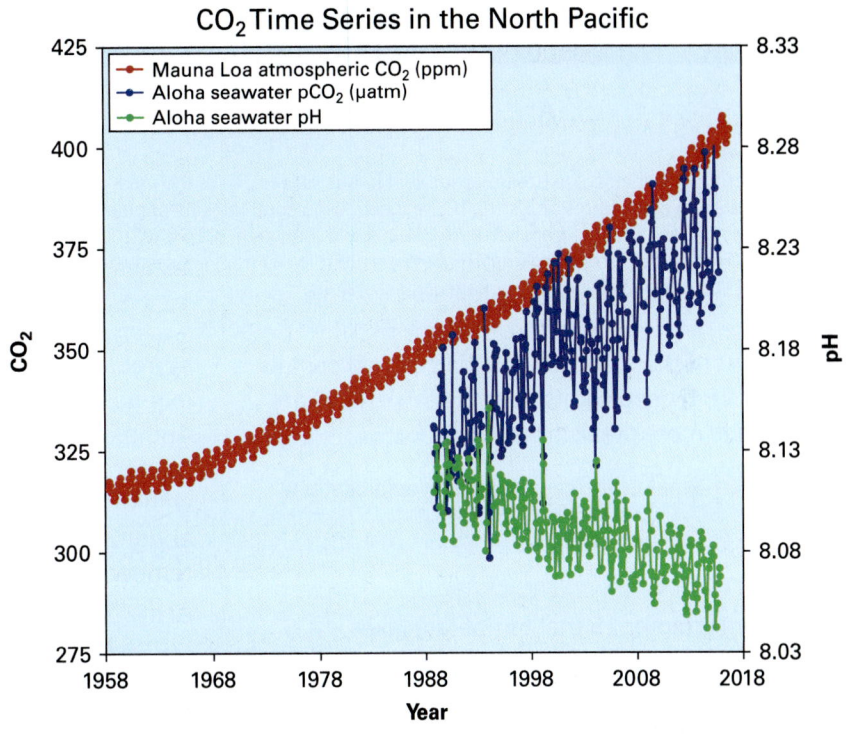

FIGURE 16-4
The graph shows concentrations of CO_2 gas in the atmosphere (red line) and dissolved CO_2 in seawater (blue line). Notice that the red and blue curves track each other. The green curve shows the measured pH from 1990 to 2018. The drop in pH from 8.1 to 8.0 represents an increase in acidity of about 50 percent. Seawater measurements were taken at the ALOHA research station, which is about 160 kilometers north of the Hawaiian Islands.

FIGURE 16-5
The pH scale is not linear. Small changes in pH correspond to large changes in acidity or alkalinity.

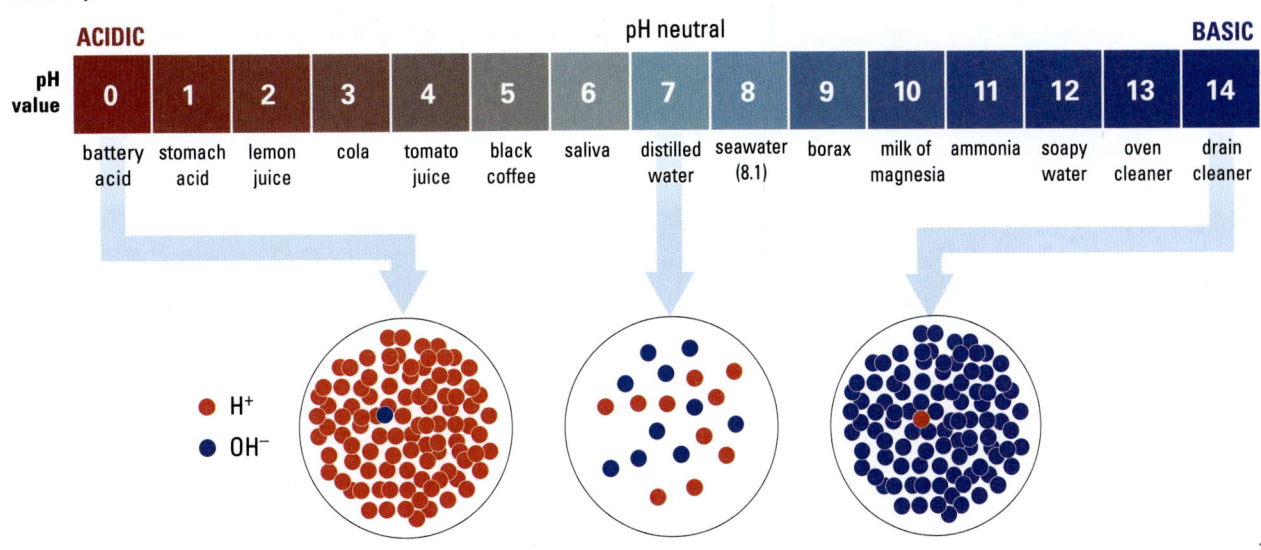

FIGURE 16-6
(A) Under normal conditions, the Great Barrier Reef in Australia is one of the most biologically diverse ecosystems on Earth. (B) Bleached, dead coral thickets appear in the Great Barrier Reef offshore from northern Queensland. Between 2016 and 2017, an estimated 50 percent of the entire Great Barrier Reef died. The vibrant ecosystem was transformed into a pile of calcium carbonate coral skeletons.

underwater thickets of bleached, calcium carbonate skeletons from dead corals (Figure 16-6). Reefs are discussed in more detail in Lesson 16.6.

checkpoint What is one cause of ocean acidification?

Temperature

Recall from Chapter 15 that temperate lakes are comfortably warm for swimming during the summer because warm water floats on the surface and does not readily mix with the deep, cooler water. Most of the global ocean develops a similar temperature layering, but because sea waves and currents stir the surface water, the warm layer can extend from the surface to a depth as great as 450 meters (Figure 16-7). Below this layer is the thermocline, where the temperature drops rapidly with depth. The thermocline extends to a depth of about 2 kilometers. Beneath the thermocline, the temperature of ocean water varies from about 1°C to 2.5°C. The cold, dense water in the ocean depths mixes very little with the surface. Thus, there are three distinct temperature zones in the ocean, as shown in Figure 16-7. This layered structure does not exist in the polar seas because cold surface water sinks, causing vertical mixing.

checkpoint Describe three temperature layers in the ocean.

FIGURE 16-7
There are three temperature layers in the ocean. The warm, upper layer exists from the surface to as deep as 450 meters. Below that, temperature cools rapidly with depth in the thermocline. Seawater in the deepest temperature layer is cold.

16.1 ASSESSMENT

1. **Recall** What are the five distinct ocean basins?
2. **Identify** What ions make up 99 percent of the ocean's dissolved compounds?
3. **Explain** What causes the ocean's salinity?
4. **Describe** What happens to seawater when carbon dioxide gas from burning fossil fuels is dissolved in ocean water?

Critical Thinking

5. **Infer** What may happen to organisms that are dependent on coral reefs as ocean acidification worsens? Explain.

16.2 TIDES AND SEA WAVES

Core Ideas and Skills
- Explain how the sun and moon cause Earth's tides.
- Describe the structure of a wave and how water waves travel.

KEY TERMS

tide	trough
tidal bulge	wavelength
spring tide	wave height
neap tide	wave orbital
tidal range	wave base
crest	

Tides

Even the most casual observer will notice that on any beach, the level of the ocean rises and falls on a cyclical basis. If the water level is low at noon, it will reach its maximum height at about 6:12 p.m. and be low again at about 12:25 a.m. These vertical displacements are called **tides**. Most coastlines experience two high tides and two low tides during an interval of about 24 hours and 50 minutes.

Tides are caused by the gravitational pull of the moon and the sun on the sea surface. Although the moon is much smaller than the sun, the moon is so much closer to Earth that its influence predominates. At any given time, one region of Earth (point A in Figure 16-8) lies directly closest to the moon. Because gravitational force is greater for objects that are closer together, the part of the ocean nearest to the moon is attracted with the strongest force. There, the ocean surface bulges upward towards the moon, forming a tidal bulge. As a particular geographic point on Earth's coastlines rotates underneath a **tidal bulge**, that coastline experiences a high tide.

But so far our explanation is incomplete. As Earth spins on its axis, a given point on Earth passes directly under the moon approximately once every 24 hours and 50 minutes, but the period between successive high tides is only 12 hours and 25 minutes. Why are there ordinarily two high tides in a day? The tide is high not only when a point on Earth is directly facing the moon, but also when it is 180° away. That is, the tidal bulge created by the gravitational effect of the moon has two ends: one end faces the moon, and one end faces directly opposite the moon on the other side of Earth.

To understand how a tidal bulge forms on the side of Earth facing away from the moon, we must consider the forces acting in the Earth–moon system. On the side of Earth facing the moon, the moon's gravitational force on the water is stronger than the inertia that keeps the water in place. The water on the near side is drawn away from Earth, resulting in high tides. The moon's gravitational force on the side of Earth facing away from the moon is weaker than the inertia that keeps the water in place (point B in Figure 16-8). The water on the far side is drawn away from Earth, also resulting in high tides. You will learn more about inertia and gravitational forces in Chapter 22.

High and low tides do not occur at the same time each day but are delayed by approximately 50 minutes every 24 hours. Earth makes one complete rotation on its axis in 24 hours. Meanwhile, the moon is revolving around Earth about every 27 days. After a point on Earth makes one complete rotation in 24 hours, that point must continue to rotate for an additional 50 minutes before it catches up with the new position of the moon in its orbit around Earth. This is why the moon rises

FIGURE 16-8
The moon's gravity causes a high-tide bulge at point A, directly facing the moon. A corresponding high tide at point B occurs at the same time on the opposite side of Earth.

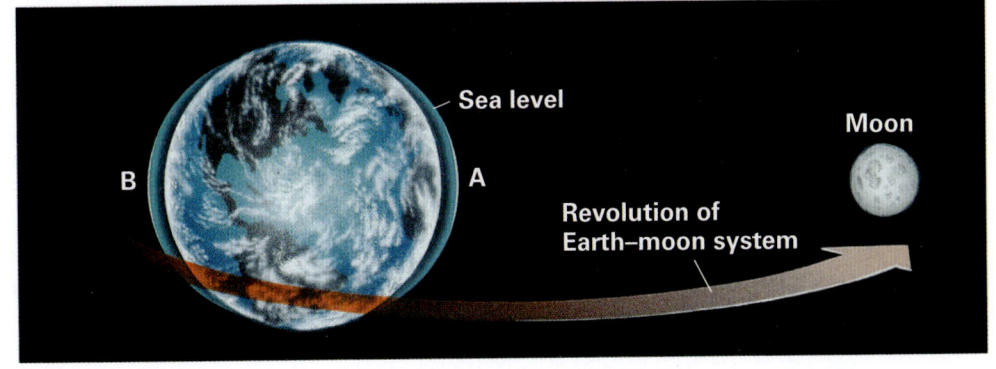

LESSON 16.2

approximately 50 minutes later each day. Likewise, the tides are approximately 50 minutes later each day (Figure 16-9).

Although the sun's gravitational pull on the oceans is smaller than the moon's, it does affect tides. When the sun and the moon are directly in line with Earth, their gravitational fields combine to create a large tidal bulge. During these times, the variation between high and low tides is large, producing **spring tides** (Figure 16-10A). When the moon is 90° out of alignment with the sun and Earth, each partially offsets the effect of the other

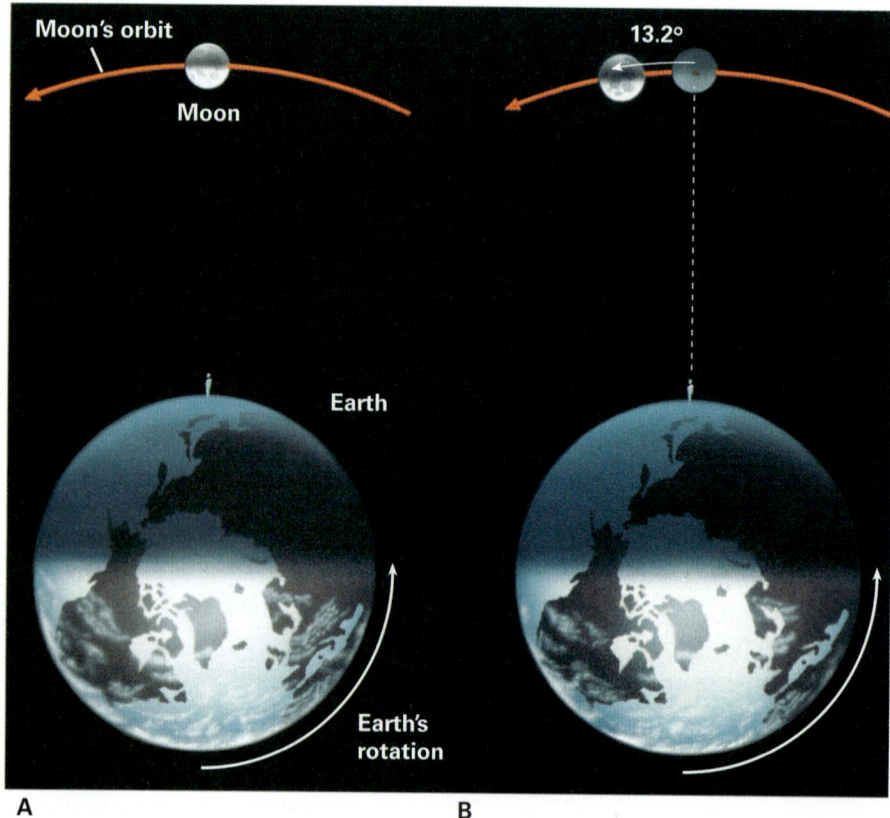

FIGURE 16-9
The moon moves 13.2° every day as a result of the difference in the time it takes Earth to complete one full rotation versus the time it takes the moon to orbit Earth. (A) The moon is directly above an observer on Earth. (B) One day later, Earth has completed one rotation, but the moon has traveled 13.2° farther than it was at that same time the previous day. Thus, Earth must now continue to rotate for another 50 minutes before the observer is again directly under the moon.

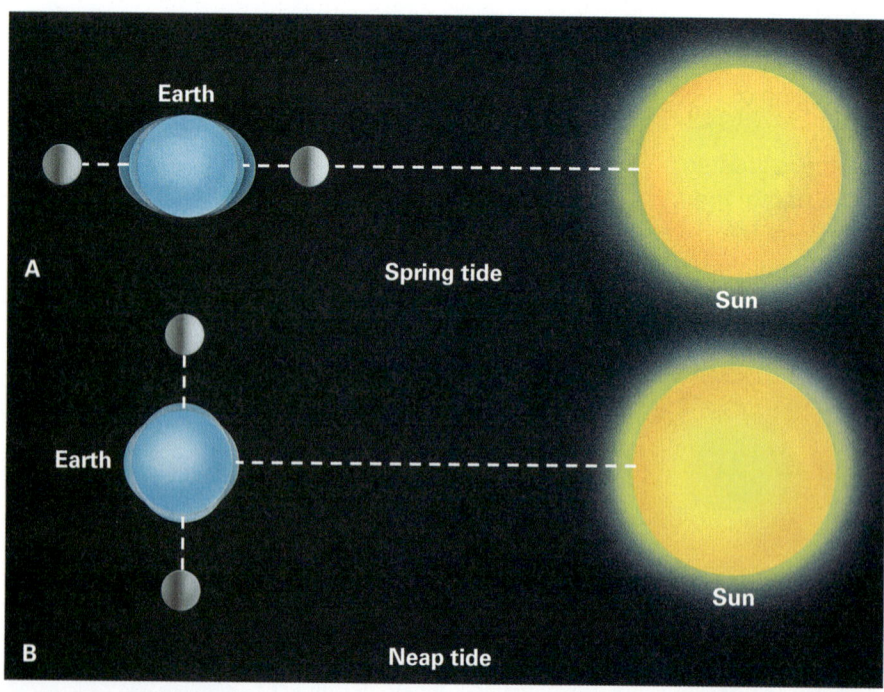

FIGURE 16-10
(A) Spring tides occur when Earth, sun, and moon are lined up. In this position, their gravitational fields combine to create a large tidal bulge. (B) Neap tides occur when the moon lies at right angles to a line drawn between the sun and Earth.

FIGURE 16-11
St. Michael's Mount is located in Cornwall, England. (A) At high tide, the walkway to the tidal island is underwater. (B) At low tide, visitors can easily traverse the walkway.

and the difference between the levels of high and low tide is smaller. These relatively small tides are called **neap tides** (Figure 16-10B).

Tidal range, which is the vertical distance between low and high tide at a given location, is strongly affected by the shape of the coastline. For example, the famous Bay of Fundy in Canada is shaped like a giant funnel. The shape of the bay concentrates the rising and falling tides. As a result, the tidal range is as much as 15 meters during a spring tide. In contrast, tides vary by less than 2 meters along a straight stretch of coast. In the open ocean, the tides average about 1 meter. Mariners consult tide tables that give the time and height of the tides in any area on any day. Pedestrians consult tide tables too. St. Michael's Mount in Cornwall, England, is accessible on foot at low tide, but not at high tide (Figure 16-11).

checkpoint Under what conditions does a spring tide occur?

Wind and Sea Waves

Most ocean waves develop when wind blows across water. Waves vary from tiny ripples to destructive giants that can topple beach houses and sink ships. In deep water, the size of a wave depends on (1) the wind speed, (2) the length of time the wind has blown, and (3) the distance that the wind has traveled. A 25-kilometer-per-hour wind blowing for two to three hours across a 15-kilometer-wide bay generates waves about 0.5 meter high. But if a storm blows at 90 kilometers per hour for several days over a distance of 3,550 kilometers, it can generate 30-meter-high waves, as tall as a ship's mast.

The highest part of a wave is called the **crest**; the lowest is the **trough** (Figure 16-12). The **wavelength** is the distance between successive crests. The **wave height** is the vertical distance from the crest to the trough.

Recall from our discussion of earthquakes in Chapter 8 that when a wave travels through rock, the energy of the wave is transmitted rapidly over large distances, but the rock itself moves only slightly. In a similar manner, a single water molecule in a water wave does not travel with the wave. While the wave moves horizontally, the

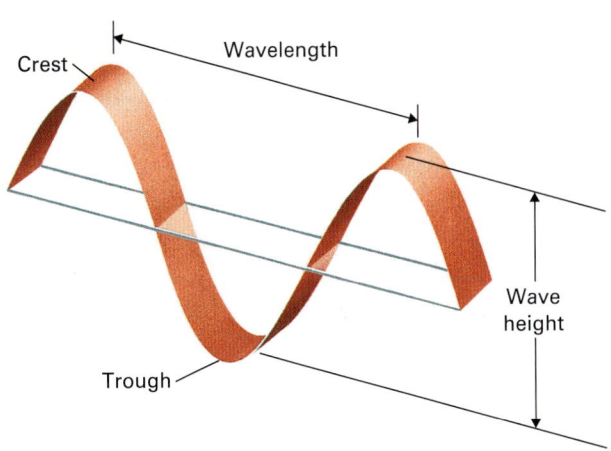

FIGURE 16-12
The parts used to describe a wave are shown here.

LESSON 16.2

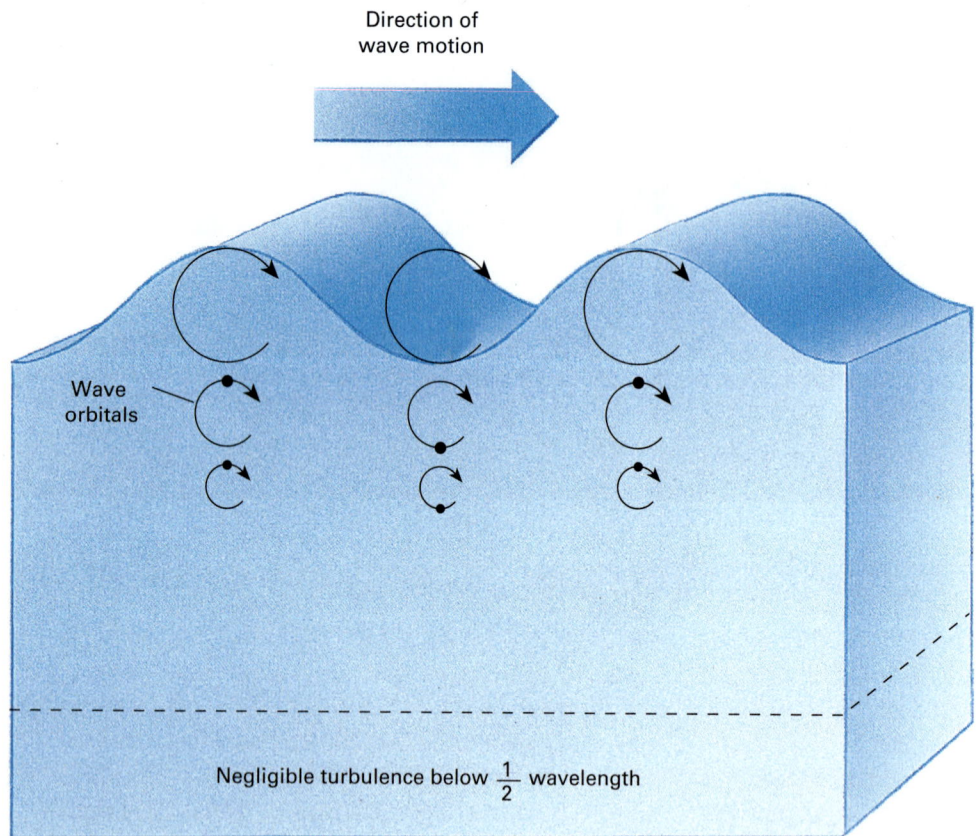

FIGURE 16-13
While a wave moves horizontally across the sea surface, the water itself moves only in small circles called wave orbitals.

water molecules move in small circles called **wave orbitals** (Figure 16-13). That is why a ball on the ocean does not travel along with the waves, but instead bobs up and down and sways back and forth as the waves pass.

The circles of water movement produced as a wave passes by become smaller with depth. At a depth equal to about half the wavelength, the disturbance becomes negligible. This depth is called **wave base**, and it varies with the size of the waves. If you dive deep enough, you can escape wave motion from even the biggest storm waves, which have a wavelength of about 500 meters and a wave base of about 250 meters. No one gets seasick in submarines because they avoid wave motion by diving below local wave base.

checkpoint Compare how a water wave moves with how the water itself moves.

16.2 ASSESSMENT

1. **Recall** What causes tides?
2. **Explain** Why does the moon have more influence on tides than the sun?
3. **Explain** What causes high and low tides?
4. **Distinguish** How are spring tides different from neap tides?
5. **Describe** On what does the size of an ocean wave depend?
6. **Identify** Name the parts of the structure of an ocean wave.

Critical Thinking

7. **Analyze** What wind conditions might cause ocean waves to be very small? What wind conditions might cause waves to be very large?

16.3 OCEAN CURRENTS

Core Ideas and Skills
- Explain what causes surface and deep-sea currents.
- Describe how ocean circulation affects Earth's climate.
- Describe the causes and effects of upwelling.

KEY TERMS

surface current
tidal current
gyre
Coriolis effect
deep-sea current
thermohaline circulation
upwelling

Surface Currents

The water in an ocean wave moves in circles, ending up where it started. In contrast, a current is a continuous flow of water in a particular direction. Although river currents are familiar and easily observed, early mariners did not recognize ocean currents. **Surface currents** are the horizontal movement of water caused by wind blowing over the sea surface. Surface currents flow in the upper 400 meters of the seas and involve about 10 percent of the water in the world's oceans.

In the mid-1700s, Benjamin Franklin lived in London, where he was deputy postmaster general for the American colonies. He noticed that mail ships took two weeks longer to sail from England to North America than did merchant ships. Franklin learned that the captains of the merchant ships had discovered a current flowing northward along the eastern coast of North America and then across the Atlantic to England. When sailing from Europe to North America, the merchant ships saved time by avoiding the current. The captains of the mail ships were unaware of this current and lost time sailing against it on their westward journeys. In 1769, Franklin and his cousin, Timothy Folger, a merchant captain, charted the current and named it the Gulf Stream (Figure 16-14).

As ship traffic increased and navigators searched for the quickest routes around the

FIGURE 16-14 ▼

A satellite image reveals sea surface temperatures in the Atlantic Ocean. The Gulf Stream is the streak of warm (dark orange) water that follows the North American eastern seaboard. It then passes eastward across the North Atlantic, south of the cold (blue) waters of the Gulf of Maine and Labrador Sea. Ireland and the southwest coast of England are visible in the upper right corner.

globe, they discovered other surface ocean currents (Figure 16-15). Ocean currents have been described as rivers in the sea. The analogy is only partially correct; ocean currents have no well-defined banks, and they carry much more water than even the largest river. The Gulf Stream is 80 kilometers wide and 650 meters deep near the east coast of Florida and moves at about 5 kilometers per hour, a moderate walking speed. As it moves northward and eastward, the current widens and slows; east of New York City, it is more than 500 kilometers wide and travels less than 8 kilometers per day.

When tides rise and fall along an open coastline, water moves in and out as a broad sheet. If the flow is channeled by a bay with a narrow entrance, the moving water funnels into a **tidal current**, a flow of ocean water caused by tides. Tidal currents can be intense where large differences exist between high and low tides and narrow constrictions occur in the shoreline. On parts of British Columbia's west coast, a diesel-powered fishing boat cannot make headway against tidal currents flowing between closely spaced islands. Fishermen must wait until the tidal current reverses direction before they can proceed. Tidal currents have no measurable effect on Earth's climate, yet their impact on coastlines and human activities may be significant.

Ocean currents (excluding tidal currents) profoundly affect Earth's climate. Every second, the Gulf Stream transports 1 million cubic meters of warm water northward past any point, warming both North America and Europe. For example, Churchill is a village in Manitoba, Canada. It is frigid and icebound for much of the year. Polar bears regularly migrate through town. Yet it is at about the same latitude as Glasgow, Scotland, which is warmed by the Gulf Stream and therefore experiences relatively mild winters. In general, the climate in western Europe is warmer than at similar latitudes in other regions not heated by tropical ocean currents.

checkpoint How are surface ocean currents and river currents different?

Forces Affecting Surface Currents

Surface currents are driven primarily by friction between wind blowing over the sea surface and the water surface itself. The winds simply drag the sea surface along in the same direction that the winds are blowing. In Figure 16-16, the green arrows show the major prevailing winds over the North and South Atlantic Oceans. The orange arrows show the elliptical surface currents, called **gyres**, in the same regions. The North Atlantic gyre circulates in

FIGURE 16-15 ▼
Wind patterns in Earth's atmosphere are the drivers of surface ocean currents. Ocean currents affect the climates of coastal regions around the world.

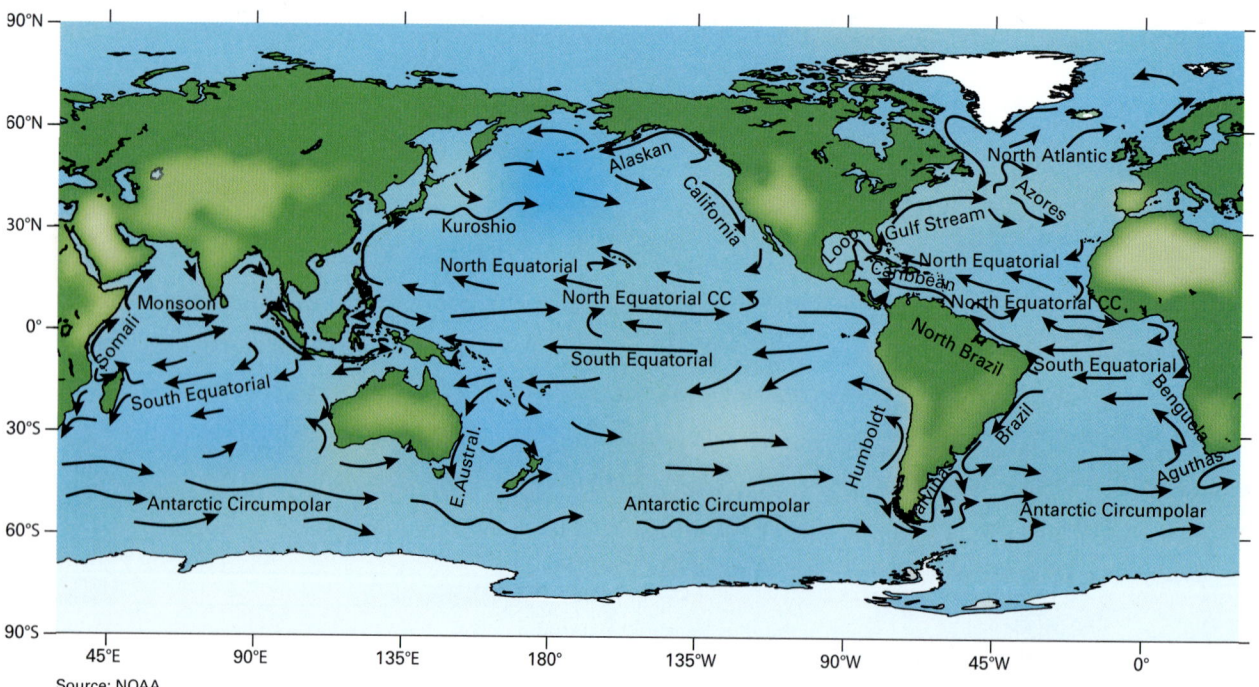

Source: NOAA

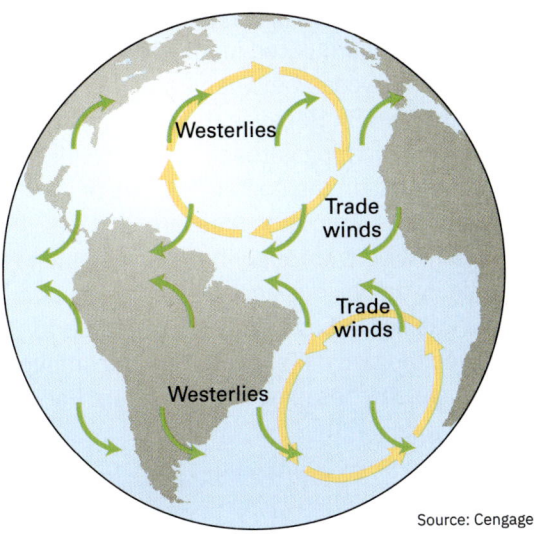

FIGURE 16-16
In this simplified model, the green arrows show major prevailing winds over the North and South Atlantic Oceans—the trade winds (easterlies) and the westerlies. The orange arrows show the surface oceanic currents in the same regions, which form the North Atlantic and South Atlantic gyres.

a clockwise direction, and the South Atlantic gyre circulates counterclockwise.

Notice that the currents do not flow in exactly the same directions as the prevailing winds. Instead, the east–west surface currents in the Northern Hemisphere are deflected to the right of the winds. Where their flow is blocked by a continent, the currents veer clockwise to continue their circuit. In the Southern Hemisphere the currents are deflected to the left of the winds and turn counterclockwise when they encounter land. These differences between prevailing wind directions and the flow directions of the surface currents suggest that forces other than wind also affect the great gyres.

One force that deflects the gyres away from the prevailing winds is the **Coriolis effect**, named for Gaspard-Gustave de Coriolis, the 19th-century French scientist who described it. To understand this effect, consider the rotating Earth: The circumference of Earth is greatest at the Equator and decreases to zero at the poles. But all parts of the planet make one complete rotation every day. Therefore, a point on the Equator must travel farther and faster than any point closer to the poles. At the Equator, all objects move eastward with a speed of about 1,600 kilometers per hour; at the poles there is no eastward movement at all and the speed is 0 kilometers per hour.

Now imagine a rocket fired from the Equator toward the North Pole. Before it was launched it was traveling eastward at 1,600 kilometers per hour with the rotating Earth. As it takes off, it is moving eastward at 1,600 kilometers per hour and northward at its launch speed. As it moves north from the Equator, it is still traveling eastward at 1,600 kilometers per hour, but points on Earth beneath it move eastward at a slower and slower speed as the rocket approaches the pole. As a result, the rocket curves toward the east, or the right. In a similar manner, a mass of water or air deflects in an easterly direction as it moves north from the Equator, as shown in Figure 16-17A.

Conversely, consider an ocean current flowing southward from the Arctic Ocean toward the

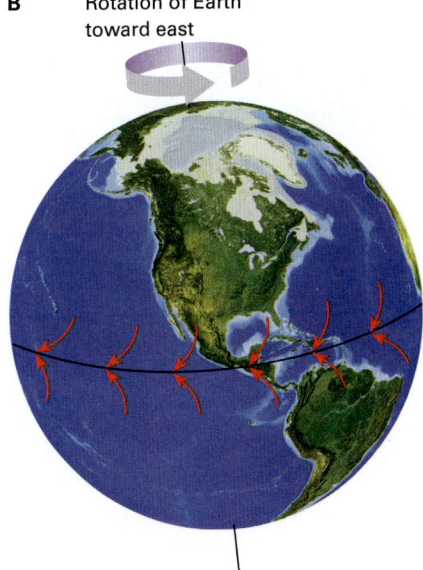

FIGURE 16-17
The Coriolis effect deflects water and wind currents. (A) Water or air moving poleward from the Equator is traveling east faster than the surface of Earth. The water or air veers to the east (turns right in the Northern Hemisphere and left in the Southern Hemisphere). (B) Water or air moving toward the Equator is traveling east slower than the surface of Earth. The water or air veers to the west (turns right in the Northern Hemisphere and left in the Southern Hemisphere).

LESSON 16.3 **511**

Equator. Since it started near the North Pole, this water moves more slowly to the east than Earth's surface near the Equator. Therefore the current veers toward the west, or to the right, as shown in Figure 16-17B. Thus, north–south currents always veer to the right in the Northern Hemisphere. In the Southern Hemisphere currents turn toward the left for the same reason.

At the sea surface, both the prevailing winds and the Coriolis effect directly influence current directions. Below the sea surface, however, the effect of the wind is indirect. Water below the surface is moved by the water above it. Hence, the effect of the wind is indirect below the surface and weakens with depth. But the Coriolis effect is as strong at depth as it is at the sea surface. As a result, each successively deeper layer is less affected by wind and more affected by the Coriolis effect. Consequently, deeper layers of water in the surface currents are deflected—to the right in the Northern Hemisphere and to the left in the Southern Hemisphere—even more than the shallowest water.

The Great Pacific Garbage Patch

The circular motion characteristic of gyres tends to trap and accumulate floating debris, especially human garbage. First discovered in the early 1990s, the Great Pacific Garbage Patch is a region of floating garbage that has accumulated in the center of the eastern Pacific Ocean between California and Hawai'i (Figure 16-18). Because the gyres move in a circular fashion, floating garbage in the oceans tends to accumulate in the center. The size of the accumulation is difficult to measure, and it is possible to sail directly through the garbage patch without noticing an obvious accumulation. The size of the Great Pacific Garbage Patch is estimated to be about twice the size of Texas. Similar but smaller floating garbage patches exist in the western Pacific and North Atlantic.

Although the name conjures up images of a floating garbage heap, these gyre-produced concentrations of garbage consist mostly of small pieces of floating or suspended plastic. Most plastics do not decompose in the ocean. Rather, they break down into smaller and smaller pieces called microplastics. Microplastic debris, tiny pieces of plastic less than 5 millimeters long, is mixed and moved about by wave and wind energy. The debris is dispersed over huge surface areas while also mixing with ocean water below the surface. At present, scientists understand very little about the effects of microplastics on the ocean ecosystem.

checkpoint What is the Coriolis effect?

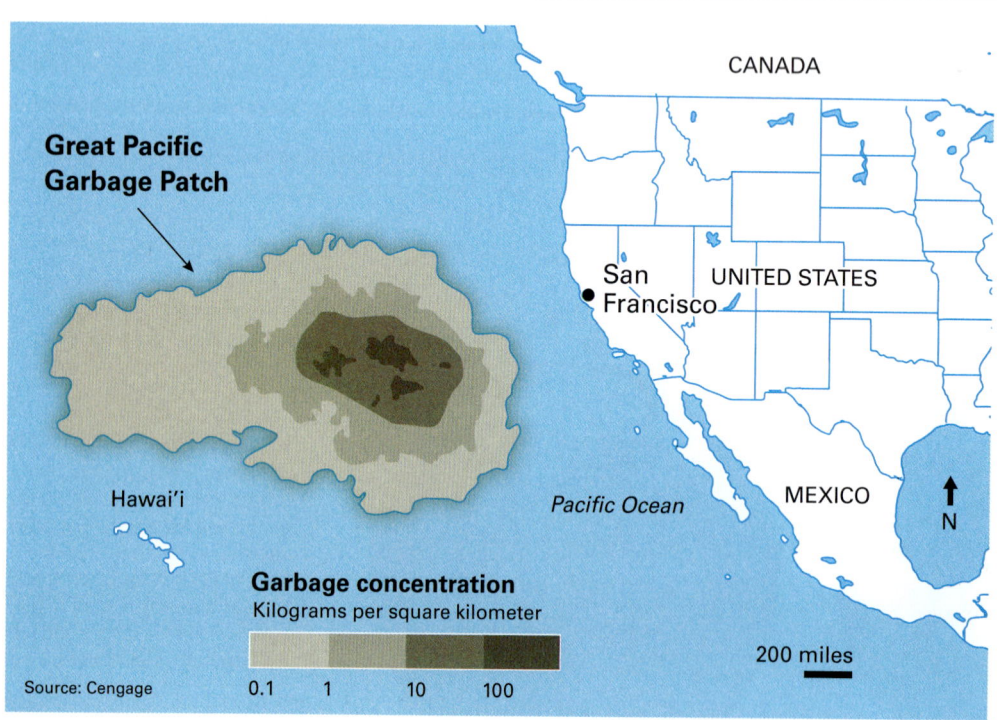

FIGURE 16-18
The Great Pacific Garbage Patch is a zone in the eastern Pacific Ocean where the North Pacific gyre concentrates floating and submerged pieces of plastic garbage. Similar floating "garbage patches" are found in the western Pacific and North Atlantic.

Deep-Sea Currents

Wind does not affect the ocean depths. Oceanographers once thought, therefore, that water of the deep ocean was almost motionless and the seafloor topography changed little over time as a result. In her 1951 book, *The Sea Around Us*, Rachael Carson wrote that the ocean depths are "a place where change comes slowly, if at all." However, in 1962, ripples and small dunes were photographed on the floor of the North Atlantic. Because flowing water forms these features, the photographs demonstrated that water must be moving in the ocean depths. More recently, oceanographers have measured **deep-sea currents** directly with flow meters and photographed moving sand and mud with underwater video cameras.

Deep-sea currents transport seawater both vertically and horizontally below a depth of 400 meters. Wind drives surface currents, but deep-sea currents are driven by differences in water density as denser water sinks and less-dense water rises. Dense water sinks and flows horizontally along the seafloor, forming a deep-sea current. Two factors cause water to become dense and sink: decreasing temperature and increasing salinity. The global deep-sea circulation shown in Figure 16-19 and Figure 16-20 is caused by these two factors and is called **thermohaline circulation** (*thermo* for temperature and *haline* for salinity).

Recall that water is most dense when it is close to freezing. Therefore, as tropical surface water moves poleward and cools, it becomes denser and

FIGURE 16-19
A cross-section of the globe shows surface and deep-sea currents in the Atlantic Ocean. This artist's rendition has simplified the currents to emphasize the major flow patterns. The numbers on the globe correspond to the numbers on the bottom-right map. This continuous exchange between surface and deep-sea currents profoundly affects climate.

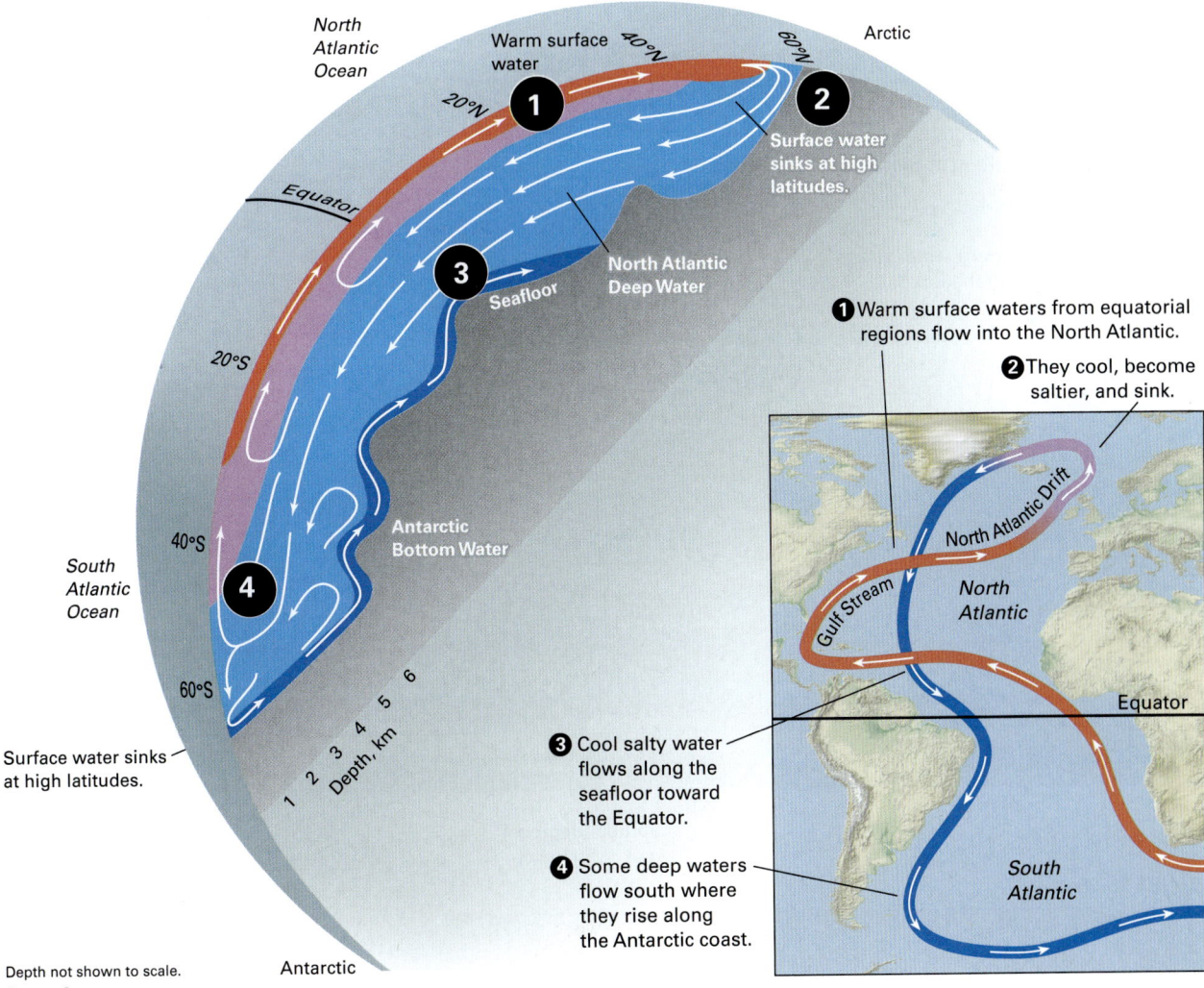

sinks. In addition, water density also increases as salinity increases, which means water sinks when it becomes saltier. Seawater can become saltier if surface water evaporates or when the surface freezes because salt does not become incorporated into the ice. Thus, Arctic and Antarctic water is dense because the water is both cold and salty.

In contrast, addition of fresh water makes seawater less salty and less dense. This effect is pronounced in enclosed bays with abundant, inflowing fresh river water. It is also evident in the polar regions when icebergs float into the ocean and melt. As we will see in Chapter 21, recent rapid melting of Greenland glaciers has introduced enough fresh water into the North Atlantic to reduce the density of surface water and alter its buoyancy.

Figure 16-20 shows that when warmer water reaches the northern part of the Atlantic Ocean near the tip of Greenland, it is cool and salty enough to sink. When the sinking water reaches the seafloor, it is deflected southward to form the North Atlantic Deep Water Current. This current flows along the seafloor all the way to Antarctica. An individual water molecule that sinks near Greenland may travel in this global conveyor belt for 500 to 2,000 years before resurfacing half a world away.

checkpoint What two factors cause some seawater to become denser, resulting in thermohaline currents?

Upwelling

If water sinks in some places, it must rise in others to maintain mass balance. This upward flow of water is called **upwelling**. Upwelling carries cold water from the ocean depths to the surface. Upwelling also brings nutrients from the deep ocean to the surface, creating rich fisheries like those along the coasts of California and Peru. Several processes can cause upwelling, both in the open oceans and along coastlines.

In Figure 16-15, note that the California Current flows southward along the coast of California. In the Southern Hemisphere, the Humboldt Current moves northward along the west coast of South America. Both currents are deflected westward—away from shore—by the Coriolis effect. As these surface currents veer away from shore, water from the ocean depths rises toward the surface along the California coast and the west coast of South America. As a result of this upwelling, surfers and swimmers along the central California coast must wear wetsuits even in August, when the water is about 15°C. By contrast, water warmed by the Gulf Stream on the mid-Atlantic coast at the same latitude is a much more comfortable 21°C.

Upwelling can also be caused by offshore winds, which blow out to sea and drag surface water away from shore. Deep water then upwells along the

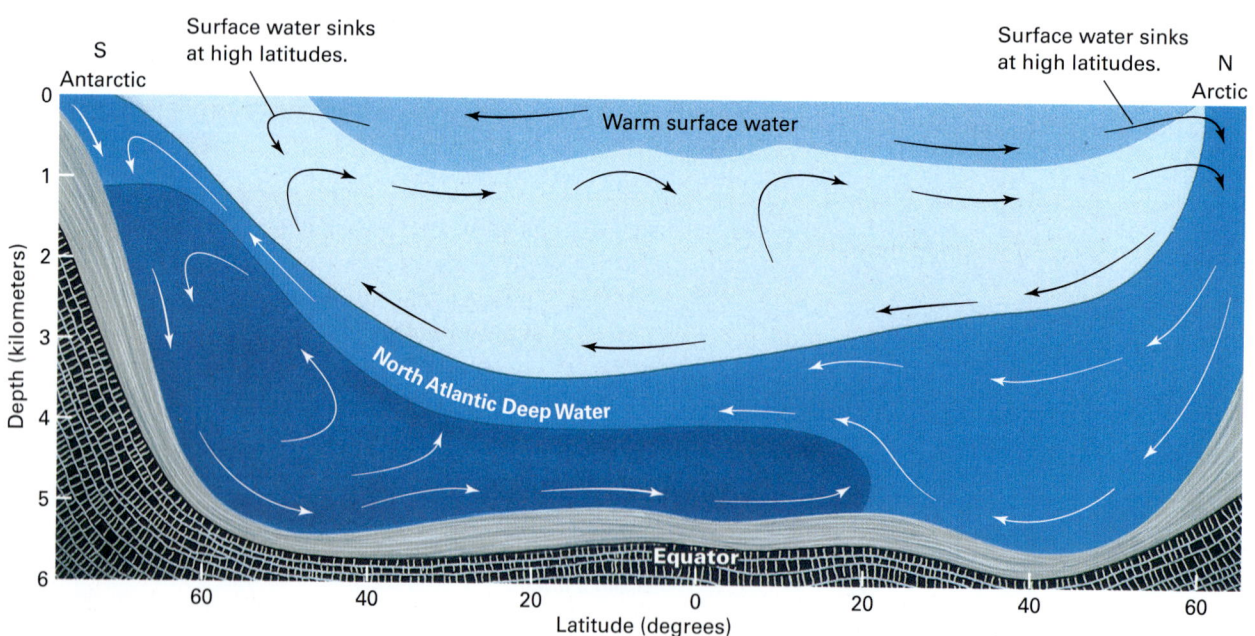

FIGURE 16-20
A north-to-south (right-to-left) profile of the Atlantic Ocean shows surface and deep-sea currents.

MINILAB Motion in the Ocean

Materials
1 tall, cylindrical container, such as a tennis ball canister
500-mL beaker
salt
food coloring (yellow and blue)
clear straw, cut in half
tablespoon measure
long wooden spoon or paint stick for stirring
high-density foam plug (0.5-in. thickness)
clothespin or tongs

CAUTION: Use caution when filling containers with water. Clean up any spilled water or food coloring immediately. Do not get food coloring on your clothing.

Procedure

1. Add about 450 mL of room-temperature water to the beaker.
2. Add 2 tablespoons of salt to the beaker and stir to dissolve it. Then add 3 drops of yellow food coloring and stir again.
3. Fill the canister half-full of water. Add 3 drops of blue food coloring and stir.
4. Insert the two halves of the straw into the holes in the foam plug as shown in the illustration. One straw should point upward and the other downward.

5. Push the plug into the canister of blue water until it reaches the surface of the water (see illustration). The plug should fit tightly; you may have to wiggle it back and forth to get it to move down.
6. Clamp the upward-facing straw with the tongs.
7. With the straw still clamped, slowly pour the yellow water into the canister, but avoid pouring directly into the hole with the downward-facing straw. Add enough yellow water so that the level is higher than the pinched straw.
8. Unclamp the straw. Record your observations.

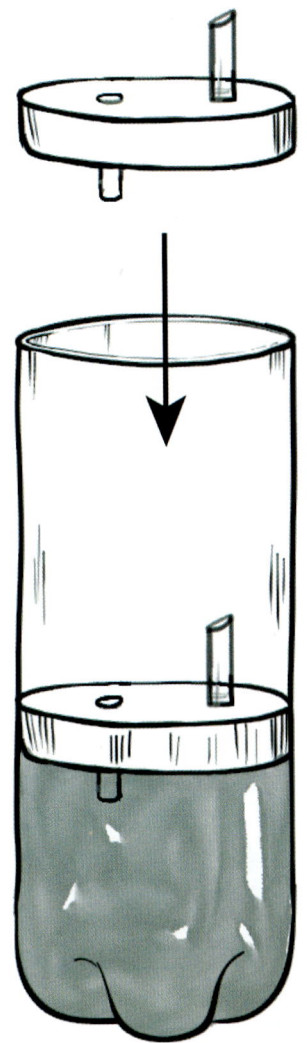

Results and Analysis

1. **Identify** What do the different types of water represent?
2. **Observe** What did you observe after the pinched straw was let go?

Critical Thinking

3. **Explain** What do your observations show about the two types of water and their properties?
4. **Connect** What is thermohaline circulation? How does this experiment model thermohaline circulation in the ocean?
5. **Predict** What do you think would happen if you performed this experiment using warm and cold water instead of salt water and fresh water? Do you think your observations would be the same? Explain your answer.

LESSON 16.3

edge of the continental shelf to replace the surface water flowing away from shore. Winds blowing parallel to shore can also create upwelling. The wind causes water to flow parallel to shore, but the Coriolis effect deflects the current, driving it away from the coast. Deep water then rises to replace the surface water.

In most years, an offshore wind drives surface water away from the coast of Peru and adjacent portions of western South America. As a result, a strong, nutrient-rich upwelling current that produces rich fisheries along that coast is created. However, about every three to seven years—in El Niño years—the offshore wind weakens and the upwelling does not occur or is much reduced. As a result, unusually warm, nutrient-poor water accumulates along the west coast of South America, displacing the cold Humboldt Current. El Niño's effects last for about a year before conditions return to normal. Many meteorologists now think that El Niño affects weather patterns for nearly three-quarters of Earth. The causes and effects of El Niño are described in Chapter 20.

Equatorial upwelling occurs in the open oceans near the Equator, where currents veer poleward because of the Coriolis effect (Figure 16-17). Hence, currents are bending both north and south, causing deeper, colder water to rise to replace the surface water.

checkpoint What causes upwelling?

16.3 ASSESSMENT

1. **Distinguish** What is the difference between a surface current and a deep-sea current?
2. **Describe** What is the path the Gulf Stream takes as it moves from the subtropics to the North Atlantic?
3. **Explain** What is upwelling and what effects does it have on waters near the ocean's surface?

Critical Thinking

4. **Infer** Describe how extreme warming at the poles might affect the "global conveyor belt" of ocean circulation.

16.4 SEACOASTS

Core Ideas and Skills

- Describe the processes of weathering and erosion on seacoasts.
- Describe how sediment is transported along seacoasts.
- Compare and contrast emergent and submergent coastlines.

KEY TERMS

surf
longshore current
beach drift
emergent coastline
submergent coastline
eustatic sea-level change

Coastlines are among the most dynamic environmental zones on Earth, where atmosphere, geosphere, hydrosphere, and biosphere all affect the local environments. Waves and currents weather, erode, transport, and deposit sediment continuously on all coastlines. In addition, the shallow waters of continental shelves are among the most productive biological ecosystems on Earth. Many organisms build hard shells or skeletons from ions made available by the chemical weathering of the continents. When the organisms die, the hard parts can be incorporated into sediment. Deposited in coastal environments, the remains make the once living organisms part of the rock cycle.

Coastlines are geologically active environments. Sea level rises and falls, flooding shallow parts of continents and stranding beaches high above the sea surface. Regions of continental and oceanic crust also rise and sink, with results similar to those caused by fluctuating sea level. Rivers deposit great quantities of sand and mud on coastal deltas. Waves and currents erode beaches and transport sand along hundreds or even thousands of kilometers of shoreline. Converging tectonic plates buckle coastal regions, creating mountain ranges, earthquakes, and volcanic eruptions.

Weathering and Erosion on the Seacoast

Waves batter most coastlines. If you walk down to the shore, you can watch turbulent water carry sand grains or even small cobbles along the beach. Even on a calm day, waves steepen as they approach

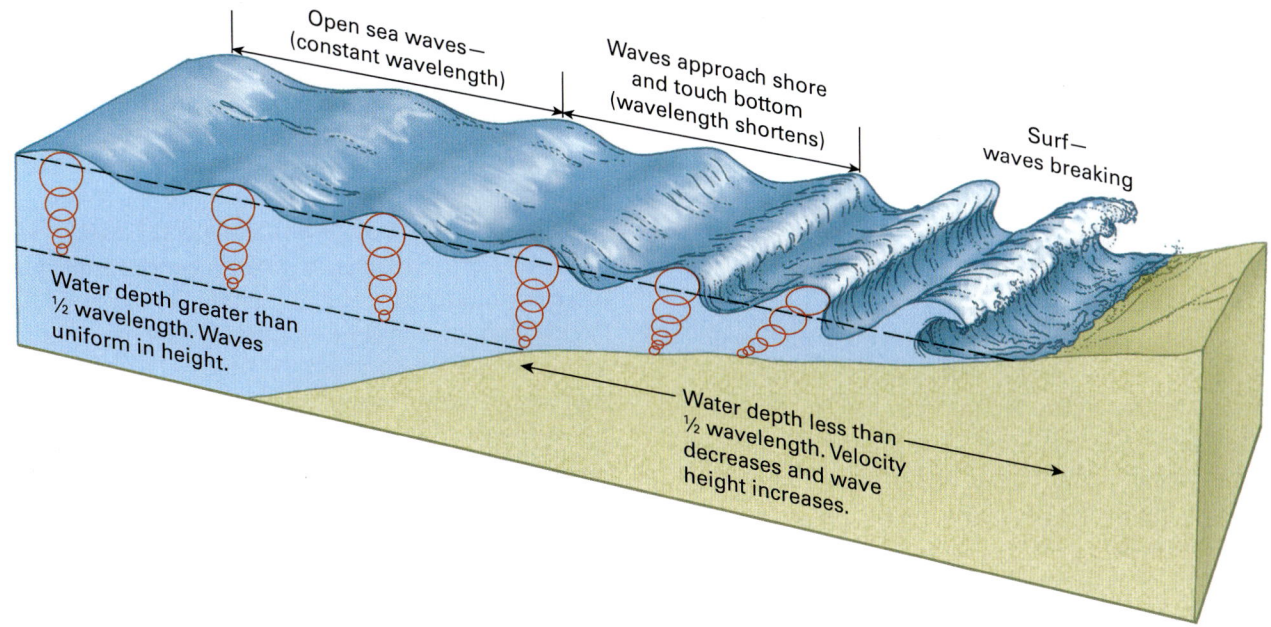

FIGURE 16-21
When a wave approaches the shore, wave orbitals at the bottom of the wave flatten out and become elliptical. The bottom of the wave drags against the seafloor. As a result, the wavelength shortens and the wave steepens until it finally breaks, creating surf.

the shoreline, crash in the surf zone, and wash up against the beach.

When a wave enters water shallower than its wave base, the bottom of the wave drags against the seafloor. This drag compresses the circular motion of the lowermost wave orbitals into ellipses. The drag also slows down the lower part of the wave. At first, the incoming wave simply builds upward from a swell to a wave as the shallowing bottom causes the incoming water to bunch up. The surface of the wave is forced upward. Eventually, however, the bottom of the wave becomes so deformed by drag, and the upper part of the growing wave gets so far ahead of the lower part, that the wave collapses forward, or breaks (Figure 16-21). Chaotic, turbulent waves breaking along a shore are called **surf**.

Most coastal erosion occurs during intense storms because storm waves are much larger and more energetic than normal waves. During windy days, 2.5-meter-high waves strike the shore with 25 times the force of the 0.5-meter-high waves seen on calm days. During a storm, a giant 9-meter-high wave strikes the shore with more than 300 times the force of a 0.5-meter-high wave. A storm wave striking a rocky cliff also drives water into cracks or crevices in the rock, compressing air in the cracks. The air and water combine to create hydraulic forces strong enough to dislodge rock fragments or even huge boulders.

Storm waves create forces as great as 25 to 30 tons per square meter. In the late 1860s, engineers built a breakwater in Wick Bay, Scotland, of car-sized rocks weighing 80 to 100 tons each. The rocks were bound together with steel rods set in concrete, and the seawall was topped by a steel-and-concrete cap weighing more than 800 tons. In 1872, a large storm broke the cap and scattered the rocks about the beach. The breakwater was rebuilt, reinforced, and strengthened. However, after a second storm hit and destroyed this wall, the project was abandoned.

Seawater also weathers and erodes coastlines by more gradual processes such as abrasion, dissolution, and salt cracking. In abrasion, waves carry large quantities of silt, sand, and gravel. Breaking waves roll this sediment back and forth over bedrock, acting like liquid sandpaper, eroding the rock. At the same time, smaller cobbles are abraded as they roll back and forth in the surf zone. In dissolution, seawater slowly dissolves rock and carries ions in solution. Salt cracking occurs because salt water soaks into bedrock. When the water evaporates, the growing salt crystals pry the rock apart. These processes are familiar from our discussions of weathering and erosion in Chapter 12.

checkpoint What causes a wave to break?

Sediment Transport Along Coastlines

Most waves approach the shore at an angle rather than head-on. When this happens, one end of the wave encounters shallow water and slows down, while the rest of the wave is still in deeper water and continues to advance at a relatively faster speed. As a result, the wave bends (Figure 16-22). This effect is called refraction.

As waves approach an irregular coast, they reach the headlands first, breaking against the point and eroding it (Figure 16-23). The waves then refract around the headland and travel parallel to its sides. Refracted waves transport sand and other sediment toward the interior of a bay. As the headlands erode and the interiors of bays fill with sand, an irregular coastline eventually smooths out.

Surfers seek such refracted waves because these waves move nearly parallel to the coast and therefore travel for a long distance before breaking. For example, the classic big-wave mecca at Waimea Bay of Oahu, Hawai'i, forms along a point of land that juts into the sea. Waves build when they strike an offshore coral reef, then refract along the point and charge into the bay.

When waves strike the coast at an angle, they form a **longshore current** that flows parallel to the shore (Figure 16-24). Longshore currents involve water from the surf zone and a little farther offshore, and they may travel for tens or even hundreds of kilometers parallel to the coastline, transporting sand for great distances. Longshore sediment transport also occurs by **beach drift**. If a wave strikes the beach obliquely, it pushes sand up and along the beach in the direction the wave is traveling. When water recedes, the sand flows straight down the beach (Figure 16-24). Thus, at the end of one complete wave cycle, the sand has moved a short distance parallel to the shoreline. The next wave transports the sand a little farther, and so on. Over time, sediment can move many kilometers down a coastline in this manner.

Longshore currents and beach drift work together to transport and deposit huge amounts of sand along a coast. Much of the sand found at Cape Hatteras, North Carolina, originated hundreds of kilometers away. The origins include the mouth of the Hudson River in New York and glacial deposits

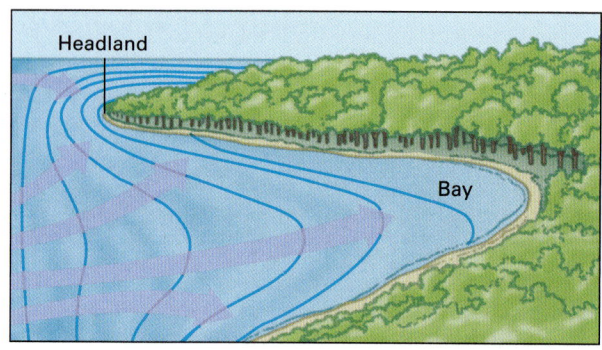

A

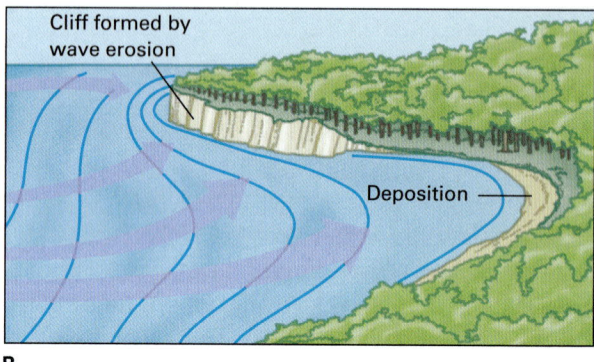

B

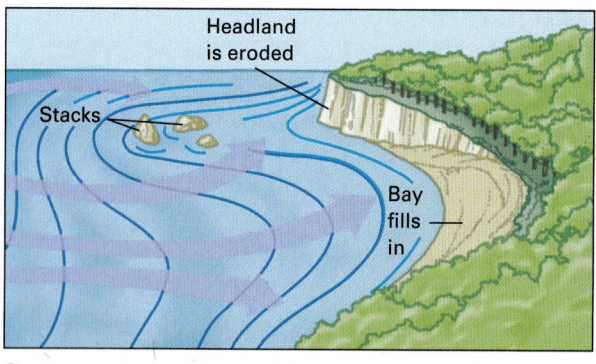

C

FIGURE 16-23
(A) When a wave strikes a headland, the shallow water causes that portion of the wave to slow down. (B) Part of the wave breaks against the headland, weathering the rock. A portion of the wave refracts, transporting sediment and depositing it inside the bay. (C) Eventually, this selective weathering, erosion, and deposition will smooth out an irregular coastline.

FIGURE 16-22
Wave refraction takes place when one end of a wave reaches shallow water and slows down.

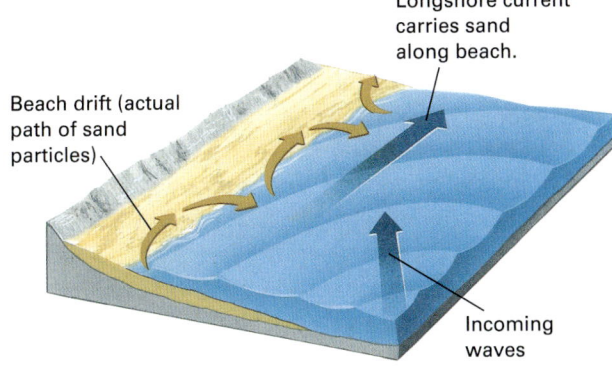

FIGURE 16-24
Longshore currents and beach drift transport sediment along a coast.

on Long Island and southern New England. At Sandy Hook, New Jersey, an average of 2,000 tons of sandy sediment a day move past any point on the beach. As a result of longshore currents, some beaches have been called rivers of sand.

checkpoint Describe how beach drift occurs.

Coastal Emergence and Submergence

Geologists have found drowned river valleys and fossils of land animals on continental shelves beneath the sea. They have also found sedimentary rocks containing fossils of fish and other marine organisms in continental interiors. As a result, we infer that sea level has changed, sometimes dramatically, throughout geologic time. An **emergent coastline** forms when a portion of a continent that was previously underwater becomes exposed as dry land (Figure 16-26). Falling sea level or rising land can cause emergence. Many emergent coastlines are sandy. In such cases, the delivery of sandy sediment to the coastline is high enough to cause the shoreline to migrate seaward. In contrast, a **submergent coastline** develops when the sea floods low-lying land and the shoreline moves inland (Figure 16-26). Submergence occurs when sea level rises or coastal land sinks. A submergent coast is commonly irregular, with many bays, sea cliffs, and headlands. The coast of Maine, with its

FIGURE 16-25
ON ASSIGNMENT How would you describe this unusual water formation? What could have caused it? Photographer Chris Gray took this photograph in Canada on the western coast of Vancouver Island. The water formation was caused by two large waves crashing into an underwater structure during a storm.

numerous fjords, inlets, and rocky bluffs, is a classic submergent coastline (Figure 16-27). Small, sandy beaches form in some protected coves, but most of the shoreline is rocky and steep because it consists of eroding bedrock and glacial sediment.

Tectonic processes can cause a coastline to rise or sink. Isostatic adjustment can also depress or elevate a portion of a coastline. About 18,000 years ago, a huge continental glacier covered most of Scandinavia, causing it to sink. As the lithosphere settled, the displaced asthenosphere flowed southward, causing the Netherlands to rise. When the ice melted, the process reversed as the asthenosphere flowed back from below the Netherlands to Scandinavia. Today, Scandinavia is rebounding and the Netherlands is sinking. During the Pleistocene Glaciation, Canada was depressed by the ice, and asthenosphere rock flowed southward. Today, the asthenosphere is flowing back north, much of Canada is rebounding, and much of the continental United States is sinking.

Sea level can also change globally. A global sea-level change, called **eustatic sea-level change**, occurs by three mechanisms: the growth or melting of glaciers, changes in water temperature, and changes in the volume of the Mid-Ocean Ridge.

During glaciation, vast amounts of water are withdrawn from the sea to form continental glaciers. This causes sea level to drop and coastlines around the world to emerge. Conversely, when glaciers melt, sea level rises globally, causing global submergence of coastlines.

Seawater expands when it is heated and contracts when it is cooled. Although this change is not noticeable in a glass of water, the volume of the oceans is so great that a small temperature change can alter sea level measurably. As a result, global warming causes sea-level rise, and global cooling leads to falling sea level.

Temperature changes and glaciation are connected. When global temperatures rise, seawater expands and glaciers melt. Thus, even

FIGURE 16-26

If sea level falls or if the land rises, the new coastline is emergent. Offshore sand is exposed to form a sandy beach. If coastal land sinks or sea level rises, the new coastline is submergent. Areas that were once land are flooded. Irregular shorelines develop, and beaches are commonly rocky.

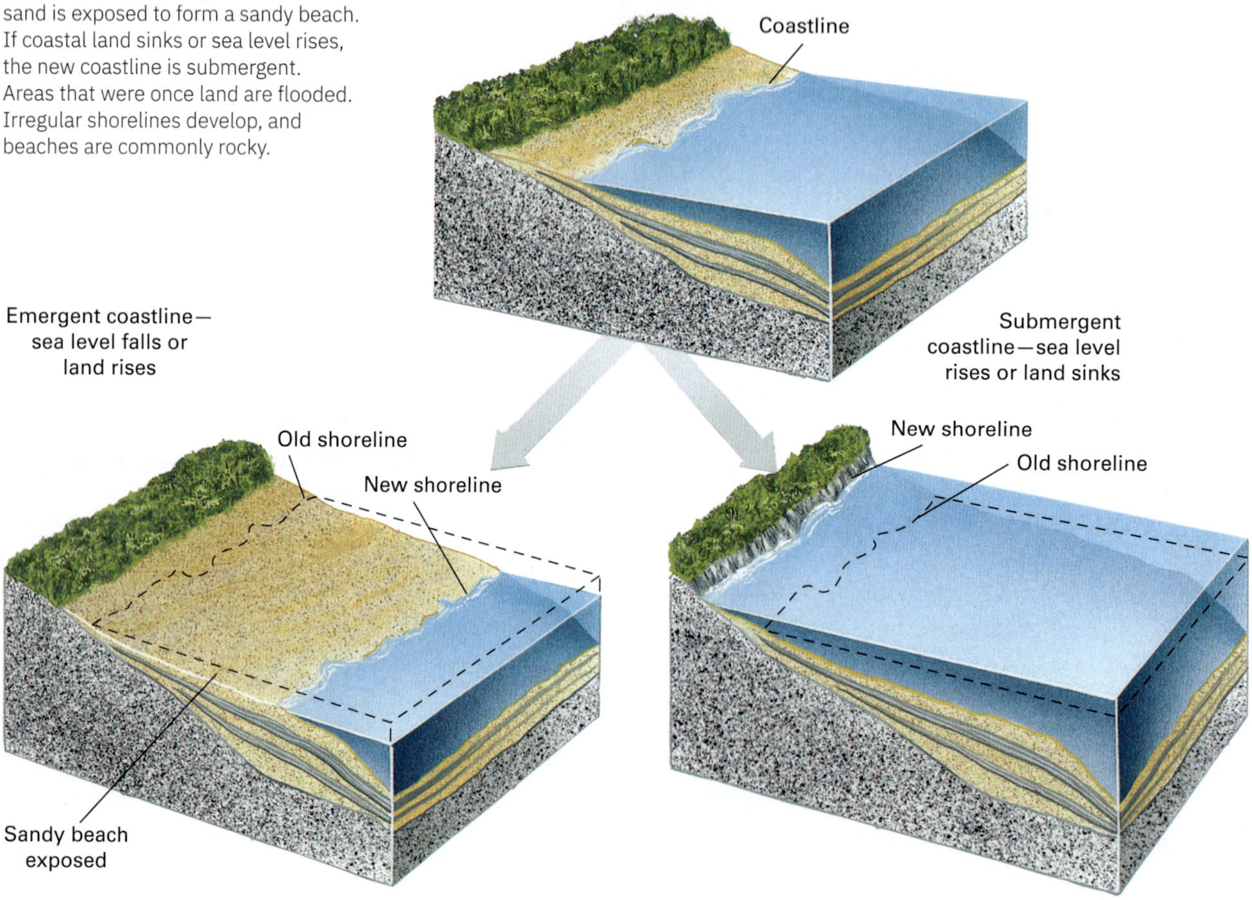

FIGURE 16-27
The Maine coast is a rocky, irregular, submergent coastline.

minor global warming can lead to a large sea-level rise. The opposite effect is also true: When temperatures fall, seawater contracts, glaciers grow, and sea level falls.

The Mid-Ocean Ridge rises highest at its spreading center, where new rock is hottest and has the lowest density. The elevation of the ridge decreases on both sides of the spreading center because the lithosphere cools, shrinks, and becomes more dense as it moves outward (Figure 11-14). Rapid seafloor spreading produces a larger volume of hot rock and pushes aside more seawater, causing global sea level to rise. If spreading later slows, a smaller volume of hot rock is produced, displacing less seawater and causing global sea level to fall. We discussed this cause of eustatic sea-level change in Chapter 11.

checkpoint What is eustatic sea-level rise?

16.4 ASSESSMENT

1. **Describe** How does wave action cause weathering and erosion on a coast?
2. **Define** What is beach drift, and what is the effect of beach drift?
3. **Compare** Compare the features of emergent and submergent coastlines.
4. **Identify** What are three causes of eustatic sea-level change?

Critical Thinking

5. **Analyze** Over time, two beaches have evolved. One beach is rocky, with many bays and sea cliffs. Another beach is sandy. Which beach is likely to be a submergent coastline? Which beach is likely to be an emergent coastline?

LESSON 16.4

16.5 BEACHES

Core Ideas and Skills
- Define different areas of a beach.
- Identify land features formed on sandy coastlines.
- Describe landforms on rocky coastlines.

KEY TERMS

spit
barrier island
groin

Beach Zones

When most people think about going to the beach, they think of gently sloping expanses of sand. However, a beach is any strip of shoreline that is washed by waves and tides. Although many beaches are sandy, others are swampy or rocky.

A beach is divided into three zones, although only two of these are exposed above the sea surface. The shoreface extends from mean low tide to mean fair-weather wave base. The shoreface typically slopes seaward and is the zone in which waves slow down, steepen, and break, expending mechanical energy against the bottom. The foreshore, also called the intertidal zone, lies between mean high and low tides. It is alternately exposed to the air at low tide and covered by water at high tide. The backshore exists above the foreshore and is usually dry but is washed over by waves during storms. It commonly slopes gently landward as a result of sediment delivered and deposited by storm waves losing energy as they wash inland. Many terrestrial plants cannot survive in salt water, so specialized, salt-resistant plants live in the backshore. The backshore can be wide or narrow, depending on its topography, tidal differences, and the frequency and intensity of storms. In a region where the land rises steeply, the backshore may be a narrow strip (Figure 16-28). In contrast, if the coast consists

FIGURE 16-28
Only small beaches with narrow backshores form on the emergent central California coastline between Monterey and Big Sur because of the mountainous topography of the coastline.

of low-lying plains and if coastal storms occur regularly, the backshore may extend several kilometers inland.

If weathering and erosion occur along all coastlines, why are some beaches sandy and others rocky? The answer lies partly in the fact that most sand is not formed by weathering and erosion at the beach itself. Instead, several processes transport sand to a seacoast. Rivers carry large quantities of sand, silt, and clay to the sea and deposit it on deltas that may cover thousands of square kilometers. In some coastal regions, glaciers deposited large quantities of sandy till along coastlines during the Pleistocene Glaciation. In tropical and subtropical latitudes, eroding reefs supply carbonate sand to nearby beaches. A sandy coastline is one with abundant sediment from one or more of these sources.

Longshore currents transport and deposit the sand along the coast. Much of the sand carried by these currents accumulates on underwater offshore bars. Thus, a great deal of sand may be stored offshore from a beach. If such a coastline emerges, this vast supply of sand becomes exposed as dry land and is available, if eroded, for building more beaches and more offshore bars. Thus, sandy beaches are abundant on emergent coastlines.

In contrast, rocky coastlines occur where sediment from any of these sources is scarce. With no abundant sources of sand, small sandy beaches may form in protected bays, but most of the coast will be rocky. On a submergent coastline, rising sea level puts the stored offshore sand even farther out to sea and below the depth of waves. As a result, submergent coastlines commonly have beaches that are rocky, not sandy.

checkpoint What is the foreshore, or intertidal zone?

Landforms Along Sandy Coastlines

Over time, deposits of sand can accumulate in various places along sandy coastlines, resulting in landforms such as spits, baymouth bars, and barrier islands. A **spit** is a small, fingerlike ridge of sand or gravel that extends outward from a beach (Figure 16-29). As sediment migrates along a coast, the spit may continue to grow. A well-developed spit may rise several meters above high-tide level and may be tens of kilometers long. A spit may eventually block the entrance to a bay, forming a baymouth bar (Figure 16-29). A spit may also extend outward into the sea, creating a trap for other moving sediment.

FIGURES 16-29B and 16-29C
Go online to view examples of (B) a spit that formed along a low-lying coast and (C) a baymouth bar.

A **barrier island** is a long, low-lying sandy island that extends parallel to the shoreline. It looks like a beach or spit and is separated from the mainland by a sheltered body of water called a lagoon. Barrier

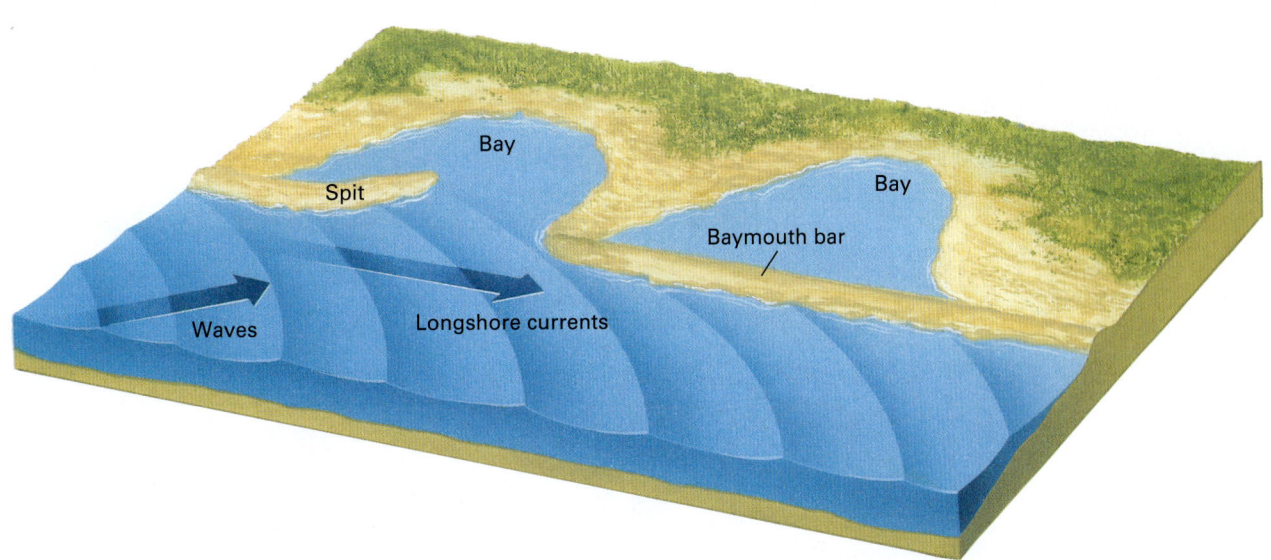

FIGURE 16-29A
Spits and baymouth bars are common features of sandy emergent coastlines with substantial longshore drift.

LESSON 16.5

islands extend along the East Coast of the United States from New York to Florida. They are so nearly continuous that a sailor in a small boat can navigate much of the coast inside the barrier island system and remain protected from the open ocean. Barrier islands also line much of the Gulf Coast, especially in Texas.

Barrier islands form in several ways. The two essential ingredients are a large supply of sand and the waves or currents to transport it. If a coast is shallow for several kilometers outward from shore, breaking storm waves may carry sand toward shore and deposit it just off the mainland as a barrier island. Alternatively, if a longshore current veers out to sea, it slows down and deposits sand where it reaches deeper water. Waves may then pile up the sand, forming a barrier island.

Other mechanisms that create barrier islands involve sea-level change. Underwater sand bars may be exposed as a coastline emerges. Alternatively, sand dunes or beaches may form barrier islands if a coastline sinks.

checkpoint What is the difference between a spit and a baymouth bar?

Development on Sandy Coastlines

The Atlantic coast of the United States is fringed with the longest chain of barrier islands in the world. Many seaside resorts are built on these islands. Developers often ignore the fact that they are impermanent and changing landforms (Figure 16-30). If the rate of erosion exceeds that of deposition for a few years in a row, a barrier island can shrink or disappear completely, leading to destruction of beach homes and resorts. In addition, barrier islands are especially vulnerable to hurricanes, which can wash over low-lying islands and move enormous amounts of sediment in a very brief time. In September 1996, Hurricane Fran flattened much of Topsail Island, a low-lying barrier island in North Carolina. Geologists were not surprised, because the homes were not only built on sand—they were built on sand that was virtually guaranteed to move.

As a second example, Long Island extends eastward from New York City and is separated from Connecticut by Long Island Sound (Figure 16-31). Longshore currents flow westward, eroding sand from glacial deposits at the eastern end of the island and depositing it to form beaches and barrier

FIGURE 16-30
Many resorts, condos, and homes are built on barrier islands, such as these on Hutchinson Island, Florida.

FIGURE 16-31 Longshore currents carry sand westward along the south shore of Long Island, creating a series of barrier islands.

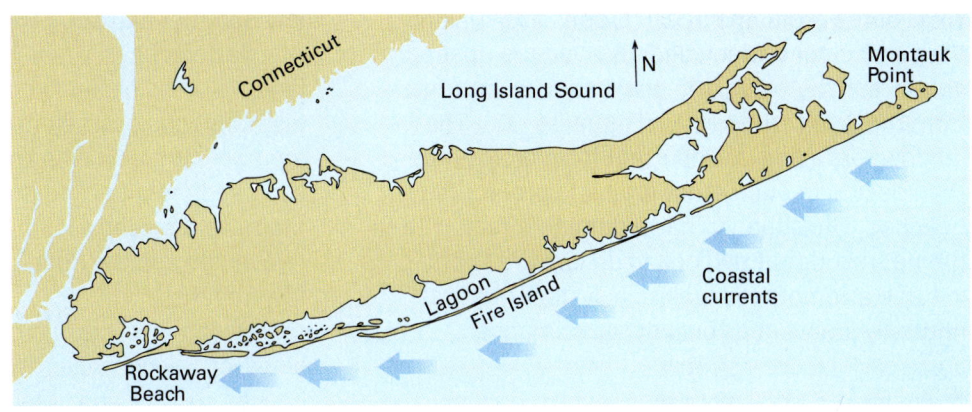

islands on the south side of the island. At any point along the beach, the currents erode and deposit sand at approximately the same rates. Geologists calculate that the supply of sand at the eastern end of the island is large enough to last a few hundred years. When the glacial deposits at the eastern end of the island become exhausted, however, the flow of sand will cease. Then the entire coastline will erode and the barrier islands and beaches will completely disappear.

Now let's narrow our time perspective and look at a Long Island beach over a season or during a single storm. Over this time frame, the rates of erosion and deposition are not equal. Thus, beaches shrink and expand with the seasons or the passage of violent gales. In the winter, violent waves and currents erode beaches, whereas sand accumulates on the beaches during the calmer summer months. In an effort to prevent these seasonal fluctuations and to protect their personal beaches, Long Island property owners have built stone barriers called **groins** from shore out into the water. The groin intercepts the steady flow of sand moving from the east and keeps that particular part of the beach from eroding. But the groin impedes the overall flow of sand (Figure 16-32).

A

Undeveloped beach; ocean currents (arrows) carry sand along the shore, simultaneously eroding and building the beach.

FIGURE 16-32
(A) Longshore currents simultaneously erode and deposit sand along an undeveloped beach. (B) A single groin or breakwater traps sand on the upstream side, resulting in erosion on the downstream side. (C) A multiple groin system multiplies the uneven distribution of sand along the entire beach.

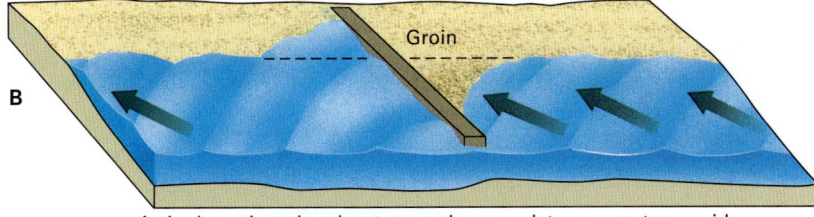

B

A single groin or breakwater; sand accumulates on upstream side and is eroded downstream.

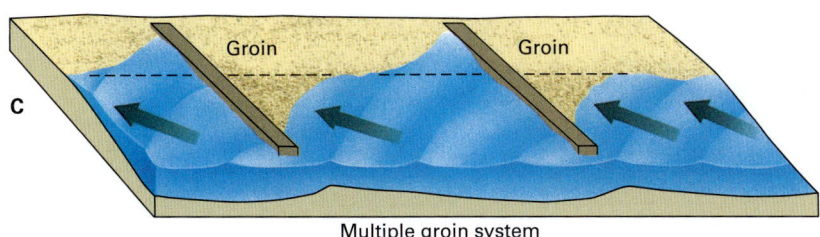

C

Multiple groin system

LESSON 16.5

West of the groin, the beach erodes as usual, but the sand is not replenished because the upstream groin traps it. As a result, beaches downcurrent from the groin erode away (Figure 16-33). The landowners living downcurrent from groins may then decide to build other groins to protect their beaches. The situation has a domino effect, with the net result that millions of dollars are spent in futile attempts to stabilize a system that was naturally stable in its own dynamic manner.

> **FIGURES 16-33A and 16-33B**
> Go online to view examples of (A) a groin and (B) a house on the downstream side of the groin.

Storms pose another problem. Hurricanes, for example, commonly strike Long Island in the late summer and fall, generating storm waves that flatten dunes, erode beaches, and transport large volumes of sand into nearby salt marshes, disrupting the ecosystem there. When the storms are over, gentler waves and longshore currents carry sediment back to the beaches and rebuild them. As the sand accumulates again, salt marshes rejuvenate and the dune grasses grow back within a few months.

This natural cycle of erosion and rebuilding, however, is inconvenient for people who build houses, resorts, and hotels on or near the shifting sands. The owner of a home or a resort hotel cannot allow the building to be flooded or washed away. Therefore, property owners often construct large seawalls along the beach (Figure 16-34). A seawall is wall or embankment built to prevent erosion of

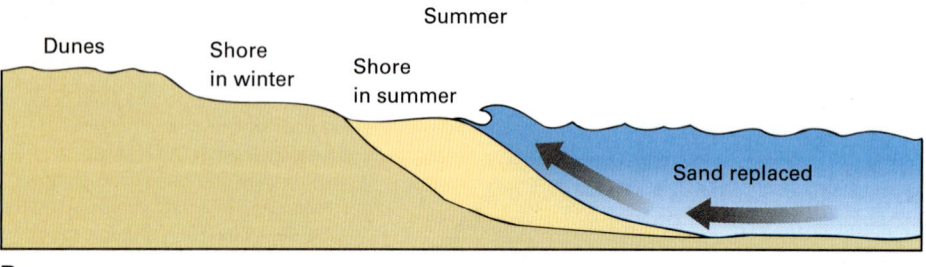

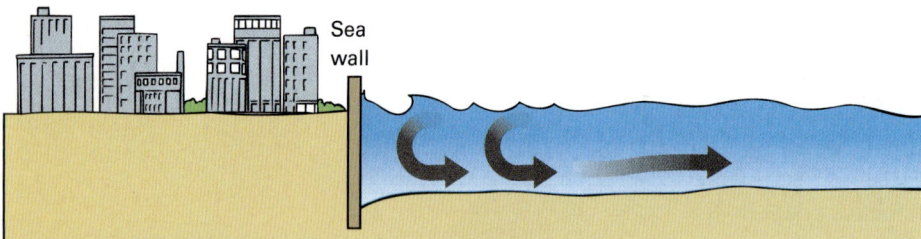

FIGURE 16-34
(A) On a natural beach, the violent winter waves often move sand out to sea. (B) The gentler summer waves push sand toward shore and rebuild the beach. (C) Wave energy concentrates against a seawall. (D) Waves will eventually destroy a seawall if it is not rebuilt.

the land and to protect land and buildings from storms or tsunamis. When a storm wave rolls across an undeveloped low-lying beach, it dissipates its energy gradually as it flows over the dunes and transports sand. A seawall interrupts this gradual absorption of wave energy. The waves crash violently against the barrier and erode sediment at its base until the wall collapses. It may seem surprising that a reinforced concrete seawall is more likely to be permanently destroyed than a beach of grasses and sand dunes, yet this is often the case.

checkpoint On beaches with multiple landowners, why does the construction of one groin often lead to other groins?

FIGURE 16-35
Waves hurl sand and gravel against solid rock, eroding cliffs and creating a wave-cut platform along the coast of Flamborough Head, Yorkshire, England.

Rocky Coastlines

A rocky coastline is one without any of the abundant sediment sources described previously. In many areas on land, bedrock is exposed or covered by only a thin layer of soil. If this type of sediment-poor terrain is submerged, and if there are no other sources of sand, the coastline is rocky. Wave-cut cliffs, sea arches, sea stacks, and fjords are landforms associated with rocky coastlines.

A wave-cut cliff forms when waves erode the headland into a steep profile. As the cliff erodes, it leaves a flat or gently sloping wave-cut platform (Figure 16-35). If waves cut a cave into a narrow headland, the cave may eventually erode all the way through the headland, forming a scenic sea arch. When an arch collapses or when the inshore part of a headland erodes faster than the tip, it leaves behind a pillar of rock called a sea stack (Figure 16-23C). As waves continue to batter the rock, eventually the sea stack crumbles.

If the sea floods a coastal U-shaped valley, a long and deep bay called a fjord is formed. Fjords are common at high latitudes, where rising sea level has flooded coastal U-shaped valleys scoured by Pleistocene glaciers. Fjords may be hundreds of meters deep, and often the cliffs at the shoreline drop straight into the sea.

checkpoint What causes some coastlines to be rocky?

16.5 ASSESSMENT

1. **Define** What are the characteristics of each of the three beach zones?
2. **Explain** What causes some coastlines to be sandy and some to be rocky?
3. **Describe** How do barrier islands form?
4. **Compare and Contrast** What are the similarities and differences between spits and groins?

   Critical Thinking
5. **Evaluate** What are some pros and cons of building seawalls to prevent shore erosion?

16.6 LIFE IN THE SEA

Core Ideas and Skills
- Identify the roles plankton play in food chains and oxygen production.
- Explain how people affect fisheries in the oceans.
- Describe conditions that are necessary for the growth of coral reefs, and how coral reefs are being harmed by humans.

KEY TERMS

plankton
phytoplankton
zooplankton

Plankton

On land, most photosynthesis is conducted by multicellular plants including mosses, ferns, grasses, shrubs, and trees. Large animals such as cattle, deer, elephants, and bison consume the plants. In contrast, most of the photosynthesis and consumption in the ocean is carried out by small organisms called **plankton**. Many plankton are single-celled and microscopic. Others are more complex and are up to a few centimeters long.

Phytoplankton conduct photosynthesis the way land-based plants do. Therefore, they are the base of the food chain for aquatic animals. Although phytoplankton are not readily visible, they are so abundant that they supply about 50 percent of the oxygen in our atmosphere. **Zooplankton** are tiny animals that feed on phytoplankton (Figure 16-36). The larger and more familiar marine plants (such as seaweed) and animals (such as fish, sharks, and whales) play a relatively small role in overall oceanic photosynthesis and consumption. However, many of these organisms are important to humans as a major source of protein.

🌐 **FIGURES 16-36A–D**
Go online to view examples of phytoplankton and zooplankton.

One major difference between terrestrial and aquatic ecosystems is that, on land, soil nutrients are abundant on the surface where light is also abundant. However, in the oceans, light is available only in the photic zone, which extends to a depth of 80 meters or more in the open ocean but is typically shallower closer to shore where the water is cloudier. Whereas light is available only in the photic zone, nutrients in the oceans are concentrated in the dark depths. Plankton live

---

DATA ANALYSIS Comparing World Fisheries

The Fisheries and Aquaculture Department of the Food and Agriculture Organization of the United Nations (FAO) conducts population surveys on 600 marine fish stocks in Earth's oceans. The oceans are divided into 19 major fishing areas, 13 of which are shown on the map (Figure 16-37). Fish species in each area are assigned a label depending on how much their populations have changed. Stocks identified as depleted are the most affected; their populations are well below historic levels and may never fully recover. Overexploited stocks are in danger of becoming depleted. Study the map and the table and use them to answer the questions that follow.

1. **Recognize Patterns** Which major fishing area shown in the table has the most overexploited and depleted species? What percentage of the world's ocean does this area represent?

2. **Identify** Which two major fishing areas have the same fish species that are overexploited or depleted? What are the species?

3. **Analyze Data** Compare the statuses and catch volumes of the two species you identified in the previous question. Use data to support your answer.

Critical Thinking

4. **Evaluate** Compare the catch volumes of the species in the table. What criteria do you think were used to classify these species as "depleted" as opposed to merely "overexploited"?

5. **Apply** People who live near the Mediterranean and Black Seas depend heavily on fish as a source of protein. These seas make up less than one percent of the world ocean, but they have more overexploited and depleted species than several other larger major fishing areas. What regulations or measures would you recommend to improve the sustainability of these fisheries?

(Continued on next page.)

DATA ANALYSIS Comparing World Fisheries

FIGURE 16-37 ▼
This map shows 13 of the world's 19 major fishing areas.

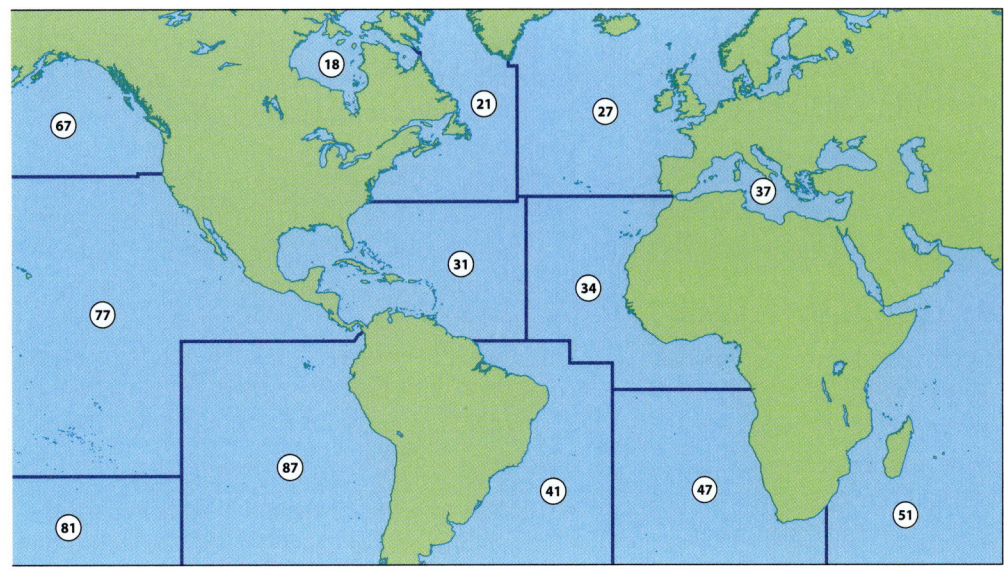

FAO Marine Major Fishing Areas
- 18 Arctic Sea
- 21 Atlantic, Northwest
- 27 Atlantic, Northeast
- 31 Atlantic, Western Central
- 34 Atlantic, Eastern Central
- 37 Mediterranean-Black Sea
- 41 Atlantic, Southwest
- 47 Atlantic, Southeast
- 51 Indian Ocean, Western
- 67 Pacific, Northeast
- 77 Pacific, Eastern Central
- 81 Pacific, Southwest
- 87 Pacific, Southeast

Source: Food and Agriculture Organization of the United Nations

TABLE 16-1 Characteristics of Five Major Fishing Areas

FAO major fishing area	Area ID number	Percent of ocean included in fishing area (by km²)	Overexploited (O) or depleted (D) fish species	Catch volumes from 2002 (tons)
Mediterranean/Black Sea	37	0.8	Atlantic Bluefin Tuna (D)	22,000
			Azov Sea Sprat (D)	27,000
			European Sprat (D)	70,000
			Pontic Shad (D)	0
Northwest Atlantic	21	1.7	Atlantic Cod (D)	55,000
			Haddock (D)	23,000
Northeast Atlantic	27	4.0	Atlantic Cod (O-D)	835,000
			Haddock (O-D)	244,000
Southwest Atlantic	41	4.8	Argentine Hake (O-D)	409,000
			Brazilian Sardinella (O)	28,000
Southeast Atlantic	47	5.1	Geelbeck Croaker (D)	0
			Red Steenbras (D)	0
			Kingklip (O)	13,000
			Bigeye Tuna (O)	19,000
			Southern Bluefin Tuna (O)	42,000
			Cunene Horse Mackerel (O)	45,000
			Southern Spiny Lobster (O)	1,000
			Perlemoen Abalone (O)	1,000

Source: Food and Agriculture Organization of the United Nations

mostly within a few meters of the sea surface, where the most light is available. When they die, their bodies sink to the bottom along with the nutrients, such as nitrogen, iron, and phosphorus, incorporated in their tissues. If the surface waters are not replenished by a fresh supply of nutrients from upwelling and surface currents, the productivity of plankton will be limited.

checkpoint Explain how phytoplankton are different from zooplankton.

World Fisheries

The shallow waters of continental shelves support large populations of marine organisms. Even many deep-sea fish spawn in shallower water within 1 or 2 kilometers of the shoreline. Shallow zones in the bays, lagoons, and estuaries of the world's continental shelves are especially hospitable to life for several reasons. They have (1) easy access to the deep sea, (2) lower salinity than the open ocean, (3) a high concentration of nutrients, (4) abundant plant life, and (5) shelter. As a result, about 99 percent of the marine fish caught every year are harvested from the shallow waters of continental shelves.

Historically, fishing pressure by humans has impacted fish populations. In one study, oceanographers documented research showing that industrial fishing fleets have caused a 90 percent reduction of large predatory fish such as tuna, marlin, swordfish, cod, halibut, and flounder. A hundred years ago, most fleets caught six to 12 fish for every 100 baited hooks. Today, despite sophisticated electronic and aerial fish-finding techniques, the catch rate has plummeted to one fish per 100 hooks.

Many countries are neither interested in nor have the capacity to tally catches from fisheries, and some nations have altered reported catch numbers for political reasons. Many wild stocks continue to decline in the absence of globally recognized catch limits designed to replenish fish stocks. Collective international agreement and adherence to a set of global catch limits are needed. Monitoring catches is important. A broad data-gathering system for determining what fish species are harvested, with the location and date of the catch, is also needed.

checkpoint What are some ways to prevent the depletion of fisheries?

Coral Reefs

Recall from Lesson 16.1 that a reef is a wave-resistant ridge or mound built by corals, oysters, algae, or other marine organisms. Because corals need sunlight and warm, clear water to thrive, coral reefs develop in shallow tropical seas where little suspended clay or silt muddies the water (Figure 16-38). As the corals die, their offspring grow on their remains.

Globally, coral reefs cover an area about the size of France. But they spread out in long, thin lines. They are extraordinarily productive ecosystems because the corals provide shelter for many invertebrates, fish, and other marine organisms. Within the past 50 years, 10 percent of the world's coral reefs have been destroyed and an additional 30 percent are in critical condition. Several factors have contributed to this destruction.

- Increased levels of carbon dioxide in Earth's atmosphere have caused higher concentrations of the gas to be dissolved in the world's oceans. This results in ocean acidification, making it more difficult for corals to live.

- Corals thrive best in a narrow temperature range. In recent years, oceanographers have compiled considerable evidence that sea surface temperatures have become warmer in recent decades. The warm water is leading to widespread die-offs.

- Silt from cities, urban roadways, farms, and improper logging smothers the delicate organisms in the reef.

- Fertilizer runoff from farms and sewage runoff from cities have added nutrients to coastal waters, feeding coral predators and lowering dissolved oxygen levels below the threshold required by corals to survive. A 2017 study by National Geographic reported that nutrient-rich runoff from the Mississippi River caused a New Jersey–size dead zone along the coastline of Louisiana and Texas (Figure 16-39).

FIGURE 16-39
Go online to view a satellite image from NASA that shows the dead zone in the Gulf of Mexico.

A dead zone has levels of dissolved oxygen too low to sustain life. The Mississippi River drainage is the largest in North America. Runoff in the

A

B

C

FIGURE 16-38
(A) Coral reefs form abundantly in clear, shallow, tropical water, as shown in this aerial view of the Great Barrier Reef, Australia. Notice the change to deep blue water on the horizon, marking the steep frontal edge of the reef. The much shallower water in the foreground consists of a network of reefs (brown-red) growing upward from the bottom, with areas of carbonate sand in between (lighter-colored areas). (B) Reefs such as the Great Barrier Reef are extremely diverse ecologically. The warm, clear, well-lit waters reveal sea fans among colorful corals and fish. (C) The surf breaks on the outer edge of a barrier reef along the southern coast of Fiji. By the time the surf reaches the top of the reef, it has dissipated.

drainage picks up fertilizers, soil particles, animal wastes, and sewage. These sources produce huge algal blooms in the northern Gulf of Mexico, altering the food chain and depleting dissolved oxygen in the water. The size and dimensions of the dead zone change with seasonal discharge from the Mississippi River, large rainfall events, as well as hurricanes and tropical storms. For some marine life, such storms make the waters in this region more hospitable, but conditions can vary from year to year.

checkpoint What four human-caused factors are involved in the decline of coral reefs?

16.6 ASSESSMENT

1. **Explain** Why are the world's fisheries located on continental shelves instead of in the open oceans?
2. **Explain** What is the photic zone, and why is it important?

Critical Thinking

3. **Synthesize** Apply what you learned in this lesson to explain why the abundant fisheries off the coast of Peru and other areas along western South America are less productive during years when nutrient-rich upwelling slows or stops, and unusually warm, nutrient-poor water accumulates along the west coast.
4. **Evaluate** What are some actions you could take to help conserve fish stocks?

TYING IT ALL TOGETHER
SEA CHANGE

In this chapter, you learned about the geography of Earth's oceans, the chemistry of seawater, and the effects of tides, sea waves, and ocean currents on coastlines and beaches. You also explored ocean ecosystems and the impact of human activity on sea life and global sea levels.

In the Case Study, you read about how various human activities are bleaching coral reefs in Hawai'i and around the world. Throughout this chapter, you considered other factors that are causing major changes throughout the world's oceans and coastlines. Sea-level rise, global warming, and ocean acidification are all aspects of climate change that can negatively affect marine ecosystems and contribute to erosion, flooding, and saltwater intrusion in coastal communities. In addition to these problems, declining fish populations, dying reefs, and collapsing food webs can have major consequences for people who depend on the sea for their food and their livelihoods.

In this activity, you will investigate one effect of human activity on Earth's oceans and coastlines in greater detail. Work with a group to complete the steps below.

1. Choose one of the following topics to research. Use reputable sources to explore your topic online.

   - Coral reef bleaching due to global warming
   - The impact of ocean acidification on fish growth and behavior
   - The effect of sea-level rise on coastal or island communities
   - The impact of ocean acidification on calcium carbonate-forming marine organisms (coral, oysters, sea urchins, sea stars, mussels, and others) due to ocean acidification

2. Create a blog post for your class blog that includes the following information:

   - An overview of the problem you investigated
   - A detailed summary of the causes and effects of your problem
   - A discussion of present and future risks posed by the problem
   - An evaluation of current efforts by scientists and engineers to address the problem
   - A proposal with steps or actions that people can or should take to help solve the problem

CHAPTER 16 SUMMARY

16.1 OCEAN GEOGRAPHY AND SEAWATER COMPOSITION

- All of Earth's oceans are connected, so Earth has a single global ocean. Geologically, there are five distinct ocean basins within the global ocean.
- Seawater contains about 3.5 percent dissolved salts. Six common ions form most of the salts in seawater. Seawater also contains dissolved gases.
- Increased absorption of carbon dioxide has led to a 50 percent increase in the acidity of seawater since the early 1990s.
- Except at the poles, the ocean has three temperature layers: a warm upper layer, the thermocline, and a cold bottom layer.

16.2 TIDES AND SEA WAVES

- Tides are caused by the gravitational pull of the moon and the sun. Two high tides and two low tides occur approximately every day.
- Ocean waves develop when wind blows across water. Wave size depends on three interrelated factors: wind speed, length of time wind has blown, and distance traveled over water.
- While a water wave moves horizontally, the water molecules move in small circles called wave orbitals.

16.3 OCEAN CURRENTS

- A current is a continuous flow of water in a particular direction.
- Surface currents are driven by wind and deflected by the Coriolis effect and landmasses.
- Deep-sea currents are driven by thermohaline circulation, or differences in seawater density. Tidal currents are caused by tides.
- Gyres are large systems of rotating surface currents.
- The Great Pacific Garbage Patch is caused by the North Pacific gyre, which concentrates plastic pollution near the center of the gyre.
- Upwelling transports cold water and nutrients from the depths to the surface.

16.4 SEACOASTS

- Most coastal erosion occurs as a result of wave action during intense storms.
- Waves also weather and erode coastlines gradually by hydraulic action, abrasion, dissolution, and salt cracking.
- The bending of a wave is called refraction. Refracted waves often form longshore currents that transport sediment along a shore.
- If land rises or sea level falls, the coastline migrates seaward and old beaches are abandoned above sea level, forming an emergent coastline.
- In contrast, a submergent coastline forms when land sinks or sea level rises.

16.5 BEACHES

- Most coastal sediment is transported to the sea by rivers. Glacial drift, reefs, and local erosion also add sand in certain areas.
- Coastal emergence may expose large amounts of sand.
- Groins or seawalls may upset the natural movement of coastal sediment and alter patterns of erosion and deposition on sandy coastlines.
- A rocky coastline is one without an abundant source of sediment.
- Wave-cut cliffs, sea arches, sea stacks, and fjords are landforms found along rocky coastlines.

16.6 LIFE IN THE SEA

- Most of the photosynthesis and nutrient consumption in the ocean is carried out by plankton.
- Phytoplankton conduct photosynthesis and are the base of the food chain for aquatic animals. Zooplankton are tiny animals that feed on the phytoplankton.
- The shallow water of a continental shelf supports large populations of marine organisms.
- Coral reefs provide ecologically diverse habitat and protection from ocean storms, but climate change and human activities have destroyed at least 10 percent of the world's coral reefs.

CHAPTER 16 ASSESSMENT

Review Key Terms

Select the key term that best fits the definition. Not all terms will be used, and no term will be used more than once.

barrier island	spring tide
beach drift	submergent coastline
Coriolis effect	surf
crest	surface current
current	thermohaline
deep-sea current	circulation
emergent coastline	tidal bulge
eustatic sea-level	tidal current
change groin	tidal range
gyre	tide
longshore current	trough
neap tide	upwelling
ocean acidification	wave base
phytoplankton	wave height
plankton	wave orbital
reef	wavelength
salinity	zooplankton
spit	

1. movement of cold water and nutrients from the ocean depths to the surface
2. the vertical distance between the crest and trough of a wave, also called amplitude
3. chaotic, turbulent waves breaking along a shore
4. a portion of a continent previously underwater that has become exposed as dry land
5. the deepening of ocean water on the sides of Earth closest to and farthest away from the moon at any given time
6. a worldwide rising or falling of ocean waters
7. a continuous flow of ocean water driven by wind and deflected by Earth's rotation
8. small, circular motion of water molecules within a sea wave
9. the total quantity of dissolved salts in a body of water, expressed as a percentage
10. the relatively small tides that occur when the moon is 90° out of alignment with the sun and Earth
11. tiny aquatic animals that feed on microscopic photosynthetic organisms
12. the process whereby the pH of seawater drops as carbon dioxide diffuses from the atmosphere into the ocean
13. a flow of ocean water caused by tides and channeled by islands or by a bay with a narrow entrance
14. larger-than-normal changes in water levels along a coast due to the combined gravitational pull of the moon and sun when they are directly in line with Earth
15. the deflection of ocean currents in response to forces generated as Earth rotates on its axis

Review Key Concepts

Answer each question on a separate sheet of paper to demonstrate your understanding of key concepts from the chapter.

16. What are Earth's five ocean basins, and why are they often collectively referred to as a "global ocean"?
17. How do increasing levels of carbon dioxide in the atmosphere affect populations of coral and shellfish?
18. Compare and contrast spring tides and neap tides. How are they similar and how do they differ?
19. What is the relationship between wind speed and ocean wave height?
20. Out on the open ocean, a wave passes under an unanchored boat. Why doesn't the wave carry the boat along with it all the way to shore?
21. Explain the connection between gyres and the Great Pacific Garbage Patch.
22. If water is too soft to abrade rock, why is abrasion by ocean waves still a major cause of weathering and erosion along coastlines?
23. What is the relationship between ocean temperatures and eustatic sea-level change?
24. Explain what a barrier island is and describe two mechanisms that can cause the formation of barrier islands.
25. How can a farmer's use of fertilizer in the northern Great Plains adversely affect a coral reef in the Gulf of Mexico?
26. How do scientists predict that sea levels will change in the future and how will this change affect humans?

Think Critically

Write a response to each question on a separate sheet of paper. Use concepts from the chapter to support your reasoning.

27. Suppose the massive ice sheet covering Greenland melted, releasing large amounts of fresh water into the sea. How would you expect this to affect thermohaline ocean circulation?

28. Suppose a friend decides to build a house on a sea cliff overlooking the ocean. What concerns would you have about this idea? Explain your thinking.

29. Explain how isostatic changes can give rise to an emergent coastline in one area and to a submergent coastline in another area.

30. A sandy beach is the site of two adjacent beachfront hotels. A longshore current moves past the first hotel, *The Sands of Time*, and then the second hotel, *The Salty Breeze*. The owner of *The Sands of Time* decides to build a groin to control sand erosion on her property. How will this groin affect the sand on the beach at *The Salty Breeze*?

31. Artificial reefs have been constructed using shipwrecked boats, decommissioned oil rigs, old tires, and construction waste materials. Discuss the potential benefits and drawbacks of artificial reefs as a substitute for natural coral reefs.

PERFORMANCE TASK

Using Computational Modeling to Assess Effects of Human Activity on Earth Systems

Earth systems can be difficult to study because of their size and complexity. Because Earth's hydrosphere, atmosphere, cryosphere, geosphere, and biosphere are so interconnected, it can be challenging to set up controlled experiments to test how changes in one variable affect another. Instead, scientists collect data and observations and look for relationships and patterns. They also develop computational models that allow them to study how various Earth systems influence one another and evaluate how human activity is affecting these relationships.

In this performance task, you will conduct research to identify a computational model developed by a scientist to study an Earth system. You will describe and analyze the model and then use it to make predictions about how human activity affects this system.

1. Describe and analyze an Earth system model.

   A Select a topic for your research. The topic should involve relationships among Earth systems and how those relationships are being modified by human activity. Some examples are given below. You may choose one of these examples or you may choose your own topic. Be sure to obtain permission from your teacher for any topic you develop on your own.

   - Ocean acidification and how fossil fuel use is accelerating it
   - Natural erosive forces that impact shorelines and how human attempts to stabilize shorelines against erosion are succeeding or failing
   - Sea-level changes and how human activities leading to global warming are affecting sea levels
   - Natural stability of fish populations and how commercial fishing affects this stability

   B Conduct research to find published results of scientific computational modeling that relate to the topic you chose.

   C Create a poster, video, or slide presentation showing how the computational model predicts outcomes for different scenarios. Scenarios must include conditions that enable you to draw conclusions about how human activity affects relationships within an Earth system. As you prepare your poster or presentation, be sure to include the following:

   - A description of the topic you chose
   - A description of the computational model you researched
   - A summary of the inputs and outputs used by the researchers to make predictions using the model
   - An analysis of the researchers' results from the modeling process

CHAPTER 17
WATER RESOURCES

An aerial photograph shows Hoover Dam on the left side of the image. Lake Mead (center) is the reservoir that resulted from the dam's construction.

The Hoover Dam, pictured on the left side of this image, is situated in the Black Canyon of the Colorado River. The dam was constructed between 1931 and 1936. The purpose of the dam was to create a freshwater reservoir, Lake Mead, from which the arid and semi-arid states of Nevada, California, and Arizona could draw. The white band of mineral deposits above the shoreline shows how much the reservoir level has varied. In 2000, Lake Mead was nearly at full capacity about 373 meters (1,221 feet) above sea level. After nearly two decades of drought, Lake Mead's water level has steadily dropped. This photo shows the lake at approximately 328 meters (1,075 feet) above sea level.

KEY QUESTIONS

17.1 What is water supply and demand?

17.2 In what ways do humans divert water for use?

17.3 Why is groundwater a valuable natural resource?

17.4 What methods can be used to remediate polluted waterways?

EXPLORERS AT WORK

PROTECTING OKAVANGO'S WATERS

WITH NATIONAL GEOGRAPHIC EXPLORER STEVE BOYES

The Okavango Delta is one of the world's largest inland deltas. Unlike a coastal delta, an inland delta forms as a river system splits into multiple streams that spread across a broad area but do not flow into an open body of water. All of the water reaching the Okavango Delta transpires or evaporates within it. During its annual winter flood, the delta's network of waterways spans as much as 20,000 square kilometers (7,700 square miles). Its mix of grassy plains, permanent marshes, and seasonal wetlands provides habitat for thousands of animal species. A million people also live in and around the Okavango. The Okavango watershed is the largest one still intact in Africa and is an invaluable water resource.

In large part due to the efforts of ornithologist Steve Boyes, the Okavango Delta was designated a UNESCO World Heritage site in 2014. The delta itself is now protected, but the rivers that supply most of its water face growing pressures from human activity. Now, Steve Boyes works to ensure their protection, too, through the National Geographic Okavango Wilderness Project (NGOWP).

Most of the water that flows into the Okavango Delta is supplied by two rivers, the Cuito and the Cubango, which originate some 800 kilometers (500 miles) away in the highlands of Angola. They converge to form the larger Okavango River, which passes through Namibia into Botswana. Floodwaters make their way from Angola to the delta through a complex process; water travels for months and arrives in pulses. Without this water, the delta would disappear. "Protecting the Cuito would help secure 45 percent of the delta's water," says Boyes, "while managing the Cubango more sustainably would help protect the rest."

To protect these crucial waterways, cooperation of all three countries is key. Boyes has worked hard to bring stakeholders together. But issues of politics, changing land use, economic priorities, and water security complicate the effort. After decades of civil war, Angola is experiencing rapid development, and more water from the Cubango is being diverted for irrigation, plumbing, and drinking water. Namibia also diverts water from the Okavango River for large-scale agricultural projects. Other industries, such as fishing, agriculture, mining, and even tourism, threaten to pollute the river. The Cuito's remote location means it is still pristine, yet it too is threatened, as human-caused fires and deforestation spur erosion along its banks. Water diversions and decreasing water quality are creating water insecurity for people living downstream.

On December 4, 2018, Boyes and his team celebrated a signing ceremony in which The National Geographic Society and the Angolan Ministries of Environment and Tourism agreed to a four-year partnership to help safeguard the Okavango River Basin headwaters. Through the partnership, the NGOWP and the Angolan government are developing ways to more sustainably manage these vital water resources.

THINKING CRITICALLY

Construct an Argument Consider the various groups of people who have a stake in how the Cuito and Cubango Rivers are managed. Develop an argument from the point of view of one of these stakeholders regarding the importance of having access to these water resources.

▲ Steve Boyes observes wildlife along a stretch of the Okavango Delta. The delta is a habitat for many birds and mammals, including elephants, rhino, and waterfowl.

◀ National Geographic Explorer Steve Boyes studies a map of the Okavango Delta.

CASE STUDY
IS SUSTAINABLE IRRIGATION POSSIBLE?

Before 1848, Los Angeles, California, was a tiny settlement with a population of approximately 1,600. The dry climate, with an average annual precipitation of just 38 centimeters (15 inches), was not naturally well-suited for farming (Figure 17-1). The Los Angeles River frequently flooded in winter, when most of the annual rainfall occurred, but diminished to a trickle in summer. Still, early Mormon settlers managed to produce a wealth of fruits and vegetables using irrigation.

Things changed quickly after the discovery of gold in California. During the California Gold Rush (1848–1855) and the years that followed, a flood of people inundated Los Angeles and the surrounding area. By 1900, the population of Los Angeles had swelled to 100,000 and by 1904, it surpassed 200,000. Unfortunately, the area's water resources simply could not keep up. Wells dried up, and the river was no longer able to supply enough water for residents and farmers.

To quench its insatiable thirst for water, the city turned to a solution 400 kilometers (250 miles) to the northeast, where the Owens River flowed out of the Sierra Nevada. Bit by bit, the city bought up the water rights from farmers and ranchers in the Owens Valley. Then the city spent millions of dollars constructing the Los Angeles Aqueduct to transport water from Owens Valley across 357 kilometers (222 miles) of mountainous, earthquake-prone terrain. On November 5, 1913, the first water poured from the aqueduct into the San Fernando Valley, located just north of Los Angeles.

But the city quickly outgrew this new source of water as well. In the 1920s, the Owens Valley dried up—as did the region's agriculture—and the Owens River could no longer supply the city's needs. By the 1930s, the Los Angeles Department of Water and Power (LADWP) had begun drilling wells into the Owens Valley and pumping its groundwater aquifer dry. Today, some four million residents are currently served by the LADWP. About one-third of the city's water comes from the Los Angeles Aqueduct while more than half comes from the California (Bay Area) and Colorado aqueducts. Groundwater represents only 10 percent of the city's water supply. The Owens Valley is now parched, with almost all of the water once used to irrigate its farms and ranches having been diverted hundreds of miles away.

Rapid population growth is not the only stressor on California's water resources. About 80 percent of California's water serves its robust agricultural industry, which produces more than half of the fruit, vegetables, and nuts grown in the United States. With water in short supply, many farms are modifying their irrigation techniques and taking steps to increase the water-holding capacity of the soil. For example, drip irrigation systems that deliver water directly to a plant's roots use far less water than traditional spray techniques because almost no water is lost to evaporation. Farmers can also conserve water by spreading mulch and compost and planting cover crops on fallow fields. Techniques such as these prevent evaporation as well as erosion and soil compaction, which can reduce the ability of the soil to hold water. With the population of California still growing, residents will need to continue to find ways to make the most of the region's limited water supply.

As You Read Explore the different sources of water on Earth and how humans make use of these sources. Consider the competing needs of consumers, agriculture, industry, and ecosystems for limited water resources. Find out where your water comes from and think about the ways you use water. Investigate steps you can take to limit water use.

FIGURE 17-1
Go online to view a map showing patterns in average precipitation amounts received per year in California.

17.1 WATER SUPPLY AND DEMAND

Core Ideas and Skills
- Distinguish between water use, water withdrawal, and water consumption.
- Describe water usage and water technology systems used in the domestic, industrial, and agricultural sectors.
- Identify regions of the world at risk for insufficient water availability.

KEY TERMS
water scarcity
consumption
withdrawal

More than two-thirds of Earth's surface is covered with water. But most of this water is salty, and most of the fresh water is frozen into the Antarctic and Greenland ice caps. Humans divert half of all the flowing water from rain and snow that falls on the continents, but even that is not enough for our burgeoning population and per capita consumption. Today, we reach—and in some cases exceed—the limits of our water resources.

Fresh Water

Currently about one-fifth of the world's population lives in a region of **water scarcity** in which the aggregate demand by all sectors exceeds availability. Under the existing climate change scenario, the United Nations has projected that by the year 2030, water scarcity will affect half the world's population. It also will displace between 24 million and 700 million people. Water scarcity will increase future levels of human poverty. It also may lead to armed conflict as the competition for the dwindling resource increases.

An important aspect of water scarcity is water quality. In addition to being physically scarce in some regions, much of the surface water and groundwater in industrial areas is heavily polluted. Pollution destroys natural habitats and makes the resource less useful for humans.

Water Withdrawal and Consumption
Rain and snow continuously replenish fresh water on land, so that the amount of available fresh water is about the same as it was two centuries ago, before large-scale colonization of North America. But the demand for water has risen dramatically, until it has approached or exceeded supply in many parts of the world (Figure 17-2). In the early 1800s, there were 1 billion people on Earth. Now there are over 7 billion. As the human population has grown, the amount of available fresh water per person has diminished. In addition, our technological society uses water at rates that would have been inconceivable to Benjamin Franklin.

Water use falls into two categories. Any process that uses water and then returns it to Earth locally is called **withdrawal**. Most of the water used by industry and homes is returned to streams or groundwater reservoirs near the place from which it was taken. For example, river water pumped through an electric generating station to cool the exhaust is returned almost immediately to the river. Water used to flush a toilet in a city is pumped to a sewage treatment plant, purified, and discharged into a nearby stream.

In contrast, a process that uses water and then returns it to Earth far from its source is called **consumption**. Irrigation water evaporates, disperses with the wind, and returns to Earth as precipitation hundreds or thousands of kilometers from its source. Industry accounts for about half of all water withdrawn in the United States. However, agriculture accounts for most of the water that is consumed and not returned to its place of origin (Figure 17-3). Globally, about two-thirds of the total water that is withdrawn is consumed.

Water use (withdrawal plus consumption) is subdivided into domestic, industrial, and agricultural categories. According to the U.S. Geological Survey (USGS), by far the largest users of water in the United States are power plants, which in 2015 accounted for about 41 percent of total water withdrawals, and irrigation for agriculture, which accounts for about 38 percent of total consumption. The specific categories for which the water is used varies by state. The arid western states consume more water for irrigation. The more populous eastern states use more water for thermoelectric power.

The United States consumes a great volume of water by global standards. However, total water withdrawals fell between 2010 and 2015 (Figure 17-4). The drop was primarily caused by decreases in withdrawals for steam-generated power, which accounted for 89 percent of the decrease, and public-supply withdrawals, which

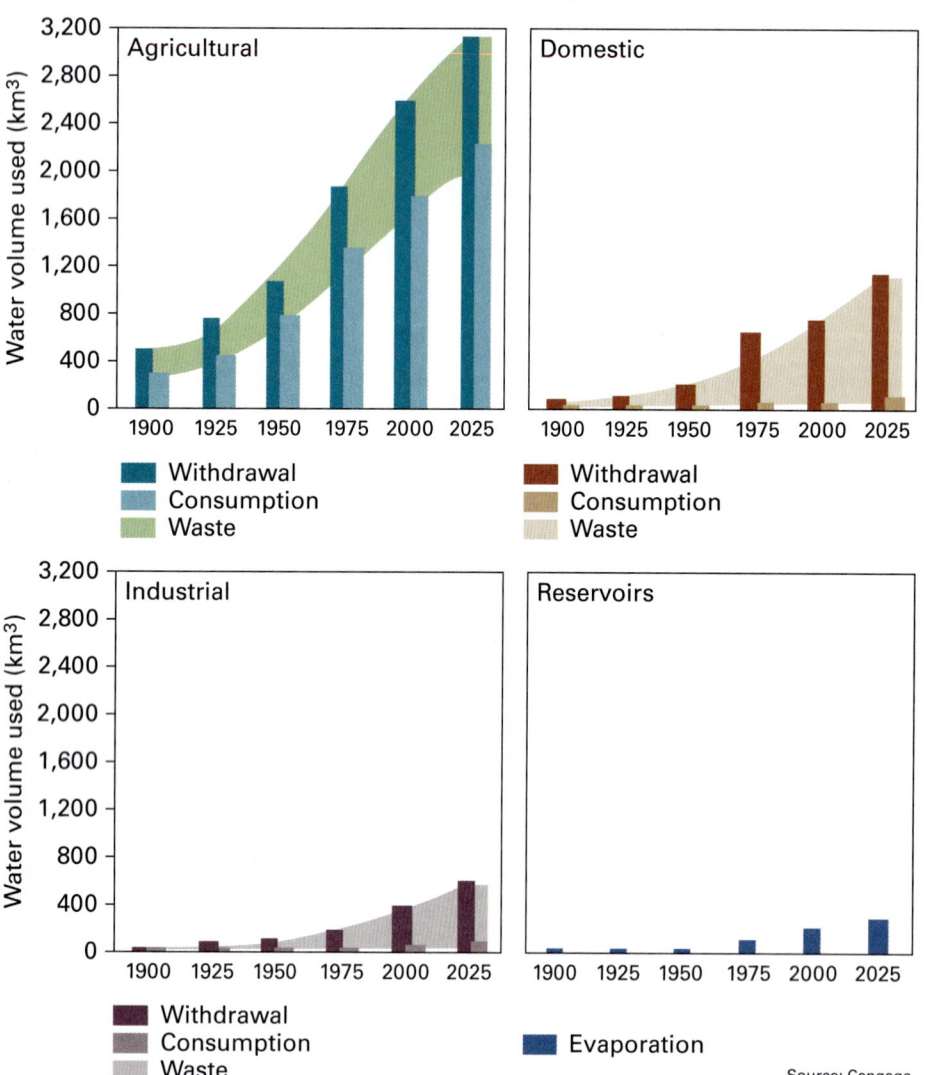

FIGURE 17-2
The graphs show actual and predicted global water use from 1900 to 2025. Between 2000 and 2054, the world's population is expected to increase by 3 billion people, so water demand will continue to rise.

Note: Domestic water consumption in developed countries (500–800 liters per person per day) is about six times greater than in developing countries (60–150 liters per person per day).

DATA ANALYSIS Global Water Use

Over the last century, global water use has increased dramatically along with the growing population of humans on Earth. Between 2000 and 2054, the world's population is expected to increase by another 3 billion people. View Figure 17-2. The graphs in this figure compare water withdrawal, consumption, and evaporation data from 1900 to 2025.

1. **Summarize** Using the graphs in Figure 17-2, summarize global water use in each of the three sectors. Which sector(s) withdraw, consume, and waste the most water?

2. **Compare** Describe the change in the rate of evaporation from reservoirs since 1900. How does this change compare to the trends shown in the first three graphs?

Computational Thinking

3. **Analyze Data** What is the relationship between consumption, withdrawal, and waste? Does it differ by sector? Explain your response by referencing the values and trends in the graphs.

4. **Generalize Patterns** Use the graphs to describe how projected increases in global population will affect global water use.

542 CHAPTER 17 WATER RESOURCES

accounted for another 9 percent of the decrease. Partially offsetting these reductions in water consumption were increases in withdrawals for irrigation and mining.

checkpoint What are the primary categories of water use?

Domestic Water Use

The average American household uses about 1,100 liters (300 gallons) per day. This is many times the amount needed to maintain a healthy life. Average consumption per household varies significantly by region. Households located in arid

FIGURE 17-3
Sprinkler irrigation systems consume water because a significant amount of water evaporates in the air instead of reaching the crops in the field.

FIGURE 17-4
This bar graph shows U.S. population (red line) and changes in groundwater (faint blue) and surface water (light blue) withdrawals from 1950 to 2015.

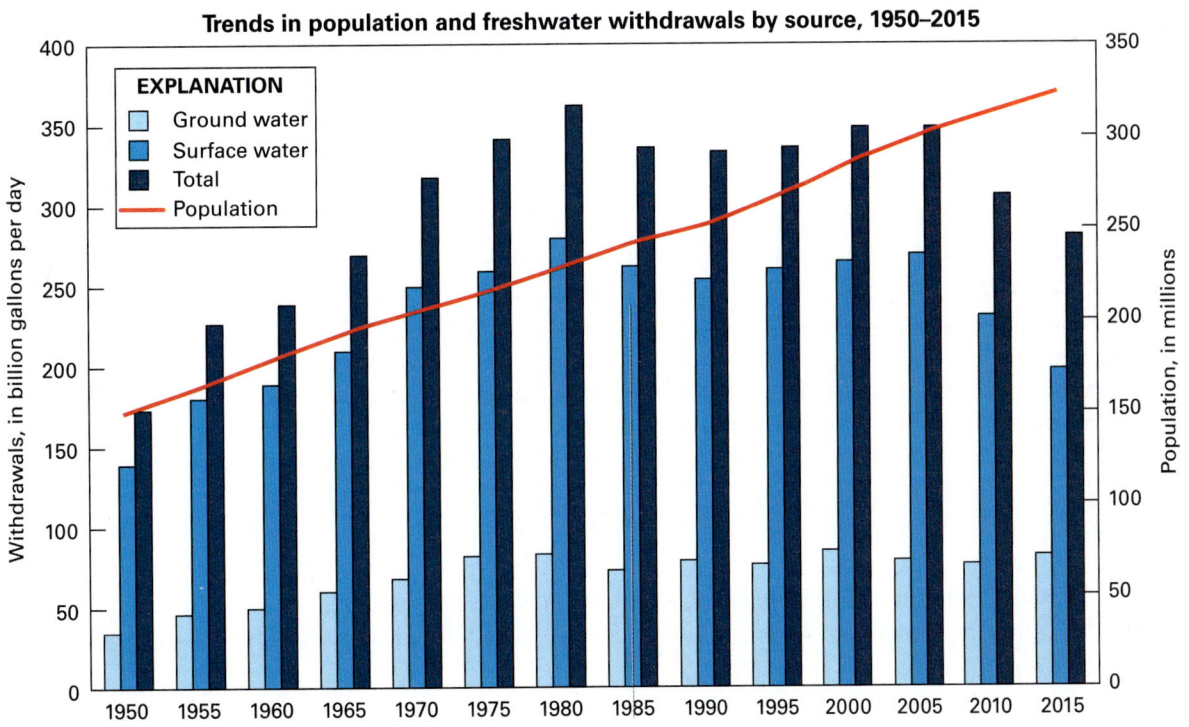

Source: USGS

LESSON 17.1 543

FIGURE 17-5 ▼
This map shows the percent population growth and the average daily water use per person over fifteen years for each state.

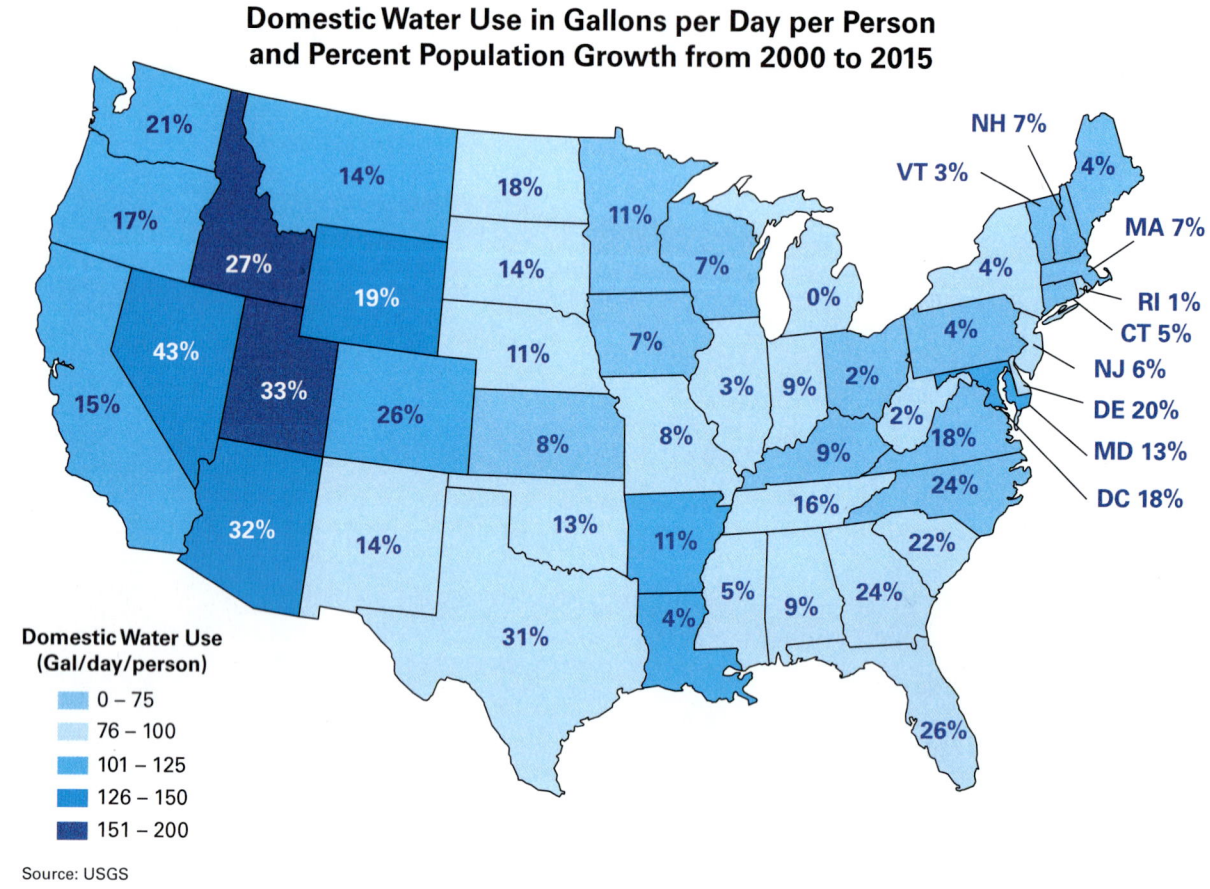

regions commonly use more than twice as much water as households located in humid regions. This is mainly because of differences in landscape irrigation— particularly lawns. These differences in average household water use among the different states are compounded by the fact that many of the states experiencing recent large population increases are also states in which the arid climate leads to higher average household consumption (Figure 17-5).

As the arid western states have become more populated, domestic water conservation practices have become more commonplace. Conservation practices led to an overall decrease in average household consumption between 2005 and 2015. Greater water awareness has reduced the average U.S. household consumption since 2005. However, it is important to keep in mind that domestic use accounts for only about 12 percent of the water used in the United States.

In contrast to the comparatively extravagant per capita water usage in the United States, the World Health Organization defines "reasonable water access" in less-developed agricultural societies as 20 liters (about 5.28 gallons) of sanitary water per person per day. This is assuming that the source is less than 1 kilometer from the home. Yet, 1.1 billion people, about one-sixth of the global population, lack even this small amount of clean water. Water access can fall short for three reasons. Either there isn't enough water, the water is unfit to drink, or the source is so far away that people must carry their water long distances from public wells or streams to their homes. The World Health Organization estimates that diarrheal illnesses caused by unsafe water account for 4.1 percent of global disease. These illnesses are responsible for the deaths of 1.8 million people each year. Most of them are children under age five.

checkpoint What are possible causes of a water access problem?

Industrial and Agricultural Water Use

Industrial Water Use The cooling systems in electric power plants (run by fossil fuels or nuclear power) account for roughly 45 percent of the water used each day in the United States. Although much of this water is returned to the stream from which it was taken, it is considerably warmer. This, as we will see, affects aquatic ecosystems. As a result, in March 2011, the U.S. Environmental Protection Agency issued proposed standards for closed-cycle cooling systems utilizing cooling towers that release heat through evaporation. Whether or not these proposed standards become reality remains to be seen, however. Besides use by power plants, all other industrial needs add an additional 10 percent to the national water demand. The quantities of water required to produce some common industrial products are shown in Figure 17-6.

Agricultural Water Use In 2010, agriculture accounted for 32.5 percent of total water use in the United States and 76 percent of all groundwater withdrawals. Figure 17-7 shows that

⊕ **FIGURE 17-8**
Go online to view a map comparing total water withdrawals by state.

FIGURE 17-6 ▼
Large quantities of water are used to produce common industrial products in the United States.

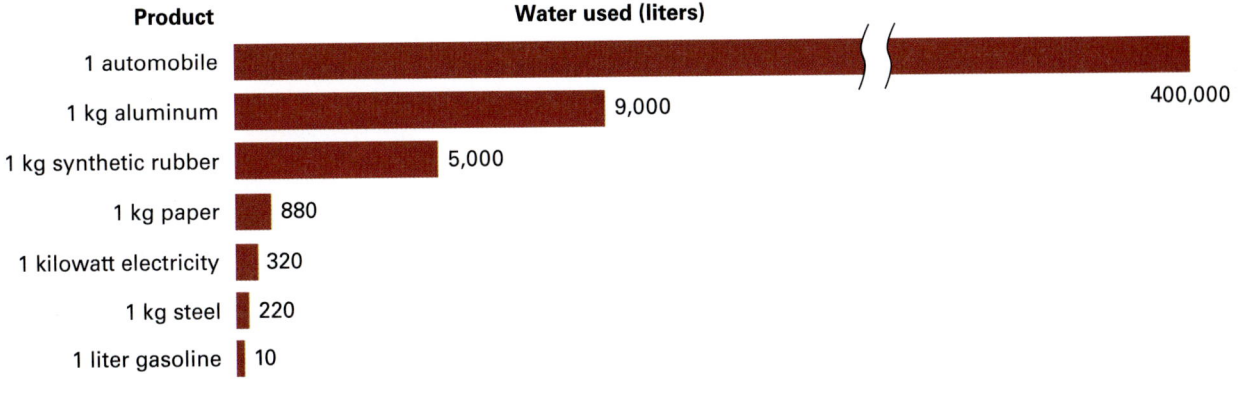

Source: Data compiled from American Water Works Association

FIGURE 17-7 ▼
Different agricultural products require vastly different amounts of irrigation water.

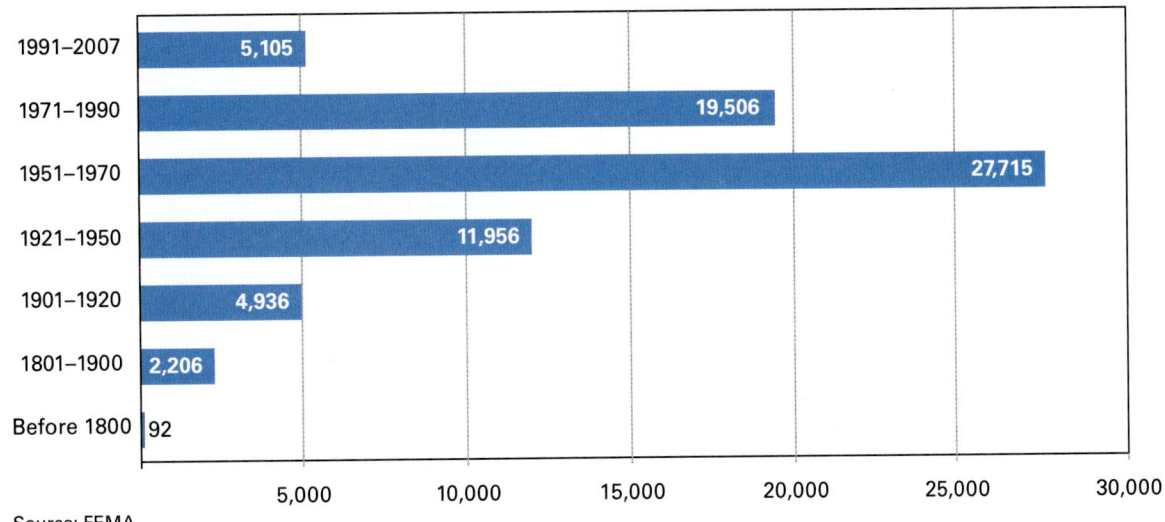

Source: FEMA

LESSON 17.1

FIGURE 17-9 ▼
Regions of the world are assigned a stress ratio based on short- or long-term water availability.

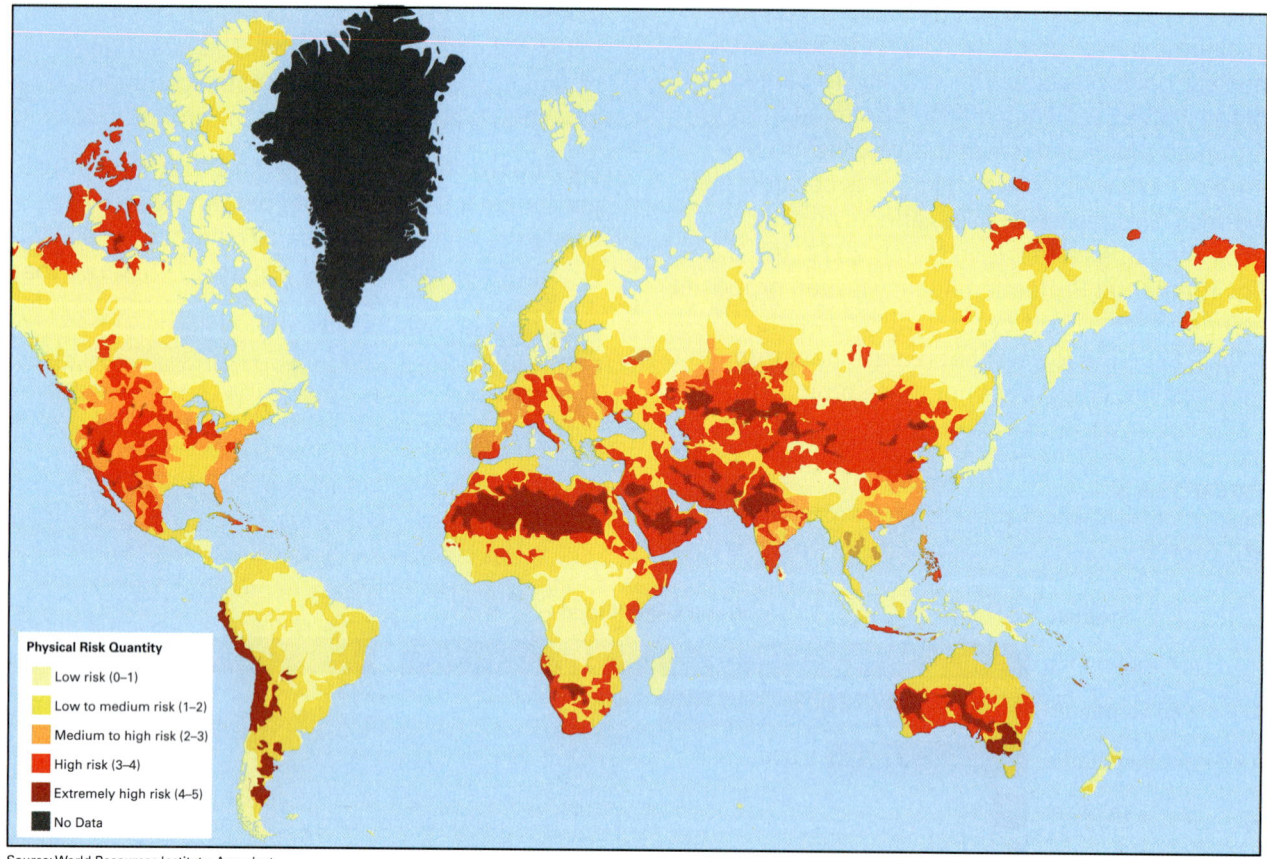

Source: World Resources Institute, Aqueduct

different agricultural products use vastly different amounts of water. California's Central Valley was once a desert. But irrigation has transformed it into an immensely productive region for growing fruits and vegetables. The productivity of the Great Plains in the central United States would decline by one-third to one-half if irrigation were to cease.

Globally, many nations rely on irrigation for more than half of their food production. Estimates vary, but roughly 70 percent of all global water consumption is used to irrigate crops. In dry regions most, if not all, farmland must be irrigated. As a result, both surface and groundwater resources are stressed (Figure 17-9), leading to a higher risk of insufficient water availability.

checkpoint What would happen to the agricultural productivity of the Great Plains without irrigation?

17.1 ASSESSMENT

1. **Distinguish** What is the difference between water withdrawal and water consumption?
2. **Describe** Why is available fresh water about the same now as it was two centuries ago?
3. **Infer** What can you infer based on the statistics in the lesson about one-sixth of the global population who lack access to clean water daily?
4. **Explain Effects** What effect has irrigation had on more naturally arid areas?

Computational Thinking

5. **Recognize Patterns** Examine Figure 17-6 and Figure 17-7 to identify a few products that require the most water consumption. What traits do the products share, if any? How are these products incorporated into everyday life?

17.2 DAMS AND DIVERSION

Core Ideas and Skills
- Identify how and why water is diverted.
- Describe harmful effects of water diversion.

KEY TERMS

diversion system

Surface Water Diversion

Much of Earth's rain falls in the wrong places or at the wrong times to be of much use to humans. For thousands of years, people eased water scarcity by building dams to store water that fell at the wrong time and by diverting water. The Tigris and Euphrates Rivers flow through deserts of the Middle East from distant mountains, but ancient Babylonians farmed the desert by irrigating the parched lowlands with diverted river water. In modern times, engineers build huge dams to store water. In 1950, there were 5,700 large dams in the world. In 2018 there were about 57,000 large dams. Almost half of the new dams were built in China. This unprecedented building boom has increased water availability in many regions and produced large amounts of hydroelectric energy, but these benefits are offset against significant human and environmental costs.

Surface water is scarce in many regions, but huge reservoirs of groundwater are accessible beneath the surface. This groundwater can be pumped to the surface for human use.

The United States receives about three times more water from precipitation than it uses. However, some of the driest regions are those that use the greatest amounts of water (Figure 17-10). For example, some of the most productive agricultural regions are located in the deserts of California, Texas, and Arizona. In these locations, crops are entirely supported by irrigation. Desert cities such as Phoenix, Los Angeles, Albuquerque, and Tucson support large populations that use hundreds of times more water than is locally available and renewable.

A **diversion system** is a pipe or a canal constructed to transport water. Many of these systems use gravity to move the water, while others use pumps to lift water uphill. Diversion systems are often augmented by dams that store water. Dams are especially useful in regions of seasonal rainfall.

FIGURE 17-10
White salts cover a vast expanse of Nevada desert. Salinization has reduced the productivity of millions of hectares of agricultural land globally.

In the arid and semiarid western United States, much of the annual precipitation occurs in winter and spring. Dams store this water for the summer irrigation season, when crops need it. In addition, the potential energy of the water in a reservoir can generate electricity, and because reservoirs can refill with water, hydroelectric power is renewable. In 2016, hydroelectric power from dams accounted for about 16 percent of total global electricity consumption. Therefore, dams are beneficial, but they also can create numerous undesirable effects.

checkpoint Why are certain areas of the United States more in need of irrigation than others?

Effects of Water Diversion

Loss of Water The reservoir formed by a dam provides more surface area for evaporation and more bottom area for seepage into bedrock than did the stream that preceded it. For example, about 270,000 cubic meters of water per year evaporate from Lake Powell, above the Glen Canyon Dam on the Colorado River. Therefore, less total water flows downstream. In addition, many canals built to carry water from a reservoir to farmland are simply ditches dug in soil or bedrock. They leak profusely.

Salinization Streams contain ions derived from weathering. When farmers use river water to irrigate cropland, the water evaporates, leaving the ions behind to form various salts in the soil. If desert or semidesert soils are irrigated for long periods of time, salinization occurs. In the United States, salinization lowers crop yields on 25 to 30 percent of irrigated farmland, more than 50,000 square kilometers, an area roughly the size of Vermont and New Hampshire combined. The problem is particularly acute in the San Joaquin and Imperial Valleys of Southern California. Globally, salinization affects between 10 and 20 percent of irrigated land, some 250,000 square kilometers or about the size of Wyoming. Areas of salinization are increasing at a rate of roughly 6 percent per year.

Silting A stream deposits its sediment in a reservoir, where the current slows. Rates of sediment accumulation in reservoirs vary with the sediment load of the dammed river. Lake Mead, shown in the Chapter Opener, lost 6 percent of its capacity in its first 35 years as a result of sediment accumulation. At such a rate, a reservoir lasts for hundreds of years, providing a considerable societal resource in terms of power generation, water storage, and recreation. However, in a few other instances, engineers have made expensive miscalculations of sedimentation rates that have greatly shortened the lifespan of reservoirs. The Tarbela Dam in Pakistan was completed in 1978 after nine years of construction. By 2005, the dam was already more than 25 percent filled with silt. The reservoir behind the Sanmenxia Dam on the Yellow River in China filled four years after the dam was finished, making both the dam and the reservoir useless.

Erosion Sediment trapping by dams can significantly affect the ecology of areas downstream that had received sediment during pre-dam floods. A classic example of the way dams affect the movement of sediment has occurred in the Colorado River within the Grand Canyon, located directly below the Glen Canyon Dam. Once the dam was built in 1963, all of the sandy sediment being transported downstream in the Colorado River was deposited in Glen Canyon Reservoir. The only sand delivered to the Colorado within the Grand Canyon came from tributaries that entered the river below the dam. Between 1963 and 1991, this reduction in sand delivery caused sandbars and beaches within the Grand Canyon to erode. This reduced the availability of prime camping spots within the park. It also greatly limited the availability of shallow, slow-flowing aquatic environments critical to the survival of fish. In addition, the lack of floods promoted the widespread establishment of one type of *Tamarix*, a non-native plant species. Growth of this plant species further disrupted the river ecosystem.

In the spring of 1996, scientists released enough water from Glen Canyon Dam to create a flood in the Grand Canyon. Water was released again in the fall of 2004 and spring of 2008. These high flow experiments were designed to mimic the effects of pre-dam flooding on the Colorado River and investigate whether such floods could help rebuild sandbars, benefit aquatic organisms, and limit the spread of *Tamarix*. The first flood in 1996 eroded sediment from existing sandbars and moved it to higher elevations. But it did not move much sand from the river bed to the sandbars as was hoped.

In response, the 2004 and 2008 floods were strategically timed to coincide with tributary floods that brought "new" sand into the Colorado. These floods succeeded in moving much of the newly delivered sand from tributaries onto the banks and sandbars in the Colorado, causing them to

FIGURE 17-11
The Glen Canyon Dam on the Colorado River enabled an invasive plant species to flourish along the riverbank. The non-native plant has negatively impacted the river ecosystem and is now widespread across many southwestern rivers.

enlarge. Research conducted on long-term study sites in the Grand Canyon showed that 75 percent of the sandbars grew over the course of the three artificial floods. In addition, the spring releases in 1996 and 2008 showed that these floods helped reduce the spread of *Tamarix*. The floods partially removed existing plants. The floods also reduced the establishment of new *Tamarix* seedlings, because the spring releases came before the plants formed seeds. Lastly, the spring floods were found to increase the population of small aquatic invertebrates, especially insects, that are an important food source for fish. The increase in shallow-water, slow-flowing environments and the rise in aquatic invertebrate population helped increase the population of rainbow trout below the dam by 800 percent.

A reservoir accumulates silt and sand. It also interrupts the supply of sediment to the seacoast, causing some coastal beaches to erode faster than new sand can be delivered by natural processes. This problem has led to expensive sand replenishment projects on some popular beaches in the eastern United States.

Risk of Disaster A dam can break, creating a disaster in the downstream floodplain. Throughout U.S. history, hundreds of dams have failed, primarily due to foundation defects, cracking and other structural failures, and inadequate maintenance. Costly property damage and devastating environmental impacts are also accompanied the loss of life. One famous event occurred in 1976 at the Teton Dam in Idaho. The dam failed because of deficiencies in the design of the dam foundation, lack of a low-fill spillway, and rapid initial filling of the reservoir. In particular, the Teton Dam was built on a foundation of highly permeable bedrock. This bedrock was insufficiently filled with grout prior to construction of the dam. Additionally, one of the dam abutments was not sealed against the bedrock.

Recreational and Aesthetic Impacts Dams are often built across narrow canyons to minimize engineering and construction costs. But when the canyon upstream of the dam is flooded, unique scenery and ecosystems are destroyed. Glen Canyon Dam on the Colorado River created Lake Powell by flooding one of the most spectacular desert canyons in the American West. Today, Glen Canyon is submerged for more than 300 kilometers above the dam. It can be seen only in old photos. Although these environments are now effectively gone, the creation of Lake Powell provides new opportunities for fishing and other water sports that did not exist prior to construction of Glen Canyon Dam.

Disputes often arise among various groups that use water in dammed reservoirs. To prevent flooding, a reservoir should be nearly empty by spring so that it can store water and fill during spring runoff. A reservoir should be drawn down slowly during summer and fall. Agricultural users also prefer to have water stored during spring runoff and supplied for irrigation during summer months. However, recreational users object to a lowering of reservoir levels during summer, when they visit reservoirs most frequently. Managers of hydroelectric dams prefer to run water through their turbines during times of peak demand for electricity. This rarely corresponds to flood-management, irrigation, or recreational schedules.

Ecological Disruptions A river is an integral part of the ecosystem that it flows through. When the river is altered, the ecosystem changes. Dams interrupt water flow during portions of the year, prevent flooding, and change the temperature of the downstream water. Dams often create unnatural daily fluctuations in stream flow. All of these changes alter populations of aquatic and riverbank species and may create specific ecological problems in individual rivers.

For example, before Egypt's Aswan Dam was built, the Nile River flooded every spring. These floods deposited nutrient-rich sediment over the floodplain and delta. When the dam and reservoir cut off the silt supply, however, erosion of the delta front by the sea caused the Nile delta to shrink. In addition, the loss of an annual supply of nutrients eliminated a source of free fertilizer for floodplain farms. The energy produced by the dam is used to manufacture commercial fertilizers. But many of the poorer farmers on the floodplain cannot afford to purchase what the river once provided for free.

Human Costs In the past 50 years, as many as 80 million people globally have been forcibly removed from their homes and farms to make way for dams and reservoirs. In many regions, waterborne diseases become more common surrounding stagnant water held behind a dam than they were near free-flowing streams that existed prior to the dam.

Construction of the Three Gorges Dam in China on the Yangtze River took 18 years to complete (Figure 17-12). It's the largest hydroelectric dam in the world. Its construction forced the relocation of about 1.4 million people. Roughly 300,000 more than originally anticipated were relocated because the rise and fall of the immense reservoir has triggered landslides along its banks. The slower flow has also caused harmful pollutants to accumulate.

Dam Removal and Hydroelectric Power Dams continue to provide an indispensable source of hydroelectric power along with water for agriculture and recreation. However, the societal risks and environmental costs of dams has significantly reduced the rate of new dam construction in the United States over the past 50 years. In fact, hundreds of dams have been removed in the United States in recent years. The dams were removed in part because of the expense required to maintain the aging structures and in part because of the rapidly growing interest in river restoration.

Few new dams are being constructed in the United States while the removal of dams is accelerating. However, this trend is largely restricted to the United States. Elsewhere around the globe, a major new initiative in hydroelectric power generation is underway in an effort to bring

FIGURE 17-12
The Three Gorges Dam was built to generate hydroelectric power and mitigate the effects of devastating floods. The dam and the reservoir are a source of significant controversy due to the forced relocation of the local population and increases in river pollution.

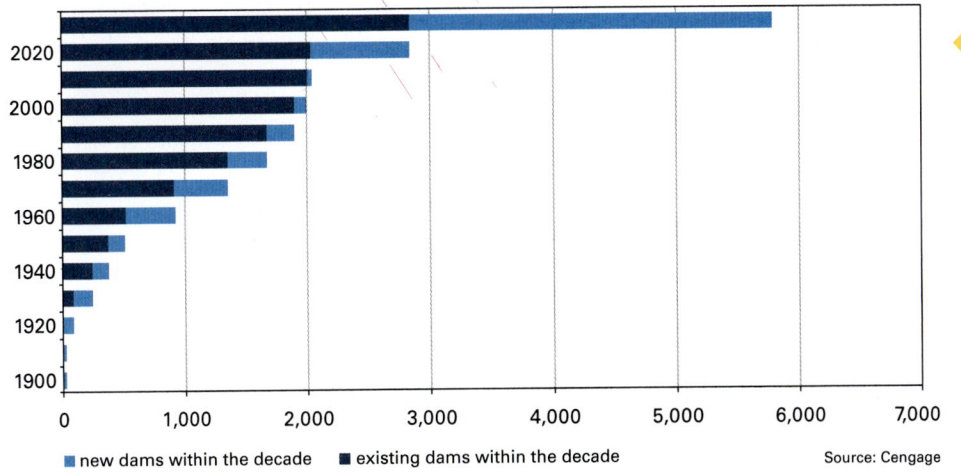

▸ **FIGURE 17-13**
A bar graph shows that construction of hydroelectric dams worldwide is accelerating and expected to nearly double in the next ten years.

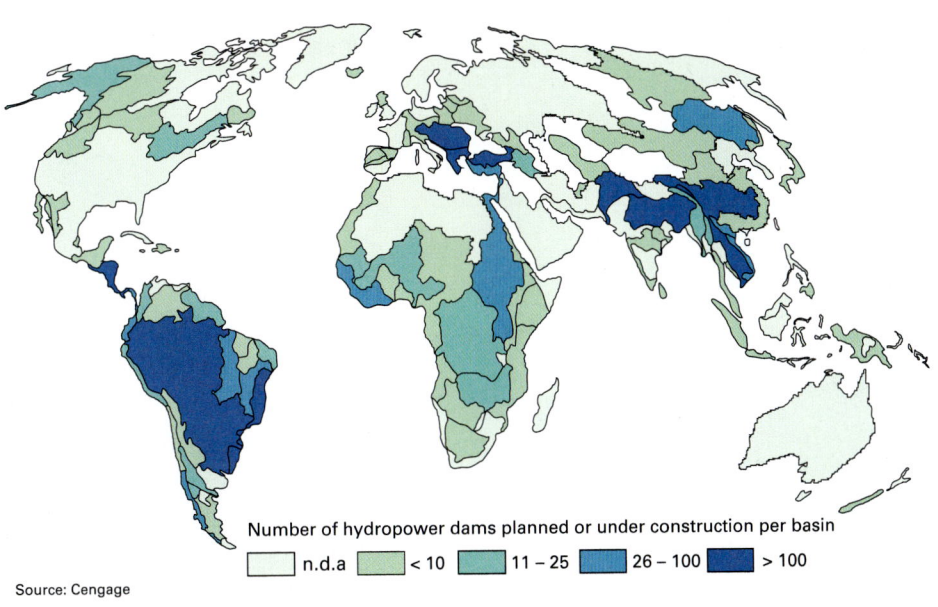

▸ **FIGURE 17-14**
This map shows the projections for hydroelectric dam projects across the globe.

reliable electricity to more people (Figures 17-13 and 17-14).

New dams are projected to increase the global capacity of hydroelectric power by 73 percent or more. But even this increase in capacity will fail to keep up with projected increase in electricity demand. The construction of new dams outside the United States is projected to reduce the number of large, free-flowing rivers by 21 percent and significantly reduce freshwater biodiversity. The construction of large dams for hydropower historically has been viewed as a sign of human progress. But the benefits are offset and may be outweighed by the negative impacts of dams on river ecosystems, the displacement of humans, and the alteration of regional water balances across watersheds.

checkpoint What caused the relocation of people in the area of the Three Gorges Dam?

17.2 ASSESSMENT

1. **Summarize** What is surface water diversion, and why has it been useful to different groups of people across time?
2. **Explain** Why does irrigation commonly lead to salinization?
3. **Summarize** What are the possible issues that can result from silting?
4. **Relate** Connect the need for hydroelectric power and the construction of new dams. How does energy consumption influence water as a resource?

Critical Thinking

5. **Critique** Design a set of guidelines or questions that could be used to review the effectiveness of dams by considering several qualities including loss of water, salinization, silting, and erosion.

LESSON 17.2 **551**

FIGURE 17-15
ON ASSIGNMENT This aerial view of the Colorado River Delta in Sonora, Mexico, was taken in 2009 by National Geographic photographer Pete McBride. The photo shows a completely dry river bed where fishing was once possible. The native people in this area once relied on the river as a food and income source.

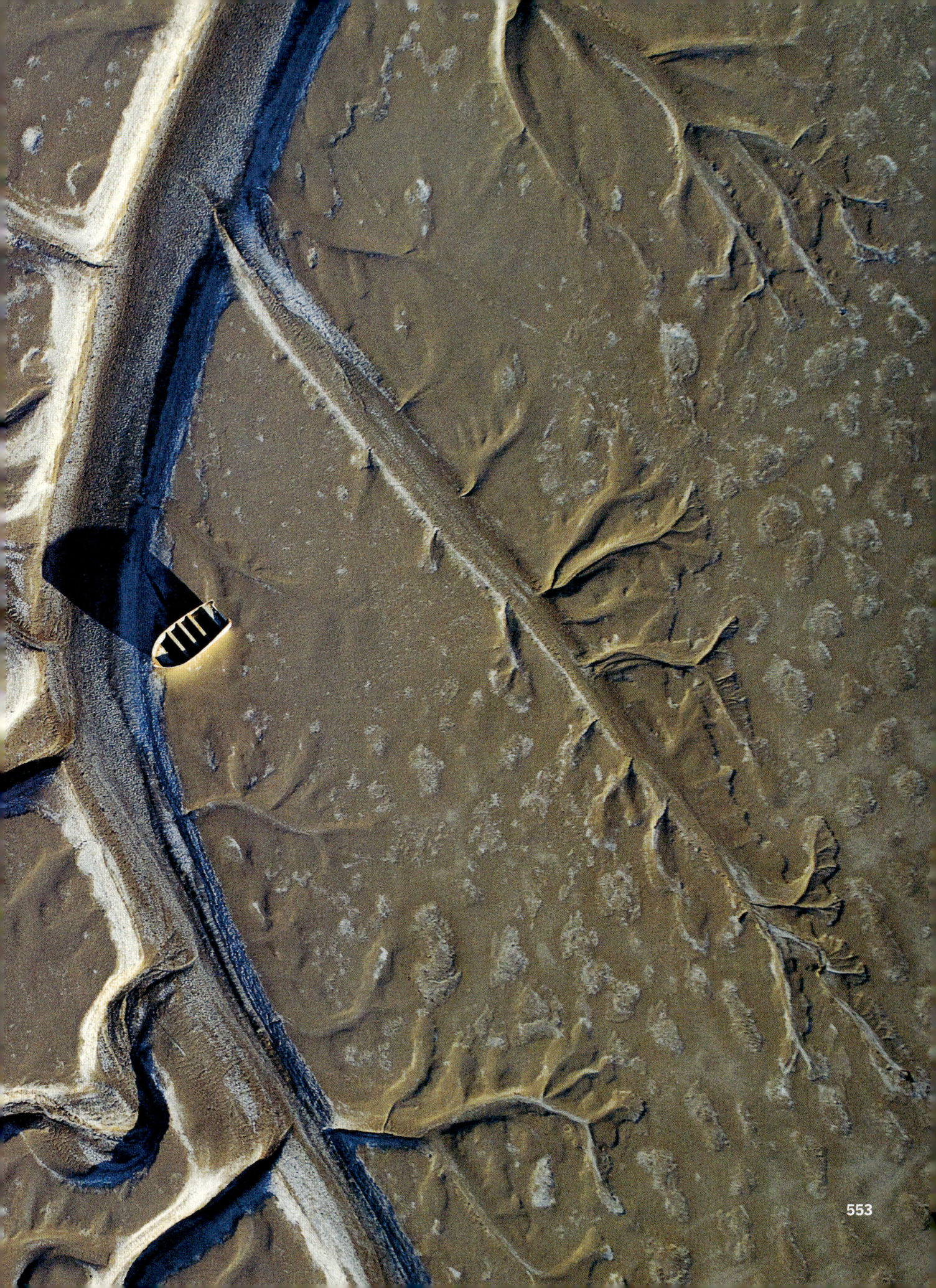

17.3 GROUNDWATER DIVERSION AND DEPLETION

Core Ideas and Skills
- Examine groundwater as a source of fresh water.
- Explain the effects of technology on groundwater reservoirs.
- Summarize projections for aquifer depletion, such as with the Ogallala aquifer.

KEY TERMS

cone of depression   saltwater intrusion
subsidence

Groundwater

Groundwater provides drinking water for more than half the population of North America and is a major source of water for irrigation and industry. Groundwater is a valuable resource because:

1. It is abundant; 60 times more fresh water exists underground than in streams and lakes combined.
2. It moves very slowly and remains available during dry periods.
3. In some regions, groundwater flows from wet environments to arid ones, providing a source of water in dry areas.

If groundwater is pumped to the surface faster than it can flow through the aquifer to the well, the water table forms a **cone of depression** surrounding the well. When the pump is turned off, groundwater flows back toward the well, and the cone of depression disappears. Typically, this occurs in a matter of days or weeks, if the aquifer has good permeability. Conversely, if water is continuously removed more rapidly than it can flow to the well through the aquifer, the water table drops (Figure 17-16).

Before the development of advanced drilling and pumping technologies, the human impact on groundwater was minimal. Today, however, deep wells and high-speed pumps can extract groundwater more rapidly than it is recharged. Where such excessive pumping is practiced, the water table falls as the groundwater reservoir becomes depleted. In some cases, the aquifer is no longer able to supply enough water to support the farms or cities that have overexploited it.

Ogallala Aquifer The Ogallala aquifer extends from the Rocky Mountains eastward from Texas to South Dakota (Figure 17-17). It consists of water-saturated porous sandstone and conglomerate shed off the southern Rocky Mountains and deposited between 6 million and 2 million years ago. The Ogallala is the world's largest known aquifer and has been heavily exploited. Today, the Ogallala aquifer supplies roughly 30 percent of all groundwater pumped for irrigation in the United States.

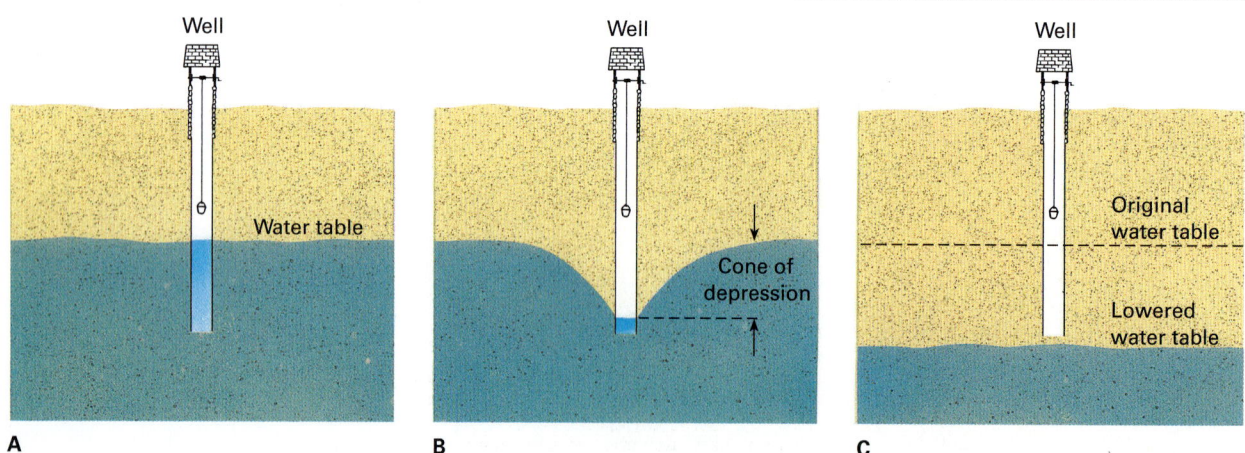

FIGURE 17-16
(A) A well is dug or drilled into an aquifer. (B) A cone of depression forms because water is withdrawn from the well at a faster rate than it can flow through the aquifer to the well and replenish the water removed. (C) If water continues to be extracted at the same rate, the water table falls.

Prior to development, the total volume of the Ogallala aquifer was about 3,800 cubic kilometers of water. Most of the aquifer's water accumulated when the last Pleistocene ice sheet melted, about 15,000 years ago. The aquifer is mostly recharged by rain and snowmelt from the southern Rocky Mountains, hundreds of kilometers to the west.

As is the case for most groundwater sources in the United States, the rate of consumption of groundwater from the Ogallala aquifer is greater than the rate of recharge (Figure 17-18).

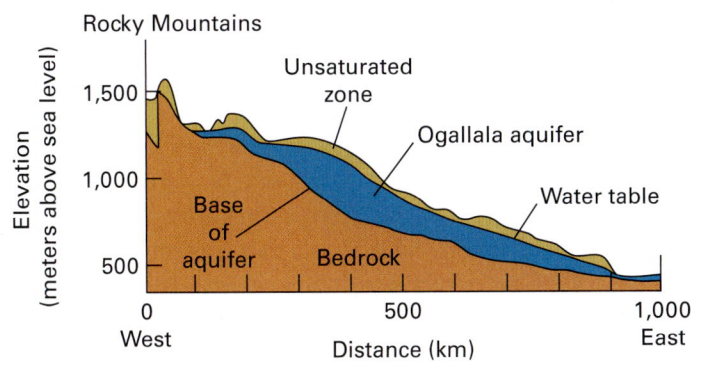

FIGURE 17-17
A cross-sectional view of the Ogallala aquifer shows that it extends from the Rocky Mountains eastward as groundwater beneath the High Plains.

FIGURE 17-18 ▼
A map shows groundwater depletion across the United States. The cross-hatched pattern in Nebraska, Colorado, and Kansas indicate multiple separate aquifers are being depleted. The Ogallala aquifer is located in the center of the map, stretching from Nebraska down through Texas. The numbers on the map report the volume of water used in excess of the recharge amount, in cubic kilometers.

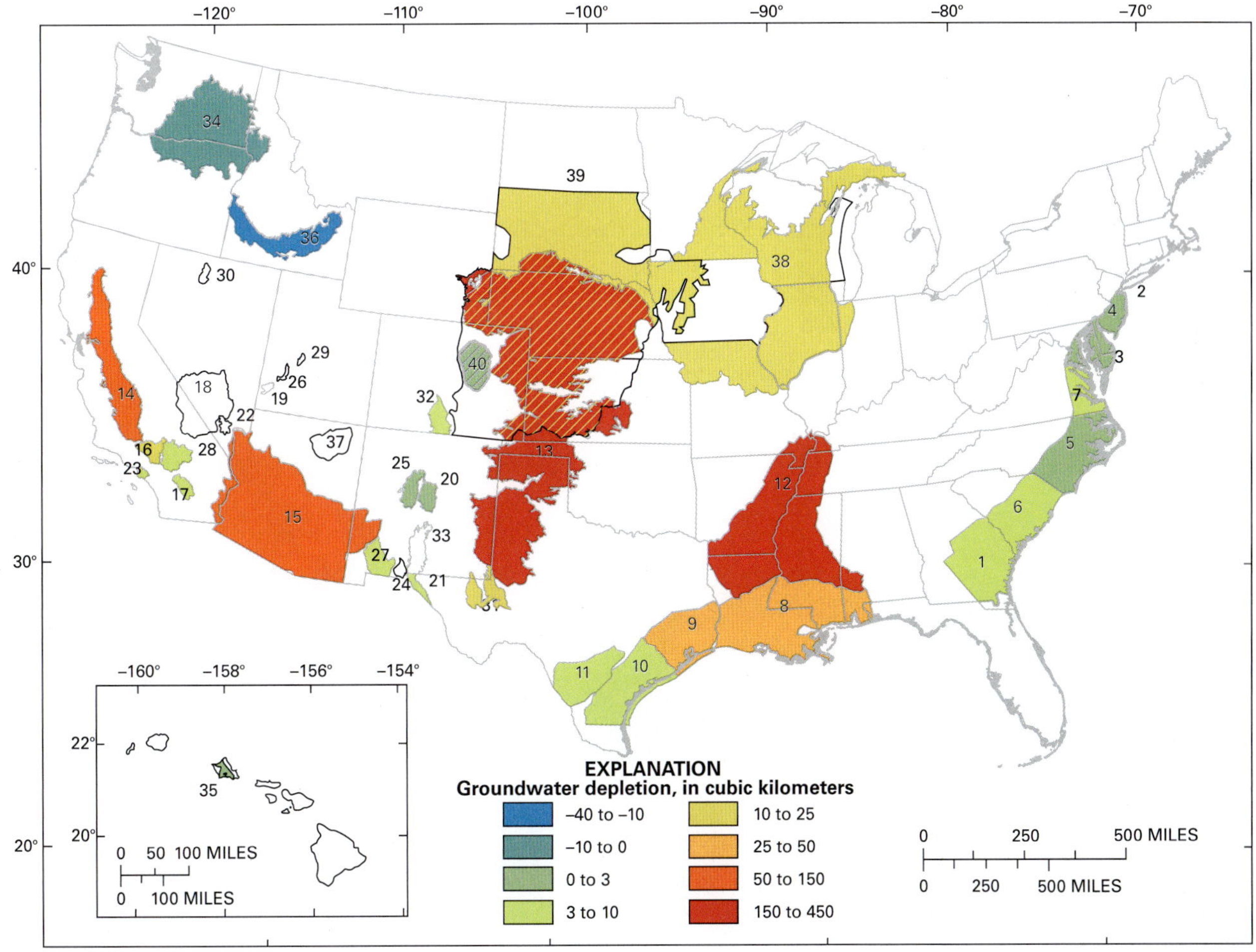

EXPLANATION
Groundwater depletion, in cubic kilometers
- −40 to −10
- −10 to 0
- 0 to 3
- 3 to 10
- 10 to 25
- 25 to 50
- 50 to 150
- 150 to 450

Source: Cengage

LESSON 17.3

Additionally, water moves very slowly through the aquifer, at an average rate of only about 15 meters per year. At this rate, groundwater takes 60,000 years to travel from the southern Rocky Mountains to the eastern edge of the aquifer. Presently, approximately 9.6 billion liters are returned to the aquifer every year through rainfall and groundwater flow. However, farmers remove nearly 10 times this volume annually. In some of the drier regions of northwest Texas and west central Kansas, wells drilled into the Ogallala aquifer are already depleted. If the present pattern of water use continues, part of the Ogallala aquifer will be consumed by the end of the century and agricultural productivity will decline by 80 percent.

Subsidence Excessive removal of groundwater can cause **subsidence**, the sinking or settling of Earth's surface. When water is withdrawn from an aquifer, pore spaces that had been filled with water collapse permanently. As a result, the volume of the aquifer decreases and the overlying ground subsides. In the aquifer underlying the San Joaquin Valley of California, the estimated reduction in water storage capacity by groundwater withdrawal is equal to about 40 percent of the total capacity of all surface reservoirs in the state.

Subsidence Excessive removal of groundwater can cause **subsidence**, the sinking or settling of Earth's surface. When water is withdrawn from an aquifer, pore spaces that had been filled with water collapse permanently. As a result, the volume of the aquifer decreases and the overlying ground subsides. In the aquifer underlying the San Joaquin Valley of California, the estimated reduction in water storage capacity by groundwater withdrawal is equal to about 40 percent of the total capacity of all surface reservoirs in the state.

Subsidence rates can reach 5 to 10 centimeters per year, depending on the rate of pumping and the nature of the aquifer. Some areas in the San Joaquin Valley have sunk approximately 2.6 meters (8.6 feet) over the past fifty years (Figure 17-19). The land surface has subsided by as much as 3 meters in the Houston–Galveston area of Texas. Subsidence problems can be particularly severe in cities. For example, Mexico City is built on an old lake bottom. Over the years, as the weight of buildings and roadways has increased and much of the groundwater has been removed, parts of the city have settled as much as 8.5 meters. Many millions of dollars have been spent to maintain the city on its unstable base. Similar problems are affecting Phoenix, Arizona, and other U.S. cities.

FIGURE 17-19
A representative from the USGS stands at the Mariposa Bypass Bridge in San Joaquin Valley, California. The marker shows land subsidence between 1965 and 2016.

Unfortunately, subsidence is irreversible. Once groundwater is withdrawn and an aquifer compacts, porosity is permanently lost. And groundwater reserves cannot be completely recharged, even if water becomes abundant again.

Saltwater Intrusion Two types of groundwater occur in coastal areas: fresh groundwater and salty groundwater that seeps in from the sea. Fresh water floats on top of salty water because it is less dense. If too much fresh water is removed from an aquifer, salty groundwater that is unfit for drinking, irrigation, or industrial use rises to the level of wells (Figure 17-20). **Saltwater intrusion** has affected much of south Florida's coastal groundwater reservoirs. Once an area is contaminated by saltwater intrusion, it dramatically reduces freshwater storage in that aquifer.

checkpoint How has drilling technology caused a negative impact on groundwater?

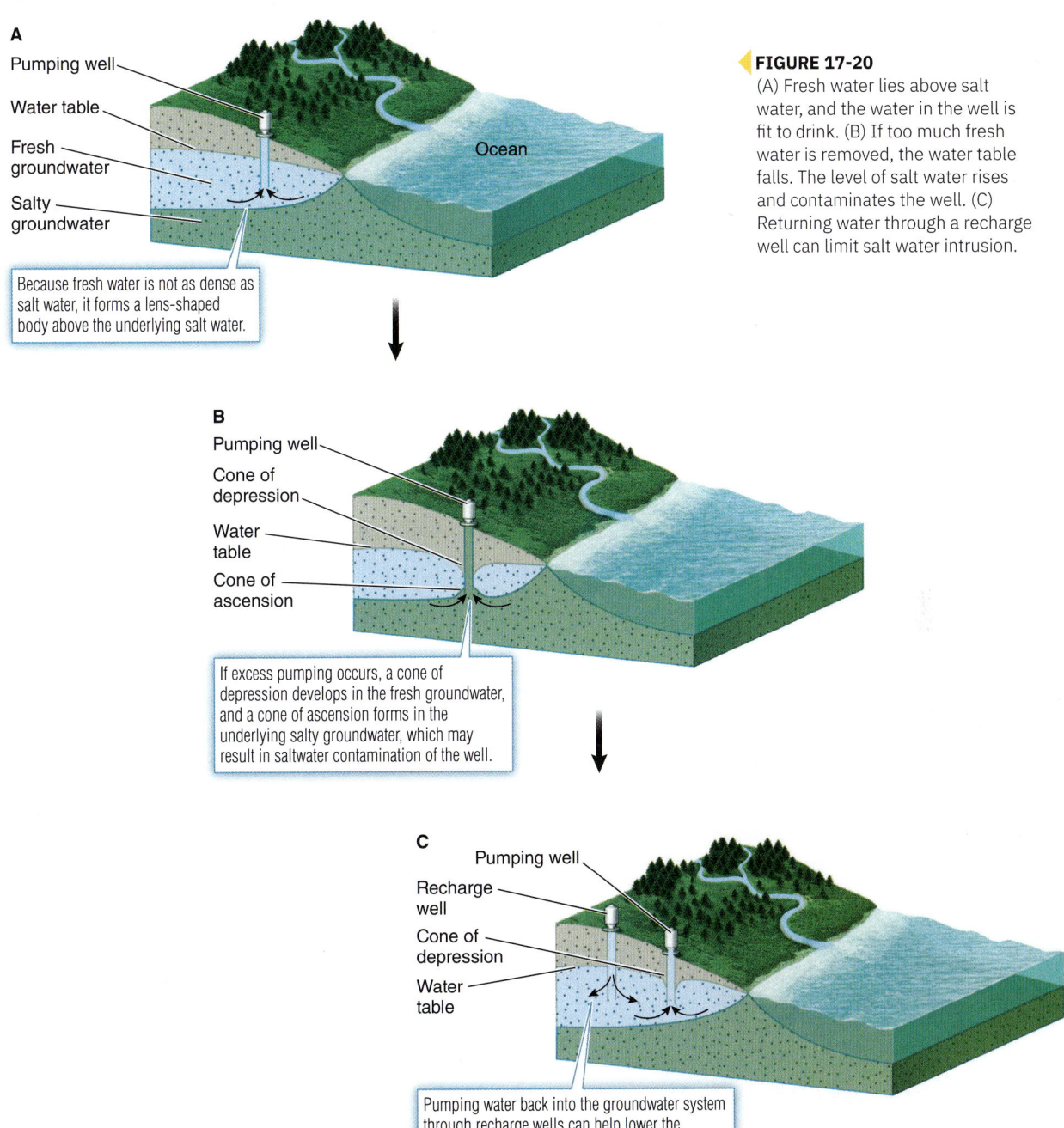

FIGURE 17-20
(A) Fresh water lies above salt water, and the water in the well is fit to drink. (B) If too much fresh water is removed, the water table falls. The level of salt water rises and contaminates the well. (C) Returning water through a recharge well can limit salt water intrusion.

LESSON 17.3

The Great American Desert

After the Civil War, an American geologist, John Wesley Powell, explored the area between the western mountains (the Sierra Nevada–Cascade Crest) and the 100th meridian. The 100th meridian is the line of longitude running through the Dakotas, Nebraska, Kansas, and Texas. Powell observed that most of this region is arid or semiarid, and he called it the Great American Desert. Despite Powell's warning about inadequate water, many Americans settled there. Parts of this region now use huge amounts of water for cities, industry, and agriculture, although they receive little precipitation. Some desert cities, including Phoenix, El Paso, Reno, and Las Vegas, rely on diversion of surface water and/or overpumping of groundwater for nearly all of their water.

Billions of dollars have been spent on projects to divert rivers and pump groundwater to serve this area of the country. In California alone, the two largest irrigation projects in the world have been built, along with 1,400 dams. More than 80 percent of the water diverted to Southern California is used to irrigate desert and near-desert cropland. The water supplied to these farms commonly irrigates crops such as cotton, rice, and alfalfa. Cotton fields require 17,500 liters of water per kilogram of cotton product. Rice and alfalfa also use great volumes of water. The irrigation water used to support cattle ranching and other livestock in California is enough to supply the needs of the entire human population of the state. Meanwhile, the U.S. government pays large subsidies to farmers in naturally wet southeastern and south-central states not to grow those same crops. Controversy over water diversion projects is not limited to the American West. New York City obtains much of its water from reservoirs in upstate New York. The New York City Bureau of Water Supply controls those reservoirs and watersheds hundreds of miles from the city. Opposing interests of local land-use groups and city water users have resulted in heated conflicts in the state.

The Colorado River

The Colorado River runs through the southwestern portion of the Great American Desert. Starting from the snowy mountains of Colorado, Wyoming, Utah, and New Mexico. There, it flows across the arid Colorado Plateau southward into Mexico, where it empties into the Gulf of California.

The river flows through a desert, so farmers, ranchers, cities, and industrial users along the entire length of the river compete for rights to use the water. In the 1920s, the Colorado River discharged 18 billion cubic meters of water per year into the Gulf of California. In 1922, the U.S. government apportioned 9 billion cubic meters for the river's upper basin (Colorado, Utah, Wyoming, and New Mexico) and the remaining 9 billion cubic meters for lower basin users in Arizona, California, and Nevada. And so, all the water was allocated. None of it was set aside for maintenance of the river ecosystem or for Mexican users south of the border. No water flowed into the sea.

In 1944 an international treaty awarded Mexico 1.5 billion cubic meters of water from the Colorado River. This was less than one tenth of the annual volume that had been flowing into the Gulf of California 20 years earlier. Half of this amount was to be taken from the Upper Colorado River Basin, and half from the Lower Basin. In 2001 the U.S. government passed legislation specifying how water surpluses within the Colorado River Basin should be shared among the states involved. But the legislation did not set aside any surplus water for Mexico. These collective actions created hard feelings among Mexicans living along the lower Colorado River toward U.S. management of river water north of the border. Finally, in 2012, the U.S. government passed legislation that included Mexico among the recipients of surplus Colorado River water stored in U.S. reservoirs. In return, Mexico agreed to cut down on its use of Colorado River water during periods of drought.

In addition to the challenge of equitably providing enough water from the Colorado River to sustain the environment and meet human demand on both sides of the international border is the challenge of maintaining the quality of water along the length of the river. The Colorado and many of its tributaries flow across sedimentary rocks. Many of these rocks contain soluble salt deposits. As a result, the Colorado is a naturally salty river. In addition, the U.S. government built 10 large dams on the Colorado, and evaporation from the reservoirs has further concentrated the salty water.

In 1961 the water flowing into Mexico contained 27,000 parts per million dissolved salts, compared with an average salinity of about 100 ppm for large rivers. For comparison, 27,000 ppm is 77 percent as salty as seawater. Mexican farmers used this water for irrigation, and their crops died. In 1973 the U.S. government built a desalinization plant to reduce the salinity of Mexico's share of the water.

There are also important questions about the possibility of a long-term drought in the Southwest. Tree ring studies have shown that more extreme and extended droughts have occurred in the past several centuries than have been experienced in modern times. With current demands on the Colorado River increasing, even a minor drought can have severe consequences. The end of 2004 marked five consecutive years of lower-than-average rainfall and decline in reservoir levels. In January 2005, Lake Powell was at 35 percent of its capacity, the lowest level since 1969, when the reservoir was being filled. Some of this water has since been replenished. But as of August 2018 the reservoir was only at 48.3 percent of capacity. Inflows over the past water year were only about 50 percent of normal, whereas outflows required by treaty were about 96 percent of normal. Clearly, maintaining a reliable supply of water for the growing population of the Colorado River Basin is one of the most challenging public policy issues of the Southwest. View Videos 17-2 and 17-3 for Colorado River footage and storytelling by National Geographic photographer, Pete McBride.

checkpoint Describe why water diversion for agriculture in southern California is necessary.

17.3 ASSESSMENT

1. **Describe** What causes a water table to drop?
2. **Summarize** What is subsidence, and why is it irreversible?
3. **Explain** Why is a coastal area susceptible to saltwater intrusion when the aquifer is overused as a source of fresh water?
4. **Conclude** Considering the different groups of people (citizens, farmers, ranchers, industries) and needs along the Colorado River, what can you conclude about water resources and use in this area? Include an example of an effect of the water usage from the Colorado River.

Computational Thinking

5. **Represent Data** Use the information about the Ogallala aquifer and the estimate of water consumption by 2020 to sketch a simple graph. The graph should show the usage based on the rates given in the lesson.

17.4 WATER POLLUTION

Core Ideas and Skills
- Compare types of water pollutants.
- Explain causes of groundwater pollution.
- Describe remediation methods for contaminated aquifers.

KEY TERMS

water pollution
biodegradable pollutant
nonbiodegradable pollutant
point source pollution
nonpoint source pollution
plume of contamination
bioremediation
chemical remediation
persistent bioaccumulative and toxic chemical (PBT)

While consumption reduces the quantity of water in a stream, lake, or groundwater reserve, **water pollution** is the reduction of the quality of the water by the introduction of impurities. A polluted stream may run full to the banks, yet it may be toxic to aquatic wildlife and unfit for human use.

Anti-Pollution Policies

In the early days of the industrial revolution, factories and sewage lines dumped untreated wastes into rivers. As the population grew and industry expanded, many U.S. waterways became fetid, foul smelling, and unhealthy. The first sewage treatment plant in the United States was built in Washington, D.C., in 1889, more than a hundred years after the Revolutionary War. Soon other cities followed suit, but few laws regulated industrial waste discharge.

In 1952 and 1969 the Cuyahoga River in Cleveland, Ohio, was so heavily polluted with oil and industrial chemicals that it caught fire. The flames and smoke spread across the water (Figure 17-21). The highly publicized photographs of the flaming river flowing through downtown Cleveland had a powerful visual impact. However, other water pollution disasters were insidious, such as that of Love Canal in New York State.

These and similar incidents made it clear that water pollution was endangering health and reducing the quality of life for people throughout the United States. As a result, Congress passed the Clean Water Act in 1972. The legislation states that:

FIGURE 17-21
Firemen attempt to put out a fire started from an oil slick on the polluted Cuyahoga River in 1952.

1. It is the national goal that the discharge of pollutants into the navigable waters be eliminated by 1985.

2. It is the national goal that, wherever attainable, an interim goal of water quality which provides for the protection and propagation of fish, shellfish, and wildlife and provides for recreation in and on the water be achieved by July 1, 1983.

3. It is the national policy that the discharge of toxic pollutants in toxic amounts be prohibited.

checkpoint Why was passing a law about discharge of waste necessary?

In 2004, the EPA reported impaired water quality for 44 percent of the nation's assessed streams, 64 percent of assessed lakes, and 30 percent of assessed estuaries. (An estuary is a former river valley that has been inundated by post-glacial sea-level rise. Chesapeake Bay is a good example.) A 2007 EPA study found that 49 percent of U.S. lakes contain fish that are contaminated with mercury. These fish are considered dangerous for human consumption. By 2008, the EPA listed more than 8,700 water bodies in 43 states, the District of Columbia and Puerto Rico, that exceeded the mercury-concentration limits established by the Clean Water Act.

Types of Pollutants

Water pollutants have different levels of toxicity to humans and the environment. In addition, some pollutants are easier and less expensive to remove or stabilize than others, depending on the physical and chemical nature of the pollutant itself. Of particular importance is whether or not a pollutant can be removed through the metabolic actions of bacteria. Bacteria either already exist or else are introduced into the polluted water as part of a cleanup operation.

Biodegradable Pollutants A biodegradable material is one that decays naturally, being consumed or destroyed in a reasonable amount of time by organisms that live in soil and water. **Biodegradable pollutants** are pollutants that will decay naturally in a reasonable amount of time. The contaminants are consumed or destroyed by organisms that live in soil or water. Examples include the following:

1. Sewage is wastewater from toilets and other household drains. It includes biodegradable organic material such as human and food wastes, soaps, and detergents. Sewage also includes some nonbiodegradable chemicals because people often flush paints, solvents, pesticides, and other chemicals down the drain.

2. Disease organisms, such as typhoid and cholera, are carried into waterways in the sewage of infected people.

3. Phosphates and nitrate fertilizers flow into surface water and groundwater mainly from agricultural runoff. Phosphate detergents as well as phosphates and nitrates from feedlots also fall into this category.

Nonbiodegradable Pollutants Many materials are not decomposed naturally by environmental chemicals or consumed by decay organisms, and therefore they are **nonbiodegradable pollutants**. Not all of these are poisons, but some are. Nonbiodegradable poisons are called **persistent bioaccumulative toxic chemicals (PBTs)**. The EPA recognizes tens of thousands of PBTs. Yet even nontoxic materials can pollute and adversely affect ecosystems. Nonbiodegradable pollutants can be roughly subdivided into three broad categories:

1. Many organic industrial compounds are particularly troublesome because they are toxic in high doses, are suspected carcinogens in low doses, and can survive for decades in aquatic systems. Examples include some pesticides; dioxin; and polychlorinated biphenyl compounds, commonly referred to as PCBs.

2. Toxic inorganic compounds include mine wastes, road salt, and dissolved ions such as those of cadmium, arsenic, and lead.

3. When sediment enters surface waters, it neither fertilizes nor poisons the aquatic system. However, the sediment muddies streams and buries aquatic habitats, thus degrading the quality of an ecosystem.

Radioactive Materials Radioactive materials include wastes from nuclear power plants, nuclear weapons, the mining of radioactive ores, and medical and scientific applications.

Heat Heat can also pollute water. In a fossil fuel or nuclear electric generator, an energy source (coal, oil, gas, or nuclear fuel) boils water to form steam. The steam runs a generator to produce electricity. The exhaust steam must then be cooled to maintain efficient operation.

The cheapest cooling agent is often river or ocean water. A 1,000-megawatt power plant heats 10 million liters of water by 35°C every hour. The warm water can kill fish directly, affect their reproductive cycles, and change the aquatic ecosystems to eliminate some native species and favor the introduction of new species. For example, if water is heated, indigenous cold-water species such as trout will die out and be replaced by carp or other warm-water fish.

Any of these categories of pollutants can be released in either of two ways. **Point source pollution** arises from a specific site such as a septic tank, a gasoline spill, or a factory. In contrast, **nonpoint source pollution** is generated over a broad area. Fertilizer and pesticide runoff from lawns and farms falls into this latter category. Figure 17-22 examines both types of pollution sources.

checkpoint What is one way that heat can pollute water?

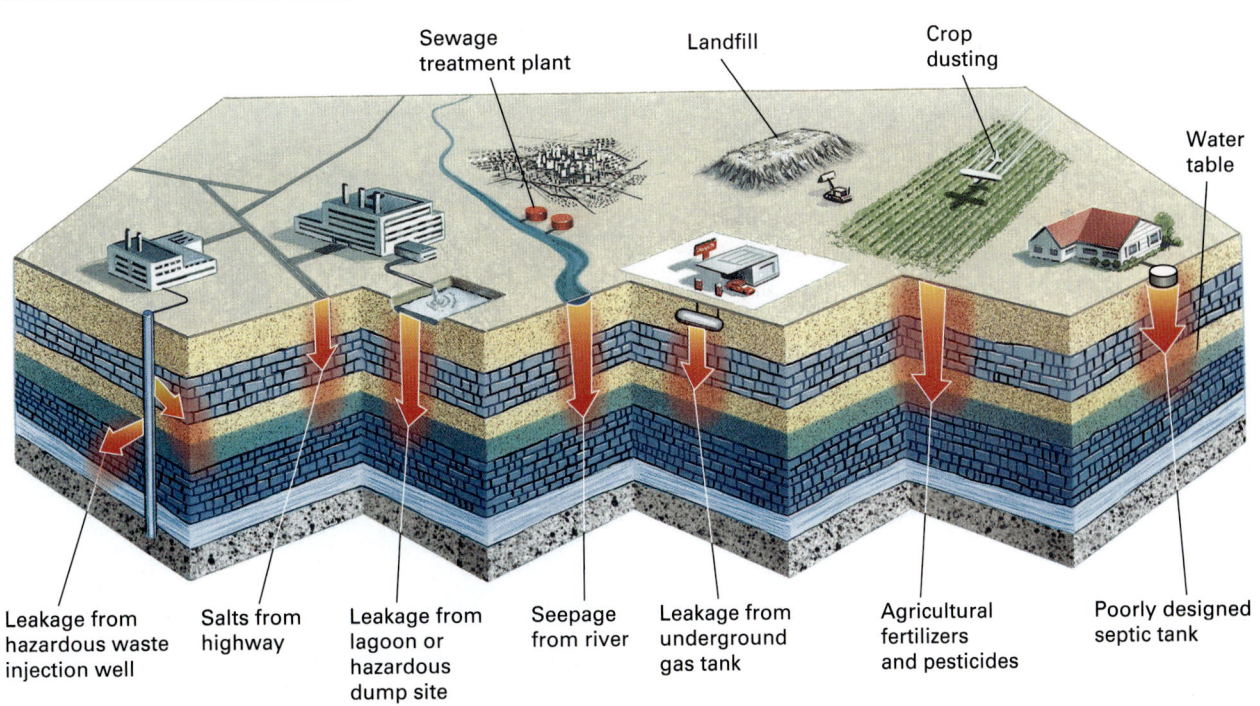

FIGURE 17-22
Pollutants from many different sources contaminate groundwater.

Polluted Waterways

An oligotrophic stream or lake contains few nutrients and clear, nearly pure water. Sewage, detergents, and agricultural fertilizers are plant nutrients. They do not poison aquatic systems; they nourish them. If humans dump one or more of these nutrients into an oligotrophic waterway, aquatic organisms multiply so quickly that they consume most of the dissolved oxygen. Many fish, such as trout and salmon, die. Other organisms, such as algae, carp, and water worms, can proliferate. Therefore, an increased supply of nutrients can transform a clear, sparkling stream or lake into a slime-covered, eutrophic waterway (Figure 17-23).

If humans release even more sewage, detergent, or fertilizer, decay organisms multiply so rapidly that they consume all the oxygen. As a result, most aquatic life dies. Then anaerobic bacteria (bacteria that live without oxygen) take over. These bacteria release noxious hydrogen sulfide gas that bubbles to the surface and smells like rotten eggs.

When raw sewage consisting of human waste is discharged into a stream, decay organisms immediately begin to break it down. At the same time, the current carries the sewage and decay organisms downstream. Over time, the organisms consume the waste and eventually die off when their food supply is consumed. Natural turbulence replenishes oxygen in the water. Then aerobic organisms become reestablished. In this way, the river cleanses itself. However, if the original source of sewage is too large, or if every town along the river dumps sewage into the same waterway, then these natural cleaning mechanisms become overwhelmed and the river remains polluted all the way to the sea. Moreover, many sewage treatment plants ultimately release the treated water directly into nearby rivers. Just because water is treated in a sewage plant does not mean that all pollutants are completely removed. And the release of treated water into surface water systems can still significantly degrade water quality.

checkpoint Why does an increase in supply of nutrients cause harm in a waterway?

Toxic Pollutants, Risk Assessment, and Cost–Benefit Analysis

Human body waste can be broken down naturally over a period of days or weeks. But many organic compounds, such as the banned pesticide DDT, persist for decades. And heavy metal contaminants such as arsenic or lead never decompose. These heavy metals may change chemical form and become benign, but the atoms never decompose.

Environmental laws in developed nations regulate the discharge of pollutants into waterways. Yet pollutants continue to enter streams and rivers. Even in situations where waterways are contaminated with very small amounts of toxic nonbiodegradable compounds, many different pollutants may be found in a single river. And even small concentrations may cause cancer or birth defects.

Scientists do not know the doses at which many compounds become harmful to human health. The uncertainty results in part from the delayed effects of some chemicals. For example, a contaminant might increase the risk of a certain type of cancer. But the cancer may not develop until 10 to 20 years after exposure. Scientists cannot perform direct toxicity experiments on humans, and they cannot wait decades to observe results. So scientists frequently attempt to assess the carcinogenic properties of a contaminant by feeding it to laboratory rats. If rats are fed very high doses, sometimes hundreds of thousands of times more concentrated than environmental pollutants, they may then contract cancer in weeks or months—rather than in years or decades. Suppose that a rat gets cancer after drinking the equivalent of 100,000 glasses of polluted well water each day for a few weeks. Can we say that the same contaminant will cause cancer in humans who drink 10 glasses a day for 20 years? No one knows. It may or may not

FIGURE 17-23
A eutrophic river supports surface algae blooms that negatively impact the aquatic ecosystem.

MINILAB Point Source Pollution Models

Materials
clear plastic food storage container
colored sugar (as used for baking decoration)
sand
small rocks
watering can
timer or stopwatch
water

Procedure
1. Construct a model of an aquifer by filling a plastic container with a layer of rocks, followed by a layer of sand on top of the rocks. Shape the layers so that they slope toward one end of the container.

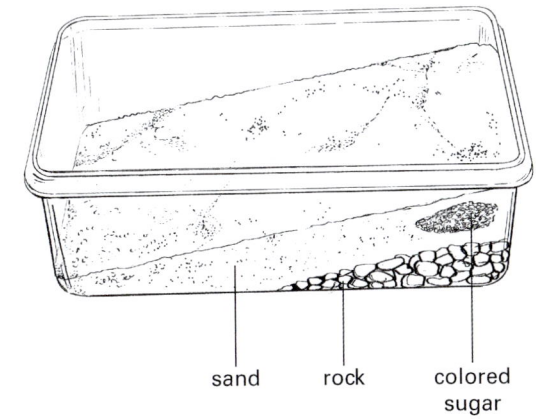

2. Add a point source of pollution to your model by burying a few tablespoons of colored sugar within the sand, as shown in the illustration.
3. Fill the watering can with room temperature tap water. Start the timer and begin sprinkling water over the sand. Observe the sand and sugar and record how long it takes for the colored sugar to dissolve and enter the rock layer.

Results and Analysis
1. **Observe** Describe how the pollution dispersed through the sand and water. What does the pattern model?
2. **Identify** What is a real-world example of a point source of pollution? Describe how it is similar to the model.
3. **Predict** Based on the results of your model, suggest a method that might prevent point source pollution from contaminating an aquifer.

Critical Thinking
4. **Critique** Describe one way in which the model differs from real-world examples of groundwater pollution.
5. **Design** Describe how you could use a similar model to represent a nonpoint source of pollution, such as fertilized farmland.

---

be legitimate to extrapolate from high doses to low doses and from one species to another.

Scientists also use epidemiological studies to assess the risk of a pollutant. For example, if the drinking water in a city is contaminated with a pesticide and a high proportion of people in the city develop an otherwise rare disease, then the scientists may infer that the pesticide caused the disease. They may conclude its presence in drinking water is a high level of risk to human health.

Because neither laboratory nor epidemiological studies can prove that low doses of a pollutant are harmful to humans, scientists are faced with a question: Should businesses and governments spend money to clean up the pollutant? Some argue that such expenditure is unnecessary until we can prove that the contaminant is harmful. Others call for the precautionary principle that it's better to be safe than sorry.

Pollution control is expensive. However, pollution is also expensive. If a contaminant causes people to sicken, the cost to society can be measured in medical bills and loss of income resulting from missed work. Many contaminants damage structures, crops, and livestock. People in polluted areas also bear the economic repercussions of diminished tourism and land values when people no longer want to visit or live in a contaminated area.

checkpoint Why is it difficult to prove that a contaminant in a water source is harmful to human health?

Groundwater Pollution
Water in a sponge saturates tiny pores and passages. To clean a contaminant from a sponge, it would be necessary to clean every pore of the sponge that came into contact with contaminated water. Because this is nearly impossible to achieve completely, no one would wash dishes with a sponge that had been used to clean a toilet.

Many different types of sources contaminate groundwater. These contaminants seep through an

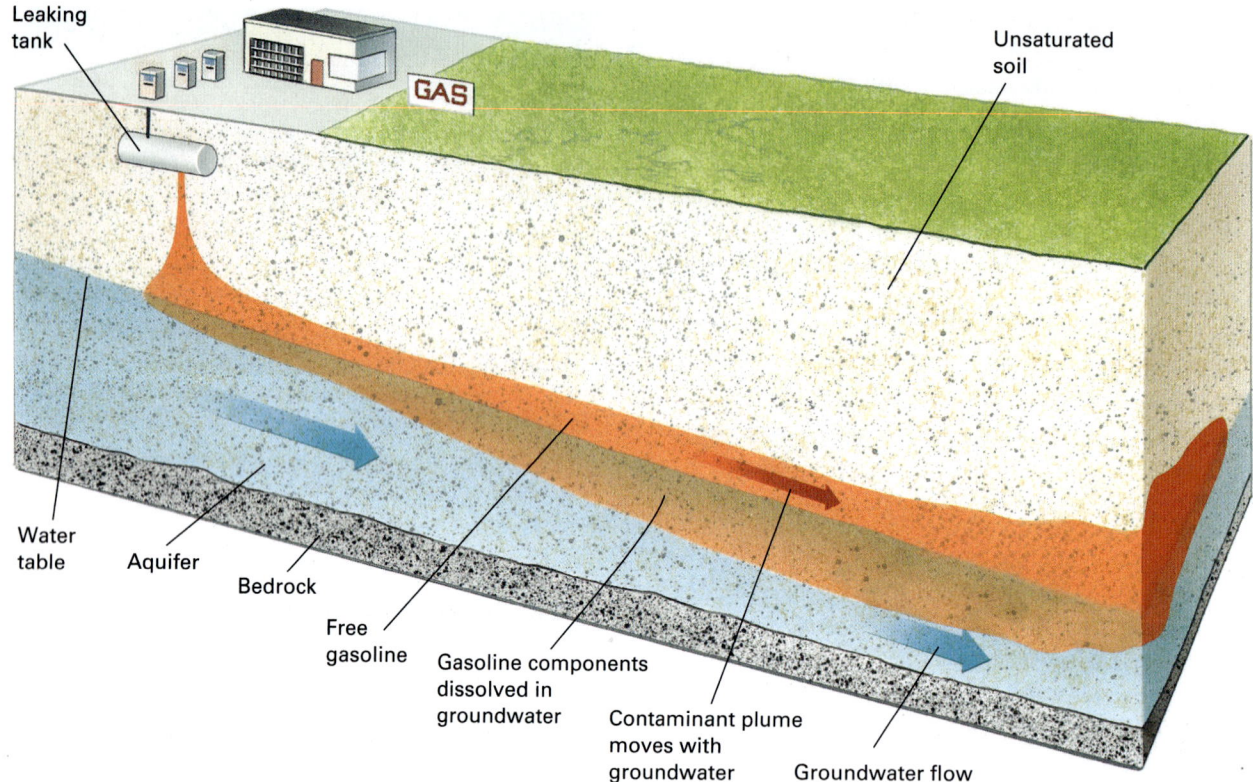

FIGURE 17-24
Pollutants disperse as a plume of contamination. Gasoline and other contaminants are less dense than water. As a result, they float and spread out on top of the water table. Soluble components may dissolve and migrate with groundwater.

aquifer in a manner analogous to the way in which contaminated water saturates a sponge. So the removal of a contaminant from an aquifer is both difficult and expensive. Once a pollutant enters an aquifer, the natural flow of groundwater disperses it, forming a **plume of contamination** (Figure 17-24). Because groundwater flows slowly, usually at a few centimeters per day, the plume spreads slowly.

Most contaminants persist in a groundwater aquifer for much longer than in a stream or lake. The rapid flow of water through streams and lakes replenishes their water quickly. But groundwater flushes much more slowly (Table 17-1). In addition, oxygen, which reacts to decompose many contaminants, is less abundant in groundwater than in surface water, thereby allowing the contaminants to persist for years. For example, in 2006 a 25,000 gallon spill of regular unleaded gasoline occurred from underground tanks at an Exxon gas station in Jacksonville, Maryland. The cleanup operation lasted for more than five years. It involved the drilling of 284 test wells to a depth of up to 187 meters (615 feet). Over 77 million gallons of polluted groundwater were treated.

Because groundwater pollution can persist for long periods of time, the following situations exist despite decades of government concern and action:

- Up to 25 percent of the usable groundwater in the United States is contaminated. About 45 percent of municipal groundwater supplies in the United States are contaminated with organic chemicals.

TABLE 17-1 Average Residence Times for Water in Various Reservoirs

Atmosphere	8 days
Rivers	16 days
Soil moisture	75 days
Seasonal snow cover	145 days
Glaciers	40 years
Large lakes	100 years
Shallow groundwater	200 years
Deep groundwater	10,000 years

- The gasoline additive Methyl *tert*-butyl ether (MTBE) is a volatile, flammable, liquid. MBTE is a frequent and widespread contaminant of underground drinking water. According to the EPA, 5 to 10 percent of drinking water in areas using reformulated gasoline shows MTBE contamination.
- Wells in 38 states contain pesticide levels high enough to threaten human health. Every major aquifer in New Jersey is contaminated.
- In Florida, where 92 percent of the population drinks groundwater, more than a thousand wells have been closed because of contamination. And more than 90 percent of the remaining wells have detectable levels of industrial or agricultural chemicals.
- In 2003, the EPA reported 436,494 confirmed releases of dangerous volatile organic compounds leaking from underground fuel storage tanks in the United States.

checkpoint If a pollutant enters an aquifer, in what manner does it disperse?

Treating a Contaminated Aquifer

The treatment, or remediation, of a contaminated aquifer commonly occurs in a series of steps:

Eliminating the Source The first step in treating an aquifer is to eliminate the pollution source so additional contaminants do not escape. If an underground tank is leaking, the remaining liquid in the tank can be pumped out and the tank dug from the ground. If a factory is discharging toxic chemicals into the groundwater, courts may issue an injunction ordering the factory to stop the discharge.

Elimination of the source prevents additional pollutants from entering the groundwater. But it does not solve the problem posed by the pollutants that have already escaped. For example, if a buried gasoline tank has leaked slowly for years, many thousands of gallons of gas may have contaminated the underlying aquifer. Once the tank has been dug up and the source eliminated, people must deal with the gasoline in the aquifer.

Monitoring When aquifer contamination is discovered, hydrogeologists monitor the contaminants to determine how far, in what direction, and how rapidly the plume is moving.

CONNECTIONS TO ENVIRONMENTAL SCIENCE

The Clean Water Act: A Modern Perspective
When it was adopted in 1972, the Clean Water Act set ambitious plans for cleaning the nation's rivers, lakes, and wetlands. Have we achieved our goals? The good news is that municipalities no longer dump raw sewage. Factories no longer discharge untreated waste directly into our waterways. The bad news is that much of our water remains polluted and most documented violations are not prosecuted. For example, a 2009 *New York Times* study reported that more than a half-million documented violations of the Clean Water Act had occurred over the prior five years. Yet less than 3 percent of these violations resulted in any fines or other punishment.

The hydrogeologists also determine whether the contaminant is becoming diluted or degraded by bacteria that live in the aquifer. The hydrogeologists may take samples from existing wells. They may drill additional wells to monitor the movement of the plume through the aquifer.

Modeling Hydrologists measure the rate at which the contaminant plume is spreading. Then they develop a computer model to predict future spread of the contaminant through the aquifer. To predict dilution effects, the model considers the local geologic structure, the permeability of the aquifer, directions of groundwater flow, and the mixing rates of groundwater.

Remediation Several processes are currently used to clean up a contaminated aquifer. For example, contaminated groundwater can be contained by building an underground barrier to isolate it from other parts of the aquifer. Or, if the trapped contaminant does not decompose by natural processes, hydrogeologists can drill wells into the contaminant plume. Then they can pump the fluid to the surface, where it is treated to destroy the pollutant.

Bioremediation uses microorganisms to decompose a contaminant. Specialized microorganisms can be developed by genetic engineers to use a particular contaminant as a source of food or energy without damaging the ecosystem. Once a specialized microorganism is developed, it is relatively inexpensive to breed it in large quantities. The microorganisms are then pumped into the contaminant plume, where they attack the pollutant. When the contaminant is gone, the microorganisms die, leaving a clean aquifer. Bioremediation can be among the cheapest of all cleanup procedures.

Chemical remediation is similar to bioremediation. It involves injecting a chemical compound into an aquifer to destroy a contaminant. The compound reacts with the contaminant, forming products that are harmless to the environment. Common compounds used in chemical remediation include oxygen and dilute acids and bases. Oxygen may react with a pollutant directly or provide an environment favorable for microorganisms, which then degrade the pollutant. Acids or bases can neutralize certain contaminants or remove pollutants from groundwater by precipitating them to solid forms.

Reclamation teams also can dig up the entire contaminated portion of an aquifer. The contaminated soil is incinerated or treated with chemical processes to destroy the pollutant. The treated soil can then be returned to fill the hole. This process is prohibitively expensive, however, and is used only in extreme cases.

checkpoint How does technology help to analyze, through modeling, a plume of contamination?

Nuclear Waste Disposal

In a nuclear reactor, radioactive uranium nuclei split into smaller nuclei, many of which are also radioactive. Most of these radioactive waste products are useless and must be disposed of without exposing people to the radioactivity. In the United States, military processing plants, commercial nuclear reactors, and numerous laboratories and hospitals generate more than 2,000 metric tons of high-level radioactive wastes every year.

Chemical reactions cannot destroy radioactive waste because radioactivity is a nuclear process and atomic nuclei are unaffected by chemical reactions. Therefore, the only feasible method for disposing of radioactive wastes is to store them in a place safe from geologic hazards and human interference. Here the wastes are allowed to decay naturally. The U.S. Department of Energy defines a permanent repository as one that will isolate radioactive wastes for 10,000 years. However, this number is totally arbitrary. It is based on the fact that 10,000 years is so far into the future that most people won't worry about pollution occurring at that time. Actually, radioactive wastes will remain harmful for 1 million years or more because of the long half-lives of some radioactive isotopes. As a result, radioactive wastes could become harmful to human or nonhuman ecosystems over the course of geologic time.

For a repository to keep radioactive waste safely isolated for long periods of time, it must meet at least three geologic criteria:

1. It must be safe from disruption by earthquakes and volcanic eruptions.
2. It must be safe from landslides, soil creep, and other forms of mass wasting.
3. It must be free from floods and seeping groundwater that might corrode containers and carry wastes into aquifers.

checkpoint What are the attributes of a permanent repository for radioactive waste?

17.4 ASSESSMENT

1. **Infer** Based on the passing of the Clean Water Act, what inference can you make about the public's and government's awareness of water pollution issues in the early 1970s?
2. **Classify** Design a simple chart to classify and organize the types of nonbiodegradable pollutants. Use the chart to help you research and answer new questions about water pollution.
3. **Summarize** Explain how sewage disposal can lead to a reduction in oxygen in a river and describe potential effects.
4. **Contrast** Compare two methods of aquifer remediation: bioremediation and chemical remediation.

Critical Thinking

5. **Apply** What research questions could you write to find out more about each method of remediation and how efficient and effective it is? What sources would you investigate to find the answers?

TYING IT ALL TOGETHER
SUSTAINABLE WATER MANAGEMENT

In this chapter, you learned about Earth's freshwater resources and how humans use water for agricultural, industrial, and domestic purposes. Although water is a renewable resource, human consumption can often outpace the natural processes that recharge aquifers and replenish rivers, lakes, and streams. Unsustainable water management practices can permanently deplete or degrade the water supply, threatening communities and ecosystems.

The Case Study for this chapter illustrated this problem, detailing the challenges faced by Los Angeles in managing its water supply. Experts recognize that sustainable water management requires conservation in all sectors. But it can be challenging to convince people to change long-established behaviors. In California, recent laws such as the Water Conservation Act (2009) and the Sustainable Groundwater Management Act (2014) have imposed regulations that require water conservation. For example, large agricultural suppliers of water are now required to submit Agricultural Water Management Plans to the state. Cities must now regulate the use of groundwater by residents, corporations, and farmers. Los Angeles has adopted several ordinances restricting domestic and industrial water use. For example, the city limits watering of home gardens and lawns to certain days for each address. It also offers financial and tax incentives, low-cost loans and grants, regulatory guidance, and professional development to help companies conserve water.

As a class, you will now continue to investigate sustainable methods of water management in various sectors. Working in groups, you will research one of the main sectors of water usage (agriculture, industry, residential) as assigned by your teacher.

Follow the steps below to guide your investigation.

1. Summarize how water usage in your assigned sector is described in the chapter.

2. Research additional information about water use in your assigned sector, focusing on data related to your state or region. Find out the amount of water used, the effects of water management on the environment, and existing conservation efforts.

3. In your group, identify any problems or challenges associated with water management in your assigned sector. Then identify possible solutions or methods for conserving water in your assigned sector.

4. Construct an argument explaining why society should prioritize sustainable water management practices for your assigned sector. Develop a list of talking points to support your argument.

5. With your group, present your argument and engage in a debate with other groups, who will be arguing why their assigned sector should be prioritized.

6. After the debate, write an op-ed article or letter to the editor. Share your findings and persuade the community to pursue specific solutions for sustainable water management in each sector.

▶ **FIGURE 17-25**
The Water Conservation Response Unit was created to enforce the Emergency Water Conservation Ordinance in Los Angeles after a severe drought in 2014. The unit educates residential and industrial customers about outdoor watering restrictions and responds to water waste complaints.

LESSON 17.4

CHAPTER 17 SUMMARY

17.1 WATER SUPPLY AND DEMAND

- Growing human populations have created demands for water that stretch, and in some cases exceed, the amount of water that is available both globally and in the United States.

- Most water used in the domestic and industrial sectors is withdrawn and then returned to streams or groundwater reservoirs near the site of withdrawal.

- Most of the water used in the agricultural section is consumed.

- U.S. water consumption falls into three categories: domestic use accounts for 10 percent; industrial use accounts for 49 percent; and agricultural use accounts for 41 percent.

17.2 DAMS AND DIVERSION

- The United States receives about three times more water from precipitation than it uses. Some of the driest regions use the greatest amounts of water.

- Water diversion projects collect and transport surface water and groundwater from places where water is available to places where it is needed.

- Dams are especially useful in regions that receive seasonal rainfall, and some dams also generate hydroelectric energy.

- Dams can create undesirable effects, including water loss, salinization, silting, erosion, disaster when a dam fails, recreational and aesthetic losses, and ecological disruptions.

17.3 GROUNDWATER DIVERSION AND DEPLETION

- Groundwater projects pump water to Earth's surface for human use, but because groundwater flows so slowly, the extraction often creates a cone of depression that may lead to a drop in the water table.

- The Ogallala aquifer illustrates groundwater depletion. Groundwater withdrawal may also cause subsidence and saltwater intrusion.

- The Great American Desert is a mostly arid region of the western United States.

- Americans have built great cities and extensive farms and ranches in the region, all supplied by irrigation systems. Diminishing water reserves and increasing costs of water-diversion projects are becoming serious challenges.

17.4 WATER POLLUTION

- Water pollution is the reduction in the quality of water by the introduction of impurities.

- In the United States, the Clean Water Act was passed in 1972 in response to serious water pollution disasters such as the Cuyahoga River fire.

- Water pollutants include biodegradable materials such as sewage, disease organisms, and fertilizers; nonbiodegradable materials such as industrial organic compounds, toxic inorganic compounds, and sediment; radioactive materials; and heat.

- Pollution can originate from both point sources and nonpoint sources.

- Biodegradable pollutants such as human sewage can promote rapid growth of decay organisms that can degrade aquatic ecosystems by depleting dissolved oxygen and lead to degradation of the ecosystem.

- Groundwater pollutants normally spread slowly into an aquifer as a plume of contamination.

- Contaminants permeate the pore spaces in an aquifer, so remediation is expensive and difficult. Techniques include elimination of the source, monitoring, modeling, bioremediation, chemical remediation, and removal of contaminated rock and soil.

- Radioactive wastes must be isolated from water resources because they are impossible to destroy, and they persist for long times.

CHAPTER 17 ASSESSMENT

Review Key Terms

Select the key term that best fits the definition. Not all terms will be used, and no term will be used more than once.

biodegradable pollutant
bioremediation
chemical remediation
cone of depression
consumption
diversion system
nonbiodegradable pollutant
nonpoint source pollution
persistent bioaccumulative toxic chemical (PBT)
plume of contamination
point source pollution
saltwater intrusion
subsidence
water pollution
water scarcity
withdrawal

1. a drop in the water level surrounding a well when groundwater is pumped to the surface faster than it can flow through the aquifer to the well
2. a pipe or canal constructed to transport water
3. scenario where aggregate demand for water by all sectors exceeds the amount of water available
4. a type of water impurity that decays naturally
5. the reduction of water quality by the introduction of impurities
6. any process that uses water and then returns it to Earth far from its source
7. method of removing of impurities from water by adding another substance that reacts with the pollutants
8. the sinking or settling of Earth's surface
9. a region of polluted groundwater that spreads gradually away from a single source of pollution
10. contamination of water that originates from a broad area
11. any process that uses water and then returns it to Earth locally
12. contamination of water that originates from a specific site
13. use of microorganisms to decompose a contaminant

Review Key Concepts

Answer each question on a separate sheet of paper to demonstrate your understanding of key concepts from the chapter.

14. Identify, describe, and differentiate between the three categories of water use.
15. What processes are responsible for replenishing surface freshwater on land?
16. Albuquerque is a large city built in a desert climate with very little water. What systems are most likely in place to meet the demand for water in Albuquerque?
17. Long-term use of irrigation can cause salinization of the soil. Why does this process have an adverse impact on crop yields?
18. People have built dams to alter water flow on Earth for centuries. What are some benefits of dams? What are some drawbacks?
19. Groundwater aquifers are important resources for communities around the world. What are some of the benefits and drawbacks of tapping groundwater for human use?
20. The figure illustrates a cone of depression. Does pumping water from a well always result in a cone of depression? Explain why or why not.

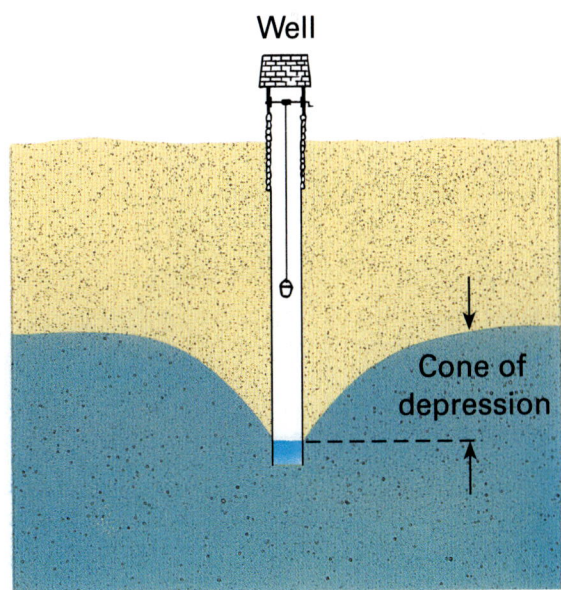

CHAPTER 17 ASSESSMENT

21. How do biodegradable pollutants differ from nonbiodegradable pollutants? Give some examples of each.

22. Explain why heat is considered a type of water pollution and give an example of this form of pollution.

23. Suppose you are an environmental inspector. You have found that an aquifer has been contaminated with pollution from a nearby factory. Describe the steps would you take to remediate the aquifer.

24. One of the chief concerns of nuclear waste disposal is the possible contamination of groundwater aquifers. Why is it important to prevent the seepage of nuclear waste into groundwater?

Think Critically

Write a response to each question on a separate sheet of paper. Use concepts from the chapter to support your reasoning.

25. During a 2015 drought, California imposed mandatory 25% water use reductions on the State Water Resources Control Board. This mandatory water use reduction applied to all water users except for agricultural users. Was this good policy? Use evidence to explain your reasoning.

26. The Grand Ethiopian Renaissance Dam is a large dam currently being constructed on the Nile River in Ethiopia. The construction has met with opposition from many sources, including other countries located downstream along the Nile. What are some of the possible problems that opponents may cite to argue against the construction of this dam?

27. Irrigation technology is often cited as one of the foundational technologies that allowed humans to build civilizations. Why is water diversion technology so important?

28. The Clean Water Act regulates pollution in the "waters of the United States." In the past, this phrase was mainly interpreted to mean large bodies of water. However, between 2015 and 2019, the Environmental Protection Agency debated expanding the definition so that regulations could be applied even to smaller streams and wetlands. Using your knowledge of water resources, identify some benefits and drawbacks of applying Clean Water Act regulations to more water in the United States.

29. Despite their dangers, many PBTs are used in industry because there are no economically viable alternatives. What are PBTs, and how should governments regulate their use?

PERFORMANCE TASK

A Local Water Problem

Many municipalities in the United States provide clean water to their residents. They have to work through many challenges to ensure that there are adequate supplies of clean water for the entire community.

In this performance task, you will work with your classmates to learn about local water pollution issues in your community and present your findings.

1. Research water pollution issues in your community.

    A With two or three partners, brainstorm to create a list of specific ways that pollutants could enter supplies of water used in your community.

    B Choose a specific type of pollutant from your list and conduct research to answer the following questions:

    - What is the origin of the pollutant? Does it primarily affect surface water or groundwater in your area?
    - What problems might the pollutant cause?
    - How is the pollutant identified and measured? Are there acceptable levels of the pollutant in water considered safe to use? How is the water treated to make it safe to use?
    - What happens to the pollutant that is removed from the water?
    - What can be done to decrease levels of the pollutant entering water supplies?

    In your research, look for public records or contact your local water board to find out how water is managed in your community.

    C After you have completed your research, create a presentation to summarize your findings. In your presentation, be sure to include graphics or illustrations to explain potential sources of a pollutant, its effects, how it is remediated, and how its levels can be decreased before it affects water supplies. Include in your presentation any relevant rules or regulations for water treatment related to the pollutant.

UNIT 6 | THE ATMOSPHERE

An artist's depiction of sea-level rise on the Miami coast shows what will likely occur during your lifetime (circa 2100). It shows the result of a 2°C (3.6°F) increase in the average global temperature. This seemingly small increase in the average global temperature is the expected upper limit resulting from carbon dioxide and other greenhouse gases that humans have already added to the atmosphere.

Chapter 18 **The Atmosphere**
Chapter 19 **Energy Balance in the Atmosphere**
Chapter 20 **Weather**
Chapter 21 **Climate and Climate Change**

Our atmosphere redistributes the sun's energy and helps regulate the temperature on Earth's surface. In the atmosphere, gases intermix while solid particles, including pollutants, move about. Weather systems also affect conditions in the atmosphere. Long-term weather patterns establish climate. Yet Earth's climate is changing rapidly. A depiction of predicted sea-level rise along a Florida coastline illustrates one of the many indirect consequences of living on a warmer planet. In this unit, you will learn about weather and climate, the composition of the atmosphere, and the balance of energy within it.

CHAPTER 18
THE ATMOSPHERE

How can humans impact the atmosphere? With extremely dry conditions in the area around Paradise, California, the stage was set for a disaster on November 8, 2018. The Camp Fire ignited around 6:30 a.m. due to faulty electrical transmission lines. Low humidity and strong winds caused the fire to race throughout the area, consuming property in and around Paradise. Normally clear air at Earth's surface became a thick blanket of wildfire smoke and ash that stretched for more than 65 kilometers (40 miles) by 10:45 a.m. This image, created using NASA's Landsat 8 satellite, shows the smoke and flames in visible and infrared light. To date, the Camp Fire is the largest and deadliest wildfire in California's history.

Although the atmosphere is a renewable resource, the activities and behaviors of humans can have negative effects on the quality of its air. Pollutants and smoke such as those produced by the wildfires in California in 2018 made breathing at ground level difficult or impossible.

As we look to the sky, Earth's atmosphere has no visible limit. It seems we have an endless supply of air surrounding Earth. In fact, Earth's atmosphere, especially the lower atmosphere that has enough oxygen for us to breathe and survive, is relatively thin. Within that space, all living things that depend on the atmosphere for survival make use of a limited resource.

KEY QUESTIONS

18.1 What was the chemical composition of Earth's early atmosphere?

18.2 How did iron and photosynthetic organisms affect the evolution of Earth's atmosphere?

18.3 How and why does atmospheric temperature and pressure change with increasing altitude?

18.4 How do some of the major sources of air pollution affect the environment and humans?

EXPLORERS AT WORK

STUDYING METHANE BUBBLES

WITH NATIONAL GEOGRAPHIC EXPLORER KATEY WALTER ANTHONY

Katey Walter Anthony crouches on the frozen surface of an Alaskan lake, peering into its icy depths. It's a situation she finds herself in frequently as part of her work with the University of Alaska, Fairbanks. Dr. Walter Anthony is an aquatic ecologist who studies how Arctic lakes such as this one may be accelerating climate change.

Walter Anthony spots what she has been looking for: patterns of white, disklike shapes trapped inside the ice. They are bubbles, frozen in the act of rising up from the lake bottom. These are not air bubbles; rather, they are bubbles of methane, a powerful greenhouse gas. In fact, one molecule of methane gas released into the atmosphere has 25 times the warming potential of one molecule of carbon dioxide gas. And if this were summer, the methane in these bubbles would indeed be released as the bubbles popped at the lake surface. The lake would literally burp out methane.

According to Walter Anthony, most lakes release methane gas, a waste product as microbes decompose organic matter in lake sediments. However, Walter Anthony has found that some Arctic lakes release much more methane gas than others—hundreds to thousands of times more. Many of these methane-bubbling "hotspots" are thermokarst lakes, which form where the land has collapsed due to an abrupt thaw in the underlying permafrost. Permafrost is a layer of permanently frozen ground just below the surface that includes solid bedrock as well as mixtures of ice, soil, gravel, sand, and mud. Permafrost covers roughly one-quarter of Earth's Northern Hemisphere. Now, it is threatened to thaw as climate warms.

As Walter Anthony explains, when permafrost thaws abruptly, carbon that has been frozen for tens of thousands of years becomes available to microbes at the bottoms of the newly formed lakes. Microbes feed on the carbon and release methane. Even small releases of methane from the soil beneath these lakes can result in further thawing of the permafrost. That can lead to the formation of more thermokarst lakes, which in turn release more methane. This creates a positive feedback loop.

Walter Anthony has analyzed gas bubble samples from hundreds of lakes in Alaska and Siberia. She hopes her research will shed light on patterns of methane emissions from lakes and how these emissions contribute to climate change.

Walter Anthony is researching ways to harness methane emissions as an alternative fuel source for people in remote areas of Alaska. These communities depend on expensive diesel fuel for power and heating. The benefits of supplying them with methane gas for fuel would be twofold. It would reduce the monetary and environmental costs of consumption associated with supplying diesel fuel. And burning methane gas only releases small amounts of greenhouse gases compared to straight emissions into the atmosphere.

THINKING CRITICALLY

Infer What would happen to the population of microbes as the climate warms and how would this impact the rate of climate warming?

Dr. Katey Walter Anthony studies methane released from lakes in the Arctic. The methane is produced by microbes that decompose organic material in the lakes.

Bubbles of methane gas released from sediments at the bottom of a lake are frozen in place during an Arctic winter.

CASE STUDY
MANURE, COLD SEEPS, AND FIRE ICE

Methane is a potent greenhouse gas that contributes to global climate change. That's one reason researchers such as Katey Walter Anthony (see Explorers at Work) are studying methane emissions in areas where permafrost melt is occurring. Although methane emissions have detrimental effects on the atmosphere, Walter Anthony and others are focused on finding ways to capture methane gas from various sources and use it as a fuel source.

Methane is considered a fossil fuel, but it burns cleaner than coal or natural gas. Using methane can reduce demand for other fossil fuels, in turn reducing the harmful effects of mining and burning those fuels.

One natural source of clean-burning methane is a substance rarely thought of as clean: animal manure. Large livestock such as cattle produce ample amounts of manure—and methane—that some farmers are working to capture and monetize. A recent collaboration in California between dairy farmers and a bioenergy company began installing machines called digesters on farms. Digesters take manure from the fields and extract the usable methane gas, removing impurities. The captured methane is then piped to locations where it can be used to produce electricity or power commercial vehicles that have engines designed to run on methane.

Another abundant source of naturally occurring methane is methane clathrate, sometimes known as "fire ice" (Figure 18-1). At the molecular level, methane clathrate consists of methane gas molecules surrounded by frozen water molecules. Essentially, it is ice with a molecule of highly flammable gas trapped inside. Fire ice forms at low temperatures and high pressures. It is often found in polar regions or deep in the ocean along tectonic plate boundaries. Some researchers have concluded that there is more energy trapped in fire ice deposits than in all of Earth's oil, coal, and natural gas deposits combined. However, mining deep-sea fire ice poses many challenges and risks. Aside from contending with the enormous pressure at such depths, mining fire ice would disrupt the ocean floor. Moreover, any accidental release of methane into the atmosphere would worsen global warming. Ongoing research into using fire ice as an energy source aims to address these issues.

Cold seeps, or methane seeps, are another potentially usable source of methane gas. Cold seeps occur along continental borders, where fluid rich in hydrocarbons, including methane, is trapped in seafloor sediments. Movements of Earth's crust can allow the fluid to seep out. Some research, analogous to Katey Walter Anthony's research on methane hotspots in thermokarst lakes, is studying ways this fluid could be collected from the seafloor and used as a substitute for natural gas.

As You Read Consider natural factors and human activities that affect the composition of Earth's atmosphere. What gases and other substances, such as particulate matter, make up different parts of the atmosphere, and how do they affect interactions with other Earth systems? Look for ways that Earth's atmosphere impacts—and is impacted by—life on Earth.

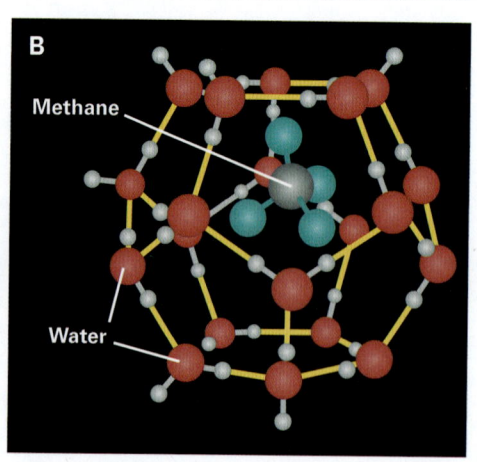

FIGURE 18-1
(A) Methane clathrate is sometimes called fire ice. (B) It is methane gas trapped inside frozen water molecules as shown by this molecular model.

18.1 EARTH'S EARLY ATMOSPHERES

Core Ideas and Skills
- Explain how oxygen can be both beneficial and detrimental to organisms.
- Describe the changes in Earth's earliest atmospheres.
- Identify the sources of volatile compounds that enriched Earth's early atmosphere.

KEY TERMS

outgassing

Nearly every multicellular organism needs oxygen to survive. If the oxygen abundance in the atmosphere were to drop below 44 percent of its current value, life on Earth as we know it would perish. If oxygen is essential to most life, would we be better off if we had an even greater supply? The answer is yes, to a limit. Even at sea level, athletes can enhance their performance by breathing a small amount of bottled oxygen. But too much oxygen is poisonous. If you breathe air that has 55 percent or more oxygen than is found at sea level, your body metabolism becomes so rapid that essential molecules and enzymes decompose. In addition, fires burn more rapidly with increased oxygen concentration. If the oxygen level in the atmosphere were to rise significantly, extremely large wildfires would engulf the planet, altering ecosystems as we know them.

Earth is the near-perfect size and distance from a stable and medium-temperature star to permit optimal atmospheric conditions for life. But Earth's atmospheric composition and temperature are not determined solely by planetary size and distance from the sun. Rather, our planet's environment is finely regulated by interactions among Earth systems. Over the past 4 billion years, the sun's output has slowly increased (although there were numerous fluctuations during this period). However, Earth's atmospheric temperature has remained relatively stable, fluctuating within a remarkably narrow range because of the interactions among Earth's systems. In this chapter we will explore these interactions and see how they led to the structure and dynamics of Earth's atmosphere today.

The First Atmospheres

Our solar system formed from a cold, diffuse cloud of interstellar gas and dust. About 99.8 percent of this cloud was composed of hydrogen and helium. Consequently, when Earth formed, its primordial, or earliest, atmosphere was composed almost entirely of these two light elements. Because Earth is relatively close to the sun and its gravitational force is relatively weak, its primordial atmosphere of hydrogen (H_2) and helium (He) rapidly boiled off into space. Table 18-1 compares the primordial and subsequent atmospheres of Earth.

TABLE 18-1 Earth's Atmosphere through Time

Event or Processes	Timeframe (Millions of Years Ago)	Atmosphere and Composition
Initial formation of planets	4,600	Primordial atmosphere Hydrogen (H_2) and helium (He)
Impacts of comets, meteoroids, and asteroids	4,500	Secondary atmosphere Carbon dioxide (CO_2), water (H_2O), and nitrogen (N_2)
Outgassing and chemical reactions	2,700 to 4,500	Predominantly hydrogen (H_2) and carbon dioxide (CO_2); some water (H_2O) and nitrogen (N_2)
Evolution of cyanobacteria; production of oxygen	2,700	Oxygen (O_2) produced by cyanobacteria is removed as quickly as it is produced
First Great Oxidation Event	2,400	Oxygen (O_2) accumulates in the atmosphere; hydrogen (H_2) becomes a trace gas
Second Great Oxidation Event	600	Atmospheric oxygen concentrations increase as biological and geological processes slow decay
Biological photosynthesis	Today	Primarily nitrogen (N_2) and oxygen (O_2), with smaller concentrations of other gases

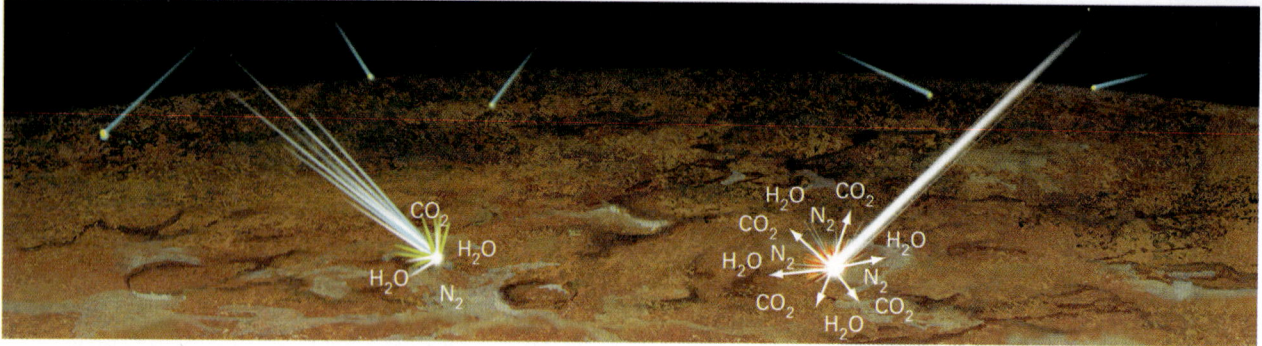

FIGURE 18-2
Bolides imported volatile compounds, such as water (H_2O) and ammonia (NH_3). These compounds reacted with Earth's materials to form water vapor, carbon dioxide (CO_2), and nitrogen gas (N_2) that formed the secondary atmosphere.

Recall that, in its infancy, the solar system was crowded with bits of rock, ice chunks, and other debris left over from the consolidation of the planets. These bolides, or bits of space debris, brought volatile compounds from outer parts of the solar system. They crashed into the planet in a near-continuous rain that lasted almost 800 million years. The result was the formation of Earth's oceans, as discussed in Chapter 11, and Earth's secondary atmosphere (Table 18-1). Ice from comet impacts vaporized into water vapor (H_2O). Carbon-bearing rocks, under the heat and pressure of impact, combined with oxygen to form carbon dioxide gas (CO_2). Ammonia (NH_3), common in the icy tail of comets, reacted to form nitrogen gas (N_2) (Figure 18-2).

checkpoint What happened to Earth's primordial atmosphere of hydrogen and helium?

The Changing Atmosphere

The secondary atmosphere composed of carbon dioxide, water, and nitrogen changed as volatiles from the Earth's mantle, such as hydrogen and methane, escaped to the surface due to volcanic activity, in a process called **outgassing**. In 1953, Stanley Miller and Harold Urey hypothesized that by 4 billion years ago, Earth's atmosphere consisted primarily of methane, ammonia, hydrogen, and water. They mixed these gases in a glass tube and fired sparks across the tube. This was to simulate lightning storms in Earth's early atmosphere. Amino acids, the building blocks of proteins, formed in the tube. The Miller–Urey model immediately became popular because scientists speculated that the first living organisms formed from these nonliving amino acids.

In the early 1970s, the idea of an atmosphere composed primarily of methane, ammonia, and hydrogen had been largely discredited and the Miller-Urey model was set aside. Scientists concluded instead that the outgassing from Earth's mantle about four and a half billion years ago, combined with reactions between gases and rocks of the geosphere, created an atmosphere rich in hydrogen (H_2) and carbon dioxide (CO_2) (Table 18-1). The enriched atmosphere also contained some water (H_2O) and nitrogen gas (N_2). Scientists hypothesize that this evolution of Earth's atmosphere set the stage for the emergence of life on the planet. Recall from Chapter 2 that life existed on Earth at least as early as 3.8 billion years ago.

checkpoint What is outgassing?

18.1 ASSESSMENT

1. **Describe** Discuss the formation and composition of Earth's primordial atmosphere.
2. **Summarize** How and why did Earth's secondary atmosphere form?
3. **Explain** How did outgassing affect the composition of the secondary atmosphere?

Critical Thinking

4. **Synthesize** The Miller-Urey model was set aside because there was insufficient evidence supporting high concentrations of the compounds the scientists used in their experiment. According to the text, what would have been the primary sources of those compounds?

18.2 LIFE, IRON, AND THE EVOLUTION OF THE MODERN ATMOSPHERE

Core Ideas and Skills

- Describe the biological, physical, and chemical processes that resulted in the two great oxidation events, approximately 2.4 billion and 600 million years ago.
- Explain the system interactions that were relevant in balancing the chemical composition of Earth's early atmospheres.
- Compare and contrast the modern atmosphere and atmospheres from Earth's past.

KEY TERMS

photosynthesis
Gaia
cyanobacteria
First Great Oxidation Event
banded iron formation
Second Great Oxidation Event

First Great Oxidation Event

The first living organisms may have developed by the accumulation of complex nonliving organic molecules formed in Earth's early environments. However, complex organic molecules become oxidized and are destroyed in an oxygen-rich environment. So, if large amounts of oxygen were present in Earth's early atmosphere, the nonliving organic molecules could not have formed.

Geochemists can determine whether a rock formed in an oxygen-rich or an oxygen-poor environment. First they analyze the mineral and chemical properties of the rock. Then they apply what is currently known about conditions needed to form that combination of minerals. Recent studies of Earth's oldest rocks indicate that the atmospheric oxygen concentration in early Precambrian time was extremely low. These results suggest that the molecules necessary for the emergence of living organisms—methane, ammonia, hydrogen, and water—would have been preserved in the primordial atmosphere.

Although life could not have arisen in an oxygen-rich environment, complex multicellular life requires an oxygen-rich atmosphere to survive. How, then, did oxygen become abundant in our atmosphere?

The world's earliest organisms probably were chemosynthesizers—organisms that get their energy from reactions with minerals such as iron and sulfur. Later, organisms survived, in part, by eating each other. But these food webs were limited because there were fewer organisms.

A crucial step in evolution occurred when early bacteria evolved the ability to use the energy in sunlight to produce organic tissue. This process, known as **photosynthesis**, is the foundation for virtually all modern life. During photosynthesis, organisms convert carbon dioxide and water to organic sugars. They release oxygen as a by-product.

In 1972, English chemist James Lovelock hypothesized that the oxygen produced by primitive organisms gradually accumulated, creating the modern atmosphere. By late Precambrian time, atmospheric oxygen concentration had reached the critical level needed to sustain efficient metabolism. As a result, multicellular organisms evolved and the biosphere as we know it was born. Lovelock was so overwhelmed by the intimate connection between living and nonliving components of Earth's systems that he likened our planet to a living creature, which he called **Gaia** (Greek for "Earth").

The Lovelock hypothesis is now about 50 years old. It remains generally accepted, but scientists are still investigating many of the details. For example, blue-green algae called **cyanobacteria** began producing oxygen 2.7 billion years ago (Figure 18-3). But significant quantities of oxygen did not begin to appear in the atmosphere until 2.4 billion years ago, in the **First Great Oxidation Event**. At that time, the concentration of atmospheric oxygen (O_2) rose abruptly (Table 18-1) and hydrogen (H_2) concentrations declined.

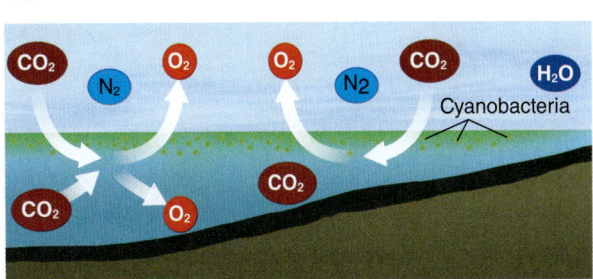

FIGURE 18-3
The oxygen content of the atmosphere slowly and steadily increased. This occurred after photosynthesizing cyanobacteria began to release oxygen 2.7 billion years ago.

The production of oxygen by cyanobacteria 2.7 billion years ago started slowly. Then, for about 700 million years, atmospheric oxygen concentration remained far below today's level. At times during this period, Earth's atmosphere may have been anoxic, meaning there was no oxygen.

These abrupt variations in the concentration of oxygen in Earth's earliest atmosphere suggest that the presence of oxygen was driven by a process involving a chemical threshold. To understand this process, we must study geosphere-atmosphere and biosphere-atmosphere system interactions.

checkpoint How can photosynthesis affect the atmosphere?

System Interactions in the Early Atmosphere

Modern hypotheses state that Earth was molten at the time of its formation. Materials separated into layers based on their density. The heaviest elements, such as iron, sank. Even though large quantities of iron sank into Earth's core when the planet formed, some remained in the mantle and crust. Today's crust, for example, is about 5 percent iron by weight. The iron that remained near the surface played an important role in concentrating oxygen in Earth's early atmosphere.

Oxygen gas dissolves in water. As a result, when early cyanobacteria released oxygen into the atmosphere, some of the gas dissolved into seawater. In an oxygen-poor environment, iron also dissolves in water. However, when oxygen concentration is above a certain threshold level, dissolved oxygen reacts with dissolved iron and precipitates rapidly as iron-oxide minerals.

Large amounts of iron dissolved in the early Precambrian Ocean when the concentration of dissolved oxygen was below the threshold level. However, when the concentration of dissolved oxygen from cyanobacteria rose above the threshold, iron that had been dissolved in the seawater precipitated rapidly as iron-oxide minerals. The minerals settled to the seafloor, forming a layer there (Figure 18-4). The precipitation of iron-oxide minerals also removed oxygen from the water. This explains why the oxygen concentration in the atmosphere didn't rise at this time in Earth's early history, even though cyanobacteria were releasing oxygen as a gas.

Roughly 90 percent of the iron ore that is mined globally comes from **banded iron formations**. These formations are chemical sedimentary rock that consist of alternating layers ("bands") of iron oxide and chert a few centimeters thick (Figure 18-5). Most of the banded iron formations on Earth are between 2.6 and 1.9 billion years old. When dissolved oxygen levels in the seas were high enough, the oxygen would combine with dissolved iron and precipitate solid iron-oxide minerals that settled and formed a layer on the seafloor. The formation of these minerals absorbed oxygen and iron from the seawater. This in turn lowered the oxygen level below the threshold at which iron-oxide minerals could accumulate. Then dissolved iron built up again in the seas from chemical weathering of exposed rock, while the oxygen was slowly replenished by photosynthesizing cyanobacteria. During that time, clay and other minerals washed from the continents

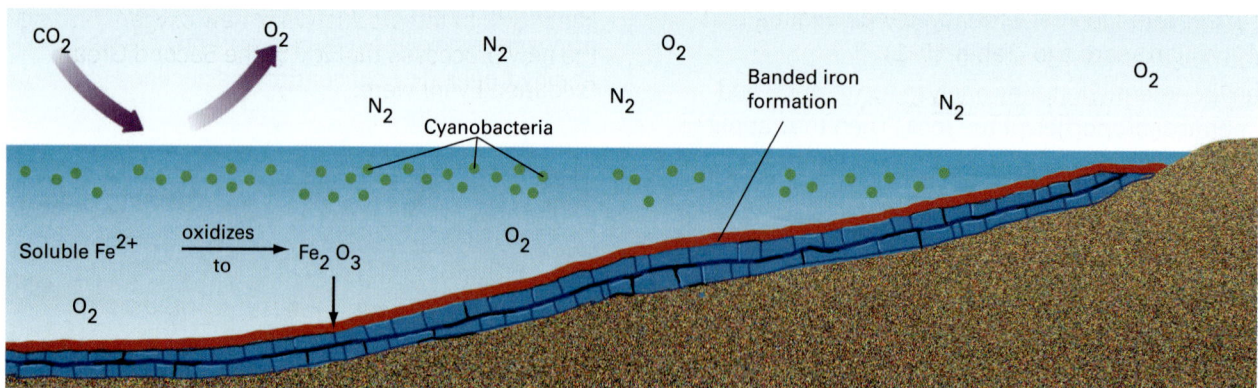

FIGURE 18-4
When the oxygen concentration in the atmosphere and in seawater reached a threshold, the oxygen combined with dissolved iron to form iron oxide minerals. The iron oxide minerals precipitated from the water as solid materials. These settled to the seafloor, forming the iron-rich layer of a banded iron formation.

and accumulated on the seafloor as they do today. These minerals formed the thin layers of silicate minerals that lie between the iron-rich layers in banded iron formations. When oxygen levels eventually rose high enough again, another layer of iron oxide minerals formed.

FIGURE 18-5
Go online to view an example of layered iron oxide minerals and chert (silica) in a banded iron formation in Michigan.

Banded iron formations contain thousands of alternating layers of iron minerals and silicates. These formations can cover tens of square kilometers. Since most banded iron formations are between 2.6 and 1.9 billion years old, the reactions that formed them must have kept the levels of dissolved oxygen close to the threshold for about 700 million years. The iron-rich rocks that support our industrial society were formed by interactions among early photosynthesizing organisms, sunlight, air, and the oceans.

In the primordial atmosphere, free oxygen reacted with hydrogen to form water. This process helped keep the oxygen concentration in the atmosphere low. However, after life evolved, bacteria in the oceans removed atmospheric hydrogen in a process that produced methane. When the hydrogen concentration decreased sufficiently, the concentration of free oxygen in the atmosphere could rise again. Evidence in the rocks indicates that the oxygen concentration in the atmosphere remained relatively low between 2.7 billion and 2.4 billion years ago. This balance was maintained by biosphere–atmosphere and geosphere–atmosphere interactions. The oxygen (O_2) concentration in the atmosphere jumped suddenly in the First Great Oxidation Event about 2.4 billion years ago (Table 18-1).

checkpoint What is a banded iron formation?

Second Great Oxidation Event

Several deposits of banded iron formed after the First Great Oxidation Event. This process continued to remove oxygen that was released during photosynthesis. The last, major banded iron layer was deposited about 1.9 billion years ago, but the oxygen concentration in the atmosphere did not increase dramatically at that time.

The sun emits significant energy in the form of high-energy ultraviolet light. These rays are energetic enough to kill evolving multicellular organisms. High-altitude oxygen absorbs ultraviolet light in a cyclical process that forms and breaks down ozone (O_3). Oxygen, largely produced by the earliest photosynthetic organisms, was necessary for life as we know it today. But as ozone in the upper atmosphere, it also protected multicellular life by filtering out harmful solar rays. The oxygen concentration could increase in the lower atmosphere, enabling the development of multicellular organisms, only after significant concentrations first accumulated in the upper atmosphere.

Multicellular plants and animals emerged in late Precambrian time. About 600 million years ago, the oxygen level in the atmosphere increased rapidly a second time, in a process called the **Second Great Oxidation Event**. What changed abruptly more than a billion years after the last banded iron layers were deposited to allow oxygen to accumulate? Scientists think that before this event, biological decay was almost as rapid as photosynthesis. The oxygen that was released into the atmosphere through photosynthesis was immediately consumed during decomposition, according to the following reactions:

Photosynthesis:
carbon dioxide (CO_2) + water (H_2O) →
 sugars + oxygen (O_2)

Decomposition:
organic matter + oxygen (O_2) →
 carbon dioxide (CO_2) + water (H_2O)

Then, beginning abruptly 600 million years ago, several new processes among Earth's spheres led to the mass burial of organic matter before it could decompose. These processes altered the balance between photosynthesis and decomposition. The removal of oxygen from the atmosphere could not keep up with the production of new oxygen. Among the new processes that led to the Second Great Oxidation Event were:

- Geochemical processes produced more nutrients. Chemical weathering released dissolved nutrients into oceans, increasing biological productivity. At the same time, sediment from physical weathering of mountains led to the rapid burial and preservation of organic matter on the seafloor, slowing decomposition rates.

- Zooplankton, photosynthetic organisms, evolved in the seas. Their dense, organic-laden wastes sank rapidly to the seafloor, accumulating in the clay sediments, removing organic carbon from the system.

MINILAB Carbon Dioxide, Photosynthesis, and Oxygen

Materials (per pair):

small piece of Elodea, approximately 5–7 cm in length
3 test tubes
test tube rack
drinking straw
25 mL distilled water, in a beaker
bright light
phenol red indicator
masking tape
indelible pen or marker

Procedures

1. Place the three test tubes in a test tube rack. Use tape and a marker to add labels: A, B, and C.
2. Add distilled water to each test tube. Add the same amount to each test tube, until the level reaches approximately 3 cm.
3. Add three drops of phenol red indicator to each test tube. (Phenol red indicator is red in neutral or basic solutions and yellow in acidic solutions.)
4. Using the straw, blow softly into test tubes B and C until you see the color change to yellow.
5. Put the piece of Elodea into test tube C.
6. Shine a bright light directly on all three test tubes.
7. Wait for about 40–60 minutes. Record what you observe.

Results and Analysis

1. **Observe** Describe the color of the water and any changes that occurred.
2. **Explain** What explains the color change in test tubes B and C in step 4?
3. **Explain** What explains your observations in step 7?

Critical Thinking

4. **Analyze** What were the independent and dependent variables in this investigation?
5. **Evaluate** Why did you only blow into two test tubes? Why was this important?
6. **Connect** How did this lab help you understand the effects of photosynthesis on water and the atmosphere?

- Simple lichens evolved on land and accelerated weathering. The weathered ions washed into the sea providing nutrients for phytoplankton. In turn, the phytoplankton fed the zooplankton.

checkpoint What is the Second Great Oxidation Event?

Composition of the Modern Atmosphere

The modern atmosphere is mostly gas but also contains droplets of liquid water and suspended particles of dust. The gaseous composition of dry air is roughly 78 percent nitrogen, 21 percent oxygen, and 1 percent other gases (Figure 18-6). Nitrogen, the most abundant gas, does not react readily with other substances. Oxygen, though, reacts chemically as fires burn, iron rusts, and plants and animals respire. Carbon dioxide, by some models, formed as much as 80 percent of Earth's early atmosphere. It is a trace gas in the modern atmosphere, with a concentration of only 0.035 percent.

The types and quantities of gases, water vapor, droplets, and dust vary with both location and altitude. In a hot, steamy jungle, air may contain 5 percent water vapor by weight. In a desert or cold polar region only a small fraction of a percent may be present.

If you sit in a house on a sunny day, you may see a sunbeam passing through a window. The visible beam is light reflected from tiny specks of suspended dust. Clay, salt, pollen, bacteria, viruses, and bits of cloth, hair, and skin are all components of dust. People travel to the seaside to enjoy the "salt air." Visitors to the Great Smoky Mountains in Tennessee view the bluish, hazy air formed by sunlight reflecting from pollen and other dust particles.

FIGURE 18-6
The modern atmosphere is composed of nitrogen, oxygen, and other gases.

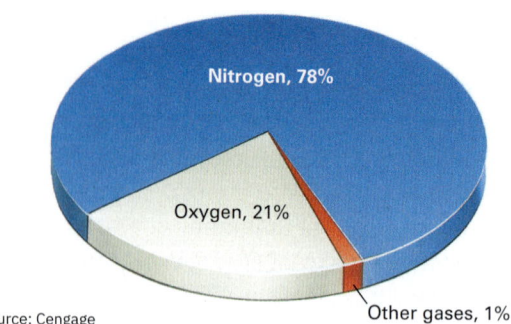

Source: Cengage

MINILAB Oxygen in the Atmosphere

Materials (per group):

test tube
glass beaker (or cup)
vinegar
water, enough to fill the cup halfway
graduated cylinder
1 g fine mesh steel wool (such as grade #00)
tape

Procedures

1. Place a small amount (about a pinch) of the steel wool in the bottom of the test tube. Press it down so that when the tube is inverted, the steel wool stays in the tube.
2. Fill the test tube with water. Then pour the water into a graduated cylinder to measure the capacity of the test tube. Record this capacity. This is the volume of air in the mostly empty test tube.
3. Pour a small amount of vinegar into the test tube, just enough to cover the steel wool. Let the vinegar stay inside the tube for one minute. Then dump the vinegar out. This will remove the coating on the steel meant to prevent rusting.
4. Fill the cup halfway with water. Turn your test tube upside down and place it, open end straight down, into the cup of water. There should be no water inside the test tube. Secure the test tube in this position with tape.
5. Observe your test tube after two days, and carefully mark the level of the water in the test tube with a small piece of tape.
6. Remove the test tube. Fill with water up to the tape and measure the volume. This is the volume of the gas when the test tube was upside down.
7. Calculate the percent change in gas in the test tube. This change is the amount of oxygen that was removed from the air inside the test tube.
8. Observe the steel wool and record any change in color.

Results and Analysis

1. **Interpret Data** Why did the water level inside the test tube rise?
2. **Analyze Data** Calculate the volume of the oxygen that was removed from the air inside the test tube by subtracting the volume of gas remaining in the test tube from the original volume of gas in the test tube.

Critical Thinking

3. **Evaluate** Was the initial environment inside the test tube oxygen-poor or oxygen-rich? What evidence from the lab supports your conclusion? Do you think the steel wool would continue to rust if you allowed it to stay in the test tube? Explain.
4. **Compare** How can you use the results of the lab to find the approximate percent of oxygen in the air? What would you expect this percent to be? Perform the calculation using your own data and explain how your results compare to your expected results.
5. **Compare** How did your results compare to other groups' results? What might account for any differences?

Within the past century, humans have altered the chemical composition of the atmosphere in many ways. We have increased the carbon dioxide concentration by burning fossil fuels and igniting wildfires. Factories release chemicals into the air—some benign, others poisonous. Smoke and soot change the clarity of the atmosphere.

checkpoint Describe the main gases and components of dust in the atmosphere.

18.2 ASSESSMENT

1. **Describe** What process first began producing oxygen about 2.7 billion years ago?
2. **Explain** Why did the oxygen concentration remain low for 300 million years after oxygen was first produced?
3. **Recall** How did banded iron formations form?
4. **Recall** List the two most abundant gases in the modern atmosphere. List one less abundant gas. List three nongaseous components of air.
5. **Describe** What are some examples of oxygen in the atmosphere reacting chemically with other substances?

Critical Thinking

6. **Apply** How might the modern atmosphere be different if little or no carbon were trapped in sediment?

18.3 ATMOSPHERIC PRESSURE AND TEMPERATURE

Core Ideas and Skills

- Explain why Earth's atmosphere is denser at its surface than far above its surface.
- Explain how a mercury barometer and an aneroid barometer measure atmospheric pressure.
- Explain how meteorologists express the measurement of atmospheric pressure.
- Identify the layers of Earth's atmosphere.
- Explain why the lower part of the troposphere is warmer than the upper part of the troposphere.
- Describe how ozone forms in the stratosphere and its significance to life on Earth's surface.
- Compare and contrast how the troposphere and the stratosphere are heated.

KEY TERMS

barometric pressure	stratopause
barometer	mesosphere
bar	mesopause
troposphere	thermosphere
tropopause	exosphere
stratosphere	

Density of Earth's Atmosphere

The molecules in a gas move about randomly. For example, at 20°C an average oxygen molecule is traveling at 425 meters per second (950 miles per hour). In the absence of gravity, or where there are temperature differences or other causes of disturbance, a gas will fill a space evenly.

Therefore, if you floated a cylinder of gas in space, the gas would disperse until there was an equal density of molecules and equal pressure throughout the cylinder. In contrast, gases that surround Earth are influenced by many factors. These factors create the complex and turbulent atmosphere that helps shape the world in which we live.

Within our atmosphere, gas molecules zoom around, but in addition, gravity pulls them downward. As a result of this downward force, more molecules concentrate near the surface of Earth than at higher elevations. Therefore, the atmosphere is denser at sea level than it is at higher elevations—and the pressure is higher. Density and pressure then decrease exponentially with elevation (Figure 18-7). At an elevation of about 5,000 meters, the atmosphere contains about half as much oxygen as it does at sea level. If you ascended in a balloon to 16 kilometers above sea level, you would be above 90 percent of the atmosphere and would need an oxygen mask to survive. At an elevation of 100 kilometers, pressure is only 3×10^{-5} that of sea level, approaching the vacuum of outer space. There is no absolute upper boundary to the atmosphere.

Atmospheric pressure is often called **barometric pressure**. It is measured with a **barometer**. A simple but accurate barometer is constructed from a glass tube that is sealed at one end. The tube is evacuated and the unsealed end placed in a dish of a liquid such as mercury. The mercury rises in the tube because atmospheric pressure depresses the level of mercury in the dish. Since there is no air in the tube to counter that pressure, the surface of the mercury in the tube rises (Figure 18-8). At sea level mercury rises approximately 76 centimeters, or 760 millimeters (about 30 inches), into an evacuated tube.

FIGURE 18-7 ▼
Atmospheric pressure decreases with altitude. One-half of the atmosphere lies below an altitude of 5,600 meters.

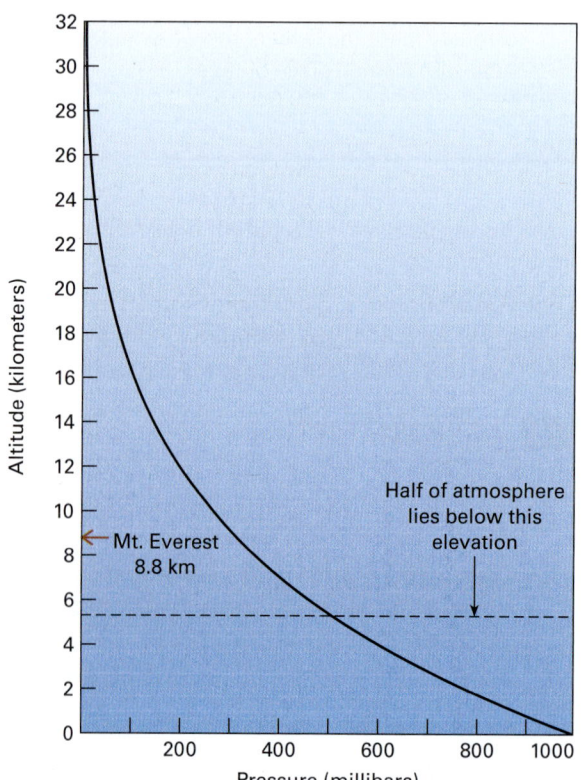

Source: Cengage

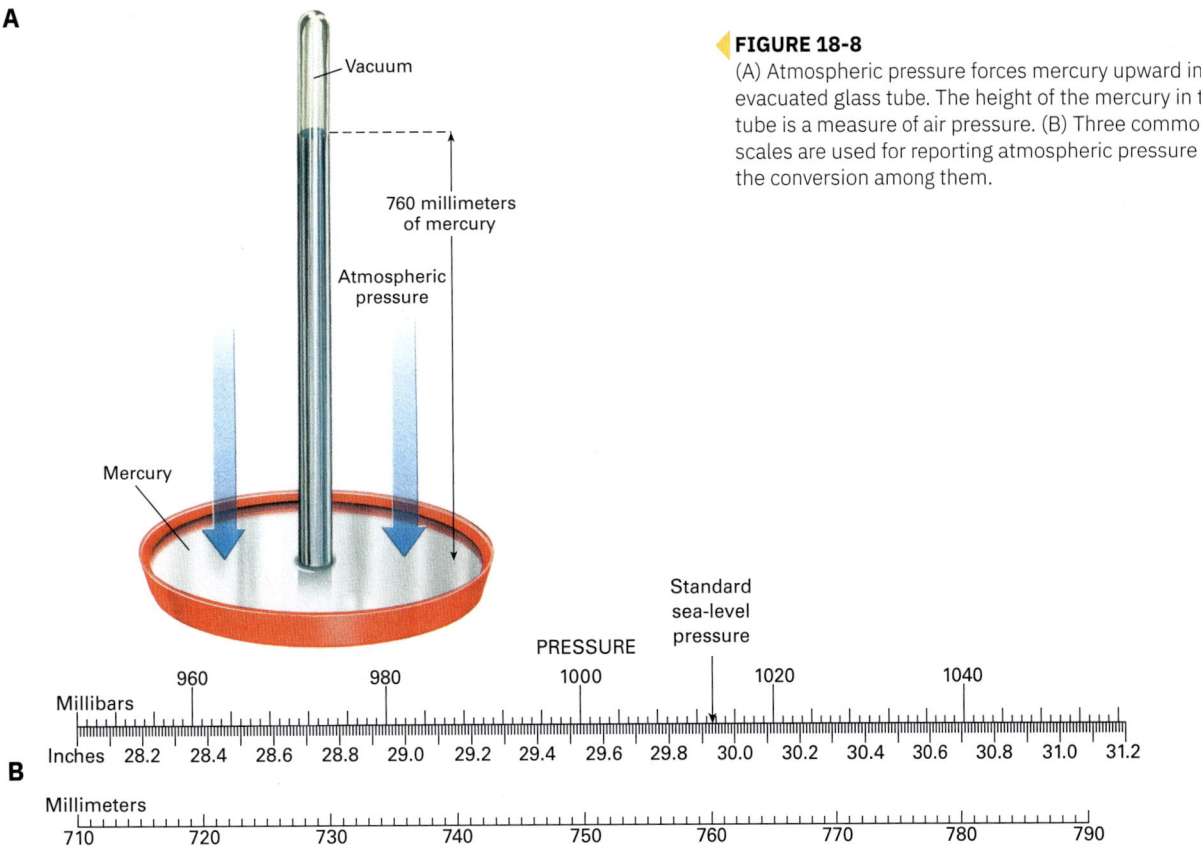

FIGURE 18-8
(A) Atmospheric pressure forces mercury upward in an evacuated glass tube. The height of the mercury in the tube is a measure of air pressure. (B) Three common scales are used for reporting atmospheric pressure and the conversion among them.

Meteorologists express pressure in inches or millimeters of mercury, referring to the height of the column of mercury in a barometer. They also express pressure in bars and millibars. A **bar** is approximately equal to sea level atmospheric pressure. A millibar is one one-thousandth of a bar.

A mercury barometer is a cumbersome device nearly a meter tall. And mercury vapor is poisonous. A safer and more portable instrument for measuring pressure, called an aneroid barometer, consists of a partially evacuated metal chamber connected to a pointer. When atmospheric pressure increases, it compresses the chamber and the pointer moves in one direction. When pressure decreases, the chamber expands, directing the pointer the other way (Figure 18-9).

Changing weather can affect barometric pressure. On a stormy day at sea level, pressure may drop to 980 millibars (28.9 inches), although barometric pressures below 900 millibars (26.6 inches) have been reported during some hurricanes. In contrast, during a period of clear, dry weather, a typical high-pressure reading may be 1,025 millibars (30.3 inches).

checkpoint Describe how an aneroid barometer works.

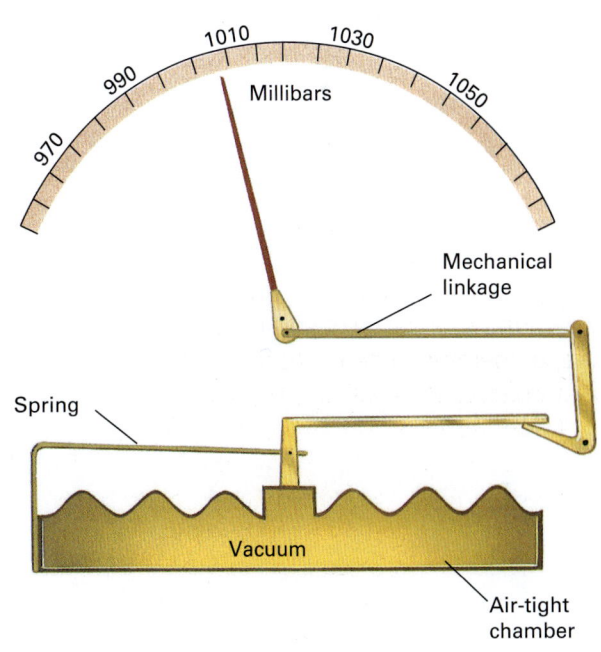

FIGURE 18-9
In an aneroid barometer, increasing air pressure compresses an airtight chamber. This causes the connected pointer to move in one direction. When the pressure decreases, the chamber expands. This causes the pointer to move in the other direction.

FIGURE 18-10
ON ASSIGNMENT National Geographic photographer Cory Richards took this photo of a mountaineer hiking up Hkakabo Razi. At a height of 5,881 meters (19,296 feet), this remote mountain is believed to be the highest in Myanmar. The atmosphere near the summit contains about half as much oxygen as it does at sea level.

Layers of Earth's Atmosphere

The temperature of the atmosphere changes with altitude (Figure 18-11). The layer of air closest to Earth—the layer we live in—is the **troposphere**. Virtually all water vapor and clouds exist in this layer, and almost all weather occurs here. Earth's surface absorbs solar energy, so the surface of the planet is warm. But, as explained earlier, continents and oceans also radiate heat, and some of this energy is absorbed by the troposphere. Lower parts of the troposphere absorb most of the heat radiating from Earth's surface; in contrast, at higher elevations in the troposphere, the atmosphere is thinner and absorbs less energy. Consequently, temperature decreases with increased elevation in the troposphere; mountaintops are generally colder than valley floors, and commercial jet airliners must heat their cabins once reaching cruising altitudes, which are generally between 10 and 11 kilometers elevation.

The top part of the troposphere is the **tropopause**. It lies at an altitude of about 10 kilometers at the Equator, although it is lower at the poles. The tropopause is the boundary between the troposphere and the **stratosphere** above. It is characterized by an abrupt cessation in the steady decline in temperature with altitude, because cold air from the upper troposphere is too dense to rise higher. As a result, little mixing of air molecules occurs across the tropopause. The tropopause forms the floor of the stratosphere, in which temperature remains constant to 35 kilometers. It then increases with altitude until, at about 50 kilometers, it is about as warm as air at Earth's surface.

The reversal in the temperature profile between the troposphere and the stratosphere occurs because the two atmospheric layers are heated by different mechanisms. As already explained, the troposphere is heated primarily from below, by Earth. The stratosphere, however, is heated primarily from above, by direct incoming solar radiation.

Oxygen molecules (O_2) in the stratosphere absorb high-energy ultraviolet (UV) rays from the sun. The radiant energy breaks the oxygen molecules apart, releasing free oxygen atoms. These free oxygen atoms then recombine to form ozone (O_3). Ozone absorbs ultraviolet energy more efficiently than oxygen does. The absorption of UV radiation by the ozone warms the upper stratosphere.

Ultraviolet radiation is energetic enough to affect organisms. Small quantities give us a suntan. But large doses cause skin cancer and cataracts of the eye, inhibit the growth of many plants, and otherwise harm living tissue. The ozone in the upper atmosphere protects life on Earth by absorbing much of this high-energy radiation before it reaches Earth's surface.

Ozone concentration declines in the upper portion of the stratosphere, and therefore at about

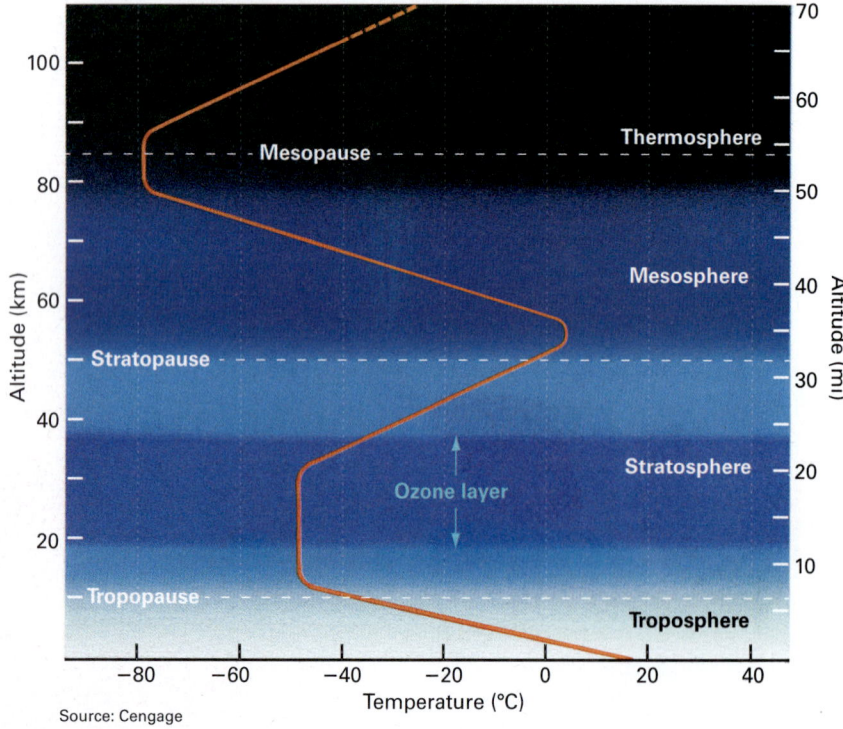

FIGURE 18-11
Atmospheric temperature varies with altitude. The atmospheric layers are zones in which different factors control the temperature.

55 kilometers above Earth, temperature once more begins to fall rapidly with elevation. This boundary between rising and falling temperature is the **stratopause**, the ceiling of the stratosphere. The second zone of declining temperature in Earth's modern atmosphere is the **mesosphere**. Little radiation is absorbed in the mesosphere, and the thin air is extremely cold. The ceiling of the mesosphere is the **mesopause**. Starting at about 80 kilometers above Earth, the temperature again remains constant and then rises rapidly in the **thermosphere**. Here the atmosphere absorbs high-energy x-rays and ultraviolet radiation from the sun. High-energy reactions strip electrons from atoms and molecules, producing ions. The temperature in the upper portion of the thermosphere is just below freezing—not extremely cold by surface standards.

The uppermost layer of the atmosphere is the **exosphere**. It is a thin layer of gas surrounding the thermosphere. In the exosphere, gas molecules are gravitationally bound to Earth, but their density is too low for the individual molecules to collide with each other, as they normally do at lower atmospheric pressure. The exosphere ultimately thins out and merges with space.

checkpoint Name the layers of Earth's atmosphere from the layer closest to Earth's surface to the layer farthest from Earth's surface.

18.3 ASSESSMENT

1. **Recall** How much of Earth's atmosphere lies below an altitude of about 5,600 meters?
2. **Explain** What is a barometer and how does it work?
3. **Contrast** How is barometric pressure typically different in stormy weather and in clear, dry weather?
4. **Contrast** What are the sources of heat in the troposphere and the stratosphere?
5. **Explain** How does ozone form in the stratosphere?

Computational Thinking

6. **Analyze** Most commercial airlines fly at altitudes between about 35,000 and 38,000 feet (10–12 km) and try to reach this height during the first ten minutes of flight. Use Figures 18-7 and 18-11 to construct an argument for why this altitude is ideal for commercial airplanes.

18.4 AIR POLLUTION

Core Ideas and Skills

- Describe the need, purpose, and impact of the U.S. Clean Air Act.
- Describe sources and effects of acid rain.
- Explain why ozone in the troposphere is harmful to living things while ozone in the stratosphere is protective to living things.
- Describe the development of the ozone hole over Antarctica and explain how chlorofluorocarbons have affected the ozone layer.
- Describe actions taken by various nations to reduce harm to the ozone layer, and how successful those actions have been.

KEY TERMS

acid precipitation
smog
particulate
aerosol
fly ash

chlorofluorocarbon (CFC)
halon
ozone hole

Ever since the first cave dwellers huddled around a smoky fire, people have introduced impurities into the air. The total quantity of these pollutants are small compared with the great mass of our atmosphere. Yet air pollution remains a significant health, ecological, and climatological problem for modern industrial society.

Sources and Types of Air Pollution

In 1948, Donora was an industrial town with a population of about 14,000, located 50 kilometers south of Pittsburgh, Pennsylvania. One large factory in town manufactured structural steel and wire, and another produced zinc and sulfuric acid. During the last week of October 1948, dense fog settled over the town. But it was no ordinary fog; the moisture contained pollutants from the two factories. After four days, visibility became so poor that people could not see well enough to drive, even at noon with their headlights on. Gradually at first, and then in increasing numbers, residents sought medical attention for nausea, shortness of breath, and constrictions in the throat and chest. Within a week, 20 people had died and about half of the town was seriously ill.

Other incidents similar to what happened in Donora occurred worldwide. In response to the growing problem, the United States enacted the Clean Air Act in 1963. As a result of the Clean Air Act and its amendments, total emissions of air pollutants have decreased and air quality across the country has improved (Figure 18-12). It is even more encouraging to note that this decrease in emissions has occurred at a time when population, energy consumption, vehicle miles traveled, and gross domestic product (GDP) have increased dramatically. Donora-type incidents in the United States have not been repeated. Smog has decreased, and rain has become less acidic. Yet some people believe that we have not gone far enough and that air pollution regulations should be strengthened further.

checkpoint What has resulted from the enactment of the Clean Air Act in 1963?

Burning Fossil Fuels

Coal is largely carbon, which, when burned completely, produces carbon dioxide. Petroleum is a mixture of hydrocarbons, compounds composed of carbon and hydrogen. When hydrocarbons burn completely, they produce carbon dioxide and water as the only combustion products. Neither is poisonous, but both are greenhouse gases. If fuels were composed purely of compounds of carbon and hydrogen, and if they always burned completely, air pollution from the burning of fossil fuels would pose little direct threat to our health (although combustion of fossil fuels would still contribute to global warming). However, fossil fuels contain impurities, and combustion is usually incomplete. As a result, other harmful products form.

Products of incomplete combustion include hydrocarbons such as benzene and methane. Benzene is a carcinogen and methane is another

FIGURE 18-12

Measurements of air pollutant concentrations in the United States between 1990 and 2015 are shown. This plot, from the U.S. EPA, shows average national concentrations of each air pollutant have fallen through the period. This reflects increased awareness of air quality and implementation of air quality measures.

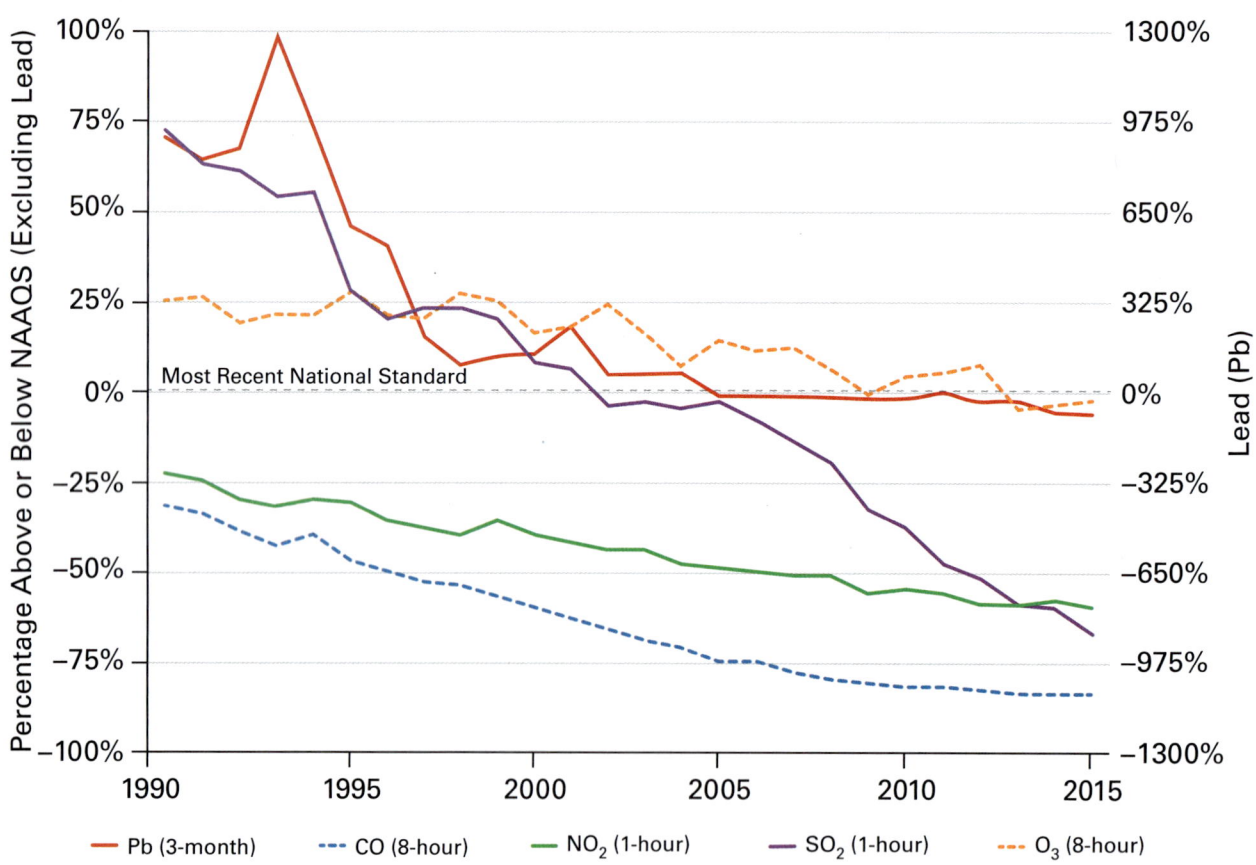

Source: U.S. EPA

DATA ANALYSIS Air Pollution

Air quality affects our health and our economy. However, many kinds of air pollutants from various sources can have specific effects on air quality. The Environmental Protection Agency (EPA) tracks the pollutants found in the air over time, as shown in the graph in Figure 18-12. Refer to the graph to answer the following questions.

1. **Collect Data** Research each pollutant shown on the graph. Identify major sources of each pollutant and explain any harmful effects on humans and the environment. Organize your findings in a data table.

2. **Synthesize** Choose an air pollutant you think is most important to reduce. Write a brief argument for reducing that pollutant. Use information from your research and from the graph to support your argument.

3. **Analyze** Why does the graph show amounts relative to the most recent national standard rather than simply the average amount in the air?

Computational Thinking

4. **Recognize Patterns** Describe the overall trend shown in the graph.

5. **Compare** Which type of pollution appears to have decreased the most relative to the national standard?

FIGURE 18-13
(A) Sources of air pollution in the United States are shown. (B) This graph represents types of air pollutants in the United States. (Although carbon dioxide is a greenhouse gas, it is not listed as a pollutant because it is not toxic.)

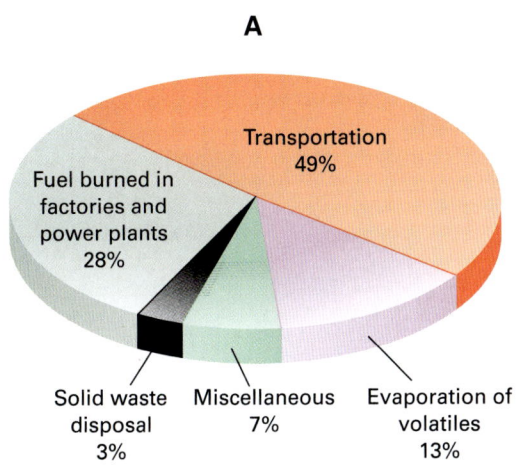

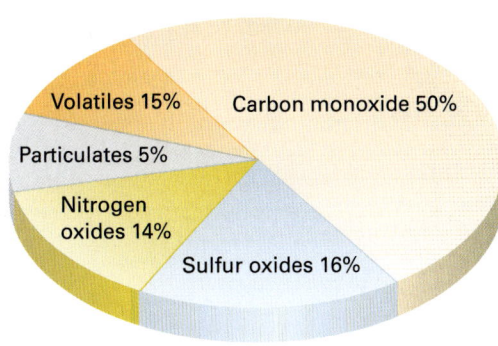

Source: Cengage

greenhouse gas. Incomplete combustion of fossil fuels releases many other pollutants. These include carbon monoxide (CO), which is colorless and odorless yet very toxic (Figure 18-13).

Additional problems arise because coal and petroleum contain impurities that create other kinds of pollution when they are burned. Small amounts of sulfur are present in coal and, to a lesser extent, in petroleum. When these fuels burn, the sulfur forms oxides, mainly sulfur dioxide (SO_2) and sulfur trioxide (SO_3). High sulfur dioxide concentrations have been associated with major air pollution disasters of the type that occurred in Donora. Today, the primary global source of sulfur dioxide pollution is coal-fired electric generators (Figure 18-13).

Nitrogen, like sulfur, is common in living tissue and therefore is found in all fossil fuels. This nitrogen, together with a small amount of atmospheric nitrogen, reacts when coal or petroleum is burned. The products are mostly nitric oxide (NO) and nitrogen dioxide (NO_2). Nitrogen dioxide is a reddish-brown gas with a strong odor. It contributes to the "browning" and odor of some polluted urban atmospheres. Automobile exhaust is the primary source of nitric oxide and nitrogen dioxide pollution. (Figure 18-13).

checkpoint What is the greatest source of air pollution in the United States?

Acid Rain

Recall that sulfur and nitrogen oxides are released when fossil fuels burn. These oxides are also released when metal ores are refined. In moist air, sulfur dioxide reacts to produce sulfuric acid. Nitrogen oxides react to form nitric and nitrous acid. These strong atmospheric acids dissolve in water droplets and fall as **acid precipitation**, also called acid rain (Figure 18-14).

Acidity is expressed on the pH scale (Figure 16-5). A solution with a pH of 7 is neutral, neither acidic nor basic. On a pH scale, numbers lower than 7 represent acidic solutions and numbers higher than 7 represent basic ones. For example, soapy water is basic and has a pH of about 10, whereas vinegar is an acid with a pH of 2.4.

Rain reacts with carbon dioxide in the atmosphere to produce a weak acid. As a result, natural rainfall has a pH of about 5.7. However, in the "bad old days" before the Clean Air Act was properly enforced, rain was much more acidic. A fog in Southern California in 1986 reached a pH of 1.7, which approaches the acidity of toilet bowl cleaners.

Consequences of Acid Rain In humans, sulfur and nitrogen oxides impair lung function, aggravating diseases such as asthma and emphysema. They also affect the heart and the liver. These oxides have been shown to increase vulnerability to viral infections such as influenza.

Acid rain can reduce nutrients in the soil that forests need, and make them more vulnerable to pests and diseases. Aquatic organisms can be affected when bodies of water become more acidic from acid rain. Many species can only survive in a narrow range of pH.

Acid rain corrodes metal and rock. Limestone and marble are especially susceptible because they dissolve rapidly in mild acid. In the United States the cost of damage and deterioration to buildings and building materials caused by acid precipitation is estimated at several billion dollars per year.

Figure 18-15A is a map of average pH for rainfall across the conterminous United States in 1992. The northeastern United States was particularly susceptible to acid rain because it was downwind from the interior rust belt, the large region of the northern interior United States in which large-scale industry had developed since the 19th century. A

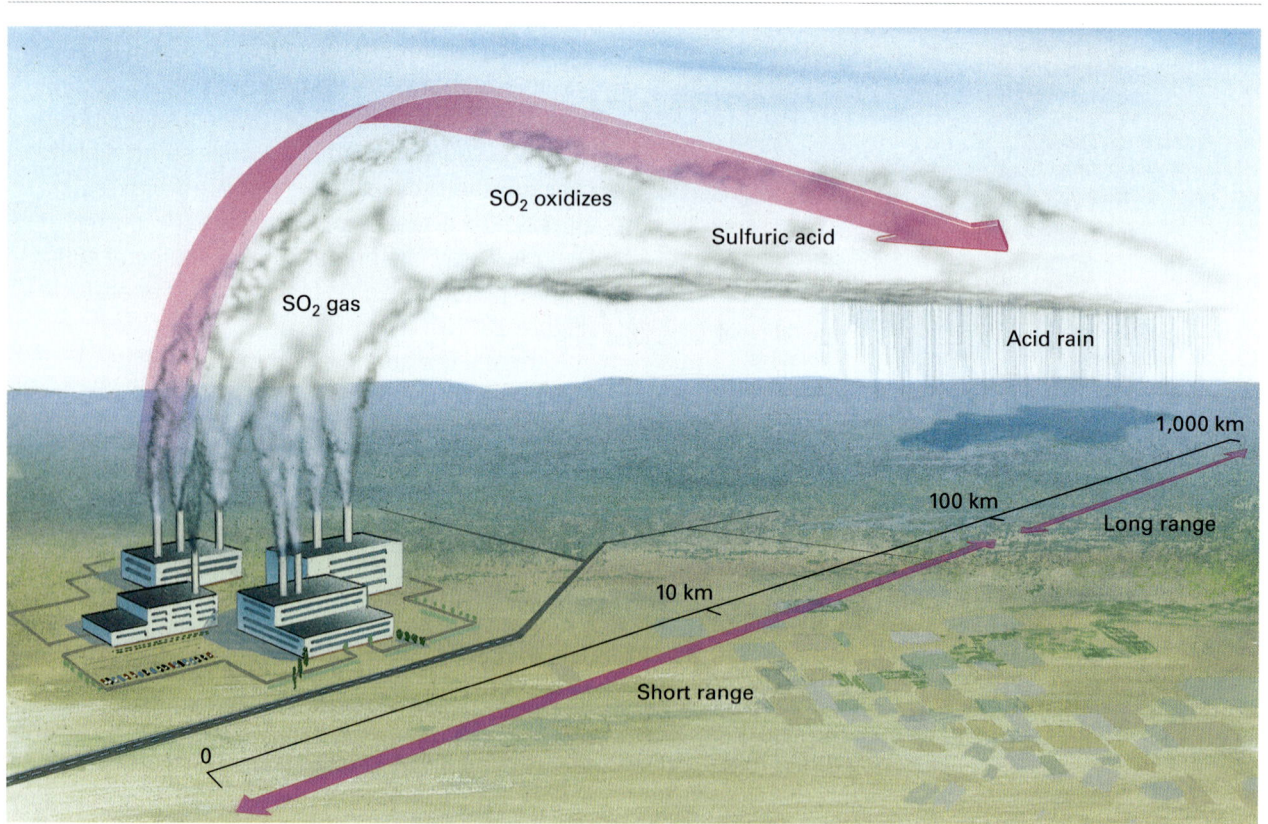

FIGURE 18-14
Acid rain develops from the addition of sulfur compounds to the atmosphere by industrial smokestacks.

FIGURE 18-15A ▼
Annual average measurements of the pH of precipitation during 1992 show that the highest levels of acidity in precipitation occurred in the northeastern United States, which is generally downwind of the industrial rust belt.

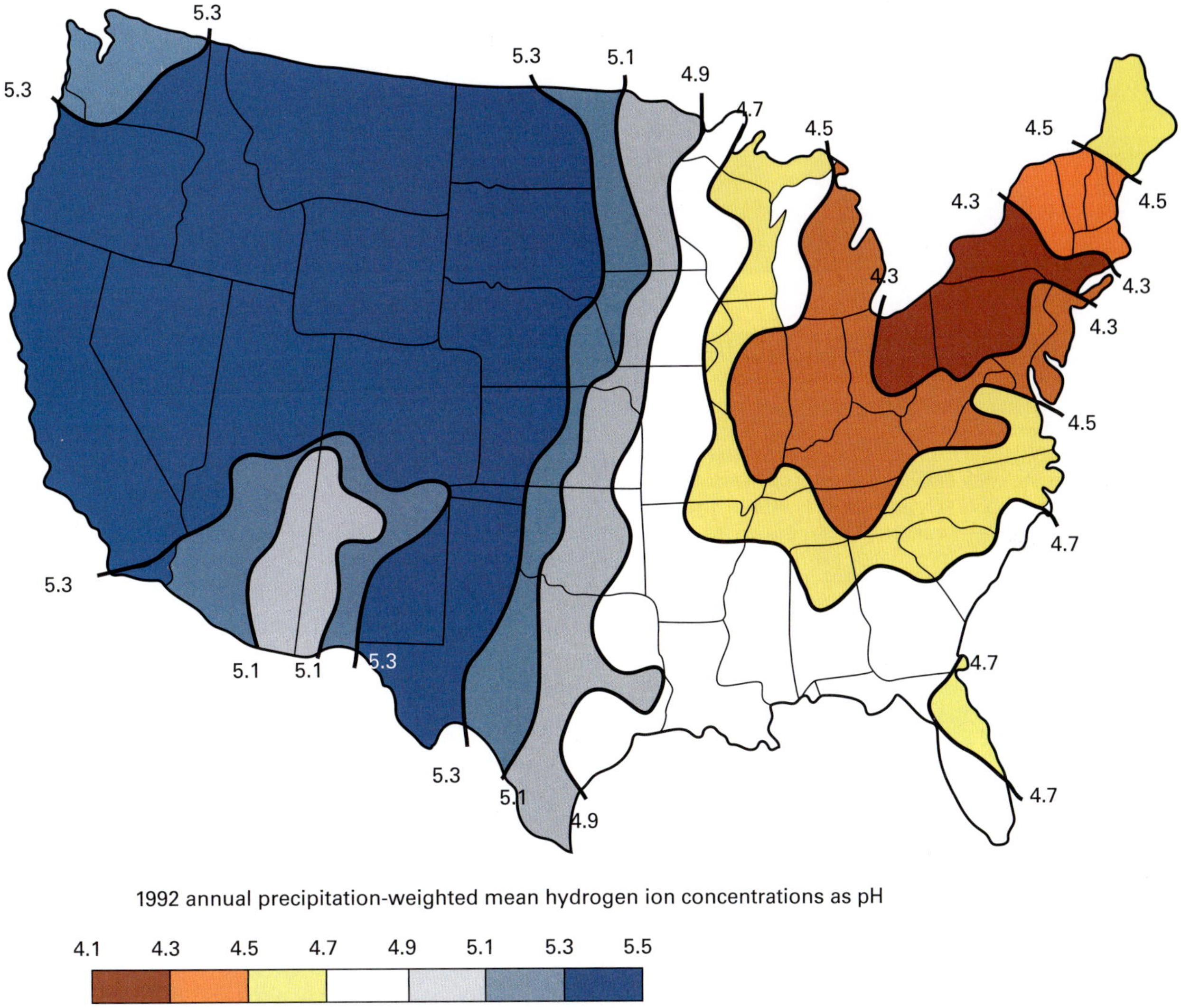

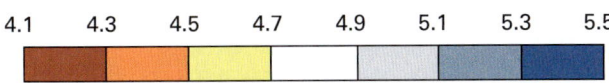

1992 annual precipitation-weighted mean hydrogen ion concentrations as pH

| 4.1 | 4.3 | 4.5 | 4.7 | 4.9 | 5.1 | 5.3 | 5.5 |

Source: Cengage

2016 study involved the analysis of 27 sites in the northeastern United States and eastern Canada over a period ranging between eight and 24 years. The results showed that long-term increases in pH occurred in soil at most sites. The study's authors attributed these trends to reversal of the effects of acid rain on North American soils.

FIGURE 18-15B
Go online to view the effects of acid rain on a North Carolina forest.

checkpoint What rocks are especially susceptible to acid rain corrosion?

Smog and Ozone in the Troposphere

Old-time photographic film was "slow" and required lots of sunlight, so many filmmakers left the polluted, overcast, industrial Northeast. Southern California, with its warm, sunny climate and little need for coal, was preferable. A district of Los Angeles called Hollywood became the center of the movie industry. Its population boomed, and after World War II automobiles became about as numerous as people. Then the quality of the air deteriorated in a strange way. People noted four different kinds of changes: a brownish haze called

LESSON 18.4

FIGURE 18-16
(A) Smog formed in the Los Angeles basin in the 1970s.
(B) Smog levels were greatly decreased in Los Angeles basin by 2018.

smog settled over the city (Figure 18-16); people felt irritation in their eyes and throats; vegetable crops became damaged; and the sidewalls of rubber tires developed cracks.

In the 1950s, air pollution experts worked mostly in the industrialized cities of the East Coast and the Midwest. When they were called to diagnose the problem in Southern California, they looked for the sources of air pollution they knew well, especially sulfur dioxide. But the smog was nothing like the pollution they were familiar with. These researchers eventually learned that incompletely burned gasoline in automobile exhaust reacts with nitrogen oxides and atmospheric oxygen in the presence of sunlight to form ozone (O_3). The ozone then reacts further with automobile exhaust to form smog (Figure 18-17).

Ozone in the stratosphere absorbs ultraviolet radiation and protects life on Earth. Yet excessive ozone in the air we breathe is harmful. Is ozone a pollutant to be eliminated, or a beneficial component of the atmosphere that we want to preserve? The answer is that it is both, depending on where it is found: ozone in the troposphere reacts with automobile exhaust to produce smog, and therefore it is a pollutant. Ozone in the stratosphere is beneficial, and the destruction of the ozone layer there creates serious problems.

Ozone irritates the respiratory system, causing loss of lung function and aggravating asthma in susceptible individuals. Ozone also increases susceptibility to heart disease. It is a suspected carcinogen. High ozone concentrations slow the

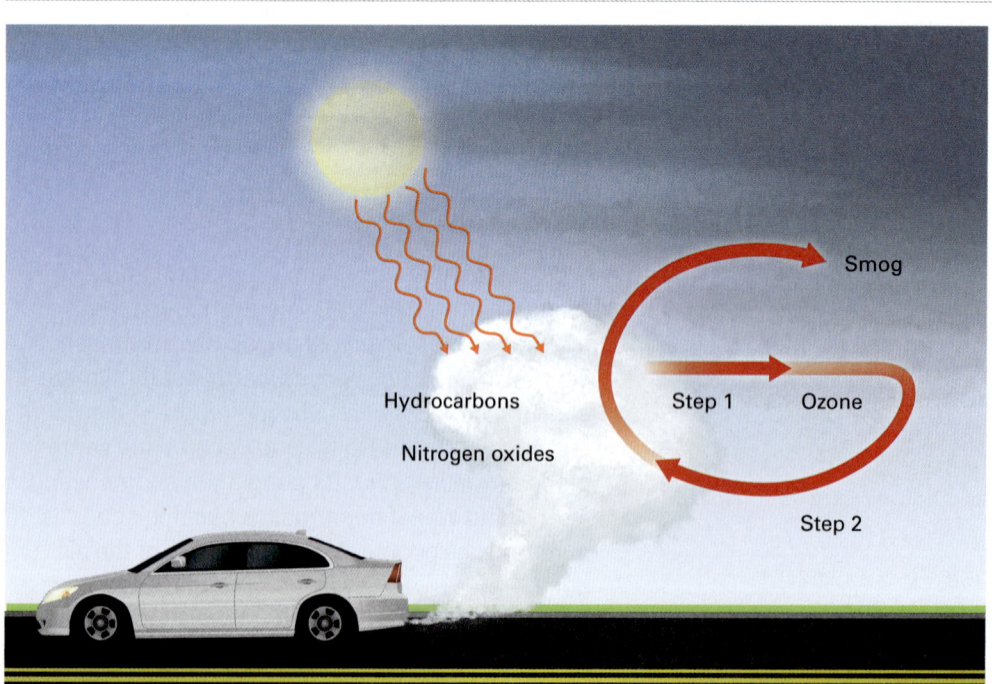

FIGURE 18-17
Smog forms in a sequential process. Step 1: Automobile exhaust reacts with air in the presence of sunlight to form ozone. Step 2: Ozone reacts with the hydrocarbons and nitrogen oxides in automobile exhaust to form smog.

growth of plants, which is a particularly serious problem in the rich, agricultural areas of California.

Ozone pollution ranks as the 33rd leading cause of human death. Air quality guidelines have been devised by the World Health Organization. According to these guidelines, as of 2018, more than 95 percent of Earth's human population lives with unsafe levels of air pollution. The *State of Global Air 2018* reported that global levels of ozone pollution increased between 1990 and 2016. Within the United States, ozone levels fell during the same period, as air pollution controls and fuel emission standards were implemented.

checkpoint What are some of the health hazards caused by ozone in the troposphere?

Toxic Volatiles

A volatile compound is one that evaporates readily and therefore easily escapes into the atmosphere. Whenever chemicals are manufactured or petroleum is refined, some volatile by-products escape into the atmosphere. When metals are extracted from ores, gases such as sulfur dioxide are released. When pesticides are sprayed onto fields and orchards, some of the spray is carried off by the wind. When you paint your house, the volatile parts of the paint evaporate into the air. As a result of all these processes, tens of thousands of volatile compounds are present in polluted air. Some are harmless, others are poisonous, and many have not been studied.

Consider the case of dioxin. Very little dioxin is intentionally manufactured. It is not an ingredient in any herbicide, pesticide, or other industrial formulation. You cannot buy dioxin at your local hardware store or pharmacy. Dioxin forms as an unwanted by-product in the production of certain chemicals and when specific chemicals are burned. For example, in the United States today, garbage incineration is the most common source of dioxin. When a compound containing chlorine, such as the plastic polyvinyl chloride (PVC), is burned, some of the chlorine reacts with organic compounds to form dioxin. The dioxin then goes up the smokestack of the incinerator, diffuses into the air, and eventually falls to Earth. Cattle eat grass lightly dusted with dioxin and store the dioxin in their fat. Humans ingest the compound mostly in meat and dairy products. The EPA estimates that the average U.S. citizen ingests about 1×10^{-10} gram (100 picograms) of dioxin in food every day. Although this is a minuscule amount, the EPA has argued that dioxin is the most toxic chemical known. Even these low background levels may cause adverse effects such as cancer, disruption of regulatory hormones, reproductive and immune system disorders, and birth defects. Others disagree, claiming there is no direct evidence to show that any of the effects of dioxins occur in humans in everyday levels.

No one knows whether very small doses of potent poisons ingested over long periods of time are harmful. Environmentalists argue that it is "better to be safe than sorry." Therefore, we should reduce ambient concentrations of volatiles such as dioxin. This argument has helped spur the movement towards increased recycling and decreasing incineration of waste plastic as a means of disposal. Others counter that because the harmful effects of compounds such as dioxin are unproven, we should not burden our economy with the costs of control.

checkpoint What are some toxic volatiles and what are their sources?

Particulates, Aerosols, and Chlorofluorocarbons

Particulates and Aerosols A **particulate**, or particle, is any small piece of solid matter, such as dust or soot. An **aerosol** is any small particle that is larger than a molecule and suspended in air. In the context of air pollution, all three terms—particulate, particle, and aerosol—are used interchangeably. Many natural processes release aerosols. Windblown silt, pollen, volcanic ash, salt spray from the oceans, and smoke and soot from wildfires are all aerosols. Industrial emissions add to these natural sources.

Smoke and soot are carcinogenic aerosols formed whenever fuels are burned. Coal always contains clay and other noncombustible minerals that have accumulated along with the organic matter in the environment in which the coal formed, commonly a swamp. When the coal burns, some of these minerals escape from the chimney as **fly ash**. The fly ash settles as gritty dust. When metals are mined, the drilling, blasting, and digging raise dust. This adds to the total load of aerosols. These aerosols can cause lung damage.

In 1988 EPA scientists noted that whenever aerosol levels rose above a certain level in Steubenville, Ohio, the number of fatalities from all causes rose. Other studies supported the

Steubenville report. The EPA proposed reductions of aerosol levels. Opponents argued that it is unfair to target all aerosols, because the term covers a range of substances from a harmless grain of salt to a deadly mix of toxic volatiles.

Chlorofluorocarbons As described in Lesson 18.3, solar energy breaks apart oxygen molecules (O_2) in the stratosphere. Free oxygen atoms (O) are released. The free oxygen atoms combine with oxygen molecules to form ozone (O_3). Ozone absorbs high-energy ultraviolet light. This absorption protects life on Earth because ultraviolet light can cause skin cancer, inhibit plant growth, and otherwise harm living tissue.

In the 1970s, scientists learned that organic compounds containing chlorine and fluorine, called **chlorofluorocarbons (CFCs)**, and compounds containing bromine and chlorine, called **halons**, rise into the upper atmosphere. Here they react with and destroy ozone (Figure 18-18). CFCs have no natural source. They were synthesized for use as cooling agents in refrigerators and air conditioners. They were also used as propellants in aerosol cans. They served as an expanding agent in foam coffee cups and some polystyrene building insulation.

In 1985 scientists observed an unusually low ozone concentration in the stratosphere over Antarctica, a phenomenon called the **ozone hole**. Since it was discovered, the low concentration of ozone over Antarctica declined further until, by 1993, it was 65 percent below normal. The ozone hole was more than 23 million square kilometers, an area almost the size of North America. Research groups also reported significant increases in ultraviolet radiation from the sun at ground level

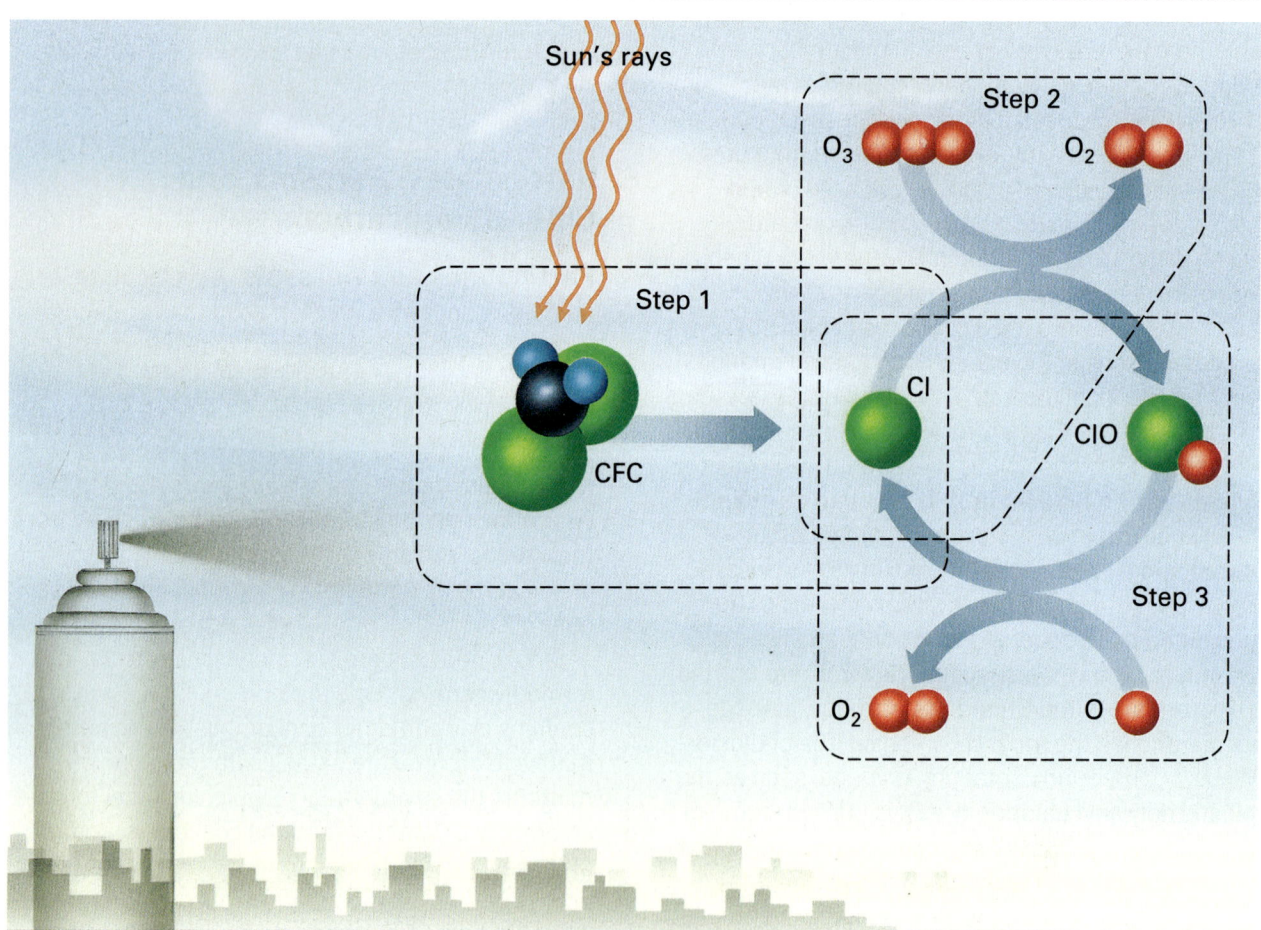

FIGURE 18-18
CFCs destroy the ozone layer in a three-step reaction. Step 1: CFCs rise into the stratosphere. Ultraviolet radiation breaks the CFC molecules apart, releasing chlorine atoms. Step 2: Chlorine atoms react with ozone, O_3, to destroy the ozone molecule and release oxygen, O_2. The extra oxygen atom combines with chlorine to produce ClO. Step 3: The ClO sheds its oxygen, producing another free chlorine atom. Therefore, chlorine is not used up in the reaction, and one chlorine atom reacts over and over again, to destroy many ozone molecules.

FIGURE 18-19

(A) Stratospheric ozone concentrations over Antarctica in 1994 are shown in this satellite image. (B) Similar data for 2004 are shown. In both images, dark purple shows the lowest ozone concentration. Ozone concentrations declined from the 1980s through the early 1990s and stabilized in the early 2000s.

A
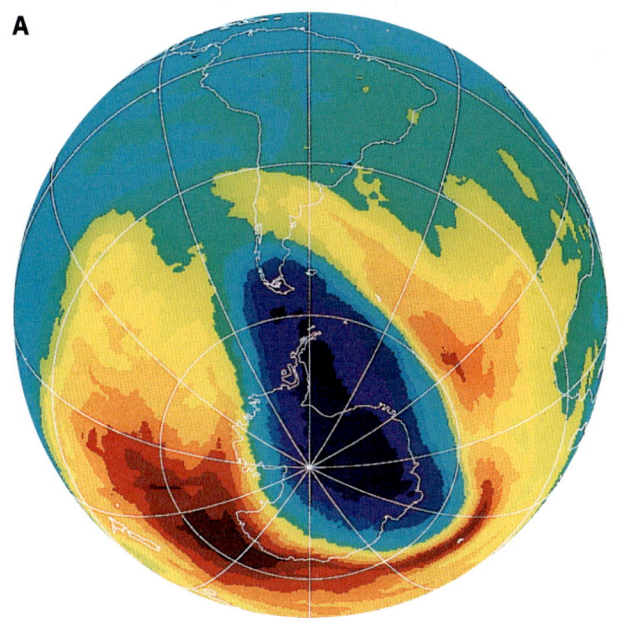

B
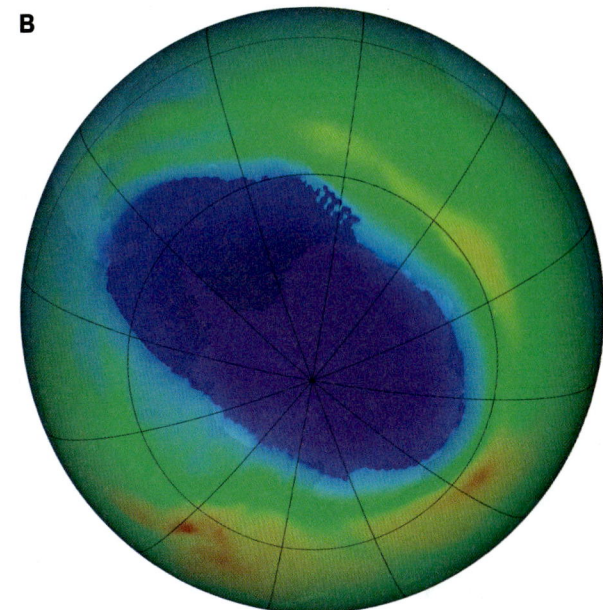

Source: NASA/Goddard Space Flight Center Scientific Visualization Studios

in the region. In addition, scientists recorded ozone depletion in the Northern Hemisphere. In March 1995, ozone concentration above the United States was 15 to 20 percent lower than during March 1979.

Data on global ozone depletion persuaded industrial nations to limit the use of CFCs and other ozone-destroying compounds. A series of international agreements were developed. Many nations of the world agreed to reduce or curtail production of compounds that destroy atmospheric ozone. Most industrialized countries stopped production of CFCs on January 1, 1996.

The international bans have had positive results (Figure 18-19). The concentration of ozone-destroying chemicals peaked in 1994 and has been declining ever since, although CFCs are projected to remain in the atmosphere for more than a hundred years. The CFCs and halons already in the atmosphere break down slowly, but at least the measured concentrations of these ozone-destroying chemicals are declining. As of 2016, eleven different CFC compounds known to be completely synthetic still occur in Earth's atmosphere.

checkpoint What caused low levels of ozone in the ozone hole?

18.4 ASSESSMENT

1. **Recall** What air pollutants are generated when coal and gasoline burn?
2. **Explain** How does acid rain affect people, forests, lakes and streams, and building materials?
3. **Summarize** Briefly discuss the health effects of aerosols.
4. **Explain** How do CFCs deplete the ozone layer?
5. **Summarize** Discuss the effects of the international ban on ozone-destroying chemicals.

Critical Thinking

6. **Evaluate** Which form of pollution do you think does the most harm? Which form of pollution do you think would be the most difficult to reduce? Explain.
7. **Evaluate** Nations cooperated with the international ban on ozone-destroying chemicals. Do you think this type of cooperation would be possible today for problems such as the burning of fossil fuels, which adds to global warming? Explain.

LESSON 18.4

TYING IT ALL TOGETHER
REDUCING AIR POLLUTION

In this chapter, you learned about the gradual transformation of Earth's primordial atmosphere into the modern, oxygen-rich atmosphere that supports life on Earth today. For much of Earth's history, changes in the atmosphere have been propelled by the gradual evolution of life. Biological processes such as photosynthesis and respiration have played a pivotal role in shaping the composition of the atmosphere.

In the Case Study, we examined one way the atmosphere continues to change today, namely through increased methane emissions. As with other drivers of atmospheric change, methane emissions arise from a complex array of factors, some natural and some linked to human activity. As a greenhouse gas, methane released into the atmosphere traps 20–30 times more heat than carbon dioxide. Scientists warn that large releases of methane now trapped in thawing permafrost could contribute to climate change. Efforts to capture methane and burn it as a cleaner alternative to fossil fuels may help curb its harmful effects, but most current technologies for methane capture are still in early stages of development.

Methane is just one of many substances that is currently altering the fragile balance of Earth's atmosphere. Addressing the negative effects of air pollution will require a multi-pronged effort to reduce emissions of harmful substances into the air.

Work with one or two other students to complete the steps.

1. Discuss the types of pollution you have learned about in this chapter. Select one type of pollution of regional or global importance to research further.

2. With your group, research the following questions about the type of pollution you chose:

   - How does this type of pollution enter the atmosphere?
   - How does this pollutant interact with or change the makeup of the atmosphere? In what ways does it alter the natural processes related to the atmosphere?
   - What are the harmful effects of this type of air pollution on human health, organisms, and/or ecosystems?
   - What efforts are currently being made to address this type of air pollution? To what extent have these measures been successful in reducing this type of pollution?
   - What still needs to be done to reduce the impact of this pollutant on Earth's systems? What possible solutions are being researched or tested?
   - How does this type of pollution affect your own community?

3. Organize your research findings into a presentation. You may wish to present to the class, create a video, or post your findings on a class blog.

FIGURE 18-20
In 1952, a "Great Smog" covered the city of London, resulting in thousands of deaths.

CHAPTER 18 SUMMARY

18.1 EARTH'S EARLY ATMOSPHERES

- Earth's primordial atmosphere was composed almost entirely of hydrogen and helium. After these light elements boiled off into space, bolides colliding with Earth brought volatile compounds that produced a secondary atmosphere. This secondary atmosphere was composed predominantly of carbon dioxide, with lesser amounts of water, nitrogen, and trace gases.

- About four billion years ago, outgassing, combined with reactions of gases with rocks of the geosphere, created an atmosphere rich in hydrogen and carbon dioxide.

18.2 LIFE, IRON, AND THE EVOLUTION OF THE MODERN ATMOSPHERE

- Photosynthesis by primitive cyanobacteria began producing oxygen about 2.7 billion years ago.

- For 300 million years, the oxygen concentration did not increase dramatically in the atmosphere because of two processes: (1) most of the early-formed oxygen reacted with dissolved iron in seawater to form banded iron formations; the oxygen concentration in the atmosphere did not rise until most of the dissolved iron in the oceans had been removed by precipitation; and (2) when life first evolved, the hydrogen concentration of the atmosphere remained high and oxygen reacted with the hydrogen to produce water; the oxygen concentration in the atmosphere rose only after the hydrogen concentration declined as bacteria converted the hydrogen to methane.

- The First Great Oxidation Event occurred about 2.4 billion years ago. After that time, further production of oxygen was roughly balanced by decomposition.

- About 600 million years ago, numerous processes caused organic carbon to be buried, leading to the Second Great Oxidation Event.

- In the modern atmosphere, dry air is roughly 78 percent nitrogen (N_2), 21 percent oxygen (O_2), and 1 percent other gases. Air also contains water vapor, dust, liquid droplets, and pollutants.

18.3 ATMOSPHERIC PRESSURE AND TEMPERATURE

- Atmospheric pressure (or barometric pressure) is the weight of the atmosphere per unit area. Pressure varies with weather and decreases with altitude.

- Atmospheric temperature decreases with altitude in the troposphere. Then the temperature rises in the stratosphere because ozone absorbs solar radiation; at an altitude of about 50 kilometers, the stratosphere is almost as warm as Earth's surface. The temperature decreases again in the mesosphere where it is extremely cold. Then in the uppermost layer, the thermosphere, temperature increases to just below freezing as high-energy radiation is absorbed.

18.4 AIR POLLUTION

- The increasing ill effects of air pollution World War II convinced lawmakers to pass legislation such as the 1963 Clean Air Act.

- Incomplete combustion of coal and petroleum produces carcinogenic hydrocarbons such as benzene, carbon monoxide, and methane, as well as nitrogen and sulfur oxides.

- Nitrogen and sulfur oxides react in the atmosphere to produce acid rain, which damages health, weathers materials, and reduces growth of crops and forests.

- Incompletely burned fuels in automobile exhaust react with nitrogen oxides in the presence of sunlight and atmospheric oxygen to form ozone. Ozone then further reacts with automobile exhaust to form smog.

- Dioxin is an example of a compound that is produced inadvertently during chemical manufacture and when certain materials are burned.

- Some scientists argue that even tiny amounts of toxic volatiles may be harmful to human health. Scientific studies show that some industrial aerosols are harmful to health.

- Chlorofluorocarbons and halons rise into the stratosphere and deplete the ozone that filters out harmful UV radiation and protects Earth.

- Ozone-destroying chemicals have been regulated, and their concentration in the troposphere is slowly diminishing. Overall, Earth's stratospheric ozone level has been increasing, but serious ozone holes remain at the poles.

CHAPTER 18 ASSESSMENT

Review Key Terms
Select the key term that best fits the definition and enter it in the blank. Not all terms will be used, and no term will be used more than once.

acid precipitation	mesosphere
aerosol	outgassing
banded iron formation	ozone hole
bar	particulate
barometer	pH scale
barometric pressure	photosynthesis
chlorofluorocarbon (CFC)	Second Great Oxidation Event
cyanobacteria	smog
First Great Oxidation Event	stratopause
fly ash	stratosphere
Gaia	thermosphere
halon	tropopause
mesopause	troposphere

1. process by which some organisms convert carbon dioxide and water to organic sugars, using sunlight as an energy source
2. Greek name for Earth used by James Lovelock, inspired by the interconnectedness of living and nonliving components of Earth's systems
3. ceiling of the atmospheric zone just below the thermosphere
4. fine, solid matter, such as dust and soot, that is suspended in the atmosphere
5. organic compounds containing chlorine and fluorine that react with and destroy ozone in the upper atmosphere
6. the atmospheric zone closest to Earth and the one we live in
7. brown haze that forms when ozone reacts with automobile exhaust
8. an area of the stratosphere centered over Antarctica that has been altered due to chlorine-containing chemicals released by humans
9. organic compounds containing chlorine and bromine that react with and destroy ozone in the upper atmosphere
10. the release of volatile substances from Earth's mantle into the atmosphere during volcanic eruptions
11. a unit of measure of atmospheric pressure
12. layer of minerals in sedimentary rock between 2.7 and 1.9 billion years old that originated as oxygen levels increased dramatically in the atmosphere
13. any small particle larger than a molecule and suspended in the atmosphere
14. the ceiling of the atmospheric zone closest to Earth
15. the time 2.4 billion years ago when oxygen levels in Earth's atmosphere increased abruptly

Review Key Concepts
Answer each question on a separate sheet of paper to demonstrate your understanding of key concepts from the chapter.

16. What was Earth's earliest atmosphere like and why didn't it last?
17. What was the source of carbon dioxide, nitrogen, and water vapor that scientists think made up Earth's primordial atmosphere?
18. What was the First Great Oxidation Event, and how did scientists find evidence of it?
19. How do scientists explain the sudden rise of oxygen in Earth's atmosphere that took place between 2.7 and 1.9 billion years ago?
20. When did the Second Great Oxidation Event occur and how do scientists explain its cause?
21. Is the air today chemically reactive? Explain.
22. Explain why barometric pressure is greatest near Earth's surface but decreases with increasing altitude.
23. Identify three mechanisms that heat Earth's atmosphere and the atmospheric zones in which these mechanisms occur.
24. Describe an event that occurred in 1948 in the United States that drew attention to air pollution as a problem that needed to be solved.
25. What is acid rain and how does it form?
26. How does ozone in the troposphere affect living things?
27. Why is loss of ozone in the stratosphere cause for concern?

Think Critically

Write a response to each question on a separate sheet of paper. Use concepts from the chapter to support your reasoning.

28. Why is it important to include a discussion of Earth's gravity in any explanation of how Earth's atmosphere formed?

29. How might the evolution and survival of organisms on Earth have been different if iron had not been a plentiful element on the planet's surface?

30. Microorganisms can be classified according to the various categories listed in the table. Construct a hypothesis about the order in which these classes of microorganisms may have evolved on Earth. Explain your reasoning.

Class	Description
obligate aerobes	require oxygenated environment to survive and can tolerate high levels of oxygen
obligate anaerobes	require anoxic environment to survive and cannot tolerate any level of oxygen
microaerophiles	require oxygenated environment to survive but can tolerate only low levels of oxygen
facultative anaerobes	can tolerate either anoxic or oxygenated environments by switching metabolisms

31. Use the two graphs to review the relationships between barometric pressure, altitude, and temperature. Explain why barometric pressure decreases steadily with decreasing altitude while temperature does not.

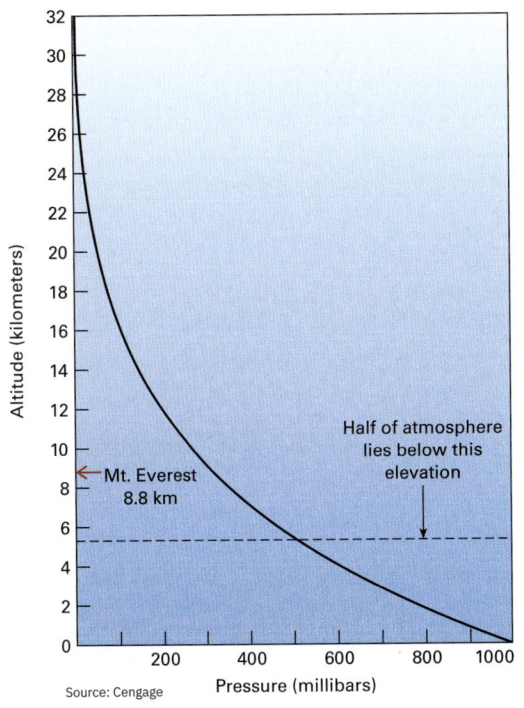
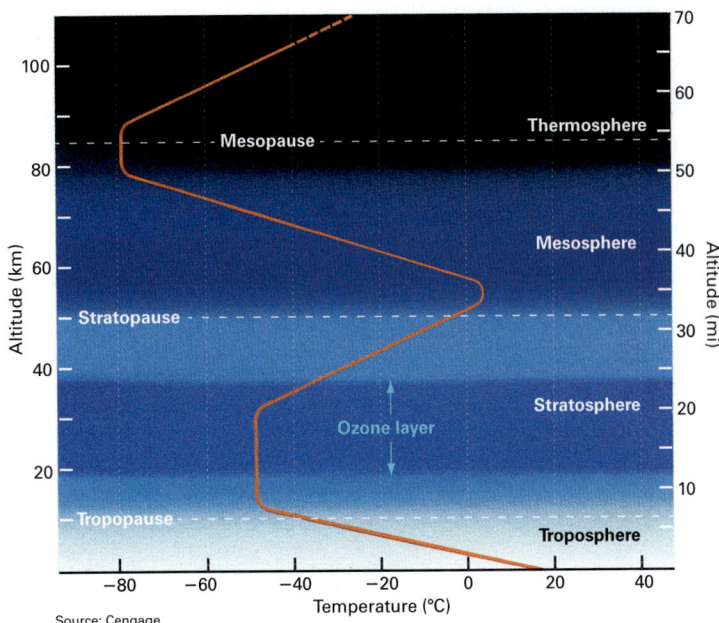

CHAPTER 18 ASSESSMENT

32. The table summarizes data collected on weather conditions and air. Air quality refers to the levels of air pollutants including particulates, aerosols, and toxic volatile compounds. Describe some relationships that you observe in the data and explain why you think they exist.

Day/Night	Sun/Clouds	Wind	Air Quality
Day	Sunny	No wind	Healthy
Day	Sunny	Light wind	Healthy
Day	Sunny	Windy	Healthy
Day	Cloudy	Windy	Moderate to healthy
Night	Less than half cloudy	Windy	Moderate to healthy
Night	Less than half cloudy	Light wind	Unhealthy
Day	Cloudy	No wind	Unhealthy
Day	Cloudy	Light wind	Unhealthy
Night	More than half cloudy	Light wind	Unhealthy
Night	More than half cloudy	Windy	Unhealthy
Night	More than half cloudy	No wind	Very unhealthy
Night	Less than half cloudy	No wind	Very unhealthy

33. The table summarizes some data related to specific halogen-containing compounds that have been banned by the Montreal Protocol. Why do you think some compounds were placed under a shorter ban schedule than others?

Compound	Montreal Protocol control measure	Lifetime of Compound in the Stratosphere (Years)	Relative Rate of Compound's Ability to Deplete Ozone
CFC-12	Phased out in 1995	100	1
CFC-113	Phased out in 1995	85	1
CFC-11	Phased out in 1995	45	1
HCFCs	35% reduction by 2004 75% reduction by 2010 90% reduction by 2015 Total phase out by 2020	1–26	0.02–0.12
Carbon tetrachloride (CCl_4)	Phased out in 1995	26	0.73
Halon-1301	Phased out in 1993	65	16
Halon-1211	Phased out in 1993	16	7.1
Methyl bromide (CH_3Br)	Phased out in 2005	0.7	0.51

Source: NOAA

PERFORMANCE TASK

Coevolution of Earth's Systems and Life on Earth

Earth's systems include the atmosphere, biosphere, geosphere, and hydrosphere. Throughout Earth's history, these systems have shaped one another in dynamic ways, influencing how each system has evolved over time.

In this performance task, you will choose a research question based on an interaction between Earth systems that affects the evolution of one, two, or three systems. You will conduct research to gather information and data relevant to your chosen question; then you will use your findings to write an evidence-based argument to answer the question.

1. Research and construct an evidence-based argument about Earth system interactions.

   A Select a question for your research. The question should be centered on a topic related to the interactions between the atmosphere, biosphere, geosphere, and/or hydrosphere in Earth's early history and how one system brought about changes in another. You may use one of the questions below or develop your own question. Be sure to obtain permission from your teacher for any questions you construct on your own.

   - When did the earliest photosynthetic organisms appear on Earth, and does this timing support the idea that photosynthesis was the driver of rising oxygen levels in the atmosphere?
   - What geoscience factors would have had to be present to enable the evolution of photosynthesis in living organisms? And does the timing of the occurrence of these factors in the geologic record support current thinking about when photosynthetic organisms first appeared?
   - Why was there a lag between when photosynthetic organisms evolved and when oxygen levels increased abruptly in Earth's atmosphere?
   - How did the rise of oxygen in the atmosphere change chemical interactions between the atmosphere and geosphere, and how did these changes affect the biosphere?
   - How did geoscience factors influence the rate of evolution of complex multicellular organisms (metazoans)?

   B Conduct research to find an answer to your question. As you search for an answer to your question, focus on the evidence collected by scientists that supports the answer.

   C Write a paper explaining the results of your research. As you write your paper, be sure to include the following things.
   - The research question you chose
   - An answer to the research question
   - Evidence to support the answer to the research question
   - An explanation of how the evidence supports the answer
   - An analysis of the quality of the evidence

CHAPTER 19
ENERGY BALANCE IN THE ATMOSPHERE

Sunlight travels about 150 million kilometers (93 million miles) in 8 minutes and 18 seconds to reach Earth. Seen from the edge of space, the atmosphere surrounds our planet like a thin blue skin.

All life on Earth depends on energy from sunlight. Our planet is but a speck in space—a minute, rocky pebble compared to the size of the sun. The sun has a mass that is about 333,000 times greater than Earth's. About 1,300,000 Earths could fit inside the sun, a roiling ball of incredibly hot, electrically charged particles that undergo nuclear fusion, putting out an unimaginable amount of energy in all directions. Without the sun, Earth would be a frozen, lifeless speck instead of the dynamic planet of water, land, air, and life it has been for several billion years. The sun makes life on Earth possible. Yet Earth only receives about one-billionth of the energy the sun radiates. So how does Earth maintain the conditions that make life possible? What is the balance between burning up from the sun's energy or freezing without it? The balance comes from complex interactions between the sun's energy and Earth's spheres.

KEY QUESTIONS

19.1 What processes influence how solar radiation affects the temperature of Earth's atmosphere?

19.2 How can you describe the balance of incoming solar radiation with the outgoing radiation from Earth's systems?

19.3 How does energy storage and transfer affect Earth's weather and climate?

19.4 How do geographic factors affect temperature?

EXPLORERS AT WORK

TAKING A CLOSER LOOK AT GLACIAL MELTING

WITH NATIONAL GEOGRAPHIC EXPLORER STEFANIE LUTZ

Stefanie Lutz is a microbiologist and ecologist who studies ways in which microorganisms adapt to extreme environments. Recently, Lutz has turned her attention to a type of green algae that thrives in the frigid water produced by melting glaciers. Lutz's research shows that these tiny organisms may play a big role in accelerating glacial melting. That, in turn, may accelerate the impacts of climate change.

Over the last 50 years, glacial melting rates have increased due to climate change. The Arctic region is being disproportionately affected; air temperatures there are rising at faster rates than anywhere else on the planet. But rising air temperature is not the only cause. Changes to the surfaces of glaciers themselves also have important effects, which is where Lutz's research comes in.

The scientific name for the algae Lutz studies is *Chlamydomonas nivalis*. It can survive in snow, and when it is present, it causes what is called red, pink, or watermelon snow. The term refers to a red pigment that *C. nivalis* produces to help protect it from high levels of light irradiation, an adaptation that is not fully understood. During spring and summer, the algae "bloom," or reproduce rapidly, as glaciers naturally begin to melt. The newly released water and nutrients spur algal growth, and patches of red snow appear on the surfaces of the glaciers. According to Lutz, the increasing distribution and duration of the algae's appearance is the problem.

Lutz explains that "snow algae" darkens the surface of a glacier, which then reflects less incoming solar radiation. This is important because glaciers and ice sheets, with their light-colored surfaces, reflect large amounts of solar radiation back into space, thus helping regulate global climate. The ratio of reflected to incoming radiation is called *albedo*. Objects with darker surfaces have a lower albedo. They absorb more solar radiation and heat up. For glaciers, says Lutz, reduced albedo leads to faster melting, which spurs more algal blooms, which in turn causes more melting—a positive feedback loop. But how significant is the impact?

Lutz headed up a team from the University of Leeds and the German Research Center for Geosciences (GFZ) that studied snow algae in the Arctic. The team collected 40 samples of red snow from 16 glaciers in Norway, Sweden, Greenland, and Iceland. Lutz's analysis showed red snow can reduce the surface albedo locally as much as 20 percent. Furthermore, this decrease in surface albedo varied little from site to site, despite variations in environmental conditions. Lutz says that means her findings could be extrapolated to predict the influence of snow algae on melt rates of glaciers throughout the entire Arctic region.

Lutz notes that current models that forecast climate change do not account for how snow algae affect glacial melting. That impact could be significant because the distribution snow algae may expand as melting increases. She hopes her research can contribute to improved climate models.

THINKING CRITICALLY

Synthesize Draw a diagram that illustrates how the presence of snow algae leads to ever-increasing glacial melt in a positive feedback loop.

Stefanie Lutz takes samples from a melting glacier in Norway. She studies how microorganisms are able to survive in the icy Arctic environment.

CASE STUDY
MELTING GLACIERS IN GREENLAND

The vast glacial ice sheets that cover most of Greenland and Antarctica hold more than 99 percent of Earth's fresh water. In Greenland, where the ice sheet averages more than 1.5 kilometers (1 mile) thick, a thin layer of surface ice normally melts each summer before refreezing again in the winter. But in recent years, satellite imagery has revealed an alarming increase in the rate of seasonal melting in Greenland (Figure 19-2). This uptick aligns closely with an upward trend in Arctic summer air temperatures.

In July 2012, an unusually warm summer, satellites showed that an unprecedented 97 percent of the surface of Greenland's ice had begun to melt. That summer, meltwater runoff flooded rivers, damaged roads, washed out bridges, and caused global sea levels to rise by 1 millimeter. Seasonal melting since 2012 has not been quite as dramatic. But the extent of melting is still considerably higher than it was in past decades, triggering a cumulative loss of glacial mass.

Satellite data on Greenland's ice cover has only been gathered since the early 1990s. But other methods are giving researchers new insights into patterns of melting going back several centuries. In 2017, a team of scientists began drilling deep into the Greenland ice sheet to extract ice cores. Using computer models to analyze ice core data, scientists can determine historical patterns of melting. The results show that seasonal melting began slowly accelerating in the mid-19th century and then spiked dramatically over the past two decades as temperatures have risen.

But factors other than temperature are also accelerating Greenland's ice loss. For example, Stefanie Lutz and other researchers have documented the presence of algae populations in the ice that darken its surface as it melts. Dust, soot, and some pollutants in the ice have a similar effect. The darkening of surface ice as it melts causes it to absorb more of the sun's energy, which in turn causes even more ice to melt.

Researchers in Greenland are conducting studies to find out what happens to all of the excess meltwater. Although some seeps down into the ice sheet and refreezes, along the coastlines much of it runs off into streams and rivers that dump straight into the ocean. Scientists predict that if all of Greenland's ice were to melt, Earth's sea level would rise by 7 meters (23 feet).

As You Read Find out more about processes and feedback mechanisms that are contributing to accelerating glacial loss in places like Greenland. What systems influence weather and climate, and what factors are responsible for changes in temperature? Consider how these factors affect other places and regions on Earth.

▶ **FIGURE 19-1**
Greenland's ice sheet is almost 2,500 meters thick near the center of the island, but it is thinning near the coast due to accelerated seasonal melting of surface ice and snow. Here, streams of meltwater plummet into a hole in the ice sheet.

🌐 **FIGURE 19-2**
Go online to see a graph showing melting rates of Greenland's ice sheet.

19.1 INCOMING SOLAR RADIATION

Core Ideas and Skills
- Explain how weather and climate are affected by many overlapping systems.
- Describe how energy is absorbed and reemitted from Earth's surface and atmosphere.
- Describe the effects that reflected sunlight has on global temperatures.
- Describe the effects of the scattering of sunlight in the atmosphere.

KEY TERMS
weather
climate
photon
electromagnetic radiation
frequency
electromagnetic spectrum
albedo

Weather, Climate, and Energy

Weather is the state of the atmosphere at a given place and time. Temperature, wind, cloudiness, humidity, and precipitation are all components of weather. The weather changes frequently, from day to day or even from hour to hour.

Climate is the characteristic weather of a region, particularly the temperature and precipitation, averaged over several decades. Miami and Los Angeles have warm climates; summers are hot and even the winters are warm. In contrast, New York and Chicago experience much greater temperature extremes. Even though winters are cool and often snowy, summers can be almost as hot as those in Miami. Seattle experiences moderate temperatures with foggy, cloudy winters. In this and the following two chapters, we discuss the atmosphere, weather, and climate. In Chapter 21 we will consider climate types in more detail and review historical and current climate change.

Many of the processes that drive energy balance in the atmosphere vary in gradual and predictable ways. But the weather and climate don't always change gradually and predictably because they are affected by many complex, overlapping system interactions and by threshold and feedback mechanisms. For example, although incoming solar radiation is most intense at the Equator and decreases predictably towards the poles, some high-latitude locations are warmer than regions closer to the Equator. (See Lesson 19.4.) On any given day, it can be warmer in Montreal than in Houston. In this and later chapters, we will explain the fundamental processes that drive weather and climate, and how they interact.

The space between Earth and the sun is nearly empty. How does sunlight travel through a vacuum? In the late 1600s and early 1700s, light was poorly understood. Isaac Newton postulated that light consists of streams of particles that he called "packets" of energy. Two other physicists, Robert Hooke and Christiaan Huygens, argued that light travels in waves. Today we know that Hooke and Huygens were correct—light behaves as a wave; but Newton was also right—light acts as if it is composed of particles. But how can light be both a wave and a particle at the same time? In a sense this is an unfair question because light is fundamentally different from familiar objects. Light is unique; it behaves as a wave *and* a particle simultaneously.

Particles of light are called **photons**. In a vacuum, photons travel only at one speed, the speed of light. The speed of light is 3×10^8 meters per second. At that rate, a photon covers the 150 million kilometers between the sun and Earth in about eight minutes. Photons appear when they are emitted by matter and disappear when they are absorbed by matter.

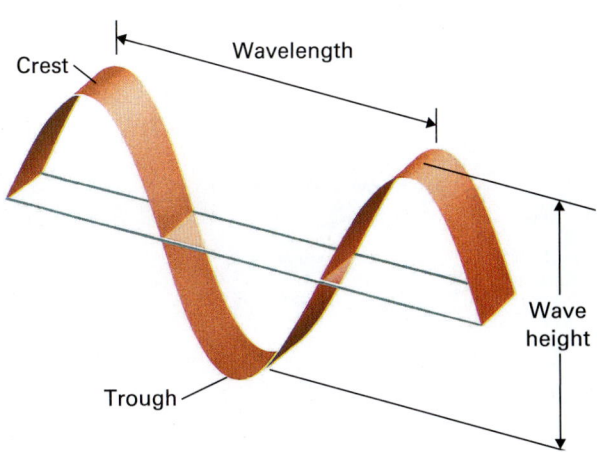

FIGURE 19-3
The terms used to describe a light wave are identical to those used for water, sound, and other types of waves.

FIGURE 19-4
Go online to view an illustration of the electromagnetic spectrum.

FIGURE 19-5
ON ASSIGNMENT National Geographic photographer Chris Hill took this photo of a rainbow at Sliabh Liag, a mountain on the Atlantic coast of County Donegal, Ireland. In a rainbow, sunlight is refracted and reflected in water droplets in the atmosphere. As sunlight enters a droplet, the light waves change direction. Different colors of light within sunlight have different frequencies, which cause them to travel at different speeds as they move through the water droplets.

FIGURE 19-6
Iron glows red, then orange, then heats to white as a blacksmith works in his shop.

As waves, light behaves with alternating electric and magnetic fields called an electromagnetic wave. **Electromagnetic radiation** is made up of electromagnetic waves. The terminology describing a light wave is the same as for mechanical waves such as sound waves or ocean waves (Figure 19-3). The wavelength of an electromagnetic wave is the distance between successive wave crests. Its **frequency** is the number of complete wave cycles, from crest to crest, that pass by a fixed point in one second. (You can think of frequency as how *frequently* the waves pass by.) Electromagnetic radiation occurs in a wide range of wavelengths and frequencies. The **electromagnetic spectrum** is the continuum of radiation of different wavelengths and frequencies (Figure 19-4). At one end of the spectrum, radiation given off by ordinary household electrical current has a long wavelength (5,000 kilometers) and low frequency (60 cycles per second). At the other end, cosmic rays from space have a short wavelength (about one-trillionth of a centimeter, or 10^{-14} meter) and very high frequency (10^{22} waves per second). Visible light is only a tiny portion (about one-millionth of 1 percent) of the electromagnetic spectrum.

checkpoint Which types of electromagnetic waves have the highest and lowest frequencies?

Absorption and Emission

What happens when sunlight strikes Earth's surface and the energy is absorbed? Each photon of light is a tiny packet of concentrated energy. (The energy of a single photon is related only to the frequency of the light, not to the surrounding temperature.) A photon's energy may be transferred to an electron in a molecule. If that happens, the electron jumps to what scientists call an excited state—it has a higher level of energy than its ground state (lowest level). But electrons do not remain in their excited states forever. They fall back to a lower energy state, initiating the emission of radiation in the process. In other words, the photon's energy was transferred to the electron and eventually the electron released that energy back in the form of a photon. All objects emit radiant energy at some wavelength.

An iron bar at room temperature emits low-energy infrared radiation. It is invisible because the energy is too low to activate the sensors in our eyes. Infrared radiation is sometimes called "heat" because we feel the warmth from objects that give off infrared radiation (whereas we do not feel ultraviolet radiation).

If you place the iron bar in a hot flame, it begins to glow with a dull, red color when it becomes hot enough. The added energy from the flame has excited electrons in the iron bar, and the excited electrons then emit visible, red, electromagnetic radiation (Figure 19-6). If you heat the bar further, it gradually changes color until it becomes white. This demonstration shows another property of emitted electromagnetic radiation: The wavelength (color) of emitted radiation is determined by the temperature of the source. With increasing temperature, the energy level of the radiation increases, the wavelength decreases, and the color changes progressively.

When the sun's radiation strikes Earth, it is absorbed by rock and soil. After the radiant energy is absorbed, the rock and soil reemit it as low-energy, infrared radiation, which has a relatively long wavelength and low frequency. Earth absorbs high-energy, visible light and emits invisible, low-energy infrared radiation.

checkpoint What causes an iron bar to glow when it is heated to a very high temperature?

Reflection

Solar radiation may be reflected from a surface instead of being absorbed and reemitted. We are familiar with the images reflected by a mirror or the surface of a still lake. Some surfaces are better reflectors than others. The reflectivity of a surface is referred to as its **albedo** (from the Latin for "whiteness") and is often expressed as a percentage. A mirror reflects nearly 100 percent of the light that strikes it and has an albedo close to 100 percent. Even some dull-looking objects are efficient reflectors. Light bounces back to your eye from the white paper of the page of a text, although very little is reflected from the black letters.

Snowfields, glaciers, and clouds have high albedos, whereas city buildings, dark pavement, and forests have low albedos (Figure 19-7). The oceans, which cover about two-thirds of Earth's surface, also have a low albedo. As a result, they absorb considerable solar energy and strongly affect Earth's radiation balance.

FIGURE 19-7
The albedos of common Earth surfaces vary greatly.

Source: Cengage

The temperature balance of the atmosphere is profoundly affected by the albedos of the hydrosphere, the geosphere, and the biosphere. If Earth's albedo were to rise by growth of glaciers or increased cloud cover, the surface of our planet would cool. Likewise, a decrease in albedo (caused by the melting of glaciers) would cause warming. Thus, the growth and shrinking of Earth's snow and ice is a classic example of a positive feedback mechanism. If Earth were to cool by a small amount, snowfields and glaciers would grow. The larger surface area of snow and ice would reflect more solar radiation back into space and lead to additional cooling—which would lead to more snow and ice, and so on.

checkpoint Give an example of a surface with a high albedo and a surface with a low albedo.

Scattering

On a clear day in the Northern Hemisphere the sun shines directly through windows on the south side of a building, but if you look through a north-facing window, you cannot see the sun. Even so, light enters through the window, and the sky outside is blue. If sunlight were only transmitted directly—as it is, for example, on the moon—a room with north-facing windows would be dark and the sky outside the window would be black. Atmospheric gases, water droplets, and dust particles scatter sunlight in all directions (Figure 19-8). It is this scattered light that illuminates a room with north-facing windows and turns the sky blue.

The amount of scattering is inversely proportional to the wavelength of light. Short-wavelength blue light scatters more than longer-wavelength red light. The sun emits light of all wavelengths. The visible wavelengths of light combine to make up white light. Consequently, in space the sun appears white. The sky appears blue from Earth's surface because the blue component of sunlight scatters more than other frequencies and therefore the atmosphere appears blue. The sun appears yellow from Earth because yellow is the color of white light with most of the blue light removed.

checkpoint How does scattering of sunlight occur in Earth's atmosphere?

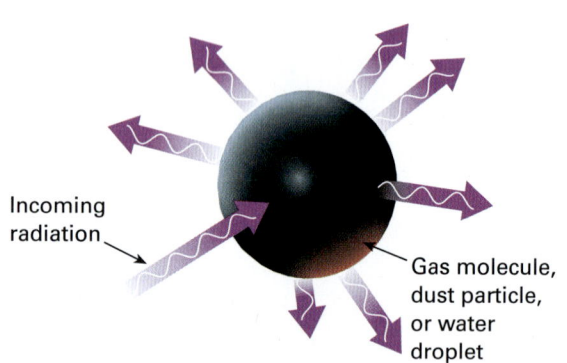

FIGURE 19-8
Atmospheric gases, water droplets, and dust scatter incoming solar radiation. When radiation scatters, its direction changes but the wavelength remains constant.

19.1 ASSESSMENT

1. **Contrast** What is the difference between weather and climate?
2. **Define** What is the electromagnetic spectrum?
3. **Explain** Why does the sky appear blue from Earth's surface?

Critical Thinking

4. **Apply** Based on what you have learned about the scattering of light, give a hypothesis for why the sky often appears yellow, orange, or red during sunrise and sunset.

Computational Thinking

5. **Analyze Data** The albedo of snow and ice is between 80 and 90 percent. This means that snow and ice absorb 10 to 20 percent of sunlight. Create a data representation, such as a stacked bar graph, that shows the percentage of sunlight that is absorbed and reflected by the different types of surfaces shown in Figure 19-7. Use this representation to compare two surfaces and construct an explanation for why one is a more efficient reflector than the other.

19.2 THE RADIATION BALANCE

Core Ideas and Skills
- Describe how solar radiation interacts with Earth's surface and atmosphere.
- Explain how the greenhouse effect works in Earth's atmosphere.

KEY TERMS

greenhouse effect

The Greenhouse Effect

Of all the solar radiation that arrives at Earth, only about 50 percent of it reaches the surface (Figure 19-9). Generally about 3 percent of that radiation is reflected by the surface and the rest is absorbed. The absorbed radiation warms rocks, soil, and water.

If Earth absorbs solar energy, why doesn't Earth's surface get hotter and hotter until the oceans boil and the rocks melt? The answer is that rocks, soil, and water reemit virtually all the energy they absorb. As explained previously, Earth's surface absorbs solar energy and then reemits the energy mostly as long-wavelength, invisible, infrared radiation. Some of this thermal energy escapes directly into space, but some is absorbed by the atmosphere. The atmosphere traps the energy reradiating from Earth and acts as an insulating blanket.

If Earth had no atmosphere, Earth's surface would cool drastically at night. Earth remains warm at night because the atmosphere absorbs and retains much of the radiation emitted by the ground. This warming process is called the **greenhouse effect** (Figure 19-10). If the atmosphere were to absorb even more of the long-wavelength radiation from Earth, the atmosphere and Earth's surface would become warmer.

FIGURE 19-9 ▼
Half of incoming solar radiation reaches Earth's surface. The atmosphere scatters, reflects, and absorbs the other half. All the radiation absorbed by Earth's surface is reradiated as long-wavelength infrared radiation.

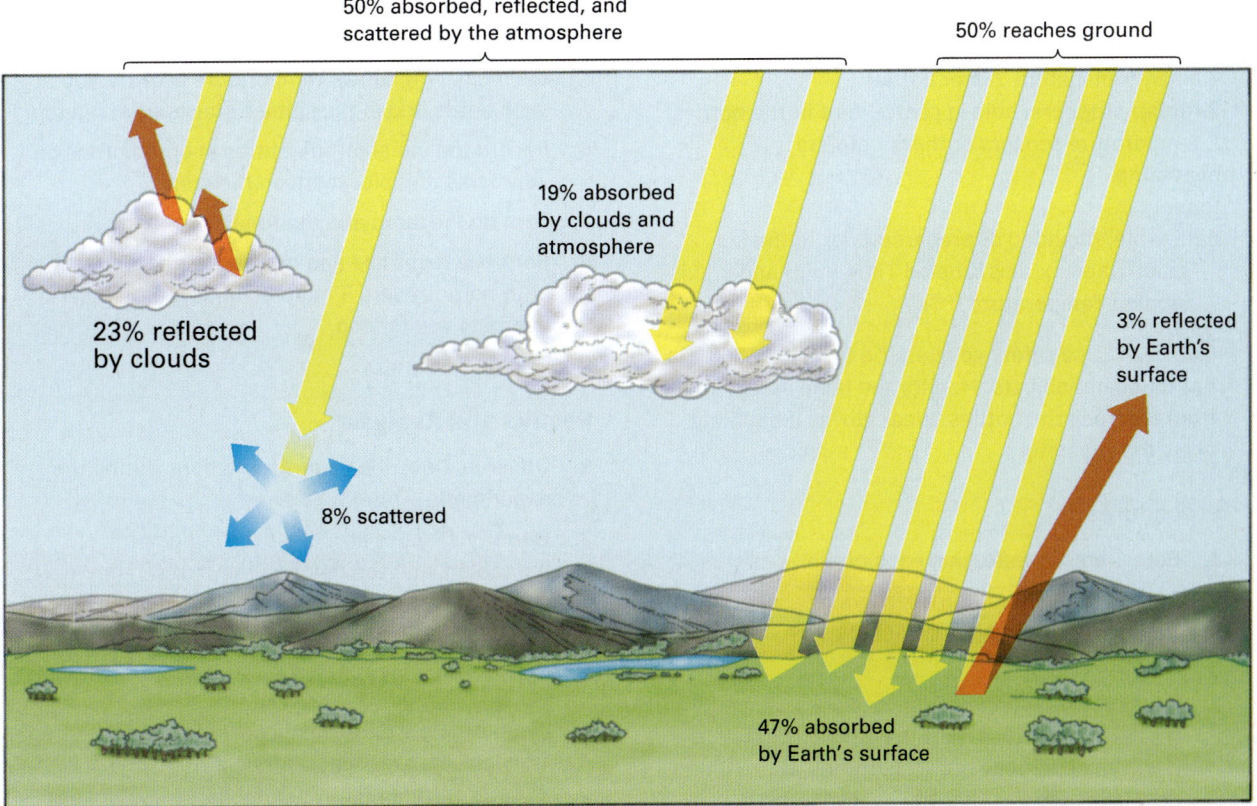

Source: Cengage

LESSON 19.2

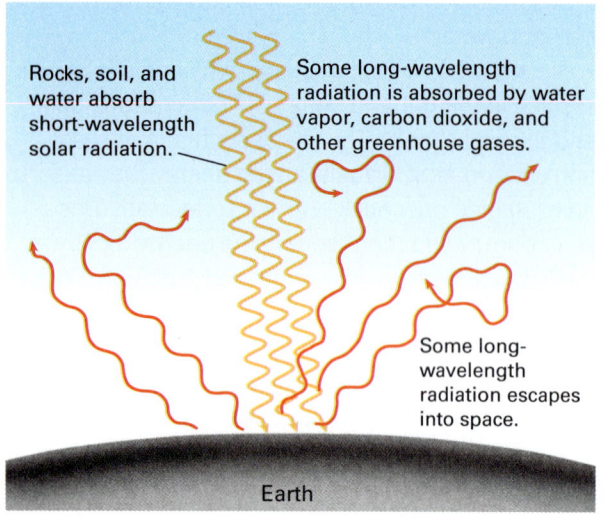

FIGURE 19-10
The greenhouse effect can be viewed as a three-step process. Step 1: Rocks, soil, and water warm up when they absorb short-wavelength solar radiation (orange lines). Step 2: Earth reradiates the energy as long-wavelength infrared radiation (red lines). Step 3: Greenhouse gases in the atmosphere absorb some of the energy, and the atmosphere becomes warmer.

Carbon dioxide is an important greenhouse gas because its abundance in the atmosphere can vary as a result of several natural and industrial processes. Methane has become important because large quantities are released by industry, agriculture, and thawing permafrost. Water is another important greenhouse gas. In fact, in humid regions during the summer, nighttime temperatures remain very warm because of the thermal energy that is retained by water vapor in the air. In arid regions, air temperatures drop noticeably once the sun goes down because there is little water vapor in the air to retain the energy. The greenhouse effect and global climate change are discussed in more detail in Chapter 21.

Dust, cloud cover, aerosols, and other particulate air pollutants also affect Earth's atmospheric temperature. These materials alter the amount of solar radiation that is absorbed or reflected. Large volcanic eruptions eject ash and dust into the troposphere, partially blocking solar radiation from reaching Earth. This can cause several years of global

---

MINILAB Simulating the Greenhouse Effect

Materials (per pair):

safety goggles
3 effervescent antacid tablets
2 liters of water
2 clear, transparent 2-liter bottles
2 rubber stoppers, hole in center, that fit the bottles
2 temperature sensors or thermometers
heat lamp
timer
data table, provided or prepared prior to the lab
  graph paper, x-axis labeled Time (min) and y-axis labeled Temperature (°C)

CAUTION: Use safety goggles. Tie back long hair. Do not touch hot objects. Monitor the lamp and keep it from touching the bottles. Wear gloves if handling glass thermometers.

Procedures

1. Pour 1 liter of water into each bottle.
2. Prepare the temperature sensors or thermometers and insert each through the hole of a stopper.
3. Position the heat lamp approximately 56 cm (2 ft.) from the bottles. Plug it in and aim it so that it will shine directly onto the bottles when turned on. Make sure the lamp shines evenly on both bottles, then turn off the lamp.
4. Drop all three effervescent tablets into one of the bottles.
5. Place the stoppers in the bottles. Position the temperature sensors or thermometers so that they are measuring the air temperature near the top of each bottle and at the same distance from the top.
6. Record the initial temperatures in both bottles on your table and plot them on the graph.
7. Turn on the lamp and the timer.
8. Continue recording and plotting temperatures in both bottles every 5 minutes until your teacher instructs you to stop.
9. Unplug the lamp.

Results and Analysis

1. **Observe** Describe your observations during the experiment.
2. **Identify** Discuss as a class the ingredients of the tablets and the reaction they triggered. What greenhouse gas was released by the dissolving antacid tablets?

Critical Thinking

3. **Evaluate** How did the experimental setup simulate the greenhouse effect?
4. **Analyze Data** Did the recorded data indicate the greenhouse effect was occurring? Explain.

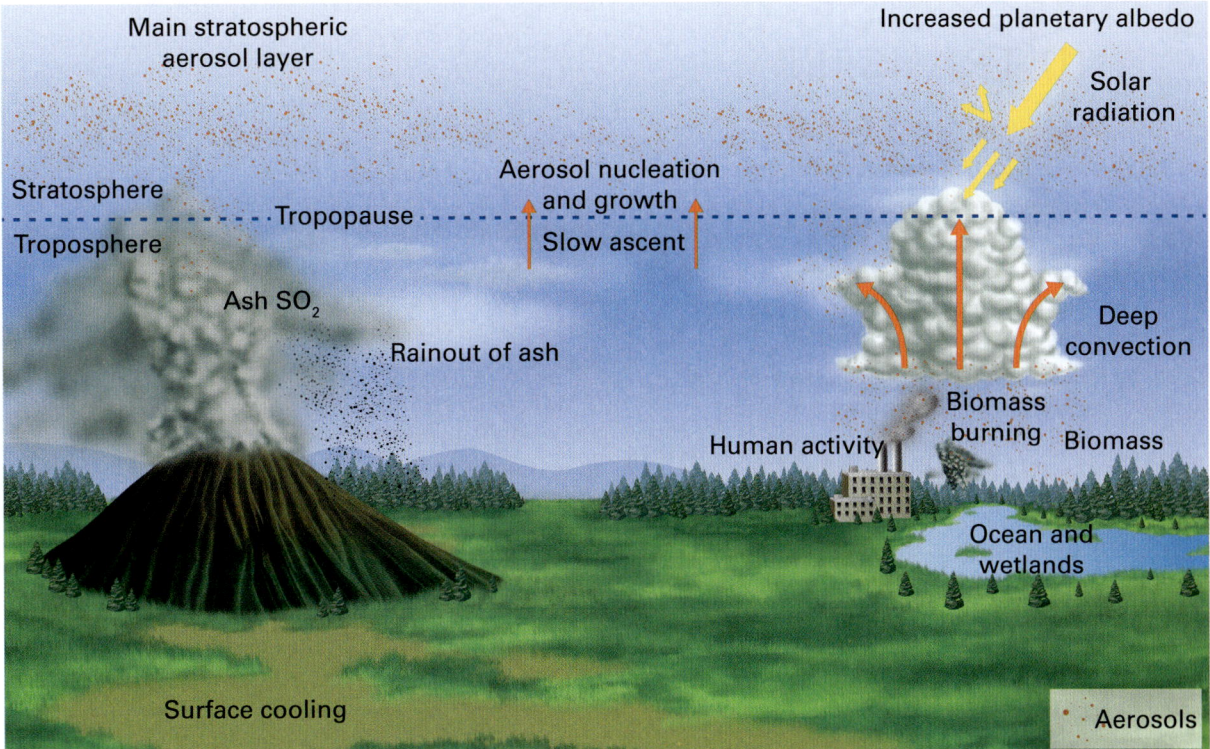

FIGURE 19-11
Large volcanic eruptions inject tons of ash into the troposphere, blocking some solar radiation. Sulfur dioxide gas from the eruption continues upward into the stratosphere. Once there, the sulfur dioxide converts to droplets of sulfuric acid, which forms a layer of aerosols that partially block solar radiation. The aerosols cause further cooling of Earth's surface. Sulfur dioxide is also introduced into the troposphere through the burning of fossil fuels and through natural environmental processes. But these aerosol sources are minor compared to that from large eruptions.

cooling as the dust settles. In addition, large volcanic eruptions in which ash clouds reach the tropopause can inject sulfur dioxide gas into the overlying stratosphere (Figure 19-11). Within about a month, the sulfur dioxide gas converts to tiny droplets of sulfuric acid about one micrometer in diameter. The droplets reflect incoming solar radiation back into space, which results in even more cooling.

The effect of human-caused aerosol and particulate pollution on global climate change is harder to interpret. Dust and aerosols close to the surface of Earth reflect incoming solar radiation but also absorb infrared radiation emitted by the ground. The net result depends on a complex balance of factors, including particle size, particle composition, and natural cloudiness. Whatever the outcome, it is certain that whenever people change the composition of air, they run the risk of altering weather and climate.

checkpoint Why are humid regions typically warmer at night than arid regions?

19.2 ASSESSMENT

1. **Identify** What are some factors that affect the amount of solar radiation absorbed by Earth's surface?
2. **Explain** If Earth's surface absorbs solar radiation, why doesn't the surface become hotter and hotter until rocks melt?
3. **Recall** What are the three important greenhouse gases in our atmosphere?
4. **Summarize** Summarize the role of the geosphere and hydrosphere in the three-step process of the greenhouse effect.

Critical Thinking

5. **Critique** A student makes a claim that aerosols from volcanic eruptions would warm Earth because the aerosols act as a blanket, keeping heat in the atmosphere. How would you revise this claim and why?

19.3 ENERGY STORAGE AND TRANSFER

Core Ideas and Skills
- Describe how convection transports heat in the air.
- Explain how different materials, such as water, change state.
- Explain how evaporation and condensation can cause changes in humidity and the temperature of land and water.
- Describe how land and water absorb and release heat at different rates and amounts, producing regional variations in weather and climate.

KEY TERMS

temperature	latent heat
heat	specific heat
conduction	advection

Heat and Temperature

All matter consists of atoms or molecules that are in constant motion. They move through space, rotate, and vibrate. The **temperature**, or the measure of the average kinetic energy (thermal energy) within a substance, is proportional to the average speed of the atoms or molecules in a sample of that substance. In a teacup full of boiling water, the water molecules are racing around rapidly, smashing into each other, spinning like boomerangs, and vibrating like spheres connected by pulsating springs. In a glass full of ice water, the water molecules are moving much more slowly. Molecules in the hot water are moving faster than those in the cold water, so the hot water has a higher temperature.

In contrast, **heat** is the energy transferred from a higher-temperature substance to a lower-temperature substance. A substance does not possess heat but does possess thermal energy. Heat and thermal energy are often used interchangeably, but thermal energy is part of the internal energy of a substance. The average energy of every molecule in a substance multiplied by the total number of molecules is called *internal energy*. It may seem counterintuitive, but there is more internal energy in a bathtub full of ice water than in a teacup full of boiling water. The average molecule in the ice water is moving slower and therefore has less thermal kinetic energy than the average molecule in the boiling water. But there are so many more molecules in the bathtub than in the teacup that the total internal energy is greater. Heat will still be transferred from the hotter teacup to the cooler bathtub full of ice because of the temperature difference.

checkpoint What is the difference between heat and temperature?

Conduction and Convection of Heat

If you place a metal frying pan on the stove, the metal handle gets hot, even though it is not in contact with the burner, because the metal conducts heat from the bottom of the pan to the handle. **Conduction** is the transport of heat by direct collisions among atoms or molecules. When the frying pan is heated from below, the metal atoms on the bottom of the pan move more rapidly. They then collide with their neighbors and transfer energy to them. Like a falling row of dominoes, energy is passed from one atom to another throughout the pan until the handle becomes hot.

Metals conduct heat rapidly and efficiently, but air is a poor conductor. To understand how air transports heat, imagine that a heater is placed in one corner of a cold room. The heated air in the corner expands, becoming less dense. This diffuse, hot air rises to the ceiling. It flows along the ceiling, cools, falls, and returns to the stove, where it is reheated (Figure 19-12A). Recall from Chapter 7 that convection is the upward and downward flow of fluid material in response to heating and cooling.

Movements of hot or cold fluids can be driven by processes other than heating or cooling and the resulting changes in density. For example, the Gulf Stream moves heat horizontally from the Caribbean region northward to the coast of western Europe. This lateral movement of heat is called **advection**. Advection is commonly confused with convection because both involve moving fluids that transfer heat. But it is the heat itself that drives convection. Advection involves a fluid that is moving for some other reason and transfers heat from one place to another in the process. For example, in a forced-air furnace, air heated by the furnace is advected into the living space by a fan and series of ducts.

Both advection and convection also occur in the atmosphere (Figure 19-12B). If air in one region is heated above the temperature of surrounding air, this warm air becomes less dense and rises—or convects—just like a hot-air balloon. Eventually the

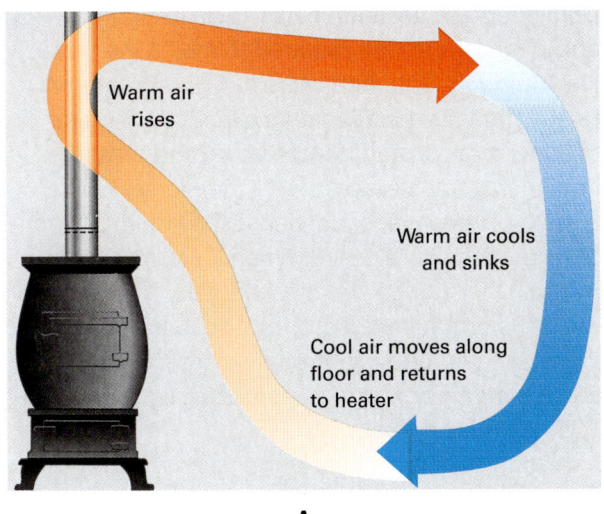
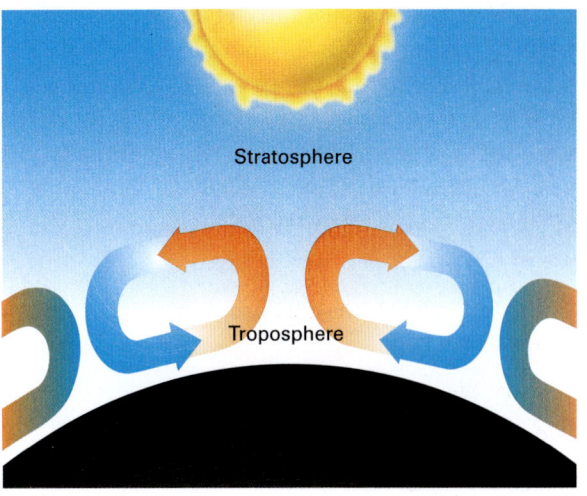

FIGURE 19-12
(A) Convection currents distribute heat throughout a room. (B) Convection also distributes heat through the atmosphere when the sun heats Earth's surface. In this case the ceiling is the boundary between the troposphere and the stratosphere.

MINILAB Modeling Atmospheric Convection and Wind

Materials (per group):

safety goggles
empty cardboard box, approximately
  20 cm × 13 cm × 6 cm (8 in. × 5 in. × 2.5 in.)
2 sheets transparent acetate,
  30.5 cm × 30.5 cm (12 in. × 12 in.)
1 sheet black construction paper,
  23 cm × 30.5 cm (9 in. × 12 in.)
scissors
cellophane tape
small candle (votive or tealight) in glass or metal
  container
lighter
incense stick

CAUTION: Use safety goggles. Tie back long hair. Do not touch hot objects. Monitor the flame at all times.

Procedures

1. Cut a rectangular panel out of one long side of the box, leaving a narrow border all around it.
2. Cut and tape construction paper to cover the back wall of the box facing the panel.
3. Cut an acetate sheet to cover the open panel. Tape it around the border.
4. Placing the box sideways, cut two holes (approximately 3 cm in diameter) on top, one near each end.
5. Cut an acetate sheet in half. Form two tubes by rolling each half. Fit the tubes into the holes.
6. Opening one end of the box, place the candle under one of the tubes. Then light the candle and close the box again.
7. Light the end of a stick of incense and hold the smoking end just above the other tube.

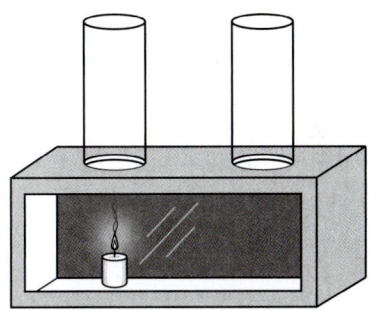

8. Observe what happens to the incense smoke.
9. Extinguish the candle and incense.

Results and Analysis

1. **Observe** Describe the behavior of the incense smoke.
2. **Connect** How did the lab model atmospheric convection and wind?

Critical Thinking

3. **Evaluate** Why was it necessary to cut two holes in the box rather than one? Explain.
4. **Design** What is the source of energy driving convection in the model?

LESSON 19.3

warm air moves laterally, causing advection of the heat as the air mass moves from one location to another. Similarly, cool, dense air from another part of the atmosphere might sink or convect downward. As it does so, it will push aside other air laterally. That displaced air will advect whatever heat it has with it as it moves. This horizontal airflow is what we call wind.

checkpoint What is the difference between convection and advection?

Changes of State

Given the proper temperature and pressure, most substances can exist in three states: solid, liquid, and gas. However, at Earth's surface, many substances commonly exist in only one state. In our experience, rock is almost always solid and molecular oxygen is almost always a gas. Water commonly exists in all three states—as solid ice, liquid, and gaseous water vapor.

Latent heat (stored energy) is the energy released or absorbed when a substance changes from one state to another. As shown in Figure 19-13, about 80 calories are required to melt 1 gram of ice at a constant temperature of 0°C. Another 540 to 600 calories are needed to evaporate 1 gram of water at a constant temperature and pressure. The energy transfers also work in reverse. When 1 gram of water vapor condenses to liquid, 540 to 600 calories are released. When 1 gram of water freezes, 80 calories are released. Sublimation is the transformation directly from solid ice to water vapor, without passing through an intermediate liquid phase. For water, this process requires about 620 calories per gram.

If you walk onto a beach after swimming, you feel cool, even on a hot day. Your skin temperature drops because water on your body is evaporating, and evaporation absorbs heat, cooling your skin. Similarly, evaporation from any body of water cools the water and air around it. Conversely, condensation releases heat. The energy released when water condenses into rain during a single hurricane can be as great as the energy released by several atomic bombs.

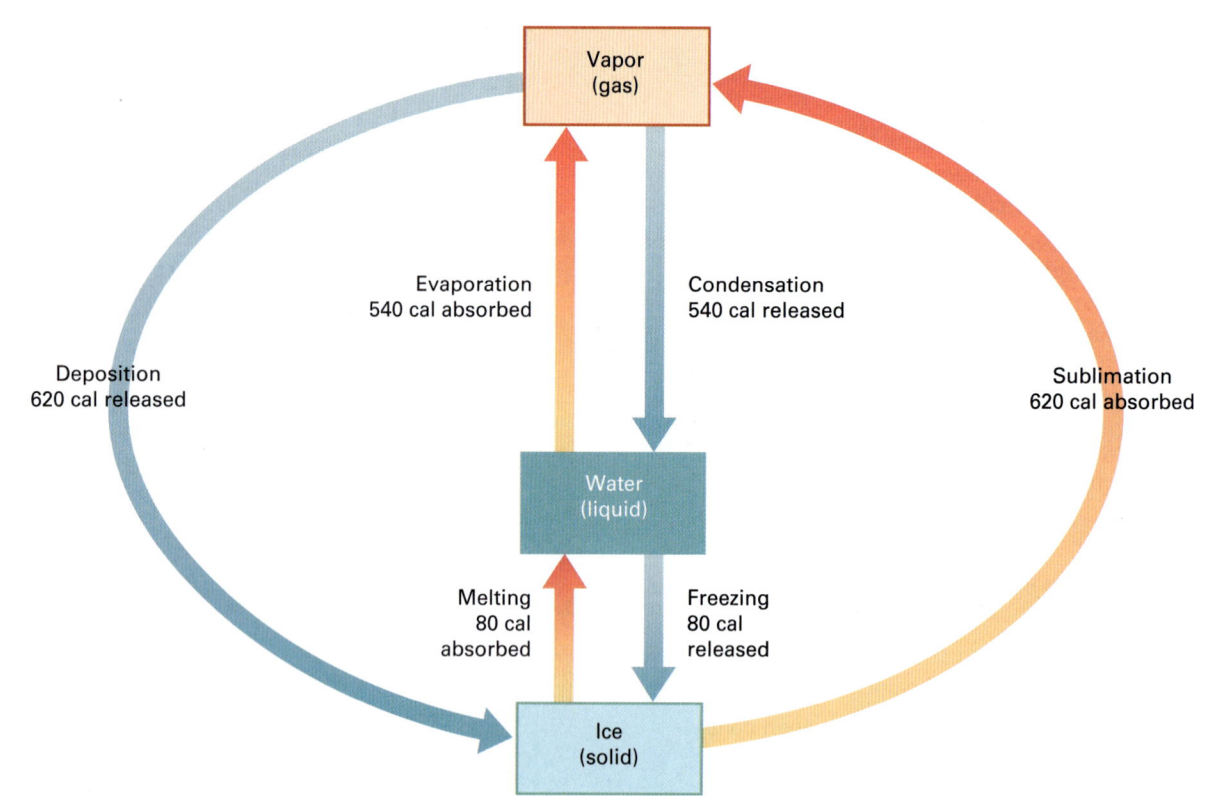

FIGURE 19-13
Water releases or absorbs latent heat as it changes among its liquid, solid, and vapor states. Red arrows show processes that absorb heat. Blue arrows show those that release heat. (Calories are given per gram at 0°C and 100°C; the values vary with temperature.)

The energy absorbed and released by water during freezing, melting, evaporation, condensation, sublimation (conversion of solid directly to gas), and deposition (conversion of gas directly to solid) is important in the atmospheric energy balance. For example, in the northern latitudes, March is usually colder than September even though equal amounts of sunlight are received in both months. In March, much of the solar energy is absorbed by melting snow. And any snow that has not melted also has a high albedo and reflects sunlight efficiently.

checkpoint What is latent heat?

Specific Heat and Temperature Change

If you place a pan of water and a rock outside on a hot summer day, the rock becomes hotter than the water. Both have received identical quantities of solar radiation. Why is the rock hotter?

1. **Specific heat** is the amount of energy needed to raise the temperature of 1 gram of material by 1°C. Specific heat is different for every substance. Water has an unusually high specific heat: the amount of energy needed to raise 1 cubic centimeter (1 milliliter) of water by 1°C is 1 calorie. In contrast, the specific heat of basalt is only about one-fifth that of water, so only about 20 percent of the energy required to heat water by 1°C is needed to heat the same volume of basalt by 1°C. Thus, if equal masses of water and rock absorb equal amounts of energy, the rock becomes hotter than the water.

2. Rock absorbs heat only at its surface, and the heat travels slowly through the rock. As a result, heat concentrates at the surface. Heat disperses more effectively through water for two reasons. First, solar radiation penetrates several meters below the surface of the water, warming it to this depth. Second, water is a fluid and transports heat by convection.

3. Evaporation is a cooling process. Water loses heat and cools by evaporation, but rock does not.

Think of the consequences of the temperature difference between rock and water. On a hot summer day you may burn your feet walking across dry sand or rock. But the surface of a lake or ocean is never burning hot. Suppose that both the ocean and the adjacent coastline are at the same temperature in spring. As summer approaches, both land and sea receive equal amounts of solar energy. But the land becomes hotter, just as rock becomes hotter than water. Along the seacoast the cool sea moderates the temperature of the land.

The interior of a continent is not cooled in this manner and is generally hotter than the coast in summer. In winter the opposite effect occurs, and inland areas are generally colder than the coastal regions. Therefore, coastal areas are commonly cooler in summer and warmer in winter than continental interiors. The coldest temperatures recorded in the Northern Hemisphere occurred in central Siberia and not at the North Pole, because Siberia is landlocked, whereas the North Pole lies in the middle of the Arctic Ocean. In summer, however, Siberia is considerably warmer than the North Pole. In fact, the average temperature in some places in Siberia ranges from −50°C in winter to 20°C in summer, the greatest range on Earth.

checkpoint Why does heat transfer differently through water than through rock?

19.3 ASSESSMENT

1. **Contrast** What is the difference between heat and temperature?
2. **Explain** What can cause air to convect or advect from one location to another?
3. **Contrast** Explain which has a bigger effect on climate, latent heat or specific heat.

Critical Thinking

4. **Apply** As water evaporates from the ocean and condenses, the energy from condensation can contribute to the formation of hurricanes. How do you think warming of the oceans would affect the energy in a hurricane?

19.4 GEOGRAPHIC FACTORS

Core Ideas and Skills
- Explain the effects of latitude, altitude, ocean currents, prevailing winds, cloud cover, and albedo on local and regional average temperature.
- Explain what causes seasons.
- Describe how isotherms connect areas of the same average temperature.

KEY TERMS

isotherm
solstice
Tropic of Cancer
Tropic of Capricorn
equinox

Latitude and Altitude

The region near the Equator is warm throughout the year, whereas polar regions are cold and ice-bound even in summer. To understand this temperature difference, consider first what happens if you hold a flashlight above a flat board. If the light is held directly overhead and the beam shines straight down, a small area is brightly lit. If the flashlight is held at an angle to the board, a larger area is illuminated. However, because the same amount of light is spread over a larger area, the intensity is reduced (Figure 19-14).

Now consider what happens when the sun shines directly over the Equator. The Equator, analogous to the flat board under a direct light, receives the most concentrated radiation. The sun strikes the rest of the globe at an angle and thus radiation is less concentrated at higher latitudes (Figure 19-15). Because the Equator receives the most concentrated solar energy, it is generally warm throughout the year. In general, the average atmospheric temperature becomes progressively cooler the closer you are to the poles. But the average temperature does not change steadily with latitude. Many other factors—such as winds, ocean currents, albedo, and proximity to the oceans—also affect the atmospheric temperature in any given region.

🌐 **FIGURE 19-15**
Go online to view an illustration of why the Equator receives the most intense solar radiation of any region on Earth.

Even though all locations at a given latitude receive equal amounts of solar radiation, some places have cooler climates than others at the same latitude. Figure 19-16 shows temperatures around Earth in January and in July. Lines called **isotherms** connect areas of the same average temperature. Note that the isotherms loop and dip across lines of latitude. For example, the January 0°C line runs through Seattle, Washington, dips southward across the center of the United States, and then swings northward to northern Norway. Such variations occur with latitude because winds and ocean

FIGURE 19-14 ▼
If a light shines from directly overhead, the radiation is concentrated on a small area. However, if the light shines at an angle, or if the surface is tilted, the radiant energy is dispersed over a larger area.

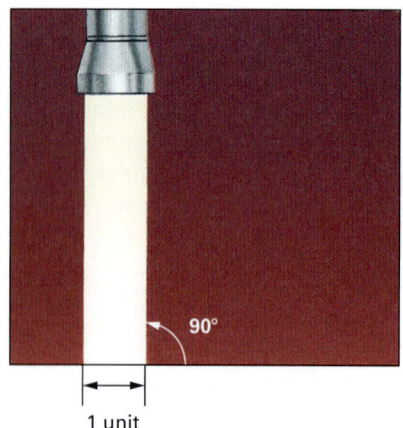

1 unit

One unit of light is concentrated over 1 unit of surface.

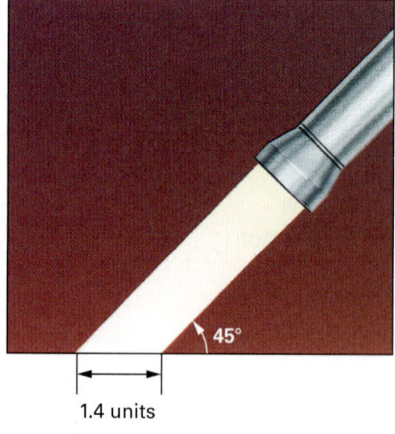

1.4 units

One unit of light is dispersed over 1.4 units of surface.

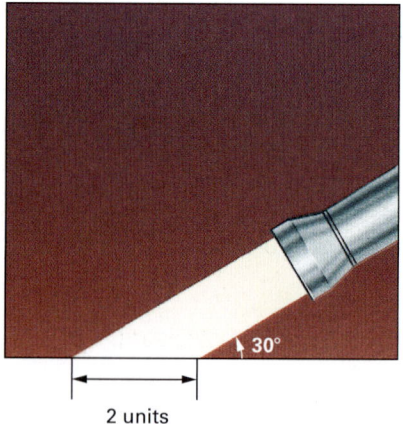

2 units

One unit of light is dispersed over 2 units of surface.

FIGURE 19-16

(A) Global temperature distributions in January are shown. Isotherm lines connect places with the same average temperatures. (B) Global temperature distributions in July are shown. Isotherm lines connect places with the same average temperatures.

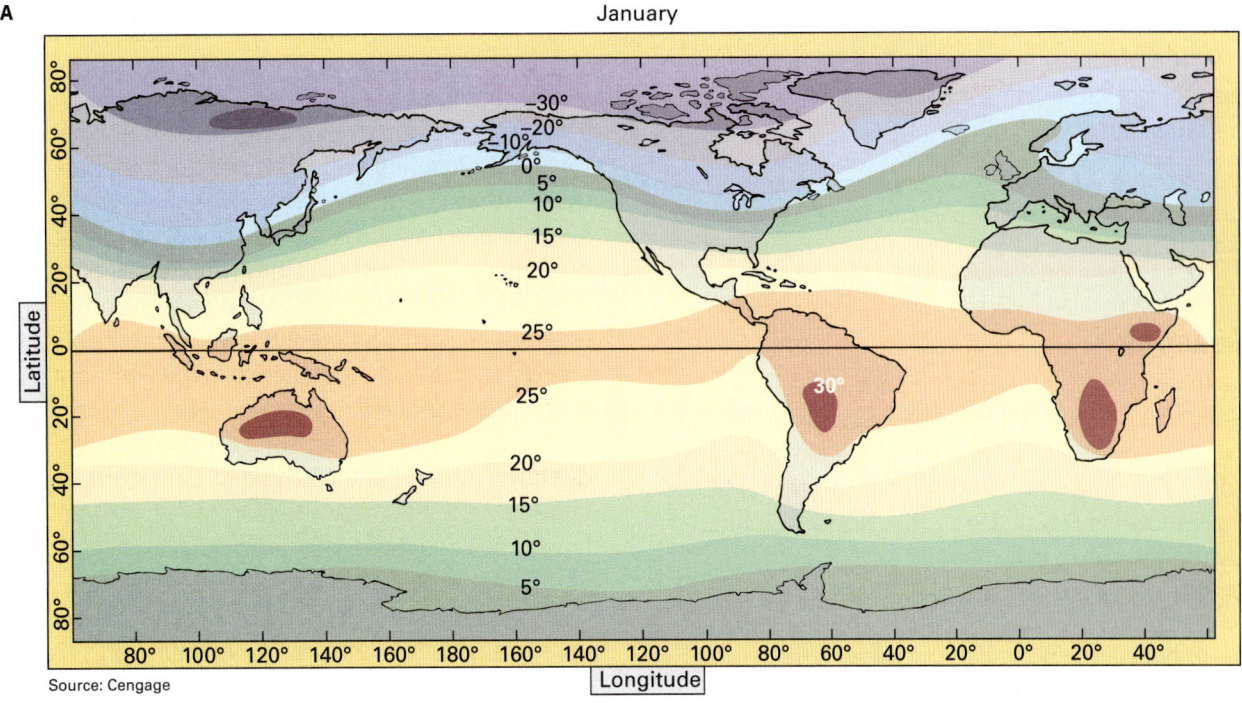

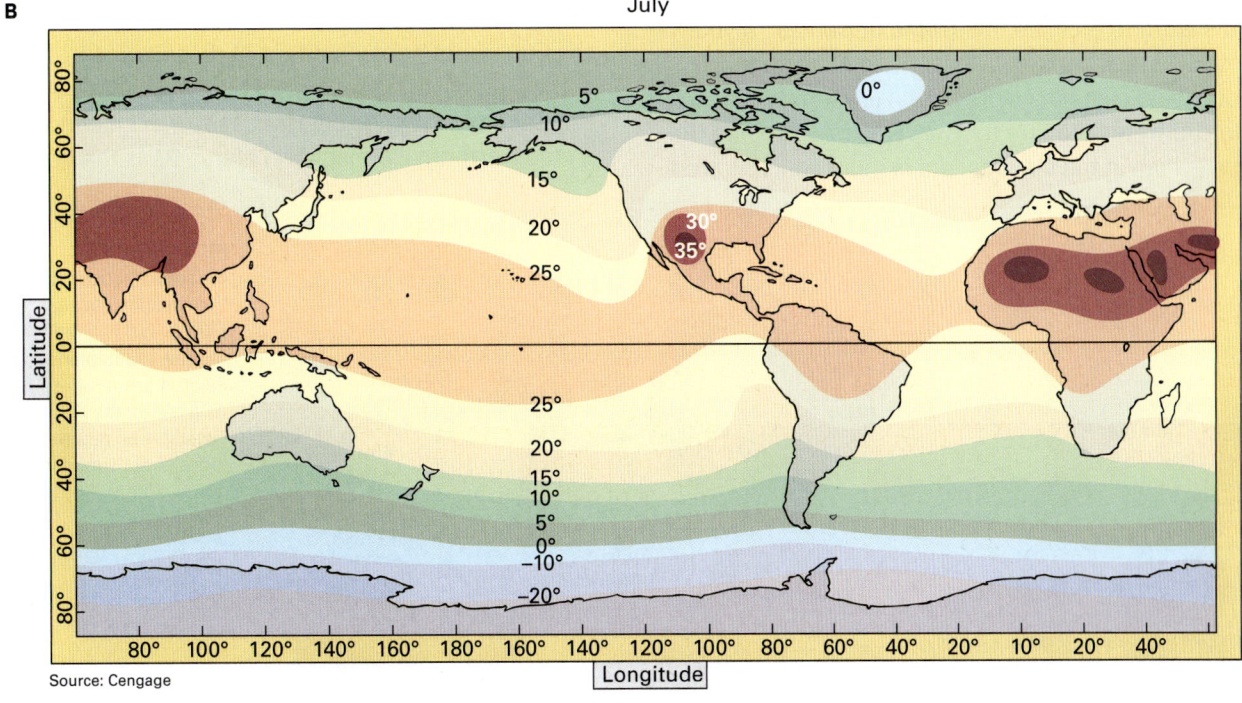

currents transport heat from one region of Earth to another. Other factors include the different energy storage properties of oceans and continents, evaporative cooling of the oceans, and the latent heat of snow and ice.

Altitude also affects average temperature. Recall from Chapter 18 that at higher elevations in the troposphere, the atmosphere is thinner and absorbs less energy. In addition, lower parts of the troposphere have absorbed much of the

LESSON 19.4 **625**

heat radiating from Earth's surface. Consequently, temperature decreases with elevation. For example, Mount Everest lies at 28° north, about the same latitude as Tampa, Florida. Yet at 8,000 meters, climbers on Everest wear heavy down clothing to keep from freezing to death. During the same season, sunbathers in Tampa lounge on the beach in swimsuits.

checkpoint Why is the Equator generally warm throughout the year as compared to areas farther north or south?

Seasons

Earth revolves around the sun once a year in a planar orbit, while simultaneously spinning on its axis once every 24 hours. This axis is tilted at 23.5° from a line drawn perpendicular to the orbital plane.

As shown in Figure 19-17, the North Pole tilts toward the sun in summer and away from it in winter. June 21 is the summer **solstice** in the Northern Hemisphere because at this time, the North Pole leans the full 23.5° toward the sun. As a result, sunlight strikes Earth from directly overhead at a latitude 23.5° north of the Equator. This latitude is called the **Tropic of Cancer**. If you stood on the Tropic of Cancer at noon on June 21, you would cast no shadow. June is warm in the Northern Hemisphere for two reasons: (1) when the sun is high in the sky, sunlight is more concentrated than it is in winter; (2) when the North Pole is tilted toward the sun, it receives 24 hours of daylight. Polar regions are called "lands of the midnight sun" because the sun never sets in the summertime (Figure 19-18). Below the Arctic Circle (66° N) the sun sets in the summer, but the days are always longer than they are in winter (Table 19-1).

When it is summer in the Northern Hemisphere, the South Pole tilts away from the sun and thus the Southern Hemisphere receives low-intensity sunlight and has short days. June 21 is the first day of winter in the Southern Hemisphere. Six months later, on December 21 or 22, the seasons are reversed. The North Pole tilts away from the sun, giving rise to the winter solstice in the Northern Hemisphere, while it is summer in the Southern Hemisphere. On this day, sunlight strikes Earth directly overhead at the **Tropic of Capricorn**, latitude 23.5° south. At the North Pole, the sun never rises and it is continuously dark, while the South Pole is bathed in continuous daylight.

On March 21 and September 22, Earth's axis lies at right angles to a line drawn between Earth and the sun. As a result, the poles are not tilted toward or away from the sun and the sun shines directly overhead at the Equator at noon. If you stood at the Equator at noon on either of these two dates, you would cast no shadow. But north or south of the Equator, a person casts a shadow even at noon. In the Northern Hemisphere, March 21 is the first

FIGURE 19-17
Weather changes with the seasons because Earth's axis is tilted relative to the plane of its orbit around the sun. As a result, the Northern Hemisphere receives more direct sunlight during summer but less during winter.

DATA ANALYSIS Hours of Sunlight per Day

How many hours of daylight does your hometown receive compared to other locations? Use a globe or a map to determine the latitude and longitude of your hometown. Compare your town to the locations shown in Table 19-1. Next, research online to find out how many hours of sunlight your town receives at the solstices and equinoxes for the current year. Note that calendar dates of the solstices and equinoxes vary slightly from year to year.

1. **Recognize Patterns** Describe the correlation between a location's distance from the Equator and the number of hours of daylight the location receives at the solstices and equinoxes.

2. **Compare** Which location in Table 19-1 is closest to your town in terms of latitude? How does the number of hours of daylight received by your hometown compare to that location?

3. **Contrast** Compare and contrast the number of hours of daylight your town or city receives with locations at the Equator and at the location in Table 19-1 farthest in latitude from your city or town.

Computational Thinking

4. **Collect Data** Create a table similar to Table 19-1 using locations in the Southern Hemisphere. Include the following locations: the Equator; Durban, South Africa; Nelson, New Zealand; Rio Gallegos, Argentina; and the South Pole. Use your online sources to determine the latitude, longitude, and daylight hours at those locations for the same dates as those you researched for your town or city.

5. **Generalize Patterns** Describe the patterns of daylight hours you observe in your table. Compare patterns in the Southern Hemisphere and Northern Hemisphere and explain why they occur.

TABLE 19-1 Hours of Sunlight per Day

Latitude	Geographic Reference	Duration of Daylight (hours:minutes)		
		Summer Solstice	Winter Solstice	Spring and Fall Equinoxes
0° N	Equator	12:00	12:00	12:00
30° N	New Orleans	13:56	10:04	12:00
40° N	Denver	14:52	9:08	12:00
50° N	Vancouver	16:18	7:42	12:00
90° N	North Pole	24:00	0:00	12:00

Source: Cengage

day of spring and September 22 is the first day of autumn, whereas the seasons are reversed in the Southern Hemisphere. On the first days of spring and autumn, every portion of the globe receives 12 hours of direct sunlight and 12 hours of darkness. For this reason, March 21 and September 22 are called the **equinoxes**, meaning "equal nights."

All areas of the globe receive the same total number of hours of sunlight every year. The North Pole and South Pole receive direct sunlight in dramatic opposition. Each of them has six months of continuous light and six months of continuous darkness. At the Equator, each day and night are close to 12 hours long throughout the year. Although the poles receive the same number of sunlight hours as do the Equatorial regions, the sunlight reaches the poles at a much lower angle and therefore delivers much less total energy per unit of surface area.

checkpoint At what latitude on Earth would you cast no shadow at noon on June 21? At what latitude on Earth would you cast no shadow at noon on December 21?

Ocean and Wind Effects

Recall from Lesson 19.3 that land heats more quickly in summer and cools more quickly in winter than ocean surfaces. As a result, continental interiors show greater seasonal extremes of temperature than coastal regions. For example, San Francisco and St. Louis both lie at approximately 38° north latitude, but St. Louis is in the middle of the continent and San Francisco lies on the Pacific coast. On average, St. Louis is 9°C cooler in winter and 9°C warmer in summer than San Francisco (Figure 19-19).

Ocean currents also play a major role in determining average temperature. Paris, France, lies at 48° north latitude, north of the U.S.–Canada border but on the other side of the Atlantic Ocean. Although winters are dark because the sun is low in the sky, the warm Gulf Stream carries enough heat northward so the average minimum temperature in Paris in January is around 6°C, comfortably above freezing. St. John's, Newfoundland, is at about the same latitude. But it is under the influence of the Labrador Current that flows from the North Pole. As a result, the average minimum January temperature in St. John's is around –3°C, considerably colder than it is in Paris (Figure 19-20).

Winds can carry cold air or warm air from region to region just as ocean currents do. Vladivostok is a Russian city on the west coast of the Pacific Ocean at about 43° north latitude. It has a near-Arctic climate with frigid winters and heavy snow. Forests are stunted by the cold, and trees are generally small.

In comparison, Portland, Oregon, lies at 45° north latitude on the east coast of the Pacific Ocean. Its temperate climate supports majestic cedar forests, and rain is more common than snow in winter (Figure 19-21). The main difference in this case is that Vladivostok is cooled by frigid Arctic winds from Siberia.

checkpoint What causes the year-round depression of temperatures in St. John's, Canada?

FIGURE 19-19 ▼
The average temperature of continental St. Louis (red line) is colder in winter and warmer in summer than that of coastal San Francisco (blue line).

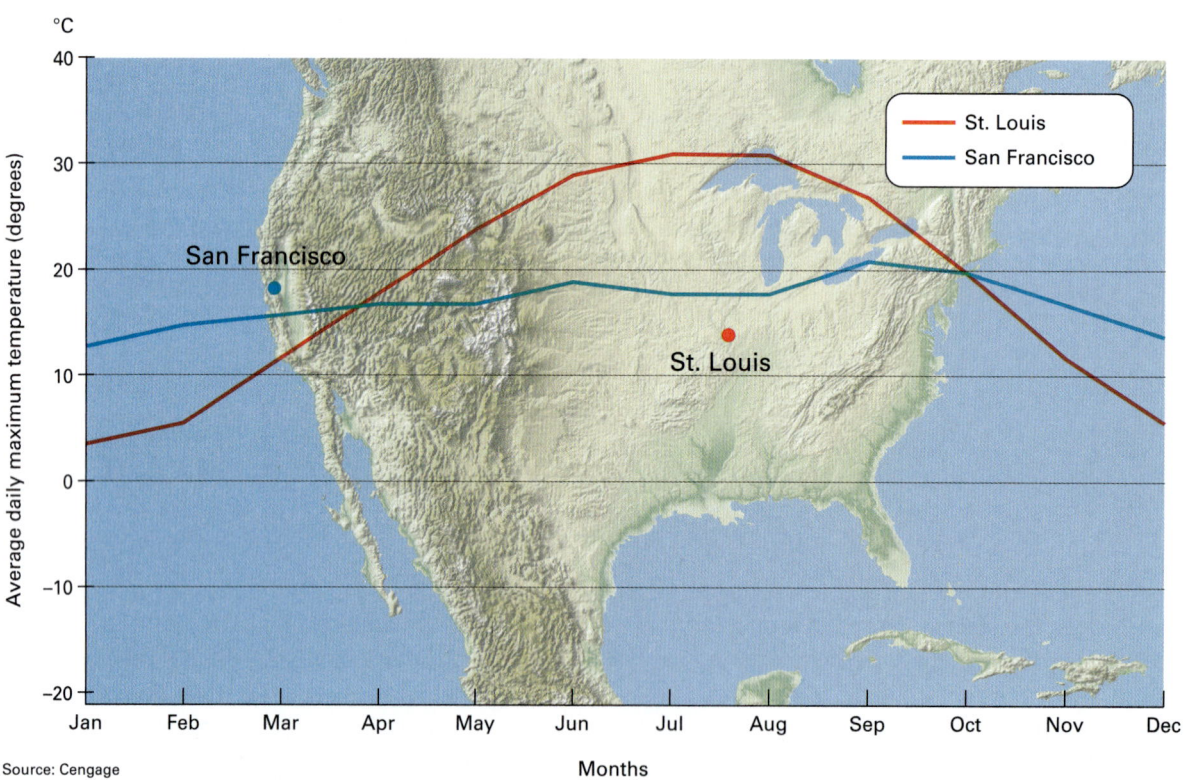

Source: Cengage

FIGURE 19-20

Paris, France, is warmed by two ocean currents, the Gulf Stream and the North Atlantic Drift. St. John's, Canada, is cooled by the Labrador Current. The cooling effect of the Labrador Current depresses the temperature of St. John's year-round.

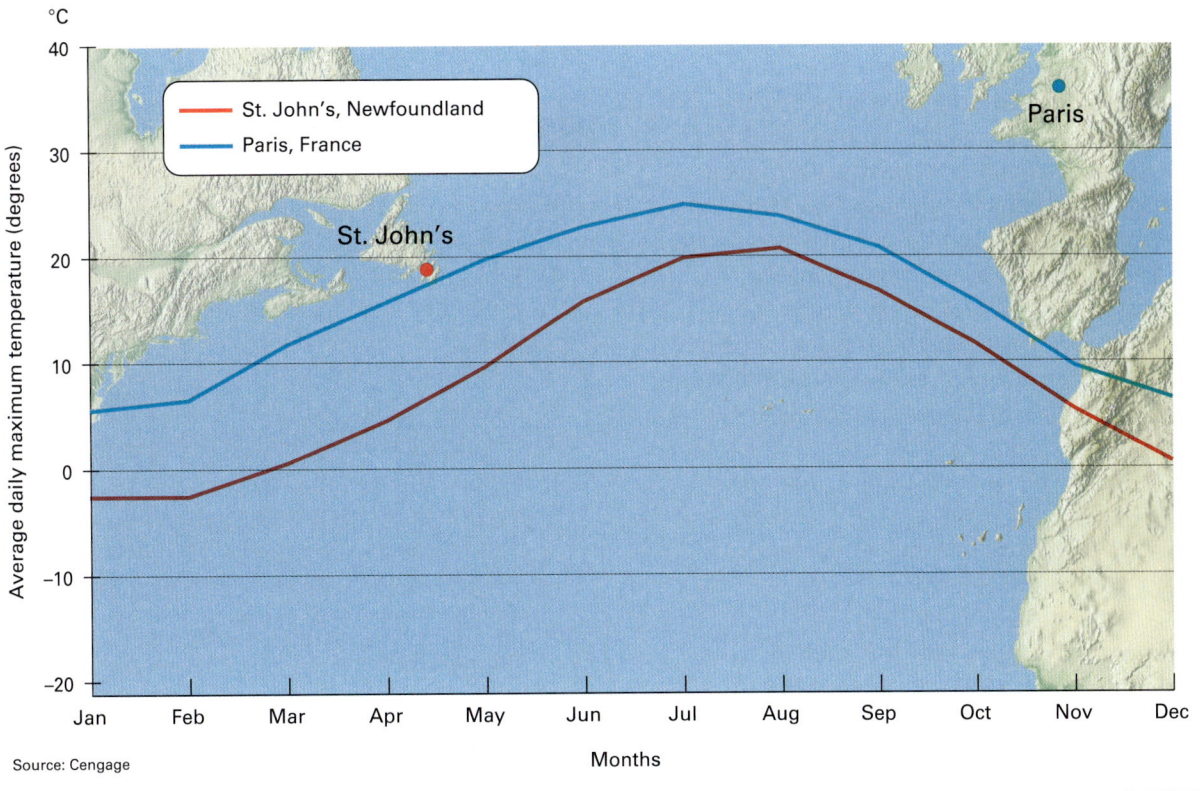

Source: Cengage

FIGURE 19-21

During summer, temperatures in Vladivostok, Russia, and Portland, Oregon, are very similar. However, frigid Arctic wind from Siberia cools Vladivostok during the winter. So in winter, the temperature in Vladivostok is significantly colder than that of Portland.

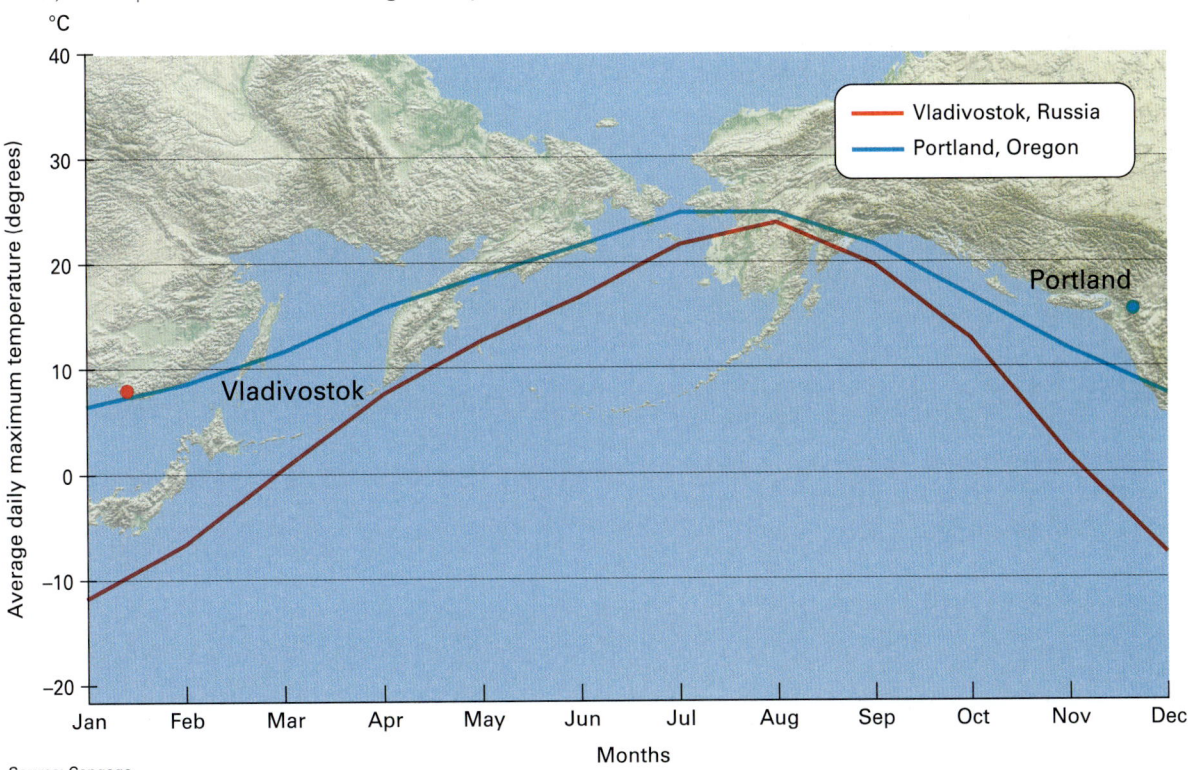

Source: Cengage

Cloud Cover and Albedo

If you are sunbathing on the beach and a cloud passes in front of the sun, you feel a sudden cooling. During the day, clouds reflect sunlight and cool the surface. In contrast, clouds have the opposite effect and warm the surface during the night. Recall that after the sun sets, Earth cools because soil, rock, and water emit long-wavelength infrared radiation into the air. Clouds act as an insulating blanket by absorbing outgoing radiation and reradiating some of it back downward (Figure 19-22). So cloudy nights are generally warmer than clear nights. Because clouds cool Earth during the day and warm it at night, cloudy regions generally have a narrower daily temperature range than regions with predominantly blue sky and clear nights.

FIGURE 19-22
Go online to view why clouds can have opposite effects on daytime and nighttime temperatures.

Snow also plays a major role in affecting regional temperature. Imagine it is raining and the temperature is 0.5°C, or just above freezing. After the rain stops, the sun comes out. The bare ground absorbs solar radiation and the surface heats up. Now imagine the temperature had dropped to −0.5°C during the storm. In this case, the precipitation would fall as snow. After the storm passed and the sun came out, the surface would be covered with a sparkling white cover with 80 to 90 percent albedo. Most of the incoming solar radiation would reflect back into the atmosphere and the surface would remain much cooler. If the surface temperature did rise above freezing, the snow would have to melt before the ground temperature would rise. But snow has a very high latent heat; it takes a lot of energy to melt the snow without raising the temperature. In this example, a seemingly insignificant one-degree initial temperature difference during a storm would cause a much larger temperature difference when the sun came out after the storm.

This type of threshold mechanism is very important in weather and climate and will be discussed in Chapters 20 and 21.

checkpoint Why are cloudy nights generally warmer than clear nights?

19.4 ASSESSMENT

1. **Describe** In general, how does temperature change with latitude?
2. **Explain** Why does the tilt of Earth's axis affect the temperatures and hours of sunlight throughout the year in the Northern Hemisphere?
3. **Describe** How do the hours of sunlight at Earth's poles change through the seasons?
4. **Explain** Although San Francisco and St. Louis are at approximately the same latitude, why is the average temperature of St. Louis cooler in winter and warmer in summer than the average temperature of San Francisco?

Critical Thinking

5. **Apply** Explain how temperatures at a high altitude at the Equator would be cooler than temperatures in a low-altitude desert area farther north.
6. **Apply** In Arizona, the cities of Flagstaff and Sedona are only about 47 kilometers apart. In Flagstaff, the average July temperature is about 19°C and the average December temperature is about 1°C. In Sedona, the average July temperature is about 27°C and the average December temperature is about 6°C. Both cities are mostly cloudy 16 percent of the year. The average annual snowfall in Flagstaff is about 255 cm, whereas in Sedona it is about 28 cm. The average elevation of Flagstaff is 2,130 m; Sedona's average elevation is 1,330 m. Neither city is near a large body of water. What are the most likely explanations for the differences in average temperatures of the two cities? Explain your reasoning.

TYING IT ALL TOGETHER
ANTARCTICA'S GLACIER RETREAT

In this chapter, you learned about processes that determine the energy balance in the atmosphere, such as solar radiation and the fate of sunlight reaching Earth. You further explored how the transport and storage of energy drives climate and weather, as well as how latitude, seasons, and geography affect climate.

In the Case Study, we considered the accelerated melting of Greenland's glaciers and how scientists are investigating its causes and effects. New technologies, collaborative data, and compiled computer models all help scientists investigate the highly complex factors that link climate change and glacial melting. Similar tools are also being used to examine glacial melting in Antarctica. Antarctica's ice cover is by far Earth's most extensive, and rates of ice loss there have tripled since 2012. In 2018, a major cooperative research effort launched to examine Antarctica's Thwaites Glacier, a massive sheet of ice in remote and dangerous territory.

In this activity, you will investigate questions researchers are attempting to answer about the Thwaites Glacier. As you complete the activity, apply your knowledge of Earth's energy balance to explain how various processes may be contributing to ice loss in Antarctica.

Work with a partner or in small groups to answer one of the research questions below. Prepare a written summary of your findings to share with others.

- Where is the Thwaites Glacier, and why are scientists particularly interested in studying ice loss there?
- What are some of the challenges of conducting research on the Thwaites Glacier, and what tools are scientists using to overcome these challenges?
- What role does albedo play in the melting of glaciers like the Thwaites Glacier?
- How could ocean energy storage affect the Thwaites Glacier?
- What geographic characteristics of the Thwaites Glacier—such as its size, location, and underlying land—influence its melting?

FIGURE 19-23
Antarctica's Thwaites Glacier is undergoing rapid changes and is the subject of major scientific research.

CHAPTER 19 SUMMARY

19.1 INCOMING SOLAR RADIATION

- Light is a form of electromagnetic radiation and exhibits properties of both waves and particles.

- Wavelength is the distance between wave crests, and frequency is the number of cycles that pass in a second. The electromagnetic spectrum is the continuum of radiation of different wavelengths and frequencies.

- A photon's energy may be transferred to an electron in a molecule. If that happens, the electron temporarily jumps to an excited energy state. The electron reemits radiation when it falls back to a lower energy state.

- All objects emit radiant energy at some wavelength.

- The sun's radiation is absorbed by rocks, soil, and water, which reemit low-energy (long-wavelength) radiation.

- Radiation reflects from many surfaces; the reflectivity of a surface is referred to as its albedo.

- Atmospheric gases, water droplets, and dust scatter incoming solar radiation.

19.2 THE RADIATION BALANCE

- The greenhouse effect is the process by which Earth's radiation balance is maintained.

- In the greenhouse effect, water vapor, carbon dioxide, methane, and other greenhouse gases absorb some of the energy reradiating from Earth and act as an insulating blanket.

19.3 ENERGY STORAGE AND TRANSFER

- Energy storage and transfer are the driving mechanisms for weather and climate.

- Temperature is a measure of thermal energy, or the average kinetic energy of atoms or molecules in a substance; temperature is proportional to the average speed of atoms or molecules in a sample.

- Heat is the transfer of thermal energy from a hotter substance to a colder one.

- Heat transfer in the atmosphere occurs by convection and advection.

- Latent heat (stored energy) is the energy released or absorbed when a substance changes from one state to another. Large quantities of stored energy are absorbed or emitted when water freezes, melts, vaporizes, or condenses.

- Oceans affect weather and climate because water has low albedo and high specific heat. Coastal areas are generally cooler in summer and warmer in winter than continental interiors at the same latitude.

19.4 GEOGRAPHIC FACTORS

- Temperature changes with geography as a result of latitude, altitude, differences in temperature between ocean and land, ocean currents, wind direction, cloud cover, and albedo.

- The general decrease in temperature from the Equator to the poles results from the decreasing intensity of solar radiation from the Equator to the poles.

- Seasonal change is caused by the tilt of Earth's axis relative to the Earth–sun plane as Earth orbits around the sun.

CHAPTER 19 ASSESSMENT

Review Key Terms
Select the key term that best fits the definition. Not all terms will be used, and no term will be used more than once.

- advection
- albedo
- climate
- conduction
- electromagnetic radiation
- electromagnetic spectrum
- equinox
- frequency
- greenhouse effect
- heat
- isotherm
- latent heat
- photon
- solstice
- specific heat
- temperature
- Tropic of Cancer
- Tropic of Capricorn
- weather

1. the warming of the atmosphere due to absorption of reemitted infrared radiation by certain gases, including carbon dioxide, water vapor, and methane
2. measure of the total kinetic energy in a substance
3. a form of energy that travels in waves with changing electric and magnetic fields
4. the characteristics of a region based on average annual temperature and average annual precipitation
5. the reflectivity of a substance
6. lines on a map that connect areas of the same average temperature
7. thermal energy transferred from one object to another
8. transport of energy by direct collisions between atoms or molecules
9. the entire range of electromagnetic radiation from very-long-wavelength to very-short-wavelength radiation
10. 23.5° North, where sunlight strikes Earth from directly overhead on June 21
11. one of two dates in the year when the sun is directly over the Equator at noon
12. particle of light
13. the amount of energy needed to raise the temperature of 1 gram of a substance by 1 degree Celsius
14. the number of complete wave cycles, from crest to crest, that pass by a fixed point in one second
15. the energy released or absorbed as a substance changes from one state to another

Review Key Concepts
Answer each question on a separate sheet of paper to demonstrate your understanding of key concepts from the chapter.

16. On the electromagnetic spectrum, what kind of electromagnetic radiation is primarily emitted by the sun, and how does it change when it is absorbed and reemitted by rocks and ground features on Earth's surface?
17. What is albedo, and how does it affect the amount of solar energy that is absorbed by Earth's surface?
18. Explain how electromagnetic radiation can be emitted by objects.
19. Are all gases in Earth's atmosphere greenhouse gases? Explain.
20. How do large volcanic eruptions influence Earth's atmospheric temperature?
21. What is advection and why is it important in the study of weather?
22. In early summer, the deep snow on a mountain begins to melt as the sun's rays become more intense. The ground temperature, however, stays very cold until all of the snow melts. Only after the snow melts does the ground temperature begin to increase. Why is warming of the ground delayed?
23. When do the summer and winter solstices occur in the Northern Hemisphere and what would a person living in the Northern Hemisphere experience as a result of these events?
24. The diagrams show how the angle of incident light affects the way it is dispersed on a surface. Explain how these diagrams can be used to model incoming energy of sunlight as it strikes different parts of Earth.

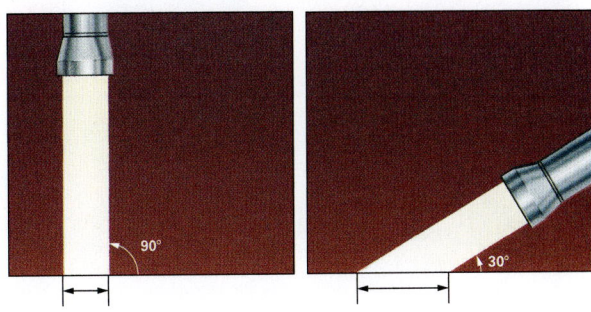

CHAPTER 19 ASSESSMENT

25. Why do regions at the Equator experience little variation in temperature over a year?
26. Explain how ocean currents affect climate.
27. Why are cloudy nights generally warmer than clear nights?

Think Critically

Write a response to each question on a separate sheet of paper. Use concepts from the chapter to support your reasoning.

28. Data concerning atmospheric gases present on Venus, Earth, and Mars are given in the table. Based on these data, which planet would you predict to have the greatest greenhouse effect and why?

Planet	Atmospheric density	Atmospheric composition
Venus	65 kg/m³	96.5% CO_2, 3.5% N_2, trace SO_2, Ar, H_2O, CO, He, Ne
Earth	1.2 kg/m³	78.1% N_2, 21.0% O_2, trace Ar, CO_2, Ne, He, CH_4, Kr, H_2, H_2O
Mars	0.020 kg/m³	95.3% CO_2, 2.7% N_2, 1.6% Ar, 0.13% O_2, 0.08% CO, trace H_2O, NO, Ne

Source: NASA

PERFORMANCE TASK

Changes in Earth's Surface Cause Changes to Earth's Systems

The energy balance in Earth's atmosphere is influenced by many factors. A change in one of these factors often causes changes in other Earth systems. These changes can create either positive or negative feedback effects.

In this performance task, you will analyze data to identify an important climate feedback system that affects Earth's energy balance.

The map shows worldwide average sea surface temperatures. The graph shows energy trapped by atmospheric water vapor as a function of latitude.

Continued on next page.

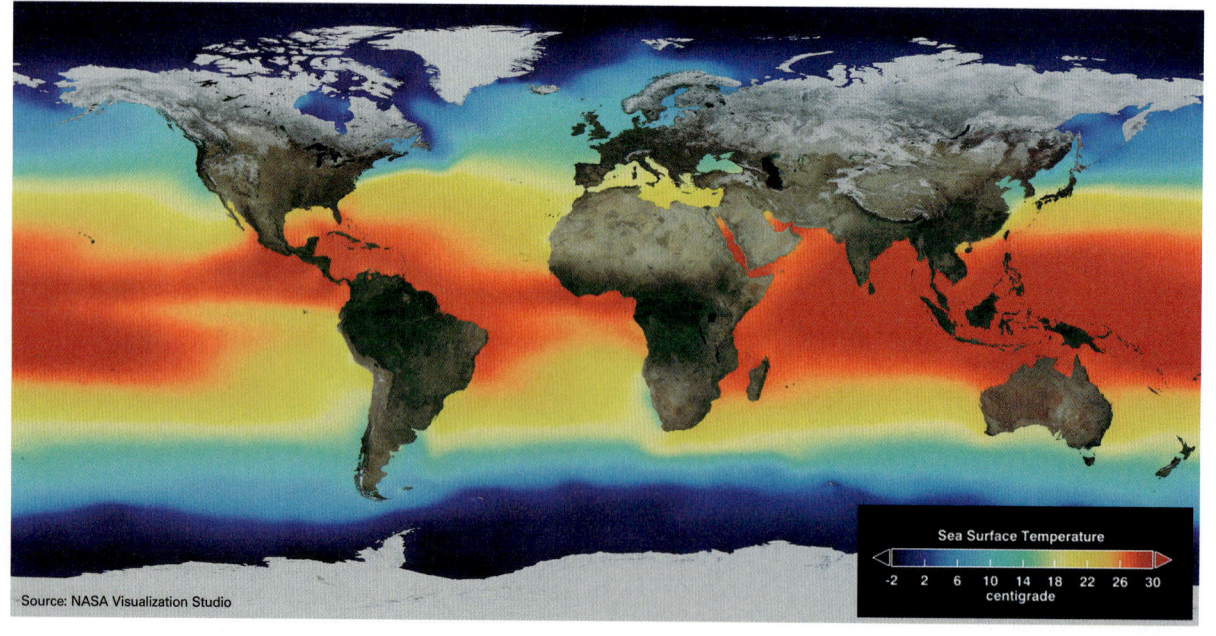

Source: NASA Visualization Studio

29. Explain how melting glaciers are part of a feedback system that has an accelerating effect on global warming.

30. The moon has no atmosphere and no clouds. Also, it takes about 27 Earth days for the moon to rotate on its axis, meaning that daytime and nighttime on the moon each last about 13 days. Based on these facts, predict what would happen to radiant energy from the sun that strikes the moon's surface. Also predict what you would expect in terms of surface temperatures on the moon. Explain your reasoning.

31. Consider an island in the middle of the ocean and a landlocked area of the same size located in the middle of a continent. Both are at the same latitude. Would you expect the island to have a warmer or colder winter climate than the landlocked area? Explain your reasoning.

32. Would North America experience solstices or equinoxes if Earth was not tilted on its axis? Explain.

33. Explain why latitude is a predictor of a region's average temperatures. Also, explain its limitations as a predictor.

PERFORMANCE TASK (continued)

1. Describe the distribution of sea surface temperatures shown in the map in terms of latitude. Use what you know about energy balance in the atmosphere to explain why any patterns you observe may occur.

2. The graph focuses specifically on water vapor as a greenhouse gas. The data show the amount of energy trapped in the atmosphere by water vapor as a function of latitude. How can you explain the large peak in the center of the graph?

3. Describe a feedback effect that could operate in the system involving water vapor and warming of Earth's atmosphere. A feedback effect is one in which a cyclical pattern develops that either positively or negatively influences an Earth system. Explain whether the feedback system you describe is positive or negative, and why. Would this feedback effect stabilize the climate or destabilize it?

4. Tropical cyclones can form over waters at temperatures of 28°C or higher. Explain how rising global carbon dioxide emissions could affect the frequency and intensity of tropical cyclones in equatorial regions.

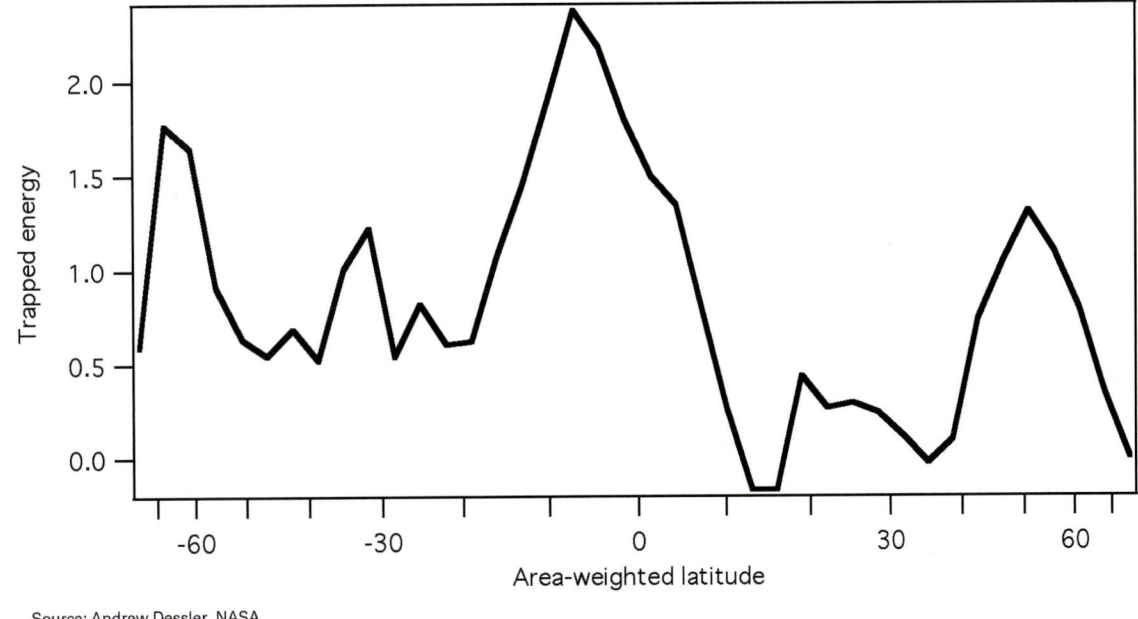

Source: Andrew Dessler, NASA

CHAPTER 20
WEATHER

Farmers pause to observe dark clouds in Sand Hills, Nebraska.

The signs for many weather events can be recognizable. Many farmers can predict, based on prior observations, when the weather will change within the hour. However, broader weather systems can be complex. Some extreme weather events, such as hurricanes and tornadoes, evolve from a series of related processes. For example, different types of clouds are produced in different storm conditions. Temperature, moisture, and movement of air all have different effects on the development of an extreme weather event. Read on to find out more about weather elements and patterns.

KEY QUESTIONS

20.1 What are the relationships between moisture and temperature in air?

20.2 How do atmospheric conditions affect cloud formation and precipitation?

20.3 How do changes in air pressure affect weather patterns?

20.4 What are the distinguishing characteristics of severe storms?

EXPLORERS AT WORK

VISUALIZING TORNADOES

WITH NATIONAL GEOGRAPHIC EXPLORER ANTON SEIMON

Anton Seimon has observed plenty of tornadoes during his more than 25 years as a scientist and geographer. In fact, he's come dangerously close to more than a few of them, having been among the first storm chasers to successfully place probes in front of oncoming twisters to collect data as they pass by. Perhaps the most memorable tornado Dr. Seimon has experienced is the one that struck near El Reno, Oklahoma, on May 31, 2013. Seimon and his wife, Tracie, were on hand to observe it from several kilometers away.

The tornado stands out to Seimon for a number of reasons. For one, it was the largest tornado ever recorded in terms of wind speed and diameter of its circulation. It killed eight people, including Seimon's good friend and colleague, National Geographic Explorer and storm chaser Tim Samaras. And finally, it was Seimon's research into that tornado that led him to develop the new technology he uses today to study the behavior of tornadoes as part of his work through Appalachian State University.

One of Seimon's goals in the wake of the tornado was to create a 3D visualization of the twister during the entire course of its progress. That meant he would need footage of the tornado. Seimon used crowdsourcing to collect hundreds of still photos and 500 gigabytes of video footage from storm chasers, science teams, and the public. Then Seimon and his team synchronized the videos and assembled them in chronological order. Seimon co-developed a new software tool called the Tornado Environment Display (TED) to present the finished product. Visitors to the El Reno Survey website can now watch a 3D visualization of the tornado.

Seimon says his visualization method has allowed him to pick apart features of the tornado, including intense microbursts that caused damage well outside the main path of the tornado vortex. He is excited about its ability to show how an evolving tornado interacts with the ground surface, because these findings reveal how moving air particles within tornadoes cause damage. This data can inform the design of tornado-resistant structures. Seimon is leading a team of storm chasers to document more tornadoes to examine wind fields right at the surface. He plans to use crowdsourcing again to assemble footage that will allow him to create 3D visualizations of those tornadoes.

Seimon's technology recently revealed something surprising. In 2018, Dr. Jana Houser, a researcher from one of the mobile radar teams that sampled the El Reno tornado, compared her data with Seimon's visualizations. Houser noticed Seimon's images clearly showed a tornado at the ground minutes before her radar detected it above the surface. Together, these data sources provided the first confirmation that tornadoes can form from the ground up, which is opposite to the long-held expectation that they grow downward from the rotating thunderstorm updraft. This new paradigm could change how forecasters go about predicting tornadoes. They usually issue tornado warnings after radar detects strong rotation above the ground, but these findings suggest forecasters also need to monitor wind activity at the ground.

THINKING CRITICALLY

Apply Anton Seimon's visualization method is allowing a more direct study of how winds interact with the landscape during a tornado. What important applications might this have for improving public safety?

Anton Seimon observes and films a supercell thunderstorm as it produces tornadoes near Carpenter, Wyoming, in June 2017. (Photograph taken by Tracie Seimon.)

Tracie Seimon captured this tornado as it tracked slowly past a wind farm at Remington, Indiana, in May 2016.

CASE STUDY
EVEN MORE POWERFUL STORMS

In Explorers at Work, you read about Anton Seimon, a storm chaser who has spent many years researching tornadoes in the Great Plains of the central United States. This region, dubbed Tornado Alley, has historically experienced more tornadoes than anywhere else in the United States. However, a recent study found that tornado frequency has actually decreased in Tornado Alley over the past 40 years while increasing in parts of the Midwest and Southeast. The region known as Dixie Alley, which includes parts of Mississippi, Louisiana, Arkansas, Tennessee, and Alabama, has begun to catch up to Tornado Alley in terms of tornado frequency.

In addition to more frequent tornadoes, states near the Gulf of Mexico have experienced another unwelcome trend in recent years: increasingly powerful hurricanes. In 2005, Hurricane Katrina, one of the most catastrophic storms in U.S. history, formed in the Atlantic Ocean, made landfall as a category 1 in Miami, and then crossed over into the Gulf of Mexico. Gaining strength over warm waters, it made a second landfall in Louisiana as a category 4 hurricane with winds of 200 kilometers per hour. In parts of New Orleans, houses were flooded to the rooftops. A record 8.5-meter storm surge inundated the Louisiana and Mississippi coastlines, flattening and flooding homes. The official death toll was 1,836 with more than 700 people still unaccounted for. Direct structural damage was estimated to cost the United States as much as $110 billion.

More than a decade later, a series of unusually powerful hurricanes battered the southeastern United States between 2016 and 2019. Hurricane Matthew (2016) struck Haiti and Cuba as a category 4 before making landfall in South Carolina as a category 1. Hurricane Harvey (2017), a category 4, caused catastrophic flooding in Houston and across much of southeast Texas, followed closely by Hurricane Irma (2017), which struck the Florida Keys as a category 4. Hurricane Maria (2017) devastated Puerto Rico after crossing the island as a category 4. Hurricane Michael (2018) made landfall on the Florida Panhandle as a category 4, the strongest recorded storm to ever hit the area. And Hurricane Dorian (2019), a category 5, was the most powerful storm to hit the Bahamas in the country's history.

This remarkable string of severe hurricanes has been attributed to the abnormally warm waters of the Atlantic Ocean and Gulf of Mexico. The higher-than-average water temperatures enabled the storms to increase in intensity much more rapidly than usual as they approached land, hindering accurate advance warnings from forecasters. With global warming leading to rising ocean temperatures, many scientists expect the trend toward more frequent destructive storms to continue.

As You Read Explore phenomena that contribute to severe weather, such as tornadoes and hurricanes, as well as more typical weather patterns. Ask questions about interactions among Earth's systems and how these affect humidity, cloud formation, precipitation, wind, weather fronts, and other elements. Find out how human activities and climate change affect and are affected by weather patterns.

FIGURE 20-1
In 2018, Hurricane Michael made landfall on the Gulf Coast of Florida as a category 4 hurricane. It was the strongest recorded storm to ever hit this area.

20.1 MOISTURE, TEMPERATURE, AND AIR

Core Ideas and Skills
- Explain the difference between absolute and relative humidity.
- Describe the causes and effects of condensation.
- Identify the processes that cool air and cause dew and frost.
- Compare the three mechanisms that cause air to rise.

KEY TERMS

absolute humidity	dew
relative humidity	frost
saturation	adiabatic temperature change
dew point	orographic lifting
supersaturation	frontal wedging
supercooling	

Humidity

Precipitation occurs only when there is moisture in the air. Therefore, to understand precipitation we must first understand how moisture collects in the atmosphere and how it behaves.

When water boils on a stove, a steamy mist rises above the pan and then disappears into the air. The water molecules have not been lost. They have simply become invisible. In the pan, water is liquid. In the mist above, the water exists as tiny droplets. These droplets then evaporate, and the invisible water vapor mixes with air. Water also evaporates into air from the seas, streams, lakes, and soil. Winds then distribute this moisture throughout the atmosphere. Therefore, all air contains some water vapor—even over the driest deserts.

Humidity is the amount of water vapor in air. **Absolute humidity** is the mass of water vapor in a given volume of air, expressed in grams per cubic meter (g/m^3).

Air can hold only a certain amount of water vapor, and warm air can hold more water vapor than cold air can hold (Figure 20-2). For example, air at 30°C can hold 30 g/m^3 of water vapor, but at 10°C, it can hold less than one-third that quantity, 9 g/m^3. **Relative humidity** is the amount of water vapor in

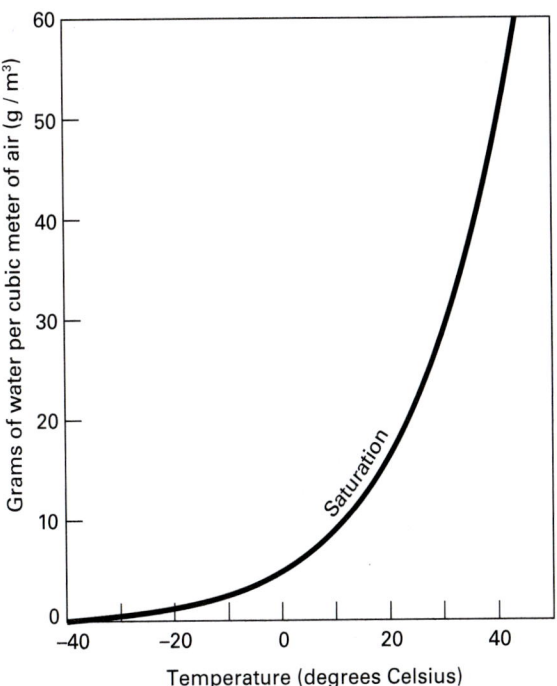

FIGURE 20-2

Warm air can hold more water vapor than cold air. As air temperature increases, the mass of water air can hold increases rapidly. Air at 30°C holds more than three times more water vapor than air at 10°C. The saturation line defines that maximum mass of water per cubic meter air can hold at any given temperature.

Source: Cengage

air relative to the maximum it can hold at a given temperature. It is expressed as a percentage:

If air contains half as much water vapor as it can hold, its relative humidity is 50 percent. Suppose that air at 25°C contains 11.5 g/m^3 of water vapor. Since air at that temperature can hold 23 g/m^3, it is carrying half of its maximum, so the relative humidity is 11.5 g ÷ 23 g × 100 = 50 percent.

Now let us take some of this air and cool it without adding or removing any water vapor. Because cold air holds less water vapor than warm air, the relative humidity increases even though the amount of water vapor remains constant. If the air cools to 12°C, and it still contains 11.5 g/m^3, the relative humidity reaches 100 percent because air at that temperature can hold only 11.5 g/m^3.

When relative humidity reaches 100 percent, the air is saturated. The temperature at which **saturation** occurs, 12°C in this example, is the **dew point**. If saturated air cools below the dew point, some water vapor may condense into liquid droplets. Special conditions can occur in the atmosphere when the relative humidity can rise above 100 percent.

LESSON 20.1   **641**

Supersaturation and Supercooling When the relative humidity reaches 100 percent (at the dew point), any further cooling results in water vapor quickly condensing onto solid surfaces such as rocks, soil, plant leaves, and airborne particles. Airborne particles such as dust, smoke, and pollen are abundant in the lower atmosphere, where they serve as tiny nuclei onto which water condenses. Water vapor condenses easily below the dew point in the lower atmosphere, where the relative humidity rarely exceeds 100 percent. However, in the clear, particulate-free air high in the troposphere, condensation occurs so slowly that for all practical purposes it does not happen. As a result, the air commonly cools below its dew point but water remains as vapor. In that case, the relative humidity rises above 100 percent, and the air reaches a point of **supersaturation**.

Similarly, liquid water does not always freeze at its freezing point. When no nuclei are available for ice crystals to form on, small droplets can remain liquid in a cloud even when the temperature is −40°C. Such water has undergone **supercooling**.

checkpoint What is the relative humidity of air at 25°C that holds 16.2 g/m^3 of water?

Condensation

Cooling and Condensation Moisture condenses to form water droplets or ice crystals when moist air cools below its dew point. Clouds are visible concentrations of this airborne water and ice. There are three atmospheric processes cool air below its dew point and cause condensation: (1) Air cools when it loses heat

FIGURE 20-3
Frost formed on the leaves of this raspberry plant when overnight temperatures cooled below the dew point. Since the dew point was lower than the freezing point of water, ice crystals formed instead of dew as water vapor condensed onto the leaves.

by radiation. (2) Air cools by contact with a cool surface such as water, ice, rock, soil, or vegetation. (3) Air cools when it rises.

Radiation Cooling The atmosphere, rocks, soil, and water absorb the sun's energy during the day and then radiate some of this energy back out toward space at night. As a result of energy lost by radiation, air, land, and water become cooler at night, and condensation may occur.

Contact Cooling You can observe condensation on a cool surface with a simple demonstration. Heat water on a stove until it boils and then hold a cool drinking glass in the clear air just above the steam. Water droplets will condense on the surface of the glass because the glass cools the hot, moist air below its dew point. The same effect occurs in a house on a cold day. Water droplets or ice crystals appear on windows as warm, moist, indoor air cools on the glass.

In some regions, the air on a typical summer evening is warm and humid. After the sun sets, plants, houses, windows, and most other objects lose heat by radiation and therefore become cool. During the night, water vapor condenses on objects that are cooler than the dew point. This condensation is called **dew**. If the dew point is below the freezing point of water, **frost** forms. Frost is not frozen dew, but ice crystals formed directly from vapor (Figure 20-3).

checkpoint Why is there a greater chance of condensation at night?

Rising Air

Cooling Radiation and contact cooling close to Earth's surface form dew and frost. However, clouds and precipitation normally form at higher elevations where the air is not cooled by direct contact with the ground. Almost all cloud formation and precipitation occur when air cools as it rises.

Air pressure and air temperature are proportionally linked. When air pressure increases, air molecules collide more often, increasing the temperature. When air pressure decreases, the temperature also decreases. Variations in temperature caused by the compression or the expansion of gas are called **adiabatic temperature changes**. An adiabatic process is one in which heat is not exchanged. During adiabatic warming, air temperature rises because compression increases the internal energy, even though no new thermal

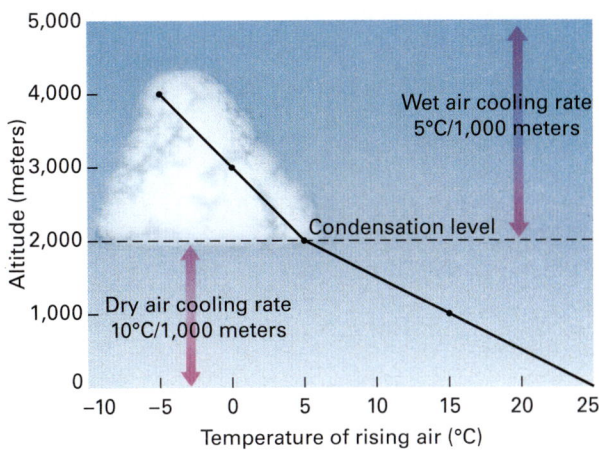

FIGURE 20-4
A rising air mass initially cools rapidly with elevation. After condensation begins and clouds start to form, it cools more slowly.

energy has been added to the air. During adiabatic cooling, air cools because the internal energy has decreased, even though no thermal energy has been lost.

Air pressure decreases with elevation. When dense surface air rises, it expands because the surrounding atmosphere is now of lower density, just as air expands when it rushes out of a punctured tire. Rising air loses energy when it expands, and therefore it cools adiabatically. How quickly air cools as it rises depends on how much moisture the air holds. Dry air cools by 10°C for every kilometer it rises. However, after condensation begins air cools by 5°C for every kilometer it rises.

The change in rate of adiabatic cooling occurs because all air contains some water vapor. As air rises and cools adiabatically, its temperature may eventually decrease to or below the dew point. Below the dew point, moisture condenses as droplets and a cloud forms. Condensing vapor releases latent heat. As the air rises through the cloud, its temperature now is affected by two opposing processes. It cools adiabatically, but at the same time it is heated by the latent heat released by condensation. However, the warming caused by latent heat is generally less than the amount of adiabatic cooling.

In contrast, sinking air becomes warmer because of adiabatic compression. Warm air can hold more water vapor than cool air can. Consequently, water does not condense from sinking, warming air. And the latent heat of condensation does not affect the rate of temperature rise.

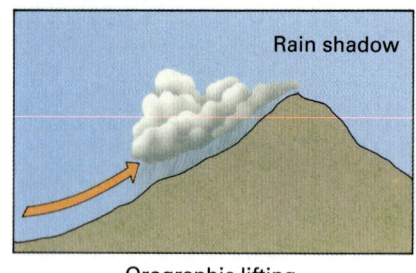
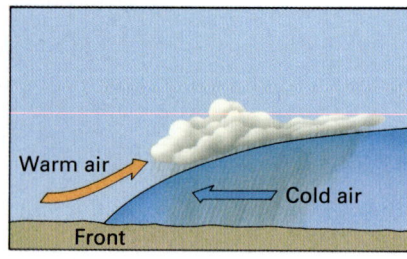
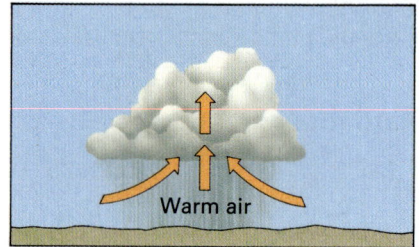

FIGURE 20-5
Three mechanisms cause air to rise and cool: (A) orographic lifting, (B) frontal wedging, and (C) convection–convergence.

When moist air rises, it cools and forms clouds. There are three mechanisms that cause air to rise (Figure 20-5): orographic lifting, frontal wedging, and convection–convergence.

Orographic Lifting When air flows over mountains, it is forced to rise by a mechanism called **orographic lifting**. This rising air frequently causes rain or snow over the mountains. Read more about this phenomenon in Lesson 20.3, Pressure, Wind, and Fronts.

Frontal Wedging A moving mass of cool, dense air may encounter a mass of warm, less-dense air. When this occurs, the cool, denser air slides under the warm air mass, forcing the warm air upward to create a weather front. This process is called **frontal wedging**.

Convection–Convergence Recall that convection is the upward, downward, and horizontal flow of fluids in response to heating and cooling. If one portion of the atmosphere becomes warmer than the surrounding air, the warm air expands, becomes less dense, and rises. A hot-air balloon rises because it contains air that is warmer and less dense than the surrounding air. If the sun heats an air mass near Earth's surface to a warmer temperature than that of surrounding air, the warm air will rise, just as the hot-air balloon rises.

On some days, clouds hang low over the land and obscure nearby hills. At other times, clouds float high in the sky, well above the mountain peaks. What factors determine the height and shape of a cloud?

Warm, moist air is unstable air because it rises rapidly and continues to rise. As water vapor condenses, towering clouds form. The vertical growth of clouds fueled by the release of latent heat promotes further instability. Air rushes along the ground to replace the rising air, generating surface winds. Most of us have experienced a violent thunderstorm on a hot summer day. Puffy clouds seem to appear out of nowhere in a blue sky. These clouds grow vertically and darken as the afternoon progresses. Suddenly, gusts of wind race across the land. Shortly thereafter, heavy rain falls. These events are all caused by unstable, rising, moist air.

In contrast, warm, dry air rises rapidly at first but then stops rising as soon as its temperature reaches that of the surrounding air. Therefore, it is unlikely to reach its dew point. Because no latent heat is released by condensation, the air does not continue to ascend, and clouds and precipitation do not form. So, warm, dry air is said to be stable air.

checkpoint How do clouds form in moist air as it rises?

20.1 ASSESSMENT

1. **Distinguish** How are relative humidity and absolute humidity determined?
2. **Compare** What is the difference between supersaturation and supercooling?
3. **Identify** Describe the set of conditions in which clouds form.
4. **Distinguish** How do frost and dew form?
5. **Explain** What are the three mechanisms that cause air to rise?

Critical Thinking

6. **Analyze Data** Suppose air at Earth's surface is at 30°C and contains 12.1 g/m^3 of water vapor, giving it a relative humidity of 40 percent. When this air rises and cools to 20°C, its relative humidity becomes 70 percent. Will water vapor start to condense as the air cools to 20°C? If so, explain why. If not, how much more water vapor would the air need to contain to become saturated at 20°C?

20.2 CLOUDS AND PRECIPITATION

Core Ideas and Skills
- Identify cloud types by their shape and structure.
- Relate cloud types to the specific meteorological conditions in which they form.
- Describe how fog forms.
- Explain what causes different forms of precipitation.
- Describe the effects of human interaction with Earth's systems.

KEY TERMS

cirrus   cumulus
stratus

FIGURE 20-6
(A) Cirrus clouds are high, wispy clouds composed of ice crystals.
(B) Stratus clouds spread out across the sky in a low, flat layer.
(C) Cumulus clouds are fluffy white clouds with flat bottoms and billowy tops.

Types of Clouds

Even a casual observer of the daily weather will notice that clouds are quite different from day to day. Different meteorological conditions create various cloud types. In turn, a look at the clouds can provide useful information about the daily weather.

Cirrus clouds are wispy clouds that look like hair blowing in the wind or feathers floating across the sky (Figure 20-6A). (*Cirrus* is Latin for "wisp of hair.") Cirrus clouds form at high altitudes, 6,000 to 15,000 meters (20,000 to 50,000 feet). The air is so cold at these elevations that cirrus clouds are composed of ice crystals rather than water droplets. High winds aloft blow them out into long, gently curved streamers.

Stratus clouds are horizontally layered, sheet-like clouds (Figure 20-6B). (*Stratus* is Latin for "layer.") They form when condensation occurs at the same elevation at which air stops rising and the clouds spread out into a broad sheet. Stratus clouds form the dark, dull-gray, overcast skies that may persist for days and bring steady rain.

Cumulus clouds are fluffy white clouds that typically display flat bottoms and billowy tops (Figure 20-6C). (*Cumulus* is Latin for "heap" or "pile.") On a hot summer day the top of a cumulus cloud may rise 10 kilometers or more above its base in cauliflower-like masses. The base of the cloud forms at the altitude at which the rising air cools below its dew point and condensation starts.

However, in this situation the rising air remains warmer than the surrounding air and therefore continues to rise. As it rises, more vapor condenses, forming the billowing columns.

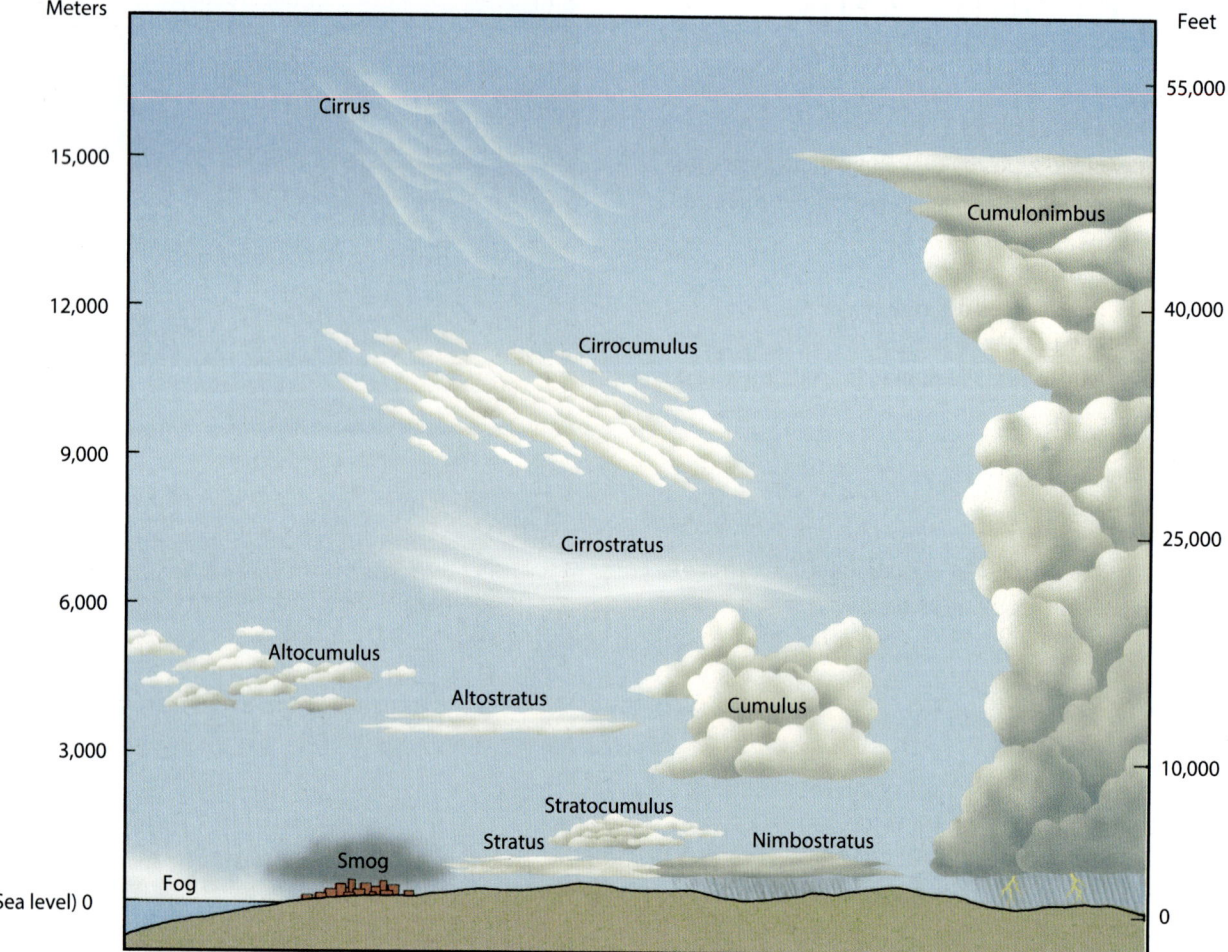

FIGURE 20-7
Cloud names are based on the shape and altitude of the clouds.

DATA ANALYSIS Record Data and Design a Game

Figure 20-7 shows the physical characteristics and altitudes of the ten types of clouds (excluding fog). Use the information in this graphic to make a four-column chart. The first three columns should contain the name of each cloud type, its physical description, and its altitude (low: less than 2,000 m; middle: about 2,000 m to 6,000 m; high: greater than 6,000 m). In the fourth column, make a detailed drawing of each cloud type. Then, use your chart to create quiz-show style questions for a game of your choice. Your questions or clues should relate hypothetical weather scenarios to cloud type.

1. **Contrast** Differentiate between stratus and nimbostratus clouds.
2. **Contrast** Differentiate between stratus and altostratus clouds.
3. **Contrast** Differentiate between stratus and cirrostratus clouds.

Computational Thinking

4. **Recognize Patterns** How can the altitude of a cloud give clues about the kind of precipitation it might produce?
5. **Recognize Patterns** What is the general relationship between a cloud type and the weather it produces?

Other types of clouds are named by combining these three basic terms (Figure 20-7). Clouds that are low and sheet-like with some vertical structure are called stratocumulus clouds. The term *nimbo* refers to a cloud that precipitates. Therefore, a cumulonimbus cloud is a towering rain cloud. If you see one, you should seek shelter, because cumulonimbus clouds commonly produce intense rain, thunder, lightning, and sometimes hail. A nimbostratus cloud is a stratus cloud from which rain or snow falls. Other prefixes are also added to cloud names. For example, *alto* is derived from the Latin root *altus*, meaning "high." An altostratus cloud is simply a high stratus cloud.

Fog Fog is a cloud that forms at or very close to ground level. Unlike other types of clouds, which result from adiabatic cooling, fog usually forms through radiation cooling or contact cooling of moist air over land or water. The air cools below its dew point, and water vapor condenses. Cities like San Francisco, California, Seattle, Washington, and Vancouver, British Columbia, all experience foggy conditions when warm, moist air blowing in from the Pacific Ocean is cooled when it comes into contact with land (Figure 20-8).

checkpoint What name would be given to a cloud that is both wispy, like feathers, and layered?

Types of Precipitation

Rain Why does rain fall from some clouds, whereas other clouds float across a blue sky on a sunny day and produce no rain? The droplets in a cloud are small, about 0.01 millimeter in diameter (about one-seventh the diameter of a human hair). In still air, such a droplet would require 48 hours to fall from a cloud 1,000 meters above Earth. But these tiny droplets never reach Earth because they evaporate before they can reach the ground.

If the air temperature in a cloud is above freezing, the tiny droplets may collide and coalesce. You can observe similar behavior in droplets sliding

FIGURE 20-8
Fog covers Vancouver, Canada.

down a window pane on a rainy day. If two droplets collide, they merge to become one large drop. If the droplets in a cloud grow large enough, they fall as drizzle (0.1 to 0.5 millimeter in diameter) or light rain (0.5 to 2 millimeters in diameter). About a million cloud droplets must combine to form an average-size raindrop.

Snow, Sleet, and Glaze When the temperature in a cloud is below freezing, the cloud is composed of ice crystals rather than water droplets. If the temperature near the ground is also below freezing, the crystals remain frozen and fall as snow. In contrast, if raindrops form in a warm cloud and fall through a layer of cold air at lower elevation, the drops freeze and fall as small spheres of ice called sleet. Sometimes the freezing zone near the ground is so thin that raindrops do not have time to freeze before they reach Earth. However, when they land on subfreezing surfaces, they form a coating of ice called glaze (Figure 20-9). Glaze can be heavy enough to break tree limbs and electrical transmission lines. It also coats highways with a dangerous, icy veneer. In the winter of 1997–1998, a sleet and glaze storm in eastern Canada and the northeastern United States caused billions of dollars in damage. The ice damaged so many electrical lines and poles that many people were without electricity for several weeks.

FIGURE 20-9
Go online to view an example of glaze forming.

Hail Occasionally, precipitation takes the form of very large ice globules called hail. Hailstones vary from 5 millimeters to a record-breaking 14 centimeters in diameter; that record breaker weighed 765 grams (more than 1.5 pounds). A 500-gram (1-pound) hailstone crashing to Earth at 160 kilometers (100 miles) per hour can shatter windows, dent car roofs, and kill people and livestock. Even small hailstones can damage crops. Hail falls only from cumulonimbus clouds. Because cumulonimbus clouds form in columns with distinct boundaries, hailstorms occur in local, well-defined areas. Therefore, one farmer may lose an entire crop while a neighbor is unaffected.

A hailstone consists of ice in concentric shells, like the layers of an onion. Ice crystals that form in a cloud can become hailstones during their descent, as supercooled water freezes onto them. The layers

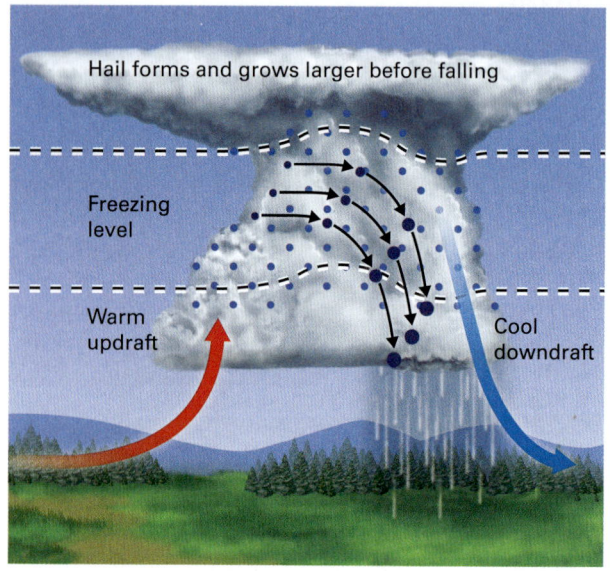

FIGURE 20-10
Hailstones form and grow as layers of supercooled water and water vapor condense on individual ice grains.

develop because different portions of a cloud have different temperatures and different amounts of supercooled water. Each layer of a hailstone forms in a different part of the cloud. Turbulent winds can also blow falling ice crystals back upward into a cloud (Figure 20-10). New layers of ice accumulate as additional water vapor condenses on the recirculating ice crystals. An individual particle may rise and fall several times until it grows so large and heavy that it drops out of the cloud.

checkpoint What causes raindrops to form glaze?

20.2 ASSESSMENT

1. **Explain** Why do cirrus clouds contain ice crystals?
2. **Describe** How is dew point connected to formation of clouds?
3. **Define** What three basic types of clouds are the foundation for all cloud types?
4. **Summarize** What causes ice crystals to fall as snow?

Critical Thinking

5. **Synthesize** What causes cumulus and cumulonimbus clouds to develop vertically whereas stratus and nimbostratus clouds develop horizontally?

20.3 PRESSURE, WIND, AND FRONTS

Core Ideas and Skills
- Explain how air pressure can affect wind speed and wind direction.
- Distinguish between different types of weather fronts.
- Explain how ocean temperature contributes to sea breezes and monsoons.
- Describe how Earth's surface can influence weather patterns.

KEY TERMS

isobar
jet stream
cyclone
anticyclone
air mass

front
occluded front
stationary front
monsoon

Wind Speed

Warm air is less dense than cold air. Therefore, warm air exerts a relatively low atmospheric pressure and cold air exerts a relatively high atmospheric pressure. Warm air rises because it is less dense than the surrounding cool air. Air rises slowly above a typical low-pressure region, at a rate of about 1 kilometer per day. In contrast, if air in the upper atmosphere cools, it becomes denser than the air beneath it and sinks.

Air must flow inward over Earth's surface toward a low-pressure region to replace a rising air mass. But a sinking air mass displaces surface air, pushing it outward from a high-pressure region. So, vertical airflow in both high- and low-pressure regions is accompanied by horizontal airflow, called wind. Winds near Earth's surface always flow away from a region of high pressure and toward a low-pressure region. Ultimately, all wind is caused by the pressure differences resulting from unequal heating of Earth's atmosphere.

Pressure Gradient Wind blows in response to differences in pressure. Imagine that you are sitting in a room and the air is still. Now you open a can of vacuum-packed coffee and hear the hissing as air rushes into the can. Because the pressure in the room is higher than that inside the coffee can, air moves from the room into the can. But if you blow up a balloon, the air inside the balloon is at higher pressure than the air in the room. If you then let some air escape the inflated balloon, air will blow from the higher-pressure interior of the balloon to the lower-pressure room.

Wind speed is determined by the magnitude of the pressure difference over distance, called the pressure gradient. Wind blows rapidly if a large pressure difference exists over a short distance. A steep pressure gradient is analogous to a steep hill. Just as a ball rolls quickly down a steep hill, wind flows rapidly across a steep pressure gradient. To create a pressure-gradient map, air pressure is measured at hundreds of different weather stations. Points of equal pressure are connected by map lines called **isobars**. (The weather map shown in Figure 1-16 has gray isolines that encircle areas of equal pressure.) A steep pressure gradient is shown by closely spaced isobars whereas a weak pressure gradient is indicated by widely spaced isobars. Pressure gradients result in forces that move air from high-pressure areas to low-pressure areas. Pressure gradients change daily, or sometimes hourly, as high- and low-pressure zones move. Therefore, maps are updated frequently.

CAREERS IN EARTH AND SPACE SCIENCES

Meteorologists: Forecasting the Weather

A meteorologist is a physical scientist who studies and observes the weather in order to forecast weather conditions. Meteorologists earn bachelor's degrees in meteorology or atmospheric science. A meteorologist can work locally or nationally.

Interpreting data, such as Doppler radar data and satellite imagery, is an important aspect of the work. Meteorologists review data along with results from computer simulations to issue warnings for severe weather, high winds, flash floods, winter storms, and other conditions. Meteorologists need to resolve conflicting data or simulation results to successfully predict and communicate weather conditions.

Meteorologists may also conduct local studies to test new techniques or technology. Findings of such studies are incorporated into a developing body of knowledge for weather forecasting.

Friction Rising and falling air generates wind both along Earth's surface and at higher elevations. Surface winds are affected by friction with Earth's surface, whereas high-altitude winds are not. As a result, wind speed normally increases with elevation. This effect was first noted during World War II. On November 24, 1944, U.S. bombers were approaching Tokyo for the first mass bombing of the Japanese capital. Flying between 8,000 and 10,000 meters (27,000 to 33,000 feet), the pilots suddenly found themselves roaring past landmarks 140 kilometers (90 miles) per hour faster than the theoretical top speed of their airplanes! Amid the confusion, most of the bombs missed their targets, and the mission was a military failure. However, this experience introduced meteorologists to **jet streams**, narrow bands of high-altitude wind. The jet stream in the Northern Hemisphere flows from west to east at speeds between 120 and 240 kilometers per hour (75 and 150 mph). As a comparison, surface winds attain such velocities only in hurricanes and tornadoes. Airplane pilots traveling from Los Angeles to New York fly with the jet stream to gain speed and save fuel, whereas pilots moving from east to west try to avoid it.

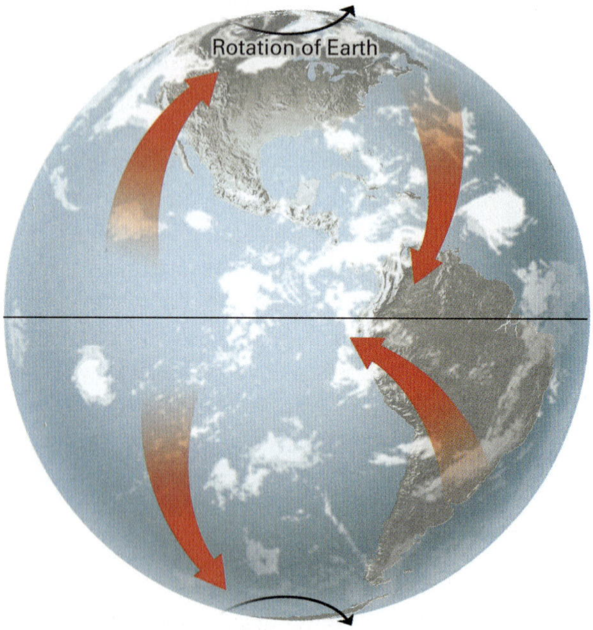

checkpoint What is the relationship between pressure gradient and wind speed?

Wind Direction

Coriolis Effect Recall from Chapter 16 that the Coriolis effect, caused by Earth's spin, deflects ocean currents. The Coriolis effect similarly deflects winds. In the Northern Hemisphere wind is deflected toward the right, and in the Southern Hemisphere, to the left (Figure 20-11). The Coriolis effect alters wind direction but not its speed.

Cyclones and Anticyclones Figure 20-12A shows the movement of air in the Northern Hemisphere as it converges toward a low-pressure area. If Earth did not spin, wind would flow directly across the isobars, as shown by the black arrows. However, Earth does spin so the Coriolis effect deflects wind to the right, as shown by the small blue arrows. This rightward deflection creates a counterclockwise vortex near the center of the low-pressure region, as shown by the large, magenta arrows. In the Southern Hemisphere (not shown) the direction is clockwise. Such a low-pressure region with its accompanying surface wind is called a cyclone. In this usage, **cyclone** means a system of inwardly directed rotating winds, not the violent storms that are sometimes called cyclones, hurricanes, or typhoons.

The opposite mechanism forms an **anticyclone** around a high-pressure region (Figure 20-12B). When descending air reaches the surface, it spreads out in all directions. In the Northern Hemisphere, the Coriolis effect deflects the diverging winds of an anticyclone to the right, forming a pinwheel pattern, with the wind spiraling clockwise. In the Southern Hemisphere (not shown), the Coriolis effect deflects winds leftward and creates wind that forms a counterclockwise spiral.

checkpoint In what direction do Northern Hemisphere cyclones and anticyclones rotate?

Fronts and Frontal Weather

An **air mass** is a large body of air with approximately uniform temperature and humidity at any given altitude. Typically, an air mass is 1,500 kilometers or more across and several kilometers thick.

FIGURE 20-11
The Coriolis effect deflects winds to the right in the Northern Hemisphere and to the left in the Southern Hemisphere. Only winds blowing due east or west are unaffected.

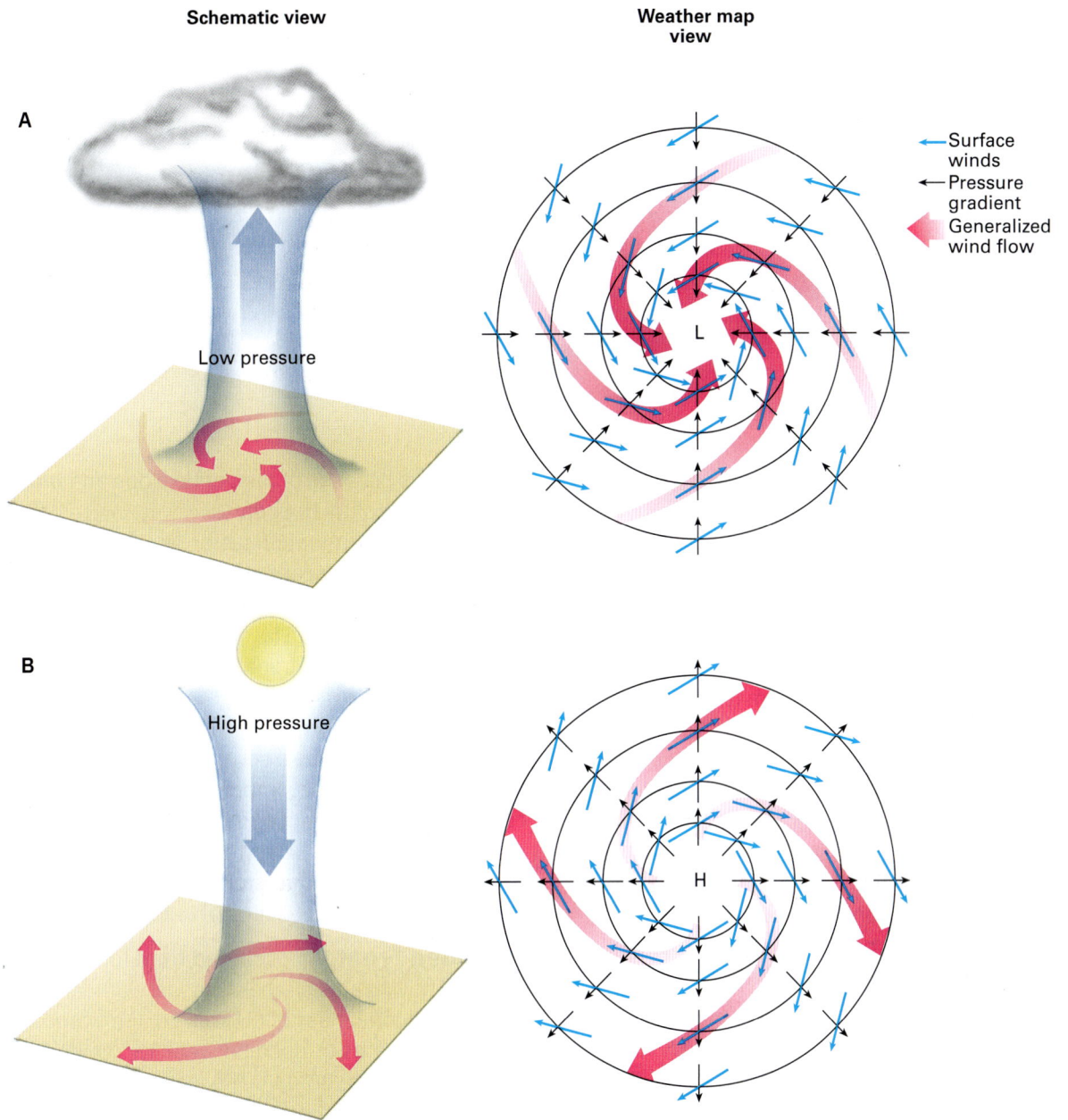

FIGURE 20-12
(A) In the Northern Hemisphere, a cyclone consists of winds spiraling counterclockwise into a low-pressure region. (B) An anticyclone in the Northern Hemisphere consists of winds spiraling clockwise out from a high-pressure zone.

Because air acquires both heat and moisture from Earth's surface, an air mass is classified by its place of origin. Temperature can be either polar (cold) or tropical (warm). Maritime air originates over water and has high moisture content, whereas continental air has low moisture content (Figure 20-13).

FIGURE 20-13
Go online to view a U.S. map that shows air masses by region.

Air masses move and collide. The boundary between a warmer air mass and a cooler one is a **front**. The term was first used during World War I (1914–1918) to describe the similarity of weather systems to armies advancing at battle lines. When two air masses collide, each may retain its integrity for days before the two mix. During a collision, one of the air masses is forced to rise, which often results in cloudiness and precipitation. Frontal weather patterns are determined by the types of air masses that collide and their relative speeds and directions.

LESSON 20.3 **651**

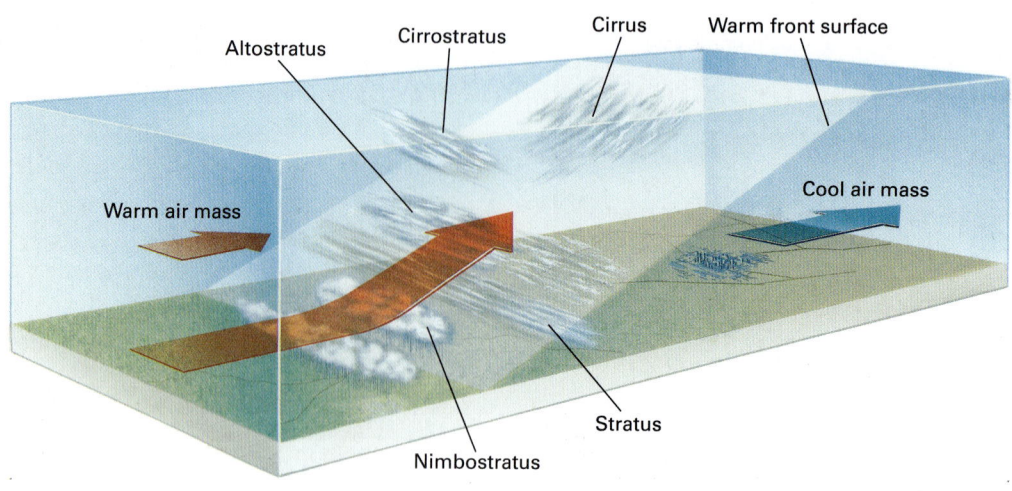

FIGURE 20-14 In a warm front, moving warm air rises gradually over cold air. High, wispy cirrus and cirrostratus clouds typically develop near the leading edge of the rising warm air, followed by descending altostratus, stratus, and nimbostratus clouds near the trailing edge of the front.

Warm Fronts and Cold Fronts Fronts are classified based on the relative movement of the warm front and cold front that are converging. A warm front forms when moving warm air collides with a stationary or slower-moving cold air mass. A cold front forms when moving cold air collides with stationary or slower-moving warm air.

In a warm front, the moving warm air rises over the denser cold air as the two masses collide. The rising warm air cools adiabatically and the cooling generates clouds and precipitation. Precipitation is generally light because the air rises slowly along the gently sloping frontal boundary. Figure 20-14 shows that a characteristic sequence of clouds accompanies a warm front. High, wispy cirrus and cirrostratus clouds develop near the leading edge of the rising warm air. These high clouds commonly precede a storm. They form as much as 1,000 kilometers ahead of an advancing band of precipitation that falls from thick, low-lying nimbostratus and stratus clouds near the trailing edge of the front. The cloudy weather may last for several days because of the gentle slope and broad extent of the frontal boundary.

A cold front forms when faster-moving cold air overtakes and displaces warm air. The dense cold air forms a blunt wedge and pushes under the warmer air (Figure 20-15). This event causes the leading edge of a cold front to be much steeper than that of a warm front. The steep contact between the two air masses causes the warm air to rise rapidly, creating a narrow band of violent weather commonly accompanied by cumulus and cumulonimbus clouds. The storm system may be only 25 to 100 kilometers in width, but within this zone downpours, thunderstorms, and violent winds are common.

Occluded Front An **occluded front** forms when a faster-moving cold air mass traps a warm air mass against a second mass of cold air. So, the warm air mass becomes trapped between two colder air masses (Figure 20-16). The faster-moving cold air mass then slides beneath the warm air,

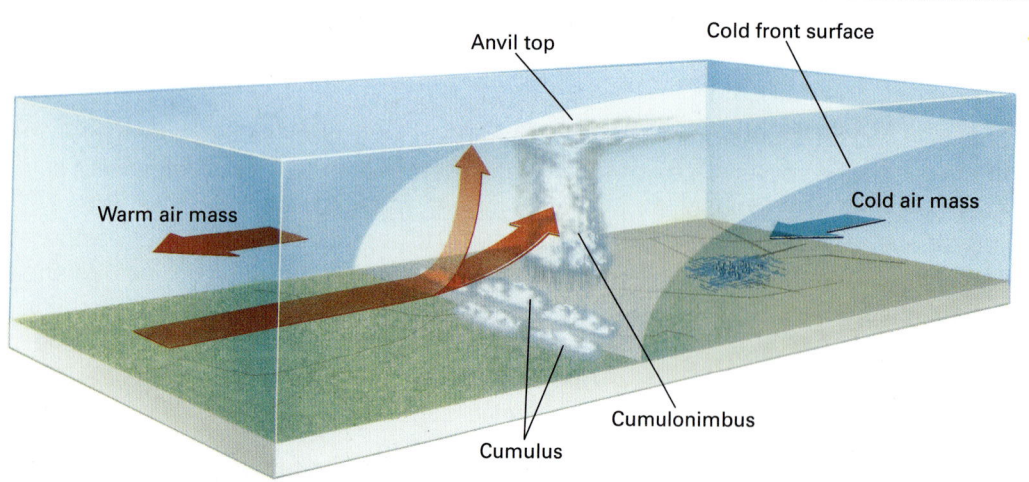

FIGURE 20-15 Correlation of sedimentary rocks from three different locations in Utah and Arizona helps geologists develop a more complete picture of geologic history in the region. The rock units are called geologic formations.

MINILAB Make a Barometer

Materials (per group):

blue food dye
clear glass bottle, empty with no label
plastic straw with no bendable neck
clay or putty
permanent marker
tape
ruler

Procedure

1. Fill the bottle halfway with room temperature water. Place two drops of food dye into the water and swirl until it is mixed.
2. Put the straw through the mouth of the bottle and down into the water. Do not allow the straw to settle on the bottom. Tape the straw in place but make sure some air can pass through the tape into the bottle. There should be some space between the straw and the bottom of the bottle.
3. Place your mouth on the straw and gently create a vacuum by slowly sucking the straw until water is three-fourths of the way to the top of the straw.
4. When the water is at the proper height in the straw, quickly remove the straw from your mouth and place your finger over the opening of the straw so that the water stays in the straw. Then quickly replace your finger with a ball of clay or putty, pressing it into the opening of the straw. Be sure the straw is sealed tightly. (If too much water leaves the straw, remove the putty and repeat steps 3 and 4.)
5. Tape the ruler to the outside of your bottle. Use a permanent marker or piece of tape to mark where the water level is in the straw on the outside of the bottle.
6. Take a measurement of the water height in the straw every day for one month. Record your data.

Results and Analysis

1. **Observe** What did you notice about the water level in the straw over the 30-day observational period?
2. **Explain** What caused the water level in the straw to change? How do you know? How does this happen?
3. **Relate** When the water level in the straw drops, what pressure does this indicate? What about when the water level rises? How do you know?
4. **Predict** How do you think the water level in the straw would behave if it were sunny and dry for an entire week? Why?

Critical Thinking

5. **Predict** How can you use your barometer to predict the weather?
6. **Critique** What factor could affect the accuracy of your barometer? Why? How could you fix this?

---

lifting it completely off the ground. Precipitation occurs along both frontal boundaries, combining the narrow band of heavy precipitation of a cold front with the wider band of lighter precipitation of a warm front. The net result is a large zone of inclement weather. A storm of this type is commonly short-lived because the warm air mass is cut off from its supply of moisture evaporating from Earth's surface.

FIGURE 20-16
Go online to view a diagram of an occluded front.

Stationary Front
A **stationary front** occurs along the boundary between two stationary air masses. Under these conditions, the front can remain over an area for several days. Warm air rises, forming conditions similar to those in a warm front. As a result, rain, drizzle, and fog may occur.

The Life Cycle of a Midlatitude Cyclone
A cyclone is a low-pressure system with rotating winds. Most cyclones in the middle latitudes of the Northern Hemisphere develop along a front between polar and tropical air masses. The storm often starts with winds blowing in opposite directions along a stationary front between the two air masses (Figure 20-17A). In the figure, a warm air mass was moving northward and was deflected to the east by the Coriolis effect. At the same time, a cold air mass traveling southward was deflected to the west.

FIGURE 20-17
Go online to view a diagram of the life cycle of a midlatitude cyclone.

In Figure 20-17B, the cold polar air continues to push southward, creating a cold front and lifting the warm air off the ground. Then, some small disturbance—a topographic feature such as a mountain range, airflow from a local storm, or perhaps a local temperature variation—deforms the straight frontal boundary, forming a wavelike kink in the front. Once the kink forms, the winds on both sides are deflected to strike the front at an angle. Therefore, a warm front forms to the east and a cold front forms to the west.

Rising warm air then forms a low-pressure region near the kink (Figure 20-17C). In the Northern Hemisphere, the Coriolis effect causes the winds to circulate counterclockwise around the kink. To the west, the cold front advances southward. To the east, the warm front advances northward. At the same time, precipitation (rain or snow) falls from the rising warm air (Figure 20-17D). Over a period of one to three days, the air rushing into the low-pressure region equalizes pressure differences, and the storm dissipates.

Many of the pinwheel-shaped storms seen on weather maps are cyclones of this type. In North America, the jet stream and other prevailing, upper-level, westerly winds generally move cyclones from west to east along the same paths, called storm tracks.

checkpoint What are the potential weather effects of a cold front?

Impact of Earth's Surface

Earth's surface features, including mountain ranges, rain forests, proximity to the sea, and uneven heating and cooling of continents, can create conditions that affect the weather of a region.

MINILAB Compare Two Coriolis Effect Models

Materials (per pair):

- light-colored, round balloon
- black permanent marker
- aluminum pie pan
- food dye
- water
- paper cup
- pushpin
- crushed ice
- rotating turntable (such as a lazy Susan)

Procedure

Part 1:

1. Blow up the balloon to full capacity and tie it off. The balloon represents Earth.
2. With the marker, draw a horizontal line around the midline of the balloon to represent the Equator.
3. Have a partner hold the balloon upright and slowly rotate it counterclockwise. As the balloon rotates, put the tip of the marker on the top of the balloon and try to draw a straight line down toward the Equator. Record your observations.
4. Repeat step 3, but this time draw a line up from the bottom of the balloon toward the Equator. Record your observations.

Part 2:

1. Place the pie pan in the middle of the turntable and fill it nearly full of water, about 1–2 cm from the top.
2. Using the pushpin, poke four small holes approximately 1 cm from the base of the paper cup.
3. Fill the cup one-half to two-thirds full of crushed ice. Place the cup in the center of the pie pan. See illustration.

4. Have a partner slowly rotate the turntable in a counterclockwise direction. As it rotates, add several drops of food coloring to the cup. Then, pour a small amount of water into the cup while continuing to rotate the turntable. Record your observations.

Results and Analysis

1. **Observe** In Part 1, what did you notice about the marker lines as they approached the Equator?
2. **Observe** In Part 2, what did you notice about the food coloring as it came out of the spinning cup?
3. **Explain** How do the two models simulate the Coriolis effect on Earth?

Critical Thinking

4. **Compare** How were the two models alike? How were they different?
5. **Critique** Which model do you think is more effective at simulating the Coriolis effect and why? Why is this the better model in your opinion? Use evidence and observations to support your claims.

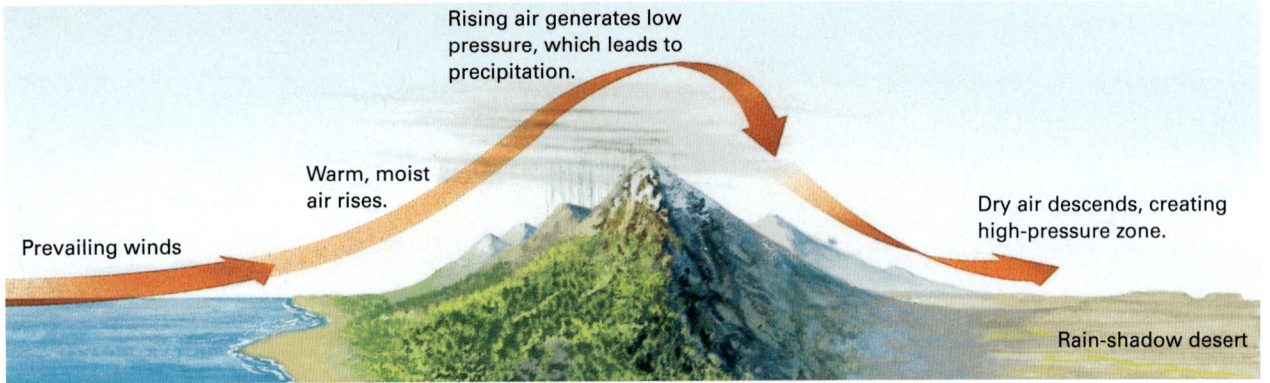

FIGURE 20-18
A rain-shadow desert forms when moist air rises over a mountain range and precipitates most of its moisture on the windward side and crest of the range. The dry, descending air on the leeward side absorbs moisture, forming a desert.

A

B

FIGURE 20-19
(A) The view from Patterson Mountain in the North Cascades National Park in western Washington state shows a lush landscape due to precipitation. (B) Lake Roosevelt National Park in eastern Washington state is on the leeward side of the Cascades and has desert conditions.

Mountain Ranges and Rain-Shadow Deserts

Orographic lifting occurs when wind forces air to rise and flows over a mountain range. As the air rises, it cools adiabatically, and water vapor may condense into clouds that produce rain or snow. These conditions create abundant precipitation on the windward side and the crest of the range. When the air passes over the crest onto the leeward (downwind) side, it sinks (Figure 20-18). This air has already lost much of its moisture. In addition, it warms adiabatically as it falls, absorbing moisture and creating a rain-shadow desert on the leeward side of the range. For example, Death Valley, California, is a rain-shadow desert and receives only 5 centimeters of rain a year, while the nearby west slope of the Sierra Nevada receives 178 centimeters of rain a year. Similarly, in the Pacific Northwest, the western side of the Cascade Mountain range receives more rain than areas on the eastern side of the range where a rain shadow takes effect (Figures 20-19).

Forests and Weather

Recall that atmospheric moisture condenses when moist air cools below its dew point. Forests cool the air. Large quantities of water evaporate from leaf surfaces in the process called transpiration. Evaporation cools the surrounding air. In addition, forests shade the soil from the hot sun, and tree roots and litter retain moisture. In an open clearing, rainwater evaporates quickly after a storm or runs off the

FIGURE 20-20
Go online to view an example of conditions during the summer monsoon season.

LESSON 20.3

FIGURE 20-21
ON ASSIGNMENT Tornado chasers wait for a tornado to develop. The vehicle in the photograph is a Doppler on Wheels truck. National Geographic photographer Carsten Peter has specialized in storm-related photography, recording images during significant weather events both beautiful and devastating in nature.

656

surface. But forest soils remain moist long after the rain dissipates. Evaporation from soil litter combines with transpiration cooling from leaf surfaces to maintain relatively cool temperatures during times when there is no rain. This results in a positive feedback mechanism: forests cool the air; cool air promotes rainfall; rainfall supports forests.

In today's tropical rain forests, local rainfall has decreased by as much as 50 percent when the forests were cut and replaced by farmland or pasture. When the rainfall decreases, wildfires become more common. More forest is destroyed, establishing a negative feedback mechanism of increasing drought, fire, and forest loss.

Sea and Land Breezes Anyone who has lived near an ocean or large lake has encountered winds blowing from water to land and from land to water. Sea and land breezes are caused by uneven heating and cooling of land and water. Land surfaces heat up faster than adjacent bodies of water and cool more quickly. If land and sea are nearly the same temperature on a summer morning, during the day the land warms and heats the air above it. Hot air then rises over the land, producing a local low-pressure area. Cooler air from the sea flows inland to replace the rising air. On a hot sunny day, winds generally blow from the sea onto land. The rising air is good for flying kites or hang-gliding but often brings afternoon thunderstorms.

At night the reverse process occurs. The land cools faster than the sea, and descending air creates a local high-pressure area over the land. Then the winds reverse, and breezes blow from the shore out toward the sea.

Monsoons A **monsoon** is a seasonal wind and weather system caused by uneven heating and cooling of continents and adjacent oceans. Just as sea and land breezes reverse direction with day and night, monsoons reverse direction with the seasons. In the summer the continents become warmer than the sea. Warm air rises over land, creating a large low-pressure area and drawing moisture-laden maritime air inland. When the moist air rises as it flows over the land, clouds form and heavy monsoon rains fall (Figure 20-20). More than half of the inhabitants of Earth depend on monsoons because the predictable, heavy, summer rains bring water to the fields of Africa and Asia. If the monsoons fail to arrive, crops cannot grow and people starve.

In winter the process is reversed. The land cools below the sea temperature, and as a result air descends over land, producing dry, continental high pressure. At the same time, air rises over the ocean and the prevailing winds blow from land to sea.

checkpoint Why would winds blow from land or shore out to sea at night?

20.3 ASSESSMENT

1. **Explain** How can air released from a balloon be used as a model for the effect of pressure gradients on wind?
2. **Explain** What happens when a warm air mass becomes trapped between two cold air masses?
3. **Explain** What are the causes for dry air on the leeward side of a mountain?
4. **Sequence** Describe the feedback loop that results in moisture and rainfall in forests.
5. **Summarize** What causes land and sea breezes to reverse direction between day and night?

Critical Thinking

6. **Design** Sketch an illustration showing how monsoonal weather systems differ between the summer and winter.

20.4 STORMS

Core Ideas and Skills
- Describe how thunderstorms develop and what causes lightning.
- Distinguish between tornadoes, tropical cyclones, hurricanes and typhoons.
- Summarize the effects of El Niño.

KEY TERMS

tornado	typhoon
tropical cyclone	storm surge
hurricane	El Niño

Thunderstorms

At any given moment, 27,000 thunderstorms are in progress over different parts of Earth. A single bolt of lightning can involve several hundred million volts of energy and for a few seconds produces as much power as a nuclear power plant. It heats the surrounding air to 25,000°C or more, much hotter than the surface of the sun. The heated air expands instantaneously to create a shock wave that we hear as thunder.

Despite their violence, thunderstorms are local systems, often too small to be included on national weather maps. A typical thunderstorm forms and then dissipates in a few hours and covers from about 10 to a few hundred square kilometers. It is not unusual to stand on a hilltop in the sunshine on a humid day and watch rain squalls and lightning a few kilometers away.

All thunderstorms develop when warm, moist air rises, forming cumulus clouds that develop into towering cumulonimbus clouds. Different local conditions cause these regions of rising air. Thunderstorms can develop as a result of convection–convergence when two moist air masses come together and the moist air rises rapidly. Central Florida is the most active thunderstorm region in the United States. As the subtropical sun heats the Florida peninsula, rising air draws moist air from both its east coast and its west coast.

Thunderstorms also form in continental interiors during the spring or summer, when afternoon sunshine heats the ground and generates cells of rising, moist air. When a moist air mass meets a mountain range, orographic lifting may force the air upward, generating mountain thunderstorms. Finally, thunderstorms commonly occur as a result of frontal wedging at cold fronts.

Lightning Lightning is an intense discharge of electricity that occurs when the buildup of static electricity overwhelms the insulating properties of air (Figure 20-22). If you walk across a carpet on a dry day, the friction between your feet and the rug shears electrons off the atoms on the rug. The electrons migrate into your body and concentrate there. If you then touch a metal doorknob, a spark consisting of many electrons jumps from your finger to the metal knob.

FIGURE 20-22
Go online to view strokes of lightning as they are released during a storm.

In 1752, Benjamin Franklin showed that lightning is an electrical spark. He suggested that charges separate within cumulonimbus clouds and build until a bolt of lightning jumps from the cloud. In the more than 250 years since Franklin, atmospheric physicists have been unable to agree upon the exact mechanism of lightning.

According to one hypothesis, friction between the intense winds and moving ice crystals in a cumulonimbus cloud generates both positive and negative electrical charges in the cloud, and the two types of charges become physically separated (Figure 20-23A). The positive charges tend to accumulate in the upper portion of the cloud. The negative charges build up in the lower reaches of the cloud. When enough charge difference accumulates, the electrical potential exceeds the insulating properties of air. A spark jumps from the cloud to the ground, from the ground to the cloud, or from one cloud to another.

FIGURE 20-23
Go online to view two examples of how lightning might form.

Another hypothesis suggests that cosmic rays bombarding the cloud from space produce ions at the top of the cloud. Other ions form on the ground as winds blow over Earth's surface. The electrical discharge occurs when the potential difference between the two groups of electrical charges exceeds, or is greater than, the insulating properties of air (Figure 20-23B).

Perhaps neither of these hypotheses is entirely correct, but rather some combination of the mechanisms causes lightning.

checkpoint From what kind of clouds do thunderstorms develop?

Tornadoes

A **tornado** is a small, short-lived, funnel-shaped storm that protrudes from the base of a cumulonimbus cloud (Figure 20-24). The base of the funnel can be from 2 meters to 3 kilometers in diameter. Some tornadoes remain suspended in air while others touch the ground. After a tornado touches ground, it may travel for a few meters to a few hundred kilometers across the surface. The funnel travels at 40 to 65 kilometers per hour, and in some cases as much as 110 kilometers per hour, but the spiraling winds within the funnel are much faster. Few direct measurements have been made of pressure and wind speed inside a tornado. However, we know that a large pressure difference occurs over a very short distance. Meteorologists estimate that winds in tornadoes may reach 500 kilometers per hour or greater. These winds rush into the narrow, low-pressure zone and then spiral upward. After a few seconds to a few hours, the tornado lifts off the ground and dissipates. Tornadoes are the most violent of all storms, yet the total destruction from tornadoes is not as great as that from hurricanes because the path of a tornado is narrow and its duration is short.

Although tornadoes can occur anywhere in the world, 75 percent of the world's twisters concentrate in the Great Plains, east of the Rocky Mountains. This area is known as Tornado Alley. (See Chapter 9 Case Study.) Approximately 700 to 1,000 tornadoes occur in the United States each year. They frequently form in the spring or early summer. At that time, continental polar (dry, cold) air from Canada collides with maritime tropical (warm, moist) air from the Gulf of Mexico. These conditions often create thunderstorms. Meteorologists cannot explain why most storms dissipate harmlessly while a few develop tornadoes. Yet tornadoes are most likely to occur when large differences in temperature and moisture exist between the two air masses and the boundary between them is sharp.

checkpoint During what time of year are tornadoes most common in the United States, and why?

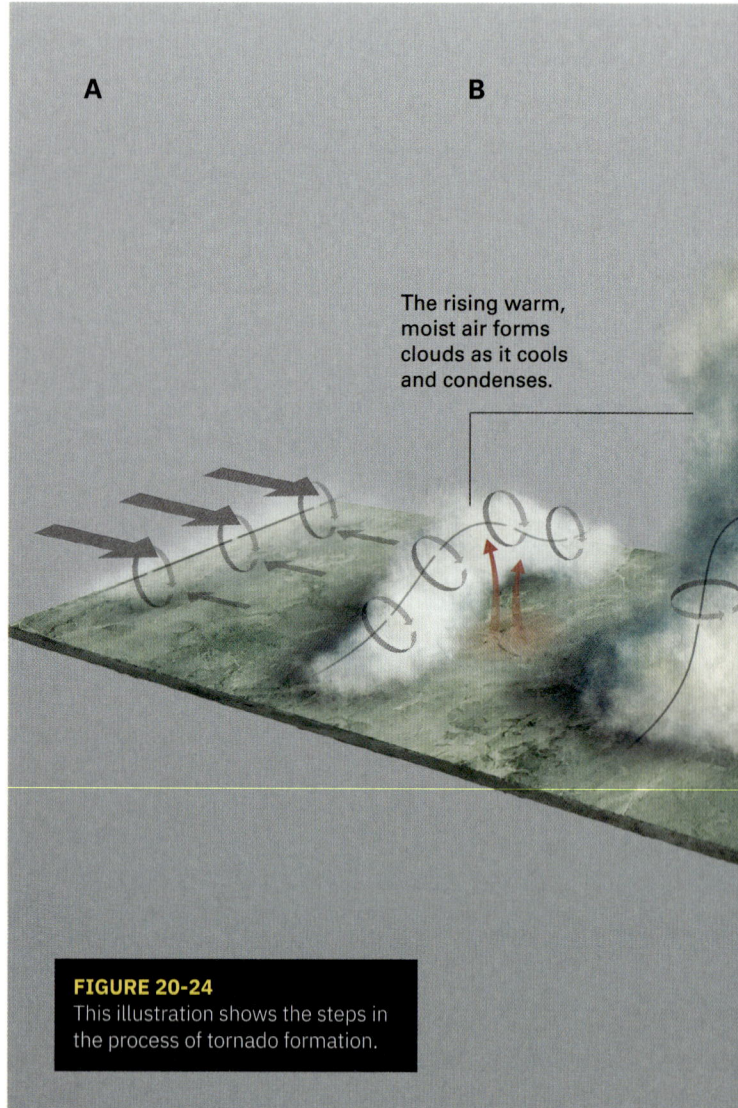

FIGURE 20-24
This illustration shows the steps in the process of tornado formation.

Tropical Cyclones

A **tropical cyclone** or tropical storm is less intense than a tornado but much larger and longer-lived (Table 20-1). Tropical cyclones are circular disturbances that average 600 kilometers in diameter and persist for days or weeks. If the wind exceeds 120 kilometers per hour, a tropical cyclone is called a **hurricane** in North America and the Caribbean, a **typhoon** in the western Pacific, and a cyclone in the Indian Ocean. Intense low pressure in the center of a hurricane can generate wind as strong as 300 kilometers per hour.

The low atmospheric pressure created by a tropical cyclone can raise the sea surface by several meters. Often, as a tropical cyclone strikes shore, strong onshore winds combine with the abnormally high water level created by low pressure to create a **storm surge** that floods coastal areas. In 2005

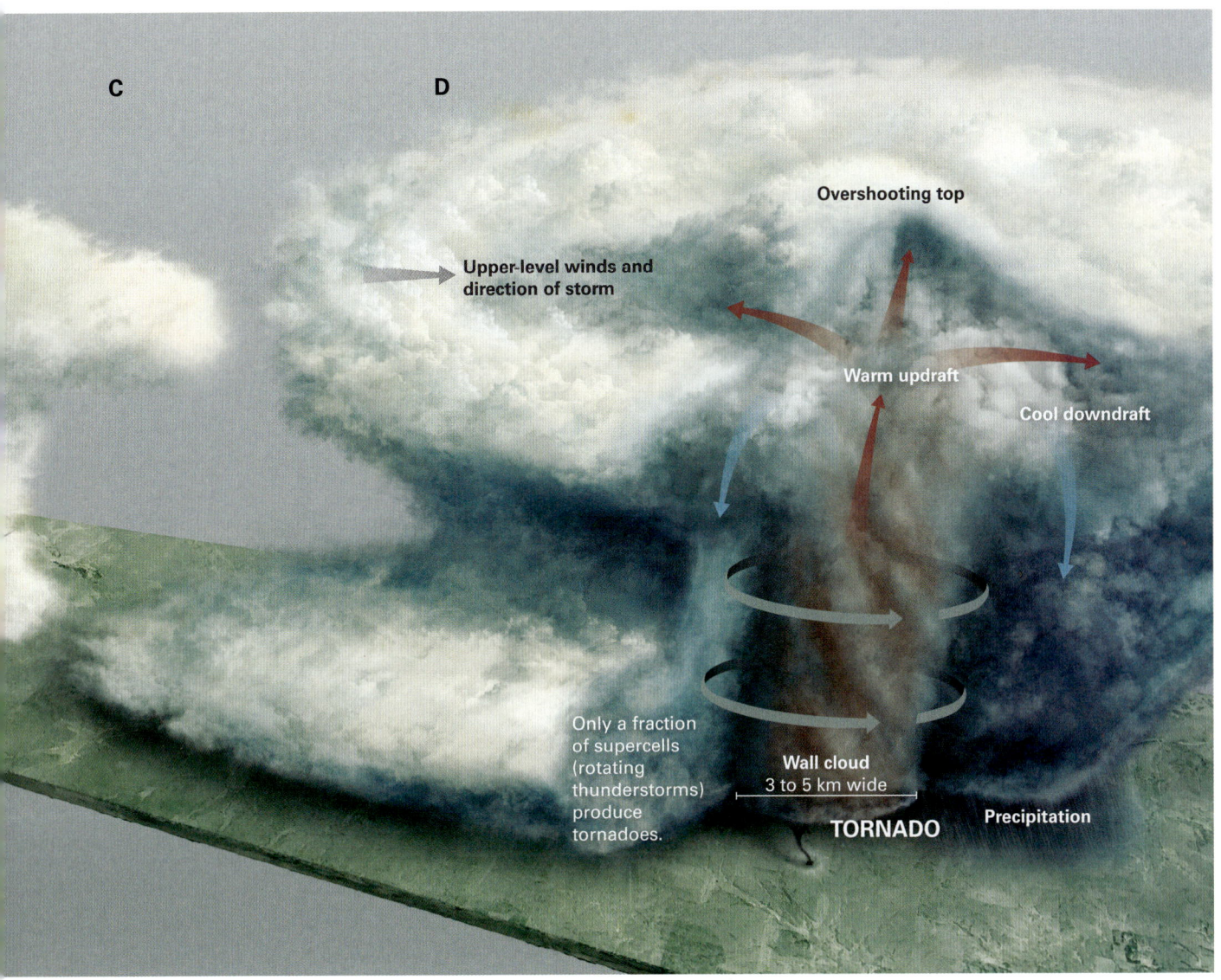

TABLE 20-1 Comparison of Tornadoes and Tropical Cyclones

Feature	Tornado	Tropical Cyclone
Diameter	2 to 3 km	400 to 800 km
Path length (distance traveled across terrain)	A few meters to hundreds of kilometers	A few hundred to a few thousand kilometers
Duration	A few seconds to a few hours	A few days to a week
Wind speed	300 to 511* km/hr	60 to 120 km/hr (tropical storm); 120 to 371† km/hr (hurricane)
Speed of motion	40 to 110 km/hr	20 to 30 km/hr
Pressure fall	20 to 200 millibars	20 to 60 millibars

*Doppler radar systems recorded a maximum wind speed of 511 km/hr during the tornado that destroyed Moore, Oklahoma, on May 3, 2013.

†On April 12, 1934, an anemometer on the summit of Mount Washington, New Hampshire, recorded a wind gust of 371 km/hr. The wind gust occurred during a hurricane.

Source: Cengage

during Hurricane Katrina, sea level rose about 8.5 meters above normal on the Gulf Coast as a result of the storm surge.

Tropical cyclones form only over warm oceans, never over cold oceans or land. This is because moist, warm air is crucial to the development of this type of storm. A midlatitude cyclone develops when a small disturbance produces a wavelike kink in a previously linear front. A similar mechanism initiates a tropical cyclone. In late summer, the sun warms tropical air. The rising hot air creates a belt of low pressure that encircles the globe over the tropics. In addition, many local low-pressure disturbances move across the tropical oceans at this time of year. If a local disturbance intersects the global tropical low, it creates a bulge in the isobars. Winds are deflected by the bulge and, directed by the Coriolis effect, begin to spiral inward. Warm, moist air rises from the low. Water vapor condenses from the rising air, and the latent heat warms the air further, which causes even more air to rise. As the low pressure becomes more intense, strong surface winds blow inward to replace the rising air. This surface air also rises, and more condensation and precipitation occur. But the additional condensation releases more heat, which continues to add energy to the storm.

The center of the storm is a region of vertical airflow, called the *eye*. In the outer, and larger, part of the eye, the air that has been rushing inward spirals upward. In the inner eye, air sinks. And so, the horizontal wind speed in the eye is reduced to near zero (Figure 20-25). Survivors who have been in the eye of a hurricane report an eerie calm. Rain stops, and the sun may even shine weakly through scattered clouds. But this is only a momentary reprieve. A typical eye is only 20 kilometers in diameter, and after it passes, the hurricane rages again in full intensity.

Therefore, a hurricane is powered by a classic feedback mechanism: the low-pressure storm causes condensation; condensation releases heat; heat powers the continued low pressure. The entire storm is pushed by prevailing winds, and its path is deflected by the Coriolis effect. A hurricane usually dissipates only after it reaches land or passes over colder water because the supply of moist, warm air is cut off. Condensing water vapor in a single tropical cyclone can release as much latent heat energy as that produced by all the electric generators in the United States in a six-month period.

The numerical "category" of tropical cyclones is based on a rating scheme called the Saffir–Simpson scale, after its developers (Table 20-2). This scale,

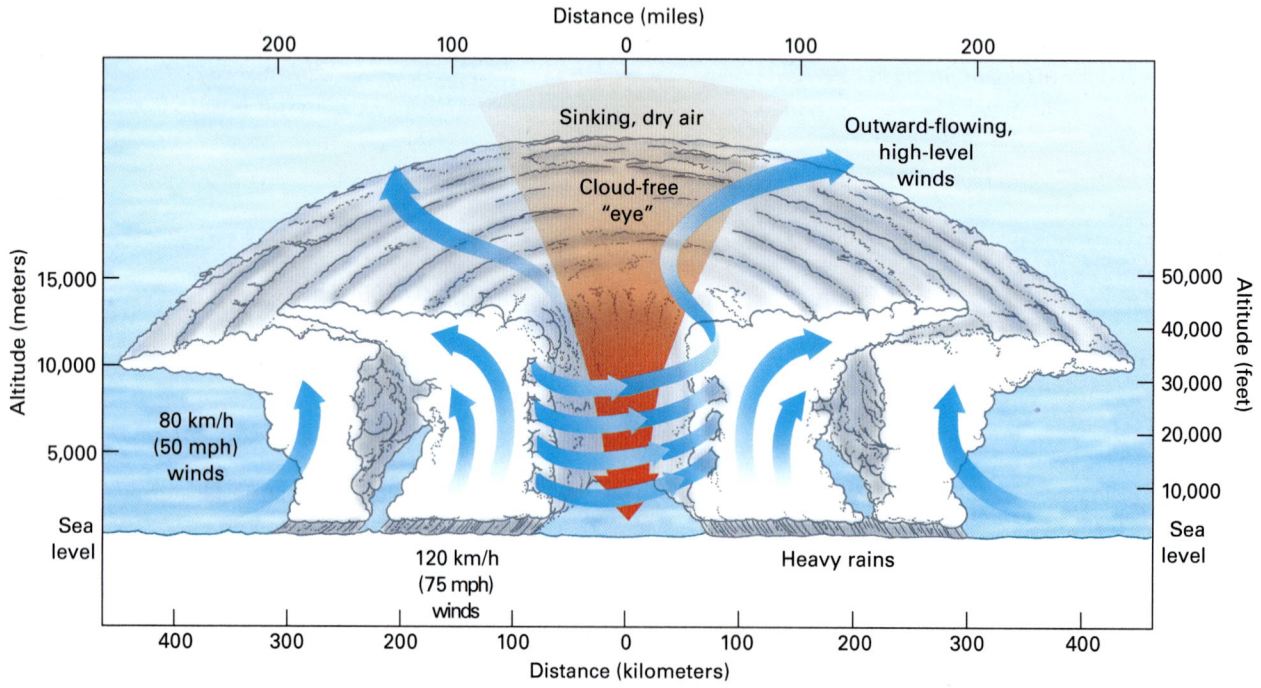

FIGURE 20-25
Surface air spirals inward toward a hurricane; it rises through the towering wall of clouds, and then flows outward above the storm.

FIGURE 20-26
This satellite image shows the presence of three hurricanes over the Atlantic in September 2017: (from left to right) Hurricane Katia, Hurricane Irma, and Hurricane Jose.

commonly mentioned in weather reports, rates the damage potential of a hurricane or other tropical storm, and typical values of atmospheric pressure, wind speed, and height of storm surge associated with storms of increasing intensity.

Hurricanes have ravaged the southeast United States, Puerto Rico, and the Caribbean, numerous times in the past and will certainly do so again (Figure 20-26). Hurricanes are not preventable. Hurricane deaths and costs result from a variety of factors, including the severity of the hurricane, building practices, and human responses.

Prior to Hurricane Katrina, the deadliest U.S. hurricanes all occurred before 1957. However, the costliest storms all occurred after 1955. With the exception of Katrina, none of the 10 deadliest storms are on the list of costliest storms, and vice versa. Why is there so little correlation between structural damage (cost) and death toll? In the last half century, structural damage has been high for three main reasons. More Americans lived near the Atlantic and Gulf coasts in the late 1990s and early 2000s than they had previously. Homes and other structures have had an increased inflation-adjusted

TABLE 20-2 The Saffir–Simpson Hurricane Damage Potential Scale

Type	Damage	Pressure (millibars)	Winds (km/h)	Storm surge (m)
Depression			>56	
Tropical storm			63 to 117	
Hurricane, Category 1	minimal	980	119 to 152	1.2 to 1.5
Hurricane, Category 2	moderate	965 to 979	154 to 179	1.8 to 2.4
Hurricane, Category 3	extensive	945 to 964	179 to 209	2.7 to 3.7
Hurricane, Category 4	extreme	920 to 944	211 to 249	4.0 to 5.5
Hurricane, Category 5	catastrophic	<920	>249	>5.5

Source: Cengage

value. Americans now own more things than ever before, so the value of their possessions is high.

But although structures are immobile, people can relocate if given enough warning. The death toll has been low in the past half century because accurate forecasting warns people of impending hurricanes. The deadliest hurricane in the United States struck Galveston, Texas, in September 1900. More than 8,000 people died because the population was caught unaware. A tropical cyclone has a sharp boundary, and even a few hundred kilometers outside that boundary, fluffy white clouds may be floating in a blue sky. Today, satellites track hurricanes, and news reports provide people with more time to evacuate.

TABLE 20-3
Go online to view a table of the deadliest storms in U.S. history.

TABLE 20-4
Go online to view a table of the costliest storms in U.S. history.

checkpoint Describe why hurricanes are considered to be powered by a positive feedback mechanism.

El Niño

Hurricanes are common in the southeastern United States and on the Gulf Coast, but they rarely strike California. Consequently, Californians were taken by surprise in late September 1997 when Hurricane Nora ravaged Baja California and then, somewhat diminished by landfall, struck San Diego and Los Angeles. The storm brought the first rain to Los Angeles after a record 219 days of drought, then spread eastward to flood parts of Arizona, where it caused the evacuation of a thousand people.

Other parts of the world also experienced unusual weather during the autumn of 1997. In Indonesia and Malaysia, fall monsoon rains normally douse fires intentionally set in late summer to clear the rain forest. The rains were delayed for two months in 1997, and as a result the fires raged out of control, filling cities with such dense smoke that visibility at times was no more than a few meters. Even an airliner crash was attributed to the smoke. Severe drought in nearby Australia caused ranchers to slaughter entire herds of cattle for lack of water and feed. At the same time, far fewer hurricanes than usual threatened Florida and the U.S. Gulf Coast. Floods soaked northern Chile's Atacama Desert, a region that commonly receives no rain at all for a decade at a time, while record snowfalls blanketed the Andes and heavy rains caused floods in Peru and Ecuador.

All of these weather anomalies have been attributed to **El Niño**, an ocean current that brings unusually warm water to the west coast of South America. But the current does not flow every year; instead, it occurs about every three to seven years, and its effects last for about a year before conditions return to normal. Although meteorologists paid little attention to the phenomenon until the El Niño year of 1982–1983, many now think that El Niño affects weather patterns for nearly three-quarters of Earth.

To understand the El Niño effects, first consider interactions between southern Pacific sea currents and weather in a normal, non–El Niño year (Figure 20-29). Normally in fall and winter, strong trade winds blow westward from South America across the Pacific Ocean. The winds drag the warm, tropical surface water away from Peru and Chile and pile it up in the western Pacific near Indonesia and Australia. In the western Pacific, the warm water forms a low mound thousands of kilometers across. The water is up to 10°C warmer and as much as 60 centimeters higher than the surface of the ocean near Peru and Chile. As the wind-driven surface water flows away from the South American coast, cold, nutrient-rich water rises from the depths to replace the surface water in the process called upwelling, discussed in Chapter 16. The nutrients support a thriving fishing industry along the coasts of Peru and Chile. Abundant moisture evaporates from the surface of a warm ocean. In a normal year, much of this water condenses to bring rain to Australia, Indonesia, and other lands in the southwestern Pacific, which are adjacent to the mound of warm water. On the eastern side of the Pacific, the cold, upwelling ocean currents cool the air above the coasts of Peru and northern Chile (Figure 20-27). This cool air becomes warmer as it flows over land. The warming lowers the relative humidity and creates the coastal Atacama Desert.

In an El Niño year, for reasons poorly understood by meteorologists, the trade winds slacken (Figure 20-28). The mound of warm water near Indonesia and Australia then flows downslope—eastward across the Pacific Ocean toward Peru and Chile.

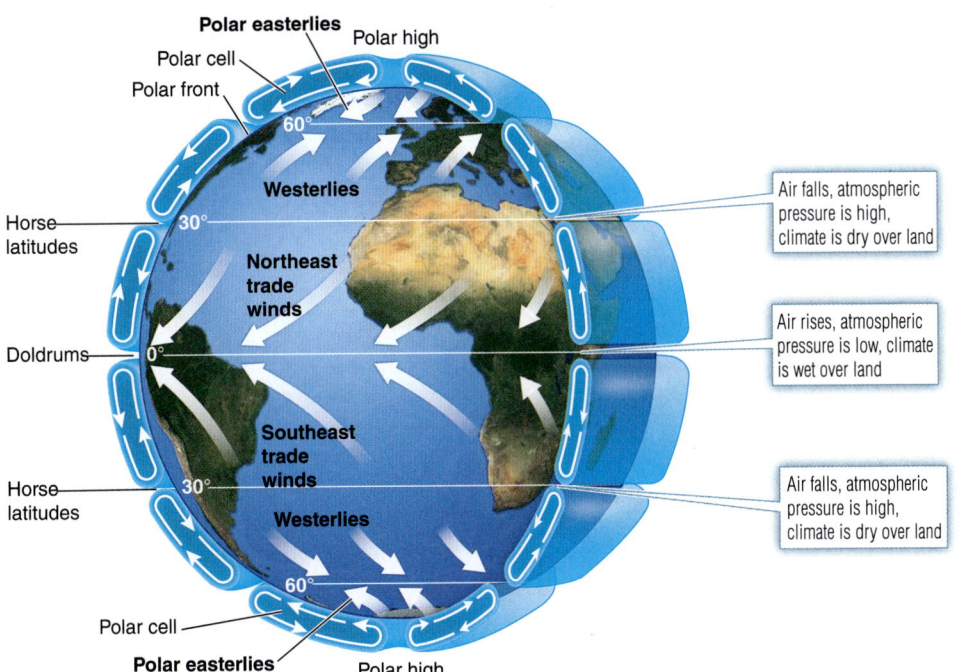

FIGURE 20-27
In a normal year, trade winds drag warm surface water westward across the Pacific near Indonesia and Australia, where the warm water causes rain. The surface flow creates upwelling of cold, deep, nutrient-rich waters along the coast of South America.

The anomalous accumulation of warm water off South America causes unusual rains in normally dry coastal regions, and heavy snowfall in the Andes. At the same time, the cooler water near Indonesia, Australia, and nearby regions causes drought.

El Niño has global effects that go far beyond regional rainfall patterns. For example, it deflects the jet stream from its normal path as it flows over North America, directing one branch northward over Canada and the other across Southern California and Arizona. Consequently, those regions receive more winter precipitation and storms than usual, while fewer storms and warmer winter temperatures affect the Pacific Northwest, the northern plains, the Ohio River valley, the mid-Atlantic states, and New England. Southern Africa experiences drought, while Ecuador, Peru, Chile, southern Brazil, and Argentina receive more rain. Globally, 2,000 deaths and more than $13 billion in damage are attributed to

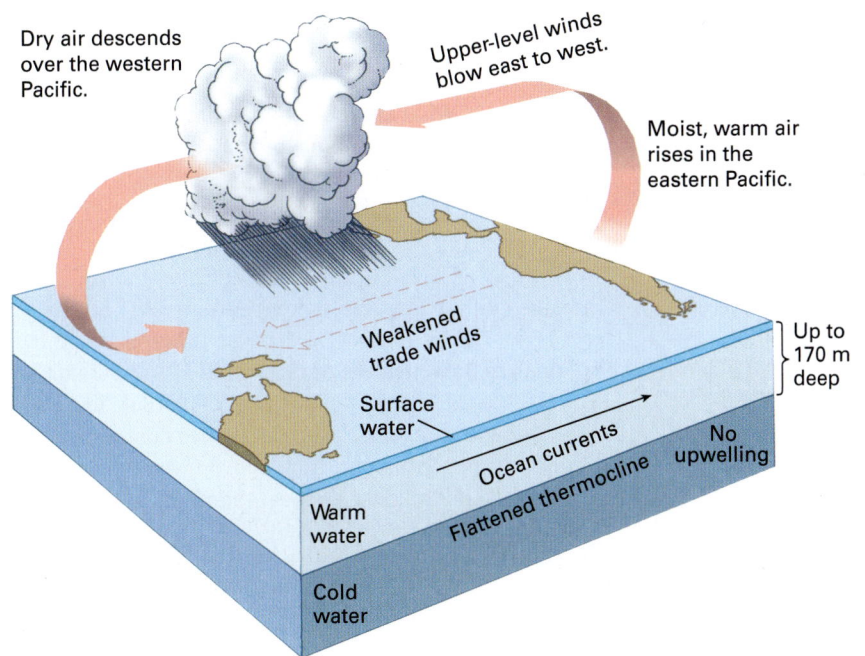

FIGURE 20-28
In an El Niño year, the trade winds weaken, and the warm water flows eastward toward South America. Moist, warm air rising in the eastern Pacific causes thunderstorms and rain along the west coast of South America and in the central Pacific. Dry air descends over the western Pacific, resulting in unusually warm, dry weather, drought, and more frequent wildfires in Australia and Indonesia.

LESSON 20.4

DATA ANALYSIS Compare Storm Statistics

Refer to Table 20-1 on page 659, which compares the general characteristics of tornadoes and tropical cyclones, two types of storms that result from intense, low-pressure centers. Visit the National Weather Service website to explore an interactive map of tornado data. Explore an interactive map of a tornado outbreak that occurred north of Jacksonville, Alabama, on March 19, 2018. Visit weather.gov. Search for "tornadoes of March 19, 2018." Choose the link for *Event Summary of Central Alabama*. Interact with the map, and familiarize yourself with the features and map symbols.

1. **Compare** Refer to Table 20-1. Which kind of storm has higher wind speeds, a tornado or a tropical cyclone? Explain your response.
2. **Compare** Refer to Table 20-1. Which kind of storm affects a larger geographical area, a tornado or a tropical cyclone? Explain your response.
3. **Identify** Follow the steps described in the activity's introduction to refer to the interactive map on the National Weather Service website. What was the maximum wind speed estimated for this storm? What was the length and width of the damage path?
4. **Compare** Based on Table 20-1, the data available through the interactive map, and the coordinating table below, Table 20-5, how did the Jacksonville tornado fall within the range of typical tornadoes? Was it particularly large, strong, or long-lasting?

Critical Thinking

5. **Synthesize** Considering the typical diameter, path length, duration, wind speed, and speed of motion, compare the extent of damage likely to be inflicted by a typical tornado versus a typical hurricane. Explain how these factors might contribute to the potential damage caused by a storm.

TABLE 20-5 Tornado Statistics From Central Alabama 2018

Rating	Estimated Maximum Wind	Injuries/ Fatalities	Damage Path Length	Maximum Path Width	Approximate Start/Point Time	Approximate End Point/ Time
EF-3	240 km/hr (150 mph)	4 injuries	55.18 km (34.29 miles)	1,800 meters (2000 yards)	SSW Silver Lakes Golf Course at 8:23 p.m. CDT	SSE Mars Hill 33.7943/ -85.3665 at 9:10 p.m. CDT

Source: NOAA

---

effects from the 1982–1983 El Niño. In the United States alone, more than 160 deaths and $2 billion in damage occurred, mostly from storm and flood damage. In southern Africa, economic losses of $1 billion and uncounted deaths due to disease and starvation have been attributed to the El Niño of 1982–1983.

checkpoint What are the observed frequency and duration patterns of El Niño?

20.4 ASSESSMENT

1. **Explain** Why do tornadoes tend to occur in the spring or early summer?
2. **Distinguish** What are the distinctions between tropical storms referred to as hurricanes, typhoons, and cyclones?
3. **Summarize** Why has the death toll from hurricanes decreased even though hurricane costs have been on the rise?

Computational Thinking

4. **Analyze Data** Use the data in Table 20-4 to create a bar graph that relates the cost of damage to hurricane category, based on the five costliest storms in U.S. history. Then explain why the data do or do not follow an expected trend.

TYING IT ALL TOGETHER
EXPLAINING WEATHER PATTERNS

In this chapter, you learned about factors that affect weather patterns, such as moisture in the air, pressure, wind, and the collision of air masses. You also explored how surface features such as mountains, forests, coastlines, and oceans contribute to weather, including the development of violent storms.

In the Case Study, we examined regions of the United States that are frequently affected by tornadoes and tropical cyclones. Storm chasers such as Anton Seimon risk their safety to learn more about tornadoes. Much like Seimon, so-called hurricane hunters collect hurricane data by flying special planes directly into tropical storms. Fitted with weather instruments, these planes are able to gather real-time hurricane data that could not be observed any other way.

Although weather often follows predictable patterns, extreme weather cannot always be predicted successfully. As Earth's climate warms and storms become more frequent and destructive in some areas, data from tornado trackers and hurricane hunters are essential to the meteorologists we rely on to help prevent tragic outcomes.

Work with a group to complete the following steps.

1. Create a graphic organizer, chart, or other visual that classifies and describes different weather events. It should include a wide range of weather events, from those that are predictable based on easily monitored elements, such as pressure, temperature, and weather fronts, to severe weather events that may develop quickly and allow only for short-range predictions. Group the events into categories and identify the defining characteristics of each category. Also include the key characteristics of each weather event and the factors that contribute to the event.

2. Select one type of weather event from your chart or visual. Assume the role of a weather specialist who is explaining the event to the general public. Present information about your selected weather event in a format of your choice. Possible formats include a page on a community website, a downloadable brochure, a newsletter, or a script for a radio spot. Regardless of the format, your final product should summarize the following:

   - Important weather traits to identify
   - Tools you will use to present and communicate information
   - Strategies for communication if the weather event's intensity increases or poses a threat
   - Warnings and tips to minimize the threats of severe weather

CHAPTER 20 SUMMARY

20.1 MOISTURE, TEMPERATURE, AND AIR

- Absolute humidity is the mass of water vapor per unit volume of air.

- Relative humidity is the amount of water vapor in the air compared to the maximum water vapor the air could hold at the current temperature. When relative humidity reaches 100 percent, the air is saturated.

- Condensation occurs when saturated air cools below its dew point.

- Three atmospheric processes can cool air below its dew point and cause condensation: (1) radiation, (2) contact with a cool surface, and (3) adiabatic cooling of rising air.

- Radiation and contact cooling cause the formation of dew and frost. Clouds and precipitation normally form as a result of the adiabatic cooling of rising air.

- Three mechanisms cause air to rise: orographic lifting, frontal wedging, and convection–convergence.

- Warm, moist air is said to be unstable because it rises rapidly, forming towering clouds and heavy rainfall. Warm, dry air is said to be stable because it does not rise to high elevations and does not lead to cloud formation and precipitation.

20.2 CLOUDS AND PRECIPITATION

- A cloud is a concentration of water droplets or ice crystals in air. The three fundamental types of clouds are cirrus, stratus, and cumulus.

- Fog is a cloud that forms at or very close to ground level. Most fog forms as a result of contact cooling or radiation cooling near Earth's surface.

- Precipitation occurs when water droplets or ice crystals in a cloud become large enough to fall.

- Temperature is a factor that determines the type of precipitation.

20.3 PRESSURE, WIND, AND FRONTS

- When air is heated, it expands and rises, creating low pressure. Cool air sinks, exerting a downward force that creates high pressure.

- Uneven heating of Earth's surface causes pressure differences, which, in turn, cause wind. Wind speed is determined by the pressure gradient.

- The Coriolis effect deflects rising air and falling air in different directions, resulting in cyclones and anticyclones that rotate in opposite directions. The effects are different in the Northern and Southern Hemispheres.

- A front is the boundary between a warmer air mass and a cooler one. When two air masses collide, the warmer air rises along the front, forming clouds and often precipitation.

- Air cools adiabatically when it rises over a mountain range, often causing precipitation.

- Winds at high altitude are not slowed by friction with Earth's surface features, resulting in jet streams.

- A positive feedback mechanism operates in forests: forests cool the air, cool air promotes rainfall, and rainfall supports forests.

- Sea breezes and monsoons occur because ocean temperature changes slowly in response to daily and seasonal changes in solar radiation and land temperature changes quickly.

20.4 STORMS

- A thunderstorm is a small, short-lived storm from a cumulonimbus cloud.

- Lightning occurs when charged particles separate within the cloud, resulting in an instantaneous rush of electrons between the positively charged and negatively charged regions.

- A tornado is a small, short-lived, funnel-shaped vortex that protrudes from the bottom of a cumulonimbus cloud.

- A tropical cyclone is a large, relatively long-lived storm that forms over warm water. These storms are powered by energy released when water vapor condenses to form clouds and rain.

- El Niño is a weather pattern caused by a weakening of trade winds that leads to changes in equatorial water currents every three to seven years.

CHAPTER 20 ASSESSMENT

Review Key Terms
Select the key term that best fits the definition. Not all terms will be used, and no term will be used more than once.

absolute humidity	jet stream
adiabatic temperature change	orographic lifting
	relative humidity
air mass	saturation
anticyclone	stationary front
cirrus	storm surge
cumulus	stratus
cyclone	supercooling
dew	supersaturation
dew point	tornado
front	tropical cyclone
frontal wedging	typhoon
frost	

1. variations caused by compression or expansion of gas rather than addition or loss of heat
2. the forced rise of air as it flows over mountains
3. amount of water vapor in air as a percentage of the maximum it can hold at a given temperature
4. condition in which a solution contains as much solute as can be dissolved at a given temperature
5. wispy clouds that look like hair blowing in the wind or feathers floating across the sky
6. process by which water remains in its liquid state even below its freezing point
7. small, short-lived, funnel-shaped storm that protrudes from the base of a cumulonimbus cloud
8. the forced sliding of cool, denser air under a warm air mass
9. temperature at which water vapor in the air starts to condense into liquid droplets
10. narrow band of high-altitude wind
11. high-pressure region that forms a system of outwardly spiraling, rotating winds
12. ice crystals formed when vapor in the air contacts cool surfaces when the dew point is below the freezing point
13. horizontally layered, sheet-like clouds

Review Key Concepts
Answer each question on a separate sheet of paper to demonstrate your understanding of key concepts from the chapter.

14. What is the difference between absolute humidity and relative humidity?
15. What three atmospheric processes are responsible for cooling an air mass? Describe each process.
16. Why does convection cause the convergence and divergence of air?
17. Describe each of the processes illustrated below. Identify the weather patterns shown and explain the mechanisms that cause air to rise.

A

B

C

CHAPTER 20 ASSESSMENT

18. What are three different types of frozen precipitation? How are the processes responsible for their formation different?
19. What is an altocumulus cloud? Does it produce precipitation? Explain how you know.
20. How do converging and diverging winds differ in the Northern Hemisphere? Describe the phenomena that cause each.
21. Do winds move faster near the surface or high in the atmosphere? Why?
22. Describe how mountain ranges and forests contribute to precipitation.
23. What are sea and land breezes? What causes them?
24. What are two hypotheses that explain the appearance of lightning in thunderstorms?
25. How are typhoons, hurricanes, and cyclones different?

Think Critically

Write a response to each question on a separate sheet of paper. Use concepts from the chapter to support your reasoning.

26. Suppose the relative humidity of a mass of air is 50 percent at 10°C. Use the graph to estimate the temperature to which the air mass would need to be cooled before reaching 100 percent relative humidity.

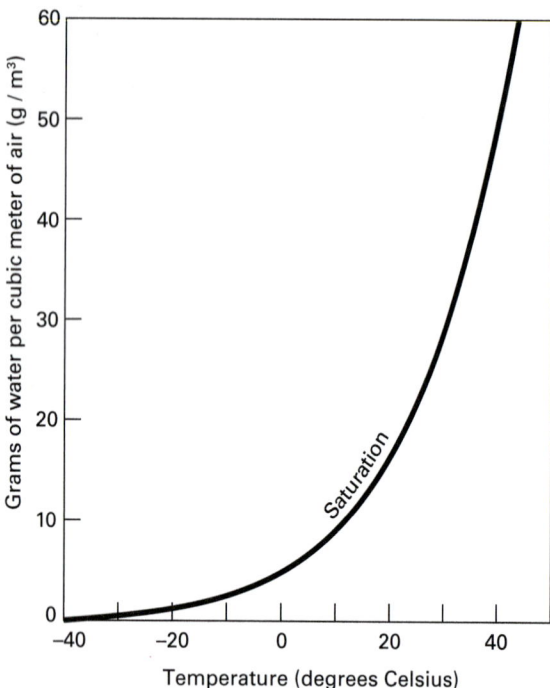

27. Is it possible for a warm air mass to slide under a cooler air mass to push the cooler air mass higher and create a weather front? Why or why not?
28. Is advection fog more likely to form during the winter or during the summer? Explain your reasoning.
20. In what direction do jet streams flow? Why?
30. La Niña is a weather phenomenon occurring every 3 to 7 years that is caused when an ocean current brings unusually cold water to the west coast of South America. How does this differ from El Niño? What weather effects are likely observed during a La Niña year?

PERFORMANCE TASK

Weather Emergency Action Plan

Severe weather events can have a significant impact on a community. Local and state governments need to work with citizens to prepare for future natural disasters.

In this performance task, you will work with other students in your class to propose suitable emergency action plans for your community.

1. Develop severe weather emergency plans for your area.
   A. Go to the National Weather Service's website, weather.gov. Collaborating with your classmates, seek out historical weather data for your community or another region identified by your teacher. Your goal is to find daily weather records for your selected area.
   B. As a class, you will research the weather for one area over the past several years. Each student in your class will be assigned to research three specific months. If there are 30 students in the class, your research will cover the past 90 months. For your assigned months, note down all severe weather events that occurred in your community, such as tornadoes, hurricanes, droughts, blizzards, or heat waves.
   C. Review the data you have collected as a class. Together with your classmates, create graphs that show how many severe weather events of each type occurred each month. Based on this data, predict what type of weather would be expected for each month of the year. In which months would you expect the types of severe weather events you identified? Design a calendar that summarizes these predictions.
   D. With your small group, choose a natural disaster that is likely to hit this area in the near future and create a portfolio of recommendations that will help the community prepare. Are there supplies that residents should keep in their home? Are there safety upgrades residents should make to their homes? Are there evacuation routes that residents should be aware of? Contact the local government to learn about town- or city-wide emergency response plans.
   E. Synthesize what you have learned in a presentation that will help others in the community prepare for weather-related emergencies.

CHAPTER 21
CLIMATE AND CLIMATE CHANGE

Coral reefs off the coast of American Samoa show the dramatic effects of bleaching caused by high ocean temperature. The photo on the left was taken in December 2014; the photo on the right is from February 2015.

The climate where you live is based on the average weather—the typical temperature throughout the year, how windy it is, and how much precipitation falls. Earth's global climate is more complex and based on interconnected systems. Global wind patterns and ocean currents connect distant regions. Scientific evidence paints a clear picture: Earth's global climate is rapidly changing. The changes we are seeing today, such as rampant coral reef bleaching, are caused in large part by human activity over the last century. Changes in our climate will have serious effects on life on this planet in the decades ahead. Mitigating the effects of climate change will require knowledge and determination to change habits and societal priorities.

KEY QUESTIONS

21.1 How does the movement of water and air impact the global climate?

21.2 What are the similarities and differences between the major climate types on Earth?

21.3 How has Earth's global climate changed over geologic time and how do we know?

21.4 What are the major changes we see in today's climate?

21.5 What are the consequences of our rapidly changing climate?

EXPLORERS AT WORK

COMPOSING STORIES OF CLIMATE CHANGE

WITH NATIONAL GEOGRAPHIC EXPLORER PAUL MILLER

Paul Miller is a unique voice for conservation. Known to his fans as DJ Spooky, That Subliminal Kid, Miller has made a name for himself as a composer, musician, performer, writer, pubic speaker, software designer, and activist. Miller applies his singular mix of talents to raising public awareness about social and environmental issues. Using art, photography, film, and music, Miller crafts multimedia experiences that address such topics as technology and society, sustainability, global culture, and climate change. Miller has performed or given talks at a range of venues, from universities and art centers to concert halls and music festivals. He has collaborated with numerous recording artists. He also is an artist-in-residence at various educational and cultural institutions, including Stanford University and the Metropolitan Museum of Art.

Over the years, Miller has created a number of symphonies that incorporate images with live music, sampled music, and field recordings of natural sounds. Those sounds have been as diverse as whale songs, rushing rivers, and cracking ice. Miller also uses computer software to convert mathematical data into digital sound patterns for his music. Miller describes his projects as acoustic portraits, and he is willing to go to extraordinary lengths to create them. For his performance piece "Heart of a Forest," Miller visited an Oregon forest during all four seasons to get the material he needed. And for the piece "Terra Nova: Sinfonia Antarctica," he spent six weeks at the bottom of the world, recording sounds with his portable studio and shooting photos and film.

"Sinfonia Antarctica" and its printed companion *Book of Ice* represent some of Miller's most ambitious projects. The symphony features a slideshow of images and video of Antarctica as well as maps, data, and historical facts. Music is performed by live musicians as well as played off Miller's laptop. *Book of Ice* features Miller's images of Antarctica as well as historical photos, original artwork, commentary from scientists, discussion of early explorations, and data related to climate change. Miller says he has been drawn to Antarctica since childhood. He is fascinated by the fact that the continent is uninhabited, has no government, and belongs to no single country. He says it is the ultimate symbol of humans' relationship with the natural world and how that relationship is impacting the planet.

"I firmly believe as we move further into the 21st century that one of the major issues that is facing us is climate change," says Miller. He also says artists have a part to play in raising public awareness. "Artists have a very important role to play in that. You can make it beautiful, you can make it interesting, you can make it immersive. The artist's role is not to say, 'Hey, this is interesting,' but to grab you by the scruff of the neck and say you can't live life without thinking a different way about this. Another world is possible." DJ Spooky's fans would certainly say he has done just that.

THINKING CRITICALLY

Evaluate How do artists such as Paul Miller add to the conversation about climate change?

Paul Miller delivers a presentation at the 2014 National Geographic Explorers Symposium.

DJ Spooky (Miller) performs "The Nauru Elegies" with Pannonia Quartet at the Grace Rainey Rogers Auditorium at the Metropolitan Museum of Art in New York.

CASE STUDY
ICE SHELF DISINTEGRATION

Along much of the Antarctic coastline, the ice sheet that covers most of the continent extends out into the sea, forming massive ice shelves. These ice shelves have begun retreating at an alarming rate. In 2000, the Ross Ice Shelf calved an iceberg roughly the size of Connecticut. Drifting north into warmer waters, the iceberg gradually broke into smaller pieces, most of which have now melted. In 2002, the Larsen ice shelf suffered major losses when a Rhode Island–size section splintered into millions of small fragments. Another Delaware–size chunk of the Larsen Ice Shelf—12 percent of its entire area—broke away in 2017.

An ice shelf is a floating slab of ice formed from glacial ice that flows out onto the sea. The world's largest ice shelves are found in Antarctica, but ice shelves also exist around Greenland and Canada's Ellesmere Island. Some ice shelves can be up to 2 kilometers thick, but their size and thickness varies over time. The flow of continental glaciers can add mass to an ice shelf, as can snow compacting on the surface and ocean water freezing along the bottom. Ice shelves can lose mass by calving icebergs, which occurs when a large chunk breaks off and floats away. Surface and undersurface melting, wind loss, and sublimation can also result in a loss of mass.

In recent decades, warmer surface temperatures and ocean waters have led to the sudden and often dramatic breakups of Antarctica's ice shelves. Warmer summers create meltwater on the surface, and warmer ocean water causes melting on the underside, leading to a thinning of the ice shelf. This thinning can lead not only to calving but also to a relatively new and more worrying process in which huge masses of the ice shelf disintegrate, breaking into many sliverlike pieces.

The melting or calving of an ice shelf does not immediately raise sea levels because the ice is already floating and displacing its weight in water. However, ice shelves serve to block the flow of continental glaciers toward the sea. When an ice shelf collapses, this relieves pressure on glacial ice, which then flows to the ocean more quickly. Since the 2002 Larsen Ice Shelf disintegration, the glaciers behind it have accelerated up to eight times the normal flow rate. With Antarctica's ice sheets continuing to collapse, scientists are concerned that the stage is set for more dramatic sea-level rise.

As You Read Consider the factors that contribute to the kind of climate warming that is affecting Antarctica's ice shelves. Think about the relationships between human activities, Earth systems, and Earth cycles, as well as their effect on climate. Also, consider the potential changes in your local environment that may result from climate change.

FIGURE 21-1
In November 2016, scientists discovered a crack in the Larsen C ice shelf in Antarctica more than 100 kilometers long. Ultimately, the crack isolated an iceberg the size of Delaware, which slowly drifted away into the Weddell Sea. This satellite image from NASA shows the iceberg in September 2017.

21.1 GLOBAL CLIMATE

Core Ideas and Skills
- Draw a diagram showing the three-cell model of global wind patterns.
- Identify where the alternating bands of high and low pressure exist on a globe.
- Distinguish between predictable wind patterns in both the Northern and Southern Hemispheres.

KEY TERMS

three-cell model
horse latitudes
trade winds
prevailing westerlies
polar easterlies
polar front
subtropical jet stream
polar jet stream

If you are planning a picnic, you hope for bright sunshine, but at the same time, you understand that your plans may be spoiled by rain. Temperature and precipitation on any day are determined by daily fluctuations of weather. But weather varies only within a relatively narrow range. You are certain that it is not going to snow on the Fourth of July in Texas, and that if you live in Montana you are not going to bask outside in shorts and a T-shirt in January.

Global Climate Stability

Over centuries, climate is stable enough that we plan our lives around it. Climate strongly influences many aspects of our lives: our outdoor recreation, the houses we live in, the clothes we wear, and the work we do. Farmers in Kansas plant wheat and never attempt to raise bananas. In winter, hotel owners in Florida prepare for an influx of tourists escaping the northern winter. Humans tend to migrate toward warm, sunny regions. In the United States for example, the populations of California and Texas have increased rapidly in the past generation, while North Dakota and Montana have seen slower growth.

Yet, such stability is only guaranteed over relatively short periods of geologic time. Dinosaur bones and coal deposits indicate that parts of Antarctica were once covered by warm, humid swamps. Ancient desert dunes lie buried beneath the fertile wheat fields of Colorado. When geologists study the climate record in a specific region, they must separate the effects of tectonic movement from global climate change. For example, the coal deposits in Antarctica do not tell us that Earth was once warm enough for coal-producing swamps to grow near the South Pole. Instead, other evidence indicates that the continent of Antarctica was once close to the Equator, as explained in our study of tectonics. Yet, even when the effects of tectonic plate movements are factored out, it is clear that global climate has changed dramatically and often abruptly over the long spans of Earth's history. Climate is regulated by many factors, including latitude, oceans and ocean currents, altitude, albedo, and wind. Let's take a closer look at global wind systems (also discussed in Chapters 16 and 19).

The sun is the ultimate energy source for winds and evaporation. The sun shines most directly at or near the Equator, warming the air near Earth's surface and causing it to rise. For a moment, consider what would happen to this air if three simplifying assumptions were met: (1) Earth's surface was uniform (such as being completely covered by water, eliminating heating differences between land and water); (2) Earth was not tilted on its axis, so at all times the Equator would receive the most direct energy from the sun and the poles would receive the least; and (3) Earth did not rotate, eliminating the Coriolis effect.

The rising equatorial air would create a permanent zone of low pressure at the surface. At high altitudes, the air would flow poleward in each hemisphere. This air would become progressively cooler and would eventually sink at either pole. Because the poles are cold year-round, they would be characterized by permanent zones of surface high pressure. At Earth's surface, the pressure-gradient force would cause surface winds to blow equatorward from each pole, completing a convection loop in each hemisphere. This simple atmospheric circulation model is known as the single-cell model.

Of course, Earth's atmospheric circulation is more complex. Because Earth rotates, the single convection cell in each hemisphere splits into three cells due to the Coriolis effect. This circulation pattern is known as the **three-cell model** (Figure 21-2). Similarities between the models include the presence of persistent equatorial low pressure and polar high pressure as a result of differential warming of Earth's surface. However, in the three-cell model the high-altitude winds do not continue to flow due north or south from the Equator. Instead, the Coriolis effect modifies their poleward movement.

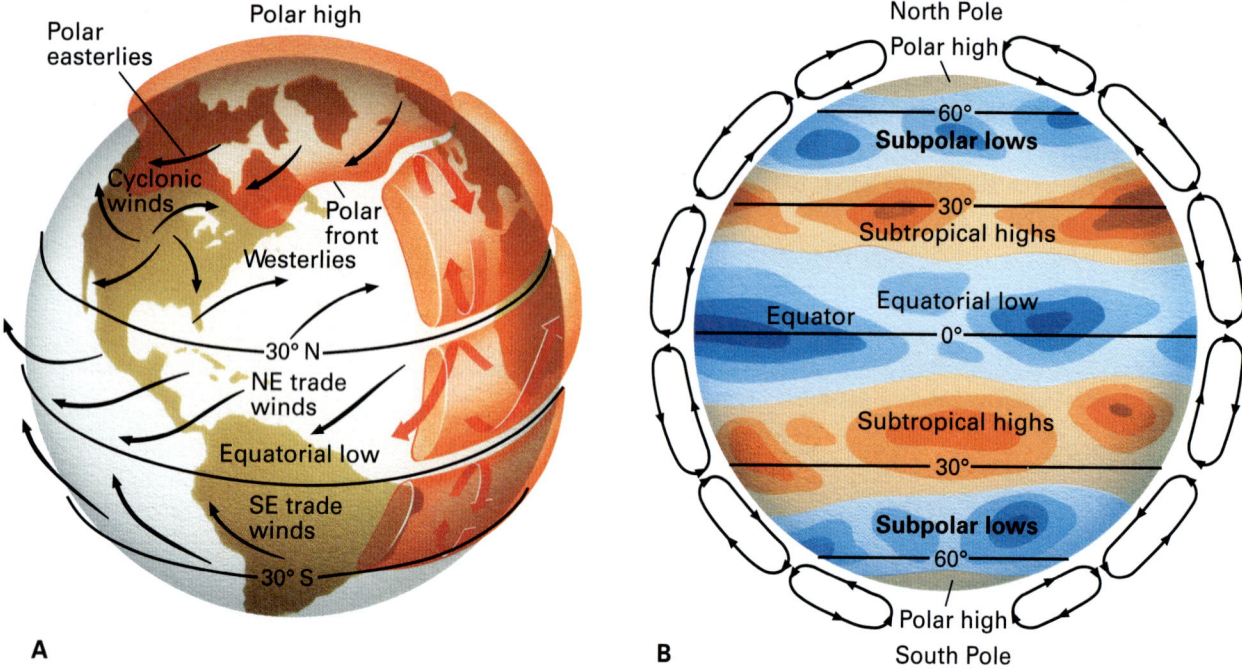

FIGURE 21-2
The three-cell model predicts observed global wind patterns. (A) Air rising at the Equator moves poleward at high elevations, falls at about 30° north and south latitudes, and returns to the Equator, forming the trade winds. The purple arrows show both upper-level and surface wind patterns. The black arrows show only surface winds. (B) High-pressure and low-pressure belts are indicated on the sphere, with surface and upper-level wind patterns shown on the edges.

The Coriolis effect causes the high-altitude winds to veer until they flow due east at about 30° north or south latitude. The air then cools enough to sink to the surface, creating subtropical high-pressure zones at 30° north and south latitudes. The descending air splits and flows over Earth's surface in two directions: toward the Equator and toward the pole. Air also sinks at each pole and flows toward lower latitudes. The surface air moving poleward from about 30° north or south converges at about 60° north or south with air flowing away from the pole. Air rises at this convergence.

The three-cell model accounts for weather and climate conditions observed at Earth's surface. At and near the equator, warm air gathers moisture from the oceans. Because the warm, moist air rises, there is little horizontal airflow. As the rising air cools, the water vapor condenses and falls as rain. Therefore, local squalls and thunderstorms are common, but steady winds are rare. This hot, still region was a serious barrier in the age of sailing ships. Mariners called the equatorial region the doldrums (from the Middle English for "dull"). The old sailing literature is filled with stories recounting the despair and hardship of being becalmed (stalled) on windless seas. On land, the frequent rains near the equatorial low-pressure zone nurture lush tropical rain forests.

In both the Northern Hemisphere and the Southern Hemisphere, high-altitude air sinks at about 30°, creating a zone of surface high pressure. This sinking air warms, hindering condensation and promoting clear, blue skies. Because the air tends to move vertically and not horizontally, few steady surface winds blow. This calm, high-pressure belt circling the globe is called the **horse latitudes**. The region was so named because sailing ships were becalmed, and horses transported as cargo often died of thirst and hunger. The warm, dry, descending air in this high-pressure zone forms many of the world's great deserts, including the Sahara in northern Africa, the Kalahari in southern Africa, and the Australian interior desert.

Descending air at the horse latitudes that flows equatorward is deflected by the Coriolis effect, so winds blow from the northeast in the Northern Hemisphere and from the southeast in the Southern Hemisphere. Sailors depended on these reliable winds and called them the **trade winds**. Deflection of descending air at the horse latitudes

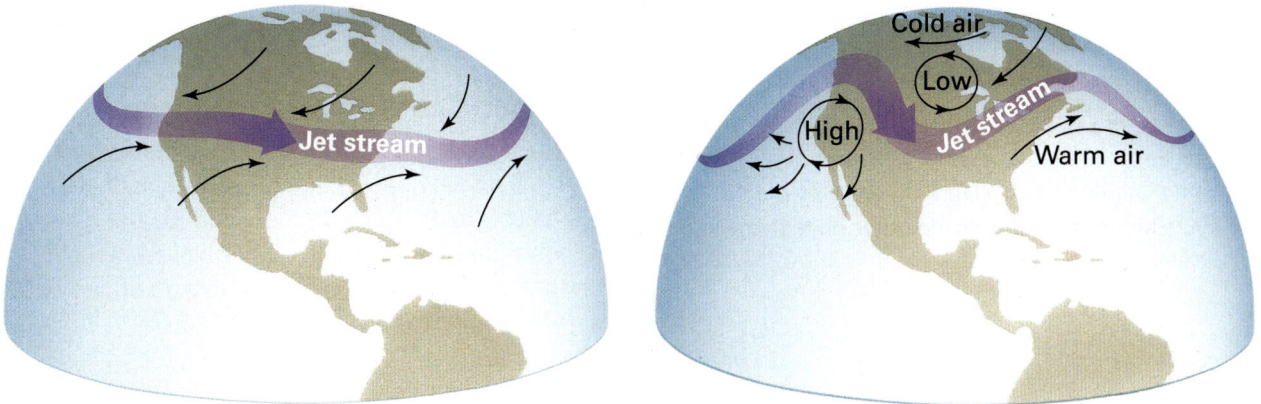

FIGURE 21-3
The polar front and the polar jet stream migrate with the seasons and with local conditions. Storms commonly occur along the jet stream.

that flows poleward forms the surface winds called the **prevailing westerlies**. They flow from the southwest in the Northern Hemisphere and from the northwest in the Southern Hemisphere. The poles are cold year-round. Air sinking at either pole and flowing toward lower latitudes forms surface winds called the **polar easterlies**. At the convergence of the prevailing westerlies and polar easterlies, air rises, forming a low-pressure boundary zone called the **polar front**.

Cyclones and anticyclones commonly develop along the polar front. These storms bring alternating rain and sunshine, conditions that are favorable for agriculture. The great wheat belts of the United States, Canada, and Russia all lie between 30° and 60° north latitude.

In the three-cell model, the three convection cells in each hemisphere are bordered by alternating bands of high and low pressure, and global winds are generated by heat-driven convection currents. But recall that assumptions for the model include that Earth has a uniform surface and is not tilted. Both of these factors modify real-world conditions. As temperatures change with the seasons, the boundaries of the pressure systems and global wind belts migrate north or south. In addition, they are distorted by surface features and local wind movement.

Recall that a jet stream is a narrow band of fast-moving, high-altitude air. Jet streams form at boundaries between Earth's climate cells as high-altitude air is deflected by the Coriolis effect. The **subtropical jet stream** flows between the trade winds and the westerlies (Figure 21-3). The **polar jet stream** forms along the polar front. When you watch a weather forecast on TV, the meteorologist may show the movement and direction of the polar jet stream as it snakes across North America. Storms commonly occur along this line because the jet stream marks the boundary between cold, polar air and the warm, moist, westerly flow that originates in the subtropics. The storms develop where the two contrasting air masses converge.

checkpoint What role does Earth's rotation have on global wind patterns?

21.1 ASSESSMENT

1. **Explain** What is the link between westerlies, polar easterlies, and storms?
2. **Infer** What are trade winds and why are they named that way?
3. **Distinguish** What are the similarities and differences between easterlies and westerlies?

Computational Thinking

4. **Analyze** The three-cell model in Figure 21-2 shows that Earth rotates but does not account for its tilt. Think about how the sun's radiation warms the planet as its axis tilts toward and away from the sun during its orbit. In reality, how does Earth's tilt affect the horse latitudes shown at 30° N and 30° S?

21.2 CLIMATE TYPES

Core Ideas and Skills
- Compare and contrast the six climate types on Earth defined by annual temperature, precipitation, and seasonal changes.
- Explain the effect that cities have on their local climate.

KEY TERMS

biome	marine west coast
tropical rain forest	temperate rain forest
tropical monsoon	humid continental
tropical savanna	subarctic
steppe	taiga
humid subtropical	tundra
Mediterranean	urban heat island effect

Earth's major climate types are classified primarily by temperature and precipitation. But an area with both wet and dry seasons has a different climate from one with moderate rainfall all year long, even though the two areas may have identical total annual precipitation. Therefore, climate types are also classified based on seasonal variations in temperature and precipitation.

Köppen Classification System

Table 21-1 describes the different climate types defined by the Köppen climate classification, used by climatologists throughout the world. Earth's climate types and subtypes based on this classification system are shown in a table and map in Appendix 2.

Although climate types are defined by temperature and precipitation, a photograph of an area can give clues about its climate type. Visual classification is possible because specific plant communities grow in specific climates. For instance, cacti grow in the desert, and trees grow where moisture is more abundant. A **biome** is a community of plants living in a large geographic area characterized by a particular climate.

The climate in any location is summarized by a climograph that records annual and seasonal temperature and precipitation. Figure 21-4 is a model climograph for Nashville, Tennessee. The recording station in Nashville is located at a latitude and longitude of 36° N and 88° W. The average annual temperature is 15.2°C, and the average annual precipitation is 119.6 cm.

checkpoint How many climate types are defined by the Köppen classification system?

TABLE 21-1 Köppen Climate Classification

Climate Type	Name	Description
A	Humid tropical	In A climates, every month is warm with a mean temperature over 18°C (64°F). The temperature difference between day and night is greater than the difference between December and June averages. There is enough moisture to support abundant plant communities.
B	Arid	B climates have a chronic water deficiency; in most months evaporation exceeds precipitation. Temperatures vary according to latitude: some B climates are hot while others are frigid.
C	Humid mesothermal	C climates occur in midlatitudes, with distinct winter and summer seasons and enough moisture to support abundant plant communities. The winters are mild. Snow may fall but snow cover does not persist, with the average temperature in the coldest month above −3°C (27°F).
D	Humid microthermal	D climates are similar to C climates, with distinct summer and winter seasons, but D climates are colder. Winters are more severe, with persistent winter snow cover and an average temperature in the coldest month below −3°C (27°F).
E	Polar	In E climates, winters are extremely cold and even the summers are cool, with the average temperature in the warmest month below 10°C (50°F).
H	Highlands	H climates, found at latitudes worldwide, are determined by elevation above sea level. Temperature changes with altitude, ranging from about −18°C (0°F) to 10°C (50°F). Precipitation tends to decrease with altitude; windward sides of mountains usually receive more precipitation than leeward sides.

Source: Cengage

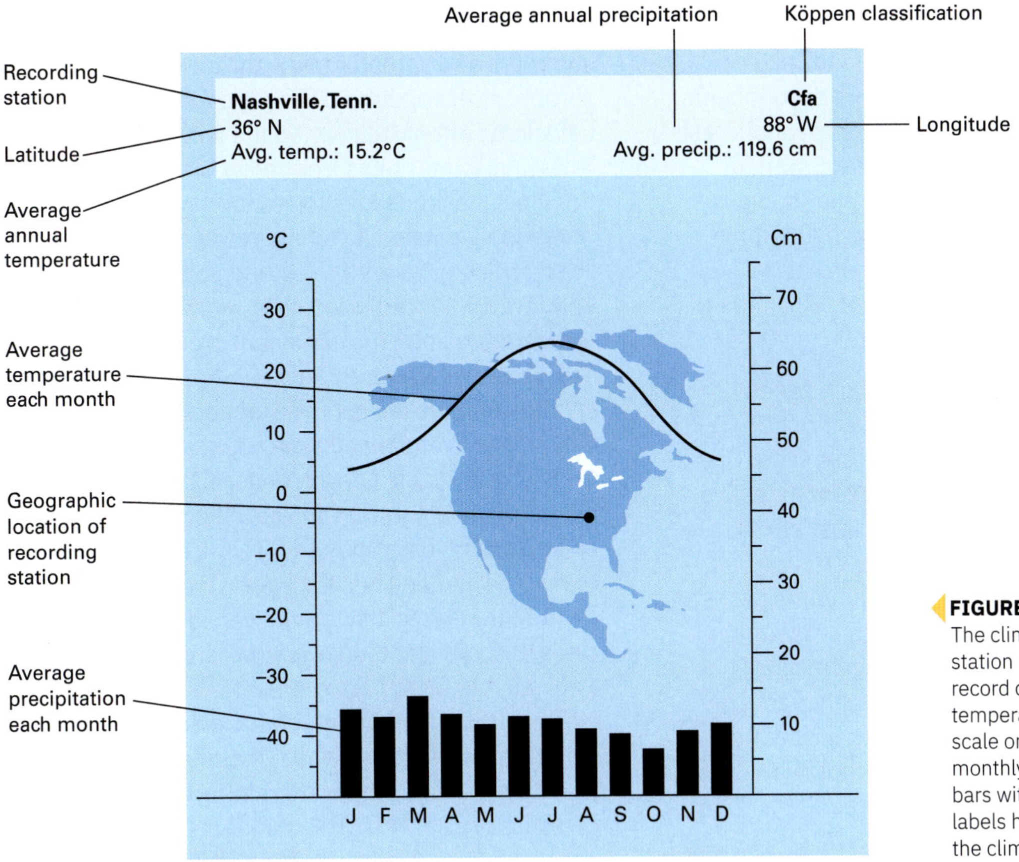

FIGURE 21-4
The climograph for a recording station in Tennessee gives a record of the average monthly temperature (curved line with scale on the left) and the average monthly precipitation (vertical bars with scale on the right). The labels highlight other features of the climograph.

Climate Subtypes

Humid Tropical Climate (A) Subtypes
The large, low-pressure zone near the Equator causes abundant rainfall, often exceeding 400 centimeters per year. This rainfall supports **tropical rain forests**. The dominant plants in a tropical rain forest are tall trees with slender trunks. These trees branch near the top, covering the forest with a dense canopy of leaves. In the densest areas the canopy blocks out most of the light, so as little as 0.1 percent of the sunlight reaches the forest floor. The ground in a tropical rain forest is soggy, the tree trunks are wet, and plants grow so fast that most nutrients are contained in vegetation instead of the soil.

The **tropical monsoon** climate and the **tropical savanna** climate (Figure 21-5) both have seasonal variations in rainfall. The monsoon climate has greater total precipitation, greater monthly variation, and a shorter dry season than the savanna. Precipitation is great enough in tropical monsoon biomes to support rain forests. The seasonal precipitation is also ideal for agriculture. Some of the great rice-growing regions in India and Southeast Asia lie in tropical monsoon climates.

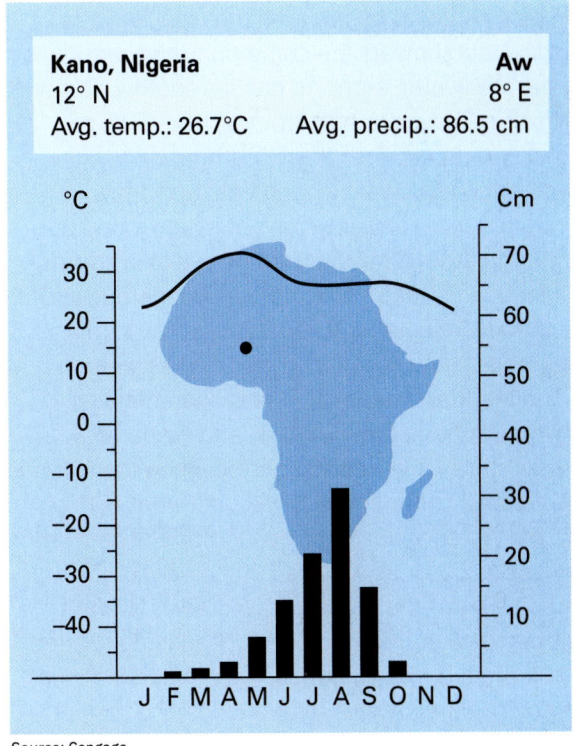

FIGURE 21-5
The climograph for Kano, Nigeria, represents data for a typical tropical savanna climate.

A tropical savanna is a grassland with scattered small trees and shrubs. Such grasslands extend over large areas, often in the interiors of continents. Rainfall is insufficient to support forests, and forest growth is prevented by frequent fires. Savannas are most extensive in Africa, where they support a rich collection of grazing animals such as zebras, wildebeest, and gazelles. Grasses sprout and grow with the seasonal rains, and the great African herds migrate with the foliage.

Arid Climate (B) Subtypes

In dry zones where the annual precipitation varies from 25 to 35 centimeters per year, the climate is semiarid. Grass-covered plains, called **steppes**, predominate. The great steppe grasslands of Central Asia fall into the dry climate category.

If the rainfall is less than 25 centimeters per year, deserts form and support only sparse vegetation. The world's largest deserts lie along the high-pressure zones of the 30° latitude, although rain-shadow deserts and coastal deserts exist at other latitudes. Some deserts are torridly hot, while Arctic deserts are frigid for much of the year.

Humid Mesothermal Climate (C) Subtypes

The southeastern United States has a **humid subtropical** climate. During the summer, conditions can be as hot and humid as in the tropics. Rain and thundershowers are common. However, during the winter, arctic air pushes southward, forming cyclonic storms. Although the average monthly temperature rarely falls below 7°C, cold fronts occasionally bring frost and snow. Precipitation is relatively constant year-round due to thunderstorms in summer and cyclonic storms in winter. This zone supports both coniferous and deciduous trees as well as valuable crops such as vegetables, cotton, tobacco, and citrus fruits.

The **Mediterranean** climate is characterized by dry summers, rainy winters, and moderate temperature. These conditions occur on the west coasts of all continents between latitudes 30° and 40°. In summer, near-desert conditions with clear skies occur as much as 90 percent of the time. In winter the prevailing westerlies bring warm, moist air from the ocean, leading to fog and rain. More than 75 percent of the annual rainfall occurs in winter. Although redwoods, the largest trees on Earth, grow in specific environments in central California, the summer heat and drought of Mediterranean climates generally slow the growth of large trees. Instead, shrubs and scattered trees dominate. Fires occur frequently during the dry summers and spread rapidly through the dense shrubbery. If vegetation is destroyed by fire, landslides often occur when the winter rains return. Torrential winter rains frequently bring extensive flooding and landslides to southern California.

Marine west coast climate types border the Mediterranean zones and extend poleward to 65°. They are influenced by ocean currents that control temperature and bring abundant precipitation. Temperature difference between the seasons is small. Summers are cool and winters are warm. For example, average monthly temperatures vary by only 15.5°C in Portland, Oregon (Figure 21-6). Seattle and other northwestern coastal cities experience rain and drizzle for days at a time. This is especially true during the winter when the warm, moist, maritime air from the Pacific flows first over cool currents close to shore and then over cool land surfaces. The total rainfall varies from moderate, 50 centimeters per year, to wet, 250 centimeters per year. The wettest climates occur where mountains interrupt the maritime air. **Temperate rain forests** grow where rainfall is greater than 100 centimeters per year and is constant throughout the year. Temperate rain forests are common along the northwest coast of North America, from Oregon to Alaska.

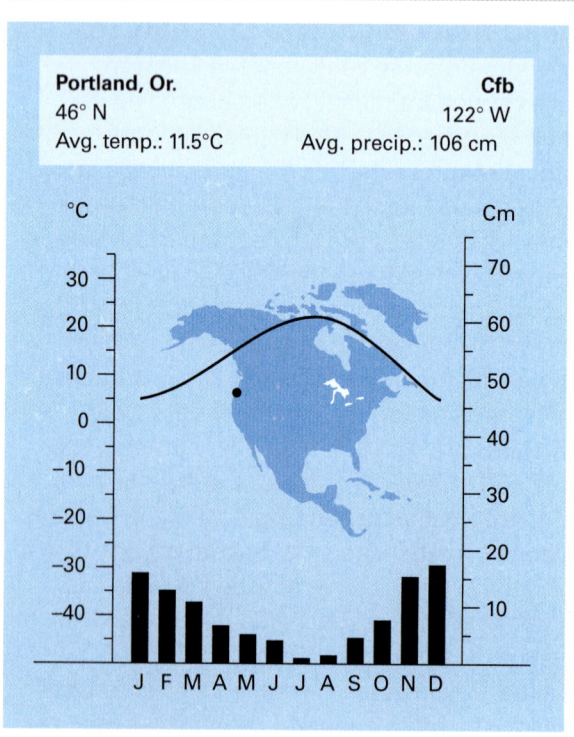

Source: Cengage

FIGURE 21-6

The climograph for Portland, Oregon, represents data for a typical marine west coast climate.

DATA ANALYSIS Compare Climographs

Climographs are useful for comparing climates from different locations. Use the climographs presented in Figure 21-7 to compare climate data at different locations around the world.

1. **Compare** Iquitos and Kozhikode both receive more than 250 centimeters of precipitation annually. In what ways are these two climates similar? In what ways are they different?

2. **Compare** Iquitos and Lima are two cities in Peru that are about 1,000 kilometers apart. Use the climographs to compare and contrast their climates. Discuss possible reasons for any similarities and differences.

Data Challenge

Go to the Data Analysis in MindTap to complete the data challenge.

FIGURE 21-7
The climograph for (A) Lima, Peru, (B) Iquitos, Peru, and (C) Kozhikode, India, are representative of the desert, the tropical rain forest, and the tropical monsoon climates, respectively.

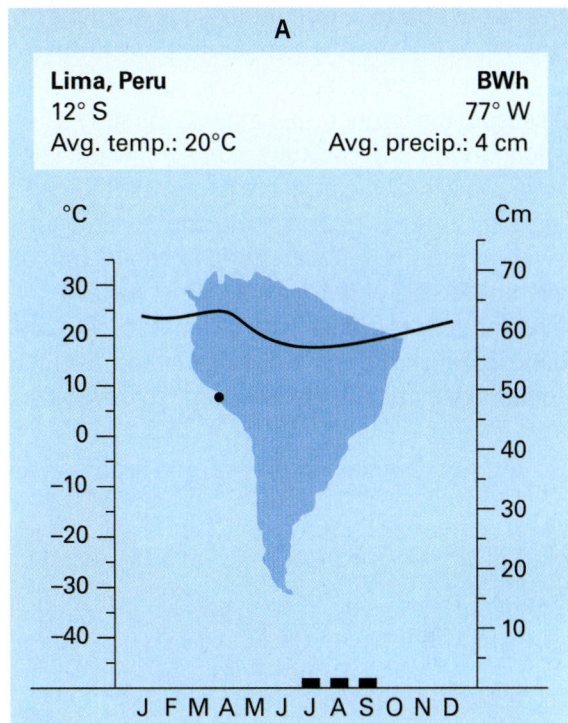

Source: Cengage

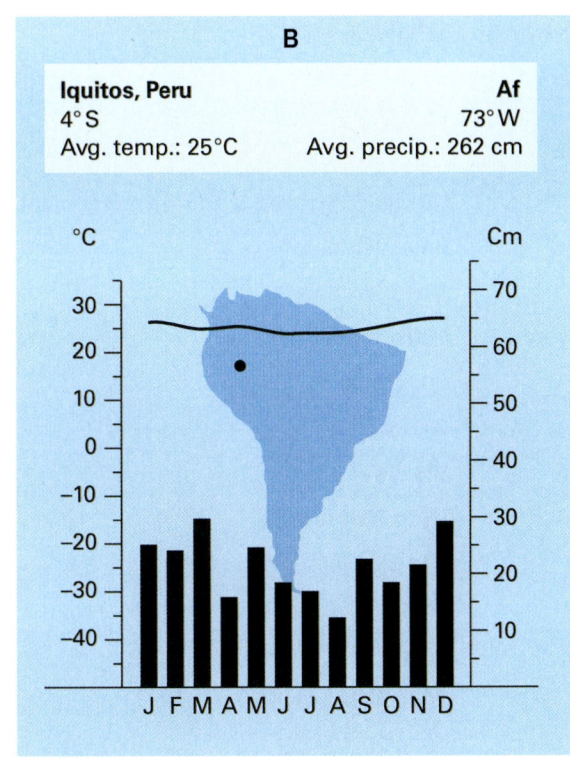

Source: Cengage

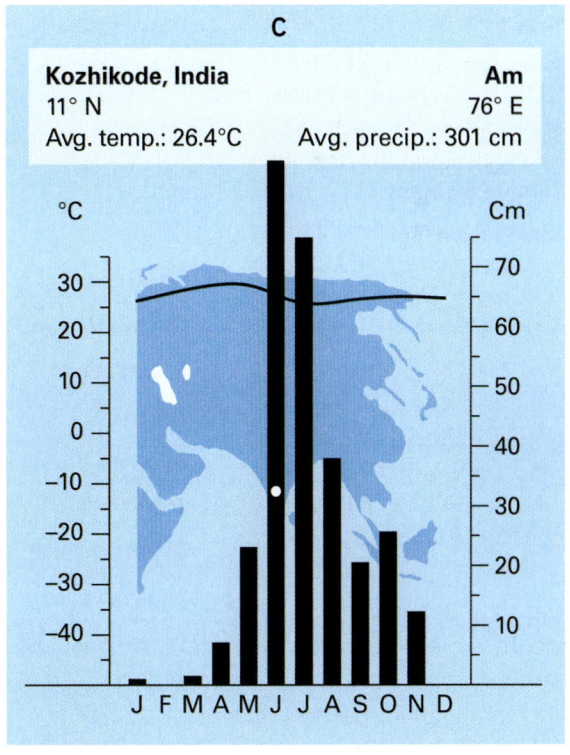

Source: Cengage

Humid Microthermal Climate (D) Subtypes

Continental interiors at midlatitudes are characterized by hot summers and cold winters, giving rise to **humid continental** climates. In the northern Great Plains the temperature can drop to −40°C in winter and soar to 38°C in summer. During one season, the temperature may vary greatly as the polar front moves northward or southward. For example, in winter the northern continental United States may experience arctic cold one day and rain a few days later. This climate supports abundant coniferous forests or grasslands in drier regions. The northernmost portion of this climate type is the **subarctic**, which supports the **taiga** biome, a forest of conifers that can survive extremely cold winters.

Polar Climate (E) Subtypes

In the Arctic and Antarctic, winters are harsh and long, and the temperature remains above freezing only during a short summer. The climate is classified as a polar climate. Trees cannot survive, and low-lying plants such as mosses, grasses, flowers, and a few small bushes cover the land. This biome is called **tundra**. A second subtype for polar climates is the ice cap in which no vegetation can survive.

Climographs for the climate types and subtypes not represented here are shown in Appendix 2.

checkpoint How are biomes useful for identifying climate types?

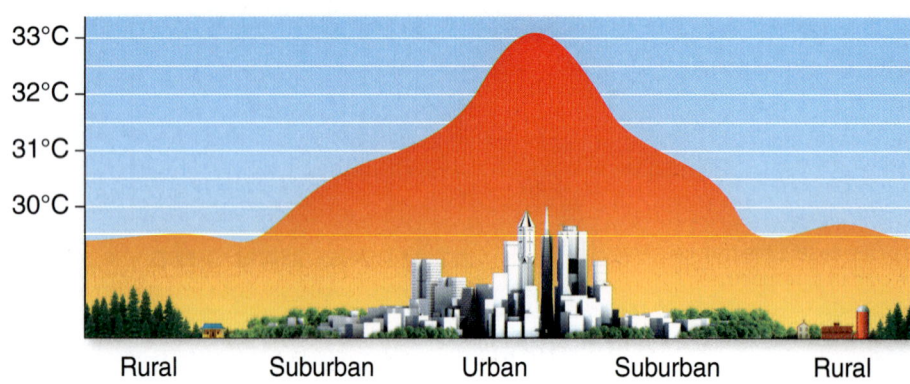

FIGURE 21-8
A graph shows typical temperature differences between rural, suburban, and urban zones due to the urban heat island effect.

Source: Cengage

TABLE 21-2 Changes in Climatic Elements Caused by Urbanization

Climate Element		Urban Compared to Rural Environment
Total Precipitation		5 to 10% more
Cloudiness	Cloud cover	5 to 10% more
	Fog, winter	100% more
	Fog, summer	30% more
Relative Humidity	Winter	2% lower
	Summer	8% lower
Radiation	Total	15 to 20% less
	Direct sunshine	5 to 15% less
Temperature	Annual mean	0.5 to 1.0°C higher
	Winter minimum	1.0 to 3.0°C higher
Wind speed	Annual mean	20 to 30% lower
	Extreme gusts	10 to 20% lower
	Calms	5 to 20% higher

Source: Cengage

Effects of Urbanization

If you travel from the center of a city toward the countryside, you may notice that the air gradually feels cooler and more refreshing as you leave the city streets and enter the green fields or hills of the outlying area. This feeling is not entirely imaginary. The climate of a city is measurably different from that of the surrounding rural regions (Table 21-2).

As shown in Figure 21-8, the temperature in rural, suburban, and urban areas can vary by almost 6 degrees Celsius. This temperature difference, called the **urban heat island effect**, is caused by numerous factors:

- Stone and concrete buildings and asphalt roadways absorb solar radiation and reradiate it as infrared waves.
- Cities are warmer because little surface water exists. As a result, little evaporative cooling occurs. In contrast, in the countryside, water collects in the soil and evaporates for days after a storm. Roots draw water from deeper in the soil. This water evaporates from leaf surfaces.
- Urban environments are warmed by the heat released when fuels are burned. In New York City in winter, the combined heat output of all the vehicles, buildings, factories, and electrical generators is 2.5 times the solar energy reaching the ground.
- Tall buildings block winds that might otherwise disperse the warm air.
- Air pollutants absorb heat emitted from the ground and produce a local greenhouse effect.

As warm air rises over a city, a local low-pressure zone develops, and rainfall is generally greater over the city than in the surrounding areas (Figure 21-9). Water condenses on dust particles, which are abundant in polluted urban air. Weather systems collide with the city buildings and linger, much as they do on the windward side of mountains. A front that might pass quickly over rural farmland remains longer over a city and releases more precipitation.

In 1600, less than one percent of the global population lived in cities. By 1950, 30 percent of the world's population was urban, and by 2000 that ratio had grown to 55 percent. Therefore, although urban climate change may not affect global climate, it affects the lives of the many people living in cities.

checkpoint What effect does the lack of surface water have on urban climates and why?

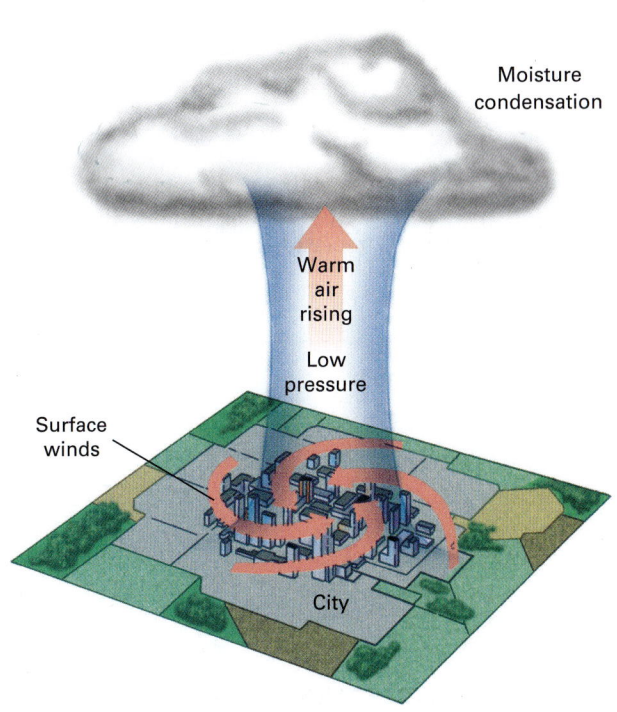

FIGURE 21-9
Warm air rising over a city creates a low-pressure zone. As a result, precipitation is greater in the city than over the surrounding countryside.

21.2 ASSESSMENT

1. **Explain** What are the two measurements that distinguish the climate types around the globe?
2. **Define** List the major climate types and at least two distinguishing characteristics for each.
3. **Compare** How is the marine west coast climate similar to and different from the tropical savanna climate?

Critical Thinking

4. **Generalize Patterns** What changes would you expect if climate change causes a region to gradually transition from a polar climate with a tundra biome to a subarctic biome?
5. **Synthesize** Look at Table 21-2 and describe how the total radiation received by an urban area compares to that of a nearby rural area. Explain which factor listed in the table is likely to contribute the most to the difference.

21.3 HISTORICAL CLIMATE CHANGE

Core Ideas and Skills
- Describe how Earth's global climate has changed over different time periods.
- Identify the types of evidence we have for historical climates of different time periods.
- Define the three basic factors that control Earth's atmospheric temperature.

To have a balanced perspective on the climate changes happening today, we must first study historical climate change. The geologic record in almost every place on Earth provides evidence that past regional climates were different from modern climates.

During the past 10,000 years, global climate has been mild and stable as compared with the previous 100,000 years. During this time, humans have developed from widely separated bands of hunter-gatherers to agrarian farmers and then moved into crowded communities in huge industrial cities. Today, with a global population of more than 7 billion, we have stressed the food-producing capabilities of the planet. If temperature or rainfall patterns were to change even slightly, crop failures could lead to famines. In addition, cities and farmlands on low-lying coasts could be flooded under rising sea level. As a result, climate change, especially human-caused climate change, is one of the most important issues we face in the 21st century.

Patterns of Climate Change

Earth's secondary atmosphere (discussed in Chapter 18) contained high concentrations of carbon dioxide (CO_2) and water vapor (H_2O). Both greenhouse gases absorb infrared radiation in the atmosphere. Astronomers have calculated that the sun gave off 20 to 30 percent less energy early in Earth's history. And yet oceans did not freeze. The high concentrations of atmospheric carbon dioxide and water vapor retained enough of the sun's radiation to warm Earth's atmosphere and surface to temperatures that kept the oceans liquid. Luckily for us, the concentration of carbon dioxide and water in the atmosphere declined only gradually as the sun warmed.

Figure 21-10 shows the long-term and short-term changes in Earth's global temperature as compared with the average temperature over the

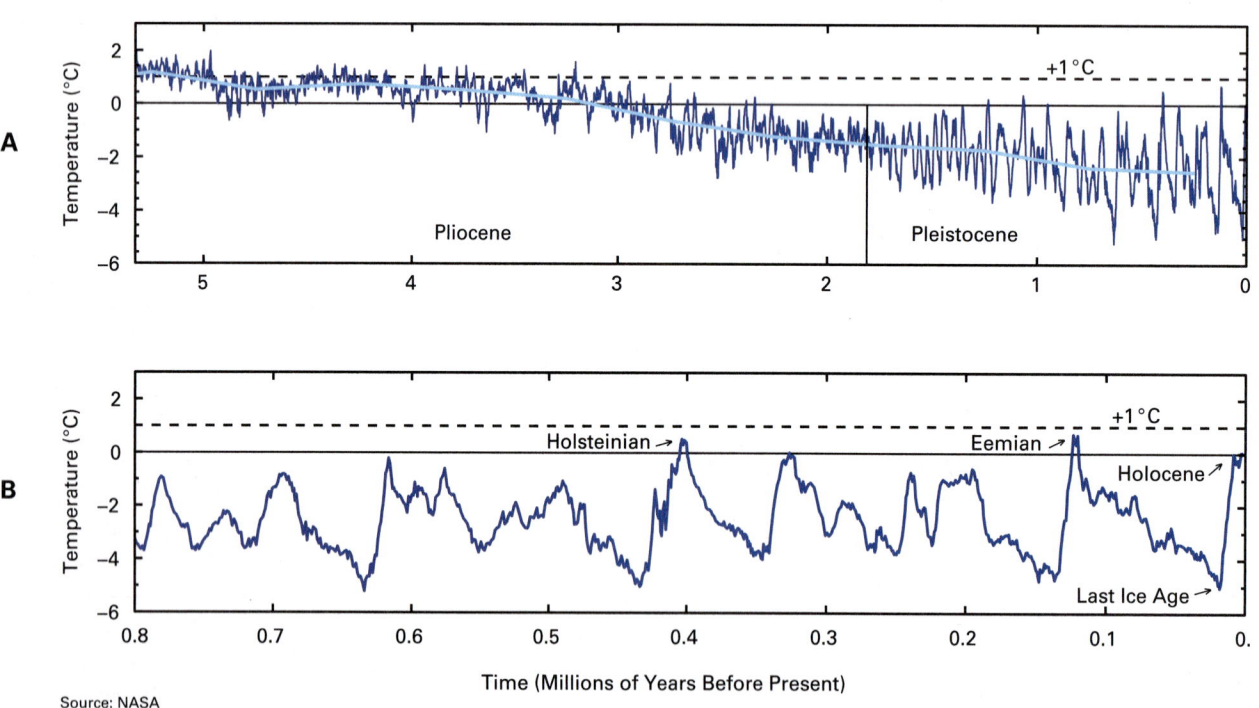

FIGURE 21-10
Two graphs show historical changes in Earth's global temperature over two timescales. The temperature estimates are based on studies of ocean sediment cores. (A) The blue line shows temperature variations from the Holocene average over the last 5.3 million years. (B) A similar graph depicts temperature fluctuations over the last 800,000 years.

last 12,000 years (Holocene epoch). Note, for example, that five million years ago, the planet was 1°C warmer than it is today, a warm trend that lasted for a few million years during the Pliocene epoch. The last two million years have witnessed the most recent ice age with the most recent deep freeze about 10,000 years ago.

Ocean sediment records from around the globe provide evidence that between 110,000 and 10,000 years ago the mean annual global temperature changed frequently and dramatically. There were two prior interglacial periods that were warmer than the Holocene: the Eemian (about 130,000 years ago) and the Holsteinian (about 400,000 years ago). Figure 21-10 shows that Earth underwent nearly a dozen rapid swings in surface temperature over the last million years. Most of the swings include very steep segments during warming episodes, suggesting that when Earth's surface temperature warmed up, it did so very quickly. Many of the cold intervals persisted for 1,000 years or more. As mentioned earlier, the past 10,000 years, during which civilization developed, have witnessed an unusually stable climate.

Figure 21-11 gives us an even more detailed look at a 128-year span from 1880 to 2008. Note that the temperature rose slowly from 1880 to 1970 and then began increasing dramatically. During the past century, people have burned large quantities of fossil fuels. When fossil fuels burn, the hydrocarbon molecules are rapidly converted to carbon dioxide and water. The burning of fossil fuels has released carbon dioxide into the atmosphere. The correlation between carbon dioxide emission and global temperature provides strong evidence that human activities have caused the current global warming. The temperature rise has been a little less than 0.8°C.

checkpoint What has been unusual about Earth's global climate during the development of civilization over the last 10,000 years?

Evidence for Climate Change

For the past 100 years, meteorologists have used instruments to measure temperature, precipitation, wind speed, and humidity. But how do we study climates from earlier periods before instruments and historical records?

Tree Rings and Pollen Growth rings in trees record variations in climate. Each year, a tree's growth is recorded as wood cells grow quickly in the spring and summer, producing a layer of thick cells called earlywood. When growth slows down during the autumn and winter months, a thin layer

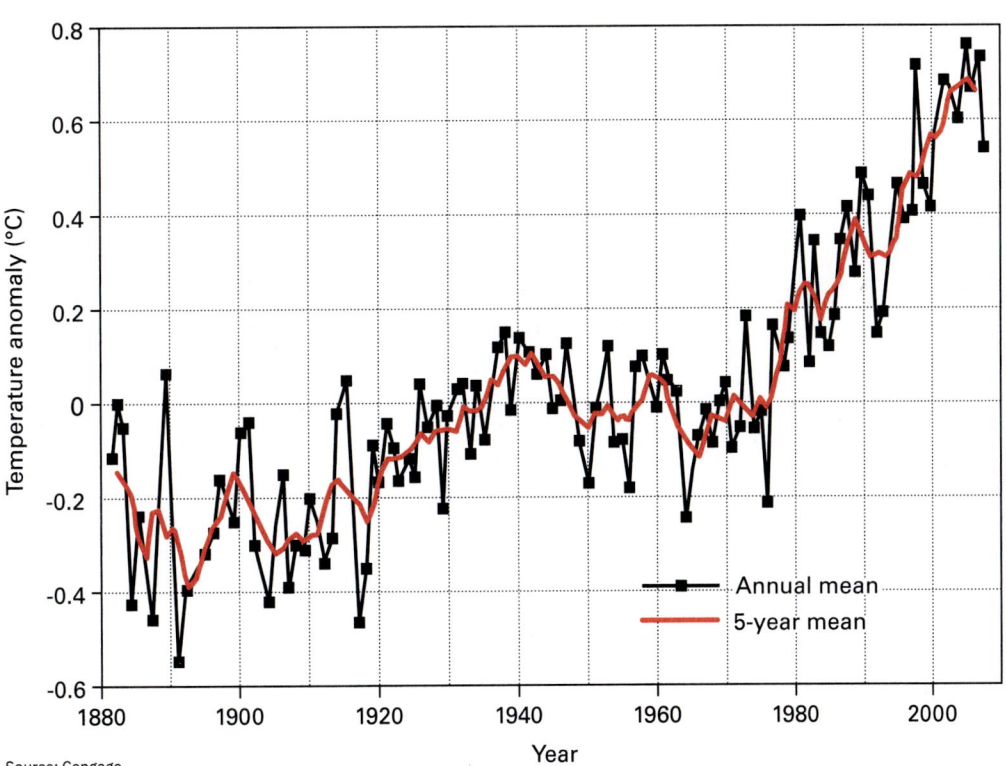

FIGURE 21-11
A graph shows the estimates of global surface temperature fluctuations over the past 130 years. The zero line represents the average temperature from 1951 to 1980. The black line shows the annual mean while the red line shows the 5-year mean.

of cells is produced called latewood. Each tree ring includes one set of earlywood and one set of latewood cells.

Trees grow slowly during a cool, dry year and more quickly in a warm, wet year. Therefore, tree rings grow wider during warmer, wetter years than during cool, dry ones. For this reason, scientists can count and measure the rings to reconstruct the history of past climate recorded in the wood. Interpretations of climate change from tree-ring data coincide well with historical data. For example, growth rings are narrow in trees that lived during the Little Ice Age between 1300 and 1870.

Plant pollen is widely distributed by wind and has a hard, waxy coating that resists decomposition. that resists decomposition. As a result, pollen grains are abundant and well-preserved in sediment in lake bottoms and bogs. For example, 11,000 years ago, spruce was the most abundant species in much of Minnesota. At that time, spruce pollen accumulated in bogs throughout the region. In modern forests, spruce dominates in colder Canadian climates but is less abundant in Minnesota. Therefore, scientists deduce that the climate in Minnesota was colder 11,000 years ago than it is at present. Pollen in younger layers of sediment shows that about 10,500 years ago, pines displaced the spruce, indicating that the temperature became warmer.

Isotope Ratios from Ice and Glacial Evidence

Oxygen consists mainly of two isotopes, abundant O-16 and rare O-18. Both isotopes are incorporated into water, H_2O. Water molecules containing O-16 are lighter and evaporate more easily than those containing O-18. Condensation works in the opposite way such that the heavier water molecules condense more readily than the lighter ones. Some of the water vapor condenses as snow, which accumulates in glaciers. The O-18/O-16 ratios in glacial ice reflect water temperature at the time the water evaporated. Because most of the atmospheric water vapor that falls as snow originated from evaporation of ocean water, scientists then use the O-18/O-16 data from glacial ice to estimate mean ocean surface temperatures. Because the sea surface and the atmosphere are in close contact, mean ocean surface temperature reflects mean global atmospheric temperature.

The oldest continuous ice core records extend to 130,000 years in Greenland, and 800,000 years in Antarctica. The age of the ice at any depth is determined by counting annual ice deposition layers or by carbon-14 dating of windblown pollen within the glacier. The oxygen isotope ratios in each layer reflect the temperature at the time the snow fell.

Erosional and depositional features created by glaciers are evidence of the growth and retreat of alpine glaciers and ice sheets. The timing of recent glacial advances can be determined by several methods. One effective technique is carbon-14 dating of organic carbon from plants preserved in glacial sediment. Unfortunately, carbon-14 dating can only be used to obtain ages spanning about 10 half-lives, from 50,000 years ago to the present.

Ocean Sediment, Rocks, and Fossils

In a technique similar to pollen studies, scientists estimate climate by studying fossils in deep-sea sediment. The dominant life-forms in the ocean are microscopic plankton that float near the sea surface. Pollen ratios change with changing climate and climate-related temperature change. Likewise, plankton species ratios change with sea-surface temperature. Fossil plankton found in sediment cores reflect sea-surface temperature.

Most hard tissues formed by animals and plants, such as shells, exoskeletons, teeth, and bone, contain oxygen. At colder temperatures, many organisms preferentially absorb the heavier isotope, causing their skeletal material to have high O-18/O-16 ratio. When temperature increases, the same organisms absorb less O-18 and the O-18/O-16 ratio decreases.

Consider foraminifera, tiny marine organisms. During a time of Pleistocene cooling and glacial growth, their shells contain an average of 2 percent more O-18 than similar shells formed during a warm interglacial interval. Just as scientists estimate paleoclimate by measuring oxygen isotope ratios in glacial ice, they can estimate ancient climate by measuring oxygen isotope ratios in fossil corals, plankton, teeth, and the remains of other organisms. Oxygen is also incorporated into soil minerals, so isotope ratios in soil and seafloor sediment also reflect paleoclimate.

Fossils are abundant in many sedimentary rocks of Cambrian age and younger. Geologists can approximate climate in ancient ecosystems by comparing fossils with modern relatives of the ancient organisms (Figure 21-12). For example, modern coral reefs grow only in tropical water. Therefore, we infer that fossil reefs also most likely formed in the tropics. Coal deposits and

FIGURE 21-12 Discovery of fossilized fern leaves indicates that a region was wet and warm in the past.

ferns formed in moist or temperate tropical environments. Cacti indicate that the region was once desert.

Looking backward even farther into the Proterozoic eon before life became abundant, it is difficult to measure climate with fossils. Geologists search for clues in rocks. Tillite is a sedimentary rock formed from glacial debris and indicates a cold climate. Lithified dunes formed in deserts or along coasts. Salt deposits require an arid climate for evaporation.

Carbonate rocks precipitate from carbon dioxide dissolved in water. Geochemists know the chemical conditions under which carbon dioxide dissolves and precipitates, so they can calculate a range of atmospheric and oceanic compositions and temperatures that would have produced limestone and other carbonate rocks. Most thick limestones formed in warm, shallow seas. Some ancient mineral deposits, such as the banded iron deposits, also reflect the chemistry of the ancient atmosphere.

checkpoint What are the major sources of evidence for historical climate change?

Causes of Climate Change

After scientists learned that Earth's climate has changed, they began to search further to understand why climates change. Despite its immense complexity, Earth's atmospheric temperature is determined by three basic factors: how much energy is input into the system, how much energy is reflected, and how much energy is absorbed.

1. **Energy Input** The sun is the primary energy source in the solar system. Energy output by the sun is input into Earth's systems. Solar output fluctuates, so in the absence of other factors, when the sun becomes hotter, Earth warms. When the sun cools, so does our planet.

2. **Energy Reflected** If the incoming solar radiation is absorbed at Earth's surface, the planet becomes warmer. If heat is reflected back into space, the planet becomes cooler. Therefore, the reflectivity, or albedo, of Earth's surface is a critical factor in determining the planet's temperature.

3. **Energy Absorbed** Heat that is reflected back into space by Earth's surface may be absorbed by gases in the atmosphere, in a process called the greenhouse effect. Because different gases retain heat differently, atmospheric composition regulates our planet's atmospheric temperature.

These three factors determine the amount of heat available to drive the planet's climate and weather engines. Feedback loops and threshold effects involving the atmosphere, the hydrosphere, the geosphere, and the biosphere then perturb the basic system to produce the resultant climate. Let's consider some natural causes of climate change.

Changes in Solar Output A star the size of our sun produces energy by hydrogen fusion over a lifetime of about 10 billion years. During this time, its energy output generally increases slowly. When Earth first formed, the sun produced only 70 percent of the energy it produces today. If sunlight were the only factor that influenced temperature, our planet would have been much cooler and Earth would have been covered with ice. Water-laden sediment lithified to form sedimentary rocks during Earth's early history, indicating that water was present at the surface. Earth's early atmosphere contained abundant greenhouse gases, but as the sun heated up, the composition of the atmosphere changed. Ultimately, processes leading to warming balanced those leading to cooling, and Earth's surface remained mostly at temperatures where liquid water was abundant.

Within the past few hundred million years, solar output has changed by only one fifty-millionth of one percent per century. This tiny variation had no measurable influence on climate change over thousands, or even millions, of years.

Changes in Earth's Atmosphere The greenhouse effect, defined in Chapter 19, is the main physical process by which Earth retains heat. Solar energy that is reflected from Earth's surface can radiate out into space or be absorbed by the

MINILAB Energy Exchanges in Atmospheric Models

Materials (per group):
48 tokens   4 cups   marking pens

Procedure

1. Label each of the cups with one of the four energy reservoirs in the model: Earth; Lower Atmosphere; Upper Atmosphere; and Sun and Space.
2. Place 18 tokens in the Sun and Space cup and place six tokens in each of the remaining cups.
3. The number of tokens in Earth and atmosphere cups represents the current temperature in each reservoir. You will transfer tokens between the cups according to the energy exchange rules in Table 21-3.
4. Day One: Start by transferring three tokens from the Sun/Space directly to Earth.
5. Transfer tokens from Earth, then from the Lower Atmosphere, and then from the Upper Atmosphere according to the rules in Table 21-3.

TABLE 21-3 Energy Exchange Rules for a Normal Atmosphere

Earth		Lower Atmosphere		Upper Atmosphere	
Token Count	Number of Tokens to Go to Lower Atmosphere	Token Count	Number of Tokens to Go to Upper Atmosphere and to Earth	Token Count	Number of Tokens to Go to Sun/Space and to Lower Atmosphere
1	0	1	0	1	0
2–8	2	2–8	1	2–8	1
9–14	3	9–14	2	9–14	2
15–19	4	15–19	3	15–19	3
20–23	5	20–23	4	20–23	4

atmosphere. Because the atmosphere has naturally changed over time, so has the global temperature and as a result, the global climate.

Tectonic activity over millions of years has affected the composition of the atmosphere and oceans. Changes in ocean currents, wind, and precipitation patterns have also altered climates. Increases in precipitation increase erosion of limestone bedrock boosting the calcium content of the oceans. Calcium reacts with carbon dioxide, removing this greenhouse gas from the atmosphere. With less ability to retain heat, global temperatures cool.

The effects of extraterrestrial impacts and volcanic eruptions on the chemistry of the atmosphere have been discussed in Chapter 5 and Chapter 19. A bolide impact about 65 million years ago caused a drastic climate change that led to a mass extinction. Freshly added aerosols and ash from massive volcanic eruptions spread across the globe, decreasing the amount of sunlight that reached the surface. When more incoming sunlight is reflected back into space, global temperatures can decrease, causing mass extinctions such as that during the Permian period.

Changes in Earth's Orbit
Over thousands of years, Earth's orbital characteristics, such as the angle between the rotational axis and orbital plane, the direction of the tilt of Earth's axis, and even the shape of its orbit around the sun, change. These changes can result in global changes on the seasons. For example, if Earth's tilt increases, the seasons become more extreme. Scientists have proposed that such changes are responsible

---

6. Repeat step 4 and step 5 for two more rounds. Then skip step 4 but repeat step 5 for three rounds.
7. Record the number of tokens in each reservoir at the end of the six rounds (one day).
8. Repeat steps 4 to 7 for Day Two, Day Three, and so on.
9. To simulate a CO_2-rich atmosphere, run the model by applying the energy exchange rules in Table 21-4.

TABLE 21-4 Energy Exchange Rules for a CO_2-Rich Atmosphere

Earth		Lower Atmosphere		Upper Atmosphere	
Token Count	Number of Tokens to Go to Lower Atmosphere	Token Count	Number of Tokens to Go to Upper Atmosphere and to Earth	Token Count	Number of Tokens to Go to Sun/Space and to Lower Atmosphere
1	0	1	0	1	0
2–8	2	2–8	0	2–8	0
9–14	3	9–14	0	9–14	0
15–19	4	15–19	1	15–19	1
20–23	5	20–23	2	20–23	2

Results and Analysis

1. **Explain** Describe what you are modeling each time you transfer three tokens from the sun to Earth.
2. **Infer** The number of tokens transferred increases with the number of tokens already in a cup. What does this imply about energy transfer?
3. **Compare** How did the outcome in the two tests of the model differ?

Critical Thinking

4. **Critique** Did the change in the model produce the intended outcome?
5. **Revise a Model** How would you revise the model? Explain the reasons for your changes.

for the Pleistocene glacial cycles described in Chapter 13. These climate changes can occur on timescales of tens to hundreds-of-thousands of years.

checkpoint What are the three basic energy-related factors that affect Earth's temperature?

21.3 ASSESSMENT

1. **Explain** Climate change studies derive their data from a wide variety of sources. Why can't these data instead be derived from a single source?
2. **Infer** How has the accuracy of evidence for climate change changed over time?
3. **Summarize** List the factors that cause global warming and global cooling.

Critical Thinking

4. **Analyze Data** Compare Figures 21-10, 5-2 (online), and 5-25. What intervals of geologic time are represented in Figure 21-10A? During these intervals, what was the overall trend in temperature? How did the number of families of organisms change, and how does this rate of change compare to the rate since the Permian–Triassic extinction?

21.4 CLIMATE CHANGE TODAY

Core Ideas and Skills
- Describe the human activities that have led to currently observed climate changes.
- Define the best-case scenario for forecasted temperature increases in the future.
- Contrast the best-case scenario with a scenario in which human activity does not change from current practices.

Human Impact on Climate Change

The amount of carbon in the atmosphere is determined by many natural factors. Within the past few hundred years, humans have become an increasingly important contributor to the carbon cycle. About 20 percent of this change results from cutting forests and the urbanization of other land surfaces. These human activities reduce the carbon uptake by plants. The remaining 80 percent is due to gases emitted by industry and agriculture. Modern industry releases four major greenhouse gases—carbon dioxide, methane, chlorofluorocarbons (CFCs), and nitrogen oxides.

People release carbon dioxide whenever they burn fossil fuels or biofuels. This release is inherent in the chemistry of combustion. Carbon in the fuel reacts with oxygen in the air to produce carbon dioxide and water. Once the carbon dioxide is released, much of it can remain in the atmosphere for centuries. Logging also frees carbon dioxide because stems and leaves are frequently burned and forest litter rots more quickly when it is disturbed by heavy machinery. The continual rise in the concentration of atmospheric carbon dioxide (Figure 2-18) has attracted considerable attention because it is the most abundant industrial greenhouse gas.

Chlorofluorocarbons (CFCs) were used widely in the 1960s and 1970s as the propellant in aerosol cans. By the early 1970s, worldwide production of the compounds had reached nearly a million tons per year. Meanwhile, evidence linking the use of CFCs to loss of ozone in the stratosphere was published. By 1978, the United States no longer permitted use of the compounds as aerosol propellants. CFCs were used as a refrigerant beyond 1978, but by 1994, production of new CFC stocks had effectively stopped worldwide. Today CFCs are only used for certain specialized uses for which an alternative has yet to be found, and the existing stock of CFCs is recycled through a set of "halon banks" designed to eliminate additional loss to the atmosphere. Although CFCs still exist in the stratosphere from their former widespread use, their concentration has decreased as the result of the Montreal Protocol, an international agreement to ban CFC production.

In addition, several other greenhouse gases are released by modern agriculture and industry. Small amounts of methane are released during some industrial processes. Larger amounts are released from the guts of cows, other animals, and termites, and from rotting that occurs in rice paddies. Nitrous oxide (N_2O), yet another greenhouse gas, is released from the manufacture and use of nitrogen fertilizer, some industrial chemical syntheses, and the exhaust of high-flying jet aircraft.

Are we sure that humans are causing climate change? Let us summarize the data:

1. Human activities release greenhouse gases.
2. The concentration of these gases in the atmosphere has risen since the beginning of the Industrial Revolution.
3. Greenhouse gases absorb infrared radiation and trap heat.
4. The atmosphere has warmed by almost 0.8°C during the last 50 years and nearly 1.5°C in the last 130 years (Figure 21-13).

Of the hottest years on record, the top five have occurred since 2014. The hottest year on record was 2016, followed by 2019. The obvious connection is that rising atmospheric concentrations of industrial greenhouse gases have caused the recent global temperature rise. Although this debate raged for a few decades, today scientists are essentially unanimous in concluding that humans are causing global warming.

checkpoint How are human activities affecting Earth's carbon cycle, and what are the results on climate?

Projected Future Climate Change

Modeling future climate change is a complex process. Predictions of how much change will occur over the next century depend largely on the emission of greenhouse gases by human activity and the sensitivity of Earth's systems to those emissions. Average annual temperature, seasonal temperature ranges, and annual precipitation patterns are relevant to these models, because climate types are largely dependent on those quantities. A broad geographical perspective is necessary because inputs to and outputs from the carbon cycle can occur anywhere on Earth.

Scientists have already provided evidence that even if all emissions resulting from human activity were to cease immediately, the average

FIGURE 21-13 ▼

The graph shows the global average surface temperature relative to the long-term average between 1880 and 2010. The vertical axis on the left shows the global average surface temperature. The black line and the vertical axis on the right show atmospheric CO_2 concentration in parts per million.

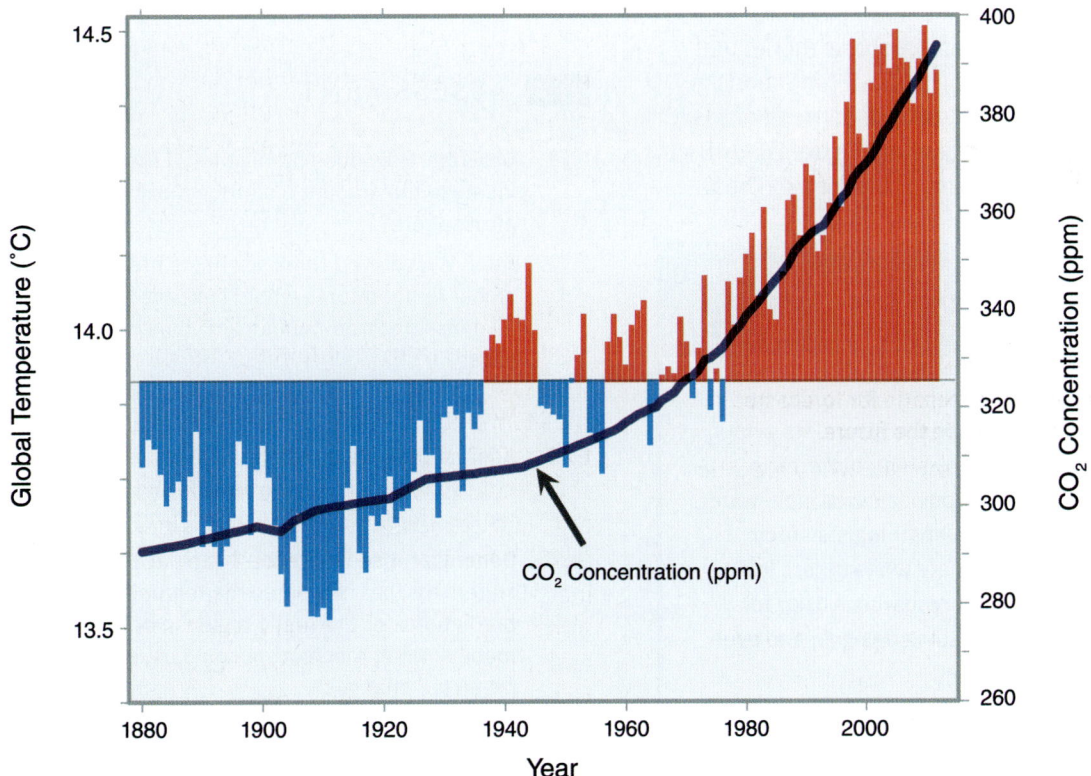

Source: U.S. Global Change Research Program, updated from Karl et al. 2009

FIGURE 21-14

Two models show different scenarios for the year 2100 based on forecasted temperature increases from the 1970–1999 average global temperature: (A) rapid emissions decreases and (B) continued emissions increases.

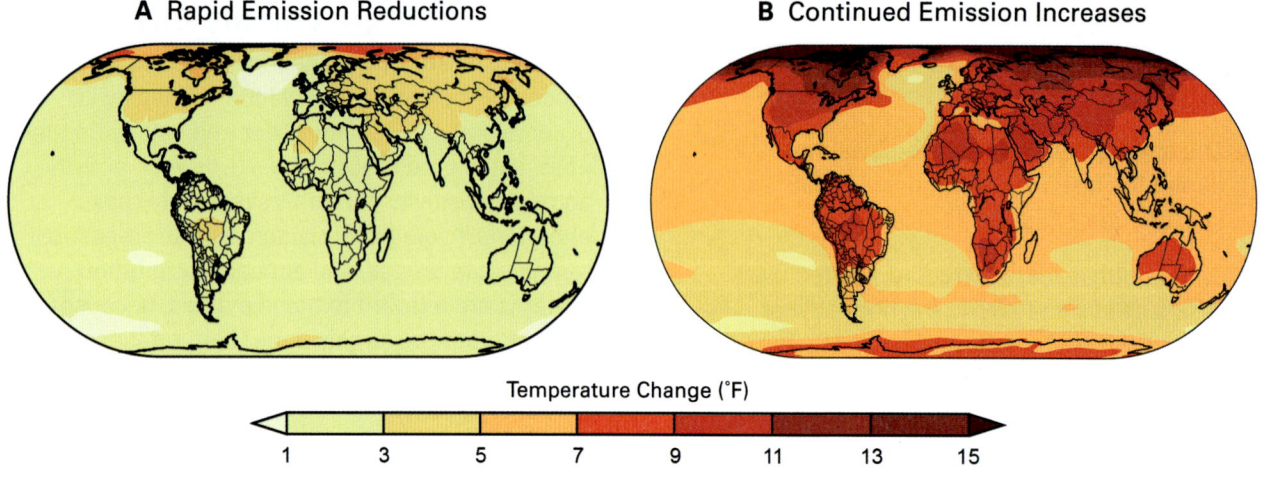

temperature will still rise in the next couple of decades. Reductions in emissions today will lead to noticeably less future warming, but those effects will take some time. Increases in emissions will result in more severe impacts than we are already experiencing today.

Figure 21-14A shows global temperature patterns predicted by an atmospheric model that assumes immediate and aggressive reductions in emissions. The average result is a warming of 1.3°C (2.4°F) by the end of this century. Figure 21-14B paints a more severe picture of the future. These predictions are based current emission levels increasing in the future. This choice leads to a warming of 4.4°C (8°F) by 2100 with the high-end forecast of a more than 6°C (11°F) increase.

checkpoint On which two factors do most predictions of climate change primarily depend?

21.4 ASSESSMENT

1. **Describe** How does burning fossil fuels increase the amount of carbon dioxide in the atmosphere?
2. **Infer** Why do you think scientists predict that global temperatures would increase even if all emissions from human activities were to cease?

Critical Thinking

3. **Analyze Data** Describe the correlation between atmospheric CO_2 concentrations and surface temperatures shown in Figure 21-13.
4. **Generalize** Use Figure 21-14 to develop a statement that compares the temperature predictions for the Arctic region in both models. What mechanism could cause these predicted changes?

CAREERS IN EARTH AND SPACE SCIENCES

Climatologist

Climatologists are atmospheric scientists who study Earth's climate. They study patterns in weather based on data from ice cores, soil, water, air, and plant life. Some climatologists study the history of Earth's climate while others look at current changes. Their research is used for agricultural planning, building designs, and even weather forecasting.

21.5 CONSEQUENCES OF CLIMATE CHANGE

Core Ideas and Skills
- Summarize the major predicted consequences of climate change.
- Identify the climate feedback loops that cause climate change to accelerate.

The consequences of climate change are mixed. A warmer global climate means a longer frost-free period in the high latitudes, a benefit for agriculture. In parts of North America, the growing season is now a week longer than it was a few decades ago. On the negative side, warmth affects plants in many ways that decrease crop yields. Heat stress, decreased soil moisture, and increased susceptibility to parasites and disease negatively affect crop growth. In a warmer world, both precipitation and evaporation are expected to increase. Computer models show that the effects would differ from region to region. Some areas will see increases in flooding while other areas will experience severe drought. The melting of ice sheets and glaciers will continue to accelerate sea-level rise. We will look at how these and other environmental consequences of climate change impact local communities and large-scale society.

Effects of Climate Change

Changing Weather Patterns In a hotter world, more soil moisture will evaporate. In the Northern Hemisphere, scientists predict that global warming will lead to increased rainfall, adding moisture to the soil. At the same time, warming will increase evaporation that removes soil moisture. Most computer models forecast a net loss of soil moisture and depletion of groundwater, despite the prediction that rainfall will increase. Drought and a decrease in soil moisture would increase demands for irrigation. But, irrigation systems are already stressing global water resources, causing both shortages and political instability in many parts of the world. As a result of all these factors, most scientists predict that higher mean global temperature would decrease global food production, perhaps dramatically.

In many regions of the world, winter precipitation accumulates as snow in the high mountains. Snowmelt feeds rivers during the summer, and this river water is used to irrigate crops. In a warmer world, more precipitation will fall as rain, and the snow that does fall is more likely to melt in early spring rather than in late summer. Also, glaciers will recede, diminishing the amount of water in long-term storage. As a result of all these factors, river levels during mid- to late summer will be lower in a warmer world than in a cooler one. Unless people build expensive dams, there will be less water available for irrigation.

Computer models predict that weather extremes such as intense rainstorms, flooding, heat waves, prolonged droughts, and violent storms—hurricanes, typhoons, and tornadoes—will become more common in a warmer, wetter world. Over the past several decades, this scenario has begun to play out in reality. In the United States, the relative number of extreme precipitation events increased between the first and second halves of the 1900s and accelerated into the new millennium (Figure 21-16). Worldwide, large floods became more frequent during the 20th century.

> **CONNECTIONS TO LIFE SCIENCE**
>
> **Threats to Amazon Basin Biodiversity**
>
> The Amazon River in South America is the largest river by discharge volume of fresh water in the world. Half the world's tropical rain forest can be found in the Amazon Basin. One in ten of all known species of plants and animals can be found there. The diversity of exotic organisms in the Amazon is without comparison anywhere else on Earth. While biodiversity is high, the population density of each species is limited, making them vulnerable to changes in the environment. Due to the impact of climate change, many native species in the Amazon are threatened or critically endangered. These include the Amazon river dolphin, the pink-throated hummingbird, the southern two-toed sloth, the black spider monkey, and the Caqueta titi monkey among many others.

🌐 **FIGURE 21-16**
Go online to view a graph that shows the alarming trends in heavy precipitation events between 1900 and 2010 in the conterminous United States.

Hurricanes are driven by warm sea surface temperatures. Many climate models predict that

FIGURE 21-15
ON ASSIGNMENT National Geographic photographer Greg Kahn documented flooding along the Chesapeake Bay. The community on Smith Island, Maryland, consists of three towns: Ewell, Rhodes Point, and Tylerton. The combined 280 residents are facing higher tides and more nuisance flooding as sea levels rise more than 3 millimeters every year.

extreme tropical storms and hurricanes like Katrina, Harvey, and Maria will become more frequent in a warmer world. We face the paradox of both more floods and more droughts. Hurricane intensity is likely to increase in a warmer world. Between 2001 and 2017, dozens of hurricanes hit the United States. Figure 21-17 shows a cost comparison of the costliest storms to hit the mainland. These include Hurricanes Katrina (2005) and Sandy (2012) in addition to the three devastating storms of 2017: Hurricanes Harvey, Irma, and Maria (see Figure 20-26).

Impact on Ecosystems According to a study published by the World Wildlife Fund in 2000, global warming could alter one-third of the world's wildlife habitats by 2100. As soil moisture decreases, trees will die and wildfires will become more common, changing forests into savannas. Plants and animals that thrive in cold temperatures will die off. Several recent studies have shown that many

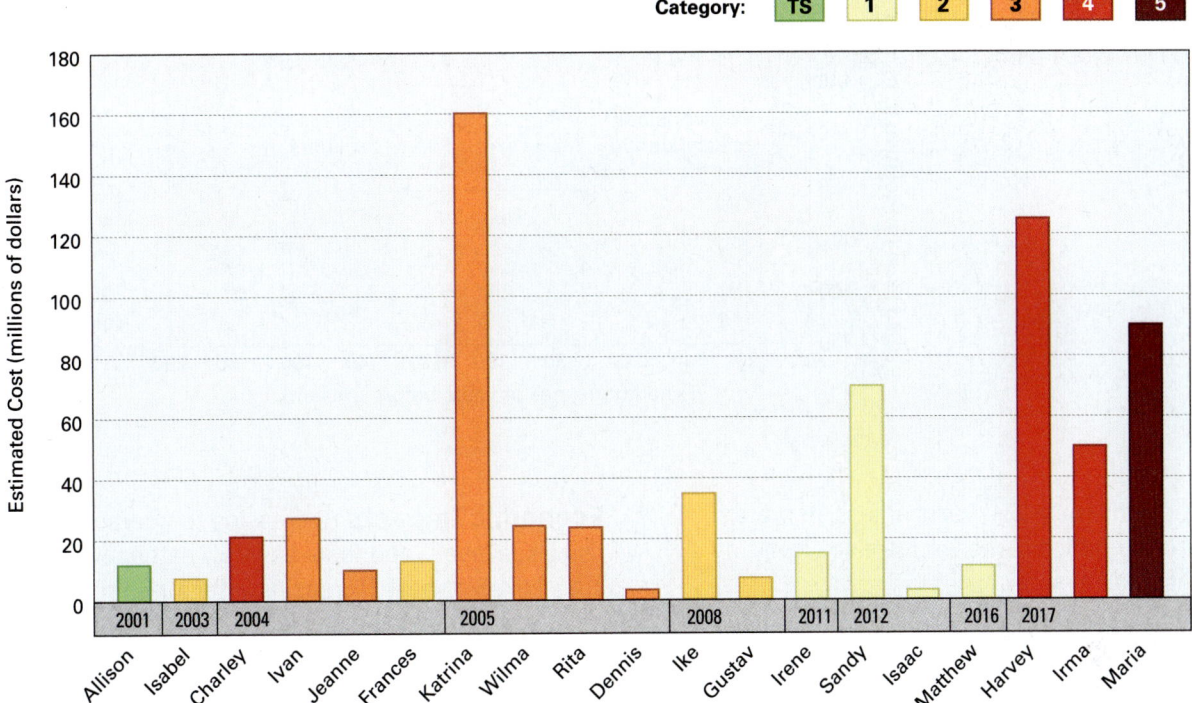

FIGURE 21-17
A bar graph shows the costliest hurricanes to hit the United States between 2000 and 2017.

Note: The reported costs are adjusted for inflation to 2017 dollars.
Source: NOAA

plant and animal pathogens thrive better in warmer temperatures than in a colder world. In a warmer world, tropical diseases such as malaria have already been spreading to higher latitudes. Wildfire frequency and size also has increased significantly in the past several decades with devastating and deadly wildfires occurring in virtually each one of the western U.S. states. Populations of frogs have declined dramatically as they have succumbed to virulent disease organisms.

In extreme cases, species that are unable to adapt to the warmer temperatures might become extinct. For example, as Arctic sea ice diminishes, polar bears are losing their hunting grounds and migration pathways. As a result, these majestic creatures are threatened. Undoubtedly, other species will flourish. Ecological systems will change, but it is important to remember that terrestrial and oceanic ecosystems have adjusted to far greater perturbations than we are experiencing today.

Sea-Level Rise and Warming
When water is warmed, it expands slightly. From 1900 to 1990, the mean global sea level rose between 1.2 and 1.7 millimeters, or roughly the thickness of a dime, annually. By 2000, the rate had increased to 3.2 millimeters per year. Oceanographers suggest that even this modest rise may be affecting coastal estuaries, wetlands, and coral reefs. However, sea-level rise is accelerating dramatically because, among other reasons, the polar ice sheets of Antarctica and Greenland are melting. In 2005, about 150 cubic kilometers of Antarctic ice melted into the oceans. To exacerbate the problem, flow of ice from Greenland into the North Atlantic Ocean has more than doubled over the past decade. If present rates continue, sea level will rise about 20 centimeters in a hundred years. Such a rise would be significant along many coastal areas such as the U.S. east coast and the west coast of Africa, but would be particularly severe in extremely low-lying areas such as the Netherlands, Bangladesh, and many Pacific Islands (Figure 21-18).

FIGURE 21-18
Go online to view a map showing the locations of glaciers and ice sheets, and the areas most vulnerable to sea-level rise.

As discussed in Chapter 13, National Geographic Explorer Erin Pettit is among the many scientists who have confirmed that the Thwaites Glacier is rapidly melting into the ocean. When it does, as

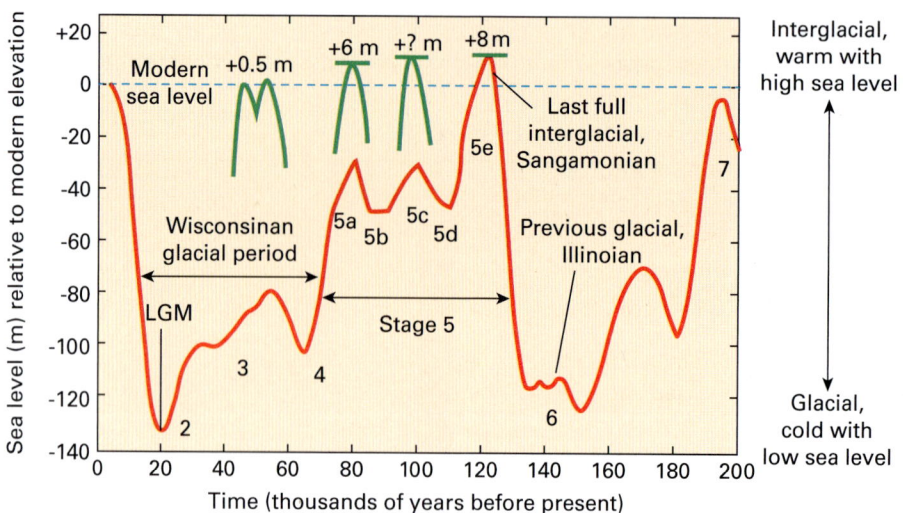

FIGURE 21-19 Sea level has fluctuated significantly over the last 200,000 years, with a maximum level 8 meters above present average 120,000 years ago.

expected in the next 250 years or less, its water volume contribution alone will raise sea levels by more than half a meter.

Sea level has risen and fallen repeatedly in the geologic past, and coastlines have emerged and submerged throughout Earth's history. During the past 40,000 years, sea level has fluctuated by about 150 meters, primarily in response to the growth and melting of glaciers (Figure 21-19). The rapid sea-level rise that started about 18,000 years ago began to level off about 7,000 years ago. By coincidence, humans began to build cities about 7,000 years ago. Civilization developed during a short time when sea level was relatively constant.

Ocean Acidification About one-third of all the carbon dioxide that enters the atmosphere dissolves in the ocean. On one hand, this buffer mechanism reduces the concentration of carbon dioxide in the atmosphere, maintaining a cooler planet. On the other hand, the dissolved carbon dioxide alters the chemistry of the oceans. When carbon dioxide dissolves in seawater, it reacts to form carbonic acid, H_2CO_3. Since preindustrial times, the oceans have become more acidic. When the oceans become more acidic, it is much harder for marine organisms to build hard shells out of calcium carbonate. For example, along the Great Barrier Reef in Australia, the calcification rate decreased by 14 percent between 1990 and 2008. Between 2016 and 2017 half of the Great Barrier Reef died outright. (See Chapter 16.) A small change in pH led directly to a large biological disruption. The resulting changes impacted marine ecosystems in ways that could in turn adversely affect ocean food webs.

Economic Impacts Increasing temperatures, rising sea levels, and more severe weather events are all likely to have increasingly negative impacts on the economy of the United States and other nations around the world. Agriculture, tourism, and other industries that depend on natural resources and stable climate conditions are especially vulnerable. For example, one study published in 2014 forecasts that some areas of the United States could see significant reductions in average crop yields by the year 2050. The same study estimates that by 2050, between $66 billion and $106 billion worth of existing coastal property will likely be below sea level.

On a global scale, rising sea level could flood coastal cities and farmlands, causing trillions of dollars worth of damage throughout the world. Wealthy, developed nations could build massive barriers to protect cities and harbors from a small sea-level rise. In regions where global sea-level rise is compounded by local tectonic sinking, dikes are already in place or planned. Portions of Holland lie below sea level, and the land is protected by a massive system of dikes. In London, where the high-tide level has risen by 1 meter in the past century, multimillion-dollar storm gates have been built on the Thames River. Venice, Italy, which is built over canals at sea level, has flooded frequently in recent years, and expensive engineering projects are underway. Rising sea level has resulted in erosion along much of the coastline along the Gulf Coast and eastern seaboard, reducing the size of nearshore wetlands needed to help decompose wastes and protect low-lying coastal areas from the effects of hurricanes.

However impressive the engineering of modern urban coastlines becomes, it is unlikely that coastal cities will ever be fully protected against a dramatic sea-level rise of several meters. Moreover, many poor countries cannot afford coastal protection even for a modest sea-level rise. For example, a 1-meter rise in sea level would flood roughly 17 percent of the land area of Bangladesh, displacing tens of millions of inhabitants.

checkpoint What major effects of climate change have already been witnessed and which are predicted?

Feedback Mechanisms

If the concentration of greenhouse gases and the mean global temperature were related linearly, a small change in one would produce a corresponding small change in the other. However, climate change does not follow linear relationships.

For example, the melting of ice is a threshold phenomenon. If the air above a glacier warms from −1.5°C to −0.5°C, the ice does not melt and the warming may cause only minimal environmental change. However, if the air warms another degree to +0.5°C, the ice warms beyond the threshold defined by its melting point. Melting ice can cause sea-level rise and a host of cascading effects.

Recall that a feedback mechanism occurs when a small initial disturbance affects another component of Earth's systems. This amplifies the original effect, which perturbs the system even more. This leads to a greater effect, and so on. There are several important climate feedback mechanisms.

Ice Melt and Albedo Sparkling snowfields and icy glaciers have high albedos, and they cool Earth by reflecting sunlight. When Earth warms a little, snow and ice covers decrease, and the albedo decreases. But when the albedo decreases, the atmosphere gets warmer, which melts more snow, which warms the atmosphere even more. Albedo is the most important factor helping to maintain a balance of heating and cooling on Earth. In the Arctic, warmer temperatures have promoted the growth of bushes that often protrude above the spring snow, trapping solar heat. Snow melts faster, providing favorable conditions for bushes to grow, which decreases the albedo even further, which warms the atmosphere, and so on.

Photosynthesis, Decomposition, and Carbon Dioxide As discussed in Chapter 18, photosynthesis removes carbon dioxide from the air by trapping carbon in plant tissue, whereas decomposition releases carbon dioxide. As the level of atmospheric carbon dioxide continues to rise, rates of photosynthesisis, and thus the storage of carbon by plants, are expected to increase. However, other factors also influence these rates. Plant growth will be limited by the availability of other plant nutrients and effects of climate change, such decreased soil moisture. The storage of carbon will likely be less than the amount of carbon released because the rate of organic decomposition in forest and grassland soils accelerates as temperatures increase. The net result is the addition of more carbon dioxide to the atmosphere, which in turn increases global warming.

Release of Methane Gas Methane is 20 times as effective in trapping heat in the atmosphere as carbon dioxide. Large deposits of methane gas are trapped both in permafrost and in sediments beneath oceans. Warming temperatures can release methane from both sources, causing a rapid and possibly catastrophic feedback warming. Vast areas of permafrost contain stores of carbon, which accumulated over millions of years. Some of this carbon already exists as methane. Additional amounts can be released quickly when the ice melts and microbes have lush, carbon-rich material to feast on. In a similar manner, methane is also trapped in seafloor sediments as frozen methane hydrates. If ocean water warms sufficiently to melt the ice, then the methane bubbles to the surface. In both these cases, initial warming from an unrelated cause, such as human release of carbon dioxide, can release the methane, which leads to additional and more rapid warming. Many scientists hypothesize that about 55 million years ago, a 100,000-year hot spell called the Paleocene-Eocene Thermal Maximum was caused by such a release of methane. This was the last time Earth was entirely without polar ice caps.

No one knows what the future will bring, but we do know that climates have changed radically in the past, from the frigid cold of Snowball Earth to the warmth of the Cretaceous. Today, by altering the atmospheric composition, we are undergoing a great and unintentional experiment with the climate system that sustains us. By making artificial changes to the natural order, we are altering not only Earth's topography but also its natural cycles in ways we may not be prepared to handle. Furthermore, global climate change has already affected Earth, its ecosystems, and the humans who call this planet home. Are we prepared to account for the consequences linked to the way we use our planet? By understanding how human activity can affect Earth's natural cycles, perhaps we can work with our environment and not against it to maintain a habitable world.

checkpoint What is the most important feedback mechanism that affects the global climate?

Mitigating the Effects of Climate Change

Reducing the threat posed by projected climate change is one of the most urgent issues we face in the 21st century. This is a global problem that will require sustained international cooperation over an extended period of time. International efforts to address this challenge began in 1988 when the United Nations established the Intergovernmental Panel on Climate Change (IPCC) to provide policymakers with regular assessments of the current scientific knowledge about climate change. So far the IPCC has issued five assessment reports, and the sixth one is expected in 2022.

The IPCC issued its Second Assessment Report in 1995, laying the groundwork for a 1997 meeting in Kyoto, Japan. Representatives from 161 nations met to negotiate a treaty to slow the rate of warming and projected climate change. The resulting Kyoto Protocol went into effect in 2005 with limited success and was allowed to expire in 2012. Then in 2015, delegates from 196 nations gathered in Paris to establish a global climate agreement. In the Paris Agreement, governments signed on to three key points:

- To accept a goal of limiting the increase in global average temperatures to less than 2°C (3.6°F) above preindustrial levels and to seek to limit the increase to 1.5°C
- To pledge to reduce greenhouse gas emissions by an amount specific to each nation
- To meet every five years to evaluate progress and increase goals

Meeting the goals of the Paris Agreement will require aggressive action both to reduce the amount of greenhouse gases emitted and to remove some of the carbon already present in the atmosphere.

Reducing Greenhouse Gas Emissions

Reducing emissions of CO_2 and other greenhouse gases is critical for avoiding the worst consequences of rising global temperatures. In 2018, the United Nations Environment Programme published the Emissions Gap Report, which compared the status of current efforts to reduce emissions with what would be required to limit the increase in global average temperatures to 1.5°C. The climate scientists who authored the report found that to avoid 2°C of warming, actions must be put in place to achieve greenhouse gas emissions in 2030 that are 25 percent lower than the 2017 levels. Limiting global warming to 1.5°C would require 2030 emission levels that are 55 percent lower than in 2017.

Carbon Capture All of the Emissions Gap Report's scenarios for limiting warming to less than 2°C assume not only a reduction in emissions but also the removal of excess CO_2 from the atmosphere. A variety of carbon capture technologies are in development, but continued investment and research will be required to make such technologies practical for large-scale use.

checkpoint Which global organization has taken the initiative to report to the world our progress in decreasing CO_2 emissions?

21.5 ASSESSMENT

1. **Explain** In what ways is climate change more than simply global warming?
2. **Contrast** Describe the opposing effects of global climate change on photosynthesis and decomposition.
3. **Identify** Where around the globe will sea-level rise have the most immediate effects?

Critical Thinking

4. **Synthesize** Describe a scenario that links the feedback mechanisms presented in this lesson, starting with the melting of a glacier.
5. **Critique** Make a list of the advantages and disadvantages of controlling the release of industrial greenhouse gases.

TYING IT ALL TOGETHER
CONSEQUENCES OF CLIMATE CHANGE

In this chapter, you learned about various factors that shape the climate in specific locations and contribute to Earth's six major climate types. You also examined how Earth's climate has changed over geologic time and considered the causes and effects of current climate change.

The Case Study highlighted the ongoing collapse of Antarctica's ice shelves as a consequence of climate warming. In this case, the connection between climate change and its consequences are obvious: as temperatures rise, ice melts to form water, and sea levels rise.

But climate change is far more complex than a simple trend in rising annual temperatures. As we discussed in this chapter, climates vary around the world not only in terms of absolute temperatures, but also in terms of precipitation and seasonal variations. In many regions, changes in precipitation and seasonal patterns may threaten human settlements more so than rising temperatures.

Work individually or with a group to complete the steps.

1. Select one of the following areas affected by climate change as a focus for your research: agriculture, biodiversity, weather extremes, water resources, sea ice and glaciers, human population, forests, sea level and coastal areas, or human health.

2. Find three or more reputable sources that discuss the effects of climate change in your selected category.

3. Make a list of small-scale and large-scale impacts. For each impact, describe whether the effects are mainly local or global.

4. For each item on your list, identify a specific location that is likely to experience an effect and explain why. Also discuss any possible secondary consequences, such as political unrest.

5. From your list, select two of the most concerning effects of climate change and prepare a short presentation about them. Explain what the anticipated effects are, why they are concerning to you, and what is being done or could be done to mitigate them. Present your findings to the class.

FIGURE 21-20
A nearly empty reservoir near Cape Town, South Africa, is still used by boaters. In 2018, the city narrowly avoided a complete loss of its water supply when a three-year drought finally relented. Weather extremes like this are one consequence of global climate change.

CHAPTER 21 SUMMARY

21.1 GLOBAL CLIMATE

- The three-cell model shows global air flow with alternating bands of high and low pressure.
- Warm air rises near the Equator, flows north and south at high altitude, cools and falls at 30° latitude.
- A portion of falling air flows toward the Equator and a portion flows poleward along the surface.
- Air rises at 60° latitude and returns to high altitude to complete the cell.
- The polar front forms where the polar easterlies and the prevailing westerlies converge.
- The jet stream blows at high altitude along the polar front and along the boundaries between airflow cells at the horse latitudes.

21.2 CLIMATE TYPES

- Climate types are classified according to annual temperature, annual precipitation, and variability in these factors.
- World climate types include humid tropical, arid, humid mesothermal, humid microthermal, polar, and highland climates.
- Urban areas are generally warmer and wetter than surrounding countryside.

21.3 HISTORICAL CLIMATE CHANGE

- Global climate has changed throughout Earth's history.
- The past 10,000 years witnessed an unusually stable climate.
- Past climates can be studied through a wide variety of observable evidence.
- Earth's atmospheric temperature is determined by energy from the sun, Earth's surface albedo, and heat retained by the atmosphere.
- When Earth first formed, the sun produced about 70 percent of the energy it produces today.
- Climate change results from the movement of continents, volcanic eruptions, and other natural changes.

21.4 CLIMATE CHANGE TODAY

- Carbon dioxide is a greenhouse gas that warms the atmosphere.
- The burning of organic carbon, including hydrocarbons, produces carbon dioxide gas.
- Industrial and agricultural processes contribute greenhouse gases to the atmosphere.
- Scientists are nearly unanimous in agreeing that human addition of greenhouse gases has altered climate in the past and continues to do so today.
- Atmospheric models predict that global temperature rise over the next few decades is expected to be one to three degrees Celsius.
- It will take time before reduction in greenhouse gas emissions will result in noticeable effects.

21.5 CONSEQUENCES OF CLIMATE CHANGE

- Global warming could lead to a longer growing season in the high latitudes, but warmth affects plants in ways that decrease crop yields, particularly through a decrease in soil moisture.
- Global warming will affect precipitation and snowmelt, cause extreme weather, alter biodiversity, melt Arctic sea ice and glaciers, lead to acidification of the oceans, and cause sea-level rise.
- Climate feedback loops include snowmelt and albedo and changes in rates of plant transpiration and photosynthesis.
- The collapse of the West Antarctic Ice Sheet could lead to rising sea level.
- Melting permafrost could release trapped methane gas, which is 20 times as effective at trapping heat as carbon dioxide. Such release could create catastrophic feedback warming.

CHAPTER 21 ASSESSMENT

Review Key Terms
Select the key term that best fits the definition. Not all terms will be used, and no term will be used more than once.

biome	subarctic
horse latitudes	subtropical jet stream
humid continental	taiga
humid subtropical	temperate rain forest
marine west coast	three-cell model
Mediterranean	trade winds
polar easterlies	tropical monsoon
polar front	tropical rain forest
polar jet stream	tropical savanna
prevailing westerlies	tundra
steppe	urban heat island effect

1. high-pressure regions at about 30 degrees north and south
2. a biome with cold temperatures and coniferous trees
3. local climate differences in areas with dense human population
4. a global pattern of airflow characterized by alternating bands of high and low pressure
5. biome found in coastal regions with moderate temperatures and high precipitation
6. climate in northernmost regions of continental interiors; severe winters and precipitation throughout the year
7. in the Northern Hemisphere, winds that originate at about 30 degrees latitude and blow from southwest to the northeast
8. semiarid grass-covered plain
9. fast-moving, high-altitude air that forms along the polar front
10. warm, wet climate with a short dry season and very heavy rainfall during the wet season; supports rain forests
11. warm climate with a winter dry season; enough rain during the summer to support grasslands and grazing animals, but not forests
12. biome found in polar climates and characterized by mosses, grasses, and a few bushes; no trees
13. climate with moderate temperatures, dry summers, and rainy winters; supports shrubs and small trees; prone to wildfires during dry summers
14. wind patterns flowing toward lower latitudes from the North Pole; deflected westward by the Coriolis effect
15. wind patterns flowing from 30 degrees latitude toward the Equator

Review Key Concepts
Answer each question on a separate sheet of paper to demonstrate your understanding of key concepts from the chapter.

16. Summarize the global patterns of high- and low-pressure areas and what causes these patterns.
17. List and describe the types of evidence that can be used to determine what the climate was like in the past.
18. Compare and contrast a tropical monsoon climate and a tropical rain forest climate.
19. Compare and contrast a marine west coast climate with a Mediterranean climate.
20. Describe an example of a feedback mechanism that leads to global climate change.
21. Summarize possible effects of global climate change and classify them as benefits or disadvantages.
22. What role does the Coriolis effect play in global wind patterns?
23. Summarize the changes in Earth's climate over the past 500 million years.
24. Describe the greenhouse effect and its impact on global temperature.
25. Summarize factors that can contribute to Earth's global climate and identify which factors are caused by human activity.
26. How can volcanoes cause both global warming and global cooling?

Think Critically
Write a response to each question on a separate sheet of paper. Use concepts from the chapter to support your reasoning.

27. Compare and contrast two methods for estimating past climates. When would it be appropriate to use each method?
28. Describe the climate you live in and identify the climate type and subtype. Provide evidence to support your answer.

CHAPTER 21 ASSESSMENT

29. Compare how a global increase in temperature of 2°C might affect a flower in an alpine meadow in the Rockies and a penguin living in Antarctica.

30. Compare and contrast the terms "global warming" and "climate change." Which term do you think is more appropriate to describe current climate trends?

PERFORMANCE TASK

Modeling Energy Flow in Earth's Climate

Energy flows to Earth from the sun. How this affects Earth's climate depends on a multitude of factors. In this performance task, you will evaluate a simple model of how energy flows in the atmosphere, link geoscience data from global climate models to the model and make an evidence-based forecast of observable climate change. Answer the questions and consider how each part of the model affects the others.

Key	Description
A	Incoming solar radiation
B	Radiation reflected off surface
C	Radiation reflected off atmosphere
D	Radiation absorbed by atmosphere
E	Radiation absorbed by surface
F	Energy stored in atmosphere

Key	Description
G	Energy stored in surface
H	Heat transferred to atmosphere
I	Heat lost to space
J	Heat returned by greenhouse effect
K	Heat radiated to space by atmosphere

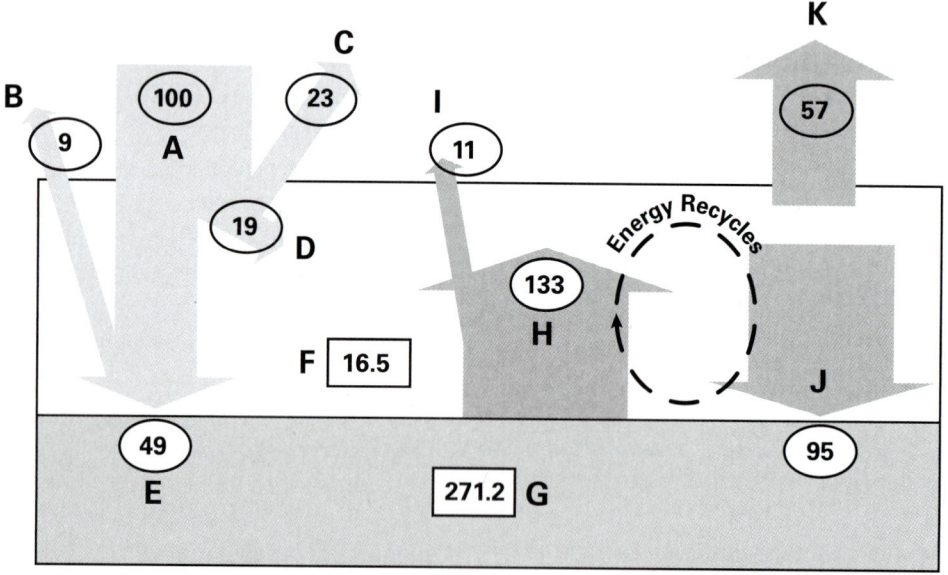

Energy Flows in the Climate System

100 energy units = 5.56x10^{24} Joules/year, total annual solar energy received at Earth

Source: Kiehl and Trenbeth, 1997. National Center for Atmospheric Research

1. List the places that solar radiation can go after coming to Earth.
2. What factors could increase the amount of light energy reflected off the land surface?
3. Suppose this model indicated a climate heat exchange in a rural area. How do you think the model would change if the area became urbanized?
4. How might a large volcanic eruption affect this model?
5. Consider a scenario in which the concentration of greenhouse gases in the atmosphere increases. What kinds of climate effects might be seen as a result?
6. Choose one RCP model and review how the concentration of atmospheric CO_2 is predicted to change by 2100. Consider the how predicted CO_2 changes would impact the energy flow model. Describe how parts of the energy flow model (including inputs, outputs, and reservoirs) will change. Give at least one example of how those changes might manifest themselves environmentally.

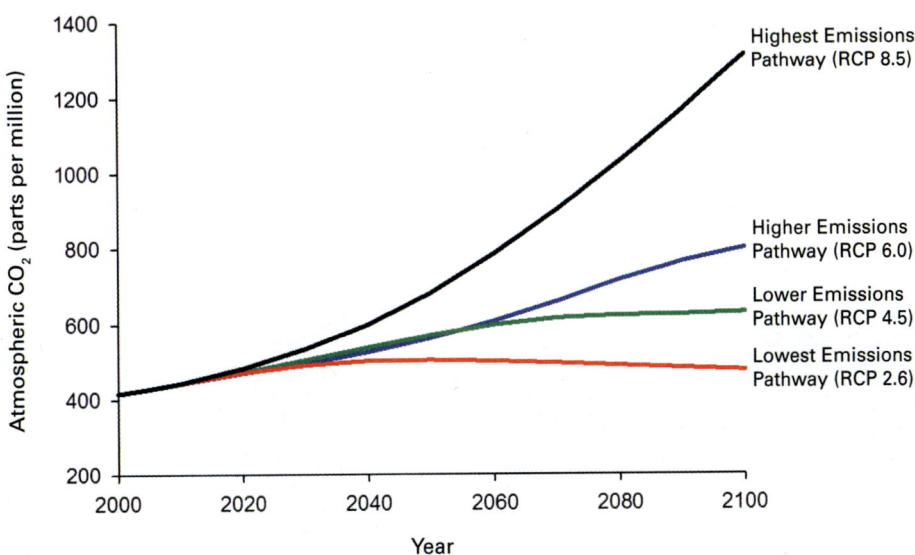

Source: International Institute for Applied Systems Analysis, Representative Concentration Pathways Database

The Projected Atmospheric Greenhouse Gas Concentrations graph shows the predicted CO_2 concentration in the atmosphere for four different Representative Concentration Pathway (RCP) global climate models. These models are based on current and future emissions from human activity.

UNIT 7 | EARTH AND SPACE

Mars

Venus

Chapter 22 **Evolving Models of the Sky**
Chapter 23 **Our Solar System**
Chapter 24 **Stars**
Chapter 25 **Galaxies and the Universe**

The study of Earth's place in space has inspired humans for thousands of years. Like other sciences, astronomy has given us an increasingly accurate picture over time as technological capabilities have progressed. Our understanding of Earth as a planet gives us the foundation for comprehending other worlds that orbit the sun. Beyond our average star lie billions of other stars in our galaxy, and beyond our galaxy, billions of other galaxies. Pulsars, spiral and elliptical galaxies, quasars, black holes, and dark matter diversify the wide repertoire of astronomical objects and phenomena that make up our incredible universe.

NASA's Cassini spacecraft snapped this scene in 2013 as Saturn and its main ring system eclipsed the sun. Labeled on the image are the locations of three of the four inner planets. The Cassini spacecraft was the first satellite to orbit the giant planet, completing nearly 300 orbits before the mission ended in 2017.

Earth and moon

CHAPTER 22
EVOLVING MODELS OF THE SKY

The Bracewell Radio Sundial at the Very Large Array in New Mexico uses the location of a shadow to measure Earth's rotation and indicate the time of day. Above the sundial, the stars trace concentric trails around the north celestial pole near the star Polaris.

Astronomy is one of the oldest sciences. Early human civilization observed the night sky and identified patterns in the stars that helped them make decisions such as when to plant crops and when to prepare for winter. Ancient astronomers performed systematic observations of the sky. They developed astronomical artifacts to aid in nautical navigation and created calendars based on the changing position of the stars and planets. The invention of the telescope allowed astronomy to develop into a modern science. Our understanding of Earth's place in the universe has changed significantly as observational astronomy has continued to become more accurate and complete.

KEY QUESTIONS

22.1 What conclusions did ancient astronomers derive from their observations of celestial objects?

22.2 How did the model of the sky change and what evidence was used to support the new models?

22.3 What role does gravity play in the motion of celestial objects and how do we know?

22.4 How have astronomical tools improved over time and what discoveries have been made possible?

EXPLORERS AT WORK

BRINGING SCIENCE DOWN TO EARTH

WITH NATIONAL GEOGRAPHIC EXPLORER BRENDAN MULLAN

Brendan Mullan has a pretty impressive résumé. He holds a Ph.D. in astrophysics and teaches college-level courses in physics and astronomy. His research interests include mind-bending concepts such as the evolution of galaxies, the birth of clusters of stars within them, and the search for extraterrestrial life. It must be pretty hard to keep up with Mullan in a conversation about his work, right? Wrong! Mullan loves nothing more than to speak about science in a way that not only makes sense to his listeners but inspires them to start asking their own scientific questions about the world around them. In fact, he says his greatest goal is to debunk the idea that science is too lofty, too complex, or too academic for the average person to care about. "We need to battle the stereotype that science is nerdy, boring, and not relevant to people's lives," Mullan says.

Mullan certainly has a gift for making science entertaining, as he demonstrated at the U.S. FameLab competition in 2012. The annual NASA-sponsored event draws scientists from around the country and challenges them to explain their research in no more than three minutes. The most effective communicator wins. That year, Mullan won for his explanation of a concept related to his research called the Fermi paradox. If life almost certainly exists elsewhere in the universe, why haven't we seen any sign of it? Mullan cast himself as a frustrated real estate agent trying to advertise Earth as a good home for aliens, none of whom were bothering to show up.

Mullan employs a variety of creative ways to make science accessible to audiences of all ages. He has put on astrobiology summer camps and frequently visits elementary and high school classrooms to give presentations such as "Everything in the Universe Is Terrible. But Here's Why That's Awesome." He also enjoys teaching physics for engineers, and astronomy and astrobiology for nonscience majors through his professorship at Park Point University. In 2015, he cofounded a nonprofit called The Wrinkled Brain Project. He helped develop an online curriculum that highlights real-world science conundrums (such as the Fermi paradox). Students learn about actual research that scientists have conducted. Then they collaborate to develop their own hypotheses to test.

Mullan says he wants to encourage critical thinking and problem solving through science inquiry, because tackling issues such as climate change and sustainable energy require those skills. On a more basic level, he simply wants to share his lifelong love of science. "I went on a school field trip to a planetarium when I was ten," explains Mullan. "The lights dimmed, all these bright pinpoints appeared overhead, and I learned about how stars are born, evolve, and die; the mystery of black holes; violent supernovae explosions. I thought it was the coolest thing in the world and decided right there I wanted to know how it all works. I was so fortunate to have access to resources like that; I want to pay it forward to the next generation. What could be more fun and meaningful than sharing the majesty of the cosmos with everyone?"

THINKING CRITICALLY

Reflect What is your view of astronomy and astronomers? What are some ways in which you find astronomy to be daunting or disconnected from your everyday life? Does reading about Brendan Mullan's approach to astronomy change your thinking in any way? If so, how?

Dr. Brendan Mullan demonstrates the ingredients that make up comets.

Mullan is cofounder and Director of Science at The Wrinkled Brain Project. He travels around the world, inspiring enthusiasm for science and scientific thinking in classrooms and at public events.

CASE STUDY
GALILEAN MOONS AND MODELS OF THE UNIVERSE

Throughout human history, people have been fascinated by the heavens. Ancient civilizations from the Sumerians, Egyptians, and Greeks to the Toltecs, Incas, and many others, considered the celestial bodies to be deities. As time went on, people observed that the sun, the moon, and the stars traveled across the sky in repeating, predictable patterns. Astronomers began measuring their movements, and many concluded that the sun, the moon, and the stars all revolved around Earth.

Much of the evidence collected by early astronomers supported this Earth-centered model of the universe, although some did not. Despite the inconsistencies, the idea that Earth was the center of the universe became deeply entrenched in European thought and religion by the Middle Ages (400–1500 C.E.). Roman Catholicism adopted an Earth-centered view of the universe to support the notion that Earth—and by extension, humanity—held a central place in God's creation. Islam adopted a similar view.

Several scholars of the Middle Ages proposed alternative models of the universe, but they were often met with resistance. One of these scholars was Galileo Galilei (1565–1642), who developed a telescope powerful enough to observe objects too distant to be seen with the naked eye. Using his telescope, Galileo tracked the paths of four bright objects near Jupiter and concluded that these bodies, now known as the Galilean moons, revolved around Jupiter. In 1610, he published his discovery and argued that Earth cannot be the sole center of motion in the universe. Unfortunately for Galileo, the Catholic Church viewed his claims as a challenge to Christian doctrine. In 1633, he was tried and convicted of heresy and sentenced to house arrest until his death in 1642.

Since Galileo's time, scientists have continued to study Jupiter and its many moons. In 1977, NASA launched the space probes Voyager 1 and Voyager 2 to fly by the giant planet. The Voyager probes sent back detailed images of Jupiter's Great Red Spot (a massive, rotating storm) and active volcanoes on the moon Io. The space probe Galileo, launched by NASA in 1989, orbited Jupiter for eight years, collecting even more detailed information. Among other discoveries, the Galileo mission revealed that the moon Europa likely has more water on its surface than Earth, and the moon Ganymede has its own magnetic field. At present, a fourth space probe named Juno is orbiting Jupiter and collecting data about the planet's atmosphere and surface. Scientists hope to use this information to continue gaining new insights about Jupiter's formation and evolution.

As You Read Learn how scientists throughout history constructed new ideas and replaced old ones about the structure of our solar system and the universe. Consider the kinds of data that were collected at various points in human history. Think about how those data helped shape the new ideas that developed as astronomy evolved as a science.

FIGURE 22-1A
Go online to view a draft of a letter written in 1609 by Galileo to Leonardo Donato, Doge of Venice, showing his notes on the motion of Jupiter's moons.

FIGURE 22-1B
An artist's rendition of the Galileo spacecraft shows how it might have looked as it arrived in 1995, becoming Jupiter's first artificial satellite.

22.1 PATTERNS IN THE SKY

Core Ideas and Skills
- Describe how astronomers organize stars seen in the night sky.
- Describe the apparent motion of stars and planets in the night sky.

KEY TERMS

constellation   retrograde motion
planet

Apparent Motion and Position of Celestial Objects

Even a casual observer notices that both the sun and the moon rise in the east and set in the west. In the midlatitudes, summer days are long, and the sun rises high in the sky. Winter days are shorter, and at midlatitudes in the Northern Hemisphere, the sun never rises very high above the southern horizon, even at noon. In contrast to the yearly cycle of seasons, the moon completes its cycle once a month. It is full and round one night, then darkens slowly until it is a thin crescent. It disappears completely after two weeks, then reappears as a sliver that grows again, completing the entire cycle in about 29.5 days.

If you had been a cowboy in the 1800s, the foreman might have told you to guard the herd at night until "the dipper holds water." This phrase refers to two features of the night sky. First, stars remain in fixed positions relative to one another. This fact led ancient peoples to identify groups of stars, which they called **constellations**. Stars are so far away that their positions relative to each other appear fixed in the sky. Ursa Major (the Big Dipper), Ursa Minor (the Little Dipper) and the Orion constellation are a few popular examples (Figure 22-2). Second, in the Northern Hemisphere, Polaris, or the polestar or North Star, is a nearly motionless fixed point in the sky, and all other stars appear to revolve around it (Figure 22-3). This motion provided the clock for the cowboy on night watch because the dipper changes its orientation, alternatively holding and then spilling water as it appears to rotate around Polaris.

Constellations appear and disappear with the seasons. For example, the Egyptians noted that Sirius, the brightest star in the sky, became visible at dawn just before the Nile began to flood. Farmers would therefore plant crops when this star first appeared, with the assurance that the high water soon would irrigate their fields.

Ancient astronomers noted several objects that appeared to be stars but were different because they changed position with respect to the stars. The ancient Greeks called these objects **planets**,

A

B

FIGURE 22-2
Three easily recognizable star patterns are shown: (A) the constellations Ursa Major, also known as the Big Dipper, and Ursa Minor, known as the Little Dipper, and (B) the Orion constellation.

FIGURE 22-3
A multihour exposure photograph of the sky shows that stars appear to rotate around one central point in the Northern Hemisphere, very near Polaris, or the North Star.

from a Greek word meaning "wanderers." For most of the year, planets appear to drift eastward with respect to the stars, but sometimes they seem to reverse direction and drift westward. This apparent reverse movement is called **retrograde motion** (Figure 22-4), a concept that we will explore further in the next lesson.

⊕ **FIGURE 22-4**
Go online to view a photograph that shows the apparent retrograde motion of Mars.

checkpoint How does the apparent motion of a celestial object help identify whether the object is a planet or a star?

22.1 ASSESSMENT

1. **Explain** Why is the apparent motion of stars an aid for telling time?
2. **Define** What is the importance or significance of the star Polaris?
3. **Explain** Why is it possible to assign a star permanently to one constellation?

Computational Thinking

4. **Evaluate** The distances to the brightest stars in the constellation Orion are given in the table. Use the data to evaluate the argument that the constellation would look different from a position in space far from Earth.

Star	Part of Constellation	Distance (10^{18} kilometers)
Betelgeuse	left shoulder	4.06
Bellatrix	right shoulder	2.31
Alnitak	left belt	7.74
Alnilam	center belt	1.28
Mintaka	right belt	8.70
Saiph	left foot	6.84
Rigel	right foot	7.34

Source: Sloan Digital Sky Survey

714 CHAPTER 22 EVOLVING MODELS OF THE SKY

22.2 EVOLVING MODELS OF THE UNIVERSE

Core Ideas and Skills
- Explain ancient astronomers' beliefs about Earth's place in the universe.
- Summarize the observations that led to the changes in the model of Earth's place in the universe.
- Explain how the sun-centered model explained the apparent strange motion of the planets.

KEY TERMS

geocentric
celestial sphere
parallax
heliocentric
astronomical unit
inertia

Geocentric Model

Greek philosopher and scientist Aristotle proposed a **geocentric**, or Earth-centered, universe. In this model, Earth is stationary and positioned at the center of the universe. A series of concentric **celestial spheres**, made of transparent crystal, surrounds Earth (Figure 22-5). The sun, the moon, the planets, and the stars are embedded in the spheres. At any one time a person can see only a portion of each sphere, but as the sphere revolves around Earth, objects appear and disappear.

⊕ **FIGURE 22-5**
Go online to view an example of a model that ancient astronomers designed to demonstrate the position of objects relative to each other in the sky.

Aristotle based his conclusions on two observations. First, a chariot driver who is not holding on will fall off his chariot when his horses start to gallop, but people do not fall off Earth. Aristotle reasoned that Earth must not be moving because people have no sensation of motion.

Aristotle's second observation was based on **parallax**, the apparent change in position of an object due to the change in position of the observer. Ancient people reasoned that if Earth moved around the sun, the stars should change

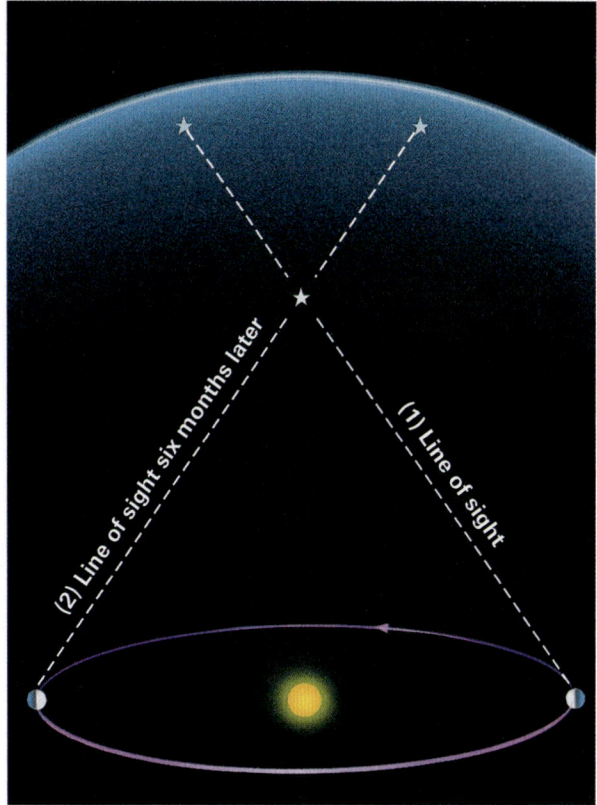

FIGURE 22-6
A nearby star appears to change positions with respect to the distant stars as Earth orbits the sun. This drawing is greatly exaggerated. In reality, the distance to the nearest star is so much greater than the diameter of Earth's orbit that the parallax angle is only a small fraction of a degree, imperceptible to the human eye.

position relative to one another (Figure 22-6). They observed no parallax shift, so they concluded that Earth must be stationary. The mistake arose not out of faulty reasoning but because stars are so far away that their parallax shift is too small to be detected with the naked eye.

Aristotle incorporated philosophical concepts, in addition to observation, into scientific theory. He argued that the gods would create only perfection in the heavens and that a sphere is a perfectly symmetrical shape. Therefore, the celestial spheres must be a natural expression of the will of the gods. The sun, the moon, the planets, and the stars must also be unblemished spheres.

Aristotle's theory, although incorrect, did explain the regular motions of the sun, the moon, and the stars. However, it failed to explain the retrograde motion of the planets. In about 150 C.E., Claudius Ptolemy modified the celestial sphere model to incorporate retrograde motion. In Ptolemy's model, each planet moves in small circles as it follows its larger orbit

LESSON 22.2   715

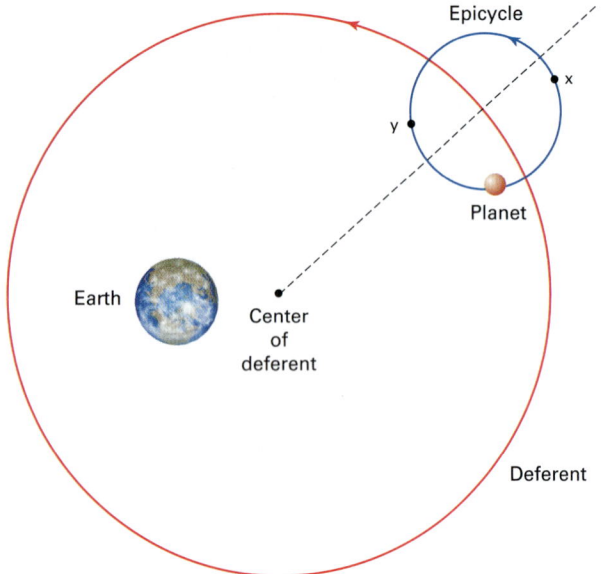

FIGURE 22-7
A diagram demonstrates Ptolemy's explanation of retrograde motion. Each planet revolves in a small orbit (epicycle) that moves along a larger orbit (deferent). When the planet is in position x, it appears to be moving eastward in the sky. When the planet is in position y, it appears to reverse direction, moving westward. Ptolemy did not realize that the planets moved in elliptical orbits. To compensate for the observed non-uniform motion of planets, he placed Earth far away from the center of the deferent.

around Earth (Figure 22-7). Ptolemy's sophisticated mathematics accurately described planetary motion and therefore his model was accepted. Even though he figured out retrograde motion, Ptolemy still retained Aristotle's erroneous idea that the sun and the planets orbit a stationary Earth.

checkpoint What apparent motion did Ptolemy's epicycles attempt to explain?

Heliocentric Model

Aristotle's and Ptolemy's ideas remained essentially unchallenged for 1,400 years. Then in the 120 years from 1530 to 1650, several Renaissance scholars changed our understanding of motion in the solar system and, in the process, revolutionized scientific thought.

Copernicus In 1530, a Polish astronomer and cleric, Nicolaus Copernicus, proposed that the sun, not Earth, is the center of the solar system and that Earth is a planet, like the other "wanderers" in the sky. Copernicus based his hypothesis on the philosophical premise that the universe must operate by the simplest possible laws. Copernicus believed Ptolemy's model was too complex and that the motions of the heavenly bodies could be explained more concisely by a **heliocentric** model with the sun at the center of the solar system.

Figure 22-8 shows how the heliocentric model explains retrograde motion. Assume that initially Earth is in position 1 and Mars is in position 1'. An observer on Earth looks past Mars (as shown by the white line) and records its position relative to more distant stars. Mars appears to be in position 1" in the night sky. After a few weeks Earth has moved to position 2, Mars has moved to 2', and Mars's position relative to the stars is indicated by 2". In the Copernican model, Earth moves faster than Mars because Earth is closer to the sun. Therefore, Earth catches up to and eventually passes Mars, as shown by positions 3–3' and 4–4'. During this passage, Mars appears to turn around and move backward, although, of course, this appearance is

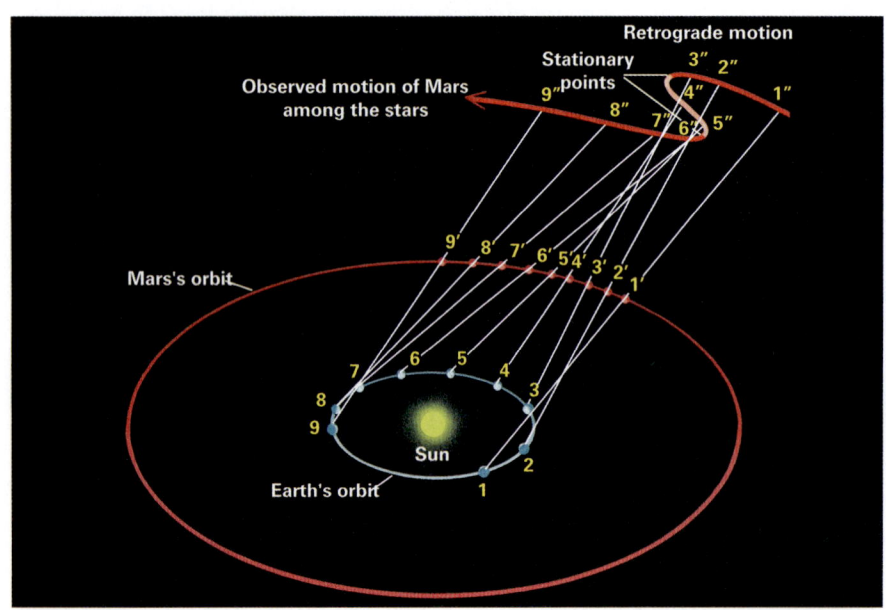

FIGURE 22-8
A model shows how the heliocentric model more simply explains the causes of the apparent retrograde motion of Mars as Earth catches up and surpasses Mars along its own orbit around the sun.

merely an illusion against the backdrop of the stars. Mars appears to reverse direction again through positions 5 and 6, thus completing one cycle of retrograde motion. In reality, Mars never reverses direction, it just appears to behave in this way as Earth catches up to it and then passes it.

Tycho and Kepler In the late 1500s, Tycho Brahe, a Danish astronomer, accurately mapped the positions and motions of all known bodies in the solar system. His maps enabled him to predict where any planet would be seen at any time in the near future, but he never explained their motions. Tycho died somewhat mysteriously in 1601. His student, Johannes Kepler, ended up with the vast amount of data that Tycho had collected.

Kepler, a mathematician, applied his reasoning skills to analyze Tycho's planetary motion data. During one analysis, Kepler sketched Mars's orbit around the sun and noted the time it took for Mars to move between two points. Then he calculated the area of the space bounded by lines connecting the two points and the sun. He found that if he repeated this process using any other two points with the same orbital duration, the areas were always the same (Figure 22-9). Astonished at this outcome, Kepler concluded that Mars must have an elliptical orbit. Up to this point, Kepler and other scientists of his time had assumed that planetary orbits were circular. However, Tycho's data showed otherwise.

⊕ FIGURE 22-9
Go online to view Kepler found sets of points corresponding to equal orbital times along Mars's orbit. He discovered that the triangular areas swept out by the planet as it passed through those points were equal.

Kepler's two discoveries—that planets move in elliptical orbits and that a line drawn between the planet and the sun will sweep across equal areas for equal durations—are now known as Kepler's first law and second law of planetary motion, respectively. Figure 22-10 illustrates both laws; these laws predict that a planet moves at a different rate of speed at different positions in its orbit. When a planet is closer to the sun, it moves at a faster rate than it does when it is farther from the sun.

CONNECTIONS TO MATHEMATICS

Applying Kepler's Laws to Venus and Mars

The orbital period of Venus is 224.7 days. Using Kepler's third law, we can find the average distance between Venus and the sun. First, we convert the orbital period of Venus in days to years:

$$224.7 \text{ days} \times \frac{1 \text{ year}}{365 \text{ years}} = 0.616 \text{ years}$$

Then, we use Kepler's third law to set the square of the orbital period, P, in years equal to the cube of the semi-major axis, a, of the planet in AU:

$$P^2 = a^3$$
$$(0.616)^2 = a^3$$

Finally, we rearrange and solve for the unknown quantity, a, which is the distance between Venus and the sun:

$$a = \sqrt[3]{(0.616)^2}$$
$$a = 0.724 \text{ AU}$$

Kepler's third law also allows us to solve for a planet's orbital period in years given its distance from the sun. For example, we would follow the steps below to solve for the orbital period of Mars given that its average distance from the sun is 1.52 AU.

$$P^2 = a^3$$
$$P^2 = (1.52)^3$$
$$P^2 = 3.51$$
$$P = \sqrt{3.51}$$
$$P = 1.87 \text{ years}$$

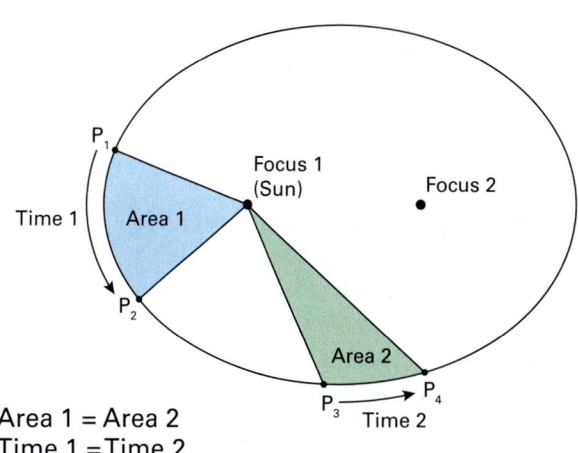

Area 1 = Area 2
Time 1 = Time 2

FIGURE 22-10 ▲
Kepler's first two laws of planetary motion are demonstrated in one diagram. The sun is at the focus of a planet's elliptical orbit. The planet moves along its orbit from point P_1 to point P_2 in the same amount of time it takes the planet to move from point P_3 to point P_4. The orbital paths define polygons of equal area for equal orbital time. This results in faster orbital speed when the planet is nearer the sun and slower orbital speed when the planet is farther from the sun.

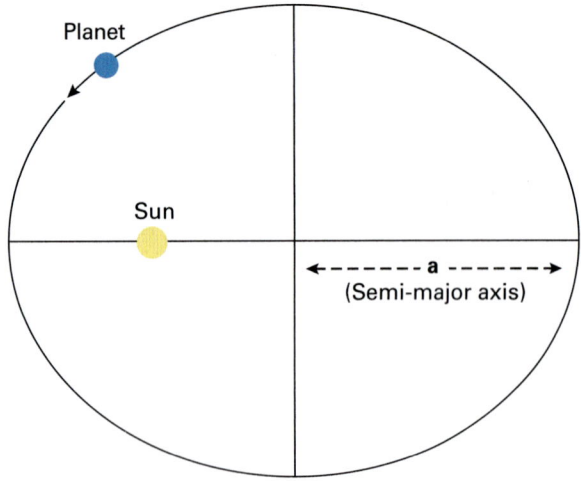

FIGURE 22-11
Kepler's third law of planetary motion states that the square of a planet's orbital period equals the cube of the semi-major axis of the planet's orbit (a).

Kepler went on to analyze more mathematical aspects of planetary orbits. For example, he examined relationships between various ellipse parameters and the period of a planet's orbit (the time required to complete one revolution around the sun). He found that the square of the orbital period, in Earth years, is directly proportional to the cube of the semi-major axis of a planet's orbit in **astronomical units** (AU) (Figure 22-11). An astronomical unit is the average distance between Earth and the sun. This relationship, known as Kepler's third law, held true for all planets that Kepler could test using Tycho's data.

checkpoint What are the two basic characteristics of planetary motion confirmed by Kepler's laws?

Galileo Galilei

Galileo Galilei was an Italian mathematician, astronomer, and physicist who made so many contributions to science that he is often called the father of modern science. Perhaps most significant was his realization that the laws of nature must be understood through observation, experimentation, and mathematical analysis. This concept freed scientists from the confines of Aristotelian dogma.

Recall that Aristotle's geocentric theory was based on two observations and one philosophical premise:

- If Earth were moving, people should fall off, but they do not.
- Aristotle was unable to observe parallax shift of the stars.
- The gods would create only an unblemished, symmetrical universe.

Galileo's experiments and observations showed that Aristotle's model was incorrect, leading him to support Copernicus's heliocentric model.

Galileo studied the motion of balls rolling across a smooth marble floor, and legend tells us he also dropped objects from the leaning Tower of Pisa. He organized the results of these experiments into laws of motion that were later expanded and quantified by Isaac Newton. One of these laws states that "an object at rest remains at rest and an object in uniform motion remains in uniform motion until forced to change." This corresponds to Newton's first law of motion, the law of **inertia**. Inertia is the tendency of an object to resist a change in motion.

According to the law of inertia, if Earth were in uniform motion, a person on its surface would be in uniform motion along with it. The person therefore would travel with Earth and would not fall off and be left behind, as Aristotle had assumed. In fact, the person could not even feel the motion.

Galileo built his first telescope in 1609 and turned it to the heavens soon afterward. He learned that the hazy white line across the sky called the Milky Way was not a cloud of light as Aristotle had proposed but a vast collection of individual stars. Next, he trained his telescope at the moon and saw hills, mountains, giant craters, and broad, flat regions on its surface, which he thought were seas. Looking at the sun, Galileo recorded dark regions, called sunspots, that appeared and then vanished.

Although these discoveries had no direct bearing on the controversy of a geocentric versus a heliocentric universe, they were important because they led Galileo to question Aristotle's views. The prevailing scientific opinion at the time was that if the Milky Way were a collection of stars, Aristotle would have known about it. Furthermore, Galileo's observations of the sun and the moon did not agree with Aristotle's philosophical assumptions that the heavenly bodies were perfectly homogeneous and unblemished. Galileo reasoned that if Aristotle was wrong about the structures of the Milky Way, the sun, and the moon, perhaps he was also wrong about the motions of these celestial bodies.

When Galileo studied Jupiter, he saw four moons orbiting the giant planet. Today we know that Jupiter has more than 60 moons, but only four are large enough to have been seen with

FIGURE 22-12

Two diagrams show the mechanics of Venus's phases. (A) Ptolemy's geocentric model showed Venus moving on an epicycle in its orbit around Earth (white dashed circle). This required that the planet would always be seen in a crescent phase, even at its furthest separation from the sun. (B) The heliocentric model explained how Venus could be seen in the different phases in which it is actually observed.

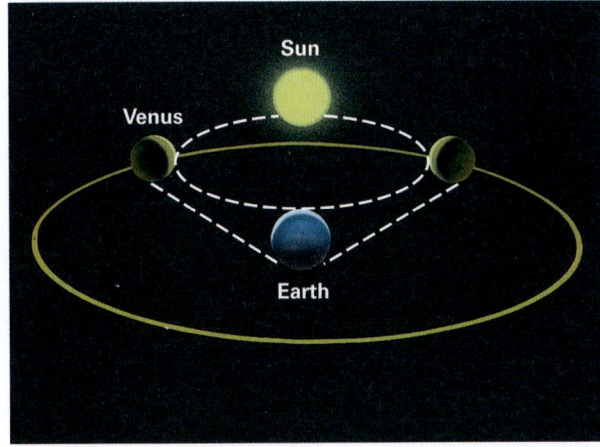

A

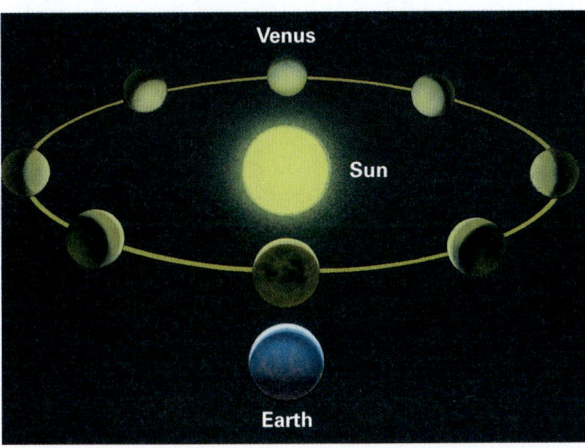

B

Galileo's telescope. According to the geocentric model, every celestial body orbits Earth. However, Jupiter's moons clearly orbited Jupiter, not Earth. The contradiction increased Galileo's doubt about Aristotelian theories.

Finally, Galileo observed that the planet Venus passes through phases as the moon does. Such cyclical phases could not be readily explained in the geocentric model, even with Ptolemy's modifications (Figure 22-12). As a result of his observations, Galileo proposed that the sun is the center of the solar system and the planets orbit around it.

checkpoint What two of Galileo's observations contradicted the geocentric model of the universe?

22.2 ASSESSMENT

1. **Explain** Why was Galileo's observations of Venus's different phases an indication that Venus does not orbit Earth?
2. **Define** What is parallax? Why didn't astronomers in Aristotle's time see the parallax of stars?
3. **Describe** As Earth and the other planets orbit around the sun, how does another planet appear to move in the sky when Earth catches up to it?

Critical Thinking

4. **Synthesize** Explain why Galileo's observations of the Milky Way, the sun, and the moon led him to question Aristotle's ideas.

22.3 THE ROLE OF GRAVITY

Core Ideas and Skills

- Explain Newton's three laws of motion and the universal law of gravitation.
- Define the role of gravity and inertia in the motion of Earth, the moon, and other objects in the solar system.
- Describe what causes lunar phases and solar and lunar eclipses.

KEY TERMS

barycenter	gibbous moon
rotation	full moon
revolution	waning
precession	solar eclipse
nutation	corona
new moon	umbra
crescent moon	penumbra
waxing	lunar eclipse

Newton's Law of Universal Gravitation

Galileo successfully described the motions in the solar system, but he never addressed the question of why the planets orbit the sun rather than fly off in straight lines into space. Aristotle had observed that an arrow shot from a bow flies in a straight

FIGURE 22-13
ON ASSIGNMENT The complete sequence of the 2017 Great American Solar Eclipse is shown in a composite photograph taken by National Geographic photographer Babak Tafreshi as the sun crossed the sky above Lake Solitude in Grand Teton National Park, Wyoming.

line but celestial bodies move in curved paths. He reasoned that arrows have an essence that compels them to move in straight lines. He also reasoned that planets and stars have an essence that compels them to move in circles. By the 1600s, Renaissance scholars such as Isaac Newton had begun questioning Aristotle's reasoning.

Isaac Newton was born in 1643, the year Galileo died. During his lifetime, Newton made important contributions to physics and developed calculus. A popular legend tells that Newton was sitting under an apple tree one day when an apple fell on his head and—presto—he discovered gravity. Of course, people knew that unsupported objects fall to the ground long before Newton was born. But he was the first to recognize that gravity is a universal force that governs all objects, including a falling apple and a flying arrow. Newton was also the first to be able to use the idea of gravity to explain the planetary motions described by Kepler and Galileo.

At the heart of Newton's work is the idea that gravity is the force that holds the universe together.

He determined that one object exerts a gravitational force on another object. Furthermore, Newton found the gravitational force that one object exerts on another is the same as the gravitational force that the second object exerts on the first. For example, the sun exerts a gravitational force on Earth that is equal in magnitude but opposite in direction to the gravitational force that Earth exerts on the sun. Using mathematical reasoning and astronomical data collected by others, Newton showed that the strength of the gravitational force between any two objects is directly proportional to the product of the masses of each object and inversely proportional to the square of the distance between them. This relationship is known as the law of universal gravitation. The variables are defined in Figure 22-14.

$$F_1 = F_2 = (6.67 \times 10^{-11} \frac{m^3}{kg\ sec^2}) \times \frac{m_1 \times m_2}{r^2}$$

Newton's law of universal gravitation explains why planets remain locked in their orbits around the sun instead of flying off into space. This same law also explains why a planet's orbit does not simply involve the planet revolving around a stationary sun. Since the sun exerts a gravitational force on a planet, and a planet exerts a gravitational force on the sun, both objects orbit around their common center of mass. This point is called the **barycenter**.

Think of a barycenter as the location where a seesaw would be perfectly balanced between two objects. As shown in Figure 22-15, if two objects were equal in mass, the barycenter would be in the middle, equidistant from the two objects. In the case of Earth and the sun, the barycenter is very close to the sun's center because the sun's mass is so much greater than Earth's mass. The

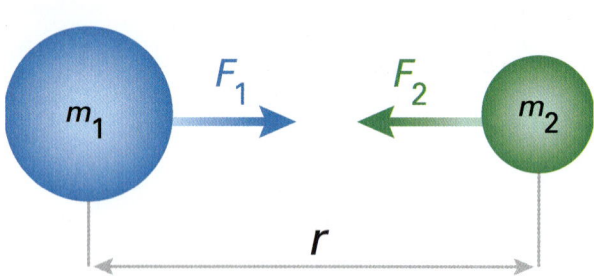

FIGURE 22-14
This diagram depicts Newton's law of universal gravitation. One object has mass represented by m_1; the other object's mass is represented by m_2. The distance between the two objects is given as r, and the number 6.67×10^{-11} is the gravitational constant, often denoted as G.

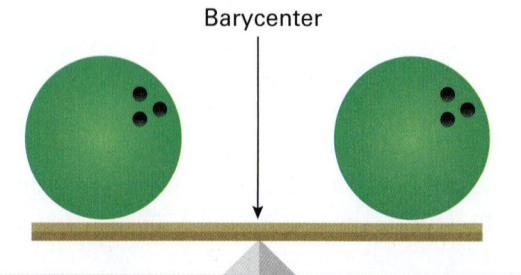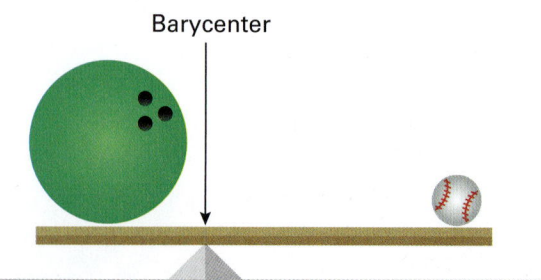

FIGURE 22-15
The barycenter is determined by the relative masses of two objects. If the masses are the same, the barycenter is equidistant between the objects. If the masses are different, the barycenter lies closer to the more massive object.

DATA ANALYSIS Quantifying the Effect of Gravity

TABLE 22-1 Data for Planets in the Solar System

Planet	Mass (10^{24} kg)	Distance from Sun (10^6 km)	Orbital period (Earth days)	Orbital velocity (km/s)
Mercury	0.33	57.9	88	47.4
Venus	4.87	108.2	224.7	35
Earth	5.97	149.6	365.2	29.8
Mars	0.642	227.9	687	24.1
Jupiter	1,898	778.6	4,331	13.1
Saturn	568	1,433.5	10,747	9.7
Uranus	86.8	2,872.5	30,589	6.8
Neptune	102	4,495.1	59,800	5.4

Source: NASA

Isaac Newton demonstrated that gravity is the force between any two objects with mass. Gravity plays an important role in the structure and motions of objects in the solar system. Consider the planetary properties summarized below.

Computational Thinking

1. **Represent Data** Use the data in the table to construct a new table showing each of the values as a ratio to Earth's values. (Note that in your table, all values for Earth will be equal to 1.) Using the data from your table, prepare three separate graphs showing the planets' properties with respect to distance from the sun. Label your graphs, including units.

2. **Pattern Recognition** Examine your graphs. Describe any trends in mass, orbital period, or orbital velocity with increasing distance from the sun.

Critical Thinking

3. **Generalize** How do you think the force of gravity may be responsible for the trends you described in your answer to question 2 above? Discuss each trend you observed.

4. **Apply** Looking only at the trends in planets Jupiter, Saturn, Uranus, and Neptune, which properties are most strongly correlated with distance from the sun? Why do you think these trends exist?

Earth-sun barycenter causes a very slight wobble in the orbital motions of Earth and the sun. Since Jupiter is so massive, a greater wobble occurs due to the Jupiter-sun barycenter, which lies outside the surface of the sun (Figure 22-16).

FIGURE 22-16
Go online to view an illustration of the barycenter between Jupiter and the sun.

According to the laws of motion introduced by Galileo and expanded by Newton, a moving body travels in a straight path unless it is acted on by an outside force. Just as a ball rolls in a straight line unless it is forced to change direction, a planet moves in a straight line unless a force is exerted on it. The gravitational attraction between the sun and a planet forces the planet to change direction and move in an elliptical orbit. Gravity is ultimately the "glue" of the universe, affecting the motions of all celestial bodies.

checkpoint What physical quantities affect the force of gravity between two objects?

MINILAB Simulating Gravity

Materials (per group):
large quilting hoop (58 cm)
sheet of flexible plastic (75 × 75 cm)
6 small marbles of the same size
ball of modeling clay, 5 cm in diameter
foam ball, 5 cm in diameter
timer
4–6 books, about the same thickness

Procedure

1. Create a 2-D model of space by securing a sheet of flexible plastic in a quilting ring. Place the ring on stacks of books so that the plastic is suspended above the table.

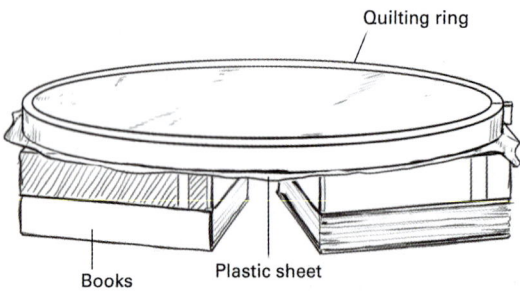

2. Drop one marble onto the plastic. Repeat, dropping the marble from the same height but several different locations relative to the ring. Record your observations after each trial.

3. Repeat step 2, this time dropping two marbles at once from different locations.

4. Repeat step 2, this time dropping the clay ball.

5. Repeat step 2 again, this time leaving the clay ball on the model and dropping a marble from different locations.

6. Repeat steps 4 and 5, this time using a foam ball instead of the clay ball.

Results and Analysis

1. **Explain** How was gravity modeled in this activity?
2. **Identify** Which object in your model exerted the greatest gravitational influence over the other objects as it encountered the plastic? Which exerted the smallest? Use evidence to support your answers.
3. **Conclude** What can you conclude from this activity about how mass and distance affect gravitational force? Explain your reasoning.

Critical Thinking

4. **Generalize** Why is mass and not size important in discussions about gravitational forces?
5. **Evaluate** Did this model have any limitations with respect to exploring ideas about orbiting planets?

Motion of Earth

By about 1700, astronomers knew that the sun is the center of our solar system and that planets revolve around it in elliptical orbits. In addition to revolving around the sun, the planets simultaneously spin on their axes. Earth spins approximately 365 times for each complete orbit around the sun. Each complete **rotation** of Earth represents one day. As Earth rotates about its axis, the sun, the moon, and the stars appear to move across the sky from east to west. We explained in Chapter 19 that Earth's axis is tilted and that this tilt, combined with Earth's orbit around the sun, produces the seasons.

Different stars and constellations are visible during different seasons. Earth's **revolution**, or orbit, around the sun is completed in approximately 365 days. This motion causes the seasonal change in our view of the night sky (Figure 22-17). Part of the sky is visible on a winter night when Earth is on one side of the sun, and a different part is visible on a summer night six months later.

FIGURE 22-17
Go online to view an illustration of the different stars and constellations that can be observed during seasonal changes.

More recent measurements have revealed several additional types of planetary motion. Earth's axis, which now points directly toward the North Star (Polaris), circles like that of a wobbling top. This circling, called **precession**, is completed once every 26,000 years (Figure 22-18). In 12,000 years, the axis will point toward Vega, and Vega will become the North Star. Because precession cycles over a 26,000-year period, Earth's axis will point toward Polaris again in thousands of years.

In addition, the moon's gravity pulls Earth slightly out of its orbit. This causes Earth's precessing axis to sway slightly, a motion called **nutation**. The

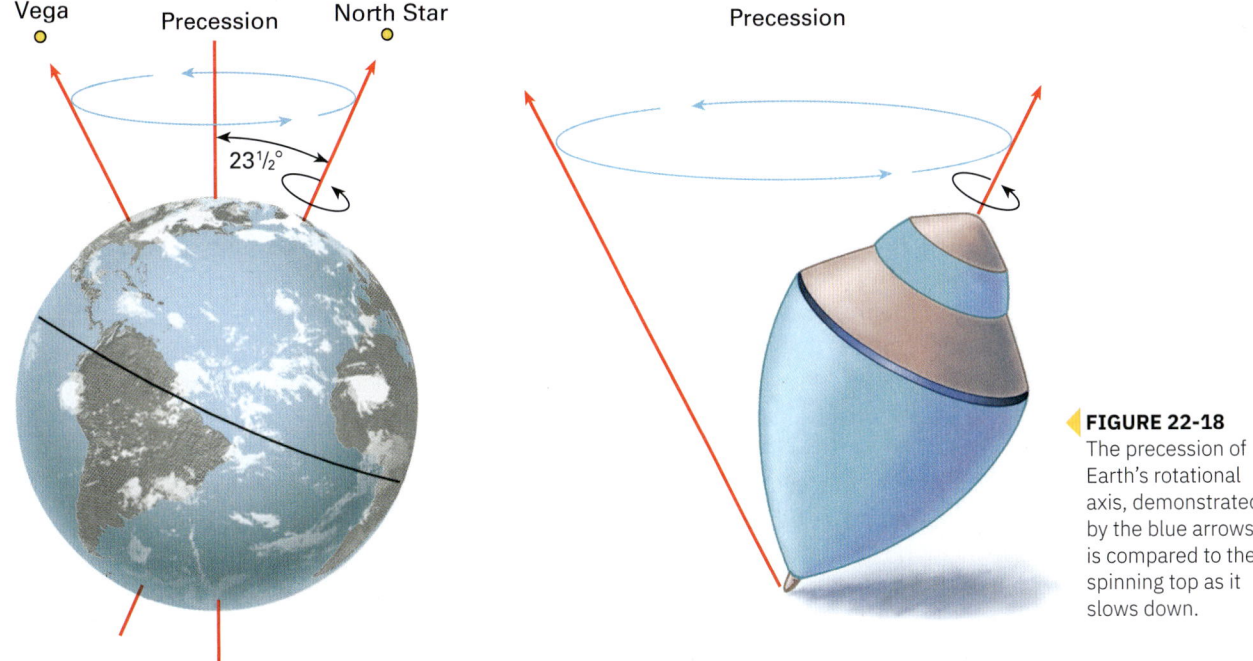

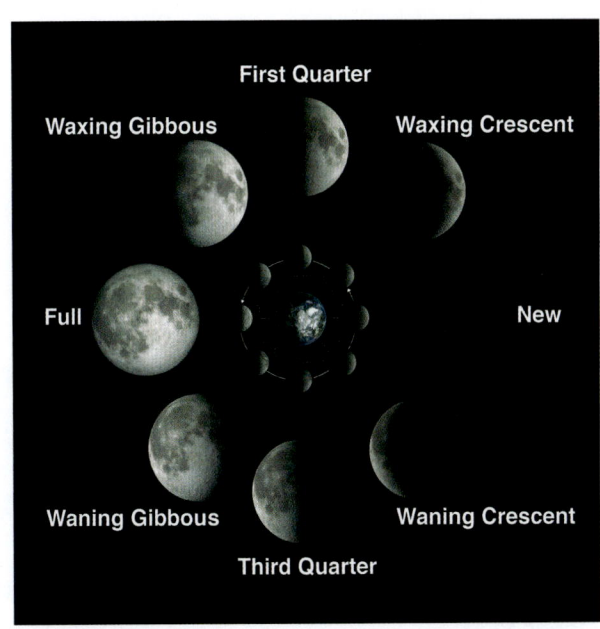

FIGURE 22-18 The precession of Earth's rotational axis, demonstrated by the blue arrows, is compared to the spinning top as it slows down.

largest component of Earth's nutation occurs with a period of about 18.6 years. Our sun, along with its solar system, orbits the center of the Milky Way galaxy. Traveling at a speed of 220 kilometers per second, the sun completes its orbit in about 200 million years. The entire Milky Way galaxy carries our sun and the planets through intergalactic space.

checkpoint Identify the five different motions of Earth and their periods.

Motion of the Moon

The gravitational attraction between the moon and Earth not only holds the two in orbit around each other but also affects the surfaces and interiors of both. In Chapter 16, you read that the moon's gravitation causes tides on Earth. In turn, Earth's gravitation pulls on the moon sufficiently to cause it to bulge. Earth's gravity attracts the bulge, so the side of the moon with the bulge always faces Earth. As a result, the moon rotates on its axis at the same rate at which it orbits Earth. Thus, we always see the same side of the moon. The moon's orbit about Earth is not circular. At its farthest distance, the moon is 406,700 kilometers (252,700 miles) from Earth. At its closest, the distance is 356,500 kilometers (221,500 miles).

To understand the phases of the moon, we must first recall that the moon does not emit its own light but reflects light from the sun. The half of the moon facing the sun is always bathed in sunlight, while the other half is always dark. The phases of the moon as we see them depend on how much of the moon's sunlit area is visible from Earth. In turn, this visible area depends on the relative positions of the sun, the moon, and Earth.

Figure 22-19 shows the lunar phases. When the moon is dark, it is called a **new moon**. The moon is dark because it is situated between Earth and the sun. The moon's unlit side faces us and we can't

FIGURE 22-19
This figure combines photographs of the moon in different phases with a diagram showing the relative positions of the moon and Earth during the 28-day cycle.

LESSON 22.3   725

see it. A few days after a new moon, a thin **crescent moon** appears. The crescent grows, and the moon is described as **waxing** or becoming full. As the moon waxes, it changes from a crescent moon to a **gibbous moon**, in which the moon is over half-full and only a thin crescent of the moon remains dark. About 14 to 15 days after a new moon, the moon appears circular and is a **full moon**. At this point, Earth is situated between the moon and the sun, so the illuminated side of the moon is facing directly toward Earth. A few evenings later, part of the disk is darkened. As the days progress, the visible portion shrinks and the moon is said to be **waning**. Eventually, only a tiny curved sliver, the waning crescent, is left. After a total cycle of about 29.5 Earth days, the moon is dark because it has returned to a new-moon position. Earth's orbit around the sun forms an elliptical plane, with the sun in the same plane and near its center. In a similar way, the moon's orbit around Earth describes another plane with Earth at its center. But the plane of the moon's orbit is tilted 5.2° with respect to Earth's orbital plane. As a result, the moon is not usually in the same plane as that of Earth and the sun (Figure 22-20A). During a new moon, the moon's shadow misses Earth; at a full moon, Earth's shadow misses the moon, as shown in Figure 22-20B.

FIGURE 22-20
Go online to view illustrations of the position of the sun and Earth during the moon's orbit.

However, on rare occasions, the new moon passes through the Earth–sun orbital plane. At these times, the moon passes directly between Earth and the sun and is in the new moon phase. When this happens, the moon's shadow falls on Earth, producing a **solar eclipse** (Figure 22-21). As the moon moves in front of the sun, an eerie darkness descends, and Earth becomes still and quiet. Birds return to their nests and stop singing.

During a total eclipse, the moon blocks out the entire surface of the sun, but the outer solar atmosphere, or **corona**—normally invisible because of the sun's brilliance—appears as a halo around the black moon. Due to the relative distances between sun, the moon, and Earth and their respective sizes, the moon's shadow is only a narrow band on Earth. The band where the sun is totally eclipsed, called the **umbra**, is never wider than 275 kilometers. In the **penumbra**, a wider band outside the umbra, only a portion of the sun is eclipsed. During a partial eclipse of the sun, the sky loses some of its brilliance but does not become dark. Viewed through a dark filter, a semicircular shadow cuts across the sun.

The moon's distance from Earth varies during its orbit, causing the moon to appear slightly smaller than the sun in the sky. Under these conditions, the moon blocks out all but a thin outer ring of the sun's disk. These eclipses are called annular eclipses. At present, roughly one third of all solar eclipses are annular but over time, these will become more commonplace. The moon is spinning away from Earth at a rate of 3.8 centimeters (1.5 inches) per year, about the same speed at which our fingernails grow.

If the moon passes through the Earth–sun orbital plane when it is full, then Earth lies directly between the sun and the moon. At these times Earth's shadow falls on the moon and the full moon temporarily darkens to produce a **lunar eclipse**. Lunar eclipses are more common and last longer than solar eclipses because Earth is larger than the moon and therefore its shadow is more likely to cover the entire lunar surface. A lunar eclipse can last a few hours.

checkpoint What is the difference between a lunar eclipse and a solar eclipse?

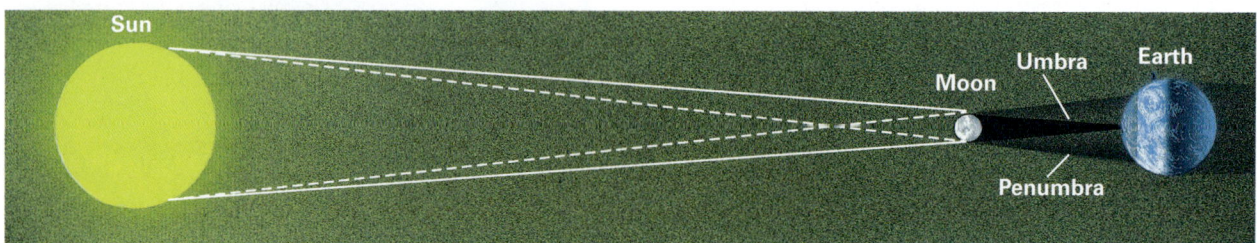

FIGURE 22-21
The mechanics of a solar eclipse require that the sun, the moon, and Earth be aligned along the same plane. The darkest part of the moon's shadow is the umbra, with the penumbra being the outer edges of the shadow.

22.3 ASSESSMENT

1. **Explain** Why does the precession of Earth affect how we define "north"?
2. **Define** What is a barycenter and why is it important for the orbit of planets?
3. **Describe** How does the appearance of the moon compare when it is waxing crescent as opposed to waning crescent? What other pairs of phases display a similar pattern?
4. **Distinguish** Identify the position of the moon in the sky relative to the sun during the four major phases: new, first quarter, full, and third quarter.

Computational Thinking

5. **Abstract Information** How would solar and lunar eclipses be different if the moon were at a distance twice as far from Earth as it is currently?

22.4 TOOLS OF MODERN ASTRONOMY

Core Ideas and Skills
- Explain the importance of telescopes to astronomy.
- Describe the advantages and disadvantages of large, ground-based telescopes.
- Describe the advantages and disadvantages of space-based telescopes.
- Summarize the information that can be obtained from the light given off by astronomical objects.

KEY TERMS

resolution	absorption spectrum
refracting telescope	emission spectrum
reflecting telescope	Doppler effect
light pollution	redshift
spectrum	blueshift

Once astronomers understood the relative motions of the sun, the moon, and the planets, they began to ask questions about the nature and composition of these bodies and to probe more deeply into space to study stars and other objects in the universe. How do we gather data about such distant objects?

Telescopes

If you stand outside at night and look at a speck of light in the sky, the information you receive is limited by several factors. For one, your eye detects only visible light, which is just one-millionth of one percent of the electromagnetic spectrum. Thus, more than 99.99 percent of the spectrum is invisible to the naked eye. In addition, the naked eye collects little light, and you may not see faint or distant objects at all. Your eye also has poor **resolution**, the degree to which details are distinguishable in an image. It may see one point of light when two actually exist. Finally, the light you see has been distorted by Earth's atmosphere. Modern astronomers attempt to overcome these difficulties with telescopes and other instruments.

A telescope is a device that collects light from a wide area and then focuses it where it can be detected. Detection devices may be simple, such as your eye, or complex, such as telescopes that stay trained on one section of the sky for several hours, collecting enough light to form an image from a weak signal. A telescope may collect light for several hours, thus seeing fainter objects than could be seen with the eye.

Galileo and many other early astronomers used a **refracting telescope**, which utilizes two lenses. The first, called the objective lens, collects light from a distant object. The second is a small magnifying lens, called the eyepiece. Light bends, or refracts, when it passes through the curved surface of the objective lens (Figure 22-22A). The bent light rays converge on the focus, forming an image of a distant object. The eyepiece then magnifies the image.

The problem with refracting telescopes is that different colors in the spectrum refract by different amounts. Therefore, if you focus the telescope to collect blue light sharply, red light will be fuzzy. As a result, most modern optical telescopes are **reflecting telescopes**. These telescopes collect light with a large curved mirror and reflect it to an eyepiece (Figure 22-22B). Modern telescopes are outfitted with both a camera and electronic detectors at the eyepiece.

As cities and suburban sprawl have grown, electric lights associated with them compete with the lights from distant stars, making them much less visible than in the countryside far from electric lights. This human-generated background light is called **light pollution,** and it reduces the distances and clarity with which telescopes operate.

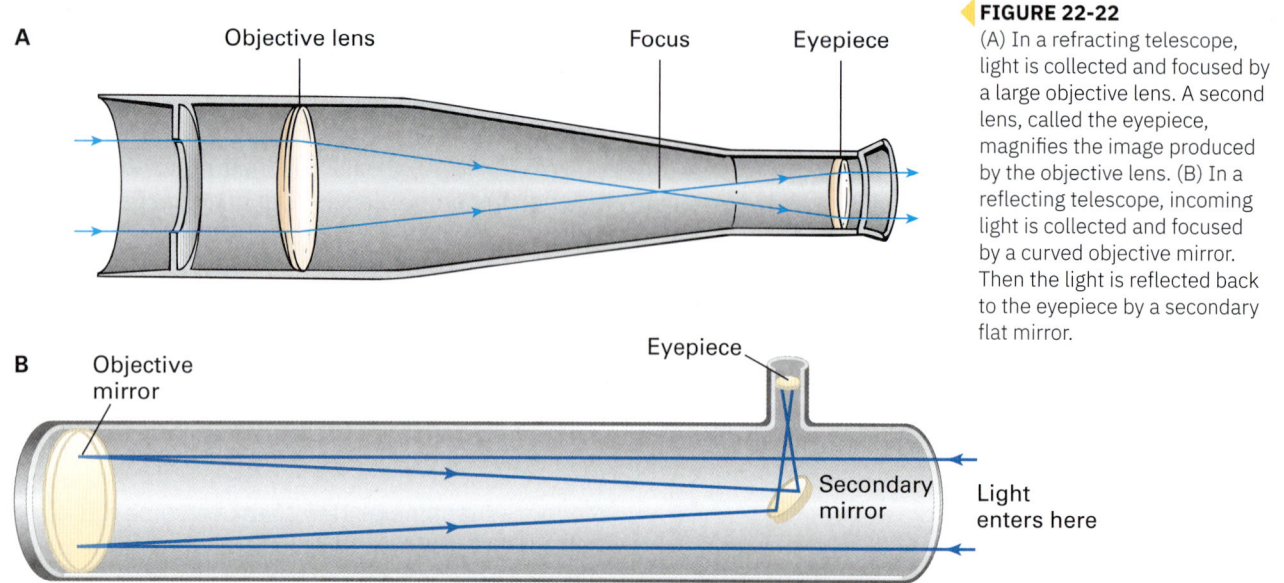

FIGURE 22-22
(A) In a refracting telescope, light is collected and focused by a large objective lens. A second lens, called the eyepiece, magnifies the image produced by the objective lens. (B) In a reflecting telescope, incoming light is collected and focused by a curved objective mirror. Then the light is reflected back to the eyepiece by a secondary flat mirror.

FIGURE 22-23
The Hubble Space Telescope orbits above Earth after repairs by astronauts on the space shuttle Columbia in 2002.

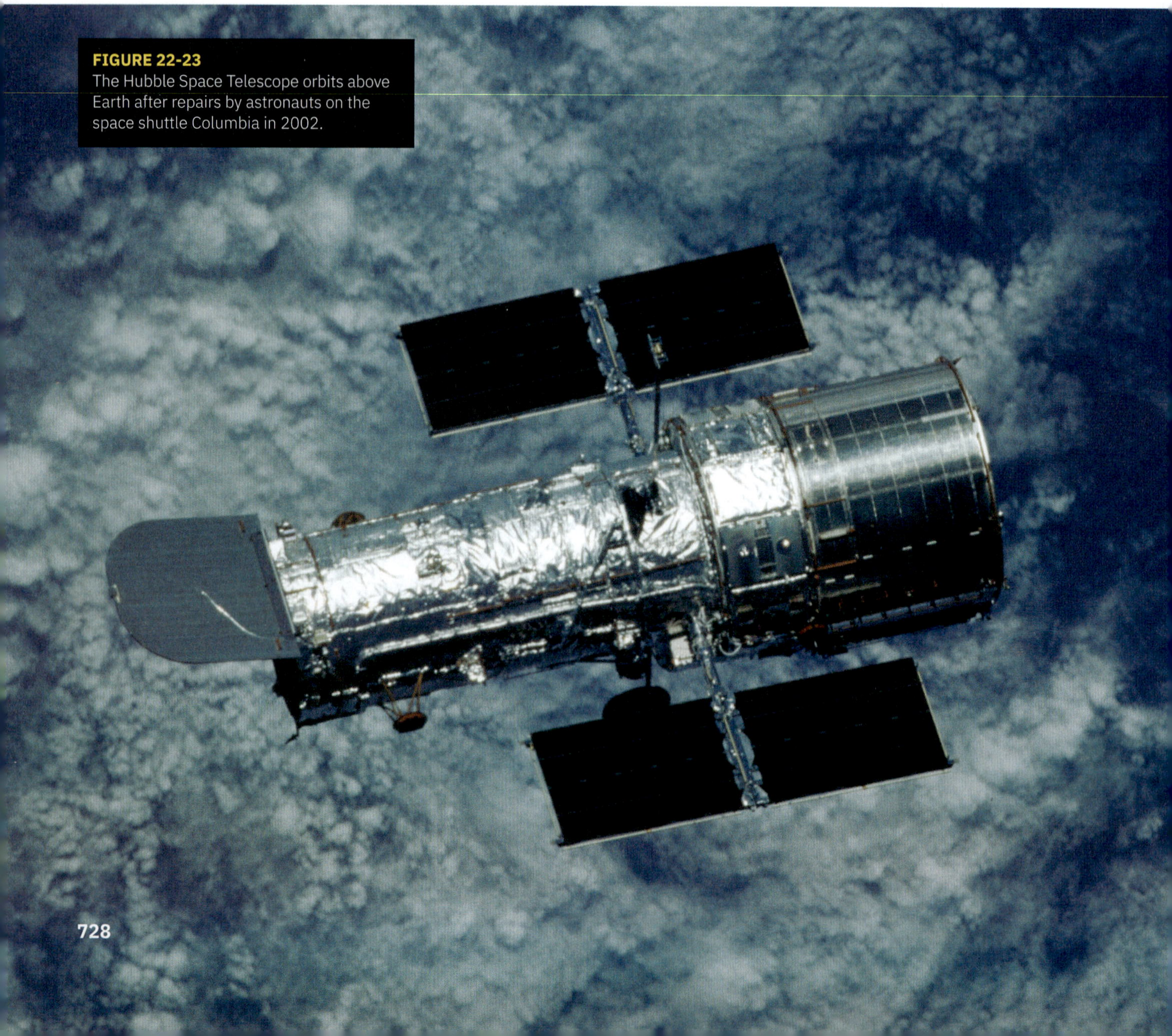

A

B

FIGURE 22-24
(A) The ground-based telescope image of galaxy M82 shows much less detail than the Hubble Space Telescope image of the same galaxy (B).

In the past few decades, astronomers have built more powerful telescopes, both on land and in space. In 1990 the Hubble Space Telescope (HST) was launched into orbit around Earth (Figure 22-23). In the vacuum of space, the HST isn't adversely affected by either light pollution or atmospheric interference. The Hubble's high-resolution images have altered our understanding of many celestial bodies (Figure 22-24). The HST is outfitted with sensors to collect visible light and radiation in other portions of the spectrum.

Projects such as Hubble are enormously expensive, but with developments in technology, astronomers can also build increasingly powerful observatories. Because a mirror much larger than 600 centimeters sags under its own weight, recent telescopes use an array of smaller mirrors to increase the mirror area. The mirrors are focused by computer, meaning that many mirrors can be focused and manipulated at once. The James Webb Space Telescope, expected to be launched in 2021, will replace the Hubble Space Telescope as the most powerful optical telescope (Figure 22-25).

FIGURE 22-25
Go online to view the team of scientists and engineers working on the James Webb Space Telescope.

checkpoint What are the main challenges of using a telescope on the ground versus one in space?

Learning from Light

Visible light is only a small portion of the electromagnetic spectrum. The wavelengths of electromagnetic radiation emitted by a star are determined by several factors, including the types of nuclear reactions that occur in the star, its chemical composition, and its temperature.

In recent years, astronomers have enhanced our knowledge of stars and other objects in space by studying many different wavelengths, from low-energy radio and infrared signals to high-energy gamma rays and x-rays. The telescopes used in these studies often do not look like conventional optical telescopes but rather are tailored to the characteristics of the wavelength being studied.

If light passes through a prism, it separates into a **spectrum**, an ordered array of colors (Figure 22-26). A rainbow is such an array, with white sunlight separated into its individual colors. Each color is formed by a band of wavelengths.

FIGURE 22-26
Go online to view how a prism works.

As light passes from the hot interior of a star through the cooler, outer layers, some wavelengths are selectively absorbed by atoms in the star's outer atmosphere. Therefore, in the continuous spectrum of starlight (Figure 22-27A), dark lines cross the band of colors. This is called an **absorption spectrum** (Figure 22-27C). Each dark line represents a wavelength that is absorbed by atoms of a particular element. Thus, an absorption spectrum enables us to determine the chemical composition of a star. This type of analysis is so effective that astronomers discovered helium in the sun 27 years before chemists detected helium on Earth. Because the brightness and width of an atom's spectral features change with temperature

LESSON 22.4

DATA ANALYSIS Evolution of the Telescope

The first telescopes had small apertures that could collect only a small amount of light. Only the very brightest and closest celestial bodies could be viewed. To see faint objects better, scientists began increasing the size of the aperture to increase the light-collecting area. At the time of Galileo, the aperture was about the size of a dime. The graph in Figure 22-30 illustrates how the light-collecting area of optical telescopes has changed over time. A larger light-collecting area increases the light sensitivity, so we can see fainter objects.

Light sensitivity is only part of the problem that telescope designers have had to overcome. Not only is it necessary to collect more light to see faint objects, but it is also necessary to be able to distinguish one distant object from another. This is a matter of resolution. Telescopes need to have adequate resolving power, also called angular resolution, so that objects can be seen as distinct objects without appearing blurry and unclear. Figure 22-31 shows how the angular resolution of optical telescopes has changed over time. Note that values decrease as you move up the vertical axis.

◀ FIGURE 22-30 The maximum aperture of telescopes has increased over time.

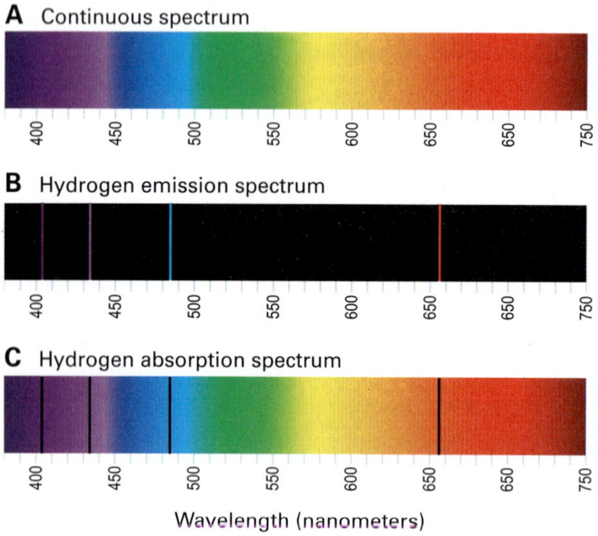

Source: Astronomical Society of the Pacific

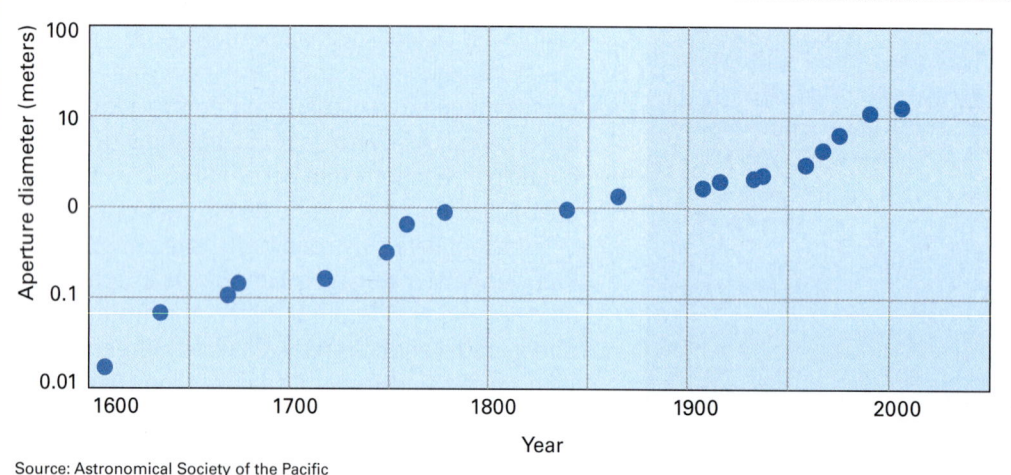

A Continuous spectrum

B Hydrogen emission spectrum

C Hydrogen absorption spectrum

Wavelength (nanometers)

FIGURE 22-27 ▲
(A) The continuous spectrum of visible light ranges from violet (short wavelength) to red (long wavelength). (B) The emission spectrum of hydrogen shows a few narrow emission lines at specific wavelengths. (C) The absorption spectrum of hydrogen shows the same few narrow lines missing from a continuous spectrum. In each spectrum, wavelengths are given in nanometers (nm) along the horizontal axis.

and pressure, spectra can also be used to determine surface temperatures and pressures of stars.

Whenever an atom absorbs radiation, it must eventually reemit it. Such an **emission spectrum** (Figure 22-27B) is often hard to detect against the bright background of starlight, but these spectra can be seen along the outer edges of some stars and in large clouds of dust and gas in space.

Have you ever stood by a train track to listen to a train speeding by, blowing its whistle? As it approaches, the pitch of the whistle sounds higher than usual. Then after it passes, the pitch lowers. This change, called the **Doppler effect**, was first explained by Austian physicist Johann Christian Doppler in 1842. He explained this change for both light waves and sound waves.

A stationary object remains in the center of the waves it generates. The waves from a moving object crowd each other in the direction of the object's motion. The object, in effect, is catching up with its own waves (Figure 22-28). If the object is moving toward you, you receive more waves

1. **Compare** How would you compare the trends in sensitivity and resolution that have been made since telescopes were first introduced in the early 1600s?
2. **Predict** What are the potential benefits to improving telescope technology?

Computational Thinking

3. **Analyze Data** How do the data in Figure 22-31 show an overall increase in resolving power of telescopes with time?
4. **Recognize Patterns** Based on the trends in the graphs above, do you think telescope technology will continue to improve? Explain your reasoning.

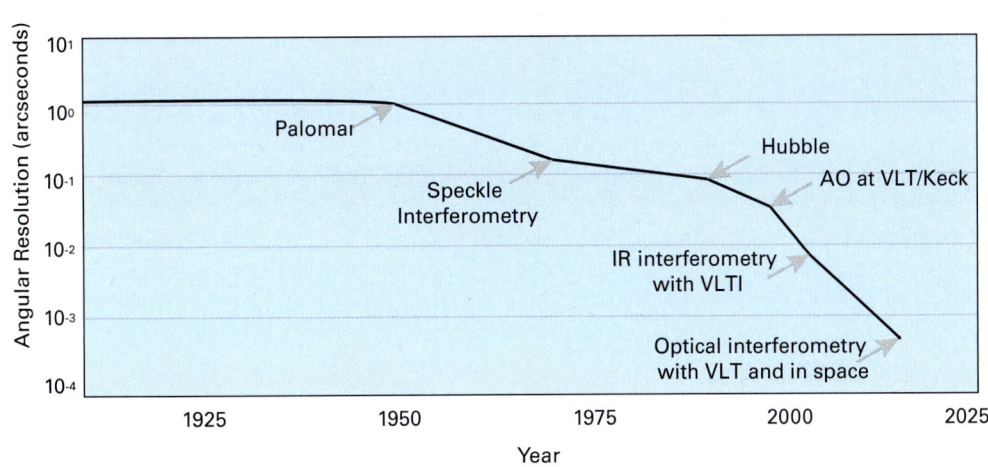

FIGURE 22-31 The angular resolution (smallest detail) of telescopes has increased over time.

per second (higher frequency) than you would if it were stationary. If it is moving away from you, you receive fewer waves per second (lower frequency).

FIGURE 22-28
Go online to view an illustration of the Doppler effect.

In the same way, the frequency of light waves changes with relative motion. The Doppler effect causes light from an object moving away from Earth to reach us at a lower frequency than it had when it was emitted. Lower-frequency light is closer to the red end of the spectrum. Thus, a Doppler shift to lower frequency is called a **redshift** (Figure 22-29). Alternatively, light from an object traveling toward us reaches Earth at higher frequency, called a **blueshift**. Using these principles, astronomers measure the relative velocities of stars, galaxies, and other celestial objects.

FIGURE 22-29
Go online to view an illustration of the Doppler shift.

checkpoint What are four pieces of information about a star that astronomers can learn by studying its absorption spectrum?

22.4 ASSESSMENT

1. **Explain** Why do astronomers prefer to use telescopes in space, and what challenges still exist with this technology?
2. **Define** What is a reflecting telescope and how is it different from a refracting telescope?
3. **Describe** What conditions cause light pollution?

Critical Thinking

4. **Apply** How would the color of light from a star moving horizontally relative to the line of sight change?

TYING IT ALL TOGETHER
GRAVITY IN EXOPLANETARY SYSTEMS

In this chapter, you considered how human understanding of the universe has evolved over the course of history. Since Galileo's time, telescopes have become exponentially more powerful, capturing views of the universe that he likely would have found difficult to imagine. Today we explore the universe using telescopes in Earth orbit and probes that travel far from our planet, collecting new data and images and sending them back to Earth.

NASA's Kepler space telescope is one example. Launched in 2009 and in operation until 2018, Kepler helped scientists identify distant exoplanets, planets that orbit stars other than Earth's sun. To detect exoplanets, Kepler used detectors that could continuously collect visual images of about 150,000 stars. These detectors could sense subtle changes in the light transmitted from a star whenever a planet passed between the star and the detector.

The tables summarize data collected by the Kepler space telescope for the star named Kepler-90 (Table 22-2) and five of the eight planets orbiting it (Table 22-3). Use the data to answer the questions.

TABLE 22-2 Characteristics of Kepler-90

Star	Mass Ratio compared to Sun	Radius Ratio compared to Sun	Surface Temperature
Kepler-90	1.2 ± 0.1	1.2 ± 0.1	6,080K (Sun 5,778K)

Source: NASA

TABLE 22-3 Characteristics of Kepler-90 Planets

Planet	Planet characteristics	Planet mass compared to Earth	Radius compared to Earth	Semi-Major axis (A.U.)	Planet orbital period (days)
Kepler-90 h	gaseous	<381	11.3	1.01	331
Kepler-90 g	gaseous	<254	8.1	0.71	211
Kepler-90 f	rocky	Not available	2.9	0.48	125
Kepler-90 c	rocky	Not available	1.3	0.074	7
Kepler-90 i	rocky	Not available	1.3	Not available	14.4

Source: NASA

1. How does Kepler-90 compare to our sun? What predictions can you make about the Kepler-90 solar system based on your comparison?

2. Based on the data provided, how does the entire solar system of Kepler-90 compare to our solar system?

3. Research the semi-major axis values (a measure of the average distance between a planet and its star) for planets in our solar system and compare them to those of the planets in the Kepler-90 system. Do the same to compare their orbital periods. What do you notice?

4. Which, if any, of the exoplanets in the Kepler-90 system might have conditions that would make them habitable by humans? Explain.

5. Assuming that the exoplanets Kepler-90 c and Kepler-90 i have the same density, how would their masses compare? Given the other data available for these two exoplanets, how do you think the semi-major axis of Kepler-90 i would compare to that of Kepler-90 c? Explain your reasoning.

CHAPTER 22 SUMMARY

22.1 PATTERNS IN THE SKY

- Constellations are groups of stars that seem to form a pattern when viewed from Earth and serve as points of reference for the positions of objects in the night sky.
- For most of the year, planets appear to drift eastward with respect to the stars, but sometimes they seem to reverse direction and drift westward. This apparent reverse movement is called retrograde motion.

22.2 EVOLVING MODELS OF THE UNIVERSE

- Aristotle proposed a geocentric, or Earth-centered, universe in which a stationary, central Earth is surrounded by celestial spheres that contain the sun, the moon, the planets, and the stars.
- Ptolemy modified the geocentric model to explain the retrograde motion of the planets. In Ptolemy's model, each planet moves in small circles within its larger orbit.
- Copernicus believed that the universe should operate in the simplest manner possible and showed that a heliocentric solar system best explains the movement of planets, including retrograde motion.
- Kepler calculated the elliptical orbits of the planets.
- Galileo used observation and experimentation to discredit the geocentric model and show that the planets revolve around the sun.

22.3 THE ROLE OF GRAVITY

- Newton proved that gravity holds the planets in elliptical orbits, keeping them from flying off into space.
- The revolution of the moon around Earth causes the phases of the moon.
- A lunar eclipse occurs when Earth lies directly between the sun and the moon.
- A solar eclipse occurs when the moon lies directly between the sun and Earth.

22.4 TOOLS OF MODERN ASTRONOMY

- Objects in space are studied with both optical telescopes and telescopes sensitive to invisible wavelengths of light.
- Instruments carried aloft by spacecraft eliminate interference by Earth's atmosphere and allow closer inspection of objects in the solar system.
- Emission and absorption spectra provide information about the chemical compositions and temperatures of stars and other objects.
- The Doppler effect causes light from an object moving away from Earth to reach us at a lower frequency than it had when it was emitted and light from an object moving toward Earth to have a higher frequency than it had when it was emitted.

CHAPTER 22 ASSESSMENT

Review Key Terms
Select the key term that best fits the definition. Not all terms will be used, and no term will be used more than once.

absorption spectrum	nutation
astronomical unit	parallax
barycenter	penumbra
blueshift	planet
celestial sphere	precession
constellation	redshift
corona	reflecting telescope
crescent moon	refracting telescope
Doppler effect	resolution
emission spectrum	retrograde motion
full moon	revolution
geocentric	rotation
gibbous moon	solar eclipse
heliocentric	spectrum
inertia	umbra
light pollution	waning
lunar eclipse	waxing
new moon	

1. the degree to which details are distinguishable in an image
2. the apparent change in position of an object due to change in position of the observer
3. the point in space representing the common center of mass for two objects and around which these objects orbit
4. the apparent reversal of direction of a planet's motion as viewed from Earth
5. the turning of a planet on its axis
6. a device that enables people to see distant images by collecting light with a large, curved mirror
7. a term referring to Earth as central in the universe
8. outer solar atmosphere that is normally not visible because of the sun's brilliance
9. tendency of an object to resist a change in motion
10. a group of stars that together form a pattern that remains constant
11. the orbiting of a celestial body around another celestial body
12. change in the frequency of light or sound waves as an object moves closer or farther away
13. a term referring to changes in the moon as visible portions of it disappear over time
14. a temporary darkening of the moon by Earth's shadow
15. the slow movement of the axis of a spinning body around another axis

Review Key Concepts
Answer each question on a separate sheet of paper to demonstrate your understanding of key concepts from the chapter.

16. What did ancient astronomers observe that allowed them to differentiate between planets and stars in the night sky?
17. Why is it that we see the same constellations that ancient astronomers observed?
18. How did Aristotle explain the motions of stars and planets in the night sky? Describe his model of the universe.
19. Explain how the concept of parallax led Aristotle and other ancient astronomers to conclude that Earth must be stationary. Explain the fault in their reasoning.
20. How did Kepler determine that planets orbit the sun in elliptical rather than circular orbits?
21. According to Newton's findings, which factors affect the amount of gravitational force between two bodies in space, and how do these factors affect the amount of gravitational force?
22. Why is one side of the moon never visible from Earth?
23. What is the arrangement of Earth, the moon, and the sun during a lunar eclipse, and what does a viewer on Earth observe during a lunar eclipse?
24. How are reflecting and refracting telescopes similar and how do they differ?
25. Compared to ground-based telescopes, what advantages does the orbiting Hubble Space Telescope have, and why?
26. How is spectroscopy used to determine the composition of a star?
27. What are some similarities and differences between a redshift and a blueshift?

Think Critically

Write a response to each question on a separate sheet of paper. Use concepts from the chapter to support your reasoning.

28. How can you use the motion of Earth to explain why we observe different constellations in the night sky during different seasons of the year?

29. Planets that are closer to the sun have faster orbital velocities than planets farther from the sun. How does this idea help explain why the planet Jupiter appears to have retrograde motion at certain times when viewed from Earth?

30. Galileo used his telescope to view Venus over many nights and observed that Venus went through phases like our moon. How do you think the apparent size of Venus from Earth might change as it proceeds through its phases? In what ways do these observations represent evidence in favor of the heliocentric model of our solar system?

31. A model of Jupiter's orbit shows it revolving around a stationary sun. Explain why this is a flawed model according to the law of universal gravitation.

32. What evidence would a scientist use to support the hypothesis that a certain star is moving away from Earth? Explain your reasoning.

PERFORMANCE TASK

Using Transit Data and Kepler's Laws to Make Predictions

Living on Earth, we have the unique opportunity to observe a type of eclipse known as a transit. A transit occurs when a planet crosses directly in front of its parent star. In our solar system, transits are only visible for planets that lie closer to the sun than Earth, so we can observe transits for Mercury and Venus, but not for other planets. Astronomers use transits as a method for detecting exoplanets—planets in other solar systems. Once they detect an exoplanet making a transit, scientists can determine the exoplanet's orbital period by measuring how much time elapses from one transit to the next.

In this performance task, you will use Kepler's third law to analyze the orbits of Mercury and Venus. You will then use both transit data and Kepler's third law to make predictions about the orbits of several exoplanets.

1. State Kepler's third law using both a mathematical equation and a written sentence.

2. The distance between Earth and the sun is 149.6 million km, or 1 AU. One Earth year is 365 days. Use this information and Kepler's third law to determine the orbital period of Mercury and Venus in both years and days.

3. Each graph shows a light curve illustrating how the transit of an exoplanet affects the observed pattern of light from its star. For each exoplanet, analyze its light curve to determine its orbital period in Earth days.

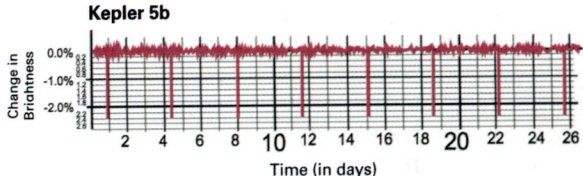

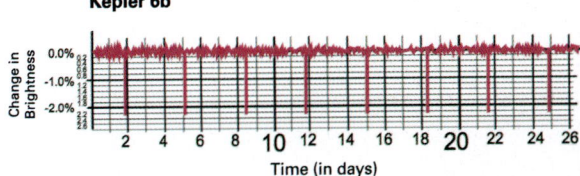

4. When applying Kepler's third law to exoplanets, a small adjustment must be made because the exoplanet's star is not the same mass as our sun. Use Kepler's third law to calculate the orbital period of Kepler-5b and Kepler-6b in both years and days. Use the adjustment equation $R^3 = T^2 \times M_{star}$ where M_{star} is the ratio of the star's mass to the sun's mass.

CHAPTER 23
OUR SOLAR SYSTEM

For thousands of years, people studied the stars and planets, but it wasn't until the past few hundred years that we found out what is really happening in space. More than 50 years ago, through tremendous feats of science, engineering, and perseverance, Americans landed on the moon. A human heart beat on the moon for the first time. We got our first glimpse of Earth as a whole planet in a vast sea of darkness and space. Since then there have been many missions to further explore the worlds in our solar system. One of the most recent, NASA's Juno spacecraft, studied Jupiter's weather, magnetic characteristics, and how the planet formed. Launched in 2011, Juno entered Jupiter's orbit in July 2016 and was scheduled to end its mission by crashing into Jupiter's clouds in 2021.

KEY QUESTIONS

23.1 What main characteristics distinguish the planets?

23.2 What are the characteristics of the terrestrial planets and Earth's moon?

23.3 How do the sizes, compositions, and atmospheres of the Jovian planets compare?

23.4 What makes each of the Jovian moons unique?

23.5 What are the characteristics of dwarf planets, minor planets, and comets?

With wildly swirling colors and textures, this image might look like an artist's rendition of Jupiter. It is actually a composite of four images taken by NASA's Juno spacecraft in May 2019 as the spacecraft orbited above the planet's northern hemisphere at altitudes as low as 8,000 kilometers above the clouds. For the first time, humans can observe close-up images of the tumultuous mix of gases and the intense, long-lasting storms occurring in Jupiter's atmosphere.

EXPLORERS AT WORK

ROVING THROUGH SPACE

WITH NATIONAL GEOGRAPHIC EXPLORER BETHANY EHLMANN

Dr. Bethany Ehlmann has always been drawn to space. She recalls being enamored with the work of astrophysicist Stephen Hawking at the age of six. By the time she entered college, Ehlmann knew she wanted to work in a space-related field. While pursing majors in Earth and planetary sciences and environmental studies, she became involved with the NASA mission team overseeing operations for two Mars rovers, Spirit and Opportunity, which were patrolling Mars for evidence of past water. That experience set the trajectory for Ehlmann's career. Today she is an accomplished planetary geologist who teaches at the California Institute of Technology and works at the Jet Propulsion Laboratory. She takes part in data analysis, mission development, and operations for various NASA missions, many of which target Mars.

For example, Dr. Ehlmann works with instrumentation aboard the Mars Reconnaissance Orbiter, a spacecraft that is circling the planet to image its landscape and atmosphere in order to detect evidence of past water and present weather. She also works with the Mars Science Laboratory, a car-size rover named Curiosity that landed on Mars in 2012. And she developed and tested instruments for use on NASA's Mars 2020 rover Perseverance. In addition to supporting Mars exploration, Ehlmann also works with teams exploring the dwarf planet Ceres, and she is helping with proposed missions to Venus, Mars's moons, asteroids, and Jupiter's moon Europa.

Ehlmann's research focuses on analyzing the mineralogy and chemistry of planetary surfaces, including Earth's surface. She makes use of everything from traditional field geologic tools such as rock hammers and hand lenses to space technology such as orbiters, rovers, landers, and spectrometers. She appreciates the technology that makes her work possible. "The rovers are our proxies on Mars," she says. "They are an extension of our curiosity and drive to explore." On Mars, Ehlmann is using Curiosity's instrumentation to study the planet's composition for clues to its environmental history. Ehlmann directs the rover to carry out such operations as using its arm to brush dust off surfaces and drilling rocks to produce powders that are analyzed with its onboard instrumentation. She's especially interested in looking for evidence of chemical processes involving water as well as evidence of organic molecules associated with life.

As Ehlmann explains, she and her team are tackling the biggest questions in Mars's history, from how a once-watery world became a cold desert to whether Mars once supported life or perhaps still does today. She says this remote study is challenging but rewarding too, as it involves collaboration with scientists in many fields, such as engineers, geobiologists, geophysicists, and atmospheric scientists.

It's work Ehlmann enjoys talking about with young audiences. She even co-authored a children's book, *Dr. E's Super Stellar Solar System: Massive Mountains! Supersize Storms! Alien Atmospheres!* By sharing her story, Ehlmann hopes to inspire the next generation of space explorers—among them, perhaps those who will take the first steps on Mars. "May we continue—multiple Mars rovers in multiple diverse places—to give more students across the country and the world the experience I had," urges Ehlmann. "Onward to Mars!"

THINKING CRITICALLY

Evaluate More missions have been launched to Mars than any other destination in the solar system. Consider the planetary characteristics that Dr. Ehlmann researches on Mars. What is the value of investing in missions to this planet?

Dr. Bethany Ehlmann uses Mars rovers to do experiments and analyze samples remotely on the surface of Mars. Her research interests include investigating evidence of past presence of liquid water on Mars and documenting physical and chemical conditions on Mars today.

Ehlmann studies Earth as well as Mars. Here Ehlmann is conducting a study of minerals in acid salt lakes of Western Australia. She studies remote-sensing data similar to that collected from spacecraft in orbit around Mars.

CASE STUDY
SMALL STEPS AND GIANT LEAPS

In 1969, as astronaut Neil Armstrong took the first human step onto the surface of the moon, he famously quipped, "That's one small step for [a] man, one giant leap for mankind." The average distance between Earth and the moon is more than 384,000 kilometers, roughly 30 times the diameter of Earth. At that distance, a trip to the moon was indeed a giant leap for mankind. Yet in the vast expanse of the universe, or even our own solar system, such a distance is hardly a step, much less a leap. More like a barely noticeable quiver.

In the five decades since the Apollo 11 mission, no astronaut has ventured any farther than Earth's moon. Yet in that time, scientists have explored the far reaches of our solar system. How? Using unmanned spacecrafts and robots that send images and data back to Earth. In Chapter 22, you read about the Voyager and Galileo probes that explored Jupiter and its moons. In this chapter, you've read about the Mars Reconnaissance Orbiter and the Mars rover Curiosity, which are exploring the surface of the red planet (see Explorers at Work). But these are merely a few of the more than 150 robotic spacecraft that have been deployed to explore our solar system.

Some of the most stunning images have come from space probes sent to orbit other planets in the solar system. For example, the Cassini spacecraft, a combined effort of NASA, the European Space Agency (ESA), and the Italian Space Agency (ISA), spent 13 years orbiting Saturn. Over the course of its mission, it traveled some 7.9 billion kilometers gathering data on the composition of Saturn's atmosphere, its famous rings, and its many moons. Among other discoveries, Cassini found that Saturn's poles are covered by gigantic hurricane-like vortexes; the moon Enceladus is covered in liquid water topped by an icy crust; and the moon Titan has rain, rivers, and seas.

More recently, since 2016, NASA's spacecraft Juno has been orbiting Jupiter. Completing one orbit about every 53 days, Juno uses advanced cameras and infrared sensors to capture three-dimensional images of parts of Jupiter not visible from Earth. For example, Juno's images of Jupiter's north and south poles, which never face toward Earth, revealed giant storms over the polar areas (Figure 23-1). Juno has also acquired close-up images of the Great Red Spot (Figure 23-16), discovered a belt of ammonia gas around Jupiter's equator, and traced changes in the planet's magnetic field.

As You Read Develop a better understanding of the size of the solar system and its planets, moons, and other bodies. Talk about the considerable resources devoted to space exploration in terms of time, effort, and technology. What do humans have to gain by studying our solar system, and what have our discoveries taught us about our own planet?

FIGURE 23-1
Jupiter's south pole never faces toward Earth, so only a space probe could take an image of the storms there. The data from three Juno orbits was combined to produce this image. The satellite passed over the planet at a height of 52,000 km. The oval-shaped storms are up to 1,000 km in diameter.

23.1 THE SOLAR SYSTEM: A BRIEF OVERVIEW

Core Ideas and Skills
- Compare and contrast the characteristics of the terrestrial and Jovian planets.

KEY TERMS

escape velocity
solar wind
terrestrial planet
Jovian planet

The Sun and the Eight Planets

The solar system formed about 4.6 billion years ago from a cold, diffuse cloud of dust and gas rotating slowly in space. The cloud was composed of about 92 percent hydrogen and 7.8 percent helium, an elemental composition similar to that of the universe. All the other elements composed only 0.2 percent of the solar nebula (Figure 23-2).

A portion of the cloud collapsed to form the sun. Here the pressure became so intense that hydrogen atoms fused together, producing energy as some of the hydrogen converted to helium. Hydrogen fusion is still the source of the sun's energy and will be discussed in Chapter 24. The remaining matter in the solar nebula formed a rotating, disk-shaped cloud of interstellar dust that eventually condensed into separate masses to produce the planets.

The formation of the planets provides an example of a feedback mechanism. Let us compare the formation of Mercury with that of Jupiter. Any object—including a rocket or a gas molecule—can escape from a planet's gravity when it reaches a speed known as the **escape velocity**. The escape velocity is proportional to the planet's mass divided by its radius. In their earliest states, all planets were composed mainly of gas. Gas particles in a planetary atmosphere are in constant motion. The higher the temperature, the higher the average speed of gas particles. Mercury, being closer to the sun, was originally hotter than Jupiter. Its gas particles were moving faster, so they were more likely to escape the planet's gravity and fly off into space. As gases escaped, the planet lost mass, so the escape velocity decreased. This made it easier for gases to escape.

In addition, the **solar wind**, a stream of charged particles radiating outward from the sun at high speed, blew even more gases away from the early planets. Combining all processes, the inner planets—Mercury, Venus, Earth, and Mars—lost most of their gases, leaving behind protoplanets composed mostly of metals and silicate rocks. These four are now called the **terrestrial planets**. In contrast, the protoplanets in the outer reaches of the solar system were so far from the sun that they were initially cool. As a result, the outer **Jovian planets**—Jupiter, Saturn, Uranus, and Neptune—retained large amounts of hydrogen, helium, and other light elements in gas form. In fact, they actually grew larger as they captured gases that escaped from the terrestrial planets. As the mass

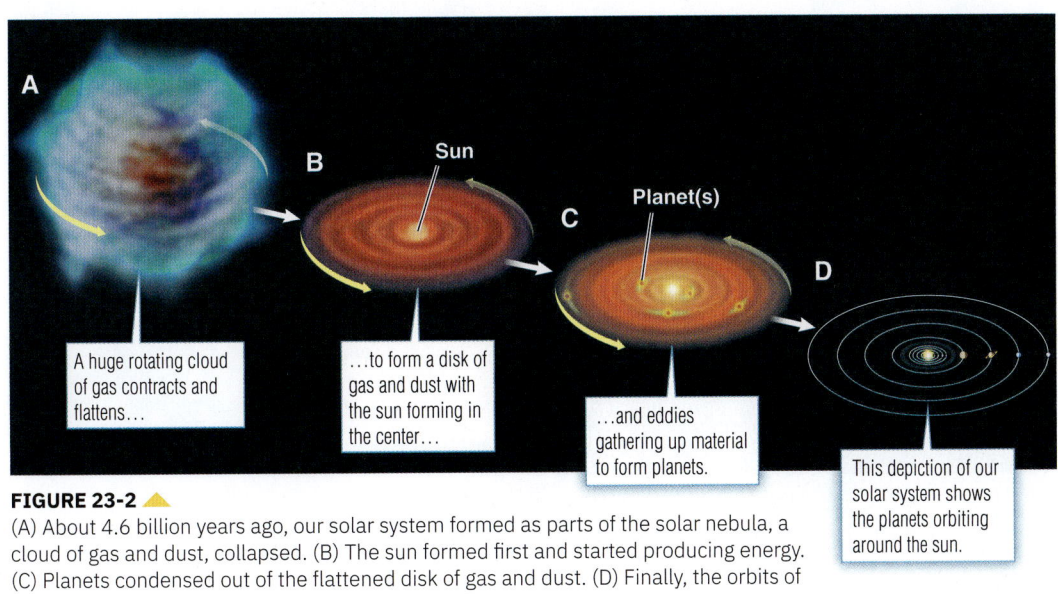

FIGURE 23-2
(A) About 4.6 billion years ago, our solar system formed as parts of the solar nebula, a cloud of gas and dust, collapsed. (B) The sun formed first and started producing energy. (C) Planets condensed out of the flattened disk of gas and dust. (D) Finally, the orbits of the planets were cleared of debris.

FIGURE 23-3
The planets and the sun of our solar system are shown drawn to scale.

TABLE 23-1 Comparison of the Eight Planets

Planet	Distance from Sun (millions of kilometers)	Radius (Earth = 1)	Mass (Earth = 1)	Density (water = 1)	Atmospheric Composition	Number of Moons*
Terrestrial Planets						
Mercury	58	0.38	0.06	5.4	none	0
Venus	108	0.95	0.82	5.2	96% CO_2, 3% N_2, 1% other	0
Earth	150	1	1	5.5	78% N_2, 20% O_2, 1% other	1
Mars	228	0.53	0.11	3.9	95% CO_2, 3% N_2, 1.5% Ar, 0.5% other	2
Jovian Planets						
Jupiter	778	11.2	318	1.3	90% H, 10% He	79
Saturn	1,427	9.4	94	0.7	96% H, 3% He, 1% other, (yellow color due to sulfur)	82
Uranus	2,871	4.0	15	1.3	83% H, 15% He, 2.5% CH_4 (blue color due to methane)	27
Neptune	4,498	3.9	17	1.7	80% H, 19% He, 1% CH_4	14

* includes moons confirmed and pending confirmation by the International Astronomical Union. Source: NASA

DATA ANALYSIS Compare Planetary Data

Table 23-1 shows characteristics of the eight planets of the solar system. Use the data in the table to compare and contrast the planets.

1. **Relate** What physical characteristics are common to the terrestrial planets? to the Jovian planets?
2. **Recognize Patterns** Describe how each of the physical characteristics of the planets are related to distance from the sun.
3. **Predict** Describe the characteristics of a hypothetical ninth planet beyond the orbit of Neptune. Assuming it follows the same general patterns as the eight planets, predict its mass, radius, density, composition, and number of moons.

Critical Thinking

4. **Explain** Why does Table 23-1 list values for radius and mass relative to Earth, but lists values for density relative to water?

Computational Thinking

5. **Generalize** What general pattern exists between the number of moons and other planetary characteristics, and what physical process is responsible for this correlation?

---

increased, the escape velocity also increased, and gas escape became more difficult. Today the Jovian planets all have relatively small, rocky or metal cores surrounded by swirling liquid and gaseous atmospheres (Figure 23-3).

Table 23-1 presents an overview of the physical properties of the eight planets in our solar system. Due to the differences in composition, the terrestrial planets are much denser than the Jovian planets. However, the Jovian planets are much larger and more massive than Earth and its neighbors.

checkpoint How are the terrestrial and Jovian planets alike and different?

23.1 ASSESSMENT

1. **Summarize** Explain the process by which the solar system we know today formed out of the solar nebula.
2. **Recall** Define *escape velocity* and how it relates to a planet's mass and radius.
3. **Contrast** How and why did the formation of terrestrial planets differ from the formation of Jovian planets?

Computational Thinking

4. **Recognize Patterns** Use the data in Table 23-1 to calculate the distances of the planets from the sun in astronomical units and describe the pattern in the distances.

23.2 THE TERRESTRIAL PLANETS AND EARTH'S MOON

Core Ideas and Skills
- Compare and contrast the atmospheres and climates of the terrestrial planets.
- Compare and contrast the geology and tectonics of the terrestrial planets.
- Describe characteristics of the moon's surface and interior.
- Summarize the discoveries made using rovers on Mars.

KEY TERMS

blob tectonics	maria

As the planets formed, the solar system was crowded with material left over from the formation of the sun, including chucks of rock and ice. This space debris crashed into the early planets, pockmarking their surfaces with millions of craters. Over geologic time, craters can be destroyed by tectonic events and by weathering and erosion. However, if these internal and surface processes do not occur, the craters remain for billions of years. Scientists learn a lot about a planet's history simply by observing crater density. With this background, let us compare the geology and the atmospheres of the four terrestrial planets.

Atmospheres and Climates

Mercury Recall that the closest planet to the sun lost essentially all of its atmosphere early during the formation of the solar system. Today it is a rocky sphere covered with craters (Figure 23-5).

FIGURE 23-4
ON ASSIGNMENT Look closely at this photograph of the sun taken by National Geographic photographer Keith Ladzinski in Colorado Springs, Colorado. A small round dot at center right is the planet Venus passing between the sun and Earth. This is a rare sight. The last transit of Venus occurred on June 5, 2012. The next Venus transit will occur on December 10, 2117.

Mercury revolves around the sun faster than any other planet. Its orbital period is about 88 Earth days. But Mercury rotates on its axis slowly, so there are only three Mercurian days every two Mercurian years. Because Mercury is so close to the sun and its days are so long, the temperature on its sunny side reaches 427°C, hot enough to melt lead. In contrast, the temperature on its dark side drops to −175°C, cold enough to freeze methane. The lack of an atmosphere is partly responsible for these extreme temperatures because there is no wind to distribute heat.

In 1991, radar images of Mercury revealed highly reflective regions at the planet's poles. Data indicated that these regions were composed of ice. Mercury's spin axis is almost perpendicular to its orbital plane around the sun, so the sun never rises or sets at the poles but remains low on the horizon throughout the year. Because the sun is low in the sky, regions inside meteorite craters are continuously shaded. With virtually no atmosphere to transport heat, the shaded regions have remained below the freezing point of water for billions of years. In 2012, NASA's MESSENGER (Mercury Surface, Space Environment, Geochemistry and Ranging) spacecraft confirmed evidence for water frozen in the craters at the north pole.

Venus The second planet from the sun closely resembles Earth in size and mass. As a result, with similar escape velocities, Venus and Earth probably had similar atmospheres early in their histories. Venus, however, is closer to the sun than Earth; therefore, it was always hotter. One hypothesis suggests that because of the higher temperatures, water never condensed out of the atmosphere— or if it did, it quickly evaporated again. Because there were no seas for carbon dioxide to dissolve into, most of the carbon dioxide also remained in the atmosphere. Water vapor and carbon dioxide combined to produce a runaway greenhouse effect, and surface temperatures became torridly hot. The hot surface temperature eventually boiled the water into space while the carbon dioxide remained behind. Today the Venusian atmosphere is 90 times denser than that of Earth's. The Venusian atmosphere is more than 97 percent carbon dioxide, with small amounts of nitrogen, helium, neon, sulfur dioxide, and other gases. Corrosive sulfuric acid aerosols float in a dense cloud layer that permanently obscures the surface. Surprisingly, Venus rotates clockwise, opposite the direction of nearly all other solar system objects. The planet rotates very slowly, taking more than 116 Earth days to complete one Venusian day. Due to severe greenhouse warming, the mean global temperature across the Venusian surface is 460°C, hotter than Mercury (Figure 23-6).

Earth Outgassing from the mantle modified the earliest atmosphere on Earth. As life evolved, complex interactions among the geosphere, the hydrosphere, and the biosphere further altered the atmosphere (Figure 23-7). Conditions became favorable for life. This complex sequence of events was discussed in detail in Chapter 18.

Mars Today the surface of Mars, the fourth planet from the sun, is frigid and dry. The surface temperature averages −56°C and never warms up enough to melt ice. At the poles, the temperature can dip to −120°C, freezing carbon dioxide to form dry ice. The atmosphere at the Martian surface is as thin as Earth's atmosphere 43 kilometers high, which for us is the outer edge of space. Although water ice exists in the Martian polar ice caps and in the soil, there is currently no liquid water on Mars.

However, abundant evidence indicates that the Martian climate was once much warmer, and that water flowed across the surface. Photographs from Mariner and Viking spacecraft show eroded crater walls and extinct stream beds and lake beds. One giant canyon, Valles Marineris, is approximately 10 times longer and six times wider than the Grand Canyon (Figure 23-8). Massive alluvial fans at the mouths of Martian canyons indicate that floods probably raced across the land at speeds up to 270 kilometers per hour.

checkpoint What evidence indicates that liquid water once flowed on Mars?

Geology and Tectonics

Mercury The first planet from the sun has a cratered surface remarkably similar to that of Earth's moon. Craters formed on all planets and their moons during intense meteorite bombardment early in the history of the solar system. Whereas tectonic activity and erosion have erased Earth's early craters, Mercury's atmosphere boiled off into space early. No wind, rain, or rivers have eroded its surface. Four-billion-year-old craters look as fresh as if they had formed yesterday.

Flat plains on Mercury are the result of lava flows that occurred when the planet's interior

FIGURE 23-5
This global mosaic of Mercury was created using images taken in 2011 by NASA's MESSENGER spacecraft, the first to ever orbit the planet.

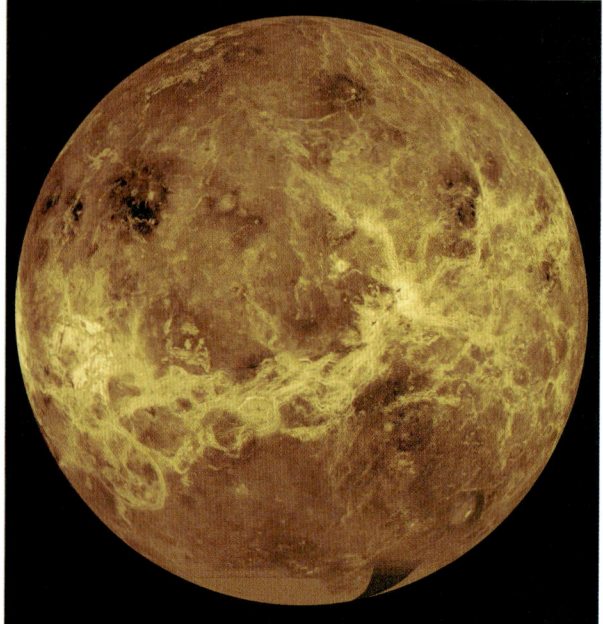

FIGURE 23-6
This radar mosaic projected onto a globe combines data from the Pioneer Venus Orbiter and the Magellan spacecraft that visited Venus in the early 1990s. The image is centered at 0 °N, 180 °E. The simulated color is based on true-color images recorded by the Soviet Venera 13 and Venera 14 spacecraft that landed and survived on the planet for about two hours in 1982.

FIGURE 23-7
This true-color image of Earth is one of a large collection of images from NASA named Blue Marble. It was produced by combining the most detailed data from the Moderate Resolution Imaging Spectroradiometer (MODIS), which is an instrument aboard two Earth-observing satellites, and topographical data from the USGS.

FIGURE 23-8
A mosaic, created from over 100 images taken by the Viking Orbiter missions, shows the entire Valles Marineris canyon system, stretching for 4,000 km across the Western Hemisphere of Mars. The dark red spots on the western limb of the planet are the 25-km-tall volcanoes called the Tharsis Montes.

was hot enough to produce magma. Mercury is so small, however, that its interior quickly cooled and hardened. Curiously, Mercury has a magnetic field. Earth's magnetic field is generated by the flow of molten metals in the rotating core. But the surface of Mercury shows no evidence of tectonic activity or of a hot interior. Therefore, astronomers cannot explain how its magnetic field is generated.

Venus Astronomers use spacecraft-based radar to penetrate the thick Venusian atmosphere. Spacecraft data can be used to infer the density of rocks near the Venusian surface, providing information about the planet's mineralogy and internal structure. The most spectacular data were obtained by the orbiting Magellan spacecraft, launched in May 1989. Magellan's detailed radar maps show that, despite the heavy bombardment of all the planets early in the history of the solar system, few craters exist on Venus. Observations indicate that the surface of Venus was reshaped. More detailed analysis shows that most of the landforms on Venus are only 300 to 500 million years old.

A leading hypothesis suggests that a catastrophic series of volcanic eruptions occurred between 300 and 500 million years ago, creating volcanic mountains and covering much of the Venusian surface with basalt flows (Figure 23-9). According to this model, rising mantle plumes generated magma that repaved the planet in a short time with a rapid series of cataclysmic volcanic eruptions. Some evidence indicates that volcanic activity has ceased permanently because the planet's interior cooled. This cooling may have occurred because radioactive elements floated toward the surface during the volcanic events, removing the mantle heat source. An alternative hypothesis contends that Venus's resurfacing occurred more slowly. Some scientists think that Venus remains volcanically active and that another repaving event may occur in the future.

FIGURE 23-9
The 8-km-tall Maat Mons volcano looms in this simulated, false-color 3-D view of the Venusian surface. Lava flows extend for hundreds of kilometers from the peak of the volcano. It is unclear whether Maat Mons is currently active. Data from the Magellan spacecraft and Soviet Venera 13 and 14 were used to construct the image.

Most Earth volcanoes form at tectonic plate boundaries or over mantle plumes, so the discovery of volcanoes on Venus led planetary geologists to look for evidence of plate tectonic activity there. Radar images from spacecraft show that 60 percent of Venus's surface consists of a flat plain. Two large and several smaller mountain chains rise from the plain (Figure 23-10). The tallest mountain is 11 kilometers high—2 kilometers higher than Mount Everest. The images also show large, crustal fractures and deep canyons. If Earth-like horizontal motion of tectonic plates caused these features, then spreading centers and subduction zones should exist on Venus. The images show no features like a mid-ocean ridge system, transform faults, or other evidence of lithospheric spreading. However, geophysicists have located 10,000 kilometers of trenchlike structures that they think are subduction zones.

FIGURE 23-10
Go online to view a detailed map of the surface features of Venus created using radar images from Pioneer Venus orbiter.

Despite the apparent existence of subduction zones, the most popular current model suggests that Venusian tectonics have been dominated by mantle plumes. In some regions the rock has melted and erupted from volcanoes by processes similar to those that formed the Hawaiian Islands. In other regions the hot, Venusian mantle plumes have lifted the crust to form nonvolcanic mountain ranges. Some geologists have proposed the term **blob tectonics** to describe Venusian tectonics because Venus is dominated by the rising and sinking of its mantle and crust. In contrast, tectonic activity on Earth causes significant horizontal movement of its plates.

Mantle plumes on Earth may initiate rifting of the lithosphere and formation of a spreading center. Why have spreading centers not developed over mantle plumes on Venus? Perhaps the surface temperature on Venus is so high that surface rocks are more plastic than those on Earth. Therefore, rock flows plastically rather than fracturing into lithospheric plates. It is also possible that Venus has a thicker lithosphere, which can move vertically but does not fracture and slide horizontally.

Earth Our planet is large enough to have retained considerable internal heat from its formation. In addition, radioactive decay of unstable isotopes in the mantle are another source of internal energy. This heat drives convection within the mantle that produces tectonic motion. In turn, as you have read in previous chapters, Earth tectonics continuously reshapes the surface of our planet and is partially responsible for the atmosphere that sustains us.

Mars The Martian surface consists of old, heavily cratered plains and younger regions that have been altered by tectonic activity. A topographical map was created using over 600 million data measurements taken by Mars Global Surveyor in 2001 (Figure 23-11). Lava flows much like those seen on Venus and the moon cover the plains. The Tharsis bulge is the largest plain, crowned by Olympus Mons (Figure 23-12), the largest volcano in the solar system. Olympus Mons is nearly three times taller than Mount Everest, with a height of 22 kilometers and a diameter of 500 kilometers. It is clearly identifiable in the upper left portion of the topographical map. Its central crater is so big that Manhattan Island would easily fit inside.

FIGURE 23-11
Go online to view a topographical map of Mars that demarcates the highlands and basins, as well as the tallest volcano in the solar system, Olympus Mons.

One hypothesis suggests that the geology of Mars is similar to that of Venus and is dominated by blob tectonics. Supporters of this concept point out that the largest volcanic mountain on Earth, Mauna Kea in the Hawaiian Islands, is

FIGURE 23-12
Olympus Mons is not only the largest volcano on Mars. It is the largest in the solar system.

limited in size because the mountain is riding on a tectonic plate. Therefore, it will drift away from the underlying hot spot before it can grow much larger. In contrast, because horizontal movement on Mars is nonexistent or very slow, Olympus Mons has remained stationary over its hot spot. As a result, it has grown bigger and bigger.

Observations reported in 2005 show very few meteorite craters in some lava flows, indicating that eruptions have occurred as recently as a few million years ago. Thus, the planet has remained hot and active until relatively recent times.

checkpoint What geologic feature is present on Earth, Venus, and Mars, but not Mercury?

Earth's Moon

Lunar Surface Most planets have small orbiting satellites called moons. The Earth's moon is close enough that we can see many of its surface features with the naked eye. In the early 1600s, Galileo studied the moon with a telescope and mapped its mountain ranges, craters, and plains. Galileo thought that the plains were oceans and called them seas, or **maria**. The word *maria* is still used today, although we now know that these regions are dry, barren, flat expanses of volcanic rock that make the lunar surface easily recognizable (Figure 23-13A).

Due to the gravitational interaction between Earth and the moon, the lunar orbital period is equal to its rotational period. This means that the same parts of the lunar crust face the Earth at all times. This results in a "near side" of the moon, the side of the moon with which we are most familiar, and a "far side" of the moon (Figure 23-13B). A distinguishing characteristic of the near side is the large percentage of area covered by maria. The moon's crust is thinner on the near side because the constant gravitational tug by our planet pulls the moon's mantle closer to the lunar surface. In the moon's past, molten lunar material easily seeped through cracks in the thin lunar crust on the side that always faces Earth. We have learned much about the moon's history by collecting data about the lunar surface using the Lunar Reconnaissance Orbiter.

Lunar Origin and Structure A Soviet orbiter named Luna-3 took the first close-up photographs of the moon in 1959. A decade later the United States landed the first of six manned Apollo spacecraft on the lunar surface. The Apollo program was designed to answer several questions about the moon: How did it form and change over its history? Was it once hot and molten like Earth? If so, does it still have a molten core, and is it tectonically active?

A

B

FIGURE 23-13
Lunar Reconnaissance Orbiter images of the moon show (A) the familiar near side, with extensive maria, and (B) the less-familiar far side.

Perhaps the single most significant discovery of the Apollo program was that much of the moon's surface consists of igneous rocks. The maria are mainly basalt flows. The highland rocks are predominantly anorthosite, a feldspar-rich igneous rock not common on Earth. Additionally, the rock of both the maria and the highlands has been crushed by meteorite impacts and then welded together by lava. Since igneous rocks form from magma, it is clear that portions of the moon were once hot and liquid. Recall that Earth was heated initially by the collisions among particles as they collapsed under the influence of gravity. Later, radioactive decay and intense meteorite bombardment heated Earth further. But what about the moon?

According to a popular hypothesis, the moon was created when an object the size of Mars or larger collided with Earth about 4.5 billion years ago, just as our planet was cooling. This massive protoplanet plowed through Earth's mantle. Silica-rich rocks from the mantle vaporized and created a cloud around Earth. The vaporized rock condensed to form the moon. So much energy was released that the moon melted to a depth of a few hundred kilometers, forming a magma ocean. Meteorite bombardment kept the moon's outermost layer molten. Eventually, meteorite bombardment slowed enough for the surface to cool. Radiometric dating of lunar samples confirms that the oldest lunar igneous rocks formed around 4.4 billion years ago. This occurred before there was enough time for radioactive decay to have melted a significant amount of lunar rock. Gravitational condensation and meteorite bombardment must have been the main causes of early melting of the moon's surface.

Swarms of meteorites bombarded the moon again between 4.2 billion and 3.9 billion years ago. In the meantime, radioactive decay was heating the lunar interior. As a result, by 3.8 billion years ago, most of the moon's interior was molten. The maria formed when lava filled meteorite craters. This episode of volcanic activity lasted approximately 700 million years. The moon and Earth had shared a similar history until this time. The moon is so much smaller that it soon cooled and has remained geologically inactive for the past 3.1 billion years.

Apollo astronauts left seismographs on the lunar surface. Initial analysis of the data indicated that the energy released by moonquakes is only one-billionth to one-trillionth that released by earthquakes on our own planet. However, in 2005, 25 years after the first reports were released, scientists realized that the computers used in the study had insufficient memory to properly record and analyze the data. Under reexamination, scientists detected numerous deep moonquakes, indicating the moon has a hot, possibly molten, core.

Lunar Water In 1998, data from the Lunar Prospector spacecraft indicated that water may be present in shadowed craters at the lunar poles. In 2009, NASA confirmed the presence of water molecules in the moon's polar regions, although in relatively small amounts. Nine years later, scientists confirmed the first direct observation of water ice in craters on both lunar poles. The data was obtained using one of two NASA instruments onboard the Chandrayaan-1 satellite, India's first lunar exploration mission.

checkpoint What are maria on the surface of the moon?

Mars Rovers

In 2004, NASA began the science phase of the Mars Exploration Rover (MER) mission after successfully landing two remote-controlled robots, Spirit and Opportunity. Each 1.5-meter-tall rover was equipped with six wheels and a drive system, communications equipment to receive and transmit data, and solar panels for power. Research was conducted with sensitive instruments: a camera, a magnifying glass, a grinding wheel to burrow into rock and expose fresh surfaces, a chemical analyzer, two remote-sensing devices to identify minerals, and a magnet to collect magnetic dust particles. One of the main goals of this mission was to look for evidence of the existence of liquid water in Mars's past.

Spirit landed in a dry area of Gusev crater and detected no evidence of water. Opportunity, however, discovered layered sedimentary rocks near its landing site in Meridiani Planum. Chemical and mineralogical analysis indicated the abundance of iron-rich minerals that usually form in the presence of water. Opportunity also found sulfate salts that had clearly precipitated from a water-rich solution. Several weeks later, Opportunity took spectacular images of ripple marks preserved on the floor of a long-dried up body of water. The rover imaged several sedimentary rock outcrops and playas, and discovered hematite and gypsum, both minerals associated with liquid water.

FIGURE 23-14
Curiosity photographed multiple images of petrified sand dunes at Mount Sharp in the center of Gale crater. These sedimentary structures, the result of a water lake on Mars, are similar to those commonly found on Earth.

Meanwhile, Spirit climbed off the lava-covered Gusev plain and found precipitated sulfur salts on the crater rim, indicating that some water had existed here. Spirit spent four years investigating the chemical composition of rocks in and around Husband Hill. It found carbonate, indicating that Mars's atmosphere had sufficient carbon dioxide at some point to precipitate those minerals, as well as pure silica deposits.

Initially planned to last only 90 days, Spirit and Opportunity are now legends in space exploration. Spirit traveled nearly 8 kilometers over six years and sent back more than 120,000 images. Opportunity survived on the red planet for more than 14 years until succumbing to power failure from a planet-wide dust storm in 2018. The rover, nicknamed "Oppy," had traveled 45 kilometers across often treacherous terrain and returned over 200,000 images.

NASA's Mars Science Laboratory (MSL) Curiosity rover, which landed on the planet in 2012, has been building upon the success and knowledge of Spirit and Opportunity. The largest and most advanced rover is roughly the size of a car, weighs nearly 4,000 kilograms, and carries a dozen instruments including cameras, spectrometers, and monitoring systems. The power system is based on radioisotope decay, so it is immune to the problems that dust storms represent for solar panels. To date, Curiosity has made many discoveries and sent back amazing images (Figure 23-14). Among the most exciting was evidence that in Mars's past, conditions could have been suitable for microbial life in and around Gale crater.

In July 2020, NASA launched the next generation rover, Mars 2020 rover Perseverance. The mission has four primary goals: Determine whether life ever arose on Mars, characterize the planet's climate, characterize its geology, and prepare for human exploration.

checkpoint What was one of the primary goals of Spirit and Opportunity?

23.2 ASSESSMENT

1. **Contrast** How are the atmospheres and climates of the terrestrial planets similar and different from Earth?
2. **Distinguish** Define blob tectonics and identify how it is different from plate tectonics.
3. **Explain** Why does the moon have a near side and a far side?

Critical Thinking

4. **Sequence** Explain how the discoveries of Spirit and Opportunity form the basis for science investigations for Curiosity, and how these all helped set up the goals for Perseverance, the Mars 2020 rover.
5. **Analyze** Compare the radar map of Venus (Figure 23-10) to the topographical map of Mars (Figure 23-11). Construct a claim, using specific evidence, that contrasts the tectonic and volcanic activity on the two terrestrial planets. For example, one line of evidence may be a comparison of the number of craters on the surface of each planet.

23.3 THE JOVIAN PLANETS

Core Ideas and Skills
- Compare and contrast the compositions and atmospheres of the Jovian planets.

> **KEY TERMS**
>
> liquid metallic hydrogen

Jupiter

The largest planet in the solar system is composed mainly of hydrogen and helium, similar to the composition of the sun. However, Jupiter did not accumulate enough mass when it formed to initiate fusion in its core, so it never became a star.

Jupiter has no solid, rocky crust but is a layered planet (Figure 23-15). Its turbulent atmosphere is a vast sea 21,000 kilometers deep, made of cold, liquid molecular hydrogen (H_2) and atomic helium (He). Beneath the hydrogen-helium sea is a layer where temperatures are as high as 30,000°C and pressures reach 100 million times the Earth's atmospheric pressure at sea level. Under these extreme conditions, molecular hydrogen dissociates to form hydrogen atoms. Pressure forces the atoms together so tightly that the electrons move freely throughout the packed nuclei, much like electrons travel freely among metal atoms. As a result, the hydrogen conducts electricity and is called **liquid metallic hydrogen**. Flow patterns in this fluid generate a magnetic field 10 times stronger than Earth's magnetic field.

Beneath the layer of metallic hydrogen, Jupiter's core is a sphere about 10 to 20 times as massive as Earth's. It is probably composed of metals and rock

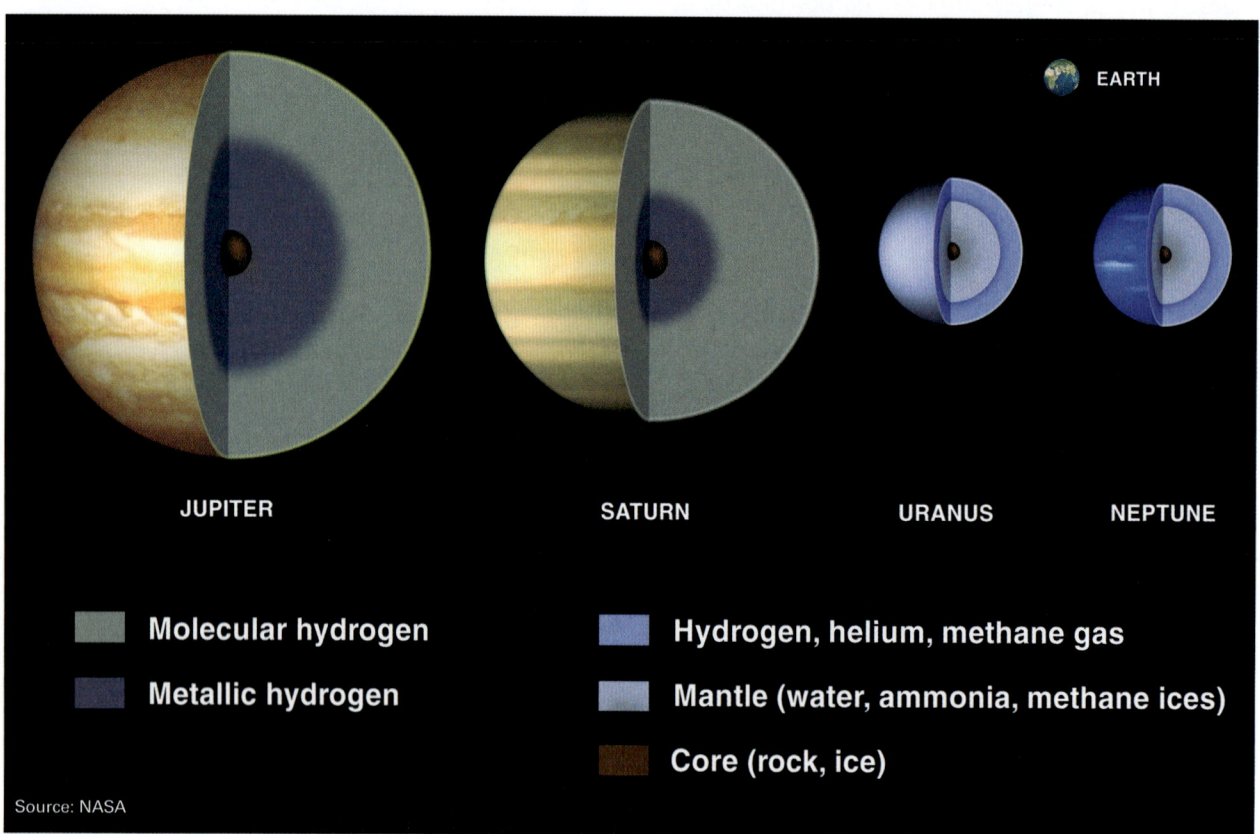

FIGURE 23-15
The Jovian inner planet structure consists of three layers. Jupiter and Saturn contain with an upper layer primarily of liquid molecular hydrogen and atomic helium, a layer of dense liquid metallic hydrogen, and a rocky metallic core. Uranus and Neptune are thought to have upper layers composed of hydrogen, helium, and methane gas above mantles of water, ammonia, and methane ice, centered on a small core of rock and ice. Earth is provided in this illustration for scale.

surrounded by lighter elements such as carbon, nitrogen, and oxygen.

More than 300 years ago, two European astronomers reported seeing what came to be called the Great Red Spot on the surface of Jupiter. Although its size, shape, and color have changed from year to year, the storm continues to this day as seen in a recent image taken by the Juno spacecraft (Figure 23-16). Measurements show that the Great Red Spot is a giant cyclone-like storm. Cyclones on Earth dissipate after days or weeks, yet this storm on Jupiter has existed for centuries. Other storm systems, rotating in linear bands around the planet, have also persisted for centuries. A recent comparison between Voyager 1 images and Juno images of the Great Red Spot have shown that the storm is rapidly shrinking in size, with a width from roughly 1.8 Earths in 1979 to 1.3 Earths in 2017. Many predict the storm could disappear within the next few decades.

Jupiter is one of the outer planets most well-studied by satellite. After a nearly six-year voyage, the Galileo spacecraft arrived at the gas giant in 1995. Once in orbit around the planet, the spacecraft dropped a probe into the outer atmosphere. As expected, data showed that the atmosphere is primarily hydrogen and helium, with smaller concentrations of ammonia, water, and methane.

The Galileo probe measured pressure, temperature, and wind speed as it fell below the visible upper layers. In the outer atmosphere, where the pressure was 0.4 bar, the probe reported a wind speed of 360 kilometers per hour with a temperature of −140°C. After falling 130 kilometers, the probe relayed that the wind speed had increased to 650 kilometers per hour, the pressure was 22 bar, and the temperature was about 150°C.

Buffeted by winds, squeezed by intense pressure, and heated beyond the tolerance of its electronics, the spacecraft stopped transmitting as it continued to fall. Scientists calculate that 40 minutes later, when the spacecraft had sunk to deeper levels of the Jovian atmosphere, where the temperature increased to 650°C and the pressure rose to 260 bar, the aluminum shell of the probe liquefied, the titanium hull melted, and streaming metallic droplets flashed into vapor as the probe vanished.

In 2018, scientists reported that the Juno satellite had gathered evidence that the jet streams that power Jupiter's long-lasting storms reach far deeper into the planet and contain more mass than previously thought. Jupiter's weather layer contains

FIGURE 23-16
This mosaic, composed of two images from the Juno spacecraft in 2019, shows the Great Red Spot in incredible detail. To the west of the spot is a sliver of red material that is being pulled off by the turbulence in the surrounding atmosphere, a phenomenon first witnessed via ground-based telescopes in 2017. The image resolution is 50 km.

about three times the mass of Earth and extends down to 3,000 km below the top of the clouds.

On Earth, wind and weather are driven by energy from the sun. But Jupiter receives only about four percent of the solar energy that Earth receives. Strong winds rip through the Jovian atmosphere far below the deepest penetration of solar light and heat. For these reasons, scientists deduce that Jovian weather is not driven by solar heat but by heat from deep within the planet itself. This heat accumulated when Jupiter first formed. As the heat slowly rises, it warms the lower atmosphere, which then transmits heat by convection to higher levels. The weather systems change very slowly because the planet's interior heat flux changes only over hundreds to thousands of years.

checkpoint What do scientists think drives weather on Jupiter?

Saturn

Saturn, known for its bright ring system (Figure 23-17), is the second-largest planet in our solar system and has the lowest density of all the planets. Such a low density implies that it is composed primarily of hydrogen and helium. Evidence suggests that Saturn's liquid molecular hydrogen layer is thicker than its inner metallic hydrogen layer, and its rock and metal core is relatively small (Figure 23-15). Saturn's atmosphere, like Jupiter's, contains dense clouds and great storm systems that envelop the planet. Winds on Saturn can reach speeds of 1,800 kilometers per hour. The temperature of Saturn's upper atmosphere is cold however, at −175°C. Lightning storms observed by Cassini have been shown to last many months. The planet also rotates very quickly, completing one rotation in just over 10 hours.

Although all the Jovian planets have one or more rings, by far the most spectacular are those of Saturn. The ring system is visible from Earth even through a small telescope. Photographs from space probes show seven major rings, each containing thousands of smaller ringlets (Figure 23-18). The entire ring system is a mere 10 to 25 meters thick, less than the length of a football field. However, the ring system is extremely wide. The innermost ring is 7,000 kilometers from Saturn's surface. The outer edge of the most distant ring is 432,000 kilometers from the planet's surface, which is a distance greater than that between Earth and our moon. Thus, the ring system measures 425,000 kilometers from the inner to outer edge. A scale model of the ring system with the thickness of 1.5 millimeters would be 30 kilometers in diameter.

FIGURE 23-18
Go online to view a color-enhanced Voyager 1 closeup image of Saturn's seven major rings and the gaps between sets of rings.

Saturn's rings are composed of dust, rock, and ice. The particles in the outer rings are only a few ten-thousandths of a centimeter in diameter (about

FIGURE 23-17
Saturn's rings cast a shadow on the planet. This view was captured by NASA's Cassini spacecraft as it orbited the giant planet on January 2, 2010.

the size of a clay particle), but the innermost rings contain chunks as large as a building. Each piece of debris orbits the planet independently.

Saturn's rings may be fragments of a moon that never formed. Alternatively, they may be the remnants of a moon, asteroid or comet that was then ripped apart by Saturn's gravitational field. If a moon were close enough to its planet, the tidal effects would be greater than the gravitational attraction holding the moon together, and it would break apart. Thus, a solid moon cannot exist too close to its planet. Images from Cassini spacecraft show that gravitational forces from Saturn's moons have herded the ring particles into intricate spirals, spokes, and twists.

checkpoint What are some ways Saturn's rings may have formed?

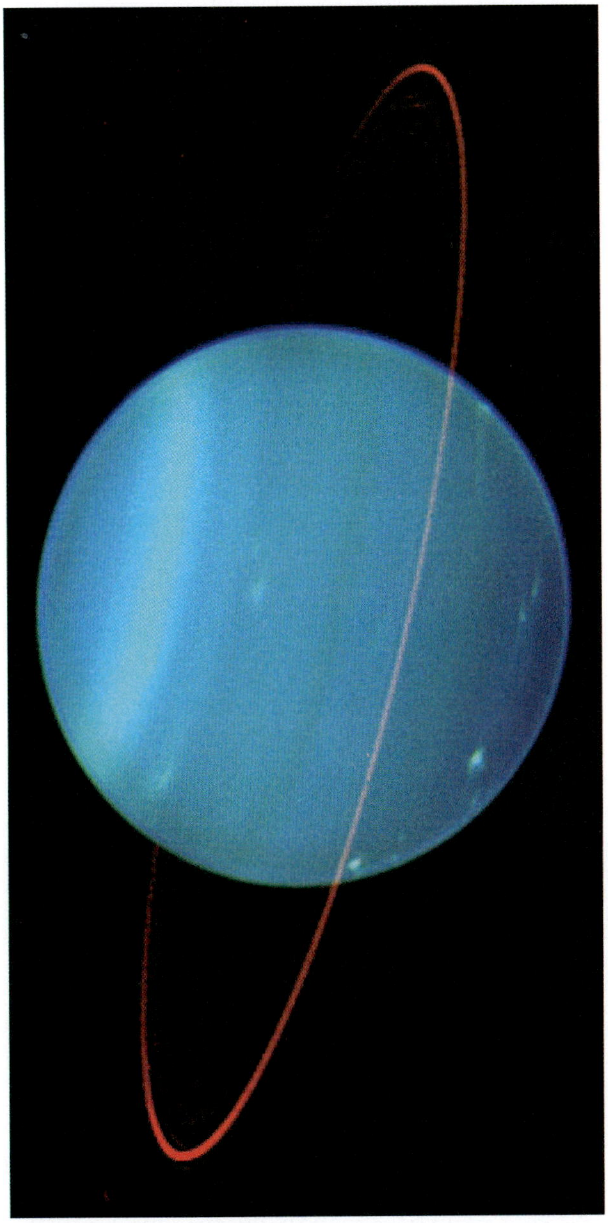

Uranus and Neptune

Uranus (Figure 23-19) and Neptune (Figure 23-20) are so distant and faint that they were unknown to ancient astronomers, who lacked telescopes. Many of the images we have of these planets come from the Voyager 2 spacecraft. Launched in the late 1970s, Voyager 2 flew by Jupiter in 1979, Saturn in 1981, Uranus in 1986, and Neptune in 1989. The journey from Earth to Neptune covered 7.1 billion kilometers and took 12 years. The craft passed within 4,800 kilometers of Neptune's cloud tops. The strength of the radio signals received from Voyager 2 measured one ten-quadrillionth (10^{-16}) of a watt. It took 38 radio antennas on four continents to collect enough radio energy to interpret the signals. Today, Voyager 2 and its twin Voyager 1, are the most distant explorers, having crossed the boundaries of the solar system into interstellar space.

FIGURE 23-19
This false-color view of Uranus was taken by the Hubble Space Telescope in August 2001. The brightness of the rings has been enhanced to make them easier to see. Its axis of rotation is tilted in relation to the orbit of Uranus, so the planet basically spins on its side.

FIGURE 23-20
This image was one of the last whole-planet images taken through the green and orange filters on the Voyager 2 narrow angle camera in 1989. Taken at a range of 7 million kilometers from the planet, the Great Dark Spot is clearly visible. In 1994, the Hubble Space Telescope viewed the planet and the storm had disappeared.

LESSON 23.3

Both Uranus and Neptune are enveloped by thick atmospheres composed primarily of hydrogen and helium, with smaller amounts of complex compounds (Figure 23-15). Beneath the atmosphere, their outer layers are molecular hydrogen, but neither world is massive enough to generate liquid metallic hydrogen. Their interiors are composed of methane, ammonia, and water, and the cores are probably a mixture of rock and metals. Uranus and Neptune are denser than Jupiter and Saturn because these outermost giants most likely contain relatively larger solid cores.

The Voyager spacecraft and the Hubble Space Telescope have revealed rapidly changing weather on both planets. On Neptune, winds of at least 1,100 kilometers per hour rip through the atmosphere, and clouds rise and fall. One region was for a time marked by a cyclonic storm system called the Great Dark Spot, similar to Jupiter's Great Red Spot. According to one controversial hypothesis, under the intense pressure near the core, methane decomposes into carbon and hydrogen and the carbon then crystallizes into diamond. Convection currents carry the heat released during the formation of diamond to the planet's surface to power the winds.

checkpoint Describe the atmospheres of Neptune and Uranus.

23.3 ASSESSMENT

1. **Describe** What makes up Jupiter's 3-layer inner structure?
2. **Infer** Why has the Great Red Spot lasted for so long?
3. **Recall** What information did the Galileo probe provide to scientists about Jupiter?
4. **Compare and Contrast** In what ways is Saturn similar to and different from Jupiter?
5. **Contrast** What main characteristics distinguish Uranus and Neptune from Jupiter and Saturn?

Computational Thinking

6. **Calculate** Jupiter receives about 4 percent the amount of sunlight that Earth receives.

   Energy from the sun spreads out in all directions in the solar system and drops in intensity as $E \times d^{-2}$ where E is the total energy flux at the source and d represents distance from the sun.

   Use this mathematical relationship to determine the amount of energy received at (a) Saturn, (b) Uranus, and (c) Neptune compared to the energy received at Earth. To simplify the calculations, determine the values as a ratio, or percentage, and use distances in AU calculated from Table 23-1. The example calculation for Jupiter has been provided:

$$\frac{\text{Energy received at Jupiter}}{\text{Energy received at Earth}} = \frac{\frac{E}{(5.2 \text{ AU})^2}}{\frac{E}{(1 \text{ AU})^2}}$$

$$= \frac{(1 \text{ AU})^2}{(5.2 \text{ AU})^2} = \frac{1}{27.04}$$

$$= 0.037 \sim 3.7\%$$

23.4 JOVIAN MOONS

Core Ideas and Skills
- Describe the distinguishing characteristics of the four Galilean moons.
- Describe the unique characteristics of the moons Titan, Miranda, and Triton.

In 1610, Galileo Galilei discovered that four tiny specks of light were objects orbiting Jupiter. By 1999, astronomers had identified 16 moons orbiting Jupiter, and now there are at least 79. More than half of these moons are relatively small, with irregular orbits. The four discovered by Galileo, referred to as the Galilean moons, are the largest and most studied: Io, Europa, Ganymede, and Callisto.

New moons orbiting the other Jovian planets are still being discovered. Among the most well-studied are Saturn's Titan, Uranus' Miranda, and Neptune's Triton.

Galilean Moons

Io The innermost moon of Jupiter is about the size of Earth's moon and slightly denser. Because it is too small to have retained heat generated during its formation or by radioactive decay, many astronomers expected it would have a cold, cratered surface. However, images beamed to Earth from Voyager 1 spacecraft in 1979 showed huge masses of gas and rock erupting to a height of 200 kilometers above the surface. This was the first evidence of active, extraterrestrial volcanism (Figure 23-21). Subsequent images from the Galileo spacecraft twenty years later showed 100 volcanoes erupting simultaneously, making Io the most active volcanic world in the solar system.

Recall that the gravitational field of our moon causes the rise and fall of ocean tides on Earth. At the same time, Earth's gravity distorts lunar rock. Thus, Earth's gravitation is responsible for deep-focus moonquakes. Jupiter is 300 times more massive than Earth, so its gravitational effects on Io are correspondingly greater. In addition, the three nearby satellites—Europa, Ganymede, and Callisto—are large enough to exert significant gravitational forces on Io, but these forces pull in directions different from that of Jupiter. This combination of oscillating and opposing gravitational forces causes so much rock distortion and frictional heating that volcanic activity is nearly continuous on Io.

Astronomers infer that meteorites bombarded Io and the other moons of Jupiter, as they did all other bodies in the solar system. Yet, the frequent lava flows on Io have obliterated all ancient landforms, giving it a smooth and nearly crater-free surface.

Europa The second closest of Jupiter's moons, Europa, is similar to Earth in that much of its interior is composed of rock and much of its surface is covered with water. One major difference is that, on Europa, the water is frozen into a vast, planetary ice crust. The Galileo spacecraft transmitted images

FIGURE 23-21
A volcanic plume erupts on Jupiter's moon, Io. Initially thought to be a cold, cratered, inactive moon, scientists discovered huge volcanic eruptions.

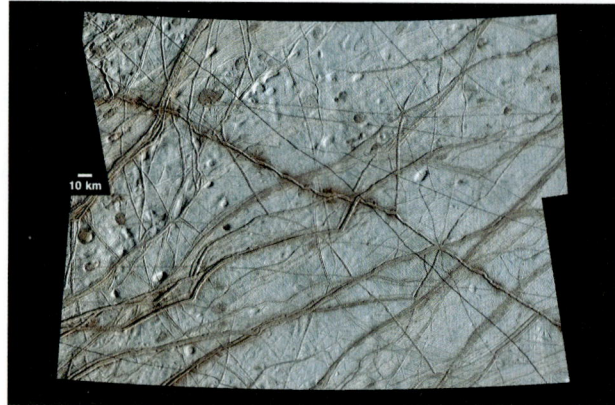

A   B

FIGURE 23-22
(A) This jumbled terrain on Europa resembles Arctic pack ice as it breaks up in the spring. Scientists estimate that in this region, the ice crust is a few kilometers thick and is floating on subsurface water. (B) A closer view of the surface of Europa reveals more detail of the smooth terrain overlaying many fractures. The environment beneath the icy surface might be favorable for life.

showing a fractured, jumbled, chaotic terrain resembling patterns created by Arctic ice on Earth.

Figure 23-22 shows areas of smooth terrain overlying an older, wrinkled surface. Data suggest that this smooth region is young ice formed when liquid water erupted to the surface and froze. Thus, pools or oceans of liquid water or a water-ice slurry probably lies beneath the surface ice. Astronomers estimate that this surface crust is 10 kilometers thick in many regions. Calculations show that the subterranean oceans are warmed by tidal effects similar to, though weaker than, those that cause Io's volcanism. Scientists speculate that the chemical and physical environment in these subterranean oceans is favorable for life.

Ganymede and Callisto The Galileo spacecraft measured a magnetic field on Ganymede, indicating that this moon has a convecting mantle and metallic core. Other measurements imply that the core is surrounded by a silicate mantle covered by a water-ice crust (Figure 23-23). The surface ice is so cold that it is brittle and behaves much like rock. Photographs show two terrains on Ganymede. One is densely cratered; the other contains fewer craters but many linear grooves. The cratered regions were formed by ancient meteorite bombardments. The grooved regions probably developed when the crust cracked and water from the warm interior flowed over the surface and froze. Lateral displacements of the grooves and ridges indicate Earthlike, horizontal tectonic activity (Figure 23-24).

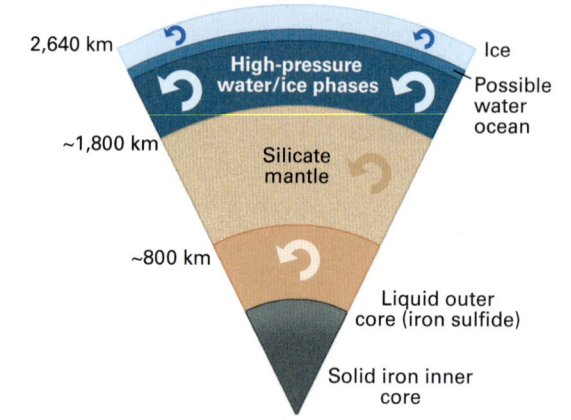

FIGURE 23-23
Recent data suggest that Ganymede has a conducting, convecting core; a silicate mantle; and surface layers consisting of ice and water.

FIGURE 23-24
Go online to view a close-up of Ganymede's young, tectonically grooved terrain.

Callisto, the outermost Galilean moon, is heavily cratered, indicating that its surface is very old. Its craters are shaped differently from those on either Ganymede or Earth's moon. Perhaps they have been modified by ice flowing slowly across its surface. Recent measurements indicate that a subterranean ocean may exist on Callisto. It is unknown if a similar feature exists on Ganymede.

checkpoint What is the most volcanically active world in the solar system and why?

Titan and Other Moons

Titan Saturn's largest moon is larger than the planet Mercury. Titan is unique because it is the only moon in the solar system with a significant atmosphere (Figure 23-25). Titan has retained its atmosphere because the moon is relatively massive and extremely cold. The major gases are nitrogen mixed with methane (CH_4) and a few trace gases. The average temperature on the surface of Titan is −178°C, and the atmospheric pressure is 1.5 times greater than that on Earth's surface. These conditions are close to the temperatures and pressures at which methane can exist as a solid, a liquid, or a gas.

Titan's atmosphere was studied directly using a probe name Huygens, dropped by the Cassini spacecraft in 2005. The probe parachuted through the outer clouds and landed on the surface after a nearly two and a half hour descent. During the 72 minutes that Huygens survived on the surface, it sent back images that showed a surprisingly Earthlike landscape, with steep-sided hills and features that looked like riverbeds, eroded hillsides, coastlines, and sandbars. These topographic features were formed by wind, tectonic activity, and flowing liquids. Due to the extreme cold on Titan, water is permanently frozen, but methane can exist in the liquid, vapor, or solid states. Evidence suggests that methane rain has fallen from the clouds and methane rivers have flowed across Titan's surface. Liquid methane may exist in reservoirs on Titan today. This situation is analogous to Earth's environment, where water can exist as liquid, gas, or solid and frequently changes among those three states.

Methane, the simplest organic compound, reacts with nitrogen and other materials in Titan's environment to form more complex organic molecules. These organic compounds do not decompose at low temperature, so the satellite's surface is likely to be covered by a tarlike organic substance. It is possible that a similar layer collected on early Earth and later underwent chemical reactions to form life. However, Titan is so cold that life probably has not occurred there.

Miranda Uranus has 27 confirmed moons and more may be discovered. Several are small and irregularly shaped, indicating that they may be debris from a collision with a smaller planet or moon. Miranda is the innermost and smallest of

▶ **FIGURE 23-25**
This composite infrared and visible-light image of Saturn's largest moon allows us to peer through the moon's thick atmosphere to its surface. The terrain shown is mostly on the Saturn-facing hemisphere of Titan. Like Earth's moon, Titan is locked in synchronous orbit and rotation such that the same side always faces its planet. The composite contains a few areas with higher resolution, called subframes. This higher resolution data, acquired with the same instruments while the spacecraft was at a closer approach, covers a smaller area of the moon.

LESSON 23.4

the five largest Uranian moons. It is composed of water ice and silicate rocks. With a diameter of only 500 kilometers, one might expect it to show little tectonic activity. However, Voyager 2 showed that its surface has three large areas covered in ridges and valleys with few craters separated by heavily cratered areas with sharp boundaries. The origins of the moon's strange structure are uncertain. Some scientists consider the possibility that the moon may have been ripped apart in a collision and then gravitationally rebound like pieces of a puzzle. Other hypotheses contend that heavy periods of meteorite bombardment caused subsurface ice to melt, rise through cracks in the rocky surface, and refreeze.

Triton Neptune has at least 13 moons. The largest is Triton, which is about 75 percent rock and 25 percent ice. Like many other planets and moons, its surface is covered by impact craters, mountains, and flat, craterless plains. While the maria on Earth's moon are blanketed by hardened lava, those on Triton are filled with ice or frozen methane. Triton, with a diameter of 2,700 kilometers, is extremely unusual because of its retrograde orbit. The moon revolves around the planet in the direction opposite to the planet's rotation. This suggests that Triton was actually captured by Neptune's gravity. It has a thin atmosphere composed of nitrogen, much of which is condensed into a frost due to the cold temperature of −235°C. Voyager 2 discovered smooth volcanic plains, frozen lava flows, and active geysers, providing evidence of tectonic activity.

checkpoint Why is life not likely to have formed on Titan?

23.4 ASSESSMENT

1. **Recall** What are the Galilean moons and what features distinguish them?
2. **Identify** Which two spacecraft used probes to study atmospheres in the outer solar system and where were the probes dropped?
3. **Compare** Which of the Jovian moons presented in the text has the greatest probability of hosting conditions favorable for life?

Critical Thinking

4. **Generalize** Describe physical and orbital characteristics that are likely to make a moon tectonically active.

23.5 DWARF PLANETS, MINOR PLANETS, AND COMETS

Core Ideas and Skills

- Describe what dwarf planets are and why this classification was created.
- Describe the characteristics of minor planets and where they exist in the solar system.
- Distinguish between meteors, meteoroids, and meteorites.
- Draw a model of a comet to demonstrate what happens as comets approach or depart the inner solar system.

KEY TERMS

dwarf planets
Kuiper belt
minor planet
asteroids
meteoroid

meteor
meteorite
chrondrule
comet

The International Astronomical Union defines three categories for objects that orbit the sun but are not planets or moons: dwarf planets, minor planets, and comets.

Dwarf Planets

Pluto has been a controversial figure in our solar system. Between its discovery in 1930 and 2006, Pluto was considered the ninth planet in our solar system. Pluto's orbit is highly elliptical, at times bringing it closer to the sun than Neptune's orbit. Its orbit is tilted 17° relative to the plane in which the eight planets orbit the sun. To better describe the peculiar characteristics of Pluto, the International Astronomical Union voted in 2006 to create a new category of objects in our solar system, the **dwarf planets**, and named Pluto its first member. Dwarf planets are roughly spherical, orbit the sun, and have not yet cleared debris from the area around their orbits. Ceres is the only known dwarf planet in the inner solar system and is the smallest object in this classification. It has merely 7 percent the mass of Pluto and orbits the sun between Mars and Jupiter, in an area known as the asteroid belt.

Pluto may be related to numerous similar icy bodies that are part of the **Kuiper belt**, a region that extends from about 30 to 50 AU from the sun. Objects in the Kuiper belt are residual planetesimals left over from the formation of the solar system. One Kuiper belt object, named Eris, is 27 percent more massive than Pluto, although its size is yet to be confirmed. Three other Kuiper belt objects found so far are at least half the size of Pluto and may have moons of their own. These bodies, while having enough gravitational force to pull themselves into spherical shapes, cannot be considered planets because they do not meet one of the International Astronomical Union's criteria: They are not large enough to dominate and gravitationally clear their orbital regions of all or most other objects.

Pluto has three moons: Nix, Hydra, and Charon. Nix and Hydra are very small, but Charon is nearly half of Pluto's diameter. By measuring the orbits of these two bodies, astronomers determined the densities of Pluto and Charon, deducing that each contain about 35 percent ice and 65 percent rock.

NASA's New Horizons spacecraft reached Pluto in 2015. Previously, our highest-resolution photographs of Pluto were of poor quality compared with those of other planets (Figure 23-26). New images revealed greater details of Pluto's rocky, icy surface. Infrared measurements showed that Pluto's surface temperature is about −220°C. Spectral analysis of Pluto's bright surface shows that it contains frozen methane (Figure 23-27). Its atmosphere is extremely thin and composed mainly of carbon monoxide, nitrogen, and some methane.

FIGURE 23-26
Go online to view (A) a pixelated ground-based image of Pluto and its moon Charon and (B) a more detailed image taken with the Hubble Space Telescope.

checkpoint What are the characteristics of a dwarf planet?

Minor Planets

The term **minor planet** refers to any object that orbits the sun but is neither a planet, dwarf planet, moon, or comet. Astronomers have discovered a wide ring between the orbits of Mars and Jupiter that contains tens of thousands of minor planets called **asteroids**. This area is commonly referred to as the asteroid belt. Ceres, the largest object in the asteroid belt, has a diameter of 930 kilometers but is actually classified as a dwarf planet, according to the International Astronomical Union.

FIGURE 23-27
NASA's New Horizons took this image of Pluto in July 2015. The colors in this image were refined from image data and are similar to the colors a human would perceive. The heart-shaped area on the lower right of the image is an icy expanse of nitrogen and methane.

The orbit of an asteroid is not permanent like that of a planet. If an asteroid passes near a planet without getting too close, the planet's gravity pulls the asteroid out of its current orbit and deflects it into a new orbit around the sun. Thus, an asteroid may change its orbit frequently and erratically.

Many asteroids orbit near Earth, and some even cross Earth's orbit. If an asteroid passes too close to a planet, it may crash into its surface. As discussed previously, ancient asteroid impacts may have caused mass extinctions on Earth. Statistical analysis shows that there is a one percent chance that an asteroid with a diameter greater than 10 kilometers will strike Earth in the next 1,000 years. Such an impact is unlikely but would possibly cause another mass extinction.

As tens of thousands of asteroids race through the solar system in changing paths, many collide and break apart, forming smaller fragments and debris. A **meteoroid** is a small asteroid or a fragment of a comet that orbits through the inner solar system. If a meteoroid travels too close to Earth, it falls into the atmosphere. Friction heats the meteoroid until it glows. To our eyes it is a fiery streak in the sky, which we call a **meteor** or "shooting star." Most meteors are barely larger than a grain of sand when they enter the atmosphere and vaporize completely during their descent. Larger ones, however, may reach Earth's surface. A meteor that strikes Earth's surface is called a **meteorite**.

Most meteorites are stony and are composed of 90 percent silicate rock and 10 percent iron and nickel. The 90:10 mass ratio of rock to metal is similar to the mass ratio of the mantle to the core in Earth. Therefore, geologists think that meteorites reflect the primordial composition of the solar system and are windows into our solar system's past. Most stony meteorites contain small grains about 1 millimeter in diameter called **chondrules**, which contain organic molecules, including amino acids, the building blocks of proteins.

Some meteorites are metallic and consist mainly of iron and nickel, the elements that make up Earth's core, while the remainder are stony-iron, containing roughly equal quantities of silicates and iron-nickel. Part of our knowledge of Earth's mantle and core comes from studying meteorites, which may be similar to the mantles and cores of other planetary bodies. Except for rocks returned from the Apollo moon missions, meteorites are the only physical samples we have from space.

checkpoint What happens when a meteoroid passes through Earth's atmosphere as it falls to the surface?

Comets

Occasionally, a glowing object appears in the sky, travels slowly around the sun in an elongated elliptical orbit, and then disappears into space. Such an object is called a **comet**, after the Greek word for "long-haired." Despite their fiery appearance, comets are cold, and their light is reflected sunlight (Figure 23-28).

Comets originate in the outer reaches of the solar system, and much of the time they travel through the cold void beyond Pluto's orbit. A comet is composed mainly of water ice mixed with frozen crystals of methane, ammonia, carbon dioxide, and other compounds. Smaller concentrations of dust particles, composed of silicate rock and metals, are mixed with the lower density frozen compounds.

When a comet is millions of kilometers from the sun, it is a ball of rock and ice without a tail. As the comet approaches the sun and is heated, some of its surface vaporizes. Solar wind blows some of the lighter particles away from the comet's head to form a long tail. At this time, the comet consists of a dense, solid nucleus, a bright outer sheath called a coma, and a long tail (Figure 23-29). Some comet tails are more than 140 million kilometers long, almost as long as the distance from Earth to the sun. As a comet orbits the sun, the solar wind constantly blows the tail so that it always extends away from the sun. By terrestrial standards, a comet tail would represent a cold laboratory vacuum—yet viewed from a celestial perspective it looks like a hot, dense, fiery arrow.

Halley's comet passed so close to Earth in 1910 that its visit was a momentous event. When the comet returned to the inner solar system in 1986, it was studied by six spacecraft as well as by several ground-based observatories. Its nucleus is a peanut-shaped mass approximately 16×8×8 kilometers, about the same size and shape as Manhattan Island. The cold, relatively dense coma of Halley's comet had a radius of about 4,500 kilometers when it passed by.

In one spectacular experiment conducted in 2005, astronomers fired Deep Impact, a

FIGURE 23-28
In March 1996, comet Hyakutake passed within 15 million kilometers of Earth. That is closer than any other comet has passed by Earth in 200 years. The comet was discovered about two months previously.

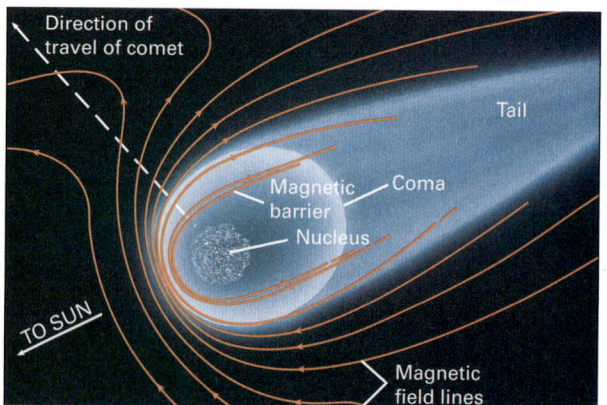

FIGURE 23-29
The comet nucleus has been enlarged several thousand times to show detail. When the comet interacts with the solar wind, magnetic field lines are generated. Gases and ions stripped from the nucleus create the characteristically shaped tail.

MINILAB Micrometeorites

Materials (per group):

3 clear plastic sandwich bags
clear plastic gallon-size bag
neodymium magnet, with hook
twist ties
kitchen sieve (opening approx. 1.5 mm, number 12 or 14 mesh)
fine sieve (opening approx. 0.4 mm, number 40 or 45 mesh)
paper plate
microscope
glass petri dish
toothpicks
protective gloves

Procedure

1. Place the magnet in one small plastic bags. Then place in a second bag so the magnet is double-bagged. Secure with twist-ties.
2. Choose a location outside, near the bottom of a downspout or where a gutter empties. Drag the plastic-covered magnet across the area.
3. Hold the magnet and the particles it has attracted over the gallon-size bag. Gently remove the smaller bags from the magnet so the sediment falls into the larger plastic bag.
4. Pass the sediment through the coarse sieve onto a paper plate. Discard the sediment that stayed in the sieve.
5. Then pass the sediment from the paper plate through the fine sieve. This time, keep the sediment that stays in the sieve. This is sediment that is neither fine nor coarse. You may rinse this material if it seems mucky.
6. Examine a sample of the medium-grained sediment through a microscope. Look for spherical objects. Use a toothpick to separate the sediments for better viewing.

Results and Analysis

1. **Observe** Describe the objects you identified as possible micrometeorites.
2. **Compare** How did the shape of micrometeorites differ from the shape of other kinds of sediment collected?

Critical Thinking

3. **Infer** Why do you think people often find micrometeorites near downspouts?
4. **Predict** Where else would you expect to find micrometeorites?

372-kilogram metal probe, into comet Tempel 1, while a spacecraft recorded the effects of the impact. The probe triggered two quick flashes. The first occurred when the collision heated the surface to thousands of degrees Celsius. The second, occurred a few milliseconds later, when the now-molten probe penetrated deeper into the nucleus and ejected a layer of volatile material. Photographs taken just prior to the collision reveal a landscape sculpted by outgassing, melting, and natural impacts—all evidence of a complex geologic history. Analysis of the material ejected at impact suggests that comet Tempel 1 was composed of fine dust, more like talcum powder than beach sand. There were no large chunks, so the comet did not have a solid ice crust. In the near-surface interior, the space probe detected organic compounds similar to those that form the basis for living organisms.

checkpoint Describe the structure of a comet.

23.5 ASSESSMENT

1. **Describe** How are Pluto's orbital characteristics different from those of the eight planets?
2. **Describe** What are the characteristics of Kuiper belt objects?
3. **Contrast** How do meteoroids, meteors, and meteorites differ?
4. **Explain** What happens to a comet as it approaches or recedes from the sun?

Computational Thinking

5. **Design an Algorithm** Consider the definitions of a planet, dwarf planet, minor planet, moon, and comet. Design an algorithm or a flowchart that uses the physical characteristics of size, mass, composition, location, orbital motion, and other qualities if necessary, to classify any object in our solar system.

TYING IT ALL TOGETHER
DISTANT WORLDS

In this chapter, you learned about worlds large and small that exist in our solar system. We explored the rocky terrestrial planets closest to the sun and the huge Jovian planets that lie far beyond the asteroid belt. We discussed the history of Earth's moon and studied some of the fascinating moons of Jupiter, Saturn, Uranus, and Neptune. We even examined some smaller bodies of the solar system such as dwarf planets, asteroids, comets, and meteoroids.

In the Case Study, we introduced a few of the missions responsible for much of what we know about our solar system. The images and data sent back by unmanned spacecraft have taught us more about distant worlds than we could ever hope to learn using even the most powerful Earth-bound telescopes. Even so, our knowledge of the solar system is still quite limited, and each new discovery breeds a host of new questions for astronomers to pursue.

Think about what sparked your curiosity as you read about our amazing solar system. What planet, moon, or other space object would you like to learn more about?

In this activity, you will select an object in the solar system to research further and then report your findings. Work individually or with a group to complete the steps.

1. Choose an object in the solar system to research. Review any information in the chapter about this object. Then, do some preliminary research to make sure there is enough outside information available to create a presentation. Consult with your teacher to make sure you have chosen an appropriate topic.

2. Develop a list of research questions about the space object. You may use the following list as a starting point, but brainstorm to generate additional research questions that are more specific to your topic.

    - What are the main characteristics of the object?
    - What was the very earliest study of this object? What technology was used to study it?
    - How have developments in technology helped people learn more about this object?
    - What are the most important discoveries that have been made about the object? Can they tell us anything about the formation or early history of the object or of Earth?
    - What hypotheses about the object are scientists currently investigating, and how?

3. Organize your research findings into a presentation. Include photos and multimedia if possible. Share your presentation with the class.

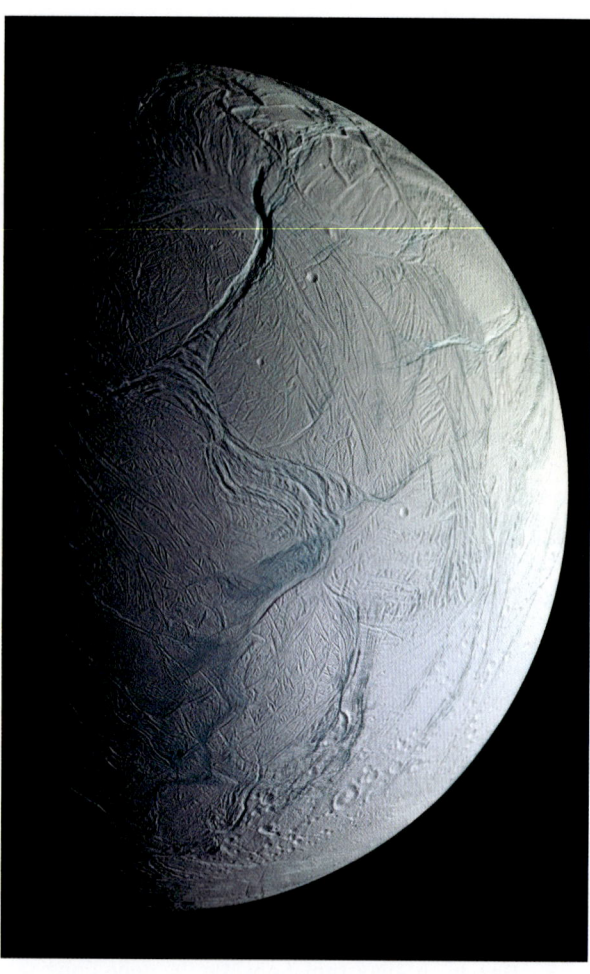

FIGURE 23-30
The Cassini mission discovered that Saturn's moon Enceladus is covered in a crust of ice under which lies an ocean of liquid water.

CHAPTER 23 SUMMARY

23.1 THE SOLAR SYSTEM: A BRIEF OVERVIEW

- The sun, eight planets, and other worlds in our solar system formed out of a cold, diffuse, rotating nebula about 4.6 billion years ago.
- The terrestrial planets are smaller, much denser, have fewer moons, and orbit closer to the sun than the Jovian planets.
- Jovian planets have dozens of moons and more moons are still being discovered.

23.2 THE TERRESTRIAL PLANETS AND EARTH'S MOON

- Mercury is the closest planet to the sun. It has virtually no atmosphere, and it rotates slowly on its axis. It experiences extremes of temperature.
- Mercury's surface is heavily cratered from meteorite bombardment that occurred early in the history of the solar system.
- Venus has a hot, dense atmosphere as a result of a runaway greenhouse effect.
- Venus's surface shows signs of recent tectonic activity called blob tectonics.
- Mars is a dry, cold planet with a thin atmosphere, but its surface bears signs of tectonic activity and ancient water erosion.
- The rovers Spirit and Opportunity detected clear signs that water once existed on Mars. Many rovers have explored the planet.
- Earth's moon probably formed from the debris of a collision between a Mars-sized body and Earth. The moon was heated by the energy released during condensation of the debris and by meteorite bombardment.
- While evidence of ancient volcanism exists, the moon is cold and mostly inactive today.
- In 2018 NASA instruments confirmed direct evidence of water ice in craters at the moon's polar regions.

23.3 THE JOVIAN PLANETS

- Jupiter, Saturn, Uranus, and Neptune are giant planets with low overall densities in comparison to terrestrial planets.
- Jupiter and Saturn have dense atmospheres, inner layers of liquid molecular hydrogen, lower zones of liquid metallic hydrogen, and cores of rock and metal.
- All the Jovian planets have one or more rings.
- Uranus and Neptune are smaller than Jupiter and Saturn and have higher density.
- Uranus and Neptune both contain an inner layer of hydrogen, helium and methane gas, and a mantle of water and ice.

23.4 THE JOVIAN MOONS

- The largest moons of Jupiter are Io, Europa, Ganymede, and Callisto, which are large spheres of rock and ice.
- Titan, the largest moon of Saturn, has an atmosphere rich in nitrogen and methane and the surface shows evidence that landscape was sculpted by wind, flowing liquids, and tectonic activity.
- Triton, Neptune's largest moon, is about 75 percent rock and 25 percent ice.

23.5 DWARF PLANETS, MINOR PLANETS, AND COMETS

- Pluto is a low density dwarf planet composed of ice and rock.
- Asteroids are small, planet-like bodies that orbit the sun between the orbits of Mars and Jupiter.
- A comet is a small body composed of water ice, frozen volatiles, dust, rock and metals that orbits the sun in an elongated, elliptical orbit.
- A meteorite is a meteoroid that falls to Earth. Most meteorites are stony, and some contain organic molecules. About 10 percent of all meteorites are metallic.

CHAPTER 23 ASSESSMENT

Review Key Terms
Select the key term that best fits the definition. Not all terms will be used, and no term will be used more than once.

asteroid	liquid metallic hydrogen
blob tectonics	maria
chondrule	meteor
comet	meteorite
dwarf planet	meteoroid
escape velocity	minor planet
Jovian planet	solar wind
Kuiper belt	terrestrial planet

1. vertical rising and sinking of the mantle and crust as opposed to the horizontal movement of plates
2. a small, dry, rocky body in orbit around the sun between Mars and Jupiter or beyond the orbit of Neptune
3. the speed that an object must attain to exit the gravitational field of a planet or other object in space
4. a celestial object that orbits the sun but is not a planet or comet
5. dry, barren, flat expanses of volcanic rock on the moon
6. a small grain, about 1 millimeter in diameter, embedded in a meteorite, consisting mostly of the silicate minerals olivine and pyroxene
7. a stream of charged particles radiating outward from the sun at high speed
8. a small interplanetary body traveling in an irregular orbit through the inner solar system
9. a small interplanetary body that does not completely vaporize and that strikes Earth's surface
10. a small interplanetary body that enters Earth's atmosphere and glows as it vaporizes
11. an outer planet with a relatively small rocky or metal core surrounded by a swirling liquid and gaseous atmosphere
12. an object composed of loosely bound rock and ice that appears to form a bright head and extended fuzzy tail when it approaches the sun
13. disk-shaped region beyond the orbit of Neptune that contains asteroids, comets, and other debris from the formation of the solar system
14. a body that orbits the sun, is not a satellite of a planet, and is massive enough to pull itself into a spherical shape but not massive enough to clear out other bodies in and near its orbit
15. any of the four Earth-like planets, which are composed primarily of metals and silicate rocks

Review Key Concepts
Answer each question on a separate sheet of paper to demonstrate your understanding of key concepts from the chapter.

16. How are the terrestrial planets different from the Jovian planets?
17. Mercury, Venus, and Mars are the three terrestrial planets besides Earth. Select one of these planets and describe its key features.
18. What evidence exists that liquid water once flowed on Mars? What technologies enabled the discovery of this evidence?
19. How is blob tectonics different from the plate tectonics that occur on Earth?
20. Based on available evidence, what is the most widely accepted hypothesis about the origin of the moon?
21. What are some characteristics of Jupiter? How does Jupiter differ from the terrestrial planets?
22. What are the four Galilean moons? What are the characteristics of each?
23. What two hypotheses have been proposed to explain the formation of Saturn's rings?
24. What is the Kuiper belt? What characteristics do Kuiper belt objects share?
25. Compare and contrast asteroids, comets, meteors, meteoroids, and meteorites.

Think Critically
Write a response to each question on a separate sheet of paper. Use concepts from the chapter to support your reasoning.

26. How would the escape velocity of an object from Earth change with its altitude? Explain your reasoning.
27. A number of organizations, such as NASA and SpaceX, are developing plans to send humans to Mars. What is the advantage of sending humans to Mars rather than relying on robots? What are the risks?

28. Compared to the other planets, scientists understand very little about Uranus and Neptune. What are some challenges that prevent scientists from studying Uranus and Neptune in more detail?

29. More than 60 moons orbit the planet Saturn along with the countless fragments of dust, rock, and ice that make up Saturn's rings. Drawing on your knowledge of the solar system, develop a hypothesis that would explain why Saturn has so many natural satellites. Explain your reasoning.

30. According to the International Astronomical Union, a planet is (1) an object that independently orbits the sun, (2) has enough mass to pull itself into a spheroidal shape, and (3) is large enough to dominate its orbit. Based on this definition, is Earth a planet? Is Earth's moon a planet? Is Pluto a planet? Explain your reasoning.

PERFORMANCE TASK

Formation and Early History of the Solar System

According to scientists, the sun and the rest of the solar system formed approximately 4.6 billion years ago. But what evidence has led scientists to this conclusion, and how do scientists go about reconstructing the formation and early history of our solar system?

In this performance task, you will work with your class to research and compare several lines of evidence. Then, you will construct your own timeline of the early history of Earth and the solar system.

1. Research and construct a timeline for the formation of the solar system.
   A. In your small group, select one of the following lines of physical evidence as a focus for your research.
      - absolute ages of ancient materials (meteorites, moon rocks, Earth's oldest minerals, etc.)
      - sizes and compositions of solar system objects (planets, moons, asteroids, etc.)
      - impact cratering of planets and moons
   B. In your small group, conduct research to identify at least five pieces of evidence from your selected category. Consulting library resources, museum archives, or internet references, investigate what each piece of evidence reveals or suggests about events and developments in the early history of Earth or the solar system.
   C. For each piece of evidence, prepare at least two slides. On the first slide, summarize the evidence. On the second slide, summarize what the evidence suggests about the formation and early development of the solar system or Earth.
   - Include any relevant images that support your findings; you may include additional slides for this purpose if needed.
   - At the end of your presentation, add a summary slide containing a timeline of all of the developments your research has uncovered.
   D. As a class, combine the timelines from all groups onto a single timeline. When complete, display your class's completed timeline and do a gallery walk to review the history of early Earth and the solar system. As you review the timeline, note any developments that are supported by multiple lines of evidence. Also note any developments in the timeline that seem contradictory.

2. Based on the class's research, what developments in the early history of Earth and the solar system are most widely accepted by the scientific community, and why?

3. Based on the class's research, what developments in the early history of Earth and the solar system are still subject to debate, and why?

CHAPTER 24
STARS

When we look up at the night sky, we see dozens of stars whose light has traveled several years to thousands of years to reach us. Our nearest star, the sun, is only eight light-minutes away and provides a majority of the energy in our solar system. It is a relatively stable, quiet, and average star compared to others we have observed. Studying the sun and other stars provides a wealth of information about how stars form, evolve, and eventually die. American astronomer Carl Sagan famously said, "The cosmos is within us. We are made of star stuff. We are a way for the universe to know itself." Thus, it is thanks to stars that we know anything about the solar system, the universe, and Earth's place in the universe.

KEY QUESTIONS

24.1 How and why do stars form?

24.2 What is the internal and external structure of the sun?

24.3 How do stars change during their lifetimes?

24.4 What are the extreme objects left after stars die?

Astronomers imaged the Westerlund 2 star cluster (red stars just right of center) in the Gum 29 star-forming region (colorful, multi-structured gas cloud left of center) using the Hubble Space Telescope. It takes light from a star along the edge of this region 30 years to travel to the other edge.

EXPLORERS AT WORK

INVESTIGATING EXOPLANETS

WITH NATIONAL GEOGRAPHIC EXPLORER MUNAZZA ALAM

Munazza Alam has stars in her eyes—and she hopes they will lead her to other planets like our own. Alam is pursuing a Ph.D. in astronomy at Harvard University. Her research focuses on exoplanets: planets orbiting stars beyond our solar system. Exoplanets are surprisingly common—there are more planets in the universe than there are stars. The question is: Are any of those distant planets capable of supporting life, as our planet does?

Alam investigates that question using data from the Hubble Space Telescope to study exoplanets' atmospheres. "My long-term goal is to find an Earth twin exoplanet with an atmosphere like ours," says Alam. With an estimated 100 billion exoplanets in our galaxy alone, the prospects are exciting. The first exoplanet was confirmed in 1995, and more than 3,000 exoplanets have been identified to date. About 2,500 more candidates await confirmation.

Directly observing exoplanets is tricky. The nearest one known is 40 trillion kilometers away. In addition, light from an exoplanet's host star makes it nearly impossible to see. Alam and other astronomers use the transit method to detect them. As an exoplanet transits, or passes in front of, its host star, the star appears to dim because the exoplanet blocks some of its light. Astronomers use this change in brightness to determine the exoplanet's size. They measure time between transits to characterize the planetary system, including how far the exoplanet is from its host star, which tells them about conditions on the planet.

Distance is critical, says Alam. Too close, and the exoplanet may be boiling hot. Too far, and the exoplanet may be frozen. Neither condition is conducive to life. "The key is to find exoplanets that orbit in the star's habitable zone," says Alam. "That is a region at a distance from the star where liquid water could exist on the surface." Early searches for exoplanets found mostly hot Jupiter-like planets because they are the brightest and biggest, but this type of planet is actually rare in the universe. Smaller planets called super-Earths, between the size of Earth and Neptune, are the most common. One of these super-Earths might just have the right atmospheric ingredients Alam is looking for—gases such as carbon dioxide, methane, and oxygen.

Transit give information about an exoplanet's atmosphere. As an exoplanet transits its host star, some starlight shines through the exoplanet's atmosphere. Various types of gas molecules in the atmosphere absorb light at different wavelengths. By looking at which wavelengths of starlight the exoplanet absorbs, Alam can determine the presence or absence of particular gases in the atmosphere. She can even tell whether clouds are present.

We may never know whether planets beyond our solar system host life. But Alam's work may help us to one day know at least whether it is possible.

THINKING CRITICALLY

Explain Scientists often must make indirect studies of objects and processes that are too complex, too dangerous, or too physically difficult to observe directly. Explain how this applies to the work that Munazza Alam does as an astrophysicist.

Astronomers Munazza Alam (center), Haley Fica (left), and Sara Camnasio (right) pose in front of one of the 6.5-meter-diameter Magellan Telescopes in Chile.

CASE STUDY
JOURNEY TO THE SUN: PARKER SOLAR PROBE

In Chapter 22, you read about several NASA space probes such as Galileo and Voyager 1 and 2 that have explored the farthest reaches of our solar system, sending us close-up images of the planets and their moons. But a more recent mission is turning NASA's attention inward to a star just 150 million kilometers (93 million miles) away: our sun. The Parker Solar Probe, launched in August 2018, set out on a seven-year journey that will bring it near the sun several times. Three of those orbits will carry the probe less than six million kilometers from the sun's surface, seven times closer than any previous spacecraft. For some of its journey, the probe will travel at speeds close to 700,000 kilometers per hour.

As you can imagine, conditions so close to the sun are extreme. The probe's path will take it inside the solar corona, the outer layer of the sun's atmosphere, where it will encounter superhot plasma. But the Parker Solar Probe is designed to withstand temperatures of up to 1,377°C (2,500°F), protected by a carbon-composite shield 11 centimeters thick.

One of the main goals of the Parker Solar Probe mission is to study the solar wind. The solar wind is a stream of charged particles that leave the sun's corona, moving with so much energy that they escape the sun's enormous gravitational pull. These high-energy particles travel into space from the corona at about 500 kilometers per second, bombarding anything in their path. Fortunately for us, Earth's magnetic field and atmosphere shield us from the solar wind. When the charged particles reach Earth's magnetic field, most are deflected around it (Figure 24-1). A few particles do enter the atmosphere, where they react with gases to produce the glowing lights we know as auroras, or the northern and southern lights.

Although Earth's magnetic field and atmosphere help protect us from the solar wind, it can still cause problems. During solar storms, disturbances on the sun can cause periods of strong solar wind that can harm satellites orbiting Earth, causing disruptions in communication networks, global positioning systems (GPS), and even electrical grids. Solar storms can also spell trouble for astronauts working outside the shield of Earth's atmosphere and magnetic field. Scientists hope the Parker Solar Probe will give us new insights into the nature of the solar corona and what causes the acceleration of particles in the solar wind.

As You Read Consider why it is important for people to study the sun. What can the sun tell us about other stars that are millions of light-years away, and why this is an important area of research?

FIGURE 24-1
The solar wind, emanating from the sun's corona, meets Earth's magnetic field, flattening it on the side facing the sun.

24.1 THE BIRTH OF STARS

Core Ideas and Skills
- Describe how interstellar gas turns into a star.
- Explain the process that begins when a gas cloud turns into a star.

KEY TERMS
nebula fusion

Birth of Stars

Over long periods of time when the universe was very young, the gravitational pull of denser regions of space drew matter inward to form huge clouds of hydrogen with traces of helium. Such a cloud of interstellar gas and dust is called a **nebula**. Within each cloud, matter further condensed into billions of smaller bodies. As the atoms in a nebula moved closer to each other under the force of gravity, they collided more and more rapidly. Parts of each cloud became very dense and hot. Under the intense heat, electrons were stripped away from their atoms, leaving a plasma of positively charged nuclei and negatively charged electrons.

Two positive protons repel each other, like two north poles of a magnet. However, if protons are hot enough, they collide with sufficient energy to overcome the repulsion and bond together. This bonding is called **fusion**. Hot protons can also fuse with hot neutrons.

Dense regions of interstellar clouds with the right temperature and pressure will begin to fuse hydrogen. A hydrogen nucleus consists of a single proton. At the right temperature, four protons (hydrogen nuclei) fuse together in a series of reactions to form a nucleus of the next-heavier element, helium, and other subatomic particles. This nuclear fusion signals the birth of a star. The process from interstellar cloud collapse to fusion ignition takes around 10 million years.

Fusion generates energy in the form of electromagnetic radiation, high-energy nuclear particles, and a tremendous amount of heat. If a dense sphere of hydrogen the size of a pinhead were to fuse completely, it would release as much energy as burning several thousand tons of coal. Inside a star, both the photons and the high-energy particles generated by fusion accelerate outward against the force of gravity. Thus, two opposing forces balance in a star. Gravity pulls particles inward, but fusion energy drives them outward (Figure 24-2). The balance between these two forces determines the diameter and density of a star of any given mass. At equilibrium, a star with an average mass has a dense inner core surrounded by a less-dense shell.

A star exists for millions, billions, or tens of billions of years. We can never observe one long enough to watch its birth, life, and death. We can, however, observe young, middle-aged, and old stars and thus piece together the story of stellar evolution.

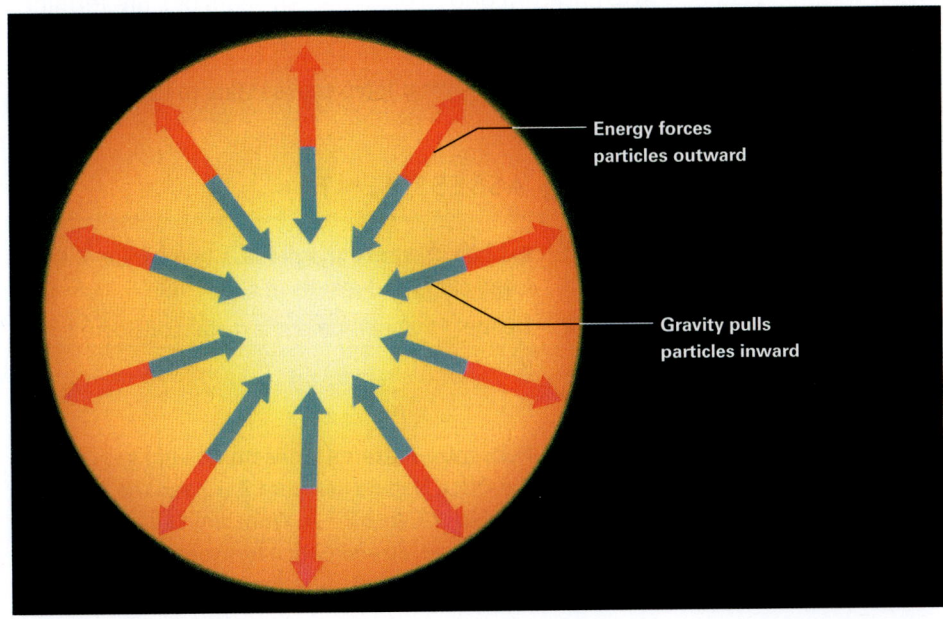

FIGURE 24-2
The diameter and density of a star are determined by two opposing forces. Gravity pulls particles inward while energy from fusion reactions in the core forces particles outward.

FIGURE 24-3
This image from NASA's Hubble Space Telescope of the Orion Nebula, the brightest spot in the sword of the Orion constellation, shows a stellar nursery where young stars are evolving.

FIGURE 24-4
NASA's Hubble Space Telescope image of Eagle Nebula shows its detailed structure. The pillar-like structures are about three light-years in length and are composed of cold gas and dust. Energy from nearby stars and shock waves from exploding ones have concentrated the gas into pillars and spurred the formation of new stars.

One of the most studied nebulae is the Orion Nebula in the constellation Orion (Figure 24-3). This nebula, like so many others, is a birthplace and nursery for newborn stars. Even though a nebula looks spectacular when viewed through a powerful telescope, the densest nebulae are only between 10^{-13} and 10^{-18} as dense as Earth's atmosphere at sea level! The Orion Nebula, which is a mere 10^{-15} as dense as Earth's atmosphere, is nearly two million AU wide. Even though it is extremely diffuse, it contains an equivalent of 200,000 times the mass of the sun. The two lightest elements, hydrogen and helium, constitute more than 99 percent of the cloud's total mass, but the nebula also contains all or most of the natural elements. In addition, astronomers have detected hundreds of different molecules in the Orion Nebula. Most are simple—such as hydrogen, water, carbon monoxide, and ammonia. But several larger, organic molecules have also been detected.

The stars in one portion of the Orion Nebula are at least 12 million years old. Stellar ages decrease progressively toward the southwest until the stars are less than a million years old. This age progression implies that star formation moved steadily from one end of the nebula to the other.

The shape and appearance of nebulae vary widely. Recent observations by the Hubble Space Telescope of the Eagle Nebula show that regions of varying density appear as towering pillars of light and shadow (Figure 24-4). Hundreds of young stars, only 300,000 to eight million years old, are forming along the dense edges of the cloud.

checkpoint What force causes stars to form?

24.1 ASSESSMENT

1. **Explain** Describe the process that begins with a gas cloud and results in a new star.
2. **Define** At what moment is a star considered born?

Critical Thinking

3. **Recognize Patterns** Consider the images of the Orion Nebula in Figure 24-3 and the Eagle Nebula in Figure 24-4. What similarities and differences do you notice between these two stellar nurseries?

24.2 THE SUN

Core Ideas and Skills
- Develop a model showing the internal and external structure of the sun.
- Describe the ways in which the sun's energy reaches or impacts Earth.

KEY TERMS

radiative zone
convective zone
photosphere
sunspot
chromosphere
spicule
prominence

The Sun Is a Star

No one has ever traveled to the sun and no human ever will. Until late 2018, with the launch of the Parker Solar Probe (Case Study), we had never sent spacecraft into the solar atmosphere. The sun's core will be forever covered and invisible to us because any spacecraft would vaporize long before it reached the solar interior. All of our information about the sun is derived from indirect evidence. Scientific reasoning is so powerful that we have a reasonably complete and accurate picture of our closest star.

By studying the orbital properties of the planets, scientists can calculate the gravitational force of the sun and hence its mass, approximately 1.99×10^{30} kilograms. Although this huge number is almost impossible to comprehend, our sun is a star of average mass. The sun's diameter is 1.4 million kilometers, or about 109 Earth diameters across.

Astronomers can determine the temperature and composition of a star by analyzing its emission and absorption spectra. Just like the iron bar example in Chapter 19, stars give off electromagnetic radiation that depends on their temperature. The color of a star yields information about its surface temperature. Bluer stars are hotter whereas red stars are cooler. In addition, the spectrum of a star provides clues about its chemical composition. The sun's visible spectrum peaks in the green and contains absorption features that give away its internal composition (Figure 24-5). From these studies, we have learned that hydrogen accounts for 92 percent of the total number of atoms making up the sun. Helium accounts for about 7.8 percent of the total number of atoms in the sun, with all remaining elements accounting for only 0.2 percent.

checkpoint What are the main elements found in the sun?

The Sun's Structure

In the late 1800s, scientists calculated that the sun emitted so much energy it could not be powered by the burning of conventional fuels. But if the sun was not a giant fire, what was it? In 1905, Albert

FIGURE 24-5
The sun's spectrum shows a series of dark absorption lines on a continuous light spectrum from red to blue.

FIGURE 24-6
ON ASSIGNMENT National Geographic photographer Babak Tafreshi caught a bright meteor streaking across the sky above a telescope at the La Silla Observatory in Chile. Such short-lived light shows are sometimes called shooting stars although they are not stars at all.

Einstein announced his theory of relativity, and shortly thereafter other scientists began to unravel the mystery of nuclear reactions. These studies showed that the sun is powered by hydrogen fusion. Every second, the sun converts four million tons of hydrogen into energy and radiates this energy into space. Yet the sun is so massive that there is no detectable change in mass from year to year, or even from century to century.

Our understanding of the solar core (Figure 24-7) is derived from calculations of hydrogen fusion reactions and gravitational forces. From these studies, scientists are reasonably certain that the core of the sun is extremely hot, more than 15 million K, and its density is 150 times that of water. The sun's core contains enough hydrogen to fuel its fusion reaction for another five billion years. When the hydrogen in the core is used up, the sun will change drastically.

Most of the energy generated by fusion in the sun's core is emitted as radiation of high-energy photons. When a photon leaves the core, atoms almost immediately absorb it in a broad region surrounding the core called the **radiative zone**. Atoms reemit photons soon after they absorb them, but the gas density in the region is so high that a photon is quickly reabsorbed again. This process of absorption and emission occurs repeatedly until the single photon reaches the **convective zone**. Warmer matter that has absorbed the photon's energy at the bottom of the convective zone moves outward while cooler mater sinks. The convective zone is much cooler on its surface than in its interior. As a result, hot gas rises, cools, and then sinks in huge convection cells. This convection transports heat to the outer layer that we see, called the **photosphere**.

If photons traveled directly from the core to the photosphere at the speed of light, they would complete the journey in about four seconds. However, the process of absorption, emission, and convection is much slower. The energy requires about a million years to make the journey from core to surface.

Even though the entire sun is gas, the photosphere is considered the first layer of the solar atmosphere because it is cool and diffuse. It is a thin surface veneer, only about 500 kilometers thick. The pressure of the photosphere is about one-hundredth that of Earth's atmosphere at sea level. Its average temperature is about 5,700 K, about the same as that in Earth's core. Fusion does not occur at this temperature. The core heats the photosphere, and the sunlight we receive on Earth comes from this thin, glowing atmosphere.

The photosphere has a granular structure. Each grain is about a thousand kilometers across, about the size of Texas. Convection currents that carry energy to the surface form the granules. If you could watch the sun's surface at close range, the granules would appear and disappear like bubbles in a pot of boiling water. The bright yellow granules

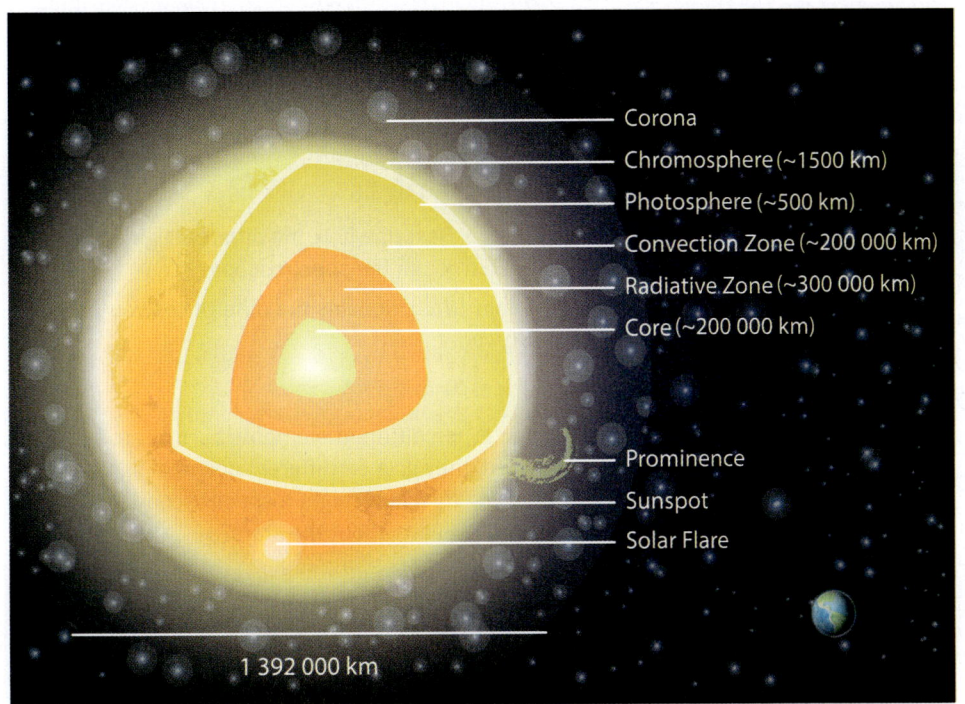

FIGURE 24-7
The sun's inner structure consists of the core, the radiative zone, and the convection zone. The photosphere, the chromosphere, and the corona make up the sun's outer atmosphere. Activity on the sun's surface includes prominences, sunspots, and solar flares, which can be as large as or larger than the entire Earth.

are formed by hot, rising gas, and the darker reddish regions are cooler areas of descending gas.

Large dark spots called **sunspots** also appear regularly on the sun's surface (Figure 24-8). A single sunspot may be small and last for only a few days. It may be as large as 150,000 kilometers in diameter and remain visible for months. Because a sunspot is about 2,000 K cooler than the surrounding area, it radiates only about half as much energy. This is why it appears dark

MINILAB Build a Spectroscope

Materials
old CD or DVD
pre-cut cardboard tube
colored pencils
scissors
tape
pencil
2 small pieces of cardstock, slightly larger than diameter of tube
different types of light sources (e.g., incandescent bulb, neon light, LED bulb, fluorescent light, candle flame)

Procedure

CAUTION: Do not point the spectroscope directly at the sun. It does not block or filter sunlight. Looking directly at the sun can cause serious eye damage.

1. Stand your cardboard tube on its end on one piece of cardstock. Trace around the tube with the pencil. Repeat with the other piece of card stock. Then cut out the circles from both pieces of card stock using scissors.

2. Use scissors to cut a thin rectangular slit (about 3 cm long) across the middle of one of the card stock circles. Cut the rectangle as narrow as you can.

3. Tape the circle with the slit at the end of the tube farthest from the pre-cut slot and viewing hole. This is the end of the tube that will point upward, or toward a light source. Tape the circle without the slit over the other end of the tube.

4. Insert the CD or DVD, shiny side up, in the diagonal cut across from the viewing hole.

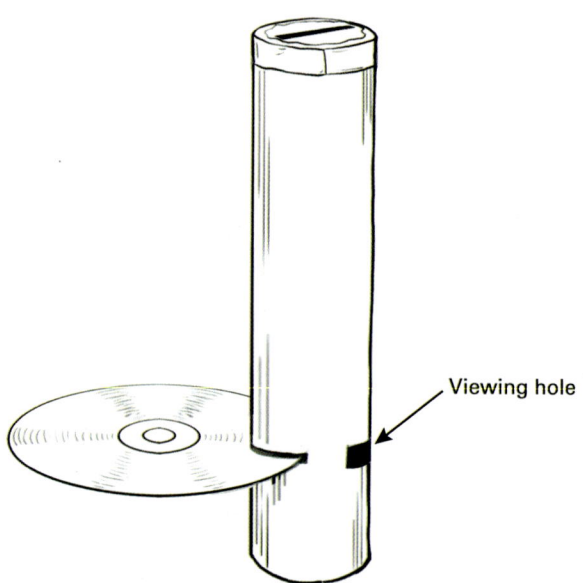

5. Use the spectroscope by pointing the top toward a light source. Look at many different kinds of light—such as an incandescent bulb, a neon light, an LED bulb, fluorescent lights, or a candle flame.

6. Use colored pencils to draw what you observe. Label your drawings and compare them with others.

Results and Analysis

1. **Observe** Describe how the different light sources looked through the spectroscope.

2. **Explain** What explains the differences you observed?

Critical Thinking

3. **Infer** When different elements burn or are heated, they give off different colors of light. Knowing this, how might a spectroscope help scientists learn about the stars?

4. **Predict** What would you expect to see if you looked at a red LED? What about sunlight reflected off a white wall? Why? Test your predictions.

compared to the rest of the photosphere. Because heat normally flows from a hot body to a cooler one, a large, cool region on the sun's surface should quickly be heated and disappear. Yet some sunspots persist for long periods.

Astronomers think the sun's magnetic field prevents material in the photosphere from mixing. The magnetic fields associated with sunspots also produce solar flares. Glowing gases are accelerated to speeds of more than 900 kilometers per hour. This sends back shock waves smashing through the solar atmosphere. The flares emit charged particles that interrupt radio communication and cause aurorae (the northern and southern lights) on Earth.

A turbulent, diffuse gas layer called the **chromosphere** lies above the photosphere. It is roughly 1,500 kilometers thick. The chromosphere can be seen with the naked eye only when the moon blocks out the photosphere during a solar eclipse. Then it appears as a narrow red fringe at the edge of the blocked-out sun. Jets of gas called **spicules** shoot upward from the chromosphere, looking like flames from a burning log. An average spicule is about 700 kilometers across and 7,000 kilometers high and lasts between 5 and 15 minutes.

Recall from Lesson 22.3 that the corona is the sun's outermost atmosphere. It is an even more diffuse region that lies beyond the chromosphere. It is normally invisible except during a full solar eclipse when the corona appears as a beautiful halo around the sun. Its density is one-billionth that of the atmosphere at Earth's surface. In a physics laboratory, that would be considered a good vacuum.

The corona is extremely hot, about two million K. How does the photosphere, at roughly 5,700 K, heat up the corona to two million K? Scientists are still investigating. According to one hypothesis, twisting magnetic fields accelerate particles in the corona. When particles are moving quickly, their temperature is high. Answering this scientific question is one of the goals of the Parker Solar Probe.

One feature found in the corona is a **prominence** (Figure 24-9). Solar prominences are red, flame-like jets of gas that rise out of the corona and travel as much as a million kilometers into space. Some prominences are held aloft for weeks or months by the sun's magnetic fields.

FIGURE 24-9
Go online to view an example of a prominence dozens of times larger than planet Earth.

The high temperature in the corona strips electrons from their atoms, reducing hydrogen and helium to bare nuclei in a sea of electrons. These nuclei and electrons are moving so rapidly that some fly off into space, forming the solar wind that extends outward toward the far reaches of the solar system.

checkpoint Name the three internal layers and three external layers of the sun.

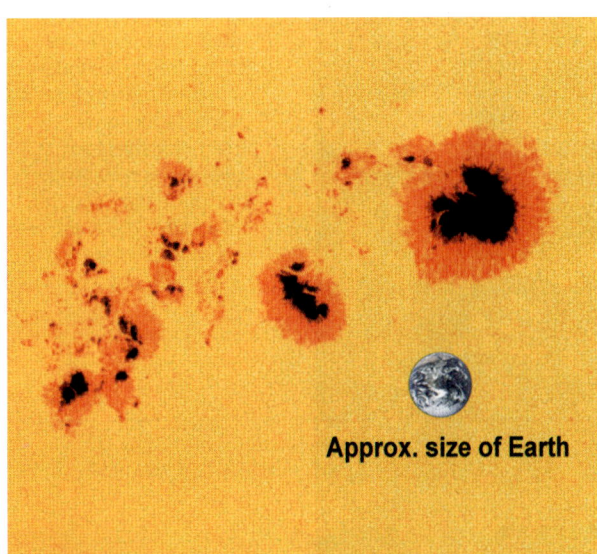

FIGURE 24-8
Sunspots on the sun's photosphere often form in groups. The largest sunspot in this photo could swallow up the entire planet Earth.

24.2 ASSESSMENT

1. **Explain** How does energy move from the sun's core to its surface?
2. **Define** What type of gas makes up the sun, and how do we know?
3. **Describe** What are the visible features that can be seen on each of the sun's external layers?

Critical Thinking

4. **Synthesize** How have scientists been able to study all three of the sun's outer layers?

Computational Thinking

5. **Represent Data** Create a pie chart that shows the sun's current chemical composition. Add labels that show how the pie chart is expected to change over millions of years.

LESSON 24.2

24.3 THE LIFE AND DEATH OF STARS

Core Ideas and Skills
- Describe what happens inside a star during its lifetime.
- Identify how stars of different mass die.

KEY TERMS

apparent brightness	solar mass
absolute brightness	planetary nebula
light-year	white dwarf
parsec	supernova
H–R diagram	population I star
main sequence	population II star
red giant	

Stellar Life

From our view on Earth, the sun is the biggest and brightest object in the sky. However, the sun looks so big and bright only because it is the closest star. When astronomers began to study stars in detail, one of their first efforts was to catalog them by their brightness. The **apparent brightness** of a star is its brightness as seen from Earth. A star can appear luminous either because it is intrinsically bright or because it is close. Think of a car headlight. The light appears to become brighter as the car approaches. The headlight bulb is not changing in true brightness; the car is just moving closer. The **absolute brightness** (also known as luminosity) is how bright a star would appear if it were a fixed distance away. Astronomers have chosen a standard distance of 10 parsecs, or 32.6 light-years, to calculate luminosities. Light-years and parsecs are common units of distance in astronomy. One **light-year** is the distance traveled by light in a year, about 9,500 trillion kilometers (9.5×10^{12} km). A **parsec** is a distance equal to 3.26 light-years. By convention, astronomers use a scale in which the luminosity, or absolute brightness, of a star is divided by the luminosity of the sun. Thus, the sun has a luminosity of 1.0. Objects brighter than the sun have a luminosity greater than 1, and objects fainter than the sun have a luminosity less than 1.

A second important visible property of a star is its color. Different stars have different colors; some are reddish, while others are yellow or blue. The color of a star is an indication of its surface temperature. Blue is the hottest; yellow is intermediate; and red is the coolest. The same relationship can be observed in a flame from a welding torch. The hottest, central portion of the flame is blue, and the cooler, outer edge is red.

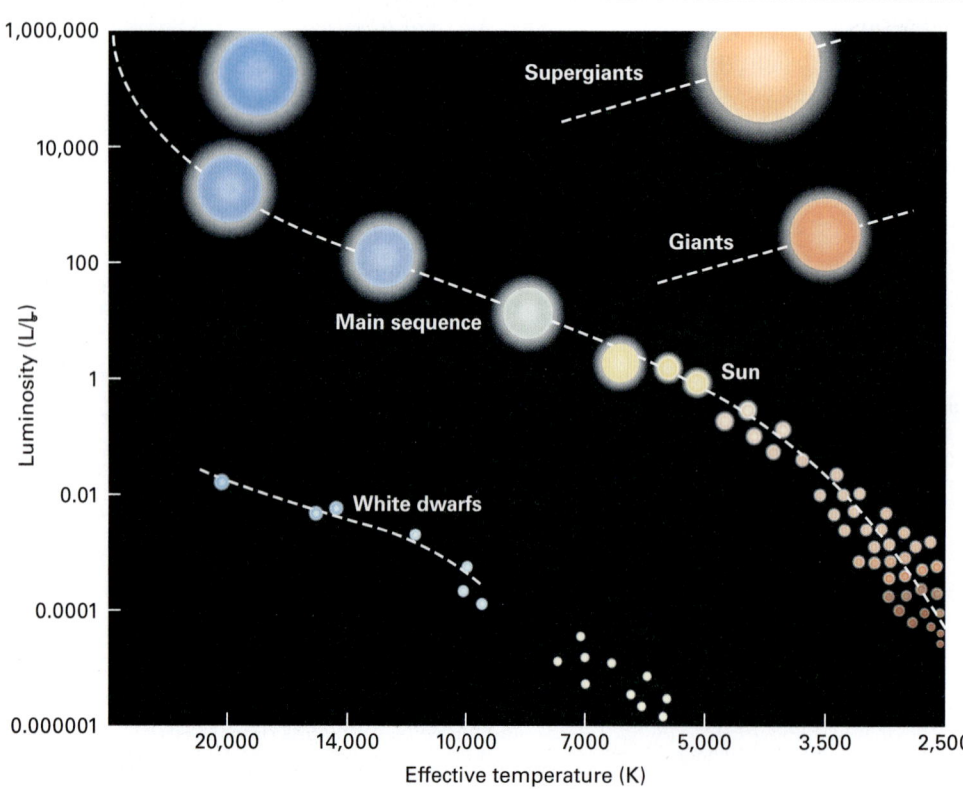

FIGURE 24-10
A Hertzsprung-Russell diagram (H–R diagram) shows that about 90 percent of all stars fall along the main sequence, the band running from upper left to lower right. Our sun is an average-size star with a surface temperature of about 5,700 K and is represented here with a luminosity of 1.

Between 1911 and 1913, Ejnar Hertzsprung and Henry Russell plotted the luminosities of stars against their temperatures. Figure 24-10 shows a Hertzsprung–Russell diagram, or **H–R diagram**. This graph uses a logarithmic scale along both the horizontal and vertical axes. Luminosity, given as a ratio to the sun's luminosity, increases from bottom to top, and temperature increases from right to left. Thus, hot, big, luminous stars are on the upper left. Cool, small, less-luminous stars are on the lower right.

The interesting observation about the H–R diagram is that about 90 percent of all stars fall along a curved band from upper left (bright and hot) to lower right (dim and cool). This band is called the **main sequence**. Thus, luminosity increases with temperature in main-sequence stars. Our sun is more or less in the middle of the main sequence.

All main-sequence stars are composed primarily of hydrogen and helium and are fueled by hydrogen fusion. The major reason for differences in temperature and luminosity among main-sequence stars is that some are more massive than others. Because the force of gravity is stronger in massive stars than in less-massive stars, hydrogen nuclei are packed more tightly and move more rapidly. As a result, fusion is more rapid and intense in massive stars, so they are hotter and more luminous (upper left of the H–R diagram). The least-massive stars, shown on the lower right, are cool and less luminous. Notice that our sun is in the yellow band about halfway along the main sequence.

Stars spend most of their lifetime on the main sequence phase, stably converting hydrogen into helium. To explain the stars that do not lie along the main sequence, we must consider the life and death of a star.

checkpoint What is the main sequence?

Low-Mass Stellar Death

In a mature star such as our sun, hydrogen nuclei fuse to form helium, but the helium nuclei do not fuse to form heavier elements. For fusion to occur, two nuclei must collide so energetically that they overcome their nuclear repulsion. Hydrogen nuclei contain only one proton, whereas helium contains two. Therefore, the repulsion between two helium nuclei is much stronger than between two hydrogen nuclei. This stronger repulsion between helium nuclei can be overcome if the nuclei are moving very rapidly. Since higher temperature causes nuclei to move more rapidly, helium fusion can occur only when a star becomes much hotter than the temperature at which hydrogen fusion occurs.

The sun is now midway through its mature phase as a main-sequence star. It has been shining for about 4.5 billion years and will go on for another 4.5 billion years. During most of its lifetime, the sun produces energy by hydrogen fusion and remains on the main sequence. Figure 24-11 shows the stages in the life cycle of a star with a mass that is similar to our sun's.

FIGURE 24-11
A series of diagrams shows the evolution of a sunlike star. (A) A star on the main sequence fuses hydrogen into helium in the core. (B) When the core has been completely converted into helium, a shell of hydrogen surrounding the core collapses and begins fusing. The star expands into a red giant. (C) The inflated star cools and collapses until helium fusion initiates in the core. (D) Helium fusion converts the core into carbon, and shells of helium and hydrogen fusion continue to produce energy, inflating the star once again.

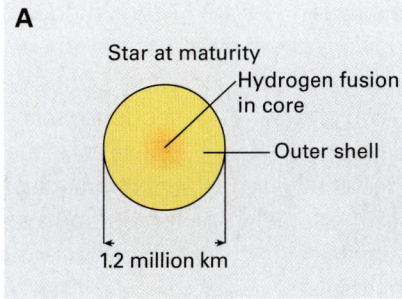
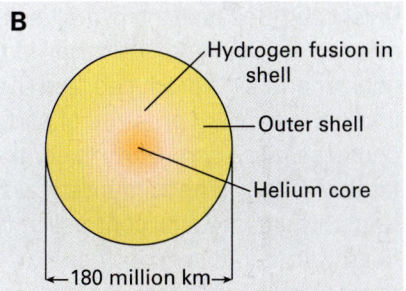
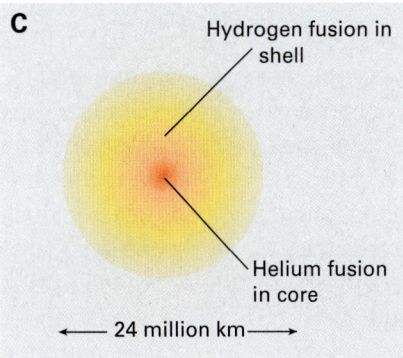
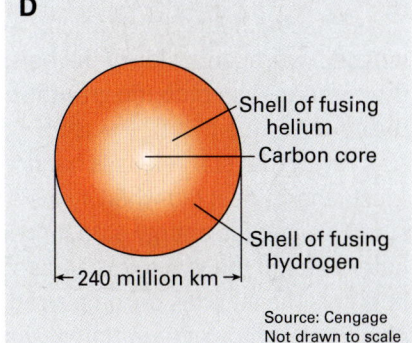

DATA ANALYSIS: Estimating the Lifetime of the Sun

Stars on the main sequence show a mathematical relationship between mass, luminosity, and lifetime. Table 24-1 shows estimated values for different types of stars, based on their masses relative to the mass of the sun. In this activity, you will use the logarithms of the values in the chart to graph the data. Pressing the logarithm (LOG) button on a calculator tells you what power of 10 the number is. This is useful for comparing very large numbers.

TABLE 24-1 Calculations for Estimating the Sun's Lifetime

Mass (relative to the sun)	Luminosity (relative to the sun)	Estimated Lifetime (years)
60	500,000	3.6×10^5
18	32,500	7.8×10^6
6	480	1.2×10^8
2	12	1.8×10^9
0.8	0.6	1.8×10^{10}
0.5	0.08	5.4×10^{10}

Computational Thinking

1. **Calculate** Using a calculator or a spreadsheet formula, calculate $\log_{10}$ of each value in the table. Record your calculations in the table. Then plot two graphs in a spreadsheet, on graph paper, or with a graphing calculator. Make one graph with the axes $\log_{10}$ of Mass and $\log_{10}$ of Luminosity. Make a second graph with the axes $\log_{10}$ of Mass and $\log_{10}$ of Lifetime.

2. **Recognize Patterns** Based on the data in the table, what overall relationship seems to exist between the mass of a star and its luminosity?

3. **Recognize Patterns** Based on the data in the table, what overall relationship exists between the mass of a star and its estimated lifetime?

4. **Use Patterns** Where along the x-axis would Earth's sun be plotted on the graphs?

5. **Use Patterns** Use the second graph to estimate the $\log_{10}$ of the lifetime of the sun. Then use that value as the exponent of 10 to estimate the lifetime of the sun.

After billions of years, the outer shell of a star such as our sun still contains large quantities of hydrogen, but most of the hydrogen in the core has fused into helium. The star's behavior now changes drastically. Because the hydrogen in the core is nearly used up, hydrogen fusion slows down, less nuclear energy is produced, and the core cools. As it cools, the outward pressure of particles and energy decreases. The core starts to contract under the force of gravity. The immense pressure resulting from this gravitational contraction causes the core to grow hotter. It seems a paradox that when the nuclear reactions decrease, the core becomes hotter, but that is what happens.

As the core heats up, the rising temperature initiates hydrogen fusion in the outer shell. The star is now heated by both the gravitational collapse of the core and the hydrogen fusion in the outer shell. As a result, the star releases hundreds of times as much energy as it did when it was a main-sequence star. This intense energy output now causes the outer parts of the star to expand and become brighter. The star has become a **red giant**. A red giant is hundreds of times larger than an ordinary star. Its core is hotter, but its surface is so large that heat escapes and the surface cools. This cool surface emits red light. (Recall that red wavelengths have the lowest energy of the visible spectrum.) These sudden changes in energy production and diameter move the star off the main sequence. Thus, a red giant is brighter and cooler than a main-sequence star of the same mass.

The core of a red giant condenses under the influence of gravity. It gets hotter until its temperature reaches 100 million K. At this temperature, helium nuclei begin to fuse to form carbon. When helium fusion starts, energy pushes outward once again, and the core expands. The star cools, its outer layers contract, and it enters a second stable phase.

Gradually, as more helium fuses to carbon, the carbon accumulates in the core just as helium did during the earlier life of the star. When the helium is used up, fusion ceases again and the carbon core contracts. This gravitational contraction causes the core to heat up again.

What happens next depends on the star's initial mass. Astronomers express the mass of a star relative to that of the sun: one **solar mass** is the mass of the sun. In a star with a mass about the same as our sun, contraction of the carbon core is not intense enough to raise its temperature sufficiently to initiate fusion of the carbon nuclei. However, gravitational contraction of the carbon core does release enough energy to blow a shell of gas out into space. This shell is called a **planetary nebula** (Figure 24-12), although it is unrelated to the formation of planets. Meanwhile, the material remaining in the star contracts until atoms are squeezed so tightly together that only the pressure exerted by the electrons prevents further compression. A dying star as massive as our sun will eventually shrink until its diameter is approximately that of Earth. Such a shrunken star no longer produces energy, and it glows solely from its residual heat produced during past eras. The star has become a **white dwarf**. It will continue to cool slowly over tens of billions of years, but it will never change diameter again. No further nuclear reactions will occur. Its gravitational force is not strong enough to overcome the strength of the electrons, so it will never contract further.

FIGURE 24-12
Go online to view the Ring Nebula, an example of what scientists think the sun will look like in about 4.5 billion years.

checkpoint At what temperature does helium fuse to form carbon?

High-Mass Stellar Death

Some stars do not die so gently. Stars greater than 1.4 solar masses continue to evolve through several stages of fusion. As helium fusion in the core ends, gravitational contraction produces enough heat to fuse carbon. Renewed fusion reactions produce increasingly heavier elements in a series of stages until iron forms in the stellar core. Iron is different from the lighter elements. When hydrogen or helium nuclei fuse, energy is released. In contrast, iron fusion requires energy and causes the stellar core to cool. When this happens, the thermal pressure that forced the stellar gases outward diminishes rapidly and the star collapses under the influence of gravity. This collapse releases large amounts of heat. Within a few seconds—a fantastically short time in the life of a star—the star's temperature reaches trillions of degrees and the star explodes in an event called a **supernova** (Figure 24-13). For a brief period, a supernova shines as brightly as hundreds of billions of normal stars and may even emit as much energy as an entire galaxy. To observers on Earth, it appears as though a new, brilliant star suddenly materialized in the sky, only to become dim and disappear to the naked eye within a few months.

On February 24, 1987, astronomer Ian Shelton was carrying out research unrelated to supernovae. When he developed one of his photographic plates, he saw a bright star where previously there was only a dim one (Figure 24-14). He walked outside,

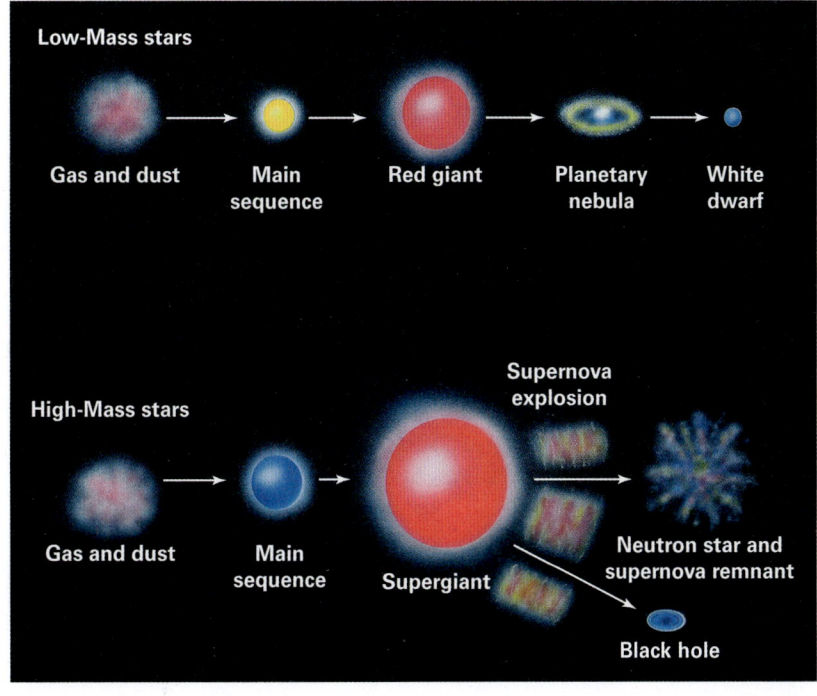

FIGURE 24-13
The evolution of a star depends on its mass. A star of low mass, such as the sun, will evolve and die as a white dwarf. A star with a large mass of at least 1.44 solar masses at the end of its life will die in a supernova explosion and form a neutron star within a supernova remnant. If the leftover neutron star has more than 2 solar masses, it collapses into a black hole.

looked into the sky, and saw the star with his naked eye. This was the first supernova explosion visible to the naked eye since 1604, five years before the invention of the telescope.

FIGURE 24-14
Go online to view an example of a supernova region imaged both before and after the explosion.

A supernova explosion is violent enough to send shock waves racing through the atmosphere of the star, fragmenting atomic nuclei and shooting subatomic particles in all directions. Many of the nuclear particles collide with sufficient energy both to fuse and to split apart. These processes form all the known elements heavier than iron. Thus, in studying the evolution of stars, scientists learned how the heavy elements originated.

Hydrogen and helium were the first elements to form when our universe was 300,000 years old. Originally, all the stars in the universe were composed entirely of these two lightest elements.

CONNECTIONS TO CHEMISTRY

Nucleosynthesis of Elements Heavier than Iron

The minimum temperature required to fuse ^{1}H into ^{4}He is 5 million K. Once this temperature is achieved in a condensed mass of gas within a nebula, a star is born. When the ^{1}H in a star's core is depleted and energy is no longer produced, the core shrinks due to gravity. This increases the temperature and allows for the fusion of ^{4}He into ^{12}C, releasing energy. The next phase is the fusion of ^{12}C into ^{16}O and then ^{16}O into ^{20}Ne. Successive nuclear reactions continue to produce elements heavier than ^{20}Ne until the core of the star is fusing ^{28}Si into ^{56}Fe. When the star's core is completely iron, additional fusion reactions require energy to proceed. Nuclear reactions cease and the star's core collapses, resulting in a supernova explosion that is hot enough to fuse elements, creating the rest of the members of the periodic table.

These old, first-generation stars are called **population II stars**, and a few still exist.

Within the cores of population II stars, hydrogen fused to helium. This changed the ratio of these two elements in the early universe. As stellar evolution continued, helium fused to carbon. Carbon fused to heavier elements, up to iron. Elements heavier than iron formed during supernova explosions of massive population II stars.

When a star dies, it blasts gas and dust into space to form a new nebula. Eventually, these nebulae condense once again into new, second-generation stars called **population I stars**. Population I stars begin life with primordial hydrogen and helium. They also contain small concentrations of heavy elements inherited from population II stars and from supernova explosions. The sun is a population I star that condensed from a nebula containing all the natural elements.

Thus, our solar system was born from the debris of dying stars. Solar systems containing Earth-like planets and living organisms could never form around population II stars. These stars do not have all the necessary elements. In studying the life cycle of stars, we also learn about the origins of the elements that make up Earth and its life-forms.

checkpoint Which stars are older, population I stars or population II stars?

24.3 ASSESSMENT

1. **Explain** Why do stars have both an apparent brightness and an absolute brightness?
2. **Define** What is a planetary nebula, and why is the name misleading?
3. **Describe** What happens to the temperature from the core of the sun to its outermost layer?
4. **Distinguish** How does the death of a 1-solar-mass star compare to the death of a 10-solar-mass star?

Computational Thinking

5. **Abstract Information** How do we know that the sun must contain elements other than hydrogen and helium?

24.4 EXTREME STELLAR REMNANTS

Core Ideas and Skills
- Describe two remnants from the death of the most massive stars and how they are related.
- Explain what causes pulsars to pulse.
- Distinguish between gamma-ray bursts and pulsars.

KEY TERMS

neutron star   black hole
pulsar   gamma-ray burst

Neutron Stars and Pulsars

In a supernova explosion, most of the matter in a star is blasted into a nebula, but a substantial fraction remains behind, compressed into a tight sphere. In the 1930s, scientists developed a hypothesis to explain what happens within this sphere. If it is between 2 and 3 solar masses, the gravitational force is so intense that the star cannot resist further compression the way a white dwarf does. Instead, the electrons and protons are squeezed together to form neutrons.

The neutrons then resist further compression and remain tightly packed. This ball of compressed neutrons is called a **neutron star**. A neutron star is extremely dense—approximately 10^{13} kg/cm^3. If the entire Earth were as dense, it would fit inside a football stadium.

The first neutron star was discovered by accident in 1967. Jocelyn Bell Burnell, a graduate student at the University of Cambridge, was studying radio emissions from distant galaxies (Figure 24-15). In one part of the sky, she detected a radio signal that switched on and off. The intermittent signal had a frequency of about one pulse every 1.33 seconds. The radio emissions Bell Burnell heard were unusual because they were sharp, regular, and spaced a little more than one second apart. If the signal were converted into sound, you would hear a "click, click, click," evenly spaced, with one click every 1.33 seconds.

At first, astronomers considered the possibility that the unusual signals might come from intelligent life, so they called the signals LGM, for "little green men." But when Bell Burnell found a similar pulsating source in a different region of

FIGURE 24-15
Jocelyn Bell Burnell poses with the radio telescope she helped construct and used to record the first data from pulsars. In 2018, Bell Burnell won the Special Breakthrough Prize in Fundamental Physics, more than 30 years after her discovery.

the sky, scientists ruled out the possibility. Once it was established that the signals did not originate from intelligent beings, their sources were called **pulsars**. But naming the source did not explain it.

The first step toward identifying the source of pulsars was to estimate their size. Not all parts of an object in space are at equal distances from Earth (Figure 24-16). If a large sphere emits a sharp burst of energy from its entire surface, some of the photons start their journey closer to Earth than others and therefore arrive sooner. The radio signal from a large source converted into sound will be heard as a prolonged "cliiiiiick" because it takes a while for all the radio waves to arrive. Alternatively, a radio signal from a small source is much sharper.

Pulsar signals are sharp, indicating that the source must be unusually small, roughly 30 kilometers in diameter. The smallest star ever recorded in the late 1960s was a white dwarf 16,000 kilometers in diameter, roughly 25 percent larger than Earth. Scientists reasoned that the

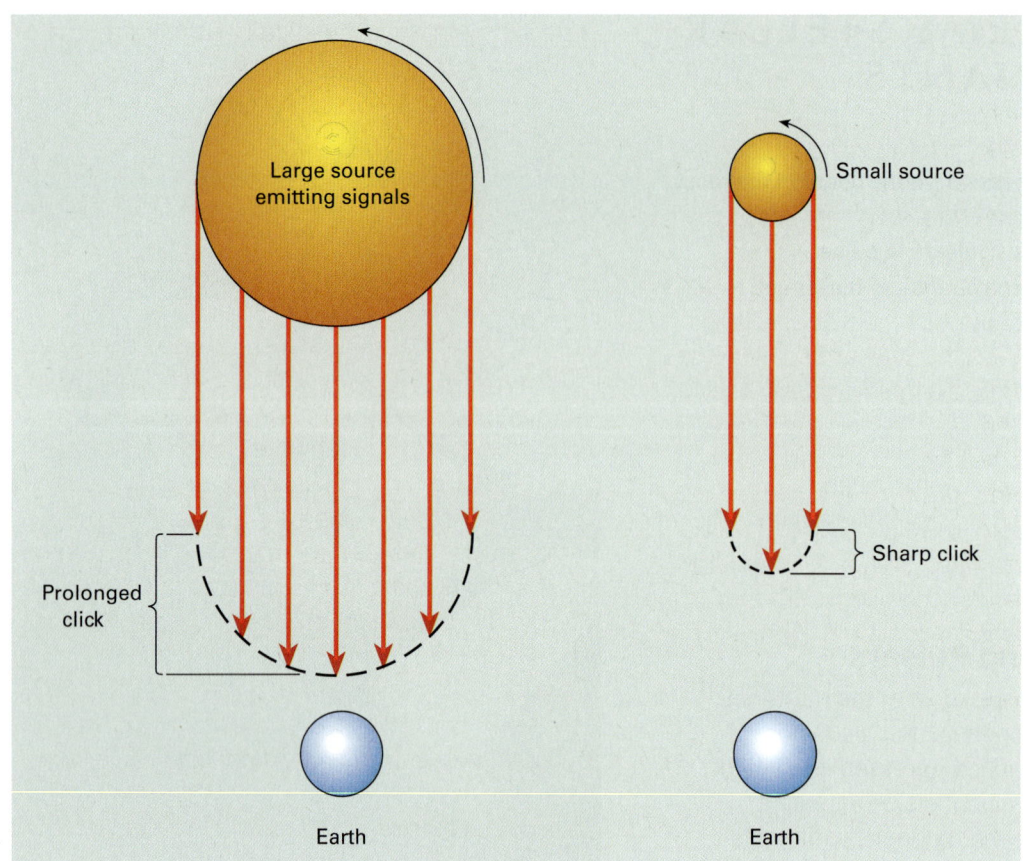

FIGURE 24-16
(A) A signal from a large rotating pulsar arrives over a longer time interval than (B) a signal from a small rotating pulsar, resulting in a prolonged click versus a sharp click.

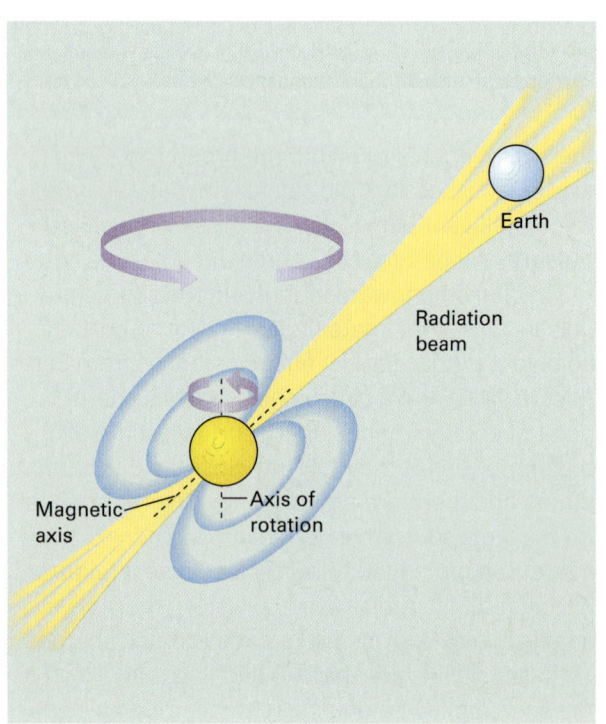

FIGURE 24-17
The radiation beam of a pulsar is detected only when it sweeps in the direction of Earth.

pulsar detected by Bell Burnell might be the long-searched-for neutron star.

Astronomers suggested that the radio signals are emitted by an electromagnetic storm on the surface of a neutron star. Thus, a pulsar is a neutron star that emits intermittent, but regular, radio signals. According to this hypothesis, as the star rotates, so does the storm center. A receiver on Earth detects one click per revolution of the star, just as a lookout on a ship sees a lighthouse beam flash periodically as the beacon rotates (Figure 24-17). If a pulse is received once every 1.33 seconds, the pulsar must rotate that rapidly. White dwarfs can again be ruled out because they are too big to rotate so rapidly. But neutron stars are small enough to rotate that fast.

Astronomers searched for an example to test the hypothesis. They focused a radio telescope on the Crab Nebula, where a supernova explosion occurred during the Middle Ages. A pulsar signal was found exactly where the supernova had occurred about 950 years ago—precisely where a neutron star should be.

checkpoint What type of telescope was used to discover pulsars?

Black Holes and Gamma-Ray Bursts

If a star with more than about 5 solar masses explodes, the remnant core remaining after the supernova explosion is more than 2 solar masses. When such a massive sphere contracts, the neutrons are not sufficiently strong to hold the star up against gravity. The star core collapses to a diameter much smaller than a neutron star and becomes a **black hole**. Such a collapse is impossible to imagine in earthly terms. A tremendous mass, perhaps a trillion, trillion, trillion kilograms, shrinks to the size of a pinhead, then continues to shrink to the size of an atom, and then even smaller. Eventually it collapses to an infinitesimally small point of infinitely high density.

Such a small point of mass creates an extremely intense gravitational field. According to Einstein's theory of relativity, gravity affects light. Starlight bends as it passes the sun. If an object is massive and dense enough, its gravitational field becomes so intense that not even light can escape. Because no light can escape such an object, it is invisible; hence the name *black hole*. If you were to shine a flashlight beam, a radar beam, or any kind of radiation at a black hole, the energy would be absorbed. The beam could never be reflected back to your eyes; therefore, you would never see it again. It would be as though the beam just vanished into space. Similarly, if matter comes too close to a black hole, it would be pulled in. No matter can move fast enough to escape.

The search for a black hole is therefore even more difficult than the search for a neutron star. How do you find an object that is invisible and can neither emit nor reflect any energy? In short, how do you find a hole in space? Although it is theoretically impossible to see a black hole, astronomers can observe the effects of its gravitational field. Many stars exist in pairs or small clusters. If two stars are close together, they orbit around each other. Even if one becomes a black hole, the two still orbit about each other, but one is visible and the other is invisible. The visible one appears to be orbiting an imaginary partner. Astronomers have studied several stars that seem to orbit in this unusual manner. In several cases, the invisible member of the pair has a mass equal to or greater than 3 solar masses. Because a normal star of 3 solar masses would be visible, and this star is not visible, it is possible that it is a black hole. However, the simple observation that a star moves around an invisible companion does not prove that a black hole exists.

If a star were orbiting a black hole, great masses of gas from the star would be pulled into the black hole, to disappear forever (Figure 24-18). As this matter started to fall into the hole, it would accelerate, just as a meteorite accelerates as it falls toward Earth. The gravitational field of a black

FIGURE 24-18
If a black hole and an ordinary star are in orbit around each other, gas from the star will eventually fall into the black hole.

LESSON 24.4

hole is so intense that particles drawn into it would collide against each other with enough energy to emit x-rays. Just as a falling meteorite glows white-hot as it enters Earth's atmosphere, anything tumbling into a black hole would glow even more energetically. It would emit x-rays. These x-rays might then be detected here on Earth. But you may ask, if light cannot escape from a black hole, how can the x-rays escape? The answer is that the x-rays are produced and escape from just outside the edge of the black hole. Thus, matter being pulled into a black hole sends off one final message before being pulled into the void from which no message can ever be sent.

Astronomers focused x-ray space telescopes on portions of the sky where a star appeared to orbit an invisible partner. In the 1980s, an orbiting x-ray telescope detected x-ray sources adjacent to stars that appeared to orbit an unseen partner. More recent observations by ground-based observatories and space telescopes such as the Hubble Space Telescope have provided evidence confirming that black holes exist.

If you could "see" electromagnetic radiation 10 million times more energetic than visible light, the universe would look different. It would appear to be a much more erratic, violent environment than we normally perceive. On average, a flash of high-frequency radiation, called a **gamma-ray burst**, comes from a random direction in space once a day. This brief pulse contains more power than we receive from the sun. Gamma-ray bursts have a duration ranging from a few milliseconds to slightly more than two seconds. In 2002, astronomers determined that the longer bursts come from the explosions of very massive stars. But the origin of the short bursts proved more elusive. In 2004, NASA scientists launched the Swift space-based observatory, equipped with multiple tools including an x-ray telescope and an ultraviolet telescope. The telescopes were aligned and coordinated for instant observations over a wide range of frequencies. From these studies, scientists determined that short gamma-ray bursts form when a neutron star is pulled into a black hole or when two neutron stars collide to form a black hole.

checkpoint What is the difference between a neutron star and a black hole?

24.4 ASSESSMENT

1. **Explain** How do black holes get their name?
2. **Describe** How do we know that a pulsar must be a neutron star and not a white dwarf?
3. **Distinguish** Compare the densities of a neutron star and a regular star.

Critical Thinking

4. **Apply** Given two groups of stars with the exact same size and color, describe two ways to determine which group of stars is older.

TYING IT ALL TOGETHER
WE ARE MADE OF STAR-STUFF

In this chapter, you learned about the birth of stars, the structure of our sun, the life and death of stars, and what happens to the material of stars after their death. In the Case Study, you read about the Parker Solar Probe and what scientists hope to learn from its harrowing expedition to the sun's corona. The sun is essential to our existence. However, the stars of the universe have been essential to human life in another way. As author and scientist Carl Sagan noted on his television show *Cosmos*, "We're made of star stuff." This might seem like a metaphor, but it is true in a literal sense. Elements that were created by fusion inside stars make up the human body and every living and nonliving thing on Earth. Without stars, neither Earth nor its organisms would exist.

In this activity, you will investigate what kinds of stars these essential elements came from and how they eventually became part of our bodies. Work individually or with a group to complete the steps below.

1. Research the following questions about the origins of elements found in the human body:

    a. What are the main elements in the human body?

    b. What types of stars produced these elements?

    c. How did these elements spread out into space and make their way to Earth and our own bodies?

2. Organize your research findings into a graphical presentation, such as a chart or a poster.

CHAPTER 24 SUMMARY

24.1 THE BIRTH OF STARS

- Stars form from condensing nebulae, clouds of gas and dust in interstellar space.
- When the condensing gases become hot enough, fusion begins.

24.2 THE SUN

- The central core of the sun has a temperature of about 15 million K and a density 150 times that of water. Hydrogen fusion occurs in this region.
- The visible surface of the sun, called the photosphere, has a temperature of about 5,700 K and a pressure of one-hundredth that of Earth's atmosphere at sea level.

24.3 THE LIFE AND DEATH OF STARS

- A star can appear luminous either because it really is bright or because it is close. The absolute brightness or luminosity of a star is how bright it would appear if it were a fixed distance of 3.26 light-years away.
- A Hertzsprung–Russell (H–R) diagram is a graph that plots stellar luminosity against temperature. Most stars fall within a band called the main sequence.
- When the hydrogen in the core of a mature star is exhausted and hydrogen fusion ends, gravity compresses the core and the temperature rises. Fusion then begins in the outer shell, and the star expands to become a red giant.
- After helium fusion in an average-mass star ends, the star releases a planetary nebula and then shrinks to become a white dwarf.
- In a more massive star, enough heat is produced to fuse heavier elements; the iron-rich core then explodes to become a supernova.

24.4 EXTREME STELLAR REMNANTS

- The remnant of a supernova can contract to become a neutron star.
- If a neutron star emits pulses of radio waves, it is called a pulsar.
- If the mass of a dying star is great enough, it collapses into a black hole.
- Gamma-ray bursts occur when two neutron stars collide or when neutron stars fall into a black hole.

CHAPTER 24 ASSESSMENT

Review Key Terms
Select the key term that best fits the definition. Not all terms will be used, and no term will be used more than once.

absolute brightness	photosphere
apparent brightness	planetary nebula
black hole	population I stars
chromosphere	population II stars
convective zone	prominence
fusion	pulsar
gamma-ray burst	radiative zone
H-R diagram	red giant
light-year	solar mass
main sequence	spicule
nebula	sunspot
neutron star	supernova
parsec	white dwarf

1. unit of distance equivalent to 32.6 light-years
2. unit of mass equivalent to the mass of the sun
3. flash of high-frequency radiation that occurs when a neutron star is pulled into a black hole or when two neutron stars collide to form a black hole
4. cloud of interstellar gas and dust
5. gaseous layer of the sun that lies outside the photosphere; from Earth, can be seen as a narrow red fringe during a solar eclipse
6. unit of distance equivalent to the distance light travels in one year
7. object in space that emits regular, intermittent radio signals
8. jet of gas that shoots up from the chromosphere of the sun
9. stars composed primarily of hydrogen and helium and fueled by hydrogen fusion
10. process that occurs when two atomic nuclei collide and bond to form a single atomic nucleus
11. the intensity of light that we would receive from a star if it were 10 parsecs away
12. late-stage star that has expanded greatly after using up most of the hydrogen in its core
13. the brightness of a star as seen from Earth
14. dense ball of electrically neutral particles that forms during a supernova explosion when electrons and protons are squeezed together by intense gravitational force
15. a broad region of the sun surrounding the core where the gas density is so high that photons emitted by the core are quickly reabsorbed again

Review Key Concepts
Answer each question on a separate sheet of paper to demonstrate your understanding of key concepts from the chapter.

16. Nuclear fusion does not occur naturally on Earth. However, in the core of a star, nuclear fusion occurs continuously. What accounts for this difference?
17. A star's lifetime is on the order of millions or billions of years. How do scientists know about the life cycle of a star?
18. What are two conditions that scientists think are necessary for a star to form from a cloud of gas and dust?
19. What evidence have scientists been able to use to determine the mass of the sun?
20. What do scientists think accounts for the occurrence and persistence of sunspots?
21. Compare and contrast spicules and prominences.
22. What is the difference between the absolute brightness of a star and its apparent brightness?
23. Based on the Hertzsprung-Russell diagram (H-R diagram), what is the relationship between temperature and luminosity for main-sequence stars?
24. What is a pulsar, and how did astronomers become aware of them even though pulsars cannot be seen directly?
25. How were astronomers able to determine the general size of the objects known as pulsars?
26. How can astronomers use x-rays and gamma rays to identify the locations of black holes?

Think Critically
Write a response to each question on a separate sheet of paper. Use concepts from the chapter to support your reasoning.

27. What would you expect to be the relationship between the density of gas and dust in a nebula and the rate of star formation within that nebula? Explain your reasoning.

CHAPTER 24 ASSESSMENT

28. The sun undergoes nuclear fusion reactions between hydrogen nuclei (protons) to produce helium nuclei. However, even though helium nuclei are building up in the sun's core, no significant quantity of helium nuclei will undergo fusion to form carbon nuclei until the sun enters the red giant stage. Why?

29. Stars with very low mass can have lifetimes of hundreds of billions of years, whereas stars with very high mass have much shorter lifetimes on the order of a billion years. How does luminosity of these stars help explain the large difference in these lifetimes?

30. Do all stars go through the same stages at the end of their lifetimes? Explain.

31. Would you expect a white dwarf to produce a black hole? Why or why not?

PERFORMANCE TASK

Modeling the Evolution of the Sun

Imagine you are writing a science fiction story about how our solar system will change over billions of years. On this time scale, scientists predict, based on a set of complex mathematical models, that the sun will evolve in dramatic ways.

In this performance task, you will use scientific predictions about the sun to construct a series of models to represent how the sun's evolution will impact the solar system over time.

The first table lists the predictions for how the sun will change over its lifetime. The second table gives the current distance of planets in our solar system.

	Predicted Changes in the Sun			
Core Processes During Stage	Radius (relative to current sun)	Surface Temperature (K)	Luminosity Relative to Current Sun (L/L_o)	Duration of Stage (Earth years)
Hydrogen fuses into helium in the core.	1.00	6,000	1	1×10^{10}
Core is helium; hydrogen fuses into helium in a shell around the core.	1.50×10^2	3,000	2,000	1×10^9
Helium fuses into carbon and oxygen in the core.	2.00×10^1	4,500	100	1×10^8
Core is carbon; hydrogen and helium fuse in shells around the core.	2.00×10^2	3,000	10,000	1×10^7
Outer layers are shed.	1.35×10^8	10,000	10,000	5×10^4
Core is exposed.	0.01	200,000	10	

Expressed in scientific notation, $1{,}000 = 1 \times 10^3$

(Continued on next page.)

PERFORMANCE TASK (continued)

Planetary Orbits	
Planet	Distance from Sun (solar radii)
Mercury	83
Venus	156
Earth	215
Mars	328
Jupiter	1,119
Saturn	2,061
Uranus	4,130
Neptune	6,463

1. Construct models to represent the evolution of the sun and how it will impact the planets in the solar system.

   A. Using data from the Predicted Changes in the Sun table, create an H-R diagram similar to Figure 24-10 for different stages in the sun's life cycle. For the horizontal axis, use a reverse scale (decreasing from left to right) for surface temperature. For the vertical axis, use a logarithmic scale for luminosity, as in Figure 24-10.

   B. Based on the descriptions of the core processes in the table, research the names of the different stages of the sun's life cycle. Then use these to label the points on your graphical model.

   C. Create a second graph to model how the size of the sun changes at different stages of its lifespan. Will the sun become larger than any of the planetary orbits at any stage?

   When you have completed your two graphical models, use them to answer the next question.

2. The sun is a main sequence star thought to be about 4.5 billion years old. Write a short narrative about the changes that will happen in the solar system as the sun ages.

The Hubble Space Telescope imaged one tiny area of sky, about one-tenth the diameter of the moon, using 800 separate exposures for a total exposure time of just over 11 days to create the Hubble Ultra Deep Field. This image shows nearly 10,000 of galaxies from large spirals, dominated by the blue light from young stars, to small ellipticals, dominated by red light from old stars. The most distant and oldest known galaxies in this image are about 13.2 billion years old while the closest and youngest formed about one billion years ago.

CHAPTER 25
GALAXIES AND THE UNIVERSE

Earth is only one planet orbiting one star among roughly 200 billion stars in our Milky Way galaxy. In turn, the Milky Way is only one galaxy of billions in the universe. Looking beyond the solar system into galactic and intergalactic space, we must stretch our minds to nearly unimaginable distances, look backward in time to events that occurred billions of years before Earth formed, and attempt to fathom energy sources powerful enough to create an entire universe.

KEY QUESTIONS

25.1 How is the matter in the universe organized?

25.2 What were the conditions in the early universe and how did they change over time?

25.3 What is the fate of the universe?

EXPLORERS AT WORK

OPENING WINDOWS ON THE UNIVERSE

WITH NATIONAL GEOGRAPHIC EXPLORER KNICOLE COLON

Peer into the night sky and one question naturally comes to mind. What's out there? Until the early 17th century, the only tools people had to study the stars were their own eyes. Italian astronomer Galileo Galilei was the first to a telescopel to study the night sky. Galileo was able to see the mountains and craters of Earth's moon as well as what appeared to be a smear of light in the sky: it was part of our own galaxy, the Milky Way. Thanks to advances in engineering and space technology, scientists now study objects and phenomena at a wide range of distances and from the incredibly small to the enormously large. Large telescopes on Earth and in space capture light from tiny interstellar dust grains in the Milky Way and interactions between multiple galaxies found in distant giant clusters. One scientist who uses multiple telescopes is astrophysicist Knicole Colon.

Dr. Colon works at NASA's Goddard Space Flight Center in Greenbelt, Maryland. The primary focus of Colon's research is the detection and characterization of exoplanets orbiting other stars in our galaxy. Colon studies both "super-Earths," rocky planets up to ten times the mass of Earth, and "hot-Jupiters," extremely hot gas giants. These planetary bodies are extremely distant and hard to detect. Consequently, Colon depends on powerful space-based technology like what is offered by NASA's Hubble Space Telescope and Spitzer Space Telescope to study them. She is looking forward to using the upcoming NASA James Webb Space Telescope as well, expected to be launched in Spring 2021.

Colon notes that most of the exoplanets discovered in our galaxy were found by NASA's Kepler space telescope. In fact, Kepler led to the confirmation of more than 2,300 exoplanets. Colon is excited about another satellite NASA launched in April 2018. The Transiting Exoplanet Survey Satellite (TESS) is hunting for exoplanets around the brightest stars close to Earth. TESS will survey the entire sky, focusing on between 30 and 300 light-years away. "Because these exoplanets are close to Earth relatively, we'll be able to better measure many of their properties, including their mass, radius, and atmosphere," explains Colon.

Space-based telescopes have also allowed astronomers like Colon to study light otherwise blocked by Earth's atmosphere. In early 2018 astronomers using the Chandra X-ray Observatory announced the discovery of as many as 2,000 possible exoplanets, the first found in another galaxy. They were detected by measuring their gravitational microlensing, or bending, of the light from extremely bright x-ray objects called quasars.

With an arsenal of powerful space research tools, Dr. Colon and other astronomers are revealing the story of how planets form and may one day even discover other Earth-like planets in our universe.

THINKING CRITICALLY

Summarize Using information from this feature, summarize in your own words the importance of technology to the study of space.

Dr. Colon visits the former Arecibo Observatory in Puerto Rico. At 305 meters in diameter, the observatory was the world's largest single-dish radio telescope until December 2020.

The Roque de los Muchachos Observatory on the island of La Palma in the Canary Islands, Spain, is home to a fleet of the most powerful telescopes on Earth, including the 10.4-meter Gran Telescopio Canarias and the William Herschel Telescope. The observatory is one of two potential sites for the planned Thirty Meter Telescope. Once constructed, the telescope will be the most advanced and powerful ground-based telescope in history.

CASE STUDY
SGR A*

Our galaxy, the Milky Way, is one of billions of galaxies in the universe. Yet it is so vast that scientists still have much to learn about it. Astronomers have long wondered what lies at the center of the Milky Way and how this central region, known as the Galactic Center, affects the galaxy's outer regions. It has taken an array of different kinds of telescopes to even begin to answer these questions. Each telescope, by focusing on a specific region of the electromagnetic spectrum, provides unique images of the Galactic Center. Like pieces of a puzzle, astronomers have assembled these data into a clearer, although as of yet incomplete, understanding of the Galactic Center.

The Galactic Center is the extremely bright area in lower-right of the images in Figure 25-1. An optical telescope—one that collects visible light—could never produce such an image. Visible light emitted from the Galactic Center is obscured by clouds of dust and gas. However, matter near the Galactic Center also emits radiation at other wavelengths that penetrate the clouds unobscured. Telescopes that detect radio waves, microwaves, infrared radiation, and x-rays each make it possible to study different wavelengths of energy emanating from the Galactic Center.

Based on data accumulated from numerous telescopes, astronomers have concluded that a supermassive black hole lies at the center of the Milky Way. Named Sgr A* (pronounced Sag A star), its mass is about four million times that of the sun. Amazingly, this supermassive black hole occupies a region of about one astronomical unit (AU). That's roughly the distance between Earth and the sun.

Scientists currently think that supermassive black holes like Sgr A* are present at the centers of most, and possibly all, large galaxies. So far, spectral data collected from other galaxies support this idea. However, astronomers are continually observing the stars, collecting more data, and reevaluating current theories. In April 2019 for example, scientists released the first image of a supermassive black hole ever captured. Produced by a network of radio telescopes known as the Event Horizon Telescope, the image depicts a black hole at the center of the galaxy Messier 87 (Figure 25-7).

As You Read Learn how astronomers accomplish the difficult task of collecting data about objects that are millions of light-years away. Think about the collaborative efforts of the many astronomers, engineers, and space scientists who contribute to our understanding of the universe. Find out more about the tools they use, such as ground-based telescopes, space telescopes, and space probes, to explore distant stars, planets, and galaxies.

FIGURES 25-1A–C
Go online to view images of the central area of our galaxy in infrared light, near-infrared light, and x-rays.

FIGURE 25-1D
A red circle on the composite of the infrared, near-infrared, and x-ray images of the center of our galaxy surrounds the bright, dense area hiding a supermassive black hole. The infrared light collected by the Spitzer Space Telescope, shows star-forming regions. The Hubble Space Telescope near-infrared light outlines the edges of gas bubbles created by young massive stars. The brightest regions in x-rays from the Chandra X-ray Observatory represent the most massive and powerful stars.

25.1 THE MILKY WAY AND OTHER GALAXIES

Core Ideas and Skills
- Explain how stars and other celestial objects in space are organized into galaxies.
- Classify galaxies by their different shapes.
- Define the main components of the Milky Way and their relative locations.
- Describe how galaxies move throughout the universe.

KEY TERMS

galaxy
elliptical galaxy
spiral galaxy
barred spiral galaxy
irregular galaxy
Hubble-Lemaître law
galactic halo
globular cluster

Galaxies

In the late 1700s French astronomer Charles Messier was studying comets. He recorded more than a hundred fuzzy objects in the sky that clearly were not stars. When these objects were studied in the 1850s with more powerful telescopes, many were observed to have spiral structures like pinwheels. They were not comets—but what were they, and how far away were they? These questions were not answered until 1924 when Edwin Hubble determined that they were farther away than even the most distant known stars. In order for us to see them at all, they must be much more luminous than a star. Hubble concluded that each object is a **galaxy**, a large volume of space composed of billions of stars and other matter held together by gravity. Today we recognize that galaxies and clusters of galaxies form the basic structure of the universe.

A galaxy with an oval shape and no spiraling arms is called an **elliptical galaxy** (Figure 25-3A). Elliptical galaxies are dominated by old, red stars that move around the center of the galaxy in unordered orbits. About one-third of the galaxies in our region of the universe are elliptical. Giant elliptical galaxies contain up to 10^{13} solar masses and may have a diameter larger than our Milky Way galaxy (which has a spiral shape, discussed next). However, most ellipticals are dwarf and contain only a few million solar masses. Recent observations indicate that dwarf ellipticals may be the most common type of galaxy.

FIGURE 25-2
ON ASSIGNMENT This stunning view of the stars, dust and gas in the Milky Way rising above Grand Teton National Park in Wyoming was photographed by award-winning National Geographic photographer and science journalist Babak Tafreshi. Tafreshi is a master of night-time photography and nightscape videos. He is founder of The World at Night program, an elite group of astrophotographers who present images to reconnect people with the night sky in natural light.

FIGURE 25-3
The portfolio of galaxy morphologies includes the elliptical galaxy (A), the spiral galaxy (B) and related barred-spiral galaxy (C), and the irregular galaxy (D).

Almost two-thirds of the bright galaxies in our region of the universe are **spiral galaxies** (Figure 25-3B). The stars, gas, and dust in a spiral are arranged in a thin disk, with arms radiating outward from a spherical center or nucleus. The stars in the outer arms rotate around the nucleus like a giant pinwheel. A typical spiral galaxy contains about 100 billion stars and are dominated by young, blue stars thanks to the high star formation rates. Nearly half of all spiral galaxies are **barred spiral galaxies** (Figure 25-3C), having a straight bar of stars, gas, and dust extending across the nucleus.

A few percent of all galaxies are lens-shaped or irregular and show no obvious pattern. One of the leading ideas about irregular galaxies is that their lack of structure and form is a result of gravitational collisions with other galaxies. Irregular galaxies exhibit knots of star formation and contain a fair number of young stars. The Large Magellanic and Small Magellanic cloud are **irregular galaxies** (Figure 25-3D). These galaxies orbit the Milky Way at distances of more than 150,000 light-years. They have already experienced several close gravitational encounters with the Milky Way and are expected to collide with our galaxy within the next two-and-a-half billion years.

Recall from Chapter 22 that astronomers measure the relative velocities of distant objects by measuring the frequency of light they emit. In 1929, five years after he had described galaxies, Hubble noted that frequency of emitted light from almost every galaxy is shifted toward the red end of the spectrum. Hubble interpreted this redshift to mean that all the galaxies are flying away from us and from each other; the universe is expanding. Moreover, he observed that the most distant galaxies are moving outward at the greatest speeds, whereas the closer ones are receding more slowly

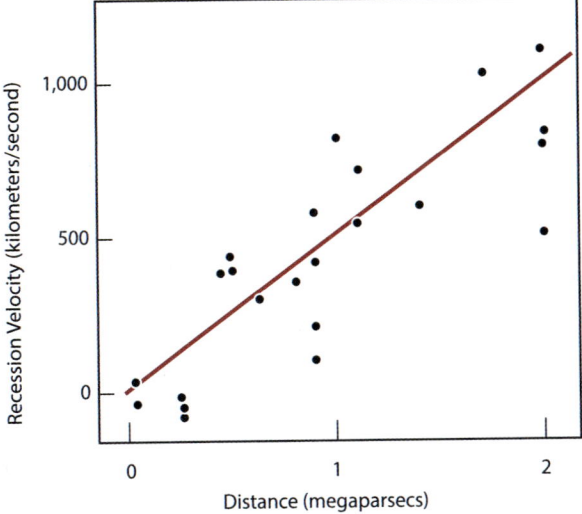

FIGURE 25-4
Edwin Hubble's original data from 1929 shows the recessional velocity of galaxies is proportional to their distance. This mathematical relationship was renamed the Hubble-Lemaître law in 2018.

(Figure 25-4). This relationship has historically been known as the Hubble Law. In late 2018 the International Astronomical Union passed a resolution to update the name of this scientific law to the **Hubble-Lemaître law** to honor the research of both Edwin Hubble and Georges Lemaître, a Belgian priest and astronomer who is thought to have first theorized that the universe is expanding.

Using the Hubble-Lemaître law, the distance from Earth to a galaxy can be calculated by measuring the galaxy's observed redshift.

checkpoint What are the three main types of galaxies?

The Milky Way

Our sun lies at the edge of a spiral arm in a barred spiral galaxy called the Milky Way. The Milky Way's galactic disk has a width of 2,000 light-years and a diameter of 200,000 light-years (Figure 25-5). The Milky Way contains about 400 billion stars. It has also swallowed up numerous independent dwarf galaxies that now rotate in the outer-spiral arms. Because the disk is relatively thin compared to its width, like a CD or DVD, an observer on Earth sees relatively few stars perpendicular to its plane. Most of the night sky contains a diffuse scattering of stars with large expanses of black space between them.

However, if you look into the plane of the disk, you see a dense band of stars. This band is commonly called the Milky Way, although astronomers use the term to describe the entire galaxy. The galactic disk rotates about its center once every 200 million years, so in the 4.6-billion-year history of Earth, we have completed about 23 rotations.

A spherical **galactic halo**—a spherical cloud of dust and gas—surrounds the Milky Way's galactic

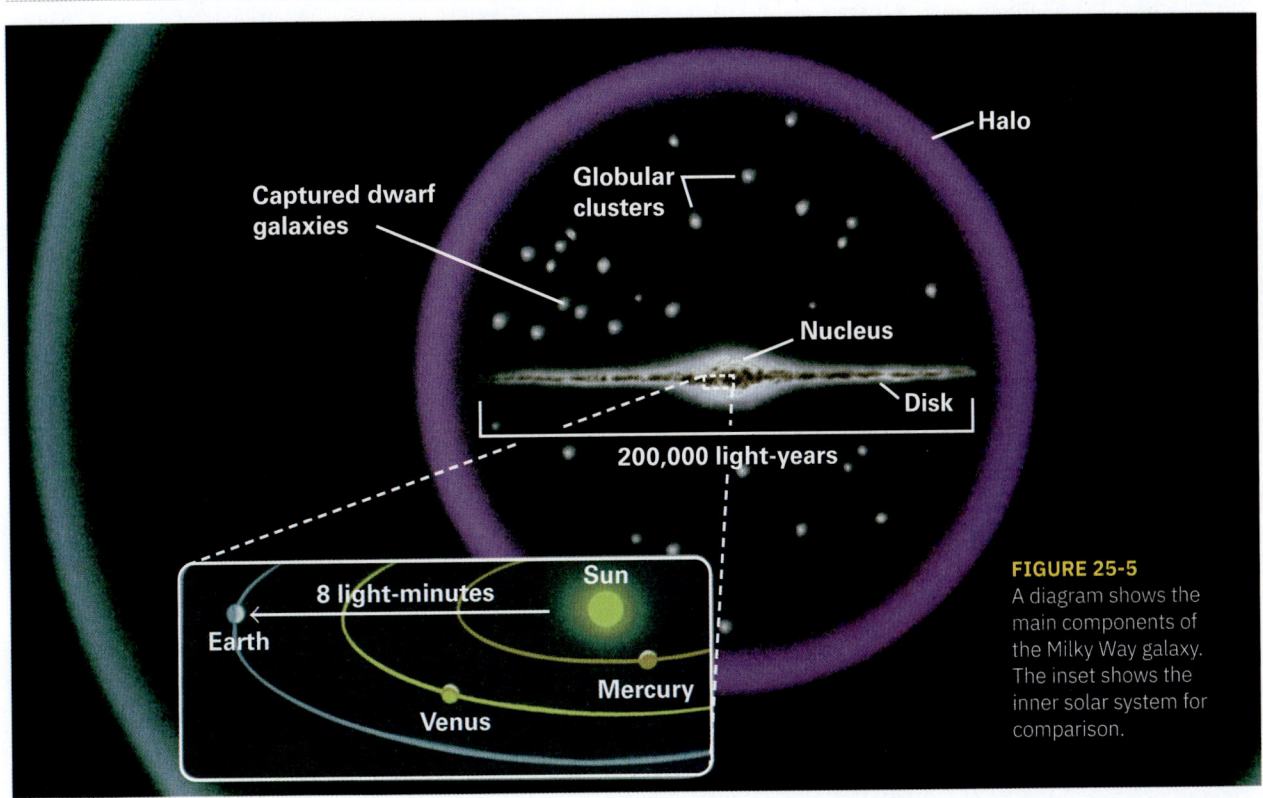

FIGURE 25-5
A diagram shows the main components of the Milky Way galaxy. The inset shows the inner solar system for comparison.

LESSON 25.1 801

disk. This halo is so large that even though it is extremely diffuse, it contains as much as 90 percent of the mass of the galaxy. Many dim and relatively old stars exist within this halo. Some are concentrated in groups of 10,000 to one million stars. Each group is called a **globular cluster** (Figure 25-6). The galactic halo and globular clusters are probably remnants of one or more protogalaxies that condensed to form the Milky Way. The spherical structure of the halo suggests that the entire galaxy was once spherical.

Photographs of other spiral galaxies show that the galactic nucleus shines much more brightly than the disk. The concentration of stars in the galactic nucleus is perhaps one million times greater than the concentration in the outer disk. If you could visit a planet orbiting one of these stars, you would never experience night, for the stars would light the planet from all directions. However, stable solar systems could not exist in this region because gravitational forces among nearby stars would rip planets from their orbits.

It is impossible to see into the nucleus of our own galaxy because visible light does not penetrate through the interstellar nebulae and dust that lie in the way. Looking into the nucleus is like trying to see a ship on a foggy day. However, just as sailors use radar to penetrate the fog, astronomers study the center of the galaxy by analyzing radio, infrared, and x-ray emissions that travel through the clouds. These studies provide a picture of the nucleus of the Milky Way.

As described in the Case Study, a cloud of dust and gas orbits the Milky Way's galactic nucleus. Infrared measurements show that this cloud is so hot that it must be heated by an energy source with 10 million to 80 million times the output of our sun. The cloud is ring-shaped, with a hole in its center, like a doughnut. Relatively recently, 10,000 to 100,000 years ago, a giant explosion blew out the center of a larger cloud and created the hole. X-ray and gamma-ray emissions tell us that matter is now accelerating inward, back into the center at a rate of one solar mass every 1,000 years.

By measuring the orbits of stars close to the galactic center, astronomers can measure the gravitational force of the galactic center, and hence its mass. Calculations show that an unseen object, about four million times as massive as our sun, lies at the heart of our galaxy. This object can only be a black hole. According to the most widely accepted hypothesis, early in the history of the galaxy, one huge star or many large ones formed in the galactic center.

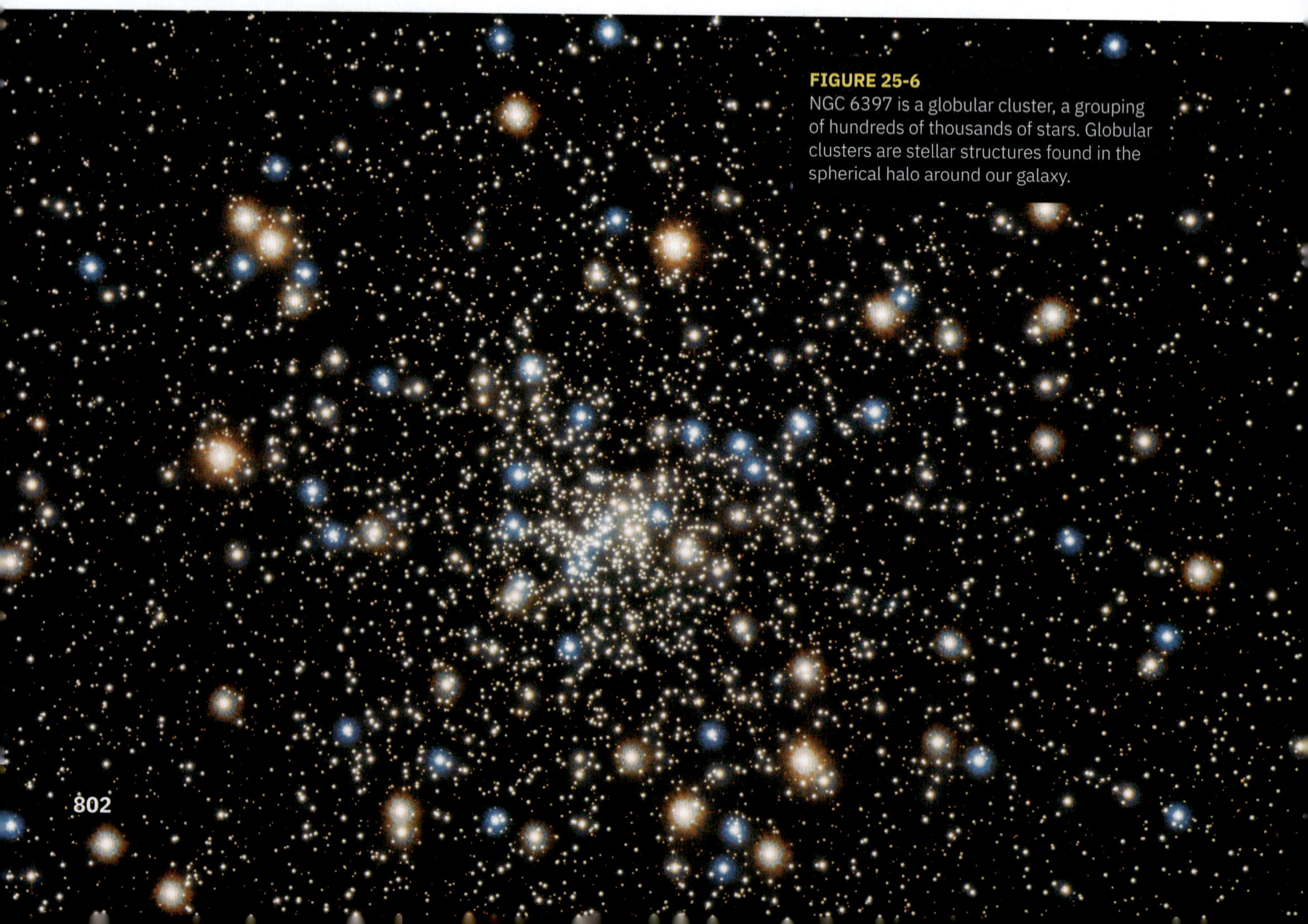

FIGURE 25-6
NGC 6397 is a globular cluster, a grouping of hundreds of thousands of stars. Globular clusters are stellar structures found in the spherical halo around our galaxy.

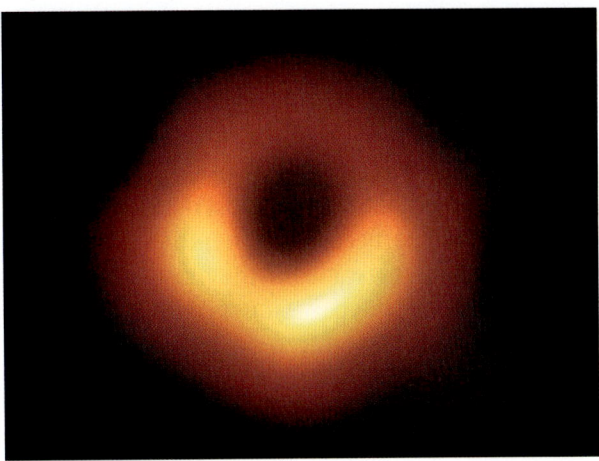

FIGURE 25-7
The Event Horizon Telescope Project produced this image of the supermassive black hole at the center of elliptical galaxy M87. The image represents the first time we have been able to image the surface, called the event horizon, of a black hole.

These stars were large enough to pass through the main sequence quickly and then collapse to become the black hole that now forms the center of our galaxy.

To date, we have been unsuccessful in imaging the black hole predicted at the center of the Milky Way. However, in April 2019 a project combining ground-based radio telescopes around the globe, the Event Horizon Telescope Project, produced the first image of a black hole (Figure 25-7). This supermassive black hole is located at the center of the giant elliptical galaxy M87.

checkpoint How is the Milky Way's spiral disk different from its halo?

25.1 ASSESSMENT

1. **Explain** What is a galaxy and how is it different from a solar system?
2. **Distinguish** How is the matter in different types of galaxies organized?
3. **Describe** How are stars, gas, and dust organized in the Milky Way galaxy?

Critical Thinking

4. **Infer** M87 is a giant elliptical galaxy whose central black hole scientists successfully imaged in 2019. Observational evidence of the motion of stars close to the center of the Milky Way suggests our galaxy also boasts a central black hole but we cannot directly image it. What inferences can be made from these two pieces of evidence?

25.2 THE BIG BANG

Core Ideas and Skills
- Summarize the big bang theory.
- Describe the observable evidence that supports the big bang theory.
- Explain how the motion of galaxies suggests that the universe is expanding.
- Describe the origin of the cosmic microwave background.

KEY TERMS

big bang
primordial nucleosynthesis
cosmic microwave background (CMB)

Big Bang Theory

In our search for the origin and history of Earth, we have looked back more than 4.6 billion years to the time when a diffuse cloud of dust and gas coalesced to form the solar system. But now, as we look into our galaxy and beyond, we ask: How did that cloud of dust and gas form? As our search for answers deepens, we finally ask: How and when did the universe begin?

Before we explore the origin of the universe, we must ask an even more fundamental question: Did it begin at all? One possibility is that the universe has always existed and there was no beginning, no start of time. An alternative hypothesis is that the universe began at a specific time and has been evolving ever since.

In 1929 Hubble observed that all galaxies are moving away from each other and that the farther galaxies are apart, the faster they move away from each other. By projecting the galactic motion backward in time, he reasoned that they must have started moving outward from a common center, and at the same time. Scientists postulated that in the beginning the entire universe was compressed into a single, infinitely dense point called a singularity. This point was so small that we cannot compare it with anything we know or can even imagine. According to modern theory, this massive point rapidly expanded in an event called the **big bang**, instantaneously creating the universe. Matter, energy, and space came into existence with this single event. It was the start of space and time.

Astronomers calculate the timing of the big bang by measuring speeds of galaxies and the distances among them. They then calculate backward in time to determine when they were all joined together in a single point. In 2013, astronomers combined a variety of techniques to determine that the universe is about 13.8 billion years old.

Let us go back to that unimaginably distant time when the universe was a trillionth of a trillionth of a billionth of a second old (Figure 25-8). At that time, the universe was only about as big as a grapefruit and the temperature was about 100 billion degrees on the Kelvin scale (100 billion K). But this primordial universe was expanding rapidly as it was propelled outward by the force of the big bang that formed it. As the universe expanded, it cooled. During the first second, the universe cooled to about 10 billion degrees, a thousand times the temperature in the center of the modern sun. At such high temperatures, atoms do not exist. Instead, the universe consisted of a plasma of radiant energy, electrons, protons, neutrons, and extremely light particles called neutrinos.

Over the next few minutes, the most exciting action came from the behavior of protons and neutrons. Recall that protons are positively charged particles. At large distances (by subatomic standards) two positive protons repel each other, like two north poles of a magnet. However, if protons are hot enough, they collide with sufficient energy to overcome the repulsion. At very close distances, two protons attract and bond together.

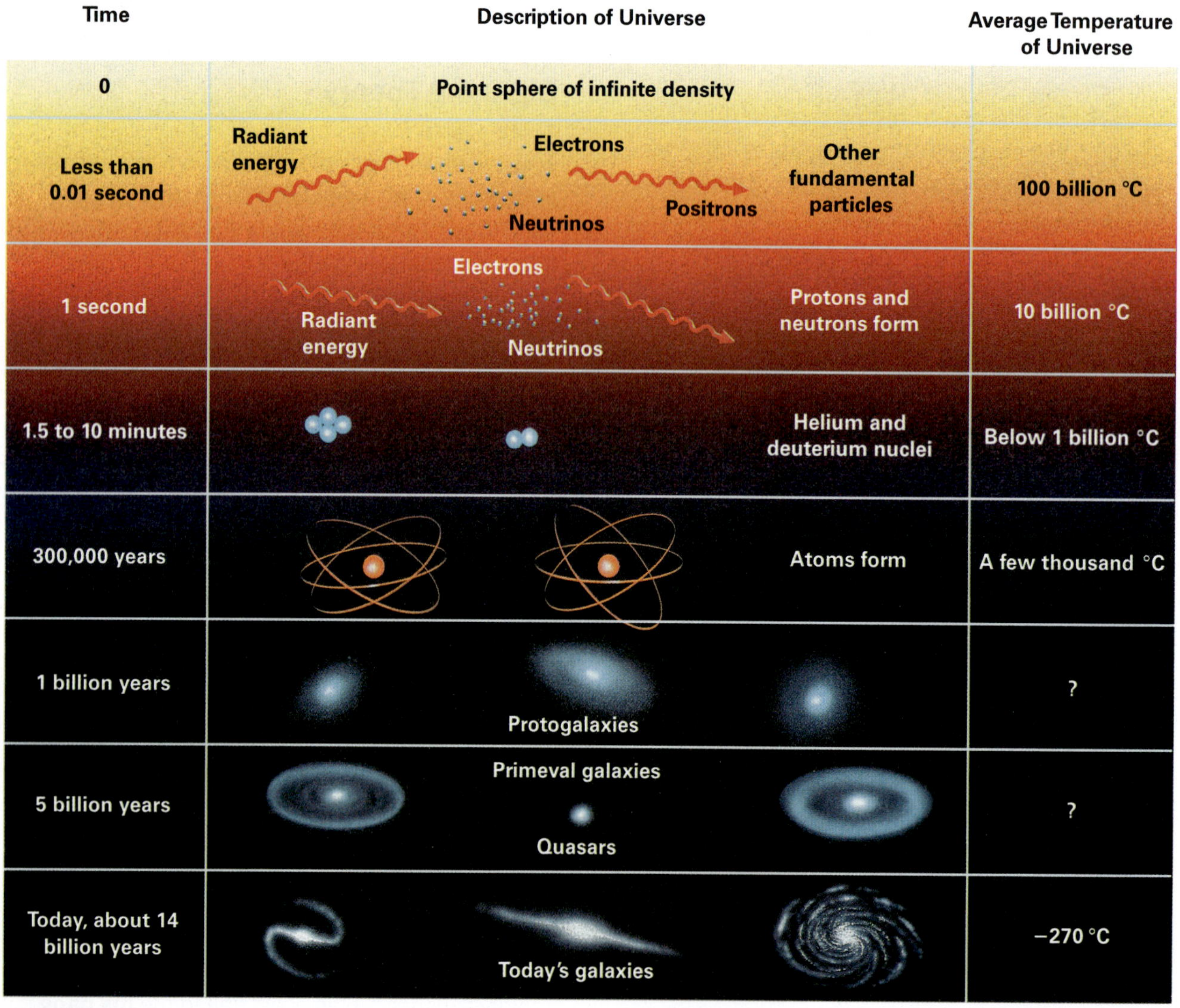

Time	Description of Universe	Average Temperature of Universe
0	Point sphere of infinite density	
Less than 0.01 second	Radiant energy, Electrons, Neutrinos, Positrons, Other fundamental particles	100 billion °C
1 second	Radiant energy, Electrons, Neutrinos, Protons and neutrons form	10 billion °C
1.5 to 10 minutes	Helium and deuterium nuclei	Below 1 billion °C
300,000 years	Atoms form	A few thousand °C
1 billion years	Protogalaxies	?
5 billion years	Primeval galaxies, Quasars	?
Today, about 14 billion years	Today's galaxies	−270 °C

FIGURE 25-8
A chart shows how scientists think the physical conditions of the universe changed over time. The universe started with extremely high temperature and contained only the most fundamental particles. Over time, the temperature decreased, and the complexity of matter increased.

This bonding is called fusion, the process that occurs to generate energy in stars. At high temperatures, protons also fuse with neutrons.

When the universe was less than 1.5 minutes old, it was so hot that helium nuclei were blasted apart almost as soon as they formed. After 10 minutes, the universe was so cool that fusion could no longer occur. The formation of helium nuclei, called the **primordial nucleosynthesis**, occurred over a time span of 8.5 minutes. Enough fusion had occurred during this time so that 6 percent of the total nuclei in the universe were helium, while 94 percent remained as hydrogen nuclei. Although primordial nucleosynthesis created a trace of lithium (the next-heaviest element after helium), it did not produce any heavier elements. There was no carbon, nitrogen, or oxygen, so life could not have evolved, and no silicon or metals, so solid planets could not have evolved. The universe was still in its infancy.

As the universe continued to expand, it finally cooled to a few thousand degrees by about 300,000 years after the big bang. At this crucial temperature, electrons became attached to the hydrogen and helium nuclei. The first atoms formed, and in a sense, the modern universe was born.

Before atoms formed, the universe had been a chaotic plasma that scattered and absorbed photons. As a result, photons could never move far in any direction. The universe was opaque. But when electrons combined with nuclei to form hydrogen and helium atoms, conditions changed radically. An atom absorbs light in only a relatively few specific wavelengths. All the rest of the light passes through unhindered. The atoms of hydrogen and helium in the early universe acted like a pair of sunglasses, filtering out certain wavelengths of light but permitting other wavelengths to pass through.

checkpoint What happened to the temperature of the universe over its first 14 billion years?

Evidence for the Big Bang

Scientists generally start their discussions of the origin of the universe when it was a trillionth of a trillionth of a billionth of a second old. How can we reach that far back in time with any degree of

---

DATA ANALYSIS Quantify the Expansion Rate of the Universe

In 1929 Edwin Hubble observed that the light coming from stars in far-off galaxies was red-shifted. Based on this observation, he proposed that the universe was expanding. Since then, a great deal of data that supports Hubble's expanding universe hypothesis has been collected. Consider the data shown in the table below.

Computational Thinking

1. **Recognize Patterns** What general trend do you observe in the data?
2. **Represent Data** Graph the data with distance on the x-axis and velocity on the y-axis. Draw a best-fit line through the data.
3. **Analyze Data** Determine the slope of the best-fit line.
4. **Abstract Information** The slope of your best-fit line can be used to determine the age of the universe, t. Use the equations below to help you solve for t. Be sure to include units in your equation.

$$\text{slope} = \frac{v}{d}$$

$$t = \frac{d}{v}$$

$1 \text{Mpc} = 3.086 \times 10^{19}$ km

Critical Thinking

5. **Evaluate** Do all the data support Hubble's hypothesis about an expanding universe? Explain why some galaxies might not appear to follow a trend that supports this idea.

Data Challenge

Go to the Data Analysis in MindTap to complete the data challenge.

TABLE 25-1 Distances and Velocities for a Set of Galaxies

Galaxy Name	Distance from Earth (Mpc)	Velocity (km/s)
NGC 7814	13.2	1,204
NGC 0157	45.7	1,769
NGC 0680	38.3	2,914
MCG-01-02-001	46.2	3,792
NGC 0252	66.5	5,176
IC 1844	85.2	6,750
MCG+01-02-004	115.0	7,576

Source: Caltech

certainty? In one series of experiments, scientists study the collision behavior of particles at very high velocities in modern particle accelerators. These results are then compared with observations of deep space. In the 90 years since Hubble's pioneering work, the experiments, observations, and calculations have led to disagreements, paradoxes, and unsolved mysteries. Yet, the preponderance of evidence is so persuasive that almost all astronomers agree with the fundamental premise of the big bang theory.

Three lines of evidence and logic support the big bang theory. The first is the expansion of the universe, discussed previously. The second is observational evidence of primordial nucleosynthesis and chemical composition. The third observation in support of the big bang theory is the discovery of the earliest radiation ever created.

Scientists have determined the expected ratio of hydrogen to helium in the modern universe based on the conditions of the primordial universe described by the big bang theory. They arrive at a number almost exactly in agreement with the observed chemical abundances in stars and galaxies. The most distant objects have few heavy elements (low metallicity) and are dominated by hydrogen and helium. Nearby galaxies are richer in elements heavier than helium (high metallicity). This observation supports the big bang theory.

Cosmic Microwave Background In the 1960s an astrophysicist at Princeton University named Robert Dicke predicted that we should be able to detect the primordial photons that began moving in a straight line as soon as atoms formed. He calculated that continued expansion of the universe would have cooled the primordial

MINILAB Big Bang Balloon

Materials

large, round balloon
string (about 35 cm)
metric ruler
felt-tip marker
graph paper
stopwatch or timer

Procedure

1. Work with a partner to create a model of an expanding universe. Start by inflating the balloon partially until it assumes a rounded shape. Pinch the end of the balloon to hold the air in, but do not tie it off.

2. Mark six locations with dots on random spots on the balloon. Number the dots 1–6.

3. Use the string and ruler to measure the distance from dot 1 to every other dot. Record each measurement in the data table under Round 1.

4. While your partner times you, take a deep breath and inflate the balloon further for 2–5 seconds. Pinch the balloon off and record the time. Then measure and record the distances in the data table under Round 2.

5. For each dot, calculate how far the dot moved (change in distance) after you inflated the balloon the second time. Record the results in your data table.

6. Calculate the speed of each dot by dividing the change in distance by the time it took to inflate the balloon in step 4. Record the results in your data table.

Results and Analysis

1. **Explain** What do the balloon and the dots represent? What does dot 1 represent? How did this exercise model the expansion of the universe?

2. **Represent Data** Create a scatter plot of your data using round 1 distance on the x-axis and speed on the y-axis.

3. **Analyze Data** How do the speeds compare for locations close to and farther away from dot 1? What does that suggest about how galaxies are moving?

Critical Thinking

4. **Predict** Suppose you are located at dot 1 in your model. What kind of Doppler shift would you expect to observe in light coming from dot 2 as the balloon inflates?

5. **Synthesize** How does your model relate to the Hubble-Lemaître law?

6. **Evaluate** What limitations does this model have with respect to the motion of galaxies in an expanding universe? Can you suggest an alternative model that might address these limitations?

photons to about 2.7K (2.7°C above absolute zero). In 1964 Arno Penzias and Robert Wilson at Bell Laboratories used the Holmdel Horn Antenna to detect a very faint photon radiation that was at 2.7 K and was uniformly distributed throughout space. Experimental observation agreed precisely with Dicke's calculation.

Walk outside at any time of day at any place on Earth and turn your palm upward to the sky. Every second, a million billion low-energy photons will strike your palm. These photons are in the microwave band of the electromagnetic spectrum. The energy is so faint and weak that you could never feel it. Yet, this radiation we call the **cosmic microwave background (CMB)** began traveling through space when the universe was only 300,000 years old. Today, the CMB is called the echo of the big bang. The prediction and discovery of the CMB provides the third convincing line of evidence to support the big bang theory.

Penzias and Wilson observed that the CMB was uniform throughout space. This measurement implied that the universe was homogeneous in its infancy. However, the modern universe is clearly not homogeneous. Matter is concentrated into stars, stars are clumped into galaxies, and galaxies are grouped into clusters containing tens of thousands of galaxies. Even the clusters group into superclusters. Most of the space between the clusters and superclusters contains no galaxies at all.

The question then arose: How and when did an initially uniform, homogeneous universe concentrate into stars, galaxies, and clusters? According to one hypothesis, the original, grapefruit-size universe had to obey laws of quantum mechanics, the same laws that describe modern atoms. Calculations based on quantum mechanics showed that the earliest universe contained tiny waves of energy, space, and time. These waves, like sound waves, contained alternating regions of higher and lower densities. If this model were correct, then these bands of varying energy densities would have created tiny temperature differences in the CMB.

In the 1980s physicists calculated that the temperature differences would be about 0.0001°C, far too small to have been detected by Penzias and Wilson's radio telescope. In 1989 astronomers launched the Cosmic Background Explorer (COBE) satellite which was capable of detecting 0.0001°C temperature differences in the CMB. As it slowly scanned the sky, COBE registered numerous fluctuations in the background temperature. After mapping the temperature of the universe for three years, COBE scientists reported that CMB varies by tiny amounts from one region to another. These data suggested that the primordial universe was not homogeneous as Penzias and Wilson had inferred. Matter and energy had concentrated into clumps during the earliest infancy of the universe, perhaps in the first billionth of a second.

In 2003 data from a second satellite, called the Wilkinson Microwave Anisotropy Probe (WMAP), confirmed and refined the COBE results. WMAP produced a high-resolution map of the CMB, with details as small as 0.2 degrees. Its data were used to determine the age of the universe—with the highest precision ever made—as 13.772 billion years. Between 2009 and 2013, the Planck satellite remapped the CMB (Figure 25-9) with three times

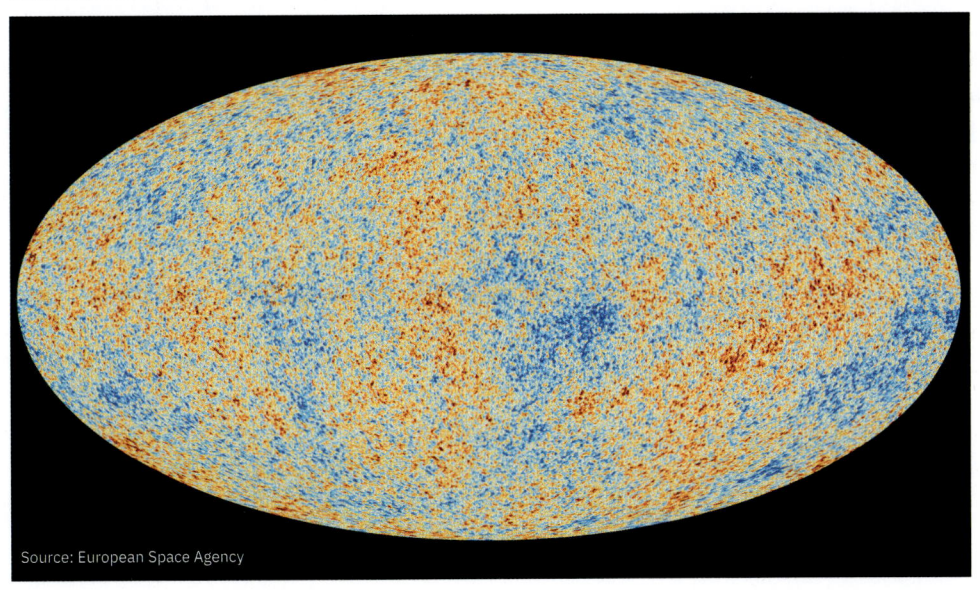

FIGURE 25-9
In 2013, observations from the Planck satellite confirmed the nearly uniform brightness of CMB at 2.73 K. The temperature fluctuations, represented by orange and blue colors on this map, are on the order of 1×10^{-5} K. The subtle temperature differences help explain how matter and energy in the universe eventually became clustered in stars, galaxies, and quasars.

Source: European Space Agency

LESSON 25.2

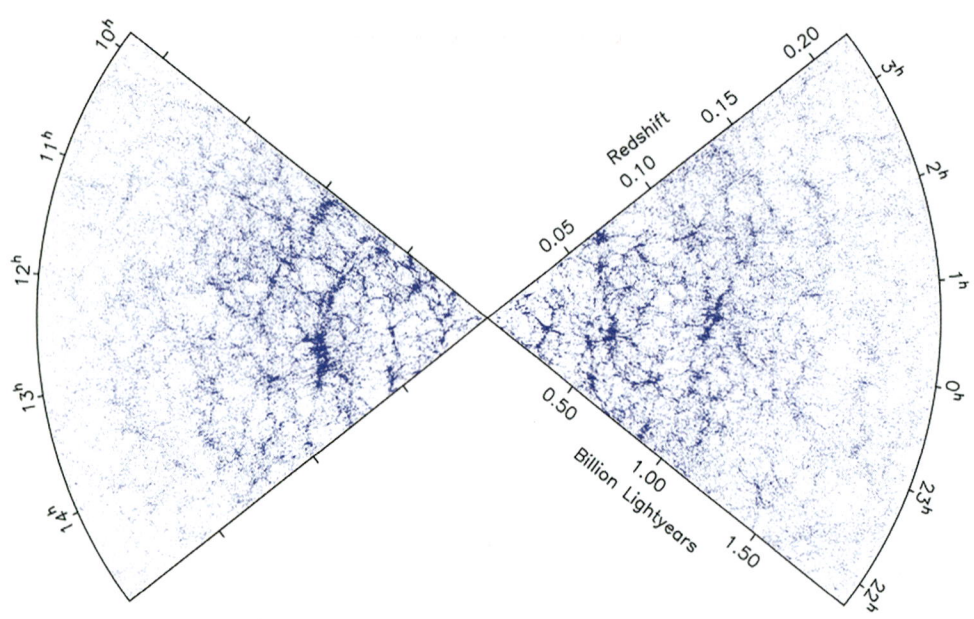

▶ **FIGURE 25-10**
This diagram represents a slice of the three-dimensional map of the visible universe created by the 2DF Galaxy Redshift Survey. The map shows the positions of more than 200,000 galaxies, covering 5 percent of the sky out to a distance of 2.5 billion light-years.

Source: 2dF Galaxy Redshift Survey Team

greater resolution than WMAP. The new data suggested that despite all the visible structure in the universe (Figure 25-10), most the universe is filled with matter we cannot see.

checkpoint How did our temperature measurements of the CMB change over time?

25.2 ASSESSMENT

1. **Identify** How old do scientists think the universe is?
2. **Explain** What was the universe's chemical composition during the first few minutes?
3. **Describe** What observational evidence supports the big bang theory?

Critical Thinking

4. **Synthesize** How has the structure of the universe changed over time?

25.3 QUASARS, DARK MATTER, AND THE FATE OF THE UNIVERSE

Core Ideas and Skills
- Describe the nature and origin of quasars.
- Explain where most of the matter in the universe is and how we know.
- Describe the possible fates of the universe.

KEY TERMS

quasar	open universe
dark matter	accelerating universe
closed universe	dark energy

Quasars

In the 1960s astronomers were studying many objects that look like stars but emit extremely large amounts of energy as radio waves. These objects were perplexing because normal stars, such as our sun, emit mostly visible and ultraviolet light. During the following decade, astronomers discovered that the spectra of many of these objects coincide with that of hydrogen, except that they have a very large redshift. Such objects are now called **quasars** (Figure 25-11).

Recall that a large redshift indicates that an object is moving away from Earth at a high speed. If quasars obey the Hubble-Lemaître law, then they are very far away. In order to be visible when they are so far away, quasars must emit tremendous amounts of energy—10 to 100 times more than an entire galaxy!

Quasars emit erratic bursts of energy. Recall from our discussion of pulsars in Chapter 24 that astronomers can estimate the diameter of a pulsar by the sharpness of a short burst of energy emitted from it. Similar techniques indicate that some quasars are several light-months in diameter. By comparison, the distance from Earth to the nearest star is four light-years. A quasar is much smaller than a galaxy but emits much more energy. Furthermore, it emits energy over a wide range of wavelengths. Most quasars are very distant from Earth. One common hypothesis for their origin is that a massive black hole, perhaps of one billion solar masses, lies at the center of a quasar and that the quasar emits energy as gas and dust accelerate into the hole.

Many quasars are eight billion to 13 billion light-years away. If an object is 10 billion light-years away, the light we see today started its journey 10 billion years ago. Therefore, we see what was happening long ago but not what is happening today. If the object blew up and disappeared nine billion years ago, we will not know about it for another billion years! When we look at close objects, we see what happened recently, but when we look at distant objects, we see what happened in the distant past. One goal of building more powerful telescopes is to study more-distant objects and therefore probe further back in time.

The oldest quasars must have formed when the universe was young. Moreover, quasars contain heavy elements, and heavy elements form only in the supernova explosions of dying stars. Therefore, quasars are second-generation structures; they formed after earlier stars were born, evolved, exploded, and died. Hence, stars must have formed and passed through their life cycles less than two billion years after the universe formed.

checkpoint How bright are quasars in comparison to galaxies?

Dark Matter

The Milky Way rotates slowly, like a giant pinwheel. Using laws of motion developed by Kepler and Newton in the 17th century, astronomers have attempted to calculate the mass of our galaxy from their knowledge of its rotational speed. However, these calculations produced a giant anomaly. The Milky Way is much more massive than we can account for if we add up all the known matter within it.

Satellite studies of the nonhomogeneity of galaxies as well as data on the CMB allow scientists to calculate the mass and gravitational force of the primordial universe. These studies indicate that matter we can see comprises only 4.6 percent of the universe! The other 95.4 percent of the universe is completely invisible and only a small fraction of that is actually considered matter, something scientists call **dark matter**.

The existence of dark matter can be inferred from the observable rotation of galaxies. Most galaxies do not have nearly enough visible mass (in the form of stars and interstellar gas) to hold their stars in orbit around their galactic centers. Only the presence of large amounts of invisible dark matter, which exerts a gravitational pull on visible matter, can explain why galaxies do not fly apart. Concentrations of dark matter can be indirectly detected by observing the phenomenon of gravitational lensing, which occurs when light from distant stars or galaxies is refracted as it passes through another galaxy on its way to Earth. By measuring how much the light bends and where it bends, astronomers can map the location of unseen dark matter.

What is dark matter? Not many years ago, many astrophysicists speculated that dark matter might

FIGURE 25-11
This image shows a bright quasar, a point source hundreds of times brighter than the full-size spiral and elliptical galaxies in the image.

be composed primarily of planet-like objects that do not radiate energy (and are therefore invisible) and black holes that cannot radiate energy. However, results from the WMAP satellite, as well as recent data from the Planck satellite, convinced most astrophysicists that the dark matter that makes up 24 percent of the universe is not made of atoms but rather an as-yet-undetected particle that exerts a conventional gravitational attraction.

The composition and nature of dark matter is still the subject of much scientific debate and investigation. A small percentage of dark matter likely consists of baryonic matter—matter made up of atoms that we simply cannot detect because they do not radiate light, such as cold intergalactic gases. Brown dwarfs, white dwarfs, black holes, and neutron stars may also account for some dark matter; however, these are relatively rare phenomena that most likely do not account for most of the dark matter known to exist. As a result, most dark matter is thought to be nonbaryonic—consisting not of atoms but of different kinds of particles that are electromagnetically neutral and only interact with baryonic matter gravitationally. These theoretical particles are often called weakly interacting massive particles (WIMPs), but

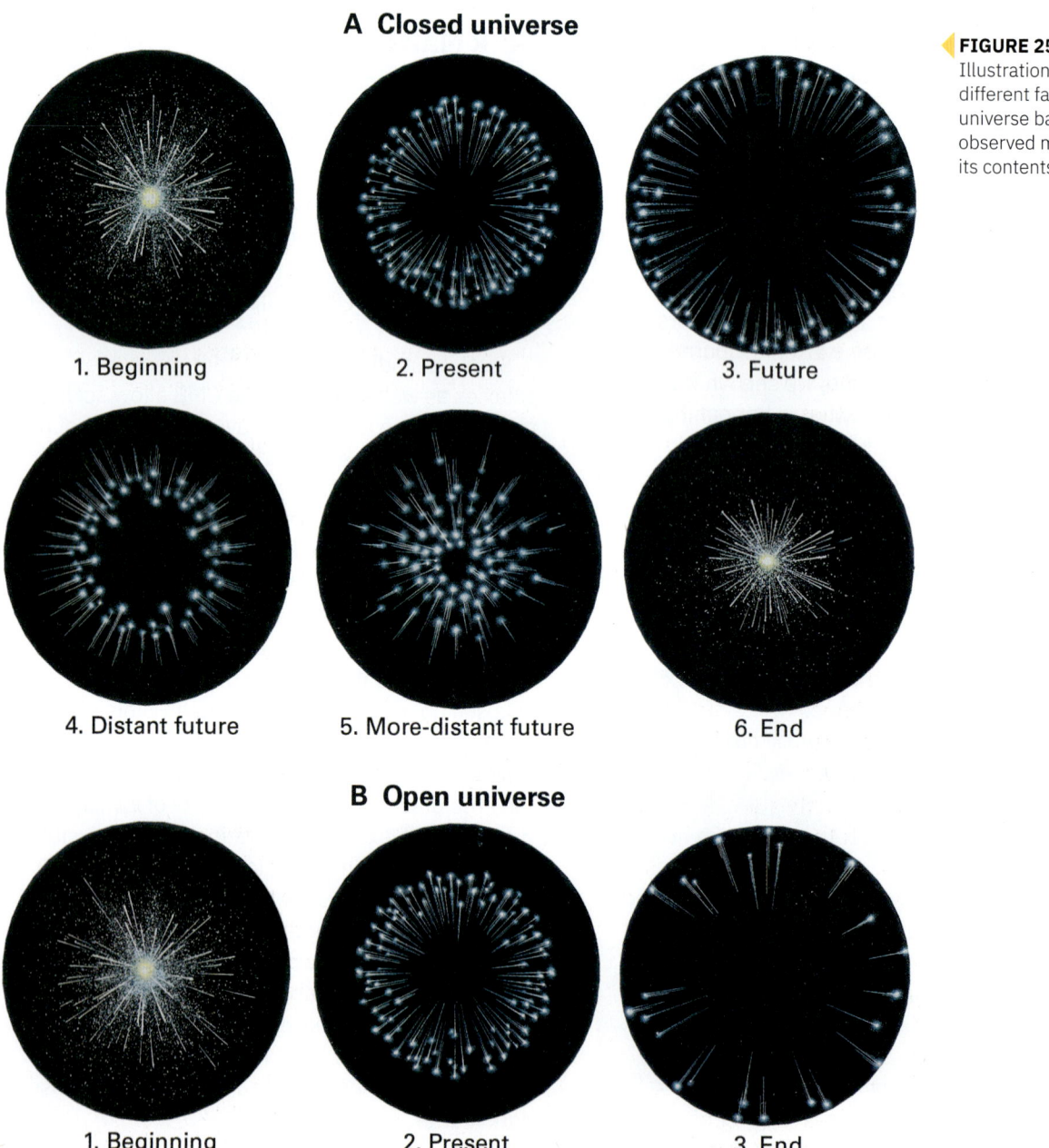

FIGURE 25-12 Illustrations show the different fates of the universe based on observed motion of its contents.

CONNECTIONS TO MODERN PHYSICS

Distorted Light and Dark Matter

In developing the theory of general relativity, Albert Einstein predicted that massive objects can change the direction of light. Gravitational lensing is the bending and distortion of light from a distant object as it passes near a massive object. More massive objects have stronger gravitational fields and cause greater distortions in the light. Just as denser glass material makes optical lenses that refract light more, greater mass results in greater bending of the light rays. Gravitational lensing happens on all scales, from small masses such as our own bodies to exoplanets to stars and larger masses such as galaxies and clusters of galaxies. Scientists have discovered quantifiable evidence for dark matter by analyzing the light from bright and extremely distant objects as it is distorted by nonvisible matter on its way to us.

---

scientists have yet to observe them. Scientists are pursuing many lines of inquiry in an effort to observe and characterize these dark matter particles, which would shed new light on the nature of the universe.

checkpoint How did astronomers discover that mass was missing from the Milky Way galaxy?

The Fate of the Universe

From the first instant of the big bang, the universe has been affected by two opposing forces. Matter is expanding outward as a result of the big bang, but at the same time the gravity of all the mass in the universe is pulling all matter back inward toward a common center. If the gravitational force of the universe is sufficient, all the galaxies will eventually slow down, reverse direction, and fall back to the center, forming another point of infinite density. This hypothesis is called a **closed universe** (Figure 25-12). Some scientists have speculated that if the universe is closed, it will collapse and then explode again to form a new universe. In turn, the new universe will expand and then collapse, creating a continuous chain of universes. However, no mathematical model explains how another big bang would occur. Instead, current models predict that in a closed system, the universe would collapse into a mammoth black hole.

Another possibility is that the gravitational force of the universe is not sufficient to stop the expansion and that the galaxies will continue to move away from each other forever. Within each galaxy, stars will eventually consume all their nuclear fuel and stop producing energy. As the stars fade and cool, the galaxies will continue to separate into the cold void. This scenario is called the **open universe**.

In the 1990s, astronomers were attempting to measure the deceleration of the universe. If they could learn how fast it was slowing down, then they reasoned that they could calculate whether the universe is open or closed. Astronomers made these measurements by studying the speed (redshift) of distant supernovae. The result was totally unexpected. The expansion of the universe is not slowing down at all. Instead, it is accelerating!

The concept of an **accelerating universe** is totally baffling. Objects accelerate only when they are acted on by a force. In this case, the force must be some sort of repulsion that acts against gravity and pushes the galaxies apart. Scientists have named this repulsive force **dark energy**. Scientists do not yet have a clear understanding of how dark energy works.

Recall that 4.6 percent of the universe is visible matter like planets, stars, galaxies and quasars. The final 71.4 percent of the universe is dark energy—the poorly understood repulsive force that is causing the modern universe to accelerate. More remains to be studied in the field of cosmology.

checkpoint What evidence makes a closed universe unlikely?

25.3 ASSESSMENT

1. **Describe** What observations distinguish quasars from other objects?
2. **Explain** What is the link between quasars and the earliest galaxies that formed?
3. **Distinguish** What are the differences between open, closed, and accelerating universes?

Critical Thinking

4. **Synthesize** As an open universe ages, describe how observations of space would change over time. Do the same for a closed universe and compare the two.

TYING IT ALL TOGETHER
QUASARS AND THE EVOLUTION OF THE UNIVERSE

In this chapter, you have learned about the origins of the universe, how its matter is organized, and various theories about its ultimate ending. But as you have learned, theories of the universe are limited by its sheer size and the challenges of gathering data from billions of light-years away.

In this chapter's Case Study, you read about ongoing research to learn about the center of the Milky Way galaxy. The evidence accumulated so far supports the idea that supermassive black holes lie at the centers of all spiral galaxies, including our own. As discussed in the chapter, astronomers have also postulated that quasars are associated with black holes. Quasars are the brightest objects in the universe, radiating more than a trillion times more energy than the sun. The source of this immense energy is thought to be the gravitational field of a supermassive black hole. Quasars are also the earliest objects in the universe that astronomers have been able to observe directly. Because they are so far away, light from quasars that we observe on Earth has taken billions of years to reach us. In other words, we are observing these objects as they existed billions of years ago, which makes them uniquely suited to studying the early history of the universe.

More than a million quasar candidates have been identified so far. In 2017, scientists discovered the quasar J1342+0928. Located more than 13 billion light-years from Earth, it was the most distant and earliest ever observed. The light coming from this quasar was emitted only 690 million years after the big bang, allowing astronomers to see farther back in time than ever before.

By studying numerous quasars like this one, astronomers are able to piece together a time line of the changes that have occurred over the history of the universe.

Work individually or with a group to complete the steps.

1. Research scientific ideas about how the universe has evolved since the big bang. As you conduct your research, take note of evidence scientists have discovered that supports any of these ideas. Include information about the types of objects present at different stages, changes in the temperature of the universe over time, and any other information you find interesting and relevant to the early history of the universe.

2. Using the information you gathered during your research, create a time line illustrating the changing nature of the universe from its formation to the present. Your time line should provide a detailed description for each time point and distinguish between those that are supported by observational evidence and those that are still theoretical or have yet to be confirmed with observations. Observational evidence should include data collected from different parts of the electromagnetic spectrum.

3. Take turns presenting your time line to the others in your class. Discuss similarities and differences between your time lines.

4. Critique your own time line based on your group discussion above. What could you do to improve your time line?

FIGURE 25-13
Scientists think that quasars, like the one depicted here in an artist's rendition, can provide clues about how the universe has evolved.

CHAPTER 25 SUMMARY

25.1 THE MILKY WAY AND OTHER GALAXIES

- Galaxies and clusters of galaxies form the basic structure of the universe.
- Commonly, galaxies are elliptical, spiral, or barred spiral, although other shapes also exist.
- The Hubble-Lemaître law states that the most distant galaxies are moving away from Earth at the greatest speeds, while closer ones are receding more slowly.
- Our sun lies in the Milky Way, a barred spiral galaxy.
- In addition to stars in the main disk, the Milky Way contains captured dwarf galaxies, diffuse clouds of interstellar gas and dust, a galactic halo, and globular clusters surrounding the disk.

25.2 THE BIG BANG

- The big bang theory states that the universe began with a cataclysmic explosion that instantly created space and time.
- Evidence for the big bang includes observations of the recessional velocity of galaxies, the chemical composition of objects in the universe, and the Cosmic Microwave Background (CMB).
- The CMB is radiation from the big bang and shows that the universe was homogeneous and uniform in all directions very early in its history.
- Tiny fluctuations in the temperature of the CMB provide evidence of the initial variations that resulted in the complex structure observed in the current universe.

25.3 QUASARS, DARK MATTER, AND THE FATE OF THE UNIVERSE

- Quasars are very far away, emit as much as a hundred times the energy of an entire galaxy, and are small compared with a galaxy.
- A black hole may lie in the center of a quasar.
- Up to 96 percent of the universe is not directly observable and a portion of this is called dark matter.
- In trying to determine whether the universe is open or closed, astronomers learned that the expansion of the universe is accelerating.

CHAPTER 25 ASSESSMENT

Review Key Terms

Select the key term that best fits the definition. Not all terms will be used, and no term will be used more than once.

accelerating universe	galaxy
barred spiral galaxy	globular cluster
big bang	Hubble-Lemaître law
closed universe	irregular galaxy
cosmic microwave background (CMB)	open universe
dark energy	primordial nucleosynthesis
dark matter	quasar
elliptical galaxy	spiral galaxy
galactic halo	

1. the hypothesis that the universe will expand forever, and the stars will eventually consume all their fuel and stop producing energy
2. the outward expansion of the universe at an increasing rate, propelled by an unknown repulsive force
3. a collection of billions of stars, along with gas and dust, held together by gravity
4. a spherical cloud of star clusters, dust, and gas surrounding the disk of a galaxy
5. the expansion of a single, infinitely dense point about 13.8 billion years ago, which produced all matter, energy, and space in the universe
6. a group of billions of stars held together by gravity in the shape of a pinwheel
7. the hypothesis that the expansion of the universe will eventually reverse itself, causing all matter to collapse back into an infinitely dense point
8. a group of billions of stars held together by gravity in the shape of an oval with no spiraling arms
9. a group of old, dim stars within the halo of a galaxy, possibly the remains of one or more protogalaxies
10. the directly proportional relationship between the velocity of a galaxy and its distance from Earth
11. the fusion of hot protons to form the first stable helium nuclei, which occurred within the first ten minutes after the big bang
12. a small, very distant object that emits enormous, erratic bursts of energy
13. invisible component of the universe that accounts for 96 percent of its mass
14. the so-called echo of the big bang, consisting of low-energy primordial photons that are distributed throughout the universe almost uniformly
15. hypothesized repulsive force that is causing the expansion of the universe to accelerate

Review Key Concepts

Answer each question on a separate sheet of paper to demonstrate your understanding of key concepts from the chapter.

16. What type of galaxy is shown?

17. What type of galaxy is shown?

18. Explain how an astronomer could use the Hubble-Lemaître law to determine the approximate distance between Earth and a newly discovered galaxy.
19. Describe the Milky Way galaxy in terms of shape, size, and composition.
20. Compare the mass of the sun to the visible mass of the Milky Way galaxy.
21. Explain why no atoms formed for the first 300,000 years after the big bang.
22. How does the current chemical composition of the universe support the big bang theory?
23. In what way do observations of quasars allow astronomers to look back in time?
24. Compare galaxies and quasars in terms of size and energy emission.

Think Critically

Write a response to each question on a separate sheet of paper. Use concepts from the chapter to support your reasoning.

25. Compare the motion of stars within a galaxy to the motion of galaxies within the universe. Explain any differences.
26. How did data from the Cosmic Background Explorer (COBE) satellite help astronomers refine the big bang theory?
27. Based on the available evidence, which do you think is more likely: the closed universe hypothesis or the open universe hypothesis? Construct an argument that demonstrates your understanding of both hypotheses.
28. How does the Doppler effect help astronomers make lower-limit estimations of the age of the universe?

PERFORMANCE TASK

Alternate Cosmology

Cosmology is the study of the nature of the universe. In this chapter, you have explored the big bang theory, which currently shapes scientists' cosmology. As you have learned, the big bang theory is supported by three main lines of observational evidence. However, scientific theories are always subject to re-examination and revision in light of new evidence.

Suppose that you are an astronomer from a planet in a distant galaxy. Civilization on your planet has developed star-gazing technology far superior to even the most powerful telescopes of Earth. Although some of your observations are in agreement with the big bang theory, others suggest that humans may need to re-examine some of their basic cosmology.

In this performance task, you will summarize observations from the two different planets and use the evidence to support distinct conclusions that are part of the big bang theory.

1. Complete a table that compares phenomena as observed from Earth and observed from your distant planet.
   A. Summarize the observations of human astronomers that support the big bang theory. Then summarize the observations of astronomers from your planet regarding the same phenomena. Observations may match those made from Earth for two of the three phenomena, but they should be different for at least one. Be creative!
   B. Add the conclusions for both sets of observations. Be sure your conclusions are based on the evidence you described in 1A.
   C. Create a brief text (1–2 pages) comparing the big bang theory with your planet's cosmology. Your text may take a form of your choosing. Some suggestions:
      - an article by an astronomer arguing against the big bang theory
      - a dialogue or debate between astronomers from both planets
      - an interview of an astronomer from your planet by a confused scientist from Earth

APPENDIX 1 REFERENCE TABLES

PERIODIC TABLE OF THE ELEMENTS

How to Read the Periodic Table

The periodic table of the elements displays all chemical elements arranged in order of increasing atomic number (the number of protons in the nucleus). Atomic number increases across a row, or period. The elements in each column, or group, share some physical and chemical properties.

Key
- 6 — Atomic number
- C — Element symbol
- Carbon — Element name
- 12.01 — Relative atomic mass

FIGURE 1 The Periodic Table

Sodium is too chemically reactive to occur in its pure form in nature, although it is a component of many minerals, including halite (table salt); sodium and calcium are the main dissolved ions in seawater.

Organisms incorporate large amounts of calcium in their bones, teeth, and shells; calcium carbonate, the primary constituent of limestone, can undergo dissolution in groundwater to form caverns with spectacular rock formations.

Zirconium, a corrosion-resistant metal, is produced from the mineral zircon; radiometric dating of uranium contained within zircon grains shows some formed 4.2–4.3 billion years ago, making them the oldest known Earth materials.

Iron, which has been used by humans for at least 5,000 years, is the heaviest element that can be formed by nuclear synthesis within stars; all elements with a higher atomic number form only in supernova explosions.

Group	1 (1A)	2 (2A)	3	4	5	6	7	8	9
1	1 H Hydrogen 1.008								
2	3 Li Lithium 6.941	4 Be Beryllium 9.012							
3	11 Na Sodium 22.99	12 Mg Magnesium 24.31							
4	19 K Potassium 39.10	20 Ca Calcium 40.08	21 Sc Scandium 44.96	22 Ti Titanium 47.88	23 V Vanadium 50.94	24 Cr Chromium 52.00	25 Mn Manganese 54.94	26 Fe Iron 55.85	27 Co Cobalt 58.93
5	37 Rb Rubidium 85.47	38 Sr Strontium 87.62	39 Y Yttrium 88.91	40 Zr Zirconium 91.22	41 Nb Niobium 92.91	42 Mo Molybdenum 95.94	43 Tc Technetium (98)	44 Ru Ruthenium 101.1	45 Rh Rhodium 102.9
6	55 Cs Cesium 132.9	56 Ba Barium 137.3	57 La Lanthanum 138.9	72 Hf Hafnium 178.5	73 Ta Tantalum 180.9	74 W Tungsten 183.9	75 Re Rhenium 186.2	76 Os Osmium 190.2	77 Ir Iridium 192.2
7	87 Fr Francium (223)	88 Ra Radium 226.0	89 Ac Actinium (227)	104 Rf Rutherfordium (261)	105 Db Dubnium (262)	106 Sg Seaborgium (263)	107 Bh Bohrium (264)	108 Hs Hassium (265)	109 Mt Meitnerium (268)

Uranium provides the energy to run electricity-generating nuclear power plants as well as to determine the ages of uranium-containing minerals via radiometric dating techniques.

58 Ce Cerium 140.1	59 Pr Praseodymium 140.9	60 Nd Neodymium 144.2	61 Pm Promethium (145)	62 Sm Samarium 150.4
90 Th Thorium 232.0	91 Pa Protactinium (231)	92 U Uranium 238.0	93 Np Neptunium (237)	94 Pu Plutonium (244)

In this periodic table, metals are shaded blue, metalloids green, and nonmetals orange. Hydrogen, shaded brown, does not belong to any group, and its properties are not similar to those of the group 1A elements below it.

Some sets of elements have their own names. Group 1A is known as the alkali metals; 2A, the alkaline-earth metals; group 7A, the halogens, and group 8A, the noble gases. The lanthanides (also called rare-earth elements) occupy the first row set off from the main table, and the actinides are below them in the second row.

Gold, a high-value metal, has an average concentration of 0.004 milligrams per kilogram in Earth's crust; one ton of seawater contains about 1 milligram of dissolved gold.

Helium, the second-most abundant element in the universe (behind hydrogen), is relatively rare on Earth, and is light enough to escape from the top of the atmosphere into space.

Nitrogen gas makes up 78.1 percent of Earth's atmosphere; certain soil bacteria can combine it with other elements into compounds needed as nutrients by plants and animals.

Oxygen is chemically reactive, readily combining with most other elements, and is the most abundant element in Earth's crust.

Silicon is the second-most abundant element in Earth's crust; in combination with oxygen, it forms silicate ions, the defining component of Earth's largest mineral group, the silicates.

Radon gas, a major health hazard, can seep into the lower levels of buildings; it is produced by the radioactive decay of radium, thorium, or uranium contained in minerals in the ground.

13 3A	14 4A	15 5A	16 6A	17 7A	18 8A
					2 He Helium 4.003
5 B Boron 10.81	6 C Carbon 12.01	7 N Nitrogen 14.01	8 O Oxygen 16.00	9 F Fluorine 19.00	10 Ne Neon 20.18
13 Al Aluminum 26.98	14 Si Silicon 28.09	15 P Phosphorus 30.97	16 S Sulfur 32.07	17 Cl Chlorine 35.45	18 Ar Argon 39.95

10	11	12						
28 Ni Nickel 58.69	29 Cu Copper 63.55	30 Zn Zinc 65.38	31 Ga Gallium 69.72	32 Ge Germanium 72.59	33 As Arsenic 74.92	34 Se Selenium 78.96	35 Br Bromine 79.90	36 Kr Krypton 83.80
46 Pd Palladium 106.4	47 Ag Silver 107.9	48 Cd Cadmium 112.4	49 In Indium 114.8	50 Sn Tin 118.7	51 Sb Antimony 121.8	52 Te Tellurium 127.6	53 I Iodine 126.9	54 Xe Xenon 131.3
78 Pt Platinum 195.1	79 Au Gold 197.0	80 Hg Mercury 200.6	81 Tl Thallium 204.4	82 Pb Lead 207.2	83 Bi Bismuth 209.0	84 Po Polonium (209)	85 At Astatine (210)	86 Rn Radon (222)
110 Ds Darmstadtium (271)	111 Rg Roentgenium (272)	112 Cn Copernicium (285)	113 Nh Nihonium (286)	114 Fl Flerovium (289)	115 Mc Moscovium (289)	116 Lv Livermorium (293)	117 Ts Tennessine (294)	118 Og Oganesson (294)

63 Eu Europium 152.0	64 Gd Gadolinium 157.3	65 Tb Terbium 158.9	66 Dy Dysprosium 162.5	67 Ho Holmium 164.9	68 Er Erbium 167.3	69 Tm Thulium 168.9	70 Yb Ytterbium 173.0	71 Lu Lutetium 175.0
95 Am Americium (243)	96 Cm Curium (247)	97 Bk Berkelium (247)	98 Cf Californium (251)	99 Es Einsteinium (252)	100 Fm Fermium (257)	101 Md Mendelevium (258)	102 No Nobelium (259)	103 Lr Lawrencium (260)

APPENDIX 1

APPENDIX 1 REFERENCE TABLES

INTERNATIONAL SYSTEM OF UNITS (SI)

The International System of Units (or SI from the French "Système International d'Unités") is used by scientists throughout the world. SI, which is based on the metric system of units, was developed to replace multiple measurement systems with a single, standardized system. Its seven basic units (Table 1) can be used to derive all other SI units. For example, a pascal (Pa), the SI unit for pressure, combines the base units for mass, length, and time: Pa = (kg/m × s^2).

The SI system uses prefixes to express the values, in multiples of 10, of quantities much larger or much smaller than a base unit, as shown in Table 2. Prefixes for large values include *mega-* (M; 10^6) and *kilo-* (k; 10^3). Prefixes for small values include *centi-* (c, 10^{-2}), *milli-* (m, 10^{-3}), *micro-* (μ, 10^{-6}), and *nano-* (n, 10^{-9}).

The SI values of some other units of measurement are shown in Table 3.

TABLE 1 SI Base Units

Measurement	Base Unit
Length	meter (m)
Mass	kilogram (kg)
Time	second (s)
Electric current	ampere (A)
Temperature	kelvin (K)
Light intensity	candela (cd)
Amount of substance	mole (mol)

TABLE 2 Common SI Units by Measurement

Measurement	Unit	Value
Length	kilometer (km)	10^3 m
	centimeter (cm)	10^{-2} m
	millimeter (mm)	10^{-3} m
	micrometer (μm)	10^{-6} m
	nanometer (nm)	10^{-9} m
Area	square kilometer (km^2)	10^6 m^2
	square meter (m^2)	1 m^2
	square centimeter (cm^2)	10^{-4} m^2
	square millimeter (mm^2)	10^{-6} m^2
Volume and Capacity	cubic meter (m^3)	1 m^3
	cubic centimeter (cm^3)	10^{-6} m^3
	kiloliter (kL)	1 m^3
	liter (L)	10^{-3} m^3
	milliliter (mL)	10^{-6} m^3

TABLE 2 Common SI Units by Measurement *(continued)*

Measurement	Unit	Value
Mass	megagram (metric ton; Mg)	10^3 kg
	gram (g)	10^{-3} kg
	milligram (mg)	10^{-6} kg
	microgram (μg)	10^{-9} kg
Energy	hertz (frequency; Hz)	1/s
	newton (force; N)	$(m \times kg)/s^2$
	joule (work, heat; J)	$(m^2 \times kg)/s^2$
	pascal (pressure; Pa)	$kg/(m \times s^2)$; N/m^2
	watt (power; W)	$(m^2 \times kg)/s^3$; J/s
	coulomb (electric charge; C)	$s \times A$
	volt (voltage; V)	$(m^2 \times kg)/(s^3 \times A)$

TABLE 3 Other Units of Measurement Expressed in SI Units

Measurement	Unit	Value in SI Units
Energy	kilowatt-hour (kW h)	3.6 MJ
	British thermal unit (BTU)	1055.06 J
Air Pressure	atmosphere	101,325 Pa
	bar	105 Pa
	millibar (mbar)	102 Pa
Astronomical Distance	astronomical unit (AU)	1.496×10^8 km
	light-year (ly)	9.461×10^{12} km
	parsec (pc)	3.086×10^{13} km

APPENDIX 1 REFERENCE TABLES

IMPORTANT MINERALS

Minerals are grouped by chemical composition (Table 4). Except for native elements, which occur as pure elements, each group is named for the anion all of its minerals contain. Two groups are characterized by a simple, single-element anion: the sulfides, S^-, and the halides, a halogen element (for example, Cl^- or F^-). Oxides may contain a simple anion (O^{2-}) or a polyatomic one such as $Cr_2O_4^{2-}$ in chromite ($FeCr_2O_4$). Minerals in the other groups always contain a polyatomic anion: sulfates, SO_4^{2-}; carbonates, CO_3^{2-}; hydroxides, OH^-; phosphates, PO_4^{3-}; and silicates, SiO_4^{4-}.

TABLE 4 Important Mineral Groups

Group	Member	Formula	Economic Use
Oxides	Hematite	Fe_2O_3	Ore of iron
	Magnetite	Fe_3O_4	Ore of iron
	Corundum	Al_2O_3	Gemstone; abrasive
	Chromite	$FeCr_2O_4$	Ore of chromium
	Ice	H_2O	Solid form of water
Sulfides	Galena	PbS	Ore of lead
	Sphalerite	ZnS	Ore of zinc
	Pyrite	FeS_2	Fool's gold
	Chalcopyrite	$CuFeS_2$	Ore of copper
	Bornite	Cu_5FeS_4	Ore of copper
	Cinnabar	HgS	Ore of mercury
Sulfates	Gypsum	$CaSO_4 \cdot 2H_2O$	Plaster
	Anhydrite	$CaSO_4$	Plaster
	Barite	$BaSO_4$	Drilling mud

TABLE 4 Important Mineral Groups *(continued)*

Group	Member	Formula	Economic Use
Native Elements	Gold	Au	Electronics; jewelry
	Copper	Cu	Electronics
	Diamond	C	Gemstone; abrasive
	Sulfur	S	Sulfa drugs; chemicals
	Graphite	C	Pencil lead; dry lubricant
	Silver	Ag	Jewelry; photography
	Platinum	Pt	Catalyst
Halides	Halite	NaCl	Common salt
	Fluorite	CaF_2	Steel making
	Sylvite	KCi	Fertilizer
Carbonates	Calcite	$CaCO_3$	Portland cement
	Dolomite	$CaMg(CO_3)_2$	Portland cement
	Aragonite	$CaCO_3$	Portland cement
Hydroxides	Limonite	$FeO(OH) \cdot nH_2O$	Ore of iron; pigments
	Bauxite	$Al(OH)_3 \cdot nH_2O$	Ore of aluminum
Phosphates	Apatite	$Ca_5(PO_4)_3(F,Cl,OH)$	Fertilizer
	Turquoise	$CuAl_6(PO_4)_4(OH)_8 \cdot 4H_2O$	Gemstone
Silicates Silicate minerals make up 92 percent of Earth's crust. For silicate minerals listed by a general name, a specific example is given.	Quartz	SiO_2	Glass; abrasive
	Feldspar (example: orthoclase)	$KAlSi_3O_8$	Glass; ceramics
	Mica (example: biotite)	$K(Mg,Fe)_3(AlSi_3O_{10})(F,OH)_2$	Paints; drilling mud
	Amphibole (example: hornblende)	$(Ca,Na)_{2-3}(Mg,Fe,Al)_5(Si,Al)_8O_{22}(OH,F)_2$	Construction materials
	Pyroxene (example: diopside)	$MgCaSi_2O_6$	Gemstone
	Olivine (example: forsterite)	Mg_2SiO_4	Refining of metals
	Clay (example: kaolinite)	$Al_2Si_2O_5(OH)_4$	Ceramics; paper

APPENDIX 1 REFERENCE TABLES

IDENTIFYING MINERALS

A mineral's hardness, a measure of its resistance to being scratched, is a useful property to test when seeking to identify an unknown mineral. A mineral with a higher hardness value on the Mohs' scale (Table 5) can scratch another mineral of lower hardness.

Other useful properties include how a mineral reflects light (its luster) and its breakage pattern (cleavage or fracture). The overall color of a mineral may be of limited help in identification, but its color in powdered form (its streak) can provide valuable information. For most minerals, identification is best made based on a combination of two or three properties (Figure 2).

TABLE 5 Mohs' Scale of Mineral Hardness

Hardness	Mineral		Common Objects with Similar Hardness
1	Talc		
2	Gypsum		Fingernail (2.5)
3	Calcite		Copper Penny (3.5)
4	Fluorite		
5	Apatite		Glass (5.5)
6	Orthoclase		Steel file (6.5)
7	Quartz		Streak plate (7)
8	Topaz		
9	Corundum		
10	Diamond		

Properties of Common Minerals

FIGURE 2 Mineral Identification Flowchart

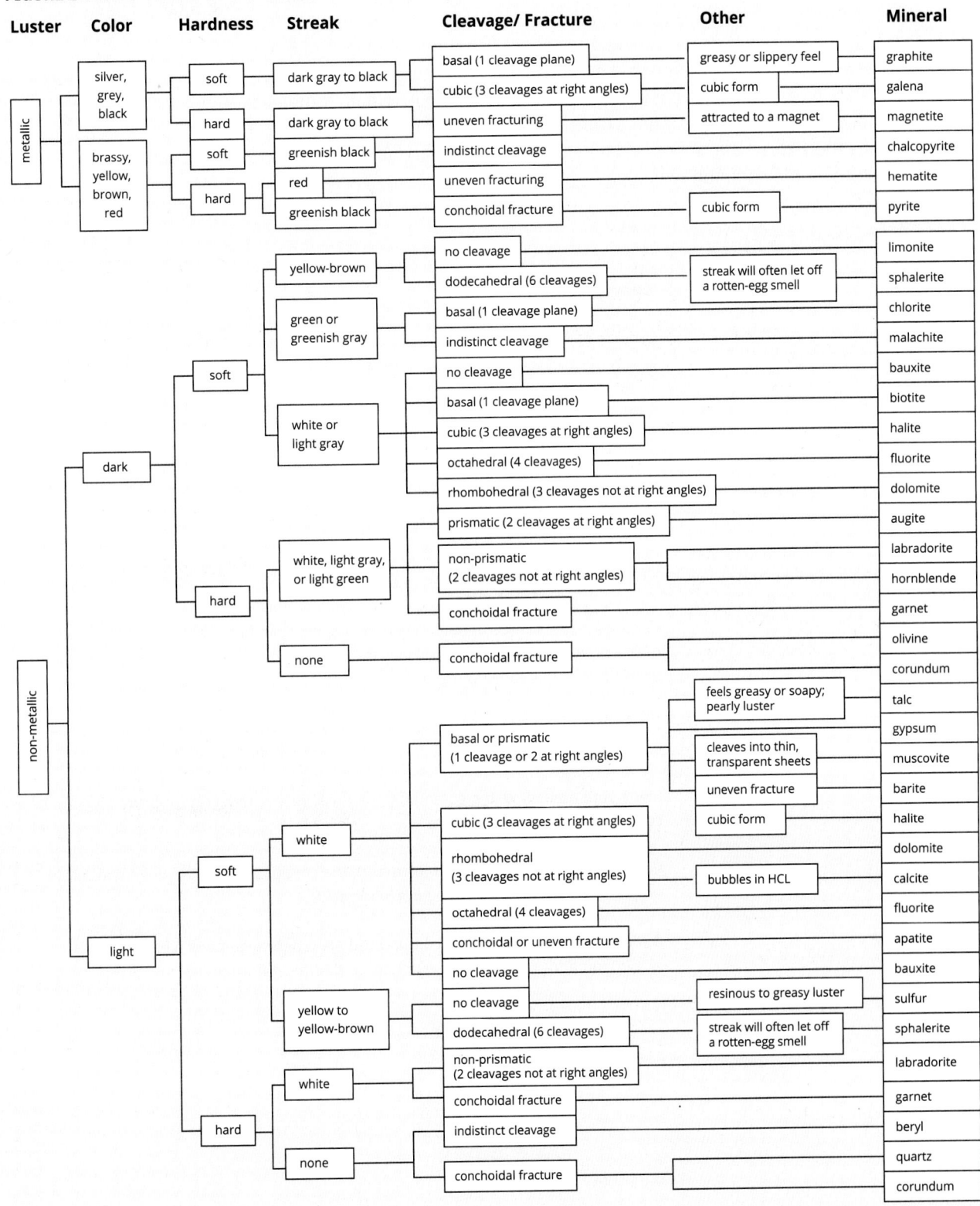

APPENDIX 1 REFERENCE TABLES

COMMON ROCKS

Sedimentary rock forms when broken fragments of rocks are lithified (compacted or cemented together) or when minerals precipitate out from solution; coal forms from plant debris. Igneous rock forms as molten rock cools and solidifies either within Earth's crust or on its surface. Metamorphic rock derives from parent rock that has undergone changes in its mineral composition and/or texture due to high temperature, high pressure, or circulating fluids. The texture of an igneous or metamorphic rock is determined by the size, shape, and arrangement of its mineral crystals. Table 6 shows characteristics of some common rock types.

TABLE 6 Some Common Rocks

Type	Name	Characteristics
Sedimentary (clastic)	Conglomerate	Consists of lithified gravel; clasts are rounded and commonly originate from a variety of rock types
	Breccia	Consists of lithified gravel; clasts are angular and commonly originate from a variety of rock types
	Sandstone	Consists of lithified sand-size grains; many sandstones consist predominantly of rounded quartz grains
	Mudstone (includes siltstone and shale)	Consists mostly of tiny clay minerals, silt, and organic particles; can vary widely in color, ranging from red to green to black
Sedimentary (bioclastic)	Limestone	Calcareous (calcium carbonate-rich) rock ranging from fine-grained rock made of microscopic fossils to rock containing abundant visible debris from the hard body parts of marine organisms (fossiliferous limestone); reacts with hydrochloric acid
Sedimentary (chemical)	Dolostone	Similar in appearance to fine-grained limestone; colors vary, but white, gray, and tan are common; reacts with hydrochloric acid when powdered
	Rock salt	Shiny, with a glassy luster; commonly colorless, yellow, red, gray, or brown; composed of precipitated halite
	Oolitic limestone	Calcareous; made of ooids that form from the precipitation of aragonite and calcite onto sand grains; reacts with hydrochloric acid
Sedimentary (organic)	Coal	Dark brown to black; made of compressed plant debris, some of which may be visible; soft and crumbly (lignite) to hard and brittle (bituminous)

TABLE 6 Some Common Rocks *(continued)*

Type	Name	Characteristics
Igneous (intrusive)	Granite	Medium- to coarse-grained felsic rock containing mostly feldspar and quartz, with some dark minerals; abundant in continental crust
	Diorite	Medium- to coarse-grained intermediate rock; may have a "salt and pepper" appearance; consists of feldspar and dark minerals such as biotite, amphibole, and/or pyroxene
	Gabbro	Medium- to coarse-grained mafic rock; dark green to black; consists of approximately equal amounts of feldspar and pyroxene; abundant deep within oceanic crust
Igneous (extrusive)	Rhyolite	Fine-grained, light-colored felsic rock; the extrusive counterpart to granite
	Andesite	Fine-grained intermediate rock; often gray or green; common in volcanic arcs; the extrusive counterpart to diorite
	Basalt	Fine-grained, dark-colored mafic rock; makes up most of the oceanic crust; the extrusive counterpart to gabbro
	Obsidian	Felsic volcanic glass; contains few to no mineral crystals; shiny, with a glassy luster; generally dark in color; fractures produce smooth, curved surfaces
	Pumice	Felsic volcanic glass containing numerous tiny vesicles; commonly light-gray to tan; some specimens may float on water
	Scoria	Mafic volcanic glass containing small vesicles; commonly reddish brown to dark gray
Metamorphic (foliated)	Slate	Finely textured, with a dull luster; tends to break along smooth, flat surfaces; resembles but is harder than its parent rock, mudstone
	Phyllite	Fine-grained, although some mineral crystals may be large enough to see with the unaided eye; fine-grained mica minerals often impart a surface sheen
	Schist	Fine- to medium-grained, displays more distinct foliation than that of phyllite; may have alternating bands of lighter and darker minerals; often shiny
	Gneiss	Medium- to coarse-grained, with distinctly separated layers of light- and dark-colored minerals; tends to break across the mineral bands rather than along them
Metamorphic (nonfoliated)	Quartzite	Composed mainly of similarly sized quartz crystals; glassy luster; may be "sugary" in appearance; often white, but can be a variety of colors
	Marble	Composed of visible calcite crystals; occurs in a variety of colors from white to black; may resemble quartzite but is softer and reacts with hydrochloric acid
	Soapstone	Soft, with a "soapy" feel and a greasy, pearly, or silky luster; commonly white, gray, or green; may have a slightly layered appearance; composed mainly of talc
	Anthracite	Glossy black; hard; the metamorphic form of coal

APPENDIX 1 REFERENCE TABLES

GEOLOGIC TIMESCALE

To organize our knowledge of the vast sweep of geologic time, scientists have developed the geologic timescale (Figure 3), based on fossil assemblage data and radiometric dating of rock.

Major events in Earth's history mark boundaries between time intervals. The largest intervals, eons, are divided into eras, which in turn are divided into periods. The periods of the Cenozoic era are divided into epochs that are further divided into ages (not included in this timescale). Note that the "ages" shown in the rightmost column provide informal descriptions and are not subdivisions of epochs.

This geologic timescale is not drawn to scale: The first 4 billion years are compressed because little is known about events early in Earth's history. Details of the timescale are frequently revised as new information is discovered.

FIGURE 3 Geologic Timescale

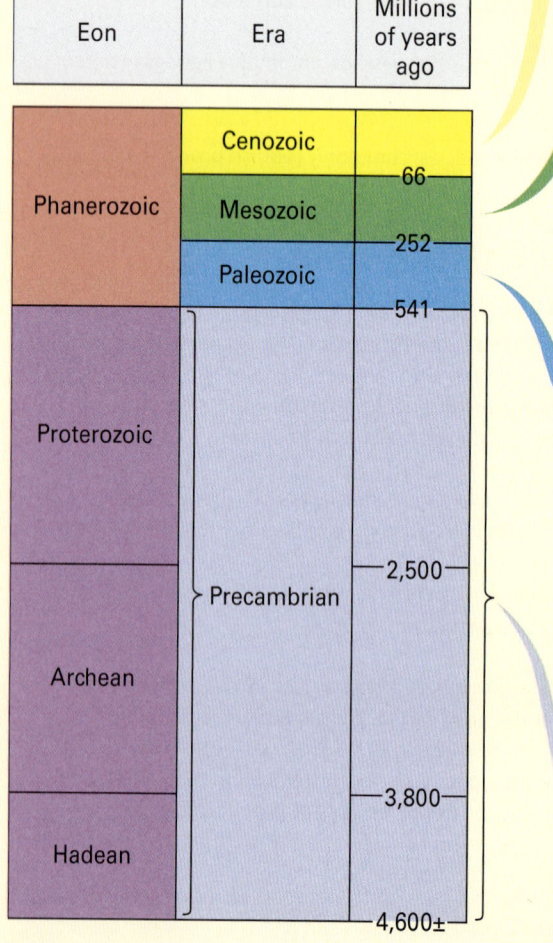

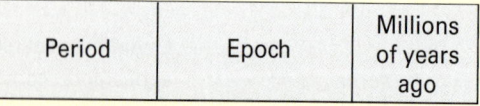

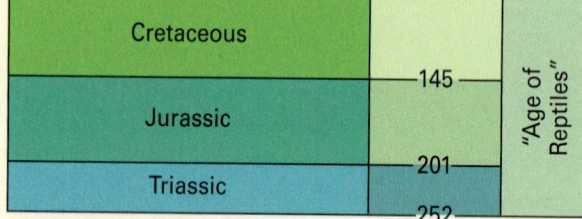

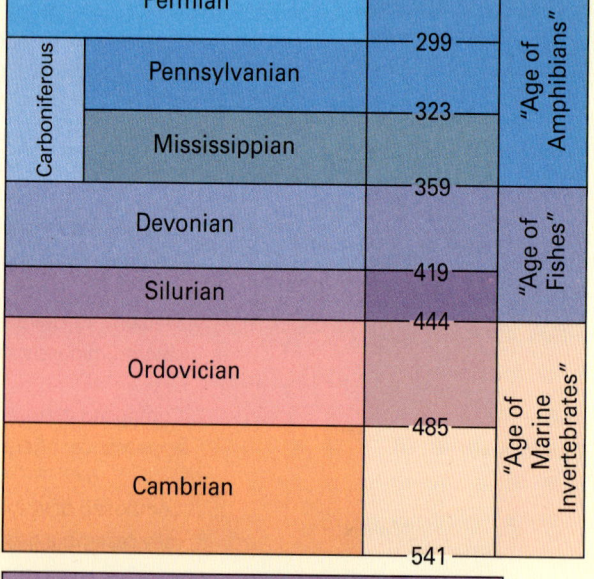

COMMON MAP SYMBOLS

Map symbols are commonly grouped into categories to enable the reader to quickly find a symbol's meaning in a map's legend. Table 7 shows geologic map symbols; Table 8 shows topographic map symbols, and Table 9 shows weather map symbols.

TABLE 7 Geologic Map Symbols

Orientation of Strata		Folds	
Horizontal strata	⊕	Monocline	
Orientation of strata tilted at 38°	⊢ 38	Anticline	
Orientation of Vertical strata	┼	Plunging anticline	
Orientation of overturned strata tilted at 60°	⚠ 60	Syncline	
Faults		Plunging syncline	
Normal fault; hachures or "D" on dropped-down side	(hachures) U/D	Dome	
Reverse or thrust fault; triangles on upper side	(triangles)	Basin	
Strike-slip fault showing relative movement	⇌		

APPENDIX 1   827

APPENDIX 1 REFERENCE TABLES

TABLE 8 Topographic Map Symbols

Contours		Water Features	
Index	—6000—	Perennial stream	
Intermediate		Intermittent stream	
Approximate or indefinite		Perennial river	
Depression		Intermittent river	
Surface Features		Perennial lake or pond	
Glacier or permanent snowfield		Intermittent lake or pond	
Gravel beach or glacial moraine	Gravel	Large rapids	
Sand or mud	Sand	Small rapids	
Disturbed surface		Spring or seep	
Levee	Levee	**Vegetation**	
Mines and Caves		Woodland	
Quarry or open-pit mine	⚒	Shrubland	
Mine shaft	◪	Mangrove	Mangrove
Mine tunnel or cave entrance	Y	Marsh or swamp	
		Submerged marsh or swamp	

828 REFERENCE TABLES

TABLE 9 Weather Map Symbols

Pressure Systems	
High	**H**
Low	**L**

Fronts	
Cold	▲▲▲
Warm	●●●
Stationary	▲●▲●
Occluded	●▲●▲

Precipitation	
Drizzle	,
Shower	▽
Rain	●
Thunderstorm	↳
Snow, ice pellets, or mix of rain and snow	✳
Fog, thick haze, or thick smoke	≡

APPENDIX 2 CLIMATE

KÖPPEN CLASSIFICATION SYSTEM

German climatologist Wladimir Köppen first published a classification system for Earth's climates in the early 20th century. The system (Table 1) defined five climate types (identified by capital letters A–E) based on the complex relationships among temperature, precipitation, and vegetation.

Köppen and colleagues modified the system several times, most notably by Rudolf Geiger and Robert Wichard Pohl, who added a subtype system (Table 2) to distinguish among specific patterns in precipitation and temperature (indicated by a second set of capital and lowercase letters) and defined a sixth climate type for highlands (H) at mountain elevations generally too high to support tree growth. This modified system is often referred to as the Köppen-Geiger-Pohl climate classification. A code composed of climate type and subtype indicates the climate classification of a region or location.

TABLE 1 Köppen Climate Classification

Climate Type	Subtype	Description	Vegetation
A Humid Tropical	Average temperatures above 18°C (64°F) with day/night difference greater than December/June difference. Precipitation supports abundant vegetation.		
	Tropical rainforest (Af)	Abundant year-round rainfall	Rainforest
	Tropical monsoon (Am)	Wet, with a short dry season	Rainforest
	Tropical savanna (Aw)	Summer rains, winter dry season	Grassland savanna
B Arid	Average temperatures range 0°C (32°F) to 22°C (72°F). In most months, evaporation exceeds precipitation.		
	Subtropical steppe (BSh)	Semiarid	Steppe grassland
	Midlatitude steppe (BSk)	Cool, dry; continental interior or rain-shadow location	Steppe grassland
	Subtropical desert (BWh)	Very warm; found at low latitude	Desert
	Midlatitude desert (BWk)	Cool, dry; continental interior or near mountain range	Desert
C Humid Mesothermal	Average temperature in the coldest month above −3°C (27°F). Mild winters are distinct from summer. Precipitation supports abundant plant communities.		
	Humid subtropical (Cfa)	Warm, wet summers with winter cyclonic storms	Forest
	Mediterranean (Cs)	Dry summer, wet winter with cyclonic storms; subtropical	Shrubby plants and oak savanna
	Marine west coast (Cfb)	Cool summer. mild winter; maritime, westerlies, cyclonic storms	Forest to temperate rainforest
D Humid Microthermal	Average temperature in the coldest month below −3°C (27°F). Cold winters are distinct from summer. Precipitation varies.		
	Humid continental (Dwa)	Midlatitude; summer temps above 22°C; drought during cold winters	Prairie and forest
	Humid continental (Dwb)	Midlatitude; summer temps below 22°C; winter drought	Prairie and forest
	Humid continental (Dfa)	High-midlatitude; summer temps above 22°C; no drought	Prairie and forest
	Humid continental (Dfb)	High-midlatitude; summer temps below 22°C; no drought	Prairie and forest
	Subarctic (Dfc)	High-latitude continental; cool summers; no drought	Forest taiga
	Subarctic (Dfd)	High-latitude continental; bitter cold winters, no drought	Forest taiga
	Subarctic (Dwc)	High-latitude continental; cool summers; winter drought	Forest taiga
	Subarctic (Dwd)	High-latitude continental; bitter cold winter drought	Forest taiga

TABLE 1 Köppen Climate Classification *(continued)*

Climate Type	Subtype	Description	Vegetation
E Polar		Average temperature in the warmest month below 10°C (50°F). Winters are extremely cold and summers are cool. Some precipitation.	
	Tundra (Et)	High latitude; short summer, harsh and long winter	Mosses, grasses, flowers; bushes, trees
	Ice cap (Ef)	High latitude; short summer, harsh and long winter	No vegetation
H Highlands		Average temperature changes with altitude, ranging from −18°C (0°F) to +10°C (50°F). Precipitation decreases with altitude.	
	Highlands	Cool to cold; mountain or high plateaus	Changes as elevation increases

Source: Cengage

TABLE 2 Köppen-Geiger-Pohl Climate Subtype Codes

Precipitation	
Code	Description
W	desert
S	steppe
f	fully humid
s	summer dry
w	winter dry
m	monsoonal

Temperature	
Code	Description
F	polar frost
T	polar tundra
h	hot arid
k	cold arid
a	hot summer
b	warm summer
c	cool summer
d	extremely continental

APPENDIX 2 CLIMATE

MAP OF KÖPPEN CLIMATE TYPES

The Köppen climate classification distinguishes global climate types and subtypes. Continual revisions to this map keep it up-to-date as climate change alters local conditions.

FIGURE 1 Köppen Climate Classification

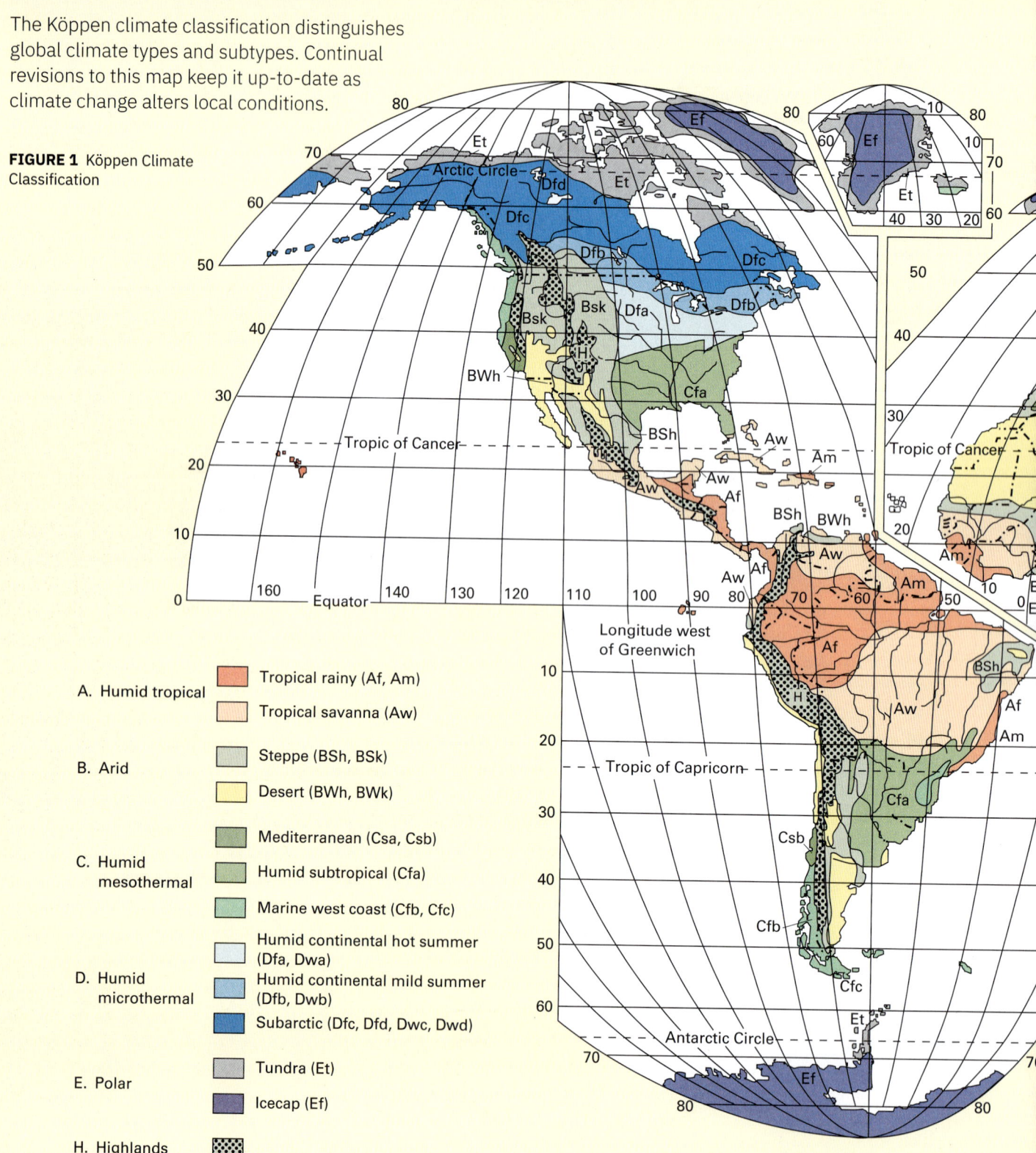

A. Humid tropical
- Tropical rainy (Af, Am)
- Tropical savanna (Aw)

B. Arid
- Steppe (BSh, BSk)
- Desert (BWh, BWk)

C. Humid mesothermal
- Mediterranean (Csa, Csb)
- Humid subtropical (Cfa)
- Marine west coast (Cfb, Cfc)

D. Humid microthermal
- Humid continental hot summer (Dfa, Dwa)
- Humid continental mild summer (Dfb, Dwb)
- Subarctic (Dfc, Dfd, Dwc, Dwd)

E. Polar
- Tundra (Et)
- Icecap (Ef)

H. Highlands

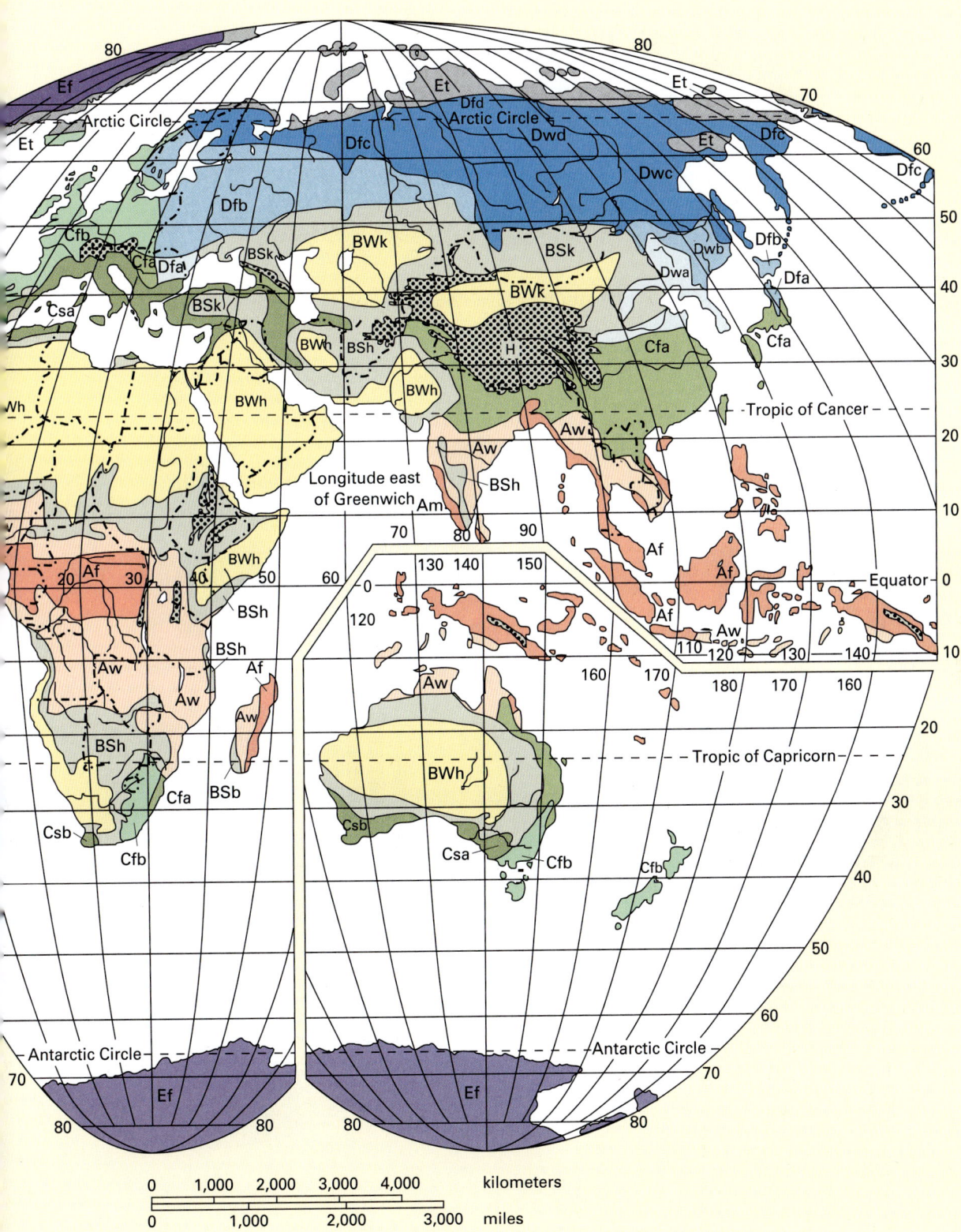

APPENDIX 2 CLIMATE

CLIMOGRAPHS

A climograph displays two types of climate data—mean monthly temperature and total monthly precipitation—for a particular location for a full year, using two vertical axes. It also shows average annual temperature and average annual precipitation. Typically, a line graph with its y-axis to the left shows temperature data, whereas a bar graph with its y-axis to the right presents precipitation data. Climographs for the same climate type display similar characteristics. Analysis of a climograph enables identification of the climate type for a particular location.

Figures 2 through 10 show climographs for locations that exhibit typical features of various climate types and subtypes. A code associated with each climograph indicates the climate type and subtype for the area. Note that not all climate types or subtypes are represented in these climographs.

FIGURE 2 Humid Tropical Climate (A) Tropical Rain Forest

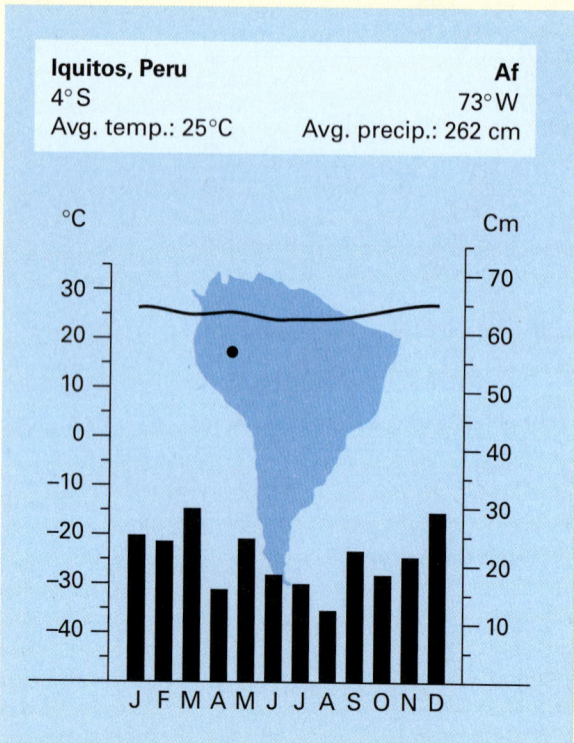

Source: Cengage

FIGURE 3 Humid Tropical Climate (A) Tropical Monsoon

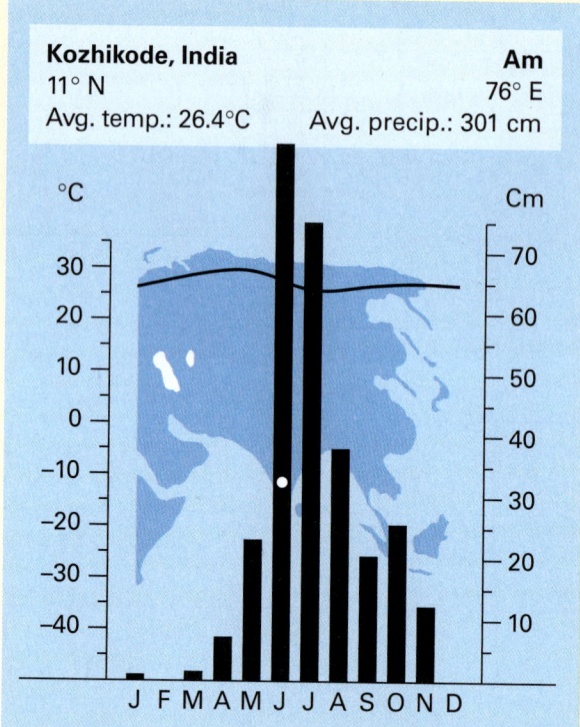

Source: Cengage

834 CLIMATE

FIGURE 4 Humid Tropical Climate (A) Tropical Savanna

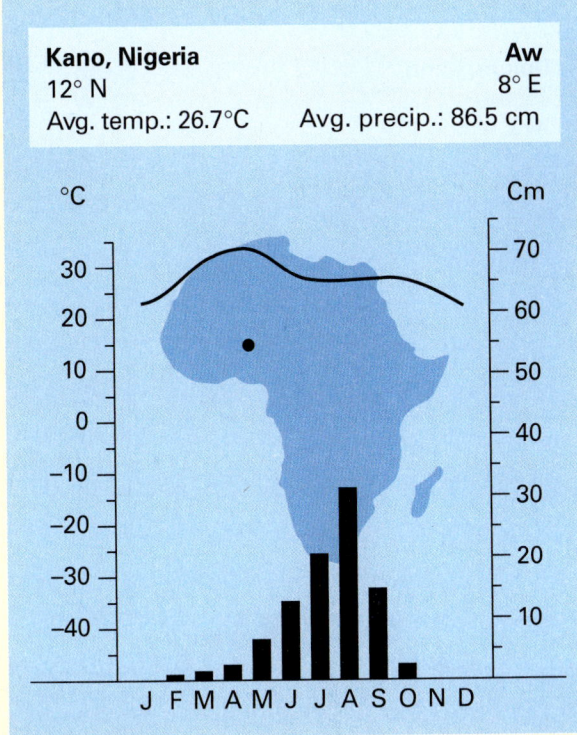

Source: Cengage

FIGURE 5 Arid Climate (B) Desert

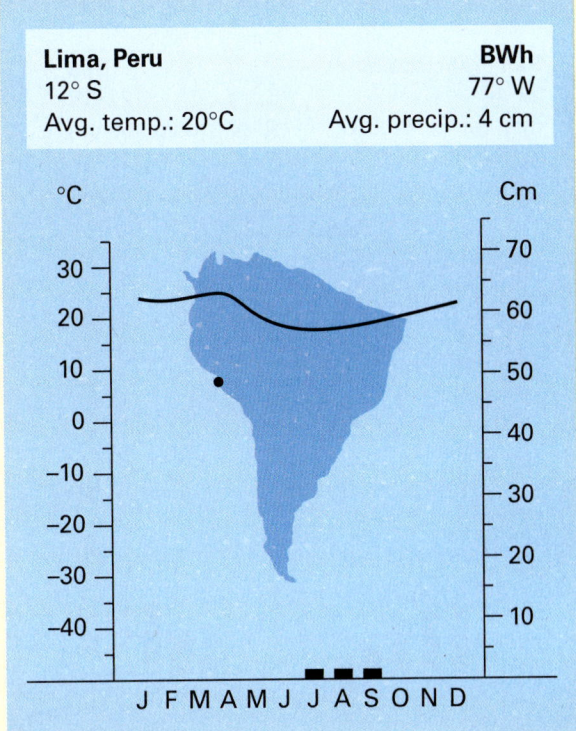

Source: Cengage

FIGURE 6 Humid Mesothermal Climate (C) Marine West Coast

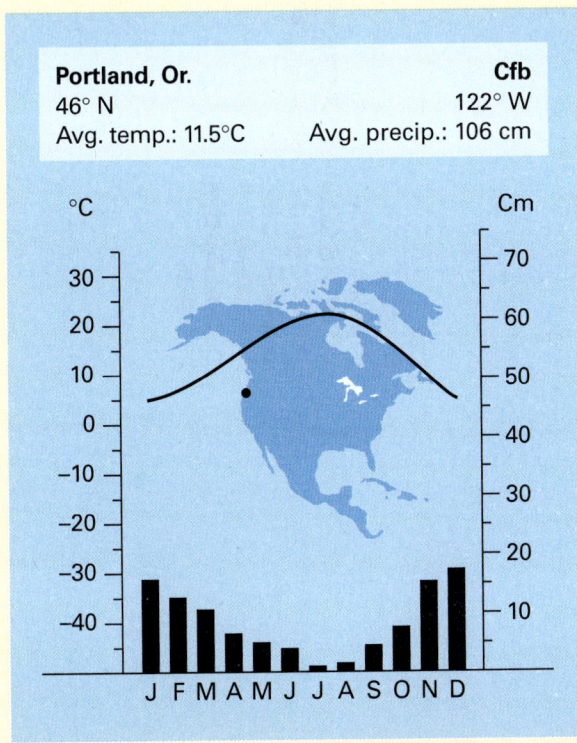

Source: Cengage

FIGURE 7 Humid Mesothermal Climate (C) Humid Subtropical

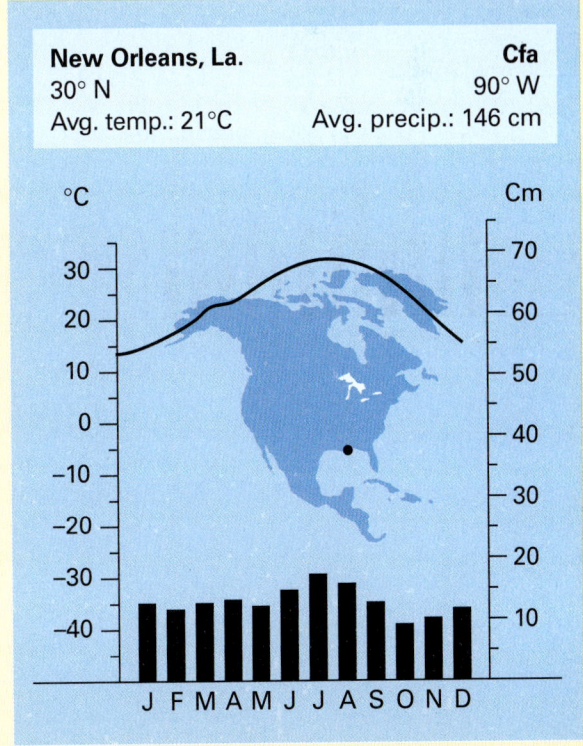

Source: Cengage

APPENDIX 2 CLIMATE

FIGURE 8 Humid Mesothermal Climate (C) Mediterranean

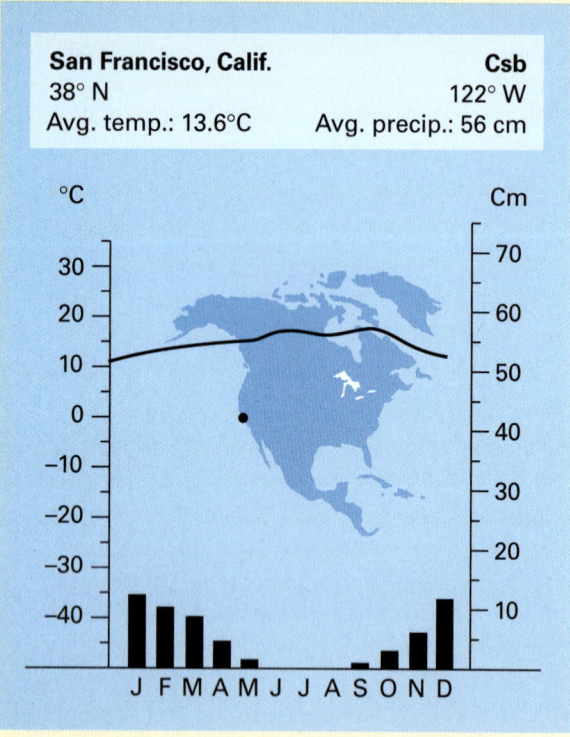

Source: Cengage

FIGURE 9 Humid Microthermal Climate (D) Humid Continental

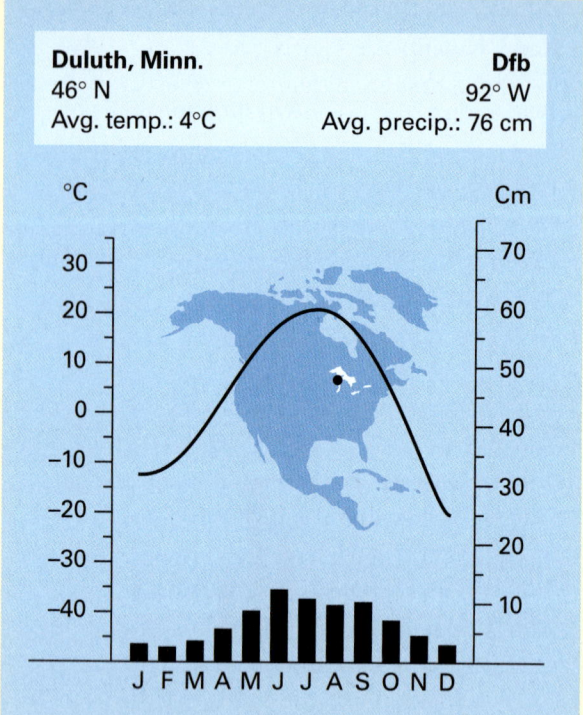

Source: Cengage

FIGURE 10 Polar (E) Tundra

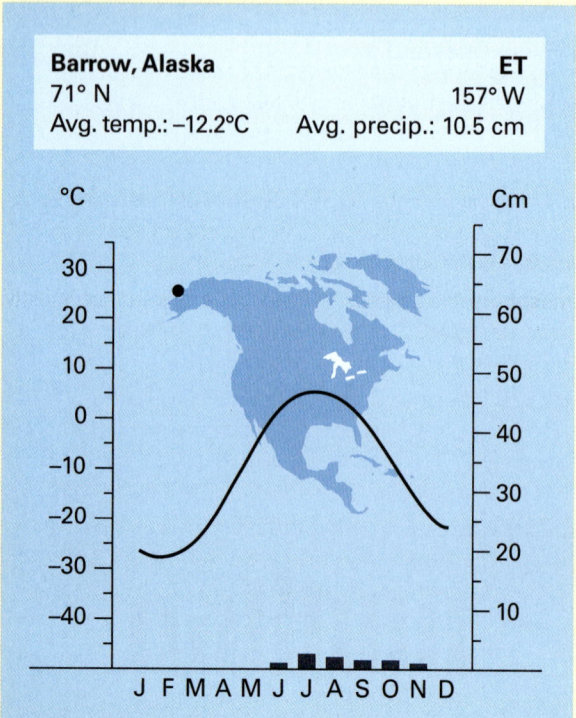

Source: Cengage

836 CLIMATE

GLOSSARY

A

absolute age the age of a geologic event based on the absolute number of years ago it occurred

absolute brightness the brightness of a star as it would appear if it were a fixed distance from Earth, also known as luminosity

absolute humidity the mass of water vapor in a given volume of air, expressed in grams per cubic meter (g/m^3)

absorption spectrum a spectrum of radiation wavelengths that are absorbed when light passes through a substance, such as a star; absorbed wavelengths appear as dark lines crossing a full-color spectrum

abyssal plain the flat, level, largely featureless parts of the ocean floor between the Mid-Ocean Ridge and the continental rise

accelerating universe a model in which an unidentified force is overpowering gravity and causing the expansion of the universe to accelerate

accreted terrane a mappable, fault-bounded landmass that originates as an island arc or a microcontinent that is later added onto a continent

acid precipitation rain, snow, fog, or mist that has become acidic after reacting with air pollutants

active continental margin a margin at a convergent plate boundary, where oceanic lithosphere sinks beneath the continent in a subduction zone

adiabatic temperature change temperature change caused by compression or expansion of gas, which occurs without gain or loss of heat

advection the transfer of heat by the horizontal flow of fluid material

aerosol in pollution terminology, a particle or particulate that is suspended in air

age the shortest unit of geologic time

air mass a large body of air that throughout has approximately the same temperature and humidity at any given altitude

albedo the reflectivity of a surface; surfaces that reflect more light have a higher albedo

alluvial fan a fan-shaped accumulation of sediment created where a steep mountain stream rapidly slows down as it reaches a relatively flat plain

alpine glacier a narrow glacier that forms at or near a mountain peak and is confined to a valley by surrounding terrain

Andean margin see *continental arc*

angle of repose the maximum slope or steepness at which loose material remains stable; if the slope becomes steeper, the material slides

angular unconformity an unconformity in which younger sediment or sedimentary rocks rest on the eroded surface of tilted or folded older rocks

anion an ion that has a negative charge

anoxic referring to the absence of oxygen

Anthropocene the current geologic age, viewed as the period during which human activity has been the dominant influence on climate and the environment

anticline a fold in rock that arches upward

anticyclone high-pressure region with its accompanying system of outwardly directed rotating winds that develop where descending air spreads over Earth's surface

apparent brightness the brightness of a star as seen from Earth

aquifer any body of rock or soil that can yield economically significant amounts of water

aquitard an area of rocks with low porosity and permeability that contains and transmits very little groundwater

arête a sharp narrow ridge of rock between adjacent valleys or between two cirques, created when two alpine glaciers moved along opposite sides of the mountain ridge and eroded both sides

artificial levee a wall built along the banks of a stream to prevent rising floodwater from spilling out of the channel onto the floodplain

ash-flow tuff a volcanic rock formed when a pyroclastic flow solidifies

asteroid small, dry rocky body in orbit around the sun between Mars and Jupiter or beyond the orbit of Neptune

asthenosphere the portion of the upper mantle just beneath the lithosphere, extending from a depth of 100–350 kilometers below the surface of Earth and consisting of weak, plastic rock where magma may form

astronomical unit the average distance between Earth and the sun; used commonly as a standard unit for describing distances in the solar system

atmosphere the gaseous layer above Earth's surface, mostly nitrogen and oxygen, with smaller amounts of argon, carbon dioxide, and other gases

atoll a circular coral reef that forms a ring of islands around a central lagoon and that is bounded on the outside by the deep water of the open sea; usually forms on top of a subsided seamount

axial surface an imaginary surface that connects all the points of maximum curvature in a fold bajada a broad, gently sloping depositional surface formed by the merging of alluvial fans from closely spaced canyons and extending outward into a desert valley

GLOSSARY

B

banded iron formation a marine chemical sedimentary rock formed mostly between 2.7 and 1.9 billion years ago and consisting of centimeter-scale interbeds of iron oxide and chert

bar unit of measurement for atmospheric pressure; one bar is roughly equal to atmospheric pressure at sea level

barometer a device used to measure barometric pressure

barometric pressure the pressure of the atmosphere at any given location and time

barred spiral galaxy a spiral galaxy with a straight bar of stars, gas, and dust extending across the nucleus

barrier island a long, narrow, low-lying island that extends parallel to the shoreline

barycenter the center of mass of two or more bodies that orbit one another and the point about which the bodies orbit

basal slip movement of a glacier in which the entire mass slides over bedrock

base level the deepest level to which a stream can erode its bed; the ultimate base level is usually sea level

batholith a large pluton, exposed across more than 100 square kilometers of Earth's surface

beach drift the gradual movement of sediment (usually sand) along a beach, parallel to the shoreline, when waves strike the beach obliquely but return to the sea directly

bed load the total mass of a stream's sediment load that is transported along the bottom or in sporadic contact with the bottom of the streambed

Benioff zone a three-dimensional zone of earthquake foci within and along the upper portion of a subducting plate; formed by release of strain as the subducting plate scrapes past the overriding plate

big bang an event 10 to 20 billion years ago, thought to mark the beginning of the universe when all matter exploded from a single infinitely dense point

bioclastic sedimentary rock rock composed of broken shell and bone fragments

biodegradable pollutant pollutants that decay naturally in a reasonable amount of time by being consumed or destroyed by organisms that live in soil or water

biodiversity the variety of living organisms on Earth or in an ecosystem

biomass energy the energy produced by the burning of organic material

biomass fuel a substance of plant or animal origin that can be burned or processed to generate energy

biome a community of plants growing in a large geographic area characterized by a particular climate

bioremediation the use of microorganisms to decompose an environmental contaminant

biosphere the zone of Earth comprising all forms of life in the sea, on land, and in the air

black hole an infinitesimally small region of space, created after the supernova explosion of a huge star, that contains matter packed so densely that light cannot escape from its intense gravitational field

blob tectonics tectonic activity dominated by rising and sinking of the mantle and crust, as opposed to the horizontal movement of plates associated with tectonic activity on Earth

blueshift a shift toward shorter (blue) wavelengths, observed in the spectrum of a distant galaxy or other object that is moving toward Earth; caused by the Doppler effect

body wave seismic waves that travel through the interior of Earth, carrying energy from the earthquake's focus to the surface

bolide a piece of space debris that typically explodes upon impact with the atmosphere

butte a flat-topped mountain, smaller and more towerlike than a mesa, characterized by steep cliff faces

C

caldera a large circular depression created by the collapse of the magma chamber after an explosive volcanic eruption

capacity the maximum quantity of sediment that a stream can transport at any one time

carbon cycle the process by which carbon moves through different parts of the biosphere, geosphere, hydrosphere and atmosphere

carbonate mineral whose chemical composition is based on the carbonate anion

carbonate rock an example of bioclastic rock with high amounts of calcite, dolomite, or both

carcinogenic a substance that has the potential to cause cancer

catastrophism the principle stating that infrequent catastrophic geologic events alter the course of Earth history, in contrast to the principle of gradualism

cation a positively charged ion

celestial sphere the hypothetical series of concentric transparent spheres surrounding Earth in Aristotle's model of the universe; Aristotle postulated that the sun, moon, planets, and stars are embedded in the spheres

chemical remediation treatment of a contaminated area by injecting it with a chemical compound that reacts with the pollutant to produce harmless products

chemical sedimentary rock rock formed by the precipitation of minerals

chemical weathering the decomposition of rock when it chemically reacts with air, water, or other agents in the environment, altering its chemical composition and mineral content

chemosynthesis a process in which bacteria produce energy from hydrogen sulfide or other inorganic compounds and thus are not dependent on photosynthesis

chlorofluorocarbon (CFC) an organic compound containing chlorine and fluorine, which rises into the upper atmosphere and destroys the ozone layer there

chondrule a small grain about 1 millimeter in diameter embedded in a meteorite, often containing amino acids or other organic molecules

chromosphere a turbulent, diffuse, gaseous layer of the sun that lies above the photosphere

cinder glassy, pyroclastic volcanic fragment 4 to 32 millimeters in size

cinder cone a small volcano, typically less than 300 meters high, made up of loose pyroclastic fragments blasted out of a central vent; usually active for only a short time

cirque a steep-walled, spoon-shaped depression eroded into a mountain peak by a glacier

cirrus wispy, high-altitude clouds composed of ice crystals

clastic sedimentary rock rock made from fragments of weathered rocks

cleavage the tendency of some minerals to break along flat surfaces, which are planes of weak bonds in the crystal; when a mineral has well-developed cleavage, sheet after sheet can be peeled from the crystal, like peeling layers from an onion

climate the characteristic weather of a region, averaged over several decades; refers to yearly cycles of temperature, wind, cloudiness, humidity, and precipitation, and not to daily variations

closed universe a model in which the gravitational force of the universe is so great that all the galaxies will eventually slow down, reverse direction, and fall back to the center, forming another point of infinite density

coal a sedimentary rock made mostly of organic carbon that will burn without refining

columnar joint regularly spaced cracks that commonly develop in lava flows, grow downward starting from the surface, and typically form five- or six-sided columns

comet an object composed of loosely bound rock and ice that appears to form a bright head and extended fuzzy tail when it approaches the sun

conduction the transport of heat by direct collision among atoms or molecules

cone of depression a cone-shaped depression in the water table, created when water is pumped out of a well more rapidly than it can flow through the aquifer to the well; if water continues to be pumped from the well at such a rate, the water table will drop

conformable a term describing sedimentary layers that were deposited continuously, without detectable interruption

constellation a group of stars that seem to form a pattern when viewed from Earth, to which astronomers have given a name for convenience in referring to the positions of objects in the night sky

constraint a factor that limits the possible solutions that could be developed to solve a problem

consumption any process that uses water and then returns it to Earth far from its source

continental arc a continental margin characterized by subduction of an oceanic plate beneath a continental plate; also called an Andean margin

continental drift the hypothesis proposed by Alfred Wegener that Earth's continents were once joined together and later split and drifted apart

continental glacier a vast, continuous mass of ice that flows outward in all directions and covers a large portion of a continent's area

continental rifting the process by which a continent is pulled apart at a divergent plate boundary

continental rise an apron of terrigenous sediment at the foot of the continental slope that merges with the deep seafloor

continental shelf a shallow, very gently sloping portion of the seafloor that extends from the shoreline to the top of the continental slope

continental slope the relatively steep submarine slope between the continental shelf and the continental rise

convection the upward and downward flow of fluid material in response to density changes produced by heating and cooling

convective zone the inner zone of the sun above the radiative zone where energy is transmitted primarily by convection

convergent boundary a plate boundary where two tectonic plates move toward each other and collide

Coriolis effect the deflection of air or water currents caused by the rotation of Earth

corona the outer atmosphere of the sun, normally invisible but appearing as a halo around a black moon during a solar eclipse

correlation the process of establishing the age relationship of rocks or geologic features from different locations on Earth; this can be done by comparing sedimentary characteristics of the layers or the types of fossils found in them

cosmic microwave background (CMB) low-energy, microwave radiation that began traveling through space when the universe was only 300,000 years old and now pervades all space in the universe

crater a bowl-like depression at the summit of a volcano, created by volcanic activity

GLOSSARY

creep form of mass wasting in which loose material moves very slowly downslope, usually at a rate of only about 1 centimeter per year and usually on land with vegetation

crescent moon a lunar phase in which the moon appears as a thin crescent

crest the highest part of a wave

crevasse a fracture or crack in the brittle upper 40 meters of a glacier, formed when the glacier flows over uneven bedrock

criteria the standards that an engineering solution must meet to be considered successful

cryosphere the part of Earth's hydrosphere that contains glaciers, ice caps, permanent snow, and all of Earth's frozen water

crystal a solid element or compound whose atoms are arranged in a regular, periodically repeated pattern

crystal face the flat surface that develops if a crystal grows freely in an uncrowded environment; under perfect conditions, the crystal that forms will be symmetrical

crystal habit the characteristic shape of an individual crystal, and the manner in which aggregates of crystals grow

cumulus fluffy white clouds with flat bottoms and billowy tops

cyanobacteria blue-green algae that were among the earliest photosynthetic life-forms on Earth

cycle a sequential process or phenomenon that returns to its beginning and then repeats itself over and over

cyclone a low-pressure region with its accompanying system of inwardly directed rotating winds; in common, nonscientific usage, the term often refers to a variety of different violent storms including hurricanes and tornados

D

dark energy one component of dark matter, comprising the poorly understood repulsive force that causes the modern universe to accelerate

dark matter mysterious invisible matter that may comprise up to 96 percent of the mass of the universe

data information collected for study or analysis

deep-sea current transport of seawater both vertically and horizontally below a depth of 400 meters, driven by differences in water density

deflation erosion by wind of fine, loose sediment from flat, dry areas

delta a fan-shaped accumulation of sediment formed where a stream enters a lake or ocean; includes a nearly flat surface that is partly onshore and partly offshore

deposition the process where sediment is deposited after being moved by erosion

desert any region that receives less than 25 centimeters (10 inches) of rain per year and consequently supports little or no vegetation

desert pavement a continuous cover of closely packed gravel- or cobble-sized clasts left behind when wind erodes smaller particles such as silt and sand

desertification a process by which semiarid land is converted to desert by human factors, natural factors, or some combination of the two

dew moisture that is condensed onto objects from the atmosphere, usually during the night, when the ground and leaf surfaces become cooler than the surrounding air

dew point the temperature at which the relative humidity of air reaches 100 percent and the air becomes saturated

dike sheet-like body of plutonic igneous rock that forms when magma flows into a rock fracture

discharge the volume of water flowing downstream over a specified period of time, usually measured in units of cubic meters per second (m^3/sec) or cubic feet per second (cfs)

disconformity a type of unconformity in which the sedimentary layers above and below the unconformity are parallel

dissolution a chemical weathering process in which mineral or rock dissolves, forming a solution

dissolved load the total mass of ions dissolved in and carried by a stream at any one time; the ions are derived from chemical weathering

divergent boundary a plate boundary where tectonic plates move apart from each other and new lithosphere is continuously forming; also called a spreading center or a rift zone

diversion system a pipe, canal, or other infrastructure that transports water from its natural place and path in the hydrologic cycle to a new place and path to serve human needs

Doppler effect the observed change in frequency of light or sound that occurs when the source of the wave is moving either toward or away from the observer

down cutting downward erosion by a stream into its bed, usually by cutting a V-shaped valley along a relatively straight path

drainage basin the region that is drained by a single stream

drift any rock or sediment transported and deposited by a glacier or by glacial meltwater

drumlin elongate hills, usually occurring in clusters, formed when a glacier flows over and reshapes a mound of till or stratified drift

dune a mound or ridge of wind-deposited sand

dwarf planet an object in orbit around the sun that is massive enough to be round or nearly round but is not massive enough to have cleared its orbital area of space debris

E

earthquake a sudden motion or trembling of Earth caused by the abrupt release of slowly accumulated elastic energy in rocks

ecosystem a complex community of individual organisms interacting with each other and with their physical environment and functioning as an ecological unit in nature

El Niño an episodic weather pattern occurring every 3 to 7 years in which the trade winds slacken in the Pacific Ocean and warm water accumulates off the coast of South America and causes unusual rains and heavy snowfall in the Andes

electromagnetic radiation radiation consisting of an oscillating electric and magnetic field, including radio waves, infrared, visible light, ultraviolet, x-rays, and gamma rays

electromagnetic spectrum the entire range of electromagnetic radiation from very-long-wavelength (low frequency) radiation to very-short-wavelength (high frequency) radiation

element a substance that cannot be broken down into other substances by ordinary chemical means

elliptical galaxy a galaxy with an oval shape and no spiraling arms; can be either giant or dwarf

eluviation the downward movement of dissolved ions and suspended matter through layers of soil by infiltrating water

emergent coastline a coastline that was previously underwater but has become exposed to air, because either the land has risen, or sea level has fallen

emission spectrum a spectrum created when radiation is directly emitted, or absorbed radiation is reemitted from its source

energy resource any naturally occurring substance or phenomenon that can be used to generate useful heat, light, and/or electricity

engineering the application of scientific and mathematical ideas to design solutions to problems

eon the longest unit of geologic time, divided into eras

epicenter the point on Earth's surface directly above the initial rupture point (focus) of an earthquake

epoch a geologic time unit that is divided into ages

equinox either of two times during the year when the sun shines directly overhead at the Equator and every area of Earth receives 12 hours of daylight and 12 hours of darkness

era a geologic time unit that is divided into periods

erosion the removal of weathered rocks that occurs when water, wind, ice, or gravity transports the material to a new location

erratic a type of boulder, usually different from bedrock in the immediate vicinity, that was transported to its present location by a glacier

escape velocity the speed that an object must attain to escape the gravitational field of a planet or other object in space

esker a long, snakelike ridge formed as the channel deposit of a stream that flowed within or beneath a melting glacier

eustatic sea-level change a global sea-level change caused by three different processes: the growth or melting of glaciers, changes in water temperature, and changes in the volume of the Mid-Ocean Ridge

eutrophic a term describing a lake containing abundant nutrients and a high level of productivity in terms of biomass

exfoliation a weathering process resulting in fracture when concentric plates or shells split away from a main rock mass like the layers of an onion

exosphere a thin layer of gas surrounding the thermosphere

experiment a procedure followed to collect data or make observations for the purpose of testing a hypothesis

extrusive igneous rock rock formed when magma solidifies above Earth's crust; usually finely crystalline; also called volcanic rock

F

fault a fracture in rock along which one side has moved relative to the other side

fault creep a continuous, slow movement of solid rock along a fault, resulting from a constant stress acting over a long time

feedback mechanism a reaction whereby a small initial perturbation in one component affects a different component of Earth's systems, which amplifies the original effect, which perturbs the system even more, which leads to an even greater effect, and so on

First Great Oxidation Event the sudden increase in Earth's atmospheric oxygen concentration from trace amounts to significant quantities that occurred about 2.4 billion years ago, probably because of a combination of biological and geochemical processes

fjord a deep, narrow, glacially carved valley on a high-latitude seacoast that was later flooded by en-croaching seas as the glaciers melted

flood basalt basaltic lava that erupts gently and in great volume from vents or fissures at Earth's surface, to cover large areas of land and form lava plateaus

floodplain the portion of a river valley adjacent to the channel; it is built upward by sediment deposited during floods and is covered by water during a flood

flow form of rapid mass wasting in which loose soil or sediment moves downslope as a slurry-like fluid, not as a consolidated mass; may occur slowly or rapidly

fly ash noncombustible minerals that escape into the atmosphere when coal burns, eventually settling as gritty dust

GLOSSARY

focus the initial rupture point of an earthquake, typically located below Earth's surface

foliation the repetitive parallel layering of minerals in a rock

forearc basin a sedimentary basin between the oceanic trench and the magmatic arc, either in an island arc or at an Andean margin

foreland basin a sedimentary basin formed by downward flexing of continental crust by the weight of thrust plates in a compressional orogen

foreshock small earthquake that precedes a large quake by a few seconds to a few weeks

fossil fuel any fuel derived from prehistoric organic remains and formed by geologic processes

fracture the manner in which minerals break, other than along planes of cleavage

frequency the number of complete wave cycles, from crest to crest, that pass by a fixed point in one second

front in meteorology, the boundary between a warmer air mass and a cooler one

frontal wedging the process where cool, dense air slides under a warm air mass, forcing the warm air upward

frost ice crystals formed directly from vapor when the dew point is below freezing

frost wedging a mechanical weathering process in which water freezes in a crack in rock, and the resulting expansion wedges the rock apart

full moon the lunar phase when the moon appears round and fully illuminated, because its entire sunlit area is visible

fusion bonding of subatomic particles that collide with sufficient energy to overcome the repulsive force between like charges

G

Gaia the term (Greek for "Earth") used by James Lovelock to refer to our planet, which he likened to a living creature due to the interconnectivity of all of Earth's systems

galactic halo a spherical cloud of star clusters, dust and gas that surrounds the galactic disk of a spiral galaxy such as the Milky Way

galaxy a large volume of space containing many billions of stars, dust and interstellar gas, held together by gravitational attraction to a central black hole

gamma ray burst a burst of high-energy radiation from space; a longer burst results from the death of a massive star, while a shorter burst forms when a neutron star is pulled into a black hole or when two neutron stars collide to form a black hole

geocentric a model that places Earth at the center of the universe

geologic formation a discrete, mappable unit of rock layers that occurs in a specific region

geologic timescale a chronological arrangement of geologic time subdivided into eons, eras, periods, epochs, and ages

geosphere the mostly solid, rocky part of Earth, consisting of the entire planet from the center of the core to the outer crust

geothermal energy the heat extracted from Earth's interior and used to generate heat and electricity at the surface

gibbous moon the lunar phase, either waxing or waning, with most of the moon lit up and only a sliver of dark

glacial striation parallel grooves and scratches in bedrock that form as rocks are dragged along at the base of a glacier

glaciation a time when alpine glaciers descend into lowland valleys and continental glaciers grow over high-latitude continents; usually used in reference to the Pleistocene Glaciation

glacier a massive, long-lasting accumulation of compacted snow and ice that forms on land and moves downslope or spreads outward under its own weight

globular cluster a concentration of older stars that shared a common origin and are held together by gravity, lying within the galactic halo of a spiral galaxy such as the Milky Way

graben a wedge-shaped block of rock that has dropped downward between two normal faults, forming a valley

gradient the steepness or vertical drop of a stream over a specific distance

gradualism a principle stating that geologic change occurs as a consequence of slow or gradual accumulation of small events, such as the slow erosion of mountains by wind and rain; more recently, scientists studying biological evolution use the term to describe a theory of evolution that proposes that species change gradually in small increments

greenhouse effect an increase in the temperature of the planet's surface caused when infrared-absorbing gases in the atmosphere trap energy from the sun

groin a narrow barrier or wall built on a beach, perpendicular to the shoreline, to trap sand transported by currents and waves

groundwater subsurface water contained in the soil and bedrock of the upper few kilometers of the geosphere, comprising about 0.63 percent of all water in the hydrosphere

guyot a flat-topped seamount, formed when the top of a sinking island, usually of volcanic origin, is eroded by wave energy

gyre a circular or elliptical current in either water or air

H

H–R diagram a graph that plots absolute stellar luminosity against temperature and shows that stars are organized into groups based on these two properties

half-life the time it takes for half of the atoms of a radioactive isotope in a sample to decay

halon a compound containing bromine and chlorine, which rises into the upper atmosphere and destroys the ozone layer there

hanging valley a small glacial valley lying high above the floor of the main valley

heat the energy transferred between two substances because of differences in their temperature

heliocentric a model that places the sun at the center of the solar system

horse latitudes a calm, high-pressure region of Earth lying at about 30° north and south latitudes, in which generally dry conditions prevail and steady winds are rare

horst in a fault zone, a block of rock between two grabens, which appears to have moved upward relative to the grabens

hot spot the hot upper mantle rock located within a plume and associated with a volcanic center that forms on the overlying lithosphere

Hubble-Lemaitre law a mathematical law that states that the velocity of a galaxy is proportional to its distance from Earth; thus, the most distant galaxies are traveling at the highest velocities

humid continental a midlatitude climate characterized by hot summers, cold winters, and precipitation throughout the year

humid subtropical a midlatitude climate characterized by hot, humid summers but cooler winters and rainfall that falls throughout the year

humus the dark, organic component of soil consisting of litter that has decomposed enough so that the origin of the individual pieces cannot be determined

hurricane a tropical storm occurring in North America or the Caribbean whose wind exceeds 120 kilometers per hour; called a typhoon in the western Pacific and a cyclone in the Indian Ocean

hydraulic fracturing also known as "fracking," the process by which fluids are injected into a dense rock in order to expand naturally occurring fractures so that trapped natural gas and oil can flow more freely

hydrocarbon any organic, combustible compound that is composed of only hydrogen and carbon atoms

hydroelectric energy the power generated by converting the energy of flowing water to electricity

hydrologic cycle the movement of water through Earth's spheres via precipitation, evaporation, and condensation

hydrolysis a chemical weathering process in which a mineral reacts with water to form a new mineral

hydrosphere all of Earth's water, which circulates among oceans, continents, glaciers, and atmosphere

hypothesis a tentative explanation for a natural phenomenon that can be tested to establish whether evidence exists to support it

I

ice shelf a thick mass of ice that floats on the ocean surface but is connected to and fed from a glacier on land

igneous rock rock formed when molten rock solidifies

index fossil a fossil that dates the layers where it is found because it came from an organism that is abundantly preserved in rocks, was widespread geographically, and existed as a species or genus for only a relatively short time

inertia the tendency of an object to resist a change in motion

interglacial period a relatively warm, ice-free time separating glaciations

intrusive igneous rock rock formed when magma solidifies within Earth's crust without erupting to the surface; usually medium to coarsely crystalline; also called plutonic rock

ion an atom with an electrical charge, either positive or negative

irregular galaxy a galaxy with stars and gas that follow no specific organized structure

island arc a gently curving chain of volcanic islands in the ocean formed by the convergence of two plates, each bearing ocean crust, and the resulting subduction of one plate beneath the other

isobar lines on a weather map connecting points of equal air pressure

isoline a line on a map that connects areas having the same elevation, depth, temperature, pressure, or other stated feature

isostasy the concept of balance between gravity and buoyancy that causes the lithosphere to float on the mantle at different elevations

isotherm line on a weather map connecting areas with the same average temperature

isotope atoms of the same element that have different numbers of neutrons

J

jet stream narrow bands of high-altitude, fast-moving wind

joint a fracture along which the rock on either side of the break does not move

Jovian planet the four outer planets with relatively small rocky or metal cores surrounded by swirling liquid and gaseous atmospheres

K

kame a small mound or ridge of stratified drift deposited by a stream that flows on top of, within, or beneath a glacier

kettle a small depression formed by a block of stagnant ice; many fill with water to become a kettle lake

GLOSSARY

kettle lake a lake that forms in a depression created by a receding glacier, filled with the water from the melting glacier

key bed a thin, widespread, easily recognized sedimentary layer that can be used for correlation because it was deposited rapidly and simultaneously over a wide area

Kuiper belt disk-shaped region beyond the orbit of Neptune that contains asteroids, comets, and other debris from the formation of the solar system

L

lake a large inland body of standing fresh water that occupies a depression in the land

latent heat stored energy; the energy released or absorbed when a substance changes from one state to another by melting, freezing, vaporization, condensation, sublimation, or deposition

lava plateau a broad plateau covering thousands of square kilometers, formed by the accumulation of many individual lava flows that occur over a short period of geologic time

law statement describing or predicting a natural phenomenon that has never been shown to be false under stated conditions

leaching the chemical dissolution of ions from soil by infiltrating water

light pollution the nighttime glow of city lights that competes with the light of distant stars and reduces the vision of a telescope

light-year the distance traveled by light in one year, approximately 9.5×10^{12} kilometers

limb either of the two sides of a fold, separated by the axial surface

liquefaction a geologic process in which a soil or sediment loses its shear strength during an earthquake and becomes a fluid

liquid metallic hydrogen a form of hydrogen under extreme temperature and pressure, which forces the atoms together so tightly that the electrons move freely throughout the packed nuclei, and as a result the hydrogen conducts electricity

lithification the process by which loose sediment is cemented or compressed together and converted into solid rock

lithosphere the 100-kilometers-thick cool, rigid, outer part of Earth, which includes the crust and the uppermost mantle and is broken into pieces called plates

litter leaves, twigs, and other plant or animal materials that have fallen to the surface of the soil but have not decomposed

loam the most fertile soil, a mixture especially rich in sand and silt with generous amounts of organic matter

loess loosely compacted deposits of windblown silt

longshore current a current generated when waves strike a shore at an angle, producing flow parallel and close to the coast

lunar eclipse a phenomenon that occurs when Earth lies directly between the sun and the moon, causing Earth's shadow to fall on the moon and darken it

luster the quality and intensity of light reflected from the surface of a mineral

M

magma molten rock generated from melting of any rock in the subsurface; cools to form igneous rock

magnetic reversal a change in Earth's magnetic field in which the north magnetic pole becomes the south magnetic pole and vice versa; has occurred on average every 500,000 years over the past 65 million years

magnetometer an instrument that measures the strength and, in some cases, the direction of a magnetic field

main sequence a band running across a Hertzsprung–Russell diagram that contains most of the stars, which are fueled by hydrogen fusion

mantle the rocky layer of Earth between the core and the crust that varies in temperature and pressure, creating some regions that are hard and firm and others that are soft and plastic

mantle plume a rising column of hot, plastic rock within the mantle

maria dry, barren, flat expanses of volcanic rock on the moon, first thought to be seas

marine west coast a midlatitude climate characterized by relatively mild but wet winters and cool summers, with little temperature difference between seasons

mass extinction a sudden, catastrophic event during which a significant percentage of all life-forms on Earth become extinct

mass wasting the downslope movement of earth material, primarily caused by gravity

meander a series of twisting curves or loops in the course of a stream

mechanical weathering the disintegration of rock into smaller pieces by physical processes without altering the chemical composition of the rock

Mediterranean a midlatitude climate characterized by dry summers, rainy winters, and moderate temperatures

Mercalli scale a scale of earthquake intensity that expresses the strength of an earthquake based on its destructive power and its effects on buildings and people; does not accurately measure the energy released by the quake

mesa a flat-topped mountain, shaped like a table, that is smaller than a plateau and larger than a butte

mesopause the ceiling of the mesosphere; the boundary between the mesosphere and the thermosphere

mesosphere the layer of air that lies above the stratopause, extending upward from about 55 kilometers to about 80 kilometers above Earth's surface

metamorphic grade a measurement of the intensity of change in a rock caused by heat, pressure, or chemical conditions

metamorphic rock rock formed when igneous, sedimentary, or other metamorphic rocks recrystallize in response to high temperature or pressure or chemical changes

metamorphism the process by which rocks change texture and mineral content in response to variations in temperature, pressure, and chemical conditions

meteor a falling meteoroid that enters Earth's atmosphere and glows as it vaporizes; colloquially called a shooting star

meteorite a meteor that does not completely vaporize and that strikes Earth's surface

meteoroid a fragment from an asteroid or comet traveling in an irregular orbit through the inner solar system

Mid-Ocean Ridge the undersea mountain chain that forms at the boundary between divergent tectonic plates within oceanic crust; it circles the planet like the seam on a baseball, forming Earth's longest mountain chain

mineral a naturally occurring inorganic solid with a definite chemical composition and a crystalline structure

mineral reserve a term to describe the known supply of ore in the ground; can be used on a local, national, or global scale

mineral resource any naturally occurring inorganic solid that is useful to humans and can be reasonably extracted at a profit now or in the future

minor planet any celestial object that orbits the sun and is not classified as a planet, dwarf planet, moon, or comet

model a physical, diagrammatic, mathematical, or other representation of a system used to describe the system or explain how it works

Mohorovičić discontinuity the boundary between the crust and the mantle, identified by a change in the velocity of seismic waves; also called the Moho

Mohs' scale a scale, based on a series of 10 fairly common minerals and numbered 1 to 10 (from softest to hardest), used to measure and express the hardness of minerals

moment magnitude an earthquake scale in which the surface area of fault movement is multiplied by an estimate of the faulted rock's strength; moment magnitude scale closely reflects the total amount of energy released

monsoon a seasonal wind and weather system caused by uneven heating and cooling of land and adjacent sea, generally blowing from the sea to the land in the summer, and from the land to the sea in winter

moraine rocks and sediment deposited by a glacier; deposited sediment can range from powdery silt to large boulders

N

native element mineral that consists of only one element and thus the element occurs in the native state (not chemically bonded to other elements); only about 20 elements occur in the native state as solids

natural gas a mixture of naturally occurring light hydrocarbons composed mainly of methane that is used for home heating and cooking and to fuel large power plants

neap tide the relatively small tide that occurs when the moon is 90° out of alignment with the sun and Earth

nebula a cloud of interstellar gas and dust; plural nebulae

negative feedback mechanism a mechanism whereby a system stays in balance; as one factor increases, it creates effects that then lead to the original factor decreasing

neutron star a small, extremely dense star, created from remnants of the supernova explosion of a large star and composed almost entirely of compressed neutrons

new moon the lunar phase during which the moon is dark when viewed from Earth, because it has moved to a position where its sunlit side faces away from Earth

nitrogen cycle processes that occur as nitrogen moves through the biosphere, atmosphere, hydrosphere, and geosphere

nonbiodegradable pollutant pollutants that do not decay naturally in a reasonable amount of time, including some industrial compounds, toxic inorganic compounds, and nontoxic sediment that muddies streams and habitats

nonconformity a type of unconformity in which layered sedimentary rocks lie on an erosion surface cut into igneous or metamorphic rocks

nonpoint source pollution pollution that is generated over a broad area, such as fertilizers and pesticides spread over agricultural fields

nonrenewable resource any resource that cannot be replenished by natural processes as quickly as it is consumed

normal fault a fault in which the hanging wall has moved downward relative to the footwall

normal magnetic polarity a magnetic orientation the same as that of Earth's current magnetic field

nuclear fuel any radioactive substance that is used to generate electricity

nutation a periodic variation in the inclination, or tilt, of the axis of a rotating object

GLOSSARY

O

occluded front a front that forms when a faster-moving cold air mass traps a warm air mass against a second mass of cold air; precipitation occurs along both frontal boundaries, resulting in a large zone of inclement weather

ocean acidification a decrease in pH of the world's oceans resulting from higher atmospheric carbon dioxide concentrations that, in turn, cause more carbon dioxide to dissolve in ocean water; once dissolved, the carbon dioxide forms carbonic acid, decreasing the pH

ocean basin the surface of Earth under the ocean

oceanic island a submarine mountain (seamount) that rises above sea level

oceanic trench a long, narrow, steep-sided depression of the seafloor formed where a subducting oceanic plate sinks into the mantle, causing the seafloor to bend downward like a flexed diving board

oligotrophic a term describing a lake that contains few nutrients and sustains relatively few living organisms

open universe a model in which the gravitational force of the universe is not sufficient to stop the current expansion and the galaxies will continue to fly apart forever

optimization the process of making the solution to a problem effective by prioritizing criteria and choosing appropriate trade-offs

ore a solid natural substance from which metals or minerals can be extracted profitably

organic sedimentary rock rock formed from the remains of living organisms

orogeny the process of mountain building; all tectonic processes associated with mountain building

orographic lifting lifting of air that occurs when air flows over a mountain

outgassing the release of volatiles from Earth's mantle and crust during volcanic eruptions

outwash sediment deposited by streams flowing from the terminus of a melting glacier

oxbow lake a crescent-shaped lake created where a meander loop is cut off from a stream and the ends of the meander become plugged with sediment

oxidation a chemical weathering process in which a mineral decomposes when it reacts with oxygen

ozone hole an unusually low ozone concentration in the stratosphere that is centered roughly over Antarctica

P

parallax the apparent change in position of an object due to the change in position of the observer

parent material the original material from which a soil forms

parsec a distance used in astronomy, equal to about 3.26 light-years

partial melting the process in which a silicate rock only partly melts as it is heated, forming magma that is more silica rich than the original rock

particulate in pollution terminology, any small piece of solid matter larger than a molecule, such as dust or soot

passive continental margin a margin where continental and oceanic crust are firmly joined together and where little tectonic activity occurs

pediment a broad, gently sloping erosional surface that forms along the front of desert mountains uphill from a bajada, usually covered by a patchy veneer of gravel only a few meters thick

peer review the process of allowing scientists to critique the experimental designs, results, and conclusions of other scientists

pelagic sediment muddy ocean sediment made up of the skeletons of tiny marine organisms

penumbra a wide band outside of the umbra, where only a portion of the sun is hidden from view during a solar eclipse

period a geologic time unit that is divided into epochs

permeability the ability of rock or soil to allow a fluid to pass through it

persistent bioaccumulative toxic chemical (PBT) nonbiodegradable toxins that are released into, and accumulate in, the environment

petroleum a complex mixture of different organic compounds that can be refined to manufacture gasoline, plastics, nylon, and other materials

pH a logarithmic scale, based on the concentration of hydrogen ions, that represents the acidity of a solution; solutions with a pH of 7 are neutral; numbers less than 7 represent acidic solutions, and numbers greater than 7 represent basic solutions

Phanerozoic eon the most recent 541 million years of geologic time, including the present, represented by rocks that contain evident and abundant fossils

photon a particle of light; the smallest particle or packet of electromagnetic energy

photosphere the surface of the sun visible from Earth; also called the solar atmosphere

photosynthesis the process by which chlorophyll-bearing plant cells convert carbon dioxide and water to organic sugars, using sunlight as an energy source; oxygen is released in the process

photovoltaic (PV) cell see *solar cell*

phytoplankton tiny plantlike plankton that live mostly within a few meters of the surface and conduct photosynthesis

pillow basalt molten basaltic lava that solidified under water, forming spheroidal lumps of basalt that resemble a stack of pillows

planet a celestial body that is in orbit around the sun, has sufficient mass for its self-gravity to result in a nearly round shape, and has cleared the neighborhood around its orbit of smaller objects

planetary nebula a nebula created when a star about the size of our sun explodes and blows a shell of gas out into space

plankton small marine and freshwater organisms that conduct most of the photosynthesis and nutrient consumption in the ocean and form the base of the marine food web

plastic flow movement of a glacier in which the ice flows as a viscous fluid

plate boundary a fracture or edge that separates two tectonic plates

plate tectonics the theory of global tectonics stating that the lithosphere is segmented into several plates that move relative to one another by floating on and sliding over the plastic asthenosphere

plateau a large, elevated area of relatively flat land

playa the dry desert lakebed of a playa lake

plume of contamination the slow-growing three-dimensional zone within an aquifer or body of surface water affected by a dispersing pollutant

pluton a body of intrusive igneous rock

plutonic rock see *intrusive igneous rock*

point source pollution pollution that arises from a specific site such as a septic tank or a factory

polar easterlies persistent polar surface winds in the Northern Hemisphere that flow from east to west

polar front the low-pressure boundary at about 60° latitude formed by warm air rising at the convergence of the polar easterlies and prevailing westerlies

polar jet stream a jet stream that flows along the polar front

population I stars a relatively young star formed from material ejected by an older, dying star; composed mainly of primordial hydrogen and helium, with a small percentage of heavier elements

population II stars an old star, with lower concentration of heavy elements than a population I star

porosity a measure of the number and size of air pockets between the particles making up a soil

positive feedback mechanism a mechanism whereby a system stays in balance; as one factor increases, it creates effects that then lead to the original factor increasing

Precambrian a term referring to all of geologic time before the Paleozoic Era, encompassing approximately the first 4 billion years, or roughly 90 percent, of Earth's history; also refers to all rocks formed during that time

precession the slow movement of the axis of a spinning body around another axis

precipitation the formation of a solid from substances previoue of volcanic ash, larger pyroclastic particles, minor lava, and hot gas that forms from collapse of an eruptive column and flows rapidly along Earth's surface

pyroclastic rock rock made up of liquid magma and solid rock fragments that were ejected explosively from a volcanic vent

Q

quasar an object, less than 1 light-year in diameter and very distant from Earth, that emits an extremely large amount of energy as radio waves

R

radiative zone the inner zone of the sun surrounding the core where energy is transmitted by absorption and emission

radiometric dating the process of measuring the absolute age of rocks, minerals, and fossils by measuring the concentrations of radioactive isotopes and their decay products

rain-shadow desert a desert formed on the downwind side of a mountain range

red giant a stage in the life of a star when hydrogen fusion occurs in a shell around a helium core; the star expands in size and is characterized by cooler surface temperatures and a red appearance

redshift a shift toward longer (red) wavelengths, observed in the spectrum of a distant galaxy or other object that is moving away from Earth; caused by the Doppler effect

reef a wave-resistant ridge or mound built by corals or other marine organisms

reflecting telescope a type of telescope that uses a mirror or mirrors to collect and focus an image, which is then reflected to the eyepiece by another mirror

refracting telescope a type of telescope that uses two lenses: one to collect light from a distant object and another to magnify the image

regolith the thin layer of loose, unconsolidated, weathered material that overlies bedrock

relative age an approach to determining the timing of geologic events based on the order in which they occurred, rather than on the absolute number of years ago in which they occurred

relative humidity the ratio of the amount of water vapor in a given volume of air divided by the maximum amount of water vapor that can be held by that air at a given temperature, expressed as a percentage

renewable resource any useful substance or phenomenon that can be replenished by natural processes faster than it is consumed

GLOSSARY

residual soil a soil formed from the weathering of bedrock below

resolution the capability to which an instrument can detect or measure detail, such as the capability of an optical telescope to define individual craters on the moon

retrograde motion the apparent motion of planets when viewed from Earth, in which they temporarily appear to move backwards (westward) with respect to the stars, before resuming their original eastward motion

reverse fault a fault in which the hanging wall has moved up relative to the footwall

revolution motion around a central point; orbiting

Richter scale a scale of earthquake magnitude that expresses the amount of energy released; calculated from the amplitude of the largest body wave on a standardized seismograph, although not a precise measure of earthquake energy

rift zone see *divergent boundary*

rock avalanche see *rockslide*

rock cycle the sequence of events in which rocks are formed, destroyed, altered, and reformed by geologic processes

rock record the rocks that currently exist on Earth and contain the record of its history

rockfall form of rapid mass wasting in which unconsolidated material falls freely or bounces down steep slopes or cliffs

rockslide a subcategory of slide mass wasting in which a segment of bedrock slides downslope along a fracture and the rock breaks into fragments and tumbles down the hillside; also called a rock avalanche

rotation turning or spinning on an axis

runoff the water that flows over Earth's surface in rivers and streams

S

salinity the total quantity of dissolved salts in seawater, expressed as a percentage

salinization a process whereby salts accumulate in soil, lowering soil fertility, usually as a result of irrigation

saltation the asymmetric jumping movement of sedimentary particles that are ejected off the bed through impact by another particle and are carried downstream by wind or water for some distance before falling back to the bed surface; most sand grains in desert environments move via saltation by wind

saltwater intrusion a condition along the coasts of oceans and inland saltwater bodies in which excessive pumping of fresh groundwater causes salty groundwater to invade an aquifer

saturation a condition in which a solution contains as much solute as can be dissolved at a given temperature; saturated air contains as much water vapor as it can hold at a given temperature and has 100% humidity

science the study of natural phenomena to better understand how the natural world works and to predict outcomes of natural events

seafloor spreading the hypothesis that segments of oceanic crust are separating at the Mid-Ocean Ridge

seamount a submarine mountain, usually of volcanic origin, that rises 1 kilometer or more above the surrounding seafloor

Second Great Oxidation Event the process, occurring about 600 million years ago, when the oxygen concentration in Earth's atmosphere increased abruptly a second time in response to interactions among chemical, physical, and biological processes

secondary wave (S wave) seismic body waves that travel slower than P waves and are the "secondary" waves to reach an observer; sometimes called shear waves due to the shearing motion in rock caused as the waves vibrate perpendicular to the direction they travel

sedimentary rock rock formed when sediment becomes compacted and cemented through the process of lithification

seismic wave an elastic wave that travels through rock, produced by an earthquake or explosion

seismograph an instrument that records seismic waves

shield volcano a large, gently sloping volcanic mountain formed by successive flows of basaltic magma

silicate mineral whose chemical elements include silicon and oxygen and whose crystal structures contain sili-cate tetrahedra; also, all rocks composed principally of silicate minerals

silicate tetrahedron the fundamental building block of all silicate minerals, a pyramid-shaped structure of a silicon ion bonded to four oxygen ions

sill a sheetlike igneous rock, parallel to the grain or layering of country rock, that forms when magma is injected between layers

sinkhole a depression on Earth's surface caused by the collapse of a cavern roof or by the dissolution of surface rocks, usually limestone

slide form of rapid mass wasting in which the rock or soil initially moves as a consolidated unit along a fracture surface

slip the distance that rocks on opposite sides of a fault have moved relative to each other

slip face the steep leeward side of a dune, typically at the angle of repose for loose sand, so that the sand flows or slips down the face, where it is deposited

slump form of mass wasting that results in the sliding of consolidated material along a curved surface

smog visible, brownish air pollution formed through chemical reactions that involve incompletely combusted automobile gasoline, nitrogen dioxide, oxygen, ozone, and sunlight

soil the upper layers of regolith that support plant growth; some Earth scientists and engineers use the terms soil and regolith interchangeably

soil erosion the degrading action that occurs when wind, ice, or water move through an area, carrying away the top layers of formed soil

soil horizon a layer of soil that is distinguishable from other layers because of differences in appearance and physical and chemical properties

soil order the highest hierarchical classification of soils by the National Resource Conservation Service; twelve soil orders are recognized

soil texture the proportion of sand, silt, and clay in a soil

solar atmosphere see *photosphere*

solar cell a device that converts sunlight to electricity, also called photovoltaic or PV cells

solar eclipse a phenomenon that occurs when the moon passes directly between Earth and the sun; the moon casts its shadow on Earth, thus blocking the sun's light

solar mass a standard unit for expressing the mass of a star relative to the mass of our sun; 1 solar mass equals the mass of the sun

solar wind a stream of charged particles radiating outward from the sun at high speed

solstice either of two times per year when the sun is farthest from the Equator and shines directly overhead at either 23.5° north or south latitude; the longest day of the year in the hemisphere receiving direct sunlight and the shortest day of the year in the other hemisphere

sonar a system detecting objects under water by emitting sound pulses and measuring their return after reflection

specific gravity the weight of a substance relative to the weight of an equal volume of water

specific heat the amount of energy required to raise the temperature of 1 gram of a substance by 1°C

spectrum an ordered array of colors; a pattern of wavelengths into which a beam of light or other electromagnetic radiation is separated

spicule a jet of gas at the edge of the sun, shooting upward from the chromosphere

spiral galaxy a galaxy characterized by arms that radiate out from the center like a pinwheel

spire individual thin column of rock, isolated from other rocks by erosion

spit a small ridge of sand or gravel extending from a beach into a body of water

spreading center see *divergent boundary*

spring tide the very high and very low tide that occurs when the sun, moon, and Earth are aligned such that the gravitational fields of the sun and moon combine to create a strong tidal bulge

stalactite an icicle-like rock structure formed by slow precipitation of calcite found on the ceiling of a limestone cave

stalagmite a cone-shaped rock structure formed by slow precipitation of calcite found on the floor of a limestone cave

stationary front a front at the boundary between two stationary air masses of different temperatures

steppe a vast, semiarid, grass-covered plain found in dry climates such as those in southeast Europe, central Asia, and parts of central North America

storm surge abnormally high coastal waters and flooding, created by a combination of strong onshore winds and the low atmospheric pressure of a storm that raises the sea surface by several meters

stratopause the ceiling of the stratosphere; the boundary between the stratosphere and the mesosphere

stratosphere the layer of air above the tropopause, extending upward to about 55 kilometers

stratovolcano a steep-sided volcano formed by an alternating series of lava flows and pyroclastic deposits and marked by repeated eruption, also known as a composite cone

stratus horizontally layered clouds that spread out into a broad sheet, usually creating dark, overcast skies

streak the color of the fine powder of a mineral, usually obtained by rubbing the mineral on an unglazed porcelain streak plate

stream a moving body of water confined in a channel and flowing downslope; a river is a large stream fed by smaller ones

strike-slip fault a vertical fault across which rocks on opposite sides move horizontally

subarctic the northernmost edge of the humid continental climate areas, lying just south of the arctic

subduction the process in which two lithospheric plates of different densities converge and the denser one sinks into the mantle beneath the other

subduction complex the complicated mass of rock consisting of deformed seafloor sediment and fragments of oceanic crust that is scraped from the upper layers of the subducting slab in a subduction zone and added to the overriding plate

submarine canyon deep, V-shaped, steep-walled troughs eroded into a continental slope, sometimes extending into the continental shelf

submarine fan a large, fan-shaped accumulation of sediment deposited on the deep seafloor, usually within and beyond the mouths of submarine canyons

submergent coastline a coastline that was previously above sea level but has been drowned, because either the land has sunk, or sea level has risen

GLOSSARY

subsidence the irreversible sinking or settling of Earth's surface

subtropical jet stream a jet stream that flows between the trade winds and the westerlies

sulfide mineral whose chemical composition is based on the sulfide anion bonded to a metal cation

sunspot a comparatively cool, dark region on the sun's surface caused by a magnetic disturbance

supercontinent a continent, such as Pangaea, consisting of all or most of Earth's continental crust joined together in a single, large landmass

supercooling a condition in which water droplets in air do not freeze even when the air cools below the freezing point

superheating heating of a substance above a phase-transition (solid to liquid or liquid to gas) without the transition occurring; high pressure can keep solid rock from melting even though it is above its melting temperature

supernova a stage in the life of a massive star where it increases dramatically in brightness because of a catastrophic explosion that ejects most of its mass

supersaturation a condition in which a solution contains more solute than a saturated solution will hold at a given temperature; supersaturated air exceeds 100 percent humidity

surf the chaotic, turbulent waves breaking along the shore

surface current horizontal flow of water in the upper 400 meters of the oceans, caused by wind blowing over the sea surface

surface wave seismic waves that radiate from the earthquake's epicenter and travel along the surface of Earth or along a boundary between layers within Earth

suspended load the total mass of a stream's sediment load that is carried within the flow by turbulence and is free from contact with the streambed

syncline a fold in rock that arches downward

system a combination of interacting parts that form a complex whole

T

taiga a biome of conifers that can survive the extreme winters of the subarctic

tectonic plate seven major and eight smaller segments of the lithosphere that float on the relatively hot, plastic mantle rock beneath and move horizontally with respect to each other

temperate rain forest a rain forest that grows in marine west coast climates where rainfall is more than 100 centimeters per year and is relatively constant throughout the year, particularly along the northwest coast of North America from Oregon to Alaska

temperature a measure of the average kinetic energy in a substance, proportional to the average speed of atoms and molecules in a sample

terminus the end, or foot, of a glacier

terrestrial planet the four inner planets composed primarily of metals and silicate rocks

terrigenous sediment sediment composed of sand, silt, and clay eroded from the continents

texture the size, shape, and arrangement of mineral grains, or crystals, in a rock

theory an explanation of a phenomenon that is generally accepted to be true based on a great deal of repeated and independent experimentation

thermocline the boundary between warm top waters and cold bottom waters in a lake in the temperate zone during the summer

thermohaline circulation the force behind deep-sea currents, created by differences in water temperature and salinity and, therefore, differences in water density

thermosphere an extremely high and diffuse region of the atmosphere lying above the mesosphere, from about 80 kilometers upward

three-cell model a model of global wind patterns that depicts three convection cells in each hemisphere, bordered by alternating bands of high and low pressure

threshold effect a reaction whereby the environment initially changes slowly (or not at all) in response to a small perturbation, but after the threshold is crossed, an additional small perturbation causes rapid change

thrust fault a type of reverse fault that is nearly horizontal, with a dip of 45 degrees or less over most of its extent

tidal bulge a bulge that forms on the surface of Earth's oceans facing directly towards and away from the moon

tidal current a current, channeled by a bay with a narrow entrance or by closely spaced islands, caused by the rise and fall of the tides

tidal range the vertical distance between low and high tide at a particular point on Earth's oceans

tide the cyclic rise and fall of ocean water caused by the gravitational force of the moon and, to a lesser extent, the sun

till glacial drift that was deposited directly by glacial ice

topography a description of the physical features of landforms in an area

topsoil the fertile, dark-colored surface soil; the combined O and A soil horizons

tornado a small, intense, short-lived, funnel-shaped storm that protrudes from the base of a cumulonimbus cloud

trade winds the winds that blow steadily toward the Equator from the northeast in the Northern Hemisphere and from the southeast in the Southern Hemisphere, between 5° and 30° north and south latitudes

trade-off a compromise in the solution to a problem where one feature, quality, or factor is favored at the cost of another

transform boundary a plate boundary where two tectonic plates slide horizontally past one another

transform fault a strike-slip fault between two offset segments of a mid-ocean ridge or along a strike-slip plate boundary

transformation a physical or chemical change that affects a soil constituent

translocation the vertical (usually downward) movement of physical or chemical soil constituents from one horizon to another

transpiration the direct evaporation of water from the surface of plants to the atmosphere

transported soil a soil formed by the weathering of regolith that is transported from somewhere else and deposited

tributary a stream that feeds water into another stream or river

Tropic of Cancer the latitude 23.5° north of the Equator; on the summer solstice in the Northern Hemisphere, sunlight strikes Earth from directly overhead at noon at this latitude

Tropic of Capricorn the latitude 23.5° south of the Equator; on the summer solstice in the Southern Hemisphere, sunlight strikes Earth from directly overhead at noon at this latitude

tropical cyclone a broad, circular storm with intense low pressure that forms over warm oceans

tropical monsoon a tropical climate with distinct wet and dry seasons, characterized by high rainfall during the wet months and a relatively short dry season

tropical rain forest a forest growing in a tropical climate with abundant, year-round rainfall; characterized by tall trees that create a dense canopy of leaves

tropical savanna a tropical climate with distinct wet and dry seasons but with relatively low annual rainfall, characterized by large areas of grassland supporting herds of grazing animals

tropopause the top of the troposphere; the boundary between the troposphere and the stratosphere

troposphere the layer of air that lies closest to Earth's surface and extends upward to about 17 kilometers

trough the lowest part of a wave

tsunami a large, destructive sea wave produced by an undersea earthquake or volcano; sometimes erroneously called a tidal wave, although it has nothing to do with astronomical tides

tundra a biome dominated by low-lying grasses, mosses, flowers, and a few small bushes that grow in the Arctic and high alpine regions

turbidity current a downslope sediment flow that is a turbulent mixture of sediment and seawater

turnover the mixing of layers of water that occurs in a lake during the breakdown of thermal stratification in spring and fall

typhoon a tropical storm occurring in the western Pacific Ocean whose wind exceeds 120 kilometers per hour; called a hurricane in North America and the Caribbean and a cyclone in the Indian Ocean

umbra the narrow band of the moon's shadow where the sun is completely blocked out during a solar eclipse

unconformity a boundary between layers of sediment or sedimentary rock that represents an interruption in the deposition of sediment, including an interruption that involves partial erosion of the older layer

underthrusting the process by which one plate subducts beneath the other in a subduction zone or during a continent–continent collision

uniformitarianism a principle stating that the geologic processes operating today also operated in the past

upwelling a rising ocean current that transports cold water and nutrients from the depths to the surface

urban heat island effect warmer temperatures in a city compared to the surrounding countryside, caused by factors in the urban environment itself

vent an opening in a volcano, typically in the crater, through which lava and rock fragments erupt

vesicle a hole in lava rock that formed when the lava solidified before bubbles of gas or water could escape

volcanic ash the smallest pyroclastic particles, less than 2 millimeters in diameter

volcanic rock see *extrusive igneous rock*

volcano a vent in Earth's crust that forms a hill or mountain from lava and rock fragments

waning becoming smaller; the 14 to 15 days after a full moon and before a new moon when the visible portion of the moon decreases every day

wash a streambed that is dry for most of the year

water pollution the reduction of the water quality by the introduction of impurities

water scarcity a situation in which the aggregate demand for water by all users, including the environment, exceeds the available supply

water table the top surface of the zone of saturation; the water table separates this zone from the zone of aeration above

GLOSSARY

wave base the depth below which surface wave energy does not reach, roughly as deep as half a single wavelength

wave height the vertical distance from the crest to the trough of a wave, also known as amplitude

wave orbital the circular motion of water that occurs as a surface wave passes by

wavelength the distance between successive wave crests or troughs

waxing becoming full; the 14 to 15 days after a new moon and before a full moon when the visible portion of the moon increases every day

weather the state of the atmosphere at a given place and time as characterized by temperature, wind, cloudiness, humidity, and precipitation

weathering the breakdown of rock at Earth's surface by chemical and physical processes

white dwarf a stage in the life of a star when fusion has halted and the star glows solely from the residual heat produced during past eras

withdrawal any process that uses water and then returns it to Earth locally

Z

zooplankton tiny animal-like plankton that live mostly within a few meters of the surface and feed on phytoplankton

GLOSARIO

A

abanico pluvial acumulación de sedimentos en forma de abanico que se genera donde un arroyo de montaña se desacelera con rapidez cuando cambia la pendiente e ingresa en una llanura relativamente plana

abanico submarino acumulación abundante de sedimentos depositados en forma de abanico en la profundidad del fondo oceánico que comúnmente se extiende cerca de la desembocadura de un cañón submarino

abultamiento de la marea inflamación abultada que se forma en la superficie de los océanos de la Tierra y se mueve en dirección a la luna o en el sentido contrario

acarreo cualquier roca o sedimento transportado y depositado en un lugar por un glaciar o por agua de deshielo de un glaciar

acidificación del océano reducción del pH de los océanos causada por una mayor concentración de dióxido de carbono en la atmósfera, lo que, a su vez, aumenta la disolución del dióxido de carbono en los océanos; una vez disuelto, el dióxido de carbono forma ácido carbónico y disminuye el pH

acuífero formación de roca o suelo que puede producir cantidades de agua económicamente significativas

acuitardo formación de roca de baja porosidad y permeabilidad que contiene o transfiere muy poca agua subterránea

advección transferencia de calor provocada por el flujo horizontal de material fluido

aerosol en términos de contaminación, partícula o conjunto de partículas suspendidas en el aire

afluente corriente de agua que vierte sus aguas en otra corriente de agua o río

agua subterránea agua contenida en el suelo y en los intersticios de las rocas a unos pocos kilómetros de profundidad en la parte superior de la geosfera; constituye aproximadamente el 0.63 por ciento de toda el agua de la hidrosfera

agujero de la capa de ozono concentración de ozono inusualmente baja que se produce en la estratósfera, aproximadamente sobre la Antártida

agujero negro región del espacio infinitesimalmente pequeña, creada luego de la supernova de una estrella gigante, que contiene material concentrado de manera tan densa que la luz queda atrapada dentro su intenso campo gravitatorio

albedo reflectividad de una superficie; las superficies que reflejan más luz tienen un albedo más alto

altura de onda distancia vertical entre la cresta y el valle de una onda, también llamada amplitud de onda

amplitud de marea distancia vertical entre la marea baja y la marea alta en un punto particular de los océanos de la Tierra

ángulo de reposo pendiente o inclinación máxima en la cual el material disperso se mantiene estable; si la pendiente se torna más empinada, el material se desliza

anión ion con carga negativa

anóxico carente de oxígeno

anticiclón región de alta presión con un sistema de vientos que rotan hacia afuera y que se generan cuando el aire descendente se esparce por la superficie terrestre

anticlinal pliegue en la roca que se arquea hacia arriba

Antropoceno era geológica actual, vista como el período en el cual la actividad humana ha sido la influencia dominante del clima y el medio ambiente

año luz distancia que recorre la luz en un año, equivalente a aproximadamente 9.5×10^{12} kilómetros

arco continental margen continental, también llamado margen andino, que se caracteriza por la subducción de una placa oceánica por debajo de una placa continental

arco de islas cadena de islas volcánicas dispuestas en forma levemente curva que se generaron por la convergencia de dos placas de corteza oceánica y la consecuente subducción de una bajo la otra

arista saliente de roca, angosta y afilada, que surge entre dos valles contiguos o entre dos circos glaciares como consecuencia de la acción de dos glaciares alpinos que se deslizaron a uno y otro lado de la saliente y erosionaron ambos laterales de la roca

arrecife cresta o montículo resistente a las olas formado por corales u otros organismos marinos

ascenso frontal proceso por el cual una masa de aire fresco y denso se desliza por debajo de una masa de aire cálido y la obliga a elevarse

astenosfera porción superior del manto, ubicada apenas debajo de la litosfera, que se extiende entre 100 y 350 kilómetros bajo la superficie de la Tierra, y está compuesta por roca frágil y flexible en la que puede formarse magma

asteroide cuerpo rocoso, pequeño y seco, que orbita alrededor del sol entre Marte y Júpiter o más allá de la órbita de Neptuno

atmósfera capa gaseosa que recubre la Tierra; está compuesta mayormente de nitrógeno y oxígeno, y pequeñas cantidades de argón, dióxido de carbono y otros gases

atmósfera solar ver *fotosfera*

atolón arrecife de coral circular que forma un anillo de islas alrededor de una laguna central y se conecta con el exterior a través de las aguas profundas del mar abierto; comúnmente se forma encina de un monte submarino

avalancha de rocas ver *desprendimiento*

B

bajada superficie depositaria extensa, de pendiente suave, formada por la fusión de abanicos aluviales generados por cañones poco distanciados que se abre en un valle desértico

bar unidad de medida en la que se registra la presión atmosférica; un bar equivale aproximadamente a la presión atmosférica al nivel del mar

GLOSARIO

baricentro centro de masa de dos o más cuerpos que orbitan uno alrededor del otro y el punto alrededor del cual orbitan

barómetro aparato utilizado para medir la presión barométrica

basalto acojinado lava basáltica fundida que se solidificó debajo del agua y formó trozos de basalto más o menos redondeados, parecidos a una pila de cojines

basalto de inundación lava basáltica que brota en gran volumen de manera no violenta de ciertas grietas o fisuras de la superficie terrestre y cubre grandes extensiones de terreno en forma de mesetas de lava

batolito plutón grande, expuesto en la superficie terrestre por más de 100 kilómetros cuadrados

Big bang fenómeno ocurrido entre diez mil y veinte mil millones de años atrás que presumiblemente marca el origen del universo, atribuido a una explosión de materia concentrada en un punto infinitamente denso

biodiversidad la variedad de organismos que viven en la Tierra o en un ecosistema

bioma comunidad de plantas que se desarrolla en una zona geográfica extensa, caracterizada por un clima particular

biorremediación uso de microorganismos para descomponer un contaminante medioambiental

biósfera zona de la Tierra que en la que se desarrollan todas las formas de vida marina, terrestre y aérea

bloque errático tipo de canto que comúnmente no pertenece al lecho de roca más próximo, sino que fue transportado a su ubicación actual por acción de un glaciar

bólido fragmento de escombros espaciales que comúnmente explota al impactar contra la atmósfera

bosque templado húmedo bosque húmedo que crece en climas oceánicos templados en donde las lluvias superan los 100 centímetros anuales y son relativamente constantes durante todo el año; tipo de bosque característico de la costa noroeste de América del Norte, desde Oregón hasta Alaska

brillo absoluto el brillo absoluto de una estrella, también conocido como luminosidad, es lo brillante que luciría si estuviera a una distancia fija de la Tierra

brillo aparente luminosidad de una estrella vista desde la Tierra

brillo calidad e intensidad de la luz que refleja la superficie de un mineral

butte montaña de cima plana, más pequeña y pronunciada que una mesa, que se caracteriza por sus laterales escarpados y empinados

C

caldera depresión circular extensa generada por el derrumbe de la cámara de magma luego de una erupción volcánica explosiva

calor específico cantidad de energía requerida para elevar en 1° la temperatura de 1 gramo de sustancia

calor latente energía almacenada; energía que se libera o absorbe cuando una sustancia cambia de estado por derretimiento, congelamiento, vaporización, condensación, sublimación o deposición

calor energía transferida entre dos sustancias por las diferencias de temperatura que presentan

cambio de temperatura adiabática cambio de temperatura causado por compresión o expansión de gas, que se produce sin aumento ni pérdida de calor

cambio eustático del nivel del mar cambio global del nivel del mar causado por tres procesos diferentes: el crecimiento o derretimiento de glaciares, cambios en la temperatura del agua y cambios en el volumen de la dorsal medio-oceánica

cancerígeno, cancerígena sustancia que puede causar cáncer

cañón submarino valle profundo, con forma de V, enmarcado por paredes empinadas erosionado en un talud continental, que ocasionalmente se extiende hasta la plataforma continental

capa freática franja superior de la zona de saturación; la capa freática separa la zona de saturación de la zona de aireación ubicada más arriba

capacidad cantidad máxima de sedimento que una corriente de agua puede transportar de una sola vez

cara de deslizamiento lado empinado de una duna, opuesto a la dirección desde la que sopla el viento, que comúnmente coincide con el ángulo de reposo de la arena suelta y por lo tanto permite que la arena fluya o se deslice por la cara de la duna, en la cual se deposita

carbón roca sedimentaria compuesta mayormente de carbono orgánico que se quemará sin refinar

carbonato mineral cuya base de composición química es el anión carbonato

carga de fondo masa total de la carga de sedimento que transporta una corriente de agua a lo largo del fondo o en contacto esporádico con el fondo de su lecho

carga disuelta masa total de iones disueltos y transportados por una corriente de agua en un momento dado; los iones son la derivación de la meteorización química

carga suspendida masa total de los sedimentos de una corriente de agua transportados por la turbulencia, que no entran en contacto con el fondo de su lecho

catastrofismo principio según el cual, a diferencia del gradualismo, los fenómenos geológicos infrecuentes alteran el curso de la historia de la Tierra

catión ion con carga positiva

célula fotovoltaica (célula PV) ver *célula solar*

célula solar dispositivo que convierte la luz solar en electricidad, conocido también como célula fotovoltaica

ceniza volante conjunto de minerales no combustibles que ingresan en la atmósfera durante la combustión del carbón y precipitan como polvo arenoso

cenizas volcánicas las partículas piroclásticas más pequeñas, de menos de 2 milímetros de diámetro

centro de la dorsal ver *límite divergente*

cianobacterias algas verdeazuladas que se encontraban entre las primeras formas de vida fotosintética de la Tierra

ciclo proceso o fenómeno secuencial que vuelve a comenzar cada vez que termina y así se repite una y otra vez

ciclo de las rocas secuencia de eventos a lo largo de la cual las rocas se forman, destruyen, alteran y vuelven a formar a partir de procesos geológicos

ciclo del carbono proceso por el cual el carbono pasa por distintas zonas de la biosfera, geosfera, hidrosfera y atmósfera

ciclo del nitrógeno procesos que se desarrollan mientras el nitrógeno circula por la biosfera, atmósfera, hidrosfera y geosfera

ciclo hidrológico circulación del agua por las distintas esferas de la Tierra a través de la precipitación, evaporación y condensación

ciclón tropical tormenta circular amplia con intensa baja presión que se forma sobre aguas cálidas

ciclón región de baja presión con un sistema de vientos que rotan hacia adentro; en un sentido amplio, no científico, el término suele referirse a una variedad de tormentas violentas, que incluyen huracanes y tornados

ciencia estudio de los fenómenos naturales para comprender mejor de qué manera funciona el mundo natural y para predecir el resultado de los eventos naturales

Cinturón de Kuiper región con forma de disco, ubicada a continuación de la órbita de Neptuno, en la que hay asteroides, cometas y escombros derivados de la formación del sistema solar

circo glacial depresión de paredes escarpadas, rodeada de cumbres, originada en una montaña por la acción erosiva de un glaciar

circulación termohalina fuerza que impulsa las corrientes marinas profundas, generada por las diferencias de temperatura y salinidad del agua y, por lo tanto, de la densidad del agua

cirro nubes altas y tenues compuestas de cristales de hielo

clima tiempo característico de una región, definido según un promedio de las condiciones meteorológicas registradas a lo largo de varias décadas; se basa en los ciclos anuales de la temperatura, humedad, viento, nubosidad, humedad y precipitaciones y no a las variaciones diarias de estos factores

clivaje propensión de ciertos minerales a quebrarse en las superficies lisas, que son planos de uniones débiles del cristal; cuando un mineral cuenta con un clivaje bien desarrollado, es posible retirar del cristal cada una de sus capas, como si se pelara una cebolla

clorofluorocarbono (CFC) compuesto orgánico con contenido de cloro y flúor que asciende a la parte superior de la atmósfera y destruye la capa de ozono

combustible de biomasa sustancia de origen animal o vegetal que puede quemarse o procesarse para generar energía

combustible fósil tipo de combustible derivado de restos orgánicos prehistóricos formado a partir de procesos geológicos

combustible nuclear cualquier sustancia radiactiva que se emplea para generar electricidad

cometa objeto compuesto por roca y hielo ligeramente fusionados que al aproximarse al sol forman una unidad compuesta por una cabeza brillante y una larga cola difusa

compensación alternativa en la solución de un problema por la cual se da prioridad o preminencia un rasgo, cualidad o factor a expensas de otro

complejo de subducción masa de roca compleja compuesta de sedimento del fondo oceánico deformado y fragmentos de corteza oceánica que se desprende de las capas superiores de la placa de subducción y se adhiere a la placa que se desliza por encima en una zona de subducción

cóndrulo grano pequeño, de aproximadamente 1 milímetro de diámetro, incrustado en un meteorito, que comúnmente contiene aminoácidos u otras moléculas orgánicas

conducción transmisión del calor generada por la colisión directa entre átomos o moléculas

conducto abertura de un volcán, comúnmente en el cráter, por la cual surgen la lava y los fragmentos de roca de la erupción

conformidad término usado para describir la condición de las capas sedimentarias que se depositaron de manera continua, aparentemente sin interrupción

cono de escoria volcán pequeño, de menos de 300 metros de altura, constituido por la acumulación de fragmentos piroclásticos sueltos expulsados por un conducto central; comúnmente activo durante poco tiempo

cono de depresión depresión en forma de cono que se produce en el nivel freático cuando el agua sale de un pozo más rápido de lo que ingresa en él a través de un acuífero; si el agua continúa saliendo del pozo a la misma velocidad, el nivel freático disminuye

constelación grupo de estrellas que, vistas desde la Tierra, dan la impresión de formar un patrón específico y a las cuales los astrónomos les han dado un nombre pertinente para determinar la posición de los objetos en el cielo nocturno

GLOSARIO

consumo cualquier proceso que involucre usar agua y devolverla a la Tierra lejos de su fuente de origen

contaminación difusa contaminación generada en zonas amplias, por ejemplo, cuando se aplican fertilizantes o pesticidas en superficies de tierra destinadas a la agricultura

contaminación hídrica disminución de la calidad del agua causada por la incorporación de impurezas en su composición

contaminación lumínica resplandor nocturno de las luces de una ciudad que compite con la luz distante de las estrellas y reduce la visión de un telescopio

contaminación puntual contaminación que surge de un lugar específico, como un tanque séptico o una fábrica industrial

contaminante biodegradable agente contaminante que se degrada de manera natural en un período de tiempo razonable al ser consumido o desintegrado por organismos que viven en el suelo o el agua

contaminante no biodegradable contaminante que no se degrada de manera natural en un lapso de tiempo razonable, como los compuestos industriales, los compuestos inorgánicos tóxicos y el sedimento no tóxico que ensucia las corrientes de agua y los hábitats

continental templado variedad de clima de latitud media caracterizado por tener veranos calurosos, inviernos fríos y precipitaciones a lo largo de todo el año

convección flujo ascendente y descendente de material fluido que resulta de los cambios de densidad causados por calentamiento y enfriamiento

corona atmósfera externa del sol, normalmente invisible, que se presenta como un halo alrededor de una luna negra durante un eclipse solar

correlación proceso por el cual se establece la relación entre diferentes lugares de la Tierra a base de la edad de las rocas o los rasgos geológicos que exhiben; esto se realiza comparando las características sedimentarias de las capas o tipos de fósiles que presentan

corriente de agua cuerpo de agua en movimiento que fluye por una pendiente contenido en un canal; un río es una corriente de agua alimentada por corrientes más pequeñas

corriente de deriva corriente marina que se genera cuando las olas golpean contra la costa en un ángulo determinado y producen un flujo paralelo y cercano a esa costa

corriente de marea corriente delimitada por una bahía con entrada angosta o islas muy poco separadas, que se genera por el ascenso y descenso de las mareas

corriente de turbidez flujo descendente de sedimento constituido por una mezcla turbulenta de sedimento y agua salada

corriente en chorro polar corriente en chorro que sopla a lo largo de un frente polar

corriente en chorro subtropical corriente en chorro que sopla entre los vientos alisios y los vientos del oeste

corriente en chorro bandas angostas de aire que se mueven a gran velocidad en las altitudes superiores

corriente marina profunda circulación vertical y horizontal de agua salada a una profundidad superior a 400 metros generada por las diferencias de densidad del agua

corriente superficial flujo de agua horizontal que se desplaza dentro de los 400 metros superiores del océano a causa del efecto del viento sobre la superficie del agua

corrimiento al azul desplazamiento hacia longitudes de onda más cortas (azules) que se observa en el espectro de una galaxia lejana u otro objeto que se mueve hacia la Tierra; es causado por el efecto Doppler

corrimiento al rojo desplazamiento hacia longitudes de onda más largas (rojas) que se observa en el espectro de una galaxia lejana u otro objeto que se aleja de la Tierra; es causado por el efecto Doppler

costa emergente costa que se encontraba bajo el agua, pero pasó a estar expuesta al aire porque se elevó la tierra o descendió el nivel del mar

costa sumergida costa que se encontraba sobre el nivel del mar, pero se hundió porque la tierra descendió o se elevó el nivel del mar

cráter depresión con forma de tazón ubicada en la cima de un volcán generada por la actividad volcánica

creciente que aumenta hasta llenarse; período de 14 a 15 días, posterior a la luna nueva y anterior a la luna llena, durante el cual la porción visible de la luna crece diariamente

cresta parte más alta de una ola

crevasse fractura o grieta que se forma en los últimos 40 metros que constituyen la zona más quebradiza de la parte superior de un glaciar a medida que el glaciar avanza sobre un lecho de roca irregular

criosfera parte de la hidrosfera de la Tierra que contiene los glaciares, casquetes polares de hielo, nieves eternas y toda el agua congelada del planeta

cristal elemento sólido o compuesto cuyos átomos forman un patrón regular y periódicamente repetido

criterios estándares que debe cumplir una solución de ingeniería para considerarse exitosa

cromosfera capa gaseosa, turbulenta y difusa del sol ubicada encima de la fotosfera

cronohorizonte capa sedimentaria delgada y extensa, que es fácil de reconocer y puede usarse para hacer correlaciones porque fue depositada de manera rápida y simultánea en una zona amplia

cuarto creciente fase lunar durante la cual la zona iluminada de la luna se ve como un semicírculo muy delgado con forma de C

cuásar objeto de menos de 1 año luz de diámetro y muy distante de la Tierra, que emite una enorme cantidad de energía

cuenca de antearco cuenca sedimentaria situada entre la fosa oceánica y el arco magmático, ya sea en un arco de islas como en un arco continental

cuenca de antepaís cuenca sedimentaria formada por la depresión de la corteza continental causada por el peso de las placas de cabalgamiento en una orogénesis compresiva

cuenca de drenaje territorio drenado por un único sistema de drenaje natural o corriente de agua

cuenca oceánica superficie terrestre que se extiende debajo del océano

cúmulo globular concentración de estrellas antiguas, con un origen común, que se mantienen unidas por acción de la gravedad y se encuentran en el halo galáctico de una galaxia espiral como la Vía Láctea

cúmulos nubes blancas, de aspecto mullido con base plana y picos ondulantes

CH

chapitel columna de roca, individual y delgada, que ha sido aislada de otras rocas por efecto de la erosión

D

datación radiométrica proceso con el cual se mide la edad absoluta de los fósiles, minerales y rocas a base de las concentraciones de isótopos radiactivos y los productos de su desintegración

datos información recogida para realizar un estudio o análisis

deflación es la erosión causada por el viento en zonas secas, de sedimentos dispersos finos

delta acumulación de sedimentos en forma de abanico que se genera donde una corriente de agua desemboca en un lago u océano; incluye una superficie casi plana que se ubica en parte sobre la costa y en parte fuera de ella

deposición proceso por el cual el sedimento se deposita en un lugar luego de ser transportado por efecto de la erosión

deriva continental hipótesis de Alfred Wegener según la cual los continentes de la Tierra alguna vez estuvieron unidos y luego se dividieron y separaron

derrumbe forma de desgaste rápido en el que material no consolidado cae libremente o rebota por la pendiente pronunciada de una montaña o acantilado

descarga volumen de agua que fluye corriente abajo a lo largo de un período de tiempo específico, comúnmente medido en metros cúbicos por segundo (m^3/seg) o pies cúbicos por segundo (pie^3/seg)

descompresión proceso de meteorización mecánica en el cual las fuerzas tectónicas elevan rocas profundamente enterradas y luego la erosión remueve la roca superpuesta y el sedimento; este proceso le quita presión al material superpuesto y hace que la roca se expanda y fracture

desertificación proceso por el cual un terreno semiárido se convierte en desierto a causa de factores humanos, factores naturales o una combinación de ambos

desgasificación liberación de elementos volátiles del manto y la corteza, que se produce durante las erupciones volcánicas

desgaste movimiento con inclinación descendente del material terrestre causado principalmente por la gravedad

desierto de sombra pluviométrica desierto que se forma del lado de sotavento de una cadena montañosa

desierto cualquier región de la Tierra que recibe menos de 25 centímetros (10 pulgadas) de lluvia annual, por lo que en consecuencia, contiene poca o ninguna vegetación

deslizamiento basal movimiento de un glaciar por medio del cual toda la masa se desliza sobre la roca madre

deslizamiento de la playa movimiento gradual del sedimento (comúnmente, arena) por la playa, en sentido paralelo a la costa, que se produce cuando las olas impactan contra la playa en forma oblicua pero regresan al mar directamente

deslizamiento forma de desgaste rápido en el cual la roca o el suelo inicialmente se mueven como unidad consolidada a lo largo de una fractura superficial

desplazamiento distancia en la que se mueven las rocas de un lado de la falla en relación con las del otro lado

desplome forma de desgaste en el cual el material consolidado cae por una superficie curva

desprendimiento subcategoría de desgaste por deslizamiento en el cual un fragmento de roca se quiebra en pedazos mientras se desliza cuesta abajo por la pendiente de una fractura y cae por la ladera de una montaña; este fenómeno también se denomina avalancha de rocas

diaclasa fractura a cuyos lados la roca no se mueve

diagrama H–R gráfico que presenta la luminosidad estelar absoluta comparada con la temperatura y muestra que las estrellas se organizan en grupos basados en estas dos propiedades

dique formación de roca plutónica ígnea que se genera cuando el magma penetra en la roca a través de una fractura

dique artificial pared construida a lo largo de la orilla de un curso de agua para impedir que el agua se desborde e inunde la llanura aledaña

disconformidad tipo de discordancia en la cual las capas sedimentarias ubicadas por encima y debajo de la disconformidad son paralelas

discontinuidad de Mohorovicic límite entre la corteza y el manto que se identifica por un cambio de la velocidad de las ondas sísmicas; también conocido como el Moho

discordancia angular discordancia que se produce cuando el sedimento más nuevo o rocas sedimentarias descansan sobre la superficie erosionada de rocas más antiguas inclinadas o plegadas

discordancia límite entre las capas de sedimento o roca sedimentaria que representa una interrupción en la deposición de sedimento, incluida la interrupción que implica erosión de la capa más antigua

disolución proceso de meteorización química en el cual un mineral o roca se disuelven y forman una solución

disyunción columnar grietas separadas de manera regular que comúnmente se producen en las emanaciones de lava, aumentan a medida que descienden por una superficie y forman columnas de entre cinco y seis caras

dorsal medio-oceánica cadena montañosa submarina que se genera dentro de la corteza oceánica entre placas tectónicas divergentes; es la cordillera más extensa de la Tierra y rodea el planeta como si fuera la costura de una pelota de béisbol

drumlin colina alargada, que comúnmente se da en grupos, formada cuando un glaciar avanza sobre un montículo de till o una acumulación de acarreo estratificado y los transforma

duna montículo o cresta de arena depositada en un lugar por el viento

E

eclipse lunar fenómeno que se produce cuando la Tierra se interpone directamente entre el sol y la luna de modo tal que su sombra se proyecta sobre la luna y la oscurece

eclipse solar fenómeno que se produce cuando la luna se interpone directamente entre la Tierra y el sol de modo tal que su sombra se proyecta sobre la Tierra y bloquea la luz solar

ecosistema comunidad compleja de organismos individuales que interactúan entre sí y con su entorno físico, y funcionan como una unidad ecológica en la naturaleza

edad unidad más breve de tiempo geológico

edad absoluta edad de un fenómeno geológico basada en la cantidad absoluta de años transcurridos desde que se produjo

edad relativa abordaje de los eventos geológicos para determinar cuándo sucedieron a base de la secuencia en la cual ocurrieron y no a base de la cantidad absoluta de años transcurridos desde que se produjeron

efecto Coriolis desviación de las corrientes de aire o agua causada por la rotación de la Tierra

efecto Doppler cambio observado en la frecuencia de la luz o sonido que se produce cuando la fuente de la onda se acerca o aleja del observador

efecto invernadero incremento de la temperatura de la superficie del planeta que se produce cuando los gases que absorben la radiación infrarroja en la atmósfera atrapan energía solar

efecto umbral situación en la cual, frente a una pequeña perturbación, el medio ambiente reacciona con un cambio lento (o, directamente, no reacciona) pero, después de traspasar un umbral, ante cualquier pequeña perturbación adicional, reacciona con un cambio rápido

El Niño patrón climático recurrente con una frecuencia de entre 3 y 7 años en el cual se debilitan los vientos alisios del Océano Pacífico y se acumula agua cálida frente a la costa de América del Sur, generando lluvias inusuales y nevadas intensas en los Andes

elemento sustancia que no puede descomponerse en otras por acción de un medio químico común

elemento nativo mineral que consta de un solo elemento que, en consecuencia, se da en estado nativo o natural (sin combinarse químicamente con otros elementos); son alrededor de 20 los elementos que se dan sólidos en estado nativo

elevación orográfica ascenso del aire que se produce cuando se encuentra con una montaña

eluviación movimiento descendente de iones disueltos y materia suspendida que se produce a través de las capas del suelo por acción del agua que se filtra por ellas

enana blanca etapa de la vida de una estrella en la que el proceso de fusión se detiene y la estrella brilla únicamente por el calor residual de eras pasadas

energía de biomasa energía producida a partir de la combustión de material orgánico

energía geotérmica calor extraído del interior de la Tierra que se usa para generar calor y electricidad en la superficie

energía hidroeléctrica energía que se genera al convertir la potencia del agua en electricidad

energía oscura uno de los componentes de la materia oscura que abarca la poco comprendida fuerza repulsiva que hace acelerar al universo moderno

eón unidad mayor de tiempo geológico, dividida en eras

eón Fanerozoico período que comprende los últimos 541 millones de años de tiempo geológico, incluido el tiempo presente, representado por rocas con abundante evidencia fósil

epicentro punto de la superficie terrestre que se encuentra directamente encima del punto inicial de ruptura (hipocentro) de un terremoto

época unidad de tiempo geológico que se divide en edades

equinoccio cualquiera de los dos momentos del año en los que el sol brilla directamente encima del ecuador y cada zona de la Tierra recibe 12 horas de luz solar y 12 horas de oscuridad

era unidad de tiempo geológico dividida en períodos

erosión remoción de roca meteorizada que se produce cuando el agua, viento, hielo o gravedad transportan el material desgastado a otro lugar

erosión del suelo degradación que se produce cuando el viento, el hielo o el agua transitan una región y arrastran con ellos las capas superiores del suelo formado

erosión vertical erosión descendente que genera una corriente de agua en su lecho y que, por lo general, resulta en un valle en forma de V que se prolonga de manera relativamente recta

escala de dureza escala basada en una serie de 10 minerales bastante comunes, numerados del 1 al 10 (del más blando al más duro), que se usa para medir y expresar la dureza de los minerales

escala de Mercalli escala de intensidad sísmica que expresa la fuerza de un terremoto a base de su poder de destrucción y efecto sobre los edificios y personas; no mide con exactitud la energía liberada durante el temblor

escala de Richter escala de magnitud sísmica que expresa la cantidad de energía liberada; se calcula a base de la amplitud de la onda sísmica interna más grande registrada en un sismógrafo estandarizado, aunque no constituye una medida exacta de la energía de un terremoto

escala de tiempo geológico distribución cronológica del tiempo geológico subdividido en eones, eras, períodos, épocas y edades

escasez de agua situación en la cual la demanda conjunta de todos los usuarios del agua, incluido el medio ambiente, supera su disponibilidad

escorrentía agua que fluye sobre la superficie terrestre por los ríos y arroyos

esfera celeste serie hipotética de esferas concéntricas transparentes que rodean Tierra, según el modelo aristotélico del universo; Aristóteles sugería que el sol, la luna, los planetas y las estrellas estaban contenidos dentro de esferas

esker cresta larga y sinuosa que se forma como depósito del canal de una corriente de agua que fluía dentro de un glaciar en proceso de deshielo o debajo de él

espectro serie ordenada de colores; patrón de longitudes de onda en el que se divide un rayo de luz u otro tipo de radiación electromagnética

espectro de absorción espectro de radiación de ciertas longitudes de onda que son absorbidas cuando la luz atraviesa una sustancia, como una estrella; al ser absorbidas, las longitudes de onda se muestran como líneas oscuras que atraviesan el espectro completo de color

espectro de emisión espectro creado cuando la radiación se emite en forma directa o cuando la radiación absorbida vuelve a emitirse desde su fuente

espectro electromagnético rango completo de radiación electromagnética que abarca desde radiación de grandes longitudes de onda (baja frecuencia) a radiación de longitud de onda muy corta (alta frecuencia)

espícula flujo de gas del borde del sol que surge de la cromosfera y se eleva

espigón barrera angosta o muralla construida en una playa, en sentido perpendicular a la costa, para contener la arena que transportan las corrientes marinas y las olas

estalactita estructura de roca, parecida a una aguja de hielo, que se forma por la precipitación lenta de calcita en el techo de una cueva de roca caliza

estalagmita estructura de roca de forma cónica que se forma por la precipitación lenta de calcita en el piso de una cueva de roca caliza

estepa llanura semiárida vasta, cubierta de pasto, característica de climas secos como los del sudeste de Europa, Asia Central y algunas zonas del centro de América del Norte

estratopausa techo de la estratosfera; límite entre la estratosfera y la mesosfera

estratos capas de nubes horizontales extensas que, por lo general, cubren el cielo y lo oscurecen

estratosfera capa de aire que se extiende sobre la tropopausa por espacio de unos 55 kilómetros hacia arriba

estratovolcán volcán de laderas empinadas, también llamado volcán compuesto, que se forma por acumulación alternada de coladas de lava y capas de depósitos piroclásticos, y se caracteriza por sus erupciones reiteradas

estrella de neutrones estrella pequeña, extremadamente densa, que se crea a partir de los restos de la explosión supernova de una estrella grande y esta compuesta casi totalmente de neutrones comprimidos

estrella de población I estrella relativamente joven formada a partir de material expulsado por una estrella moribunda, más vieja; su composición incluye principalmente hidrógeno y helio primordial, y un pequeño porcentaje de elementos más pesados

estrella de población II estrella vieja, con menor concentración de elementos pesados que una estrella de población I

estriación glaciar conjunto de estrías o surcos paralelos que se forman en la roca a medida que las rocas son arrastradas a lo largo de la base de un glaciar

eutrófico término usado para describir un lago con abundantes nutrientes y un alto nivel de productividad en términos de biomasa

evaluación por los pares proceso en el cual se permite a ciertos científicos realizar una crítica de los diseños experimentales, resultados y conclusiones de otros científicos

exfoliación proceso de meteorización por el que se desprenden distintas capas o placas concéntricas de una masa de roca, como si fuera una cebolla, y se producen fracturas

exosfera capa de gas delgada que rodea la termosfera

expansión del fondo oceánico hipótesis según la cual los segmentos de la corteza oceánica se encuentran en proceso de separación en la dorsal medio-oceánica

experimento procedimiento llevado a cabo para recoger datos o realizar observaciones con el propósito de poner a prueba una hipótesis

extinción masiva evento catastrófico repentino durante el cual se extingue un porcentaje significativo de todas las formas de vida terrestre

extracción cualquier proceso que emplea agua y la devuelve a la Tierra de manera local

GLOSARIO

F

faceta superficie plana que se forma si un cristal se desarrolla libremente en un ambiente no saturado; en condiciones óptimas, el cristal se forma de manera simétrica

falla fractura a lo largo de la cual la roca de un lado se ha movido con relación a la roca del otro lado

falla choque-deslizamiento falla vertical a cuyos lados las rocas se mueven de en dirección horizontal

falla de cabalgamiento tipo de falla inversa, casi horizontal, que presenta una inclinación de 45 grados, o menos, en toda su extensión

falla inversa falla en la cual el bloque superior se eleva respecto del bloque inferior

falla normal falla en la cual el bloque superior desciende respecto del bloque inferior

falla reptante movimiento lento y continuo de la roca sólida que se produce a lo largo de una falla y que es resultado de un esfuerzo ejercido de manera constante durante mucho tiempo

falla transformante falla de choque-deslizamiento que se produce entre dos segmentos de una dorsal medio-oceánica o a lo largo de un límite de placas de choque-deslizamiento

fiordo valle profundo y angosto tallado por un glaciar sobre una costa de altas latitudes que fue posteriormente invadido por aguas marinas a medida que los glaciares se derritieron

fitoplancton plancton de plantas diminutas que en su mayoría viven a pocos metros de la superficie del agua y llevan a cabo la fotosíntesis

flanco cualquiera de los dos lados de un pliegue separado del otro por la superficie axial

flujo forma de desgaste rápido en el cual el suelo suelto o sedimento se desplaza cuesta abajo como fluido espeso y no como masa consolidada; puede producirse de manera lenta o rápida

flujo piroclástico mezcla incandescente de cenizas volcánicas, partículas piroclásticas más grandes, lava menor y gas caliente sumamente destructiva que se forma al colapsar una columna eruptiva y fluye rápidamente sobre la superficie terrestre

flujo plástico movimiento de un glaciar en el cual el hielo fluye como un líquido viscoso

foliación disposición de los minerales en capas paralelas repetidas en el interior de una roca

Fondo Cósmico de Microondas (CMB) radiación de microondas de baja energía que comenzaron a viajar por el espacio cuando el universo tenía apenas 300,000 años y ahora se propagan por todo el espacio del universo

formación de hierro bandeado formación química de roca sedimentaria marina en su mayoría originada entre 2,700 y 1,900 millones de años, que presenta una estructura de bandas de escala centimétrica intercaladas, compuestas de óxido de hierro y chert

formación geológica unidad diferenciada de capas de roca que se da en una región específica

fosa oceánica depresión submarina estrecha y alargada del fondo oceánico que se forma cuando una placa oceánica en proceso de subducción se hunde en el manto y hace que el fondo oceánico se flexione como un trampolín

fósil guía fósil que permite datar las capas que lo rodean porque provenía de un organismo abundantemente preservado en la roca, se encontraba ampliamente extendido en términos geográficos y existió como especie o género apenas durante un tiempo relativamente corto

fotón partícula de luz; paquete o partícula más pequeña de energía electromagnética

fotosfera superficie del sol visible desde la Tierra, también conocida como atmósfera solar

fotosíntesis proceso por el cual, usando el sol como fuente de energía, las células vegetales con clorofila transforman el dióxido de carbono y el agua en azúcares orgánicos; durante el proceso se libera oxígeno

fractura manera en que se quebrantan los minerales, distinto a la fracturación que se produce a lo largo de los planos de clivaje

fracturación hidráulica proceso por el cual se inyectan fluidos en una roca densa para expandir sus fracturas naturales y permitir que el gas y petróleo que normalmente contienen fluya con mayor facilidad; también conocido como *fracking*

frecuencia número de ciclos de onda completos, de cresta a cresta, que pasan por un punto fijo en un segundo

frente de glaciar límite inferior o pie de un glaciar

frente de oclusión frente que se forma cuando una masa de aire frío se desplaza a mayor velocidad que una masa de aire cálido y la encierra contra una segunda masa de aire frío; en estos casos, se producen precipitaciones en ambos límites frontales y se genera una extensa zona de mal tiempo

frente estacionario frente que se genera en el límite entre dos masas de aire estacionario de diferente temperatura

frente polar franja de baja presión, ubicada a aproximadamente 60° de latitud, que se genera por el aumento de aire cálido en la zona de convergencia de vientos polares del este y los vientos del oeste

frente en meteorología, límite entre una masa de aire cálido y una masa de aire fresco

fusión parcial proceso por el cual, al calentarse, una roca de silicato se funde solo parcialmente y forma magma con mayor cantidad de sílice que la roca original

fusión por descompresión derretimiento de la roca causado por disminución de la presión, expansión del volumen y fusión, que comúnmente se produce en la astenosfera

fusión unión de partículas subatómicas que colisionan con suficiente energía como para vencer la fuerza repulsiva entre cargas iguales

G

Gaia término usado en griego para nombrar a la Tierra y con el cual James Lovelock se refirió a nuestro planeta, al que además comparó con un ser viviente por la interconectividad que existe entre todos sus sistemas

galaxia espacio de gran volumen que contiene miles de millones de estrellas, polvo y gas interestelar que se mantienen juntos por efecto de la atracción gravitacional que ejerce un agujero negro central

galaxia elíptica galaxia ovalada, gigante o enana, sin brazos en forma de espiral

galaxia espiral galaxia caracterizada por brazos que proyectan a partir de un centro, como las aspas de un molinillo.

galaxia espiral barrada galaxia espiral con una banda recta de estrellas, gas y polvo que atraviesa el núcleo de lado a lado

galaxia irregular galaxia con estrellas y gas que no presenta ninguna estructura organizada específica

gas natural combinación de hidrocarburos naturales livianos compuestos principalmente de metano, que se usa para calentar las viviendas y cocinar alimentos y poner en funcionamiento grandes plantas de energía eléctrica

gelifracción proceso de meteorización mecánica por el cual el agua que se congela dentro de las grietas de una roca se expande y la quiebra

geocéntrico clase de modelo que ubica a la Tierra como centro del universo

geosfera parte más sólida y rocosa de la Tierra que abarca todo el planeta, desde el centro de su núcleo hasta la corteza externa

gigante roja etapa de la vida de una estrella en la que el hidrógeno se fusiona en alrededor del núcleo de helio y la estrella expande su tamaño, desarrolla temperaturas superficiales más frescas y adopta una apariencia rojiza

giro oceánico corriente circular o elíptica que se genera en el agua o el aire

glaciación período en el que los glaciares alpinos descendieron a los valles de las tierras bajas y los glaciares continentales se crecieron en los continentes de altas latitudes; nombre comúnmente usado para referirse a la glaciación del Pleistoceno

glaciar alpino glaciar angosto que se forma en un pico montañoso o cerca de él y queda confinado en un valle por el terreno circundante

glaciar continental masa de hielo grande y continua que se extiende en todas las direcciones y cubre una vasta porción de la superficie continental de la Tierra

glaciar masa de hielo duradera, originada por acumulación de nieve compactada, que se forma en la tierra y se desplaza en forma descendente o se extiende de manera lateral por acción de su propio peso

glacis continental superficie de sedimento terrígeno ubicada en la base del talud continental, que se fusiona con el fondo oceánico

graben bloque de roca con forma de cuña que al caer entre dos fallas normales forma un valle

gradiente inclinación o pendiente vertical de una corriente de agua que se da a lo largo de una distancia específica

grado metamórfico medida de la intensidad de cambio que se produce en una roca a causa del calor, la presión o ciertas condiciones químicas

gradualismo principio según el cual el cambio geológico ocurre durante un largo período de tiempo como consecuencia de la acumulación lenta o gradual de sucesos casi imperceptibles, como la erosión paulatina de las montañas por efecto del viento o la lluvia; en tiempos más recientes, los científicos que estudian la evolución biológica han usado el término gradualismo para describir una teoría de evolución que propone que las especies cambian en forma gradual, poco a poco

Gran Oxidación aumento repentino de la concentración del oxígeno en la atmósfera terrestre que tuvo lugar hace aproximadamente 2,400 millones de años y por el cual la cantidad de oxígeno pasó de ser de ínfima a significativa, debido a una combinación de procesos biológicos y geoquímicos

gravedad específica peso de una sustancia relativo al peso de igual volumen de agua

guyot monte submarino de cumbre plana que se forma cuando la energía de las olas erosiona la parte superior de una isla de origen volcánico que está en proceso de hundimiento

H

hábito cristalino forma característica de un cristal individual y manera en la que se desarrollan los agregados cristalinos

halo galáctico nube esférica conformada por estrellas, polvo y gas, que rodea el disco galáctico de una galaxia espiral como la Vía Láctea

halón compuesto con contenido de cloro y bromo que asciende a la parte superior de la atmósfera y destruye la capa de ozono

helada cristales de hielo que se forman directamente del vapor cuando el punto de rocío alcanza una temperatura bajo cero

heliocéntrico modelo que ubica al sol en el centro del sistema solar

hidrocarburo cualquier compuesto combustible orgánico conformado exclusivamente por átomos de carbono e hidrógeno

GLOSARIO

hidrógeno metálico líquido tipo de hidrógeno que se da en condiciones extremas de temperatura y presión que obligan a los átomos a unirse de manera tan apretada que los electrones se mueven libremente por los núcleos comprimidos y hacen que el hidrógeno se comporte como conductor de energía

hidrólisis proceso de meteorización química en el cual un mineral reacciona con el agua y forma un nuevo mineral

hidrosfera totalidad del agua que circula por los océanos, continentes, glaciares y atmósfera de la Tierra

hipocentro punto inicial de ruptura de un terremoto, comúnmente localizado debajo de la superficie de la Tierra

hipótesis explicación tentativa de un fenómeno natural que puede ponerse a prueba para determinar si existe evidencia que lo avale

horizonte edáfico capa del suelo que se distingue de las demás porque presenta una apariencia distinta y, tiene propiedades físicas y químicas diferentes

humedad absoluta masa de vapor de agua por unidad de volumen de aire; se expresa en gramos por metro cúbico (g/m^3)

humedad relativa proporción que surge de dividir la cantidad de vapor de agua presente en un volumen determinado de aire por la máxima cantidad de vapor de agua que puede contener ese aire a una temperatura dada

humus material orgánico oscuro del suelo, compuesto por elementos en tal estado de descomposición que no se puede determinar el origen de cada fragmento individual

huracán tormenta tropical característica de América del Norte y el Caribe cuyos vientos superan los 120 kilómetros por hora; en el oeste del Pacífico recibe el nombre de tifón y en el océano Índico se llaman ciclón

I

inconformidad tipo de discordancia por el cual las rocas sedimentarias descansan sobre una superficie de erosión formada en la roca ígnea o metamórfica

inercia tendencia de un objeto a resistir un cambio de movimiento

infracabalgamiento proceso por el cual una placa se desliza por debajo de otra en una zona de subducción o durante una colisión continente-continente

ingeniería aplicación de ideas científicas y matemáticas para diseñar soluciones a distintos problemas

intrusión de agua salada situación que se produce en la costa de los océanos y en los cuerpos de agua salada continentales por la que la extracción excesiva de agua dulce subterránea hace que el agua salada subterránea invada un acuífero

inversión magnética cambio en el campo magnético de la Tierra en el cual polo norte magnético se convierte en el polo sur magnético y viceversa; en promedio, se ha repetido cada 500,000 años a lo largo de los últimos 65 millones de años

inversión fenómeno en el cual las capas de agua de un lago se mezclan durante el colapso de la estratificación térmica de primavera y otoño

ion átomo con carga eléctrica positiva o negativa

isla barrera isla larga, estrecha y baja dispuesta en forma paralela a la costa

isla de calor urbano situación que se da en las zonas urbanas, donde las temperaturas de la ciudad son más cálidas que las de sus alrededores no urbanos como consecuencia de factores asociados específicamente con el entorno urbano

isla oceánica montaña sumergida (monte submarino) que se eleva por encima del nivel del mar

isobara línea que conecta puntos de igual presión atmosférica sobre un mapa meteorológico

isolínea línea que en un mapa conecta zonas que presentan la misma altura, profundidad, temperatura, presión o algún otro rasgo determinado

isostasia concepto de equilibrio entre la gravedad y la flotabilidad por el cual la litosfera flota sobre el manto a distintas alturas

isoterma línea que en un mapa conecta zonas con la misma temperatura promedio

isótopo átomos de un mismo elemento que presentan diferentes números de neutrones

K

kame pequeño montículo o colina de material de acarreo estratificado, que fue depositado en un lugar por una corriente de agua que corre encima, dentro o debajo de un glaciar

L

lago cuerpo de agua dulce de tamaño considerable que ocupa una depresión del terreno aislada del mar

lago de marmita gigante lago que se forma en una depresión originada por un glaciar en retirada cuya agua es la que llena el lago

laguna desértica cuenca seca de una laguna que se forma en el desierto en periodos de lluvia

latitud alta subtropical región de calma y alta presión ubicada aproximadamente entre los 30° de latitud norte y sur de la Tierra, en la que suelen predominar las condiciones secas y son infrecuentes los vientos constantes

ley enunciación que describe o predice un fenómeno natural que, en las condiciones establecidas, no ha podido demostrarse que sea falso

ley Hubble-Lemaître ley de la matemática según la cual la velocidad de una galaxia es proporcional a su distancia de la Tierra; por lo tanto, las galaxias más lejanas son las que viajan a mayor velocidad

licuefacción proceso geológico por el cual el suelo o algún sedimento pierden su resistencia al esfuerzo cortante durante un terremoto y se convierten en líquido

límite convergente tipo de límite hacia el cual se aproximan dos placas tectónicas que, al encontrarse, colisionan

límite de placa fractura o borde que separa dos placas tectónicas

límite divergente tipo de límite en el que las placas tectónicas se separan y distancian y, como consecuencia, se genera permanentemente litosfera nueva; también se denomina punto de expansión o centro de la dorsal

límite transformante límite por el que dos placas tectónicas se ponen en contacto mientras se deslizan horizontalmente en direcciones opuestas

litificación proceso por el cual el sedimento suelto se compacta o comprime y se convierte en roca maciza

litosfera capa rígida y fría, de 100 kilómetros de espesor, que recubre la Tierra e incluye la corteza y el manto superior, y se divide en distintas láminas sólidas llamadas placas

lixiviación disolución química de los iones del suelo por infiltración de agua

loess depósitos de limo acarreado por el viento que se encuentran ligeramente compactados

longitud de onda distancia entre las sucesivas crestas o valles de onda

luna gibosa fase lunar, tanto cuarto creciente como menguante, durante la cual la mayor parte de la luna se ve iluminada y apenas una porción se encuentra en penumbra

luna llena fase lunar en la cual la luna se ve redonda porque su cara visible está enteramente expuesta a la luz del sol

luna nueva fase lunar en la que la luna se ve oscura porque al moverse oculta a la Tierra su cara iluminada por el sol

LL

llanura abisal extensión de terreno llano de las zonas del fondo marino ubicadas entre la dorsal medio-oceánica y el glacis continental

llanura de inundación zona de un valle fluvial aledaña al canal formada por el sedimento que queda depositado sobre su superficie durante las inundaciones y que es sucesivamente cubierta por el agua de cada inundación

M

magma roca derretida que resulta de la fusión de cualquier roca del subsuelo; al enfriarse da origen a la roca ígnea

magnetómetro instrumento que mide la fuerza y, en algunos casos, la dirección de un campo magnético

magnitud de momento sísmico escala sísmica en la cual el área superficial del movimiento de una falla se multiplica por un valor aproximado de la resistencia de la roca quebrada; la escala de magnitud de momento refleja la cantidad de energía liberada

mancha solar región comparativamente fresca y oscura de la superficie solar que se genera por perturbación magnética

manto capa rocosa de la Tierra, situado entre el núcleo y la corteza, cuyas condiciones de temperatura y presión son cambiantes y generan algunas regiones duras y firmes y otras blandas y elásticas

mar lunar extensión plana, seca e inhóspita de roca volcánica de la superficie lunar que los científicos originalmente creían que era un mar

marea ascenso y descenso cíclico del agua del océano causado por la fuerza gravitacional de la luna y, en menor medida, del sol

marea de tormenta inundaciones y elevación anormal del nivel del mar en la costa que se producen por la combinación de fuertes vientos en la tierra con la baja presión atmosférica de una tormenta que eleva la superficie del mar por varios metros

marea muerta marea relativamente pequeña que se produce cuando la luna se encuentra 90° desplazada de su plano de alineación con el sol y la Tierra

marea viva marea muy alta y muy baja que se produce cuando el sol, la luna y la Tierra están alineados de manera tal que los campos gravitatorios del sol y la luna se combinan para generar un fuerte alargamiento de la marea

margen andino ver *arco continental*

margen continental activo margen localizado en un límite de placa convergente en el cual la litósfera se hunde debajo del continente en una zona de subducción

margen continental pasivo margen en el que la corteza continental se une firmemente a la corteza oceánica y en el cual se genera poca actividad tectónica

marmita de gigante pequeña depresión formada por un bloque de hielo estancado; en algunos casos, la cavidad se llena de agua y se convierte en un lago

masa de aire cuerpo de aire de tamaño considerable que mantiene aproximadamente la misma temperatura y humedad a una altitud dada

masa solar unidad estándar usada para expresar la masa de una estrella en comparación con la masa del sol; 1 masa solar equivale a la masa total del sol

materia oscura materia misteriosa invisible que puede abarcar hasta el 96 por ciento de la masa del universo

material pyroclástico fragmento volcánico piroclástico vítreo de 4 a 32 milímetros de tamaño

material parental material original con el que se forma un suelo

meandro serie de curvas sinuosas que define el curso de una corriente de agua

GLOSARIO

meandro abandonado lago con forma de herradura que se genera en el lugar donde la curva de un meandro queda separada de la corriente de agua porque los extremos del meandro están bloqueados con sedimento

mecanismo de retroalimentación reacción por la cual una pequeña perturbación inicial ocurrida en un componente afecta a otro componente de los sistemas de la Tierra y amplifica el efecto original, lo que tiene un impacto aún mayor en el sistema y conduce a un efecto todavía mayor

mecanismo de retroalimentación negativa mecanismo por el cual un sistema conserva el equilibrio; a medida que un factor aumenta, genera efectos que luego harán que el factor original disminuya

mecanismo de retroalimentación positiva mecanismo por el cual un sistema conserva el equilibrio; a medida que un factor aumenta, genera efectos que luego harán que el factor original aumente

mediterráneo variedad de clima de latitud media caracterizado por tener veranos secos, inviernos lluviosos y temperaturas moderadas

mena sustancia natural sólida de la cual pueden extraerse metales o minerales de manera rentable

menguante que mengua o reduce su tamaño; período de 14 a 15 días, posterior a la luna llena y anterior a la luna nueva, durante el cual la porción visible de la luna disminuye diariamente

mesa montaña de techo plano, con forma de mesa, más pequeña que una meseta y más grande que un butte

meseta extensión de terreno elevado y relativamente plano

meseta de lava meseta amplia, formada por la acumulación de múltiples corrientes de lava individuales ocurridas en un corto período de tiempo geológico, que cubre miles de kilómetros cuadrados

mesopausa techo de la mesosfera; límite entre la mesosfera y la termosfera

mesosfera capa de aire ubicada encima de la estratopausa, que se extiende hacia arriba desde aproximadamente 55 kilómetros a aproximadamente 80 kilómetros sobre la superficie de la Tierra

metamorfismo proceso por el cambian la textura y el contenido mineral de las rocas como consecuencia de variaciones en sus condiciones químicas, temperatura y presión

meteorito meteoro que no se volatiliza completamente y al caer impacta sobre la superficie terrestre

meteorización descomposición de la roca de la superficie terrestre causada por procesos químicos y físicos

meteorización mecánica desintegración de la roca en fragmentos pequeños producida por procesos físicos que no alteran su composición química

meteorización química descomposición que sufre la roca al reaccionar con el aire, agua u otros agentes del medio ambiente que alteran su composición química y contenido mineral

meteoro meteorito que al caer ingresa en la atmósfera terrestre y resplandece al volatilizarse; coloquialmente suele llamarse estrella fugaz

meteoroide fragmento de un asteroide o cometa que describe una órbita irregular mientras se desplaza por el interior del sistema solar

método de la raya procedimiento que consiste en trazar con un trozo de mineral una línea sobre una placa de porcelana de tonalidad mate para observar el color del polvo que deja el mineral y analizar sus características

mineral sustancia natural inorgánica sólida, de composición química definida y estructura cristalina

modelo representación física, diagramática, matemática o de otro tipo que sirve para describir o explicar cómo funciona un sistema

modelo de tres celdas modelo de los patrones globales de circulación del viento que muestra tres celdas de convección en cada hemisferio, bordeadas por bandas alternadas de alta presión y baja presión

monte submarino montaña sumergida en el mar, comúnmente de origen volcánico, que se eleva al menos 1 kilómetro por encima del fondo oceánico que la rodea

monzón viento estacional y sistema meteorológico generado por el calentamiento y enfriamiento irregular de la tierra y el mar adyacente que, en general, sopla desde el mar hacia el continente durante el verano y desde el continente hacia el mar durante el invierno

morrena cúmulo de rocas y sedimentos derivados de la deposición de un glaciar; el sedimento depositado puede ser limo pulverulento o cantos de gran tamaño

movimiento orbital circular movimiento circular del agua que se genera cuando una ola superficial se desplaza horizontalmente

movimiento retrógrado movimiento aparente de los planetas, los cuales, vistos desde la Tierra, parecen retroceder (hacia el oeste) con respecto a las estrellas antes de retomar su movimiento original (hacia el este)

N

nebulosa nube de polvo y gas interestelar

nebulosa planetaria nebulosa que se genera cuando una estrella de tamaño similar al sol estalla y lanza una envoltura de gas al espacio

nivel de base del oleaje profundidad por debajo de la cual la energía superficial de las olas no llega a penetrar; aproximadamente equivale a la mitad de una longitud de onda

nivel de base nivel más profundo hasta donde una masa de agua puede erosionar su lecho; el máximo nivel de base, comúnmente, es el nivel del mar

nucleosíntesis primordial formación de núcleos de helio que se produjo por fusión nuclear durante los primeros 8.5 minutos de vida del universo

nutación variación periódica en la inclinación del eje de un objeto que rota

O

oceánico templado variedad de clima de latitud media caracterizado por tener inviernos húmedos pero relativamente templados y veranos frescos, con escasa diferencia térmica entre las estaciones

oligotrófico término con el que se describe un lago que contiene pocos nutrientes y sustenta a relativamente pocos organismos vivos

onda primaria (onda P) onda sísmica que viaja más rápido que otras y, por eso, es la primera en llegar al observador; se forma por la alternancia de compresión y expansión de la roca en paralelo a la dirección en que se propaga la onda

onda secundaria (onda S) onda sísmica que viaja más lentamente que una onda primaria y, por eso, es la segunda en llegar al observador; a veces llamada onda de cizalla por el movimiento de corte que experimentan las rocas dado que las ondas vibran en sentido perpendicular a la dirección en la que se propagan

onda sísmica interna onda que se propaga por el interior de la Tierra transportando energía desde el foco sísmico a la superficie

onda sísmica onda elástica que se propaga por la roca como consecuencia de un terremoto o explosión

onda superficial onda sísmica que se genera en el epicentro del terremoto y se propaga por la superficie terrestre o a lo largo de alguno de los límites entre las capas internas de la Tierra

optimización proceso por el cual se da prioridad a determinados criterios y se eligen los sacrificios apropiados para hacer que la solución a un problema resulte efectiva

orden del suelo clasificación jerárquica más alta de los tipos de suelo, establecida por el Servicio de Conservación de Recursos Naturales de los EE. UU., que reconoce doce órdenes

orogénesis proceso por el cual se originan las montañas; todos los procesos tectónicos relacionados con las montañas

oxidación proceso de meteorización química por el cual un mineral se descompone al reaccionar con el oxígeno

P

paralaje cambio aparente de la posición de un objeto debido al cambio de posición del observador del objeto

parsec unidad de distancia usada en astronomía, equivalente a alrededor de 3.26 años luz

partículas término del campo de la contaminación que alude a los fragmentos de materia sólida en suspensión de mayor tamaño que una molécula, como el polvo u hollín

pavimento desértico superficie continua de grava, o clastos, muy compacta que se forma cuando el viento erosiona partículas más pequeñas, como el limo y la arena

pedimento superficie de erosión ancha y levemente inclinada que se forma a lo largo de la parte frontal de las montañas desérticas en su ascenso desde una bajada y que en algunas zonas aisladas comúnmente se halla cubierta de grava de unos pocos metros de espesor

penumbra franja ancha fuera de la umbra en la que solo una porción del sol queda oculta durante un eclipse solar

período unidad de tiempo geológico que se divide en épocas

período interglaciar lapso de tiempo que se da entre las glaciaciones y se caracteriza por ser relativamente templado y no presentar formación de hielo

permeabilidad capacidad de la roca o el suelo de permitir que un líquido los penetre

petróleo mezcla compleja de diferentes compuestos orgánicos que pueden refinarse para producir gasolina, plástico, nailon y otros materiales

pH escala logarítmica basada en la concentración de iones de hidrógeno que representa la acidez de una solución; las soluciones con pH 7 son neutras; los valores inferiores a 7 representan soluciones ácidas mientras que los superiores a 7 representan soluciones básicas

pilar tectónico en una zona de fallas, bloque de roca ubicado entre dos grabenes que parece haberse desplazado hacia arriba en relación con ellos

placa tectónica uno de los siete segmentos mayores u ocho segmentos menores de la litosfera que flotan encima de un manto plástico y relativamente caliente de roca y se mueven en sentido horizontal

plancton organismos pequeños de agua salada y dulce, encargados de la mayor parte de la fotosíntesis y el consumo de nutrientes en el océano, que constituyen la base de la cadena alimentaria de los animales acuáticos

planeta cuerpo celeste que gira alrededor del sol, tiene suficiente masa como para que su propia gravedad le permita tener una forma bastante redonda y ha limpiado su órbita de objetos más pequeños

planeta enano objeto que gira alrededor del sol y tiene la suficiente masa como para ser redondo o casi redondo, pero no tanta como para limpiar su órbita de escombros espaciales

planeta joviano nombre que recibe cualquiera de los cuatro planetas exteriores con núcleos metálicos o rocosos relativamente pequeños rodeados por atmósferas líquidas y gaseosas

planeta menor objeto celeste que orbita alrededor del sol y no califica como planeta, planeta enano o cometa

GLOSARIO 865

GLOSARIO

planeta terrestre nombre que recibe cualquiera de los cuatro planetas internos compuestos principalmente de metales y silicatos

plataforma continental porción poco profunda del fondo oceánico, que se extiende con una pendiente suave desde la costa hasta la parte superior del talud continental

plataforma de hielo masa gruesa de hielo que flota en la superficie del océano pero que está conectada con la tierra a través de un glaciar, del cual se alimenta

pluma de contaminación zona tridimensional de desarrollo lento dentro de un acuífero o cuerpo de agua superficial afectada por un contaminante en dispersión

pluma del manto columna de roca blanda y caliente que asciende por el interior del manto

plutón masa de roca ígnea intrusiva

polaridad magnética normal orientación magnética coincidente con la del campo magnético terrestre actual

porosidad medida del número y tamaño de los huecos de aire entre las partículas que conforman un determinado suelo

Precámbrico término con el que se denomina al tiempo geológico anterior a la era Paleozoica, que abarca aproximadamente los primeros 4,000 millones de años, o alrededor del 90 por ciento, de la historia de la Tierra; el término también se emplea para referirse a todas las rocas que se formaron durante ese tiempo

precesión movimiento lento del eje de un cuerpo que mientras rota gira también alrededor de otro eje

precipitación formación de un sólido a partir de sustancias previamente disueltas en un líquido

precipitación ácida lluvia, nieve, niebla o bruma que se volvió ácida luego de reaccionar con los contaminantes del aire

premonitor terremoto menor que precede a un sismo de grandes proporciones por algunos segundos o semanas

presión barométrica presión de la atmósfera en un lugar y momento dado

principio de horizontalidad original principio según el cual la mayor parte del sedimento se deposita en lechos casi horizontales y, por lo tanto, la mayoría de las rocas sedimentarias al formarse presentan capas de sedimento casi horizontales

principio de inclusión principio según el cual primero debe existir una roca como pieza completa para que de ella luego puedan desprenderse fragmentos que se incorporen a otra pieza completa de roca

principio de relaciones de corte principio obvio según el cual una roca o rasgo primero debe existir para que luego algo pueda sucederle; por lo tanto, si un dique de basalto atraviesa una capa de arenisca, el basalto es más joven

principio de sucesión faunística principio según el cual las especies se suceden en el tiempo conforme a un orden definido de modo que la edad relativa de las rocas sedimentarias fosilíferas puede determinarse a base de su contenido fósil

principio de superposición principio según el cual, en las capas de roca sedimentaria o de sedimento inalterado, la edad disminuye progresivamente de abajo hacia arriba de modo que las capas más nuevas siempre se acumulan sobre las más viejas

protuberancia solar chorro de gas rojizo, similar a una llama, que sobresale de la corona del sol

púlsar estrella de neutrones que emite pulsos de ondas de radio

punto caliente roca caliente de la parte superior del manto que se ubica dentro de una pluma y se asocia con un centro volcánico que se forma en la litosfera

punto de rocío temperatura a la cual la humedad relativa del aire alcanza el ciento por ciento y el aire se satura

Q

quimiosíntesis proceso en que las bacterias producen energía a partir del ácido sulfhídrico u otros compuestos inorgánicos y, en consecuencia, no depende de la fotosíntesis

R

radiación electromagnética radiación que consiste en un campo magnético y eléctrico oscilante que incluye ondas de radio, luz infrarroja, luz visible, y rayos ultravioletas, rayos X y rayos gamma

ráfaga de rayos gamma brote de radiación de alta energía proveniente del espacio; las ráfagas más prolongadas se generan como consecuencia de la extinción de una estrella gigante mientras que las ráfagas más breves se producen cuando una estrella de neutrones es atraída hacia un agujero negro o cuando dos estrellas de neutrones colisionan y forman un agujero negro

rambla lecho fluvial que permanece seco durante la mayor parte del año

recurso energético sustancia o fenómeno que se da en la naturaleza y puede usarse para generar calor, luz y/o electricidad de manera productiva

recurso mineral cualquier tipo de sólido natural inorgánico útil para los seres humanos, que puede extraerse de manera razonable para que obtener un beneficio económico en el presente o en el futuro

recurso no renovable cualquier recurso que no puede volver a generarse mediante procesos naturales con la misma rapidez con la que se lo consume

recurso renovable cualquier sustancia o fenómeno útil que puede volver a generarse mediante procesos naturales más rápido de lo que se lo consume o emplea

registro de las rocas rocas que actualmente existen en la Tierra y permiten constatar su historia

regolito capa delgada de material meteorizado no consolidado que cubre el lecho de roca

remediación química tratamiento de una zona contaminada que consiste en inyectar un compuesto químico que reacciona con el contaminante para generar productos inocuos

reptación forma de desgaste en la cual un material suelto se desplaza muy lentamente cuesta abajo, por lo general, a razón de 1 centímetro por año y sobre la tierra con vegetación

reserva mineral término con el cual se describe la cantidad de mena alojada en el suelo; puede referirse a las reservas locales, nacionales o mundiales

residuos hojas, ramas y otros tipos de material vegetal o animal que han sido depositados sobre la superficie del suelo pero que aún no se han descompuesto

resolución capacidad de un instrumento para detectar o medir hasta los más mínimos detalles, como la que ofrece un telescopio óptico al definir individualmente los cráteres de la luna

restinga pequeña punta o lengua de arena o grava que se prolonga desde la costa hasta un cuerpo de agua en el que se adentra

restricción factor que limita las soluciones que podrían desarrollarse para resolver un problema

revolución movimiento alrededor de un punto central; desplazamiento en órbita

roca carbonática ejemplo de roca bioclástica con cantidades elevadas de calcita, dolomita o ambas

roca ígnea extrusiva roca cristalina, también llamada roca volcánica, que se forma cuando el magma se solidifica sobre la corteza terrestre

roca ígnea intrusiva roca de grano grueso a grano medio, también conocida como roca plutónica, que se forma cuando el magma se solidifica dentro de la corteza terrestre, sin salir a la superficie

roca ígnea roca formada a partir de la solidificación de roca fundida

roca metamórfica tipo de roca que se forma cuando las rocas ígneas, sedimentarias u otros tipos de roca metamórficas se recristalizan en respuesta a cambios químicos, de temperatura o de presión

roca piroclástica roca compuesta de magma líquido y fragmentos de roca sólida que fueron expulsados de manera explosiva por un conducto volcánico

roca plutónica ver *roca ígnea intrusiva*

roca sedimentaria roca que se forma cuando el sedimento se compacta y comprime por efecto del proceso litificación

roca sedimentaria bioclástica roca compuesta de conchas marinas partidas y fragmentos de hueso

roca sedimentaria clástica roca compuesta de fragmentos de roca meteorizada

roca sedimentaria orgánica roca formada a partir de los restos de organismos vivos

roca sedimentaria química tipo de roca que se forma por la precipitación de minerales

roca volcánica ver *roca ígnea extrusiva*

rocío humedad de la atmósfera que se condensa sobre los objetos, comúnmente durante la noche, cuando el suelo y la superficie foliar se enfrían más que el aire a su alrededor

rompiente zona caótica y turbulenta de la costa en la que desarman o terminan las olas

rotación movimiento que consiste en girar en torno de un eje

S

salinidad cantidad total de sales disueltas en el agua salada expresada como porcentaje

salinización proceso por el cual las sales se acumulan en el suelo, comúnmente a causa de la irrigación, y disminuyen su fertilidad

saltación salto asimétrico de las partículas sedimentarias que son expulsadas de un lecho por el impacto de otra partícula y son transportadas una distancia corta por el viento o el agua antes de volver a reposar en la superficie del lecho; la mayoría de los granos de arena de los ambientes desérticos se mueven por saltación eólica

saturación situación que se da cuando una solución contiene todos los solutos que pueden disolverse a una temperatura determinada; el aire saturado contiene tanto vapor de agua como puede absorber a una temperatura determinada y alcanza el 100% de humedad

secuencia principal banda del diagrama Hertzsprung–Russell en la cual se encuentran la mayoría de las estrellas alimentadas por fusión de hidrógeno

sedimento de lavado sedimento depositado por las corrientes de agua que fluyen desde el frente de un glaciar en proceso de deshielo

sedimento pelágico sedimento oceánico terroso compuesto por los esqueletos de microorganismos marinos

sedimento terrígeno sedimento compuesto de arena, limo y arcilla que es producto de la erosión de los continentes

segmentación continental proceso por el cual un continente se separa en un límite de placas divergente

Segunda Gran Oxidación aumento abrupto de la concentración del oxígeno en la atmósfera terrestre que por segunda vez tuvo lugar hace aproximadamente 600 millones de años en respuesta a la interacción de distintos procesos químicos, físicos y biológicos

selva tropical bosque que crece en un clima tropical, con abundantes lluvias a lo largo de todo el año, y que se caracteriza por la presencia de árboles altos que generan un techo de follaje muy tupido

GLOSARIO

silicato mineral cuya composición química incluye el silicio y el oxígeno, y cuyas estructuras cristalinas contienen tetraedros de silicato; término con el que también se nombran todas las rocas compuestas principalmente de minerales de silicato

sill roca ígnea, dispuesta como una lámina de manera paralela al grano o a las capas de roca nativa, que se forma cuando el magma penetra entre las capas rocosas

sinclinal pliegue de una roca que se arquea hacia abajo

sismógrafo instrumento que registra las ondas sísmicas

sistema combinación de partes que al interactuar forman un todo complejo

sistema de desviación conducto, canal o cualquier otro tipo de infraestructura que transporta agua desde su ubicación y cauce natural en el ciclo hidrológico a una ubicación y cauce distintos para satisfacer las necesidades humanas

smog contaminación visible del aire, caracterizada por su color pardusco, que se forma por reacciones químicas relacionadas con la combustión incompleta de gasolina, dióxido de nitrógeno, oxígeno, ozono y luz solar

sobresaturación situación que se da cuando una solución contiene más solutos de los que contendría una solución saturada a una temperatura determinada; el aire sobresaturado supera el 100 por ciento de humedad

solsticio cualquiera de los dos momentos del año en los que el sol se encuentra más lejos del ecuador y brilla directamente encima de los 23.5° de latitud norte o sur; el día más largo del año en el hemisferio que recibe directamente la luz solar y el más corto en el hemisferio opuesto

sonar sistema que emite pulsos sonoros y mide su eco de retorno para detectar la presencia de objetos bajo del agua

subártico extremo norte de las zonas de clima continental templado, ubicado apenas al sur del ártico

subducción proceso por el cual dos placas litosféricas de diferente densidad convergen y la más densa se hunde en el manto y desliza por debajo de la otra

subenfriamiento situación en la que las gotas de agua del aire no se congelan aún cuando el aire alcanza temperaturas bajo cero

subsidencia hundimiento o asentamiento irreversible de la superficie terrestre

subtropical húmedo variedad de clima de latitud media caracterizado por tener veranos calurosos y húmedos, inviernos más frescos, y lluvias a lo largo de todo el año

suelo conjunto de las capas superiores del regolito que sustentan el crecimiento de las plantas; algunos científicos e ingenieros utilizan indistintamente los términos suelo y regolito

suelo franco el tipo de suelo más fértil, caracterizado por contener una mezcla particularmente rica en arena y limo, y cantidades abundantes de materia orgánica

suelo residual suelo formado a partir de la meteorización del lecho de roca que tiene debajo

suelo superficial capa fértil, de tonalidad oscura de la parte superior del suelo en la que se combinan los horizontes O y A del suelo

suelo transportado suelo que se forma por la meteorización del regolito que es trasladado a otro lugar, en el cual se deposita

sulfuro mineral cuya composición química se basa en la unión del anión de sulfuro con un catión metálico

sumidero depresión que se forma en la superficie terrestre por el colapso del techo de una cueva o por disolución de roca superficial, como la roca caliza

supercalentamiento calentamiento de una sustancia por encima de la fase de transición (de estado sólido a líquido o de líquido a gaseoso) sin que se produzca la transición; la alta presión puede evitar que la roca sólida se derrita aún cuando se encuentra por encima de la temperatura de fusión

supercontinente continente, como la Pangea, que está compuesto por casi la totalidad de la corteza continental de la Tierra, unida en una sola masa de tierra

superficie axial superficie imaginaria que conecta todos los puntos de máxima curvatura en un pliegue

supernova etapa de la vida de una estrella gigante durante la cual su brillo aumenta enormemente a causa de una explosión catastrófica en la que expulsa la mayoría de su masa

surgencia corriente ascendente que transporta agua fría y nutrientes desde la profundidad del mar hasta su superficie

sustancia química persistente, bioacumulable y tóxica (PBT) toxina no biodegradable que se libera y acumula en el medio ambiente, bioacumulable y persistente

T

taiga bioma de coníferas capaces de sobrevivir los inviernos extremos del subártico

talud continental pendiente submarina relativamente pronunciada que se extiende desde la plataforma continental hasta el glacis continental

tectónica de placas teoría de la tectónica global, según la cual, la litosfera se segmenta en diversas placas que flotan o se deslizan respectivamente en la astenosfera plástica

tectónica sin subducción actividad tectónica regida por la elevación y el hundimiento del manto y la corteza terrestre, en contraposición al movimiento horizontal de las placas comúnmente asociado con la actividad tectónica de la Tierra

telescopio reflector variedad de telescopio óptico que emplea uno o varios espejos para recoger y enfocar una imagen, la cual luego se refleja en el ocular mediante otro espejo

telescopio refractor variedad de telescopio que emplea dos lentes: una para recoger luz de un objeto distante y otra para ampliar la imagen

temperatura medida de la energía cinética promedio de una sustancia, proporcional a la velocidad promedio de los átomos y moléculas de una muestra

teoría explicación de un fenómeno, generalmente aceptada como válida, que se basa en la repetición de experimentos independientes

termoclina capa de agua que se forma entre las aguas cálidas de la superficie y las aguas frías del fondo de un lago de una región templada durante el verano

termosfera región extremadamente alta y difusa de la atmósfera que se extiende sobre la mesosfera desde aproximadamente los 80 kilómetros hacia arriba

terremoto movimiento o temblor repentino de la Tierra causado por la liberación abrupta de energía elástica acumulada en las rocas

terreno adicionado por acreción masa continental delimitada por fallas que se origina en un arco insular o un microcontinente, que más tarde se une a un continente por acreción

tetraedro de silicato estructura piramidal que forma un ion de silicio al unirse a cuatro iones de oxígeno y constituye el componente fundamental de todos los minerales de silicato

textura del suelo proporción de arena, limo y arcilla que presenta un determinado suelo

textura tamaño, forma y disposición de los granos del mineral, o cristales, de una roca

tiempo condiciones de la atmósfera en un momento y lugar dados, determinadas por la temperatura, el viento, la nubosidad, la humedad y la precipitación que se registran

tifón tormenta tropical que se produce en el oeste del Pacífico cuyos vientos superan los 120 kilómetros por hora; en América del Norte y el Caribe recibe el nombre de huracán y en el océano Índico se llama ciclón

till acarreo glaciar depositado directamente por el hielo de un glaciar

toba de ceniza roca volcánica que se forma cuando el flujo piroclástico se solidifica

topografía descripción de las características físicas de los accidentes geográficos de una zona

tornado tormenta pequeña, intensa, de corta duración y forma de embudo, que se proyecta desde la base de una nube cumulonimbos

transformación cambio físico o químico que afecta a un componente del suelo

translocación movimiento vertical (usualmente hacia abajo) de los componentes físicos o químicos del suelo de un horizonte a otro

transpiración proceso por el cual el agua se evapora directamente de la superficie de las plantas y se transfiere a la atmósfera

tropical de sabana variedad de clima tropical con estaciones húmedas y secas bien diferenciadas, pero con lluvias anuales relativamente escasas, caracterizado por el desarrollo de extensos pastizales y animales de pasteoreo

tropical monzónico variedad de clima tropical con estaciones húmedas y secas bien diferenciadas, caracterizado por abundantes lluvias durante los meses húmedos y una estación seca relativamente corta

trópico de Cáncer paralelo ubicado a 23.5° de latitud, al norte del ecuador; en el hemisferio norte, durante el solsticio de verano, la luz solar impacta sobre la Tierra directamente sobre esta latitud justo al mediodía

trópico de Capricornio paralelo ubicado a 23.5° de latitud, al sur del ecuador; en el hemisferio sur, durante el solsticio de verano, la luz solar impacta sobre la Tierra directamente sobre esta latitud justo al mediodía

tropopausa parte superior de la troposfera; límite entre la troposfera y la estratosfera

troposfera capa de aire más cercana a la superficie de la Tierra que se extiende unos 17 kilómetros hacia arriba

tsunami ola marina grande y destructiva generada por un volcán o terremoto submarino, que, a diferencia de las olas oceánicas, no tiene relación con el viento ni las mareas astronómicas

tundra bioma característico del Ártico y las regiones alpinas altas en el que predominan los pastos cortos, el musgo, las flores y algunos arbustos pequeños

U

umbra banda angosta de la sombra lunar en la cual el sol se encuentra totalmente bloqueado durante un eclipse

unidad astronómica distancia promedio entre la Tierra y el sol; unidad de medida comúnmente usada como estándar para expresar las distancias en el sistema solar

uniformitarismo principio según el cual los procesos geológicos actuales operan de la misma manera que lo hacían en el pasado

universo abierto modelo en el cual la fuerza gravitacional del universo no es suficiente para detener la expansión en curso, por lo cual, las galaxias continuarán distanciándose para siempre

universo cerrado modelo en el cual la fuerza gravitacional del universo es tan intensa que todas las galaxias finalmente se desaceleran, revierten su dirección y vuelven al centro para formar otro punto de densidad infinita

universo en expansión acelerada modelo en el que una fuerza no identificada domina la gravedad y hace que la expansión del universo se acelere

GLOSARIO

V

valle depresión o parte más baja de una onda

valle colgante pequeño valle glaciar ubicado a mucha más altura que la base del valle principal

velocidad de escape velocidad que debe alcanzar un objeto para escapar del campo gravitacional de un planeta u otro objeto del espacio

vesícula orificio que se forma en una roca de lava cuando la lava se solidifica antes de que puedan escapar de ella las burbujas de gas o agua

vida media tiempo que tardan en desintegrarse la mitad de los átomos de una muestra de isótopo radiactivo

viento solar corriente de partículas cargadas que son irradiadas desde el sol a gran velocidad

vientos alisios vientos que soplan de manera constante hacia el ecuador desde el noreste en el hemisferio norte y desde el sudeste en el hemisferio sur, entre los 5° y los 30° de latitud norte y sur

vientos del oeste vientos que soplan de manera constante hacia los polos desde el sudoeste en el hemisferio norte y desde el noroeste en el hemisferio sur, entre los 30° y 60° de latitud norte y sur

vientos polares del este vientos polares de superficie persistentes, característicos del hemisferio norte, que soplan de este a oeste

volcán conducto que se abre en la corteza terrestre por el que surgen los fragmentos de roca y la lava que determinan su forma de colina o montaña

volcán en escudo montaña volcánica de gran tamaño y pendiente suave que se formó por acumulación de sucesivas coladas de magma basáltico

Z

zona convectiva zona interna del sol ubicada encima de la zona radiactiva, donde la energía se transmite principalmente por convección

zona de Benioff zona tridimensional ubicada dentro y a lo largo de la porción superior de una placa de subducción en la cual se agrupan focos de terremotos; se forma a partir de la liberación de tensión que se produce cuando la placa descendente de la zona de subducción raspa la placa que se desliza por encima de ella

zona radiativa zona interna del sol que rodea al núcleo en la que energía se irradia por absorción y emisión

zooplancton plancton diminutos que en su mayoría viven a pocos metros de la superficie del agua y se alimentan, parecido a los animales, de fitoplancton

INDEX

An italicized *f* indicates a figure.
An italicized *t* indicates a table.
Boldface type indicates a figure or table in the online edition.

A

Aa lava, 276, 277*f*
Abrasion
 deserts, wind abrasion in, 428–429
 by ocean waves, 517
 of rocks, 361, 361*f*–363*f*, **361*f***, 428–429
 by waves, 517
Absolute age, 134, 140, 144–147, 155
Absolute brightness (star), 780, 790
Absolute humidity, 641, 668
Abyssal plains, 340, 349
Accelerating universe, 811, 813
Accreted terranes, 337, 337*f*
Acid, minerals dissolving in, 79
Acid mines, 87, 87*f*
Acid rain, 365, **365*f***, 594–595, 594*f*, 601
Acidification, of oceans, 500, 502, 503*f*, 698
Acids, pH, 365, 502, 503*f*, 594–595, 595*f*, 698
Active continental margins, 346
Additions (soil), 447–448
Adiabatic cooling, 558, 643, 665, 668
Adiabatic process, 643
Adiabatic temperature changes, 643
Adirondack Mountains (New York), 151
Advection, 620, 621*f*, 622
Aerosols, as pollutants, 597–598
African plate, 206*f*, 222
Agate, 73
Age spectrum, 146
Ages, 151
Agriculture
 climate change and, 698
 cover crops, 463
 Dust Bowl (United States), 459–460, 461, 461*f*
 erosion of cropland, 442, **442*f***
 irrigation, 461, 463, 540, 543*f*, 546, 547, 558, 568, 695
 monoculture, 463
 optimizing food production, 469
 perennial crops, 440, 441*f*, 464
 smallholder farms, 456*f*–457*f*
 soil depletion and degradation, 464, 466
 soil quality, 459
 soil types, 450–451, 451*f*, 452*t*
 sustainable food systems, 440, 441*f*
 terracing, 463, 463*f*
Air, 41
 convection–convergence, 644, 659, 668
 cooling and condensation, 642–643, 668
 frontal wedging, 644, 668
 heating of, 668
 humidity, 641, 641*f*
 orthographic lifting, 644, 644*f*
 rising air, 643, 649, 650, 668, 702
 sinking air, 649, 650, 668, 702

Air mass, weather, 650, 651, **651*f***
Air pollution, 591–599, 601
 acid mines and heavy metals, 87, 87*f*
 acid rain, 365, **365*f***, 594–595, 594*f*, 601
 chlorofluorocarbons (CFCs), 598–599, 598*f*, 601, 692
 fossil fuel burning, 592–593
 global climate change and, 619
 greenhouse gases, 592, 618, 686, 692, 702
 ozone hole, 598–599, 599*f*, 601
 particulates and aerosols, 597–598
 reducing, 600
 smog, 595–596, 596*f*
 sources and types, 591–592, 593*f*
 toxic volatile substances, 561, 563, 597, 601
 in United States, 591–592, 592*f*
Alam, Munazza, 770, 771*f*
Albedo, 615–616, 615*f*, 699, 702
Aleutian Islands (Alaska and Russia), 209, 217, 244*t*, 304, 336
Algae, 608
Algal blooms, 481, 608
Ali, Saleem, 160, 161*f*
Alkali feldspar, in common rock types, 80*t*
Alluvial fans, 378, 378*f*, 423, 423*f*, 745
Alluvium, 378
Alnilam (star), 714*t*
Alnitak (star), 714*t*
Alpine glaciers, 399–400, 400*f*, 403, 403*f*, 410, 414, 416, 433
Alps (mountains in Europe), 138, 220, 315, 317, 416, 472*f*–473*f*
Altai Mountains (Asia), 420
Altitude, average temperature and, 624–625
Altostratus clouds, 646*f*, 647
Altyn Emel National Park (Kazakhstan), 361*f*
Aluminum
 bauxite, 80*t*, 167, 167*f*
 in Earth's crust, 73, 73*t*, 164*t*
Alvarez, Luis, 15, 15*f*, 130
Alvarez, Walter, 14–15, 15*f*, 130
Alvord Playa (Oregon), 422*f*
Amazon rain forest (South America), 70
Amazon River (South America), 369, 369–370, 369*f*
Amethyst, 83
Amphibole, 80*t*, 81, 81*f*, 85, 86, 91
Ancohuma (peak in South America), 307*f*
Andarko Basin (United States), geothermal resource maps, 27, 28, 28*f*
Andean margins, 317
Andes (mountains in South America), 208*t*, 209, **209*f***, 217, 269, 280, 304, 305, 307, 307*f*, 309*f*, 317, 664
Andesite, 107
Andisols, 453, 454, **454*f***
Aneroid barometer, 587, 587*f*

Angle of repose, 386–387, 387*f*, 388*f*
Angular unconformity, 140, 140*f*, 141
Anhydrite, 80*t*
Animals
 biodiversity, 42, 50–51, 52*f*–53*f*, 54, 61
 biosphere support for, 41–42
 climate change and, 697
 coastal wetlands, 483
 dinosaurs, 14–16, 44
 ecosystems, 43, 43*f*, 51
 endemic animals, 480
 freshwater biodiversity, 474, 475*f*
 invasive species, 495
 mass extinctions, 126, 129–133, 155
 mountain habitats, 314
 National Geographic Photo Ark, 52*f*–53*f*
 nitrogen and, 58
 plastic pollution in mercury cycle, 68
 rock weathering caused by, 364
 on St. Matthew Island (Alaska), 50, 51*f*
 threshold effects, 49, 49*f*, 50, 54
 trace fossils, 109*f*, 110
Anions, 73
Anker, Conrad, 294
Anoxia, 133
Antarctic Circle, 20*f*
Antarctic Circumpolar Current, 220
Antarctic Ocean, 404, 501
Antarctic plate, 206*f*
Anthracite, 172*t*
Anthropocene Period, 54
Anticline, 298, **298*f***
Anticyclones, 650, 651*f*, 679
Apatite, 76*t*, 80*t*, 83
Apollo program (NASA), 10, 749–750
Appalachian Mountains (United States), 152, 220, 296, 296*f*, 305, 306, 315, 316, 317, 355, 398
Apparent brightness (stars), 780
Aquifers, 488, 568
 confined, 489*f*
 remediation of, 565–566
 See also Ogallala aquifer
Aquitards, 488
Arabian Desert (Middle East), 429
Arabian plate, 196, **196*f***, 206*f*, 222
Aragonite, 80*t*, 111
Archean eon, 45*f*, 150*f*, 151
Arctic Circle, 20*f*
Arctic Ocean, 34, 35*f*, 416, 501
Arecibo Observatory (Puerto Rico), 797*f*
Arête, 408, 408*f*
Arid climate, 682
Aristotle, 715, 718, 719, 722, 733
Aristotle's model of the universe, 715, 716, 718, 733
Armstrong, Neil, 10, 740
Arsenic, toxicity, 85, 87
Artificial levees, 380–381, 380*f*, 381*f*, 389, 389*f*
Artisanal small-scale gold mining (ASGM), 70, 90, 90*f*
Asbestos, 85–86, 86*f*, 91

Ash, volcanic, 44, 60, 114, 131, 277
Ash-flow tuff, 281*f*, 282, 289
Assimilation, 58
Asteroid belt, 761
Asteroids, 14–16, 761, 765
Asthenosphere, 267, 520
 in Earth's layers, 199*t*, 200, 200*f*, 201*f*, 223, 259
 plate tectonics, 40*f*, 199*t*, 200–201, 200*f*, 201*f*, 205, 205*f*, 206, 206*f*, 208, 217, 223
 seismic waves in, 254
Astronomical units (AU), 718
Astronomy, 10, 10*t*, 709
 early astronomy, 712, 713, 715, **715*f***
 telescopes, 709, 712, 718, 727–729, 728*f*, 729*f*, **729*f***
 tools, 727–731, 733
Aswan Dam (Egypt), 550
Asymmetric folds, 298, 298*f*
Atacama Desert (South America), 417, 420, 420*f*
Atlantic Ocean, 326, 501, 509*f*, 513*f*, 514, 514*f*, 529*t*
Atlantic Ocean basin, 327
Atmosphere, 33, 37–38, 37*f*, 38*f*, 41, 42, 44, 61, 573*f*, 574–605
 air pollution, 591–599
 barometric pressure, 586–587, 586*f*, 587*f*, 601
 carbon in, 56, 56*f*, 57*f*, 57*t*, 61
 carbon cycle, 56*f*, 57*f*, 61
 carbon dioxide in, 56, 57*f*, 61, 128, 584, 618, 699
 climate change and, 690, 693, 693*f*
 composition of modern atmosphere, 584–585, 584*f*, 601
 composition through time, 579*t*
 density of, 586–587
 Earth's early atmospheres, 579–580, 579*t*, 580*f*
 energy balance in, 606–635
 energy exchange in, 690*t*, 691*t*
 evolution of modern atmosphere, 581–585, 601
 First Great Oxidation Event, 579*t*, 581–582, 583, 601
 global climate stability, 677
 global temperature rise, 572*f*, 694*f*, 695
 greenhouse effect, 617–619, 618*f*, 632, 690
 heat advection and convection, 620, 621*f*, 622
 of Jupiter, 752*f*, 765
 layers of, 590–591, 590*f*
 of Mars, 745
 of Mercury, 744, 745, 765
 methane in, 56
 Miller–Urey model, 580
 of Neptune, 752*f*, 756, 765
 organisms and, 42
 outgassing, 580, 601
 oxygen level in, 579, 581, 601
 ozone in, 583
 ozone hole, 598–599, 599*f*, 601
 rock cycle and, 101
 of Saturn, 752*f*, 765
 Second Great Oxidation Event, 579*t*, 583–584, 601

INDEX

secondary atmosphere, 686
solar radiation, 611, 614–616, 632
tectonic movements and, 220, 223
temperature of, 572*f*, 590–591, 590*f*, 601, 693
temperature balance of, 616
three-cell model, 677–678, 678*f*, 679, 702
of Titan, 759, 765
of Uranus, 752*f*, 756, 765
of Venus, 745, 765
water in, 41, 41*f*, 44, 55
Atmospheric cooling, volcanic eruptions and, 131
Atmospheric pressure, 586–587, 586*f*, 587*f*, 601
Atmospheric science, 10*t*
Atmospheric warming, volcanic eruptions and, 131
Atolls, 338, 349
Atomic structure, 144
Atoms, isotopes, 144–147, 145*f*, 146*t*, 147*t*, 155
Auto industry, energy efficiency, 182–183
Autonomous underwater vehicle (AUV), 348*f*
Axial surface (of fold), 298

B

Backshore, 522, 522*f*
Bacteria, 42, 333, 349
Baffin Island (Canada), 401*f*
Bajadas, 423–424, 423*f*
Ballinger, Adrian, 294
Banded iron formations, 167, 167*f*, 582–583, 582*f*, **583***f*, 601
Banff National Park (Canada), 361*f*, 479*f*
Banks (streams), 370
Bar (unit of measure), 587
Barchan dunes, 430, 430*f*
Barite, 80*t*
Barometer, 586, 587, 587*f*, 653
Barometric pressure, 586–587, 586*f*, 587*f*, 601
Barred spiral galaxy, 800, 800*f*, 801
Barrier islands, 523, 524–526, 524*f*–526*f*
Barycenter, 722, 722*f*, **723***f*
Baryonic matter, 810
Basal slip (glaciers), 404, 405, 433
Basalt, 81, 81*f*, 82*t*, 104–105, 107, 121, 136, 136*f*, 165, 203, 268, 270*f*, 289
basaltic lava flow, 264
basaltic magma, 207, 268, 271, 271*t*, 308, 332, 346
basaltic oceanic crust, 340–341
mid-ocean ridges, 212
pillow basalt, 340, **341***f*
specific heat, 622
Basalt plateau, 278*t*
Base level (streams), 373–375, 373*f*, 374*f*, **374***f*
Basement rock, 106
Basin and Range Province, 217, 217*f*, 232, 249, 268, 301, 301*f*
Basins, 217*f*, 299, 299*f*, 476
Batholith, 275, 275*f*, 289

Battery-powered vehicles, 182–183
Bauxite, 80*t*, 167, 167*f*
Baymouth bars, 523, 523*f*
Beach drift, 518
Beaches, 522–527, 533
beach zones, 522–523, 523*f*
erosion of, 524–526, 525*f*, 526*f*, **526***f*
groins, 525–526, 525*f*, 526*f*, 533
rock record, 142, 143*f*
seawalls, 526–527, 526*f*, 533
Beaufort Sea (Alaska), 343
Bed (streams), 370
Bed load (stream), 372–373
Bedded chert, 114, 114*f*
Bedding, 109, 109*f*
Bedrock, 106, 108, 108*f*, 246, 362*f*–363*f*
Belarus, 237*f*
Bell Burnell, Jocelyn, 785, 785*f*
Bell Island (Russia), **481***f*
Bellatrix (star), 714*t*
Bengal Fan (Indian Ocean), 345
Benioff zone, 243, 243*f*
Benzene, 592
Betelgeuse (star), 714*t*
Bicarbonate, 61, 66*f*–67*f*
Big bang, 803, 804, 813
Big bang theory, 803–806, 813, 815
Big Dipper (constellation), 713, 713*f*
Big Sur (California), 522*f*
Bingham Canyon mine (Utah), 168*f*, 169
Bioclastic limestone, 112, 112*f*
Bioclastic sedimentary rocks, 110, 111–113, 112*f*
Biodegradable pollutants, 560, 568
Biodiversity, 42, 50–51, 52*f*–53*f*, 54
Amazon Basin biodiversity, 695
extinction of species, 51
freshwater biodiversity, 474, 475*f*
National Geographic Photo Ark, 52*f*–53*f*
population growth and, 61
soil biodiversity, 448, 448*f*
wetlands, 483, 484*f*
Biofuel, 180
Biomass energy, 180
Biome, 680
Biosphere, 33, 37*f*, 38, 38*f*, 44, 61
carbon in, 48, 56, 56*f*, 58
tectonic processes and, 221, 223
Biotite, 71
Bituminous coal, 172*t*
Black, Ira, 426*f*
Black holes, 783*f*, 787–788, 787*f*, 790, 798, 802, 803, 809, 810, 812, 813
Black lung disease, 85
Black smokers, 166, 166*f*, 333, 333*f*, **333***f*
Blackfoot River (Montana), 370, 370*f*, 371, 371*f*
Bladensburg National Park (Australia), **428***f*
Blob tectonics, 748
Blowout (deserts), 430
Blue-green algae, 581

Blue Ridge Mountains (United States), 296, 296*f*
Blue stars, 775, 780
Blueshift, 731
Blyde River (South Africa), 373*f*
Body waves, 234, 234*f*, 254–255, 255*f*, 259
Bogs, 482
Boiling River of the Amazon (Peru), 6, 7*f*
Bolides, 324, 580, 580*f*, 691
Borax, 164, 422, 422*f*
Bornite, 80*t*
Bowman, Katlin, 68, 69*f*
Bracewell Radio Sundial, 708*f*
Brahe, Tycho, 717
Braided streams, 376, 376*f*, 412*f*
Braidplain, 412
Breakwater, 525*f*
Bromine, 598
Brown dwarfs (stars), 810
Bryce Amphitheater (Utah), 352*f*, 353*f*
Bryce Canyon National Park (Utah), 352*f*, 353*f*
Bugaboo Mountain Range (Canada), **403***f*
Buildings
building materials, 152, 153
earthquake damage and, 246–247, 259
energy consumption in, 181–182, 181*f*
Bullivant, Nic, 161*f*
Burial metamorphism, 117, 121
Buttes, 424–425, 425*f*

C

Calcite, 73, 76*t*, 78*f*, 80*t*, 83, 111, 112, 365, 489
Calcium, in Earth's crust, 73, 73*t*
Calcium feldspar, 270
Calcium ions
bioclastic limestone formation and, 111–112, 112*f*
hot springs, 66*f*–67*f*
Caldera, 278*t*, 282, 282*f*, 289
California Current, 514
Calisti, Marcello, 322, 323*f*
Callisto (moon of Jupiter), 747, 765
Cambrian period, 45*f*, 150*f*, 153, 197*f*
Camnasio, Sara, 771*f*
Camp Fire (California), 574*f*–575*f*
Canadian Shield (Canada), 147*t*
Cancer, 86
See also Carcinogens
Cape Fold Belt (South Africa), 197*f*, 198
Cape Hatteras (North Carolina), 518
Carbon
energy balance, 699
fixation, 56
in the geosphere, 56, 58
in the hydrosphere, 58, 61
isotopes of, 144–145
in permafrost, 699
Carbon-14 dating, 145–147, 145*f*, 146*t*, 151, 155
Carbon capture, 700
Carbon cycle, 44, 56, 56*f*, 57*f*, 58, 61
cement production and, 63

climate change and marine carbon cycles, 34, 35*f*
fossil fuels, 56*f*, 58, 63
human impact on, 63
modeling of, 63
Carbon dioxide, 61
in the atmosphere, 56, 57*f*, 61, 128, 584, 618, 699
in the biosphere, 56, 56*f*, 58
in the Carboniferous period, 128
climate change and, 51
Earth's carbon dioxide budget, 131–133, 132*f*
emissions, 687, 700
fixation, 56
in hot springs, 66*f*–67*f*
in the hydrosphere, 58, 61
in seawater, 58, 533
in volcanic gas, 131
Carbon monoxide, 593, 593*f*
Carbonates
in carbon cycle, 61
carbonate platforms, 343, 343*f*
in hot springs, 66*f*–67*f*
marine organisms and, 502–503
as minerals, 80*t*, 81–82, 91, 111–112, 689
Carbonic acid, 502
Carboniferous period, 45*f*, 128, 150*f*
Carcinogens, 85, 86
Careers in Earth and Space Sciences
climatologist, 694
earthquake engineer, 253
engineering geologist, 220
geobiologist, 51
limnologist, 481, 481*f*, **481***f*
marine geophysicist, 334
meteorologists, 649
museum conservator, 136
paleontologist, 115
Carmichael, Sarah, 126, 127*f*
Carnotite, 79
Carpenter, Frank, 128
Carson, Rachel, 513
Cartography, 19
See also Maps and mapping
Cascade Mountains (North America), 107, 107*f*, 217, 304, 655*f*
Case Study, 8, 36, 70, 98, 128, 162, 196, 230, 266, 296, 324, 358, 398, 442, 476, 500, 540, 578, 610, 640, 676, 712, 740, 772, 798
Coral Reef Killers, 500, 500*f*, **500***f*
Devils Tower National Monument, Wyoming, 98, 98*f*
Even More Powerful Storms, 640, 640*f*
Flooding from Hurricane Katrina, 358, 358*f*
Galilean Moons and Models of the Universe, 712, 712*f*, **712***f*
Giant Dragonflies?, 128, 128*f*
Glacier to Watch, 398, 398*f*
Great Lakes Ecosystem, 476, 476*f*
Ice Shelf Disintegration, 676, 676*f*

Is Sustainable Irrigation Possible?, 540, **540f**
Journey to the Sun: Parker Solar Probe, 772, 772f
Manure, Cold Seeps, and Fire Ice, 578, 578f
Mapping Geothermal Energy, 8, 8f, **8f**
Melting Glaciers in Greenland, 610, 610f, **610f**
Mexico City Earthquakes, 1985 and 2017, 230, 230f, **230f**
Probing the Depths, 324, 324f, **324f**
Rising Atmospheric and Ocean Mercury Concentrations Linked to Small-Scale Gold Mining, 70, 70f
SGR A*, 798, 798f, **798f**
Small Steps and Giant Leaps, 740, 740f
Soil Erosion and Human Activities, 442, 442f, **442f**
Splitting of a Continent, 196, 196f, **196f**
Story of a Mountain Chain, 296, 296f
Strategically Important Minerals, 162, 162f, **162f**
Volcanic Activity: Past, Present, and Future, 266, 266f
Yellowstone Volcano: Major Eruptions Can Affect Earth's Systems, 36, 36f
Cassini spacecraft, 706f–707f, 740, 754f, 759, 764f
Catastrophic events, mass extinctions, 126, 129–133, 155
Catastrophism, 46, 47, 47f, 61
Cations, 73
Causation, correlation and, 11, 12
Caverns, 489–491, 490f, 491f, 493
Celestial spheres, 715
Cell phones, minerals for components, 90
Cement production, 28, 63
Cenozoic era, 45f, 150f, 152, 217f
Central Valley (California), 546
Ceres (dwarf planet), 738, 761
Chalcopyrite, 80t
Chalk, 113, 113f, 152–153
Chalky soils, 451, 451f, 452t
Chandra X-ray Observatory, 796, 798f
Change, 61
  catastrophism, 46, 47, 47f, 61
  Earth and, 44–50, 353f
  gradualism, 46–47, 61
  oceans, 532
  rates of change, 44–46
  threshold effects, 48–50, 49f, 50f, 61
  uniformitarianism, 46
Changes of state, energy balance, 622–623
Charon (moon of Pluto), 761, **761f**
Chemical contaminants, in water, 562
Chemical elements, 73
  CHNOPS (acronym), 55
  in Earth's crust, 73, 73t, 80, 91
  electrical charge, 73
  elements heavier than iron, 784
  isotopes, 144–145, 155
  in minerals, 73, 74, 91
  most abundant, 54–55, 73, 73t
  primordial nucleosynthesis, 805, 806
  radioactive isotopes, 144–147, 145f, 146t, 147t, 155
  role of, 73
  symbols, 73
  toxic elements in minerals, 85
Chemical sedimentary rocks, 113–114, 114f
Chemical weathering, 360, 360f, 365–368, 390
  acting together with mechanical weathering, 366–368, 367f
  dissolution, 365, **365f**, 517
  hydrolysis, 366, 367
  oxidation, 369
  soil formation and, 446
Chemosynthesis, 333, 349
Chernobyl accident (Ukraine), 177
Chert, 114, 114f
Chesapeake Bay (United States), 696f
Chesapeake Bay watershed (United States), 466
Chlorine, 598, 598f
Chlorofluorocarbons (CFCs), 598–599, 598f, 601, 692
CHNOPS (acronym), 55
Chondrules, 762
Chromite, 80t
Chromosphere (sun), 779
Chrysotile, in common rock types, 85, 85f, 86
Churchill (Canada), 510
Cinder cone volcanoes, 278t, 279–280
Cinnabar, 80t
Cirques, 407, 407f
Cirrostratus clouds, 652, 652f
Cirrus clouds, 645, 645f, 646f, 652, 652f
Cities, climate and urbanization, 684–685, 684t, 684f, 685f, 702
Clastic sedimentary rocks, 110–111, 110f, 110t
Clasts, 136
Clay, 80t, 81, 81f, 91, 110, 110t, 367–368
Clay soils, 450, 451, 451f, 452t
Clean Water Act, 559–560, 565, 568, 592, 601
Cleavage (minerals), 78, 78f
Cleavage planes, 78, 78f
Climate, 611, 672–705
  acid rain, 365, **365f**
  arid climate, 682
  average temperature, 624–626, 625f, 628, 628f, 629f, 630
  climate maps, 23f
  climatologist, 694
  climograph, 680, 680f–683f
  clouds, 630, **630f**
  energy balance and, 611, 614
  energy flow, 704–705
  equatorial region, 678
  glaciation, 414–415, 414f
  global climate stability, 677–679, 678f, 679f
  global temperature rise, 572f, 694f, 695
  hours of sunlight per day, 627t
  humid continental climate, 684
  humid mesothermal climate, 682
  humid microthermal climate, 684
  humid subtropical climate, 682
  humid tropical climate, 680t, 681–682
  of Jupiter, 753–754
  Köppen classification, 680, 680t
  latitude and altitude, 622–624, 622f
  marine west coast climate, 682, 682f
  of Mars, 745, 765
  Mediterranean climate, 682
  of Mercury, 744
  of Neptune, 756
  ocean currents and, 510
  oceans and, 327, 510, 632
  plate tectonics and, 220–221, 221f, 223
  polar climate, 684
  of Saturn, 754
  seasons, 616–617, 616f, 632
  snow and, 630
  soil formation and, 444, 444f
  subarctic climate, 684
  tropical monsoon climate, 681
  tropical rain forest climate, 681
  tropical savanna climate, 681–682, 681f
  types, 680–684, 680t, 702
  urbanization and, 684–685, 684t, 684f, 685f, 702
  of Venus, 745
  volcanic eruptions and, 283, 284, 285f, 287
  See also Weather
Climate change
  air pollution and, 619
  carbon cycle, 44, 56, 56f, 57f, 58, 61
  causes of, 689–692
  consequences of, 695–700, 701, 702
  economic impacts of, 698–699
  ecosystems and, 696–697
  effects of, 695–700
  emissions and, 51, 694, 694f
  energy production and, 185
  evidence for, 687–689
  feedback loops, 50, 699, 702
  future projections, 694, 694f, 702
  global cooling in Permian period, 131, 132f, 133
  historical climate change, 686–689, 689f, 702
  human activities and, 619, 692–693, 702
  marine carbon cycles and, 34, 35f
  mean annual temperature, 51
  mitigating effects of, 700
  mountains and, 316
  ocean acidification, 698
  pathogens, 697
  precipitation, 695, **695f**
  projected future climate change, 693–694, 694f
  sand dunes and, 394
  sea-level rise and sea warming, 697–698, **697f**, 698f
  tree rings as evidence of, 687–688
  volcanic eruptions and, 60, 131, 223, 691
  "volcanic winter," 60
  weather patterns and, 695–696, **695f**, 696f
Climate maps, 23f
Climograph, 680, 680f–683f
Closed universe, 810f, 811, 813
Clouds, 630, **630f**, 636f–637f, 644, 668, 684t
  types, 645, 645f–647f, 647
Coal, 172, 172t, 173f
  carbon content of, 172, 173f
  combustion of, 592, 593, 597, 601, 692
  end-use efficiencies of, 181f
  formation of, 72, 113, 113f, 172, 173f
  grades of, 172t, 173f
  mercury pollution and, 89, 172
  mining, 85, 168f, 169, 170f–171f, 187
Coal mining, 85, 168f, 169, 170f–171f, 187
Coastal deserts, 418, 420, 420f
Coastal emergence, 519, 520f, 533
Coastal submergence, 519, 520f, 533
Coastlines (seacoasts), 516–521, 533
  beach drift, 518
  coastal emergence and submergence, 519, 520f, 533
  erosion of, 698
  landforms along sandy coastlines, 523–524
  longshore currents, 518–519, 519f, 523, 525f
  marine life at, 528–531, 533
  rocky coastlines, 527, 527f, 533
  sediment transport along, 518–519, 518f, 519f
  surf, 517, 517f
  weathering and erosion, 516–517, 517f
  See also Beaches
Cocos plate, 258
Cold seeps, 578
Colon, Knicole, 796, 797f
Color (minerals), 76–77
Colorado Plateau (United States, 424–425, 424f
Colorado River (North America), 377, 377f, 424, 536f–537f, 547, 548–549, 549f, 558–559
Colorado River delta (North America), 552f–553f
Columbia Glacier (Alaska), 416
Columbia River (United States), 347, 377, 377f
Columbia River basalt, 277, 279f
Columbia River plateau, 277
Columnar joints, 276, 277f
Combustion, of fossil fuels, 592–593, 601, 692
Comets, 325, 762–763, 763f, 765
Commercially important minerals, 82–83, 83f
Composite cone volcano, 278t, 280, 280f
Condensation, 37f, 55, 55f, 61, 642–643, 668
Cone of depression, 554

INDEX **873**

Conformity, 139, 155
Conglomerate, 111, 111*f*, 135, 135*f*
Connections
  to art, 48, 48*f*, 99
  to chemistry, 784
  to environmental sciences, 183, 565
  to history, 89, 283, 422, 422*f*
  to human geography, 309, 309*f*
  to life science, 483, 695
  to modern physics, 811
Conservation, environmental. *See* Environmental conservation
Conservation of matter and energy, 44
Constellations, 713, 713*f*, 733
Constraints (engineering), 16
Contact cooling, 643, 668
Contact metamorphism, 117, 121
Continental arcs, 307, 317
Continental collision zones, mountain building, 305
Continental crust, 40*f*, 199–200, 199*t*, 326, 340
  convergence of, 209
  granitic, 343
  rifting in, 207–208
Continental Divide (North America), 377
Continental drift, 199, 223
Continental drift hypothesis, 197–199, 197*f*, 223, 330*f*
Continental glaciers, 401, 410, 433
  erosion by, 408
  melting of, 334
Continental margins, 345–347, 346*f*, 347*f*, 349
  active continental margins, 346
  continental rise, 344, 349
  continental shelf, 342–344, 342*f*, 343*f*, 349
  continental slope, 342*f*, 344, 344*f*, 349
  passive continental margins, 345, 349
  submarine canyons, 344–345, 345*f*, 349
  submarine fan, 344–345, 345*f*
Continental rifting, 207–208, 217, 222, 270
Continental rise, 344, 349
Continental shelf, 342–344, 342*f*, 343*f*, 349
Continental slope, 342*f*, 344, 344*f*, 349
Continents
  accreted terranes, 337, 337*f*
  continental shelf, 342–343, 342*f*, 343*f*, 349
  formation of, 289
  movement of, 131–132, 223, 297
  subduction at a continental margins, 307–308, 307*f*–309*f*, 310
  See also Supercontinents
Contour interval, 24, 24*f*
Contour lines (isolines), 22, 23*f*, 24, 24*f*
Convection, 620, 621*f*
  convection–convergence in weather, 644, 659, 668
  tectonic plate movement, 205*f*, 211–212, 211*f*, 213, 223

Convective zone (sun), 777
Convergence
  of continental crust, 209
  convection–convergence in weather, 644, 659, 668
  of oceanic crust, 209
Convergent plate boundaries, 205, 206*f*, 208–209, 208*t*, 220, 223, 302, 346, 347
  earthquakes at, 243–244, 243*f*
  island arcs and, 336
  locations, 244*t*
Cooling
  adiabatic cooling, 558, 643, 668
  of air, 642–643, 668
  atmospheric cooling, 131
  contact cooling, 643, 668
  of lava, 276
  of magma, 104, 106, 274
  radiation cooling, 643, 668
  of seawater, 520
  supercooling, 642
Coordinate system, 20, 29
Copernican model of the universe, 716, 733
Copernicus, Nicolaus, 716, 733
Copper, 79, 80*t*, 82, 168*f*, 169
Coral and coral reefs, 112, 138, 199, 338–339, 498, 499*f*, 500, 500*f*, **500*f***, 530–531, 531*f*, 533, 672*f*–673*f*, 698
Coral Reef Task Force, 500
Core (of Earth), 37*f*, 39, 40*f*, 199*t*, 200*f*, 201*f*, 203, 205*f*, 223, 254–256, 255*f*
Core (of the sun), 777, 777*f*, 782, 790
Coriolis effect
  ocean currents and, 511, 511*f*, 512, 514, 516, 533
  wind and, 650, 650*f*, 654, 662, 668, 677, 678, 678*f*, 679
Corona (sun), 726, 779
Correlation
  age of rocks, 141–142, 142*f*, 144, 155
  causation and, 11, 12
Corundum, 76*t*, 77, 80*t*
Cosmic Background Explorer (COBE) satellite, 807
Cosmic microwave background (CMB), 806–808, 813
Cosmology, big bang theory, 803–806, 813, 815
Cover crops, 463
Crab Nebula, 786
Crandall, Dwight, 283
Crater Lake (Oregon), 282, 288, 479
Craters, 279, 359
Creep (landslides), 384, 384*f*, 385*t*
Crescent moon, 725*f*, 726
Crest (wave), 507, 507*f*
Cretaceous period, 34, 45*f*, 150*f*, 335
  mass extinction, 129, 130–131, 131*f*
  mountain building, 309*f*
  origin of name, 152–153
Cretaceous Western Interior Seaway (North America), 34, 152
Crevasses, 402–403, 402*f*, 403*f*, **403*f***, 433
Criteria (engineering), 16

Crop failure, volcanic eruption and, 283
Crosscutting concepts, 18, 18*t*, 29
Crosscutting relationships, 135, 135*f*
Crude oil, 173, 173*f*
Crust (of Earth), 199–200, 199*t*, 200*f*, 206
  chemical elements in, 73, 73*t*, 80
  composition of, 326
  continental crust, 40*f*, 199–200, 199*t*, 207–208, 209, 326, 340
  crust–mantle boundary, 253–254, 259
  density of, 255
  formation of, 359
  metal concentrations in, 164*t*
  metamorphism, 118
  minerals in, 81, 81*f*, 91, 106
  oceanic crust, 199, 199*f*, 209, 215, 259, 268, 306, 312, 313, 326, 340–341, 341*f*, 349
Cryosphere, 40–41, 41*f*
Crystal face, 74–75
Crystal habit, 77, 77*f*
Crystals, mineral structure, 74–75, 74*f*, 75*f*, 91, 104, 105*t*, 106
Cubango River (Africa), 538
Cuito River (Africa), 538
Cumulonimbus clouds, 646*f*, 647, 652, 652*f*, 659, 668
Cumulus clouds, 645, 645*f*, 646*f*, 652, 652*f*, 659
Curiosity (Mars rover), 738, 751, 751*f*
Cutler, Joe, 474, 475*f*
Cuyahoga River (Ohio), 559, 560*f*
Cyanobacteria, 65
Cycles, 37*f*, 44, 54–59, 61
  carbon cycle, 44, 56, 56*f*, 57*f*, 58, 61, 63
  humans and Earth cycles, 63
  nitrogen cycle, 44, 58, 59*f*, 61
  rock cycle, 44, 96, 99–101, 100*f*, 147
  water cycle, 44, 55–56, 55*f*, 61, 477–478, 478*f*, 492–493
Cyclones, 650, 651*f*, 679, 682
  hurricanes, 358, 358*f*, 382, 389, 389*f*, 640, 640*f*, 660, 695
  midlatitude cyclone, 653–654, 653*f*, 662
  Saffir-Simpson scale, 662, 663*t*
  storm surge, 660, 662
  strength of, 662, 663*t*
  tropical cyclones, 660, 661*t*, 662, 664, 668
Cynognathus fossils, 198*f*

D

DAGGER Group, 126
Dams, 547–551, 552*f*–553*f*, 568
  ecological disruptions, 550
  erosion and, 548–549
  human costs, 550
  hydroelectric power from, 179, 550–551, 551*f*
  recreational and aesthetic impacts, 549–550
  removal of, 550
  risk of disaster, 549
  sediment movement, 548
Dark energy, 811

Dark matter, 809–810, 813
Darwin, Charles, 339
Data
  correlation and causation, 11, 12
  nature of science, 10, 13
  variables, 13
Data Analysis activity, 12, 26, 57, 82, 118–119, 147, 184, 201, 239, 244, 286, 314, 327, 371, 415, 419, 448, 465, 485, 528–529, 542, 593, 627, 646, 666, 683, 723, 730–731, 743, 782, 805
  Abundance and Distribution of Silicate Minerals, 82, 82*t*
  Air Pollution, 593
  Analyze Patterns in U.S. Energy Consumption, 184, 184*f*
  Analyze Patterns of P and S Waves, 239, 240*f*
  Blackfoot River Discharge Over Time, 371, 371*f*
  Compare Climographs, 683, 683*f*
  Compare Elevations, 314, 314*t*
  Compare Planetary Data, 743
  Compare Seafloor Data, 327, 327*f*
  Compare Storm Statistics, 666, 666*t*
  Compare Wetland Gains and Losses, 485, 485*f*
  Comparing World Fisheries, 528–529, 529*f*, 529*t*
  Develop a Topographic Profile, 26, 26*f*
  Distinguish Between Correlation and Causation, 12, 12*t*
  Effects on Soil Biodiversity, 448, 448*f*
  Estimating the Lifetime of the Sun, 782, 782*t*
  Exploring Earthworm Bioremediation, 465, 465*f*
  Exploring Volcanic Risk in the United States, 286, 286*f*
  Evolution of the Telescope, 730–731, 730*f*–731*f*
  Global Water Use, 542
  Graphing Glacial Melt, 415
  Graphing the Carbon Cycle, 57, 57*f*, 57*t*
  Hours of Sunlight per Day, 627
  Interpret Temperature and Pressure Graphs, 201, 201*f*
  Map Earthquakes in Real Time, 244, 244*t*
  Metamorphism of Mudstone, 118–119, 118*f*–119*f*
  Quantify the Expansion Rate of the Universe, 805, 805*t*
  Quantifying the Effect of Gravity, 723, 723*t*
  Think About Radiometric Testing, 147, 147*t*
  Record Data and Design a Game, 646
  Which Is the Desert?, 419, 419*f*
Daughter isotopes, 145, 145*f*, 155
Dead Sea (Middle East), 422, 422*f*
Dead zone, 530–531, **531*f***
Death Valley (California), 108, 108*f*, 221*f*, 378*f*, 424*f*, 425, 655

Death Valley National Park (California), 422*f*
Deccan Traps (India), 14, 16
Decompression melting, 268
Deep Impact probe, 762–763
Deep-sea currents, 131–133, 132*f*, 513–514, 513*f*, 514*f*, 533
Deflation, 428
Degrees of latitude, 20, 29
Degrees of longitude, 20, 29
Delta front, 379, 379*f*
Delta top, 378, 379, 379*f*
Deltas, 117, 358, 378–379, 379*f*
Denali (mountain in Alaska), 292*f*–293*f*, 314, 314*t*, 315, 399
Denitrification, 58, 59*f*
Dependent variable, 13
Deposition
  sedimentary rocks, 108
  in streams, 377–378, 378*f*, 390
Desert lakes, 421–422
  Great Salt Lake (Utah), 114
Desert pavement, 428, 428*f*, **428*f***, 433
Desertification, 459–460, 461, 461*f*, 462*f*, 467
Deserts, 395, 417–431, 682
  alluvial fans, 423, 423*f*
  bajadas, 423–424, 423*f*
  buttes, 424–425, 425*f*
  coastal deserts, 418, 420, 420*f*
  deflation, 428
  desert biome, 419
  desert pavement, 428, 428*f*, **428*f***, 433
  dunes, 108, 394, 428, 429–431, 429*f*, 430*f*, 433, 497, 688
  erosion in, 421
  evaporative mineral deposits, 166
  formation of, 432
  fossils in, 138
  groundwater, 557
  interior deserts, 418, 420
  latitude and, 417–418, 417*f*, 418*f*
  loess, 430–431, 430*f*, 433
  major deserts of the world, 418*f*
  mesas, 424–425, 425*f*
  mountains and, 418
  pediments, 424
  plateaus, 424–425
  rain-shadow deserts, 418, 419*f*, 655, 655*f*
  rainfall in, 417, 433
  sand dunes, 108, 394, 428, 429–431, 429*f*, 430*f*, 433
  sand transport in, 429
  spires, 424–425, 425*f*
  streams and lakes in, 421–422
  washes, 421, 421*f*
  water in, 421–425, 433
  wind abrasion in, 428–429
  wind and, 428–436
  wind erosion, 428, 432
Devils Postpile National Monument (California), 277*f*
Devils Tower National Monument (Wyoming), 98, 98*f*, 120, 120*f*
Devonian period, 45*f*, 126, 150*f*
Dew, 643
Dew point, 641

Diamond, Jared, 54
Diamonds, 72, 76*t*, 80*t*, 83, 84*f*
Dicke, Robert, 806–807
Dikes, 135, 135*f*, 275–276, 276*f*, **276*f***, 289, 698
Dinosaurs
  in geologic time, 44
  mass extinction, 14–16, 129
Dioxin, 597, 601
Discharge (streams), 369–371, 369*f*, 370*f*, 371
Disconformities, 139–140, 139*f*, 140*f*, 155
Dissolution, 365, **365*f***, 517
Dissolved load (stream), 372
Distributary channels, 378–379, 382, 390
Divergent plate boundaries, 205, 206–207, 206*f*, 207*f*, 208*t*, 217, 223, 243–244, 243*f*
Diversion system, 547
Dixon Waterfowl Refuge (Illinois), 483
DJ Spooky. *See* Miller, Paul
Dolomite, 80*t*, 109*f*, 111, 373–374, 373*f*
Domes, 299, 299*f*
Domestic water use, 543–544, 544*f*
Donato, Leonardo, 712
Doppler effect, 730, **731*f***, 733
Doppler on Wheels truck, 656*f*–657*f*
Doppler shift, 731
Downcutting, 373, 373*f*
Dragonflies, giant dragonfly fossils, 128, 128*f*, 129
Drainage basins (streams), 377, 377*f*
Drainage divides (streams), 377
Dredging, 328
Drift, 409, 433
Drilling, of seafloor, 328
Drought, 695
Drumin fields, 411
Drumlins, 411–412, 413, 433
Dunes, 108, 394, 428, 429–431, 429*f*, 430*f*, 433, 497, 688
Durdle Door (England), 152*f*–153*f*
Dust Bowl (United States), 459–460, 461, 461*f*
Dust storms, 459–460, 461*f*
Dwarf elliptical galaxies, 799
Dwarf galaxies, 813
Dwarf planets, 760–761

E

Eagle Nebula, 774, 774*f*
Earth
  age of, 61, 133, 340
  average temperature of, 624–626, 625*f*, 628, 629*f*, 630, 686, 686*f*
  changes in, 44–50, 353*f*
  core. *See* Core (of Earth)
  crust. *See* Crust (of Earth)
  cycles, 37*f*, 44, 54–59
  density of, 255
  distribution of water on, 477–478, 477*f*
  early atmospheres, 579–580, 579*t*, 580*f*, 601
  formation of, 39, 197, 256
  formation of life on, 61
  as Gaia, 581

geologic structure, 37*f*, 39–40, 39*f*, 297–303, 750
geologic time, 44–46, 45*f*, 124–157
glacial retreat, 415–416, 416*f*, 429, 631
global temperature of, 572*f*, 686–687, 686*f*, 687*f*
global warming, 520–521, 700
interior of, 99, 253–256, 259
internal heat of, 101, 197
latitude and longitude, 19, 20, 20*f*
layers of, 199–201, 199*f*, 200*f*, 201*f*, 203, 223, 225*t*
magnetic field, 203–204, 203*f*, 204*f*, 256, 256*f*, 257*f*
mantle. *See* Mantle
map projections, 19–21, 20*f*, 21*f*, 29
maps and mapping, 19–27, 29
mass extinctions, 126, 129–133, 155
motion in universe, 724–725
nutation of, 724
ocean formation from primordial solar system, 325–326
oldest rock in, 147*t*
orbital changes and climate change, 691–692
origin of life, 581
ozone depletion, 599
Pangaea, 132, 132*f*, 198, 214, 214*f*, 296, 345–346, 346*f*
Pangaea Proxima, 192*f*–193*f*, 214
as planet, 707, 709, 723*t*, 743*t*, 745, 746*f*, 748, 793*t*
population, 440, 685
primordial atmospheres, 579–580, 579*t*, 580*f*
primordial Earth, 325
resource depletion, 51
revolution of, 724
rock cycle, 44, 96, 99–101, 100*f*, 147
rotation of, 724, 725*f*
Snowball Earth, 415, 699
solar energy, 44
sources of heat, 33, 37–42, 44
spheres, 33, 37–42, 44
systems, 43–44, 60
tectonic plates, 37*f*, 40, 40*f*
temperature of, 39, 51, 115, 601
two-dimensional maps of, 20–21, 20*f*
volcanic eruptions affecting systems, 36, 44, 60
water cycle, 477, 478*f*, 488*f*, 492
Earth-centered (geocentric) model of the universe, 712, 715–716, 715*f*, **715*f***, 716*f*, 719, 719*f*
Earth science
  branches of, 10, 10*t*
  geologic time, 44–46, 45*f*
  GPS use in, 21–22, 29
  models in, 13
  nature of, 10
  See also Geology
Earthflow, 383
Earthquake engineer, 253
Earthquakes, 47, 227–261
  anatomy of, 231–233, 232*f*
  Benioff zone, 243, 243*f*

body waves, 234, 234*f*, 254–255, 255*f*, 259
building damage from, 246–247, 259
earthquake waves, 234–236, 234*f*, 235*f*, **235*f***, 237*f*, 239, 240, 253, 259
epicenter, 234, 259
faults, 231–232, 232*f*, 242
foreshocks, 252
India (2005), 228, 229*f*
landslides in, 246, 388, 390
locating source, 239–240, 240*f*
loss of life in, 247
mapping, 244
Mercalli scale, 237
Mexico City (1985 and 2017), 230, 230*f*, **230*f***, 246, 258
Missouri earthquakes (1811 and 1812), 244, 249
moment magnitude, 237, 239, 261
Philippines (2012), 226*f*–227*f*
plate boundaries and, 241–245, 242*f*, 243*f*, 259
in plate interiors, 244–245, 245*f*
plate movement, 232
plate tectonics, 220, 232
predicting, 249, 252, 259
preparing for, 228, 229*f*, 230
research, 228, 229*f*, 324
Richter scale, 237, 239
San Francisco earthquake (1906), 242, 242*f*, 243
seismic waves, 234–236, 234*f*, 235*f*, **235*f***, 237*f*, 239, 240, 253, 254, 259
seismograph, 236, 236*f*, 237*f*, 238, 238*f*, 239, 252
strength of, 237–239
Sumatra-Andaman earthquake, 47, 247–248, 247*f*
Tohoku (2011), 236, 237*f*, 248–249, 248*f*–251*f*, 324, 324*f*, 348
travel-time curve, 239–240, 240*f*
tsunamis, 247–249, 248*f*, 249*f*
vulnerability to, 258
Earth's crust. *See* Crust (of Earth)
East Africa, plate tectonics, 194, 195*f*, 196, 196*f*, **196*f***, 197, 197*f*
East African Rift, 194, 196, 196*f*, 208*f*, 217, 244*t*, 268, 301, 304
East Pacific Rise, 208*t*
Eastern Hemisphere, 20
Echinoderms, 112
Echo sounder, to map seafloor, 328–329, 331
Eclipses
  Great American Solar Eclipse, 720*f*–721*f*
  lunar eclipse, 726, 733
  solar eclipse, 726, 726*f*, 733
Ecosystems, 43, 43*f*, 51, 696–697
Edwards, Jason, 362*f*
Eemian interglacial period, 687*f*, 688, 688*f*
Ehlmann, Bethany, 738, 739*f*
Einstein, Albert, 775, 777, 811
El Niño, 516, 664–666, 665*f*, 668
Elastic deformation, of rocks, 231, 232*f*, 317

INDEX **875**

INDEX

Electric cars, 182–183
Electrical charge, of chemical elements, 73
Electrical storm, 239
Electricity
 from geothermal energy, 8, 8f, 16, 17, 17f, 27–28
 lightning, 239, 263f, 659, **659f**, 668
 See also Power plants
Electromagnetic radiation, 614, 632, 729
Electromagnetic spectrum, **611f**, 614, 632, 729
Elements. *See* Chemical elements
Elliptical galaxy, 799, 800f
Eluviation, 447
Emergence, coastal, 519, 520f, 533
Emission reduction, 694, 694f, 700
Emissions, climate change and, 51, 694, 694f
Emissions Gap Report, 700
Empirical evidence, 13
Enceladus (moon of Saturn), 740, 764f
End moraines, 410, 410f
Endemic animals, 480
Energy, 18t, 29
 carbon in, 56–58, 56f, 58f
 climate change and, 689
 conservation of, 44
 dark energy, 811
 latent heat, 622, 622f, 632
 specific heat, 623
Energy balance, 606–635
 analysis, 189, 189t
 changes of state, 622–623
 cloud cover, 630
 energy storage and transfer, 620–623, 632
 geographic factors, 624–630, 632
 heat and temperature, 620
 ocean and wind effects, 628, 628f, 629f, 630
 radiation balance, 617–619, 617f, 632, 668
 weather and climate, 611, 614
Energy conservation, 181–185, 187
Energy consumption, 181–182, 181f, 184, 184f
Energy efficiency, 181, 181f
Energy exchange, 690t, 691t
Energy resources, 163
 end-use efficiencies of, 181f
 environmental conservation, 181–185, 187
 future of, 183–185
 nonrenewable resources, 172–177, 187
 renewable resources, 177–180, 187
Energy return on investment (EROI), 189
Energy storage and transfer, 620–623, 632
Engare Sero Footprint Project (Tanzania), 356, 357f
Engineering, 16–19, 29
 criteria and constraints, 16
 defining engineering problems, 16–17, 16f
 design process, 16–17, 16f, 29

developing and optimizing solutions, 16f, 17
models, 12
practices, 11f
prototypes, 12–13
See also Science
Engineering geologist, 220
Entisols, 454, **454f**
Environmental conservation, energy resources, 181–185, 187
Environmental science, 10t
Eocene epoch, 45f, 150–151, 154
Eons, 45f, 150, 150f
Epicenter (earthquake), 234, 259
Epochs, 45f, 150f, 151
Equator, 20f
Equilibrium line altitude (ELA), 403, 404
Equinoxes, 627, 627t
Eras, 45f, 150–151, 150f
EROI. *See* Energy return on investment
Erosion, 359, 359f, 395, 437
 of barrier islands, 524–526, 525f, 526f, **526f**
 of beaches, 524–526, 525f, 526f, **526f**
 of coastlines, 698
 by continental glaciers, 408
 of cropland, 442, **442f**
 dams and, 548–549
 in deserts, 421
 fossilized human footprints, 356
 glacial erosional landforms, 405, 406f–408f, 407–408, **407f**
 by glaciers, 361, 362f–363f, 398, 405–408, 405f–408f, **407f**, 412, 433
 by groundwater, 489
 ice erosion, 405, 432
 lateral erosion, 373–374, 373f, 425
 minerals containing toxic elements, 85
 of Mississippi Delta, 358
 in the rock cycle, 108
 by sand, 361, 361f
 of soil, 442, 442f, **442f**, 458, 458f, 459f, 460, 467
 from storms, 525
 by streams, 372–373, 372f, 373–374, 373f, 377, 390
 wind erosion, 428, 432
Erosional landforms, glaciers, 405, 406f–408f, 407–408, **407f**
Erratics, 409, **409f**
Erta Ale (volcano in Ethiopia), 217
Erwin, Douglas, 129
Escape velocity, 741, 743
Eskers, 412, 433
Euphrates River (Middle East), 547
Eurasian plate, 206f, 210f, 228, 317
Europa (moon of Jupiter), 712, 738, 757–758, 758f, 765
Eustatic sea-level change, 520
Eutrophic lake, 480f, 481, 562, 562f
Euxinia, 133
Evaporation, 477–478, 622, 623
 carbon cycle, 56f, 61
 chemical sedimentary rock formation, 114
 water cycle, 37f, 55, 55f, 61, 478f

Event Horizon Telescope, 798, 803, 803f
Everglades (Florida), 483f
Evidence, 10, 13
Evolution, theory of, 136
Exoplanets, 732, 732t, 770, 771f, 796
Exosphere, 591
Expanding universe, 800, 801, 803, 805, 806
Experiment, 13
Explorers at Work, 34, 68, 96, 126, 160, 194, 228, 264, 294, 322, 356, 396, 440, 474, 498, 538, 576, 608, 638, 674, 710, 738, 770, 796
 Ancient Coral Reefs, 498, 499f
 Antarctica, 396, 397f
 Bioinspiration, 322, 323f
 Bringing Science Down to Earth, 710, 711f
 Constructing Rock Records, 126, 127f
 Everest Summit, 294, 295f
 Exoplanets, 770, 771f
 Footprints of Early Humans, 356, 357f
 Freshwater Biodiversity, 474, 475f
 Glacial Melting, 608, 609f
 Lava Flows, 264, 265f
 Managing Global Mineral Resources, 160, 161f
 Mapping Earthquake-Causing Faults, 228, 229f
 Marine Carbon Cycles and Climate Change, 34, 35f
 Measuring Tectonic Movements, 194, 195f
 Methane Bubble, 576, 576f
 Opening the Windows on the Universe, 796, 797f
 Protecting Okavango's Waters, 538, 539f
 Roving Through Space, 738, 739f
 Soil Resources and Sustainable Food Systems, 440, 441f
 Stories of Climate Change, 674, 675f
 Tracking Mercury, 68, 69f
 Visualizing Tornadoes, 638, 639f
 Volcanic Activity, 96, 97f
 World Is My Classroom, 6, 7f
Extinction
 of dinosaurs, 14–16
 of species, 51
Extraterrestrial impacts, as cause of mass extinction, 130–131, 131f
Extrusive igneous rocks, 104–105, 105f, 121

F

Farallon plate, 309f, 310
Fault creep, 259
Fault zone, 299, 299f
Faults, 231–232, 232f, 242, 299, 299f–303f, 301–303, 317, 334, 334f
Faunal succession, principle of, 136, 137f, 138f
Feedback mechanisms, 50, 58, 60, 61, 699, 702, 741
Feldspar, 71, 80t, 81, 81f, 85, 91, 106, 270, 366, 367

Felsic magma, 269
Felsic rocks, 106, 107
Fermi paradox, 710
Fertilizers, as pollutants, 560, 561
Fica, Haley, 771f
Field guides, 93
Finger Lakes (New York), 478
Fire, fracturing of rocks by, 360
Fire ice, 578, 578f
Firn, 399, 399f
First Great Oxidation Event, 579t, 581–582, 583, 601
Fish, coastal wetlands, 483
Fisheries, 528–529, 529t, 530, 664
Fissure eruptions, 277
Fissure flows, 277
Fissures, 279
Fjords, 408, 527, 533
Flamborough Head (England), 527f
Flathead River (Montana), 375f
Flint (chert), 113
Flood basalt, 277
Flood control, 380–382, 380f, **380f**, 381f, 389, 389f
Floodplains, 379–380, 482
Floods and flooding, 379–380, 390
 Chesapeake Bay (Maryland, Virginia), 696f
 climate change and, 698
 flood control, 380–382, 380f, **380f**, 381f
 floodplains, 379–380, 382
 from Hurricane Katrina, 358, 358f, 382, 389, 389f, 662
 storm surge, 660, 662
 wetlands to control flooding, 483
Flow (landslides), 383, 383f, 385t
Fluorescence, of minerals, 79
Fluorescent bulbs, 182
Fluorite, 76t, 78f, 80t
Fly ash, as pollutants, 597–598
Focus (earthquake), 234
Fog, 646, 647f, 649, 668
Folds, 297–299, 297f–299f, **298f**, 302, 317
Folger, Timothy, 509
Foliated metamorphic rock, 116, 116f
Foliation, 116–117, 116f
Food
 coastal wetlands, 483
 fisheries, 528–529, 529t, 530
 Midwest Wetlands Initiative, 483
 optimizing food production, 469
 sustainable food systems, 440, 441f
Fool's gold, 77
Footwall, 299
Foreland basin, 308, 308f
Foreshocks, 252
Foreshore, 522
Forests
 forest fires, 360
 tropical rain forests, 417, 417f, 681
 weather and, 655, 658, 668
Fossil fuels, 172–175, 187
 carbon cycle, 56f, 58, 63
 coal, 172, 172t, 173f, 189t
 combustion of, 592–593, 601, 692
 natural gas, 71, 174, 181f, 187, 189t

876 INDEX

nontraditional fossil fuel reservoirs, 174
peat, 113, 172*t*, 443–444
petroleum, 173–174, 173*f*–175*fV*, 187
Fossils
 ancient ocean sediments in Mongolia, 126, 127*f*
 climate change, evidence for in, 688–689
 coral reefs, 138, 498, 499*f*
 in geologic formation, 138
 geologic timescale, 45*f*, **129*f***, 150–153, 150*f*, 155
 giant dragonflies, 128, 128*f*, 129
 at Grand Canyon (United States), 9*f*
 human footprints, 356, 357*f*
 index fossils, 142, 143*f*, 155
 interpreting geologic history from, 138
 mass extinctions, 126, 129–133, 155
 Meganeuropsis fossils, 128, 128*f*, 129
 Mesosaurus fossils, 198*f*
 Mid-Ocean Ridge, 204
 on Mount Everest (Asia), 99
 paleontologists, 115, 138
 Paleozoic era, 151
 Pangaea and, 198–199, 198*f*
 Phanerozoic eon, 151–152, 155
 Precambrian era, 151, 155
 relative age of, 136, 138*f*, 155
 scientific illustration, 48, 48*f*
 snail fossils, 154*f*
 stromatolites, 65*f*
 trace fossils, 109*f*, 110
 trilobites, 136, 137*f*
Fracking (hydraulic fracturing), 175, 175*f*
Fracturing (minerals), 78, 79*f*, 231, 360, 360*f*, 367–368, 367*f*
Franciscan Assemblage, 309*f*
Franklin, Benjamin, 509, 659
French Revolution, volcanic eruption and, 283
Frequency (waves), 614
Fresh water, 41, 41*f*, 472–495
 agricultural use, 542*f*, 545, 545*f*
 algal bloom, 481
 biodiversity, 474, 475*f*
 domestic use, 542*f*, 543–544, 544*f*
 groundwater, 487–491
 industrial use, 542*f*, 545, 545*f*
 polluted waterways, 562
 springs, 489
 supply and demand, 541, 542*f*, 543, 543*f*, 544*f*
 wetlands, 482–486, 483*f*
 See also Lakes; Rivers; Streams
Fringing reef, 339*f*
Frontal wedging, 644, 668
Fronts, weather, 651–654, 668
Frost, 642*f*, 643
Frost wedging, 360, 361*f*
Fukushima Daiichi nuclear accident (Japan), 177, 324
Full moon, 725*f*, 726
Fusion, in stars, 773, 783, 789, 790, 805

G

Gabbro, 80*t*, 107
Gaia, 581
Galactic center (nucleus), 798, 798*f*, **798*f***, 802
Galactic disk, 801, 802
Galactic halo, 801–802, 813
Galaxies, 794*f*–795*f*, 799–803, 813
 clusters of, 813
 recessional velocity of, 801*f*, 813
Galena, 80*t*, 82, **82*f***
Galileo Galilei, 13, 712, 712*f*, **712*f***, 718–719, 723, 727, 733, 749, 796
Galileo probe, 752
Gamma-ray bursts, 788, 790
Ganges River (India), 345
Ganymede (moon of Jupiter), 757, 758, 758*f*, **758*f***, 765
Garnet, 83
Gas giants, 796
Gases
 in atmosphere, 41, 61, 584, 584*f*
 methane bubble, 576, 576*f*
 in seawater, 502–504, 503*f*
 in soil, 443, 448, 449
 in the sun, 779
 volcanic gases, 44
Gems, 72, 83
Geocentric model of the universe, 715–716, 715*f*, **715*f***, 716*f*, 719, 719*f*
Geographic information systems (GIS), 22, 29
Geologic change. *See* Change
Geologic formations, 138, 142*f*
Geologic maps, 23*f*
Geologic structures, 297
 basins, 299, 299*f*
 domes, 299, 299*f*
 faults, 231–232, 232*f*, 242, 299, 299*f*–303*f*, 301–303, 317, 334, 334*f*
 folds, 297–299, 297*f*–299*f*, **298*f***, 302, 317
 graben, 299, 300*f*, 301, 301*f*
 horst, 301, 301*f*
 mountains, 297–319
Geologic time, 44–46, 45*f*, 124–157
 absolute age, 134, 144–147, 155
 age spectrum, 146
 correlation, 141–142, 142*f*, 144, 155
 fossils, interpreting geologic history from, 138
 geologic timescale, 45*f*, **129*f***, 150–153, 150*f*, 155
 glaciation, 414–415, 414*f*
 global climate stability, 677–679, 678*f*, 679*f*
 Grand Canyon (United States), 9*f*, 39*f*, 128, 139, **139*f***, 141*f*, 142*f*
 index fossils, 142, 143*f*, 155
 interglacial periods, 414
 key beds, 142, 143*f*, 144, **144*f***, 155
 mass extinctions, 126, 129–133, 155
 radiometric dating, 145–147, 145*f*, 146*t*, 147, 147*t*, 151, 155
 relative age, 134–136, 134*f*–136*f*, 155

rock record, 139–144, 155
 supercontinents, 132, 132*f*
 unconformities, 139–141, 139*f*–141*f*, 155
Geologic timescale, 45*f*, **129*f***, 150–153, 150*f*, 155
Geology, 10*t*, 750
 See also Earth science
Geomorphology, 378
Geoscience
 geobiology, 51
 geochemist, 126
Geosphere, 33, 37–40, 37*f*–40*f*, 41, 42, 61
 carbon in, 56, 58
 water in, 477
 See also Minerals; Rocks
Geothermal energy, 8, 179, 179*f*
 Amazon River (South America), 6, 7*f*
 cement curing, 28
 drilling, 8, 8*f*
 electricity from, 8, 8*f*
 hot springs, 66*f*–67*f*
 in Iceland, 5, 6, 7*f*, 8, 16, 17, 17*f*, 27–28
 power plants, 8, 8*f*, 16, 17, 17*f*, 27–28
 resource maps, 8, 8*f*, **8*f***, 27, 27*f*
 resource planning, 27–28, 28*f*
 in the United States, 8, 8*f*, **8*f***, 27–28, 28*f*
 uses, 28
Gibbous moon, 725*f*, 726
GIS. *See* Geographic information systems
Glacial erosion, 361, 362*f*–363*f*, 398, 405–408, 405*f*–408*f*, **407*f***, 412, 433
Glacial ice, 215, 409, 688
Glacier Bay (Alaska), 404
Glacier National Park (Montana), **276*f***, 302, 302*f*, 416, 416*f*
Glacier Peak (Washington), **144*f***, 151
Glaciers, 40, 41, 395, 399–416, 433
 alpine glaciers, 399–400, 400*f*, 403, 403*f*, 410, 414, 416, 433
 Antarctica, 220, 362*f*–363*f*, 394*f*–397*f*, 396, 398, 398*f*, 404, 416
 arête, 408, 408*f*
 basal slip, 404, 405, 433
 cirques, 407, 407*f*
 continental glaciers, 334, 401, 408, 410, 433
 crevasses, 402–403, 402*f*, 403*f*, **403*f***, 433
 downhill flow, 399
 drift, 409, 433
 drumlins, 411–412, 413, 433
 equilibrium line altitude (ELA), 403, 404
 erosion by, 361, 362*f*–363*f*, 405–408, 405*f*–408*f*, **407*f***, 412, 433
 erosional landforms, 405, 406*f*–408*f*, 407–408, **407*f***, 688
 erratics, 409, **409*f***
 eskers, 412, 433
 firn, 399, 399*f*
 fjords, 408
 formation of, 399

glacial deposits, 199, 409–413, 433
 glacial meltwater, 412–413, 412*f*, 413*f*
 glacial striations, 405, 405*f*
 glaciation, periods of, 220, 414–415, 414*f*, 433
 hanging valley, 408
 in Himalaya (Asia), 305*f*
 ice erosion, 405, 432
 in Iceland, 5
 isotope ratios as evidence of climate change, 688
 kame, 412, 433
 kettle lakes, 412, 479
 lake formation from, 478, 493
 locations of, **697*f***
 loess, 413, 430–431, 430*f*, 433
 mass balance, 403–404
 melting of, 215, 396, 398, 398*f*, 412–413, 412*f*, 413*f*, 415, 416, 479, 514, 608, 610, 610*f*, **610*f***, 697–698, 699
 moraines, 409–410, 410*f*, 411, 413, 414, 425, 433
 movement of, 401–404, 433
 outwash, 412, 433
 paternoster lakes, 407, 407*f*
 plastic flow, 404–405, 405*f*, 433
 retreat of, 404, 415–416, 416*f*, 429, 433, 631, 695
 sandy till, 523
 Snowball Earth, 415, 699
 subglacial streams, 412
 terminus, 403, 409–410, 410*f*
 till, 408, 409*f*, 412
 tributary glacier, 408
 U-shaped valleys, 406*f*, 407
 V-shaped valleys, 405
 zone of ablation, 433
 zone of accumulation, 403, 433
Glasgow (Scotland), 510
Glaze (weather), 648, **648*f***
Glen Canyon Dam (Arizona), 548–549, 549*f*
Global positioning system (GPS), 21–22, **21*f***, 29
Global warming, 520–521, 700, 702
Globular clusters, 802, 802*f*, 813
Glossopteris fossils, 198*f*
Glover, Jerry, 440, 441*f*, 442
Gneiss, 117, 117*f*, 118, 119*f*, 121, 124
Gobi Desert (Asia), 420, 431
Gokyo Ri (Nepal), 210*f*
Gold
 in Earth's crust, 164*t*
 extraction with mercury, 88
 in native form, 74, 82
 mining, 68, 70, 70*f*, 90, 90*f*, 163
 ore formation, 166
 placer deposits, 166, 167*f*
 in seawater, 502
 specific gravity, 79
 uses, 80*t*
Gold Rush (California), 166
Golden Gate Bridge (California), 46*f*, 310
Goma Volcano Observatory (GVO), 288
Gondwana, 198, 198*f*, 312*f*
Gorner Glacier (Switzerland), 400, 400*f*
Graben, 299, 300*f*, 301, 301*f*, 425
Graded beds, 109–110, 110*f*

INDEX **877**

INDEX

Gradualism, 46–47, 61
Gran Telescopio Canarias, 797*f*
Grand Canyon (United States), 9*f*, 39*f*, 128, 139, **139*f***, 141*f*, 142*f*, 421*f*, 424, 548
Grand Prismatic Spring (Yellowstone National Park, Wyoming), 33, 33*f*
Grand Teton National Park (Wyoming), 720*f*–721*f*, 799*f*
Granite, 71, 71*f*, 75, 75*f*, 80*t*, 86, 106–107, 121, 268, 270
  Baffin Island (Canada), 39*f*
  chemical weathering of, 366*f*
  fracturing by exfoliation, 367–368, 367*f*
  weathering of, 367
Granitic continental crust, 343
Granitic magma, 270, 271, 271*t*, 275, 282, 282*f*, 289
Granitic plutons, 274
Granitic water, hydrothermal water and, 165
Graphite, uses, 80*t*
Grassland ecosystem, 129
Gravel, 110, 111, 153, 449–450
Gravitational lensing, 809, 811
  microlensing, 796
Gravity, 719–726, 733
  Newton's law of universal gravitation, 719, 722–723, 722*f*, 723*t*, 733
  star formation, 773, 773*f*
Gray, Chris, 519*f*
The Great American Desert, 558, 568
Great American Solar Eclipse, 720*f*–721*f*
Great Barrier Reef (Australia), 503, 531*f*, 698
Great Basin (Nevada), 424*f*, 425
Great Dark Spot (Neptune), 756
Great Lakes (North America), 215, 476, 476*f*, 478, 492, 492*f*
Great Lakes Basin (North America), 476, 476*f*, 492, 492*f*
Great Lakes Restoration Initiative (GLRI), 495
"The Great Nonconformity," 141
Great Pacific Garbage Patch, 183, 512, 512*f*, 533
Great Rare Earth Rock (monument in China), 161*f*
Great Red Spot (Jupiter), 712, 740, 753
Great Rift Valley (Africa), 222
Great Salt Lake (Utah), 114, 422
Great Smoky Mountains (Tennessee), 584
Greenhouse effect, 617–619, 618*f*, 632, 690
Greenhouse gases, 592, 618, 686, 692, 702
Groins, 525–526, 525*f*, 526*f*, 533
Gros Morne National Park (Canada), 124*f*–125*f*
Ground moraines, 410
Groundwater, 40, 41, 477, 487–491, 493, 554–559, 568
  aquifers, 488, 489*f*, 554–556, 555*f*, 565, 568
  cone of depression, 554
  depletion of, 695
  erosion by, 489
  hydrothermal water and, 165
  mining by-products, 87
  movement of, 488, 489*f*–491*f*, 491, 493
  permeability of, 487
  pollution of, 563–566, 564*f*, 564*t*, 568
  porosity of, 487
  saltwater in, 366
  saltwater intrusion, 557, 557*f*
  subsidence, 556–557
  water table, 487–488, 488*t*
Groundwater resources, 487
Gulf of Aden, 222
Gulf of Mexico, **531*f***
Gulf Stream, 509–510, 509*f*, 620
Guyot, 338, **338*f***
Gypsum, 76*t*, 80*t*, 82–83
Gypsum Springs formation, 98, 98*f*
Gyres, 510–511, 512, 533

H

Hadean eon, 45*f*, 150*f*, 151
Hadron collider (Europe), 162*f*
Hail, 648, 648*f*
Hailstones, 648, 648*f*
Half-life, 145, 145*f*, 146, 146*t*, 155
Halide minerals, 80*t*
Halite, 74–75, 74*f*, 80*t*, 108, 108*f*
Halley's comet, 762
Halons, 598, 601
Hanging valleys, 408
Hanging walls, 299
Hard coal, 172*t*
Hardness, minerals, 76, 76*t*
Hawaiian Island-Emperor Seamount chain (Pacific Ocean), 337, 338*f*, 339
Hayward Fault, 233, **233*f***
Hazardous rocks and minerals, 85–87, 91
  acid mines and heavy metals, 87, 87*f*
  asbestos, 85–86, 86*f*, 91
  carcinogens, 85, 86, 91
  environmental hazards, 85
  erosion and, 85
  mercury. See Mercury
  radon, 86–87, 91
Headlands, 518, 518*f*
Heat, 614, 620, 632
  advection, 620, 621*f*, 622
  change of state, 622
  climate change and, 689–690
  conduction of, 620
  convection, 620, 621*f*
  as groundwater pollutant, 561, 561*f*
  rock cycle and, 101
Heat transfer, 632
Heating, of seawater, 520
Heezen, Bruce, 330*f*
Heliocentric model of the universe, 716–718, 716*f*
Helium, 729, 784, 790, 805
Hematite, 77, 80*t*
Hermit formation, 128
Hertzsprung, Ejnar, 781
Hertzsprung–Russell diagram (H–R diagram), 780*f*, 781, 790
Hill, Chris, 612*f*
Himalaya (mountains in Asia), 208*t*, 210*f*, 220, 228, 243, 244*t*, 294, 295*f*, 296, 305*f*, 311–315, 311*f*, 314*t*, 316, 317
Hkakboo Razi (Mountain in Myanmar), 588*f*–589*f*
Hokkaido earthquake (2018), 386*f*
Holdren, John, 54
Holmdel Horn Antenna, 807
Holocene epoch, 45*f*, 150*f*, 414, 414*f*, 687
Holsteinian interglacial period, 687*f*, 688, 688*f*
Hooke, Robert, 611
Hoover Dam (Nevada), 536*f*–537*f*
Horse latitudes, 678
Horseshoe Falls (Canada), 354*f*
Horsts, 301, 301*f*, 425
"Hot-Jupiters," 796
Hot spots, 212, 223
Houser, Jana, 638
Hubble, Edwin, 799, 800, 801, 801*f*, 803, 805
Hubble Law, 801
Hubble-Lemaître law, 801, 809, 813
Hubble Space Telescope (HST), 728, 728*f*, **761*f***, 769*f*, 770, 774, 774*f*, 795*f*, 796, 798*f*
Hubble Ultra Deep Field, 795*f*
Hudson Bay (Canada), 41, 215
Hulett sandstone, 98
Human activities
  chemical composition of atmosphere, 585
  climate change and, 619, 692–693, 702
  coral reefs and, 500, 530
  global climate change and, 619
  oceans and, 532
  soil erosion and, 442, 442*f*, **442*f***, 467
  soil quality and, 459, 459*f*
  water pollution and, 562
Human health
  acid mines and heavy metals, 87, 87*f*
  asbestosis, 86
  black lung disease, 85
  carcinogens, 85, 86
  hazards of asbestos, 85–86, 85*f*, 86*f*, 91
  hazards of radon, 86–87, 91
  health hazards from rocks and minerals, 85–89
  industrial mineral extraction hazards, 85–86
  mercury, 68, 70, 85, 88–89, 88*f*, 89*f*, 91
  silicosis, 85
Humans
  carbon cycle impact, 63
  Earth systems and, 50–54, 61
  fossilized footprints, 356, 357*f*
  fossils of, 138
  health. See Human health
  mass extinctions and human activity, 155
  minerals and history, 71
  nitrogen cycle impact, 58, 61
  origins of, 133
  population threshold effects, 54, 61
  poverty, 54
Humboldt Current, 514
Humid continental climate, 684
Humid mesothermal climate, 682
Humid microthermal climate, 684
Humid subtropical climate, 682
Humid tropical climate, 680*t*, 681–682
Humidity, 641, 641*f*, 668
Humus, 443
Hurricanes, 640, 660, 662–664, 662*f*, 663*f*
  barrier islands and, 524
  categories of strength, 663*t*
  climate change and, 695–696
  costliest storms, **664*f***, 696
  damage from, 663
  deadliest storms, 663, **664*f***
  death toll, 664
  Dorian (2019), 640
  Fran (1996), 524
  Harvey (2017), 640, 696
  Irma (2017), 640, 663*f*, 696
  Jose (2017), 663*f*
  Katia (2017), 663*f*
  Katrina (2005), 358, 358*f*, 382, 389, 389*f*, 640, 662, 696
  Maria (2017), 640, 696
  Matthew (2016), 640
  Michael (2018), 640, 640*f*
  Nora (1997), 664
  Saffir–Simpson scale, 662, 663*t*
  Sandy (2012), 696
  strength of, 662, 663*t*
Hutchinson Island (Florida), 524*f*
Hutton, James, 46
Huygens, Christiaan, 611
Huygens space probe, 759
Hyakutake (comet), 762*f*
Hybrid cars, 182
Hydra (moon of Pluto), 761
Hydraulic fracturing (fracking), 175, 175*f*
Hydrocarbons, 592
Hydroelectric power, 179, 550–551, 551*f*
Hydrogen, 730*f*, 752, 784
Hydrogen fusion, 741, 781, 782
Hydrogen sulfide, acid mines, 87
Hydrologic cycle. See Water cycle
Hydrology, 10*t*
Hydrosphere, 33, 37, 37*f*, 38*f*, 40–41, 41*f*, 42, 44, 61, 471
  carbon in, 56, 58, 61
  fresh water, 41, 41*f*
  glaciers, 40, 41
  groundwater, 40, 41
  ice, 40–41
  plate tectonics and, 223
  rock cycle and, 101
  See also Fresh water; Oceans; Rivers; Streams; Water; Water resources
Hydrothermal explosions, 36
Hydrothermal metamorphism, 118, 121
Hydrothermal processes, 165–166, 165*f*, 187
Hydroxide minerals, 80*t*
Hypothesis, 13

I

Ice
  age of, 688
  Arctic sea ice, 416, 433
  as cryosphere, 40–41, 41*f*
  on Earth, 399, 401
  glacial ice, 409
  ice erosion, 405, 432
  ice shelf, 401, 416, 676, 676*f*, 701

melting of, 699
See also Glaciers
Ice age, 414*f*
Ice caps, 684
Ice erosion, 405, 432
Ice sheets, 401, 415, 610, 610*f*, 610*f*, 697, 697*f*
Ice shelf, 401, 416, 676, 676*f*, 701
Icebergs, 215, 215*f*, 404, 404*f*, 676, 676*f*
Idaho, 277
Igneous intrusions, 135, 135*f*
Igneous rocks, 100, 104–107, 121
  common rocks, 106–107, 121
  cooling, 165
  crystal sizes of, 104, 105*t*, 106
  intrusive rocks, 105*f*, 106, 121
  extrusive rocks, 104–105, 105*f*, 121
  features of, 104
  formation of, 99, 104–107
  magmatic processes, 165, 187
  plutonic rock, 105*f*, 106, 121, 269, 270, 288, 289, 314
  pressure-release fracturing, 360, 368
  spheroidal weathering, 368, **368*f***
  textures of, 104–105, 105*f*
  types, 104
  volcanic rocks, 104–105, 105*f*, 121
Illampu peak (Andes), 307*f*
Imago Mundi (map), 19, **19*f***
Incandescent lighting, 182, 182*f*
Included fragments, principle of, 136, 136*f*, 147
Independent variable, 13
Index fossils, 142, 143*f*, 155
Indian-Australian plate, 206*f*
Indian Ocean, 345, 501
Indian plate, 210*f*, 228, 312, 313*f*, 317
Indo-Australian plate, 247
Industrial minerals, 82–83, 83*f*
Industrial processes, hazardous minerals and, 85, 90, 90*f*
Industrial Revolution, 71
Industry, energy efficiency, 182
Inertia, 718
Infrared radiation, 614–615
Inorganic compounds, 72
Instability, of systems, 18*t*, 29
Interglacial periods, 414, 687, 687*f*, 688, 688*f*
Intergovernmental Panel on Climate Change (IPCC), 185, 700
Interior deserts, 418, 420
Internal energy, 620
International date line, 19–20
International Thwaites Glacier Collaboration, 396
Intrusive igneous rocks, 105*f*, 106, 121
Invasive species, 495, 548, 549
Io (moon of Jupiter), 712, 757, 757*f*, 765
Ions, 73
  sedimentary rock formation, 111
  weathering of rock and, 99, 108, 108*f*
Iridium, in clay, 15, 130
Iron

banded iron formations, 167, 167*f*, 582–583, 582*f*, **583*f***, 601
commercial importance, 82
early atmosphere, 582
in Earth's crust, 73, 73*t*, 164*t*
heating of, 614–615, 614*f*
mineral deposits, 167, 167*f*
rust formation, 360, **360*f***, 366
Irregular galaxy, 800, 800*f*
Irrigation, 461, 463, 540, 543*f*, 546, 547, 558, 568, 695
Island arcs, 306–307, 306*f*, 317, 336–337, 336*f*, 337*f*, 349
Islands
  barrier islands, 523, 524–526, 524*f*–526*f*
  volcanic islands, 306, 337–338, 338*f*, 349
Isobars, 649
Isolines (contour lines), 22, 23*f*, 24, 24*f*
Isostasy, 214–216, 215*f*, 223, 305
Isotactic adjustment, 215, 215*f*, 223
Isotherms, 624
Isotope ratios, as evidence of climate change, 688
Isotopes, 144–147, 145*f*, 146*t*, 147*t*, 155
Isua supracrustal belt (ISB, Greenland), 147*t*
Izu-Ogasawara Trench (Pacific Ocean), 327*t*

J

J1342+0928 (quasar), 812
James Webb Space Telescope, 729, 729*f*, **729*f***, 796
Jasper, 83
Jeju Island (South Korea), 158*f*–159*f*
Jet stream, 650, 668, 679, 679*f*, 702
Ji, Kang-Hyeun, 194
Joe Creek (Canada), 375*f*
Joint, 317
Jovian planets, 741, 742*f*, 743, 743*t*, 752–756, 765
  See also Jupiter; Neptune; Saturn; Uranus
Juan de Fuca plate, 347
Juno space probe, 712, 736*f*–737*f*, 740, 740*f*, 753
Jupiter (planet), 757, 765
  atmosphere of, 752*f*, 765
  barycenter of, **723*f***
  climate of, 753–754
  formation of, 741
  Galileo and, 712
  geology of, 752–753
  Great Red Spot, 712, 740, 753, 753*f*
  images of, 753, 753*f*
  moons of, 712, 712*f*, **712*f***, 718–719, 757–760, 765, 757*f*
  north and south poles, 740, 740*f*
  properties of, 723*t*, 743*t*, 793*t*
  scale drawing of, 742*f*
Jurassic period, 45*f*, 150*f*, 152*f*, 153

K

K–Pg extinction, 14–15

Kahn, Greg, 696*f*
Kame, 412, 433
Kelly, Patrick, 102*f*–103*f*
Kepler, Johannes, 717–718, 717*f*, **717*f***, 733, 809
Kepler-90 (star), 732, 732*t*
Kepler space telescope, 732, 796
Kermadec Trench (Pacific Ocean), 327*t*
Kettles, 412, 479
Key beds, 142, 143*f*, 144, **144*f***, 155
Kīlauea volcano (Hawai'i), 213*f*, 218*f*–219*f*, 277, 279, 279*f*, **279*f***, 288
Kinetic energy, 620, 632
Klein, Dave, 49
Komatiite, 107
Köppen classification, 680, 680*t*
Kuiper belt, 760–761
Kuril-Kamchatka Trench (Pacific Ocean), 327*t*
Kyoto Protocol, 700

L

La Brea Tar Pits (California), 174, 174*f*
La Sila Observatory (Chile), 776*f*
Ladzinski, Keith, 744*f*
Lake Baikal (Russia), 479–480
Lake Bermin (Cameroon), 474
Lake Malawi (Africa), 222
Lake Mead (United States), 536*f*–537*f*, 548
Lake Okeechobee (Florida), 479
Lake Powell (United States), 548, 549
Lake Roosevelt National Park (Washington), 655*f*
Lake Turkana (Africa), 222
Lake turnover, 480, 481–482
Lakes, 478–482, 493
  deep lakes, 480
  desert lakes, 114, 421–422
  eutrophic lakes, 480*f*, 481
  formation of, 479–480
  glacial lakes, 412, 479
  lava lakes, 217, 266
  methane bubbles, 576, 576*f*
  oligotrophic lakes, 480–481, 562
  oxbow lakes, 376*f*, 479
  paternoster lakes, 407, 407*f*
  playa lakes, 422, 422*f*
  thermocline, 481
  turnover, 480, 481–482
  volcanic crater lakes, 474
Laki (volcanic fissure in Iceland), 283, 285, 287
Landforms
  along sandy coastlines, 523–524
  deposited by glacial meltwater, 412–413, 412*f*, 413*f*
  glacial erosional landforms, 405, 406*f*–408*f*, 407–408, 407*f*
Landslides, 246, 382–388, 383*f*, 390
  angle of repose, 386–387, 387*f*, 388*f*
  creep, 384, 384*f*, 385*t*
  debris flow, 383
  earthflow, 383
  in earthquakes, 246, 388, 390
  effects of, 385–386
  flow, 383, 383*f*, 385*t*

mudflow, 383
orientation of rock layers and, 386
risk factors for, 386–388
rockfall, 383*f*, 384, 385*t*
rockslides, 383–384, 383*f*
sand and, 387, 388*f*
slides, 383, 383*f*, 385*t*
slump, 383*f*, 384–385, **385*f***, 385*t*
steepness of slope and, 386
type of rock and, 386
unconsolidated materials, 386–387, 387*f*
volcanoes and, 388, 390
water and vegetation and, 387–388, 388*f*
Lapidary art, 99
Large Magellanic Cloud (galaxy), 800
Larsen Ice Shelf (Antarctica), 676, 676*f*
Last Interglacial, 498
Latent heat, 622, 622*f*, 632
Lateral erosion, of streams, 373–374, 373*f*, 425
Lateral moraines, 410*f*, 411, 425
Latitude, 19, 20, 20*f*
  average temperature and, 624–626, 625*f*
  deserts and, 417–418, 417*f*, 418*f*
  horse latitudes, 678
Laurasia, 198, 198*f*
Laurentian Channel (Canada), 344*f*
Lava, 104, 213*f*, 276, 276*f*
  aa lava, 276, 277*f*
  cooling of, 276
  lava flows, 264, 265*f*
  lava lakes, 217, 266
  lava plateau, 277, 289
  pahoehoe lava, 276, 277*f*, 279
  solidification of, 104–105
Lava lakes, 217, 266
Lava plateau, 277, 289
Lava ponds, 218*f*–219*f*
Lava tubes, 218*f*–219*f*
Laze, 279
Leaching, 447
Lead, 79, 82, 85, 87
Lead-206 and -207, 147*t*
Levees, 380–381, 380*f*, 381*f*, 389, 389*f*
Lewis Overthrust (Montana), 302, 302*f*
Libyan Desert, 361*f*
Lichens, 583
LIDAR (light detection and ranging), 24
Life
  atmospheric support for, 41
  biodiversity, 42, 50–51, 52*f*–53*f*, 54, 61
  biosphere support for, 41–42, 56
  evolution of, 44, 583, 601
  formation of, 61
  geobiologist, 51
  mass extinctions, 126, 129–133, 155
  origins of, 581
  species, 42
  theory of evolution, 136
  types of, 41–42
  See also Animals; Biosphere; Human health; Humans; Marine life; Plants

INDEX

Light, 632
  absorption and emission, 614–615
  gravitational lensing, 811
  in oceans, 528
  photons, 611, 614, 632
  properties of, 729–731
  reflection of, 615–616, 615f
  scattering of, 616, 616f
  spectrum of, 729, 730f
  sunlight, 606f–607f, 611, 612f–613f, 624, 627t
  waves, 611, 611f
Light pollution, 728
Light-year, 780
Lighting technologies, 181–182, 181f, 182f
Lightning, 239, 263f, 659, **659f**, 668
Lignite, 172t
Limbs, 298, 298f
Limestone, 73, 98, 121, 365
  bioclastic, 112
  as building material, 365
  carbonate platforms, 343, 343f
  dissolution of, 489
  folding and tilting of, 134, **134f**
  formation of, 108, 112, 112f
  marble formation from, 116, 117, 121
  oolitic limestone, 112, 112f
  Paleozoic limestone, 152, 152f
  uses of, 152, 163
Limnologist, 481, 481f, **481f**
Limonite, 80t
Liquefaction, 246
Liquid metallic hydrogen, 752
Lithification, 99, 112
Lithium, 183
Lithosphere, 37f, 40, 40f, 94f, 259
  continental lithosphere, 208–209, 215f
  downslope sliding of, 211–212, 212f
  in Earth's layers, 199t, 200, 200f, 201f, 223, 225t
  Hudson Bay (Canada), 41
  isostasy, 214–216, 215f, 223, 305
  mountain building, 220, 305
  oceanic lithosphere, 209, 215f
  plate tectonics, 205, 205f, 206, 206f, 207, 208, 209
  tectonic plate movement, 211–212, 212f, 214–216, 215f
Lithospheric plates, 205, 207, 207f, 208
Little Dipper (constellation), 713, 713f
Little Ice Age, 688
Liutkus-Pierce, Cynthia, 356, 357f
Loam, 450, 452t
Loamy soil, 451, 451f, 452t, 453
Loess, 413, 430–431, 430f, 433, 454–455, 455f, 467
Loktak Lake (India), 481
Long Island Sound (United States), 524–525, 525f
Long Valley (California), 282
Longitude, 19, 20, 20f
Longitudinal dunes, 430, 430f
Longshore currents, 518–519, 519f, 523, 525f
Los Angeles Aqueduct (California), 540
Los Angeles River (California), 540
Love waves, 235–236, 235f
Lovelock, James, 581
Luminosity (stars), 781, 790
Luna-3 (Soviet orbiter), 749
Lunar eclipses, 726
Lunar phases, 725–726, 725f
Lunar Reconnaissance Orbiter, 749, 749f
Luster (minerals), 76f, 77
Lutz, Stephanie, 608, 609f
Lyell, Charles, 154
Lystrosaurus fossils, 198f

M

M87 (giant elliptical galaxy), 803
Maat Mons (volcano on Venus), 747f
Mafic rocks, 106
Magellan spacecraft, 25, 25f, 27, 746f, 747, 747f
Magellan Telescope (Chile), 771f
Magma, 5, 44, **60f**, 96, 100, 101, 267–271, 274, 289
  basaltic magma, 207, 268, 271, 271t, 308, 332, 346
  behavior of, 271, 274
  cooling of, 104, 106, 274
  formation of, 99, 104, 267–269, 267f–269f
  lava, 104, 213f, 217, 264, 265f, 276, 276f
  mantle plumes, 212, 223, 269, 269f
  metamorphism, 118
  plate tectonics, 208–209
  plutons, 274–275, 274f–276f, 288, 289
  spreading centers, 206, 217, 268–269, 268f, 335
  types of, 269–271, 271t
  volcanic formations, 207
  in volcanoes, 105, 105f, 106, 107
  See also Basaltic magma; Granitic magma
Magnesium, in Earth's crust, 73, 73t
Magnetic field
  of Earth, 203–204, 203f, 204f, 256, 256f, 257f
  normal magnetic polarity, 203
  of sun, 779
Magnetic reversal, 203
Magnetism
  in minerals, 79
  in seafloor rocks, 203–204, 203f, 204f
Magnetite, 79, 80t
Magnetometers, for seafloor research, 329
Magnifying lens (telescopes), 726, 727f
Main sequence (stars), 781, 783f, 790
Malthus, Thomas, 51
Manganese, 186, 186f
Mantle, 37f, 39, 40f, 199t, 200, 200f, 206, 223, 259
  convection, 205f, 211–212, 211f
  crust–mantle boundary, 253–254, 259
  hot spots, 212, 223
  mantle–core boundary, 254
  mantle plumes, 212, 223
  structure of, 254, 254f, 255
Mantle convection cells, 40f, 205f
Mantle plumes, 212, 223, 269, 269f, 748
Mantle rock, 206
Map projections, 19–21, 20f, 21f, 29
Maps and mapping, 19–27, 29
  ancient mapping, 19
  coordinate system, 20, 29
  earthquakes, 244
  geographic information systems (GIS), 22, 29
  geothermal resource maps, 8, 8f, **8f**, 27, 27f
  global positioning system (GPS), 21–22, **21f**, 29
  history of, 19, 29
  interpreting maps, 31
  isolines (contour lines), 22, 23f, 24, 24f
  latitude and longitude, 19, 20, 20f
  map projections, 19–21, 20f, 21f, 29
  maps as models, 22, 23f, 29
  maps as representations of data, 19
  Mercator projections, 20–21, 20f, 29
  Mid-Atlantic Ridge, 330f
  Mollweide projections, 21, 21f, 29
  planets and space objects, 24, 25f, 27
  scale of a map, 22
  seafloor mapping, 24–25, **25f**
  technology and, 21–22, 24
  topographic maps, 23f, 24–27, 24f–26f, 31
  types of maps, 23f, 29
  Winkel tripel projections, 21, 21f, 29
Marble, 116, 117, 121, 163
Mariana Trench (Pacific Ocean), 326, 326f, 327f, 337, 347
Marine geophysicist, 334
Marine life
  algae, 608
  biosphere support for, 41
  blue-green algae, 581
  carbon dioxide budget, 131
  carbonate platforms, 343, 343f
  carbonates and, 502–503
  chemosynthetic bacteria, 333, 349
  coastal wetlands, 483
  coastlines, 528–531, 533
  cyanobacteria, 581, 581f
  foraminifera, 341f, 688
  fossils of, 137f
  limestone formation, 112, 112f
  mass extinction, 126, 127f
  microplastics and, 68, 183
  in pelagic sediment, 340
  Permian Ocean, 131, 133, 151, 287
  plankton, 113, 113f, 480, 528, **528f**, 530, 533, 688
  plastic pollution in mercury cycle, 68
  pollutants and, 562
  rock cycle and, 99–100
  shell formation, 99
  tubeworms, 333, 333f
  zooplankton, 528, **528f**, 533, 583
  zooxanthellae algae, 500
Marine organisms, acid rain and, 594
Marine reefs, 112
Marine west coast climate, 682, 682f
Mariner spacecraft, 745
Mars (planet), 745, 748–749
  atmosphere of, 745
  climate of, 745, 765
  geology of, 748
  images of, 746f
  plate tectonics, 748–749
  properties of, 723t, 743t, 793t
  retrograde motion, 714, **714f**
  scale drawing of, 742f
  topographical map of, **748f**
  volcanoes on, 748, 748f, **748f**
Mars Exploration Rover (MER) mission, 750
Mars Global Surveyor, 748
Mars Reconnaissance Orbiter, 738, 740
Mars rovers, 738, 739f, 750–751, 765
Marshes, 482
MARUM-SEAL (robotic vehicle), 324, 348
Marum Volcano (Vanuatu), 278f
Mass balance (glaciers), 403–404
Mass extinctions, 126, 129–133, 155
  Cretaceous period, 129, 130–131, 131f
  extraterrestrial impacts as cause, 130–131, 131f
  hypotheses for, 129–130
  Permian period, 129, 131, 151, 287
  volcanic eruptions and, 131
Mass wasting, 382–388, 390
  See also Landslides
Matter, 18t, 29
  baryonic matter, 810
  conservation of, 44
  dark matter, 809–810, 813
  in Earth's spheres, 37f, 44
  specific heat, 623
Matterhorn (mountain in Switzerland), 407–408, 408f
Matthews, Drummond, 203
Mauna Kea (volcano in Hawai'i), 279
Mauna Loa Observatory (Hawai'i), 57f
McBride, Pete, 552f–553f
Meanders, 373–374, 373f
Mechanical weathering, 359–364, 390
  acting together with chemical weathering, 366–368, 367f
  frost wedging, 360, 361f
  by plants and animals, 364, 364f
  pressure-release fracturing, 360, 360f
  processes causing, 359–364
  thermal expansion and contraction, 360
  See also Erosion
Medial moraines, 410f, 411
Mediterranean climate, 682
Mediterranean Sea, 529t
Melting
  of continental glacier, 334
  decompression melting, 268

of glaciers, 215, 396, 398, 398f, 412–413, 412f, 413f, 415, 416, 479, 514, 608, 610, 610f, **610f**, 697–698, 699
of ice, 699
of ice sheets, 697
of ice shelf, 676
island arc formation and, **336f**
of minerals, 267, 267f
partial melting, 270
of permafrost, 699, 702
pressure-release melting, 268, 268f, 269f
of rocks, 267–268, 267f
of snow, 702
of two or more minerals, 270
Mercalli scale, 237
Mercator projections, 20–21, 20f, 29
Mercury (chemical element), 68, 70
in Earth's crust, 164t
gaseous mercury, 68
gold extraction with, 68, 70, 70f, 88
methylmercury, 68
as native element, 88
as pollutant, 68, 91, 172
toxic mercury in seawater, 68, 69f
toxicity, 70, 85, 88–89, 88f, 89, 89f, 91
Mercury (planet), 744–745, 765
atmosphere if, 744, 745
atmosphere of, 744, 745, 765
formation of, 741
geology of, 745
images of, 745, 746f
properties of, 723t, 743t, 793t
scale drawing of, 742f
Mercury barometer, 587
Mercury cycle, 68, 70
Mertz Glacier (Antarctica), 404
Mesas, 424–425, 425f
Mesopause, 590f, 591
Mesosphere, 590f, 591
Mesozoic era, 45f, 150f, 152, 310
MESSENGER spacecraft, 745, 746f
Messier, Charles, 799
Metal processing, hazardous minerals and, 85
Metals
commercially important metals, 82
heat conduction by, 620
natural concentrations in Earth's crust, 164t
ore deposits, 164–167, 164t
ores, 82, 163, 164, 187
specific gravity of, 79
Metamorphic grade, 115, 119f
Metamorphic rocks, 115–119, 121
features of, 115–117, 116f
foliation, 116–117, 116f
formation of, 100, 115–117, 116f
metamorphism, 115–118
pressure-release fracturing, 360, 368
Metamorphism, 115–119, 118f–119f, 121
Meteor Crater (Arizona), 47f
Meteorites, 761–762

as cause of mass extinction, 16, 130–131, 131f
formation of, 256
Meteoroids, 761
Meteorologists, 649
Meteors, 761
impact crater, 15f, 16, 47f, 130, 131f
Methane, 61, 568, 578f, 592, 600, 618, 692, 699–700, 702
on Pluto, 761, 761f
on Titan, 759
Methane bubble, 576, 576f
Methane clathrate, 578, 578f
Methane seeps, 578
Methylmercury, 68
Mica, 78, 78f, 80t, 81, 81f, 91, 116, 116f
Microcline, cleavage plane, 78f
Microplastics, 68, 183, 512
Mid-Atlantic Ridge, 203, 203f, 207, 207f, 208t, 244t, 330f, 521
Mid-Ocean Ridge, 203–204, 204f, 207, 208t, 212, 217, 243–244, 268, 301, 304, 326, 331–334, 331f, 333f, 335, 340, 341, 349
Mid-ocean-ridge basalt (MORB), 96
Mid-ocean ridges, 220, 243–244
Midewin National Tallgrass Prairie (Illinois), 483
Midlatitude cyclones, 653–654, **653f**, 662
Midwest Wetlands Initiative, 483
Migmatite, 118, 119f
Milky Way (galaxy), 718, 799f, 801–803, 801f, 809, 813
galactic disk, 801
galactic halo, 801–802, 813
globular clusters, 802, 802f, 813
Miller, Paul, 674, 675f
Miller–Urey model, 580
Milling, hazardous minerals and, 85
Mineral deposits
evaporative deposits in deserts, 166
formation, 165–167
hydrothermal vein deposits, 165, 165f
sedimentary processes, 166, 187
See also Mining
Mineral reserves, 169, 187
Mineral resources, 187
banded iron formations, 167, 167f, 582–583, 582f, **583f**
prehistoric use of, 163
rare earth metals, 162
types of, 163–164
Minerals, 66–93, 71–72, 91, 267
chemical composition of, 73–74, 91
chemical formula of, 73–74
cleavage of, 78, 78f
color of, 76–77
commercially important minerals, 82–83, 83f
crystal habit of, 77, 77f
crystalline structure of, 74–75, 74f, 75f, 91, 104, 105t, 106
dissolving in acid, 79
in Earth's crust, 81, 81f
field guide, 93

fluorescence of, 79
fracture of, 78, 79f
gems, 72, 83
global coordination of mineral development, 160
groups, 74
hardness of, 76, 76t
hazardous rocks and minerals, 85–89, 91
hazards from mining and extraction, 85–86, 89, 90, 90f, 91
hydrothermal processes, 165–166, 165f, 187
identification of, 75, 91
industrial minerals, 82–83, 83f
inorganic, 73
luster of, 76f, 77
magmatic processes, 165, 187
magnetism of, 79
melting of, 267, 267f
Mohs' scale, 76, 76t
naturally occurring, 71–72
nature of, 72
as nonrenewable resources, 164
observing, 71
ore, 82, 153
phosphorescence of, 79
physical properties of, 75–79, 91
radioactivity of, 79
rock-forming minerals, 74, 79, 80–82, 91
scratch test, 75
specific gravity of, 78
stable minerals, 115
streak, 77
toxicity of, 71
See also Geosphere; Mineral deposits; Mineral reserves; Rocks
Minilab activity, 14, 42, 71, 101, 146, 180, 213, 216, 238, 271, 310, 339, 364, 431, 452, 482, 486, 515, 563, 584, 585, 618, 621, 653, 690, 763
Atmospheric Convection and Wind, 621
Biofuel, 180
Big bang model, 806
Carbon Dioxide, Photosynthesis, and Oxygen, 583
Convection, 213
Coriolis Effect Models, 654
Energy Exchanges in Atmospheric Models, 690
Fault-Block Mountain Models, 310
Greenhouse Effect Simulation, 618
Hawaiian Hot Spot, 339
Isostasy, 216
Lake Turnover, 482
Make a Barometer, 653
Micrometeorites, 763
Modeling a Cloud in a Bottle, 42
Modeling Soil Erosion, 460
Motion in the Ocean, 515
Observing Minerals, 71
Oxygen in the Atmosphere, 585
Point Source Pollution Models, 563
Radioactivity and Half-Life, 146, 146t
Rock Cycle Model, 101
Role of Wetlands, 486

Scientific Law Versus Theory, 14
Seismograph, 238, 238f
Simulating Gravity, 724
Simulating Wind Erosion, 431
Soda Bottle Volcano, 271
Spectroscope, 778
Weathering by Plant Roots, 364
Weathering of Iron, 368
What Is in the Soil?, 452
Mining
acid mines, 87, 87f
of coal, 168f, 169, 170f–171f
of copper, 168f, 169
hazardous minerals and, 85–90
hazards from, 85–86, 89, 90, 90f, 91
of manganese nodules, 186, 186f
open-pit mines, 168f, 169
of rare earth metals, 162
of stone, 163
surface mines, 168f, 169
underground mines, 168f, 169, 170f–171f
Minor planets, 761–762
Mintaka (star), 714t
Miocene epoch, 45f, 150f, 154
Miranda (moon of Uranus), 759–760
Mississippi River (United States), 369, 370, 377, 377f, 380f, 393, 530
Mississippi River Delta, 358, 382, 389, 389f
Mississippian period, 45f, 150f
Missouri earthquakes (1811 and 1812), 244, 249
Missouri River (United States), 390
Models, 12
maps as models, 22, 23f, 29
in science and engineering, 12–13
Moderate Resolution Imagine Spectroradiometer (MODIS), 746
Moho (Mohorovičić discontinuity), 254, 259
Mohorovičić, Andrija, 253
Mohs' scale, 76, 76t
Mollweide projections, 21, 21f, 29
Moment magnitude, 237, 239, 261
Monoculture, 463
Monsoons, **655f**, 658, 664, 668
Monument Valley (United States), 426f–427f
Moon (Earth's moon), 748–749, 765
bombarded by meteorites, 750
distance from Earth, 725, 740, 742f
formation of, 750
images of, 749f
maria, 749
motion of, 725–726, 725f, **726f**
orbit of, 725–726, 725f, **726f**
origin and structure, 749–750
phases of, 725–726, 725f, 733
quakes on, 750
tides and, 505–507, 505f, 506f, 533
water on, 750, 765
Moraines, 409–410, 410f, 411, 413, 414, 425, 433
MORB. *See* Mid-ocean-ridge basalt
Morley, Lawrence, 203

INDEX **881**

INDEX

Motion, Newton's laws of, 14, 718, 723, 809
Mount Cayambe (South America), 399
Mount Everest (Asia), 99, 294, 295f, 314t, 315
Mount Hood (Oregon), 280f
Mount Kenya (Africa), 304, 399
Mount Kilimanjaro (Tanzania), 304
Mount McKinley. See Denali
Mount Nyiragongo (Congo), 97f, 266, 272f–273f, 288
Mount Pinatubo (Philippines), 285, 285f, 287
Mount Rainier (Washington), 107, 107f, 280
Mount Sakurajima (Japan), 263f
Mount St. Helens (Washington), 280, 283, 284f
Mount Stephen (British Columbia), 137f
Mount Vesuvius eruption, 266, 266f, 288
Mount Waialeale (Hawai'i), 417
Mountain building, 220, 223, 304–306
  foreland basin, 308, 308f
  island arcs, 306–307, 306f, 317, 336–337, 336f, 337f, 349
  subduction at a continental margins, 307–308, 307f–309f, 310
  underthrusting, 305, 306
Mountaineering, 309f, 360, 399, 403
Mountains, 297–319
  deserts and, 418
  elevations of, 314t
  formation of, 220, 223, 304–306, 317
  graben, 299, 300f, 301, 301f, 425
  horsts, 301, 301f, 425
  island arcs, 306–307, 306f, 317, 336–337, 336f, 337f, 349
  orogeny, 304
  tectonic stress and, 297
  weather and, 655, 655f
  world mountain chains, 316f
  See also Mountain building
Mouth bars, 378, 379f
Mud cracks, 109, 109f
Mudflats, 482
Mudflow, 383
Mudstone, 80t, 86, 98, 111, 111f, 117–119, 117f–119f, 121
Mullan, Brendan, 710, 711f
Mullineaux, Don, 283
Multnomah Falls (Washington), 288
Muscovite, 81f
Museum conservator, 136
Myakka River State Park (Florida), 484

N

National Geographic Explorers, 6, 34, 52, 68, 96, 126, 160, 194, 228, 264, 294, 322, 356, 396, 440, 474, 498, 538, 576, 608, 638, 674, 738, 770, 796
  Alam, Munazza, 770, 771f
  Ali, Saleem, 160, 161f
  Bowman, Katlin, 68, 69f
  Boyes, Steve, 538, 539f

Calisti, Marcello, 322, 323f
Carmichael, Sarah, 126, 127f
Colon, Knicole, 796, 797f
Cutler, Joe, 474, 475f
Ehlmann, Bethany, 738, 739f
Glover, Jerry, 440, 441f
Liutkus-Pierce, Cynthia, 356, 357f
Lutz, Stephanie, 608, 609f
Miller, Paul (DJ Spooky), 674, 675f
Mullan, Brendan, 710, 711f
Pettit, Erin, 396, 397f, 398, 697
Quanbeck, Caroline, 498, 499f
Richards, Cory, 294, 295f
Ruzo, Andrés, 6, 7f, 8
Sartore, Joel, 52f
Seimon, Anton, 638, 639f, 640
Shah, Afroz Ahmad, 228, 229f
Sims, Kenneth Warren, 96, 97f
Soldati, Arianna, 264, 265f
Stamps, D. Sarah, 194, 195f, 196
Tessin, Allyson, 34, 35f
Walter Anthony, Katey, 576, 577f
National Geographic Okavango Wilderness Project (NGOWP), 538
National Geographic Photo Ark, 52f–53f
Native element minerals, 79, 80t, 88
Natural gas, 71, 174, 181f, 187, 189t
Natural resources, 159, 163
  See also Energy resources; Mineral resources
Nazca plate, 209, **209f**
Neap tides, 506f, 507
Nebulae, 773, 774
Negative correlation, 11
Negative feedback mechanism, 50
Neodymium–142, 147t
Neogene period, 45f, 150f
Neptune (planet), 755–756
  atmosphere of, 752f, 756, 765
  climate of, 756
  Great Dark Spot, 756
  images of, 755f
  moons of, 760, 765
  properties of, 723t, 743t, 793t
  scale drawing of, 742f
Neutron stars, 783, 783f, 785, 790, 810
New Horizons spacecraft, 761, 761f
New Madrid Fault zone (United States), 244, 245, 249
New moon, 725–726, 725f
Newton, Isaac, 611, 722, 809
  law of universal gravitation, 719, 722–723, 722f, 723f, 733
  laws of motion, 14, 718, 723, 809
NGC 6397 (globular cluster), 802f
Niagara Falls (North America), 354f–355f, 373–374, 373f
Nigge, Klaus, 484f
Nile delta (Africa), 550
Nile River (Africa), 550
Nimbostratus clouds, 646f, 647, 652, 652f
Niobium, 162, **162f**
Nishinoshima (volcano in Japan), 191f
Nitrates, 58, 59f

Nitrification, 58, 59f
Nitrites, 58
Nitrogen, 584, 584f, 593, 601
  isotope (nitrogen-14), 145, 146t
  gas, 58, 58f
Nitrogen cycle, 44, 58, 59f, 61
Nitrogen fixation, 58, 59f
Nitrogen mineralization, 58, 59f
Nitrogen oxides, 58, 59f, 593, 601, 692
Nitrogen uptake, 58, 59f
Nix (moon of Pluto), 761
Nodular chert, 114, 114f
Nonbiodegradable pollutants, 561
Nonconformity, 140–141, 155
Nonmetallic mineral resources, 163, 163f
Nonpoint source pollution, 561
Nonrenewable energy resources, 172–177, 187
  coal, 172, 172t, 173f, 189t
  end-use efficiencies of, 181f
  fossil fuels, 172–175, 187
  future of, 183–185
  hydraulic fracturing (fracking), 175, 175f
  natural gas, 71, 174, 181f, 187, 189t
  nontraditional fossil fuel reservoirs, 174
  nuclear fuels, 175, 176f, 177, 187, 189t
  petroleum, 173–174, 173f–175f
Nonrenewable resources, 164
Nontraditional fossil fuel reservoirs, 174
Normal faults, 299, 300f, 301, 317
Normal magnetic polarity, 203
North American Atlantic shelf, 342
North American plate, 206f, 241–243, 241f, 245, 258, 347
North Atlantic gyre, 510–511
North Cascades National Park (Washington), 655f
North magnetic pole, 256, 257f
North Pacific gyre, 533
North Pole, 20f
North Sea, 343
North Star, 713, 714f, 724
Nubian plate, 196, **196f**, 222
Nuclear energy
  nuclear accidents, 176f, 177
  nuclear fuels, 175, 176f, 177, 187
  reactors, 175, 176f, 177
  waste, 177, 561, 568
Nuclear waste disposal, 566
Nucleus (comet), 762, 762f
Nuna (supercontinent), 214

O

O-18/O-16 ratios, as evidence of climate change, 688
Objective lens (telescopes), 726, 727f
Obsidian, 106, 106f
Obsidian Cliff (Yellowstone National Park), 107f
Occluded fronts, 652–653, 653f
Ocean acidification, 500, 502, 503f
Ocean basins, 326–327, 326f, 501, 501f
Ocean currents, 509–516, 533

  climate and, 510, 632
  Coriolis effect, 511, 511f, 512, 514, 516, 533, 650, 650f, 654
  deep-sea currents, 131–133, 132f, 513–514, 513f, 514f
  El Niño, 516, 664–666, 665f, 668
  Gulf Stream, 509–510, 509f
  gyres, 510–511, 512, 533
  heat transport and, 327
  longshore currents, 518–519, 519f, 523, 525f
  surface currents, 509–512, 533
  thermohaline circulation, 513–514, 513f, 514f
  tidal current, 510
  upwelling, 514, 514f, 516, 533
  wind and, 510f, 513, 516, 628
Ocean sediment, historic changes, 687
Ocean waves, 507–508, 507f, 508f, 533
Oceanic crust, 199, 199t, 215, 259, 268, 306, 326
  basaltic oceanic crust, 340–341
  convergence of, 209
  layers of, 340–341, 340f, 349
  subduction of, 312, 313
Oceanic islands, 337, 349
Oceanic lithosphere, 40f
Oceanic trenches, 40f, 205f, 327t, 336, 346–347, 347f, 349
Oceans, 325–351, 472, 497
  acidification of, 500, 502, 503f, 698
  albedo of, 615
  atolls, 338, 349
  average sea surface temperatures, 634, 634f
  biosphere, 41–42
  carbon cycle, 34, 35f, 56f
  change in, 532
  climate and, 327, 510, 632
  coastlines, 516–521
  continental margins, 345–347, 346f, 347f, 349
  continental rise, 344, 349
  continental shelf, 342–344, 342f, 343f, 349
  continental slope, 342f, 344, 344f, 349
  coral reefs, 112, 138, 199, 338–339, 400, 400f, 498, 499f, **500f**, 530–531, 531f, 533, 672f–673f, 698
  crust, 199, 199t, 209, 215, 259, 268
  deep water currents, 131–133, 132f, 513–514, 513f, 514f, 533
  deltas, 117, 358, 378–379, 379f
  depth of, 326, 326f, 327t
  fisheries, 528–529, 529t, 530
  formation from primordial solar system, 325–326
  guyot, 338, **338f**
  human activities and, 532
  hydrosphere, 40, 41f, 61
  island arcs, 306–307, 306f, 317, 336–337, 336f, 337f, 349
  light in, 528
  lithosphere, 40f, 209
  mid-ocean-ridge basalt (MORB), 96

Mid-Ocean Ridge, 203–204, 204f, 207, 208t, 212, 217, 243–244, 268, 301, 304, 326, 331–334, 331f, 333f, 335, 340, 341, 349
   mining manganese nodules, 186, 186f
   ocean basins, 326–327, 326f, 501, 501f
   ocean-floor sediment, 340–341
   oceanic islands, 337, 349
   origin of, 325–326
   photic zone, 528
   plastic pollution in, 183, 183f
   rifting in, 207, 207f
   river deltas, 117, 358
   rock cycle in, 96
   sea-level changes, 334–335, 335f, 433, 697–698, **697f**, 698f
   seacoasts, 516–521
   seafloor mapping, 24–25
   seafloor spreading, 203–204, 203f, 204f, 223, 521
   seamounts, 337, 349
   submarine canyons, 344–345, 345f, 349
   submarine fans, 344–345, 345f
   surf, 517, 517f
   tectonic movements and, 220
   temperature of water, 504, 504f, 509f, 514, 533, 634, 634f, 697
   tides, 505–507, 505f, 506f, 533
   trenches, 40f, 205f, 327t, 336, 346–347, 347f, 349
   turbidity currents, 344–345, 345f, 349
   underwater robots for exploration, 322, 323f, 348, 348f
   waves, 507–508, 507f, 508f, 533
   wind, 507, 628
   See also Marine life; Ocean currents; Oceanic crust; Seafloor; Seawater
Ogallala aquifer (United States), 554–556, 555f, 568
Oil, 71
   exploration, drilling, 328
   refineries, 173, 173f
   seeps, 174, 174f
   wells, 173, 173f, 175f
Okavango Delta (Africa), 538, 539f
Okavango River (Africa), 538
Okavango watershed (Africa), 538
Ol Doinyo Lengai (volcano in Tanzania), 195, 195f, 357, 357f
Oligocene epoch, 45f, 150f
Oligotrophic lakes, 480–481, 562
Olivine, 78, 80t, 81f, 83, 104, 270
Olympus Mons (volcano on Mars), 748, 748f, **748f**
On Assignment, photographs, 52f–53f, 84f, 102f–103f, 146f–147f, 170f–171f, 218f–219f, 250f–251f, 272f–273f, 304f–305f, 330f, 362f–363f, 426f–427f, 456f–457f, 484f, 519f, 612f–613f, 696f, 720f–721f, 776f, 799f
Ooids, 112, 112f
Open-pit mines, 168f, 169

Open universe, 810f, 811, 813
Opportunity (Mars rover), 738, 750, 765
Optical telescopes, 798
Optimization, 17
Ordovician period, 45f, 150f
Ore, 82, 163, 164, 187
Ore deposits, 164–167, 164t
Organic compounds, 72
Organic matter
   in soil, 443
   soil formation from, 443–444
Organic sedimentary rocks, 110, 113, 113f
Organisms
   Earth's spheres and, 42
   ecosystem, 43, 43f
   in soil, 443, 444
Original horizontality, 134, **134f**
Orion Nebula, 774, 774f
Orogen, 296, 304
Orogeny, 304, 319
Orographic lifting, 644, 644f, 668
Orthoclase, 76t, 81
Osprey, mercury levels in blood of, 88, 89f
Out of Eden Walk, 304f
Outgassing, 580, 601
Outwash (glaciers), 412, 433
Owens River (California), 540
Oxbow lake, 376f, 479
Oxidation, chemical weathering by, 369
Oxide minerals, 80t
Oxygen
   in atmosphere, 128, 581, 584, 584V, 601
   in Earth's crust, 73, 73t, 80
   isotope ratios as evidence of climate change, 688
   production by cyanobacteria, 581–582
Ozone, 583, 595–597, 596f
Ozone depletion, 599
Ozone hole, 598–599, 599f

P

P waves (primary seismic waves), 234–235, 234f, 237f, 239, 240, 253, 254–255, 254f, 259
Pacific Ocean, 326–327, 336, 337, 501, 664
Pacific plate, 206f, 209, 241–243, 241f, 248, 258
Pahoehoe lava, 276, 277f, 279
Paleocene-Eocene Thermal Maximum, 699
Paleocene epoch, 45f, 150f
Paleogene period, 45f, 150f
Paleontologists, 115, 138
Paleontology, geologic timescale, 45f, **129f**, 150–153, 150f, 155
Paleozoic era, 45f, 150f, 151–152, 152f, 197f
Pangaea (supercontinent), 132, 132f, 198, 214, 214f, 296, 345–346, 346f
Pangaea Proxima (supercontinent), 192f–193f, 214
Parabolic dunes, 430, 430f
Parallax, 715
Parent isotopes, 145, 145f, 155

Parent rock (parent material), 106, 269, 359, 443–444
Paris Agreement, 700
Paris Basin (France), 154, 154f
Parker Solar Probe, 772, 775, 789
Parsec, 780
Partial melting, 270
Particle accelerator, 162f
Particulates, as pollutants, 597–598
Particulates and aerosols, as pollutants, 597–598
Passive continental margins, 345, 349
Passive solar house, 178
Paternoster lakes, 407, 407f
Pathogens, climate change and, 697
Patterns, 18t, 29
Patterson Mountain (Washington), 655f
PBTs. *See* Persistent bioaccumulative toxic chemicals
Peat, 113, 172t, 187, 443–444
Peaty soil, 451, 451f, 452t
Peer review, 13
Pelagic sediment, 340, 341f
Penfield, Glen, 130
Pennsylvanian period, 45f, 150f
Penumbra, 726
Penzias, Arno, 807
Perennial crops, 440, 441f, 464
Performance Task activity, 31, 63, 93, 123, 157, 189, 225, 261, 291, 319, 351, 437, 469, 495, 535, 571, 605, 634–635, 671, 704, 735, 767, 792, 815
   Alternate Cosmology, 815
   Changes in Earth's Surface and Changes in Earth's Systems, 634–635, 634f–635f
   Coevolution of Earth's Systems and Life on Earth, 605
   Computational Modeling of Human Effects on Earth Systems, 535
   Early History of Solar System, 767
   Energy Balance of Power Generation System, 189, 189t
   Erosion Model, 437
   Evolution of the sun, 792
   Great Lakes Invasive Species, 495
   Interpreting Maps, 31
   Local Water Problem, 571
   Mineral Field Guide, 93
   Mississippi and Missouri Rivers, 393
   Modeling Carbon Cycle, 63
   Modeling Energy Flow in Earth's Climate, 704–705
   Optimizing Food Production and Soil Conservation, 469
   Present-Day Mountain Building, 319
   Regional Geologic History, 157
   Release of Energy in Earthquakes, 261
   Rocks in U.S. National Parks, 123
   Scientific Illustration, 225
   Seafloor Features Model, 351

   Transit Data and Kepler's Laws, 735
   Volcanic Effects, 291
   Weather Emergency Action Plan, 671
Peridotite, 107, 200, 269, 270, 289
Periods, 45f, 150f, 151
Permafrost, 699, 702
Permeability, of groundwater, 487
Permian mass extinction, 129, 131, 151, 287
Permian Ocean, 132, 132f, 133
Permian period, 45f, 129, 131–133, 132f, 150f, 151, 287
Perseverance (Mars rover), 738, 751
Persistent bioaccumulative toxic chemicals (PBTs), 561
Pesticides, 597
Peter, Carsten, 218f–219f, 272f–273f, 656f–657f
Petroleum, 173–174, 173f–175f, 187
   combustion of, 592, 593, 601, 692
   end-use efficiencies of, 181f
   formation of, 174
Petroleum reserves, 343
Pettit, Erin, 396, 397f, 398, 697
pH, 365, 450, 502, 503f, 594–595, 595f, 698
Phanerozoic eon, 45f, 150f, 151–153, 155
Philippine Trench (Philippine Sea), 327t
Phosphate minerals, 80t, 83
Phosphorescence, of minerals, 79
Photic zone, 528
Photons, 611, 614, 632
Photosphere (sun), 777, 790
Photosynthesis, 581, 583, 601, 699
Photosynthetic organisms, 58
Photovoltaic cells (PV, solar cells), 178, 178f
Phyllite, 118, 119f
Physical weathering. *See* Mechanical weathering
Phytoplankton, 533
Pillow basalt, 340, **341f**
Pine Island Glacier (West Antarctica), 404
Pioneer Venus Orbiter, 746f, **748f**
Pistol butt trees, 384, 385f
Piston coring, 328, 328f
Placer deposits, 166, 167f
Plagioclase, in common rock types, 82t
Plagioclase feldspar, 81, 81f
Planck satellite, 807–808, 807f
Planetary astronomers, 10
Planetary nebulae, 783, 783f
Planetesimals, 38–39
Planets, 713–714
   barycenter, 722, 722f, **723f**
   dwarf planets, 760–761
   exoplanets, 732, 732t, 796
   formation of, 38–39, 325, 741, 765
   Jovian planets, 741, 742f, 743, 743t
   Kepler–90 planets, 732, 732t
   Kepler's discoveries, 717, 717f, **717f**, 718
   mapping, 25, 27

INDEX

minor planets, 761–762
movement of, 733
orbits of, 733
properties of, 723*t*, 743*t*, 793*t*
retrograde motion, 714, 714*f*, 733
scale drawing of, 742*f*
terrestrial planets, 741, 742*f*, 743, 743*t*
See under names of individual planets
Plankton, 113, 113*f*, 480, 528, **528*f***, 530, 533, 688
Plants
 biodiversity, 42, 50–51, 52*f*–53*f*, 54
 biome, 680
 biosphere support for, 41
 climate change and, 699
 desert biome, 419
 early Earth origins, 583
 invasive species, 495, 548, 549
 in lakes, 480
 landslides and, 387–388, 388*f*
 mass extinctions, 126, 129–133, 155
 mountain habitats, 314
 perennial grains, 440, 441*f*
 photosynthesis, 581, 583, 601, 699
 rock weathering caused by, 364, 364*f*, 368
 salt-resistant, 522
 soil biodiversity, 448, 448*f*
 for soil erosion prevention, 458, 459*f*
 soil from, 444
 sustainable food systems, 440, 441*f*
 threshold effects, 49, 49*f*, 50, 54
 trace fossils, 109*f*, 110
 transpiration, 477
Plastic deformation, of rocks, 231, 231*f*, 297, 297*f*, 317
Plastic flow (glaciers), 404–405, 405*f*, 433
Plastics
 energy conservation and, 183
 global consumption, 183
 Great Pacific Garbage Patch, 183, 512, 512*f*, 533
 in mercury cycle, 68
 microplastics, 68, 183, 512
 in ocean, 183
 single-use plastics, 183, 183*f*
Plate boundaries, 205, 205*f*
 convergent plate boundaries, 205, 206*f*, 208–209, 208*t*, 220, 223, 243–244, 243*f*, 244*t*, 302, 336, 346, 347
 divergent plate boundaries, 205, 206–207, 206*f*, 207*f*, 208*t*, 217, 223
 earthquakes and, 241–245, 259
 faults and, 302–303
 mountain building and, 305
 transform plate boundaries, 205, 206*f*, 208*t*, 209, 223
 types, 244*t*
Plate interiors, earthquakes in, 244–245
Plate movement, 46–47, 211–215, 217, 223, 231
 atmosphere and, 220, 223
 basins, 299, 299*f*
 convection, 205*f*, 211–212, 211*f*, 213, 223
 domes, 299, 299*f*
 downslope sliding of, 211–212, 212*f*
 earthquakes, 232
 faults, 299, 299*f*–303*f*, 301–303
 folds, 297–299, 297*f*–299*f*, **298*f***, 302
 isostasy, 214–216, 215*f*, 223, 305
 measuring movement, 194, 195*f*
 rates, 232
 spreading centers, 206, 217, 268–269, 268*f*, 335
 stress in rocks, 297
 supercontinents, 132, 132*f*, 192*f*–193*f*, 214, 214*f*, 223
 underthrusting, 305, 306
Plate tectonics, 40, 192–225
 biosphere and, 221
 blob tectonics, 748
 climate and, 220–221, 221*f*
 continental drift hypothesis, 197–199, 197*f*, 223, 330*f*
 earthquakes, 220
 hot spots, 212, 223
 mantle plumes, 212, 223, 269, 269*f*
 of Mars, 748–749
 mountain building, 220, 223
 Pangaea, 132, 132*f*, 198, 214, 214*f*, 296, 345–346, 346*f*
 Pangaea Proxima, 192*f*–193*f*, 214
 plate boundaries, 205–209, 205*f*–207*f*, 208*t*, 223
 plate movement, 211–215
 rifting, 207–208, 207*f*, 217, 222
 seafloor spreading hypothesis, 203–204, 203*f*, 204*f*, 223
 subduction, 206, 207*f*, 208, 209, 220, 259, 269, 270*f*, 274
 supercontinents, 132, 132*f*, 192*f*–193*f*, 198, 214, 214*f*, 223
 theory of, 205–209, 223
 of Venus, 748
 volcanoes, 217, 217*f*, 223
 See also Tectonic plates
Plateaus, 424–425
Platinum, uses, 80*t*
Playas, 422, 422*f*
Playa lakes, 422, 422*f*
Pleistocene epoch, 45*f*, 150*f*, 335, 410, 414, 414*f*, 415, 431, 433
Pleistocene Glaciation, 414
Pliocene epoch, 45*f*, 150*f*, 154, 414*f*
Plume of contamination, 564, 564*f*
Pluto (dwarf planet), 760–761, 765
 images of, **761*f***
 moons of, 761
Plutonic rocks, 105*f*, 106, 121, 269, 314
Plutons, 274–275, 274*f*–276*f*, 288, 289
Point bars, 375*f*, 376
Polar climate, 684
Polar easterlies, 679
Polar front, 678*f*, 679, 702
Polar jet stream, 679
Polaris (North Star), 713, 714*f*, 724
Pollen, as evidence of climate change, 688
Pollution
 acid mines and heavy metals, 87, 87*f*
 in Chesapeake Bay Watershed, combatting, 466
 from coal power plants, 172
 light pollution, 728
 nonpoint source pollution, 561
 point source pollution, 561
 polycyclic aromatic hydrocarbons (PAHs) in soil, 465
 wetlands to control pollution, 483, 485
 See also Air pollution; Water pollution
Polycyclic aromatic hydrocarbons (PAHs), 465
Polyvinyl chloride (PVC), 597
Population
 biodiversity, 42, 50–51, 52*f*–53*f*, 54, 61
 global population, 686
 limiting factors, 51, 61
 Malthus on, 51
 threshold effects and, 49, 50, 50*f*, 54, 61
 uncontrolled human growth, 61
 urban population, 685
Population I stars, 784
Population II stars, 784
Porosity
 of groundwater, 487
 sedimentary rocks, pore space, 110, 110*f*, 111
 of soil, 449, 450*f*
Porphyry, 105*f*
PoseiDRONE (soft robot), 322, 323*f*
Positive correlation, 11
Positive feedback mechanisms, 50, 668
Potassium, in Earth's crust, 73, 73*t*
 in Earth's crust, 73, 73*t*
 isotope (potassium-40), 146*t*
 feldspars, structure of, 81*f*
Power plants
 biomass energy, 180
 coal-fired, 172, 189*t*
 energy balance analysis, 189, 189*t*
 energy return on investment (EROI), 189
 geothermal energy, 179, 179*f*
 hydroelectric power, 179, 550–551, 551*f*
 natural gas, 174, 189*t*
 nuclear reactors, 175, 176*f*, 177, 189*t*
 solar energy, 178, 178*f*, 189*t*
 wind energy, 158*f*–159*f*, 178, 189*t*
Precambrian era, 45*f*, 150*f*, 151, 155, 197*f*, 244, 415, 571
Precession, 724, 725*f*
Precious gems, 83, 84*f*
Precipitation, 647–648, 668
 climate change and, 695, **695*f***
 rock cycle, 100, 108, 110, 112
 snow, 399, 399*f*, 630, 648, 702
 in United States, 547, 548
 warm front, 652
 water cycle, 37*f*, 55, 55*f*, 61, 478*f*
 See also Rain; Snow; Rock cycle
Precipitation of minerals, 100, 108, 110, 112, 165–167
Pressure, plastic deformation and, 297
Pressure gradient, weather and, 649, 668
Pressure-release fracturing, 360, 360*f*, 368
Pressure-release melting, 268, 268*f*, 269*f*
Prevailing westerlies, 679
Primary seismic waves. *See* P waves
Prime meridian, 19, 20, 20*f*
Primordial nucleosynthesis, 805, 806
Primordial universe, 804, 804*f*, 809
Principle of crosscutting relationships, 135, 135*f*
Principle of faunal succession, 136, 137*f*, 138*f*
Principle of included fragments, 136, 136*f*, 147
Principle of original horizontality, 134, **134*f***
Principle of superposition, 134, 134*f*, 136
Prism, 729, **729*f***
Prodeltas, 378, 379*f*
Proterozoic eon, 45*f*, 150*f*, 151, 688
Prototypes, 12–13
Ptarmigan Cirque (Canada), 407*f*
Ptolemy's model of the universe, 715–716, 716*f*, 719*f*, 733
Pulsars, 785–786, 786*f*
Pumping, of groundwater, 554
Purcell sill, **276*f***
Pyrite, 76*f*, 77, 80*t*
Pyroclastic flows, 36, 276, 281–282, 281*f*, **281*f***, 289
Pyroxene, 80*t*, 81, 81*f*, 91, 107

Q

Quanbeck, Caroline, 498, 499*f*, 500
Quarries, 163
Quartz, 71
 chemical composition, 74, 80, 80*t*
 chert, 114, 114*f*
 cleavage, 78
 color, 77
 in common rock types, 80*t*, 106
 conchoidal fracture, 78, 79*f*
 crystal habit, 77, 77*f*
 in Earth's crust, 81, 81*f*, 91
 hardness, 76*t*, 111
 structure of, 81*f*
 toxicity of dust, 85
 uses, 80*t*, 91
 varieties, 83
 weathering of, 366
Quasars, 808–809, 809*f*, 812, 812*f*, 813
Quaternary period, 45*f*, 150*f*

R

Radiation, 632, 684, 684*t*
 balance, 617–619, 617*f*, 632
 cooling, 643, 668

earliest. *See* Cosmic Microwave Background
Radiative zone (sun), 777
Radio telescopes, 785*f*, 798, 807
Radioactive materials, as pollutants, 561, 566, 568
Radioactivity, 144
  isotopes, 145–147, 145*f*, 146*t*, 147*t*, 155
  of minerals, 79, 86–87
  radiometric dating, 145–147, 145*f*, 146*t*, 147*t*, 151, 155
  radon, 86–87
Radiometric dating, 145–147, 145*f*, 146*t*, 147*t*, 151, 155
Radon, 86–87, 91
Rain, 647–648
  acid rain, 365, **365*f***, 594–595, 594*f*, 601
  in California, **540*f***
  in deserts, 417, 433
  El Niño and, 516, 664–666, 665*f*, 668
  groundwater and, 493
Rain forests, 68, 70, 417, 417*f*, 681, 682
Rain-shadow deserts, 418, 419*f*, 655, 655*f*
Rainwater, 365
Rare earth metals, 162
Rayleigh waves, 235–236, 235*f*
Recent epoch, 45*f*
Recessional moraines, 410
Recessional velocity, 801*f*, 813
Recumbent folds, 298, 298*f*
Red giants, 782, 783*f*, 790
Red Sea (Africa, Asia), 207, 222
Red stars, 775, 780, 799
Redshift, 731, 800, 805, 809
Reef limestone, 112, 112*f*
Reefs, 112, 503, 523
  See also Coral and coral reefs
Reflection, of light, 615–616, 615*f*
Refracting telescopes, 726, 727*f*
Refraction, of waves, 518, 518*f*, 533
Regional metamorphism, 117–118
Regolith, 443
Relative age, 134, 155
Relative humidity, 641, 642, 668, 684*t*
Relativity, theory of, 777, 787
Relief maps, 23*f*
Remote sensing, for seafloor sampling, 328–329, 329*f*, 349
Renewable energy resources, 177–180, 187
  biofuel, 180
  biomass energy, 180
  future of, 183–185
  geothermal energy, 179, 179*f*
  hydroelectric energy, 179, 550–551, 551*f*
  solar energy, 178, 178*f*, 189*t*
  wind energy, 158*f*–159*f*, 178, 189*t*
Reservoirs, 547*f*, 548, 549, 564*t*, 701*f*
Residual soil, 444
Resolution, of telescope, 727
Resource maps, 23*f*
Resource planning, geothermal energy, 27–28, 28*f*

Retrograde motion, 714, 714*f*, **714*f***, 716, 733
Reverse fault, 301, 301*f*, 317
Revolution, of Earth, 724
Reykjadalur Valley (Iceland), 5, 6, 10
Rhyolite, 106, 106*f*, 107
Rhyolite magma, 207
Richards, Cory, 294, 295*f*, 588*f*–589*f*
Richardson, Jim, 456*f*–457*f*, 464, 464*f*
Richter scale, 237, 239
Ridge push, 212
Rift valleys, 222, 332, 332*f*
Rifting, 207–208, 207*f*, 217, 222, 345–346, 346*f*
Rigel (star), 714*t*
Ring Nebula, **783*f***
Ring of Fire (Pacific Ocean), 241, 241*f*, 269
Rings (of Jovian planets), 754, 765
Rings (of Saturn), 754–755, 754*f*, **754*f***
Rio Grande (United States), 377, 377*f*
Rio Tinto (Spain), 87*f*
Ripple marks, 47*f*, 108–109, 109*f*
Rising air, weather and, 643, 649, 650, 668, 702
Risk assessment, volcanoes, 283, 286, 286*f*
Rivers, 369
  beaches on, 523
  dead zone, 530–531, **531*f***
  deltas, 117, 358, 378–379, 379*f*
  in deserts, 421
  flood control, 380–382, 380*f*, **380*f***, 381*f*
  floodplains, 109, 109*f*
  rock record, 141–142, 143*f*
  tributaries, 369, 377
  See also Streams
Robots
  underwater exploration, 322, 323*f*, 348, 348*f*
  space exploration, 738, 740, 750
Rock cycle, 44, 96, 99–101, 100*f*, 121, 147
  Earth's systems and, 101
  lithification, 99, 112
  precipitation, 100, 108, 110, 112
  weathering, 99, 108, 187
Rock deformation
  basins, 299, 299*f*
  domes, 299, 299*f*
  faults, 231–232, 232*f*, 242, 299, 299*f*–303*f*, 301–303, 317, 334, 334*f*
  folds, 297–299, 297*f*–299*f*, **298*f***, 302
Rock dredge, 328
Rock-forming minerals, 74, 79, 80–82, 91
  carbonates, 80*t*, 81–82, 91, 111–112, 112*f*
  silicates, 79, 80*f*, 80*t*, 81*f*, 82, 82*f*, 91
Rock salt, 108
Rockfall, 383*f*, 384, 385*t*
Rocks, 71–72, 91, 94–123, 94*f*–95*f*, 99
  abrasion of, 361, 361*f*–363*f*, **361*f***, 428–429

absolute age, 134, 140, 144–147, 155
age spectrum, 146
characteristics of, 297, 317
chemical weathering, 360, 360*f*, 365–368
correlation, 141–142, 142*f*, 144, 155
dating, 144–147, 155
earthquakes. *See* Earthquakes
elastic deformation of, 231, 232*f*, 317
erratics, 409, **409*f***
fracture, 231
frost wedging, 360, 361*f*
geologic timescale, 45*f*, **129*f***, 150–153, 150*f*, 155
hazardous rocks and minerals, 85–89
heated by fire, 360
in Himalaya (Asia), 314
hydrothermal processes, 165–166, 165*f*, 187
igneous rocks, 104–107, 121
lapidary art, 99
magmatic processes, 165, 187
mechanical weathering, 359–364
melting of, 104, 267, 267–268, 267*f*
metamorphic rocks, 115–119, 121
oldest rock on Earth, 147*t*
Pangaea, and 198
Phanerozoic eon, 151–152, 155
plastic deformation of, 231, 231*f*, 297, 297*f*, 317
Precambrian era, 151, 155
pressure-release fracturing, 360, 360*f*, 368
principle of crosscutting relationships, 135, 135*f*
principle of faunal succession, 136, 137*f*, 138*f*
principle of included fragments, 136, 136*f*, 147
principle of original horizontality, 134, **134*f***
principle of superposition, 134, 134*f*, 136
radiometric dating, 145–147, 145*f*, 146*t*, 147*t*, 151, 155
rock record, 141, 155
sedimentary processes, 166, 187
sedimentary rocks, 108–114, 121
specific heat, 622
stress, 297, 317
talus, 360, 361*f*
texture, 100
thermal expansion and contraction, 360
volcanic rocks, 104–105, 105*f*, 106, 107, 107*f*
weathering, 99, 108, 167, 167*f*, 187
  See also Geologic time; Geosphere; Rock cycle
Rockslides, 383–384, 383*f*
Rocky coastlines, 527, 527*f*, 533
Rocky Mountains (North America), 120, 138, 151, 152, 355, 373*f*, 375*f*, 377
Rodinia (supercontinent), 214

Roque de los Muchachos Observatory (Canary Islands), 797*t*
Ross Ice Shelf (Antarctica), 676
Rub al-Khali (sand dune), 429
Rubidium-87, 145, 146*t*, 147*t*
Ruby, 77
Runoff, 477
Russell, Henry, 781
Rust, formation of, 360, **360*f***, 366
Ruzo, Andrés, 6, 7*f*, 8

S

S waves (secondary seismic waves), 235–236, 235*f*, **235*f***, 237*f*, 239, 240, 254, 255*f*, 259
Saffir–Simpson scale, 662, 663*t*
Sagan, Carl, 769, 789
Sahara Desert (Africa), 429
Saint Lawrence River (Canada), 344*f*
Saiph (star), 714*t*
Salinity, of seawater, 502, 502*f*
Salinization
  of soil, 461–462, 462*f*
  of surface water, 547*f*, 548
  in the United States, 548
Salopek, Paul, 304*f*–305*f*
Salt, crystalline structure of, 74–75, 74*f*
Salt beds, 167
Salt cracking, 366–367, 367*f*
Salt-resistant plants, 522
Salt water, 472
Saltation, 428, 429
Salts, in seawater, 533
Saltwater, intrusion into groundwater, 557, 557*f*
Samaras, Tim, 638
Samarium-146, 147*t*
San Andreas Fault (United States), 208*t*, 209, 232, 232*f*, 233, 233*f*, 241–242, 242*f*, 243*f*, 244*t*, 259, 302, 334
San Francisco earthquake (1906), 242, 242*f*, 243
San Joaquin Valley (California), 556, 556*f*
Sand, 110, 110*t*, 152, 153
  erosion by, 361, 361*f*
  landslides and, 387, 388*f*
  porosity of, 449
  saltation, 429
  in soil, 450
  transport in deserts, 429
  windblown sand, 361, 429
Sand dunes, 108, 394, 428, 429–431, 429*f*, 430*f*, 433, 497, 688
Sandstone, 98, 99, 121, 163, 367*f*
  disconformities, 139–140, 139*f*, 140*f*
  downcutting, 373*f*
  formation, 46, 111, 111*f*
  Pangaea, 199
  ripple marks, 47*f*, 108–109, 109*f*
  rock record, 141–142
  silicate minerals in, 80*t*
Sandy clay soil, 450, 452*t*
Sandy Hook (New Jersey), 519
Sandy loam, 451*f*, 452*t*
Sandy sediment, earthquakes and, 246
Sandy soil, 451, 451*f*, 452*t*

INDEX **885**

INDEX

Sanmenxia Dam (China), 548
Sapphire, 77, 83, **83f**
Saria, Elifuraha, 194
Sartore, Joel, 52f
Satellite altimetry, seafloor mapping, 24–25, **25f**
Satellite-based microwave radar, 329
Saturation, 641
Saturn (planet), 706f–707f, 754–755, 765
   atmosphere of, 752f, 765
   climate of, 754
   image of, 706f–707f
   moons of, 740, 759, 759f, 764f, 765
   properties of, 723t, 743t, 793t
   rings of, 754–755, 754f, **754f**
   scale drawing of, 742f
Scattering, of light, 616, 616f
Schist, 118, 119f, 121
Science, 9–16, 29
   branches of, 29
   correlation and causation, 11, 12
   crosscutting concepts, 18, 18t, 29
   data, 10, 13
   empirical evidence, 13
   experiments, 13
   goal of, 16
   hypothesis, 13
   laws, 14, 29
   models, 12–13
   natural laws, 9
   nature of, 9–10, 9t, 29
   peer review, 13
   practices, 11f
   scientific inquiry, 10, 11–12, 11f, 12, 13, 29
   scientific knowledge, 9–10, 9t
   scientific laws, 14, 29
   scientific theory, 13–15, 29
   simulations, 13
   See also Engineering
Science and engineering practices, 11, 11f
Science on Wheels program, 264
Scientific illustration, 48, 48f, 225
Scotland, 161f
Scratch test (minerals), 75
Sea breezes, 658, 668
Sea cliffs, weathering of, 366–367, 367f
Sea ice, 416, 433
Sea level, global warming and, 520–521
Sea-level changes, 334–335, 335f, 433, 520–521, 697–698, **697f**, 698f
Sea life. *See* Marine life
Sea waves, 507–508, 507f, 508f
Seacoasts. *See* Coastlines
Seafloor, 320–351
   abyssal plains, 340, 349
   age of, 335, 340, 349
   characteristics of, 326
   continental shelf, 342–344, 342f, 343f, 349
   depth of, 326, 326f, 327t
   dredging, 328
   drilling, 328
   earthquake zone research, 324
   features of, 338–341, 351
   island arc formation, 306, 306f, 317, 336–337, 336f, 337f, 349

   magnetic orientation of seafloor rocks, 203–204, 203f, 204f
   maps and mapping, , 24–25, **25f**, 320f–321f, 330f
   mid-ocean-ridge basalt (MORB), 96
   Mid-Ocean Ridge, 203–204, 204f, 207, 208t, 212, 217, 243–244, 268, 301, 304, 326, 331–334, 331f, 333f, 335, 340, 341, 349
   mining of manganese nodules, 186, 186f
   mountain building, 306
   ocean-floor sediment, 340–341
   ore formation, 166
   piston coring, 328, 328f
   plate tectonics, 203–204, 203f, 204f
   research, 324
   rift valleys, 222, 332, 332f
   robots for exploration, 322, 323f, 348, 348f
   rocks, 209
   sampling methods, 328, 328f, 329f, 349
   spreading of, 203–204, 203f, 204f, 223, 335, 335f, 521
   subduction complex, 306, 306f
   submarine canyons, 344–345, 345f, 349
   submarine fans, 344–345, 345f
Seafloor spreading hypothesis, 203–204, 203f, 204f, 223
Seamounts, 337, 349
Seasons
   stars and constellations and, 724, **724f**
   weather, 616–617, 616f, 632
Seawalls, 526–527, 526f, 533
Seawater
   acidification of, 500, 502, 503f, 698
   carbon dioxide in, 58, 533
   composition of, 533
   cooling of, 520
   dissolved gases in, 502–504, 503f
   heating of, 520
   mercury in, 68, 69f
   methane in, 699
   salinity of, 502, 502f
   temperature of, 504, 504f, 509f, 514, 533, 697
Second Great Oxidation Event, 579t, 583–584, 601
Secondary atmosphere, 686
Secondary seismic waves. *See* S waves
Sediment, 108, 121
   continental shelf, 342–343, 342f, 343f, 349
   deposition of, 359, 359f, 369–382
   drift, 409
   erosion. *See* Erosion
   fossilized human footprints, 356, 357f
   lakes, 478
   ocean-floor sediment, 340–341
   pelagic sediment, 340, 341f
   regolith, 443
   in streams and rivers, 372–373, 373f
   terrigenous sediment, 340, 343
   transport of, 372

   turbidity currents, 344–345, 345f, 349
   wind abrasion, 428–429
Sedimentary rocks, 95f, 108–114, 121
   angular unconformity, 140, 140f, 141
   banded iron formations, 167, 167f
   bedding, 109, 109f
   bioclastic rocks, 110, 112f
   Cenozoic era, 152
   chemical sedimentary rocks, 113–114, 114f
   clastic rocks, 110–111, 110f, 110t
   deposition, 108
   disconformities, 139–140, 139f, 140f, 155
   features of, 108–110, 108f–109f
   folding and tilting, 134, **134f**
   folds, 298f
   forearc basin, 306
   formation of, 99, 100, 108
   geologic formations, 142f
   graded beds, 109–110
   index fossils, 142, 143f, 155
   island arc building, 306, 306f, 317, 336–337, 336f, 337f, 349
   key beds, 142, 143f, 144, **144f**, 155
   landslides and, 386, 387f
   Mesozoic era, 152
   nonconformity, 140–141, 155
   organic rocks, 110
   organic sedimentary rocks, 113, 113f
   Paleozoic era, 151–152, 152f
   Pangaea, 199
   Phanerozoic eon, 151–152, 155
   pore space, 110, 110f, 111
   porosity of, 487
   principle of crosscutting relationships, 135, 135f
   principle of faunal succession, 136, 137f, 138f
   principle of included fragments, 136, 136f, 147
   principle of original horizontality, 134, **134f**
   principle of superposition, 134, 134f, 136
   rippling, 47f, 108
   rock record, 141–142, 142f
   stratification, 109, **109f**
   trace fossils, 109f, 110
   unconformities, 139–141, 139f–141f, 155
   in the United States, 152
Seimon, Anton, 638, 639f, 640
Seismic reflection profiler, 329, 329f
Seismic waves, 234–236, 234f, 235f, **235f**, 237f, 239, 240, 253, 254, 259
Seismograph, 236, 236f, 237f, 238, 238f, 239, 252
Seismology, 259
Semiprecious gems, 83
Serpentinite, 341, **341f**
Seven Rila Lakes region (Bulgaria), **407f**
Sewage, as pollutant, 560, 562
Sewage treatment, 559, 562

Seymour, Brett, 149f
SGR A* (black hole), 798
Shadow zone, 254
Shah, Afroz Ahmad, 228, 229f
Shark Bay (Australia), 65f
Shells, 73
Shield volcanoes, 278t, 279, 279f, 289
Shoreface, 522
Sierra de la Ventana Fold Belt (South America), 197f, 198
Sierra Nevada (mountains in California), 99, 274f, 275f, 288, 309f, 360f
Silicon
   in Earth's crust, 73, 73t, 80
   magma behavior and silica, 271, 289
   silica dust, 85
   silicate minerals, 79, 80–81, 80f, 80t, 81f, 82, 82f, 91, 114, 270
   silicate tetrahedron, 80, 80f, 81f, 91
   silicosis, 85
Sill, 276, **276f**, 289
Silt, 110, 110t, 449
Silting, 548
Silty soil, 451, 451f, 452f
Silurian period, 45f, 150f
Silver, 74, 79, 80t, 82, 88
SILVER project, 322, 323f
Sims, Kenneth, 96, 97f, 272f
Singularity, 803
Sinkholes, 491, 491f, 493
Sinking air, weather and, 649, 650, 668, 702
Sirius (star), 713
660-kilometer discontinuity, 254
Skerry, Brian, 149f
Sky, models of, 708–734, 708f–709f
   See also Universe
Slab pull, 212
Slate, 118, 119f, 121, 163
Sleet, 648
Sliabh Lang (mountain in Ireland), 512f–513f
Slides, 383, 383f
Slip (of rocks), 299
Slip face (dunes), 429, 429f, 430f
Sloughs, 482
Slump (landslides), 383f, 384–385, **385f**, 385t
Slurry, 383
Small Magellanic Cloud (galaxy), 800
Smallholder farms, 456f–457f
Smith Island (Maryland), 696f
Smith Rock State Park (Oregon), 281f
Smog, 595–596, 596f, 600f
Smoke, 597–598, 664
Snail fossils, 154f
Snow, 399, 399f, 630, 648, 702
Snow algae, 608
Snowball Earth, 415, 699
Sodium, in Earth's crust, 73, 73t
Soft robots, 322, 323f
Soil
   additions, 447–448
   biodiversity, 448, 448f
   bioremediation, 354, 465
   changes in, 447–449, 447f
   classification of, 449–455, 452t, 467

components of, 443, 449–450, 450f
conservation of, 469
degradation of, 458
desertification, 459–460, 461, 461f, 467
eluviation, 447
formation of, 443–449, 462, 467
gases in, 443, 448, 449
horizons, 446–447, 449, 454f, 467
human impacts on soil quality, 459, 459f
humus, 443
leaching, 447
liquefaction, 246
loess, 430–431, 430f, 433, 454–455, 455f, 467
organisms in, 443, 444
parent material, 443–444
porosity of, 449, 450f
residual soil, 444
salinization of, 461–462, 462f
soil depletion and pollution, combatting, 466
soil orders, 453, 453f, **453f**, 467
sustainability of soil resources, 462–464, 463f, 464f, **464f**
terracing, 463, 463f
texture of, 449
topsoil, 447
transformations, 447, 449
transported soil, 444
translocation, 447, 449, 467
types, 450–451, 451f, 452t, 453, 467
Soil Conservation Service (SCS), 464
Soil depletion, combating, 466
Soil erosion, 442, 442f, **442f**, 458, 458f, 459f, 460, 467
Dust Bowl (United States), 459–460, 461, 461f
dust storms, 459–460, 461f
prevention strategies, 463, 464, 467
Soil orders, 453, 453f, **453f**, 467
Soil resources, 438–469
sustainability of, 462–464, 463f, 464f, **464f**
sustainable food systems, 440, 441f
Soil texture, 449
Solar cells (photovoltaic cells), 178, 178f
Solar eclipse, 726, 726f
Great American Solar Eclipse, 720f–721f
Solar energy, 44, 178, 178f, 189t
Solar heating systems, 178
Solar mass, 783
Solar physicists, 10
Solar radiation, 611, 614–616, 632
absorption and emission, 614–615
albedo, 615–616, 615f, 699, 702
climate change and, 689
early Earth solar output, 690
greenhouse effect, 617–619, 618f, 632, 690
latitude and altitude, 622–624, 622f, **624f**

radiation balance, 617–619, 617f, 632
reflection of, 615–616, 615f
scattering of, 616, 616f
volcanic eruptions blocking, 618–619, 619f
Solar system, 736–767
asteroids, 14–16, 761, 765
bolides, 324, 580, 580f, 691
chondrules, 762
comets, 325, 762–763, 763f, 765
dwarf planets, 760–761
formation of, 38, 325, 359, 579–580, 741, 741f, 765, 767, 784
Kuiper belt, 760–761
meteorites, 16, 130–131, 131f, 256, 761–762
meteoroids, 761
meteors, 761
minor planets, 761–762
motion of Earth, 724–725
nutation, 725
ocean formation from primordial solar system, 325–326, 325f
precession, 724, 725f
primordial solar system, 744
solar wind, 741, 772, 772f
space debris, 744
See also Planets; Stars; Sun
Solar wind, 741, 772, 772f
Soldati, Arianna, 264, 265f
Solstices, 626, 626f
Somali plate, 196, **196f**, 222
Sonar, 24, 328–329
Soot, as pollutants, 597–598
Soufrière Hills volcano (Montserrat), **281f**
South American plate, 206f
South magnetic pole, 256, 257f
South Pole, 20f
Space-based telescopes, 796
Space probes, 712, 738, 740
Space science, 706f–707f
branches of, 10, 10t
galaxies and the universe, 794–815
mapping used in, 24, 25f, 27
Spearfish formation, 98, 98f
Species, 42, 51, 52f–53f, 126, 155
Specific gravity, minerals, 78
Specific heat, 623
Spectrum
absorption, 729, 730f
emission, 730, 733
of light, 729, 730f
of the sun, 775f
Sphalerite, 80t
Spicules, 779
Spiral galaxy, 800, 800f, 802, 812
Spires, 424–425, 425f
Spirit (Mars rover), 738, 750, 751, 765
Spits, 523, **523f**
Spitzer Space Telescope, 796, 798f
Spreading centers, 206, 217, 268–269, 268f, 335
Spreading ridges, 40f, 205f
Spring tides, 506–507, 506f
Springs, 489, 493
St. Matthew Island (Alaska), 50, 51f

St. Michael's Mount (England), 507, 507f
Stalactites, 490
Stalagmites, 490–491
Stamps, D. Sarah, 194, 195f, 196
Stanmeyer, John, 304f
Stars, 707f–708f, 768–793
absolute brightness, 780, 790
apparent brightness, 780
birth of, 773–774, 773f, 774f, 790, 804, 805
blue stars, 775, 780
brown dwarfs, 810
color of, 780
death of, 781–784, 790
evolution of, 783f
extreme stellar remnants, 785–788, 790
globular cluster, 801
high-mass stellar death, 783–784
life cycle of, 781, 781f, 790
low-mass stellar death, 781–783
luminosity, 781, 790
main sequence, 781, 783f, 790
neutron stars, 783, 783f, 785, 790, 810
parallax shift, 715, **715f**
population I stars, 784
population II stars, 784
properties of, 780–781
pulsars, 785–786, 786f
red giants, 782, 783f, 790
red stars, 775, 780, 799
stellar life, 780–781
supernovas, 783–784, 783f, **784f**, 785, 790, 809
white dwarf, 783, 783f, 810
State of Global Air 2018, 597
Static electricity, lightning, 239, 263f, 659, **659f**, 668
Stationary fronts, 653
Steppes, 682
Steubenville (Ohio), 597
Stockade Beaver mudstone, 98
Storm surge, 660, 662
Storms, 659–664, 668
beach erosion from, 524–526, 525f, 526f, **526f**
climate change and, 695
costliest storms, **664f**
cyclones, 650, 651f
deadliest storms, 663, **664f**
hurricanes, 358, 358f, 382, 389, 389f, 640, 640f, **644f**, 660, 662, 662f, 663–664, 663f, 695–696
midlatitude cyclone, 653–654, 653f
monsoons, **655f**, 658, 664, 668
storm surge, 660, 662
thunderstorms, 659, 668, 682
tornadoes, 638, 639f, 640, 656f–657f, 660, 660f–661f, 661t, 666t, 668, 695
tropical cyclones, 660, 661t, 662, 668
typhoons, 660, 695
Stratification, sedimentary rocks, 109, **109f**
Stratocumulus clouds, 646f, 647
Stratopause, 590f, 591
Stratosphere, 590, 590f, 596, 598
Stratovolcanoes, 278f, 280, 280f, 289

Stratus clouds, 645, 645f, 646f, 652, 652f
Streak (minerals), 77
Streams, 369–382, 478
alluvial fans, 378, 378f, 423, 423f
banks, 370
base level, 373–375, 373f, 374f, **374f**, 433
bed, 370
bed load, 372–373
braided streams, 376, 376f, 412f
capacity of, 372
channel characteristics, 370
deltas, 117, 358, 378–379, 379f
deposition, 377–378, 378f, 390
in the desert, 421–422
discharge, 369–371, 369f, 370f, 371
dissolved load, 372
distributary channels, 378–379, 382, 390
downcutting, 373, 373f
drainage basins, 377, 377f
drainage divides, 377
erosion and sediment transport, 369–382, 372–373, 372f, 373–374, 373f, 377, 390
flood control, 380–382, 380f, **380f**, 381f
floods, 379–380, 390
gradient, 369
lateral erosion of, 373–374, 373f
meanders, 373–374, 373f
mouth bar, 378, 379f
oxbow lake, 376f, 479
path of stream channel, 375–376, 376f, 375f, 376f
point bars, 375f, 376
polluted waterways, 562
subglacial streams, 412
suspended load, 372
tributaries, 369, 377
velocity, 369
See also Rivers
Strike-slip fault, 242, 243f, 259, 302, 317
Stromatolites, 65f
Strontium-87, 147t
Subarctic climate, 684
Subduction, 206f, 207f, 208, 209, 220, 269, 270f, 274
at a continental margins, 307–308, 307f–309f, 310
Himalaya (Asia), formation of, 312–313, 312f
of oceanic crust, 312, 313
on west coast of South America, 307–308, 308f
Subduction complex, 306, 306f
Subduction zone, 217, 259, 269, 270f, 324, 324f, 336
Subglacial streams, 412
Submarine canyons, 344–345, 345f, 349
Submarine fans, 344–345, 345f
Submergence, coastal, 519, 520f, 533
Subsidence, 556–557
Subtropical jet stream, 679, 679f
Sulfur
acid rain, 365, **365f**, 601
in native form, 83, 83f
and oxidation, 366

INDEX **887**

INDEX

sulfate minerals, 80t
sulfide minerals, 79, 80t, 82, 87, 91, 166
sulfur dioxide in volcanic gas, 131
sulfur oxides as pollutants, 593–594, 594f, 601
sulfuric acid in acid mines, 87
uses, 80t
Sumatra-Andaman earthquake, 47, 247–248, 247f
Summer solstice, 626, 626f, 627t
Sun, 607f, 632, 775–779, 789, 813
  climate change and, 689
  core, 777, 777f, 782, 790
  corona, 779
  early Earth, 686, 690
  evolution of, 792
  formation of, 741
  global climate stability, 677
  helium in, 729
  lifetime of, 781, 782, 782f
  magnetic field of, 779
  orbit of, 725
  Parker Solar Probe, 772, 775, 789
  predicted changes in, 79f
  prominence, 779, 779f
  structure of, 775, 777–779
Sundance formation, 98, 98f
Sunlight, 606f–607f, 611, 612f–613f
  hours of sunlight per day, 627t
  latitude and, 624
  See also Solar radiation
Sunspots, 778–779, 779f
"Super-Earths," 796
Superconductors, 162f
Supercontinents, 214, 214f, 223
  Nuna, 214
  Pangaea, 132, 132f, 198, 214, 214f, 296, 345–346, 346f
  Pangaea Proxima, 192f–193f, 214
  Rodinia, 214
Supercooling, 642
Supergiants, 783f
Superheating, 268
Supernovas, 783–784, 783f, **784f**, 785, 790, 809
Superposition, principle of, 134, 134f, 136
Supersaturation, 642
Surf, 517, 517f
Surface currents, 509–512, 533
Surface mines, 168f, 169
Surface processes, 355
  See also Erosion; Landslides; Stream erosion and deposition; Weathering
Surface water
  currents, 509–512, 533
  distribution of on Earth, 477–478, 477f
  diversion, 437–438
  runoff, 477
  salinization of, 547f, 548
Surface waves, 234, 259
Suspended load (stream), 372
Sustainability
  mineral resources, 160, 161f
  food systems, 440, 441f, 469
  of soil resources, 462–464, 463f, 464f, **464f**
  water management, 567

Sustainable Groundwater Management Act (California, 2014), 567
Swamps, 482, 483f
Swarr, Gretchen, 69f
Swift Observatory, 788
Sylvite, 80t
Symmetric folds, 298, 298f
Syncline, 298
Synthetic gems, 72
Systems, 18t, 34, 43–44
  changes in, 44–50, 634–635, 634f–635f
  ecosystem, 43, 43f, 51
  feedback mechanisms, 50, 58, 60, 61
  humans and Earth systems, 50–54, 61, 63, 535
  interconnectedness, 60
  rock cycle and, 101

T

Tafreshi, Babak, 720f–721f, 776f, 799f
Taiga biome, 684
Taklamakan Desert (Asia), 431
Talc, hardness, 76t
Tallgrass prairies, 442, 442f
Talus, 360, 361f
Tarbela Dam (Pakistan), 548
Taylor Glacier (Antarctica), 397f
Tectonic plates, 37f, 40, 40f, 194, 195f
  See also Plate boundaries; Plate movement; Plate tectonics
Tectonic stress (tectonic forces), 297
  basins, 299, 299f
  coastlines and, 520
  domes, 299, 299f
  earthquakes, 232
  faults, 231–232, 232f, 242, 299, 299f–303f, 301–303, 317, 334, 334f
  folds, 46f, 297–299, 297f–299f, **298f**, 302, 317
  folding and tilting, 134, **134f**, 140
  joints, 299, 302
  metamorphism, 116, 116f
  mountain building, 405, 418, 432
  uplifting, 140, 275, 360, 424
  slip, 299
Tedesco, Dario, 272f
Telescopes, 709, 712, 718, 727–729, 796
  angular resolution of, 731f
  aperture of, 730, 730f
  evolution of, 730
  optical telescopes, 798
  radio telescopes, 785f, 798, 807
  reflecting telescopes, 728
  refracting telescopes, 726, 728f
  space-based telescopes, 796
  x-ray space telescopes, 788
Tempel 1 (comet), 763
Temperate rain forests, 682
Temperature, 620, 632, 668, 684t
  of atmosphere, 590–591, 590f, 601
  average temperature, 624–626, 625f, 628, 628f, 630, 630f, 632, 668
  climate change, 51
  clouds and, 630, **630f**

  of Earth, 39, 51, 115, 601
  global temperature, 572f, 625f, 687, 687f, 694, 694f
  of oceans, 504, 504f, 509f, 514, 533
  plastic deformation and, 297
Terminal moraines, 410, 413, 414
Terminus, 403, 409–410, 410f
Terracing, 463, 463f
Terrestrial planets, 741, 742f, 743, 743t, 744–751, 765
  See also Earth; Mars; Mercury; Venus
Terrrigenous sediment, 340, 343
Tertiary period, 45f, 150f, 154
Tessin, Allyson, 34, 35f
Teton Dam (Idaho), 549
Tharp, Marie, 330f
Tharsis (on Mars), 748
Tharsis Montes (shield volcanoes on Mars), 746f
Theories, 13–15, 29
Thermal expansion and contraction, 360
Thermal pools, 5, 66f–67f
Thermal rivers, 6, 7f
Thermocline, 481
Thermohaline circulation, 513–514, 513f, 514f
Thermosphere, 590f, 591
Thingvellir National Park (Iceland), 202f
Thirty Meter Telescope (planned), 797f
Three-cell model, 677–678, 678f, 679, 702
Three Gorges Dam (China), 550, 550f
Three Mile Island accident (Pennsylvania), 176f, 177
Threshold effects, 48–50, 49f, 50f, 54, 61
Thrust faults, 302, 302f, 317
Thunder, 239
Thunderbird Glacier (Montana), 416
Thunderstorms, 659, 668, 682
Thwaites Glacier (Antarctica), 396, 398, 631, 631f, 697–698
Tian Shan (China), 138
Tibetan Plateau (Asia), 220, 315, 420
Tidal bulge, 505
Tidal currents, 510
Tidal range, 407
Tidal waves, 247
Tides, 505–507, 505f, 506f, 533
Tien Shan Mountains (Asia), 420
Tigereye, 83
Tigris River (Middle East), 547
Till, 408, 409f, 412
Tillite, 688
Time
  geologic. See Geologic time
  soil formation and, 445–446, 445f
Titan (moon of Saturn), 740, 759, 759f, 765
Tohoku earthquake (2011, Japan), 236, 237f, 248f–251f, 324, 324f, 348
Tonga Trench (Pacific Ocean), 327t
Topaz, 76t, 83, **83f**
Topographic maps, 23f, 24–27, 24f–26f, 31
Topographic profile, 26

Topography, 24
Topsail Island (North Carolina), 524
Topsoil, 447
Tornado Alley, 640, 660
Tornado Environment Display (TED) Jana, 638
Tornadoes, 638, 639f, 656f–657f, 660, 660f–661f, 661t, 668, 695
Trace fossils, 109f, 110
Trade-offs, 16
Trade winds, 511f, 664, 665f, 678
Transarctic Mountains (Antarctica), 316f
Transform faults, 334
Transform plate boundaries, 205, 206f, 208t, 209, 223, 241–242, 242f–244f
Transformations (soil), 447, 449
Transiting Exoplanet Survey Satellite (TESS), 796
Transits, 744f, 770
Translocation, 447, 449, 467
Transpiration, 477
Transported soil, 444
Transverse dunes, 430, 430f
Travel-time curve, 239–240, 240f
Travertine, 67, 67f
Trees
  landslides and, 387–388, 388f
  pistol butt, 384, 385f
  rain forests, 68, 70, 417, 417f, 681, 682
  tree rings as evidence of climate change, 687–688
Trenches (oceanic), 40f, 205f, 327t, 336, 346–347, 347f, 349
Triassic period, 45f, 150f
Tributaries, 369, 377
Tributary glacier, 408
Trilobites, 136, 137f, 138f
Triton (moon of Neptune), 760, 765
Tropic of Cancer, 626
Tropic of Capricorn, 626
Tropical cyclones, 660, 661t, 662, 664, 668
Tropical rain forests, 68, 417, 417f
Tropopause, 590, 590f
Troposphere, 590, 590f, 595–597, 596f
Trough (wave), 507, 507f
Tsunamis, 247–249, 248f, 249f
  Tohoku earthquake (Japan), 324, 324f
Tundra, 314, 482, 684
Tungsten, 164
Turbidity currents, 344–345, 345f, 349
Turnover, 480, 481–482
Turquoise, 80t, 83
Turretellid gastropods, 154f
2DF Galaxy Redshift Survey, 808, 808f
Tying It All Together activity, 27, 60, 90, 120, 154, 186, 222, 258, 288, 316, 348, 389, 432, 466, 492, 532, 567, 600, 631, 667, 701, 732, 764, 789, 812
  Antarctica's Glacier Retreat, 631
  Communities Near Volcanoes, 288

Consequences of Climate Change, 701
Costs of Mining, 90, 90f
Devils Tower Formation, 120, 120f
Distant Worlds, 764
Earthquake Vulnerability, 258
Flood Control and Management, 389
Geothermal Energy Resource Planning, 27–28, 28f
Gravity in Exoplanetary Systems, 732
Great Lakes Connections, 492, 492f
Ice and Wind Erosion, 432
Mining Manganese Nodules, 186, 186f
Mountain Chains of Change, 316
Quasars and the Evolution of the Universe, 812, 812f
Reconstructing Earth's Past, 154
Reducing Air Pollution, 600
Rifting Apart, 222
Robotic Exploration of the Seafloor, 348, 348f
Sea Change, 532
Soil Depletion and Pollution in Chesapeake Bay Watershed, 466
Super Disruption: Natural Disasters, 60
Sustainable Water Management, 567
We Are Made of Star-Stuff, 789
Weather Patterns, 667
Typhoons, 660, 695

U

U-shaped valleys, glaciers 406f, 407, 527
Ultramafic rocks, 106
Ultraviolet radiation, 44
Umbra, 726
Unconformities, 139–141, 139f–141f, 155
Underground mines, 168f, 169, 170f–171f
Underthrusting, 305, 306
Unfoliated metamorphic rock, 116, 116f
Uniformitarianism, 46
United Nations, 54, 70, 125, 160, 185, 266, 528, 529f, 541, 700
Universe
 accelerating universe, 811, 813
 Aristotle's model of, 715, 716, 718, 733
 Big bang theory, 803–806, 813, 815
 closed universe, 810f, 811, 813
 Copernican model of, 716, 733
 cosmic microwave background (CMB), 806–808, 813
 dark energy, 811
 dark matter, 809–810, 813
 deceleration of, 811
 expanding universe, 800, 801, 803, 805, 806
 fate of, 810f, 811
 geocentric model of, 715–716, 715f, **715f**, 716f, 719, 719f
 heliocentric model of, 716–718, 716f
 models of, 712, 715–718, 733
 open universe, 810f, 811, 813
 primordial nucleosynthesis, 805, 806
 primordial universe, 804, 804f, 809
 Ptolemy's model of, 715–716, 716f, 719f, 733
 singularity, 803
Upwelling, 514, 514f, 516, 533
Ural Mountains (Europe and Asia), 315, 317
Uranium, 86, 145, 146t, 164t, 174, 175, 177, 181f, 189t
Uranus (planet), 765
 atmosphere of, 752f, 756
 images of, 755f
 moons of, 759–760
 properties of, 723t, 743t, 793t
 scale drawing of, 742f
Urban heat island effect, 684
Urbanization, climate and, 684–685, 684f, 684f, 685f, 702
Ursa Major (constellation), 713, 713f
Ursa Minor (constellation), 713, 713f

V

V-shaped valleys, glaciers, 405
Valles Marineris (canyon on Mars), 745
Vancouver Island (Canada), 519f
Vannuatu Island (South Pacific), 278f
Vehicles. See Auto industry
Venera spacecraft, 746f
Venice (Italy), 698
Venus (planet), 745, 746
 atmosphere of, 745, 765
 blob tectonics of, 748
 climate of, 745
 geology of, 748
 images of, 746f, 747f, 748, **748f**
 phases of, 719f
 properties of, 723t, 743t, 793t
 scale drawing of, 742f
 surface of, 25, 27
 transit, 744f
 volcanic eruptions on, 747, 747f
Very Large Array (New Mexico), 708f
Vesicles, 276
Victoria Falls (Zimbabwe), 470f–471f
Victoria Valley (Antarctica), 394f–395f
Viking Orbiter missions, 745, 746f
Vine, Frederick, 203
Volatile compounds, as pollutants, 597
Volcanic arcs, 107
Volcanic crater lakes, 474
Volcanic eruptions, 281–285, 287, 289
 ash, 44, 60, 114, 131
 climate change and, 60, 131, 223, 691
 fissure flows, 277
 gases, 44
 global climate and, 283, 284, 285f, 287
 greenhouse effect and, 618
 notable volcanic eruptions, 266, 266f, 288
 magma, 5, 44, **60f**, 101, 105, 105f, 106, 107
 mass extinctions and, 131
 predicting, 283, 289
 pyroclastic flows, 36, 276, 281–282, 281f, **281f**, 289
 research, 96, 97f
 risk assessment, 283, 286, 286f
 systems affected by, 36, 44, 60
 of Vesuvius, 266, 266f, 288
Volcanic formations, 274–279, 278t, 288
 batholith, 275, 275f, 289
 dikes, 275–276, 276f, **276f**, 289
 plutons, 274–275, 274f–276f, 288, 289
 sills, 276, **276f**, 289
Volcanic Threat Assessment scores, 286
"Volcanic winter," 60
Volcanoes, 191, 262–291
 ash-flow tuff, 281f, 282, 289
 black smokers, 333, 333f, **333f**
 caldera, 278t, 282, 282f, 289
 cinder cone volcanoes, 278t, 279–280
 composite cone volcanoes, 278t, 280, 280f
 crater, 279
 fissures, 279
 formation of, 207, 218f–219f
 haze, 279
 islands, 306, 337–338, **338f**, 349
 island arc formation, 306
 lava, 104–105, 213f, 217, 264, 265f, 276, 276f
 magma in, 105, 105f, 106, 107
 on Mars, 746f, 748, 748f, **748f**
 Mid-Ocean Ridge, 203–204, 204f, 207, 208t, 212, 217, 243–244, 268, 301, 304, 326, 331–334, 331f, 333f, 335, 340, 341, 349
 oceanic islands, 337, 349
 plate tectonics, 217, 217f, 223
 research, 96, 97f, 194, 195f
 rock formation, 102f–103f, 104–107, 105f, 107f
 seamounts, 337, 349
 shield volcanoes, 278t, 279, 279f, 289
 stratovolcanoes, 278t, 280, 280f, 289
 subduction and, 313
 types of, 279–280
 vents, 279
 on Venus, 27
Voyager space probes, 712, 753, **754f**, 755–757, 755f, 760

W

Walter Anthony, Katey, 576, 577f, 578
Waning moon, 725f, 726
Warm front, 652, 652f
Washes, 421, 421f
Waste
 from mining, 169
 from nuclear power plants, 177
 radioactive waste, 177
Water, 471, 641, 641f
 in atmosphere, 41, 41f, 44, 55
 in biosphere, 56
 change of state, 622
 condensation of, 642
 in deserts, 421–425, 433
 distribution of on Earth, 477–478, 477f
 evaporation of, 477–478, 622, 623
 freezing of, 642
 frost wedging, 360, 361f
 in geosphere, 477
 as greenhouse gas, 618, 632
 in hydrosphere, 33, 37, 37f, 38f, 40–41, 41f, 42, 44
 hydrothermal metamorphism, 118, 121
 hydrothermal water, 165
 latent heat, 622, 622f
 lunar water, 750, 765
 magma behavior and, 274, 289
 in oceans, 327
 rock cycle and, 101
 sedimentary rocks and, 108
 in soil, 443
 specific heat, 622
 wells, 554, 554f, 565
 See also Fresh water; Groundwater; Hydrosphere
Water Conservation Act (California, 2009), 567
Water cycle, 44, 55–56, 55f, 61, 101, 477–478, 478f, 488f, 492–493
Water pollution, 559–566
 acid mines and heavy metals, 87, 87f
 anti-pollution policies, 559–560
 aquifers, 565–566
 environmental laws, 562
 groundwater, 563–566, 564f, 564t
 nonpoint source pollution, 561
 nuclear waste disposal, 566
 plume of contamination, 564, 564f
 point source pollution, 561
 risk assessment, 563
 toxic pollutants, 562, 563
 types of pollutants, 560–561
 waterways, 562
Water resources, 536–571
 California, 540
 dams and diversion, 536f–537f, 547–548
 groundwater, 554–559
 See also Fresh water; Groundwater; Water pollution
Water supply and demand, 541–546, 568
 agricultural use, 545–546, 545f
 consumption, 541, 542f, 543, 543f
 domestic water use, 543–544, 544f
 fresh water, 541, 542f, 543, 543f, 544f
 global consumption figures, 542f
 industrial use, 545–546, 545f, 546f
 water scarcity, 541
 withdrawal and consumption, 541, 543f, **545f**
 in United States, 541, 543, 543f, 545–546, 545f, **545f**, 568
Water table, 487–488, 488t
Water vapor, 477
Watersheds, 492, 492f
Wave base, 508
Wave-cut platform, 527, 527f, 533

Wave height, 507, 507f
Wave orbitals, 508, 508f, 517f
Wavelength, 507, 507f, 632
Waves, 507–508, 507f, 508f, 533
   coastal erosion and, 517
   of light, 611
   refraction of, 518, 518f, 533
   storms, 517, 517f
   surf, 517, 517f
Waxing moon, 725f, 726
Weakly interacting massive particles (WIMPs), 810–811
Weather, 636–659
   anticyclones, 650, 651f, 679
   barometric pressure and, 587, 601
   convection–convergence, 644, 658, 668
   Coriolis effect and, 650, 650f, 654, 668
   cyclones, 650, 651f, 679, 682
   dew point, 641
   dust storms, 459–460, 461f
   Earth's surface features and, 654–655, 655f, 658
   El Niño, 516, 664–666, 665f, 668
   energy balance and, 611, 614
   fog, 646, 647f, 649, 668
   forests and, 655, 658, 668
   frontal wedging, 644, 668
   fronts and frontal weather, 650–654, 668
   frost, 642f, 643
   glaze, 648, **648f**
   hail, 648, 648f
   humidity, 641, 641f
   lightning, 239, 263f, 659, **659f**, 668
   maps, 23f
   meteorologists, 649
   midlatitude cyclone, 653–654, **653f**
   monsoons, **655f**, 658, 664, 668
   mountains and, 655, 655f
   orographic lifting, 644, 644f, 668
   pressure gradient and, 649, 668
   rain-shadow deserts, 655, 655f
   relative humidity, 641, 642, 684t
   rising air, 643, 649, 650, 668, 702
   sinking air, 649, 650, 668, 702
   sleet, 648
   snow, 399, 399f, 630, 648, 702
   storm surge, 660, 662
   storms, 659–664, 668
   thunderstorms, 659, 668, 682
   tropical cyclones, 660, 661t, 662, 668
   typhoons, 660, 695
   weather patterns, 667, 695–696, 695f–696f
   wind speed, 649, 668
   See also Climate; Climate change; Clouds; Hurricanes; Precipitation; Rain; Tornadoes
Weathering, 99, 108, 167, 167f, 187, 359, 359f, **359f**, 390
   chemical weathering, 360, 360f, 365–368, 390
   mechanical weathering, 359–364, 390
   by plant roots, 368
   of seacoast, 517

Wegener, Alfred, 197–199, 197f, 223
Wells, 554, 554f, 565
West Antarctic Ice Sheet, 404, 702
Westerlund 2 (star cluster), 768f–769f
Western Aleutians Rift, 208t
Wetlands, 482–486, 483f, 493
   biodiversity, 483, 484f
   degradation of, 485–486, 485f
   density in the U.S., 485f
   food webs, 483
   importance of, 486
   Midwest Wetlands Initiative, 483
   properties of, 482
Whewell, William, 46
White Cliffs of Dover (England), 113, 113f
White dwarfs, 783, 783f, 810
Wildfires, 574f–575f, 697
Wilkinson Microwave Anisotropy Probe (WMAP), 807, 810
William Herschel Telescope, 797f
Wilson, Robert, 807
Wind, 41, 649
   abrasion 428-429
   average temperature and, 628, 628f, 629f, 630
   Coriolis effect, 650, 650f, 654, 662, 668, 677, 678, 678f, 679
   deserts and, 428–436
   direction of, 650
   El Niño, 516, 664–666, 665f, 668
   energy, 158f–159f, 178, 189t
   erosion, 428, 432
   friction and, 650
   at high altitude, 668
   jet stream, 650, 668, 679, 679f, 702
   ocean currents and, 510f, 513, 516, 628
   polar easterlies, 679
   polar jet stream, 679
   pressure gradient and, 649
   prevailing westerlies, 679
   speed of, 649, 668, 684t
   subtropical jet stream, 679, 679f
   three-cell model, 677–678, 678f, 679, 702
   in tornadoes, 660
   trade winds, 511f, 664, 665f, 678
Winkel tripel projections, 21, 21f, 29
Winter Park sinkhole (Florida), 491, 491f
Winter solstice, 626, 626f, 627t
Withdrawal (water). See Water withdrawal and consumption
WMAP satellite, 807, 810
Wolinsky, Cary, 84f
The World at Night Program, 799f
Wright Brothers, 13
The Wrinkled Brain Project, 710, 711f

X

X-ray space telescopes, 788
X-rays, 796, 798, 798f, 801
   black holes, 787

Y

Yamashita, Michael, 250f–251f

Yangtze River (China), 550, 550f
Yellow River (China), 372f, 381, 548
Yellowstone Lake (Wyoming), 148f–149f
Yellowstone National Park (Wyoming)
   geologic features, 33, 33f, 36f, 107f
   hydrothermal vents, 148f–149f
   underwater spire, 148f–149f
   volcanic eruptions, 36, 36f, 60, **60**, 107f, 269, 282
Yellowstone River (Montana), 369
Yellowstone Volcano (Wyoming), 36, 60
Yosemite National Park (California), 274, 274f
Yosemite Valley (California), 408
Yucatán (Mexico), 15f, 16, 130–131, 131f

Z

Zambezi River (Zimbabwe), 470f–471f
Zinc, 82, 87, 164t
Zone of ablation, 433
Zone of accumulation, 403, 433
Zooplankton, 528, **528f**, 533, 583

ACKNOWLEDGMENTS

National Geographic Learning gratefully acknowledges the contributions of the following National Geographic Explorers and affiliates to our program:

Munazza Alam, National Geographic Grantee
Saleem Ali, National Geographic Emerging Explorer
Ira Black, National Geographic Photographer
Katlin Bowman, National Geographic Grantee
Steve Boyes, National Geographic Fellow
Marcello Calisti, National Geographic Grantee
Sarah Carmichael, National Geographic Grantee
Knicole Colón, National Geographic Grantee
Joe Cutler, National Geographic Explorer
Jason Edwards, National Geographic Photographer
Bethany Ehlmann, National Geographic Emerging Explorer
Jerry Glover, National Geographic Emerging Explorer
Chris Hill, National Geographic Photographer
Greg Kahn, National Geographic Photographer
Patrick Kelly, National Geographic Photographer
Robb Kendrick, National Geographic Photographer
Keith Ladzinski, National Geographic Photographer
Cynthia Liutkus-Pierce, National Geographic Grantee
Stefanie Lutz, National Geographic Grantee
Pete McBride National Geographic Grantee
Paul Miller, National Geographic Emerging Explorer
Brendan Mullan, National Geographic Emerging Explorer
Klaus Nigge, National Geographic Photographer
Carsten Peter, National Geographic Grantee
Erin Pettit, National Geographic Emerging Explorer
Alicia Pressel, National Geographic Emerging Explorer
Caroline Quanbeck, National Geographic Grantee
Jim Richardson, National Geographic Photographer
Andrés Ruzo, National Geographic Explorer
Joel Sartore, National Geographic Fellow
Brian Skerry, National Geographic Fellow
Anton Seimon, National Geographic Grantee
Afroz Ahmad Shah, National Geographic Grantee
Kenneth Warren Sims, National Geographic Grantee
Arianna Soldati, National Geographic Grantee
D. Sarah Stamps, National Geographic Grantee
John Stanmeyer, National Geographic Grantee
Babak Tafreshi, National Geographic Photographer
Allyson Tessin, National Geographic Grantee
Katey Walter Anthony, National Geographic Emerging Explorer
Cary Wolinsky, National Geographic Photographer
Michael Yamashita, National Geographic Photographer

Photo Credits:

i ESA/NASA; **iv** Robbie Shone/National Geographic Image Collection; **v** Frans Lanting/National Geographic Image Collection; **vii** © Hydrographic and Oceanographic Department, Japan Coast Guard; **viii** © Jacqueline Kehoe; **ix** Dea/V. Giannella/De Agostini/Getty Images; **xi** © Climate Central/climatecentral.org; **xii** NASA/JPL-Caltech/SSI; **xiv** Courtesy of Marc S. Hendrix; **xv** (t) Courtesy of Graham R. Thompson, (b) Courtesy of Jonathan Turk; **xvi** (cl1) © Munazza Alam, (cl2) © Maria Ali, (c) © Katlin Bowman, (cr1) Sora Devore/National Geographic Image Collection, (cr2) © Marcello Calisti, (bl1) © Felix Kunze, (bl2) Courtesy of Knicole Colon, (bc) © Joe Cutler, (br1) Courtesy of Caltech, (br2) Courtney Rader/National Geographic Image Collection; **xvii** (tl1) © Marie Freeman, (tl2) Stefanie Lutz/National Geographic Image Collection, (tc) © Rebecca Drobis, (tr1) Rebecca Emily Drobis/National Geographic Image Collection, (tr2) Courtesy of Erin Pettit, (cl1) © Caroline Quanbeck, (cl2) © Sofía Ruzo, (c) © Tracie Seimon, (cr1) Courtesy of Afroz Ahmad Shah, (cr2) © Maura Hanning, (bl1) © Arianna Soldati, (bl2) © Sarah Stamps, (bc) © Timothy Gallagher, (br) Rebecca Hale/National Geographic Image Collection; **2–3** (spread) Robbie Shone/National Geographic Image Collection; **4–5** (spread) Raul Touzon/National Geographic Image Collection; **7** (t) © Parker Young, (bl) © Steve Winter; **9** Bill Hatcher/National Geographic Image Collection; **15** Lawrence Berkeley National Laboratory; **17** Snorri Gunnarsson/Alamy Stock Photo; **19** Andriy Popov/Alamy Stock Photo; **22** US Government Accountability Office; **23** (tl) National Geographic Cartographic Division/National Geographic Image Collection, (tr) NREL, (cl) Department of Conservation and Natural Resources, Bureau of Geological Survey, (cr) Pilvitus/Shutterstock.com, (bl) Andrew Garcia-Phillips/National Geographic Image Collection, (br) Contemporary Images/Alamy Stock Photo; **25** NASA/JPL/USGS; **32–33** (spread) Milko Marchetti/500px/Getty Images; **35** (t) © Mark Zindorf, (bl) © Stephen Widdicomb; **39** (bl) Robert Harding Picture Library/National Geographic Image Collection, (br) Gordon Wiltsie/National Geographic Image Collection; **43** Yva Momatiuk and John Eastcott/Minden Pictures; **46** (t) Chintla/Shutterstock.com, (tr) Courtesy of Marc S. Hendrix; **47** (tr) Action Sports Photography/Shutterstock.com, (b) Courtesy of Tom Foster; **48** Chase Studio/Science Source; **49** Mark Reid/U.S. Geological Survey; **52–53** (spread) Joel Sartore, National Geographic Photo ARK/National Geographic Image Collection; **56** Data taken from U. Siegenthaler and J. L. Sarmiento, "Atmospheric Carbon Dioxide and the Ocean," Nature 365 (September 9, 1993): 119; **57** ESRL/NOAA; **64–65** (spread) Frans Lanting/National Geographic Image Collection; **66–67** (spread) © Andrea Loriga; **69** (t) (b) © Gretchen Swarr; **70** Randy Olson/National Geographic Image Collection; **71** (tr) Courtesy of Marc S. Hendrix, (br) Tabor Chichakly/Shutterstock.com; **75** Courtesy of Marc S. Hendrix; **76** Nyura/Shutterstock.com; **77** (t) Sebastian Janicki/Shutterstock.com, (cr) yul38885/Shutterstock.com; **78** (tl) vvoe/Shutterstock.com, (bl) Mark A. Schneider/Science Source, (bc) Fernando Sanchez Cortes/Shutterstock.com, (br) Juraj Kovac/Shutterstock.com; **79** © Breck P. Kent; **83** Konstantin Nikolaevich/Shutterstock.com; **84** Cary Wolinsky/National Geographic Image Collection; **85**

ACKNOWLEDGMENTS

Bildagentur Zoonar GmbH/Shutterstock.com; **87** Jose Arcos Aguilar/Shutterstock.com; **88** Maureen and Mike Mansfield Library Digital Collections; **89** Autumn's Memories/Shutterstock.com; **90** Alexander Loos/EyeEm/Getty Images; **94–95** (spread) Andrew Coleman/National Geographic Image Collection; **97** (t) Courtesy of Kenneth Warren Sims, (b) © Carsten Peter; **98** Rex Wholster/Alamy Stock Photo; **104–105** (spread) Patrick Kelley/National Geographic Image Collection; **106** (bl) Bragin Alexey/Shutterstock.com, (bc) Tyler Boyes/Shutterstock.com, (br) Only Fabrizio/Shutterstock.com; **107** (tl) Courtesy of James St. John (Geology, Ohio State University at Newark), (cl) Galyna Andrushko/Shutterstock.com; **108** (tr1) Courtesy of Marc S. Hendrix, (tr2) tobkatrina/Shutterstock.com; **109** (cl) Doug McLean/Shutterstock.com, (cr) (br) Courtesy of Marc S. Hendrix, (bl) Courtesy of James St. John (Geology, Ohio State University at Newark); **111** (tr) Fokin Oleg/Shutterstock.com, (cr1) Sci Stk Photog/Science Source, (cr2) Siim Sepp/Alamy Stock Photo; **112** (tl) Arjen de Ruiter/Shutterstock.com, (cl) Jen Watson/Shutterstock.com, (bl) Sementer/Shutterstock.com; **113** (t) Peter Cripps/Shutterstock.com, (cr1) www.sandatlas.org/Shutterstock.com, (cr2) Susan E. Degginger/Alamy Stock Photo; **114** (tl) Weldon Schloneger/Shutterstock.com, (tr) Courtesy of Marc S. Hendrix; **117** (tl) Courtesy of Marc S. Hendrix, (tr) Courtesy of Graham R. Thompson/Jonathan Turk; **119** (cl1) Anfisa Kameneva/EyeEm/Getty Images, (cl2) Bortel Pavel-Pavelmidi/Shutterstock.com, (c) Marafona/Shutterstock.com, (cr1) Scott Camazine/Science Source, (cr2) Jack Barr/Alamy Stock Photo; **120** (bl) Aaron Huey/National Geographic Image Collection, (br) Richard Cummins/Lonely Planet Images/Getty Images; **122** vvoe/Shutterstock.com; **123** sonsam/Shutterstock.com; **124–125** (spread) David Doubilet/National Geographic Image Collection; **127** (t) (b) © Felix Kunze; **128** © Lucas Lima/Studio 252MYA; **135** Courtesy of Marc S. Hendrix; **137** Peter Essick/National Geographic Image Collection; **139** Courtesy of Marc S. Hendrix; **141** Paul Racenet/Shutterstock.com; **148–149** (spread) Brian Skerry/National Geographic Image Collection; **152** (tl) Clint Farlinger/Alamy Stock Photo; **152–153** (spread) Riska Parakeet/Shutterstock.com; **154** © Shanan Peters/University of Wisconsin-Madison; **158–159** (spread) geophoto/Multi-bits/ImaZinS/Getty Images; **161** (t) © Richard Paradis, (b) © Saleem Ali; **162** Luca Locatelli/National Geographic Image Collection; **163** Lester Lefkowitz/Stone/Getty Images; **166** NOAA/Science Source; **167** (tr) The Natural History Museum, London/Science Source, (cr) Courtesy of Marc S. Hendrix; **168** (t) Monty Rakusen/Cultura/Getty Images, (b) Michael Melford/National Geographic Image Collection; **170–171** (spread) Robb Kendrick/National Geographic Image Collection; **173** Courtesy of Marc S. Hendrix; **174** Iris Schneider/Los Angeles Times/Getty Images; **176** Andrew Harrer/Bloomberg/Getty Images; **178** Xinhua/Cao Yang/Xinhua News Agency/Getty Images; **179** Michael Peuckert/imageBroker/Getty Images; **182** (bl) Vladimir Gjorgiev/Shutterstock.com, (br) Pokpak Stock/Shutterstock.com; **183** Rich Carey/Shutterstock.com; **186** (bl) SPL/Science Source, (br) Courtesy of Marc S. Hendrix; **190–191** (spread) © Hydrographic and Oceanographic Department, Japan Coast Guard; **192–193** (spread) Art: Charles Preppernau. Source: C. R. Scotese, Paleomap Project/National Geographic Image Collection; **195** (t) (b) © Sarah Stamps; **196** guenterguni/E+/Getty Images; **202** Alex Mustard/Nature Picture Library/Getty Images; **207** © Planetary Visions Ltd; **210** Feng Wei Photography/Moment/Getty Images; **213** Mario Tama/Getty Images News/Getty Images; **214** (tl) (cl1) (cl2) (bl) Art: Charles Preppernau. Source: C. R. Scotese, Paleomap Project/National Geographic Image Collection; **218–219** (spread) Carsten Peter/National Geographic Image Collection; **221** Bill Perry/Shutterstock.com; **226–227** (spread) Erik De Castro/Reuters; **229** (t) © Afroz Shah, (b) Courtesy of Afroz Ahmad Shah; **230** Mario Vazquez/AFP/Getty Images; **231** Courtesy of Graham R. Thompson/Jonathan Turk; **232** Stocktrek/Stockbyte/Getty Images; **235** Encyclopaedia Britannica/Universal Images Group/Getty Images; **242** (tl) NOAA/NGDC, University of Colorado at Boulder, (tr) Library of Congress, Prints & Photographs Division, Reproduction number LC-USZ62-47591 (b&w film copy neg. of half stereo); **248** NOAA Center for Tsunami Research; **250–251** (spread) Michael S. Yamashita/National Geographic Magazines/Getty images; **252** USGS; **262–263** (spread) The Asahi Shimbun/Getty Images; **265** (t) (b) © Arianna Soldati; **266** Quintanilla/Shutterstock.com; **272–273** (spread) Carsten Peter/National Geographic Image Collection; **274** Thomas Roche/Moment/Getty Images; **277** (tl) Bruce Omori/Paradise Helicopters/EPA-EFE/Shutterstock.com, (tr) Manfred Thürig/Alamy Stock Photo, (bl) LucynaKoch/E+/Getty Images, (br) Courtesy of Graham R Thompson/Jonathan Turk; **278** Ronald Karpilo/Alamy Stock Photo; **279** Wildnerdpix/Shutterstock.com; **280** tusharkoley/Shutterstock.com; **281** Peter Unger/Lonely Planet Images/Getty Images; **284** Images & Volcans/Science Source; **291** Massimo Lama/500Px Plus/Getty Images; **292–293** (spread) Menno Boermans/Aurora Open/Getty Images; **295** (t) Mark Stone/National Geographic Image Collection, (b) Cory Richards/National Geographic Image Collection; **296** Cory Richards/National Geographic Image Collection; **297** J M Barres/AGE Fotostock; **300** (tr) Fletcher & Baylis/Science Source, (b) Doug Steley A/Alamy Stock Photo; **301** Courtesy of James St. John (Geology, Ohio State University at Newark); **302** Paul E. Carrara/USGS; **303** Ted Kinsman/Science Source; **304** © John Stanmeyer; **305** Jeffrey Kargel, USGS/NASA JPL/AGU; **307** SeppFriedhuber/E+/Getty Images; **309** © Jeff Heimsath; **311** Andy Bardon/National Geographic Image Collection; **320–321** (spread) NG Maps/National Geographic Image Collection; **323** (t) (bl) © Marcello Calisti; **324** © Jamstec; **330** Heinrich C. Berann/National Geographic Image Collection; **333** NOAA Okeanos Explorer Program, Galapagos Rift Expedition 2011; **341** SPL/Science Source; **343** Liam

Gumley, Space Science and Engineering Center, University of Wisconsin-Madison; **348** © MARUM–Center for Marine Environmental Sciences; University of Bremen; V.Diekamp; **352–353** (spread) © Jacqueline Kehoe; **354–355** The New York Power Authority; **357** (t) © Sarah Carmichal, (b) © Cynthia Liutkus-Pierce; **358** DigitalGlobe/Getty Images; **360** Courtesy of Graham R. Thompson/Jonathan Turk; **361** (tr) Courtesy of Marc S. Hendrix, (bl) Maxim Petrichuk/Shutterstock.com, (br) Denis Burdin/Shutterstock.com; **362–363** (spread) Jason Edwards/National Geographic Image Collection; **366** Courtesy of Marc S. Hendrix; **367** (tl) Courtesy of Marc S. Hendrix, (br) Courtesy of Graham R. Thompson/Jonathan Turk; **369** Planet Observer/Universal Images Group/Getty Images; **372** zhuda/Shutterstock.com; **373** (tr) Karel Gallas/Shutterstock.com, (b) Peter Unger/Lonely Planet Images/Getty Images; **375** (tr) (bl) (br) Courtesy of Graham R. Thompson/Jonathan Turk, **376** Daniel Bosma/Moment/Getty Images; **378** Dr. Marli Miller/Visuals Unlimited/Getty Images; **385** Courtesy of Graham R. Thompson/Jonathan Turk; **386** JIJI Press/AFP/Getty Images; **388** Courtesy of Marc S. Hendrix; **389** Prisma by Dukas Presseagentur GmbH/Alamy Stock Photo; **391** imageBroker/Alamy Stock Photo; **392** Masami Goto/Minden Pictures; **393** © Bruce Dale/National Geographic Image Collection; **394–395** (spread) © Dr Charlie Bristow; **397** (t) (b) courtesy of Erin Pettit; **398** NASA/J. Yungel; **400** Patrick Poendl/Shutterstock.com, **404** Michael Melford/National Geographic Image Collection; **405** Courtesy of Graham R. Thompson/Jonat han Turk; **406** ptnphoto/Shutterstock.com; **407** Brandon Smith/Alamy Stock Photo; **408** Chinnaphong Mungsiri/Moment/Getty Images; **409** Courtesy of Marc S. Hendrix; **410** © Steve sheriff; **411** Chris Hill/National Geographic Image Collection; **412** Jason Edwards/National Geographic Image Collection; **416** (tl) Morton Elrod, courtesy of GNP Archives/USGS, (tr) D. Farge/G. Pederson/USGS; **420** abriendomundo/iStock/Getty Images; **421** Mike Buchheit/Shutterstock.com; **422** (tl) Olesya Baron/Shutterstock.com, (tr) Gary Weathers/Tetra images/Getty Images, (bl) Pete Ryan/National Geographic Image Collection, (br) Apic/Hulton Archive/Getty Images; **423** (tl) Courtesy of Marc S. Hendrix, (b) Courtesy of Graham R. Thompson/Jonathan Turk; **426–427** (spread) Ira Block/National Geographic Image Collection; **438–439** (spread) Jim Richardson/National Geographic Image Collection; **441** Jim Richardson/National Geographic Image Collection; **442** stevegeer/E+/Getty Images; **445–446** Courtesy of Marc S. Hendrix; **453** NRCS/USDA; **454** (cl) Rachel Blaser/Shutterstock.com, (br) soils.usda.gov; **455** Roman Khomlyak/Shutterstock.com; **456–457** (spread) Jim Richardson/National Geographic Image Collection; **458** Universal Images Group/Getty Images; **459** Construction Photography/Avalon/Hulton Archive/Getty Images; **461** USDA; **462** (t) NRCS/USDA, (bl) Eduardo Pucheta/Alamy Stock Photo; **463** Chan Srithaweeporn/Moment/Getty Images; **464** Jim Richardson/National Geographic Image Collection; **470–471** (spread) Dea/V. Giannella/De Agostini/Getty Images; **472–473** (spread) Cyril Ruoso/Minden Pictures; **475** © Sebastien Lavoué; **476** NASA image by Jeff Schmaltz, MODIS Rapid Response Team, Goddard Space Flight Center; **479** Thomas Barrat/Shutterstock.com; **480** Maksim Oleynik/Shutterstock.com; **481** Pallava Bagla/Corbis News/Getty Images; **483** (t) Tomasz Szymanski/Shutterstock.com, (tr) akphotoc/Shutterstock.com; **484** Klaus Nigge/National Geographic Image Collection; **490** Robbie Shone/National Geographic Image Collection; **491** (tl) AP images, (tr) © Red Huber, Orlando Sentinel; **494** David Boyer/National Geographic Image Collection; **496–497** (spread) © Jon Frank; **499** (t) (b) © Caroline Quanbeck; **500** Oxford Scientific/Getty Images; **501** AridOcean/Shutterstock.com; **504** (tl) David Doubilet/National Geographic Image Collection, (tr) Kyodo News/Getty Images; **507** (tl) (tr) Jim Richardson/National Geographic Image Collection; **509** NOAA; **519** Chris Gray/National Geographic My Shot/National Geographic Image Collection; **521** Sean Pavone/Shutterstock.com; **522** Oscity/Shutterstock.com; **524** FloridaStock/Shutterstock.com; **527** simonbradfield/iStock/Getty Images; **531** (tl) JC Photo/Shutterstock.com, (tr) Debra James/Shutterstock.com, (cl) Courtesy of Marc S. Hendrix; **536–537** (spread) U.S. Bureau of Reclamation; **539** (t) (b) Courtesy of Steve Boyes; **543** northlight/Shutterstock.com; **547** Courtesy of Graham R Thompsone/Jonathan Turk; **549** © Jon Waterman; **550** Imagine China/Yichang/Hubei/China/Newscom; **552–553** (spread) Pete McBride/National Geographic Image Collection; **556** USGS; **560** Bettmann/Getty Images; **562** goran cakmazovic/Shutterstock.com; **567** Christian Science Monitor/Getty Images; **572–573** (spread) © Climate Central/climatecentral.org; **574–575** (spread) NASA Earth Observatory; **577** (t) © University of Alaska Fairbanks, (b) © TH Culhane; **578** Mark Thiessen/National Geographic Image Collection; **588–589** (spread) Cory Richards/National Geographic Image Collection; **596** (tl) nik wheeler/Corbis NX/Getty Images, (cl) Steve Proehl/Corbis/Getty Images; **599** (tl) NASA/GRIN/GSFC, (tr) NASA/Goddard Space Flight Center Scientific Visualization Studio; **600** TopFoto/The Image Works; **606–607** (spread) NASA; **609** Stefanie Lutz/National Geographic Image Collection; **610** Ginny Catania/National Geographic Image Collection; **612–613** (spread) Chris Hill/National Geographic Image Collection; **614** grafvision/Shutterstock.com; **631** NASA; **636–637** (spread) © Jim Richardson/National Geographic Image Collection; **639** (t) (b) Courtesy of Anton Seimon; **640** NOAA; **642** Tim Laman/National Geographic Image Collection; **645** (tr) Marc Moritsch/National Geographic Image Collection, (cr) elen_studio/Shutterstock.com, (br) Keenpress/National Geographic Image Collection; **647** Vancouver Fog/Moment/Getty Images; **655** (cl) Marina Poushkina/Shutterstock.com, (bl) Zack Frank/Shutterstock.com; **656–657** (spread) Carsten Peter/National Geographic Image Collection; **660–661** (spread) Nick Kaloterakis/National Geographic Image Collection; **663** NOAA;

ACKNOWLEDGMENTS

672–673 (spread) © Richard Vevers/The Ocean Agency; **675** (t) Sora Devore/National Geographic Image Collection, (b) Richard Perry/The New York Times/Redux; **676** NASA Earth Observatory; **689** © James St. John; **693** Kenneth E. Kunkel, Cooperative Institute for Climate and Satellites-NC; **694** (tl) (tr) Kenneth E. Kunkel/NOAA NCDC/CICS-NC; **696** © 2013 Greg Kahn; **701** Per-Anders Pettersson/Getty Images News/Getty Images; **706–707** (spread) NASA/JPL-Caltech/SSI; **708–709** (spread) © NRAO/AUI/NSF; Miles Lucas; **711** (t) (b) Courtesy of Brendan Mullan; **712** Comstock Images/Stockbyte/Getty Images; **713** (bl) (br) Gerard Lodriguss/Science Source; **714** Mike Ver Sprill/Shutterstock.com; **720–721** (spread) Babak Tafreshi/National Geographic Image Collection; **725** NASA/Bill Dunford; **728** NASA; **729** (tl) Science Source; (tr) NASA, ESA, A. Goobar (Stockholm University), and the Hubble Heritage Team (STScI/AURA); **730** Designua/Shutterstock.com; **736–737** (spread) NASA/JPL-Caltech/SwRI/MSSS/Kevin M. Gill; **739** (t) (b) Courtesy of Bethany Ehlmann; **740** NASA/JPL-Caltech/SwRI/MSSS/Betsy Asher Hall/Gervasio Robles; **742** (t) (c) NASA/JPL, (tr) NASA, ESA and Erich Karkoschka (University of Arizona), (cl) NASA, (b) Keith Ladzinski/National Geographic Image Collection; **744** Keith Ladzinski/National Geographic Image Collection; **746** (tl) NASA/Johns Hopkins University Applied Physics Laboratory/Carnegie Institution of Washington, (tr) NASA/JPL, (cl) Nasa Images, (cr) NASA/JPL-Caltech; **747** NASA/JPL; **748** SPL/Science Source; **749** (bl) (br) NASA/Goddard Space Flight Center/Arizona State University; **751** NASA/JPL-Caltech/MSSS; **752** NASA/Lunar and Planetary Institute; **753** NASA/JPL-Caltech/SwRI/MSSS/Image processing by Kevin M. Gill, © CC BY; **754** NASA/JPL-Caltech/Space Science Institute; **755** (cl) California Association for Research in Astronomy/Science Source, (br) NASA/JPL; **757** NASA/JPL/University of Arizona; **758** (tl) NASA/JPL-Caltech/SETI Institute, (tr) NASA/JPL-Caltech/University of Arizona; **759** NASA/JPL/University of Arizona/University of Idaho; **761** NASA/Johns Hopkins University Applied Physics Laboratory/Southwest Research Institute; **762** SPL/Science Source; **764** NASA/JPL/Space Science Institute; **768–769** (spread) NASA, ESA, the Hubble Heritage Team (STScI/AURA), A. Nota (ESA/STScI), and the Westerlund 2 Science Team; **771** © Jacqueline Faherty; **772** SOHO (ESA & NASA); **774** (tl) NASA,ESA, M. Robberto (Space Telescope Science Institute/ESA) and the Hubble Space Telescope Orion Treasury Project Team, (tr) NASA; **775** N.A.Sharp, NOAO/NSO/Kitt Peak FTS/AURA/NSF; **776** Babak Tafreshi/National Geographic Image Collection; **777** Siberian Art/Shutterstock.com; **779** NASA/SDO; **785** Daily Herald Archive/SSPL/Getty Images; **787** Stocktrek Images, Inc./Alamy Stock Photo; **794–795** (spread) NASA, ESA, H. Teplitz and M. Rafelski (IPAC/Caltech), A. Koekemoer (STScI), R. Windhorst (Arizona State University), and Z. Levay (STScI); **797** (t) (b) © Knicole Colon; **798** NASA, ESA, SSC, CXC, and STScI; **799** Babak Tafreshi/National Geographic Image Collection;

800 (tl) NASA, ESA, R.M. Crockett (University of Oxford, U.K.), S. Kaviraj (Imperial College London and University of Oxford, U.K.), J. Silk (University of Oxford), M. Mutchler (Space Telescope Science Institute, Baltimore), R. O'Connell (University of Virginia, Charlottesville), and the WFC3 Scientific Oversight Committee, (tr) Credit for Hubble Image: NASA, ESA, K. Kuntz (JHU), F. Bresolin (University of Hawaii), J. Trauger (Jet Propulsion Lab), J. Mould (NOAO), Y.-H. Chu (University of Illinois, Urbana), and STScI, (cl) NASA, ESA, and The Hubble Heritage Team (STScI/AURA);Acknowledgment: P. Knezek (WIYN), (cr) NASA, ESA, A. Aloisi (STScI/ESA), and The Hubble Heritage (STScI/AURA)-ESA/Hubble Collaboration; **802** NASA, ESA, and T. Brown and S. Casertano (STScI) Acknowledgement: NASA, ESA, and J. Anderson (STScI); **803** Event Horizon Telescope Collaboration; **807** ESA/Planck Collaboration; **808** © 2dF Galaxy Redshift Survey Team; **809** NASA & ESA; **812** Science Source; **814** (cr) NASA, ESA, and the LEGUS team, (br) NASA, ESA, and The Hubble Heritage Team (STScI/AURA);Acknowledgment: P. Goudfrooij (STScI).

Illustration:

All illustrations are owned by © Cengage.